CORRECTIONS

Other Titles of Interest

CORRECTIONS

A Text/Reader

Mary Stohr
Boise State University

Anthony Walsh
Boise State University

Craig Hemmens
Boise State University

Los Angeles • London • New Delhi • Singapore • Washington DC

For information:

SAGE Publications, Inc.
2455 Teller Road
Thousand Oaks, California 91320
E-mail: order@sagepub.com

SAGE Publications Ltd.
1 Oliver's Yard
55 City Road
London EC1Y 1SP
United Kingdom

SAGE Publications India Pvt. Ltd.
B 1/I 1 Mohan Cooperative Industrial Area
Mathura Road, New Delhi 110 044
India

SAGE Publications Asia-Pacific Pte Ltd
33 Pekin Street #02-01
Far East Square
Singapore 048763

Printed in the United States of America

Library of Congress Cataloging-in-Publication Data

Corrections : a text/reader / [edited by] Mary Stohr, Anthony Walsh, Craig Hemmens.
    p. cm.
Includes bibliographical references and index.
ISBN 978-1-4129-3773-3 (pbk.)
  1. Corrections—United States. 2. Criminal justice, Administration of—United States.
3. Punishment—United States. I. Stohr, Mary K. II. Walsh, Anthony, 1941- III. Hemmens, Craig.

HV9275.C633 2009
364.60973—dc22   2008026114

This book is printed on acid-free paper.

08 09 10 11 12   10 9 8 7 6 5 4 3 2 1

Acquisitions Editor:	Jerry Westby
Editorial Assistant:	Eve Oettinger
Production Editor:	Catherine M. Chilton
Copy Editor:	Karen E. Taylor
Typesetter:	C&M Digitals (P) Ltd.
Proofreader:	Doris Hus
Indexer:	Hyde Park Publishing Services LLC
Cover Designer:	Edgar Abarca
Marketing Manager:	Christy Guilbault

Brief Contents

Detailed Table of Contents

Section V. Treatment Programming and Rehabilitation Research   283

INTRODUCTION   283

READINGS   296

Section XI. Corrections in the 21st Century 703

Foreword

You hold in your hands a book that we think is something new. It is billed as a "text/reader." What that means is we have taken the two most commonly used types of books, the textbook and the reader, and blended them in a way that we anticipate will appeal to both students and faculty.

Our experience as teachers and scholars has been that textbooks for the core classes in criminal justice (or any other social science discipline) leave many students and professors cold. The textbooks are huge, crammed with photographs, charts, highlighted material, and all sorts of pedagogical devices intended to increase student interest. Too often, though, these books end up creating a sort of sensory overload for students and suffer from a focus on "bells and whistles," such as fancy graphics, at the expense of coverage of the most current research on the subject matter.

Readers, on the other hand, are typically comprised of recent and classic research articles on the subject matter. They generally suffer, however, from an absence of meaningful explanatory material. Articles are simply lined up and presented to the students, with little or no context or explanation. Students, particularly undergraduate students, are often confused and overwhelmed.

This text/reader represents our attempt to take the best of both the textbook and reader approaches. This book comprises research articles on corrections. This text/reader is intended to serve either as a supplement to a core textbook or as a stand-alone text. The book includes a combination of previously published articles and textual material introducing these articles and providing some structure and context for the selected readings. The book is divided into a number of sections. The sections of the book track the typical content and structure of a textbook on the subject. Each section of the book has an introductory chapter that serves to introduce, explain, and provide context for the readings that follow. The readings are a selection of the best recent research that has appeared in academic journals, as well as some classic readings. The articles are edited as necessary to make them accessible to students. This variety of research and perspectives will provide the student with a grasp of the development of research, as well as an understanding of the current status of research in the subject area. This approach gives the student the opportunity to learn the basics (in the textbook-like introductory portion of each section) and to read some of the most interesting research on the subject.

There is also an introductory chapter explaining the organization and content of the book and providing context for the articles that follow. This introductory chapter provides a framework for the text and articles that follow, as well as introducing relevant themes, issues, and concepts. This will assist the student in understanding the articles.

Each section will include a summary of the material covered. There will also be a selection of discussion questions, placed after each reading. These summaries and discussion questions should facilitate student thought and class discussion of the material.

It is our belief that this method of presenting the material will be more interesting for both students and faculty. We acknowledge that this approach may be viewed by some as more challenging than the traditional textbook. To that we say, "Yes! It is!" But we believe that, if we raise the bar, our students will rise to the challenge. Research shows that students and faculty often find textbooks boring to read. It is our belief that many criminal justice instructors would welcome the opportunity to teach without having to rely on a "standard" textbook that covers only the most basic information and that lacks both depth of coverage and an attention to current research. This book provides an alternative for instructors who want to get more out of the basic criminal justice courses and curriculum than one can get from a typical textbook that is aimed at the lowest common denominator and filled with flashy but often useless features that merely serve to drive up its cost. This book is intended for instructors who want to go beyond the ordinary, standard coverage provided in textbooks.

We also believe students will find this approach more interesting. They are given the opportunity to read current, cutting-edge research on the subject, while also being provided with background and context for this research.

We hope that this unconventional approach will be more interesting, and thus make learning and teaching more fun, and hopefully more useful as well. Students need not only content knowledge but also an understanding of the academic skills specific to their discipline. Criminal justice is a fascinating subject, and the topic deserves to be presented in an interesting manner. We hope you will agree.

Craig Hemmens, JD, PhD
Department of Criminal Justice
Boise State University

Preface

⊠ Why This Book?

There are a number of excellent corrections textbooks and readers available to students and professors, so why this one? The reason is that stand-alone textbooks and readers (often assigned as expensive additions to textbooks) have a pedagogical fault that we seek to rectify with the present book. Textbooks focus on providing a broad overview of the subject and lack depth, while readers often feature in-depth articles about a single topic with little or no text to unify the readings and little in the way of pedagogy. This book provides more in the way of text and pedagogy, and it will use recent research-based articles to help students understand corrections. This book is unique in that it is a hybrid text/reader offering the best of both worlds. It includes a collection of articles on corrections that have previously appeared in a number of leading criminal justice or criminology journals; these articles are accompanied by original textual material that serves to explain and synthesize the readings. We have selected some of the best recent research and literature reviews and assembled them into this text/reader for an undergraduate or graduate corrections class.

Journal articles selected for inclusion have been chosen based primarily on how they add to and complement the textual material and how interesting we perceive them to be for students. In our opinion, these articles are the best contemporary work on the issues they address. However, journal articles are written for professional audiences, not for students, and thus often contain quantitative material students are not expected to understand. They also often contain concepts and hairsplitting arguments over minutia that tend to turn students glassy-eyed. Mindful of this, we edited and abridged the articles contained in this text/reader to make them as student friendly as possible. We have done this without doing injustice to the core points raised by the authors or detracting from the authors' key findings and conclusions. Those wishing to read these articles (and others) in their entirety are able to do so by accessing the Sage Web site provided for users of this book.

This book can serve as a supplemental reader or the primary text for an undergraduate course in corrections, or as the primary text for a graduate course. In a graduate course in corrections, it would serve as both an introduction to the extant literature and a sourcebook for additional reading, as well as a springboard for enhanced class discussion. When used as a supplement to an undergraduate course, this book can serve to provide greater depth than the standard textbook. It is important to note that the readings and the introductory textual material in this book provide a comprehensive survey of the current state of the existing scientific literature in virtually all areas of corrections, as well as giving a history of how we got to this point in each topic area.

⊠ Structure of the Book

The structure of this book mirrors that found in standard textbooks on corrections. We begin with an overview of the history of corrections. The next two sections focus on how inmates and staff experience corrections. We then examine current legal issues in corrections before moving to an examination of treatment programming. We then turn to probation, parole, and community corrections. After this, we review the literature on jails, an often under-examined aspect of corrections. Then, we focus on two corrections populations that require special attention: women and juveniles. Last, we examine what the future of corrections holds.

1. Introduction
2. History of American Corrections
3. Corrections Experience for Inmates
4. Corrections Experience for Staff
5. Legal Issues in Corrections
6. Treatment Programming and Rehabilitation Research
7. Probation and Community Corrections
8. Parole and Prisoner Reentry
9. Jails
10. Gender
11. Juvenile Justice/Corrections System
12. Corrections in the 21st Century

⊠ Ancillaries

To enhance the use of this text/reader and to assist those using this book as a core text, we have developed high-quality ancillaries for instructors and students.

Instructors' Resources on CD. Various instructor's materials are available. For each section, this material includes PowerPoint slides, Web resources, a complete set of test questions, and other helpful resources.

Student Study Site. This comprehensive student study site features chapter outlines that students can print for class, flashcards, interactive quizzes, Web exercises, and more.

◿ Acknowledgments

We would first of all like to thank executive editor Jerry Westby. Jerry's faith in and commitment to the project helped make this book a reality. We also would like to thank Jerry's developmental editors, Denise Simon and Eve Oettinger, who helped shepherd the book through the review process and whose gentle prodding ensured that deadlines would be met. Our copy editor, Karen Taylor, made sure that we didn't leave in any really embarrassing sentence fragments or dangling participles. We also would like to express our gratitude to Erin Conley, who wrote the "How to Read a Research Article" guide for students. The authors would also like to thank Brian Innacchione for his assistance with research and fact checking for this book.

We are also very grateful to the reviewers who took the time to review early drafts of our work and who provided us with helpful suggestions for improving both the introductory material and the edited readings. Their comments undoubtedly made the book better than it otherwise would have been. Heartfelt thanks to the following experts: Gaylene Armstrong, Southern Illinois University; Rosemary Gido, Indiana University of Pennsylvania; Jodi Lane, University of Florida; Steve Owen, Radford University; Bob Moore, Central Washington University; Brian Colwell, San Jose State University and Stanford University; Deborah Baskin, California State University, Los Angeles; Kristy Holtfreter, Florida State University; Elizabeth C. McMullan, University of North Florida; Robert Hanser, University of Louisiana at Monroe and Ellis College of New York Institute of Technology; Marie Griffin, Arizona State University; and Bruce Bikle, California State University, Sacramento.

◿ Dedication

Mary Stohr: I would like to note that this book has presented the opportunity to work with two of the most scholarly and driven men I know: Tony Walsh and Craig Hemmens. Luckily they are also the most affable and understanding of colleagues, and I love them, respect them, and thank them for all of these inestimable qualities. I would also like to acknowledge Craig, as my husband, and our daughter Emily Rose Stohr-Gillmore, for their ceaseless love and support; I could do nothing well without it and them.

Anthony Walsh: I would like to dedicate this book to my ever-pleasant, ever-gorgeous wife, Grace Jean, for the love and support she has provided me over the years.

Craig Hemmens: I would like to dedicate this book, all the books in this series, and everything of value that I have ever done to my father, George Hemmens, who showed me the way; to James Marquart and Rolando Del Carmen, who taught me how; and to Mary and Emily, for giving me something I love even more than my work.

INTRODUCTION TO THE BOOK

The Philosophical and Ideological Underpinnings of Corrections

☒ What Is Corrections?

Corrections is a generic term covering a wide variety of functions carried out by government (and, increasingly, private) agencies having to do with the punishment, treatment, supervision, and management of individuals who have been accused of or convicted of criminal offenses. These functions are implemented in prisons, jails, and other secure institutions, as well as in community-based correctional agencies such as probation and parole departments. Corrections is also a field of academic study of the theories, missions, policies, systems, programs, and personnel that fall under the correctional rubric, as well as of the behaviors and experiences of those who make the institution of corrections an unfortunate necessity, its unwilling customers. As the term implies, the whole correctional enterprise exists to "correct," "amend," or "put right" the criminal behavior of its clientele. This is a difficult task as most of them are possessed by voracious appetites for

1

acquiring by force or fraud things that are not theirs, and many often behave as if sobriety is a difficult state for them to tolerate.

The more cynical among us might consider the phrase *correctional process* as a euphemism for what is more aptly termed the *punishment process*, and we would be right (Logan & Gaes, 1993). The correctional enterprise is preeminently about punishment, but, if something positive results from that punishment (such as cessation of criminal behavior), it is a bonus. Earlier scholars were more honest, calling what we now call *corrections* by the name *penology*, which means the study of punishment for crime. No matter what we call our prisons, jails, and other systems of formal social control, we are compelling people to do what they do not want to do, and such arm-twisting is experienced by them as punitive regardless of what we call it.

▲ **Photo I.1**   An interior view of Sing Sing Prison in Ossining, New York

▧ The Origins of Punishment

Ever since humans first devised rules of conduct they have wanted to break them. Human are not very good at obeying rules; the straight and narrow road does not come naturally. In the earliest days of our lives, parents chastise and scold us for doing things that are in our immediate interest to do—throwing temper tantrums, stealing others' property, hitting our siblings, biting the cat's tail, and so on. Later, teachers may paddle us, peers ostracize us, and employers fire us if we don't behave according to the rules. We have to learn to be good children and good citizens, and we only learn these lessons when we realize that our wants and needs are inextricably bound with the wants and needs of others. We also learn that our wants and needs are best realized through cooperation with others who want the same things. Punishment, then, is a form of social control used to try to achieve peace, harmony, and predictability in social relationships. Punishment by the state is formal social control exercised against those who have not learned to behave well via informal social control methods.

Can you imagine a society in which punishment did not exist? What would such a society be like; could it survive? If you cannot realistically imagine such a society, you are not alone, for the desire to punish those who have harmed us or otherwise cheated on the social contract is as old as the species itself. Punishment (referred to as *moralistic* or *retaliatory aggression*) aimed at discouraging cheats is also observed in every social species of animal, leading evolutionary biologists to conclude that punishment is an evolutionarily stable strategy designed by natural selection both for the emergence and the maintenance of cooperative behavior (Alcock, 1998; Clutton-Brock & Parker, 1996; Fehr & Gachter, 2002; Walsh, 2000). Imaging of the human brain via positron emission tomography (PET) and functional magnetic resonance imaging (fMRI) provides hard evidence that positive feelings accompany the punishment of those who have wronged us and that punishing others reduces the negative feelings evoked when we are wronged. Neuroimaging studies such as these show that, when subjects punished cheats, even at a cost to themselves, they had significantly increased blood flow to areas of the brain that respond to reward, suggesting that punishing those who have wronged us provides emotional relief and reward for the punisher (de Quervain, et al., 2004; Fehr & Gachter, 2002). These studies strongly imply that we are hard wired to "get even," as suggested by the popular saying "revenge is sweet."

Sociologists will note the similarity of the evolutionary argument with Emile Durkheim's (1893/1964) contention that crime and punishment are central to social life. Durkheim considers crime as normal in the sense that it exists in every society and that criminal behavior is in everyone's behavioral repertoire. Punishing criminals maintains solidarity, in part, because the rituals of punishment reaffirm the justness of the social norms, particularly those concerning cooperation among society's members. Punishment is functional because it defines the boundaries of acceptable behavior and allows citizens to express their moral outrage. Durkheim recognized the inborn nature of the punishment urge and that punishment serves an expiatory role, but he also recognized that we can temper the urge with sympathy. He observed that, over the course of social evolution, humankind had largely moved from retributive justice (characterized by cruel and vengeful punishments) to restitutive justice (characterized by reparation).

Repressive justice is driven by the natural passion for punitive revenge that "ceases only when exhausted . . . only after it has destroyed" (Durkheim, 1893/1964, p. 86). Durkheim goes on to claim that restitutive justice is driven by simple deterrence and is more humanistic and tolerant, although it is still "at least in part, a work of vengeance" because it is still "an expiation" (1893/1964, pp. 88–89). Both forms of justice satisfy the human urge for social regularity by punishing villains, but repressive justice oversteps its adaptive usefulness and becomes socially destructive. For Durkheim, restitutive responses to wrongdoers offer a balance between calming moral outrage, on the one hand, and exciting the emotions of empathy and sympathy, on the other.

◪ A Short History of Punishment

The earliest known written code of punishment is the Code of Hammurabi, created about 1780 BC. This code expressed the well-known concept of *lex talionis* (the law of equal retaliation), which is further enunciated in the Mosaic Code as "an eye for an eye, a tooth for a tooth . . . a life for a life." These laws codified the natural inclination of individuals harmed by another to seek revenge, but they also recognize that personal revenge must be restrained if society is not to be fractured by a cycle of tit-for-tat blood feuds. Thus, although the urge for vengeance is an *adaptation* (if we didn't retaliate against those who steal our bananas they would have no reason not to steal them again in the future), it is not *adaptive* in social groups much larger than the hunter-gatherer bands that characterized the social life throughout the vast majority of our evolutionary history. Blood feuds perpetuate and expand the injustice that "righteous" revenge was supposed to assuage. As Susan Jacoby (1983) put it,

> The struggle to contain revenge has been conducted at the highest level of moral and civic awareness at each stage in the development of civilization. The self-conscious nature of the effort is expectable in view of the persistent state of tension between uncontrolled vengeance as destroyer and controlled vengeance as an unavoidable component of justice. (p. 13)

"Controlled vengeance" is about the state taking responsibility for punishing wrongdoers from the individuals who were wronged. Nevertheless, early state-controlled punishment was typically as uncontrolled and vengeful as that which any grieving parents might inflict on the murderer of their child. Prior to the eighteenth century, all human beings were considered born sinners because of the Christian legacy of original sin. Cruel tortures used on criminals to literally "beat the devil out of them" were justified by the need to save sinners' souls. Earthly pain was temporary and certainly preferable to the eternity of torment awaiting sinners who died unrepentant. Punishment was often barbaric regardless of whether punishers bothered to justify it with such arguments and regardless of whether such justifications were believed.

The practice of brutal punishment began to wane in the 18th century with the beginning of a period historians call the *Enlightenment*, which was essentially a major shift in the way people began to view the world and their place in it. This new worldview questioned traditional religious and political values and began to embrace humanism, rationalism, and a belief in the primacy of the natural world. The Enlightenment also ushered in the beginnings of a belief in the dignity and worth of all individuals, a view that would eventually find expression in the law and in the treatment of criminal offenders.

Perhaps the first person to apply Enlightenment thinking to crime and punishment was English playwright, author, and judge Henry Fielding (1707–1754). Fielding's book, *Inquiry into the Causes of the Late Increase of Robbers* (1751/1967), provided his thoughts on the causes of robbery (which was being fueled by London's gin epidemic in much the same way as crack fueled robbery in American cities in the 1980s), called for a "safety net" for the poor (free housing and food) as a crime prevention strategy, and campaigned for alternative punishments to hanging. Many of his suggestions were implemented and were remarkably successful by most accounts (Sherman, 2005).

⬚ The Emergence of the Classical School

Enlightenment ideas eventually led to a school of penology that has come to be known as the *classical school*. More than a decade after Fielding's book, Italian nobleman and professor of law Cesare Bonesana, Marchese di Beccaria (1738–1794) published what was to become the manifesto for the reform of judicial and penal systems throughout Europe—*Dei Delitti e della Pene* (On Crimes and Punishment [1764/1963]). The book was an impassioned plea to humanize and rationalize the law and to make punishment just and reasonable. Beccaria did not question the need for punishment, but he believed that laws should be designed to preserve public safety and order, not to avenge crime. He also took issue with the common practice of secret accusations, arguing that such practices led to general deceit and alienation in society. He argued that accused persons should be able to confront their accusers, to know the charges brought against them, and to have the benefit of a public trial before an impartial judge as soon as possible after arrest and indictment.

If offenders were found guilty, punishment should be proportionate to the harm done to society, should be identical for identical crimes, and should be applied without reference to the social status of either the offender or the victim. Beccaria championed the abolition of the death penalty (not necessarily on humanitarian grounds, though, but because he felt that a life of penal servitude would be more of a deterrent), and he believed that punishments should only minimally exceed the level of damage done to society. Punishment, however, must be certain and swift to make a lasting impression on the criminal and to deter others. To ensure a rational and fair penal structure, society should decree in written criminal codes punishments for specific crimes and severely curtail the discretionary powers of judges. The judge's task was to determine guilt or innocence and then to impose the legislatively prescribed punishment if the accused was found guilty.

Beccaria's work was so influential that many of his recommended reforms were implemented in a number of European countries within his lifetime (Durant & Durant, 1967, p. 321). Such radical change over such a short time across many different cultures suggests that Beccaria's rational reform ideas tapped into and broadened the scope of emotions such as sympathy and empathy among the political and intellectual elite of Enlightenment Europe. Alexis de Tocqueville (1838/1956, Book III, Chapter 1) noticed the diffusion of these emotions across the social classes, beginning in the Enlightenment with the spreading of egalitarian attitudes, and he attributed the "mildness" of the American criminal justice system to the country's democratic spirit. We tend to feel empathy for those whom we view as being "like us," and empathy often leads to sympathy, which may translate the vicarious experiencing of the pains of others into an active concern for their welfare. A number of vignette studies have shown that people tend to recommend more lenient punishment for criminals whom they perceive to be similar to themselves (reviewed in Miller & Vidmar, 1981). With cognition and emotion gelled into the Enlightenment ideal of the basic unity of humanity, people began to expand their circle of others whom they saw as "like them" and their concept of social commonality and justice became both more refined and more diffuse (Walsh & Hemmens, 2000).

Another prominent figure was British lawyer and philosopher, Jeremy Bentham (1748–1832). His major work, *Principles of Morals and Legislation* (1789/1948), is essentially a philosophy of social control based on the principle of utility, which prescribes "the greatest happiness for the greatest number." The principle posits that a human action should be judged moral or immoral by its effect on the happiness of the community. The proper function of the legislature is thus to promulgate laws aimed at maximizing the pleasure and minimizing the pain of the largest number in society—"the greatest good for the greatest number" (Bentham, 1789/1948, p. 151).

If legislators are to legislate according to the principle of utility, they must understand human motivation, which for Bentham was easily summed up: "Nature has placed mankind under the governance of two sovereign masters, pain and pleasure. It is for them alone to point out what we ought to do, as well as to determine what we shall do" (Bentham, 1789/1948, p. 125). This was essentially the Enlightenment concept of human nature, which characterized individuals as hedonistic, rational, and endowed with free will. The classical explanation of criminal behavior, and how to prevent it, can be derived from these three assumptions about human nature.

Bentham devoted a great deal of energy (and his own money) to arguing for the development of prisons as punitive substitutes for torture, execution, or transportation. He designed a prison in the 1790s called the *panopticon* ("all seeing"), which was to be a circular "inspection house" enabling guards to constantly see their charges, thus requiring fewer staff. Because prisoners could always be seen without seeing who was watching or when they were being watched, the belief was that the perception of constant scrutiny would develop into self-monitoring. Bentham felt that prisoners could be put to useful work and thus pay for their own keep, with the hoped-for added benefit that they would acquire the habit of honest labor. Unfortunately, what was once considered humane and progressive is now best known in terms of Michael Foucault's (1991) pejorative "panopticism," which he defines as modern society's way of spying on, controlling, and disciplining all its members, not just its incarcerated criminals.

The final important figure of the era is English Christian activist and sheriff John Howard (1726–1790). Howard had been imprisoned by the French for a short period during one of the many Anglo-French wars, and he never forgot the experience. As sheriff of the county of Bedfordshire, he was able to tour and inspect many English gaols (jails) and prisons, and was highly critical of the wretched conditions found in them. His 1777 book *The State of Prisons in England and Wales, with an Account of some Foreign Prisons* shocked his readers with descriptions of the horrible conditions of English prisons and helped influence the British Parliament to pass penal reform legislation.

▲ **Photo I.2**   The Stateville Penitentiary was built along the principles of Jeremy Bentham's panopticon, a model for a prison in which the inmate would always be watched. This is the interior of Cell House C, with a guard tower in the center. Photographed in 1954.

The British government established new and better facilities, housing inmates in sanitary cells and providing them with adequate food and clothing under the Penitentiary Act of 1779.

Howard coined the term *penitentiary* and considered this institution a place of penitence and contemplation. He also, like Bentham, put great stock in hard labor as a way to reform criminals. Perhaps his greatest legacy is the Howard League for Penal Reform in Britain and the John Howard Society in Canada. According to an online pamphlet, the aim of the Howard League enunciated at its inception was the "promotion of the most efficient means of penal treatment and crime prevention" and of "a reformatory and radically preventive treatment of offenders."

⬚ The Emergence of Positivism: Should Punishment Fit the Offender or the Offense?

Just as classicism arose from the 18th century humanism of the Enlightenment, positivism arose a century later from the 19th century spirit of science. Classical thinkers were "armchair" philosophers in the manner of the thinkers of classical Greece (hence the term *classical*), while positivist thinkers took upon themselves the methods of empirical science, from which more "positive" conclusions could be drawn (hence the term *positivism*). An essential tenet of positivism is that human actions have causes and that these causes are to be found in the uniformities that typically precede those actions. The search for the causes of human behavior led positivists to dismiss the classical notion that humans are free agents who are alone responsible for their behavior.

Early positivism went to extremes to espouse a hard form of determinism (such as Cesare Lombroso's "born criminal"), which modern thinkers see as untenable. Nevertheless, positivism slowly moved the criminal justice system away from a singular concentration on the criminal act as the sole determinant of the type of punishment to be meted out and toward an appraisal of the characteristics and circumstances of the offender as an additional determinant. Because human actions have causes, many of which are involuntary, the concept of legal responsibility was called into question. For instance, Italian lawyer Raffael Garofalo (1885/1968) believed that, because human action is often evoked by circumstances beyond human control, the only thing to be considered at sentencing was the offenders' "peculiarities" or risk factors for crime.

Garofalo's only concern when recommending individualizing sentencing, however, was the danger offenders posed to society, and his proposed sentences ranged from execution for what he called the *extreme criminal* (whom we might call psychopaths today) to transportation to penal colonies for *impulsive* criminals to simply changing the law to deal with what he called *endemic* criminals (what we might call today those who commit some of the so-called victimless crimes). German criminal lawyer Franz von Liszt, on the other hand, campaigned for customized sentencing based on the rehabilitative potential of offenders, which was to be based on what scientists discovered about the causes of crime (Sherman, 2005). This ideal of sentences tailored to the characteristics of individuals meant that judges (or, for the early positivists, preferably behavioral scientists) were to enjoy wide sentencing discretion, which militated against another tenet of classical thinking—predetermined statutory sentences imposed on all who commit the same crime without any consideration at all for individual differences.

⊠ The Objectives of Punishment

Philosophers, legal scholars, and criminologists have traditionally identified four major objectives or justifications for the practice of punishing criminals: retribution, deterrence, rehabilitation, and incapacitation. Criminal justice scholars have recently added a fifth purpose to the list: reintegration. Before we discuss these objectives, we must emphasize that all theories and systems of punishment are underlain by conceptions of basic human nature. The view of human nature on which the law relies today is the same view enunciated by classical thinkers Becarria and Bentham, namely, that human beings are hedonistic, rational, and possessors of free will.

Hedonism is a doctrine that maintains that all life goals are desirable only as means to the end of achieving pleasure or avoiding pain. Pleasure is intrinsically desirable, pain is intrinsically undesirable, and we all seek to maximize the former and minimize the latter. We pursue these goals in rational ways, that is, in ways that are consistent with logic. People are said to behave rationally when we observe a logical "fit" between the goals they strive for and the means they use to achieve them. The goal of human activity is self-interest, and self-interest governs our behavior whether in prosocial or antisocial directions.

Hedonism and rationality are combined in the concept of the *hedonistic calculus,* a method by which individuals are assumed to weigh logically the anticipated benefits of a given course of action against its possible costs. If, on balance, the consequences of a contemplated action are thought to enhance pleasure or minimize pain, then individuals will pursue it; if not, they will not. If people miscalculate, as they frequently do, it is because they are ignorant of the full range of consequences of a given course of action, not because they are irrational.

The final assumption about human nature is that humans enjoy a free will, which enables them to purposely and deliberately choose to follow a calculated course of action. If people seek to increase their pleasures illegally, they do so freely and with full knowledge of the wrongness of their acts. Because criminals know what is right and what is wrong and choose the latter, society has a perfectly legitimate right to punish those who harm it.

Retribution

Retribution is the justification for punishment underlined by the concept of *lex talionis.* It is a "just deserts" model that demands that punishments match the degree of harm criminals have inflicted on their victims, i.e., what they justly deserve. Those who commit minor crimes deserve minor punishments, and those who commit more serious crimes deserve more serious punishments. This is perhaps the most honestly stated justification for punishment because it both taps into our most primal urges and posits no secondary purpose for it, such as the reform of the criminal. California is among the states that have explicitly embraced this justification in their criminal codes (California Penal Code Sec. 1170a): "The Legislature finds and declares that the purpose of imprisonment for a crime is punishment" (cited in Barker, 2006, p. 12). This model of punishment avers that, regardless of any secondary purpose punishment might serve, it is simply right to punish criminals because justice demands it.

Some scholars consider retribution to be nothing more than primitive revenge and therefore morally wrong (Tutu, 1999). However, retribution as presently conceived is not Durkheimian revenge "that ceases only when exhausted." Rather, it is constrained revenge curbed by proportionality and imposed by neutral parties bound by laws mandating respect for the rights of the individuals on whom it is imposed. Logan and Gaes (1993) go so far as to claim that only legal retributive punishment "is an affirmation of the autonomy, responsibility, and dignity of the individual" (p. 252). By holding offenders responsible and blameworthy for their actions, in other words, we are treating them as free moral agents, not as mindless rag dolls being blown hither and thither by the winds of malevolent forces in the environment.

Deterrence

A more complex justification for punishment is deterrence, i.e., the prevention of crime by the threat of punishment. The principle that people respond to incentives and are deterred by the threat of punishment is the philosophical foundation behind all systems of criminal law. Deterrence may be either specific or general.

Specific deterrence refers to the effect of punishment on the future behavior of persons who experience the punishment. For specific deterrence to work, it is necessary that a previously punished person make a conscious connection between an intended criminal act and the punishment suffered as a result of similar acts committed in the past. Unfortunately, such connections, if made, rarely have the desired effect, either because memories of the previous consequences are insufficiently potent or are discounted.

Committing further crimes after being punished for one is called *recidivism* ("falling back" into criminal behavior), which is a lot more common among ex-inmates than rehabilitation. Nationwide, about 33% of released prisoners recidivate within the first six months after release, 44% within the first year, 54% by the second year, and 67.5% by the third year (Robinson, 2005, p. 222), and these are just the ones who are caught. Offenders matched for seriousness of crime who are placed on probation, however, appear to recidivate at significantly lower rates (Austin & Irwin, 2001). Nevertheless, among ex-inmates who do desist, many cite the fear of additional punishment as a major factor (reviewed in Wright, 1999).

As the classical scholars of criminology remind us, the effect of punishment on future behavior depends on its certainty, celerity (swiftness), and severity. In other words, there must be a relatively high degree of certainty that punishment will follow a criminal act, the punishment must be administered very soon after the act, and it must be quite harsh. As we see from Figure 1.1, the probability of getting caught is very low, especially for property crimes—so much for certainty. Factoring out the immorality of the enterprise, burglary, for instance, appears to be a very rational career option for the capable criminal.

If a person is caught, the wheels of justice grind excruciatingly slowly, with many months passing between the act and the imposition of punishment—so much for celerity. This leaves the law with severity as the only element it can realistically manipulate (it can increase or decrease statutory penalties almost at will), but severity of punishment is, unfortunately, the least effective element when it comes to deterrence (National Center for Policy Analysis, 1998a). Studies from the United States and the United Kingdom find substantial negative correlations (as one factor goes up the other goes down) between the likelihood of conviction (a measure of certainty) and crime rates but much weaker ones (albeit in the same direction) between the severity of punishment and crime rates (Langan & Farrington, 1998).

The effect of punishment on future behavior also depends on the *contrast effect*, which is the distinction between the circumstances of the possible punishment and the

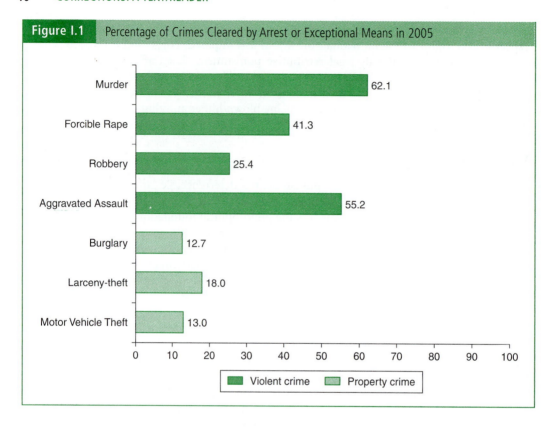

Figure I.1 Percentage of Crimes Cleared by Arrest or Exceptional Means in 2005

usual life experience of the person who may be punished. For people with little or nothing to lose, an arrest may be perceived as little more than an inconvenient occupational hazard, an opportunity for a little rest and recreation, and a chance to renew old friendships, but for those who enjoy a loving family and the security of a valued career, the prospect of incarceration is a nightmarish contrast. Like so many other things in life, deterrence works least for those who need it the most (Austin & Irwin, 2001).

General deterrence refers to the preventive effect of the threat of punishment on the general population; it is thus aimed at *potential* offenders. Punishing offenders serves as an example to the rest of us of what might happen if we violate the law. As Radzinowicz and King put it, "People are not sent to prison primarily for their own good, or even in the hope that they will be cured of crime. . . . It is used as a warning and deterrent to others" (1979, p. 296). The existence of a system of punishment for law violators deters a large but unknown number of individuals who might commit crimes if no such system existed.

What is the bottom line on the effectiveness of deterrence? Are we putting too much faith in the ability of criminals and would-be criminals to calculate the cost-benefit ratio of engaging in crime? Although many violent crimes are committed in the heat of passion, there is quite a bit of evidence underscoring the notion that individuals do (subconsciously at least) calculate the ratio of expected pleasures to possible pains when contemplating a course of action. Nobel Prize–winning economist Gary Becker (1997) is a major adherent of the position. He dismisses the idea that criminals lack the knowledge and the foresight to take punitive probabilities into consideration when deciding whether or not to continue committing crimes. He says that "Interviews of young people in high crime areas who do engage in crime show an amazing understanding of what punishments are, what young people can get away with, how to behave when going before a judge" (p. 20). Becker also compared crime rates in Great Britain and the United States and demonstrated that crime rates rose in the former as its penal philosophy became more and more lenient and fell in the United States as its penal philosophy became more and more punitive.

More general reviews of deterrence research indicate that legal sanctions do have "substantial deterrent effect" (Nagin, 1998, p. 16; see also Wright, 1999), and some researchers have claimed that increased incarceration rates account for about 25% of the variance in the decline in violent crime over the last decade or so (Spelman, 2000; Rosenfeld, 2000). Of course, this leaves 75% of the variance to be explained by other factors, such as an improved economy. Unfortunately, even for the 25% figure, we cannot determine whether we are witnessing a *deterrent* effect (has violent crime declined because more would-be violent people have perceived a greater punitive threat?) or an *incapacitation* effect (has violent crime declined because more violent people are behind bars and thus not at liberty to commit violent crimes on the outside?).

Incapacitation

Incapacitation refers to the inability of criminals to victimize people outside prison walls while they are locked up. Its rationale is aptly summarized in James Q. Wilson's (1975) trenchant remark: "Wicked people exist. Nothing avails except to set them apart from innocent people" (p. 391). The incapacitation justification probably originated with Enrico Ferri's concept of *social defense.* Ferri was one of the early positivists who dismissed the classical ideas about human nature as myths. To determine punishment, he argued, we must subordinate notions of culpability, moral responsibility, and intent to an assessment of offenders' strength of resistance to criminal impulses, with the express purpose of averting future danger to society. He believed that moral insensibility and lack of foresight, underscored by low intelligence, were the criminal's most marked characteristics: The criminal has "defective resistance to criminal tendencies and temptations, due to that ill-balanced impulsiveness which characterizes children and savages" (Ferri, 1897/1917, p. 11).

Ferri's social defense asserts that the purpose of punishment is not to deter or to rehabilitate but to defend society from criminal predation. Ferri reasoned that the characteristics of criminals prevented them from basing their behavior on rational calculus principles. So how could their behavior be deterred, and how could born criminals be rehabilitated? Given the assumptions of early biological positivism, the only reasonable rationale for punishing offenders was to incapacitate them for as long as possible so that they no longer posed a threat to the peace and security of society.

It probably goes without saying that incapacitation "works," at least while criminals are incarcerated. Elliot Currie (1999) uses robbery rates to illustrate this point. He states that in 1995 there were 135,000 inmates in state and federal institutions whose most serious crime was robbery and that each robber, on average, commits five robberies per year. Had these robbers been left on the streets, they would have been responsible for an additional 135,000 × 5 or 675,000 robberies on top of the 580,000 actual robberies reported to the police in 1995. Similarly (Wright, 1999) estimated that imprisonment averted almost seven million offenses in 1990. The incapacitation effect is more starkly driven home by a study of the offenses of 39 convicted murderers. This study examined offenses committed *after* these 39 men had served their time for murder and were released from prison. It was found that they had 122 arrests for serious violent crimes (including additional murders), 218 arrests for serious property crimes, and 863 "other" arrests between them (DeLisi, 2005, p. 165). Had these men remained behind bars, much social harm in the form of pain, suffering, and economic loss would have been averted.

Rehabilitation

The term *rehabilitation* means to restore or return to constructive or healthy activity. Whereas deterrence and incapacitation are primarily justified philosophically on classical grounds, rehabilitation is primarily a positivist concept. The rehabilitative goal is based on a medical model that used to view criminal behavior as a moral sickness requiring treatment. More recently, the model views criminality as "faulty thinking" and criminals as in need of programming rather than treatment. Although the goal of rehabilitation is the same as that of deterrence, rehabilitation aims to achieve this goal by changing offenders' attitudes so that they come to accept that their behavior was wrong rather than by deterring crime using the threat of further punishment. The difficulty with this justification of punishment is that it is asking convicted criminals to return to a state most of them have never experienced, the state of being fit to function in society (habilitation). Habilitation basically means to change oneself, and, used in this sense, it means to change oneself into a law-abiding person. One might argue that to attempt to habilitate people while they are in prison is like attempting to dry out alcoholics by locking them up in a brewery. Consequently, most such efforts are conducted in community settings (probation or parole). Because this justification for punishment has its own section in this book devoted to it alone, it will not be discussed further now.

Reintegration

The goal of reintegration is to use the time criminals are under correctional supervision, either in institutions or in the community, to prepare them to reenter the free community as well equipped to do so as possible. This goal is also known as *reentry* or *restoration*. In effect, reintegration is not much different from rehabilitation, but it is more pragmatic, focusing on concrete programs such as job training rather than attitude change. There are many challenges associated with this process, so much so that, like rehabilitation, it warrants a section to itself and will be discussed in much more detail there.

Table 1.1 is a summary of the key elements (e.g., justification and strategy) of the five punishment philosophies or perspectives discussed. The commonality that they all share to various extents is, of course, the prevention of crime.

⊠ Is the United States Soft on Crime?

One of the most frequently heard criticisms of the criminal justice system in the United States is that it is soft on crime. If we define hardness or softness in terms of incarceration

Table I.1	Summary of Key Elements of Different Punishment Perspectives				
	Retribution	**Deterrence**	**Incapacitation**	**Rehabilitation**	**Reintegration**
Justification	Moral	Prevention of further crime	Risk control	Offenders have deficiencies	Offenders have deficiencies
Strategy	None: Offenders simply deserve to be punished	Make punishment more certain, swift, and severe	Offenders cannot offend while in prison Reduce opportunity	Treatment to reduce offenders' inclination to re-offend	Concrete programming to make for successful reentry into society
Focus of Perspective	The offense and just deserts	Actual and potential offenders	Actual offenders	Needs of offenders	Needs of offenders
Image of Offenders	Free agents whose humanity we affirm by holding them accountable	Rational beings who engage in cost/benefit calculations	Not to be trusted but to be constrained	Good people who have gone astray Will respond to treatment	Ordinary folk who require and will respond to concrete help

rates, Figure I.2 showing the comparative incarceration rates for selected countries in 2004 conveys the opposite message, as does the U.S. retention of the death penalty, which has been eschewed by other "civilized" nations. Only the Russian incarceration rate comes close to that of the United States, and, when we consider the rates of only modern Western nations, that of England and Wales, which is five times lower than the U.S. rate, is the closest.

However, if we define hardness/softness in terms of alternative punishments or the conditions of confinement, then the "soft" (humane?) criticism is valid. For instance, although China is listed as having an incarceration rate more than five times lower than that of the United States, it is the world's leader in the proportion of its criminals it executes each year. The death penalty may be applied for nearly 70 different offenses, including murder, rape, economic crimes committed by high-level officials, and "hooliganism." According to Amnesty International, there were at least 3,400 confirmed executions in China in 2004, but sources inside China put the true figure at around 10,000 (Amnesty International, 2005). Adjusting for population size differences, the official figure (3,400) is still over 13 times more than the number of executions that occurred in the United States in the same year. Additionally, punishment in many Islamic countries often includes barbaric corporal punishments for offenses considered relatively minor in the West (e.g., petty theft) or for acts no longer criminalized in the West, such as drinking alcohol or homosexuality (Walsh & Hemmens, 2000).

Another problem with assessing the hardness or softness of the American criminal justice system based on incarceration rates is that these are calculated per 100,000 *citizens* not per 100,000 *criminals*. If the United States has more criminals than these other countries, then perhaps the greater incarceration rate is justified. Of course, no one knows how many crooks any country has, but we can get a rough estimate from a country's crime rates. For instance, the U.S. homicide rate is about five times that of England and Wales,

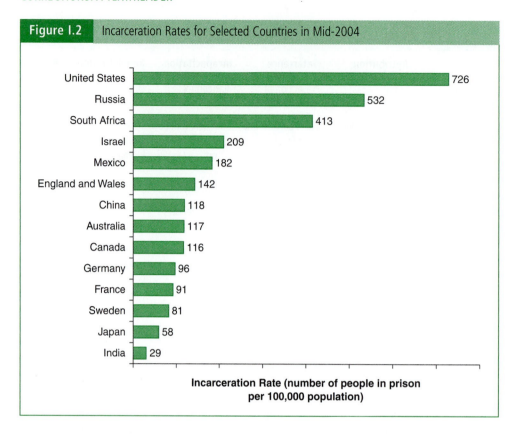

Figure I.2   Incarceration Rates for Selected Countries in Mid-2004

Country	Rate
United States	726
Russia	532
South Africa	413
Israel	209
Mexico	182
England and Wales	142
China	118
Australia	117
Canada	116
Germany	96
France	91
Sweden	81
Japan	58
India	29

Incarceration Rate (number of people in prison per 100,000 population)

which roughly matches its five times greater incarceration rate. However, when it comes to property crimes, Americans are in about the middle of the pack of nations in terms of the probability of being victimized (less than in England and Wales, incidentally). This fact notwithstanding, burglars serve an average of 16.2 months in prison in the United States compared with 6.8 months in Britain and 5.3 months in Canada (Mauer, 2005), which makes the United States harder on crime than its closest cultural relatives and suggests that we may be overusing incarceration to address our crime problem. So is the United States softer or harder on crime than other countries? The answer obviously depends on how we conceptualize and measure the concepts of hardness and softness and with which countries we compare ourselves with.

⊠ Summary

- Corrections is a social function designed to punish, supervise, deter, and possibly rehabilitate criminals. It is also the study of these functions. The urge to punish those who have wronged us is probably an evolutionary adaptation because it is observed universally, at all times, and in all social species. Neuroimaging studies have shown that, when subjects punish wrongdoers, their brains register rewarding effects.

- Although it is natural to want to exact revenge when people mistreat or steal from us, it has long been recognized that to allow individuals to pursue this goal is to invite a series of tit-for-tat feuds that may fracture a community. The state has thus taken over responsibility for punishment, although it was often just as passionately barbaric in its form of punishment as any wronged individual might be. Over the course of social evolution, however, the state has moved to more restitutive forms of punishment, which serve to assuage the community's moral outrage but temper it with sympathy.

- Much of the credit for the shift away from retributive punishment must go to the great classical thinkers such as Fielding, Beccaria, Bentham, and Howard, all of whom were imbued with the humanistic spirit of the Enlightenment period. The concept of the individual (hedonistic, rational, and possessing free will) held by these men led them to view punishment as primarily for deterrent purposes, as, ideally, only just exceeding the "pleasure" (gains) of crime, and as applicable equally to all who have committed the same crime regardless of any individual differences.

- Opposing classical notions of punishment are those of the positivists who rose to prominence during the 19th century and who were influenced by the spirit of science. Positivists rejected the philosophical underpinnings regarding human nature of the classicists and declared that punishment should fit the offender rather than the crime.

- The objectives of punishment are retribution, deterrence, incapacitation, rehabilitation, and reintegration, all of which have come and gone out of favor, and come back again, over the years.

- Retribution is simply just deserts—getting the punishment you deserve, no other justification needed.

- Deterrence is the assumption that people are prevented from committing crime by the threat of punishment.

- Incapacitation means that criminals who are incarcerated cannot commit further crimes against the innocent.

- Rehabilitation centers on efforts to socialize offenders in prosocial directions while they are under correctional supervision so that they won't commit further crimes.

- Reintegration refers to efforts to provide offenders with concrete skills they can use that will provide them with a stake in conformity.

- The United States leads the world in the proportion of its citizens that it has in prison. Whether this is indicative of hardness (more time for more people) or softness (imprisonment as an alternative to execution or mutilation) depends on how we view hardness versus softness and with which countries we compare the United States.

KEY TERMS

Classical explanation of criminal behavior	Hedonistic calculus
Corrections	Penitentiary
Deterrence	Positivism
Enlightenment	Punishment
Hedonism	Repressive justice

Restitutive justice

Retribution

Specific deterrence

General deterrence

Incapacitation

Rehabilitation

Reintegration

DISCUSSION QUESTIONS

1. Discuss the implications for a society that decides to eliminate all sorts of punishment in favor of forgiveness.

2. Why do we take pleasure in the punishment of wrongdoers? Is it a good or bad thing that we take pleasure in punishment, and what evolutionary or social purpose might it serve?

3. Discuss the assumptions about human nature held by the classical thinkers. Are we rational, seekers of pleasure, and free moral agents? If so, does it make sense to try to rehabilitate criminals? What about punishment? Is this always carried out in a rational manner?

4. Discuss the assumptions underlying positivism in terms of the treatment of offenders. Do they support Garofalo or von Liszt in terms of the meaning these assumptions have for punishment?

5. Which justification for punishment do you favor? Is it the one that you think "works" best in terms of preventing crime, or do you favor it because it fits your ideology?

6. What is your position on the hardness or softness of the U. S. stance on crime? We are tougher than other democracies; is that OK with you? We are also softer than more authoritarian countries; is that OK with you?

INTERNET SITES

American Correctional Association: www.aca.org

American Jail Association: www.aja.org

American Probation and Parole Association: www.appa-net.org

Bureau of Justice Statistics (information available on all manner of criminal justice topics): www.ojp.usdoj.gov/bjs

National Criminal Justice Reference Service: www.ncjrs.gov

Office of Justice Research (information available on all manner of criminal justice topics, specifically probation and parole here): www.ojp.usdoj.gov/bjs/pub/pdf/ppus05.pdf

Pew Charitable Trust (Corrections and Public Safety): www.pewtrusts.org/our_work_category.aspx?id=72

Vera Institute of Justice (information available on a number of corrections and other justice related topics): www.vera.org

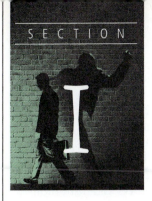

HISTORY OF AMERICAN CORRECTIONS

Introduction: Not Much Is New Under the Corrections Sun

The history of corrections is riddled with the best of intentions and the worst of abuses. Correctional techniques and facilities (e.g., galley slavery, transportation, jails and prisons, community corrections) were created, in part, to remove the riffraff—both poor and criminal—from urban streets or at least to control and shape them. Prisons and community corrections were also created to avert the use of more violent or coercive responses to such folk. In this section, the focus is on the recurring themes that run through and define the American experience with corrections.

It is somewhat ironic that one of the best early analyses of American prisons and jails was completed by two French visitors to our country—Gustave de Beaumont and Alexis de Tocqueville—while our country was in its relative infancy in 1831 and at the virtual birthing of prisons themselves (Beaumont & Tocqueville, 1833/1964). These two astute French observers came to American shores to investigate the relatively young democracy and its newly minted prisons, among many other social phenomena. One hundred seventy-three years later, Norman Johnston (article included here) traces the history of one of the prisons the French visitors wrote about: the Eastern State Penitentiary. The irony is that, as outsiders and social critics, Beaumont and Tocqueville could so clearly see what others (namely, Americans who were thought to have "invented prisons" and who worked in them) were blind to. Leonard Orland (1995) and Johnston, writing from a twentieth- and a twenty-first-century perspective, respectively, rather than a continental one, have the knowledge that comes from temporal distance, which allows them to see the context for such a social reform—where correctional practices and institutions, as social derivatives, came from, and what they represented—throughout the centuries.

Nichole Hahn Rafter (article included here), in a chapter from her classic work on women's imprisonment, provides yet another "take" on the nature of incarceration in some of our historic prisons in New York, Ohio, and Tennessee. Women have always represented only a small fraction of the correctional population in either prisons or jails, and their experience with incarceration, as shaped by societal expectations of and for them, can be wholly different from that of men. Some of the themes that run through the practice of corrections apply to the women and girls as well, but with a twist. As literal outsiders to what was the "norm" for inmates of prisons and jails and as a group whose rights and abilities were socially controlled on the outside more than those of men and boys, women prisoners and the female experience in corrections history are worth studying.

What is clear from all of these writings is that what was *intended* when prisons, jails, and reformatories were conceived and how they *actually operated* then and more recently are often two very different things (Rothman, 1980; Welch, 2005). So it was in these early facilities, and so it is now. As social critics ourselves, we can use these writings on the history of corrections to identify a series of "themes" that will run through this entire book. Such themes will reinforce the tried, yet true, maxim that "Those who do not know their history are doomed to repeat it" (Santayana, 1905, p. 284). Too often, we do not know or understand our history of corrections, and, as a consequence, we are forever repeating it.

In this brief introductory chapter, and as a fitting prelude to the works of the authors, we will identify some themes that appear in their and others' writings on correctional institutions, programs, and practices, themes that are illustrative of where we have been and where we are likely to go.

⬛ Themes: Truths That Underlie Correctional Practice

Some themes have been almost eerily constant, vis-à-vis corrections, over decades and even centuries. Of these, a few are obvious, such as the influence that money or its lack exerts over virtually all correctional policy decisions. Other themes are less apparent but no less potent in their effect on correctional operations. For instance, there appears to be an evolving sense of compassion or humanity that, though not always clear in the short term or in practice, policy, or statute, has underpinned reform-based decisions about corrections, at least in theory, throughout its history in this country. The creation of the prison, with a philosophy of penitence (hence penitentiary) along with retribution, was a grand reform itself, and, as such, it represented, in theory at least, a major improvement over the brutality of punishment that characterized early English law and practice (Orland, 1995). Some social critics do note, however, that the prison and the expanded use of other such social institutions also served as "social control" mechanisms to remove punishment from public view while making the state appear more just (Foucault, 1979; Welch, 2005). Therefore, this is not to argue that such grand reforms in their idealistic form, such as prisons, were not constructed primarily out of the need to control but rather that there were philanthropic, religious, and other forces aligned that also influenced their creation and design, if not so much their operation (Hirsch, 1992). Also of note, the social control function becomes most apparent when less powerful populations such as the poor, minorities, or women are involved, as will be discussed in the following chapters.

Other than the influence of money and a growing sense of compassion or humanity in correctional operation, the following themes are also apparent in our history: how to use labor and technology (which are hard to decouple from monetary considerations); a decided religious influence; the intersection of class, race, and gender in shaping one's experience in corrections; architecture as it is intermingled with supervision; methods of control; overcrowding; and, finally, the fact that good intentions do not always translate into effective practice. Though far from exhaustive, this list contains some of the most salient issues that become apparent streams of influence as one reviews the history of corrections. Some of the larger philosophical issues, such as conceptions of right and wrong and whether it is best to engage in retribution or rehabilitation using correctional sanctions (or both or neither or incapacitation, deterrence, and reintegration), are also obviously associated with correctional change and operation. The authors deal with these broader matters either directly or indirectly as they underpin many of the identified themes.

Money, Money, Money, Labor, and Technology

When Beaumont and Tocqueville arrived in the United States, the country was literally in its first prison-building boom. We built it, and they came. But what they "came" to, in terms of prisons and jails, was very different from state to state and locality to locality. As our French documenters remark, each state, each city and county in each state had, and still has, its own laws and practices that determine how their local prison or jail is operated. From the Frenchmen's perspective, this lack of unity of ideas and implementation was both an advantage for Americans and a curse. It was an advantage in that it led to the fermentation of new ideas and the ultimate experimentation with some of them, and a curse in that it led to a lack of uniformity such that some states and municipalities were becoming more humane in their correctional practices while, virtually "next door," other states and municipalities were engaged in what the visitors termed the "ancient abuses."

What shaped these differences in part was money, or a lack of it, for such luxuries as jails or prisons. Rather than punishing offenders with the whip or stocks or even executions, states, large cities, and counties (in the case of jails) were faced with building, maintaining, and staffing correctional institutions with public funds (Orland, 1995; see also Rafter reading). As Beaumont and Tocqueville remarked, such funding, if not supplemented with some sort of income from inmate labor, could be ruinous to the treasury!

The cost of building and operating prisons and jails was, and is, prohibitively expensive (see readings by Johnston and by Rafter; Irwin & Austin, 1994; Rothman, 1980). Part of this expense, then and now, is for technological wonders, for example, the plumbing and heating systems built into early facilities and our surveillance equipment of today. But just raising the capital to build and maintain a facility takes a commitment of public capital and lots of it. For instance, as indicated in Table 1.1, the Georgia Department of Corrections (2007) estimates that it costs $31,675 to build a minimum security bed and then an additional $14,016 to operate it. Notably, those costs increase if the bed is in max-imum security and decrease substantially if the offender is on probation. When compared to the prison expenditures of other states, say Washington, the Georgia costs are cheap. The Washington State Department of Corrections estimates it costs on average $26,736 per offender per year to institutionalize an inmate or about $7,000 more than the average for maximum security inmates in Georgia. But then Southern states have some of the lowest incarceration and supervision costs (though they have the highest regional incar-ceration rates) as compared to other regions of the country, with the Northeast having the highest incarceration and supervision costs (though they have the lowest regional incar-ceration rates) (JFA Institute, 2007). Notably, these costs are averages for all adults, and, typically, costs are much higher, because of decreased economies of scale and different needs, for adult women and for juveniles.

In contrast to current costs to incarcerate, correctional costs in the past were miti-gated by the labor of inmates. Galley slavery, transportation, and the earliest jails and pris-ons (the Auburn, New York model) all used labor as an intrinsic part of the punishment. Those who labored at the oars, though they might die there from abuse and lack of clean food or water, provided the labor necessary for the ship's power before the technology of sails rendered such work unnecessary (Orland, 1995; Welch, 2004). Transportation to the new colonies, a practice in use for over 400 years, also had the double benefit of remov-ing the "criminal class" from English or French streets and providing the essential labor-ers needed for settlements (Feeley, 1991; Orland, 1995). Early bridewells and workhouses and prison-like structures (e.g., the Newgate Connecticut prison that was a copper mine) also confined the poor, the criminal, and the displaced (from the rural to urban areas of Europe and the colonies) in places where, if needed, their labor might be exploited (Phelps, 1996).

Moreover, as labor was formally restricted in some early prisons, maintenance could not be done "officially" by inmates (e.g., in the early Pennsylvania prisons). This was

Table 1.1	Costs of Corrections in Georgia (2006)	
		Capital Outlay ($ per unit or prisoner)
Initial cost		
Minimum security bed		$31,675
Medium security bed		$60,700
Maximum security bed		$109,400
Operating costs		
Maximum security bed		$18,582
Medium and minimum security beds		$14,016
Prerelease, parole, and probation work and treatment centers		$12,000–21,582
Regular probation		$475
Intensive probation		$1,241

another costly aspect of the prison. Also, the labor that could be produced in the individual cells of the Pennsylvania prisons did not compare to the output of the factory-like prisons (e.g., New York's Auburn and Sing Sing prisons) that other states were developing in the 19th century (Rothman, 1980). It is for these reasons that the Pennsylvania system was not much copied inside the United States, and, when it was reproduced in other countries, individual cells were unlikely to include their own exercise yards. It is also, perhaps, that jails and prisons have almost always used inmate labor to supplement staff and sometimes in lieu of staff (Marquart & Roebuck, 1985). However, it is worth noting that the labor of women in our early prisons, as documented by Rafter, was even more restricted than men's because of sex role expectations but also because women were more likely to be left unsupervised and secluded so that they could not perform labor with the male inmates.

One of the reasons that the Auburn and Sing Sing model of prisons spread was the congregate labor systems these prisons employed. Not only did prison administrators appreciate the remedial effects of labor, as did the Pennsylvania Prison Society, but they noted that it reduced expenses for the prisons. Maintenance of the prison could be done by inmates, and congregate work, albeit in silence, as prisoners were not allowed to communicate with one another in Auburn and Sing Sing, allowed the inmates to produce more goods for sale. Of course, the factory prisons of the North in the late 1800s and early 1900s and the convict lease system of the South during that same period survived because the cost of incarceration could be offset almost entirely by inmate labor (Pollock, 2004; Seiter, 2002). In the factory prisons, inmates labored together to produce goods, sometimes for the state and at other times for private contractors. Under the convict contract or lease system, inmates' labor was sold by the prison to farmers or other contractors. As the supply of prisoners, mostly in the form of ex-slaves, was plentiful during the post–Civil War period, their lives were treated cheaply, and they were often not fed or clothed or sheltered adequately.

Another related example of the importance of money and labor in shaping corrections would be the plantation prisons of the post–Civil War period, which acted as proxies for slavery by keeping poor ex-slaves essentially enslaved (Oshinsky, 1996). It is not surprising, then, that we see the same thing today with offenders sentenced to probation or work release, who must refund monies to the state from their wages from outside work to pay for their housing and supervision (Georgia Department of Corrections, 2007). Although one cannot help but notice that a stay in a work release or in most prisons or jails these days is, by all accounts, much less physically trying than it once was, the scale of our current use of corrections (expanding caseloads and populations) is very troublesome (JFA Institute, 2007).

In sum, despite guards' use of the lash to maintain control,

▲ **Photo 1.1**   Prison cells in Alcatraz, San Francisco, California.

Beaumont and Tocqueville found early American prisons to be vast improvements over the jails that preceded them. As Kerle (2003) notes in his book on jails, colonial jails charged inmates fees for all life's necessities while incarcerating inmates without regard to gender, age, or type of crime. So the early prisons may have represented some improvement, but perhaps only marginally in some cases, and when those prisons were operated like the southern lease system, not at all.

A Greater Compassion or Humanity in Correctional Operation

Appearing in some articles in this and other sections of this book is a sense that the development of jails and then prisons and probation and parole encompassed major reforms spurred by a need to clean up the corrupted and corruptive influences of corrections and punishment as they then existed. For instance, John Howard, an English ex-sheriff and an advocate for jail reform from the 1770s to the 1790s, called for the elimination of the fee systems in jails, whereby inmates paid to be incarcerated; argued for the provision of decent food and clothing for inmates; and lobbied for the frequent inspection of jails so that they would provide decent facilities and work to reform their inmates (Krebs, 1978; Orland, 1995; Radzinowicz, 1978; see also the discussion of Howard in the introduction to this book). Such a call for reform must be viewed in light of the efforts by other Enlightenment-influenced leaders and philosophers, such as William Penn, a Quaker who founded the Pennsylvania colony and worked to eliminate harsh punishments by substituting imprisonment for whippings and executions (Johnston). English philosopher Jeremy Bentham (1811/1930), and Italian philosopher Cesare Beccaria (1764/1963), who were the great eighteenth-century advocates of deterrence theory and who also argued for less severe punishment for more minor offenses, were also very influential (see also the discussion of Bentham and Beccaria in the introduction to this book). In fact, Orland (1995, p. 10) argues that the combination of the Enlightenment belief in "right reason" (or that some humans are rational and not innately depraved) and the need to deal with burgeoning urban populations led to the creation of prisons and the need for more jails.

Prisons, such as the Eastern Pennsylvania Prison, the Western Pennsylvania Prison that followed it, and Auburn and Sing Sing, were created to "alleviate the miseries" of offenders. As Conover (2001) comments in his book on Sing Sing, "The Quakers' goals were prevention of further harm to society, deterrence, and, by the early nineteenth century, encouragement of prisoners to engage in 'penitent reflection,' which could result in their personal reformation" (p. 173). Eastern, Johnston argues, was explicitly created by its founders, who were influenced by Enlightenment thinkers and the Quakers, to rehabilitate inmates and, absent that, to at least deter them from further criminality. Although the goal for Auburn and Sing Sing, where physical punishments were inextricably tied to operation and control, was decidedly not rehabilitation, those prisons certainly provided a much safer and more decent incarceration experience than had the houses of correction that came before them or that existed in other states and municipalities.

Since these early prisons, there have been multiple waves of reform aimed at reducing the miseries prisoners suffer while incarcerated. Probation, as it was originally practiced by its founder John Augustus, was just such a philanthropic endeavor (Augustus, 1852/1972). As Augustus himself recounts, his volunteer work as a probation officer began in 1841, when he was 57. He bailed out a drunkard before the Boston courts and then proceeded to help him. Until his death at 75, Augustus followed this procedure with the other 5,000 or so men, women, and children whom he subsequently bailed out and saved from jail or prison time. But Augustus's work did not extend just to bailing out offenders; he would literally take these mostly minor offenders and help them find employment, housing, and aid for their families. He was known to house several probationers in his own home with his family over the years when he could not find alternate housing.

As the Harvard professor Sheldon Glueck wrote in his introduction to the 1939 (xxiii) reprinted version of *A Report of the Labors of John Augustus, for the Last Ten Years, In Aid of the Unfortunate*, "These rehabilitative efforts by John Augustus were inspired not only by a strong humanitarian impulse but by a definite view that 'the object of the law is to reform criminals, and to prevent crime and not to punish maliciously, or from a spirit of revenge'" (p. 23). In other words, Augustus and reformers like him were compelled to act and to change the criminal justice system out of an impulse to improve it and the lot of those less fortunate.

Efforts to reform the conditions of women in male prisons or the women's sections of early male prisons also appear to have been inspired by some reformers with humane impulses. For instance, the "ladies" that Rafter mentions who visited women incarcerated at the Eastern Pennsylvania Prison, and the countless other men and women who decried the conditions for men and women in America's early prisons (Dorothea Dix is mentioned often by Rafter, for instance) would appear to have been compelled to do good by a genuine desire to reform not just the institutions but the people they held.

The first major prison reform occurred approximately 50 years after the first New York and Pennsylvania prisons were built and was doubtless the result of all of those calls for change: The 1870 American Prison Congress was held in Cincinnati, Ohio with the express purpose of trying to recapture some of the idealism promised with the creation of prisons (Rothman, 1980). Despite their promises of reform and attempts at preventing "contamination," the early prisons had become, by the 1860s, warehouses without hope or resources. All of the themes mentioned in this section, save the desire for reform and that was remedied with the next round of reforms to follow the Congress, applied to the operation of the 19th century prisons: they were overcrowded, underfunded, brutal facilities where people would spend time doing little that was productive or likely to prepare them to reintegrate into the larger community.

Appropriately enough, then, the Declaration of Principles that emanated from the American Prison Congress was nothing short of revolutionary at the

▲ **Photo 1.2** Prison cells at Elmira Reformatory. Photographed in 1955.

time and provided a blueprint for the prisons we see today (Rothman, 1980). Some of those principles were concerned with the grand purposes of prisons—to achieve reform—while others were related to their daily operations, for example, training of staff, eliminating contract labor, and the treatment of the insane (American Correctional Association, 1983). As a result of these principles, a spirit of reform in corrections was again energized, and the Elmira Reformatory was founded in 1876 (Rothman, 1980). The reformatory would encompass all of the rehabilitation focus and graduated reward system (termed the *marks system,* as in you behave and you earn marks that entitle you to privileges) promoted by reformers, along with trained staff and uncrowded facilities. Unfortunately, and as before, this attempt at reform was thwarted when the funding was not forthcoming and the inmates did not conform as expected. The staff soon resorted to violence to keep control. It should not be forgotten, however, that, even on its worst day, the Elmira prison was likely no worse than the old Auburn and Sing Sing prisons and probably much more humane.

Similar sorts of reforms followed with the creation of probation and parole in the latter part of the 19th century and the early 20th century. The idea here, too, was to reduce the use of incarceration and to help the offender to transition more smoothly back into the community. Doubtless, the intent was good, and, although the execution of this reform was less than satisfactory, it did represent an improvement over the correctional practices that preceded it (Rothman, 1980).

The Intersection of Class, Race, and Gender

Some of the earliest descriptions of criminal law and depictions of correctional practice make clear that *who one was* demographically (class, race or ethnicity, gender) determined to a large extent *how one was handled* (Orland, 1995; Oshinsky, 1996; Reiman, 1998). Throughout history, in the earliest of legal codes (e.g., the Code of Hammurabi or the Justinian Code), in the English "Black Laws" (punishing with death the killing of deer in the king's forest or even the "blacking" of one's face with mud or charcoal to go into that forest), and in the differential treatment of those who enter the criminal justice system today, if one is rich, of the powerful race or ethnicity, and male, one is treated substantially differently than one would be if these were not accurate descriptors (Orland, 1995; Oshinsky,1996; Reiman, 1998; Thompson, 1975).

The well-to-do were less subjected to physical punishments in the Middle Ages and more likely to "pay" for their crimes literally, if at all. However, the poor were subjected to all manner of abuse and violent punishments meted out by the Crown and its functionaries (Orland, 1995). When incarcerated in English and Irish jails of the 1700s and 1800s, those who could pay were housed in comfortable quarters with plenty to eat and were even able to visit with friends and relatives (Kerle, 2003; Stohr & Cooper, 2007). Such differential treatment by class was so institutionalized that facilities were constructed with this difference in mind: separate rooms were reserved for the wealthy inmates, and begging windows were built in next to busy thoroughfares for the poor. Simply put, and historically speaking, being rich has meant that either an offender's punishment was much lighter, more comfortable, or nonexistent. As Reiman (1998) notes in his book *The Rich Get Richer and the Poor Get Prison,* such distinctions, though less obvious, continue today in our focus on "street crimes," which tend to be largely perpetrated by poor and minority group members, rather than on corporate and white collar crime, which tends to be the purview of the middle and upper classes and which, he argues, actually results in greater loss of life and property.

As indicated by Reiman's (1998) analysis, race and ethnicity have also served to differentiate correctional practice. In the frontier days, murder was approved practice against those American Indians who allegedly transgressed against white settlers (Blalock 1967; Kitano, 1997; Stannard, 1992). Never mind that the whites had taken their land and committed offenses against them. Similarly, there were separate laws governing Chinese immigrants laboring in western mining towns or on the railroads, regarding their right to citizenship and to own property (Blalock 1967; Kitano, 1997). Mexicans in the American Southwest were treated more severely by the criminal justice system than the white settlers who followed them and settled on their land (Moore & Pachon 1985; Weyr 1988). African Americans saddled with the legacy of slavery and the barbaric treatment they suffered as property under its practices had no protection under the law (*Dred Scott v. Sanford*, 1856). Even after the Civil War, however, African American men and women were incarcerated differently than their fellows; in fact, some "prison farms" in the deep south were nothing more than plantation slavery continued (Oshinsky, 1996). Even outside of the south, male and female minority group inmates would often be separated from whites and treated in a more discriminatory fashion in terms of assignments, punishments, and programming (Hawkes, 1998; Joseph & Taylor, 2003). Unfortunately, some of these "traditions" of discrimination still bedevil us.

The most prominent driver of this discrimination exists in the drug war, which has spurred the greatest incarceration rate increase in jails and prisons in American history, along with the concomitant rise in probation and parole caseloads (Chesney-Lind, 2001; Joseph, Henriques, & Richards-Ekeh, 2003; Pollock, 2004; Young & Adams-Fuller, 2006; Zimring & Hawkins, 1995). Couple these rising incarceration and correctional supervision rates with the use of "three strikes" laws and other forms of mandatory sentences and you have the resulting "harsh justice"—the increased use of corrections—that sets the United States apart from the rest of the Western and civilized world. We incarcerate and use correctional supervision at higher rates than other comparable countries (Ruddell, 2004; Whitman, 2003; see also a related discussion in the introduction to this book).

A related theme is that of differential treatment for incarcerated women and girls in corrections (see the Rafter article featured in this section). Some of the earliest correctional institutions (e.g., the bridewells or workhouses) did not always separate the genders or divide their inmates by age. But some of the earliest criminal statutes did provide punishments for women in particular (e.g., punishing the village "scold" or the ability of a man to discipline his wife with a stick no bigger than his thumb or the burning of witches), punishments that appeared to be aimed at forcing women to conform to a certain reduced role in social, political, and communal affairs (Anderson, 2006; Pollock, 2002). Much like African American slaves, women and girls throughout history have been legally defined as the property of their male relatives, and it was only over a gradual period of reform, which lasted several hundred years and included much struggle, that they gained the same rights and liberties in law as men and boys (Stohr, 2000). Of course, the degree to which women, or less powerful groups such as the poor and minorities, can exercise their rights and liberties in law, and thus over correctional operation, has depended to some extent on that intersection of class, race, and gender (Joseph & Taylor, 2003).

Religious Influence

As becomes clear when one reviews the history of the Walnut Street Jail and the Western and Eastern Pennsylvania prisons, many U.S. correctional facilities would not have been created and operated in the manner that they were if not for the influence of the Quakers (Orland, 1995; see also Beaumont & Tocqueville and Johnston Readings). From the architecture to the supervision to the type of activities allowed, all aspects of these prisons were, in part, shaped by the need to provide inmates with the opportunity to reform via contemplation of the Bible and the desire to isolate them from the corrupting influence of other criminals.

In fact, the history of corrections is replete with instances of correctional institutions and practices being shaped by religious influences. The Catholic Church constructed and operated a "prison-like" existence for offenders in monasteries (Welch, 2004). Perhaps the most ubiquitous programming in prisons or jails from their inception till now, however, has been the religious outreach provided by pastors and priests and rabbis in the local community. Today, more than ever, we see such faith-based initiatives promoted for both the community and corrections (Sipes & Young, 2006).

Architecture as It Is Associated With Supervision

Jeremy Bentham was one of the first to argue cogently that architecture and supervision were complementary. In recommending to the British Parliament in 1843 the creation of his "panopticon" (an architecturally rounded prison with a central guard station in the middle and a glass ceiling), he touted the ability of the officers to supervise inmates more efficiently (Foucault, 1979). Although approved by parliament, the panopticon was never funded, so Bentham was never able to test his marriage of architecture and supervision idea, though others did (e.g., the Stateville prison of Illinois; Jacobs, 1977).

Much like the panopticon, Sing Sing, Auburn, and the Pennsylvania prisons were architecturally shaped to fit a certain supervision style (see Beaumont & Tocqueville and Johnston in this section; Orland, 1995). That style was to be removed and indirect in most cases, with restricted interactions between staff and inmates even in the congregate, but silent, New York prisons. This "ideal" of restricted contact, however, was corrupted itself when prisons confronted the reality of overcrowding.

More recently, in the last 25 years, we have seen this important connection between architecture and supervision represented in the architecture of "New Generation" jails (Zupan, 1991). In such facilities, the podular, or rounded, architecture allows officers in the living units a greater opportunity to supervise visually and physically —or directly— what is occurring. Notably, the podular architecture, coupled with direct supervision, has spread to numerous jails and prisons across the country (see Section VIII for a fuller discussion of such jails).

Overcrowding

Today, given what we know about the tendency of prisons, jails, and other correctional facilities to be overused, or at least well used, we are probably not surprised to learn that the prisons of the early 1800s, like the jails built before them and since, were prone to overcrowding. Almost from the very beginning, the Walnut Street Jail was overcrowded, even after it was remodeled as, arguably, the first prison, and this overcrowding spurred the building of the Western and then the Eastern Pennsylvania prisons. Likewise, once the Auburn prison became crowded, the building of Sing Sing was virtually preordained. So we see today the apparently insatiable need for prison and jail space leading to the current incarceration of over two million in those institutions, with no serious abatement in prison and jail growth in the foreseeable future (Bureau of Justice Statistics, 1997, 2006b; see also Rafter below).

But what is often not noted by correctional commentators, probably because of the lesser strictures on liberty for offenders and the lesser relative cost for the state, is that the explosive growth in jail and prison populations has been exceeded by the growth of community corrections populations, particularly in the area of probation. For instance, from 1990 to 2005 there was an average increase of 53.5% in probation populations (along with a 9.1% increase in parole populations) as compared to 25.2% and 12.3% increases in populations, consecutively, for prisons and jails (Glaze & Bonczar, 2006; see also a discussion of probation and parole in other sections of this book). What these increases signify is that caseloads and not just correctional institutions get "overcrowded," and this overcrowding complicates the ability of correctional personnel to "manage" offenders. These recent data and what we know about bridewells, poorhouses, jails, and early prisons all indicate that, if an institution is built, it will be filled and usually beyond capacity—necessitating in some policy makers minds the need to build still more institutions.

Good Intentions Do Not Always Translate Into Effective Practice

It has become an accepted truism of public policy making that what you plan for may not be what you get once programs are in place (Pressman & Wildavsky, 1984; Rothman, 1980). In his classic work *Conscience and Convenience*, Rothman (1980) describes the many reasons a disjunction often occurs between even the best of intentions and how programs, and in this case, institutions, actually "work." Some of those reasons have to do with the other themes mentioned here, particularly money or the lack of it. Other reasons have to do with the politics of the time and whether a particular correctional practice or institution fits the culture it emanates from, which in turn is influenced by religious forces and perceptions of morality.

Maconochie, a progressive prison warden who was charged with operating the brutal Norfolk Island penal colony from 1840 to 1844, instituted some of the earliest prison reforms, including the mark system (a program akin to good time), a form of indeterminate sentencing (whereby inmates could reduce the severity of their sentences by behaving appropriately), and decent, nonviolent responses to most inmates' misbehavior (Morris, 2002; Orland, 1995). But because his proposals for change were a poor fit with the politics and perceptions of morality of his time, Maconochie was removed from his post as warden, and his reform efforts were abandoned for a time until prison reform became politically popular some 30 years later.

Beaumont and Tocqueville note, as does Johnston, the good intentions that served as the conceptual rampart for the construction of America's early prisons. Then they explain why those intentions were never fully matched by actual practice. Similarly, Rothman (1980) finds in his review of major criminal justice reforms of the latter part of the 19th and first half of the 20th centuries—the Elmira prison of 1870, the creation and operation of the juvenile court and probation and parole, and reform efforts in prisons—that many missteps can occur between the "intent" of a program and its actual operation.

Thirty some years ago—in 1974—Martinson claimed that almost nothing works in correctional rehabilitation, and, though this assessment is generally thought to be an exaggeration, he had hit on an essential truth: some programs clearly were not "working" or, if they were, they were not being evaluated correctly so that anyone could tell (Cullen & Gilbert, 1982; Martinson, 1974; Palmer, 1983). In the last five to ten years, belief in rehabilitation programming has returned, though some rightly note that we never abandoned it as a central purpose of corrections (e.g., see Cullen & Gilbert, 1982). The lesson we can take away from these earlier attempts at reform and apply to today's efforts—some of which you will read about in this book—is not that we should never try to improve corrections or its operation but that each such effort should be approached with a healthy degree of caution and grounded in empirically derived findings, rather than in the political or populist fads of the moment (Rothman, 1980).

⬛ Summary: Knowing Where We Have Been Helps Us Determine Where We Should Go

- Correctional institutions, whether jails or prisons or, more recently, community corrections, have been shaped by several themes throughout their history. These themes, though apparently constant, are products of their times. For instance, the Eastern Penitentiary would not be built today as a *general use prison* because it would be considered cruel to isolate inmates from other human contact. Yet this kind of isolation, sometimes even with the tiny cells, is seen as beneficial by those today who build and operate super-max prisons for *special uses* to control incorrigible inmates (Kluger, 2007).

- Whether there is an overall movement toward greater compassion and humanity in corrections is debatable. Certainly, the current willingness to use correctional punishment and very long sentences for some offenses would appear to contradict this idea. However, the whole move toward a greater use of treatment (beginning in the late 1990s and continuing today), though certainly motivated by a need to reduce the warehousing costs of "get tough" policies of the 1980s, 1990s, and today, is certainly also supported by old-time "Enlightenment" beliefs in "right reason" and in the basic humanity and dignity of most offenders. (See other sections of this book for a fuller discussion of the trend toward treatment.)

- Class, race, and gender are ostensibly not as prominent in law and practice today as they were previously, at least in relation to determining how corrections operates. Yet some would argue that the failure to focus criminal justice energies on corporate crimes and an overenthusiasm for prosecuting low-level drug offenders has the predictable effect of differential punishment in corrections by class and race, if not by gender.

- Likewise, religious programming, at least in large institutions, is much more diverse in content than it was when prisons and jails were first conceived. But its basic thrust is still reform or repentance through contact with and assistance by a higher power. Moreover, the current focus on faith-based initiatives ensures that a religious influence in corrections will remain prominent.

- Good intentions sometimes lead to outstanding practice; at other times, they do not. Programming today is doubtless created with the best of intentions, and, when analyzed, it *sometimes* lives up to its promise.

- In the following chapters, we will see themes such as those mentioned here dealt with again and again by the authors. That they reappear and reappear and then reappear again does not mean, however, that we cannot make and have not made any progress in corrections. There is no question that, on the whole, the vast majority of jails and prisons in this country today are much better than those that served us for most of the last 170 years. However, the unprecedented *use* of correctional sanctions in our country could be regarded by some as overly harsh and thus a regressive trend. These themes presented here merely represent conundrums (e.g., how much money or compassion or religious influence, is the "right" amount), and, as such, we are constantly called upon to address them.

KEY TERMS

Congregate but silent labor systems

Convict lease system

Factory prisons

Plantation prisons

Prison with a philosophy of penitence

Silent and penitent systems

INTERNET SITES

American Correctional Association: www.aca.org

American Jail Association: www.aja.org

American Probation and Parole Association: www.appa-net.org

Bureau of Justice Statistics (information available on all manner of criminal justice topics): www.ojp.usdoj.gov/bjs

National Criminal Justice Reference Service: www.ncjrs.gov

Office of Justice Research (information available on all manner of criminal justice topics, specifically probation and parole here): www.ojp.usdoj.gov/bjs/pub/pdf/ppus05.pdf

Pew Charitable Trust (Corrections and Public Safety): http://www.pewtrusts.org/our_work_category.aspx?id=72

Vera Institute of Justice (information available on a number of corrections and other justice related topics): www.vera.org

READING

This book chapter excerpt by Gustave de Beaumont (a prosecutor) and Alexis de Tocqueville (a lawyer) is of great historical interest because these two French aristocrats came to the United States in 1831 purposely to observe and report upon America's experiment with the penitentiary system. Beaumont and Tocqueville studied the Cherry Hill Prison in Philadelphia and the Auburn Prison in New York as well as some others. They found these prisons somewhat different from older American prisons and European prisons; for example, prisoners were kept in isolation so that they could not corrupt one another, and prisoners were required to work throughout their sentences. The biggest innovation was that attempts were made to reform prisoners morally and spiritually (hence the term "penitentiary").

An Historical Outline of the Penitentiary System

Gustave de Beaumont and Alexis de Tocqueville

Though the penitentiary system in the United States is a new institution, its origin must be traced back to times already long gone by. The first idea of a reform in the American prisons, belongs to a religious sect in Pennsylvania. The Quakers, who abhor all shedding of blood, had always protested against the barbarous laws which the colonies inherited from their mother country. In 1786, their voice succeeded in finding due attention, and from this period, punishment of death, mutilation and the whip were successively abolished in almost all cases by the Legislature of Pennsylvania. A less cruel fate awaited the convicts from this period. The punishment of imprisonment was substituted for corporal punishment, and the law authorized the courts to inflict solitary confinement in a cell during day and night, upon those guilty of capital crimes. It was then that the Walnut Street prison was established in Philadelphia. Here the convicts were classed according to the nature of their crimes, and separate cells were constructed for those whom the courts of justice had sentenced to absolute isolation. These cells also served to curb the resistance of individuals, unwilling to submit to the discipline of the prison. The solitary prisoners did not work.

This innovation was good but incomplete. The impossibility of subjecting criminals to a useful classification, has since been acknowledged, and solitary confinement without labor has been condemned by experience. It is nevertheless just to say, that the trial of this theory has not been made long enough to be decisive. The authority given to the judges of Pennsylvania, by the law of April 5, 1790, and of March 22, to send criminals to the prison in Walnut Street, who formerly would have been sent to the

SOURCE: "An Historical Outline of the Penitentiary System," pages 37–52; originally published in *On the Penitentiary System in the United States and Its Application in France* by Gustave de Beaumont and Alexis de Tocqueville. © 1964 by Southern Illinois University Press.

different county jails, soon produced in this prison such a crowd of convicts, that the difficulty of classification increased in the same degree as the cells became insufficient.

To say the truth there did not yet exist a penitentiary system in the United States. If it be asked why this name was given to the system of imprisonment which had been established, we would answer, that then as well as now, the abolition of the punishment of death was confounded in America, with the penitentiary system. People said—*instead of killing the guilty, our laws put them in prison; hence we have a penitentiary system.*

The conclusion was not correct. It is very true that the punishment of death applied to the greater part of crimes, is irreconcilable with a system of imprisonment; but this punishment abolished, the penitentiary system does not yet necessarily exist; it is further necessary, that the criminal whose life has been spared, be placed in a prison, whose discipline renders him better. Because, if the system, instead of reforming, should only tend to corrupt him still more, this would not be any longer a penitentiary system, but only a bad system of imprisonment.

This mistake of the Americans has for a long time been shared in France. In 1794, the Duke de la Rochefoucauld-Liancourt published an interesting notice on the prison of Philadelphia: he declared that this city had an excellent prison system, and all the world repeated it. However, the Walnut Street prison could produce none of the effects which are expected from this system. It had two principal faults: it corrupted by contamination those who worked together. It corrupted by indolence, the individuals who were plunged into solitude.

The true merit of its founders was the abolition of the sanguinary laws of Pennsylvania, and by introducing a new system of imprisonment, the direction of public attention to this important point. Unfortunately that which in this innovation deserved praise, was not immediately distinguished from that which was untenable.

Solitude applied to the criminal, in order to conduct him to reformation by reflection, rests upon a philosophical and true conception. But the authors of this theory had not yet founded its application upon those means which alone could render it practical and salutary. Yet their mistake was not immediately perceived, and the success of Walnut Street prison boasted of in the United States still more than in Europe, biased public opinion in favor of its faults, as well as its advantages.

The first state which showed itself zealous to imitate Pennsylvania, was that of New York, which in 1797, adopted both new penal laws and a new prison system.

Solitary confinement without labor, was admitted here as in Philadelphia, but, as in Walnut Street, it was reserved for those who especially were sentenced to undergo it by the courts of justice, and for those who opposed the established order of the prison. Solitary confinement, therefore, was not the ordinary system of the establishment; it awaited only those great criminals who, before the reform of the penal laws, would have been condemned to death. Those who were guilty of lesser offenses were put indiscriminately together in the prison. They, different from the inmates of the solitary cells, had to work during the day, and the only disciplinary punishment which their keeper had a right to inflict, in case of breach of the order of the prison, was solitary confinement, with bread and water.

The Walnut Street prison was imitated by others: Maryland, Massachusetts, Maine, New Jersey, Virginia, etc., adopted successively, the principle of solitary confinement, applied only to a certain class of criminals in each of these states. The reform of criminal laws preceded that of the prisons.

Nowhere was this system of imprisonment crowned with the hoped-for success. In general it was ruinous to the public treasury; it never effected the reformation of the prisoners. Every year the legislature of each state voted considerable funds towards the support of the penitentiaries, and the continued return of the same individuals into the prisons, proved the inefficiency of the system to which they were submitted.

Such results seem to prove the insufficiency of the whole system; however instead of accusing the theory itself, its execution was attacked. It was believed that the whole evil resulted from the paucity of cells, and the crowding of the prisoners; and that the system, such as it was established, would be fertile in happy results, if some new buildings were added to the prisons already existing. New expenses therefore, and new efforts were made.

Such was the origin of the Auburn prison [1816]. This prison, which has become so celebrated since, was at first founded upon a plan essentially erroneous. It limited itself to some classifications, and each of these cells was destined to receive two convicts: it was of all combinations the most unfortunate; it would have been better to throw together fifty criminals in the same room, than to separate them two by two. This inconvenience was soon felt, and in 1819 the Legislature of the State of New York, ordered the erection of a new building at Auburn (the northern wing) in order to increase the number of solitary cells. However, it must be observed, that no idea as yet existed of the system which has prevailed since. It was not intended to subject all the convicts to the system of cells, but its application was only to be made to a greater number. At the same time the same theories produced the same trials in Philadelphia, where the little success of the Walnut Street prison would have convinced the inhabitants of Pennsylvania of its inefficiency, if the latter, like the citizens of the State of New York, had not been led to seek in the faults of execution, a motive for allowing the principle to be correct.

In 1817, the Legislature of Pennsylvania decreed the erection of the penitentiary at Pittsburgh, for the western counties, and in 1821, that of the penitentiary of Cherry Hill, for the city of Philadelphia and the eastern counties. The principles to be followed in the construction of these two establishments were, however, not entirely the same as those on which the Walnut Street prison had been erected. In the latter, classification formed the predominant system, to which solitary confinement was but secondary. In the new prisons the classifications were abandoned, and a solitary cell was to be prepared for each convict. The criminal was not to leave his cell day or night, and all labor was denied to him in his solitude. Thus absolute solitary confinement, which in Walnut Street was but accidental, was now to become the foundation of the system adopted for Pittsburgh and Cherry Hill. The experiment which was to be made, promised to be decisive; no expense was spared to construct these new establishments worthy of their object, and the edifices which were elevated, resembled prisons less than palaces.

In the meantime, before even the laws which ordered their erection, were executed, the Auburn prison had been tried in the State of New York. Lively debates ensued on this occasion, in the legislature, and the public was impatient to know the result of the new trials, which had just been made. The northern wing having been nearly finished in 1821, eighty prisoners were placed there, and a separate cell was given to each. This trial, from which so happy a result had been anticipated, was fatal to the greater part of the convicts. In order to reform them, they had been submitted to complete isolation; but this absolute solitude, if nothing interrupts it, is beyond the strength of man; it destroys the criminal without intermission and without pity; it does not reform, it kills.

The unfortunates on whom this experiment was made fell into a state of depression, so manifest, that their keepers were struck with it; their lives seemed in danger, if they remained longer in this situation; five of them had already succumbed during a single year; their moral state was not less alarming; one of them had become insane; another, in a fit of despair, had embraced the opportunity when the keeper brought him something, to precipitate himself from his cell, running the almost certain chance of a mortal fall.

Upon similar effects the system was finally judged. The Governor of the State of New York pardoned twenty-six of those in solitary confinement; the others to whom this favor was not extended, were allowed to leave the cells during day, and to work in the common workshops of the prison. From this period, (1823) the system of unmodified isolation ceased entirely to be practiced at Auburn. Proofs were soon afforded that this system, fatal to the health of the criminals, was likewise inefficient in producing their reform. Of twenty-six convicts, pardoned by the governor, fourteen returned a short time after into the prison, in consequence of new offenses.

This experiment, so fatal to those who were selected to undergo it, was of a nature to endanger the success of the penitentiary system altogether. After the melancholy effects of isolation, it was to be feared that the whole principle would be rejected: it would have been a natural reaction. The Americans were wiser: the idea was not given up, that the solitude, which causes the criminal to reflect, exercises a beneficial influence; and the problem was, to find the means by which the evil effect of total solitude could be avoided without giving up its advantages. It was believed that this end could be attained, by leaving the convicts in their cells during night, and by making them work during the day, in the common workshops, obliging them at the same time to observe absolute silence. Messrs. Allen, Hopkins, and Tibbits, who, in 1824, were directed by the Legislature of New York to inspect the Auburn prison, found this new discipline established in that prison. They praised it much in their report, and the Legislature sanctioned this new system by its formal approbation.

Here an obscurity exists which it has not been in our power to dissipate. We see the renowned Auburn system suddenly spring up, and proceed from the ingenious combination of two elements, which seem at first glance incompatible, isolation and reunion. But that which we do not clearly see, is the creator of this system, of which nevertheless some one must necessarily have formed the first idea.

Does the State of New York owe it to Governor Clinton, whose name in the United States is connected with so many useful and beneficial enterprises? Does the honor belong to Mr. Cray, one of the directors of Auburn, to whom Judge Powers, who himself was at the head of that establishment, seems to attribute the merit? Lastly, Mr. Elam Lynds, who has contributed so much to put the new system into practice, does the glory also of the invention belong to him? We shall not attempt to solve this question, interesting to the persons whom we have mentioned, and the country to which they belong, but of little importance to us. In fine, does not experience teach us that there are innovations, the honor of which belongs to nobody in particular, because they are the effects of simultaneous efforts, and of the progress of time?

The establishment of Auburn has, since its commencement, obtained extraordinary success. It soon excited public attention in the highest degree. A remarkable revolution took place at that time in the opinions of many. The direction of a prison, formerly confided to obscure keepers, was now sought for by persons of high standing, and Mr. Elam Lynds, formerly a captain in the army of the United States, and Judge Powers, a magistrate of rare merit, were seen, with honor to themselves, filling the office of directors of Auburn.

However, the adoption of the system of cells for all convicts in the state of New York, rendered the Auburn prison insufficient, as it contained but 550 cells after all the successive additions which it had received. The want of a new prison, therefore, was felt. It was then that the plan of Sing Sing was resolved upon by the legislature (1825) and the way in which it was executed is of a kind that deserves to be reported.

Mr. Elam Lynds, who had made his trials at Auburn, of which he was the superintendent, left this establishment; took one hundred convicts, accustomed to obey, with him, led them to the place where the projected prison was to be erected; there, encamped on the bank of the

Hudson, without a place to receive, and without walls to lock up his dangerous companions; he sets them to work, making of every one a mason or a carpenter, and having no other means to keep them in obedience, than the firmness of his character and the energy of his will.

During several years, the convicts, whose number was gradually increased, were at work in building their own prison, and at present the penitentiary of Sing Sing contains one thousand cells, all of which have been built by their criminal inmates. At the same time (1825) an establishment of another nature was reared in the city of New York, but which occupies not a less important place among the improvements, the history of which we attempt to trace. We mean the house of refuge, founded for juvenile offenders.

There exists no establishment, the usefulness of which, experience has warranted in a higher degree. It is well known that most of those individuals on whom the criminal law inflicts punishments, have been unfortunate before they became guilty. Misfortune is particularly dangerous for those whom it befalls in a tender age; and it is very rare that an orphan without inheritance and without friends, or a child abandoned by its parents, avoids the snares laid for his inexperience, and does not pass within a short time from misery to crime. Affected by the fate of juvenile delinquents, several charitable individuals of the city of New York conceived the plan of a house of refuge, destined to serve as an asylum, and to procure for them an education and the means of existence, which fortune had refused. Thirty thousand dollars were the produce of a first subscription. Thus by the sole power of a charitable association, an establishment eminently useful, was founded, which, perhaps, is still more important than the penitentiaries, because the latter punish crime, while the house of refuge tends to prevent it.

The experiment made at Auburn in the state of New York (the fatal effects of isolation without labor) did not prevent Pennsylvania from continuing the trial of solitary confinement, and in the year 1827, the penitentiary of Pittsburgh began to receive prisoners. Each one was shut up, day and night, in a cell, in which no labor was allowed to him. This solitude, which in principle was to be absolute, was not such in fact. The construction of this penitentiary is so defective, that it is very easy to hear in one cell what is going on in another; so that each prisoner found in the communication with his neighbor a daily recreation, i.e., an opportunity of inevitable corruption. As these criminals did not work, we may say that their sole occupation consisted in mutual corruption. This prison, therefore, was worse than even that of Walnut Street, because, owing to the communication with each other, the prisoners at Pittsburgh were as little occupied with their reformation, as those at Walnut Street. And while the latter indemnified society in a degree by the produce of their labor, the others spent their whole time in idleness, injurious to themselves, and burdensome to the public treasury.

The bad success of this establishment proved nothing against the system which had called it into existence, because defects in the construction of the prison, rendered the execution of the system impossible. Nevertheless, the advocates of the theories on which it was founded, began to grow cool. This impression became still more general in Pennsylvania, when the melancholy effects caused by solitude without labor in the Auburn prison, became known, as well as the happy success of the new discipline, founded on isolation by night, with common labor during the day.

Warned by such striking results, Pennsylvania was fearful she had pursued a dangerous course. She felt the necessity of submitting to a new investigation the question of solitary imprisonment without labor, practiced at Pittsburgh and introduced into the penitentiary of Cherry Hill, the construction of which was already much advanced.

The legislature of this state, therefore, appointed a committee in order to examine which was the better system of imprisonment. Messrs. Charles Shaler, Edward King, and

T. I. Wharton, commissioners charged with this mission, have exhibited, in a very remarkable report, the different systems then in practice (December 20, 1827), and they conclude the discussion by recommending the new Auburn discipline, which they pronounce the best. The authority of this inquiry had a powerful effect on public opinion. It however met with powerful opposition: Roberts Vaux, in Pennsylvania and Edward Livingston, in Louisiana, continued to support the system of complete solitude for criminals. The latter, whose writings are imbued with so elevated a philosophy, had prepared a criminal code, and a code of Prison Discipline for Louisiana, his native state. His profound theories, little understood by those for whom they were destined, had more success in Pennsylvania, for which they had not been intended. In this superior work, Mr. Livingston admitted, for most cases, the principle of *labor of the convicts*. Altogether, he showed himself less the advocate of the Pittsburgh prison, than the adversary of the Auburn system. He acknowledged the good discipline of the latter, but powerfully opposed himself to corporal punishment used to maintain it. Mr. Livingston, and those who supported the same doctrines, had to combat a powerful fact: this was the uncertainty of their theories, not yet tested, and the proven success of the system they attacked. Auburn went on prospering: everywhere its wonderful effects were praised, and they were found traced each year with great spirit, in a work justly celebrated in America, and which has essentially co-operated to bring public opinion in the United States, on the penitentiary system, to that point where it now is. We mean the annual publications of the Prison Discipline Society at Boston. These annual reports—the work of Mr. Louis Dwight, give a decided preference to the Auburn system.

All the states of the Union were attentive witnesses of the controversy respecting the two systems. In this fortunate country, which has neither troublesome neighbors, who disturb it from without, nor internal dissensions which distract it within, nothing more is necessary, in order to excite public attention in the highest degree, than an essay on some principle of social economy. As the existence of society is not put in jeopardy, the question is not how to live, but how to improve.

Pennsylvania was, perhaps, more than any other state, interested in the controversy. The rival of New York, it was natural she should show herself jealous to retain, in every respect, the rank to which her advanced civilization entitles her among the most enlightened states of the Union. She adopted a system which at once agreed with the austerity of her manners, and her philanthropical sensibility. She rejected solitude without labor, the fatal effects of which experience had proved everywhere, and she retained the absolute separation of the prisoners—a severe punishment, which, in order to be inflicted, needs not the support of corporal chastisement.

The penitentiary of Cherry Hill, founded on these principles, is therefore a combination of Pittsburgh and Auburn. Isolation during night and day, has been retained from the Pittsburgh system: and, into the solitary cell, the labor of Auburn has been introduced. This revolution in the prison discipline of Pennsylvania, was immediately followed by a general reform of her criminal laws. All punishments were made milder; the severity of solitary imprisonment permitted an abridgment of its duration; capital punishment was abolished in all cases, except that of premeditated murder.

While the states of New York and Pennsylvania made important reforms in their laws, and each adopted a different system of imprisonment, the other states of the Union did not remain inactive, in presence of the grand spectacle before them.

Since the year 1825, the plan of a new prison on the Auburn model, has been adopted by the legislature of Connecticut; and the penitentiary at Wethersfield has succeeded the old prison of Newgate. In spite of the weight which Pennsylvania threw into the balance, in favor of absolute solitude with labor, the Auburn system, i.e., common labor during the day, with isolation during night, continued to obtain a preference. Massachusetts, Maryland, Tennessee, Kentucky, Maine, and Vermont, have gradually adopted the Auburn plan, and have taken the Auburn prison as a model for those which they have caused to be erected.

Several states have not stopped here, but have also founded houses of refuge for juvenile offenders, as an addition, in some measure, to the penitentiary system, in imitation of New York. These latter establishments have been founded in Boston in 1826, and in Philadelphia in 1828. There is every indication that Baltimore also, will soon have its house of refuge.

It is easy to foresee, that the impulse of reform given by New York and Pennsylvania, will not remain confined to the states mentioned above. From the happy rivalship which exists among all the states of the Union, each state follows the reforms which have been effected by the others, and shows itself impatient to imitate them. It would be wrong to judge all the United States by the picture which we have presented of the improvements adopted by some of them.

Accustomed as we are to see our central government attract everything, and propel in the various provinces all the parts of the administration in a uniform direction, we sometimes suppose that the same is the case in other countries; and comparing the centralization of government at Washington with that at Paris, the different states of the Union to our departments, we are tempted to believe that innovations made in one state, take, of necessity, place in the others. There is, however, nothing like in the United States.

These states, united by the federal tie into one family, are in respect to everything which concerns their common interests, subjected to one single authority. But besides these general interests, they preserve their entire individual independence, and each of them is sovereign master to rule itself according to its own pleasure. We have spoken of nine states which have adopted a new system of prisons;

there are fifteen more which have as yet made no change.

In these latter, the ancient system prevails in its whole force; the crowding of prisoners, confusion of crimes, ages, and sometimes sexes, mixture of indicted and convicted prisoners, of criminals and debtors, guilty persons and witnesses; considerable mortality; frequent escapes; absence of all discipline, no silence which leads the criminals to reflection; no labor which accustoms them to an honest mode of subsistence; insalubrity of the place which destroys health; ignism of the conversations which corrupt; idleness that depraves; the assemblage, in one word, of all vices and all immoralities—such is the picture offered by the prisons which have not yet entered into the way of reform.

By the side of one state, the penitentiaries of which might serve as a model, we find another, whose jails present the example of everything which ought to be avoided. Thus the State of New York is without contradiction one of the most advanced in the path of reform, while New Jersey, which is separated from it but by a river, has retained all the vices of the ancient system.

Ohio, which possesses a penal code remarkable for the mildness and humanity of its provisions, has barbarous prisons. We have deeply sighed when at Cincinnati, visiting the prison. We found half of the imprisoned charged with irons, and the rest plunged into an infected dungeon; and are unable to describe the painful impression which we experienced, when, examining the prison of New Orleans, we found men together with hogs, in the midst of all odors and nuisances. In locking up the criminals, nobody thinks of rendering them better, but only of taming their malice; they are put in chains like ferocious beasts; and instead of being corrected, they are rendered brutal.

If it is true that the penitentiary system is entirely unknown in that part which we mentioned, it is equally true that this system is incomplete in those states even where it is in vigor. Thus at New York, at Philadelphia, and

Boston, there are new prisons for convicts, whose punishment exceeds one or two years' imprisonment; but establishments of a similar nature do not exist to receive individuals who are sentenced for a shorter time, or who are indicted only. In respect to the latter, nothing has been changed; disorder, confusion, mixture of different ages and moral characters, all vices of the old system still exist for them: we have seen in the house of arrest in New York (Bridewell) more than fifty indicted persons in one room. These arrested persons are precisely those for whom well-regulated prisons ought to have been built. It is easy in fact to conceive, that he who has not yet been pronounced guilty, and he who has committed but a crime or misdemeanor comparatively slight, ought to be surrounded by much greater protection than such as are more advanced in crime, and whose guilt has been acknowledged.

Arrested persons are sometimes innocent and always supposed to be so. How is it that we should suffer them to find in the prison a corruption which they did not bring with them? If they are guilty, why place them first in a house of arrest, fitted to corrupt them still more, except to reform them afterwards in a penitentiary, to which they will be sent after their conviction? There is evidently a deficiency in a prison system which offers anomalies of this kind. These shocking contradictions proceed chiefly from the want of unison in the various parts of government in the United States.

The larger prisons (state prisons) corresponding to our *maisons centrales*, belong to the state, which directs them; after these follow the county jails, directed by the county; and at last the prisons of the city, superintended by the city itself.

The various branches of government in the United States being almost as independent of each other, as the states themselves, it results that they hardly ever act uniformly and simultaneously. While one makes a useful reform in the circle of its powers, the other remains inactive, and attached to ancient abuses.

We shall see below, how this independence of the individual parts, which is injurious to the uniform action of all their powers, has nevertheless a beneficial influence, by giving to each a more prompt and energetic progress in the direction which it follows freely and uncompelled.

We shall say nothing more of the defective parts in the prison system in the United States. If at some future period France shall imitate the penitentiaries of America, the most important thing for her will be to know those which may serve as models. The new establishments then, will form the only object of our further inquiry.

We have seen, in the preceding remarks, that few states have as yet changed entirely their system of imprisonment; the number of those which have modified their penal laws is still less. Several among them yet possess part of the barbarous laws which they have received from England.

We shall not speak of the Southern states, where slavery still exists. In every place where one-half of the community is cruelly oppressed by the other, we must expect to find in the law of the oppressor, a weapon always ready to strike nature which revolts or humanity that complains. Punishment of death and stripes—these form the whole penal code for the slaves. But if we throw a glance at those states even which have abolished slavery, and which are most advanced in civilization, we shall see this civilization uniting itself, in some, with penal laws full of mildness, and in others, with all the rigor of a code of Draco.

Let us but compare the laws of Pennsylvania with those of New England, which is, perhaps, the most enlightened part of the American Union. In Massachusetts, there are ten different crimes punished by death—among others, rape and burglary. Maine, Rhode Island, and Connecticut, count the same number of capital crimes. Among these laws, some contain the most degrading punishments, such as the pillory; others revolting cruelties, as branding and mutilation. There are also some which order fines equal to confiscations. While we find the remains of barbarism in some states, with an old population, there are others, which, risen since yesterday, have banished from their laws all cruel punishments not called for by the interest of society. Thus, Ohio, which certainly is not as enlightened as New England, has a penal code much more humane than those of Massachusetts or Connecticut.

Close by a state where the reform of the penal laws seems to have arrived at its summit, we find another, the criminal laws of which are stamped with all the brutalities of the ancient system. It is thus that the States of Delaware and New Jersey, so far behind in the path of improvement, border on Pennsylvania, which, in this respect, marches at the head of all others.

We should forget the object of our report were we to dwell any longer on this point. We were obliged to present a sketch of the penal legislation of the United States, because it exercises a necessary influence on the question before us. In fact it is easy to conceive to what point the punishments which degrade the guilty, are incompatible with a penitentiary system, the object of which is to reform them. How can we hope to awaken the moral sense of an individual who carries on his body the indelible sign of infamy, when the mutilation of his limbs reminds others incessantly of his crime, or the sign imprinted on his forehead, perpetuates its memory?

Must we not ardently wish, that the last traces of such barbarism should disappear from all the United States, and particularly from those which have adopted the penitentiary system, with which they are irreconcilable, and whose existence renders them still more shocking? Besides, let us not blame these people for advancing slowly on the path of innovation. Ought not similar changes to be the work of time, and of public opinion? There are in the United States a certain number of philosophical minds, who, full of theories and systems, are impatient to put them into practice; and if they had the power themselves to

make the law of the land, they would efface with one dash, all the old customs, and supplant them by the creations of their genius, and the decrees of their wisdom. Whether right or wrong the people do not move so quickly. They consent to changes, but they wish to see them progressive and partial. This prudent and reserved reform, effected by a whole nation, all of whose customs are practical, is, perhaps, more beneficial than the precipitated trials which would result, had the enthusiasm of ardent minds and enticing theories free play.

Whatever may be the difficulties yet to be overcome, we do not hesitate to declare that the cause of reform and of progress in the United States, seem to us certain and safe. Slavery, the shame of a free nation, is expelled every day from some districts over which it held its sway; and those persons themselves who possess most slaves, are convinced that slavery will not last much longer. Every day punishments which wound humanity, become supplanted by milder ones; and in the most civilized states of the north, where these punishments continue in the written laws, their application has become so rare that they are to be considered as fallen into disuse. The impulse of improvement is given. Those states which have as yet done nothing, are conscious of their deficiency; they envy those which have preceded them in this career, and are impatient to imitate them.

Finally, it is a fact worth remarking, that the modification of the penal laws and that of prison discipline, are two reforms intimately associated with each other, and never separated in the United States. Our special task is not to enlarge on the first; the second alone shall fix our attention. The various states in which we have found a penitentiary system, pursue all the same end: the amelioration of the prison discipline. But they employ different means to arrive at their object. These different means have formed the subject of our inquiry.

DISCUSSION QUESTIONS

1. Discuss the relative benefits and drawbacks of the "Pennsylvania" versus the "New York" models of early prisons. What did Beaumont and Tocqueville think of them and why? Which type of prison would you rather work in, or be incarcerated in, and why?

2. What was it about America that made its approach to prisons different from that of Europe, according to Beaumont and Tocqueville?

3. In what ways has prison reform changed since the era of Beaumont and Tocqueville?

❖

READING

Norman Johnston's article traces both the history and worldwide influence of what he calls "the world's most influential prison." This prison system, which completely separated prisoners from each other during their entire sentence and created a unique architecture, was developed and instituted on a large scale at Eastern State Penitentiary in Philadelphia in 1829. Although not followed in other U.S. prison systems, the so-called Pennsylvania system was adopted, along with its architecture, in most of Europe, South America, and large parts of Asia until early in the 20th century. This article considers the successes and failures of the system and assesses its place in the history of corrections.

The World's Most Influential Prison

Success or Failure?

Norman Johnston

In 1822, on the outskirts of Philadelphia, construction began on the new Eastern State Penitentiary, which was to be not only one of the largest and most expensive structures in the country at the time but also, in both its architecture and its program, the most influential prison ever built.

To understand how Philadelphia and Pennsylvania became the center of prison reform worldwide, it is necessary to look briefly at the early development of penal practices in William Penn's colony. Penn, who himself had been confined in England for his Quaker beliefs, abolished the severe criminal

SOURCE: Johnston, N. (2004). The world's most influential prison: Success or failure? *The Prison Journal, 84*(4 supplemental), 20–40. Reprinted with permission of Sage Publications, Inc.

code, instituted by the Duke of York, that was in effect in other parts of British North America. Upon Penn's death, conservative elements in the colony and in England reintroduced many of the more sanguinary punishments. As late as 1780, punishments such as the pillory and hanging were carried out in public. An account of an execution that year related how two prisoners "were taken out amidst a crowd of spectators—they walked after a cart in which were two coffins and a ladder, etc., each had a rope about his neck and their arms tied behin [*sic*] them . . . they were both hanged in the commons of this city [Philadelphia] abt. 1 o'clock" (Teeters, 1955, p. 15). In spite of these practices, liberal thinkers and reformers never abandoned their concerns for prisoners and continued to be influenced by the Enlightenment ideals emanating from Europe, especially those calling for imprisonment in place of corporal and capital punishment. The investigations of John Howard and Elizabeth Fry, as well as the reforms instituted in a few exemplary prisons such as the San Michele House of Correction for juveniles in Rome and the prison in Ghent, Belgium, were also well known to the Philadelphia reformers.

Overcrowding and mingling of men, women, and boys in the Old Stone Jail at Third and High (Market) streets in Philadelphia prompted the construction of the Walnut Street Jail. It was built opposite the State House, later to be Independence Hall, and opened in 1776. After the peace of 1783, a group of prominent citizens led by Benjamin Franklin, Benjamin Rush, and others organized a movement to reform the harsh penal code of 1718. Their efforts resulted in the new law of 1786 that substituted public labor for the previous severe punishments. But reaction against the public display of convicts on the streets of the city and the disgraceful conditions in the Walnut Street Jail led to the formation in 1787 of the Philadelphia Society for Alleviating the Miseries of Public Prisons, a

name it retained for 100 years. Members of the society were appalled by what they learned about the new Walnut Street Jail. Garnish was common. It was a practice in which inmates, when entering the prison, were shaken down by other prisoners for their money, which was then used to buy rum and other drinks available at inflated prices from the jailer. New prisoners lacking money had to relinquish some of their own clothing, resulting in some being nearly naked in the jail. There was no separation of men from women or hardened offenders from others, and the press reported that some women had themselves arrested and confined for fictitious debts to consort with male prisoners. Riotous and disorderly behavior and escapes were common.

As the result of lobbying by the Pennsylvania Prison Society, the legislature was urged to use solitary labor to effect reform. They asked for particulars, and, in December of 1788, the society prepared an account of their investigations of conditions at the jail and recommended solitary confinement at hard labor. An act of 1790 brought about sweeping reforms in the prison and authorized a "penitentiary house" to be built in the yard of Walnut Street Jail to carry out solitary confinement with labor for "hardened and atrocious offenders." There is some evidence that few criminals received such sentences and that this little cellblock with 16 cells was used primarily for infractions of prison rules. However, following 1790, the jail, now a state prison, became a showplace, with separation of different sorts of prisoners even though the main building of the prison had only common sleeping rooms. Workshops were constructed and provided useful trade instruction, and the old idleness and abuses seemed to have been eliminated.

Walnut Street Jail had been built in response to overcrowding in the old jail. Now with Walnut Street a state prison and with the population of Philadelphia increasing rapidly, it too became intolerably crowded. The Society continued to urge the creation of large penitentiaries for the more efficient handling of prisoners. Partially as the result of their efforts, money was appropriated for a state penitentiary to be built at Allegheny, now part of Pittsburgh. The Society continued to remain convinced that, in spite of the small-scale isolation cellblock at Walnut Street, that prison would never prove the value of the system of separate confinement. Only a larger structure built specifically to separate inmates from one another would be needed. New York State, responding to developments in Philadelphia, had constructed individual cells in a portion of their new prison at Auburn, but because of poor architectural design and an insufficient internal regimen, the first serious use of separate confinement in the United States lasted only from 1821 to 1823.

Conditions at the Walnut Street Jail worsened, resulting in further efforts to get a penitentiary for Philadelphia and the eastern part of the state to house felons. Both the Pennsylvania Prison Society and the Board of Inspectors of Walnut Street Jail sent memorials to the legislature. Authorizing legislation was finally passed on March 20, 1821, and the governor appointed 11 building commissioners. Among them was Samuel Wood, later to be the first warden of the prison. All but three of the building commissioners were either members of the Pennsylvania Prison Society or had served on the board of inspectors of the Walnut Street Jail. The commissioners first met on March 20, 1821. A competition for plans produced only two designs that merited serious consideration: one by the well-known Philadelphia architect, William Strickland, the designer of the disastrous first Western Penitentiary,[1] and the other by John Haviland, a relative newcomer to the United States. The board immediately split into two factions favoring either the circular plan of Strickland or Haviland's radial design, then common in England where he had trained. Haviland's plan was accepted, but Strickland was appointed supervising architect—which at that time included purchasing building materials, hiring workers, and acting in the capacity of a general contractor. A few months later, while Strickland was in Europe, his rival Haviland was hired to build the prison.

The building commissioners not only oversaw the construction of the new prison but also were intent upon solving some of the persistent problems they had observed at Walnut Street Jail and at other contemporary prisons in the United States and Europe. These problems included the obvious influence of bad associations; idleness that led to disorder and violence; overcrowding that had plagued each prison built in Philadelphia and in other U.S. cities; poor supervision by sometimes venal and untrained personnel; abysmal health conditions of the inmates; and, of course, the questionable rehabilitative value of such incarceration. The reformers had gradually developed the idea that the key to true reform was complete isolation of inmates from one another, providing them with the right mix of solitude for reflection and perhaps reading and some vocational training or useful work. And, if all of this did not always work, the dread of the experience of generally extroverted, sociable criminals being isolated from contact with their fellow prisoners for years on end would deter them from further crimes, whatever their earlier motivations had been. This meant that a whole new kind of prison would have to be designed so that the entire sentence would be served in a cell alone. The inmates would only be visited occasionally by a guard to bring work materials, food, or fresh bedding or clothing. The inmate would work, sleep, learn, and worship in the solitary cell and exercise in the personal exercise yard, which was attached to the cell.

The idea of solitary confinement as an answer to the evils of congregate imprisonment had been tried earlier in England but had been abandoned because of inadequate buildings and the lack of a carefully worked out internal regimen. What had developed extensively in the little county prisons in Ireland and

England—and to very limited extent in France—was a radial layout of the prison structure to give the governor (warden) better control over the movements and activities of both the inmates and the guards. What evolved was a series of structures, sometimes with hub-and-spoke designs and sometimes with half-circular designs, with a governor's house in the center and cell buildings sometimes attached but often a short distance away from the center. The aim was what was termed inspection at the time, but there must have been little opportunity to observe the activities in the prison through the center building's windows.

Writers sometimes have left the impression that Eastern State Penitentiary's architect, John Haviland, created his radial plan through sheer innovation and that it was a truly fortuitous invention. But as historians of technology have observed, inventions and innovations are seldom that simple. An invention usually consists of a combination of elements already available to the inventor. This certainly was the case with Haviland's building. His original plan for the prison was remarkably similar to plans and layouts for mental hospitals being built or proposed in England during his apprenticeship there (Johnston, 1994, p. 34).

Constructing a Model Prison

From among 23 sites, the building commissioners selected a farm on the northern outskirts of Philadelphia. On what had been a cherry orchard, the prison, commonly known as Cherry Hill even abroad, was begun in May of 1822, and the original seven wings were completed in 1836, although prisoners were received in 1829 while construction was still going on. No one had ever tried to design quarters for 24-hour single occupancy of large numbers of inmates. The technology of indoor plumbing and large-scale central heating was very rudimentary at the time. Eschewing the use of a toilet bucket, common in almost every prison, some into the mid-20th century, Haviland provided a cell flush toilet years before they were available in the White House and central heating before the U.S. Capitol had it. Showers, apparently the first in the country (and where the inmates were taken individually about every 2 or 3 weeks), were in place before those installed shortly thereafter in a first-class Boston hotel. Although the heating, ventilation, and plumbing were far from perfect, Eastern State Penitentiary was clearly using cutting edge technology. Because the inmate was not to leave his or her cell, it also had to serve as a workshop. This resulted in large cells, even by 21st century standards, that were 8 feet wide and from 12 to 16 feet long, most with an attached exercise yard. All of these features did not come cheap. The initial appropriation of the state legislature barely covered the cost of constructing the perimeter wall and an elaborate Gothic front building. The overall cost per cell of $1,800, compared with Connecticut's $150 per cell, caused a minor scandal.[2] But in spite of its critics, Eastern State Penitentiary became Philadelphia's pride and joy, a cause for celebration before it was even completed. As an attraction for travelers, Cherry Hill was said to rank with Niagara Falls and the U.S. Capitol.

Keeping inmates absolutely out of contact with other inmates during their entire sentences was easier to put forth as a correctional principle than it was to carry out.[3] These precautions proved both expensive and difficult. Prisoners were not to learn each other's name or to ever see one another's face. Entering prisoners were led to their cells with hoods over their faces. This was also done on any occasion when they were taken out of their cell to the dispensary or to the showers. Numbers rather than names were used during their entire sentence. The separate regimen required that each prisoner have not only his or her own cell but also his or her own exercise yard, which was attached to the cell. Prisoners were let out in their yards at different times so that no conversations could take place from one yard to the next. Because inmates could not leave their cells to empty slop buckets, as was customary at other prisons, cell plumbing was necessary. For the same reason, food had to be delivered to each cell by guards. Eastern State Penitentiary also had to restrict the prisoners to in-cell work such as cigar making, shoe-making, and textile production. As inmates could not gather in a church for worship, services were conducted at the end of each cell-block. The wooden cell doors were opened so the inmates could listen to the sermon through the iron latticework inner door, but a canvas was strung the length of the corridor so that inmates could not see one another.

Prisoners in the first years of the penitentiary were inside a bubble which no outside, uncontrolled influences could penetrate, and they were not allowed visits from their family nor were they permitted newspapers. They were sealed off from the outside world like unwilling monks. Their human contacts consisted of infrequent visits from official visitors from the Pennsylvania Prison Society or from a minister, a trade instructor, or the guard who brought the meals and work materials. All of these extreme measures to insulate inmates from the corrupting influence of fellow prisoners were at the same time intended to allow the prisoners, in their solitude, to reflect on the error of their ways and to be exposed to moral guidance, appropriate reading, and to regular work habits through some instruction in a trade.

⊠ How Separate Was Separate Confinement?

Serious scandals surfaced 5 years after the prison opened. A joint state legislative committee held hearings. It became clear that the separate system, as portrayed to the public and to the proponents of the rival Auburn system, was far from the reality of everyday life inside Cherry Hill. For a variety of reasons, some inmates were out of their cells, walking around without masks and unescorted by an officer,

talking to one another, having the freedom of the prison yard. From the beginning, inmates were sometimes used as helpers for the carpenters and stonemasons constructing the prison. Inmates performed some maintenance services such as keeping the stoves in the cell-blocks supplied with fuel and, in the case of female prisoners, working in the kitchen and the laundry. Although these flagrant deviations from the publicly presented picture of the Pennsylvania or separate system of prisoner treatment that the investigation revealed undoubtedly resulted in tightening the rules, the costs of using hired labor for all the housekeeping and maintenance made it tempting to continue using inmates as workers. Prisoners classified as "invalid" continued to be allowed to work in the gardens maintained in the spaces between the original seven cellblocks. It is clear that guard supervision would have been insufficient to prevent those inmates from communicating with each other and even perhaps with prisoners in their exercise yards.

The planners of Eastern State Penitentiary intended that the inmates, confined to their cells and released to their exercise yards when their neighbors were not, would not be able to communicate with each other. Such was not always the case at Cherry Hill, just as it has not been the case in other prisons past and present where such contact has been forbidden. Some inmates were able to communicate through the sewer pipes during periods when they were empty, and some developed rapping codes that were tapped on the walls or heating pipes.[4] Prisoners threw notes weighted with pebbles over their exercise yards, making contact with prisoners even two cells away. The investigation of 1834 and 1835 revealed that an inmate housed in a two-story cellblock, while out in the exercise yard, could speak with the occupant of the cell directly above through the ventilation flue or, standing on a workbench or stool, could communicate through the skylight with neighbors. These skylights were permanently closed in 1852.

But the dirty secret of Pennsylvania's separate system was that over-capacity inmate numbers, due to the extremely rapid growth of Philadelphia, resulted in double celling, although the state still maintained a public posture of a system based on strict separation. As early as 1841, the visiting committee from the Pennsylvania Prison Society reported with disapproval that some cells contained two convicts. Prior to a major building program in 1876, 795 inmates occupied 585 cells. More than half of the prisoners in a penitentiary organized around the principle of separation were sharing cells. Before the turn of the century, the prison's population approached 1,400, with as many as four inmates occupying one cell.

The Pennsylvania system, fiercely defended by its local partisans against the rival Auburn system, never maintained strict seclusion for all of its inmates and, without public acknowledgment, seriously eroded throughout the 19th century. The reality within its walls was finally recognized with a repeal of relevant legislation in 1913.

⚑ Eastern State Penitentiary's Influence in the United States

The idea of quarantining criminals one from the other during their imprisonment seemed at the time a sensible solution to the problems of disorder, bad associations, and other problems common in the prisons of the late 18th and early 19th centuries. To the Philadelphia reformers, separate confinement appeared to solve these problems nicely. But there was a catch; in fact there were several catches. The system turned out to be very costly. The physical plant was more expensive to build. Eastern's cells, because they were workrooms, were large, and initially each had an attached exercise yard. Prison labor, limited to production in individual cells and unavailable for routine maintenance work around the prison, was underutilized in a country that had a chronic shortage of labor in the 19th century.

Some early attempts to institute the Pennsylvania system in other state's prisons, such as New York's, were soon abandoned. They too found the system to be expensive not only to build but also to run. This came at a time when some states were claiming that their prisons were self-supporting and even making a profit because of inmate factory work that was producing for a civilian market without restrictions on prison-made goods. Cherry Hill was forced to confine itself to in-cell production and, in theory, to hire outside labor for maintenance tasks that, elsewhere, were done by inmates. As machinery was introduced into manufacturing consumer goods such as shoes, Pennsylvania law forbade the use of machines in prison labor. In the 19th century United States, with a chronic shortage of laborers, the underutilization of prisoners' labor and the expense of the system doomed its widespread adoption around the country.

There were also disquieting rumors of mental illness brought on by the relative isolation inmates experienced in the Philadelphia prison. In New York and Massachusetts, an alternative architecture and program, known as the Auburn system, was developed. Here, prisoners slept in tiny cells. Those at Auburn and Sing Sing prisons were 7 feet 6 inches by 3 feet 6 inches and 6 feet 3 and a half inches, respectively; Kingston Penitentiary in Canada, designed and built by Auburn's master builder, contained sleeping rooms 6 feet 6 inches by 2 feet 6 inches. Sing Sing's original single cellblock housed 1,000 inmates on five tiers of cells in what was once described as a human filing system. Warden Elam Lynds at Auburn, a former professional soldier, instituted a machine-like, meticulous routine for the prisoners, designed to regulate every aspect of their behavior at all times. Iron discipline and harsh punishments further maintained order. To avoid the expenses linked to the Pennsylvania system while still preventing inmates from communicating, the so-called silent system was created. Inmates were never to speak unless spoken to by an officer and were marched with military precision to and from workshops and mess halls. Ultimately, other U.S. prisons in the 19th and early 20th centuries followed this Auburn-Sing Sing regimen along with its associated cellblock construction of multitiered inside cells.

⌖ The United States' Second Cultural Export

Construction of Cherry Hill had created an international sensation. Ideas on penal treatment began to influence European reformers as early as 1796.[5] The experiments in penal treatment pioneered in Pennsylvania and New York came to the attention again of Europeans in the 1820s now seriously searching for a system of penal treatment which would be effective. Philadelphia, a small city at that time in a relatively unpopulated country, came to be recognized as the world center of prison reform. In the process of evaluating the competing U.S. systems, government delegations and notables from all over the world came to the United States and particularly to Philadelphia to view Eastern State Penitentiary. Heads of government during this time often took an active interest in prison reform: Frederick William IV, King of Prussia, visited new prisons in England, and King Oscar of Sweden published a book in 1840, translated into several languages, discussing noteworthy prisons, including Philadelphia's Walnut Street Jail and the Eastern State Penitentiary, and even including a radial prison plan. Another influence on world prison reform was the International Penal and Penitentiary Congresses (IPPC), held periodically from 1846. Beginning with the first congress in Germany, the IPPC favored the Pennsylvania system and its architecture. As a consequence of these and other professional meetings and books, the system of separation that was first tried in a limited way at the Walnut Street Prison and later at Eastern State Penitentiary, along with the latter's distinctive architecture, was to become the second major cultural export to the rest of the world.[6]

The first delegation to visit the yet-unfinished Cherry Hill was headed by Sir William Crawford, whose report to the British government favored the Pennsylvania system and its architecture. In one of those strange twists of history and cultural diffusion, the radial plan originally developed in British and European prisons was not widely used for large prisons in Europe until after it had been transplanted to the United States by Haviland. Pentonville prison in London, opened in 1842 and still in use, became the template for a total rebuilding of English prisons. All were more or less variations of the Eastern State or Haviland's Trenton prison plans and operated with some form of separate confinement into the 20th century. Having visited the Philadelphia prison or its offspring, Pentonville, and constructed so-called model prisons in their capitals, all other countries in Europe followed England's early lead. In the case of Belgium, Spain, and Germany, their entire prison systems were rebuilt, using Eastern State and Pentonville models of multiple wings radiating from a central rotunda and the separate system. In countries such as Russia, France, and Italy, regime changes, wars, and chronic shortages of funds resulted in limited construction of new prisons in the 19th century. One of the first effects of Westernization in Japan was a new concern for prison reform. British colonial prisons were studied, and officials were sent to the United States and Europe. The result was the enthusiastic adoption of the use of separate confinement and the acceptance of the radial prison plan. The first of over 36 new radial prisons was opened in 1879 at Miyagi. Such plans continued into the 1930s. Prison reform came later to China. By 1918, the entire prison system had been rebuilt, resulting in about 40 radial prisons, most of which are still in use.[7] In Europe, separate confinement as it had developed in Philadelphia continued long after its de facto abandonment at Eastern State Penitentiary.[8]

⊠ Fall From Grace

As large prison factories became common-place in U.S. prisons, Eastern State Penitentiary continued to maintain the public image of separation of inmates from one another, ultimately using the term individual treatment instead of separation, until the fiction of separation was finally ended by law in 1913 and a congregate regimen, already a reality, was publicly recognized. For years, the Pennsylvania Prison Society and Eastern State had resisted reforms and developments occurring elsewhere. When the system of separation was publicly abandoned, the prison gradually modernized by introducing new programs and becoming, in contrast to its past, not the center of new and daring efforts to reform but just another U.S. prison with aging, albeit unique, buildings. Outmoded and expensive to maintain, it closed as a state penitentiary in January of 1970, when its last remaining inmates were moved to Graterford prison, which was built as Eastern's replacement and opened over 40 years earlier. In 1988, a task force was formed to prevent developers from destroying the prison, and it reopened 11 years ago as a national historic site.

⊠ Success or Failure?

Granted that Eastern State Penitentiary was an enormously important milestone in the history of prisons and corrections, but the question must be asked, did it work? The answer of course depends on what one's expectations are for imprisonment: rehabilitation, punishment, safe containment of lawbreakers, or profit from prisoners' labor or at least not be an undue financial burden on the government.

Concerning earlier prison experiments, the matter of the effective reform of prisoners is much harder to resolve without the benefit of a national fingerprint file, computerized records, and long-term studies that occasionally are carried out today on recidivism.

Anecdotal success stories always have been gathered by supporters as proof of a prison's effectiveness. Failures were known only if the former inmate was sentenced in the same state or returned to the same prison. Comparable recidivism rates for specific prisons were never available. What is clear, however, is that the founders of Eastern State regarded rehabilitation of its inmates to be their primary goal. In the mid-19th century, this was in dramatic contrast to the opposing system developed in New York State by Warden Elam Lynds.

The goal of punishment by imprisonment would seem to be a given, although in the 20th century complaints that particular prisons were not punitive enough surfaced regularly. As a matter of fact, such complaints had appeared from time to time in the past when prisoners were first given free schooling or recreational facilities or when the exterior of the prison seemed too costly and attractive. Eastern State, devoid of the severe physical punishments of other contemporary prisons and institutions in the 19th century, but characterized by a high degree of solitude, was accused of mental rather than physical cruelty, and there seems little doubt that it was so experienced by most of its inmates. Other prisons, although they used isolation as one form of punishment, relied heavily on frequent and severe physical punishments. Severe physical punishments, including whipping, survived in some U.S. prisons and juvenile institutions even into the second half of the 20th century.

Was Eastern State Penitentiary successful in containing its inmates and preventing escapes? Existing evidence makes comparisons with other prisons impossible, but it appears that, compared with earlier prisons, the Philadelphia prison was relatively secure. When escape attempts were temporarily successful, carelessness of guards was usually the cause.

On the matter of cost, Eastern State Penitentiary clearly was a loser. The warden was responsible for purchases of raw material and marketing of products produced by inmates in their cells. Long after products such as shoes were machine-made, the prison legally was required to make them by hand. Maintenance tasks around the prison, on principle, had to be carried out by paid staff. These internal operational costs, as well as the initial outlay for the more elaborate prison structure, meant that the system was expensive both to build and to operate at a time when other prisons were claiming self-support and even a profit.

Although Eastern State Penitentiary may not have met its founders' and supporters' sanguine and idealistic expectations, it should not be forgotten that the system clearly did remedy many of the intractable ills of the prisons it displaced and, in fact, those of the prisons which followed in other states. The system of relative isolation for all practical purposes eliminated the disorder, exploitation, and corruption that were implicit in congregate housing. Riots and disorders were rare while the separate system survived. Inmates could no longer exploit other inmates, and prison officials no longer tolerated shake-downs, alcohol, and mixing of male and female prisoners. When both the silent system and the separate system were abandoned in the 20th century, and when a freer mingling of inmates in prison became common, the old evils—inmate exploitation of other prisoners, violence at the hands of both guards and fellow prisoners, and a return to the full effects of the criminal subculture or inmate community with its attendant assimilation of values, attitudes, and behaviors inimical to adequate adjustment outside prison—returned.[9]

The success of several other issues connected with Eastern State's system of separation is more ambiguous. One was the matter of the health of prisoners. Prisons had never been healthy places. But the reformers were determined that the new prison be a healthy environment with features well beyond those of any other prison of the time—shower baths and individual toilets and a water tap in each cell. When the prison was built, the technology for providing adequate heating and ventilating of large buildings was primitive and experimental. Inmate cells at Eastern were cold and damp in

the winter and hot and damp in the summer. Nevertheless, the inmates most certainly suffered less from the cold and stale air in their cells than did other prisoners of that period. Were Cherry Hill inmates physically healthier than those in earlier prisons? Undoubtedly. Were they healthier than inmates elsewhere at the time? As with recidivism rates, the evidence does not exist to make such a judgment.

One of the persistent claims of opponents of the Philadelphia prison's regimen in both the United States and abroad was that it adversely affected the mental health of prisoners. It must be remembered that all sorts of mentally fragile and all but the most dramatically mentally ill passed through the criminal justice system in the 19th century largely unremarked and undetected. However, the system of cellular isolation, though not complete, undoubtedly resulted in some mental breakdowns that would not have occurred in other prison settings. The elimination of the Pennsylvania system, of course, was not the end of isolation in prison. It continues to be used for prison rule infractions and as a model for the new supermax prison. It is the reincarnation of Eastern State Penitentiary without its humanitarian ideals.

What then should be the final judgment of the bold new social experiment in corrections carried out in Philadelphia beginning in 1829? Cellular isolation had earlier been tried briefly in several English prisons, but it was without the carefully laid out details used in Philadelphia and without an architecture enabling the system to be carried out. Haviland's genius was not in his radial plan itself, which he merely adapted from many examples in England, but rather in the care and lavish attention to details in the cells and in the use of cutting edge heating and plumbing, all with consequent heavy outlays of public funds. The system of 24-hour cellular separation diminished the need for the radial layout. Ironically, the practical utility of Haviland's plan was increased once the Pennsylvania system was abandoned and inmates circulated about the prison. However, the Eastern State Penitentiary plan became firmly linked with the Pennsylvania system in the minds of reformers, and everywhere in the world where that separate system was tried, radial style prisons were built. Although the Philadelphia prison was always mentioned, it was the half-circle array of cellblocks without attached exercise yards of Haviland's Trenton State Prison that was most frequently duplicated abroad and in the United States. Foreign reformers never mentioned the Trenton prison.

A mid-20th century Harvard psychiatrist was once characterized by one of his colleagues as knowing in his research where the big fish were but not having the right bait (i.e., being aware of a problem that needs attention but lacking the techniques to find its solution).[10] The Philadelphia reformers knew clearly what was wrong with prisons. They thought they had a solution that would be both humane and effective in bringing about law-abiding behavior, but they observed, perhaps more realistically, that should rehabilitation not take place, the experience of enforced isolation would be enough to deter further crime, even without a change of heart. Although their experiment in modifying criminal behavior may not have been particularly successful, they created a system of prison architecture and a coherent philosophy of treatment that influenced prisons around the world into the early 20th century and beyond. Although some prisons have enjoyed greater notoriety or association with historic events—the Bastille, Newgate, Alcatraz, and Devil's Island—no other single prison has ever come close to wielding such influence in the field of corrections. But more than a prototype of architecture and a strategy for rehabilitation, Eastern State Penitentiary was a remarkable symbol of the optimism, energy, and good intentions of that period of the 19th century during which Philadelphia and the rest of the country believed that anything was possible in the new republic.

✎ Notes

1. Strickland's prison has the dubious distinction of being the shortest-lived permanent prison structure ever built, being razed after less than 7 years of operation.

2. Gustave de Beaumont and Alexis de Tocqueville, in their classic and even-handed description of the competing systems they found in the United States, *On the Penitentiary System in the United States*, characterized Cherry Hill as "truly a palace," complaining that "each prisoner enjoyed all the comforts of life."

3. Although official reports did not discuss it, female prisoners were frequently used in the kitchen and laundry and probably in other tasks outside their cells so that the strict regimen of separate confinement was not usually applied to them.

4. Frederick Wines, a 19th century penologist, described such codes used in European prisons in the early 20th century as consisting of a combination of sharp and dull raps for each letter of the alphabet. He mentions a Russian fortress prison where cell floors and walls were covered with felt and where a wire netting, covered with paper, was strung 5 inches from the wall, all to prevent communication by rapping.

5. An influential Frenchman, François Duc de la Rochefoucauld-Liancourt, became interested in the reforms at the Walnut Street Prison in Philadelphia. His book, *On the Prisons of Philadelphia*, was published in four subsequent editions in French and in Spanish, bringing the Philadelphia reforms to a larger European audience.

6. Although U.S. architecture had represented British and Continental influences, Eastern State Penitentiary has been characterized as the first U.S. architectural contribution exported to Europe. It must be considered the U.S. building most widely imitated in Europe and Asia in the 19th century. No other U.S. building form, until the modern skyscraper, played such a seminal role.

7. In the large cells originally intended for work and sleeping for one inmate, I have observed arrangements for sleeping ten on two sleeping platforms in a cell.

8. When, in 1958, I visited the large departmental prison at Fresnes, outside Paris, the director expressed to me his concern that prisoners could see each other from one wing to the next. They were still being fed in their cells and exercised in individual yards.

9. The concept of prisonization was first put forward by Donald Clemmer (1940) in his pioneering study of the inmate social system in the Illinois State Penitentiary at Menard.

10. The researcher was William H. Sheldon, and his work, attempting to link body build and behavior, especially with criminality, was prominent but controversial in the mid-20th century. See his *Varieties of Human Physique* (1940).

References

Beaumont, G., & Tocqueville. A. (1833). *On the penitentiary system in the United States* (F. Lieber, Trans.). Philadelphia: Carey, Lee, and Blanchard.

Clemmer, D. (1940). *The prison community.* Boston: Christopher.

Johnston, N. (1955). John Haviland (1792–1852). *Journal of Criminal Law, Criminology and Police Science, 45,* 509–519.

Johnston, N. (1994), *Eastern State Penitentiary: Crucible of good intentions.* Philadelphia: Philadelphia Museum of Art.

Johnston, N. (2000). *Forms of constraint: A history of prison architecture.* Urbana, IL: University of Illinois Press.

Sheldon, W. (1940). *Varieties of human physique.* New York: Harper and Row.

Teeters, N. (1955). *The cradle of the penitentiary.* Philadelphia: Pennsylvania Prison Society.

Teeters, N., & Shearer, J. (1957). *The prison at Philadelphia: Cherry Hill.* New York: Columbia University Press.

Vaux, R. (1872). *Brief sketch of the origin and history of the State Penitentiary for the Eastern District of Pennsylvania, at Philadelphia.* Philadelphia: McLaughlin Brothers.

DISCUSSION QUESTIONS

1. Why did Pennsylvania become the center of prison reform in the 1800s?

2. What made Eastern State Penitentiary different from other prisons of its era? In what ways did it change how prisons were designed and operated?

3. What were some of the successes of the Eastern State system? Some failures?

❖

READING

In this book chapter excerpt, Nichole Hahn Rafter examines gender differences in prisons at various locations in the United States during the 19th and early 20th centuries, highlighting the atrocities that the women had to endure and the reforms that took place. Although women inmates were separated from the men inmates, they were often the sexual targets of male guards and inmates. Rafter notes that female criminals were regarded as worse than male criminals because criminal behavior was so atypical of their sex. She notes that the first prison reform involving female offenders was placing matrons in women's prisons with the hope that matrons would serves as middle-class models for female convicts. The overall ideal, as it was for male offenders, was to reform female convicts by inculcating moral values and by educating them. Rafter introduces us to the Mount Pleasant Female Prison, the first all-female prison in the United States.

"Much and Unfortunately Neglected"

Women in Early and Mid-Nineteenth-Century Prisons

Nicole Hahn Rafter

The early nineteenth century witnessed the emergence of one of the most dramatic innovations in the history of punishment: the penitentiary, a fortress-like institution designed to subject prisoners to total control. Although historians argue about why the penitentiary came into existence at that time, most agree about the nature of convict life within such prisons. According to the usual picture, penitentiaries were designed to isolate inmates from the moral contamination of other felons; unlike the very first state prisons, in which several inmates were held together in one room, penitentiaries separated convicts into individual cells. Some held prisoners in perpetual solitary confinement, while others herded them together during the day for labor. But in both "separate" and congregate penitentiaries, speech and even eye contact were forbidden. In congregate institutions especially, strict routines governed every activity. Convicts with shaven heads and identical striped uniforms rose with the morning bell, marched in lockstep to their meals and workshops, and returned in the evening to their cells. Officials scorned idleness as corrupting; they scheduled every moment of their charges' lives, mainly for the labor that, in congregate penitentiaries, was expected to be financially profitable to the institution as well as morally profitable to the prisoners. "The doctrines of separation, obedience, and labor," writes David Rothman in a typical description, "became the trinity around which officials organized the penitentiary."[1]

Few historians of the penitentiary have noted that women as well as men inhabited these gloomy institutions. Had they investigated the treatment of incarcerated women, they would have found that in nearly every respect, it contradicted the usual picture of penitentiary discipline. Women were punished in nineteenth-century prisons, but few officials tried to transform them into obedient citizens through seclusion and rigorous routines.[2]

Even after states replaced their original prisons with penitentiaries, many continued to hold a number of women en masse in old-fashioned large cells, inside penitentiary walls but away from the men's cellblocks. Women did not receive the supposed benefits of unbroken silence and individual isolation. Exempted from the most extreme forms of regimentation, they encountered other sorts of deprivation. Descriptions of women's conditions in Jacksonian prisons emphasize the intolerable noise and congestion of their quarters. Whereas male prisoners were closely supervised, women seldom had a matron. Often idleness rather than hard labor was their curse. As time went on and women were transferred from large rooms to individual cells, their treatment became more like that of men. But in general, female convicts in nineteenth-century prisons experienced lower levels of surveillance, discipline, and care than their male counterparts. Describing the situation of women in his institution as "inhuman—barbarous—unworthy of the age," one penitentiary

---

SOURCE: Nicole Hahn Rafter. Chapter 2: "Much and Unfortunately Neglected": Women in Early and Mid-Nineteenth Century Prisons. In *Partial Justice: Women, Prisons, and Social Control, 1800–1935.* © by Nicole Hahn Rafter. Reprinted with permission of University Press of New England.

chaplain concluded that "To be a *male* convict in this prison, would be quite tolerable; but to be a *female* convict for any protracted term, would be worse than death."[3]

To illustrate these differences between the sexes, we will look in some detail at the conditions of women incarcerated from the late eighteenth through the mid-nineteenth centuries in the prisons of New York, Ohio, and Tennessee—the three states that this study will use throughout to exemplify penal practices in the Northeast, Midwest, and South. (Because the southern prison system developed more slowly than that of the Northeast and Midwest, several late nineteenth-century changes are used to illustrate southern practices; because the prison system of the West developed later still, no western representative is included here.) These examples indicate diversity in regional styles of incarceration, but they also show that women imprisoned in the various regions had much in common. Their conditions of confinement, while outwardly resembling those of men held in the same institutions, were often inferior.

◪ Differential Punishment: Women in the Prisons of New York, Ohio, and Tennessee

New York's Newgate prison, opened in 1797 in the Greenwich Village section of New York City, was the first state institution established to hold felons only. In it, as in other prisons of the pre-penitentiary era, there were no marked differences in the handling of the sexes. Evenhandedness was not to remain the rule for long, but in Newgate and other original state prisons, officials had little alternative. Like Newgate's male convicts, women were lodged in chambers "sufficient for the accommodation of eight persons." Their quarters were separated, for women resided in a north wing with "a courtyard entirely distinct from that of

the men"; yet the institution's small size did not permit women to be isolated from the mainstream of prison life, as they were later isolated in penitentiaries. The women had no matron, but as several lived together in each room, they could protect one another from lascivious turnkeys. Thus they were less exposed to sexual attack than women later held in individual penitentiary cells. Newgate's women were required to wash and sew, while the males were assigned to shoemaking and other manufactures. However, because profit making had not assumed the importance that it later did in penitentiary management, the women's apparently lower productivity did not yet furnish an excuse for inequitable care.[4]

Treatment of both male and female felons changed radically when, about 1820, the penitentiary system was inaugurated at New York's new Auburn State Prison. Auburn's disciplinary methods—the individual cell, lockstep, prohibition on prisoner communication, harsh punishments for rule infractions, hard labor—captured the imagination of penologists throughout the western world and soon became staples in penal regimens. With the advent of penitentiary discipline, New York closed the outmoded Newgate, transferring that prison's male inmates to Auburn. But women did not receive the benefits thought to accrue to penitentiary discipline for another decade. Instead they became pawns in a heated dispute between Auburn and another New York state penitentiary that opened somewhat later at Sing Sing. Neither wanted the women, who were shunned as a particularly difficult type of prisoner; each made strenuous efforts to ensure that females would be sent to the other location. While the men's prisons engaged in this squabble, women formerly incarcerated at Newgate, along with others subsequently committed from the New York City area, were held at the city's Bellevue Penitentiary.

At Bellevue, standards sank far below those established at the state penitentiaries, for aside from semiannual visits by the inspectors of Sing Sing (who technically had custody of these women), the female prisoners almost wholly lacked supervision. No matron was hired to attend to their needs and maintain order. Visiting state officials lamented the wretchedness of conditions at Bellevue: the impossibility of separating old from young, and hardened criminals from novices; the women's "constant and unrestrained intercourse" (a fault of special seriousness at a time when most penologists endorsed the silent system); the poor quality and quantity of the food; the lack of a matron; and the absence of proper sanitary and security precautions (during a cholera epidemic, eight women died and eleven escaped). But these complaints had little effect on Bellevue's officials, who also actively sought to avoid responsibility for the state's female convicts.[5]

While these conditions prevailed for New York City–area women, courts in the western part of the state began, in 1825, to commit females to Auburn. There, however, they were housed not in cellblocks but in a third-floor attic above the penitentiary's kitchen. Like their Bellevue counterparts, they suffered extreme neglect. Until a matron was hired in 1832, women at Auburn had no supervision. Once a day a steward delivered food and removed the waste, but otherwise prisoners were left to their own devices. Their lack of protection from one another, and the psychological strain of being forced to share an overcrowded, unventilated space, sharply distinguished their care from that of men in the nearby cellblocks. Visiting in the early 1830s, Harriet Martineau reported a scene of almost complete chaos:

> The arrangements for the women were extremely bad. . . . The women were all in one large room, sewing. The attempt to enforce silence was soon given up as hopeless; and the gabble of tongues among the few who were there was enough to paralyze any matron. . . . There was an engine

in sight which made me doubt the evidence of my own eyes; stocks of a terrible construction; a chair, with a fastening for the head and for all the limbs. Any lunatic asylum ought to be ashamed of such an instrument. The governor [warden] liked it no better than we; but he pleaded that it was his only means of keeping his refractory female prisoners quiet while he was allowed only one room to put them all into.[6]

Reports such as Martineau's—together with the scandal that ensued when one Auburn inmate became pregnant, was flogged while five months into her pregnancy, and later died—finally forced New York to construct regular quarters for its female felons. This was the Mount Pleasant Female Prison, to which Auburn and Bellevue inmates were transferred in 1839. Nearly two decades had passed since New York had indicated, at Auburn and Sing Sing, that individual cells, close supervision, and reformational discipline were desirable for prisoners.[7]

The development of Ohio's prison system paralleled that of New York's. Like Newgate, Ohio's first prisons resembled large houses; they were relatively small buildings, and their cells opened off central corridors. In 1834, Ohio abandoned this nonsecure type of structure, substituting a penitentiary patterned after Auburn. Like Auburn and Bellevue, the Ohio penitentiary segregated female prisoners into separate quarters. But in two respects, the penal practices of Ohio diverged from those of New York. First, Ohio had much lower standards for handling prisoners of both sexes: throughout the nineteenth century, observers ranked the Ohio penitentiary as one of the worst prisons in the country. Diseases ravaged the population, administrative corruption flourished, and, as Dorothea Dix put it in an indictment scathing even for her, the institution was "so totally deficient of the means of moral and mental culture . . . that little remains to be said, after stating the fact." Second, Ohio's prisoner population was smaller than New York's, reflecting its smaller general population. Whereas New York's relatively large number of female convicts pushed the state into creating a separate women's prison at mid-century, Ohio was able to wait until the early twentieth century to do so.[8]

The Ohio penitentiary developed a novel method of sequestering females, that of building a Women's Annex adjacent to the institution but outside its front wall. Constructed in 1837, the annex was one of the earliest extramural structures in the country designed specifically for female state prisoners. Originally the annex consisted of eleven two-person rooms and a yard. Crowding later necessitated construction of additional cells, but because the annex had been jammed between the perimeter wall and the street, it could not be expanded, and the women's quarters became increasingly cramped. Like the men's section, the annex sometimes fell into such disrepair that it was impossible to keep out the elements. The wings of the men's prison, "which have leaked for years," were recovered in 1850 with cement. "The female prison has been served in the same way," according to the annual report, but "Much more needs to be done by way of improvements." Similar observations about the miserable state of the annex were made for the next sixty years.[9]

As in New York, segregation of women led to their neglect. The men's section was patrolled by guards, but until 1846, the annex had no matron. Thereafter, owing to underfunding and political turmoil in the central administration, supervision remained sporadic and inadequate at best. As a result, discipline was often more lax for female than male prisoners. Lack of discipline was accompanied by absence of other forms of attention and control. Unguarded, the women were vulnerable to unwelcome sexual advances by male officials. Pandemonium sometimes prevailed. Gerrish Barrett, a representative of the

Boston Prison Discipline Society who visited the Ohio penitentiary in the mid-1840s, reported that although there were only nine women, they gave more trouble than the five hundred male convicts. "The women fight, scratch, pull hair, curse, swear and yell, and to bring them to order a keeper has frequently to go among them with a horsewhip." That there was some accuracy in Barrett's description is indicated by Dorothea Dix's independent observation of about the same period: "There was no matron in the woman's wing at the time I was there, . . . and they were not slow to exercise their good and evil gifts on each other." Later in the century, former prisoner Sarah Victor wrote, "[T]he knives had all been taken from the female department, to prevent some refractory prisoners from cutting each other, which they had done, in a terrible manner, at times. . . ."[10]

Even at the Ohio penitentiary, strict rules were sometimes imposed on women. Sarah Victor reported that in the early 1870s the "discipline of the prison was very strict. . . . the prisoners not being allowed to speak to each other. . . ." And on occasion, women were punished severely. Victor describes one woman beaten so terribly "that she was black-and-blue all over her body." She herself, when she arrived at the penitentiary about 1870, was kept in solitary confinement for five months. In 1880, a new matron alluded with awe to a brutal penalty used by her predecessor; she hoped never to resort to it herself. This was probably the "hummingbird," a form of punishment that forced the naked offender to sit, blind-folded, in a tub of water while steam pipes were made to shriek and electric current was applied to the body.[11]

At times, then, women in the annex did taste the bitterness of penitentiary discipline. At others, they were practically free of control. But because leniency often went hand in hand with anarchy, it was not necessarily preferable to the austerities of a solitary cell and close surveillance.

Tennessee's first prison, opened in Nashville in 1831, also adopted the Auburn system of convict discipline: inmates were brought together during the day for silent labor (for the women, mainly sewing), and at night they were locked in individual cells. For its first ten or fifteen years of operation, the Tennessee State Penitentiary was a relatively progressive institution. More than Ohio's prison, it attempted to approximate New York's standards of care. By 1845, however, decline had set in. Concern with maximizing profits from prison labor, combined with increasing preoccupation with the issue of slavery, corroded the quality of convict care. Tennessee started exhibiting the indifference to prisoner health and safety that came to characterize penal treatment throughout the South. On the eve of the Civil War the institution lacked an adequate water supply, and the warden was forced to inquire of the legislature, "What shall be done with the excrement arising in the prison in the future? You are aware it has been deposited on a vacant lot adjoining the prison property for the last fifteen years." The Civil War destroyed any possibility that Tennessee might have returned to northern standards. As in other Southern states, the conflict severely damaged the prison system, and prisoners were virtually forgotten. After it was over, . . . the penitentiary filled with newly freed blacks and began to replicate the techniques of slavery.[12]

Few women were held in Tennessee's penitentiary before the Civil War. The first male prisoners had arrived in 1831; the first woman, sentenced to the institution in 1840, had been preceded by 453 men. In outward respects, they received care similar to that of men. Apparently they were not even isolated in a separate section of the prison. In the early 1840s, Governor James K. Polk appealed for "suitable apartments" in which to segregate the women, but his recommendation went unheeded for four decades. So egalitarian was the penitentiary in its treatment of the sexes that, after the Civil War, it sent women to labor alongside men in coal mines and on railroads.[13]

But despite surface similarities between the care of female and male prisoners, women held at the Tennessee penitentiary in fact experienced disadvantages. Because of their low commitment rate, the institution's failure to hire a matron, and its post-war policy of assigning some women to work-gangs, the women were nearly (sometimes entirely) alone among the men, isolated not by a policy of segregation (as in the North) but by their small numbers and sex. The issue of privacy was always a problem for the women. Sexual exploitation through rape and forced prostitution was constantly present as a threat, if not an actuality. Thus the women's experience of incarceration must have been one of greater loneliness, vulnerability, anxiety, and humiliation than that of men at the Tennessee penitentiary.[14]

Not until the 1880s did the Nashville institution begin to separate women from the main population of prisoners. At this point it segregated women in an upper story of the entrance building, much as Auburn had isolated women above its kitchen fifty years before. Tennessee's slowness in sequestering women was due to several factors: the minuscule size of its female population until after the Civil War; the war itself, which totally disrupted prison management; and the indifference to prisoner care that formed part of the southern penal tradition. As in other states, once women were segregated, aspects of their care deteriorated. They came to suffer from extreme overcrowding: in 1894, forty-five women were incarcerated in a wing of sixteen cells. In their separate female department, women had no employment or other programs, not even a place to exercise. And until a matron was hired at the century's end, they were almost entirely cut off from prison personnel. These conditions were not worse than those of Tennessee's male prisoners, many of whom experienced forced labor as well as desperate crowding. But isolation from the men meant a change in the sources of misery for women, just as it had in New York and Ohio: earlier, women suffered from superficial egalitarianism; later, from segregation.

✉ Stages in the Segregation of Female Prisoners and Sources of Unequal Treatment

As these examples indicate, there were phases in the process by which female prisoners were separated from the main, male populations. Patterns differed from state to state, but broadly speaking, there were three stages in the separation process. During the first, women formed part of the general prison population; they were confined in large rooms or individual cells but not further isolated. During the second, they were removed to separate quarters within or attached to the men's section—to the kitchen attic at Auburn, the annex at the Ohio penitentiary, and the upper floor of the entrance building in Tennessee. In the third stage, women were relocated to an even more isolated building on or near the main prison grounds. New York's Mount Pleasant Female Prison illustrates this third stage, as does the Woman's Building erected in the northwest corner of the yard of the new penitentiary opened by Tennessee in 1898. The evolution of separate quarters for women tended to be most rapid in the Northeast, a bit slower in the Midwest, slower still in the South, and most laggardly of all in the West. (In the late 1970s, a few western states still remained in the second stage, holding their few female felons in small units adjacent to their prisons for men.[15])

After the separation process began, female prisoners tended, in Francis Lieber's phrase, to be "much and unfortunately neglected." Their care paralleled that of men in the same institution, but the superficial resemblances concealed important discrepancies. Women were sometimes treated less severely. However, for each exemption they paid a corresponding price. Less discipline meant less supervision, and hence less protection from one another and male officers. When they lacked individual cells, women could not be classified by seriousness of offense or behavior in prison. In congregate cells they had more opportunity for conversation and companionship than men, but they suffered from lack of privacy. And in nearly every state, separation meant less access to fresh air, exercise yards, and prison staff.[16]

One source of differential treatment was the simple fact that so few women were sentenced to state prisons. In the late eighteenth and nineteenth centuries, women rarely made up more than 10 percent of a prison's population, often far less. Visiting in the early 1830s, William Crawford found only ninety-seven women in the penitentiaries of the seven most populous states and the District of Columbia. Fifteen years later, Dorothea Dix counted a grand total of 167 females in prisons from Maine through Virginia.[17] Beaumont and Tocqueville remarked of female inmates,

> It is because they occupy little space in the prison, that they have been neglected. It is the same with most evils of society, a remedy for which is ardently sought if they are important; if they are not alarming they are overlooked.[18]

Officials resented the extra demands that a few female prisoners placed on their resources. Chaplains and physicians found it bothersome to visit the female department after making their usual rounds. And to hire a matron to supervise a handful of women seemed an unwarranted expense.

Several nineteenth-century penologists addressed the question of why male prisoners so greatly outnumbered female ones. Francis Lieber found the answer in a lower female rate of crime. "In all countries women commit less [*sic*] crimes than men, but in none is the disproportion of criminals of the two sexes so great as in ours,"[19] This lower crime rate Lieber in turn attributed mainly to women's social conditions:

> Women commit fewer crimes from three causes chiefly: (1) because they are, according to their destiny and the

consequent place they occupy in civil society, less exposed to temptation or to inducement to crime; their ambition is not so much excited, and they are naturally more satisfied with a dependant [*sic*] situation; (2) they have not the courage or strength necessary to commit a number of the crimes which largely swell the lists of male convicts, such as burglary, robbery, and forcible murder; (3) according to their position in society they cannot easily commit certain crimes, such as bigamy, forgery, false arrest, abuse of official power, revolt, etc.[20]

William Crawford, on the other hand, suspected that chivalry, together with a need for female services and lack of cell space for women, worked to keep their numbers low:

> Few circumstances . . . impress a visitor more forcibly than the small number of females to be found in the penitentiaries of the United States. . . . I fear, however, that the criminal calendars do not convey a correct idea of the extent of crime among the female population: at least I have been assured that from the general sense which exists of the value of female services, particularly in those parts of the country which have been but recently settled, there prevails a strong indisposition to prosecute, especially if the offender be not a woman of colour. Magistrates are also reluctant to commit women from the circumstance of there not being any suitable prisons for their reception. With the exception of Pennsylvania and Connecticut, there is not a single State in which the treatment of female prisoners is not entirely neglected.[21]

In all likelihood, both Lieber and Crawford were partially correct. Today, too, far fewer women than men are committed to prison, because of lower rates of offending. Women are socialized to be more passive and nurturing than men, and as a result, they commit less crime. The same was true in the past. And in the nineteenth century, judges probably were reluctant to incarcerate women, especially those who seemed to remain within the bounds of respectability. As Sing Sing's chaplain put it in 1841, courts evidently refused "to send any female to the State prison save the *vilest of the vile.*"[22]

But it was not only their small numbers that made female convicts seem insignificant. To many officials, women were by definition more troublesome and less able than men. For example, women—the outsiders—were identified as the source of sexual trouble. The proximity of women was thought to drive men to the unhealthy practice of masturbation; the presence of women led to scandals when officers were discovered fostering prostitution or fathering children. One former inmate, a male, complained of Newgate's women that "The utmost vulgarity, obscenity, and wantonness, characterizes their language, their habits and their manners[,] . . . agonizing . . . every fibre of delicacy and virtue." Women alone seemed to be at fault in such matters. Enoch Wines and Theodore Dwight, national authorities on prison management, argued for entirely separate prisons for women "on moral grounds, because where the two sexes are confined in the same building or the same enclosure, the very fact of this contiguity has an exciting and bad effect, and leads to endless attempts to communicate, which are not unfrequently, against all probability, successful." Wines and Dwight found, moreover, that "where prisoners of different sexes are confined in the same building or enclosure it is often necessary to impede light and ventilation by half closing windows, and by putting doors across passages which would otherwise be left open, thus violating the laws of hygiene and obstructing an important condition of health." Such barriers were usually placed on the smaller, women's units. Thus women bore the burden for separation of the sexes, just as they often bore the blame for sexual disturbances in the first place.[23]

Neglect was also a result of female prisoners' apparent inability to earn their keep. Administrators had little interest in inmates who could not turn a profit. "The product of women's labor in the State prisons," Dix observed, "fails to meet the expenses of their department." Of the female department at Connecticut's Wethersfield prison, Crawford reported: "The directors have stated their conviction that no contract can be made for the profitable employment of this part of the establishment, after paying the expense of its support and management." In an era when prisoners were expected at least to support their institution, and sometimes to contribute significantly to state revenues as well, these were serious charges against female convicts. However, those who complained of the financial liability of female departments ignored the fact that women were simply not assigned to or contracted for high profit tasks. Provided with less lucrative work, they were in fact prevented from earning as much as males.[24]

Finally, differential treatment of women stemmed from the common belief that a female criminal was far worse than any male, depraved beyond redemption. (This conviction prevailed until the reformatory movement began.) In 1833 Francis Lieber stressed that "the injury done to society by a criminal woman, is in most cases much greater than that suffered from a male criminal," a phenomenon he attributed to woman's role as guardian of society's morals: beginning from a higher elevation, woman had further to fall. "[A] woman once renouncing honesty and virtue, passes over to the most hideous crimes which women commit, with greater ease than a man proceeds from his first offense to the blackest crimes committed by his sex." Thus, Lieber maintained, "a woman, when she commits a crime, acts more in contradiction to her whole moral organization, i.e., must be more depraved, must have sunk already deeper than

a man." Similarly, Mary Carpenter, the English authority on women and crime, wrote that "female convicts are, as a class, even more morally degraded than men." We hear echoes of this belief in the complete corruption of the female criminal in Gerrish Barrett's report that the nine women in Ohio's annex gave more trouble than the penitentiary's five hundred males. A similar conviction informed the first European treatise on female crime, in which Lombroso explained that the female offender outdistanced the male in primitive depravity. At bottom the conception was rooted in the archetype of the Dark Lady—dangerous, strong, erotic, evil—a direct contrast to the obedient, domestic, chaste, and somewhat childlike Fair Lady of popular imagery.[25]

Gender-based perceptions—that female prisoners were the source of sexual mischief; that they could not earn as much as men; that they had gone beyond the pale of redemption—combined with the problem of smaller numbers to create a situation in which women's needs were slighted. The same factors operated to ensure that, once the double standard of care developed in informal practice, it continued and, in some institutions, intensified as the years went by.

✉ Matrons, Lady Visitors, and Women's Prison Reform

Toward the end of the nineteenth century, women themselves led the movement to establish separate prisons for their sex, and women directed these institutions as well. To what degree were women involved in the care and supervision of female convicts earlier in the century? The answer to this question helps complete the picture of treatment of female prisoners in the early period. In addition, it indicates the limits on female supervision in predominantly male institutions, particularly as a means for promoting change.

By the mid-nineteenth century, a number of states had hired matrons to supervise their female convicts. As the female populations expanded, need for matrons became ever more obvious, and one legislature or prison administration after another grudgingly established the office of matron. Quite early, moreover, far-sighted penologists began to formulate a theory about matrons that later played an important ideological role in women's prison reform. According to this theory, matrons were necessary because female prisoners by nature needed special treatment that only other women could provide. This idea appeared, in embryonic form, in Lieber's observation of the early 1830s that

> a matron [is] necessary for the special superintendence of the female prisoners; she is quite indispensable if the Auburn system is applied to women as well as men; she alone can enforce the order of this system, while it is nearly impossible for male keepers. The whole spirit of opposition in womankind is raised against him [sic]. Besides, the moral management of female convicts must differ from that of male criminals.[26]

Over the next three decades this belief was expanded and refined into the notion that female supervisors were not only necessary but also possibly reformative agents. Carefully selected matrons, went the new argument, could provide role models and thus effect positive change in their charges. In the late 1860s, Wines and Dwight endorsed an English authority's belief that:

> It is especially important . . . that female officers should be distinguished for modesty of demeanor, and the exercise of domestic virtues, and that they should possess that intimate knowledge of household employment, which will enable them to teach the ignorant and neglected female prisoner how to economise her

means, so as to guard her from the temptations caused by waste and extravagance.[27]

Matrons who exhibited characteristics of middle-class homemakers might inspire female criminals to become respectable women.

But although these new conceptions and justifications of the matron's role were in the air, they had little effect on practice before 1870. Matrons in the early prisons and penitentiaries seldom performed more than a custodial function. In many institutions, there was but one matron to supervise the entire female population. Usually she lived in the institution (sometimes in the women's section itself) and was on duty twenty-four hours a day, six and one-half days a week. Little biographical information is available on these early matrons, but they appear often to have been older women, widowed, forced by economic hardship into such unpleasant work. In other instances, they were wives of wardens. In any case, as working-class women they were unlikely to provide the middle-class role models that progressive penologists recommended. Moreover, even if some early matrons had had the energy for reform, they had no authority with which to realize such an ambition. Hired by the warden, they could also be fired by him if they strayed too far from prison tradition. Mary Weed, who ran Philadelphia's Walnut Street Jail between 1793 and 1796, was an exception to this rule. But elsewhere, women involved in prison administration were subordinate to men.[28]

Another route through which women became involved in prison operations before 1870 was lay visiting. Inspired by religious principles, some women passed through the dreary gateways of penitentiaries to succor and uplift female criminals. Most notable among this group were the lady visitors at Philadelphia's Eastern Penitentiary. Eastern was famous in the United States and Europe for its system of perpetual solitary confinement. Inmates were kept in total isolation from one another on the theory that they would thus reflect and repent instead of picking up new criminal habits. Despite the spartan aspects of their care, Eastern's prisoners were well supervised, and they were permitted to have selected visitors. At a time when its female population numbered as low as twenty, the institution hired a matron to care for their needs. Moreover, according to Dorothea Dix, although they were "chiefly employed in making and repairing apparel," the women had "full time for the use of books, and the lessons which are assigned weekly by the ladies who visit the prison to give instruction."[29] These visitors belonged to the Association of Women Friends. Dix reported that the Quaker women made

> visits every Monday afternoon throughout the year; and you may see them there seriously and perseveringly engaged in their merciful vocation. Their care extends to the convicts after the expiration of sentences. These ladies read the scriptures, furnish suitable books for the prisoners, give instruction in reading, writing, and arithmetic; and what is of great value, because reaching them through a direct influence, instruct them by conversation, suited to their capacity.[30]

The efforts of the Philadelphian ladies to comfort and educate female inmates echoed, if but faintly, events already underway in England, where about 1815 Elizabeth Fry, another Quaker, had begun similar work. Such efforts were soon mirrored in several New York State developments. At Ossining, New York, two women took over management of the Mount Pleasant Female Prison in the mid-1840s and introduced some of the techniques of reform that later became staples of women's prison discipline. At almost the same time, middle-class women in New York City organized a women's branch of the reformist Prison Association of New York. Under the leadership of yet another Quaker, Abigail Hopper Gibbons, this Female Department of

the Prison Association established a halfway house where discharged female prisoners could receive shelter, moral training, and help in finding positions as domestic servants. On both sides of the Atlantic, these early manifestations of concern by women outside the walls foreshadowed the tremendously strong current of reform that eventually swept some women out of men's prisons entirely and into female-run institutions of their own.[31]

But true reform, in the sense of structural change in the care of prisoners, could not be accomplished by these lady visitors, high-status matrons, and prison association members. Such activists supplemented the matron's role and contributed to its redefinition, but at mid-century they did not actually challenge the conditions under which women were confined. Despite prisons' growing willingness to employ matrons and the unquestionably meliorative efforts by outsiders in Philadelphia, Ossining, and New York City, profound change in the condition of female prisoners could not occur until reformers identified what was later recognized as the source of the problem: the fact that women were held in institutions designed for men. Once the problem was defined, the solution seemed obvious—a fourth stage in the process of separation of female from male prisoners, in which the former would be removed to prisons of their own.

⧉ The Mount Pleasant Female Prison

This fourth stage did not occur until the reformatory movement began in the last quarter of the nineteenth century, but it was anticipated by New York's establishment of the Mount Pleasant Female Prison at Ossining in 1835. The founding of Mount Pleasant was a milestone in the evolution of women's prisons. This institution was the first women's prison in the United States: it was deliberately established by an act of the legislature,[32] in contrast to other women's units of the period, which developed haphazardly as appendages to men's prisons. Although Mount Pleasant was both geographically close to and administratively dependent on the Sing Sing prison for men, it existed apart, with its own buildings and staff. Moreover, during the years in which Mount Pleasant was administered by two innovative women, it became the site for experiments that forecast the great reformatory movement just ahead.

Construction having taken several years, Mount Pleasant was not ready to receive prisoners until 1839. Even before the prison officially opened, women at Bellevue were transferred to Sing Sing, so anxious was New York City to pass them on. They were held in a cellblock of the men's institution until the new prison was ready, at which point they and the women from Auburn were transferred to Mount Pleasant. A New York law of 1841 instructed that, thereafter, all women sentenced to a state prison should be sent to the female department at Sing Sing.[33]

The women's prison was situated on the hill behind Sing Sing, separated from the men's quarters by a roadway and overlooking the Hudson River. According to a description of the late 1860s, it was "a handsome building . . . with a Doric portico of imposing proportions."[34] The inside was modeled on the Auburn plan, with three tiers of twenty-four cells each. In the west end of the building, from which the view was best, the matrons had their quarters. At the east end, within the prison area, stood an elevated platform used for chapel services and lectures. Below it was a nursery. In addition to the main building this prison included a workshop and two large, separate cells for punishment, each cell with its own yard. The men's section of Sing Sing was as yet unenclosed, but a high wall was built around the women's complex to minimize communication between the sexes. More cell space was needed within a few years. The women's prison, however, could not easily be expanded or remodeled. "Poorly designed and difficult to alter," W. D. Lewis has observed, "the Greek temple overlooking the Hudson

was an example of penny-wisdom and pound-foolishness."[35]

Ultimate authority for management of Mount Pleasant lay with the board of inspectors of Sing Sing, but daily administration was left to a matron to whom was delegated the same authority over government and discipline as that given the principal keeper of the men's section. Several assistant matrons helped her with these tasks. During most of the years of its operation, the prison's matrons were at best unremarkable, at times disastrous. Two, however, were outstanding: Eliza Farnham, who served as chief matron from 1844 to 1847, and Georgiana Bruce, a former resident of Brook Farm, the Utopian community, who assisted Farnham during her first year. Their experiments with reformational techniques were the most radical and ambitious efforts of the time to improve criminals morally.

Georgiana Bruce's account of the background to Farnham's appointment gives an idea of how tumultuous life could be in an inadequately supervised female department.

[Under the previous matron] there had been a sort of rebellion among the convicts, or among some of the most daring, who had deliberately refused to conform to the rules of the prison, or to perform the duties assigned them. They tyrannized over and maltreated the weaker and more docile of their fellows, and made night hideous by singing blasphemous and obscene songs.

The matron, a respectable, but incompetent person, had finally been attacked and the clothes torn from her body. A well-meaning, tightskulled little chaplain had prayed frantically for the rebels,—prayed to them also. They made a feint of yielding, then turned the prison into a pandemonium again. . . . The Board, on making a visit to the prison, had been met by shouts of derision and insolent defiance, and they had to make a hasty exit to escape the kids [wooden food tubs] flung at them by the rioters.

By a most fortunate chance Judge Edmunds [Edmonds; board member and prison reformer] heard of Mrs. Farnham, and one interview with her convinced him that he had found the person he was in search of, and she was shortly engaged as a matron.[36]

Farnham was only twenty-eight years old at the time, and Bruce but twenty-five; yet through a combination of firmness and kindness, they rapidly reestablished order at Mount Pleasant. A phrenologist, Farnham believed that if she could stimulate her charges with positive influences, their criminal tendencies would be overcome. To this end she introduced a program of education, instructing the women each morning on "the more interesting persons in the history of our country, in . . . astronomy, in geography, and also in . . . the elements of physiology and physical education." She added novels such as *Oliver Twist* to the prison's library (which at the time of her arrival consisted solely of seventy-five copies of *Call to the Unconverted Sinner*), and she permitted inmates to take books to their cells. "The wayward creatures," Bruce reported, "found by degrees that their prison was turned into a school, and they lost the inclination to make trouble." Farnham was a strict disciplinarian, yet she tried to keep rules to a minimum. She modified the rule of total silence, permitting inmates "to talk in a low tone to each other half an hour every afternoon, providing that they had conformed to the rule of silence during the remainder of the day." In another departure from contemporary practice, she and Bruce attempted to alleviate the grimness of the prison environment by introducing flowers, music, and visitors from the outside. Significantly, Bruce referred to the prison as "our reformatory."[37]

Such innovations, though widely endorsed toward the end of the century, shocked many of Farnham's contemporaries. Conservatives such as Sing Sing's chaplain considered novel-reading irreligious. Moreover, Farnham's relaxation of the silent rule sowed dissension at the neighboring men's prison, where the rule still prevailed. Farnham's opponents publicly attacked her and her reforms. She fought back but eventually lost the struggle, resigning in 1847.

Like most other female convicts of the mid-nineteenth century, those at Mount Pleasant worked long hours at tasks considered appropriate to women—in this case button-making and hat-trimming, as well as sewing clothes for male prisoners. Other than work, their program was minimal, consisting in 1841 of only a Sabbath school taught by lady visitors. By 1843 (the chaotic year preceding Farnham's appointment), even the Sabbath school had been discontinued. With Farnham's arrival began a brief period of programs. Convicts continued to work, but Farnham made time for religious observances and instruction. This enraged her critics, who charged that the women's prison should be earning higher profits. Farnham retorted that women in the female prison, like their counterparts outside the wall, were paid much less than men. But this logic did not appease her profit-minded opponents.[38]

Mount Pleasant seems to have been the first state prison to include a nursery. Babies could occasionally be found in other nineteenth-century custodial institutions: in an act of 1843, for example, the Tennessee legislature instructed the penitentiary to "receive with Pricilla Childress, a convict from the county of Giles, her infant child," and in 1869 New Jersey officials complained that a black woman who had been incarcerated for years had recently given birth to a mulatto child, fathered by a guard. But only the records of Mount Pleasant make reference to separate accommodations for infants. Before special arrangements were made for the care of newborns, infant mortality rates ran high. According to the Prison Association of New York, for instance, before establishment of the Mount Pleasant nursery, every child born at Sing Sing had died. Women's reformatories that opened in the late nineteenth and early twentieth centuries usually allowed women to keep their babies, and because these institutions received mainly young sex delinquents, their infant populations were often sizeable. Babies posed a greater problem for custodial prisons that, like Mount Pleasant, received felons; the serious offenses and lengthy sentences of female felons made prison administrators reluctant to admit infants with their mothers. But Mount Pleasant, because it held the largest population of female convicts in the country and because it was founded explicitly for the care of women, came to terms with the problem posed by infants. The presence of a nursery at Mount Pleasant meant that the difference between the incarceration experiences of men and women was greater at Sing Sing than at other penitentiaries. Moreover, women did not automatically lose their babies or have to devise ways to care for them in single cells. The special needs of female prisoners were being recognized.[39]

Disobedient women were punished less brutally than men at Sing Sing, yet their chastisements were severe. One punishment was gagging, which Dorothea Dix found "shocking and extremely objectionable." Judge Edmonds reported that at Sing Sing, "The gag has been sometimes applied, but it has been only among the females that it has been rendered *absolutely* necessary!" On the other hand, he found, "In the women's prison, the lash is never used. There the punishments are confinement to their own cells in the main dormitory, or in separate cells, with reduction of food," and, of course, gagging.[40] Farnham preferred kindness to punishment as a means to achieve order, but even she could react harshly, meting out long periods of solitary confinement, cropping women's hair, and using the gag and straitjacket. A list of violations and punishments in 1846 included:

—Noise and violence in her room at noon. Shower bath.[41]

—Disobedience and noise in her room. Twelve days in solitary confinement in outer cell.

—Noise in her room at night. Straight [*sic*] jacket for the night, and *bread and water for one week.*

—For rushing from her cell when the door was open . . . and repeating it many times . . . a chain six feet in length was made fast to the wall and locked upon her wrist.[42]

Administration of corporal punishment had not yet become a matter about which even reform-minded authorities felt embarrassment.

Overcrowding at Mount Pleasant—or rather New York's refusal to create space for its growing numbers of female convicts—eventually led to the institution's demise. By 1865, with nearly two hundred prisoners, Mount Pleasant was close to double its capacity. That year the legislature ruled that women from the seventh and eighth judicial districts should be sent to local institutions instead of to the women's prison. A law of 1877 ordering transfer of all Mount Pleasant's inmates to a county penitentiary emptied the prison entirely. For more than a decade thereafter, New York held its female state prisoners in local institutions.[43]

As the phenomenon of Mount Pleasant demonstrates, the seeds of the women's reformatory had been sown by mid-century. At Mount Pleasant, women were confined apart from men and supervised by other women. Under Farnham and Bruce, female convicts—for perhaps the first time—were encouraged by prison administrators to reform. And the techniques introduced by Farnham—education, example, sympathy—later became crucial to the reformatory program. But as Mount Pleasant also indicates, these seeds could not survive in the harsh environment of the penitentiary. Education could have little effect in an institution whose main interest was profits; role models and sympathetic understanding were incompatible with straitjackets and bars. Most importantly, no thoroughgoing change in the treatment of women could take place until female prisoners and their matrons were freed from second-class status in institutions that insisted on male authority and precedence.

Until about 1870 the custodial institution was the only type of penal unit for women. It received its fullest articulation at Mount Pleasant, but all other female departments exhibited its traits. The custodial model was a masculine model: derived from men's prisons, it adopted their characteristics—retributive purpose, high-security architecture, a male-dominated authority structure, programs that stressed earnings, and harsh discipline. In comparison to women's reformatories, women's custodial institutions treated inmates like men. But as we have seen, this did not mean that women's care and experience of incarceration were identical to those of males. Probably lonelier and certainly more vulnerable to sexual exploitation, easier to ignore because so few in number, and viewed with distaste by prison officials, women in custodial units were treated as the dregs of the state prisoner population.

⬙ Notes

1. Rothman, *Discovery of the Asylum*, p. 105. On the debate about the origins of the penitentiary, see Michael Ignatieff, "State, Civil Society, and Total Institutions: A Critique of Recent Social Histories of Punishment," in Michael Tonry and Norval Morris, eds., *Crime and Justice: An Annual Review of Research*, vol. 3 (Chicago: University of Chicago Press, 1981), pp. 153–92.

2. For exceptions to the rule of disregard of women within penitentiaries, see W. David Lewis, *From Newgate to Dannemora: The Rise of the Penitentiary in New York, 1796–1848* (Ithaca: Cornell University Press, 1965); and McKelvey, *American Prisons*.

3. New York, Auburn State Prison, *AR 1832:* 17 (emphasis in original).

4. New York, Inspectors of State Prisons, *An Account of the State Prison or Penitentiary House, in the City of New-York* (New York: Isaac Colling and Son, 1801), p. 18 (both quotations).

5. For the complaints, see New York Committee on State Prisons, *Report of the Committee on State Prisons, relative to a prison for female convicts* (New York Sen. Doc. No. 68, 1835), p. 1 (no classification, "constant . . . intercourse"); New York, Mount Pleasant State Prison, *AR 1835:* 5 (food), *AR 1836:* 5 (need for a matron); New York Committee on State Prisons, *Report of the Committee on State Prisons* (New York Sen. Doc. No. 32, 1833), p. 3 (cholera and ensuing events). New York, Mount Pleasant State Prison, *AR 1836:* 5, notes the desire of New York City officials to rid themselves of the women at Bellevue.

6. Harriet Martineau, *Retrospect of Western Travel*, vol. 1 (London: Saunders and Otley, 1838), pp. 124–25.

7. On the unfortunate pregnancy of Rachel Welch, see Lewis, *From Newgate to Dannemora*, pp. 94–95.

8. On the two Ohio state prisons that preceded its penitentiary, see Clara Belle Hicks, "The History of Penal Institutions in Ohio to 1850," *Ohio State Archeological and Historical Society Publications* 33 (1924): 359–426; Jacob H. Studer, *Columbus, Ohio: Its History, Resources, and Progress* (Columbus: n.p., 1873); and George H. Twiss, ed., "Journal of Cyrus P. Bradley," *Ohio Archeological and Historical Publications* 15 (1906): 240–42.

According to Orlando F. Lewis, *The Development of American Prisons and Prison Customs, 1776–1845* (orig. 1922; repr. Montclair, N.J.: Patterson Smith, 1967), p. 262, Ohio's prison of 1818 included five underground, unheated cells, and it was here that the women were kept. No other historian of the building refers to women in the underground cells, however, and according to William Crawford's plan of the institution *(Report on the Penitentiaries of the United States,* [orig. 1835; repr. Montclair, N.J.: Patterson Smith, 1969], Plan 18), its women resided in a small building inside the yard.

9. Ohio Penitentiary, *AR 1850:* 133.

10. Gerrish Barrett as quoted by Lewis, *Development of American Prisons and Prison Customs,* p. 263; D. L. Dix, *Remarks on Prisons and Prison Discipline in the United States*, 2d ed. (orig. 1845; repr. Montclair, N.J.: Patterson Smith, 1967), p. 48; Sarah Maria Victor, *The Life Story of Sarah M. Victor for Sixty Years. Convicted of Murdering Her Brother, Sentenced to be Hung, Had Sentence Commuted, Passed Nineteen Years in Prison, Yet is Innocent* (Cleveland: Williams Publishing Co., 1887), p. 317.

11. Dan J. Morgan, *Historical Lights and Shadows of the Ohio State Penitentiary and Horrors of the Death*

Trap (Columbus: Champlin Printing Company, 1895), p. 91; Victor, *The Life Story,* pp. 298, 326, 327; Ohio Penitentiary, *AR 1880:* 91. The humming-bird is described in Victor, *The Life Story,* pp. 324–35.

12. Tennessee State Penitentiary, *BR 1859:* 232. On the Tennessee penitentiary in the early and mid-nineteenth century, see Jesse Crawford Crowe, "The Origin and Development of Tennessee's Prison Problem, 1831–1871," *Tennessee Historical Quarterly* 15 (2) (June 1956): 111–35; and E. Bruce Thompson, "Reforms in the Penal System of Tennessee, 1820–1850," *Tennessee Historical Quarterly* 1 (4) (December 1942): 291–308.

13. Tennessee State Archives, *Convict Record Book 1831–1874,* Record Group 25, Ser. 12, v. 86; Messages of Governor James K. Polk, Tennessee General Assembly, *House Journal,* 24th Assembly, 1st sess., 1841–1842: 23, as quoted in Crowe, "Origin and Development of Tennessee's Prison Problem," p. 117.

14. For a hint of forced prostitution at the Tennessee penitentiary late in the century, see Tennessee, *Acts and Resolutions 1897,* Ch. 125, sec. 28, making it a misdemeanor for any prison officer "to hire or let any female convict to any person on the outside as cook, washerwoman, or for any other purpose."

15. Montana and Utah. Utah opened two work-release facilities for women in the late 1970s but apparently continued to send women to the state prison until near the end of their terms. See American Correctional Association, *Directory 1980* (College Park, Md.: American Correctional Association, 1980), pp. 139, 233. Cf. Joan Potter, "In Prison, Women are Different," *Corrections Magazine* (December 1978): 15 ("Montana's 12 women are divided between a separate Life Skills Center in Billings and a coed facility in Missoula").

16. Francis Lieber, "Translator's Preface" to Gustave de Beaumont and Alexis de Tocqueville, *On the Penitentiary System in the United States and Its Application in France* (orig. 1833; repr. Carbondale: Southern Illinois University Press, 1964), p. 8.

17. Crawford, *Report on the Penitentiaries,* Appendix; Dix, *Remarks on Prisons,* pp. 107–8.

18. Beaumont and Tocqueville, *On the Penitentiary System in the United States,* p. 72.

19. Lieber, "Translator's Preface," p. 8.

20. Ibid., p. 12.

21. Crawford, *Report on the Penitentiaries,* pp. 26–27.

22. New York, Mount Pleasant State Prison, *AR 1841:* 28 (emphasis as in original). For reviews of the literature on current patterns of female crime, current explanations of female crime patterns, and the evidence for and against judicial chivalry, see Nicolette Parisi, "Exploring Female Crime Patterns: Problems and Prospects" and "Are Females Treated Differently? A Review of the Theories and Evidence on Sentencing and Parole Decisions," both in Nicole Hahn Rafter and Elizabeth A. Stanko, eds., *Judge, Lawyer, Victim, Thief: Women, Gender Roles, and Criminal Justice* (Boston: Northeastern University Press, 1982).

23. W. A. Coffey, *Inside Out, or an Interior View of the New-York State Prison* (New York, 1823), p. 61, as quoted in Lewis, *From Newgate to Dannemora,* p. 38; E. C. Wines and Theodore W. Dwight, *Report on the Prisons and Reformatories of the United States and Canada* (Albany: van Benthuysen & Sons, 1867), p. 71.

24. Dix, *Remarks on Prisons,* p. 108; Crawford, *Report on the Penitentiaries,* Appendix, p. 68.

25. Lieber, "Translator's Preface," pp. 9–11; Mary Carpenter, *Our Convicts,* vol. 1 (orig. 1864; repr. Montclair, N.J.: Patterson Smith, 1969), p. 207; Caesar Lombroso and William Ferrero, *The Female Offender* (first English ed. 1895; repr. New York: D. Appleton & Company, 1915). On the troublesome women in Ohio, see note 10 and accompanying text. On the archetype of the Dark Lady and its influence, see Paula Blanchard, *Margaret Fuller: From Transcendentalism to Revolution* (New York: Delacorte Press, 1978), pp. 193–94; and Rafter and Stanko, *Judge, Lawyer, Victim, Thief,* pp. 2–7 and chapter 11. Lewis's "The Ordeal of the Unredeemables" (*From Newgate to Dannemora,* chapter 7) also deals with nineteenth-century conceptions of female criminals and their effects upon prison treatment.

26. Lieber, "Translator's Preface," p. 13.

27. Wines and Dwight, *Report on the Prisons,* pp. 123–24.

28. On Mary Weed, see Negley K. Teeters, *The Cradle of the Penitentiary: The Walnut Street Jail at Philadelphia, 1773–1835* (Philadelphia: Pennsylvania Prison Society, 1955), pp. 47, 61.

29. Dix, *Remarks on Prisons,* p. 107.

30. Ibid., pp. 62–63. For background on the Association of Women Friends, see Teeters, *Cradle of the Penitentiary,* p. 107.

31. See Elizabeth Fry, *Memoir of the Life of Elizabeth Fry with Extracts from her Journal and Letters,* two volumes edited by two of her daughters (Philadelphia: J. W. Moore, 1847); and, for a discussion of the Female Department of the Prison Association of New York, see Freedman, *Their Sisters' Keepers,* pp. 28–35.

32. New York, *Laws of 1835*, Ch. 104.

33. New York, *Laws of 1841*, Ch. 200, sec. 3.

34. Wines and Dwight, *Report on the Prisons*, p. 107.

35. Lewis, *From Newgate to Dannemora*, p. 177.

36. Georgiana Bruce Kirby, *Years of Experience: An Autobiographical Narrative* (orig. 1887; repr. New York: AMS Press, 1971), pp. 190–91.

37. New York, Mount Pleasant State Prison, *AR of the Inspectors 1846* (New York Sen. Doc. No. 16, 1846): Appendix D, p. 94 (first quotation); Kirby, *Years of Experience*, pp. 193, 199, 218.

38. New York, Mount Pleasant State Prison, *AR of the Inspectors 1846*: Appendix D, p. 88 (Farnham's retort).

39. Tennessee, *Acts of the General Assembly 1843–44*, Resolution No. 16; New Jersey, *Report of the Commissioners to Examine the Various Systems of Prison Discipline and Propose an Improved Plan* (Trenton: The True American Office, 1869), p. 5; Prison Association of New York, *AR 1846*: 48.

40. Dix, *Remarks on Prisons*, pp. 13–14 (includes the Edmonds report; emphasis as in original).

41. Despite its mild name, the shower bath was one of the prison's cruelest punishments. The prisoner was bombarded by a powerful stream of water until close to drowning.

42. New York, Mount Pleasant State Prison, *Report of the Inspectors of the Mount Pleasant State Prison in answer to a resolution of the Assembly* (New York Ass. Doc. No. 139, 1846), Appendix C, pp. 113–14.

43. Clifford M. Young, *Women's Prisons Past and Present and Other New York State Prison History* (Elmira Reformatory: The Summary Press, 1932), p. 13 (ruling of 1865); New York, *Laws of 1877*, Ch. 172, secs. 1 and 2.

DISCUSSION QUESTIONS

1. Relate how race, class, and gender have affected the operation of correctional institutions in the past. Can you think of how they might influence correctional operation these days?

2. How did prison conditions for women vary by state in the 1800s? How were they similar? How did they differ from the conditions experienced by male prisoners?

3. Why were women inmates treated differently from male inmates during this period? How has this changed?

❖

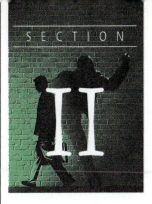

SECTION

II

THE CORRECTIONS EXPERIENCE FOR OFFENDERS

Introduction

I t has become axiomatic to say that correctional programs and institutions are overcrowded, underfunded, and unfocused these days. As the drug war rages on and mandatory sentencing has its effect, probation and parole caseloads and incarceration rates spiral past any semblance of control (Bureau of Justice Statistics, 1997, 2006b; Glaze & Bonczar, 2006; Pollock, 2004; Ruddell, 2004). As a consequence, though spending on corrections has steadily, and steeply, climbed over the last few years, it is nearly impossible for most states and localities to meet the needs for programs, staff, and institutions. So they do not. Thus, the corrections experience for offenders is shaped by shortages.

But, as has been discussed already in this text, this has always been somewhat true. If it is built, or in the case of probation and parole, offered, then they will come—because, as with all corrections sentences, they are forced to. A case in point: almost immediately after the first American prisons were built, the Walnut Street Jail (1790), the Auburn Prison (1819), the Western Pennsylvania Prison (1826), and the Eastern Pennsylvania Prison (1829), they were full, and, within a few years, they were expanded or new prisons were under construction (Augustus, 1852/1972; Conover, 2001; Johnson, 2002; Lewis, 1922; Pollock, 2004).

To say that crowding and corrections have always been linked, of course, is not to dismiss the negative effects of overfilling or to argue that it might not be worse than ever now. Certainly, the U.S. incarceration rate has never been so high, nor can we easily dismiss the monetary or social costs of corrections (Clear, 1994; Harrison & Beck, 2005; Zimring & Hawkins, 1991).

In this section, we discuss the nature of the correctional experience for incarcerated individuals, an experience that, as of 2004, 1 in 138 U. S. residents have (Harrison & Beck, 2005). Of course, the odds of incarceration increase if you are male and a minority, though the number of incarcerated women has increased faster proportionately than it has for males in recent years. Certainly, as illustrated by the research presented in this section, *how* people experience corrections is influenced to some extent by the reality of crowding and the concomitant lack of resources.

⊠ Jails and Prisons

Jails were the first type of correctional institution created. The form they took varied by the sophistication and size of the community and its relative remoteness. When some sort of legal representative holds the persons, either a sheriff or a designee, so that they will appear at trial, then wherever they are placed is by definition a form of a jail. When need be, accused offenders or political opponents were held in caves, shacks, barns, pits, castles, dungeons, the sheriff's own house, or they were just tied to a tree or rock (Hirsh, 1992; Irwin, 1985; Kerle, 1998; Zupan, 1991).

The history of adult and juvenile American jails, per se, has been intertwined with that of prisons. In fact, the two modes of confinement were virtually indistinguishable both in design and theory before the nineteenth century when separate structures called "penitentiaries" were created in Pennsylvania and New York. More than 80 years ago, Joseph Fishman articulated the differences between jails and prisons in his book *Crucibles of Crime: The Shocking Story of the American Jail* (1923/1969) and these distinctions remain somewhat true today:

> In the pure sense, there is a difference, and a vast difference, between these two types of institutions [prisons and jails]. Men and women are confined in prison strictly speaking, after they have been convicted of an offense of a grave character which involves moral turpitude. They are confined in jails after they have been convicted of misdemeanors, such as being drunk and disorderly, petit larceny, working on Sunday, etc. Also (and here is the great difference) jails are used for the confinement of men and women who have been arrested, charged with committing crimes or misdemeanors, but who have not yet been tried. As a general rule, persons are sentenced to jail for comparatively short periods of time from thirty days to a year. They are sentenced to prison for periods of from one year to life. (pp. 15–16)

These differences that Fishman delineated in 1923 became clearer once the penitentiaries were developed for the convicted and sentenced "long term" offenders and the correctional purposes of jails and prisons diverged (Hirsh, 1992). Prisons were and are used for long-term and convicted offenders who were to be simultaneously punished, deterred, and reformed while being isolated (incapacitated) from the community. Conversely, jails were and are used to hold pretrial detainees and short-term, sentenced offenders with diverse justifications for incarceration (Goldfarb, 1975). Despite these distinctions, however, the common history shared by jails and prisons ensures that there are parallels between the two types of correctional institutions in both the managerial, architectural, and theoretical spheres. Juvenile facilities, work releases, halfway houses, and probation and parole developed in reaction to the problems inherent in jails and prisons; they were a form of reform. Ironically, jails, and then prisons, themselves were developed when the reformist spirit prevailed and before it inevitably ebbed in populist politics (Hirsh, 1992; Lewis, 1922; Sullivan, 1990).

⊠ The Promise of Reform

As rehabilitation as a justification for incarceration gained primacy in America's prisons, jails, and juvenile institutions, correctional facilities with a more open architecture were constructed (Rothman, 1980). Dormitories, cottages, and less institutionalized settings were thought to complement treatment purposes effectively. In time, "reformatories" geared primarily toward juveniles, young adult male offenders, and women were developed in the latter half of the nineteenth century and into the twentieth; they tended to reflect a distinctly rehabilitative purpose. Developing in tandem with the reformatory era was the indeterminate sentence, probation parole, and education and vocational training programs (Rotman, 1990; Sullivan, 1990). Such reformatory style facilities eventually came to influence adult male institutions as well, and reformatories remain in widespread use in some form in many states today. Because they tended to be more open architecturally, direct and constant supervision by staff, to prevent abuse, is required in such facilities.

In the latter half of the nineteenth century, there was a sense developing that the great American prison experiment had failed. The conditions in some prisons were notoriously unhealthy, brutal, and in some disarray. Prisons were overcrowded, and the Auburn style of formal discipline was shredding. At the inception of their creation, prisons had been meant to punish and reform. The incontrovertible evidence was, however, that punishment did occur but reform rarely did in America's prisons. Hence, in the last decades of the nineteenth century, the National Prison Association, the predecessor of the American Correctional Association, was created, and the idea of the *reformatory* was born (Roberts, 1997).

This reformatory, as epitomized by the Elmira, New York prison (1877), was conceptualized as operating on the precepts of the developing medical, psychological, and social work fields (Lewis, 1922; Rothman, 1980; Sullivan, 1990). Geared toward younger male inmates, who were believed to be more amenable to treatment, the Elmira reformatory, in theory, was to use a reward system, such as "good time" and early release to parole, as a means of motivating the positive behavior and ultimate reform of inmates. The staff was to be composed of professionals with at least some college training who could mentor these young men and model prosocial behavior. The use of corporeal punishment was prohibited.

The reality of the implementation of these ideals was less than the promise of reform for a number of reasons, as was discussed in Section I (Rothman, 1980). Chief among these was the failure of the state to fund the reform fully, thus resulting in fewer programs and less than professional staff and a descent into the renewed use of the strap to maintain order. Despite these failures in implementation, and perhaps in conceptualization, the medical model of corrections was more concretely represented in Elmira and in other reform initiatives, leading to

- ◆ good time,
- ◆ parole and probation,
- ◆ rehabilitation programming, and
- ◆ indeterminate sentencing.

By the early decades of the twentieth century, all states had adopted all or a portion of these programming components of the reformatory model for their correctional systems. Of course, as with the Elmira experiment, formal adoption of a model does not mean that it is implemented perfectly or funded adequately. Moreover, as our discussion of treatment programming in Section V of this book will illustrate, the implementation of effective programming takes more than just full funding.

⊠ Total Institutions, Importation, Prisonization, and the Pains of Imprisonment

In addition to the evident patterns and trends that appear in corrections historically and that continually crop up in more modern correctional institutions and practices, the *operation* of corrections, particularly of prisons and jails, shapes the offender's experience. One central component of that operation is the totality of it.

Erving Goffman (1961) coined the term *total institution* to describe the situation of inmates of mental hospitals (but also of prisoners) in this country in the 1950s. For one year he served as a staff member (athletic director's assistant) and did ethnographic research in a federal mental health hospital in Washington, DC. While avoiding sociable contact with staff, he immersed himself in the inmate world, or as much of it as he could without being admitted to the hospital, and what he observed allowed him to learn much. Goffman (1961) defines a total institution as "[a] place of residence and work where a large number of like-situated individuals, cut off from the wider society for an appreciable period of time, together lead an enclosed, formally administered round of life" (p. xiii). Another key component of this social world is that there are clear social strata in such institutions dividing the "inmates" and the "staff" (Goffman, 1961, p. 7). There are formal prohibitions against even minor social interactions between these two groups, and all of the formal power resides with one group (the staff) over the other group (the inmates).

This definition is directly applicable to prisons, even today, though it more aptly describes both the prisons and jails of the past (Jacobs, 1977). For jail and prison inmates, the institution is where they live, and often work, with people like themselves in terms of not only their criminal involvement but also, largely, their social class. Though there is some ability to visit with others, the mode and manner of this contact with the outside world is quite limited in prisons and, to some extent, even more so in jails, and it is also dependent on the security status of the institution (e.g., whether it is a work release facility or a maximum security prison). The formal rules of prisons and jails also closely control inmate behavior and movement. Another key formal prohibition of total institutions governs interactions between staff and inmates. Simply put, staff members are to restrict such interactions to business alone and are to parcel out information only as absolutely necessary. As Goffman (1961) put it, "Social mobility between the two strata is grossly restricted; social distance is typically great and often formally prescribed" (p. 7).

How do these aspects of total institutions affect the lives of inmates of jails and prisons today? In the 1950s, Goffman (1961) believed that total institutions had the effect of debilitating their inmates. As he saw it, upon entrance into the institution, the inmate might become *mortified* or suffer from the loss of the many roles he or she occupied in the wider world (Goffman, 1961; Sykes, 1958). Instead, only the role of inmate is available, and that role is formally powerless and dependant. Additionally, though each person entering a jail or prison *imports* their own culture from the outside, to some extent they are likely to become *prisonized* or socialized into the inmate culture of the institution (Clemmer, 2001). Couple this mortification and subsequent role displacement with prisonization into the contingent inmate subculture, and you have the potential for the new inmate to experience a life in turmoil while adjusting to the institution and some difficulty when reentering the community.

Part and parcel of this inmate world are what Gresham Sykes (1958) describes as the *pains of imprisonment,* such pains include "the deprivation of liberty, the deprivation of goods and services, the deprivation of heterosexual relationships, the deprivation of autonomy, and the deprivation of security"(pp. 63–83). To this list we would add *deprivation of contact with family members, particularly their children,* a severe pain that many inmates experience when, as an artifact of their incarceration, they are unable to have regular interactions with their own children or to have any control over their children's environment on the outside (Gray, Mays, & Stohr, 1995; Stohr & Mays, 1993).

Sykes (1958) argued that the first five of these pains, though not physically brutalizing, have the cumulative effect of destroying the psyche of the inmate. In order to avoid this destruction, inmates in prisons (he did research in a maximum security prison) may be motivated to engage in deviance while incarcerated as a means of alleviating their pain. So bullying other inmates, involvement in gangs, buying items through the underground economy, and homosexual acts might all be motivated, in fact, by the need for some autonomy, liberty, security, goods and services, and sexual gratification (Johnson, 2002). Extrapolating from this point, the extent to which female inmates form pseudofamilial relationships may be a means of alleviating the pain experienced due to the separation from children and other close family members (Owen, 1998).

▲ **Photo 2.1**   A prisoner on Death Row walks in the corridor on April 16, 1997 at the Ellis Unit in Huntsville, Texas. Texas has more than 500 prisoners on death row. The state leads all records in executing people around the United States.

⊠ The Correctional Subculture

Subcultures, or environments with their own norms, values, beliefs, and even language, tend to solidify when people are isolated from the larger culture and when members have continual contact with each other for an extended time. In other words, it would appear that total institutions provide the perfect admixture for an inmate subculture to form. Accordingly, the degree to which a correctional environment fits the definition of a total institution will determine the extent to which a client subculture exists. Not surprisingly, then, most correctional research on inmate subcultures has tended to focus on prison inmates and specifically maximum-security prison inmates (see Hemmens & Marquart and Lutze & Murphy in this section; Clemmer, 2001; Irwin & Cressey, 1962; Owen, 1998; Sykes, 1958). This is not to say, of course, that those in a jail or a minimum security prison do not have distinguishable "norms, values, beliefs, and language" that set them apart from the wider community, but it is much less likely that they do than is the case for inmates in a maximum security prison. By definition, the longer people are in an institution, associating with others like them, and the more "total" the institution is in its restrictions on liberty and contact with "outsiders," the more subject inmates are to the pains of imprisonment. This observation would appear to suggest that the more a correctional environment fits the descriptors of a total institution, the more likely that a definable inmate subculture exists and that those correctional clients are "prisonized" or have adopted the subculture.

Indicators of such a subculture, as identified by prison researchers, include defined roles for inmates and prescribed values and behavior patterns (Clemmer, 2001; Owen, 1998). For instance, Clemmer (2001) broadly defines criminal subcultural values as including "the notion that criminals should not betray each other to the police, should be reliable, wily but trustworthy, coolheaded, etc." (p. 87). Though all criminals might have some exposure to these values, they are more likely to be reinforced and adhered to in a prison setting for the reasons noted above. This criminal subculture becomes a "convict subculture," according to Clemmer (2001), when such inmates "[s]eek positions of power, influence, and sources of information" so that they might get the goods and services they desire to alleviate those pains of imprisonment. Similarly, Owen (1998) notes that some women would engage in a version of this subculture although it might be tempered by the relationships they had and the goods and services they needed. Notably, both Clemmer and Owen, however, found that a significant portion of inmates in the male and female prisons they studied were not at all interested in being involved in the convict subculture or the "mix" of behavior that can lead to trouble in prisons. Rather, these inmates either chose not to connect to the inmate subculture or they held on to more traditional and legitimate values from the larger culture.

⊠ Mature Coping

In his research on corrections, Johnson (2002) noticed that, despite the mortification, prisonization, and pains experienced to different degrees by incarcerated individuals, some were able to adjust prosocially, even to grow, in a prison setting. Though the exception rather than the rule, these inmates, he noted, developed another means of adjusting. This alternative means of handling incarceration, or supervision in the case of probationers and parolees, is *mature coping*. As identified and defined by Johnson (2002), "Mature coping means, in essence, dealing with life's problems like a responsive and responsible human being, one who seeks autonomy without violating the rights of others, security without resort to deception or violence, and relatedness to others as the finest and fullest expression of human identity" (p. 83). As indicated by this definition, offenders need to learn how to be adults with some autonomy in an environment where, formally, they have little power (although the informal reality may be different—more about this in Section III of this book) and their status is almost subhuman by wider community standards. Moreover, they must accomplish this feat without doing violence to others—though Johnson (2002) allows that violence in self-defense may be necessary—and they need to exercise consideration of others in their environment.

Johnson (2002) notes that mature coping is relatively rare among the inmate population for a number of reasons. He argues that inmates are typically immature in their social relations to begin with, which, of course, is one of the reasons they are in prison in the first place (Conover, 2001). Because of impoverishment, poor or absent or abusive parenting, mental illness, schools that fail them or that they fail, offenders enter the criminal justice system with a number of social, psychological, and economic deficits. They are often not used to voluntarily taking responsibility for their actions as one would expect of "mature" individuals. Nor are they typically expected to "[e]mpathize with and assist others in need," especially in a prison or jail environment (Johnson, 2002, p. 93).

Second, Johnson (2002) argues that, for inmates to maturely cope it is helpful if they are incarcerated in what he terms a *decent prison*. Such a facility does not necessarily have more programming, staffing, or amenities than the norm, though he thinks it might be helpful if they did; rather, such institutions or programs are relatively free of violence and include some opportunities so that inmates might find a *niche* for involvement. In order for inmates to find this niche, however, decent prisons need to include some opportunities for inmates to act autonomously.

Being secure from violence, like autonomy, is basic to human development. In fact, if the security need is not fulfilled, it will preoccupy offenders and motivate them to engage in behaviors (e.g., bullying or gang activity) that they normally might avoid if they were not feeling continually threatened (Johnson, 2002; Maslow, 1998). Then, assuming that the offender perceives that he or she is relatively safe, there need to be prosocial activities, including work, school, athletic, church, treatment, or art programs that provide some sort of means for positive self-value reinforcement. Such places are termed *niches* by Johnson (2002), and the opportunities they afford provide redress for the mortification and pains that offenders, particularly those who are incarcerated, experience.

⬛ How Corrections Is Experienced Depends

Offender Demographics

In the readings that follow, the authors either directly or indirectly discuss the correctional topics mentioned earlier. Their research gives these facts and concepts real and current application. In their research, they more fully illustrate the point that all offender adjustment and types of institutions are not monolithic. The research by Hemmens and Marquart and Lutze and Murphy makes plain that demographics such as gender, race or ethnicity, social class, and age can all affect how offenders perceive their experience with corrections. Furthermore, the research by Pizarro and Stenius and Richards and Ross indicates that a powerless status and the type of institution or correctional program offenders are exposed to also shape their reality and how they adjust.

In terms of gender, the history and practice of incarcerating women and girls provides a context for current practice. Initially, women and girls were not incarcerated separately from men and boys. As their numbers grew and a sense of moral propriety developed, however, women and girls were allotted their own separate institutions (see the Rafter article in Section I). To some extent, their separate treatment was a boon as their institutions tended to be softer structurally (e.g., fewer fences, a more cottage-like atmosphere). Separate institutions, particularly when staffed by women, also provided female prisoners protection from sexual abuse by male staff (Pollock 2002). However, cultural pressures to create "ladies" and "conformity" with stereotypical female roles were fierce in such institutions (see the Rafter article in Section I). As a consequence, historically, women and girls have tended to be incarcerated and supervised in the community for lesser offenses than those for which men and boys are "punished" (Chesney-Lind & Shelden, 1992) because females, by definition, are *double deviants*; they have been deviant criminally (like men), but they also have deviated from societal expectations for their gender. Moreover, the drug war of the last 20-plus years has tended to capture and incarcerate low-level female offenders, resulting in what some researchers regard as an over-incarceration of women (Chesney-Lind, 2002).

The correctional experience for women and girls has also been defined by their separation from family and friends. Perhaps because of this separation, some of them have tended to form pseudo family units as a means of adjustment to the pain they experience (Owen, 1998; Pollock, 2002).

Lutze and Murphy note that the gendered nature of some correctional institutions—in male institutions this means there are more "masculinized" values placed on behavior—may reward or promote aggression and impede the ability of inmates to transition effectively out of prison. In contrast, in female institutions where "feminized" values are placed on behavior, the inmate may be rewarded for more submissive and nurturance-related behaviors (Britton, 2003). Although Britton (2003) in her book *At Work in the Iron Cage: The Prison as Gendered Organization* is primarily focused on how gender affects work in prisons, she does note that organization and culture tend to shape inmate behavior so that it conforms to sex role stereotypes about what it means to be a woman.

As regards age and race, and as discussed in Section I, scholars note that inmates have historically been incarcerated in different types of institutions because of their race or their age and treated differently by criminal justice and correctional authorities (see Hemmens & Marquart in this section; Carroll, 1974; MacKenzie, 1987; Maitland & Sluder, 1996; Oshinsky, 1996; Wright, 1989; Young, 1994). Moreover, the research indicates that different racial and age groups might adjust in dissimilar ways, though there is some dispute in the literature on racial differences at least (see Hemmens & Marquart in this section; Wright, 1989). We do know that younger inmates will tend to engage in more deviant behavior while incarcerated, perhaps because of their inability or unwillingness to "maturely cope."

Powerlessness and Inmate Status

Power, as defined by the political scientist Dahl (1957), is essentially *the ability to get others to do what they otherwise wouldn't*. Formally, at least, and to differing degrees depending on their role, the status of their subjects, and the type of institution or program, probation and parole officers and correctional officers in jails and prisons have the power to get their clients to do what they otherwise would not in terms of behavior or work or programming. As Richards and Ross note in their article, this power to control and classify in corrections is somewhat absolute and can be debilitating. Welch (2005) notes that such control is "ironic" in that, although a lack of social control in the larger community is usually believed to be a cause of crime, this control might also create crime or greater violence or deviance in a correctional environment.

Sykes (1958) also noted about 50 years ago that the formal roles of staff and inmates do not always reflect the informal reality. In the maximum security institution he studied, Sykes (1958) found that this formal power was in fact tempered by the fact that staff members were grossly outnumbered by inmates and needed inmate compliance to get work done (e.g., guards could not easily or efficiently or safely use force to get inmates to make their beds, attend programs, leave other inmates alone, etc.). Also, the regular human contact between staff and inmates led them to form friend-like relationships that moderated the formal power

▲ **Photo 2.2**   Johnny Lee Wilson, convicted of murder, stands handcuffed in the hallway of the Jefferson City Correctional Centre, Missouri.

of staff. For these reasons, Sykes (1958) claimed that inmate and staff relationships would naturally incline toward degrees of corruption, some more serious than others.

Johnson (2002) also recognizes that these informal relationships between staff and inmates do exist (as will be discussed more fully in another section of this book), albeit not as a "corrupted" relationship per se but as a real world representation of prisons and prison work. For Johnson (2002), such relationships mean that some staff members will engage in human service work to facilitate the adjustment of offenders or inmates.

Whether corrections employees are inclined to be human service workers or not is likely shaped by their own proclivities, the professional culture of their institution or program, the type of institution they work in, and the sort of offender or inmate they work with. In a number of sections in this book, we will touch more on how the type of facility shapes staff behavior. It is sufficient to say here that probationers likely have a different experience with corrections, because of their status and the degree of liberty, than do jail or prison inmates, or even parolees. But we should also note here that between jurisdictions (even in the same state), between the adult and the juvenile system, between large and small and medium jails, and between different prison security levels, the experience of offenders and inmates is likely to vary greatly. Less restrictive environments are *possibly* more preferred by offenders, though there is some evidence that intensive community supervision is not necessarily prized over prison, particularly among those who have already "done time" (Crouch, 1993).

⌦ Summary

- Though individual experiences, or individual perceptions and adjustment, might vary, the issues identified by philosophers and correctional researchers of yesteryear are still pertinent to our understanding of corrections today. To varying degrees, offenders and inmates experience pains related to their status, and, as humans, we can be sure that they will behave in either prosocial or antisocial ways to ameliorate that pain.

- Despite the formal roles of staff and correctional clients in "total institutions," it is likely that such roles are "corrupted" or at least operate differently than one would think. In other words, staff power is not truly absolute, though it can be used for good or ill.

- Mature coping is one way that correctional clients can fruitfully "adjust" and perhaps reform in the corrections environment.

- Extralegal factors such as gender, race, age, and class likely color the view and experience of correctional clients. Correctional institutions and programs are likely still engaged in the social control of their clients and can serve to facilitate their reform or disable it.

- The good news is that we can direct offender adjustment in ways that are likely to yield a calmer and more secure environment for all offenders and inmates and that might actually lead to their reform (and ours?).

KEY TERMS		
Decent prison	Mortification	Reformatories
Double deviants	Pains of imprisonment	Subcultures
Jails	Power	Total institution
Mature coping	Prisons	

INTERNET SITES

American Correctional Association: www.aca.org

American Friends Service Committee (a Quaker organization interested in correctional reform): www.afsc.org

American Jail Association: www.aja.org

American Probation and Parole Association: www.appa-net.org

Bureau of Justice Statistics (information available on all manner of criminal justice topics): www.ojp.usdoj.gov/bjs

National Criminal Justice Reference Service: www.ncjrs.gov

Office of Justice Programs, Bureau of Justice Statistics (periodic statistical reports on all manner of criminal justice topics, e.g., HIV in prisons and jails, probation and parole, and profiles of prisoners): www.ojp.usdoj.gov/bjs/periodic.htm

Pew Charitable Trust (Corrections and Public Safety): www.pewtrusts.org/our_work_category.aspx?id=72

Vera Institute (information available on a number of corrections and other justice related topics): www.vera.org

How to Read a Research Article

As you travel through your criminal justice or criminology studies, you will soon learn that some of the best-known and emerging explanations of crime and criminal behavior come from research articles in academic journals. This book has research articles throughout, but you may be asking yourself, "How do I read a research article?" It is my hope to answer this question with a quick summary of the key elements of any research article, followed by the questions you should be answering as you read through the assigned sections.

Every research article published in a social science journal will have the following elements: (1) introduction, (2) literature review, (3) methodology, (4) results, and (5) discussion or conclusion.

In the introduction, you will find an overview of the purpose of the research. Within the introduction, you will also find the hypothesis or hypotheses. A hypothesis is most easily defined as an educated statement or guess. In most hypotheses, you will find that the format usually followed is this: If X, Y will occur. For example, a simple hypothesis might be "If the price of gas increases, more people will ride bikes." This is a testable statement that the researcher wants to address in his or her study. Usually, authors will state the hypothesis directly, but not always. Therefore, you must be aware of what the author is actually testing in the research project. If you are unable to find the hypothesis, ask yourself two questions: what is being tested and/or manipulated, and what are the expected results?

The next section of the research article is the literature review. At times, the literature review will be separated from the text in its own section, and, at other times, it will be found within the introduction. In any case, the literature review is an examination of what other researchers have already produced in terms of the research question or hypothesis. For example, returning to my hypothesis on the relationship between gas prices and bike riding, we may find that five researchers have previously conducted studies on the effects of increasing gas prices. In the literature review, I will discuss their findings, and then discuss what my study will add to the existing research. The literature review may also be used as a platform of support for my hypothesis. For example, one researcher might have already determined that an increase in gas causes more people to roller-blade to work. I can use this study as evidence to support my hypothesis that increased gas prices will lead to more bike riding.

The methods used in the research design are found in the next section of the research article. In the methodology section, you will discover who and what was studied, how many subjects were studied, the research tool (e.g., interview, survey, observation), how long the subjects were studied, and how the data that was collected was processed. The methods section is usually very concise, with every step of the research project recorded. Concise but complete recording of steps is important because a major goal of the researcher is "reliability," or being able to do the research over again in the same way and achieving the same the results.

The results section is an analysis of the researcher's findings. If the researcher conducted a quantitative study (using numbers or statistics to explain the research), you will find statistical tables and analyses that explain whether or not the researcher's hypothesis is supported. If the researcher conducted a qualitative study (nonnumerical research for the purpose of theory construction), the results will usually be displayed as a theoretical analysis or interpretation of the research question.

Finally, the research article will conclude with a discussion and summary of the study. In the discussion, you will usually find the hypothesis restated and perhaps a small explanation of why this is the hypothesis. You will also find a brief overview of the methodology and results. Finally, the section will end with a discussion of the implications of the research and of what future research is still needed.

Now that you know the key elements of a research article, let us examine a sample article from your text.

▧ Ultramasculine Prison Environments and Inmates' Adjustment

It's Time to Move Beyond the "Boys Will Be Boys" Paradigm

Faith E. Lutze and David W. Murphy

1. What is the thesis or main idea from this article?

   ◆ The thesis or main idea is found in the introduction of this article. Lutze and Murphy first state that "few studies have considered the influence of gender-stereotyped environments on all-male correctional populations" (p. 82). This statement is then followed with the explanation of this study in the conclusion of the introduction: "In this study we explore the influence of ultramasculine prison environments on inmates' adjustment to prison" (p. 82). Thus, from the introduction, we learn that the main idea of this article is to fill in the space where prior research has failed to completely address the effect of ultramasculine prisons on male inmates and the process of rehabilitation.

2. What are the hypotheses?

   ◆ The hypotheses are found in the middle of the article on page 85. In a section titled "Hypotheses," Lutze and Murphy state and explain their two hypotheses. Hypothesis 1: "The gendered or ultramasculine nature of the prison environment influences inmates' adjustment and perceptions of the environment" (p. 85). Hypothesis 2: "Shock incarceration programs are more gendered or more ultramasculine than traditional prisons and influence inmates' adjustment and perceptions of the institutional environment differently than traditional prisons" (p. 86).

3. Is there any prior literature related to the hypotheses?

   ◆ Lutze and Murphy state in the section "Hypotheses," "This study builds on prior research by exploring the relationship between inmates' adjustment and the gendered nature of the prison environment in a male boot camp prison" (p. 85). The previous section, "Gender and Environmental Effects on Behavior," presents the prior literature to which the authors are referring.

Throughout this section, the authors cite numerous studies that analyze the gendered nature of ultramasculine prisons. The authors then provide their own hypotheses to add to the already existing research.

4. What methods are used to support the hypotheses?

    ◆ Lutze and Murphy outline their methods in the section titled "Methodology." Here, the authors state, "on the basis of survey evaluations of two all-male groups of inmates, we compare the differences in the perception of the prison environment present in a shock incarceration program and in a traditional minimum-security prison" (p. 86). We know that the authors are utilizing surveys and that this is a comparative study conducted on male inmates. The authors continue the "Methodology" section by describing in detail the setting, the subjects, the procedure, the sample, and the measures.

5. Is this a qualitative study or quantitative study?

    ◆ To determine whether or not a study is qualitative or quantitative, you must look at the results. Are Lutze and Murphy using numbers to support their hypotheses (quantitative) or are they developing a non-numerical theoretical argument (qualitative)? Because Lutze and Murphy utilize statistics in this study, we can safely conclude that this is a quantitative study.

6. What are the results, and how does the author present the results?

    ◆ The results are presented in both the "Findings" and the "Discussion" sections. The "Findings" section discusses the statistical results, whereas the "Discussion" section analyzes the statistical results. If the reader has no prior statistical knowledge, it is best to uncover the results in the "Discussion" section. In this section of the Lutze and Murphy article, the authors state that "these findings indicate that perceptions of gender are important for inmates' adjustment to the institution. In addition, they show that perceived differences in the gendered nature of the prison environment do not influence inmates' patterns of adjustment differently in the two prison populations" (p. 91). The authors then provide a detailed discussion of the findings, as well as the implications of the findings.

7. Do you believe that the authors provided a persuasive argument? Why or why not?

    ◆ This answer is ultimately up to the reader, but, looking at this article, I believe that it is safe to assume that the readers will agree that Lutze and Murphy present a persuasive argument. Let us return to the first major premise: The gendered or ultramasculine nature of the prison environment influences inmates' adjustment and perceptions of the environment. This proposition is supported with a statistical analysis based on a comparative study of inmates' attitudes and perceptions.

    ◆ The second premise is this: Shock incarceration programs are more gendered or more ultramasculine than traditional prisons and influence inmates' adjustment and perceptions of the institutional environment differently than traditional prisons. Although the findings showed a mixed statistical support for this proposition, the authors discuss the limitations of the study that may have led to the statistical conclusion. Lutze and Murphy argue that the "failure to find dramatic or consistent differences between prisons may be related to the general concept that prison environments are designed by men for men, and thus tend to be gendered in similar ways regardless of their design or pragmatic intent" (p. 92). In other words, the authors realize that other variables are at work and future research is necessary. This limitation does not take away from the persuasiveness of the argument but rather acknowledges the many variables that affect inmates' attitudes and perceptions.

8. Who is the intended audience of this article?

   ◆ As you read any article, ask yourself, to whom is the author wanting to speak? After you read this article, you will see that Lutze and Murphy are writing for not only students but also professors, criminologists, psychologists, and criminal justice personnel. The target audience may most easily be identified if you ask yourself, who will benefit from reading this article?

9. What does the article add to your knowledge of the subject?

   ◆ Again, this answer is for the reader to answer independently, but it is safe to assume that the reader will find that this article adds an analysis of the gendered effect of ultramasculine prisons on the male prisoner. As Lutze and Murphy argue, most literature regarding ultramasculine or gendered prisons focuses on the effect on female inmates. The authors of this article add to the existing literature the effect of gendered prisons on male inmates and how this effect will ultimately affect the rehabilitation of male inmates.

10. What are the implications for criminal justice policy that can be derived from this article?

   ◆ Many policy implications can be derived from this article, but the most important implication is a reevaluation of gendered and ultramasculine prisons. Does this type of prison environment assist or inhibit the rehabilitation of male inmates? After reading this article, you will see that the effect of ultramasucline prisons on male inmates' attitudes and perceptions requires further research and evaluation so that the criminal justice system can better fulfill its function of changing behavior and preventing future criminality.

Now that we have gone through the elements of a research article, it is your turn to continue through your text, reading the various articles and answering the same questions. You may find that some articles are easier to follow than others, but do not be dissuaded. Remember that each article will follow the same format: introduction, literature review, methods, results, and discussion. If you have any problems, refer to this introduction for guidance.

Faith Lutze and David Murphy examine the concept of shock incarceration in this article. Shock incarceration is supposed to function by "shocking" people sent to them, by sending home the message that, yes, there are prisons and that they are not nice places. Shock programs have become increasingly popular as an alternative to traditional prisons because short stays are obviously cheaper than long stays. Critics of such programs state that they are characterized by ultramasculine environments, which may lead to a number of negative outcomes for inmates. Lutze and Murphy's study compares inmates in a shock incarceration program with inmates in a traditional minimum-security prison on the degree to which they perceive their environments as masculine and on how these perceptions relate to institutional adjustment. The authors find that inmates who describe their environment as possessing ultramasculine attributes were more likely to report negative patterns of adjustment.

Ultramasculine Prison Environments and Inmates' Adjustment

It's Time to Move Beyond the "Boys Will Be Boys" Paradigm

Faith E. Lutze and David W. Murphy

Gender is universally recognized as one of the most fundamental determinants of human behavior. Criminologists consistently propose that it is one of the strongest predictors of criminal involvement (Messerschmidt 1993:1).[1] Decades of evidence indicate that men are far more likely than women to engage in nearly all types of illegal behavior, especially violent crime (Belknap 1996; Chesney-Lind and Shelden 1998; Cullen, Golden, and Cullen 1979). Although gender has been acknowledged as an important influence on behavior and is commonly used in criminological studies about female offenders and their treatment by the criminal justice system (Belknap 1996; Daly and Chesney-Lind 1988), the influence of gender on behavior is often ignored in criminal justice research on or about men (Collier 1998; Newburn and Stanko 1994).

Understanding how behaviors are influenced is a prerequisite to creating appropriate correctional institutions and effective correctional programs. Although many studies have considered gender differences in criminal offending (see Adler 1975; Belknap 1996; Chesney-Lind and Shelden 1998; Messerschmidt 1993; Simon and Landis 1991; Steffensmeier 1995) and the influences of gender-biased

SOURCE: Ultramasculine Prison Environments and Inmates' Adjustment: It's Time to Move Beyond the "Boys Will Be Boys" Paradigm," Faith E. Lutze and David W. Murphy, *Justice Quarterly* 16(4): 709-733 (1999), Taylor & Francis, Ltd., http://www.informaworld.com, reprinted by permission of Taylor & Francis, Ltd.

sex-role stereotypes on women in prison (Pollock-Byrne 1990; Rafter 1995), few studies have considered the influence of gender-stereotyped environments on all-male correctional populations (Morash and Rucker 1990; Sim 1994; Wright 1991b). With the relatively recent advent of boot camp prisons, also known as shock incarceration programs, concerns have been raised about subjecting women to a military model of corrections designed for men (MacKenzie and Donaldson 1996; Marcus-Mendoza, Klein-Saffran, and Lutze 1998; also see Hannah-Moffat 1995). Yet few observers have voiced concerns about the wisdom of subjecting men to an ultramasculine model of corrections that emphasizes aggressive interactions and male sex-role stereotypes (Morash and Rucker 1990).

In this study we explore the influence of ultramasculine prison environments on inmates' adjustment to prison. We begin by examining the meanings and significance of gender and the influence of social environments on gender formation. We further explore this influence by comparing inmates of a shock incarceration environment with those of a traditional minimum-security prison.

◙ Definitions of Masculinity and Femininity

Despite the lack of general agreement on the specific definition of masculinity, certain elements of the concept appear in virtually all related discussions.[2] First, it is assumed that a general relationship exists between sex (i.e., being male or female) and gender (i.e., being masculine or feminine) (Vetterling-Braggin 1982). Second, specific behavioral characteristics and personality traits tend to be associated with masculinity, specifically in contrast to femininity (Deaux 1987). These terms largely have emerged from traditional stereotypes linking femininity to "weakness and passion" and masculinity to "power and rationality"

(Money 1987:13). Although the definitions of these terms continue to evolve, they still reflect similar stereotypes today.[3]

The words associated with masculine characteristics carry a sense of superiority, which is unmistakable in contrast with the terms used to describe feminine characteristics. For instance, *sissified* carries profoundly derogatory implications; being *powerful*, however, is commonly considered desirable. The influence of gender labeling is evident in instances where men and women depart from sex-role stereotypes (see Collier 1998 for a discussion). For instance, men who fail at particular tasks are often labeled with derogatory feminine terms such as being a sissy, while women who demonstrate power are often referred to as bitch. Thus sex-role labeling tends to reinforce behavioral patterns, social networks, and existing institutions of power and rewards.

Subscribing to masculine sex-role stereotypes (i.e., valuing masculinity and/or devaluing femininity), however, has been associated with a number of negative attitudes and behaviors. Although masculine attributes are not inherently bad, they have the potential of being detrimental when alternative behaviors are ignored because of stereotypical beliefs that "real" men can act only in one way (see Basow 1986; Collier 1998; Thompson, Grisanti, and Pleck 1985; Toch 1998). For instance, independence may be considered a positive attribute in some circumstances, but sometimes help from others is needed to overcome personal problems such as drug and alcohol abuse or financial difficulties. Also, although authoritative action and assertiveness may be necessary at times, they may be counterproductive if they are the only means considered acceptable for men in communicating with others. Such behavior may cause conflict in the home and in public places (Thompson et al. 1985), and may cause crime. Refusal to accept less stereotypical masculine responses inhibits the solving of problems that may be related to criminal behavior.

Consequently an overreliance on masculine sex-role stereotypes may hinder prosocial behavior.

Gender and Environmental Effects on Behavior

Historically, the apparent difference between males' and females' criminal tendencies were attributed to sex (i.e., biological traits). Modern criminological theorists, however, subscribe to the general assumption that masculinity and femininity are gender-linked characteristics which are the function of socialization and the environment (see Belknap 1996). According to accepted theories of masculinity, males engage primarily in certain behaviors because those behaviors represent their masculinity effectively. In other words, masculinity is *learned*, and males use it as a means of establishing their gender identity (Franklin 1984).

Particular forms of masculinity, and the appropriate behaviors used to express them, are established not at the individual level but within a larger social group or network. Masculinities therefore are constructed, maintained, and restructured according to the relationships that exist within various social networks in a given environment (Franklin 1984). Consequently, "different types of masculinity exist in the school, the youth group, the street, the family, and the workplace," but each is subject to the influences of the patriarchal culture in which they exist (Chatterbaugh 1990; Messerschmidt 1993:84) Moreover, these types of masculinity exist in relatively specific and identifiable contexts; yet they develop systematically and are ever-changing, always unfinished products (Chatterbaugh 1990; Messerschmidt 1993). Therefore the behaviors in which men engage depend on what type of masculinity that exists in their particular environment or social setting. These settings serve as the venues in which males find support for their self-identity and in which they ultimately contribute to the identification of others (Franklin 1984).

The Gendered Nature of Prison

Understanding how support for masculine behavior is influenced by social networks and given environments is important to the study of correctional environments. Whether prisons are meant to punish or to rehabilitate, their purpose is to change behavior and prevent future criminality. Correctional environments that support overreliance on male sex-role stereotypes may inadvertently support behaviors and attitudes that inhibit prosocial behavior by rewarding aggression and hindering the transition from prison to a law-abiding lifestyle (see Abbott 1981; Bernard and McCleary 1996; Sim 1994).

Traditional Prison Environments

Research on traditional prison environments tends to show that the institutional environment, or prison climate, influences inmate behavior and adjustment. Gender, however, is rarely considered as an environmental attribute. When the prison is considered as gendered or as reinforcing sex-role stereotypes, it usually is in regard to women inmates rather than men. Concern about the reinforcement of gender-biased policy and environments in women's prisons has long been considered detrimental to women and their success after prison. Even when issues of equality are discussed, especially when programs designed for men are provided to women, the appropriateness of subjecting women to such environments is questioned because women's needs are different from men's and because of the gendered nature of programs designed to fit men's needs. Yet in spite of this history of concern about the gendered nature of women's prisons, and the long-standing gender disparity in the commission of violent crime, sex-role stereotyped prison environments for men have been called into question only recently.

Unfortunately, little research has been conducted on how masculine environments affect inmates in traditional male prisons. The literature largely provides graphic accounts of survival in maximum-security institutions permeated with personal and institutional violence and aggression (see Abbott 1981; Bernard and McCleary 1996; Sykes 1958). It appears that masculinity, along with violence and aggression, are valued in the traditional prison culture (Lockwood 1982; Toch 1998). Wright (1991b) argues that the establishment and reinforcement of a masculine identity legitimizes an inmate's use of violence and aggression. Moreover, violence appears to be most common among those inmates who have negative perceptions of (and difficulty in adjusting to) the prison environment and the effects of incarceration (Wright 1991b). Furthermore, Wilson (1986) concludes that male and female inmates differ in their adherence to the inmate code: Male inmates are less trusting of the staff than are female inmates. Wilson attributes this difference to the prison setting, not to gender differences.

Most discussions of masculine prison environments have been linked to violence in maximum-security institutions, where serious, often violent offenders are incarcerated. Few studies have considered how masculine sex-role stereotypes may be replicated or perpetuated in minimum-security institutions, where violent exchanges between inmates are less common and where the institutional environment may be considered safer and more predictable than in maximum-security institutions.

Shock Incarceration Environments

Shock incarceration programs appeared in the early 1980s as an alternative to traditional prisons and as a means of getting tough with young male offenders. Boot camp prisons mirror military boot camps in their common use of "strict discipline, physical training, drill and ceremony, military bearing and courtesy, physical labor, and summary punishment for minor misconduct" (Morash and Rucker 1990:205).

The purpose of a strict, disciplined correctional environment is to scare offenders away from future criminality and to transfer prosocial behavior to the street. Soon after boot camp prisons developed, however, the severity of such an environment was strongly criticized, and its impact on inmates was questioned (Morash and Rucker 1990; Sechrest 1989).

Because of concern about the principles of military boot camps, which include the promotion of aggression, toughness, and intimidation, Morash and Rucker (1990:206) asked perhaps the most pertinent question: "Why would a method that has been developed to prepare people to go to war, and as a tool to manage legal violence, be considered as having such potential in deterring or rehabilitating offenders?" Specifically, Morash and Rucker (1990) argue that ultramasculine sex-role stereotyped environments may promote aggression and competitiveness as well as leading to isolation and helplessness; each of these outcomes may inhibit prosocial adjustment in some males. Thus an important aspect of the boot camp environment may be the extent to which it is gendered. A gendered boot camp environment generally values what is known as *ultramasculinity* (Morash and Rucker 1990). Such a setting supports notions of male forcefulness and strength of will and informally rewards bravado, aggression, and toughness (see Karner 1998). Morash and Rucker state that these beliefs "rest on the assumption that forceful control is to be valued" and that to succeed is to have masculine characteristics—characteristics that may be a cause of crime (1990:215).

For instance, many boot camp programs emphasize the complete dominance of staff members over inmates, confrontational modes of expression, physical fitness, and repeated verbal insults that degrade minorities, women, and homosexuals (Lutze and Brody 1999). Examples of confrontation often are displayed proudly to the media by shock administrators. Numerous newspapers have depicted staff members at boot camp prisons "in the face" of offenders, yelling commands or insults and administering discipline through demanding physical exercise (see Bohlen 1989; Brodus 1991; McDermand 1993; Stobbe 1993). Confrontation, dominance, and control are displayed without much consideration of the idea that other forms of communication (e.g., personal or group sessions, private discussions/counseling) may be more effective in initiating positive responses. Considering a less gender-stereotyped response to men in prison is important in view of recent research on male victimization. According to research (Newburn and Stanko 1994), and personal accounts (Abbott 1981; Bernard and McCleary 1996), some men victimize others to regain their "manhood" and a sense of control in their lives.

In spite of early concerns about the abusive nature of boot camp prisons, most of the work on shock incarceration programs suggests that such prison environments have a positive effect on inmates' attitudes and adjustment (Burton et al, 1993; MacKenzie and Shaw 1990; MacKenzie and Souryal 1995; McCorkle 1995) and do not tend to increase aggression (Lutze 1996a, 1996b).[4] More recent research on the nuances of inmates' attitudinal change and adjustment has yielded mixed results, however (see Lutze 1996b, 1998; Lutze and Marenin 1997). Although researchers have not yet considered empirically whether the environment in boot camp prisons is any more ultramasculine than in traditional prisons, emerging evidence suggests that the environments of boot camp prisons in fact are gendered and may influence inmates' adjustment accordingly. For instance, MacKenzie and Shaw (1990) report that inmates were well-adjusted overall and remained so over time, except for an increase in conflict. This outcome may indicate the confrontational nature of boot camp prisons and the gender-limited means of dealing with stress among inmates and between inmates and staff. Lutze (1996a, 1996b) also reports that although shock incarceration inmates were better-adjusted and held more positive attitudes toward the program and staff than inmates in a traditional prison, inmates in the shock incarceration program reported

greater feelings of isolation and helplessness over time than did the comparison group. These findings tend to support Morash and Rucker's (1990) argument that inmates in boot camp prison may be forced to deal with stress in predominantly masculine ways (i.e., conflict) or by withdrawing (i.e., isolation and helplessness).

As further evidence that shock incarceration programs may be ultramasculine, Lutze (1998) discovered that such programs supported rehabilitation by providing a safe, controlled environment for inmates' participation in programs, but did not offer any greater emotional feedback and support than a traditional prison. Lutze (1998:561) concludes that there is a need to explore how fully "shock incarceration programs, or other correctional settings, can be developed to increase both the external controls that inhibit negative inmate behaviors and the support and emotional feedback that promote psychological and emotional change." Psychological and emotional support may be associated with modes of coping not fostered in an ultramasculine environment.

▧ Hypotheses

This study builds on prior research by exploring the relationship between inmates' adjustment and the gendered nature of the environment in a male boot camp prison. We test the following hypotheses:

Hypothesis 1. The gendered or ultramasculine nature of the prison environment influences inmates' adjustment and perceptions of the environment.

We expect that all prison environments are gendered, and that the extent of gendering will influence how inmates adjust to prison and how they perceive their correctional environment. We expect that the more inmates define their environment as possessing strongly masculine attributes, the more they will experience feelings of isolation, helplessness, anxiety, assertive interactions, and conflict with staff members and other inmates. We also expect that inmates will perceive their environment as being less safe, and as providing less support and emotional feedback. In addition, we believe that they will report higher levels of coercion.

Hypothesis 2. Shock incarceration programs are more gendered or more ultramasculine than traditional prisons, and influence inmates' adjustment and perceptions of the institutional environment differently than traditional prisons.

On the basis of Morash and Rucker's (1990) analysis of boot camp prisons, we expect that shock incarceration programs will emphasize ultramasculine attributes. Given these observations, we expect that shock incarceration inmates will experience more feelings of isolation, helplessness, anxiety, assertive interactions, and conflict with staff members and other inmates. They will also be likely to report less safety, support, and emotional feedback and to report being subjected to more coercion.

In spite of the heavily military environment prevailing in most of these programs, a few also include types of rehabilitation not typically geared toward males. The presence of these features may help to neutralize the otherwise ultramasculine environment at these institutions. Some shock programs emphasize family relationships, family planning, antidrug and alcohol consumption, and personal freedom. Many of these programs are not given as much attention or credibility in traditional male institutions. Because of the dual existence of an ultramasculine environment and "feminine-like" programs in some shock incarceration facilities, it may be difficult to determine how inmates define the gendered nature of their environment.

⊠ Methodology

On the basis of survey evaluations of two all-male groups of inmates, we compare differences in the perception of the prison environment present in a shock incarceration program and in a traditional minimum-security prison. We use these comparisons to examine the influence of the prison environment on variables measuring inmates' adjustment.

Research Setting

The subjects compared in this study are a group of inmates housed at the Intensive Confinement Center (ICC) in Lewisburg, Pennsylvania and a group of inmates in the Federal Prison Camp (FPC) at Allenwood, Pennsylvania. ICC inmates typically spend up to 17 hours a day for six months in the program, which emphasizes physical activity, work, and treatment programs such as education and drug counseling. The ICC also closely regulates inmates' appearance, demeanor, and activity, in accordance with strict militaristic guidelines. All staff members are male except for one teacher, one drug and alcohol specialist,[5] and two secretaries, who have limited contact with inmates.

The public areas of the ICC may be characterized as a loud, tightly structured, highly disciplined environment. Officers commonly yell orders and inmates yell responses in acknowledgment or in request for further instruction. The staff totally dominates the inmates. Inmates must request permission to proceed past each staff member they encounter, to enter the cafeteria, and to speak with staff members. Their physical appearance is under constant scrutiny, including presentation of uniforms, length of hair, amount of facial hair, posture, and body weight. Inmates who fail to conform to the rigid expectations of the program are quickly subjected to verbal reprimands and discipline in the form of physical exercise. Inmates who show weakness or who do not measure up in their athletic ability are often called sissies, girls, or mama's boys. In the private areas of the institution (classrooms and work areas), staff members still maintain a strictly disciplined environment, though communication is somewhat more relaxed. Inmates are allowed to speak with staff members more openly and without formality. Strict rules still apply to appearance and posture, however. Treatment staff members (teachers, counselors, case managers) still may discipline inmates for their behavior, and they share their progress reports with the custody staff.

FPC Allenwood, on the other hand, is a traditional minimum-security institution in which inmates develop their own programs. These programs typically include work; they also may include treatment and recreation. The characteristics of offenders sentenced to institutions such as FPC Allenwood are generally similar to those sentenced to the ICC. For example, inmates in both institutions tend to be less violent, to have less serious records, and to be serving shorter sentences than inmates in more secure traditional prisons. Although most of the staff at FPC Allenwood are males, women serve as correctional officers, teachers, hospital staff members, secretaries, and work supervisors.

The public areas of FPC Allenwood are less disciplined and less structured than in the ICC. Although FPC inmates are required to observe general rules and regulations governing inmate behavior, their movement through the institution is less restricted and much more relaxed than in the ICC. Although staff members maintain control of inmates, the general staff-inmate interaction is not confrontational.

Research Procedure

In this study we use data originally generated from a larger study in which self-report questionnaires were administered to the subjects. The first of two questionnaires was given within two weeks of an inmate's arrival at his institution (Time 1); inmates completed the second questionnaire approximately six months later (Time 2). All demographic data were gathered during the first phase; inmates' evaluations of their environment as well as their attitudes and personal adjustment to confinement were collected during the second phase. The first author administered confidential surveys to inmates in small groups, away from direct observation by the staff.

Sample

Intensive Confinement Center. All inmates arriving at the ICC between December 1993 and October 1994 were surveyed for this study. Of the 334 ICC inmates completing the Time 1 questionnaire, 271 (81%) also completed the Time 2 questionnaire.[6] Of the 63 inmates (19%) who were not included in the second phase, 23 (7%) withdrew voluntarily from the ICC program, 19 (6%) were removed formally by the prison staff for misconduct, 10 (3%) were released prior to graduating from the program, 9 (3%) were unable to participate because of medical conditions, and 2 (1%) inmates' sentences were commuted by the court. This study includes only those ICC inmates who completed the program by participating at both Time 1 and Time 2.[7]

FPC Allenwood. The comparison group used in this study consists of inmates arriving at FPC Allenwood between March 1994 and January 1995. A total of 170 of these inmates were selected to participate in the study. Comparison group selections were made according to age (less than 45 years) and sentence length (less than 60 months). Of the original 170 inmates, 106 (62%) participated fully in the study by completing questionnaires at Time 2. Of the 64 (38%) inmates who did not complete the study, 54 (32%) were released from the institution before Time 2, and 10 (6%) declined to participate at Time 2. This study includes only those FPC inmates for whom Time 1 and Time 2 measures were collected.[8]

Measures

The measures used here relate to three specific areas: demographic information, institutional environment, and adjustment to prison.

Demographics. Subjects were asked to provide information on a variety of background characteristics including age, race, marital status, occupation, education, prior arrests, and offense leading to current confinement.

Environment. The environment of each institution is measured with the Gendered Environment Scale, Wright's (1985) Prison Environment Inventory (PEI) and the Coercion Scale. We developed the Gendered Environment Scale to explore the extent to which the environment is gendered; it measures the extent to which the prison environment supports male sex-role stereotypes (see Appendix Table Al). Composed of 11 Likert-type questions, it is based on components of the male sex role specified by Brannon and Juni (1984; also see Thompson et al. 1985). Gender stereotyping in this study relates to the overgeneralization of feminine attributes, definition of feminine attributes as negative, the use of feminine descriptors as derogatory labels for men, the portrayal of men as better than women, and support for assertive interactions between men.[9]

The PEI is used to measure how inmates define their environment (Wright 1985). The PEI consists of 80 Likert-type questions that address Toch's (1977) eight situational variables: privacy, safety, structure, support, emotional feedback, social stimulation, activity, and freedom. Only the scales for safety, support, and emotional feedback are used in this study. The PEI safety scale measures the extent to which the institutional environment allows inmates to feel safe from being attacked by others, or from being robbed. The support scale measures the extent to which the environment provides inmates with reliable assistance from people with programs that advocate opportunities for self-improvement. The emotional feedback scale measures the extent to which the environment allows inmates to develop intimate relationships with people who care about them and the extent to which inmates are allowed to show their emotions without being ridiculed. This scale includes interactions with staff members and with other inmates.

Adjustment. We use six measures to assess inmates' adjustment. Isolation is measured with the Isolation Scale. Isolation, as measured by this scale, relates to the inmates' feelings that few, if any, people can be trusted and that it is better to stay away from others if one is to complete his sentence successfully. The Isolation Scale consists of six items such as "There are no *real* friends in prison" and "In the institution I keep pretty much to myself."

The state version of the State-Trait Anxiety Inventory (Speilberger, Gorsuch, and Lushene 1970) is used to measure stress. This 20-item Likert-type scale, which has been used in prior prison research (see MacKenzie, et al. 1987), consists of items such as "I feel calm" and "I am presently worrying over possible misfortunes."

An inmate's feeling of helplessness is measured by the Victim Scale, taken from MacKenzie and Shaw's (1990) 37-item Attitude Toward IMPACT Scale. (For the remainder of this study, we refer to this scale as inmates' "helplessness.") Items consist of statements such as "I do not think I can take this anymore" and "This place is unfair."

Conflict is measured by two Guttman scales indicating how many times an inmate has been involved in a conflict either with another inmate or with the staff (Shoemaker and Hillery 1980). The two scales each consist of seven items that measure conflict on a continuum ranging from verbal conflict to physical violence; inmates are asked if they have been in a variety of situations involving other inmates or staff members. For example, they are asked whether they have ever been in "[a] discussion in which some disagreement occurred" and "[a] situation in which some physical force was used on someone."

Assertiveness is measured by the Assertive Interactions Scale (Goodstein and Hepburn 1985). This scale consists of nine questions relating to the likelihood that an inmate will assert himself in a difficult situation: for example, "I try to stay out of trouble but nobody is going to push me around and get away with it."

⊠ Findings

Demographic Characteristics

We compared inmates at the ICC and the FPC to see whether the groups differed on demographic characteristics at Time 1 of the study (see Table 1). The groups were statistically similar in most characteristics, except that ICC inmates tended to be younger, serving shorter sentences, younger at the age of first arrest, and single.[10] These characteristics are related primarily to age and sentence; therefore we introduce these variables as controls in all multilevel analyses.[11]

Table 1	Demographic Characteristics of the Sample					
	ICC (N = 271)		Allenwood (N = 106)			
Variable	Mean	SD	Mean	SD	F	Sig.
Age	26.2	5.2	33.2	7.0	109.1	.01
Education	12.3	1.9	12.7	2.0	3.1	NS
Sentence Length (Months)	24.4	7.8	30.7	17.4	23.4	.01
Felony Convictions	.25	.89	.36	1.04	1.1	NS
Age at First Arrest	23.5	5.7	28.8	8.3	48.7	.01
Juvenile Time	.23	1.0	.08	.54	1.9	NS
	%	n	%	n	Chi-Square	Sig.
Race						
African-American	25.9	70	19.8	21	4.42	NS
Hispanic	17.8	48	12.3	13		
White	51.5	139	61.3	65		
Other	4.8	13	6.6	7		
Total	100.0	270	100.0	106		
Marital Status						
Married	40.1	107	46.2	49	24.76	.01
Single	51.7	138	29.2	31		
Divorced	8.2	22	24.6	26		
Total	100.0	267	100.0	106		
Employment						
Full-time	73.5	197	71.5	75	.18	NS
Part-time	12.7	34	13.3	14		
Unemployed	13.8	37	15.2	16		
Total	100.0	268	100.0	105		

Influence of Perceptions of a Gendered Environment on Inmate Adjustment

We use an independent-sample t-test to test for the difference between the two prisons regarding the presence of gendered environments. ICC inmates ($M = 35.1$) reported the environment to be significantly more gendered than did FPC inmates ($M = 31.7$).[12]

We use linear regression (OLS) to test the relationship between the inmates' assessment of the gendered nature of the environment and how they adjust to prison. Each of the adjustment measures (assertive interactions, isolation, helplessness, stress, conflict) is regressed on the nature of the gendered environment, age, sentence, and prison setting. We pool the two prison populations for these analyses.

The gendered nature of the environment relates significantly to each of the adjustment measures (see Table 2). As inmates defined the environment as possessing more masculine

| Table 2 | Relationship Between Gendered Environment and Inmates' Adjustment |

	Assertive Interaction		Isolation		Helplessness		Anxiety	
	B	t	B	t	B	t	B	t
Constant	28.042	12.81	13.869	7.79	13.098	6.37	42.978	8.43
Gender	.208	4.75***	.108	3.02**	.277	6.73***	.418	4.12***
Prison	−3.652	−5.62***	1.021	1.91	1.032	1.68	3.363	2.20*
Age	−.118	−2.74**	.066	1.87	−.020	−.47	.118	1.17
Sentence	.005	−.217	.030	1.70	.015	.73	−.040	−.780
R^2	.117		.048		.150		.083	
F (df)	11.320(4,346)***		4.375(4,348)**		15.147(4,346)***		7.588(4,340)***	
N	347		349		347		341	

	Inmate-Staff Conflict		Inmate-Inmate Conflict	
	B	t	B	t
Constant	1.307	1.95	2.267	4.512
Gender	.029	2.19*	.030	2.98**
Prison	.389	1.95*	.215	1.43
Age	−.030	−2.28*	−.014	−1.46
Sentence	.018	2.70**	−.0001	−.02
R^2	.077		.060	
F (df)	7.070(4,342)***		5.486(4,348)***	
N	343		349	

* $p < .05$; ** $p < .01$; *** $p < .001$

attributes, they were more likely to report greater levels of assertive interactions, isolation, feelings of helplessness, stress, and conflict with staff members and other inmates. These relationships between adjustment and the gendered nature of the environment did not differ significantly by prison setting.[13]

Perceptions of Gendered Environment on Safety, Emotional Feedback, Support, and Coercion

We use linear regression to test the relationship between the inmates' assessment of the gendered nature of the environment and their perception of safety, emotional feedback, support, and coercion. Each of the environmental measures is regressed on the gendered nature of the environment, age, sentence, and prison setting.

The gendered nature of the environment significantly influences inmates' perceptions of other environmental attributes (see Table 3). Inmates defining their environment as more masculine were more likely to report lower levels of safety, lower levels of emotional feedback and support, and greater levels of coercion. The

relationship between the gendered nature of the environment and environmental attributes did not differ significantly by prison setting.[14]

▨ Discussion

These findings indicate that perceptions of gender are important for inmates' adjustment to the institution. In addition, they show that perceived differences in the gendered nature of the prison environment do not influence inmates' patterns of adjustment differently in the two prison populations.

Gender Matters: The Influence of Environment on Inmates' Adjustment

Regardless of institutional setting, inmates who defined the environment as possessing ultramasculine attributes were more likely to report greater levels of assertiveness, isolation, helplessness, stress, and conflict with other inmates and with the staff. They were also more likely to perceive the environment as

| Table 3 | Relationship Between Gendered Environment and Inmates' Perceptions of the Environment |

	Emotional Feedback		Support		Safety		Coercion	
	B	t	B	t	B	t	B	t
Constant	38.219	18.53	47.497	19.86	40.979	16.925	3.722	3.27
Gender	−.211	−5.08***	−.233	−4.85***	−.191	−3.95**	.224	9.76***
Prison	.640	1.04	−.868	−1.21	6.714	9.342***	1.712	5.01***
Age	−.069	−1.70	−.167	−3.55***	.011	.230	.013	.573
Sentence	.005	.231	.014	.590	−.016	−.668	.014	1.27
R^2	.081		.102		.255		.316	
F (df)	7.443 (4,340)***		9.654 (4,345)***		29.132 (4,344)***		39.85 (4,349)***	
N	341		346		345		350	

**p < .01; *** p < .001.

providing less safety, support, and emotional feedback, and as more coercive.

These findings support the notion that gendered prison environments may be directly related to adjustment in ways that inhibit rehabilitation. Prior research on inmates' adjustment shows that safety, support, emotional feedback, and positive interactions with others are important to prosocial change (see Goodstein and Wright 1989; Lutze 1998). Yet it appears that these very attributes are compromised in environments that emphasize male sex-role stereotypes. It is disconcerting that ultramasculine environments may be magnified in boot camp prisons which are designed to create a more positive correctional environment; this fact calls into question new programs that basically do "more of the same" in housing male inmates.

These findings suggest that Morash and Rucker.(1990) are justified in their concern about creating an environment in shock incarceration programs which actually may promote behavior related to ultramasculine stereotypes, which in turn are believed to be correlated with criminality (see Karner 1998 for a similar discussion related to the military). In this study we offer a possible explanation why inmates in prior studies of shock incarceration programs reported greater levels of conflict (MacKenzie and Shaw 1990), isolation, and helplessness (Lutze 1996b), and reported similar levels of support and emotional feedback than those in traditional prisons (Lutze 1998).

Prison as Support for Masculine Social Groups

We find mixed support for the hypothesis that inmates will perceive boot camp prison to be more masculine than traditional prison, and thus will follow different patterns of adjustment. Although ICC inmates perceived their environment as significantly more masculine than did inmates in traditional prison (in support of Morash and Rucker 1990), inmates at the ICC did not differ overall from traditional inmates in their patterns of adjustment.

A failure to find dramatic or consistent differences between prisons may be related to the general concept that prison environments are designed by men for men, and thus tend to be gendered in similar ways regardless of their design or programmatic intent (Hannah-Moffat 1995; Sim 1994). The ICC may merely reinvent the masculine nature of traditional prisons, thus generating similar outcomes related to adjustment.

The types of behavior in which men engage depend on the type of masculinity that exists in their particular environment or social setting. These environments or social settings serve as the venues in which males find support for their self-identity; ultimately they contribute to the identification of others (Franklin 1984). Therefore, in prison, just as in free society, masculinity is reinforced through social groups and networks. The social setting of prison is an arena in which ultramasculine sex-role stereotypes are promoted and must be confronted, whether or not the individual inmate or staff member subscribes to such beliefs or behavior (see Collier 1998; Toch 1998).

Male inmates and staff members may find it difficult to provide higher levels of support and emotional feedback in programs designed to accomplish rehabilitation because more personal, more caring forms of support are not perceived as acceptable masculine forms of communication. It may be that the prison environment is not "safe" enough to enable an inmate to depart from traditional male paradigms of communication. For instance, because of a lack of trust between all actors in the prison and a fear of being perceived as weak, many men may not transcend the traditional male boundaries, which inhibit the provision of emotional support. Immersion in an ultramasculine setting may hinder the opportunity to incorporate gender-diverse modes of coping; such modes may support personal changes that may enhance success after release.

Patterns of adjustment such as increased isolation, helplessness, and stress also may have negative consequences. For instance, inmates who do not contend effectively with isolation or helplessness may continue, after release, to separate themselves from others who may be helpful in providing services that support law-abiding behavior. On the other hand, inmates who internalize repeated examples of confrontation and assertive interaction as acceptable behavior for gaining control of others may experience problems after release by overasserting their position and refusing to compromise.

It may be that sex-role stereotypes are replicated when the gendered nature of the prison environment is not considered in the creation of new programs such as shock incarceration. Thus prisons are similar to each other, not only because all total institutions function to control and influence inmates' behavior in similar ways (see Goffman 1959), but also because their environments are similarly gendered; thus they reinforce behavior related to sex-role stereotypes that create division between inmates and staff. Reinforcement of ultramasculine sex-role stereotypes actually may contribute to behaviors that perpetuate criminal behavior.

Although this study contributes to the literature on gender and on inmates' adjustment to prison, it also raises many questions that need to be addressed through future research. First, masculinity is not a unidimensional construct. Individuals differ in their adherence to different dimensions of masculinity, depending on demographic characteristics, personality, socialization, and social setting (see Messerschmidt 1993). Consequently, future studies should consider how different types of individuals interact with others (male and female) and with the environment (coercive versus noncoercive settings).

Second, more complex measures should be developed and implemented to assess the gendered nature of the environment and how inmates adjust to different social climates. The environmental and adjustment measures used in this study were derived on the basis of male populations; therefore they also may be gendered so as to influence outcomes.

Finally, it is unknown how individual inmates are affected by ultramasculine prison environments. Are weaker inmates victimized, or do they become the victimizers (see Bernard and McCleary 1996; Newburn and Stanko 1994)? Are aggressive individuals rewarded, and do they become more aggressive as a result (Morash and Rucker 1990)? To more fully understand the existence of ultramasculine environments and how individuals interact with them, researchers must explore these questions more fully.

⊠ Conclusion

Our findings support the notion that an inmate's ability to undergo prosocial adjustment (the goal of boot camp prisons) may be inhibited if the environment emphasizes ultramasculine values. Such environments, whether located in new, creative prison programs or in traditional prisons, are limited in their ability to provide the freedom and support that inmates need to pursue rehabilitation wholeheartedly. Ultramasculine environments also may inhibit staff members from providing full support to inmates who wish to seek personal change, because they are similarly inhibited by sex-role stereotypes. It is time for criminal justice scholars and criminologists to stop viewing gender bias as applicable only to the evaluation of women, and to begin exploring how sex-role stereotypes influence men negatively and inhibit treatment attempts to change antisocial male behavior.

⊠ Notes

1. Theorists in psychology, feminist studies, masculine studies, and education have also analyzed the influence of gender on behavior (see Sapiro 1990 for a review).

2. As with many concepts encountered in the social sciences, a single, agreed upon definition of masculinity does not exist. This is quite evident when we examine the variety of ways in which masculinity

has been operationalized across studies (Baker and Chopik 1982; Norland, Wessell and Shover 1981; Wilkinson 1985) and defined across disciplines (e.g., psychology, feminist studies, etc.). Wilkinson (1985) explains that masculinity exists on multiple dimensions, and leaves conceptualization to the discretion of individual researchers. Consequently the validity of any analysis of masculinity depends on the appropriateness of the definition and the measurement.

3. Common adjectives used to describe masculine and feminine characteristics include the following:

> Masculine: virile, manly, mannish, gentlemanly, strong, vigorous, brawny, muscular, broad-shouldered, powerful, *forceful*, macho, *red-blooded, two fisted* (Webster's New World Thesaurus 1997; emphasis added).

> Feminine: effeminate, epicene, *sissified, sissyish*, unmanly, womanish (Roget's II 1997; emphasis added).

4. Research on aggression and boot camp prisons has focused on behavior within the institution as opposed to behavior after release. Studies on recidivism among boot camp prison graduates do not indicate increases through new violent offenses (see MacKenzie 1991; MacKenzie and Shaw 1993).

5. The female drug and alcohol specialist was hired toward the end of the period of study.

6. This study includes only those who completed questionnaires at both Time 1 and Time 2, because data related to the environment were collected only at Time 2. Although dropouts did not differ significantly from completers for either group on most of the demographic variables, it is not known how dropouts may have defined their environment in terms of being gendered. Thus the results should be interpreted cautiously.

7. On the variables listed in Table 1, ICC inmates who dropped out of the study did not differ statistically from ICC inmates who completed the study.

8. On the demographic variables listed in Table 1, Allenwood inmates who dropped out of the study did not differ statistically from Allenwood inmates who completed the study, except on sentence length. Dropouts (M = 24.7, sd = 17.4) tended to have slightly shorter sentences than those who completed the study (M = 30.7, sd = 19.1, p < .05). These differences should not affect the generalizability of study results.

9. Results of a varimax factor analysis for the Gendered Environment Scale produced five factors (see Lutze 1996a for detailed results). Each of the factors relates to masculine attributes defined by prior research (see Brannon and Juni 1985), and they relate conceptually to what we call the gendered environment. Therefore we combine them into a single scale for the purposes of this study. Items remaining in the scale were determined on the basis of their corrected interitem correlations and on how they affect Cronbach's coefficient alpha for the scale. Responses were measured on a five-point Likert scale ranging from "strongly agree" (= 5) to "strongly disagree" (= 1).

10. ICC inmates differed statistically from FPC inmates in the following manner: age (ICC: M = 26.2, sd = 5.2 vs. FPC: M = 33.2, sd = 7.0, p < .01), sentence length, in months (ICC: M = 24.4, sd = 7.8 vs. FPC: M = 30.7, sd = 17.4, p < .01), age at first arrest (ICC: M = 23.5, sd = 5.7 vs. FPC: M = 28.8, sd = 8.3, p < .01), and marital status (chi square = 24.8, p < .01).

11. Statistical correlations showed that age, sentence length, age at first arrest, and marital status are highly correlated. Therefore the use of age as a single control is justified.

12. The results of the t-test are t = 5.11; df = 359; p < .001.

13. The interaction term (prison x gendered environment) was introduced into each of the regression models in Table 2. No interaction was significant. The interaction term for assertive interactions, however, approached significance (p = .067). As the gendered nature of the environment increased, both the ICC and FPC inmates reported that they would be more likely to assert themselves in a difficult situation. The relationship between gendered environment and assertive interactions, however, was stronger for the FPC inmates.

14. The interaction term (prison x gendered environment) was introduced into each of the regression models in Table 3. No interaction was significant. The interaction term for safety (p = .079), however, approached significance. As the gendered nature of the environment increased, both ICC and FPC inmates described the environment as less safe. The relationship between gendered environment and safety, however, was stronger for the FPC inmates.

◪ References

Abbott, J. (1981). *In the Belly of the Beast: Letters from Prison.* New York: Vintage Books.

Adler, F. (1975). *Sisters in Crime: The Rise of the New Female Criminal.* New York: McGraw-Hill Book Company.

Ajdukovic, D. (1990). "Psychosocial Climate in Correctional Institutions: Which Attributes Describe It?" *Environment and Behavior* 22:420–432.

Baker, P. and K. Chopik. (1982). "Abandoning the Great Dichotomy: Sex vs. Masculinity-Femininity as Predictors of Delinquency." *Sociological Inquiry* 52:349–357.

Basow, S. (1986). *Gender Stereotypes: Traditions and Alternatives.* Monterey, CA: Brooks/Coles Publishing Company.

Belknap, J. (1996). *The Invisible Woman: Gender, Crime and Justice.* New York: Wadsworth Publishing Co.

Bernard, T. and R. McCleary. (eds.) (1996). *Life Without Parole: Living in Prison Today.* Los Angeles: Roxbury Publishing Company.

Bohlen, C. (1989). "Expansion Sought for Shock Program." *New York Times,* June 8, pp. A20.

Brannon, R. and S. Juni. (1984). "A Scale for Measuring Attitudes About Masculinity." *Psychological Documents* 14(1):6.

Brodus, M. (1991). "Guard Combines Toughness, Concern." *The Flint Journal,* Flint, Michigan.

Burton, V., J. Marquart, S. Cuvelier, L. Alarid, and R. Hunter. (1993). "A Study of Attitudinal Change Among Boot Camp Participants." *Federal Probation* 57:46–52.

Chatterbaugh, K. (1990). Contemporary Perspectives on Masculinity: Men, Women, and Politics in Modern Society. Boulder: Westview Press.

Chesney-Lind, M. and R. Shelden. (1998). *Girls, Delinquency, and Juvenile Justice* (2nd ed.). Cincinnati, OH: Wadsworth Publishing Company.

Collier, R. 1998. *Masculinities, Crime and Criminology.* London: Sage Publications.

Cullen, F., K. Golden, and J. Cullen. (1979). "Sex and Delinquency: A Partial Test of the Masculinity Hypothesis." *Criminology* 17:301–310.

Daly, K. and M. Chesney-Lind. (1988). "Feminism and Criminology." *Justice Quarterly* 5:497–538.

Deaux, K. (1987). "Psychological Constructions of Masculinity and Femininity." Pp. 289–303 in *Masculinity / Femininity: Basic Perspectives,* edited by J. Reinisch, L. Rosenblum, and S. Sanders. New York: Oxford University Press.

Eaton, M. (1993). *Women After Prison.* Philadelphia, PA: Open University Press.

Flanagan, T. and K. Maguire. (1991). *Bureau of Justice Statistics Sourcebook of Criminal Justice Statistics 1991.* Albany, NY: The Hindelang Criminal Justice Research Center.

Franklin, C. (1984). *The Changing Definition of Masculinity.* New York: Plenum Press.

Goffman, E. (1959). *Asylums: Essays on the Social Situation of Mental Patients and Other Inmates.* Chicago: Aldine Publishing Company.

Goodstein, L. and J. Hepburn. (1985). *Determinate Sentencing and Imprisonment: A Failure of Reform.* Cincinnati, OH: Anderson Publishing Company.

Goodstein, L. and K. Wright. (1989). "Inmate Adjustment to Prison." Pp. 229–251 in *The American Prison: Issues in Research and Policy,* edited by L. Goodstein and D. MacKenzie. New York: Plenum Press.

Hannah-Moffat, K. (1995). "Feminine Fortresses: Woman-Centered Prisons?" *The Prison Journal* 75:135–164.

Karner, T. (1998). "Engendering Violent Men: Oral Histories of Military Masculinity." Pp. 197–232 in *Masculinities and Violence,* edited by L. Bowker. London: Sage Publications.

Lockwood, D. (1982). "Reducing Prison Sexual Violence." Pp. 257–265 in *The Pains of Imprisonment,* edited by R. Johnson and H. Toch. Newbury Park, CA: Sage Publications.

Lutze, F. (1998). "Do Boot Camp Prisons Possess a More Rehabilitative Environment Than Traditional Prison? A Survey of Inmates." *Justice Quarterly* 15:547–563.

——. (1996a). "Does Shock Incarceration Provide a Supportive Environment for the Rehabilitation of Offenders? A Study of the Impact of a Shock Incarceration Program on Inmate Adjustment and Attitudinal Change." Doctoral dissertation, The Pennsylvania State University.

——. (1996b). "The Influence of a Shock Incarceration Program on Inmate Adjustment and Attitudinal Change." A paper presented at the annual meetings of the Academy of Criminal Justice Sciences, Las Vegas, Nevada.

Lutze, F. and D. Brody. (1999). "Mental Abuse as Cruel and Unusual Punishment: Do Boot Camp Prisons Violate the Eighth Amendment?" *Crime and Delinquency* 45:242–255.

Lutze, F. and O. Marenin. (1997). "The Effectiveness of a Shock Incarceration Program and a Minimum Security Prison in Changing Attitudes Toward Drugs." *Journal of Contemporary Criminal Justice* 12:114–138.

MacKenzie, D. (1990). "Boot Camp Prisons: Components, Evaluations, and Empirical Issues." *Federal Probation* 54:44–52.

——. (1991). "The Parole Performance of Offenders Released from Shock Incarceration (Boot Camp Prisons): A Survival Time Analysis." *Journal of Quantitative Criminology* 7:213–236.

MacKenzie, D. and H. Donaldson. (1996). "Boot Camp for Women Offenders." *Criminal Justice Review* 21:21–43.

MacKenzie, D., L. Goodstein, and D. Blouin. (1987). "Prison Control and Prisoner Adjustment: An Empirical Test of a Proposed Model." *Journal of Applied Social Psychology* 16:109–228.

MacKenzie, D., L. Gould, L. Riechers, and J. Shaw. (1989). "Shock Incarceration: Rehabilitation or Retribution?" *Journal of Offender Counseling, Services & Rehabilitation* 14:25–40.

MacKenzie, D., and J. Shaw. (1990). "Inmate Adjustment and Change During Shock Incarceration: The Impact of Correctional Boot Camp Programs." *Justice Quarterly* 7:125–150.

——. (1993). "The Impact of Shock Incarceration on Technical Violations and New Criminal Activities." *Justice Quarterly* 10:463–487.

MacKenzie, D. and C. Souryal. (1991). "Rehabilitation, Recidivism Reduction Outrank Punishment as Main Goals." *Corrections Today* 53:90–96.

——. (1995). "Inmates' Attitude Change During Incarceration: A Comparison of Boot Camp Prison with Traditional Prison." *Justice Quarterly* 12:325–354.

Marcus-Mendoza, S., J. Klein-Saffran, and F. Lutze. (1998). "A Feminist Examination of Boot Camp Prison Programs for Women." *Women and Therapy* 12(1):173–185.

McCorkle, R. (1995). "Correctional Boot Camps and Change in Attitude: Is all This Shouting Necessary? A Research Note." *Justice Quarterly* 12:365–375.

McDermand, D. (1993). "Hard Time: Intensive Program for Female Prisoners Teaches Responsibility." *Bryan-College Station Eagle*, October 3, p. Dl. College Station, Texas.

Messerschmidt, J. (1993). *Masculinities and Crime: Critique and Reconceptualization of Theory.* Lanham, MD: Rowman and Littlefield Publishers, Inc.

Money, J. (1987). "Propaedeutics of Diecioua G-I/R: Theoretical Foundations for Understanding Dimorphic Gender-Identity/Role." Pp. 13–28 in *Masculinity/Femininity: Basic Perspectives*, edited by J. Reinisch, L. Rosenblum, and S. Sanders. New York: Oxford University Press.

Moos, R. (1968). "The Assessment of the Social Climates of Correctional Institutions." *Journal of Research Crime and Delinquency* 5:173–188.

——. (1975). *Evaluating Correctional and Community Settings.* New York: Wiley.

Morash, M. and L. Rucker. (1990). "A Critical Look at the Idea of Boot Camp as a Correctional Reform." *Crime and Delinquency* 36:204–222.

Newburn, T. and E. Stanko. (1994). *Just Doing Business? Men, Masculinities and Crime.* New York: Rutledge.

Norland, S., R. Wessell, and N. Shover. (1981). "Masculinity and Delinquency." *Criminology* 19:421–33.

Parent, D. (1989). *Shock Incarceration: An Overview of Existing Programs.* Washington, DC: U.S. Department of Justice.

Pollock-Byrne, J. (1990). *Women, Prison, & Crime.* Pacific Grove, CA: Brooks/Cole Publishing Company.

Rafter, N. (1995). *Partial Justice: Women, Prisons, and Social Control* (2nd ed.). London: Transaction Publishers.

Sapiro, V. (1990). *Women in American Society: An Introduction to Women's Studies* (2nd ed.). London: Mayfield Publishing Company.

Sechrest, D. (1989). "Prison 'Boot Camps' Do Not Measure Up." *Federal Probation* 53:15–20.

Sim, J. (1994). "Tougher Than the Rest? Men in Prison." Pp. 100–152 in *Just Doing Business? Men, Masculinities and Crime*, edited by T. Newburn and E. Stanko. New York: Rutledge.

Simon, R. and J. Landis. (1991). *The Crimes Women Commit: The Punishments They Receive.* Lexington. Mass: Lexington Books.

Shoemaker, D. and A. Hillery, Jr. (1980). "Violence and Commitment in Custodial Settings." *Criminology* 18:94–102.

Speilberger, C., R. Gorsuch, and R. Lushene. (1970). *Manual for the State-Trait Anxiety Inventory.* Palo Alto: Consulting Psychologists Press.

Steffensmeier, D. (1995). "Trends in Female Crime: It's Still a Man's World." Pp. 89–104 in *The Criminal Justice System and Women: Offenders, Victims, and Workers*, edited by B. Price and N. Sokoloff. New York: McGraw-Hill, Inc.

Stobbe, M. (1993). "Camp Is No Picnic for Young Men in Trouble." *The Flint Journal*, August 30, pp. A1:A6. Flint, Michigan.

Stratton, J. and L. Lanza-Kaduce. (1980). Project Report, Mt. Pleasant Medium Security Unit. Unpublished report.

Sykes, G. (1958). *The Society of Captives: A Study of a Maximum Security Prison.* Princeton, NJ: Princeton University Press.

Thomas, C. and C. Foster. (1972). "Prisonization in the Inmate Contraculture." *Social Problems* 20:229–329.

Thompson, E. Jr., C. Grisanti, and J. Pleck. (1985). "Attitudes Toward the Male Role and Their Correlates." *Sex Roles* 13:413–427.

Toch, H. (1977). *Living in Prison: The Ecology of Survival.* New York: The Free Press.

——. (1998). "Hypermasculinity and Prison Violence." Pp. 168–178 in *Masculinities and Violence*, edited by L. Bowker. London: Sage Publications.

Vetterling-Braggin, M. (1982). *"Femininity," "Masculinity," and "Androgyny."* New Jersey: Rowman and Littlefield.

Wilkinson, K. (1985(. "An Investigation of the Contribution of Masculinity to Delinquent Behavior." *Sociological Focus* 18:249–263.

Wilson, T. (1986). "Gender Differences in the Inmate Code." *Canadian Journal of Criminology* 24:397–405.

Wright, K. (1985). "Developing the Prison Environment Inventory." *Journal of Research in Crime and Delinquency* 22:257–277.

——. (1991a). "A Study of Individual, Environmental, and Interactive Effects in Explaining Adjustment to Prison." *Justice Quarterly* 8:217–242.

——. (1991b). "Violent and Victimized in the Male Prison." *Journal of Offender Rehabilitation* 16:1–25.

Wright, K. and L. Goodstein. (1989). "Correctional Environments." Pp. 253–270 in *The American Prison: Issues in Research and Policy*, edited by L. Goodstein and D. MacKenzie. New York: Plenum Press.

Zamble, E. and F. Porporino. (1988). *Coping, Behavior, and Adaption in Prison Inmates*. Kingston, Ontario: Queens University.

DISCUSSION QUESTIONS

1. Define what a total institution is and how it might vary by type of correctional arrangement (e.g., probation, parole, jail, prison) and inmate status.

2. Explain why the power of staff in corrections is not absolute? How might this power be used to improve the correctional environment and to destroy it?

3. According to Lutze and Murphy, how do inmates perceive minimum security and shock incarceration facilities and operations?

❖

READING

In this article Craig Hemmens and James Marquart survey the correctional research literature on studies of inmate adjustment patterns to prison life. Early studies assumed inmates were part of a monolithic homogeneous whole, though later research suggested factors such as race, age, and socioeconomic status affect inmate adjustment to prison life. Their own research focuses on the relationship between perceptions of one aspect of the institutional experience, inmate-staff relations, and age and race or ethnicity. A survey of recently released Texas inmates revealed that race and age have a major impact on inmate perceptions of staff.

Friend or Foe?

Race, Age, and Inmate Perceptions of Inmate-Staff Relations

Craig Hemmens and James W. Marquart

 Introduction

The prison experience has historically been meant to be unpleasant, and prisoners have been expected to suffer to some degree. There is, however, some dispute in the correctional research literature as to whether prison has any impact whatsoever on today's inmates. There is a general consensus in the correctional literature that institutionalization is a dehumanizing and demoralizing experience (Goffman, 1961; Irwin, 1980), but some research suggests that some inmates prefer prison to probation (Crouch, 1993), and that some inmates fare better in prison than others (Goodstein & Wright, 1989; Goodstein & MacKenzie, 1984; Wright, 1991). Some research indicates sociodemographic characteristics such as race and age may affect adjustment to prison, as well as perceptions of the institutional experience.

This article builds upon prior research on inmate adjustment patterns. A total of 775 recently released Texas inmates, or "exmates," were surveyed about several aspects of the inmate-staff relationship. Their responses were examined and compared with a host of sociodemographic and criminal history variables. This article isolates and examines the effect of race and age on inmate perceptions of inmate-staff relations.

◪ Prior Research

There is a substantial body of literature on inmate adjustment to prison (much of it focusing on race and age) as well as on inmate-staff relations. Each of these is examined in turn.

Inmate Adjustment

Research on inmate adjustment to incarceration dates from the pioneering study by Clemmer (1940) in which he developed the concept of "prisonization" to explain how prisoners become assimilated into the informal social structure of the prison. Some subsequent researchers built upon Clemmer's work, focusing on the inmate subculture that developed around shared "pains of imprisonment" (Sykes, 1958), which unified the inmate population and created a subculture based on a set of norms and values in opposition to those espoused by the prison staff. This portrayal of prison life became known

as the "deprivation" model, and received mixed empirical support.

Some researchers decried the focus on the so-called pains of imprisonment and institutional factors as the key to understanding inmate adjustment. They examined, instead, attributes and experiences that inmates brought with them to prison. They also argued that inmates brought into prison values learned on the street, in the inner city (Carroll, 1974; 1982). This was referred to as the "importation model." Researchers have studied a variety of extra-prison variables, including race, age, socioeconomic status, and criminal history, a number of which have been found to be related to inmate adjustment patterns. Race and age are prominent among these variables. The focus of this article is on the impact of race and age on inmate perceptions of one aspect of the institutional experience, inmate-staff relations.

Race and Inmate Adjustment

Several studies have examined the racial heterogeneity of the inmate population. Earlier studies made passing reference to racial differences and their potential impact on the prisoner subculture (Sykes, 1958), and the picture generally painted was one of inmate homogeneity. Regardless of whether such a picture was accurate then, indications are that it certainly is not accurate today. Carroll (1974) and Jacobs (1974) both detailed the changing nature of the inmate population, focusing particularly on racial differences.

Jacobs's description of Stateville prison portrayed Black inmates as much more cohesive and unified than White inmates. He argued that this racially based group unity was a direct consequence of the Black inmates' shared preprison experiences with racism and discrimination in their lives in general and the criminal justice system in particular (Jacobs, 1976, 1977). Carroll posited that Blacks were successful in

SOURCE: Hemmens, C., & Marquart, J. W. (2000). Race, age, and inmate perceptions of inmate-staff relations. *Journal of Criminal Justice, 28*, 297–312.

adjusting to prison not only because of their shared history of discrimination on the basis of race, but also because so many of them came to prison from the urban ghetto, where "making it" on the streets required a greater degree of toughness (Carroll, 1982).

Goodstein and MacKenzie (1984) found that White and Black inmates did vary on the degree of prisonization and time spent in the criminal justice system. White inmates with multiple convictions were more highly prisonized than Whites with one conviction, while Black inmates remained at the same level of prisonization regardless of the number of convictions (Goodstein & MacKenzie, 1984). Black inmates were also more "radicalized" than were White inmates, regardless of the length of confinement. These findings suggest there are significant differences between Black and White inmates, possibly because Blacks have suffered discrimination at the hands of criminal justice actors at all levels of the system.

Several studies of racial differences in prison indicated that Black inmates are significantly more likely than White inmates to be involved in conflicts with both staff and other inmates. Fuller and Orsagh (1977) found that Black inmates were more likely than White inmates to be aggressors. Other studies have also found that interracial conflict most often involves a Black aggressor and a White victim (Bowker, 1980; Lockwood, 1980; Wooden & Parker, 1982). The evidence of racial differences on aggressive behavior is mixed, however. A number of other studies have found little or no support for the hypothesis that non-White inmates are in fact more aggressive or violent than are White inmates, when controlling for other factors such as age, number of prior arrests, and drug and alcohol dependency (Ellis et al., 1974; Goodstein & MacKenzie, 1984; Wright, 1988; Zink, 1957;). In addition, while a number of studies have found that Blacks are much more likely to be involved in conduct that results in official condemnation by the prison administration (Flanagan, 1983; Ramirez, 1983), it has been

suggested that racial discrimination on the part of prison administrators or correctional officers may account for this differential (Flanagan, 1983; Howard et al., 1994; Poole & Regoli, 1980; Wright, 1988).

Other studies of racial differences in adaptation to prison indicated that White inmates may suffer from higher levels of stress and fear than Black inmates. One study found that White inmates are more likely than Black inmates to injure themselves intentionally (Wright, 1988). Other studies showed that White inmates had more psychological problems, including breakdowns (Johnson, 1987) and depression. There may be, however, alternative explanations for the difference in adjustment by race. Goodstein and MacKenzie (1984) found no differences in the level of anxiety or the likelihood of depression between Black and White inmates.

Age and Inmate Adjustment

Age is frequently cited as an important explanatory variable in criminal justice (Gottfredson & Hirschi, 1991). A number of studies of violence in prison suggested that age is an important factor in inmate adjustment patterns. Age has been closely linked to the likelihood of aggressive behavior in prison. As inmates become older, there is a linear decline in the number of aggressive acts toward other inmates and/or correctional staff (Ekland-Olson et al., 1983; Porporino & Zamble, 1984), which mirrors the age-crime relationship in the free world (Gottfredson & Hirschi, 1991; Nagin & Farrington, 1992).

A study of age and aggressive behavior in prison by MacKenzie (1987) revealed a slightly more complex picture, however. Rather than a linear decline with age, she found that the rate of aggressive behavior rose until the late twenties, then declined. In addition, interpersonal conflicts with other inmates remained high for a longer period of time than did interpersonal conflicts with correctional officers. Recent research suggests that age is related not only to the likelihood of being involved in violent activity in prison, but also to perceptions of prison as safe or dangerous. Hemmens and Marquart (1999b) found that younger inmates were more likely to perceive prison as a dangerous place than were older inmates.

Inmate-Staff Relations

Living in prison means losing control over much of one's life. Personal autonomy is replaced by the requirement that the individual obey the commands of correctional staff. Inmate-staff relations thus comprise a crucial aspect of the institutional experience, and have been the subject of correctional research since the 1930s. Early researchers assumed that inmates had different norms and values than guards, and that guards and inmates distrusted one another. This research on the prison experience also tended to dichotomize the prison setting, with inmates on one side, guards on the other (Clemmer, 1940; Sykes, 1958).

It was taken for granted by these early researchers that inmates would have different norms and values than guards, and that guards and inmates would distrust one another. Later research has suggested that such a picture is overly simplistic. The Ramirez (1983) research indicated that staff and inmates share similar attitudes concerning a variety of issues. The Marquart (1986a, 1986b) participant observation research indicated that correctional officers who are involved in daily interaction with inmates can establish a rapport with and an understanding of inmates.

Historically, correctional officers have been White males with relatively low education levels, drawn in large part from rural areas, where many prisons were first located (Clemmer, 1940; Jacobs, 1977; Lombardo, 1982). Only recently has the presence of minorities and females in the correctional officer pool become noticeable. Studies of the impact on corrections of the changing correctional officer work force have produced mixed findings. Some studies indicated that minority

officers have more punitive attitudes towards inmates than White officers (Jacobs & Kraft, 1978), while other studies found just the opposite (Jurik, 1985) or no difference (Crouch & Alpert, 1982).

In regards to female correctional officers, several studies indicated that female officers do not hold substantially different attitudes towards inmates than male officers (Cullen et al., 1985; Jurik, 1985, 1988; Jurik & Halemba, 1984). In addition, research suggests that while sexist attitudes still exist among inmates and male officers (Crouch, 1985; Zimmer, 1986), in general both inmates and male officers believe female correctional officers can do their jobs as well as male officers (Kissel & Katsampes, 1980; Walters, 1993).

The focus of the research presented in this article was on inmate attitudes towards and perceptions of correctional officers, a relatively understudied aspect of the inmate-staff equation. Much of the prior research has focused on staff attitudes toward inmates. The composition of the Texas correctional officer force has moved from what was characterized by Marquart (1986b) as a "good old boy" system dominated by White males from rural areas, to one with a substantial number of minority and female correctional officers, many from more urban areas. How have inmates perceived this shift in the demographic makeup of the Texas correctional officer force? This research sought to answer this question by examining inmate perceptions of female correctional officers, inmate-staff relations, and the use of force by correctional staff.

▧ Methods

The data for this study were obtained from a survey, administered over a six-week period, to 775 men released from incarceration in the Texas Department of Corrections—Institutional Division (TDCJ-ID). These former inmates, or "exmates," were interviewed at the bus station in Huntsville, located two blocks west of the Walls Unit. There are over one hundred prisons in TDCJ-ID, though virtually all inmates are processed and released through the Walls Unit. According to TDCJ-ID data, 1,900 inmates were released during the interview period, and 775 inmates submitted to interviews. Assuming that virtually all of these 1,900 inmates passed through the bus station on their way home, as required by law, this generated a response rate of 41 percent. No data were collected on precisely why inmates refused to participate. Care was taken to create a research design that would yield accurate responses and generalizable results, though there are areas of concern, which may limit the usefulness of this study and which should be noted. These include the issues of selection and response bias.

Statistical Procedures

The data analysis included two steps. First, responses to individual items were examined using analysis of variance (ANOVA). Second, a logistical regression was conducted, incorporating all of the sociodemographic variables collected, to determine which individual variables remained in the equation when controlling for the effects of other variables.

Analysis of Variance

Analysis of variance (or ANOVA) is a statistical procedure by which the ratio of variance between group means is examined to determine the likelihood that a difference in mean scores occurred by chance alone. In this research, exmate responses on each of the individual survey items were compared on the basis of sociodemographic and criminal history variables. Sociodemographic variables included race, age, education level; criminal history variables included age at first arrest, number of prior incarcerations, and number of years spent in prison. ANOVA was used to determine whether the difference in the mean scores of the various groups (race/ethnicity and four age categories) occurred through chance variation

or whether there is a statistically significant relationship between the sociodemographic/criminal history characteristics and responses to the survey items.

It should be noted that while information concerning a host of sociodemographic and criminal history variables was collected, in this article the focus is exclusively on race and age. This is because, in most instances, the other variables failed to reveal a statistically significant relationship between the variable and perceptions (Hemmens, 1998). These have therefore been excluded from the presentation of results.

The possible range of scores on each item was from 1 (strongly agree) to 4 (strongly disagree). A lower mean score indicated that an exmate group tended to agree with the statement; a higher mean score indicated that an exmate group tended to disagree with the statement. Items were reverse scored when necessary to achieve consistency of interpretation.

Logistic Regression

Logistic regression allows the researcher to perform a regression-like analysis of data when the dependent variable is dichotomous rather than continuous. Logistic regression was chosen for these data because the narrow range of answers (1 to 4 on a modified Likert scale) tended to produce very clear response patterns—exmates either agreed or disagreed with most statements. Exmate responses were recoded into two categories (agree/disagree), and logistic regression was performed.

▨ Findings

Sociodemographic Characteristics of the Exmate Sample

First, the demographic characteristics of the exmate sample were analyzed and compared with state and national data. Descriptive statistics for the sociodemographic and criminal history characteristics of the 775 male exmates who comprised the sample are summarized in Table 1. Sample characteristics were similar to national level data regarding sociodemographic characteristics of male inmates in 1994 (Gilliard & Beck, 1996).

Blacks made up the largest racial/ethnic group in the exmate sample, comprising almost one-half (48 percent) of all respondents. Whites accounted for approximately one-third (33.7 percent) of all respondents, while Hispanics made up just 17.2 percent of the sample. The racial/ethnic composition of the sample was similar to that of inmates nationally. National statistics indicate 45.6 percent of inmates are White, and 48.2 percent are Black/minority (Gilliard & Beck, 1996).

The average age of the exmate sample was 33 years. White exmates were slightly older, with a mean age of 33.8 years, compared with a mean age of 32.7 years for Black exmates, and 32.2 years for Hispanic exmates. The difference in mean ages was not statistically significant. Precise national statistics on age are not kept, but estimates of the number of inmates by age category compiled by the U.S. Bureau of Justice Statistics suggest the mean age of the exmate sample is in line with national statistics. Approximately 55 percent of all inmates nationally are age 32 or younger, while 45 percent are age 33 or older (Snell, 1995). In addition, the median age of inmates on admission (a figure that is available), was 29 years in 1992 (Perkins, 1994).

The mean years of education completed for the exmate sample was just less than eleven years, or less than a high school degree (twelve years). Over one-half (55.4 percent) of all the exmates had not completed high school. Slightly over one-quarter (28.8 percent) of the exmates had a high school degree or GED, while 15 percent had at least some college experience. National statistics on inmate education levels indicate that roughly 40 percent of all male inmates do not have so much as a high school degree, while slightly less than 47 percent have a high school degree, and slightly more than 13 percent have at

least some college experience (Snell, 1995). The median education level nationally of inmates at admission in 1992 was eleven years (Perkins, 1994). This closely mirrors the exmate sample. Comparison of the exmate and national samples revealed exmates as a whole were somewhat less educated than inmates nationally, with a greater percentage of exmates having failed to graduate from high school.

Perceptions of Inmate-Staff Relations

Next, the responses to questions dealing with inmate-staff relations were examined. The exmates were divided into three racial groups: White, Black, and Hispanic. Responses by racial/ethnic group were compared to determine whether exmates of different racial/ethnic groups held different perceptions of inmate-staff relations.

Four age-group categories were also created: 19–29 years, 30–39 years, 40–49 years, and 50 and older. Responses by age category were compared to determine whether exmates of different ages held different perceptions of inmate-staff relations. Age categories were created to highlight age differences. Tracking responses by individual years tended to mask differences as the changes between one year and the next were very slight.

Exmates were asked their level of agreement with eight statements regarding correctional staff. ANOVA was then conducted to determine statistically significant differences based on race and age. The eight statements and the responses to each statement are displayed in Table 2.

Exmates appeared to share similar feelings about correctional officers and inmate-staff relations. Mean scores on the eight questions regarding perceptions of correctional staff did not often vary significantly based on most of the selected sociodemographic and criminal history characteristics,

Table 1	Sociodemographic and Criminal History Characteristics of the Exmate Sample		
Characteristic	**Frequency**	**Percent**	**Mean**
Race/ethnicity			
White	261	33.7	
Black	363	46.8	
Hispanic	133	17.2	
Other	17	2.2	
Age			32.98
19–29	289	37.3	
30–39	326	42.1	
40–49	119	15.4	
50–72	36	4.6	
Education level			10.96
No high school degree	429	55.4	
High school degree	223	28.8	
Some college	87	11.2	
College degree	28	3.6	
Prior incarcerations			2.44
0	351	45.3	
1	61	7.8	
2	215	27.7	
3 or more	148	19.1	
Years in prison			6.08
1–3	112	26.9	
4–5	141	33.8	
6–9	164	39.3	
Age at first arrest			19.37
Under 17	379	48.9	
18	88	11.4	
19–21	111	14.3	
22–29	128	16.5	
30 or older	56	7.2	

such as education level, socioeconomic status, and criminal history. The results for these items were thus excluded for ease of presentation here. Only two variables revealed statistically significant results. These were exmate age and race (discussed following).

Table 2	Perceptions of Inmate-Staff Relations Using Analysis of Variance

I had very few problems with guards in TDC.

Race/Ethnicity	White	Black	Hispanic
N	260	363	132
Mean	2.408	2.532	2.561
SD	.867	.867	.822

F ratio = 2.055; F probability = .1288

Age	19–29	30–39	40–49	50+
N	289	325	118	28
Mean	2.640	2.498***	2.246**	2.143[a]
SD	.887	.891	.703	.524

F ratio = 7.744; F probability = .000

Female guards do their job as well as male guards.

Race/Ethnicity	White	Black	Hispanic
N	259	359	130
Mean	2.398	2.320	2.346
SD	.792	.723	.723

F ratio = .8188; F probability = .4413

Age	19–29	30–39	40–49	50+
N	287	318	119	28
Mean	2.317	2.393	2.387	2.143
SD	.758	.749	.678	.705

F ratio = 1.370; F probability = .2506

There are enough guards to provide safety and security for inmates.

Race/Ethnicity	White	Black	Hispanic
N	259	362	130
Mean	2.568	2.547	2.592
SD	.776	.773	.723

F ratio = .1785; F probability = .8366

Age	19–29	30–39	40–49	50+
N	287	322	119	28
Mean	2.711	2.528*	2.437**	2.214[a]
SD	.787	.745	.721	.568

F ratio = 7.085; F probability = .0001

The quality of new guards entering TDC today is as good as it ever was.

Race/Ethnicity	White	Black	Hispanic
N	224	333	125
Mean	2.808	2.796	2.856
SD	.711	.690	.631

F ratio = .352; F probability = .7034

Age	19–29	30–39	40–49	50+
N	260	295	105	28
Mean	2.869	2.820	2.781	2.464[a,b]
SD	.691	.669	.679	.637

F ratio = 3.145; F probability = .0247

TDC staff often act unfairly towards inmates.

Race/Ethnicity	White	Black	Hispanic
N	255	359	132
Mean	2.835	2.894	2.924
SD	.696	.701	.561

F ratio = .917; F probability = .4003

Age	19–29	30–39	40–49	50+
N	286	319	119	27
Mean	2.923	2.871	2.849	2.593
SD	.671	.722	.591	.752

F ratio = 2.097; F probability = .0992

Overall, they treated me pretty good in TDC.

Race/Ethnicity	White	Black	Hispanic
N	260	360	133
Mean	2.300	2.386	2.300
SD	.623	.711	.623

F ratio = 1.3431; F probability = .2617

(Continued)

Table 2	(Continued)			
Age	**19–29**	**30–39**	**40–49**	**50+**
N	288	324	118	28
Mean	2.493	2.318*	2.220**	2.036[a]
SD	.703	.654	.572	.429
F ratio = 8.483; F probability = .000				

Most guards in TDC treat inmates like they are less than human.

Race/Ethnicity	White	Black	Hispanic
N	260	362	133
Mean	2.873	3.008	2.985
SD	.737	.700	.627
F ratio = 2.936; F probability = .0537			

Age	**19–29**	**30–39**	**40–49**	**50+**
N	288	325	119	28
Mean	3.101	2.932*	2.773**	2.714[a]
SD	.699	.691	.657	.810
F ratio = 8.180; F probability = .000				

Prison guards often use too much force on inmates.

Race/Ethnicity	White	Black	Hispanic
N	255	358	133
Mean	2.576	2.821*	2.752
SD	.784	.746	.690
F ratio = 8.317; F probability = .0003			

Age	**19–29**	**30–39**	**40–49**	**50+**
N	285	322	116	28
Mean	2.853	2.683*	2.612**	2.571
SD	.778	.723	.695	.742
F ratio = 4.437; F probability = .0042				

NOTE: In this table, the following notations indicate statistical significance at the .05 level between groups: 1,2 (*), 1,3 (**), 2,3 (***), 1,4 ([a]), 2,4 ([b]). TDC = Texas Department of Corrections.

Age and Perceptions

These results indicate that younger exmates tend to have more problems with correctional staff, believe staff treat inmates in an inhumane fashion, feel new guards are less qualified than in the past, and feel that they were not well-treated in prison. They are also more likely to believe there are not enough guards to ensure inmate safety, and that correctional officers use more force than is necessary.

On the item "I had very few problems with guards in TDC" the difference in the mean score for exmates in the 19–29 age category and exmates age 30 or older was statistically significant at the .05 level. In addition, the difference in mean scores for exmates in the 30–39 and 40–49 age category was statistically significant at the .05 level. This indicates that younger inmates had more problems with correctional staff, and that, as exmates age, the likelihood of trouble with staff declines steadily, in a linear fashion. This finding is in accord with research that indicated younger inmates generally have more disciplinary infractions than do older inmates (Ellis, et al., 1974; Howard, et al., 1994; Zink, 1957).

Related to this finding is the observation that younger exmates were more likely to agree with the statement "Prison guards often use too much force on inmates" (this item was reverse coded). The difference in mean scores for exmates in the 19–29 age category and exmates in the older categories was statistically significant at the .05 level, with exmates in the 19–29 age category more likely to agree with the statement. A possible explanation for this finding is that younger inmates are more likely to be involved in altercations with staff, and therefore be the subject of use of force by correctional officers.

In accord with the finding that younger exmates believe staff use excessive force, younger exmates also were more likely to believe that guards treat inmates in an inhumane fashion. On the item "Most guards in TDC treat inmates like they are less than human," the difference in the mean score for exmates in the 19–29 age category and the other three categories was statistically significant at the .05 level (this item was reverse coded). On the item "Overall, they treated me pretty good in TDC," the difference in the mean score for exmates in the 19–29 age category and the other three categories was statistically significant at the .05 level. Younger exmates were clearly more likely than were older exmates to feel they were not well treated. This may be because correctional staff are harder on younger prisoners, or because younger prisoners are less willing to accept what happens to them in prison as proper, or because younger prisoners are involved in more disciplinary infractions and resent being punished by correctional staff.

The differences in mean scores for the various age categories on the item "TDC staff often act unfairly towards inmates" were not statistically significant, although younger exmates did tend to agree with the statement more than their older counterparts. Taken together, responses to these three items suggest that younger exmates think they are treated *poorly*, but not *inappropriately*. It is unclear whether this is a reflection of low self-esteem or simply a statistical artifact. In addition, it is worth noting that age is related to perceptions in the free world as well. Younger citizens have less positive attitudes towards police than do older citizens (Huang & Vaughn, 1996; Walker, 1992).

Given the tendency of younger exmates to feel that they are poorly treated by correctional staff, they were nonetheless more likely to feel that there are not enough correctional officers to ensure safety and security in prisons. The difference in the mean score of the 19–29 age group and each of the older age groups was statistically significant at the .05 level. In addition, as exmate age increased, belief that there were enough guards increased steadily, which suggests, perhaps, that older exmates think there are too many officers getting into their business.

Related to the perception of whether there were enough correctional officers was the perception of the quality of new officers. The mean scores for exmates in the 19–29, 30–39 age categories, and exmates in the 50 and older age category were statistically significant at the .05 level, with the younger exmates tending to disagree with the statement "The quality of new guards entering TDC today is as good as it ever was." This finding suggests that the oldest exmates think new officers are not as qualified as officers hired in the past, an interesting finding given that TDCJ-ID preservice training has increased rather than decreased in recent years.

Exmates were also asked whether they believed that female correctional staff do their jobs as well as male officers. The mean scores for the four age groups did not reveal any statistically significant differences. In this case, a finding of no difference is perhaps as interesting as a finding of some difference. Exmates were fairly unified in their estimation of the ability of female guards vis-à-vis male guards. The mean scores for the age categories fall around 2.3, indicating a slight tendency towards agreement with the statement "Female guards do their jobs as well as male guards." This finding is in accord with prior research on inmate attitudes (Kissel & Katsampes, 1980; Walters, 1993). Inmates seem to hold more positive opinions of female correctional officers than do many male staff. Prior research indicated male officers often resent the introduction of female officers into their male world (Owen, 1985; Zimmer, 1986), and frequently denigrate the job performance of female officers (Jurik, 1985; Zupan, 1992).

Race and Perceptions

The relationship between race and perception of inmate-staff relations turned up as statistically significant on only one item: "Prison guards often use too much force on inmates." On this item (which was reverse coded), the mean score for Black exmates was 2.8, while for White exmates it was 2.6. Black exmates, therefore, were more likely to agree with the statement than White exmates. The Hispanic exmate group had a mean score close to the Black exmates, but this was not statistically significant relative to White exmates. This finding is in accord with research conducted on attitudes of nonincarcerated populations. Black citizens, as compared with White citizens, appear to trust the police less (Cole, 1999; Lasley, 1994) and believe that police use excessive force more often (Huang & Vaughn, 1996).

Black and White exmates had different perceptions of use of force by correctional officers, though they did not have substantially different perceptions of whether correctional officers acted unfairly, or treated inmates as "less than human," which might be expected.

Logistic Regression Analysis

Logistic regression was conducted, to determine the impact of the various sociodemographic and criminal history variables when controlling for the effect of other variables. Table 3 displays the results of the logistic regression equations for the eight items regarding exmate perceptions of inmate-staff relations. Race and age appeared as significant variables in several items. They were also the only variables to remain in any equation.

These results suggest race/ethnicity is related to perceptions of inmate-staff relations, a fact obscured by the ANOVA procedure. Race/ethnicity and age both play important roles in the logistic regression models for the items regarding perceptions of inmate-staff relations. Differences between White and Black exmates remained in all but two of the eight equations, indicating White and Black exmates have very different perceptions of the inmate-staff relationship. Differences between White and Hispanic exmates remained in several equations as well, suggesting that race is a powerful predictor of perceptions of this aspect of the institutional experience.

White exmates were somewhat more likely than Black exmates to agree with the statement "Overall they treated me pretty good in TDC." White exmates were almost twice as likely as Black or Hispanic exmates to disagree with the statement "Most guards in TDC treat inmates like they are less than

Table 3	Logistic Regression on Perception of Inmate-Staff Items				
Group	**B**	**Wald**	**Σ**	**R**	**Odds Ratio**
I had very few problems with guards in TDC.					
Age (B/W)	−.3412	11.467	.0007	−.1063	1.41
Age (H/W)	−.3137	5.320	.0211	−.0799	1.39
Age (H/B)	−.4962	16.672	.0000	−.1494	1.64
Female guards do their jobs as well as male guards.					
Race (B/W)	.4125	5.879	.0153	.0698	1.51
There are enough guards to provide safety and security for inmates.					
Race (B/W)	−.3412	11.465	.0007	−.1055	1.40
Age (H/W)	−.3599	8.960	.0028	−.1150	1.43
Age(H/B)	−.3976	11.436	.0007	−.1198	1.50
The quality of new guards entering TDC today is as good as it ever was.					
Age(H/B)	−.2878	5.323	.0210	−.0776	1.33
TDC staff often act unfairly towards inmates.					
Race (H/W)	−.6414	5.868	.0154	−.0939	1.90
Overall, they treated me pretty good in TDC.					
Race (B/W)	−.3944	4.608	.0318	−.0550	1.48
Age (B/W)	−.4496	14.812	.0001	−.1308	1.57
Age (H/W)	−.3691	6.280	.0122	−.0955	1.45
Age(H/B)	−.5681	18.573	.0000	−.1637	1.76
Most guards in TDC treat inmates like they are less than human.					
Race (B/W)	−.6260	10.984	.0009	−.1135	1.87
Age (B/W)	−.4003	13.253	.0003	−.1271	1.49
Race (H/W)	−.7344	8.023	.0046	−.1142	2.08
Age(H/B)	−.4918	13.060	.0003	−.1511	1.64
Prison guards often use too much force on inmates.					
Race (B/W)	−.8832	25.778	.0000	−.1713	2.42
Age (B/W)	−.2550	6.0157	.0142	−.0704	1.29
Race (H/W)	−.6960	9.756	.0018	−.1216	2.00
Age(H/B)	−.3512	8.434	.0037	−.1030	1.42

Note: TDC = Texas Department of Corrections; B = Black; W = White; H = Hispanic.

human." White exmates were more than twice as likely as Black or Hispanic exmates to disagree with the statement "Prison guards often use too much force on inmates." In addition, White exmates were 1.5 times more likely than Black exmates to agree with the statement "Female guards do their jobs as well as male guards," and 1.5 times more likely than Black exmates to agree with the statement "There are enough guards to provide safety and security for the inmates." Finally, White exmates were almost twice as likely as Hispanic exmates to disagree with the statement "TDC staff often act unfairly towards inmates." All of this suggests that White exmates have a higher opinion of correctional staff than minority exmates.

Age also plays an important role in shaping perceptions of inmate-staff relations. As they get older, exmates were more likely to agree with the statement "I had very few problems with the guards in TDC." For each increase in age category, exmates were approximately 1.5 times more likely to agree with the statement. The same was true in regard to the statement "Overall, they treated me pretty good in TDC." Older exmates were also more likely to disagree with the statement "Most guards in TDC treat inmates like they are less than human." Older exmates were also more likely to disagree with the statement "Prison guards often use too much force on inmates." These findings suggest that older exmates are significantly more likely than younger exmates to have positive opinions of correctional staff, feeling that staff treat inmates decently. This corresponds with the ANOVA findings.

⊠ Conclusion

Examination of the exmate responses by race and age revealed some significant relationships between race, age, and perceptions of the institutional experience. Other factors (such as criminal history variables) consistently failed to show up as statistically significant in the ANOVA and do not remain in the logistic

regression model, though two factors consistently showed up in both the ANOVA and logit procedures: race/ethnicity and age. Age remained in virtually every equation, while racial differences between White and Black exmates remained in several equations as well.

Regarding age, younger exmates also differed from older exmates in their perceptions of correctional staff and inmate-staff relations. Younger exmates reported having more problems with staff, and were more likely to believe that correctional staff treats them poorly: younger exmates tended to agree that staff often acted unfairly, treated inmates as less than human, and used too much force on inmates. All of this suggests that age is closely related to perceptions of staff behavior and inmate-staff interactions. Whether younger exmates feel this way because they are more often engaged in activity that is likely to be the subject of staff reprisals, or because staff treat older exmates with more care and respect, or because younger exmates come into prison with different beliefs about how they should be treated by staff is unclear. Further research on these differences in perception is clearly warranted.

Regarding race/ethnicity, the logistic regression analysis indicated that both Black and Hispanic exmates are more likely than White exmates to feel correctional staff treat inmates unfairly and/or poorly, and use excessive force. This finding is not altogether surprising, given the long history of poor race relations in Texas prisons (Marquart, 1986a, 1986b). It also mirrors research findings in the free world, which indicate that minority populations have a more negative perception of the criminal justice system in general and the police in particular than do Whites (Cole, 1999; Huang & Vaughn, 1996; Walker, 1992).

These findings are in accord with prior research, which found that individual, extra-institutional variables such as race play a major role in inmate adjustment patterns. In addition, this research added to the literature a clearer picture of the impact of age, an understudied variable, on inmate perceptions. Younger and older

inmates differ in their perceptions of prison and this difference is often linear in nature, but particularly pronounced when the youngest and oldest inmate age groups are compared.

References

Babbie, E. (1995). *The Practice of Social Research* 6th ed. Belmont, CA: Wadsworth Publishing Company.

Berk, B. (1968). Organizational goals and inmate organization. *Am J Soc 71*, 522–534.

Bowker, L. H. (1980). *Prison Victimization.* New York: Elsevier.

Carroll, L. (1974). *Hacks, Blacks, and Cons.* Lexington, MA: Lexington.

Carroll, L. (1982). Race, ethnicity, and the social order of the prison. In R. Johnson & H. Toch (Eds.), *The Pains of Imprisonment* (pp. 181–203). Beverly Hills, CA: Sage Publications.

Clemmer, D. (1940). *The Prison Community.* New York: Holt, Rinehart and Winston.

Cole, D. (1999). *No Equal Justice: Race and Class in the American Criminal Justice System.* New York: The New Press.

Crouch, B. M. (1985). Pandora's box: women in men's prisons. *J Crim Just 13*, 535–548.

Crouch, B. M. (1993). Is incarceration really worse? analysis of offenders' preferences for prison over probation. *Just Quart 10*, 67–88.

Crouch, B. M., & Alpert, G. P. (1982). Prison guards' attitudes towards components of the criminal justice system. *Criminology 18*, 227–236.

Cullen, F., Link, B., Wolfe, N., & Frank, J. (1985). The social dimensions of correctional officer stress. *Just Quart 2*, 505–533.

Ekland-Olson, S., Supancic, M., Campbell, J., & Lenihan, K. J. (1983). Postrelease depression and the importance of familial support. *Criminology 21*, 253–275.

Ellis, D., Grasmick, H. G., & Gilman, B. (1974). Violence in prisons: a sociological analysis. *Am J Soc 80*, 16–43.

Faine, J. R. (1973). A self-consistency approach to prisonization. *Soc Quart 14*, 576–588.

Flanagan, T. J. (1983). Correlates of institutional misconduct among state prisoners. *Criminology 21*, 29–39.

Fowler, F. J. (1993). *Survey Research Methods.* Newbury Park, CA: Sage Publications.

Fuller, D. A., & Orsagh, T. (1977). Violence and victimization within a state prison system. *Crim Just Rev 3*, 35–55.

Garabedian, P. G. (1963). Social roles and processes of socialization in the prison community. *Social Probs 11*, 139–152.

Gilliard, D. K., & Beck, A. J. (1996). *Prison and Jail Inmates, 1995.* Washington, DC: U.S. Department of Justice.

Goffman, E. (1961). *Asylums.* Chicago, IL: Aldine de Gruyter.

Goodstein, L., & MacKenzie, D. L. (1984). Racial differences in adjustment patterns of prison inmates—prisonization, conflict, stress, and control. In D. Georges-Abeyie (Ed.), *The Criminal Justice System and Blacks* (pp. 271–306). New York: Clark Boardman Company.

Goodstein, L., &. Wright, K. N. (1989). Inmate adjustment to prison. In L. Goodstein & D. L. MacKenzie (Eds.), *The American Prison: Issues in Research and Policy* (pp. 229–251). New York: Plenum Press.

Gottfredson, D. M., & Hirschi, T. (1991). *A General Theory of Crime.* Palo Alto, CA: Stanford University Press.

Grusky, O. (1959). Organizational goals and the behavior of informal leaders. *Am J Soc 83*, 59–67.

Hemmens, C. (1998). *Life in the Joint and Beyond: An Examination of Inmate Attitudes and Perceptions of Prison, Parole, and Self at the Time of Release.* Ph.D. diss. College of Criminal Justice, Sam Houston State University: Huntsville, TX.

Hemmens, C, & Marquart, J. W. (1998). Fear and loathing in the joint: the impact of race and age on inmate support for prison AIDS policies. *Prison J 78*, 133–151.

Hemmens, C, & Marquart, J. W. (1999a). The impact of inmate characteristics on perceptions of race relations in prison. *Int J Offender Ther & Comp Crim 43*, 230–247.

Hemmens, C, & Marquart, J. W. (1999b). Straight time: inmates' perceptions of violence and victimization in the prison environment. *J Offender Rehab 28*, 1–21.

Howard, C, Winfree, L. T., Mays, G. L., Stohr, M. K., & Classon, D. L. (1994). Processing inmate disciplinary infractions in a federal correctional institution: legal and extralegal correlates of prison-based legal decisions. *Prison J 73*, 5–31.

Huang, W., & Vaughn, M. S. (1996). Support and confidence: public attitudes toward the police. In T. J. Flanagan & D. Longmire (Eds.), *Americans View Crime and Justice: A National Public Opinion Survey* (pp. 98–119). Thousand Oaks, CA: Sage Publications.

Irwin, J. (1980). *Prisons in Turmoil*. Boston, MA: Little, Brown & Company.

Irwin, J., & Cressey, D. R. (1962). Thieves, convicts and the inmate culture. *Soc Probs 10*, 142–155.

Iverson, C., & Norpoth, C. (1987). *Analysis of Variance*. Newbury Park, CA: Sage Publications.

Jacobs, J. B. (1974). Street gangs behind bars. *Social Probs 21*, 395–409.

Jacobs, J. B. (1976). Stratification and conflict among prison inmates. *J Crim Law & Crim 66*, 476–482.

Jacobs, J. B. (1977). *Stateville*. Chicago, IL: University of Chicago Press.

Jacobs, J. B., & Kraft, L. (1978). Integrating the keepers: a comparison of Black and White prison guards in Illinois. *Social Probs 25*, 304–318.

Johnson, R. (1987). *Hard Time: Understanding and Reforming the Prison*. Belmont, CA: Wadsworth Publishing Company.

Jurik, N. (1985). Individual and organizational determinants of correctional officer attitudes towards inmates. *Criminology 23*, 523–539.

Jurik, N. (1988). Striking a balance: female correctional officers, gender role stereotypes, and male prisons. *Soc Inquiry 58*, 291–305.

Jurik, N., & Halemba, G. J. (1984). Gender, working conditions, and the job satisfaction of women in a nontraditional occupation: female correctional officers in men's prisons. *Soc Quart 25*, 551–566.

Kachigan, S. K. (1991). *Multivariate Statistical Analysis: A Conceptual Introduction*. New York: Radius Press.

Kissel, P. J., & Katsampes, P. L. (1980). The impact of woman correctional officers on the functioning of institutions housing male inmates, *J Offender Counseling, Ser & Rehab 4*, 213–231.

Lasley, J. R. (1994). The impact of the Rodney King incident on citizen attitudes toward police. *Policing & Society 3*, 245–255.

Lawson, D. P., Segrin, C, & Ward, T. D. (1996). The relationship between prisonization and social skills among prison inmates. *Prison J 76*, 293–309.

Lockwood, D. (1980). *Prison Sexual Violence*. New York: Elsevier.

Lombardo, L. X. (1982). *Guards Imprisoned*. Cincinnati, IL: Anderson Publishing Co.

MacKenzie, D. L. (1987). Age and adjustment to prison: interactions with attitudes and anxiety. *Crim Just & Behav 14*, 427–447.

Marquart, J. W. (1986a). Doing research in prison: the strengths and weaknesses of full participation as a guard. *Just Quart 3*, 15–32.

Marquart, J. W. (1986b). Prison guards and the use of physical coercion as a mechanism of prisoner control. *Criminology 24*, 347–366.

Maxfield, M. G., &. Babbie, E. (1995). *Research Methods for Criminal Justice and Criminology*. Belmont, CA: Wadsworth Publishing Company.

Menard, S. (1995). *Applied Logistic Regression Analysis*. Newbury Park, CA: Sage Publications.

Nagin, D. S., & Fanington, D. P. (1992). The onset and persistence of offending. *Criminology 30*, 501–523.

Owen, B. (1985). Race and gender relations among prison workers. *Crime & Del 31*, 147–159.

Perkins, C. (1994). *National Corrections Reporting Program, 1992*. Washington, DC: U.S. Department of Justice.

Poole, E. D., & Regoli, R. M. (1980). Race, institutional rule-breaking, and disciplinary response: a study of discretionary decision making in prison. *Law & Society Rev 14*, 931–946.

Porporino, F., & Zamble, E. (1984). Coping with imprisonment. *Can J Crim 25*, 403–421.

Ramirez, J. (1983). Race and the apprehension of inmate misconduct. *J Crim Just 11*, 413–427.

Schrag, C. (1961). Some foundations for a theory of correction in the prison. In D. Cressey (Ed.), *The Prison* (pp. 309–358). New York: Holt, Rinehart, and Winston.

Snell, T. L. (1995). *Correctional Populations in the United States, 1992*. Washington, DC: U.S. Department of Justice.

Street, D. (1970). The inmate group in custodial and treatment settings. *Am Soc Rev 33*, 40–55.

Sykes, G. M. (1958). *The Society of Captives: A Study of a Maximum Security Prison*. Princeton, NJ: Princeton University Press.

Thomas, C. W. (1970). Toward a more inclusive model of the inmate contraculture. *Criminology 8*, 251–262.

Thomas, C. W. (1977). Theoretical perspectives on prisonization: a comparison of the importation and deprivation models. *J Crim Law & Crim 68*, 135–145.

Tittle, C. R. (1968). Inmate organization: sex differentiation and the influence of criminal subcultures. *Am Soc Rev 30*, 492–504.

Walker, S. (1992). *The Police in America*. New York: McGraw-Hill.

Walters, S. (1993). Changing the guard: male correctional officers' attitudes toward women as coworkers. *J Offender Rehab 20*, 47–60.

Wellford, C. (1967). Factors associated with adoption of the inmate code: a study of normative socialization. *J Crim Law, Crim & Police Sci 58*, 197–203.

Wheeler, S. (1961). Socialization in correctional communities. *Am Soc Rev 26*, 697–712.

Wilson, T. P. (1968). Patterns of management and adaptations to organizational roles: a study of prison inmates. *Am J Soc 71*, 146–157.

Wood, B. S., Wilson, G. G., Jessor, R., & Bogan, J. B. (1968). Trouble-making behavior in a correctional institution: relationship to inmates' definition of their situation. *Am J Orthopsych 36*, 795–802.

Wooden, S., & Parker, J. (1982). *Men Behind Bars: Sexual Exploitation in Prison.* New York: Plenum Press.

Wright, K. N. (1988). The relationship of risk, needs, and personality classification systems and prison adjustment. *Crim Just & Behav 15*, 454–471.

Wright, K. N. (1989). Race and economic marginality in explaining prison adjustment. *J Res Crime & Del 26*, 67–89.

Wright, K. N. (1991). A study of individual, environmental, and interactive effects in explaining adjustment to prison. *Just Quart 8*, 217–242.

Zimmer, L. (1986). *Women Guarding Men.* Chicago, IL: University of Chicago Press.

Zink, T. M. (1957). Are prison troublemakers different? *J Crim Law & Crim 48*, 433–434.

Zupan, L. (1992). The progress of women correctional officers in all-male Prisons. In I. Moyer (Ed.), *The Changing Roles of Women in the Criminal Justice System* (pp. 145–174). Prospect Heights, IL: Waveland.

DISCUSSION QUESTIONS

1. Inmate subcultures are thought to be related to the concepts of prisonization, importation, and the pains of imprisonment. Note how and why this might be so.

2. The extra legal characteristics of correctional clients are likely to shape how they view and experience corrections. Discuss how race and age might affect how one adjusts in a correctional environment.

3. According to Hemmens and Marquart, what demographic factors most influence inmate perceptions of staff? What is their explanation for this?

READING

In this article, Stephen Richards and Jeffrey Ross examine inmates' perspectives on prison policy or procedure, particularly their opinions about prison classification schemes. Classification involves decisions about where convicts spend their time in prison based on various risk factors (dangerousness, type of crime, escape risk, etc.), and, of course, inmates themselves have little voice in making these correctional decisions, although how they are classified and in which security level they are confined will have a huge impact on their lives in prison. The authors find little right and much wrong with current classification practices. They do not like classification according to individual characteristics without attention being paid to what they call "bigger structural issues," such as poverty, discrimination, and the drug war.

A Convict Perspective on the Classification of Prisoners

Stephen C. Richards and Jeffrey Ian Ross

▨ Introduction

Convicts are rarely asked to comment on prison policy or procedure. They have little voice in correctional decisions. This essay attempts to give the men and women who live in cages a voice in how they are classified. This is no small issue for convicts. Although prisoners may not be considered "stakeholders" (as Berk et al. point out), they may stake their very lives on how they are classified and in which security level they are confined.

Typically, new prisoners enter prison systems through "reception centers" or in what the Federal Bureau of Prisons (FBOP) calls "Receiving and Departure." Although intake procedures differ, they never receive a pleasant welcome. The prisoners arrive scared and worn, wearing handcuffs, belly chains, and dragging leg irons. Standing in line, they are ordered to strip, searched, sprayed or dusted with delousing chemical, issued clothes, and ordered to submit to a battery of token medical and psychological examinations administered by guards pretending to be medical staff. Convicts call this "kicking the tires." If it's not flat, don't fix it. It makes no difference if the prisoner is HIV positive, ready to have another stroke, or near death. The line marches on.

The new prisoners may spend weeks or months at the "reception center," housed in cells or dormitories. Eventually, they are ordered to a classification meeting where an officer announces what has already been decided: their official security level and prison assignment. Some time later, they are transported to their new home: a penitentiary, correctional institution, or camp.

Prisoner classification is reviewed once or more per year, depending on the system. In the FBOP, this is called "team meeting." A prisoner with a major disciplinary report may be reclassified in the blink of an eye and transferred to administrative segregation (the hole) or cuffed up and transported to a high security prison. To no surprise, being reclassified to lower security takes more time, is rarely initiated by staff, and may require repeated requests by the prisoner. In the upside down world of prison, you fall up fast and climb back down ever so slow.

⊠ The Problems With State-Sponsored Prison Research

Fair warning, beware any research that discusses men and women as "offenders" or "inmates." This is the official language used by prosecutors, judges, jailers, prison administrators, and the media to degrade and dehumanize. Even persons with better intentions use these words because they are so rarely challenged. Still, the words we use are important.

It is no surprise that most prison research reflects the language and special interests of the prison bureaucracy. After all, the government funds the research and therefore sets the agenda, limits the parameters, and decides if the final report will collect dust on a shelf or be read and used to inform new policy and procedure. Nevertheless, correctional administrators must be reminded that public taxes pay for their prison budgets, their personal salaries, and the research (Ross, 2002). The public would be horrified to know that their tax money is being spent on human warehouses, where little attention is paid to rehabilitation, treatment, or providing prisoners with the opportunity to better prepare themselves for a law-abiding life (Ross and Richards, 2002).

Regardless of whether the research is state-sponsored, statistical analyses typically mean that researchers do not have to get their "hands dirty" by interacting with convicts or ex-convicts to have a better contextual understanding of their findings. Simply analyzing "inmate files" and observing classification hearings does not explore the full dimensions of the problem under study. Ethnographic or qualitative research can be employed to get a better understanding of the real issues involved (Ross and Richards, 2002, 2003).

SOURCE: Richards, S. C., & Ross, J. I. (2003). A convict perspective on the classification of prisoners. *Criminology & Public Policy*, 2(2), 243–252. © American Society of Criminology.

⊠ Experiments With Prisoners

Social science has used mice, pigeons, and monkeys as experimental subjects. When using human beings, we need to give them a voice. At the very least, it would be interesting to know what the prisoners thought of the alternative classification system. After all, they are the experimental subjects that will reap the benefits or suffer the consequences of the changes proposed by the research. Interviewing the convicts might also raise important questions. For example why do prison systems now house so many prisoners in maximum-security penitentiaries, super-maximum control units, administrative segregation detention, and protective custody? Is this overuse of high-security incarceration the result of increased rates of prison violence, disciplinary violations, prisoner refusal to program, or the reclassification of prisoners designed to fill new maximum-security penitentiaries? Is this the result of overcrowding, the lack of constructive prison programs, or the failure to "do corrections"? How have mandatory minimum sentencing, the implementation of longer sentences, three strikes legislation (Austin and Irwin, 2001:184–218), and the "rising tide of parole violations" (Austin and Irwin, 2001:143–159) contributed to prisoners doing more time in prison? How has the reclassification of prisoners created the "perpetual incarceration machine" (Richards and Jones, 1997, 2003), where prisoners are recycled from prison to parole and back to prison? The system feeds, getting larger, on its own failure to properly prepare prisoners for reentry and legal citizenship. These are the questions that prisoners might suggest need to be addressed before alternative schemes for classification are created.

⊠ Statistical Analysis Does Not Solve the Real Puzzle

There is an implicit belief that better data and statistical analysis will somehow improve things for prisoners and correctional staff alike. The problem is that convicts and guards are different constituencies with competing concerns. The prisoners want less restrictive classification (minimum or medium-security), where they might have better living conditions (more time out of cell, less restrictive family visits, better access to programs, and less violence). In comparison, prison staff may want prisoners to be housed in more restrictive environments (maximum-security, control units, segregation) where they are "locked in" and have little freedom of movement, thus giving the guards more control and less exposure possibly to assault and injury. Statistical analysis does not solve the real puzzle: How does prisoner classification, which decides where individual prisoners will be designated to live, impact the day-to-day routine of prisoners and staff? How is prison classification reflected in the design and construction of new facilities and the remodeling of existing institutions?

A second problem, implied by the discussion above, is that "inmate files" (which usually include presentencing investigation reports, criminal offenses, institutional reports) should not be the sole determinate of classification decisions. Although evaluating prisoners individually is one important criterion, it fails to look at the bigger issues, such as the growing incarceration of minorities and women, conditions of confinement, and problems with reentry (Austin et al., 2001).

For example, using disciplinary reports as the primary criteria for reclassification of prisoners may lead to the construction of more maximum-security prisons. It costs more to house prisoners in high-security institutions. Prisoners that serve time in these institutions suffer more deterioration and are less prepared for release. Do we want correctional departments to spend more tax dollars on concrete and steel or rehabilitation programs?

⊠ FBOP Prisoner Classification

To further illustrate the complexities of classification, we provide the following discussion of the FBOP. Notice, that the FBOP had six levels and now five levels, compared to only four levels for the California Department of Corrections. The FBOP uses an "inmate classification system" as a means to segregate, punish, and reward prisoners. This is a "classification ladder" with maximum security at the top and minimum security at the bottom. Ideally, if the FBOP operated to facilitate rehabilitation, prisoners would work their way down the ladder with good conduct and program participation. As they completed their sentences and got "short" (which means a year to release), they would be moved to minimum-security camps or community custody. Unfortunately, most men and women move up the ladder from minimum to medium, or medium to maximum, rather than down. Few medium- and maximum-security prisoners ever make it to the camps.

The classification designations have changed over the years to accommodate the growth in FBOP prisons and population. The old system had six security levels, with 6–5 being maximum, 4–2 being medium, and 1 being minimum. USP Marion (the first super-maximum penitentiary) was the only level 6 institution. U.S. penitentiaries were level 5 (e.g., USP Atlanta, USP Leavenworth, USP Lewisburg, USP Lom Poc); the federal correctional institutions ranged from 4 to 2 (e.g., FCI Talladega, FCI Sandstone, FCI Oxford, etc.), and the federal prison camps were 1. Security levels 6–2 are "in" custody, which means inside the fence or wall. Level 1 is "out" custody, which means federal camps that do not have serious security fences. Level 1community custody refers to prisoners in camps that were eligible for community programs—work assignments or furloughs.

In the 1990s, the FBOP collapsed these six security designations into five: high, medium high, medium low, minimum, and administrative. The BOP prisoner population is approximately 10% high (USP), 25% high medium (FCI), 35% low medium (FCI), and 25% minimum (FPC), with the rest not assigned a security level; many of these men and women are in administrative facilities (medical or detention), transit, or held in local jails or private prisons. "Administrative" refers to Administrative Detention Max (ADX) Florence (CO) (the highest security prison in the country), FTC Oklahoma City (a medium-security transport prison), and the federal medical centers (which may be maximum, medium, or minimum security).

⊠ The Central Inmate Monitoring System

There are additional variables that may not appear on official classification forms. Some of these categories are unique to the FBOP. Prisoners complain these labels adversely affect their ability to reach low-level security prisons, despite good conduct records and short time to do on a given sentence. FBOP staff must check the Central Inmate Monitoring System (CIMS) before any prisoner is reassigned to a new cellblock, dormitory, or prison. Convicts may not know they have been singled out for such attention.

CIMS is a computer system that tracks nine special categories of prisoners: (1) "Witness Security" prisoners are government informers that have testified, are testifying, or will testify in court cases; (2) "Special Security" prisoners are prison snitches cooperating in internal investigations; (3) "Sophisticated Criminal Activity" prisoners are those inmates identified as being involved in large-scale criminal conspiracies, for example, organized crime, drugs, or white collar. They may be men or women who were targets of the federal Racketeer Influenced and Corrupt Organizations (RICO) or Continuing Criminal Enterprise (CCE) prosecution, which carry life sentences (Richards, 1998:133). Many of these convicts are suspected of being connected to major drug-smuggling organizations, or they refused to plead guilty, cooperate, and inform on other persons. (4) "Threats to Government' Officials" prisoners have been convicted of writing letters, making phone calls, or issuing verbal remarks that convey the intent to do bodily harm to public officials; (5) "Broad Publicity" prisoners are those inmates involved in high-profile cases; (6) "State Prisoners" are inmates serving state sentences that were transferred into the fed system because they were "difficult"; (7) "Separation" prisoners are those who have been moved to another institution because they are government witnesses, institutional snitches, gang leaders, or persons in danger of being killed or killing someone else; (8) "Special Supervision" prisoners are police, judges, and politicians that are provided protective privilege (Richards and Avey, 2000). These men and women are usually designated to camps (they may not live long in a penitentiary). (9) "Disruptive Groups" prisoners may include members of organizations, such as street or prison gangs and political groups (i.e., Black Panthers, Communists). The point is that classification includes additional variables that may not be amenable to statistical number crunching. Some of these variables may not even be known to the research team or the prisoner.

◪ Classification May Be Used for Unofficial Purposes

Officially, prison systems design classification systems as a means to designate prisoners to different security levels. Typically, the hardcore violent convicts serving long sentences are assigned to maximum security, the incorrigible prisoners serving medium-length sentences, are sentenced to medium-security prisons, and the relative lightweight men serving short sentences are sentenced to minimum-security camps, farms, or community facilities.

Women prisoners are also subject to "classification." Still, women make up less than 10% of the correctional population. They are usually confined in one or a few institutions in each state. These prisons may hold women prisoners classified for different security levels in various sections of the same institution. Exceptions include the large states and the FBOP where women with different security levels may be imprisoned in separate institutions. In any case, the dramatic increase in the incarceration of women may result in the

further differentiation of women's prisons. We predict there will be future studies of classification systems for women prisoners.

Classification may load up high-security prisons with minorities. African-American, Hispanic, Latino, and Chicano prisoners are more likely to be "young," gang affiliated, and collect bad conduct "tickets." This is readily apparent to most observers of prison, including DOC and FBOP administrators. The FBOP and many states have struggled for years with schemes to "racially balance" institutions. Like school busing programs, they bus prisoners from prison to prison trying to somehow racially integrate prisons as dictated by some policy directive addressing the problem. The public does not like to read in the newspaper that maximum-security prisons are mainly occupied by underclass minorities, whereas minimum-security prisons are reserved for middle- and working-class European-Americans.

Depending on the prison system (budget, number of institutions, population counts, level of disorder), prisoners are shuffled from one institution to another. These transfers may or may not reflect official classification schemes. When a given prison is bursting at the seams, with men sleeping in hallways, three to a one-man cell, or on bunk beds arranged in recreational areas of classrooms converted into make-shift dormitories, the "correctional fairy" (Jon Marc Taylor) waves his magic wand, tears up official policy, and transfers bus loads of prisoners to whichever facility has empty beds. "Population Over Ride" is commonly used.

Classification consists of reviewing any "disciplinary actions" and "demonstration of positive participation in an inmate program." However, convicts will tell you prison guards issue "write-ups," what are called "shots" in the FBOP, "115s" in the California Department of Corrections, or simply "tickets" in many prison systems (disciplinary reports), every chance they get. Prisoners housed in overcrowded cellblocks or dormitories may collect minor "tickets" for petty infractions or major tickets for defending themselves against predatory or aggressive individuals. Many prisoners claim disciplinary committees rule against prisoners without due process. There have been a number of studies that suggest prison staff disproportionately find minority prisoners guilty in disciplinary hearings. Write a letter to a newspaper, call a congressional office, or complain about staff or the lack of medical services, and you collect tickets, get dragged to the "hole," and are reclassified and shipped out to the penitentiary or super-maximum.

Using prisoner participation in prison programs as a second measure has similar problems. Convicts will tell you prison activities (work, vocational training, education) include custodial duties (washing dishes, mopping floors, cleaning bathrooms), duties that masquerade as vocational training (cooking in the kitchen, mowing lawns, painting and repair), and token education programs (ABE, GED). Few of these activities elicit prisoner enthusiasm or are considered real opportunities to learn new skills. "Positive program participation" is usually defined by prison staff as the convict showed up, did not refuse direct orders, and made a good show of pretending to work or study. In many institutions, what programs exist are considered token, as they serve few prisoners, while the rest wait years to participate. Then again, correctional authorities have to have some institutional programs, at least to silence naïve academics and give themselves something to brag about in year-end reports and to the news media. Of course when push comes to shove and the correctional budgets are cut, what programs do exist are the first to get the axe.

A more important problem, briefly alluded to in the article by Berk et al. is that the prison system may no longer expect prisoners to participate in programs, as the programs no longer exist. Most U.S. prison systems do not pretend to provide vocational or educational programming. Prison administrators limit their responsibility to operating orderly institutions, trying to control contraband and violence and prevent escapes. The most efficient way although incredibly expensive and destructive is to build high-security institutions and fill them with reclassified prisoners.

Conclusion: Placing a Finger in the Dike

Redesigning classification reminds us of the old "placing a finger in the dike" story. In the United States massive numbers of people are incarcerated on a daily basis. And there is a belief, sometimes unstated, that better classification procedures will to a greater or lesser extent minimize our problems with incarceration; at the very least, it may save the taxpayer the increased costs of housing prisoners in more restrictive settings. Nevertheless, as long as classification of prisoners is based entirely on out-dated measures of individual behavior (criminal offense, institutional conduct, gang affiliation), without references to the bigger structural issues (poverty, racial discrimination, drug war) that have created the boom in prison population, or prison programming that could lower the rate of disciplinary reports and predictable parole failure, very little will change. Meanwhile, the little boy has his finger stuck in the hole, whereas many states are awash in the budgetary debt rushing over the wall from the construction and operation of new prisons.

Perhaps we can expect no more from research sponsored by the government with such limited vision. At best, the research will result in policy review that merely tinkers with how prisoners are classified. And so it goes, across the country, millions of Americans live in cages, academics do studies that appear like they are rearranging the chairs on the Titanic when it is sinking, departments of corrections talk of policy reforms, prison conditions worsen, and the taxpayers drown in red ink. Maybe it is time to close some prisons, send men and women home to their families, and spend the public dollars saved on economic and community development (Clear and Cadora; 2003)? Research on prisons needs to explore these wider contexts and implications.

✉ References

Austin, James, Marino A. Bruce, Leo Carroll, Patricia L. Mc Call, and Stephen C. Richards. 2001 The Use of Incarceration in the United States. American Society of Criminology National Policy Committee. *Critical Criminology: An International Journal 10*:17–41.

Austin, James and John Irwin. 2001 *It's About Time.* Belmont, Calif.: Wadsworth.

Clear, Todd and Eric Cadora. 2003 *Community Justice.* Belmont, Calif.: Wadsworth.

McCleary, Richard. 1992 *Dangerous Men: The Sociology of Parole.* New York: Harrow and Heston.

Richards, Stephen C. 1998 Critical and Radical Perspectives on Community Punishment: Lesson from the Darkness. In Jeffrey Ian Ross (ed.), *Cutting the Edge: Current Perspectives in Radical/Critical Criminology and Criminal Justice.* New York: Praeger.

Richards, Stephen C. and Michael J. Avey. 2000 Controlling State Crime in the United States of America: What Can We Do About the Thug State? In Jeffrey Ian Ross (ed.), *Varieties of State Crime and Its Control.* Monsey, New York: Criminal Justice Press.

Richards, Stephen C. and Richard S. Jones. 1997 Perpetual Incarceration Machine: Structural Impediments to Post-Prison Success. *The Journal of Contemporary Criminal Justice 13*:4–22.

Richards, Stephen C. and Richard S. Jones. 2003 Beating The Perpetual Incarceration Machine. In Shadd Maruna and Russ Immarigeon (eds.), *Ex-Convict Reentry and Desistance from Crime.* Albany: State University of New York Press.

Ross, Jeffrey Ian. 2002 Grants-R-Us: Inside a Federal Grant-Making Research Agency. *American Behavioral Scientist 43*: 1704–1723.

Ross, Jeffrey Ian and Stephen C. Richards. 2002 *Behind Bars: Surviving Prison.* Indianapolis, Ind.: Alpha.

Ross, Jeffrey Ian and Stephen C. Richards. 2003 *Convict Criminology.* Belmont, Calif.: Wadsworth.

DISCUSSION QUESTIONS

1. The authors believe inmates should have a voice in how they are classified. Do you agree? Why or why not?

2. Why would inmates care about the classification process?

3. What factors most affect the inmate classification process, according to the authors? What other factors do they believe should play a significant role?

READING

In this article, Jesenia Pizarro and Vanja Stenius take an in-depth look at the history and function of supermax prisons and their effects on inmates. Supermax prisons are a fairly new development in corrections. Pizarro and Stenius explore the roots of these institutions, explain how they operate, and examine their potential effects on inmate populations. They conclude that supermax facilities have the potential to damage inmates' mental health while failing to meet their purported goals (e.g., deterring inmates in the general prison population from committing criminal acts inside prison), resulting in added problems for correctional administrators and increased economic costs to public budgets without apparent benefits.

Supermax Prisons

Their Rise, Current Practices, and Effect on Inmates

Jesenia Pizarro and Vanja M. K. Stenius

The United States has built the largest prison system in the world (Currie, 1998), and the prison population has skyrocketed in the past decade. Between 1973 and the beginning of the 1990s, the number of prisoners increased by 332%, and the incarceration rate per 100,000 citizens increased by over 200% (Bureau of Justice Statistics, 1993; Clear, 1994). The growth in the prison population brought with it an increase in young, more violent inmates as well as court rulings affecting the powers of guards and administrators. The combination of these factors pushed many corrections practitioners and scholars to try to develop more effective ways to manage penal institutions and to ensure prison safety. In doing so, a number of new approaches in corrections emerged in recent years, one of which is the super-maximum, or "supermax," prison. The National Institute of Corrections (NIC) (1997) defined supermax prisons as

> free-standing facilities, or a distinct unit within a facility, that provides for the management and secure control of inmates who have been officially designated as exhibiting violent or seriously disruptive behavior while incarcerated. Such inmates have been determined to be a threat to safety and security in traditional high security facilities, and their behavior can be controlled only by separation, restricted movement, and limited direct access to staff and other inmates. (p.1)

The advent of supermax institutions has not been without controversy. Opponents argue that supermax institutions violate prisoners' rights, contribute to inmates' psychological problems, and are extremely costly (Fellner & Mariner, 1997). Proponents claim that the "toughening" of the inmate population, increased gang activity, and difficulties associated with maintaining order in severely crowded prisons necessitate supermax facilities (Riveland, 1999). This article explores the roots of these controversial institutions, explains how they operate, and examines their potential effects on inmate populations. Although limited, the extant empirical research on supermax facilities demonstrates that these institutions have the potential to damage inmates' mental health while failing to meet their purported goals (e.g., deterring inmates in the general prison population from committing criminal acts inside prison), resulting in added problems for correctional administrators and increased economic costs to public budgets without apparent benefits.

▨ Supermax Prisons: Their Rise, Current Practices, and Legal Issues

The Origins and Rise of Supermax Prisons

Supermax institutions separate the most serious and chronic troublemakers from the general prison population (Henningsen, Johnson, & Wells, 1999). These institutions house inmates in solitary confinement, with minimal contact with other humans and virtually no educational, religious, or other programs. Their general purpose is to increase control over inmates who are known to be

SOURCE: Pizarro, J., & Stenius, V.M.K. (2004). Supermax prisons: Their rise, current practices, and effect on inmates. *The Prison Journal* 84(2), 248–264. Reprinted with permission of Sage Publications, Inc.

violent, assaultive, major escape risks, or likely to promote disturbances in the general prison population (National Institute of Corrections, 1997; Riveland, 1999). The rationale behind supermax facilities is to segregate the most dangerous inmates to protect prison staff members and inmate populations. Furthermore, proponents of supermax facilities assert that the threat of the harshness of supermax prisons deters other inmates from committing criminal acts inside the walls (Fellner & Mariner, 1997).

Correctional scholars and practitioners alike consider order and safety to be very important in managing prisons (Dilulio, 1987; Logan, 1992; Reisig, 1998; Riveland, 1999; Useem & Reisig, 1999). This is why prisons have historically had "jails within prisons" to securely house violent and disruptive inmates (Barnes, 1972; Riveland, 1999). Some assert that Alcatraz, which was the home of the most publicized disobedient inmates of the early and mid-1900s, paved the way for modern-day supermax prisons (King, 1999). Alcatraz followed a "concentration model," which refers to the creation of specific units or facilities to manage specific types of troublesome inmates (King, 1999; Riveland, 1999). In 1963, the Bureau of Prisons (BOP) decided to close Alcatraz and replace it with a new, special, high-security prison in Marion, Illinois; however, the prison was not completed by the time of Alcatraz's closure. As a result, the BOP dispersed Alcatraz's inmates to facilities throughout the federal prison system. As time elapsed, practitioners noticed that the new dispersion approach appeared to "work." As a result, the BOP embraced the dispersion approach. This approach is generally referred to as the "dispersion model" because inmates who are considered troublemakers are spread throughout the system to prevent them from enticing others into collective misconduct (Riveland, 1999).

In the early 1970s, the level of assaults and violence directed toward staff members and other inmates escalated (Bureau of Prisons, 1973b; King, 1999). This increase in violence prompted the BOP to begin sending

troublesome prisoners to the high-security prison in Marion, which was originally intended to replace Alcatraz, and to once again embrace the concentration model. In 1972, the BOP built the H unit at Marion, which was designed to separate offenders whose behavior seriously disrupted the orderly operation of the institution from the general prison population. The mission of the H unit ironically fell within the purview of "reform." It was designed to assist individuals in changing their attitudes and behaviors to facilitate their return to the general prison population (Bureau of Prisons, 1973a; King, 1999).

The escalating violence in BOP institutions continued into the mid-1970s, with an increase of 45.5% in assaults on prison staff members by inmates (Henderson, 1979; King, 1999). As a result, in 1979, the BOP recommended the addition of a new administrative maximum-level unit to the classification system for prisons. Later that year, Marion became the first Level 6 (super-maximum-security) prison (Bureau of Prisons, 1979; King, 1999). Its mission was to provide long-term segregation within a controlled setting for prisoners throughout the federal system who threatened or injured other inmates or staff members, possessed deadly weapons or dangerous drugs, disrupted the orderly operation of a prison, or escaped or attempted to escape (Bureau of Prisons, 1979; King, 1999).

Violence at Marion escalated during the early 1980s. From 1980 to 1983, there were 14 escape attempts, 10 group disturbances, 54 serious assaults on inmates, 28 assaults on staff members, and eight prisoners and two corrections officers killed by inmates in its supermax unit (King, 1999). These incidents led to a complete lockdown of Marion during the fall of 1983. The warden and correctional officers at Marion claimed that this act reduced assaults and made the environment safer in the prison (Fellner & Mariner, 1997).

As a result of such claims, many states followed in Marion's footsteps. The NIC (1997) reported that as of 1997, approximately

34 jurisdictions in the United States operated 1 or more supermax facilities or were in the process of opening one. As of 1997, over 55 supermax facilities or units were operating nationwide (National Institute of Corrections, 1997). At the end of 1998, about 20,000 prisoners, or 1.8% of all those serving sentences of 1 year or more in state and federal prisons, were housed in such facilities (King, 1999).

The Operation of Supermax Facilities

In a survey distributed to correctional institutions nationwide, the NIC (1997) found that jurisdictions vary considerably in the operation and management of supermax facilities. Nevertheless, they found that all supermax prisons share certain defining features. For example, inmates are confined in their cells for 22 to 23 hours a day (Fellner & Mariner, 1997; NIC, 1997; Riveland, 1999). These institutions limit human contact to instances when medical staff members, clergy members, or counselors stop in front of inmates' cells during routine rounds. Physical contact is limited to being touched through security doors by correctional officers while being put in restraints or having restraints removed. Most verbal communication occurs through intercom systems (Riveland, 1999).

Placement of Inmates in a Supermax Prison. In most jurisdictions, admission into a supermax facility or unit does not depend on a formal disciplinary hearing but is rather based on the criminal and behavioral history of an inmate while incarcerated (Committee to End the Marion Lockdown, 1992; Riveland, 1999). The inmates in these institutions are not those who committed the worst crimes in society but those whom correctional staff members deem as threats to the safety, security, or orderly operation of the facilities in which they are housed (National Institute of Corrections, 1997; Riveland, 1999). Placement in a supermax institution is not a penalty but an administrative decision based on a pattern of dangerousness or

unconfirmed but reliable evidence of pending disruption (e.g., a prisoner is a leader of a gang or other radical movement) (Committee to End the Marion Lockdown, 1992; Riveland, 1999).

Inmate Programming. Jurisdictions vary in the extent of the programs and activities they offer to their inmates. Some jurisdictions allow inmates to have televisions in their cells and provide education and self-help programs through intra-institutional cable television. Other jurisdictions provide inmates with instructors that assist inmates through cell-front visits. During these visits, instructors stand in front of inmates' cells and talk to them through openings in the cell doors. Other jurisdictions, however, provide no programs to inmates.

The amount of exercise allowed to inmates in supermax facilities is generally limited to 3 to 7 hours per week in indoor spaces or small, secure, attached outdoor spaces within the facilities (National Institute of Corrections, 1997; Riveland, 1999). Inmates exercise one at a time, and at least two correctional officers escort inmates to and from the exercise spaces. Group exercise occurs only in transition programs (programs designed to reintegrate inmates into the general prison populations or society), which only some facilities provide.

Visitation privileges also vary from facility to facility (National Institute of Corrections, 1997; Riveland, 1999). Some institutions allow only 1 hour of visitation per month, whereas others allow several hours per month. Even so, inmates typically have no direct contact with visitors; their visitation consists of video visiting, which means that inmates and visitors communicate and see each other through a 13-inch, black-and-white television (Chacon, 2000). In other jurisdictions, inmates sit in small cubicles separated by clear partitions from their visitors and communicate through intercoms (National Institute of Corrections, 1997).

Physical Coercion. Correctional officers in supermax institutions may use proportionate and reasonable force to subdue inmates when dangerous situations erupt. For example, guards are allowed to conduct cell extractions—the forceful removal of prisoners from their cells—when inmates refuse to come out of their cells or cover the glass windows in their cell doors (Fellner & Mariner, 1997). Correctional administrators justify cell extractions as a procedure for reducing the harm to staff members that could occur with less intrusive means. Quick-response teams carry out this procedure. In most facilities, teams consist of five correctional officers wearing body armor, helmets with visors, neck support, and heavy leather gloves. Each member is responsible for subduing a specific part of an inmate's body. Other correctional staff members, including a supervising sergeant, an officer with a video camera who records the extraction, and a medical assistant, accompany the extraction team. Guards usually administer chemical sprays (mace and pepper spray) into inmates' cells through openings in the doors prior to the extraction, rush in when the cell door opens, gain control of inmates, and then place them in restraints (Fellner & Mariner, 1997).

Supermax institutions also use the four-point restraint as a security method (Fellner & Mariner, 1997). This technique uses the leather restraints with which inmates' beds are equipped to immobilize prisoners by strapping and holding their arms and legs secure. This procedure may be used only if offenders present themselves as imminent threats to themselves or others (Fellner & Mariner, 1997).

Release From Supermax Prisons. In most jurisdictions, the criteria for release from supermax facilities are not published or revealed to prisoners. In fact, only 23 jurisdictions have written criteria under which inmates can earn transfer from supermax prisons (National Institute of Corrections, 1997). Furthermore, the amount of time inmates serve in supermax facilities also varies across jurisdictions (Riveland, 1999). Some jurisdictions have determinate periods to be served, but most have indeterminate placement. The amount of time served may depend on the

perceived risk an inmate presents, any behavioral changes that may take place, the amount of time left on his or her sentence, changes in his or her physical and psychological conditions, and his or her willingness to renounce allegiance to gangs (Riveland, 1999). An inmate may be either returned to the general prison population or, if his or her court-ordered sentence is up, released into the community.

The Cost of Running Supermax Institutions

Riveland (1999) noted that "in most jurisdictions, operating costs for extended control facilities are generally among the highest when compared to those of other prisons" (p. 21). For example, in 1999, the average daily cost for inmates at the Colorado State Penitentiary (a supermax facility) was $88.72, whereas the cost at the maximum-security facility (the Colorado Correctional Center) was $50.82. When compared annually, the average cost to house an inmate in the supermax facility in Colorado was $32,383, whereas in the maximum-security facility, it was $18,549 (Rosten, 1999). In addition, the construction of supermax facilities is very costly because of the need for high-security components. These institutions are composed of high-security doors, fortified walls, and sophisticated electronic systems. Although construction costs are high, the cost of staffing these facilities is even higher because correctional officers provide services to inmates and perform maintenance work within the facilities (Riveland, 1999).

Legal and Ethical Issues With Supermax Prisons

The overall constitutionality of supermax prisons remains unclear (Riveland, 1999). The Eighth Amendment, which prohibits cruel and unusual punishment, requires that prisoners be afforded a minimum standard of living (Law Information Institute, 2001). Many argue that the living conditions and treatment provided to inmates in supermax facilities do not meet the standards of the Eighth Amendment (Fellner & Mariner, 1997). Federal court judges, however, have repeatedly ruled that prolonged segregation is cruel and unusual punishment only for the mentally ill (Rogers, 1993). U.S. district courts maintain that, although the conditions in these institutions are horrible, they are necessary for security reasons and therefore do not violate inmates' constitutional rights (Henningsen et al., 1999).

Consequences of Supermax Confinement

Pains of Imprisonment

In his classic work *The Society of Captives*, Gresham Sykes (1958) asserted that life in a maximum-security prison is a painful experience that influences inmates' behavior and psychological well-being. In addition to restricting inmates' behavior and autonomy, incarceration punishes them emotionally and psychologically through what Sykes called the "pains of imprisonment." These include the feelings of deprivation and frustration caused by the (a) loss of liberty, (b) loss of autonomy, (c) lack of heterosexual relationships, (d) deprivation of goods and services, and (e) lack of personal security and safety. Inmates in supermax facilities suffer these pains in addition to almost complete isolation, although personal security and safety may be greater for inmates in supermax facilities than for those in general populations because they do not have contact with other inmates. The addition of isolation, however, suggests that the pains of imprisonment in supermax facilities are more severe than those in maximum-security prisons. Consequently, any negative emotional or psychological reactions to imprisonment should be greater in supermax facilities than in lower security facilities.

The Effect of Supermax Incarceration on Inmates' Health

A major concern voiced by critics of supermax facilities is their potential effect on inmates' mental health because of isolation and the lack of activity. Early U.S. experiments with isolation in Pennsylvania and New York in the 1800s demonstrated the severe impact that isolation has on inmates' psychological and physical health (Toch, 2001). As a result, prison administrators quickly abandoned solitary confinement as a general correctional tool and used isolation as only a temporary form of punishment.

Although the conditions in prisons today are certainly quite different from those in the first penitentiaries, the impact of isolation on inmates' psyches is likely to be quite similar. Despite the increased use of modern supermax facilities, no research to date has directly examined the effect of supermax confinement on inmates' psychological and physical health. Inferences about the impact of these facilities on inmates' mental and physical health are based primarily on research examining the effects of temporary solitary confinement or administrative segregation within regular prisons. Although this research is informative, differences in the scope of restrictions and deprivations, as well as the duration of the isolation, must be considered (see Bonta & Gendreau, 1990). For example, spending a specified number of days in isolation is quite different from serving the remainder of one's sentence, possibly years, in a supermax facility. Similarly, spending 23 hours a day in isolation with no activities is not comparable to spending 23 hours a day in isolation with meaningful activities.

Isolation research supports the notion that greater levels of deprivation contribute to more psychological and emotional problems (Brodsky & Scogin, 1988; Grassian, 1983; Grassian & Friedman, 1986; Miller, 1994; Scott & Gendreau, 1969). As inmates face greater restrictions and social deprivations, their levels of social withdrawal increase (Miller, 1994; Scott & Gendreau, 1969). Scott and Gendreau (1969) argued that increasing inmates' restrictions by limiting human contact, autonomy, goods, or services requires more intense activity programming to counteract the adverse effects of these restrictions. Imposing more restrictions without appropriate activity programming is detrimental to inmates' health and rehabilitative prognoses. Potentially beneficial programming includes educational, recreational, and psychological services. More recent studies support these contentions in that increasing restrictions, namely through segregation, tends to result in such forms of psychological distress as depression, hostility, severe anger, sleep disturbances, physical symptoms, and anxiety (Brodsky & Scogin, 1988; Miller, 1994). Although the types of restrictions and outcomes measured vary across studies, the general consensus is that increasing the level of restrictions increases the risk for psychological and emotional problems. The extent of the effects of these restrictions depends not only on the nature of the confinement and deprivations but also on inmates' characteristics (Grassian & Friedman, 1986). Given the high rates of mental health problems within the inmate population, the potential for adverse effects is especially high (Ditton, 1999).

Overall, the research suggests that solitary confinement has potentially serious psychiatric risks (Brodsky & Scogin, 1988; Grassian, 1983; Grassian & Friedman, 1986; Korn, 1988; Kupers, 1999; Miller, 1994). Isolation can produce emotional damage, declines in mental functioning, depersonalization, hallucination, and delusion (Brodsky & Scogin, 1988; Grassian, 1983; Korn, 1988; Kupers, 1999; Miller, 1994; Scott & Gendreau, 1969). Inmates in isolation, whether for the purpose of protective custody or punishment, suffer from numerous psychological and physical symptoms, such as perceptual changes, affective disturbances (notably depression), difficulties in thinking, concentration and memory problems, and problems with impulse control (Brodsky & Scogin, 1988; Grassian, 1983; Grassian & Friedman, 1986; Miller, 1994).

Interviews with inmates in high-security facilities have demonstrated similar findings. In particular, Korn (1988) found that women living in a high-security unit experienced claustrophobia, chronic rage reaction, depression, hallucinatory symptoms, defensive psychological withdrawal, and apathy. Korn attributed these problems to factors such as depersonalization, the denial of individuality, the denial of personal initiative, and humiliation. Similarly, Kupers (1999) argued that inmates placed in an environment as stressful as that in a supermax prison begin to lose touch with reality and exhibit symptoms of psychiatric decomposition. He indicated that the majority of the inmates he had interviewed in administrative segregation units had difficulty concentrating, heightened anxiety, intermittent disorientation, and a tendency to strike out at people.

Since isolation was abandoned as an effective means of reforming offenders, it has primarily been used as a means of punishment within correctional institutions or as an administrative tool to protect individual inmates or others in the general prison population. As such, the goal of using solitary confinement is generally to induce behavioral change within an institution. Although most research does not support the claim that isolation results in desirable behavior modification, a couple of studies have supported this assertion (Suedfeld & Roy, 1975; Suedfeld, Ramirez, Deaton, & Baker-Brown, 1982). Suedfeld and Roy (1975) found that "short-term" segregation is an effective tool for dealing with disruptive inmates because it aids in modifying their behavior and produces beneficial psychological and behavioral effects (e.g., inmates become more pleasant, optimistic, self-confident, and compliant with institution rules). In addition, Suedfeld et al. (1982) found no support for the claim that solitary confinement is adverse, stressful, or damaging. It is important to note, however, that Suedfeld et al. conducted this research on simulated solitary confinement units with inmates who volunteered to take part in the experiment. Accordingly, the implications of this research for supermax facilities are limited given the differences in the duration of the confinement and status of the inmates. Isolation in supermax facilities is not short term, nor are inmates there on a voluntary basis.

In sum, the vast majority of research suggests that inmates placed in restricted environments, such as in solitary confinement, for prolonged periods of time tend to develop psychological problems. Most, if not all, of these studies, however, are weak methodologically. For example, using inmates who volunteer to be placed in solitary confinement could lead to erroneous results because the inmates know that the situation is not real and that they can get out of the situation whenever they want to (Suedfeld et al., 1982). The studies that examined inmates involuntarily placed in segregation failed to administer pretests or to look at the inmates' past psychological and behavioral records (Brodsky & Scogin, 1988; Korn, 1988; Kupers, 1999; Miller, 1994; Suedfeld & Roy, 1975). In the absence of information on inmates' presegregation psychological status, it is difficult to make valid assessments of changes in status, because inmates could had been suffering from psychological problems before being placed in isolation. Finally, some of the studies drew inferences on the basis of inmates under special circumstances, such as class-action suits against jurisdictions for the treatment they received in isolation (Grassian, 1983; Grassian & Friedman, 1986), which makes their results difficult to generalize to other populations. Making general inferences from studies using small sample sizes is similarly problematic (Korn, 1988; Grassian, 1983; Grassian & Friedman, 1986; Suedfeld & Roy, 1975).

Despite these problems and limitations, the research suggests that inmates placed in supermax facilities are likely to suffer some form of psychological distress. Although the available research is limited in its applicability to supermax facilities and flawed methodologically, the research suggests that solitary confinement has a detrimental impact on individuals' mental health, although the extent and specific nature of this impact are unclear. As Robins (1978) noted,

> In the long run, the best evidence for the truth of any observation lies in its replicability across studies. The more the populations studied differ, the wider the historical eras they span; the more the details of the methods vary, the more convincing becomes the replication. (p. 611)

Although none of the studies is perfect, taken together, they suggest that solitary confinement negatively influences individuals' psychological and/or emotional well-being. The implications for supermax facilities are not clear, apart from purporting that confinement within these facilities is likely to contribute to the development of mental health problems and/or exacerbate any existing problems. Given the differences in the lengths and conditions of confinement between research participants, with relatively short periods of confinement, and supermax inmates, one would expect a greater detrimental impact among the latter population than the reviewed studies suggest. Similarly, the effects should be larger for inmates housed in supermax units that have more restrictions and less, or no, programming. The research suggests that inmates housed in supermax facilities for longer periods of time, without programming and with more restrictions on human contact, should be the most adversely affected by supermax confinement; however, this is a function of inmates' characteristics.

Supermax Prisons as a Deterrent

Prison administrators assert that supermax prisons serve as a general deterrent within the correctional population—that their presence curbs violence and disturbances within penal institutions. General deterrence may occur as individuals observe the imposition of the threatened punishment on others or solely by the knowledge that a given behavior carries a given punishment. This theory asserts that if punishment is distributed with certainty, adequate (and appropriate) severity, and celerity, rates of offending should be low (Beccaria, 1764/1994; Bentham, 1789/1992; Zimring & Hawkins, 1973).

For deterrence strategies to be effective, offenders must not only be aware of the sanctions but also believe that they will get caught and punished with the threatened sanctions. What is important in the efficacy of sanctions as deterrents is not their actual certainty or severity but individuals' perceptions of their certainty and severity (Paternoster, 1987). It is unlikely that supermax facilities serve as a deterrent because of the certainty of punishment; placement in these facilities is relatively rare and often based on administrative decisions using risk factors over which inmates have little control (Riveland, 1999; Toch, 2001). The perceived certainty of placement in supermax facilities is likely to be low and become increasingly so as inmates engage in and observe disruptive or violent behavior that does not result in placement in a supermax institution. "Experiential effects" suggest that threatening inmates with placement in supermax institutions for specified behavior and then failing to follow through may actually increase problematic behavior (Claster, 1967; Jensen, 1969; Paternoster, 1987). Additionally, increasing the severity of punishment has generally been found to be a less effective means of achieving deterrence than increasing its certainty (Zimring & Hawkins, 1973). The argument that the severity of supermax confinement acts as a deterrent does not find support in the deterrence literature, especially if inmates question the certainty of such confinement for violent or disruptive behavior.

Furthermore, Sherman (1993) argued that individuals abstain from offending according to four key concepts in emotional responses to the sanctioning experience: legitimacy, social bond, shame, and pride. Legitimacy is the perceived degree of respectfulness and procedural fairness of an enforcing agent by an individual. Social bond is conceptualized as the relationship an offender has with a sanctioning agent. Shame is whether an offender acknowledges or bypasses a sanction. Finally, pride is how an offender feels in the aftermath of a sanction. If an offender perceives sanctioning as illegitimate or unfair, has social weak bonds with a sanctioning agent and the community that the agent represents, and denies his or her shame, then the sanctioning could cause future involvement in crime. Moreover, Sherman pointed out that the level of deterrence achieved through sanctions varies as a function of the offender. The effects depend on an offender's personality type, social bonds, and perceptions of legitimacy. As a result, sanctions may deter crime among some groups but increase crime in others.

Arguably, supermax facilities are not a deterrent for institutional misconduct, because inmates generally neither have bonds with sanctioning agents nor believe that they will be treated fairly. Because the deterrence perspective targets only those who would otherwise engage in the proscribed behavior, the threat of placement in supermax facilities is unlikely to serve as a deterrent. Disruptive and violent inmates may be less likely than other inmates to be concerned about the consequences of their actions, to have bonds with sanctioning agents, or to feel shame or pride over their behavior. If supermax facilities are effective in deterring only inmates who would otherwise not engage in misconduct, then they do not add any deterrent value.

In addition, Barak-Glantz (1983) argued that solitary confinement plays a minimal role in deterring inmates' behaviors. Existing empirical evidence does not suggest that the placement of problematic inmates in supermax prisons decreases prison violence. Research in the area of deterrence indicates that, in most cases, deterrence as a correctional policy does not work (Clear, 1994; Cullen, 1995; Paternoster, 1987; Sherman, 1993). Deterrence research in conjunction with theory and empirical evidence on inmates' behavior suggests that supermax facilities are unlikely to be effective as a general deterrent for violence and disturbances in prisons.

Conclusion and Policy Implications

Since the 1970s, the U.S. correctional system has undergone dramatic changes. Prison populations have skyrocketed in response to changing sentencing policies and crime rates, which has contributed to numerous problems within facilities, such as overcrowding and violence. In the face of inmate violence, lawsuits, federal oversight, and other problems, prison administrators have sought—and continue to seek—means of addressing these issues. Supermax facilities present one solution whose growing popularity has made them "one of the most dramatic features of the great American experiment with mass incarceration during the last quarter of the 20th century" (King, 1999, p. 163). As of 1998, 1.8% of all those serving sentences of 1 year or more in state and federal prisons were housed in such facilities (King, 1999). This number is likely to increase, because many practitioners have classified supermax facilities as an effective tool in the management of problematic prisoners despite the lack of empirical evidence demonstrating such effectiveness (King, 1999).

Research on supermax facilities and solitary confinement within prisons is limited and generally lacking in sound methodology, which makes it difficult to draw clear conclusions regarding what effect these facilities may have on inmates' behavior and mental health. Given that existing isolation research examines the effects of deprivations that are arguably less restrictive and shorter in duration than supermax confinement, one expects that the risk for psychological harm and other detrimental impacts is greater in supermax facilities than the research suggests. In terms of controlling behavior, the available research does not support the assertion that supermax facilities are effective management tools for controlling violence and disturbances within prisons. Although no research has looked at the deterrent effect of supermax facilities on behavior within prisons, deterrence research suggests that supermax facilities are not effective in reducing violence or disturbances within the general population. In conjunction with the potential for a detrimental impact on the mental health of inmates placed in supermax prisons, the implications of existing deterrence literature suggest that supermax prisons should not be used for their current purpose.

In addition to the lack of apparent benefit from placing inmates in supermax facilities, doing so imposes costs on society. Foremost is the expense of operating supermax facilities. Advocates claim that the cost is worthwhile because these facilities serve as a general deterrent and ensure security in the general prison population, but these assertions are not empirically supported, making it difficult to justify the costs of constructing and operating these facilities. The impact of solitary confinement and the lack of activities or programming on inmates' psychological well-being presents additional costs to society by necessitating psychiatric care within institutions and potentially leading to more disruptive behavior and violence against staff members.

The costs associated with supermax facilities are not limited to incarceration costs. If these inmates have been abused, treated violently, and confined in dehumanizing conditions that threaten their mental health, then they may leave prison angry, dangerous, and far less capable of leading law-abiding lives than when they entered prison (Fellner & Mariner, 1997). It is probable that inmates who have spent prolonged periods in solitary confinement have a more difficult time adjusting to life outside of prison, especially given the potential for the development or exacerbation of psychological problems. Supermax inmates may be more likely than comparable inmates serving sentences in regular institutions to recidivate (or to escalate their offending). Furthermore, the presence of psychological problems means that the release of these individuals into society poses additional burdens on communities trying to deal effectively with mentally ill offenders.

An additional question arises as to whether it is worthwhile to place someone in a supermax facility for the sake of reducing violence and disturbances within prisons (which research suggests is not accomplished), only to release that individual into society as less capable of normal social functioning than when he or she was sent to prison. Research in this area is sorely lacking, but given the increasing popularity of supermax facilities, the implications of supermax confinement need careful consideration, because most of the inmates housed in such facilities are returned either to the general prison population or to society. Do the benefits, if any, of placing inmates in supermax facilities outweigh or justify the costs? Available research suggests that they do not; in which case, one must ask why supermax prisons are so popular and whether they are justifiable. Policy makers and prison administrators need to consider why they favor supermax institutions and carefully weigh the consequences of expanding the use of supermax prisons.

Nearly 200 years ago, prison administrators abandoned the first experiments with solitary isolation as a practice, not just a temporary punishment, solely on the basis of its detrimental effects on inmates. Although conditions within institutions are certainly different today, one must ask why this practice, which was once dropped because it was deemed inhumane, is once again justifiable. If it cannot be justified on the basis of current purposes, then the aims of supermax facilities need reconsideration. Current research, which is certainly limited, suggests that it is difficult to justify them for utilitarian purposes (e.g., effective inmate management, deterrence). It is, however, possible to justify them on punitive grounds. Their increasing popularity may mirror the increased punitiveness seen in sentencing across the United States since the 1970s, in which case the arguments surrounding their use are quite different, focusing more on theoretical and moral justifications for their existence, as opposed to their ability to help correctional administrators manage the inmate population.

Regardless of the rationale for using, or not using, supermax facilities, more research is needed to better understand their use, the impact that they have on inmates while in the facilities as well as after release, and ultimately the implications for affected communities. In the absence of more empirical evidence, the conclusions that can be drawn regarding supermax facilities, although informative, are limited. We hope that the conclusions presented here demonstrate just how much is not known about the impact of these increasingly popular facilities as well as point out some areas in need of further exploration, both empirically and philosophically.

References

Barak-Glantz, I. L. (1983). Who is in the "HOLE?" *Criminal Justice Review, 8,* 29–37.

Barnes, H. E. (1972). *The story of punishment: A record of man's inhumanity* (2nd ed.). Montclair, NJ: Patterson Smith.

Beccaria, C. (1994). On crimes and punishment. In J. E. Jacoby (Ed.), *Classics of criminology* (2nd ed., pp. 277–286). Prospect Heights, IL: Waveland. (Original work published 1764)

Bentham, J. (1992). Punishment and deterrence. In A. von Hirsch & A. Ashworth (Eds.), *Principled sentencing* (pp. 62–66). Boston: Northeastern University Press. (Original work published 1789)

Bonta, J., &. Gendreau, P. (1990). Reexamining the cruel and unusual punishment of prison life. *Law and Human Behavior, 14,* 347–372.

Brodsky, S., & Scogin, F. (1988). Inmates in protective custody: First data on emotional effects. *Forensic Reports, 1,* 267–280.

Bureau of Justice Statistics. (1993). *Survey of state prison inmates, 1991.* Washington, DC: U.S. Department of Justice.

Bureau of Prisons. (1973a). *Control unit policy statement* (M 17300, 90). Washington, DC: U.S. Department of Justice, National Institute of Corrections.

Bureau of Prisons. (1973b). *Policy statement* (5212.1). Washington, DC: U.S. Department of Justice, National Institute of Corrections.

Bureau of Prisons. (1979). *Program statement* (5212.3). Washington, DC: U.S. Department of Justice, National Institute of Corrections.

Chacon, D. J. (2000). *A new home for problem prisoners.* Retrieved April 4, 2001, from http://www.super maxed.com/newhome.htm

Claster, D. (1967). Comparison of risk perception between delinquents and non-delinquents. *Journal of Criminal Law, Criminology, and Police Science, 58,* 80–86.

Clear, T. R. (1994). *Harm in American penology: Offenders, victims, and their communities.* Albany: State University of New York Press.

Committee to End the Marion Lockdown. (1992). *From Alcatraz to Marion to Florence—Control unit prison in the United States.* Retrieved April 18, 2001, from http://www-unix.oit.umass.edu/kastor/ceml.html

Cullen, F. T. (1995). Assessing penal harm movement. *Journal of Research of Crime and Delinquency, 32,* 338–358.

Currie, E. (1998). *Crime and punishment in America: Why the solutions to America's most stubborn social crisis have not worked and what will.* New York: Henry Holt.

Dilulio, J. J. (1987). *Governing prisons: A comparative study of correctional management.* New York: Free Press.

Ditton, P. M. (1999). *Mental health and treatment of inmates and probationers* (Bureau of Justice Statistics special report). Washington, DC: U.S. Department of Justice, Office of Justice Programs.

Fellner, J., & Mariner, J. (1997). *Cold storage: Super-maximum security confinement in Indiana.* New York: Human Rights Watch.

Grassian, S., (1983). Psychopathological effects of solitary confinement. *American Journal of Psychiatry, 140,* 1450–1454.

Grassian, S., & Friedman, N. (1986). Effects of sensory deprivation in psychiatric seclusion and solitary confinement. *International Journal of Law and Psychiatry, 8,* 49–65.

Henderson, J. D. (1979). *Marion task force report* (International report for the federal Bureau of Prisons). Washington, DC: U.S. Department of Justice.

Henningsen, R. J., Johnson, W. W., & Wells, T. (1999). Supermax prisons: Panacea or desperation? *Corrections Management Quarterly, 3,* 53–59.

Jensen, G. F. (1969). Crime doesn't pay: Correlates of a shared misunderstanding. *Social Problems, 17,* 189–201.

King, R. D. (1999). The rise and rise of supermax: An American solution in search of a problem? *Punishment and Society, 1,* 163–186.

Korn, R. (1988). The effects of confinement in the high security unit in Lexington. *Social Justice, 15,* 8–19.

Kupers, T. A. (1999). *Prison madness: The mental health crisis behind bars and what we must do about it.* San Francisco: Jossey-Bass.

Law Information Institute. (2001). *Prisons and prisoner's rights: An overview.* Retrieved April 4, 2001, from http://www4.law.cornell.edu/cgi-binhtm

Logan, C. H. (1992). Well-kept: Comparing quality of confinement in private and public prisons. *Journal of Criminal Law and Criminology, 83,* 577–613.

Miller, H. A. (1994). Reexamining psychological distress in the current conditions of segregation. *Journal of Correctional Health Care, 1,* 39–50.

National Institute of Corrections. (1997). *Supermax housing: A survey of current practices, special issues in corrections.* Longmont, CO: National Institute of Corrections Information Center.

Paternoster, R. (1987). The deterrent effect of perceived certainty and severity of punishment: A review of the evidence and issues. *Justice Quarterly, 4,* 173–217.

Reisig, M. D. (1998). Rates of disorder in higher-custody state prisons: A comparative analysis of managerial practices. *Crime and Delinquency, 4,* 229–244.

Riveland, C. (1999). *Supermax prisons: Overview and general considerations.* Washington, DC: U.S. Department of Justice, National Institute of Corrections.

Robins, L. (1978). Sturdy childhood predictors of adult antisocial behavior: Replications from longitudinal studies. *Psychological Medicine, 8,* 611–622.

Rogers, R. (1993). Solitary confinement. *International Journal of Offender Therapy and Comparative Criminology, 37,* 339–349.

Rosten, K. L. (1999). *Statistical report fiscal year 1999.* Colorado Springs: Colorado Department of Corrections, Office of Planning and Analysis.

Scott, G., & Gendreau, P. (1969). Psychiatric implications of sensory deprivation in a maximum security prison. *Canadian Psychiatric Association Journal, 14,* 337–341.

Sherman, L. W. (1993). Defiance, deterrence, and irrelevance: A theory of the criminal sanction. *Journal of Research in Crime and Delinquency, 30,* 441–474.

Suedfeld, P., Ramirez, C., Deaton, J., & Baker-Brown, G. (1982). Reaction and attributes of prisoners in solitary confinement. *Criminal Justice and Behavior, 9,* 303–340.

Suedfeld, P., & Roy, C. (1975). Using social isolation to change the behavior of disruptive inmates.

International Journal of Offender Therapy and Comparative Criminology, 19, 90–99.

Sykes, G. M: (1958). *The society of captives.* Princeton, NJ: Princeton University Press.

Toch, H. (2001). The future of supermax confinement. *The Prison Journal, 81,* 376–388.

Useem, B., & Reisig, M.D. (1999). Collective action in prisons: Protest, disturbances, and riots. *Criminology, 37,* 735–760.

Zimring, F. E., & G. J. Hawkins. (1973). *Deterrence: The legal threat in crime control.* Chicago: University of Chicago Press.

DISCUSSION QUESTIONS

1. Why did supermax prisons develop?

2. Are supermax prisons doing what they were originally designed to do, or has their purpose changed over time?

3. What are the benefits and detriments (to staff and inmates) of supermax prisons?

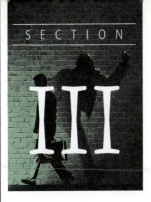

THE CORRECTIONS EXPERIENCE FOR STAFF

Introduction: What Do You Want to Be When You Grow Up?

I t would be the rare child who would reply "correctional officer" when asked the perennial question, usually by the adult relative, "What do you want to be when you grow up?" The typical answers might predictably include "police officer," "lawyer," "engineer," "doctor," or "computer programmer," but never "correctional officer" or "probation officer" or "juvenile counselor" or anything else in corrections. Why is that? The answer to this question tells us much about the status of correctional work as a profession and the regard in which these jobs are held by the larger community.

Part of the answer to this question is that the public does not think that correctional work is a "profession." Despite a century of effort by some determined correctional administrators, corrections organizations (such as the American Correctional Association, the American Jail Association, and the American Probation and Parole Association), and some academicians, many correctional jobs are not structured like professions. A profession is typified by four things: prior educational attainment, formal training on the job, pay and benefits that are commensurate with the work, and the ability to exercise discretion. Yet most jobs in corrections still do not adequately meet the first three of these criteria for professional status.

◪ How Corrections Could Become More Professional

First of all, and unfortunately, most correctional institutions and programs do not have prior educational requirements that would elevate personnel to the level of "professionals." Though it is true that many probation and parole officers (also known as community corrections officers in some states) must have a college degree or at least some college to qualify for the job, most jails, prisons, and even juvenile institutions, even those with a greater emphasis on rehabilitative programming, do not require such a qualification from applicants.

Yet, the oft-cited Stanford Prison Experiment provides a powerful argument for the value of formal education and training for correctional staff. In this 1971 experiment volunteer students, with no training as officers and only their own expectations and beliefs to guide them, were divided into officers and inmates in a makeshift "prison" (Haney, Banks, & Zimbardo, 1981). The "officers" were outfitted in uniforms, including reflective sunglasses, and given nightsticks. The "inmates" were given sack-like attire. Neither "officers" nor "inmates" were told of any rules or policies to guide or restrict their behavior. Predictably, a few of these "officers" or "guards" really engaged in verbal and psychological abuse of the "inmates." In the end, about a third of the "officers" engaged in the abuse and others stood by while it was going on. The experiment was stopped after a few days, but it is often referenced as an example of how correctional work, and the subcultures that develop as part of the job, can foster corrupt behavior by officers.

The problem was that the "officers" were never given any education or training in corrections work. They were to exercise their discretion in controlling the inmates, but it was a discretion that was not necessarily anchored to any history or knowledge or "best practices" in corrections. Rather the "choices" were shaped by the movies and popular press depictions of corrections that tended to reinforce the stereotypes of the institutions or programs and of work. Not knowing how best to *get people to do what they otherwise wouldn't* (Dahl, 1961), the "guards" used what knowledge of corrections they had, even if it was all wrong.

The Abu Ghraib scandal, in which prisoners were tortured by untrained "correctional officers" in the Iraqi prison of that same name, tends to reinforce the lessons of the Stanford Prison Experiment, as contrived as the experiment's circumstances were. Simply put, the lesson could be that some people will not act professionally, or even decently, especially when they have no education or training in that profession.

Second, correctional work often does not resemble other professions because the formal training provided for many new hires, including the number of hours required and the quality of that training, does not approach that of other professions, which may schedule months of training (e.g., police departments with an average number of 749 hours for new recruits [Hickman & Reeves, 2006]) or extensive internships lasting months or years (e.g., teaching, social workers, doctors). Moreover, in some professions, the requisite college or professional degree is geared toward the work itself (e.g., computer programming, law school, or a masters in social work). Yet, when a college degree is *required* for a job in corrections, it is rarely just specifically a criminal justice degree; usually, the requirement is for a more general social sciences degree, which may include no classes on corrections or the criminal justice system at all.

The typical correctional job has lesser requirements for formal training or structured experience. For instance, in a *Corrections Compendium* (2003) survey, the researchers found that 31 of the reporting United States agencies required at least 200 hours of pre-service training for those destined to work in a correctional institution. Likewise, in a quick survey of 150 directors and staff trainers, with responses received from 13 states in April 2004, the Juvenile Justice Trainers Association found that about 140 to 180 hours of preservice, academy-like training is required for most new hires in juvenile facilities (Collins, 2004).

Of course, these deficits in training and in formal education and knowledge base leave correctional workers less suited to perform their job in anything approaching a historical or contemporary research-based context. Because they have not studied corrections or been provided with sufficient training, they may not understand the reason some practices are undertaken or why others are abandoned. They do not have the requisite tools to suggest changes or the background in research to know whether something "works" or not. Their ability to be and develop as a professional is limited. So when they use their "discretion" (defined here as *the ability to make choices and to act or not act on them*), they could be making ill-informed choices that are not based on knowledge or experience and are overly influenced by personal ideology, politics, or the media (Merlo & Benekos, 2000).

Third, children (and their parents) may not view correctional work as a desirable career choice because they do not understand it. The truth is that few people outside of correctional work (or academe) probably know how institutions and community supervision actually operate. Nor do they hold the roles of staff working in those agencies in very high regard. Students are acculturated by a media preoccupied with violence that tends to depict correctional institutions as dark, corrupted places peopled by abusive or, at a minimum, cynical and distant "guards" (Conover, 2001; Johnson, 2002; O'Sullivan, 2006). In the movies and television specials, prisons are almost always maximum security, old, and noisy; jails are crowded and huge monstrosities; juvenile facilities are depressing and havens for child predators. Perhaps as discouraging, community corrections, which is arguably—based on our criteria here—the most professionalized sector of corrections, is rarely depicted in the mass media at all.

Unfortunately, the mass media is not alone in misleading the public and students of criminal justice and criminology about corrections and correctional work. Academics have also tended to focus much of their attention on only the biggest, and the "baddest," of correctional institutions and programs and the labor of their staff. Maximum security institutions, and to a lesser extent metropolitan city jails, have been showcased, though they are not the norm for most corrections in this country (e.g., see Conover, 2001; Hassine, 1996; Jacobs, 1977; Johnson, 2002; Morris, 2002; Sykes, 1958). Research on these institutions tends to focus on the negative, on what the staff or institutions are doing wrong, rather than on what is working well. Of course, it is understandable that the "negative" shines through when these particular institutions, and their type, are at the center of attention: there is much that is amiss in such places.

Predictably, work in probation and parole, much like the study of jail staff, receives short shrift by academics who tend, like the media, to be preoccupied with what is "sexy," violent, and controversial. Given these depictions by the media and academics, it is hard to discern the truth about correctional institutions and programs and work in them, because it is clear that the work is underappreciated, little understood, and hampered by misguided perceptions of it.

▲ **Photo 3.1** Correctional officers bring an inmate to his cell on death row at Ellis Unit in Huntsville, Texas.

✑ Some Truths About Corrections and Correctional Work

Some Correctional Facilities Have Not Been, and Are Not, Operating Well

Of course, as serious students of correctional history and current operation know, there is some truth to the media depictions and the academic critiques of corrections. Some whole correctional institutions in the past have been declared unconstitutional in their operation (e.g., Arkansas, Alabama) because they had horrific conditions for inmates and staff (e.g., see the United States Supreme Court case regarding Arkansas, the *Holt v. Sarver*, 1970 case, or a discussion of Arkansas prisons by ex-warden Murton in Murton & Hyams, 1976). Some wardens have been authoritarian dictators of their own correctional fiefdoms and programs (Bergner, 1998; Jacobs, 1977; Rothman, 1980). There are corrupt and abusive staff working in some institutions even today (e.g., read news accounts of "gladiator" fighting instigated by correctional officers at Corcoran prison in California in the 1990s [Arax & Gladstone, 1998]). Correctional institutions typically are overcrowded, and there is not enough discussion between the affected organizations about how to reduce it (Davis, Applegate, & Otto, 2004; Harrison & Beck, 2005). Programming is limited or poorly conceptualized, funded and delivered in some correctional settings though not in others (Andrews, Zinger, Hoge, Bonta, Gendreau, & Cullen, 2001; Knight, Simpson, & Hiller, 2004). In short, some of the negative depictions of and disparaging comments about correctional institutions, programming, and personnel is merited, but it does not represent the whole or even most of the truth (Applegate, Cullen & Fisher, 2001; Gibbons & Katzenbach, 2006; Johnson, 2002; Lombardo, 2001).

The Misleading Depiction of Corrections

The media tends to depict maximum security prisons so often that one might be forgiven for believing that they represent the normative experience for staff and offenders or inmates in corrections. Yet nothing could be further from the truth. About half as many people are incarcerated in a jail on any given day (713,990) as in prisons (1,421,911). However, most accused people who experience "corrections" do so within the confines of a jail because a jail stay averages a week or less, so the absolute number of people in jail per year far exceeds the number in prison. Even among prisons, maximum security facilities are in a clear minority, representing only about one fifth of the prisons currently in operation (Bureau of Justice Statistics, 2003). This means that, among prisons, most are operated at the minimum or medium security level.

Of those who stay in the correctional system because of a convicted status, almost three times as many are on probation rather than prison (see Figure III.1). In fact, as Figure III.1 illustrates, on any given day in America, there are about one million more probationers than there are all the people on parole, in jail, or in prison combined.

But if this is true, then why do the media and academics tend to focus on those maximum security prisons? Part of the explanation is that, when prisons were first created for adult males in the states, they were of the maximum security variety (Jacobs, 1977; Pollock, 2004). Because the incarceration of women and children was much rarer, their institutions (or sections of institutions) were sometimes of the maximum security variety and sometimes not (Irwin, 2005; Pollock, 2002; Rothman, 1980; see also the Rafter and Young articles

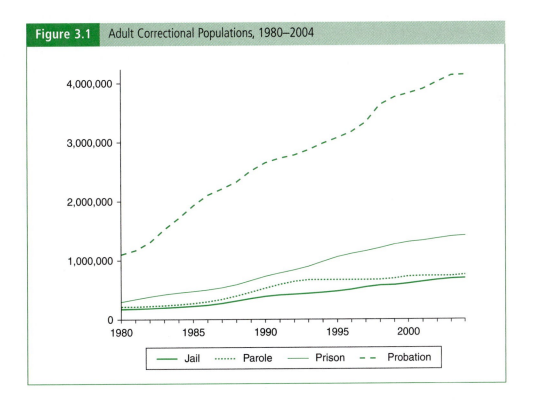

Figure 3.1   Adult Correctional Populations, 1980–2004

in this book). But the incarcerated adult male population was by far the most numerous, and virtually all of their prisons were what we would regard as at the maximum security level. Thus, a possible explanation for the first adult male prisons being maximum security is that, when your incarcerated population is small, your focus should be on the most serious and violent of offenders. Therefore, if you build a prison, it should be to fit these offenders, even if the number of less serious offenders is greater. Some might argue that this focus on a higher security level is understandable given the possible consequences of serious and violent offenders escaping and wreaking further havoc and misery in communities. Therefore, the historic focus on maximum security institutions by the media and academics makes some sense. (Also, it was not until the 1970s and 1980s that states started to build more medium and minimum security institutions to fit the vast majority of inmates coming into the system.) But these past correctional circumstances do not explain the current preoccupation with maximum security institutions.

Why the Continued Fascination With Maximum Security and How That Preoccupation Hinders a True Understanding of Corrections

The explanation for the continued and current focus on maximum security prisons probably rests, in part, with the prurient interests of the producers and consumers of correctional information. Given the popularity of certain television programs, movies, and video games, it would appear that people are attracted to viewing and knowing about violence and the subcultures that produce it, which maximum security institutions tend to have in abundance. They tend to incarcerate those with the most disruptive behavioral problems in any given prison system—though not always those with the most violent history of offending in the free world—who might be adequately housed in medium or even minimum security prisons (Hensley, Tewksbury, & Wright, 2006; Ireland & Ireland, 2006). They also tend to have younger, less experienced staff who might be more willing to engage in violent encounters with inmates. Also, as mentioned in Section II, maximum security institutions likely foster the development of subcultures (their inmates have a greater tendency to use violence or to be disruptive, and they tend to have longer sentences), which, in turn, tends to support the violence and gangs. The combination of more violence-prone inmates and staff, with a highly developed inmate subculture, all in an institution that is the most isolated from the public because of its security level, may provide the perfect admixture for violence—which would, of course, explain the fascination with maximum security institutions by the media, academics, and the general public.

But this very fascination with violent institutions or, more accurately, with those institutions containing inmates that are more prone to violence—and with the maximum security prison in particular—can serve to impede researchers and the public from gaining an understanding of what corrections is truly like. This is one of the reasons we still do not know how much violence inmates experience in jails or most other prisons. The depiction of the more violent-prone inmate in a max, and the reciprocal violence of staff, may have caused many in the community to dismiss the need to make correctional institutions more humane or to fund programming. Instead, the public might be more inclined to allow people who are perceived as more violent and thus a greater threat to public safety (as a greater percentage of maximum security inmates are) to be warehoused for longer and longer periods of time.

Members of the public may also have been less inclined to support the increased professionalization of staff when they thought the staff role was limited merely to the use of brute force, a role termed *guard* in old-style "Big House" prisons and defined as a *hack* by current scholars (Farkas, 1999, 2001; Johnson, 2002). If a correctional officer is viewed by the public as a hack, or as a violent, cynical, and alienated keeper of inmates in a no-hope warehouse prison, then there would appear to be little need to encourage education or to provide the training and pay that would elevate corrections staff to "professional status." Yet, there is reason to believe that most officers in prisons, jails, juvenile facilities, and in community corrections actually regularly engage in "human service" work, which serves as an alternative, more developed and more positive role for correctional workers (Johnson, 2002). They provide "goods and services"; serve as "advocates" for inmates, when appropriate; and assist them with their "adjustment" and use "helping networks" (Farkas, 1999, 2001; Johnson, 2002, p. 242–259). When the public does not know about, and the correctional organization does not recognize or sanction, the alternate "human service" work performed by correctional or juvenile justice officers and by probation and parole officers, then, again, there is no need to provide the training and pay that would be commensurate with that role.

Management of Correctional Institutions and Programs

In addition to exploding inmate and offender populations, the number of employees in corrections has grown astronomically in the last 20 years, albeit these employees are often undereducated, inadequately trained, and poorly paid for their work. From 1982 to 2003, there has been a 300% increase in federal, state, and local government funding of

▲ **Photo 3.2** A Texas Department of Corrections officer watches over inmates going to work in the fields outside the prison in Huntsville, Texas.

all justice functions; that is, police, courts, and corrections (Hughes, 2006, p. 2). During that same time period, just the expenditures for corrections increased by 423%! Not surprisingly, employment in corrections more than doubled from 300,000 to over 748,000. Given these increases in not just inmate but also staff populations, one can appreciate the organizational problems that develop for correctional managers seeking to hire, train, and retain the best employees.

Perhaps in part because of this recognized need, the management of correctional institutions and programs has shifted over the years as efforts to professionalize, democratize, and routinize corrections have had some success. Usually, for a profession, compensation that is commensurate with job requirements and skills is a clear indication of the value ascribed to it. It is true, however, that it would be difficult to regard the staff members of some correctional institutions or programs as "professionals" because they do not meet the educational, training, or pay requirements of people in a "profession." However, the truth is that there are a number of correctional institutions and programs that have made much progress in this area, though the path to professionalization and the creation of a work environment that is conducive to employee growth and greater welfare has not always been clear in corrections.

So if we were to imagine a continuum of correctional professionalism across the field, we could probably generalize that community corrections officers and their work are more professionalized—as that is defined as more education, training, and pay—than are prison and jail correctional officers. Next along that continuum would come correctional officers who work in public prisons and then those who work in jails. Of course, these generalizations do not hold true for every institution or every locality. For instance, in some larger cities, jail officers might be paid and trained as well as or better than prison officers or even adult probation and parole officers. However, in general, our imaginary continuum of correctional professionalism does typify the field and means that attempts to professionalize corrections need more effort in some areas than in others. Students of criminal justice tend to recognize this difference: when asked what they want to do when they graduate, they are more likely to identify a community corrections (probation and parole) job as desirable over work in a prison or jail.

⬙ Organizational and Individual Level Factors That Affect the Correctional Workplace

In the articles contained in this section of the book, the authors discuss just these factors and how they affect correctional work settings. What they find is that organizations can be reconfigured to suit workers so that they are more satisfied with, less stressed by, and more challenged by their work. The key is in the support of a professional role for officers and in the evolving conception of an organization that is more responsive to its dynamic environment and more supportive of its employees and clients. The research discussed in this section would indicate that the extent to which a correctional—or any—organization and its employees attend to these matters determines ultimate success on a number of levels.

In their article, Stohr, Lovrich, Menke, and Zupan review how different jails attend to the issues of employee investment and how that affects stress, turnover, and job satisfaction.

As regards job satisfaction, role ambiguity (or confusion about one's role on the job), whether because of a lack of managerial direction (read training) and reinforcement, is an indicator of a lack of a conception of a professional role. This finding was reinforced by the meta-analysis of work-related stress conducted by Dowden and Tellier. These authors also find, after a review of 20 studies of work-related stress in corrections, that role difficulties affected stress, as did the dangerousness of the job, the ability to participate in the decisions that affect one's work, job satisfaction, and commitment, along with a propensity to leave or not (turnover).

As the correctional workplace has diversified, so has the importance of race as it might affect workers' perceptions. The article by Camp, Steiger, Wright, Saylor, and Gilman explores whether employees in different racial groups perceive that their job opportunities are shaped by their race and by the race of others in their workplace.

As much as race, the change in the gender composition of the employee workplace has had an effect on correctional worker attitudes, perceptions, and behavior. As regards sexual harassment in the correctional workplace, the research indicates that it is not restricted to female victims and male offenders, nor is it limited to just staff. In the article by Marquart, Barnhill, and Balshaw-Biddle, the authors find that female officers are much more involved as aggressors, particularly in the more minor versions of the sexual harassment of male inmates, than are their male counterparts.

▧ Summary

The correctional experience for staff is fraught with challenges and much promise. It involves a diverse role that encompasses work with juvenile and adult inmates in institutions and offenders in the community. It is not as narrow a role as is commonly perceived by the public, and it includes many opportunities to effectuate a just incarceration or community supervision experience for inmates.

- Though many have worked long and hard to "professionalize" correctional work, there is every indication that most jobs in this area do not meet standard professional criteria.
- When correctional workers do not receive the requisite professional training and education they need to do their job in an appropriate manner, both clients and the community are likely to suffer.
- The focus on maximum security institutions by the media and academics has misrepresented the reality of correctional work and operation in this country. Most correctional institutions are not of the maximum security variety, most employees do not work in such facilities, and most inmates or offenders or clients are not housed there. Rather, maximum security institutions, even among prisons, represent only one fifth of the incarcerated population. By far most people under correctional supervision are under the least restrictive type of programs or on probation. This misrepresentation of correctional operation may have served to warp the public view of correctional work, making professional standards seem less necessary. Another possible corollary of this is that the public could think that the potential of inmates, offenders, and clients to reform is less than it might be because

they are seen as more "violent and hardened"—or as more typically max inmates—than they may in fact be.

♦ Many who labor in corrections undertake the "human service" role as that is presented: providing goods and services, advocating for inmates or offenders or clients, and assisting in their adjustment (Johnson, 2002).

♦ The research presented here would indicate that there are organizational factors that can be manipulated to improve this experience for staff so that they, and their organization, can realize their promise. Namely, better pay and benefits and training can be used to foster the greater development of professional attributes, such as education, which will promote a consequent reduction in role ambiguity, low job satisfaction, turnover, and stress. This research also indicates that the organization can do much to reduce the problems associated with the greater diversification of its staff. It can be open in its promotion practices so that false impressions regarding unfair advantage are not perpetuated and do not have a demoralizing effect on the workforce. It is also within the correctional organization's power to prevent most sexual harassment, whether practiced on staff or inmates. In short, the promise of a positive correctional experience for staff is achievable and, as that perception seeps into the public consciousness, we are much more likely to hear "correctional officer" in response to that question, "What do you want to be when you grow up?"

KEY TERMS

Discretion	Human service officer
Hack	Profession

INTERNET SITES

American Correctional Association: www.aca.org

American Jail Association: www.aja.org

American Probation and Parole Association: www.appa-net.org

Bureau of Justice Statistics (information available on all manner of criminal justice topics): www.ojp.usdoj.gov/bjs

National Criminal Justice Reference Service: www.ncjrs.gov

Office of Justice Programs, Bureau of Justice Statistics (periodic statistical reports on all manner of criminal justice topics, e.g., HIV in prisons and jails, probation and parole, and profiles of prisoners): www.ojp.usdoj.gov/bjs/periodic.htm

Pew Charitable Trust (Corrections and Safety): www.pewtrusts.org/our_work_category.aspx?id=72

Vera Institute (information available on a number of corrections and other justice related topics): www.vera.org

READING

In this article, Mary Stohr and her colleagues study staff in five modular direct-supervision jails to compare two different management models: the control and employee investment models. The authors analyze variables such as job satisfaction, job enrichment, workplace stress, role conflict, organizational commitment, training, and staff turnover. They found only limited support for the benefits of the control model in contemporary jails, and they find some evidence that investment in personnel is beneficial to the jail organization.

Staff Management in Correctional Institutions

Comparing DiIulio's "Control Model" and "Employee Investment Model" Outcomes in Five Jails

**Mary K. Stohr, Nicholas P. Lovrich, Jr.,
Ben A. Menke, and Linda L. Zupan**

The evidence presented in this five-jail comparative study adds support to the belief shared increasingly by correctional facility managers that correctional institutions are not well served by maintaining "control-oriented" staff management practices. We believe that the staff has the potential to contribute to the correctional environment of the twenty-first century, and that some intensive development of that potential is warranted. We are not arguing here that participatory management and investment in staff development will be the cure-all for correctional institutions. Certain contemporary social, political, and economic conditions have produced public policies that present perhaps insurmountable challenges to the managers and staff in United States jails; participatory management is no easy remedy for these societal problems. We believe, however, that prisons and jails, their management, and their staffs (and, we suspect, their inmates) can benefit substantially by investing in their staffs and by facilitating some empowerment of employees among lower levels of the staff.

Correctional scholars have long recognized that jails occupy the lowest level of status in relation to the nation's other social institutions, including prisons (Fishman 1923; Goldfarb 1975; Irwin 1985). Thus the movement toward an employee investment model has not been clear or linear. Unflattering appellations for jails are justified, for the most part: too many jails have been little more than "dirty little

Source: Staff Management in Correctional Institutions: Comparing DiIulio's "Control Model" and "Employee Investment Model" Outcomes in Five Jails; Mary K. Stohr, Nicholas P. Lovrich, Ben Menke, and Linda L. Zupan. *Justice Quarterly* 11(3): 471–497 (1994); Taylor & Francis, Ltd., http://www.informaworld.com, reprinted by permission of Taylor & Francis, Ltd.

holes" that hold hapless inhabitants, garner few resources and scant attention from the larger political or social community, and treat their staff much as they treat their inmates.

During the 1980s, however, jails received uncharacteristically close attention from scholars, judges, and the news media. In recent years they have become the unlikely birthplace for a correctional management model that is transforming the operation of correctional facilities across the country. The impetus for this transformation probably came from several sources, including broad social and political changes external to and including sponsored research, technical support offered by the Jails Division of the National Institute of Corrections, the development of a separate "jail" magazine and national organization (*American Jails* and the American Jail Association), and the concomitant calls for professionalization of correctional work. In addition, a significant cadre of academics have recognized that jails represent a critical criminal justice institution worthy of careful study, as do the efforts of dedicated and thoughtful jail managers and staffs who are wresting jails from their dark Victorian moorings and pushing them into the limelight of the twenty-first century.

The principal harbinger of change has been and continues to be the evolving concept of a "new generation" or podular direct-supervision jail. This type of facility was conceived for federal jails in the 1970s; the idea was adopted by numerous counties in the 1980s and continues to gain ground in the early 1990s. These jails, now numbering approximately 80 across the country (American Jail Association 1991), combine podular architecture and placement of direct-supervision staff members, theoretically to achieve greater safety for staff and inmates and to provide inmates with a more humane confinement (Gettinger 1984; Manning et al. 1988; Menke et al. 1986; Gettinger 1984; Zupan 1991).[1] As a corollary to this change, the "new generation" philosophy has come to encompass much

more than reconfigured buildings and close contact between staff and inmates. Once free of some of the restrictions of long-dominant traditional corrections control models (e.g., linear architecture and remote or intermittent supervision), some jail managers and their staffs have moved toward participatory management practices and investments in human resource development that were largely unthinkable only a decade ago.

The need for an organizational structure that is more flexible than the conventional monocratic hierarchy is recognized increasingly by some jail employees because corrections officers now play a key role in the workings of podular direct-supervision jails (Fuqua 1991). Correctional staff members may be hired and trained with an eye to their fitness for the more demanding role played by a correctional officer in a podular direct-supervision jail. Once placed in a module, these officers enjoy considerable autonomy as unit supervisors and social leaders in that artificial environment. It follows that such employees would develop strong problem-solving skills and would not fit easily into a rigid organizational structure that provides them with little opportunity to influence broader organizational policy affecting their work and workplace (Conroy 1989; Fuqua 1991).

We explore here the evidence of the outcomes associated with such change in podular direct-supervision jail management. First, however, it is appropriate to discuss the still predominant corrections management model and its shortcomings for the workforce of 2000.

The Control Model of Corrections

Although the soil from which the control model grew in a general sense derived from ubiquitous social-political arrangements and institutions prevalent in western European cultures (e.g., class structure, hierarchical family and business relations), it came to be represented more consciously and more purposefully in the United States in the Pennsylvania and Auburn prison systems of the early nineteenth century. Inmate isolation, separation of staff from inmates, the rigidity of roles in the facility, the inhospitality of large secured buildings, the prevalence of force-based subcultures, and the totality of this confinement—as described by contemporary writers on prisons, jails, and related social institutions—first were evidenced in such facilities (Bartollas 1990; Bartollas and Conrad 1992; Clear and Cole 1990; Duffee 1989; Goffman 1961; Hawkins and Alpert 1989; Murton 1976; Rothman 1980).

In *Governing Prisons: A Comparative Study of Correctional Management*, DiIulio (1987) borrows from and builds on this long-dominant nineteenth-century correctional management model. He explicitly rejects Sykes's (1958) argument that the keepers are necessarily kept in such facilities, and he disagrees with Murton's (1976) and Jacobs's (1977) view that some amount of control over institutional governance ought to be accorded to prison inmates. DiIulio disagrees with Jacobs about the relevance of broader social change for greater involvement and participation in issues of governance in correctional settings. He maintains instead that prison management practices emphasizing *order first* and the provision of some services and amenities as *earned benefits* (as opposed to rightful claims) are required to ensure smooth operation and to provide safe environments for inmates and staff alike.

Correct management, according to DiIulio, means not only tight control over procedural and security matters but also modeling of the "straight" way of life (e.g., dressing neatly, speaking respectfully). Prison staff management is also tight-fisted, demonstrating that respect for rules and for authority is expected of everyone, staff and inmates alike. DiIulio believes that the more "governing" of a prison, the better; the ultimate ends for correctional staff and inmates are increased order, finer amenities, and more helpful services. The existence of order, for

DiIulio, is determined by the level of assaults and violence, the regularity of administrative routines, and the predictability of institutional programming. Services are apparent when the prison offers regular work, schooling, and other programmed activities. Amenities are those small but valued things one does not expect but is pleased to discover in a prison, such as hot coffee, tasty food, or clean sheets.

In his study of three state prison systems (Texas, Michigan, and California), DiIulio describes three distinct models for inmate management, which correspond directly to the three state prison systems in question: the *control model* (Texas, until and including the early 1980s), the *responsibility model* (Michigan), and the *consensual model* (California). In the control model the inmates' activities and time are tightly monitored and controlled: "inmate movement was monitored too closely, rules were enforced too tightly, inmates were kept too busy, the punishments for misbehavior were too swift and certain, and the rewards to be earned in sentence reduction and other privileges were too great for inmates to commit many infractions" (1987:108).

DiIulio appears to favor the control model for several reasons. His studies lead him to observe that prisons governed by this approach experience less violence and less staff turnover, inspire greater staff morale in the form of "attachment" to the organization, provide inmates with more worthwhile self-development programs, and maintain a safer and more civilized environment for staff and inmates alike.

According to DiIulio, several factors are central to the effective operation of the control model: 1) a formal bureaucratic organizational structure (including hierarchy, specialized labor, published rules and regulations, retention and promotion based on merit, and "impersonal relations") (1987:237); 2) administrators who are devoted to "hands-on" leadership, in which they are involved in the facility's daily operation (e.g., visiting the cell blocks at least once a day) and are capable of garnering resources and playing the necessary political games; prison wardens must "manage both behind the walls and beyond them" (1987:241); and 3) a need for groups such as correctional officers, judges, unions, politicians, and the media to move beyond criticism so as to reach a consensus on the conditions required to maintain the proper working of the prison.

The Control Model: Adherents and Critics

John DiIulio received some of his graduate training from James Q. Wilson at Harvard University. DiIulio's debt to Wilson extends not only to graduate training but also to Wilson's conception of the "rightness" of community order, whether a neighborhood, a public agency, or a prison. This strong focus on maintenance of order as a means of enhancing the community's quality of life and of decreasing "lawlessness" made a marked debut in contemporary criminal justice literature in Wilson and Killing's (1985) "Broken Windows: The Police and Neighborhood Safety." Once high levels of order are achieved in the neighborhood or the prison, say the authors, culture and productive business in the neighborhood and amenities and desired services in the prison can flourish. Conversely, if order is lacking, Wilson and Kelling—and DiIulio as well, in spirit—believe that little else is possible, whether in a rich communal life or in a well-managed prison.

The critique of this control perspective, advocated for community policing by Wilson and Kelling (and inherited by DiIulio), is multidimensional. Irwin (1985) criticizes the "uses" to which such control perspectives have been put and the diversionary tactics to which this perspective often leads in practice. He charges that the police, in the name of maintaining order, have been known to harass people or to harbor an "us versus them" mentality, which is reinforced by control strategies that focus on getting the "disreputables" off the street.

Among other critiques, Irwin also observes that the emphasis on street crimes and community appearances, as engendered by the focus on order maintenance, is merely a means of political diversion. Rather than concentrating on corporate or white-collar criminals, who tend to be wealthy and established and yet are responsible for more loss of life and money than are street criminals, the police choose to pursue those who are least able to protect and defend themselves—those who live and interact primarily on the street. Such a critique of a community control model also might be fruitfully applied to the control model for jails and prisons. This approach would tend to rigidify the officer and inmate roles and the attendant hierarchy, reinforcing the "us versus them" mentality, and would lead to forsaking the qualitative issues (e.g., true order, services, and amenities) for the appearance of order.

Walker (1984) criticizes what he regards as a selective revisionist view of police history, as presented by Wilson and Kelling. He charges that the idyllic past of policing to which Wilson and Kelling would have us return is essentially a myth. The police of yesteryear (before the automobile and the professionalization movement removed them from "real" contact with people) were not particularly interested in fighting crime as they walked their beat. Instead they were often involved in graft and corruption, or in doing as little work as possible. The salience of this "revisionist history" critique by Walker is reinforced by a similar critique of DiIulio offered by Crouch and Marquart (1990).

Crouch and Marquart question DiIulio's contention that the Texas prison system of the 1960s, 1970s, and early 1980s (the prototype control system of DiIulio's analysis) was safe, orderly, or predictable for inmates. In fact, these authors state that under the old "control" system touted by DiIulio, inmates perceived Texas prisons as unsafe.

This conclusion moves our understanding of prereform Texas prisons beyond that offered by DiIulio (1987).

His account (influenced strongly by the TDC of the 1960s and early 1970s under Director Beto) presents a control model of Texas prisons that offered all prisoners a safe, predictable environment. Yet the data presented here reveal that although in the "old days" powerful directors may have been able to limit official violence and to maintain a level of order envied by prison managers and researchers, most TDC prisoners did not share this perception of safety and order in the years before the court-ordered reforms. (Crouch and Marquart 1990:121)

"New Generation" Jails, the Control Model, and Beyond

We wish here to add another element to the emerging critique of the control model. Specifically, we address the staff management component of this model by comparing outcomes of the applications of different staff management approaches in five podular direct-supervision "new generation" jails. We acknowledge fully that much of what DiIulio describes as the control model for older-style Texas prisons and what Wilson and Kelling advocate for neighborhood policing has become part of podular direct-supervision jail operations. Such jails rely heavily on order maintenance that focuses on minor infractions, the "broken windows," to avoid the development of major types of disorder. The procedures are routinized, and staff and inmates are expected to adhere closely to the routines. Authority is centralized in the correctional officer within the pod. Civility is expected in the living unit; those who transgress are removed to a less pleasant living environment. The provision of many services and amenities is contingent on maintenance of order.

Podular direct-supervision jails also resemble DiIulio's model in that they are highly bureaucratized, like most correctional facilities, and typically operate in a paramilitary manner. There is a clear line of command: staff members report to and take orders from their immediate supervisors, who then report up the line. Procedures and rules are standardized, and considerable documentation of activities and events is required. Correctional and support staff members tend to specialize in particular areas as a function of needs and/or training (e.g., booking staff, module correctional staff, cooks, and nurses).

Some podular direct-supervision jails, however (as well as more "traditional" jails and prisons), diverge considerably from the control model, as advocated by DiIulio. In these jails the correctional officer in the module is trained not only to use supervisory rule enforcement skills, but also to employ proactive interpersonal skills in his or her relations with inmates. Rather than exerting full control and applying impersonal rule enforcement, the officer in many "new generation" jails is urged to develop more of a guidance role, in which "personalization of supervision" and "leadership skills" are exercised. Furthermore, the decentralized organizational structure of podular direct-supervision jails, despite the bureaucratic nature of the jail authority structure, has led many jurisdictions to develop some means for systematic input and problem solving by lower-level staff members. Mechanisms for facilitating input might include the use of quality circles or employee teams (empowering officers and managers) who apply insight and problem-solving skills to jail problems (Chandler Halford 1992; Conroy 1989; Fuqua 1991; Sigurdson, Wayson, and Funke 1990).

We hypothesize that insofar as jails provide for parity (with law enforcement) in pay and status for jail staff members, invest heavily in training and staff development, and use participative management practices, they will experience *less workplace stress, less staff turnover, less role conflict, greater organizational commitment*, and *greater job satisfaction* than jails which do not feature these policies and practices. We were unable to classify the jails beforehand as to their adherence to our "employee investment model" versus DiIulio's "control model." We are not arguing here that any of the jails conformed to an ideal-typical representation of either model (as presented in Table 1), nor are we arguing that these models are at opposite poles of a management continuum. Rather we believe that our model encompasses the "best" of the control model and improves on it in a dialectic synthesis (see Table 1).

We propose that those jails which tend to have characteristics related to our model also tend to reap the identifiable benefits cited above.

◈ Methods: Multijail Design, Staff Surveys, and Organizational Profiles

Site Selection and Visitation

We selected study sites containing medium-sized, urban-area facilities with at least one year of operation to ensure that generalizations made about podular direct-supervision jails were applicable across a wide array of

geographic settings. Podular design facilities with direct supervision initially were selected *at random* in the south, the northeast, the southwest, the northwest, and the mountain states.

At least two researchers visited each of the six jails included in the study. During the site visits the researchers spoke with the chief administrative officer on the premises, talked to personnel officers and those responsible for coordinating the collection of the staff survey data, toured the facility, spoke with inmates, correctional staff, and support staff, and formulated a general impression of each jail.

Research Design and Instruments: The Employee Survey and the Organizational Profile

To collect systematic data on staff members' perceptions and attitudes, we constructed a survey that included both some original measures and some standardized indicators widely used in personnel research. The standardized measures were employed to compare podular direct-supervision jail employees with known

Table 1	Attributes of the Employee Investment Model

- Similarities to the Control Model
  - o   Bureaucratic structure
  - o   Civil service rules and protections
  - o   Basic training in security and control techniques
  - o   Proactive leadership by managers
  - o   "Broken windows" attentiveness by staff

- Employee Investment
  - o   Investment in staff training that prepares them to be leaders and supervisors in their living unit. Such training might include courses in problem solving, interpersonal and communication skills, participatory management, ethics, and leadership.

- Pay equivalent to that of law enforcement personnel

- Many opportunities (both formal and informal) for input by staff

- Involvement of staff (as much as is possible and reasonable) in decision making that affects their workplace

- Valuing of input by staff

national norms. Survey items included questions, issues, and attributes that have been associated with stress and turnover in other social service workplaces. The surveys also contained several conventional questions on background and job status. We wished to assess whether outcomes such as measures of an enriched job, higher job satisfaction, lower stress, turnover, and role conflict were related to organizational attributes associated with an employee investment model (e.g., pay parity, developmental training, and opportunities for input).

We included the Job Diagnostic Survey (JDS) developed by Hackman and Oldham (1974) as an important outcome indicator to measure the presence of enriching characteristics of a job. Use of the JDS requires that respondents estimate, on a seven-point scale, the degree to which certain attributes presently apply to their jobs. The instrument includes questions designed to assess the degree of task identification, autonomy, skill variety, task significance, and feedback perceived to exist in one's work; these characteristics represent the attributes of an "enriched" job.[2] According to Hackman and Oldham (1974), these five "core job dimensions" translate into three "critical psychological states," which result in high internal work motivation, high quality of work performance, high satisfaction with the work, and low absenteeism and turnover. Skill variety, task identity, and task significance combine to constitute the critical psychological state of "experienced meaningfulness of the work," autonomy measures "experienced responsibility for outcomes of the work," and feedback measures the equally important element "knowledge of the actual results of the work activities." Each of the job characteristic scores is weighted to reflect the relative importance of each dimension; they are combined mathematically to create a single "motivating potential" score. The motivating potential scores range from 0 to 343, reflecting a range from total absence of motivating potential to total fulfillment.[3]

The Role Conflict Instrument developed for this research is based on work by Rizzo, House, and Lirtzman (1970) and is used to measure the extent to which an employee experiences role conflict on the job. Of 13 items that loaded on the "role conflict" factor in the Rizzo et al. study, we selected only the eight highest-loading items. On a seven-point Likert-type scale (very false to very true, with "undecided" as midpoint), the respondents were asked to rate the degree to which they might encounter role conflict stemming from decisions they had to make about how to do their work. For instance, they were asked to indicate the extent to which "I have to do things that should be done differently" or "I receive incompatible requests from two or more people" on the job. The higher their agreement with such statements, the greater the role conflict they experienced.

The first subscale measure (labeled Person) contains three items and is believed to assess the level of conflict between a person's "internal standards or values and the defined role behavior" (Rizzo et al. 1970:155). The second subscale measure (labeled Resource) contains three items and is believed to assess the level of conflict "between the focal person and defined role behavior" (Rizzo et al. 1970:155). The third and fourth subscale measures (labeled Policy and Request) contain two items and one item respectively and are believed to assess the "conflicting expectations and organizational demands in the form of incompatible policies" and "conflicting requests from others" (Rizzo et al. 1970:155).[4]

The Job Descriptive Index (JDI) devised by Smith, Kendall, and Hulin (1969) is a widely used measure of job satisfaction. Satisfaction is measured as it relates to five areas of work life: character of work, level of pay, promotion opportunity, supervision quality, and regard for other people on the job. For instance, employees are asked to indicate whether adjectives such as "fascinating" or "tiresome" describe their present job ("undecided" also was an option). Each of the five subscales contained eight to 18 items. A summary job satisfaction measure is created by summing the five index scores; low index scores indicate low job satisfaction.[5]

The standardized Brief Symptom Inventory instrument (the SCL-53), developed by Derogatis (1977), represents a measure of symptoms of physiological, psychological, and behavioral difficulties that are known to be related to stress. Nine subscales are formed by adding together all nonzero items constituting each dimension, and then dividing the total by the number of items in that subscale.[6]

This instrument contains nine separate subscales pertaining to psychosomatic symptomatology. Because jails screen routinely for gross psychological disorientation (such as is assessed by the phobic anxiety, paranoid ideation, and psychoticism subscales of this instrument), the most highly differentiating subscales in this analysis should be the somatization, obsessive-compulsive, interpersonal sensitivity, depression, anxiety, and hostility subscales.[7]

We also included the O'Reilly and Chatman (1986) measure of organizational commitment and attachment. In the use of this instrument, respondents must estimate the degree to which they agree or disagree, on a seven-point scale, with 12 statements related to congruence in values, rewards, ownership of the job, and pride in the job. The greater their agreement with such statements as "If the values of this organization were different, I would not be as attached to this organization," the higher their organizational attachment.

The survey instrument was administered to the population of full-time noncontract and nonvolunteer employees at six facilities by a "contact person" at each location, usually a personnel specialist or chief administrative officer. Each contact person also was responsible for collecting the surveys and mailing them back to us for coding and analysis. Each respondent was provided with an individual envelope in which to seal the survey in order to provide privacy and ensure frankness in responses. This method yielded a 76 percent return rate in the County 1 jail (110 usable questionnaires out of 145 administered). The same method, however, was less successful at three other jails: it produced return rates of 33 percent from County 2 (179 administered and 59 collected), 35 percent at County 5 (220 administered and 78 collected), and virtually zero from the County 4 jail. Even after we sent a mail survey to the County 4 staff, the return rate was still only 24 percent (37 of 157 questionnaires returned). Because of this return rate and because the County 4 jail was not fully operational as a podular direct-supervision jail at the time of the site visit and data collection (as we originally believed it would be), we were reluctant to put much credence in the findings from this jail. Accordingly we dropped it from this analysis.

To avoid such response rate problems in County 3, we arranged for the surveys to be administered directly by the contact person rather than being placed in mailboxes. When each survey was returned, we crossed the respondent's name off his roster. This method of distribution and collection clearly threatens to produce a response bias. Yet on the basis of some of the negative comments on the surveys regarding certain aspects of the workplace (such comments were common in all the jails), it is difficult to believe that many of the respondents at County 3 were inordinately concerned that their responses would be reviewed by jail administrators at this site. The return rate at County 3 was 79 percent (200 administered and 157 collected).

At the County 6 jail the surveys were distributed in a similar manner, and yielded an 83 percent return rate (166 administered and 138 returned). In County 6 the surveys were distributed at the end of training sessions conducted each Thursday in rotation to one-fourth of the correctional staff. Those present were required to fill out the survey before leaving the room. Support staff members at the County 6 jail were given surveys by their supervisors; they were similarly urged to complete the questionnaires and return them in a sealed envelope addressed to the research team. Again, this distribution method represents a compromise of the validity of the responses. As the analysis indicates, however, the respondents at this site were not highly enthusiastic about their organizational environment in comparison with respondents at the other jails. Consequently it would appear that County 6 jail employees were not discouraged from being candid in their responses, either explicitly or implicitly, by jail personnel and/or the method of collection.

The Organizational Profile Form

We developed the organizational profile form to identify personnel processes and document circumstances known to be related to turnover, stress, and inmate management. In addition, the form measured workload level as it related to jail overcrowding. It was quite evident that all of the facilities surveyed were operating beyond capacity, some by enormous amounts in particular living units.

Characteristics and Practices of Jails

Table 2 displays the characteristics of each of the five jails in question. The findings displayed in this table suggest some degree of variance among facilities on several background factors. For example, most of the County 2 and 5 employees tend to be concentrated in a relatively young age bracket, whereas employees in the remaining facilities tend to reflect greater age disparities. Not surprisingly, these differences also are evident for the "length of employment" variable: employees from jails in County 2 (and, to a lesser degree, in County 5) tend to be more recent hires than employees of the other jails. Given these differences, one might conclude that the relatively low return rates in the County 2 and 5 jails yielded biased information. When we examine the other variables on this table, however (e.g., gender, ethnicity, education), these

cross-jurisdictional differences tend to dissipate and/or disappear.

We also explored employees' comments on the type, amount, importance, and usefulness of training received; these findings are only summarized here because of space constraints. Most employees in the County 1 and 2 jails reported that they had received "moderate to a great deal of" training (69.7% and 86.4% respectively); this finding is in contrast to responses in the County 3 (46.5%) and County 6 (56.6%) jails (and, to a lesser degree, in County 5, with 32.1%), who were much more likely to report that they had received "not much to moderate" amounts of training. Somewhat in line with these findings, those employees who reported the greatest amount of training (Counties 1 and 2) were more likely to rate the training they received as moderately useful to very useful (60.4%, 67.7%, and 75.6% respectively) than were those who had received less training.

In all of the jails, the most highly regarded type of training was that which involved "Interpersonal Skills" (IPS). The employees (with the sole exception of County 3) regarded "Problem Solving" as the second most valuable type of training. Training in "policies and procedures," "observation skills," "crisis management," "report writing," or "self-defense" received third-place rankings. Ironically, however, the types of training that were rated most useful were reportedly offered to only a few of the employees as part of their in-service training. A review of the amount and type of training provided in each jail facility is central to the hypothesis presented here, because training represents an investment in employees. The types of training valued by employees (e.g., the top two choices, IPS and problem solving) suggest that they recognize that the skills they need to operate effectively at work are more complex than the traditional control and security techniques.

We also collected and analyzed information about the amount of inmate-generated disruption of employees. As a result, of inmates' action, 45 percent and 41 percent of

Table 2	Distribution of Employees' Demographic Characteristics by County Facility				
County	1	2	3	5	6
N	110	59	157	78	138
Age					
Under 24					
25 to 34	41.3%	73.0%	33.8%	52.0%	39.4%
35 to 49	44.9	22.1	44.0	36.4	49.0
50+	13.8	5.1	22.3	11.7	11.7
Gender					
Female	39.8	22.6	34.0	20.8	44.2
Male	60.2	77.4	66.0	79.2	55.8
Ethnicity					
Black	1.9	11.3	.7	9.7	3.1
Hispanic	0.0	3.8	2.0	2.8	1.6
White	91.3	83.0	94.0	83.3	91.5
Other	6.7	1.9	3.3	4.2	3.9
Education					
High school or less	17.6	10.2	17.9	35.9	20.3
Some college	68.5	66.1	65.4	38.5	56.6
BA or more	11.2	23.7	15.4	25.6	22.5
Other	2.8	0.0	1.3	0.0	.7
Position					
CO	74.1	74.6	43.9	57.7	65.2
CO supervisor	6.5	6.8	10.1	12.8	4.4
Treatment	4.6	11.9	.6	6.4	6.7
Clerical	12.0	13.6	14.0	12.8	5.2
Support	2.8	0.0	17.2	10.3	8.9
Other	0.0	0.0	14.0	0.0	9.6
Length of Employment (years)					
Less than 1	19.3	40.7	20.5	26.9	14.6
1 to 5	60.5	47.4	75.0	46.1	65.7
6 to 11	9.2	10.2	1.3	19.3	15.3
12+	11.0	1.7	3.2	7.7	4.4

the inmate contact staff in County 1 and County 6 respectively received minor injuries. In the County 1, 2, and 6 jails, 64.5 percent, 50.8 percent, and 56.5 percent respectively were the subject of inmates' grievances. Very small percentages of the staff in any of the facilities, however, suffered inmate violence requiring medical care; fewer yet were involved in a hostage situation. Except for County 1, fewer than 6 percent of the respondents were named in a lawsuit by an inmate. County 3 jails had the highest number of inmate management staff members who stated that they had not been the object of an inmate's action (67.5%). This finding may reflect the fact that a higher proportion of the staff at this jail indicated that writing reports on inmates and placing them in administrative lockdown or physical restraints are "not applicable" to their job. Employees at all jails were more likely to choose the "not applicable" response if their job did not bring them into daily contact with inmates. This is especially true in County 3, where a much smaller percentage of the respondents than elsewhere were correctional staff members (officers or supervisors).

When we sum the three middle responses in each column for each jail for the "reports written," "placed in administrative lockdown" and "placed in physical restraints" variables, a consistent pattern emerges. The highest number of reports written per staff member and the greatest number of inmates placed in administrative lockdown and in physical restraints tend to occur in County 2, County 1, and (on two of the variables, "reports written" and "placed in administrative lockdown") in County 6. Again, this finding merely may reflect the greater percentage of County 3 employees who indicated that these activities are "not applicable" to their job.

According to DiIulio (1987), facilities that emphasize policies, procedures, and swift attention to small disruptions are likely to experience less violence than those which allow minor disruptions to go unpunished.

This hypothesis is not borne out by data for two of the jails (County 1 and County 5), which reported high levels of reports written and placements in administrative lockdown and physical restraints, as well as high rates of minor injuries and more serious injuries requiring medical attention. The hypothesis is supported, however, by the data from County 2. Employees in County 1 and County 5 jails also reported high percentages of "inmate grievances" (along with County 2) and high incidences of "inmate suits."

⬥ Findings

Job Diagnostic Survey

The results reported in Table 3 suggest that there are few significant cross-jail differences on job characteristics, as measured by the Job Diagnostic Survey. The students' t-test is used to produce the findings shown here and in the other relevant tables. In ascending order, County 1, 3, and 6 facilities had low means on the measures of task identity, autonomy, and feedback from job (only the County 1 mean differed significantly from the County 5 mean on this last variable). This same pattern is reflected in the scores for motivating potential: County 1, 3, and 6 facilities have the lowest means, but only the County 1 mean differs significantly from means for all the other counties (except County 3).

Job Satisfaction

The results displayed in Table 4 suggest that employees' satisfaction is noticeably higher in three of the jails than in the others. Although we found no significant differences on work satisfaction, and few on the quality of supervision or of co-workers on the job, the county facilities show discernible differences on the remaining variables and on the overall satisfaction measure. That is, staff members at Counties 2, 3, and 5 jails tend to have higher

Table 3	Comparison of Job Diagnostic Survey Scale Mean Scores Across County Facility				
County	1	2	3	5	6
N	110	59	157	78	138
Skill Variety (norm = 5.18)					
Mean	4.4	4.4	4.5	4.4	4.7
SD	1.4	1.2	1.6	1.4	1.4
Task Identification (norm = 5.09)					
Mean	3.7	4.2[a]	4.1[a]	4.7[a,c]	4.3[a]
SD	1.3	1.1	1.7	1.5	1.4
Task Significance (norm = 6.06)					
Mean	5.5	5.5	5.7	5.8	5.7
SD	1.2	1.3	1.2	1.2	1.1
Autonomy (norm = 5.04)					
Mean	4.6	5.2[a]	4.4[b]	5.2[a,c]	4.8[c,d]
SD	1.2	1.2	1.5	1.1	1.3
Feedback from Job (norm = 5.12)					
Mean	4.3	4.5	4.5	4.7[a]	4.6
SD	1.3	1.1	1.4	1.4	1.4
Motivating Potential (norm = 140)					
Mean	99.2	119.0[a]	110.5	131.9[a]	117.5[a]
SD	59.4	59.6	79.79	71.4	60.8

NOTE: The higher the mean, the greater the presence of the characteristic in the job. Norms lifted for public-sector employees are based on numerous studies conducted in state and local government agencies (see Hackman and Oldham 1974).

a. Significant difference (at .05) between County 1 and Counties 2, 3, 5, and 6.

b. Significant difference (at .05) between County 2 and Counties 3, 5, and 6.

c. Significant difference (at .05) between County 3 and Counties 5 and 6.

d. Significant difference (at .05) between Counties 5 and 6.

and significantly different means on the job satisfaction measures; they tend to be more satisfied with pay levels and opportunity for promotion (except for County 3), and with respect to the overall satisfaction measure.

Role Conflict

In our analysis (which is only summarized here because of space constraints), we discovered that employees at two of the jails (Counties 1 and 6) report significantly higher levels of role conflict than do those at most of the other jails. The distinction between these two jails and the others on role conflict is both notable and consistent on virtually all of the subscales, particularly between County 2 (with the least reported role conflict) and Counties 1 and 6 (with the most reported role conflict).

Table 4	Comparison of Job Descriptive Index Mean Scores Across County Facility				
County	1	2	3	5	6
N	110	59	157	78	138
Character of Work (norm = 34.5)					
Mean	25.3	24.9	26.5	27.2	26.9
SD	10.5	11.2	11.9	11.4	10.4
Level of Pay (norm = 25.0)					
Mean	20.6	28.2[a]	33.0[a,b]	16.6[a,b,c]	23.6[a,b,c,d]
SD	11.0	12.0	10.4	10.6	12.0
Opportunity for Promotion (norm = 17.4)					
Mean	9.5	34.9[a]	19.8[a,b]	23.2[a,b]	11.6[b,c,d]
SD	8.2	15.8	15.31	15.8	11.6
Quality of Supervision (norm = 39.3)					
Mean	31.4	33.7	34.3	34.6	28.78[b,c,d]
SD	14.3	14.3	14.3	14.3	15.4
People on the Job (norm = 40.3)					
Mean	38.5	42.2	35.5b	32.1[a,b]	34.3[a,b]
SD	11.9	11.1	12.7	14.5	14.4
Overall Satisfaction (norm = 156.5)					
Mean	125.4	163.9[a]	148.9[a,b]	133.74[b,c]	125.2[b,c]
SD	35.6	44.1	43.4	44.5	41.9

NOTES: Higher means indicate higher levels of satisfaction in the job. Norms listed for public-sector employees are based on 10 separate studies conducted in state and local government agencies from 1980 to 1992 (Hackman and Oldham 1974).

a. Significant difference (at .05) between County 1 and Counties 2, 3, 5, and 6.

b. Significant difference (at .05) between County 2 and Counties 3, 5, and 6.

c. Significant difference (at .05) between County 3 and Counties 5 and G.

d. Significant difference (at .05) between Counties 5 and 6.

Stress

The findings presented in Table 5 indicate that employees at the County 1 and County 5 facilities suffer the highest levels of stress symptomatology. More important, however, the overall pattern of survey findings is unsettling. The findings are startling indeed when the means generated from this analysis are compared with the symptom profile means that Derogatis (1977) constructed when he aggregated the responses to this instrument using representative nonpatient, outpatient, and inpatient psychiatric groups. On all the subscales, staff members in each of the five jails report symptoms that are above the profile score for nonpatients. Staff members in two jails in particular report disturbingly high symptoms related to stress. On five of the subscales (obsessive-compulsive, anxiety, hostility,

Table 5	Comparison of Brief Symptom Inventory Scales Across County Facilities				
County	1	2	3	5	6
N	110	59	157	78	138
Somatization (symptom profile = .36)					
Mean	.37	.12[a]	.23[b]	.29[b]	.28[b]
SD	.55	.25	.42	.55	.44
Obsessive-Compulsive (symptom profile = .39)					
Mean	.71	.38[a]	.52[a]	.68[b]	.61[b]
SD	.59	.49	.55	.81	.60
Interpersonal Sensitivity (symptom profile = .29)					
Mean	.45	.31	.33	.51	.40
SD	.58	.56	.54'	.80	.52
Depression (symptom profile = .36)					
Mean	.46	.19[a]	.33[a]	.45[b]	.37[b]
SD	.55	.36	.46	.69	.51
Anxiety (symptom profile = .30)					
Mean	.53	.27[a]	.33[a]	.52[b,c]	.39[a]
SD	.58	.32	.44	.55	.42
Hostility (symptom profile = .30)					
Mean	.62	.33[a]	.39[a]	.61[b,c]	.44[a]
SD	.58	.38	.50	.73	.47
Phobic Anxiety (symptom profile = .13)					
Mean	.20	.12	.11[a]	.19	.14
SD	.39	.26	.31	.43	.33
Paranoid Ideation (symptom profile = .34)					
Mean	.63	.34[a]	.54[b]	.77[b,c]	.55[b,d]
SD	.62	.51	.65	.85	.59
Psychoticism (symptom profile = .14)					
Mean	.30	.21	.29	.35	.25
SD	.45	.44	.48	.56	.38

NOTE: The higher the mean, the higher the level of stress symptoms.

a. Significant difference (at .05) between County 1 and Counties 2, 3, 5, and 6.

b. Significant difference (at .05) between County 2 and Counties 3, 5, and 6.

c. Significant difference (at .05) between County 3 and Counties 5 and 6.

d. Significant difference (at .05) between Counties 5 and 6.

phobic anxiety, and psychoticism) County 1 employees report symptom means that approach or surpass twice the threshold score on the Derogatis profile chart. On six of the subscales (interpersonal sensitivity, anxiety, hostility, phobic anxiety, paranoid ideation, and psychoticism) County 5 employees report similarly extreme symptom means. Although it is widely recognized that correctional staff members work in a stressful environment, it is surprising that the magnitude of this stress is so apparent, particularly among the County 1 and County 5 employees.

The large standard deviations relative to the means indicate a wide individual-level variation in responses. A close inspection of the several frequency distributions for these subscales (data not shown) confirms that there are a few extreme cases, which cause the variance observed. This is particularly true in Counties 1 and 5, where five or six subjects for each subscale record high values, whereas only one or two subjects in each of the other counties record such high scores.

A different picture appears when we review the percentage of subjects in each jail who exceed the profile score. From these findings it is clear that a substantial proportion of employees—not only a handful of troubled souls—are reporting scores beyond normal levels of stress tolerance. For instance, the summary percentage (across all nine stress indicators) of persons exceeding the profile scores is highest in Counties 6 and 1 (23.9% and 22.6% respectively), followed by Counties 6, 3, and 2 (18.31%, 13.78%, and 9.28%).

Crowding and Personnel Practices

The level of crowding is gauged by the "reported levels" data, which was gathered from the organizational profile form and from "census levels" data provided by the U.S. Department of Justice (1988). Although the Reported Level data for one facility (County 1) are unavailable, it is clear that the remaining jails were either near or well over acceptable levels of crowding. In fact, for all the jails providing information, crowding apparently increased or remained virtually stable (County 2) at very high levels. County 2 and County 3 facilities reported the highest levels of jail crowding in both the Census Levels and the Reported Levels data. One would expect, given these findings, that County 2 and County 3 staff members would experience greater stress than those of other facilities; this is not the case, however.

Some additional information about events and actors at the surveyed jails may provide a richer organizational profile, a fuller picture of its operation character, and a clearer view of practices with respect to employees. Within a year of this study, for instance, Counties 1 and 5 suffered various organizational ills such as disruption among leaders, high turnover, and low pay (at least in County 5).[8] Moreover, County 5 jail employees reported at the time of this research that they had few, if any, opportunities for input regarding jail operations.

At the County 6 facility the amount of formal training provided for all staff members, particularly correctional officers, was relatively low, although the jail manager had established a comprehensive informal training program for all employees that was geared toward building organizational cohesion and teamwork. County 6 also had one of the lowest overall turnover rates (Stohr, Self, and Lovrich 1992).[9]

The County 2 jail has long occupied a "prototype" position nationwide among podular direct-supervision jails. This jail was among the first to experiment at the county level with the combination of podular design and direct-supervision practices. We believe that this "model" status was still justified at the time of our research. The jail was severely overcrowded during the study period: the number of inmates and staff members—75 to 2 respectively—had doubled in most living units, and a number of officers complained about administrative decisions and arrangements. Even so, County 2 correctional officers had the highest pay,[10] the most training, and the second lowest turnover rate of all the jails studied here. These jail officers not only enjoy parity in pay with the county's deputy sheriffs, but also occupy a parallel status in the personnel system. Moreover, a new jail was scheduled to open within months of this research to provide some relief of the overcrowding. The "hands-on" management style and the many resulting opportunities for staff input, the regular weekly pod inspections, and the generally upbeat professional attitudes of most staff members also made it clear that the ingredients for a well-run jail had been combined successfully.

On the basis of our observations, these data, and related research (e.g., Stohr et al. 1992), County 3 may be similar to County 2 in its attempt to reconcile organizational with individual needs and thereby to manage the stress and maintain morale of the jail staff. Although overtime was running quite high for the jail's correctional officers (some officers were making so much from overtime that their salaries exceeded that of the facility's lieutenant), the jail staff members nonetheless maintained a strong attachment to their work and their jail organization. The correctional officers enjoy parity in pay with sheriff's deputies in this county as well.[11] This jail also was impressive for its open leadership style, its informal opportunities for input, its upbeat attitudes as commonly expressed by employees, and the focus of at least one employee (with the approval of upper management) on making employment in the jail fun. These factors would seem to make a considerable difference in the stress level and the sense of organizational commitment among staff members in this facility.

Assessing the Payoff of Model Personnel Practices

It is possible to use a combination of information derived from organizational profiles, archival records, qualitative assessments of site visits, and survey results to construct a "ranking" of the several jails studied here. Observations of all of the jails allow us to judge, on solid grounds, how closely each jail approaches "model" personnel and management practices. Practices such as those promoted in a total quality management (TQM) program (Walters 1992) or in Osborne and Gaebler (1992) *Reinventing Government* constitute the elements of model personnel and management practices. In the context of jail administration, we considered the following criteria for ranking: 1) parity in pay and jail staff status in relation to equivalent positions in law enforcement, 2) high investment in training and staff development, 3) use of participative management in organizational problem solving, and 4) commitment to empowerment of employees (devolution of decisional authority) as a goal of staff development. Contemporary literature on public-sector management practices and public personnel administration suggests that the approximation of such model practices would be associated with positive outcomes such as higher levels of job satisfaction, lower levels of role conflict, lower workplace stress, higher levels of organizational commitment, and lower turnover rates. Contrary to this view, however, DiIulio argues on the basis of his prison research that correctional institutions are managed most effectively by use of a "broken windows" philosophy (Wilson and Kelling 1982), whereby strict control, rule governed behavior, and predictability are the primary goals of inmate and staff management and of correctional staff personnel systems that affect discipline and rewards (DiIulio 1987:95, 175–79, and 236–42). If the latter argument holds for jails as well as for prisons, it is likely that the "model" practices advocated generally for public-sector organizations would be inappropriate for contemporary new generation jails.

What observations on this question can be made from the multijail study reported here? Table 6 sets forth findings based on a comparison of *rankings* along several outcome dimensions that pertain to employees' morale and well-being. The five jails for which management and personnel policy data are sufficient for classification are listed by rank from most proximate to model practices to least proximate. To draw accurate conclusions as to how such practices are related to management outcomes, we compare the rankings for six separate dimensions of well-being on the jail staff against the ranking of facilities with respect to approximation of model personnel and management practices.[12] The dimensions of interest are those we investigated earlier: job satisfaction (JDI), Rizzo et al.'s role conflict measure, the job enrichment level ("motivating potential score") of Hackman and Oldham, the level of psychosomatic stress symptoms (the Derogatis index), the organizational commitment level

as gauged by the O'Reilly and Chatman measure, and turnover. Table 6 displays the Spearman's rho statistic as a measure of the degree of association between approximation of model personnel and management practices and level of well-being among jail staff members.

The findings reported in Table 6 show that concrete and demonstrable benefits accrue to the implementation of model personnel and management practices in these five jails. Although such practices have only slight effects on role conflict and job enrichment, their positive impact is quite apparent in job

Table 6	Comparison of Rankings of Jails on Approximation of Model Personnel and Management Practices and Measures of Morale and Organizational Commitment					
	Indicators of Staff Morale and Commitment: Rank Orders					
	Approximation to Model Personnel and Management Practices: Rank Order (most to least model)					
County	**Job Sat (SD1)[a]**	**Role Conflict[b]**	**Job Motive[c]**	**Work Stress[d]**	**Organ Ident[e]**	**Negative Turnover[f]**
County 2 (1)	1	2	2	1	1	1
County 3 (2)	2	3	4	2	2	3
County 6 (3)	5	5	3	3	5	2
County 5 (4)	3	1	1	5	3	5
County 1 (5)	4	4	5	4	4	4
Spearman's rho	.70	.20	.30	.90	.70	.80
(stat. signif.)	(.09)	(.37)	(.31)	(.02)	(.09)	(.05)

a. Composite job satisfaction, an "overall satisfaction" indicator: Additive index (sum of means) for satisfaction with work task, pay, opportunity for promotion, supervision, and co-workers. The higher the satisfaction, the lower the rank number.

b. Composite role conflict: Additive index (sum of means) for person, resource, policy, and request sources of role conflict. The higher the role conflict, the higher the rank number.

c. Motivating potential score (MPS): Motivation in the job; additive and multiplicative index composed or perceptions of one's job in these dimensions: task identity, autonomy, skill variety, task significance, and feedback. The higher the motivating potential score, the lower the rank number.

d. Composite psychosomatic: Additive index (sum of means) for nine categories of stress symptoms: somatization, obsessive-compulsive, interpersonal sensitivity, depression, anxiety, hostility, phobic anxiety, paranoid ideation, and psychoticism. The higher the stress, the higher the rank number.

e. Composite organizational commitment and attachment: Additive index. The higher the commitment, the lower the rank number.

f. Negative turnover: Voluntary turnover. The higher the turnover, the lower the rank number.

satisfaction, psychosomatic stress symptoms, organizational identification, and turnover. The findings reported in this table suggest that the faith in model personnel and management policies, featuring investment in development of employees and use of participative management practices throughout the public sector, is not misplaced. Such policies and practices seem as appropriate for new generation jails as for other public agencies (Golembiewski 1985). From these findings it appears that the complete application of DiIulio's prison-based "control" management model to jails, at least to the podular direct-supervision facilities increasingly coming into operation across the country (Cronin 1992), would be unwise.

◼ Conclusions

In our opinion, the evidence presented here suggests that continued adherence to the correctional control model (in this case, in podular direct-supervision jails) would be imprudent. We recognize, however, that our alternative, the employee investment model, is not a comprehensive remedy for the personnel ills of correctional facilities. Moreover, we acknowledge that in two jails (County 3 and County 6) it is likely that data collection methods were not completely voluntary; thus it is possible that some of the profile information provided by personnel at those jails could be misleading or could misrepresent the true circumstances. We have no reason to believe, however, that these data were falsely reported; in fact, the other data and our own site visits tend to confirm the profile data. In social science reality, no model's "fit" is perfect or predictable (e.g., see Table 6), nor are experimental conditions achieved. The outcomes we expected, however, such as reduced stress and turnover, greater job satisfaction, motivation in the job, and organizational commitment, were realized with some consistency, in part and in general, for the two jails (County 2 and County 3) most closely approximating our employee investment model.

This multijail comparison represents a partial replication of the study method employed in DiIulio's work on prisons. Although our findings are at odds with DiIulio's in most respects, we share with him a great enthusiasm for multifacility comparison employing a combination of quantitative and qualitative research techniques. We also recognize that the alternative model we have suggested is not a polar opposite of DiIulio's. Rather the employee investment model we propose encompasses some focus on control and security, but also includes a healthy recognition of the value of staff resources and the unique contribution (e.g., in knowledge) that staff can bring to the correctional workplace, if only they are allowed.

In addition to providing managers of podular direct-supervision jails with some useful information on contemporary jail issues, we hope that this study will increase the evidence of the need for such research on a broader range of correctional facilities and programs. We also hope that these findings will encourage those corrections administrators who are seeking to reform their organizations for effective operation in concert with the needs and diversity of their workforce.

◼ Notes

1. Local county jails embraced and refined the concept of podular direct "new generation" jails in the 1980s (Gettinger 1984), and the evolution of these jails (often fashioned to local needs and sensibilities) continues in the 1990s. These jails have "podular" self-contained living units: the inmates sleep, eat, visit, and take recreation inside the pod. Pods are staffed "directly" by correctional officers 24 hours a day—that is, officers are located among inmates. The officers in such jails are called on to interact with inmates face-to-face; no bars or steel doors divide inmates from correctional officers. When operated well, this correctional arrangement can be both challenging and rewarding for the staff (Nelson 1986).

2. The complete version of the Job Diagnostic Survey includes six additional sections that also

measure the enriching characteristics of the job, along with measures of affective reactions to the job and the presence of three associated psychological states. Because of time and space limitations, we used only the two sections designed to assess the presence of enriching job characteristics.

3. The motivating potential score is created by adding the three task-related subscales and dividing them by 3 to create a task-related mean variable. This newly created variable is then multiplied first by the autonomy subscale and then by the feedback subscale; the product represents the motivating potential score. We assessed the reliability of the subscales and the summary "motivating potential score" scale by computing standardized Cronbach's alphas. The alphas for the five subscales ranged from .65 to .74, indicating acceptable levels of reliability.

4. The standardized alphas for these items ranged from .63 to .73.

5. We computed the standardized alpha for each of the subscales and the summary satisfaction measure to test reliability. The alphas for the subscales ranged from .77 to .90; the overall satisfaction measure had an alpha of .71.

6. The BSI is a shorter version of the SCL-90-R. In a study of 565 psychiatric outpatients, Derogatis (1977:33) found that the BSI correlated very highly with the larger form on each of the subscales (correlations ranged from .92 to .99); hence we employed the shorter version in this survey.

7. We computed a standardized Cronbach's alpha for each of the subscales to assess scale reliability for this population of subjects. The alphas for the subscales ranged from .68 for psychoticism to .83 for the obsessive-compulsive scale.

8. The average monthly salaries for correctional officers in Counties 5 and 1 at the time of this research were $1,596 and $1,833 respectively.

9. The average monthly salary for correctional officers in County 6 at the time of this research was $2,047.

10. The average monthly salary for correctional officers in County 2 at the time of this research was $2,963.

11. The average monthly salary for correctional officers in County 3 at the time of this research was $2,228.

12. Model conformance was ranked on the basis of a given jail's tendency to approximate investment in employees (i.e., pay parity, developmental training, opportunities for input).

▧ References

American Jail Association. (1991). *Who's Who in Jail Management.* Hagerstown, MD: American Jail Association.

Bartollas, C. (1990). "The Prison Disorder Personified." In J. W. Murphy, and J.E. Dison (eds.), *Are Prisons Any Better? Twenty Years of Correctional Reform,* pp. 11–22. Newbury Park, CA: Sage.

Bartollas, C. and J. P. Conrad. (1992). *Introduction to Corrections.* New York: Harper Collins.

Chandler Hallford, S. (1992). "Thoughts on Jail Management." *American Jails* 5(6): 11–13.

Clear, T.R. and G.F. Cole. (1990). *American Corrections.* Pacific Grove, CA: Brooks/Cole.

Conroy, R. W. (1989). "Santa Clara County Direct Supervision Jail." *American Jails* 3(3): 59–67.

Cronin, M. (1992). "Gilded Cages: New Designs for Jails and Prisons Are Showing Positive Results. The Question Is, Can We Afford Them?" *Time,* May 25, pp. 52–54.

Crouch, B. M. and J. W. Marquart. (1990). "Resolving the Paradox of Reform Litigation, Prisoner Violence and Perceptions of Risk." *Justice Quarterly* 7: 103–23.

Derogatis, L. R. (1977). *The SCL-90 Administration, Scoring and Procedures Manual—I for the Revised Version and Other Instruments of the Psychopathology Rating Scale Series.* Baltimore: Johns Hopkins University School of Medicine.

DiIulio, J. J. (1987). *Governing Prisons: A Comparative Study of Correctional Management.* New York: Free Press.

Duffee, D. E. (1989). *Corrections: Practice and Policy.* New York: Random House.

Fishman, J. F. (1923). *Crucibles of Crime: The Shocking Story of the American Jail.* Montclair, NJ: Patterson Smith.

Fuqua, J.W. (1991). "New Generation Jails: Old Generation Management." *American Jails* 5(1): 80–83.

Gettinger, S.H. (1984). *New Generation Jails: An Innovative Approach to an Age-Old Problem.* Boulder: National Institute of Corrections.

Goffman, E. (1961). *Asylums: Essays on the Social Situation of Mental Patients and Other Inmates.* Garden City, NY: Anchor.

Goldfarb, R. (1975). *Jails: The Ultimate Ghetto.* Garden City, NY: Anchor.

Golembiewski, R. T. (1985). *Humanizing Public Organizations: Perspectives on Doing Better-Than-Average When Average Ain't at All Bad.* Mt. Airy, MD: Lomond Publications.

Hackman, J. R. and G. R. Oldham. (1974). "The Job Diagnostic Survey: An Instrument for the Diagnosis of Jobs and the Evaluation of Job Redesign Projects." Technical report, Department of Administrative Services, Yale University.

Hawkins, R. and G. P. Alpert. (1989). *American Prison Systems: Punishment and Justice*. Englewood Cliffs, NJ: Prentice-Hall.

Irwin, J. (1985). *The Jail: Managing the Underclass in American Society*. Berkeley: University of California Press.

Jacobs, J. B. (1977). *Stateville: The Penitentiary in Mass Society*. Chicago: The University of Chicago Press.

Manning, D., B. A. Menke, L. L. Zupan, M. K. Stohr-Gillmore, M.W. Stohr-Gillmore, and N.P. Lovrich. (1988). "Performance Appraisal in New Generation Jail Facilities." Final report for the National Institute of Corrections.

Menke, B. A., L. L. Zupan, N. P. Lovrich, and D. Manning. (1986). "Model Selection Process for Podular Direct-Supervision New Generation Jails." Final report for the National Institute of Corrections.

Murton, T. O. (1976). *The Dilemma of Prison Reform*. New York: Praeger.

Nelson, W. R. (1986). "Can Cost Savings Be Achieved by Designing Jails for Direct Supervision Inmate Management?" In J. Farbstein and R. Wener (eds.), *Proceedings of the First Annual Symposium on New Generation Jails*, pp. 13–20. Washington, DC: National Institute of Corrections.

Nelson, W. R., M. OToole, B. Krauth, and C. G. Whitmore. (1984). *Direct Supervision Models*. Washington, DC: National Institute of Corrections.

O'Reilly, C. and J. Chatman. (1986). "Organizational Commitment and Psychological Attachment: The Effects of Compliance, Identification, and Internalization on Prosocial Behavior." *Journal of Applied Psychology* 71(3): 492–99.

Osborne, D. and T. Gaebler. (1992). *Reinventing Government: How the Entrepreneurial Spirit Is Transforming the Public Sector from Schoolhouse to Statehouse, City Hall to the Pentagon*. Reading, MA: Addison-Wesley.

Rizzo, J. R., R .J. House, and S. I. Lirtzman. (1970). "Role Conflict and Ambiguity in Complex Organizations." *Administrative Science Quarterly* 15(2): 150–63.

Rothman, D. J. (1980). *Conscience and Convenience: The Asylum and Its Alternatives in Progressive America*. Glenview, IL: Scott, Foresman.

Sigurdson, H. R., B. Wayson, and G. Funke. (1990). "Empowering Middle Managers of Direct Supervision Jails." *American Jails* 3(4): 52–85.

Smith, P. C., L. M. Kendall, and C. L. Hulin. (1969). *The Measurement of Satisfaction in Work and Retirement*. Chicago: Rand McNally.

Stohr, M. K., R. L. Self, and N. P. Lovrich. (1992). "Staff Turnover in New Generation Jails: An Investigation of Its Causes and Prevention." *Journal of Criminal Justice* 20(5): 455–78.

Sykes, G. M. (1958). *The Society of Captives* Princeton, NJ: Princeton University Press.

U.S. Department of Justice. (1988). *Census of Local Jails*, 1988, Vols. 2, 4, and 5. Washington, DC: Bureau of Justice Statistics.

Walker, S. (1984). "'Broken Windows' and Fractured History: The Use and Misuse of History in Recent Police Patrol Analysis." *Justice Quarterly* 1: 57–90.

Walters, J. (1992). "The Cult of Total Quality." *Governing* (May): 38–42.

Wilson, J. Q. and G. L. Kelling. (1982). "Broken Windows: The Police and Neighborhood Safety." *Atlantic Monthly*, March, pp. 29–38.

Zupan, L. L. (1991). *Jails: Reform and the New Generation Philosophy*. Cincinnati: Anderson.

DISCUSSION QUESTIONS

1. According to the authors, is the control model or the employee investment model a better approach to jail management? Why?

2. Explain how the correctional organization can provide the right environment to reduce stress, turnover, and the harassment of its staff.

3. Explain and discuss the "hack" versus the "human service" role for staff. Which role do you think the public typically ascribes to correctional staff? Which role do you think is most commonly undertaken by staff?

READING

In this article, Camp and his colleagues explore the effectiveness of affirmative action programs in establishing perceptions of a "level playing field" for historically disadvantaged groups. Their research uses both outcome and attitudinal data of correctional officers employed by the Federal Bureau of Prisons, an affirmative action employer. Black and white correctional officers provide evaluations of their own opportunities for job advancement, with black and white officers exhibiting wide disagreement about opportunities available for minorities. The authors examine the processes by which discrepancies between black and white evaluations of minority opportunities arise by testing two competing hypotheses to explain the disagreement between blacks and whites. The first hypothesis, the denial of minority opportunity hypothesis, holds that minorities underestimate minority opportunities relative to their own opportunities. The second hypothesis, the denial of majority opportunity hypothesis, maintains that nonminorities overestimate minority opportunities. The authors conclude that white correctional officers tend to overestimate minority opportunities.

Affirmative Action and the "Level Playing Field"

Comparing Perceptions of Own and Minority Job Advancement Opportunities

Scott D. Camp, Thomas L. Steiger, Kevin N. Wright, William G. Saylor, and Evan Gilman

Social science research helps us to understand the sources of opposition to and support for proactive remedies (e.g., affirmative action). Social science research also illuminates the perceptual differences about and support for affirmative action among women, racial minorities, and White men (Kluegel & Bobo, 1993; Kluegel & Smith, 1986; Taylor, 1995). However, social science research has not been particularly informative regarding people's attitudes and perceptions about particular affirmative action programs and/or their experiences with affirmative action programs. In particular, while there is some research that

examines whether people feel personally discriminated against (Crosby, 1984; Crosby, Pufall, Snyder, O'Connell, & Whalen, 1989), there is little research on how people view their job promotion opportunities in organizations that practice affirmative action (for an exception, see Camp & Steiger, 1995).

It is important to understand and track perceptions regarding the effects of affirmative action and the resulting diversity (Cox, 1994). For example, it is useful to know how minority group members perceive their own opportunities for promotion in comparison to majority group members, especially when controlling

SOURCE: Camp, S., Steiger, T.L., Wright, K.N., Saylor, W.G., & Gilman, E. (1997). Affirmative action and the "level playing field": Comparing perceptions of own and minority job advancement opportunities. *The Prison Journal* 77(3), 313–334. Reprinted with permission of Sage Publications, Inc.

for objective evidence regarding promotion opportunities for Whites and minorities. Likewise, it is helpful to know how majority and minority group members perceive opportunities for minorities (considered as a group) in organizations, especially in relation to the respondents' evaluations of their own, personal opportunities. This is important because most existing research on affirmative action asks respondents to compare group opportunities— Black versus White, for example. But do people make the same judgments about opportunities when they are asked about groups as opposed to personal experiences? As we argue later, there are strong reasons to believe this is not the case, and existing research has been remiss in not addressing this issue.

From a practical viewpoint, there are legal, training, and policy consequences that result from employee perceptions of promotion opportunities. For example, if an organization has a positive, objective track record regarding the promotion of minority members and women, yet minority members perceive there to be a problem, it is not unreasonable to expect problems to arise for the organization from these perceptions. Minority members would probably be expected to file job grievances or even bring legal action because of their perceptions. At the very least, there would probably be suspicions in the workplace about job promotion opportunities, although the perceptions are incongruent with the actual facts in this hypothetical case. Such suspicions could lead to lower morale, lower levels of worker cooperation, and a host of other workplace problems. On the flip side, if research shows that majority members have an exaggerated view of the opportunities available to minority members, then we would also expect organizational problems such as those mentioned above. However, in this case the causes and appropriate remedies would obviously differ.

In short, we need to understand not only whether affirmative action creates more equitable outcomes in employment and promotion opportunities, the area in which most research has been conducted to date, we also need to understand how the consequences of affirmative action policies are perceived by employees. Where racial and gender differences in perceptions continue to exist in organizations committed to affirmative action, it is important to understand the sources of those differences if effective policy and training are to be designed and implemented by the organization to address possible misperceptions about promotion opportunities. At the very least, we think it is reductionist to assume that perceptions follow in a simple manner from experiences in organizations.

⧅ Literature Review

Affirmative action differs from previous legislative attempts to end gender and racial employment discrimination. Title VII of the Civil Rights Act, passed by Congress in 1964, established the explicit prohibition of racial, sexual, ethnic, or religious discrimination. Shortly after the passage of the Civil Rights Act, President Lyndon Johnson issued Executive Order 11246, establishing what is now known as affirmative action. Affirmative action and "equal opportunity," as the Civil Rights Act came to be known, differ in at least five ways (Crosby, 1994, pp. 18–21). For our purposes, it is sufficient to note that affirmative action is much more proactive in rectifying discrimination, whereas equal opportunity is more reactive and serves to redress documented instances of discrimination.

Support among the public for affirmative action varies. Broad cross-sectional studies of the United States population, such as the studies done by Kluegel and his colleagues (Kluegel & Bobo, 1993; Kluegel & Smith, 1986), suggest that women and racial minorities are more supportive of affirmative action than White males. Kluegel and Bobo (1993) found that proactive measures such as race targeting diminished White support for opportunity-enhancing policies by about 22% on average.

Using data similar to that of Kluegel and Bobo (1993), Taylor (1995) examined the contextual effects of working in organizations with and without affirmative action programs. Taylor found no evidence to support the contention that experience with race targeting creates White resentment or polarization. Indeed, Taylor (1995) suggested that experience with affirmative action for White men may generate support for the policy as a whole. Of course, Taylor's study does not contradict directly the work of Kluegel (Kluegel & Bobo, 1993; Kluegel & Smith, 1986), since she compares only White workers whose employers do and do not practice affirmative action. Taylor's research does show that experience with affirmative action does not necessarily make White males any more negative about affirmative action, and in some cases it may make them more supportive.

For our purposes, the value of the research by Kluegel is limited because the studies do not control for the respondents' experiences with affirmative action. Taylor, on the other hand, does control for respondents' experiences with affirmative action, but Taylor does not provide comparisons between different racial groups. More limiting is that none of the studies examines how respondents' perceptions of opportunities for job advancement are affected by affirmative action, an outcome we feel has been surprisingly overlooked. When outcomes are examined at all, the outcomes in question are generally objective in nature, as in looking at racial or gender representativeness within the workforce or promotions awarded. Even with objective outcomes, the number of studies is relatively small, and the results are generally mixed (see Blum, 1990; Hanna, 1988; Thomas, 1991; for studies on the effects of affirmative action in the federal government, see Benokraitis & Gilbert, 1989; DiPrete & Soule, 1986; Kellough, 1989).

The only studies that deal with perceptions of opportunities among respondents are those by Kluegel and Smith (1986) and Camp and Steiger (1995). As Kluegel and Smith argued, it

is better to ask people about their opportunities than to ask them about discrimination. Discrimination is a "hot button" issue in American society; one that may evoke excessive ideological imagery among respondents that is borrowed from the media and other social institutions. Perceptions of opportunities, on the other hand, are probably more rooted in personal experience (Kluegel & Smith, 1986, p. 56). We agree. The job advancement opportunities items analyzed here were designed for a similar purpose, that is, to assess differences in staff perceptions about opportunities with as little ideological contamination as possible (Saylor, personal communication, 1996).

Although Americans in general perceive personal opportunity, there are important group differences. According to Kluegel and Smith (1986), "Blacks have a less favorable assessment of opportunity in general and of their personal opportunity than do whites, and women's beliefs about these aspects of opportunity differ from men in the same direction" (p. 64). Further analysis of these differences shows that the greatest differences lie between people at different levels of socioeconomic status (p. 68); that Blacks perceive their opportunity is much more limited by discrimination than do Whites (p. 200); and that Black-White differences are greater on issues related to Black opportunity, whereas male-female differences are less on issues related to women's opportunity (p. 240).

Camp and Steiger (1995) also found group differences in perceptions of opportunity in a case study of workers of one organization, but their results are somewhat different than those of Kluegel and Smith (1986). Much of the difference between the findings of Camp and Steiger and Kluegel and Smith is probably due to the implicit control for exposure to affirmative action programs in the Camp and Steiger study. The respondents studied by Camp and Steiger were all employed by the same affirmative action employer, the Federal Bureau of Prisons (BOP). We do not have any information about respondents' experiences with affirmative action in the Kluegel and Smith study. Regardless, unlike Kluegel and Smith, Camp and Steiger found no difference between males and females in their assessments of personal job advancement opportunities either within the specific prison of employment or within the BOP overall. Similar to Kluegel and Smith, men rated opportunities for minorities more favorably than did women. But, again, unlike Kluegel and Smith, Camp and Steiger found that Blacks, more strongly than Whites, agreed that they personally have job advancement opportunities. As did Kluegel and Smith, Camp and Steiger found that Whites, more strongly than Blacks, agreed that minorities have job advancement opportunities.

The results presented by Camp and Steiger (1995) are intriguing. When asked about their own opportunities, there is very little difference between the average ratings provided by Blacks and Whites, and no difference between men and women (see Appendix B). These findings alone seem to imply that, at the BOP anyway, there is recognition among minority and majority members of the fairly equal opportunities for job advancement available to minority and majority group members.[1] However, when asked to assess opportunities available to minorities as a group, there is a large discrepancy between Blacks and Whites and a similar gap between men and women. Taken together, these findings imply that majority and minority group members think about their own opportunities for promotion in a similar fashion, but they do not extend that same reasoning when thinking about the opportunities available to minorities as a group at the BOP.

Camp and Steiger (1995) did not explore the nature of the difference between Black and White officers in their responses to questions about minority job advancement opportunities. We believe that such an examination is crucial. In particular, we think it is necessary to examine the empirical nature of the discrepancies between majority and minority evaluations of own and minority opportunities. Why do the perceptions of Blacks and Whites and men and women regarding their own opportunities for advancement accurately reflect the objective data on promotions presented below, while the perceptions regarding minority opportunity demonstrate wide disagreement between majority and minority group members? We analyze the same data used in the study by Camp and Steiger to address these issues.

⬣ Hypotheses and Method

One possible explanation for the discrepancy in attitudes comes from the work of Crosby on the denial of personal disadvantage (Crosby, 1984; Crosby et al., 1989). In her research, Crosby found that racial minorities and women are willing to indicate on surveys that the groups they are part of face discrimination and suffer from disadvantage, but the same respondents are unwilling to acknowledge that the disadvantage extends to them personally. The reasons given for this phenomenon are largely psychological. Although the data analyzed here and by Camp and Steiger (1995) pertain to perceptions of opportunity rather than discrimination and disadvantage, we can still use the logic provided by Crosby to derive a denial of minority opportunity hypothesis.[2]

> *Hypothesis 1* The lower evaluations of minority opportunities for job advancement provided by women and minorities are due to women and minorities not recognizing the opportunities available to minorities. Thus, women and minorities provide evaluations of minority opportunity that are lower than their evaluations of their own opportunities.

Evidence against Hypothesis 1 would come in the form of women and minority members providing comparable evaluations of their own and minority opportunities.

Another possibility for the reversal in attitudes about job advancement opportunities is provided by the research on national opinions toward affirmative action. National surveys find that women and minorities more strongly favor affirmative action programs. It seems likely that items inquiring into minority job advancement opportunities tap into these opinions about affirmative action, especially among Whites and males who typically hold more negative views. As such, rather than minorities failing to recognize minority opportunities, the discrepancy may be due to historically favored groups (Whites and males) overestimating the opportunities for job advancement available to minorities, and, thus, underestimating the opportunities available to majority members. This is the denial of majority opportunity hypothesis on the part of formerly privileged groups.

> *Hypothesis 2* The lower evaluations of minority opportunities for job advancement provided by women and minorities are due to overestimates of minority opportunity provided by White and male correctional officers. That is, women and minority correctional officers recognize that minorities as a group share job opportunities comparable to their own personal opportunities, but White and male correctional officers exaggerate the small advantages going to minorities.

Counterevidence to Hypothesis 2 comes primarily in the form of women and minority members providing lower evaluations of minority opportunities than their own opportunities. We already know from the work of Camp and Steiger (1995) that minorities assess their own opportunities in a slightly more favorable fashion than White officers. As will be shown below, there is slight empirical justification for these assessments. As such, it is expected that White officers should provide slightly higher evaluations of minority opportunities than their own opportunities. However, these evaluations should be comparable to the

evaluations of minority opportunities provided by minority members, and we know from Camp and Steiger that the evaluations provided by majority members are much higher. If minorities and women provide comparable evaluations of their own and minority opportunities, then the large discrepancy must arise from majority males providing exaggerated evaluations of minority opportunities, and Hypothesis 2 is supported.

It is also possible that both processes outlined in the first two hypotheses are at work. As such, we are left with a mixed hypothesis.

> *Hypothesis 3* The lower evaluations of minority opportunities for job advancement provided by women and minorities are due to a combination of the denial of minority opportunities by women and minority correctional officers and the denial of majority opportunities by male and White correctional officers.

The data used in this study come from two general sources. First, the major source of data analyzed, that on perceptions of job opportunities, is taken from the 1994 administration of the Prison Social Climate Survey (PSCS). The PSCS has been given annually since 1988 to a stratified proportional probability sample of BOP staff working at field locations. In 1994, the PSCS was administered to 6,903 staff working in 74 distinct BOP prisons; 6,004 staff responded for a response rate of 86.98%. In this analysis, we select only the 1,242 respondents who self-identified as nonsupervisory correctional officers and who completed the items in the survey about job opportunities. Correctional officers constitute a large group of individuals with similar job responsibilities.

The data on promotions, which make up Table 1 and provide the objective context for job advancement opportunities at the BOP, are taken from operational databases used to generate statistical information reported in the

Executive Staff Module (Muth, 1995) of the BOP's Key Indicators/Strategic Support System (KI/SSS). KI/SSS is an interactive management information system that is pressed onto CD-ROM monthly and distributed to BOP managers (Gilman, 1991; Saylor, 1988).

In this analysis, we use the same measures analyzed by Camp and Steiger (1995) as dependent variables, but we use them in a different manner. Whereas Camp and Steiger simply look at the four job opportunity measures as dependent measures in isolation from one another, we use the measures in a comparative fashion. We compute our dependent measures by comparing the Likert-type category that respondents choose to assess their own opportunities for job advancement to the Likert-type category that they select to rate minority opportunities for job advancement.[3] This gives three possibilities: respondents can rate minority opportunity as higher than their own, lower than their own, or the same as their own.

We compute dependent measures for the items about job advancement opportunities at the specific institution where respondents are employed and for job advancement opportunities within the BOP, respectively. We do not expect there to be major differences in the performance of the two measures. Within the BOP, promotion opportunities have been linked historically with transfers to other locations. Our concern here is not with whether respondents perceive more opportunity at one level of the organization than the other, although this historical pattern does explain why the survey instrument contains questions about opportunities at both the institution and overall organization level. The institutional comparison score is derived by comparing the values provided for the following two items: (a) institution has opportunity for me measure (INOPPME) and (b) the institution has opportunity for minorities (INOPPMIN).[4] Likewise, the difference in perceptions of opportunity within the overall organization is

Table 1	Organizational Data on Promotions for 1992, 1993, 1994, by Sex and Race		
	1992	**1993**	**1994**
Total active staff (count)	23,145	24,244	25,505
Total promotions (count)	7,099	4,767	4,696
% staff receiving a promotion	31%	20%	16%
Females			
% of total promotions	31%	31%	28%
% of total staff	27%	27%	26%
Equity ratio[a]	1.16	1.17	1.06
% females promoted	35%	23%	17%
Males			
% of total promotions	69%	69%	72%
% of total staff	73%	73%	74%
Equity ratio[a]	0.94	0.94	0.98
% males promoted	29%	18%	16%
Blacks			
% of total promotions	19%	16%	21%
% of total staff	18%	18%	18%
Equity ratio[a]	1.04	0.88	1.13
% Blacks promoted	32%	17%	18%
Whites			
% of total promotions	68%	71%	64%
% of total staff	71%	71%	70%
Equity ratio[a]	0.96	1.01	0.92
% Whites promoted	29%	20%	15%
Other race			
% of total promotions	13%	13%	15%
% of total staff	11%	11%	12%
Equity ratio[a]	1.21	1.14	1.25
% other promoted	37%	22%	20%

a. The equity ratio is computed as the ratio of the percentage of total promotions going to staff in a given category to the percentage of total staff in the respective category.

obtained by comparing the following items: (a) the BOP has opportunity for me (BOPOPPME) and (b) the BOP has opportunity for minorities (BOPOPMIN).

We perform multinomial logistic regression analysis on the difference measures described above to assess Hypotheses 1 through 3. These models allow us to assess in a multivariate sense whether respondents rate minority opportunities as being lower than their own opportunities, comparable to their own opportunities, or higher than their own opportunities.

In addition to entering race and sex into models of the difference measures to evaluate Hypotheses 1 through 3, we control for several other individual-level effects that we feel are related to evaluations of job advancement opportunity. In particular, we control for education, Hispanic ethnicity, age, tenure at the BOP, and whether the respondent has ever accepted a transfer from one BOP facility to another. We are not confident enough about the effects of these variables to generate formal subhypotheses, nor are we substantively interested in the effects in this analysis, but we do generally expect them to behave in specific ways. We expect higher levels of education to mediate against misperceptions of relative job advancement opportunities. We expect Hispanic ethnicity to generally behave in a manner similar to race, given the same general barriers faced by Hispanics and members of racial minority groups. We expect higher values for age and tenure to lead respondents to exaggerate the differences in opportunity between minorities and nonminorities. And, we expect the greater experience with the BOP obtained by accepting a transfer from one BOP facility to another to narrow the gap between perceptions of own and minority job advancement opportunities.

Initially, we included aggregate and organizational measures in the models. In preliminary ordinary least squares results (not presented here), we did not find any effects for the aggregate and organizational measures we considered, including factors such as the security level of the institution. In fact, when modeling the actual magnitude of the difference between own and minority opportunity, we found little variation in institution means for the difference measures when the effects of the individual-level variables were controlled. Therefore, we concluded that these contextual effects have little impact on the determination of perceptions, and we have thus focused on individual-level variables.

⊠ Results

Table 1 shows the promotion rates for all staff, females, males, and the different racial groups within the BOP for the years 1992 through 1994. It is important to note that these values are for BOP staff in all types of jobs, and not just for correctional officers alone.[4] Table 1 also reports a simple equity measure for females and males and the different racial groups. The equity measure is defined as the ratio of the percentage of promotions going to the group in question divided by the respective percentage of BOP staff in the group in question.[5] For example, females received 28% of the promotions in 1994 and constituted, on average, 26% of BOP staff. The ratio of these two percentages, 1.06, is the equity measure for females in 1994, demonstrating that females received more of the promotions in 1994 than would be expected simply from their representation among BOP staff.

As can be seen in the results for the equity measures reported in Table 1, females receive a disproportionate share of promotions in 1992, 1993, and 1994. The equity scores for females in 1992 and 1993 are higher than the equity score in 1994, suggesting a slight drop-off in 1994. Males, on the other hand, share the converse of the female situation. That is, males receive a smaller percentage of the promotions as a group in 1992, 1993, and 1994 relative to their numbers.

Table 1 also presents findings for promotions aggregated by race. The results show that Blacks have a slightly higher equity rate for 1992 and a clearly higher equity rate for 1994. In 1993, Blacks received promotions at a rate lower than what would be expected from their representation among BOP staff. Other staff, on the other hand, had clearly higher equity scores for all of the years. The White equity scores are lower than 1 in 1992 and 1994, and the equity score is effectively 1 in 1993.

Given these results, it seems reasonable to assume that the BOP has been aggressively pursuing affirmative action during the years leading up to and including the administration of the 1994 PSCS.

Table 2 presents the bivariate results between sex and minority opportunity within the BOP, and Table 3 provides results for the relationship between sex and minority opportunity within the respondents' institutions. As can be seen, males are more likely than females to rate minority opportunities higher than their own opportunities, both at the levels of the BOP and the institution, although the difference is not statistically significant for the institution. In evaluations of opportunities within the BOP, more than 43% of males rate minority opportunities within the BOP as being higher, whereas only 28.4% of females do. Females are more likely than males to rate minority opportunities as being the same as their own.

Table 2	Bivariate Relationship Between Sex and Perceptions of Minority Opportunities Within the Overall Organization			
	Minority Opportunities Lower Than Own	**Minority Opportunities Same as Own**	**Minority Opportunities Higher Than Own**	**Total**
Males	11.9%	44.8%	43.3%	1,037
Females	13.9%	57.7%	28.4%	201
Total	151	581	506	1,238

$\chi^2 = 15.73$

Probability $> \chi^2 = .00038$

Table 3	Bivariate Relationship Between Sex and Perceptions of Minority Opportunities at the Specific Institution of Employment			
	Minority Opportunities Lower Than Own	**Minority Opportunities Same as Own**	**Minority Opportunities Higher Than Own**	**Total**
Males	12.9%	27.7%	59.4%	1,035
Females	18.1%	30.7%	51.3%	199
Total	169	348	717	1,234

$\chi^2 = 5.77$

Probability $> \chi^2 = .05593$

Tables 4 and 5 provide the bivariate results between race and the difference variables at the BOP and institution levels. As can be seen, the relationship between race and differences in opportunity is stronger than the relationship between sex and differences in opportunity. White correctional officers are much more likely than Black correctional officers (or those of other races) to evaluate the opportunities of minorities higher than their own opportunities. This is especially true for the difference in opportunities within the BOP.

The bivariate results presented in Tables 4 and 5, and to a lesser extent Tables 2 and 3, generally support Hypothesis 2. That is, the difference between minority and nonminority, and male and female, evaluations of minority opportunities seems to be caused more by males and especially Whites overestimating the opportunities minorities have for job advancement. There is less of a tendency for minority members and women to report minority opportunities as being less than their own, although indeed there are some instances of reporting in this direction, especially for evaluations by Black correctional officers (and those of other races) of opportunities within the institution (see Table 5). The question now is, Does this partial support for Hypothesis 2 hold up in multivariate analyses?

Table 4	Bivariate Relationship Between Race and Perceptions of Minority Opportunities Within the Overall Organization			
	Minority Opportunities Lower Than Own	**Minority Opportunities Same as Own**	**Minority Opportunities Higher Than Own**	**Total**
White	7.5%	39.9%	52.6%	859
Black	26.1%	60.7%	13.2%	234
Other	17.4%	65.1%	17.4%	149
Total	151	582	509	1,242

$\chi^2 = 177.69$
Probability $> \chi^2 = .00001$

Table 5	Bivariate Relationship Between Race and Perceptions of Minority Opportunities at the Specific Institution of Employment			
	Minority Opportunities Lower Than Own	**Minority Opportunities Same as Own**	**Minority Opportunities Higher Than Own**	**Total**
White	5.7%	22.8%	71.5%	859
Black	35.1%	39.0%	26.0%	231
Other	26.5%	41.5%	32.0%	147
Total	169	347	721	1,237

$\chi^2 = 247.24$
Probability $> \chi^2 = .00001$

Table 6 presents the results for the baseline logistic model predicting differences in evaluations of minority and own opportunities in the BOP. The R^2 and adjusted R^2 measures suggest a reasonable fit to the data analyzed with this model, with values of .202 and .235, respectively. The effects of the educational comparisons, age, and sex are not statistically significant at the conventional cutoff of $\alpha = .05$.

The effects for race presented in Table 6 are generally consistent with Hypothesis 2. That is, for a typical White, non-Hispanic, female correctional officer, the highest probabilities are that she indicates that minority officers have greater opportunities for promotion than she has ($p_3 = .460$) or the opportunities are the same ($p_2 = .471$) (see Table 7).[6] The

typical Black, non-Hispanic, female officer, on the other hand, is much less likely ($p_3 = .107$) to see the opportunities of minorities as being greater than her own opportunities. In fact, there is somewhat of a tendency to see minority opportunities as lower than one's own for Black, female officers ($p_2 = .345$), although the predominant response is to see the opportunities in the same light ($p_1 = .548$). Non-Hispanic, female officers of other races respond in between the patterns noted for Black and White female officers, although the model predicts that the overwhelming majority see their opportunities as being the same as minority opportunities ($p_2 = .573$).

Table 8 presents the results for the baseline logistic model predicting differences in

Table 6	Logistic Regression Model of Probabilities That Staff See Minority Opportunities as Being Higher, the Same, or Lower Than Their Own Opportunities Within the Overall Organization			
Variable	**Estimate β_i**	**Odds Ratio**	**Wald Chi-Square**	**Probability > Chi-Square**
Intercep1	−0.1621		0.71	.3989
Intercep2	2.6074		147.99	.0001
Some college	0.1643	1.179	1.38	.2396
College degree	0.1662	1.181	0.82	.3646
Graduate degree	0.4549	1.576	2.94	.0865
Black	−1.9649	0.140	132.73	.0001
Other	−0.6940	0.500	7.13	.0076
Hispanic	−1.0597	0.347	17.53	.0001
Male	0.3088	1.362	3.43	.0640
Age (log)	−0.7505	0.472	3.13	.0770
Tenure (log)	0.6062	1.833	51.31	.0001
Transfer (yes)	−0.4081	0.665	5.84	.0157

$R^2 = .2016$

Adjusted $R^2 = .2346$

Table 7	Probabilities Associated With Comparisons of Minority and Own Opportunities Within the BOP for Typical Female Correctional Officers			
Probability	**Description**	**White**	**Black**	**Other**
p_1	Minority lower	.069	.345	.129
p_2	Minority same	.471	.548	.573
p_3	Minority higher	.460	.107	.298

opportunities at the institution. The R^2 and adjusted R^2 measures suggest a reasonable fit to the data analyzed with this model, with values of .237 and .279, respectively. The effects of two of the educational comparisons (officers with some college education and officers with a bachelor's degree compared to officers with a high school education), transfer status, and sex are not statistically significant. The finding for sex is especially surprising given our theoretical expectations and previous analysis.

The effects for race presented in Table 8 are also generally consistent with Hypothesis 2. The typical White, female, non-Hispanic correctional officer clearly sees the opportunities for minorities at the institution as being higher than her own opportunities ($p_3 = .678$). The typical Black, female, non-Hispanic officer, on the other hand, sees minority opportunities as being lower ($p_1 = .430$) or the same ($p_2 = .394$). The typical female, non-Hispanic officer of another race, somewhat surprisingly, sees minority opportunities as being either higher ($p_3 = .420$) or the same ($p_2 = .398$) (see Table 9).

Table 8	Logistic Regression Model of Probabilities That Staff See Minority Opportunities as Being Higher, the Same, or Lower Than Their Own Opportunities at Their Own Institution of Employment			
Variable	**Estimate β_i**	**Odds Ratio**	**Wald Chi-Square**	**Probability > Chi-Square**
Intercep1	0.7458		13.17	.0003
Intercep2	2.5722		133.44	.0001
Some college	0.2581	1.295	2.96	.0852
College degree	0.2898	1.336	2.13	.1440
Graduate degree	0.7553	2.128	6.79	.0092
Black	−2.2891	0.101	185.65	.0001
Other	−1.0694	0.343	16.87	.0001
Hispanic	−1.0042	0.366	15.85	.0001
Male	0.2020	1.224	1.31	.2527
Age (log)	−1.3571	0.253	8.90	.0027
Tenure (log)	0.5889	1.802	45.75	.0001
Transfer (yes)	−0.2198	0.803	1.48	.2235

$R^2 = .2365$

Adjusted $R^2 = .2791$

Table 9	Probabilities Associated With Comparisons of Minority and Own Opportunities at the Institution for Typical Female Correctional Officers			
Probability	**Description**	**White**	**Black**	**Other**
p_1	Minority lower	.071	.430	.182
p_2	Minority same	.251	.394	.398
p_3	Minority higher	.678	.176	.420

⬛ Discussion

The results of the multinomial logistic regression analysis are most consistent with Hypothesis 2. Hypothesis 2 states that the differences between minorities and nonminorities, and women and men, in perceptions of own and minority opportunity are caused by nonminorities and males exaggerating the opportunities available to minorities. Perhaps Whites accurately perceive the marginal underrepresentation of Whites in the promotion pool. If so, Whites should rate minority opportunities as higher than their own, but those evaluations should be generally consistent with the evaluations given by minorities. We know from Camp and Steiger (1995, p. 271) that there is great disagreement between Black and White officers on perceptions of minority opportunity. On a 7-point Likert-type scale from *strongly disagree* to *strongly agree*, Camp and Steiger reported that Whites are 1.65 units higher on average than Blacks for assessments of opportunities for minorities within the institution and .93 units higher on assessments of opportunities for minorities within the overall BOP when multivariate controls similar to those used here are introduced. Plus, we know from the logistic regression analysis reported here that minorities, for the most part, rate minority opportunities as being consistent with their own (generally equal) opportunity. Therefore, it is not an underestimation

of minority opportunities by Black officers that generally causes the disagreement, it is an overestimation of minority opportunities by White officers.

Support for Hypothesis 2 regarding the group findings for race is not unequivocal. In particular, there is some evidence for the denial of minority opportunity hypothesis among Black correctional officers, especially when they are evaluating opportunities for job advancement within their current institutions of employment. Even so, the most typical pattern is for minorities to see the group minority opportunities as being comparable to or higher than their own opportunities, which is consistent with Hypothesis 2. For Whites, the general pattern is to see minority opportunities as being higher than personal opportunities. As such, we tentatively conclude that Hypothesis 2 is better supported in this analysis than Hypothesis 1 or Hypothesis 3.

Because we do not find statistically significant sex effects for our difference measures in the multivariate models, we conclude that Hypotheses 1 through 3 are irrelevant for sex, although only at the Federal Bureau of Prisons. The statistically significant relationships between sex and the difference measures reported in Tables 2 and 3 disappear when controls for other characteristics of the correctional officers are added.

Although not the specific focus of this analysis, the results for the other individual-level variables are interesting. Education does not play a role in mediating perceptions of job advancement opportunity. In fact, given the educational comparisons we make, we find only one educational comparison to be statistically significant. In the model of differences in opportunity within the respondent's own institution, respondents with an advanced college degree are more likely to see minority opportunities as being more favorable than their own. This finding could very well represent the frustrations of correctional officers with advanced degrees as reported by Jurik, Halemba, Musheno, and Boyle (1987).

Regarding the comparisons between Hispanic and non-Hispanic correctional officers, we did indeed find that Hispanic correctional officers provide more equitable evaluations of the differences between minority opportunities and their own opportunities. This finding is much as we expected.

Age and tenure did not affect the differences in minority and own evaluations exactly as we expected. Age obtains a statistically significant effect only in the model of differences in opportunity within the respondent's own institution. Even here, the effect is not what we expected. Although we expected older workers to provide less favorable evaluations of the differences between minority opportunities and their own opportunities, we found the opposite. Increasing age works to lower the difference. Tenure did behave as we expected. Officers with more tenure are more likely to rate minority opportunities as higher than own.

Finally, whether a correctional officer had ever transferred from one BOP facility to another was statistically important only in the model of differences in opportunity within the overall organization. Officers who had transferred within the BOP were less likely to evaluate minority opportunities as being higher than their own, as expected. Again, we suspect this finding is due to the broader understanding of the operations of the BOP as officers gain experience at more than one BOP facility.

✎ Conclusions

What are the implications of this study for theory and practice? For theory, we see from the research of Camp and Steiger (1995) that there is a general correspondence between the perceptions of personal job advancement opportunity and objective conditions, at least for racial groups. Black workers at the BOP may enjoy a slight advantage in the aggregate promotion rate, and, as Camp and Steiger noted, this is reflected in the perceptions of Black and White workers. It is probably more accurate to say that the playing field has been leveled for all races rather than to say it favors any particular race (given the slight differences between racial groups). Regardless, the attitudinal results suggest that workers are fairly objective in evaluating their own opportunities for job advancement and are not influenced, on average, by the racial group to which they belong. In the case study of correctional officers at the BOP, it appears that perceptions of own opportunities that Black and White correctional officers provide reflect the slight advantages that minorities enjoy.

The same cannot be said for evaluations of minority opportunities. These results suggest that Black and White correctional officers continue to view their places of employment quite differently. From the previous research of Camp and Steiger (1995), we were motivated to investigate why there appears to be such a large amount of disagreement between Black and White workers about minority opportunities for job advancement. Although the results do not totally rule out the notion that Black workers deny minority opportunities for job advancement, the results presented here more strongly suggest that the differences between Black and White evaluations of minority opportunities are due to a failure of White correctional officers to recognize the opportunities that Whites have. White officers appear to exaggerate the opportunities for minority advancement relative to their own opportunities.

What cannot be determined from this study is why White officers exaggerate the opportunities available to minorities. It could be that the exaggeration is introduced by general feelings about affirmative action, in line with the trends reported by Kluegel and Smith (1986), or it could be that the feelings arise from the officers' perceptions of their experiences with affirmative action at the BOP. If forced to choose, we would guess that questions asking specifically about minority opportunities are "loaded" in the sense that they invoke responses based on more than direct experience with affirmative action, but we have no means of substantiating this point at present. Rather, we base the judgment on the proper correspondence between objective conditions and perceptions of advancement opportunity when the questions are asked about personal opportunities, which probably more directly tap into respondents' experiences with affirmative action at the BOP.

From a policy or practice viewpoint, the results presented here suggest that efforts to address the discrepancies between Black and White perceptions of minority opportunities need to be addressed mostly to White workers, at least in the case of the BOP. What should be involved in these efforts is a much more difficult proposition to tackle. The difficulty arises from not knowing whether the exaggerated views of Whites are imported into the organization from outside influences or are generated from experiences with affirmative action policy at the BOP. Regardless, these results suggest that the discrepancy between Black and White opportunities for job advancement is predicated on misperceptions by White officers. We feel this is an important starting point for future research.

Clearly, more research is needed to address the theoretical and policy issues raised here. In particular, research is badly needed on the responses of formerly privileged groups, namely White males, as employment opportunities approach greater equity. Of course, the diminishing support for and increasing hostility toward affirmative action may undermine proactive efforts at establishing level playing fields for all. Additionally, we need more information about which perceptions are more instrumental in affecting organizational outcomes, perceptions of respondents' own opportunities or perceptions of minority opportunities.

In addition, further research is needed to ensure that promotion opportunities are available to minorities at the BOP when controls are introduced for factors that influence promotion rates. In particular, it is necessary to examine the movement of minorities and women into supervisory and managerial positions in the BOP. Despite the limitations noted, we believe this research begins to address, in a meaningful manner, the experiences of workers with affirmative action. We need to see more of this type of research to ensure adequate theoretical understanding of the operation of affirmative action in the workplace and to ensure the practical success of affirmative action policy.

⬙ Notes

1. We demonstrate in this analysis that, objectively, majority and minority group members and men and women have fairly comparable promotion opportunities at the Federal Bureau of Prisons (BOP), at least as measured by aggregate promotion data for 1992, 1993, and 1994.

2. Keep in mind that the hypotheses are derived for an agency in which there is an affirmative action program. As shown below, minorities and women tend to be slightly overrepresented at the aggregate level in terms of receiving promotions.

3. We chose this method over computing a simple difference score between the respective measures because it simplifies the assumptions we have to make about the metric of the scales for the different measures. All we assume with this approach is that the respondents answer the respective questions while taking into account their previous responses. Because the items are located together in the survey instrument, this is not an unreasonable assumption.

4. The promotion rate is defined as the total number of promotions that occur over the course of the year for each aggregate category in question divided by the total number of promotions for the year. Because federal regulations generally prohibit an individual from receiving more than one promotion in a 12-month period, these data are not biased by "star" performers who receive multiple promotions during one year.

5. The percentage of staff in the category in question is computed by taking an average of monthly data on the number of individuals in the aggregate category of question and dividing it by an average of monthly data on the total number of individuals in active staff status.

6. *Typical* hereafter refers to a correctional officer with a high school education, mean values on age and tenure, and no location transfer within the BOP.

⬛ References

Benokraitis, N. V., & Gilbert, M. K. (1989). Women in federal government employment. In F. A. Blanchard & F. J. Crosby (Eds.), *Affirmative action in perspective* (pp. 65–80). New York: Springer-Verlag.

Blum, D. E. (1990). Ten years later, questions about Minnesota sex-bias settlement. *Chronicle of Higher Education, 36*(39), A13–A15.

Camp, S. D., & Steiger, T. L. (1995). Gender and racial differences in perceptions of career opportunities and the work environment in a traditionally male occupation: Correctional workers in the Federal Bureau of Prisons. In N. A. Jackson (Ed.), *Contemporary issues in criminal justice: Shaping tomorrow's system* (pp. 258–277). New York: McGraw-Hill.

Cox, T., Jr. (1994). *Cultural diversity in organizations: Theory, research & practice.* San Francisco: Berrett-Koehler.

Crosby, F. (1984). The denial of personal discrimination. *American Behavioral Scientist, 27*(3), 371–386.

Crosby, F. (1994). Understanding affirmative action. *Basic and Applied Psychology, 15*(1-2), 13–41.

Crosby, F., Pufall, A., Snyder, R. C., O' Connell, M., & Whalen, P. (1989). The denial of personal disadvantage among you, me, and all the other ostriches. In M. Crawford & M. Gentry (Eds.), *Gender and thought: Psychological perspectives* (pp. 79–99). New York: Springer-Verlag.

DiPrete, T., & Soule, W. T. (1986). The organization of career lines: Equal employment opportunity and status attainment in a federal bureaucracy. *American Sociological Review, 51*(3), 295–309.

Gilman, E. (1991). Implementing key indicators. *Federal Prisons Journal, 2*(3), 48–56.

Hanna, C. (1988). The organizational context for affirmative action for women faculty. *Journal of Higher Education, 59*, 390–411.

Jurik, N. C, Halemba, G. J., Musheno, M. C., & Boyle, B. V. (1987). Educational attainment, job satisfaction, and the professionalization of correctional officers. *Work and Occupations, 14*(1), 106–125.

Kellough, J. E. (1989). *Federal equal employment opportunity policy and numerical goals and timetables: An impact assessment.* New York: Praeger.

Kluegel, J. R., & Bobo, L. (1993). Opposition to race-targeting: Self-interest, stratification ideology, or racial attitudes? *American Sociological Review, 58*(4), 443–464.

Kluegel, J. R., & Smith, E. R. (1986). *Beliefs about inequality: Americans' view of what is and what should be.* Hawthorne, NY: Aldine de Gruyter.

Muth, W. R. (1995). *Implementing a management information tool for executives in the federal prison system.* Washington, DC: Federal Bureau of Prisons.

Nagelkerke, N. J. D. (1991). A note on a general definition of the coefficient of determination. *Biometrika, 78*, 691–692.

Saylor, W. G. (1984). *Surveying prison environments.* Washington, DC: Federal Bureau of Prisons.

Saylor, W. G. (1988). *Developing a strategic support system: Putting social science research into practice to improve prison management.* Washington, DC: Federal Bureau of Prisons.

Taylor, M. C. (1995). White backlash to affirmative action: Peril or myth? *Social Forces, 73*(4), 1385–1414.

Thomas, R. R. (1991). *Beyond race and gender.* New York: AMACOM.

DISCUSSION QUESTIONS

1. According to the authors, how do minority and nonminority correctional officers view affirmative action policies?

2. According to this research are employee perceptions and reality in alignment, or do they diverge?

3. In what ways to black and white correctional officers view their workplace differently? How do you explain these different perspectives?

READING

Whenever opposite-sex individuals are placed in common situations for extended periods, attractions inevitably occur. In prison settings, correctional officer-inmate attractions, say Marquart, Barnhill, and Balshaw-Biddle, can prove fatal. Prison employees are trained to maintain their distance from prisoners and to do their job professionally without personal entanglements or abuse of prisoners in any way. Efforts are made to ensure that those who enter the correctional service serve honorably; these efforts, however, do not prevent some prison employees from ending their careers in disgrace. The authors examine boundary violations among 508 Texas state prison security staff members disciplined between January 1, 1995 and December 31, 1998. They find that some employees engage in career-ending infractions whereby they are suspended and/or reprimanded; some are terminated for engaging in inappropriate relationships (or boundary violations) with inmates.

Fatal Attraction

An Analysis of Employee Boundary Violations in a Southern Prison System, 1995–1998

James W. Marquart, Madeline B. Barnhill, and Kathy Balshaw-Biddle

Over 40 years ago, Gresham Sykes (1958:42) stated, "[F]ar from being omnipotent rulers who have crushed all signs of rebellion against their regime, the custodians are engaged in a continuous struggle to maintain order—and it is a struggle in which the custodians frequently fail."

Prison employees maintain only a theoretical dominance over inmates. Three institutional influences corrupt their authority. First, the staff cannot physically coerce the prisoners into total submission, and therefore must make deals with inmates wherein they let some things slide to secure compliance elsewhere. This has been termed the "norm of reciprocity," whereby employees exercise power over inmates by giving them "freedom" in return

for good behavior. Second, the custodians rely on prisoners to perform numerous institutional chores (e.g., bookkeeping); such reliance cedes power to the kept. Third, the prison staff works closely with the kept; this situation increases the desire to get along with the inmates. Pressures to get along blur the boundary between employees and prisoners. As a result of their constant interaction with inmates, staff members often redefine inmates as "people in prison" rather than as dangerous offenders.[1] In the present paper, we focus on this latter aspect of the corruption of authority.

Proximity fosters personal bonds between prison employees and inmates, encouraging favoritism and selective rule enforcement (Crouch and Marquart 1989). Worse, such

SOURCE: Fatal Attraction: An Analysis of Employee Boundary Violations in a Southern Prison System, 1995–1998; James W. Marquart, Madeline B. Barnhill, and Kathy Balshaw-Biddle, *Justice Quarterly* 18(4): 877–910 (2001); Taylor & Francis, Ltd., http://www.informaworld.com, reprinted by permission of Taylor & Francis, Ltd.

proximity can facilitate romantic relationships, "consensual" love affairs, or even criminal sexual involvement (Baro 1997; Hanson 1999). In general, correctional occupational deviance refers to inappropriate work-related activities in which correctional employees participate (Kappeler, Sluder, and Alpert 1994:22). We define actions that blur, minimize, or disrupt the professional distance between correctional staff members and prisoners as boundary violations; such activities represent one form of correctional occupational deviance. Boundaries are violated when the managerial desire to maintain formality and distance between keepers and kept erodes in the face of the operational realities of institutional life (i.e., the norm of reciprocity). Correctional staff members' violation of traditional boundaries through the formation of friendships, romantic relationships, or sexual involvement with inmates constitutes a major compromise of correctional work ethics and values. Such violations represent an extension of Sykes's (1958) concept of the corruption of authority.

Before the advent of sex-segregated prisons in 1870, female prisoners often were housed in separate wings of male prisons; accommodations set the stage for sexual victimization by male custodians (Dobash, Dobash, and Gutteridge 1986; Rafter 1985). Even sex segregation, however, did not eliminate sexual encounters between staff and prisoners. Zedna (1995:310) reported that "undue intimacy" between female prisoners and their female warders occurred with "surprising frequency" in Victorian prisons. The National Women's Law Center (1996) suggested that correctional employee sexual abuse of prisoners is widespread. United Nations researchers contend that guards' sexual involvement "is common in women's prisons" in America (Day 1998; Olson 1999). As Amnesty International (1999:1) noted, "[M]any women in prisons and jails in the USA are victims of sexual abuse by staff." Sexual misconduct, primarily by male staff members against female prisoners, has prompted lawsuits against 23 prison systems and jails (Hanson

1999). In light of this problem, 42 states, the District of Columbia, and the federal government have enacted laws prohibiting staff members' sexual misconduct with inmates (National Institute of Corrections 2000).

These reports imply that sexual misconduct involves only male staff and female prisoners. Inmate sexual contact, however, is only one form of occupational deviance resulting from prison personnel's close association with, and proximity to, inmates. Other forms of boundary violations, such as excessive familiarity, friendships, and romantic relationships, are probably much more common than overt sexual contact. Three other gender combinations also are possible in contemporary prisons: male staff-male prisoner, female staff-male prisoner, and female staff-female prisoner.

Until quite recently, prisons and jails were sex-segregated by both correctional staff and inmates (Jacobs 1983; McDonald 1997). Sex segregation, however, severely restricted women's employment and promotional opportunities in corrections (Jacobs 1983). Title VII of the Civil Rights Act of 1964 banned sexual discrimination and gave women a foundation for seeking legal remedies (Martin and Jurik 1996). Female correctional employees filed lawsuits[2] and won the right to work in male prisons (Pogrebin and Poole 1997). In 1998, women constituted 22 percent of the American correctional officer workforce. Currently, eight female officers in 10 work in male prisons. Men accounted for 78 percent of the correctional officer workforce in 1998, but only 4 percent were assigned to female institutions (Camp and Camp 1998).

Sexual integration of the correctional workforce has expanded the employment possibilities for women. Ironically, it also has facilitated female prison staff members' opportunity to engage in occupational deviance, especially boundary violations (forming personal relationships with male prisoners). We currently know little about gender differences in boundary violations among prison staff members: prison administrators customarily have regarded this matter as taboo (Mooney 1995). Sykes's (1958) discussion of the corruption of authority centered on male officers and male prisoners, and modern prisons are far different places than he described decades ago. Despite recent media attention to male prison employees' sexual exploitation of female inmates in Florida, Georgia, and Michigan (Rivera 1999), as well as anecdotes (Mooney 1995) and testimony from court cases (Elliot 1999), the research literature on deviance in prison organizations lacks systematic inquiry into prison boundary violations (Stewart 1998).

Building on Sykes's (1958) concept of the corruption of authority, we develop the rudiments of a general theory of boundary violations in prisons. We employ Goffman's (1974) model of social frames to examine gender differences in a group of prison staff members investigated for establishing personal relationships with prisoners. Our perspective helps us to understand gender differences in boundary violations, and illustrates how the processes that pull male and female prison employees into boundary violations may differ (Steffensmeier and Allan 1996).

◼ Boundaries and Social Frames

Boundaries demarcate areas where we are welcome or forbidden. Epstein (1994:15) states, "The concept of personal boundaries employs a spatial metaphor that helps us describe and define our relationships with other beings and objects in the external world." Social and psychological boundaries demarcate space and protect our individuality. Boundaries and their maintenance are essential to the survival of human communities; without them, social organization would be impossible (Erikson 1966).

Social Frames and Boundary Maintenance

Erving Goffman (1974) defined social frames as principles of organization that govern social events and our subjective involvement in those events. At any level of human discourse, whether loving intimacy, group processes, business dealings, or sporting events, we tend to maintain a basic set of expectations and rules to comprehend reality. Goffman (1974) stressed that nearly every interpersonal situation included frames, or "ways of doing," that delineate the purpose and meaning of the relationship. Social frames organize involvement by providing participants with a sense of what is going on, as well as limiting how deeply the individual is to be carried into the activity. Involvement and a sense of purpose are critical aspects in fiduciary relationships, wherein one party has a professional obligation to care for or provide service to another (Friedman and Boumil 1995).

Conceptualizing the Custodial Frame

In keeping with Goffman's (1974) notion of social frames, we conceptualize the custodial frame as a set of expectations that organize and give meaning to staff-inmate interactions. It is commonly understood that prison systems differ greatly as to how far inmates can and should be trusted, the causes of criminality, the role of treatment, and the proper scope of prisoners' rights. Prison personnel share the "keeper" philosophy, which guides staff behavior (DiIulio 1987). In their encounters with prisoners, prison employees are expected and trained to be firm but fair, nonabusive, impersonal, dispassionate, and nonconfrontational. These interaction rules are gender-neutral; they constitute the "custodial frame" that legitimizes and enforces the boundary between superordinates (staff) and subordinates (inmates). The custodial frame also facilitates cohesion among the staff members.

Throughout the history of prisons, employees have been forbidden from becoming personally involved with prisoners. The "no friendship" rule in prisons is akin to the incest taboo. Contemporary prisons ban interaction beyond the custodial frame, and have established disciplinary procedures to punish employees who engage in inappropriate relationships with prisoners (National Institute of Corrections 2000). The custodial frame is maintained through physical barriers such as uniforms, grooming standards, and a variety of duty posts (e.g., gun towers, perimeter patrol cars, catwalks) and shift work. At the social level, formal modes of address (Sir, Mrs., Ms., Mr., Officer, Boss) have been institutionalized to enforce a polite but firm impersonal boundary between prison employees and prisoners. Inmates are expected to be polite, docile, nonthreatening/nonconfrontational, and impersonal in their interactions with staff.

Sometimes, however, the well-established and historic boundary or custodial frame between staff and inmates is broken. Staff members may denigrate inmates, ignore inmates' requests for assistance, physically abuse inmates, or bring contraband such as drugs or weapons to prisoners (Crouch and Marquart 1980, 1989; Marquart 1986). In his discussion of the corruption of authority, Sykes (1958) described how prison staff members bend the rules to secure inmates' compliance. The custodial frame is not absolute, nor can it govern every staff-inmate interaction.

Breaking the Frame

Breaking a social frame constitutes a boundary violation, a well-researched topic in psychology, psychiatry, and health care. This body of research on power imbalances posits that boundary violations encompass an array of activities and behaviors. Most important for the present paper is Strom-Gottfried's (1999) research on the violation of therapeutic boundaries by social workers or therapists with their clients. She reviewed 894 ethics complaints submitted to the National Association of Social Workers over an 11-year period.

Strom-Gottfried uncovered three types of ethical transgressions: general blurring of roles (i.e., discussing personal information with

clients, meeting with clients after hours); dual or overlapping relationships (i.e., dating, pursuing joint hobbies and social activities with clients); and sexual contact. One hundred and forty-seven cases went to formal hearings and resulted in affirmative findings of boundary violations. Of these cases, 107 involved some form of sexual violation, 62 entailed dual relationships, and 70 involved general breaches. Moreover, Strom-Gottfried found that each boundary violation, or frame break, consisted of a series of steps or a sequence of actions committed by the clinician. Further, a minority of cases involved a "slippery slope": clinicians began with a general blurring of roles, which progressed into sexual contact (Strom-Gottfried 1999).

In the present paper we apply to the prison Strom-Gottfried's (1999) typology of boundary violations among social workers and their clients. Our goal is to extend current theorizing to inform future research. We strongly suggest that the typology of boundary violations presented here is not exhaustive, but is a first pass at an unexplored problem in prisons.

Applying the Boundary Violation Perspective to Penal Organizations

In our application of Strom-Gottfried's (1999) continuum to the prison, general boundary violations constituted "unserious" frame-breaks committed by employees who accepted from inmates, or exchanged with inmates, items such as soft drinks, food, and craftwork or materials, or wrote letters to prisoners whom they had known before incarceration. Dual relationships occurred when employees blurred the boundary between themselves and inmates, and established romantic relationships with prisoners through excessive flirting and disclosure of personal information. These employees pursued an inmate's attention or affection, and fell in love. Staff members who engaged in dual relationships committed one or more of the following behaviors:

Discussed their personal life in detail with a prisoner (including sexual life, social life, marital status, spouse and/or children, experiences with domestic abuse);

Exchanged letters and/or personal photographs (including nude photos) with a prisoner;

Exchanged erotica (pornographic poems and letters) with an inmate;

Placed money in an inmate's trust fund;

Used aliases and post office boxes in nearby towns or cities to hide the relationship;

Contacted an inmate's family to relay information about the prisoner;

Established a relationship with an inmate in prison and then lived with the inmate upon his or her release from prison;

Moved in with an inmate's family member;

Provided a cellular phone to a prisoner, or took collect calls to facilitate off-duty conversations with an inmate;

Engaged in on-the-job subterfuge to hide the relationship.

We realize that placing money in an inmate's trust fund and/or exchanging "dirty" letters certainly stretch the definition of dual relationships. In our view, however, dual relationships encompass diverse behaviors up to, but not including, sexual contact. It is possible that the exchange of erotica may constitute another behavioral type suggesting eventual sexual contact. Additional research in other prisons (based on more cases) may further clarify the differences among dual relationships.

Staff-inmate sexual contact is the most serious boundary violation. In keeping with the research literature on sexual contact between patients and therapists (Wincze et al. 1996), we defined employee-inmate sexual contact as vaginal and/or anal intercourse, fondling, masturbation, genital exposure, and/or any oral-genital contact. Sexual boundary violations had to be observed or corroborated by a third party (another employee), and/or the employee under investigation must have admitted to the act (in written statements or under oath during a polygraph session).[3] Momentary kisses, hugs, and hand holding were not considered sexual contacts. For example, one employee admitted to investigators that she loved inmate X and had delivered 30 letters and 11 photos of herself to this inmate. She admittedly had kissed inmate X, had deposited $700 in his trust fund, wanted to have a "quickie" with him, and "fantasized about having sex" with him. These incidents were not considered sexual contact. Our criteria for sexual contact yielded 42 incidents (involving 18 males and 24 females) of reported employee-inmate sexual contact.[4] Additional acts of sexual contact in the same period probably went unreported (Day 1998).

Some scholars and practitioners will argue that, because the relationship between staff and inmates is coercive, the kept cannot refuse a sexual relationship with employees. We suggest, however, that under certain circumstances it may be possible for the keeper and the kept to establish a consensual sexual relationship. The organizational links (e.g., a culture of silence) and the legal implications (e.g., consent) of sexual contact between prison employees and inmates are complex, and merit additional scholarly inquiry.

What kinds of employees had sexual contact with inmates? Where did this behavior occur? Where were the supervisors when the behavior took place? Answers to these questions will have important policy implications for prison management. Rather than simply developing a "kinds of people" explanation, we also examined the situational aspects of correctional employees' framebreaks. To this end, our research was guided by the research conducted over the past 25 years by Phillip Zimbardo and his colleagues, who emphasized the surprising and ironic power of institutional environments over the participants. Their research illustrates clearly how situational dynamics can distort an individual's judgment so far as to defy all individual expectations (Haney and Zimbardo 1998:710).

In keeping with this line of research, we expect to find that certain situations and places in the prison will be conducive to staff-inmate framebreaks. We expect to find that such framebreaks will be more prevalent where supervisors and colleagues are absent. The lack of authority figures to exert formal control over an employee's contact with prisoners should be conducive to boundary violations. These violations will be more prevalent in unstructured employee-prisoner situations (i.e., choir practice, recreational activities) than in structured work-intensive situations such as supervising inmates on industrial work details (Osgood et al. 1996). Unstructured activities involving staff and inmates (one-on-one situations away from the supervisors' view) set the stage for blurring the line between staff and inmates, disrupting formal relations, and weakening the custodial frame.

▧ Methodology

Data for this inquiry were obtained from the personnel files of 549 employees of the Texas Department of Criminal Justice, Institutional Division (TDCJ-ID). These employees were investigated by Internal Affairs and were formally disciplined between January 1, 1995 and December 31, 1998: They were punished for violating Rule 42 of the "Employees' General Rules of Conduct."[5] Originally the data set consisted of security officers, parole staff, health services staff, clerical staff, teachers, chaplains, and other members of the support staff. We focused on the prison staff members (security personnel, mail room clerks, secretaries) who had daily interactions with inmates. We eliminated 41 employees (parole officers and health services staff)[6] to reduce the data set to 508 "investigated" prison employees.[7]

Data and Contents of Personnel Files

Personnel files included employee applications, background investigations, and in-service work records. The employment application and background investigation contained demographic, educational, employment, military, and criminal history information. In-service records contained prison unit assignments, academy test scores, promotions, annual performance evaluations, initial interview and evaluation scores, records of sick and injured leave, disciplinary actions, dispositions of complaints, and all official correspondence or records. Each file contained the case and investigation history surrounding the circumstances and actors and/or actions leading to the employee's investigation for violating Rule 42.

We content-analyzed each case history, searching for themes in the employees' version of the events (87 data categories emerged from this examination). We developed a coding instrument to capture relevant variables. (Employees' and inmates' names used in this paper are pseudonyms.) We readily acknowledge that our analysis is employee centered and that we have lost a great deal by not incorporating the inmates' perspective. The inmate's voice is critical but it lies beyond the present paper's scope.

Sample of "Successful" Employees

At the request of human resources administrators, we conducted a comparative analysis of "successful" and "unsuccessful" employees to determine, for recruiting purposes, whether significant demographic differences existed between the two groups. We were provided with a random sample of 585 prison employees hired between 1995 and 1998, who were still employed as of December 31, 1999. These individuals had undergone no Internal Affairs investigations and were considered successful employees. In the comparative analysis we explored differences between investigated and noninvestigated employees, based on the same data elements as gathered from the employment application and background investigation and from in-service records.

Interviews With Supervisors and Internal Affairs Investigators

Finally, to obtain personal insights into inappropriate employee-inmate relationships, we interviewed 12 employee supervisors (lieutenants, majors, and wardens) from five Huntsville-area prisons, as well as four internal affairs investigators. We asked the supervisors and investigators to explain the process by which employees become involved in inappropriate inmate relationships, why and how such relationships develop, and whether or not improved hiring standards and training could deter such relationships and improve retention of employees. We also asked the interviewees about the impact of recent organizational growth on staff recruitment, retention, and inappropriate relationships. The 20-question interviews averaged 60 minutes each; we carefully noted and later content-analyzed all responses.

Background Characteristics of the Investigated Employees

At the time of the investigation, the average age of the 508 prison employees was 36 years (31 at initial employment), and their tenure averaged slightly less than four years (46 months on the job). Fifty-four percent were Anglo, 30 percent were African-American, and 16 percent were Hispanic. Over two-thirds (68 percent) had a high school diploma. Thirty-eight percent reportedly were attending college, and three subjects claimed to be criminal justice majors. Most subjects (85 percent) had no prior military experience, and eighty-six percent reported performing service work or manual labor before their prison employment.

Seventy-seven percent of the investigated employees were females. This skewed proportion (or sex ratio) must be interpreted cautiously, however. In 1978, Texas prisons were sexually segregated workplaces for security staff members. In 1988, 17 percent of all female security staff worked in male units. By 1998, this proportion had increased to 80 percent. On the other hand, 95 percent of all male security officers worked in male prisons (Camp and Camp 1998). Thus a larger proportion of female employees were exposed to male prisoners.

Raw percentages can be misleading, however. Table 1 presents employment data on male and female security officers between 1995 and 1998. In 1995, 3,941 females worked in male prisons. In the same year, 98 female security employees (in our data set) were investigated for engaging in inappropriate relationships with male prisoners (a rate of 2.4 per 100 officers). In comparison, 348 male officers worked in female prisons; 12 were investigated for engaging in inappropriate relationships with female prisoners (a rate of 3.5 per 100 officers). Our analysis, although preliminary, shows that the overall rates of inappropriate relationships per 100 officers between 1995 and 1998, for male and female employees in sexually integrated settings were roughly the same.

Yet despite the similarity in rates of boundary violations, we suggest that the risk for deviance is far greater for female than for male employees in sexually integrated prisons. A male-dominated work environment, especially a prison, is characterized by a sexual ambiance and by the expression of male sexuality (Carroll 1974). Women employed in male-dominated settings are more likely to experience social-sexual behaviors (e.g., invitations to sex, innuendo, harassment), a point underscored by research on sex and the workplace (Statham 1996). We strongly suggest that sex-role spillover (the carryover into the workplace of gender-based expectations about behavior) also helps to explain the higher proportion of employee boundary violations by females (Gutek 1985:149). It was highly likely that male prisoners perceived the women security officers as females first and employees second. Thus male prisoners, in keeping with sex-role behaviors, may have planned and initiated relationships and sexual interaction. (Research on the inmates' version of the events will help to clarify this process. Male prisoners also may "hit on" female employees to advertise and validate their heterosexuality to themselves and their peers.)

The sexual ambiance, sex-role behaviors, and expectations surrounding male employees in female prisons are different and deserve separate analysis. Although the rates of deviance may appear on the surface to be similar, the dynamics involved in male and female employees' boundary violations are probably very different, making any comparison difficult and subject to interpretational error.

Boundary violations occurred quickly, often within the first year of employment. Fifty-seven employees were investigated for engaging in inappropriate relationships with inmates after less than one month of employment, 161 employees were investigated after one to 12 months of employment, 99 between 13 and 24 months, and 64 between 25 and 36 months. In all, 381 subjects were investigated within 36 months of employment.

Most (71 percent) investigated employees were hired between 1994 and 1998, in part because of rapid ID growth in the mid-1990s. During that period many existing criminal laws were enhanced, and new laws mandating longer sentences were enacted. Texas spent $1 billion on prison construction in the early 1990s, and thousands of individuals were hired

to manage the influx of new prisoners (Robison 1991). Between 1995 and 1998, the ID prison population increased from 118,386 to 140,718, and the number of prisons jumped from 69 to 108. Between 1995 and 1998, TDCJ-ID also increased its annual operating budget from $1.5 to $2 billion, increased the number of employees from 35,458 to 39,418, and increased its total security force from 18,920 to 27,509 (Camp and Camp 1995–1998). Roughly equal numbers of subjects were assigned to "old" prisons (1989 and earlier) and to "new" prisons (1990 and later). Fifty-five percent were terminated or dismissed, 23 percent resigned, and 31 percent received disciplinary probation.[8] Employees who resigned did so after they were confronted with the prospect of an investigation. Those who resigned were allowed to leave the prison service without any negative comments or blemishes in their personnel file. We also realize that some unknown number of employees simply quit to avoid formal investigation when they felt that they were under suspicion.

Eligible applicants undergo an interview and take five pre-employment tests.[9] A final evaluation score is computed from the combination of pre-employment tests, along with nine dimensions evaluated by interviewers. Applicants with high school diplomas, no felony convictions, and evaluation scores of 60 and above are offered positions. Human resource managers review applicants with

Table 1		Employment Figures for Security Staff Members by Year and Rate of Disciplinary Actions							
Year	Total	Males	Females	Females in Male Institutions	Number of Females Disciplined	Rate of Female Violations	Males in Female Institutions	Number of Males Disciplined	Rate of Male Violations
1995	18,920	14,067	4,853	3,941	98	2.40	348	12	3.50
1996	22,846	16,655	6,191	5,332	115	2.10	542	12	2.40
1997	26,894	19,229	7,665	6,317	100	1.60	540	9	1.60
1998	27,509	19,009	8,500	6,616	54	.80	571	4	.70

scores of 59 and below; depending on the factors that caused the low score and on the needs for facility personnel, some applicants may be offered employment. The average evaluation score for investigated employees was 63; 119 (23 percent), however, scored 59 or less, and one scored 38.

In summary, the "average" employee investigated for establishing an inappropriate relationship with an inmate was a 31-year-old white female with a high school diploma and a history of service and nonmilitary work, who was employed by the prison system in the mid-1990s as a uniformed security officer. These employees typically established an inappropriate inmate relationship within 36 months of employment.

⊠ Findings

In this section we compare the "investigated" employees with the "successful" employees, and apply Strom-Gottfried's (1999) continuum of boundary violations to the data from the Texas prison system. We then discuss the importance of situational dynamics as a possible explanation for inappropriate relationships in prisons.

Investigated Versus Non-investigated Employees

As stated above, prison human resource administrators wanted to know, for recruiting and training purposes, whether significant demographic differences existed between successful and unsuccessful employees. Accordingly we gathered relevant background and ID employment data from a random sample of 585 noninvestigated employees to compare with the investigated employees (see Table 2). A comparative analysis uncovered several important differences between the two groups. Investigated employees were significantly more likely than nondisciplined

employees to be Anglos and females. In addition, they were significantly more likely than successful employees to have a general equivalency degree instead of a high school diploma, were significantly less likely to have military experience, and scored significantly lower on the pre-employment application (with average scores of 63 versus 68). Another important and statistically significant difference involved prior disciplinary problems: 45 percent of the investigated employees had at least one disciplinary write-up, compared with 31 percent of the noninvestigated employee group.

Analyses of Three Kinds of Boundary Violations

Using Strom-Gottfried's (1999) categorization scheme, we found that 38 employees (8 percent) had committed general boundary violations, 428 (80 percent) had engaged in dual relationships, and 42 (12 percent) had had sexual contact with prisoners (see Table 3). A significant relationship emerged between gender and the type of boundary violation: female employees were more likely than males to commit a boundary violation, and 80 percent of non-sexual dual relationships involved female employees. Minor breaks typically resulted in disciplinary probation, while more serious boundary violations led to severe punishment. Employees investigated for dual relationships were significantly younger (average age 30) than employees investigated for the other boundary violations. In 1998 the average age of an ID prisoner was 35; this age differential (35 to 30) may be related in some way to dual relationships.

General boundary violations. Among the 38 employees who committed general boundary violations (e.g., accepted or exchanged food products or craft work/materials with prisoners, or wrote letters to prisoners) most were Anglo (53 percent) and female (63 percent), had high school diplomas, and were 36 years

	Table 2	Investigated Staff Members Compared With Sample of Successful Staff Members, 1995–1998	

Variable	Investigated Staff (N = 508)	Successful Staff (N = 585)
Race/Ethnicity*		
White	54%	51%
African-American	30	27
Hispanic	16	22
Gender***		
Male	23%	68%
Female	77	32
Education***		
GED	32%	14%
High school diploma	68	86
Marital Status		
Married	50%	52%
Single	50	48
Military Experience***		
Yes	15%	26%
No	85	74
Employment History		
Service	86%	93%
Supervisory	14	7
Prior Disciplinary Violations***		
Yes	45%	31%
No	55	69
Application Score***		
59 or less	24%	11%
60 and above	76	89
Average Application Score***	63	68
Average Age at Employment	31	31

* $p < .05$; *** $p < .001$

old when hired by the ID. Most (79 percent) were hired between 1994 and 1998, and were investigated (66 percent) between 1997 and 1998. Slightly more than three-quarters (76 percent) were punished with disciplinary probation and received a second chance. All admitted their offense and expressed some degree of remorse.

Table 3	Investigated Staff Members' Background Characteristics, 1995–1998		
	Type of Boundary Violation		
Variables	**General Boundary Violations (n = 38)**	**Dual Relationships (n = 428)**	**Sexual Involvement (n = 42)**
Race/Ethnicity			
African-American	29%	30%	36%
White	53	55	50
Hispanic	18	15	14
Gender***			
Male	37%	20%	40%
Female	63	80	60
Education			
GED	32%	32%	31%
High school diploma	68	68	69
Marital Status			
Married	43%	49%	66%
Single	57	51	34
Year Hired			
1977–1987	8%	5%	4%
1988–1993	13	25	17
1994–1998	79	70	79
Year Investigated			
1995	21%	26%	19%
1996	13	31	45
1997	32	27	24
1998	34	16	12
Case Outcome**			
Terminated/dismissed	16%	57%	62%
Probation	76	19	5
Resigned	8	24	33
Unit Opened 1989 and Before	37%	48%	45%
Unit Opened 1990 and After	63	52	55
Application Score			
59 or less	24%	24%	17%
60 and above	76	76	83
Mean Application Score	64	63	65
Average Age at Employment*	36	30	34
Average Age at Termination	38	35	39
Mean Number of Months on Job	35	57	36

* $p < .05$; ** $p < .01$; *** $p < .001$

The following vignettes (taken from the case files) illustrate common themes involved in general boundary violations:

Case 1. In April 1998, Officer Daly admitted that she corresponded with Inmate Y and submitted a written statement that she was writing the inmate. She also stated that she had known Y before his current incarceration and did not realize that corresponding with former inmate friends was a violation of policy. She freely admitted that they were friends and that she wrote all the letters.

Case 2. In a written statement, Officer Jones admitted that he corresponded with Inmate A. He wrote: "I have known A since we was kids. His mom used to baby sit me. We have been friends before I became a CO [correctional officer]. I wasn't thinking when I wrote him. I did not mean to break policy. I thought I would write to give an old friend some advice."

Case 3. In July 1997, a unit investigation found that Officer Bailey deposited $20 in Inmate B's trust fund in return for a leather watchband. When confronted, Bailey admitted to the actions and further stated, "I deeply regret this and apologize for my actions and want disciplinary actions. I didn't think this was a big deal."

Most (63 percent) general boundary violations occurred in "newer" prisons. This finding suggests four possible explanations. First, it is plausible that the new units were flooded with rookies who failed, through ineffective training and supervision, to internalize the boundaries between themselves and inmates. When new employees enter a new prison, they may not know the organizational cultural taboos. The rapid expansion of the workforce may have weakened the organizational culture, as well as overwhelming the stitutional norms that govern inmate-staff relationships. White (1995) states:

> During periods of organizational turbulence, there is a weakening of organizational culture and values. The organization loses its power to shape, monitor, and self-correct boundary problems within worker-client relationships. Weak organizational cultures lose the capacity to define boundaries of appropriateness in service relationships. Weak organizational cultures exert little influence or control on individual practitioners. Rapid staff turnover or growth opens up the possibility of new workers and emergent subcultures that deviate from an organization's historical values. Turbulence within organizational systems, just as in family systems, marks a period of great vulnerability for role boundary violations. (p. 189)

Second, organizational pressure to recruit staff may have led to the erosion of standards, compromising hiring practices between 1995 and 1998. Similarly, the rapid expansion of police forces in New York, Washington, Miami, Houston, and Detroit led to careless screening and hasty training, which contributed to increases in police corruption and brutality (Krauss 1994). The recent Board of Inquiry into the Rampart area corruption incident in the Los Angeles Police Department attributed the criminal activity of several officers to accelerated hiring practices (in the late 1980s and early 1990s) and to the erosion of hiring standards (Board of Inquiry 2000).

The interviewees maintained that massive prison expansion eroded hiring standards: in the rush to build prisons and attract "warm bodies" to staff the new prisons, hiring standards declined. As one supervisor stated:

The hiring standards were loosened up to staff the units. They had to hire whoever they could. Look, this prison system is so big now that I believe we have exhausted the labor pool of decent applicants. They hired anybody that was warm and could walk. Inappropriate relationships? The reason for these things is due to a lack of standards. It only stands to reason that when you lower the standards you get a lower-quality employee. Poor employees translate into management problems.

Third, it is possible that new units were staffed with new or inexperienced supervisors, who enforced employee rules and regulations aggressively. General boundary violations may have been the form of misconduct most visible for remediation.

Fourth, some of the post-1990 prison units are new-generation prisons, with an open design, which could reduce the supervisors' span of control and facilitate staff-inmate deviance. In short, new prison architecture may be associated with inappropriate relationships.

All four explanations underscore the unintended consequences of organizational growth on employees' misconduct, and require further research. In addition, organizational growth represents a serious historical effect that could confound the internal validity of the present inquiry (Cook and Campbell 1979). The data presented here clarify the effects of system expansion on personnel recruitment and boundary violations.

Dual relationships. As stated above, the analysis found 428 employees who broke the custodial frame by becoming personally involved with an inmate. Each of these employees was both a friend and a supervisor of an inmate. Dual relationships commonly involved Anglos (55 percent), females (80 percent),

security officers (91 percent), and male prisoners. These relationships involved one employee and one prisoner, and appeared to be consensual. Most of the employees had high school diplomas (68 percent), were hired between 1994 and 1998 (70 percent), and were investigated in 1996 and 1997 (58 percent); more than half were terminated (57 percent). Their average age was 30 at employment, and they were on the job for roughly five years at the time of investigation.

We failed to determine whether inmates or employees initiated these custodial frame-breaks. The employees' accounts of the events, however, suggest that inmates initiated the relationship for a variety of reasons such as companionship, loneliness, money, desire for favoritism, sex, contraband, boredom, and competition with other inmates. It is possible, for example, that male inmates engaged in "fox hunting" and manipulated the employees to violate policy (Sapp 1997). The employees' version of the events revealed no instances in which the inmate forced the staff member into a relationship. The staff members allowed the relationships to develop despite their training.

Dual or overlapping relationships involved a process or series of steps. Glances were exchanged; notes, photographs, and rings were passed; smiles and friendly conversations were offered. Typically an inmate engaged an employee in small talk (e.g., on the weather or current events). If an employee reciprocated, then other topics (such as personal likes and dislikes or home and social life) were broached incrementally until the employee disclosed sensitive personal information (e.g., marital status, friends, and off-duty interests). In other cases, an inmate passed a "hook letter" (see below) to an employee and waited for a response. Employees who corresponded with the prisoners were approached and engaged in additional small talk. Sometimes an employee requested a duty assignment near his or her inmate "friend."

Dear Officer Jones:

Commit thy works unto the Lord, and thy thoughts shall be established. (Proverbs 16:3).

Giving all praises an [*sic*] honor to the Lord and Savior, Jesus Christ. I'm praying that this small note becomes a long lasting friendship based on truth and honesty. To let you know that I am a honest man from the heart. Just searching for someone to relate to on verious [*sic*] issues of life. Most of all, life itself and what it truly means. To let you know a few things about me; I'm 36 years of age and I have a 30 yr. sentence. This is time that has no true meaning. I'm from Houston's North side. I'm a quiet type of guy and I stay to myself most of the time. So if this grabs you or even suits your fancy. Let me know, and of course let me know what turns you on and makes you happy. Just maybe I'm that friend you've been looking for.

"Looking to hear from you!"

Ricky

Dual relationships were typified by "lovesickness" (Gonsiorek 1995)—situations in which employees became infatuated and then fell in love with inmates. These employees desired a soulmate. The two typically exchanged letters detailing their life histories, goals, aspirations, and devotion to each other. Many letters contained erotic themes and a desire for marriage when the inmate was paroled. Dual or overlapping boundary violations often involved romantic love and idealization of the inmate. Some employee-inmate correspondence revealed that female employees were drawn to male offenders to "straighten them up" or to tame a "rowdy man." In other cases, the employees were so consumed with romantic love that they identified (like Patty Hearst) with the prisoners rather than with the staff. The following vignettes from employees' letters or cards to inmates illustrate the intense emotions at work in dual relationships:

Dear Jay, As I lay here alone, listening to music, all I can think of is you. Do you realize that in such a little time we developed a love that is undescribable. In my heart I have so many feelings, we have a bond that will last forever. I need you at home with me so that I can love you right, you're my dreams, love, and you being locked up is the ultimate test of our friendship and love. (Female employee to male inmate)

I love you, I need you. I want you. Forget the past, I only wanna hear about the future. I dream about you every night. Please don't leave. I would marry you now. (Female employee to male inmate)

I love you as always. I have you in my thoughts and dreams. Girl you will always have a special place in my heart. I love you, and your whole body. God just made you so beautiful—I can't resist you. Oh yeah, never argue with the officers for It's bad news. (Male employee to female inmate)

I love you and I love this feeling also, the desire, the anxiety and excitement. I can't wait to get to work. (Female employee to male inmate)

Look darling, We've been through a lot together. We've risked a lot together and Today we came together to be man and wife. I am not only your friend and lover, or girlfriend anymore. I am your partner in life. I will always be there for you until the end of time. This is my solemn vow to you. PS. They are watching all the female officers because of that shit with that boss on the first shift, be careful. (Female employee to male inmate)

Remember I love you. My heart is yours forever. I know there is an age difference between us, but that does not matter to me because I love you, I live for the time that we can be together. (Male employee to female inmate)

These letters voiced obsession, excitement, romantic love, and a desire for a life together beyond the penitentiary. These employees were eager to come to work and to overlook the prisoner's criminal past. In eight dual relationships, the employee fell in love with an inmate gang member. Many married employees caught up in dual relationships appeared to be personally vulnerable.

Employees investigated for this violation often went to great lengths to keep the relationship secret. When we examined employee-inmate messages and letters, we often found employees instructing inmates to "keep a low profile," "lay low, they're [security staff] watching us," or "keep away for now, let things cool down." Security staff members often accidentally discovered the relationship during cell/property searches, after finding the employee's artifacts (e.g., love letters, poems, cards, photos) in the inmate's possession. In four incidents, inmate property searches unearthed an employee's cellular telephone. In one case, an inmate informed unit supervisors of an employee-inmate dual relationship "because everybody knows about it and it was causing friction on the wing." In two cases, employee-inmate relationships were discovered when the employee's name was found boldly tattooed, one on an inmate's stomach and the other on an inmate's leg. In six cases, the employee's spouse uncovered love letters in the employee's possession and informed unit supervisors. In four cases, investigated employees spoke of their relationships to their colleagues, who then informed investigators.

Where were the supervisors? The Mollen Commission, which investigated police wrongdoing in the New York City Police Department, reported that supervisors had a "willful blindness" to police misconduct (Commission to Investigate 1994). The supervisors did not want the bad publicity and scandal that would follow if it was discovered that employees under their watch were engaged in misconduct. As a result, they looked the other way and refused to take action against officer wrongdoing. It is possible that the prison supervisors in Texas knew of dual relationships but practiced "willful blindness" to avoid scandal and embarrassment to the agency.

Sexual contact. Forty-two acts of employee-inmate sexual contact occurred between 1995 and 1998, accounting for 8 percent of all boundary violations. Similarly, Strom-Gottfried (1999) found that over an 11-year period, 12 percent of all ethics complaints submitted by clients to the National Association of Social Workers dealt with sexual violations. Half of the prison employees were Anglos, 60 percent were women, and 79 percent were hired between 1994 and 1998. Over two-thirds had high school diplomas (69 percent), and nearly all were security staff members (90 percent) in their early thirties (average age 34) at the time of employment. These employees averaged three years on the job when they were investigated for sexual contact. Two-thirds were married; we had no information on the quality of the spousal relationship, however. Forty incidents involved heterosexual contact; two (5 percent) were same-sex pairings (male-male). A similar percentage (4 to 5 percent) was found in the literature on sexual contact between male therapists and male clients (Gonsiorek 1995). Most of the employees investigated for sexual contact were terminated (62 percent) or resigned (33 percent) from employment. The National Institute of Corrections (2000:10–11) recently reported that in fiscal year 1998, more than 115 officers were discharged for substantiated incidents of sexual contact with inmates, and 20 were prosecuted. (Data were taken from 37 agencies.)

The following vignettes illustrate "typical" acts of employee-inmate sexual contact:

Case 1. In a written statement, Officer Flanigan stated that I was working the night shift and doing my rounds in a dormitory and I saw inmate J masturbating in her bunk. I walked up to her and told her to stop but she unzipped my pants and rubbed my penis and placed it in her mouth. Well after that she demanded $40 from me and then $80. I didn't pay and she turned me in. I admit to the whole thing and I know I was wrong.

Case 2. In a written statement, Officer Adams admitted that she loved inmate D and was involved in a personal relationship with D going on two years. Adams stated that she got to know D through casual conversations, which eventually grew into the present relationship. Adams also stated that she was in love with D and had discussed marriage and life together after he was released. Adams admitted that the two had engaged in acts of sexual intercourse in the chapel. She also admitted that she placed $350 in D's trust fund.

Case 3. In written statements, Lieutenant Tan and Sergeant Green stated that they witnessed Officer Hays [female employee] stroking and fondling inmate H's penis in the dayroom. Officer Hays was also kissing inmate H and stroking his buttocks. When confronted by unit supervisors, Officer Hays, who was married and had two children, admitted the acts were true.

Case 4. In a written statement, Officer Toms observed Officer Funworth [female employee] being fondled by inmate A. The inmate had his hands under the shirt of Officer F and was fondling her breasts and she did nothing to stop it. When confronted, inmate A and Officer F admitted the act and Officer F resigned in lieu of termination.

Sexual contact typically occurred away from supervisory staff, a finding that parallels the police literature (Kappeler et al. 1994).

Kitchen, chapel, closets, laundry, library, bathrooms, and offices were the primary locales for sexual contact. Discovery by coworkers was usually accidental. In eight incidents, unit supervisors staked out the employee by capturing the incident on film, on microphone, and/or through "wired" inmates. In the other cases, suspected employees under investigation for dual relationships admitted to the relationship and sexual contact when confronted by investigators.

The employees' version of sexual contact illustrates the strong gender differences in this form of occupational deviance. Male employees typically were sexual predators. The male prison officers' uniform (as a symbol of power and authority) and their custodial access to female prisoners were conducive to exploitation (Kraska and Kappeler 1995). Female prison employees' motivation for sexual contact with prisoners involved love and postprison commitments. We found no evidence that female employees used sex as a mechanism (a commodity or a quid pro quo) to exploit inmates for personal gain. The inmates' perspective, when examined, will support or negate the female employees' interpretation of the situation.

Comparison of Boundary Violations by Situations

Four general types of situation led to official punishment. The employees did not simply violate policy in a vacuum; various situational dynamics must be accounted for or explained. Table 4 presents the findings from a comparative analysis of the three boundary violations by the four situations. We found a statistically significant relationship between boundary violations and situations.

Rescue situations. In these situations, an employee felt sorry for a prisoner and then broke the rules to aid or assist him or her. We found only five rescue situations, but they involved unusual contexts worthy of separate treatment. In one instance, an employee placed

Table 4	Boundary Violations Compared With Situations, 1995–1998

| | Type of Boundary Violation*** | | | |
Type of Situation	General Boundary Violations	Dual Relationships	Sexual Acts	Total
Rescuer	.00% (0)	1.00% (4)	2.00% (1)	5
Näiveté/Accidents	84.00% (32)	16.00% (67)	2.00% (1)	100
Lovesick	13.00% (5)	75.00% (326)	60.00% (25)	356
Predators	3.00% (1)	7.00% (31)	36.00% (15)	47
Count	38	428	42	508

NOTE: Numbers are given in parentheses.
*** $p < .001$

money in the inmate's trust fund after the inmate told the employee of his dire situation, "now that he had AIDS." In another instance, a male employee wanted to "save" an inmate from a lesbian relationship, and counseled her on religious proscriptions against homosexuality. Rescue situations typically involved employees who violated organizational policies to save inmates from a "life-threatening" situation. Rescuers expressed no remorse about their actions and insisted that they were providing "therapy" to the inmate.

Naïveté or accidents. One hundred employees were involved in situations that we defined loosely as naïveté or accidents: situations in which the employees appeared to be naïve or socially unaware about the professional relationship between themselves and inmates (Gabbard and Lester 1995). The employees apparently were poorly prepared for correctional work, and/or had failed to internalize (or understand) the custodial frame. We found several employees who tried to place money in an inmate's trust fund after purchasing craftwork from the inmate. The hallmark of these naïve situations or accidents was remorse; the employees all expressed profound self-reproach for not knowing the rules, and regretted their transgression. Further, 69 of these situations occurred in newer

prison units; this finding suggests a link between expansion, organizational culture, and the internalization of professional boundaries by employees and supervisors. Additional preservice training may be needed to prepare new employees to work in new units.

Lovesickness. We found 356 employees enmeshed in lovesick situations. These involved "romantic idealism," in which the employee disregarded the offender's personal history, placed the inmate on a pedestal, and fantasized about life together when the inmate was released. Their love was regarded as so extraordinary that employee policies and ethical codes were irrelevant (Gabbard 1995). In dual or overlapping boundary violations and lovesick situations, romantic love and idealization of the inmate were often combined.

Seventy-five percent of all dual/overlapping relationships involved lovesick situations. Additional data analysis found that 268 (75 percent) of these relationships involved female employees: The employee perceived the inmate as a person, a boyfriend, or a soulmate. These employees actively fostered a relationship with an inmate, typified by the letters presented above. Some married employees had experienced a catastrophic traumatic event before or during employment: their correspondence to inmates mentioned domestic violence, sexual frustration, marital strife or discord, boredom, ruptured dreams, or separation from their spouses. Others felt at a loss because their children had left the nest, or someone close to them had passed away. This personal vulnerability then was sensed by the inmates, who seized upon the employees' troubles to manipulate the situation.

Personal vulnerability coupled with institutional vulnerability (proximity to and manipulation by inmates) fostered or enhanced employees' deviance. Gabbard and Lester (1995) noted a similar phenomenon among therapists who fell in love with their patients, in which infatuation occurred in connection with extreme stress in the analyst's life (e.g., divorce, separation, disillusionment with their own career or marriage). The prison employees willingly accepted the inmates' attention and violated policy.

The combination of dual relationships and lovesick situations unfolded in duty posts such as the laundry, commissary, issue room, kitchen, library, chapel, dormitories, and mailroom. These situations often involved one employee (in an out-of-the-way location) and one or more prisoners. We realize that prison employees, regardless of gender, are assigned to a variety of duty posts on the basis of need. Lovesick situations, however, develop in institutional areas where supervisory contact is minimal and where employees and offenders can interact with each other as males and females, eroding the boundary between employee and offender. Not all female employees who work (for example) in ID stewards' departments succumb to prisoners' advances; our data simply suggest that specific hot spots exist in the units and bear close supervision.

Subsequent data analysis led us to conclude that the combination of dual relationships and lovesick situations led to sexual encounters: 60 percent of the sexual contacts involved lovesick situations. These data suggest a slippery slope, in which some employees (at least those who were caught) who develop a romantic relationship with an inmate have sexual contact. The progression from nonsexual to sexual contact has been found as well among psychotherapist-patient relationships (Stake and Oliver 1991; Strasburger, Jorgenson, and Sutherland 1992).

Predators. We found 47 predatory situations in which employees sought out and manipulated prisoners, and thought they could get away with it for personal or private gain (Friedman and Boumil 1995). In these situations, the employee actively preyed on an inmate for money, property, and/or sex. One employee, for example, established relationships with offenders and then cased their possessions and stole their property (e.g., watches, necklaces, rings) under the cover of cell searches. Male employees were more likely to become involved in predatory situations.

Kraska and Kappeler's (1995; see also Horswell 2000; Vaughn 1999) demonstrated that police sexual violations ranged from unobtrusive behavior (e.g., viewing photographs of victims) to criminal behavior (e.g., sexual assault). We found two primary types of correctional employee-inmate sexual contact.

First, as stated above, 25 cases (60 percent) involved lovesick situations: 22 employees were females and three were males. Similarly, Gartrell and colleagues (1986) found that 65 percent of psychiatrists who were sexually involved with a patient described themselves as in love with the patient. Among the prison employees, we found many letters filled with declarations of love, infatuation, lifetime commitment, and plans beyond the penitentiary. Some employees and inmates even exchanged vows and rings, and referred to themselves as married couples.

The employees' stories suggests that staff members took their first small steps down the slippery slope of boundary erosion, and ended up much farther down than they ever could have imagined (Strasburger et al. 1992). Our findings support those of Strom-Gottfried (1999), who found that clinicians who "break frame" often proceed from minor to major transgressions. How many dual relationships culminate in sexual contact is unknown.

In addition, we found 15 cases (all male employees and female prisoners) of employee-inmate sexual contact that involved predatory situations. In 14 of these cases, male employees used their position and power to procure sexual favors from female prisoners, such as sex in exchange for preferential treatment or marriage upon release. Sex becomes a commodity, and an underground economy evolves between male staff and female prisoners; this situation encourages female prisoners to prostitute themselves in exchange for preferential treatment and to avoid punishment or harm. Such behavior institutionalizes a form of exploitation and sexual slavery (Jolin 1994). The predatory situations we researched paralleled Sapp's (1997:146) discussion of "sexual shakedowns" committed by police officers.

Currently it is unknown whether or not these 14 employees have been criminally prosecuted or have been served with civil lawsuits. All were terminated from prison employment. In one case of sexual contact, a male employee cohabited with a female parolee, whom he met while she was incarcerated. Upon questioning, both admitted to sexual activity. Under current Texas law, it is a second-degree felony (sexual assault) for a public servant to coerce another individual to participate in sexual activity (Texas Penal Code 2000). The data suggest that male prison employees, like police officers, may be engaging in sexual activity with female prisoners simply because they can. The presence of numerous potential partners in routine encounters with inmates provides the opportunity for seduction (Crank 1998:145).

Female prisoners, like women in the wider community, are at risk of victimization by coercive sexual strategies simply because of their gender (Finkelhor and Asdigian 1996; Pogrebin and Poole 1997). The data displayed in Table 4 illustrate that employees' boundary violations were linked conceptually to certain situations. Our findings indicate the need to view each boundary violation and the specific situation as part of a process rather than as an isolated deviant episode. For example, employees might begin by asking inmates about their life and how they arrived in prison. An inmate might respond to this concern by writing the employee a note of thanks. If the employee reciprocates, the inmate receives a "mixed message" and assumes that the employee wants a relationship.

Inappropriate employee-inmate relationships often involved a mixture of situations, behaviors, emotions, needs, and human desires. Breaking the custodial frame entails an array of actions that must be considered in combination with various situational determinants. Our analyses of boundary violations and situational contexts illustrate the complexity of such inappropriate relationships.

Additional analyses (not reported in Table 4) revealed a significant finding when we examined the relationship between incidents and the average number of months on the job at time of punishment. Employees involved in predatory situations averaged nearly 11 years of tenure at punishment. Predator situations are probably difficult to detect because employees take great care to cover their actions. It also appears that rescue situations occur quickly, illustrating how readily prisoners can manipulate receptive employees. Oddly, employees caught in naïveté or accident situations averaged 44 months of tenure at punishment. We expected these situations to unfold more quickly and cannot account for this finding, unless it relates to the difficulty of detection and the need for thorough investigation. Employees caught in lovesick situations averaged 37 months at punishment.

Why would seemingly normal employees become romantically (and even sexually) involved with prisoners? Haney, Banks, and Zimbardo (1973) posed a similar question when they asked why bright, emotionally stable young male college students (most of whom were in the peace movement) would abuse their fellow students in a prisonlike environment. Their answer to this question emphasized the surprising power of institutional environments over the participants (Haney and Zimbardo 1998:710). As Haney and colleagues (1973:90) suggested,

"[T]he abnormality here [the mock prison] resided in the psychological nature of the situation and not in those who passed through it." Over the past 27 years, these researchers have generated a body of research that clearly illustrates how "mind altering social psychological dynamics" can bend and twist human behavior (Haney and Zimbardo 1998:710). If one accepts the situational explanation, then complex issues such as consent become difficult to define and comprehend. The inmates' version must be researched if we are to fully ascertain how and why these dual relationships develop.

◪ Discussion and Conclusions

This paper represents the first study of boundary violations between correctional employees and inmates. It is based on four years of data, 1995–1998, a period of substantial organizational expansion that could affect the validity of our findings (Cook and Campbell 1979). Also, our formulation of the continuum of boundary violations and situational dynamics is incomplete. Clearly the types of boundary violations overlap, and it is not possible to arrange a linear relationship among these behaviors (Strom-Gottfried 1999). We suspect that each combination of boundary violation and situation contains subsets of other violations, as in the case of dual relationships and lovesick situations.

With these warnings in mind, we produced several important findings with implications for correctional personnel management, training, theory, and research on occupational deviance in penal settings.

First, the most important finding from our analysis was that inappropriate employee-inmate relationships must be understood as boundary violations. Custodial framebreaks involved various behaviors and activities. In this way, Strom-Gottfried's (1999) conceptualization of boundary violations (general, dual/overlapping, and sexual) enhances our understanding of this phenomenon in prisons. Dual/overlapping relationships were the most frequent form of deviance; typically they involved female employees who became romantically involved with prisoners. Boundary violations, especially dual relationships, involved a process or sequence of behaviors that transpired over weeks and months.

Second, our comparative analysis found that employees who broke the custodial frame were significantly more likely than not to be Anglos and females who had a GED and no military experience. Investigated employees received significantly lower pre-employment evaluation scores than successful employees; lower scores were risk markers for boundary violations. Investigated employees also were more likely than successful employees to have a history of rule violations.

Third, inappropriate employee-inmate relationships unfolded in specific situations. The most common context involved lovesick female employees who, through manipulation by inmates or their own volition, negated the role differences between themselves and the inmate. These employees viewed the inmate as a soulmate, a boyfriend or girlfriend, a pen pal lover, or a future spouse. The most important aspect of these situations was the employee's recasting of the offender as a nonprisoner or a "person who just happens to be in prison." In some cases, the employee identified with the prisoner. Ironically, inmates, rather than being powerless in their dealings with the staff, exert some control over interested employees. Naïveté/accident situations often occurred at newer prison units; thus it is possible that organizational expansion may precipitate these situations as a result of an unsettled organizational culture.

Fourth, most custodial framebreaks occurred within 36 months of employment. The first 12 months of tenure represent an at-risk period in which some employees (e.g., those with low application scores and with general equivalency diplomas) have not bonded to the organization, and perceive offenders to be "just other people." Perhaps seasoned employees could conduct additional on-the-job training, stressing the pitfalls and consequences of boundary violations to strengthen the new employees' ties to the organization. The interviewees strongly suggested that prison organizations should use "field training officers" (like their police counterparts) to guide and mentor new employees in the critical first few months on the job. Under the rubric of employee survival skills, the Texas prison system recently has expanded its pre-service training from one hour on "games inmates play" to nearly eight hours. This change in training underscores the agency's sensitivity to employees' vulnerability, and to the need for tools to forestall further problems.

Fifth, sexual violations represented 8 percent (or 42 cases) of all violations, a figure paralleling the research on sexual contact in other occupations. Most sexual contacts involved lovesick female employees; these situations suggest the escalation of an employee-inmate relationship from nonsexual to sexual contact. We found 14 male employees who used their position to procure sexual favors from female prisoners. Sexual contact is the most serious violation of professional conduct: currently, in 42 states, legislation prohibits sexual contact between correctional employees and inmates. Since 1996, 20 departments of corrections have developed or revised their policies regarding sexual misconduct (National Institute of Corrections 2000).

Sixth, research is needed on the effects of sex-role spillover in prisons. Females employed in male prisons and males employed in female prisons work in nontraditional jobs, and sex-role spillover can be studied empirically as an underlying factor contributing to employees' boundary violations. Such research also will clarify the changing and complex nature of the corruption of authority.

The deviant employees studied here violated policy voluntarily. To suggest that there was something wrong with them initially implies that improved recruitment, proper screening measures, psychological tests, and incentives (i.e., merit raises) can eliminate inappropriate personal contact between employees and inmates. Osgood et al. (1996: 639), however, suggest that "most people have the potential for at least occasionally succumbing to an opportunity for deviant behavior." Future prison researchers, like their police counterparts (Sherman 1980), must examine the individual, situational, and organizational correlates of boundary violations in prisons. Such violations, and the situations in which they unfold, are an ironic aspect of the pathology of imprisonment, as well as an eternal defect of total power.

Notes

1. We interviewed one of the first women employed as a security officer in a male Texas prison. She said she interacted with a notorious male prisoner who worked in the front office with her. Shortly after the inmate was paroled, he was killed while robbing a 7-11. Upon hearing of the inmate's death, the officer became upset. She remembered saying to a coworker "Who would want to kill Bill [an alias]? He was just trying to make it." This employee assumed that Bill was working in the store, not that he was trying to rob it. This story illustrates how staff members sometimes see inmates as people who happen to be in prison rather than as dangerous felons.

2. See Coble v. Texas Department of Corrections (1983); Dothard v. Rawlinson (1977); Griffin v. Michigan Department of Corrections (1982); Grummett v. Rushen (1985).

3. Eighty-eight employees had alleged or inferred sexual contacts, and 32 engaged in nonsexual contacts (e.g., hugging, hand holding).

4. Studies of psychologists have yielded rates of sex misconduct between 5 percent and 12 percent for males, and 1 percent and 3 percent for females (Stake and Oliver 1991). Research on male psychotherapists revealed that 7 percent had reported sexual contact with female clients (Pope and Bouhoutsos 1986). Kardener, Fuller, and Mensch (1973) found that between 5 percent and 13 percent of therapists had engaged in erotic behavior with their patients. We found that 5 percent of the female officers and 4 percent of the males engaged in sexual misconduct with inmates.

5. Rule 42, Employee-Inmate/Client Relationships (association or correspondence): Employees are prohibited from continuing or establishing any personal relationships with inmates/clients or with family members of inmates/clients which jeopardizes, or has the potential to jeopardize, the security of the Agency or which compromises the effectiveness of the employee. Employees are required to report to agency officials any previous or current relationships between: (1) the employee with an inmate/client; (2) the employee with a family member of an inmate/client; (3) a family member of the employee with an inmate/client; or (4) a family member of the employee with a family member of an inmate/client.

6. Parole/health services staff members work for separate agencies. We lacked access to these records.

7. We began in 1995 because these personnel files represented the most recent and most accessible files for individual review, coding, and analysis.

8. Internal Affairs investigators informed us that employees who resigned would have been terminated in most cases. Therefore we decided to keep resignation in the database.

9. The criteria are as follows: (1) eligible to work in the United States; (2) at least 18 years old, with no age maximum; (3) high school diploma or GED; (4) males age 18 to 25 must be registered with the Selective Service; (6) not on active duty in the military; (6) no felony or convictions for drug-related offense; (7) no convictions for offense involving domestic violence; (8) no Class A or B misdemeanor convictions within the last 5 years; (9) not on probation for any criminal offense or pending charges; and (10) pass TDCJ pre-employment test and drug test.

◪ References

Amnesty International. (1999). *Sexual Abuse, International Women's Day, 1999.* London, UK: Amnesty International.

Baro, A. (1997). "Spheres of Consent: An Analysis of the Sexual Abuse and Sexual Exploitation of Women Incarcerated in the State of Hawaii." *Women and Criminal Justice* 8:61–84.

Board of Inquiry. (2000). "Rampart Area Corruption Incident, Los Angeles Police Department." Bernard C. Parks, Chief of Police, Executive Summary, March 1.

Camp, C. and G. Camp. (1995–1998). *Corrections Yearbook, 1995–1998.* Middletown, CT: Criminal Justice Institute, Inc.

Carroll, L. (1974). *Hacks, Blacks, and Cons: Race Relations in a Maximum Security Prison.* Lexington, MA: Lexington Books.

Commission to Investigate. (1994). "The Report of the Commission to Investigate Allegations of Police Corruption and the Anti-Corruption Procedures of the New York City Police Department." New York: City of New York.

Cook, T. and D. Campbell. (1979). *Quasi-Experimentation: Design and Analysis Issues for Field Settings.* Boston, MA: Houghton Mifflin.

Crank, J. (1998). *Understanding Police Culture.* Cincinnati, OH: Anderson.

Crouch, B.M. and J. Marquart. (1980). "On Becoming a Prison Guard." Pp. 63–106 in *The Keepers: Prison Guards and Contemporary Corrections,* edited by B.M. Crouch. Springfield, IL: Thomas.

——. (1989). *An Appeal to Justice: Litigated Reform of Texas Prisons.* Austin, TX: University of Texas Press.

Day, A. (1998). "Cruel and Unusual Punishment of Female Inmates: The Need for Redress Under 42 U.S.C. 1983." *Santa Clara Law Review* 38:555–87.

DiIulio, J. (1987). *Governing Prisons.* New York: Free Press.

Dobash, R., R. Dobash, and S. Gutteridge. (1986). *The Imprisonment of Women.* Oxford: Basil Blackwell.

Elliot, J. (1999). "Former Parole Officer Liable in Civil Suit," *Texas Lawyer,* October 4, pp. 9–12.

Epstein, R. 1994. *Keeping Boundaries.* Washington, DC: American Psychiatric Press.

Erikson, K. (1966). *Wayward Puritans.* New York: Wiley.

Eschholz, S. and M. Vaughn. (2001). "Police Sexual Violence and Rape Myths: Civil Liability Under Section 1983." *Journal of Criminal Justice* 29:389–405.

Finkelhor, D. and N. Asdigian. (1996). "Risk Factors for Youth Victimization." *Violence and Victims* 11:3–19.

Friedman, J. and M. Boumil. (1995). *Betrayal of Trust: Sex and Power in Professional Relationships.* Westport, CT: Praeger.

Gabbard, G. (1995). "The Early History of Boundary Violation in Psychoanalysis." *Journal of the American Psychoanalytic Association* 43:1115–36.

Gabbard, G. and E. Lester. (1995). *Boundaries and Boundary Violation in Psychoanalysis.* New York: Basic Books.

Gartrell, N., J. Herman, S. Olarte, M. Feldstein, and R. Localio. (1986). "Psychiatrist-Patient Sexual Contact: Results of a National Survey." *American Journal of Psychiatry* 14:690–94.

Geier, A. (2000). "Smith Says She Had Sex With a Guard." Associated Press. Retrieved August 31, 2000 (http://www.Charleston.net).

Goffman, E. (1974). *Frame Analysis*. Boston, MA: Northeastern University Press.

Gonsiorek, D. (1995). *Breach of Trust: Sexual Exploitation by Health Care Professionals and Clergy*. Thousand Oaks, CA: Sage.

Gutek, B. (1985). *Sex and the Workplace*. San Francisco, CA: Jossey-Bass.

Haney, C, W. Banks, and P. Zimbardo. (1973). "Interpersonal Dynamics in a Simulated Prison." *International Journal of Criminology and Penology* 1:69–97.

Haney, C. and P. Zimbardo. (1998). "The Past and Future of U.S. Prison Policy." *American Psychologist* 53:709–27.

Hanson, E. (1999). "Prisoner Who Said Deputy Raped Her Wins Suit." *Houston Chronicle*, October 27, Section 2, p. 3.

Horswell, C. (2000). "Patrol Man Resigns Amid Sex Scandal." *Houston Chronicle*, February 17, p. 35a.

Jacobs, J. (1983). "Female Guards in Men's Prison." Pp. 178–201 in *New Perspectives on Prisons and Imprisonment*, edited by J. Jacobs. Ithaca, NY: Cornell University Press.

Jolin, A (1994). "On the Backs of Working Prostitutes: Feminist Theory and Prostitution Policy." *Crime and Delinquency* 40:69–83.

Kappeler, V., R. Sluder, and G. Alpert. (1994). *Forces of Deviance*. Prospect Heights, IL: Waveland.

Kardensr, S., M. Fuller, and I. Mensch. (1973). "A Survey of Physicians' Attitudes and Practices Regarding Erotic and Non-Erotic Contact With Patients." *American Journal of Psychiatry* 130:1077–81.

Kraska, P. and V. Kappeler. (1995). "To Serve and Pursue: Exploring Police Sexual Violence Against Women." *Justice Quarterly* 12:85–111.

Krauss, C. (1994). "The Perils of Police Hiring." *New York Times*, September 18, Section 4, p. 3.

Marquart, J. (1986). "The Use of Physical Force by Prison Guards: Individuals, Situations, and Organizations." *Criminology* 24:347–66.

Martin, S. and N. Jurik. (1996). *Doing Justice, Doing Gender*. Thousand Oaks, CA: Sage.

McDonald, M. (1997). "A Multidimensional Look at the Gender Crisis in the Correctional System." *Law and Inequality* 15:505–46.

Mooney, T. (1995). "In Prison, Sex, Abuse Prevalent." *Providence Journal-Bulletin*, April 3, p. A4.

National Institute of Corrections. (2000). *Sexual Misconduct in Prisons: Law, Remedies, and Incidence*. Longmont, CO: Information Center, U.S. Department of Justice.

National Women's Law Center. (1996). *Fifty-State Survey on State Criminal Laws Prohibiting the Sexual Abuse of Female Prisoners*. Washington, DC: National Women's Law Center.

Olson, E. (1999). "UN Panel Is Told of Rights Violations at U.S. Women's Prisons." *New York Times*, March 31, p. A16.

Osgood, D., J.K. Wilson, P.M. O'Malley, J.G. Bachman, and L.D. Johnston. (1996). "Routine Activities and Individual Deviant Behavior." *American Sociological Review* 61:635–55.

Phillips, R. (1999). "Jail Sex Scandal Nets Twelve Indictments." Retrieved December 17, 1999 (APBnews.com).

Pogrebin, M. and E. Poole. (1997). "The Sexualized Work Environment: A Look at Women Jail Officers." *Prison Journal* 77:41–57.

Pope, K. and J. Bouhoutsos. (1986). *Sexual Intimacy Between Therapists and Patients*. New York: Praeger.

Rafter, N. (1985). *Partial Justice*. Boston, MA: Northeastern University Press.

Rivera, G. (1999). "Women in Prison: Nowhere to Hide." Retrieved September 10, 1999 (www.msnbc.com .news/geraldorivera_front.asp).

Robison, C. (1991). "Prison OK'd by Voters May Never Be Built." *Houston Chronicle*, November 7, p. A21.

Rutter, P. (1989). *Sex in the Forbidden Zone*. Los Angeles, CA: Jeremy Tarcher.

Sapp, A. (1997). "Police Officer Sexual Misconduct." Pp. 139–51 in *Crime and Justice in America*, edited by P. Cromwell and R. Dunham. Upper Saddle River, NJ: Prentice-Hall.

Sherman, L. (1980). "Causes of Police Behavior: The Current State of Quantitative Research." *Journal of Research in Crime and Delinquency* 17:69–100.

Siegal, N. (1999). "Stopping Abuse in Prison." *The Progressive* 63:31–33.

Stake, J. and J. Oliver. (1991). "Sexual Contact and Touching Between Therapist and Client." *Professional Psychology* 22:297–307.

Statham, A. (1996). *The Rise of Marginal Voices: Gender Balance in the Work Place*. Lanham, MD: University Press of America.

Steffensmeier, D. and E. Allan. (1996). "Gender and Crime: Toward a Gendered Theory of Female Offending." *Annual Review of Sociology* 22:459–87.

Stewart, C. (1998). "Management Response to Sexual Misconduct Between Staff and Inmates." *Corrections Management Quarterly* 2:81–88.

Strasburger, L., L. Jorgenson, and P. Sutherland. (1992). "The Prevention of Psychotherapist Sexual

Misconduct: Avoiding the Slippery Slope." *American Journal of Psychotherapy* 4:544–54.

Strom-Gottfried, S. (1999). "Professional Boundaries: An Analysis of Violations by Social Workers." *Families in Society* 80:439–49.

Sykes, G. (1958). *Society of Captives.* Princeton, NJ: Princeton University Press.

Texas Penal Code, 14th Edition. 2000. Eagan, MN: West Publishing Company.

Vaughn, M. (1999). "Police Sexual Violence: Civil Liability Under State Tort Law." *Crime and Delinquency* 45:334–57.

"Virginia Governor Orders Prison Sex-Abuse Inquiry." (1999). *New York Times,* October 13, pp. A22.

White, W. (1995). "A Systems Perspective on Sexual Exploitation of Clients by Professional Helpers." Pp. 220–44 in *Breach of Trust: Sexual Exploitation by Health Care Professionals and Clergy,* edited by J. Gonsiorek. Thousand Oaks, CA: Sage.

Wincze, J., J. Parsons, J. Richards, and S. Bailey. (1996). "A Comparative Survey of Therapist Sexual Involvement Between an American State and an Australian State." *Professional Psychology* 27: 289–94.

Zedna, L. (1995). "Wayward Sisters: The Prison for Women." Pp. 295–324 in *The Oxford History of the Prison,* edited by N. Morris and D. Rothman. New York: Oxford University Press.

Cases Cited

Coble v. Texas Department of Corrections, 568 F. Supp. 410 (1983)

Dothard v. Rawlinson, 433 U.S. 321 (1977)

Griffin v. Michigan Department of Corrections, 654 F. Supp. 690 (1982)

Grummett v. Rushen, 779 F. 2d 491 (1985)

DISCUSSION QUESTIONS

1. Why do correctional officers view inmates differently than people in the free world view them?

2. What factors lead to corruption of authority in the corrections workplace?

3. What constitutes a boundary violation, according to the authors of this study?

READING

Craig Dowden and Claude Tellier's meta-analysis examines the predictors of job stress in correctional officers. Twenty studies were selected for inclusion, producing 191 individual effect size estimates. The authors findings revealed that work attitudes (i.e., participation in decision making, job satisfaction, commitment, and turnover intention) and specific correctional officer problems (i.e., perceived dangerousness and role difficulties) generated the strongest predictive relationships with job stress overall. Furthermore, both favorable (i.e., human service/rehabilitation orientation and counseling) and unfavorable (i.e., punitiveness, custody orientation, social distance, and corruption) correctional officer attitudes yielded moderate relationships with job stress, with the country of study emerging as a critical moderating variable. The weakest correlates of job stress were demographic variables and job characteristics (e.g., security level). The authors conclude by offering the implications of their findings and by providing directions for future research.

Predicting Work-Related Stress in Correctional Officers

A Meta-Analysis

Craig Dowden and Claude Tellier

⊠ Introduction

Several recent reviews of the research literature exploring job stress in correctional officers discussed the real or perceived high-stress nature of this type of work (Finn, 1998; Huckabee, 1992; Schaufeli & Peeters, 2000). There was, however, a paucity of empirical evidence supporting this contention and, more importantly, no study demonstrated that this line of work was more stressful than other types of occupations (Huckabee, 1992). Nonetheless, the ubiquitous assumption that correctional work is indeed stressful has increased the amount of research attention given to this topic over the past two decades.

Despite the lack of direct comparative evidence, several studies illustrated the multiple consequences of pursuing a career as a correctional officer using more proximal outcomes. Cheek and Miller (1983) revealed that the average rates of divorce and stress-related illnesses (i.e., heart disease, hypertension, and ulcers) for correctional officers were unusually high, while another study reported that the average life span of correctional officers (fifty-nine years) was sixteen years lower than the national average (Cheek, 1984). The increased health problems found in correctional personnel were also documented by Adwell and Miller (1985) who found that correctional officers were more likely to suffer from heart attacks, high blood pressure, and ulcers than

members of the general public. Finally, negative health outcomes were documented in countries other than the United States. For example, one study that compared Swedish prison staff to individuals in other occupations revealed that a significantly higher proportion of correctional officers reported heart disease, diabetes, asthma or bronchitis, and hypertension (Harenstam, Palm, & Theorell. 1988).

The adverse impact of working as a correctional officer manifested in other ways than those listed above. For example, a study of Israeli prison officers found that feelings of occupational tedium, emotional exhaustion, and negative attitudes toward self and others were so widespread that 50 percent of the officers left within the first eighteen months of taking the job (Shamir & Drory, 1982). In Canada, several studies reported similar findings. More specifically, Karlinsky (1979) reported that job-related stress was the most frequently mentioned job concern and the largest source of job dissatisfaction while, more recently, Kelloway, Desmarais, and Barling (2000), using absenteeism as a proxy for stress, found that correctional employees who worked in institutions evidenced more absenteeism than those who worked in other locations. More importantly, across the two years for which complete data was available, the front line correctional personnel (CX) group experienced the highest absenteeism rates. Finn (1998) also highlighted the high

SOURCE: Dowden, C., and Tellier, C. (2004). Predicting Work-Related Stress in Correctional Officers: A Meta-Analysis. *Journal of Criminal Justice, 32*(1), 31–47.

sick leave and turnover rates within his review of the correctional officer job stress literature.

Defining Stress

Despite the widespread use of the word in both academic and nonacademic publications, there was a noticeable lack of consensus with regard to what actually constitutes stress. This situation evolved, in no small part, as a result of the various ways in which stress was operationalized (Parker & DeCotis, 1983). For example, stress has been "treated as a stimulus, a response, an environmental characteristic, an individual attribute, and an interaction between an individual and his or her environment" (p. 161). Although there is still some disagreement over whether stress is the antecedent or the result of various strains, the latter conceptualization was used in the present review as this definition fits well within the broader confines of the job stress literature for correctional officers. Furthermore, for the purposes of this review, the factors that contributed to the stress response in the work place were viewed as job stressors.

One final point should also be made regarding the construct of stress. Selye (1982), one of the leading experts in this field of research, differentiated both positive and negative forms of stress, labeled *eustress* and *distress* respectively. For the purposes of this review, the negative side of stress was the sole focus.

Predictors of Stress in Correctional Officers

Research focusing specifically on correctional personnel and their attitudes has become an increasingly important and prevalent area of scientific inquiry over the past twenty years (Schaufeli & Peeters, 2000). This was not always the case, however. Over twenty years ago, Jacobs (1978) reported that although there were many studies published on the attitude and demographic characteristics of inmates, there was a noticeable absence of similar research that examined correctional personnel. One area that was neglected was work stress. This was especially surprising given the importance of the correctional officer in terms of providing for the welfare of inmates, as well as their integral role in the attainment of institutional goals (Long, Shouksmith, Voges, & Roache, 1986).

Fortunately, research focusing on work stress in correctional personnel increased in the 1980s (Triplett, Mullings, & Scarborough, 1999), and received more detailed attention over the past decade. Three comprehensive literature reviews specifically focusing on stress in correctional personnel were published (Finn, 1998; Huckabee, 1992; Schaufeli & Peeters, 2000), and research recently examined gender (Gross, Larson, Urban, & Zupan, 1994; Hurst & Hurst, 1997; Triplett et al., 1999; Van Voorhis, Cullen, Link, & Wolfe, 1991) and ethnic differences (Van Voorhis et al., 1991; Wright & Savior, 1992) in the work stress experienced by correctional officers.

The Need for a Meta-Analytic Review

Despite increased attention, very little consensus currently existed regarding the underlying sources of stress for correctional officers. Most important for the present review, each summary of the job stress literature for correctional officers lamented the lack of systematic and consistent findings in the field. Huckabee concluded his review of this literature by stating "at this point, there seems to be no clear consensus as to which factors can be consistently correlated with stress in corrections" (1992, p. 484) while Finn (1998) stated that his review "confirmed that there is little reliable empirical evidence that identifies the severity and sources of stress for correctional officers" (p. 72).

Given the current disparate and inconclusive findings in this area, a meta-analytic review provides the best mechanism to appropriately and systematically aggregate the results to date while simultaneously highlighting areas in need of further research. This approach enables researchers to have a much clearer and more precise description of the "state-of-the-art" and allows finer detailed analyses than possible through the traditional narrative review.

A meta-analysis is a statistical technique that aggregates the results of a group of independently conducted studies in order to form a conclusion (Glass, McGaw, & Smith, 1981). Meta-analytic reviews essentially replaced traditional narrative reviews given that narrative reviews were typically viewed as more subjective and open to reviewer bias (Rosenthal, 1991; Wolf. 1986). Furthermore, and perhaps more importantly, as highlighted by Glass et al. (1981), individual studies were no more "comprehensible without statistical analysis than the hundreds of data points in one study" (p. 12). A meta-analysis addresses each of these concerns within its methodological framework.

Meta-analyses were conducted in diverse areas of research including criminal justice, education, medicine, and other social science's (Lipsey & Wilson, 1993). In the field of criminal justice in particular, meta-analytic techniques were used to consolidate the prediction (Bonta, Law, & Hanson, 1998; Dowden & Brown, 2001; Gendreau, Little, & Goggin, 1996; Hanson & Bussiere, 1998) and treatment literatures (Andrews et al., 1990; Dowden & Andrews, 1999, 2000; Izzo & Ross, 1990; Losel, 1995; Whitehead & Lab, 1989).

Despite the research attention given to the area of correctional officer stress in terms of primary studies and traditional narrative literature reviews, a meta-analysis has yet to be conducted on this topic. Therefore, the purpose of the present article is to provide the first meta-analytic examination of the strongest predictors of correctional officer stress. Furthermore, the meta-analytic approach taken will enable a more systematic exploration of the questions under review than previously possible.

▨ Method

Sample of Studies

A literature search was conducted to identify published studies between January 1950 and January 2001 using the PsycInfo and National Criminal Justice Reference Service (NCJRS) computerized databases. Key search terms included: correctional officers, correctional personnel, attitudes, job stress, work stress, stress, burnout, stressors, and distress. Furthermore, the reference section of each included article was surveyed to identify potential unpublished and published studies.

As mentioned previously, Schaufeli and Peeters (2000) conducted the most recent narrative literature review of job stress and correctional officers. The forty-three studies identified in their review were also examined to determine their suitability for inclusion. It should be noted that these search techniques resulted in approximately 150 studies being identified as potential candidates for inclusion.

Inclusion Criteria

In order to be included, studies were required to meet the following criteria: (1) the study included an outcome measure of job stress, that was operationalized according to the definition presented earlier, and (2) sufficient statistical information was available to allow the reported statistic to be converted into an effect size using the procedures outlined by Rosenthal (1991). Overall, twenty studies met the criteria and were included in the meta-analysis. Although the number of studies selected for inclusion in this review might seem small based on the number of initial studies, an important reason for this was that the primary focus of the initial literature search was to gather studies exploring correctional officer attitudes. Of the studies that met this broad selection criterion, however, a much smaller number included job stress as one of their outcome measures. Several additional reasons also contributed to this small number of eligible studies and these were more specific to the correctional officer stress literature and will be discussed below.

As stated previously, for this review, work stress was defined as any psychological work-related discomfort or anxiety. Studies that utilized physiological measures of stress (i.e., Lasky, Gordon, & Srebalus, 1986; Pollack & Sigler, 1998, etc.) were excluded to maintain consistency of outcome measures as well as to control for the lack of systematic study present in the field of correctional officer stress (Huckabee, 1992). The rationale for the former decision was based on the review by Cacciopo, Klein, Bemtson, and Hatfield (1993), which showed that self-report and physiological measures tapped into different aspects of the affective experiences of individuals and that these measurement scales were often uncorrelated. Further evidence for this decision came from a study conducted by Pollack and Sigler (1998) who found extremely low levels of work stress in correctional officers in the northern region of Ontario. The authors openly questioned whether these findings resulted from the type of physiological pathology stress measure they employed by pointing out "although the findings reflect lower levels of stress for Canadian correctional officers than reported in other studies of stress in correctional officers, the measures in other studies of correctional officers are not sufficiently similar to permit direct comparison" (p. 123). The decision to exclude studies using physiological outcome measures seemed appropriate based on these findings.

One further set of studies was excluded due to problematic outcome measures. More specifically, studies using measures of burnout such as the Maslach Burnout Inventory (Maslach & Jackson, 1981) were excluded. Although stress and burnout were often used interchangeably (Farber, 1983) in both the mainstream and academic literatures, burnout was the protracted consequence of unabated job stress. Therefore, the two terms, although related, should not be considered synonymous.

Thus, to preserve the integrity of the results of this meta-analysis, it was felt that combining the two outcomes could confound the interpretation of the results.

Finally, studies that examined the predictors of work-related stress for probation and parole officers were also excluded. Once again, this decision was made to maintain the homogeneity of the study group as well as to enhance external generalizability. Furthermore, the work environments for these two occupational groups were assumed to be sufficiently dissimilar to warrant this decision.

It should also be noted that several studies included in the meta-analysis were conducted by the same author or authors (i.e., Cullen, Link, Wolfe, & Frank, 1985; Lindquist & Whitehead, 1986a, 1986b, 1986c; Saylor & Wright, 1992; Van Voorhis et al., 1991; Wright & Saylor, 1992). In these cases, the authors often reported identical or slightly altered correlations across different studies that reported on the same sample of correctional officers. To ensure that each contributing effect size was based on a unique sample of individuals, only one effect size per predictor category was extracted from sets of studies that were clearly based on the same sample of correctional officers.

Predictor Categories

For the present meta-analysis, predictor categories were based on the most frequently reported stressors outlined within the correctional officer job stress literature. Further, predictor categories were mutually exclusive such that an effect size could only contribute once to each predictor category. Predictor categories were operationalized as follows:

- ◆ *Demographic information*: age, gender (female), ethnicity (minority), educational level, and marital status.
- ◆ *Shift* (evening).
- ◆ *Security level* (maximum).
- ◆ *Contact hours with inmates*.

- *Commitment:* the degree to which an individual wishes to remain a member of the organization and identifies with its goals and overall mission (Mowday, Steers, & Porter, 1979).
- *Turnover intention:* the degree to which the correctional officers had thought of quitting their job or a measure of their intentions to leave their current position.
- *Job satisfaction:* the degree to which the correctional officer was satisfied with his or her job.
- *Dangerousness:* the perceived dangerousness of the correctional environment by the correctional officer. In one study (Tellier & Robinson, 1995), this variable was operationalized by inverse scoring the correctional officers' satisfaction level with their personal safety.
- *Role problems:* the vast majority of studies on correctional officer stress generally failed to recognize the different forms of role problems encountered by correctional officers (Philliber, 1987). Although the present review attempted to divide these problems into role conflict and role ambiguity, only one study (Triplett et al., 1999) used role ambiguity. The remaining studies included measures of role conflict and a generic role problems measure adapted from Poole and Regoli (1980).
- *Favorable correctional attitudes:* this variable was composed of positive treatment/human services orientation and the counseling subscale of Klofas and Toch (1982). For the analysis of the composite variable, if more than one scale was used in a study, the average correlation between the scales was entered into the analysis.
- *Unfavorable correctional attitudes:* this variable was composed of punitiveness, custody orientation, and the social distance and corruption subscales of the Klofas and Toch (1982) measure of correctional orientation. Once again, for the analysis of the composite variable, if more than one scale was used in the primary study, the average correlation between the scales was entered into the analysis.

Although social support was thought to be a particularly important predictor variable, only two studies incorporated this measure and thus further analyses were deemed inappropriate. Furthermore, it should be noted that several of the above predictor categories were combined into super-ordinate categories where appropriate.

Effect Size Calculations

The primary unit of analysis in a meta-analysis is the effect size, which reflects the strength of the relationship between two variables (i.e., a predictor variable and job stress). Although several effect size estimates were available, the one chosen for the present review was the Pearson r correlation coefficient. Both unweighted (Mr) and weighted (Mz^+) effect sizes were reported in the research literature and while the weighted effect size adjusted the size of the correlation based on the size of the sample contributing to it, the former did not. Thus, studies based on larger samples of offenders were afforded more weight than studies based on smaller samples. Generally, the weighted effect size was considered to be more appropriate and was utilized in the present meta-analysis.

The relationship between each predictor category and job stress was extracted from the raw statistics reported in each study. If a statistic other than a Pearson r was reported (e.g., t, F, or chi-square), it was transformed into a Pearson r using the formulae provided by Rosenthal (1991). Further, if a study reported a nonsignificant relationship between a particular predictor category and job stress and evidence indicated that appropriate statistical analyses was conducted to support this contention, the effect size was recorded as .00.

Fail Safe N

One of the criticisms against meta-analytic reviews was referred to as the file drawer problem (Rosenthal, 1991), and was based on the fact that studies that rejected the null hypothesis (i.e., produce statistically significant results) were more likely to be published while the remaining null studies were kept in file drawers. Therefore, if meta-analysts exclusively drew their sample of studies from the published literature, there was an inherent bias in that the stronger and significant effects were more likely to be included in the review. This, in turn, could lead to a skewed conclusion of the magnitude of the relationship involved.

There are two ways meta-analysts can confront this problem. The first solution is rather straightforward and involves a comparison of the mean effect sizes generated between published and unpublished studies. If the published studies had a significantly higher mean effect size than unpublished studies, then it is known that publication bias was an important moderating role in the relationship studies. Although an attempt was made to track down unpublished works, very few were found and so conducting such a comparative analysis was not practical, especially when considering the relatively small number of effect sizes in each predictor category.

A second solution, which was developed by Rosenthal (1991), addressed this problem at a statistical level and was commonly referred to as the fail safe N (N_{fs}). This statistic corresponded to "the number of additional studies required in a meta-analysis that would be necessary to reverse the overall probability obtained from our combined test to a value higher than our critical value for statistical significance" (Wolf, 1986, p. 38) and provided the readers with further validation of the external generalizability of the findings. In other words, if only a few additional studies were required to change the direction of the conclusions, the findings should be viewed as cautionary whereas a fail safe N of

several hundred made the findings considerably more robust. Unfortunately, a standard criterion did not currently exist in the literature that indicated the exact number of studies required to yield more reliable conclusions. This decision was generally left up to the discretion of the researcher by attending to the number of studies involved in the meta-analysis, as well as the magnitude of the corresponding effect size.

⬛ Results and Discussion

Study Descriptives

In the present meta-analysis, twenty unique studies yielded 191 individual effect sizes examining the relationship between various predictor categories and measures of work-related stress. The characteristics of these studies are presented in Table 1.

Notably, there was a rather constant focus on the area of correctional officer stress over the past fifteen years. Over 75 percent of the effect sizes were derived from studies composed of predominantly male correctional officers and almost 75 percent were based on predominantly Caucasian samples. Clearly, most studies were conducted in the United States (63 percent) and came from the published academic literature (71 percent). Interestingly, the three unpublished papers came from Canada. Finally, the most commonly used measure of job stress in this series of studies was developed by Cullen, Link et al. (1985), followed by the single-item Smith and Ward (1983) scale. This finding was undoubtedly influenced by the fact that several authors published multiple journal articles based on the same sample of correctional officers with the same job stress measure (i.e., Lindquist & Whitehead, 1986a, 1986b, 1986c).

To facilitate organization of the results, the various stressors were aggregated into composite level categories and will be discussed accordingly. Seven separate broad categories

Table 1	Characteristics of the Studies Contributing to the Meta-analysis

Variable	Freq (percent) (k = 191)
Publication source	
Published article	135 (70.7 percent)
Unpublished	56 (29.3 percent)
Gender	
Male	17 (8.9 percent)
Female	17 (8.9 percent)
Predominantly male	127 (66.5 percent)
Mixed	15 (7.9 percent)
Not reported	15 (7.9 percent)
Ethnicity	
Caucasian majority (> 60 percent)	141 (73.8 percent)
Black majority (> 60 percent)	16 (8.4 percent)
Other minority (> 60 percent)	17 (8.9 percent)
Mixed	17 (8.9 percent)
Country	
United States	120 (62.8 percent)
Canada	69 (36.1 percent)
England	2 (1.0 percent)
Type of stress measure	
Cullen, Link et al. (1985)	94 (49.2 percent)
Smith and Ward (1983)	31 (16.2 percent)
Prison social climate survey	12 (6.3 percent)
Occupational envt scale	3 (1.6 percent)
Police stress survey	28 (14.7 percent)
Other	8 (4.2 percent)
Year of publication	(k = 20)
1985–1989	7
1990–1995	6
1996–2001	7

were created and included: demographic characteristics, work attitudes, specific correctional officer concerns, job characteristics, positive correctional attitudes, negative correctional attitudes, and support.

Demographics

The meta-analytic findings for demographic factors and their role in predicting correctional officer stress are presented in Table 2.

Clearly, the impact of demographic variables on correctional stress is relatively unimpressive when considering the composite mean effect size of .00 (k = 69). Furthermore, specific variables such as gender, marital status, education, and number of children are particularly weak, especially considering that their confidence intervals include zero. Although age had a significant impact on predicting work-related stress in correctional personnel, the magnitude of the effect was quite low and the statistical significance undoubtedly resulted from the large sample size.

Although country of origin was a significant moderator variable, examination of the outlier analysis revealed this was most likely as a result of the outlier study (Saylor & Wright, 1992) and not due to any substantive differences between countries.

The failure of correctional officer age to have a more meaningful impact on job stress may be due to an artifact of this relationship. More specifically, when correlating age with job stress, the size of the correlation (or in this case, effect size) is reliant on the degree of the linear relationship between the two variables (i.e., the stronger the linear relationship, the higher the correlation). Previous studies, however, indicated a curvilinear relationship between age and job stress in correctional officers (Launay & Fielding, 1989; Patterson, 1992). Thus, if this type of curvilinear trend existed, the deflation in effect size magnitude might be accounted for with this type of relationship.

Finally, the significant mean effect size for ethnicity (– .07) was noteworthy. More specifically, correctional officers who were minorities experienced significantly less job stress than Caucasian officers. This finding was

Table 2	Unweighted (*Mr*) and Weighted Mean Effect Sizes (*Mz⁺*) for Demographics				
Predictor (k)	*N*	*Mr*	*Mz⁺ᵃ*	*CI*	*Q*
Gender (female) (15)	14,813	.00	.00	−.01 to .02	36.02*
−Blau, Light, and Chamlin (1986) excluded (14)	12,133	.00	−.01	−.03 to .01	7.46 ns
Ethnicity minority (11)	7,994	−.01	−.07*	−.10 to −.05	31.46*
Canada (2)	1,338	.01	.00 ns	−.06 to .06	.27 ns
U.S.A. (9)	6,650	−.02	−.09*	−.11 to −.06	22.29*
−Blau et al. (1986) excl'd	4,012	.00	−.05*	−.08 to −.02	4.89 ns
Age (18)	13,189	−.02	0.03*	.01 to .05	295.69*
Canada (6)	5,648	−.05	−.04*	−.06 to −.01	4.87 ns
U.S.A. (12)	7,541	.00	.08*	.06 to .10	246.60*
−Saylor and Wright (1992) excl'd	4,216	−.04	−.07*	−.10 to −.04	21.79 (<50%)
Marital status (married) (7)	3,641	.03	.06*	−.00 to .06	1.66 ns
Education level (13)	7,452	−.02	−.04*	−.00 to .06	11.31 ns
Number of children (5)	906	.00	.04 ns	−.03 to .12	3.86 ns
Demographics (69)	47,995	−.01	.00	−.01 to .01	.380.00*

k = number of effect sizes per predictor category.

N = number of subjects per predictor category.

CI = confidence intervals about *Mz⁺*; ᵃ*Mz⁺* values are weighted according to sample size. *Q* = the test statistic for effect size heterogeneity referenced in the text. This value reflects the amount of variability present in the categories for a particular category. Larger values are indicative of greater heterogeneity. This value is compared to the χ^2 distribution (df = 1) to determine significance.

* *p* < .05

important as it related directly to a debated area in this literature regarding the role of ethnicity in the etiology of correctional officer stress. This idea, commonly expressed as the "identification" hypothesis (Jacobs & Kraft, 1978), was supported by largely anecdotal evidence in the past and stated that minority officers (such as Blacks and Hispanics) were less alienated and concomitantly felt less strain when interacting with the inmate population due to the disproportionate representation of minorities in the prison system in the United States (Britton, 1997).

The meta-analytic findings provide preliminary support for this proposition, although much additional empirical work is required. Further support for this proposition arguably comes from the moderator analysis where correctional officers in United States prisons revealed significantly lower on-the-job stress (−.09) whereas, in Canada, a similar relationship did not exist (.00). One potential explanation may be that in the United States minorities are more disproportionately represented in prisons than in Canada. Unfortunately, the nature of meta-analytic work precludes investigating such a hypothesis. Nonetheless, future work should focus on the interaction between the ethnic composition of correctional officers and the inmates whom they supervise.

Work Attitudes

The impact of work-related attitudes on correctional officer stress displayed a much stronger relationship than that found for demographic factors (−.37, k = 28). Table 3 presents these findings across four specific variables including participation in decision-making, job satisfaction, commitment, and turnover. In each case, the results were in the expected direction and the magnitude of the findings across the three positive work attitudes was quite consistent.

Overall, participation in decision-making, job satisfaction, and commitment had significant negative impacts on correctional officer stress. In other words, correctional personnel who participated more in decision-making, had higher job satisfaction, and expressed higher commitment to the organization were significantly less likely to express job-related stress. In contrast, correctional officers who expressed the intention of leaving their job evidenced significantly higher levels of work-related stress than those who did not express such sentiments.

The positive relationship between turnover intentions and job stress was also important, as turnover intention was found to be a strong predictor of employee turnover (Steel & Ovalle, 1984). This relationship is even more important as turnover of correctional personnel has tremendous direct and indirect costs for the organizations (Stohr, Self, & Lovrich, 1992). Arguably, one of the most important consequences of employee turnover is the lowering of employee morale, which may lead to further turnover (Stohr et al., 1992). This cyclical pattern may have far reaching negative implications for correctional officers, as well as for institutional and inmate management.

One final point should also be mentioned. The moderator analysis focusing on turnover intentions revealed that ethnicity once again played a significant role in this relationship.

Table 3	Unweighted (*Mr*) and Weighted Mean Effect Sizes (*Mz⁺*) for Work Attitudes				
Predictor (k)	**N**	**Mr**	**Mz⁺**	**CI**	**Q**
Participation (4)	866	−.37	−.36*	−.43 to −.30	37.16*
Lindquist and Whitehead (1986a, 1986b, 1986c) excluded	625	−.48	−.49*	−.57 to .41	3.85 ns
Job satisfaction (11)	5,294	−.41	−.46*	−.49 to −.43	32.47*
Canada (6)	4,332	−.48	−.49*	−.52 to −.46	6.77 ns
U.S.A. (5)	962	−.33	−.32*	−.38 to −.26	3.04 ns
Commitment (6)	5,655	−.34	−.33*	−.36 to −.30	3.44 ns
Turnover (7)	1,880	.19	.25*	.20 to .29	51.82*
Ethnic majority (5)	1,310	.15	.15*	.10 to .22	1.04 ns
Ethnic minority/mixed (2)	570	.29	.46*	—	15.25*
Work attitudes (28)	13,695	−.34	−.37	−.39 to −.36	124.88

* $p < .05$

More specifically, although Caucasian (or predominantly Caucasian) groups of correctional officers yielded a moderately significantly positive relationship between job stress and turnover intentions (.15), the mean effect size present in ethnic minority/mixed groups was substantially larger (.46).

Specific Correctional Officer Problems

A separate category of problems specific to the correctional officer was created and included perceived dangerousness and role difficulties such as role conflict, role ambiguity, or some comparable measure (i.e., role problems scale of Poole & Regoli, 1980). Although it is true that these factors are found in other occupations in law enforcement such as police work, these are salient outcomes for correctional officers as they are ever-present in their daily work routines and were thus analyzed separately (Table 4).

In terms of the specific occupational problems facing correctional officers, the perceived dangerousness of the position (.48) far outweighs role difficulties (.21) in terms of its contribution to job stress. Interestingly, perceived dangerousness was associated with significantly higher self-reported stress levels in Canadian prison workers than it was in the United States. This difference may be explained by the work of Philliber (1987) who suggested that the attitudes of correctional personnel might be intricately linked to the attitudes of the surrounding community. Although the relevance of this assertion to the present finding does not easily lend itself to cogent interpretation, one plausible explanation is that cultural differences between the two countries (i.e., gun control and more exposure to violent crime) may heighten or decrease an individual's sensitivity to dangerous situations.

Job Characteristics

An equally important area of concern when discussing correctional officer stress is the role of job characteristics. The four job characteristics included in this meta-analytic review were shift, security level, contact hours, and years of experience. Table 5 presents the meta-analytic findings for each of these predictors.

Table 4	Unweighted (*Mr*) and Weighted Mean Effect Sizes (*Mz⁺*) for Specific Correctional Officer Problems				
Predictor (k)	**N**	**Mr**	**Mz+**	**CI**	**Q**
Danger (10)	6,061	.40	.48*	.46 to .51	47.13*
Canada (5)	5,436	.52	.51*	.48 to .54	2.57 ns
U.S.A. (5)	625	.29	.28*	.20 to .36	14.67*
–Bazemore and Dicker (1994) excluded	516	.37	.34*	.25 to .43	3.09 ns
Role difficulty (8)	1,154	.21	.20*	.14 to .26	26.03 ns
Role conflict (4)	610	.12	.13*	.05 to .21	9.75*
–Triplett et al. (1999) excluded	408	.16	.21*	.11 to .31	2.66 ns
Role problems (3)	342	.37	.38*	.27 to .48	0.24 ns
CO specific problems (18)	7,557	.32	.44*	.42 to .46	152.51*

* $p < .05$

Table 5	Unweighted (*Mr*) and Weighted Mean Effect Sizes (*Mz⁺*) for Job Characteristics				
Predictor (k)	**N**	**Mr**	**Mz⁺**	**CI**	**Q**
Shift (evening) (3)	3,589	.06	−.01	−.04 to .02	3.34 ns
Security level (higher) (9)	12,062	.09	.04*	.03 to .06	46.88*
Canada (5)	5,497	.10	.09*	.07 to .12	8.43 ns
U.S.A. (4)	6,565	.07	.01	−.01 to .03	15.63*
−Saylor and Wright (1992) excluded	3,240	.10	.05*	.01 to .08	3.71 ns
Contact hours (5)	6,805	.10	−.02	−.04 to −.00	19.56*
−Saylor and Wright (1992) and Britton (1997) excluded	501	.18	.13	−.07 to .33	3.79 ns
CO experience (18)	13,565	.06	.03*	.01 to .04	191.22*
Canada (6)	5,686	.05	.05*	.03 to .08	7.63 ns
U.S.A. (12)	7,879	.07	.06*	.04 to .09	176.43*
−Britton (1997) excluded	4,554	.09	.13*	.10 to .16	10.16 ns
Job characteristics	36,021	.07	.02*	.01 to .03	260.99

* $p < .05$

Very little empirical support was found for the role of job characteristics in the development of stress in correctional personnel. Furthermore, although the magnitude of the effect was significant, this was clearly a result of the enormous sample size. Several key variables such as shift, security level, and years of experience failed to have much of an impact.

Although type of shift (i.e., day, evening, and night) failed to have a large impact on correctional officer stress, this might be a direct result of how the variable was operationalized (i.e., evening) in these three studies. Support for this interpretation came from Cheek and Miller (1983), who found a nonlinear trend between shift and job-related strain as the highest average stress score was found on the evening (i.e., approximately 2:30 p.m. to 10:30 p.m.) shift. Since this hypothesis is based on a single study, further analysis is warranted on several different samples of officers.

These results also challenge the conclusions put forward by Huckabee (1992) regarding the integral role of maximum-security institutions in correctional officer stress. More specifically, Huckabee (1992) stated "it appears from the extant research that stress is greatest in the tense and dangerous world of the maximum-security institution" (p. 482). The findings from the present meta-analytic review fail to support this assertion. Although job stress is certainly higher in maximum-security prisons, the absolute magnitude of the effect is relatively small (.04). Thus, the environmental differences between the different security levels appear not to be particularly important for job stress.

Although the initial findings for amount of contact hours with inmates yielded a negative relationship (i.e., correctional officers who spent more time with inmates experienced less stress), the removal of two outlier studies revealed an opposite conclusion. These contradictory findings were further compounded by the fact that the outlier studies that were removed were the largest in terms of sample size, and thus, extreme caution must be exercised in drawing any firm conclusions from the data in this predictor category. Notably, the two studies (Britton, 1997: Saylor & Wright, 1992) that were identified as outliers were based on national studies of correctional workers who worked at institutions under the jurisdiction of the Federal Bureau of Prisons in the United States. Whether these discrepancies can be explained based on differences in the nature of the work or the use of a problematic outcome measure (i.e., the two studies used the job-related stress subscale of the Prison Social Climate Survey) must be examined more completely in future studies. Support for this latter interpretation can be found in the present meta-analysis as in several of the outlier analyses for various predictor categories, the study by Saylor and Wright (1992) was identified as an outlier and its removal generally restored effect size homogeneity to acceptable levels.

The findings also lend support to the commonly voiced position that age and years of experience are somewhat confounded as the same trends were found for both of these variables. More specifically, as was the case for age in the demographics category, years of

experience yielded the highest effect size heterogeneity and the mean effect size for years of experience was identical in magnitude to that found for age (.03). Consistent with the findings for contact hours, removal of Saylor and Wright (1992) yielded a substantial increase in effect size magnitude (from .06 to .13). The relative instability of these results, however, precludes the formation of any solid conclusions. Thus, whether years of experience is a weak or moderate predictor of work-related stress remains very much in need of an answer.

⬛ Professional Orientation

Favorable Correctional Attitudes

Considerable attention was focused on both the predictors of professional orientation (i.e., rehabilitation-focused versus punitive/custody-oriented) as well as its role in predicting various work-related outcomes. The prevailing research suggests that these variables may affect the job stress experienced by correctional officers (Liou, 1995). The findings regarding positive correctional attitudes (defined as supportive of human service/rehabilitation and counseling) are presented in Table 6.

| Table 6 | Unweighted (Mr) and Weighted Mean Effect Sizes (Mz⁺) for Positive Correctional Attitudes |

Predictor (k)	N	Mr	Mz⁺	CI	Q
Human services/rehab (10)	6,534	−.09	−.15	−.18 to −.13	46.54*
Canada (7)	6,161	−.17	−.17*	−.19 to −.13	15.78*
−Lariviere (2001) excl'd	5,011	−.19	−.19*	−.22 to −.16	3.85 ns
U.S.A. (3)	373	.11	.09 ns	−.01 to .19	8.11*
−Cullen, Link et al. (1985) excluded	218	.20	.20*	—	—
Counseling (3)	948	−.14	−.15*	−.21 to −.08	17.65*
−Whitehead el al. (1987) excluded	690	−.24	−.23*	—	—
Positive attitudes	7,482	−.10	−.15*	−.17 to −.13	64.19

* p < .05

Overall, officers who possess a human service/rehabilitation orientation experienced considerably less job stress than those who did not endorse such a position. In addition, as shown in Table 6, the various scales used to measure this construct produced similar results. One interesting and critical finding, however, emerged in the moderator analyses and has important implications for future research in this area. Table 6 reveals that the country where the study was conducted had an enormous impact on the overall conclusions. In particular, correctional officers who possessed human service/rehabilitation orientation within Canadian prisons experienced significantly less stress than those officers who did not whereas in the United States, the opposite was found.

This finding is important, as correctional officers who endorse positive correctional attitudes are valued within an institutional setting, particularly within Canada where rehabilitation is a major component of the overall correctional philosophy (Simourd, 1997). This substantial discrepancy between countries might be explained by the somewhat disparate correctional orientations of prisons in both countries. More specifically, as previously mentioned, although rehabilitation was the primary focus of Canadian correctional institutions (Simourd, 1997), researchers in the United States highlighted the duality of their custody and treatment goals (Hepburn & Albonetti, 1980). Therefore, Canadian correctional officers who support a human service/rehabilitation orientation should reasonably be expected to experience less stress as their views align more strongly with those of their employer. It may be that correctional personnel working in institutions in the United States are more vulnerable to stress due to the competing ideologies of punitive/custody and rehabilitation. In other words, correctional officers who espouse a human service/rehabilitation orientation may be in direct conflict with the predominant or equal emphasis placed on the custody/control functions of the correctional officer. This interpretation is speculative at this point and additional research is needed to explore this distinction. Regardless of the position taken, the magnitude of this difference and the subsequent reversal in the direction of the relationship caused by country origin makes this an essential area for future research.

The moderator analysis conducted within the counseling dimension also provided additional support for the above interpretation. More specifically, the study identified as an outlier was the only one involving American correctional officers (the other two were Canadian studies). Clearly, the interaction between professional orientation and the philosophy of the prison must be subjected to further empirical scrutiny.

Unfavorable Correctional Attitudes

The findings related to more negative or undesirable correctional attitudes are shown in Table 7. As can be seen, these findings are in the opposite direction to those found for rehabilitation/human service. More specifically, the results indicate that individuals who support custodial/punitive statements are more likely to experience elevated levels of stress compared to those who fail to endorse such statements. A particularly interesting finding was that the mean effect size for the composite score of negative correctional attitudes was identical to that reported for positive correctional attitudes (.15), but in the opposite direction. This raises the question of whether these constructs are mirror images of each other.

These findings provide even more striking illustrations of between-country differences in terms of the role of professional orientation in evoking job stress in correctional officers. For example, the findings concerning punitiveness demonstrate significant between-country

Table 7			Unweighted (*Mr*) and Weighted Mean Effect Sizes (*Mz+*) for Negative Correctional Attitudes		
Predictor (K)	**N**	**Mr**	**Mz+**	**CI**	**Q**
Custody (3)	4,015	.20	.09*	.06 to .12	26.01*
−Saylor and Wright (1992) excluded	690	.28	.26*	—	—
Punitive orientation (9)	6,519	.18	.20*	.17 to .22	62.45*
Canada (7)	6,152	.22	.21*	.18 to .24	41.38*
−Lariviere (2001) excl'd	5,005	.25	.25*	.22 to .28	6.65 ns
U.S.A. (2)	367	.03	−.02 ns	—	—
Social distance (3)	948	.14	.14*	.07 to .20	3.92 ns
Corruption (3)	948	.14	.15*	.09 to .22	8.51*
−Whitehead et al. (1987) excluded	690	.20	.21*	—	—
Negative attitudes	12,430	.17	.15*	.13 to .17	100.89*

* $p < .05$

differences. More specifically, Canadian correctional personnel who endorsed these sentiments experienced significantly higher levels of job stress than those who did not. On the other hand, a punitive orientation in correctional officers in the United States did not impact job stress, and in fact, was weakly negatively associated. Moreover, those individuals who scored positively on custody and corruption scales experienced significantly greater job stress when employed in prisons in Canada as compared to the United States. This finding was supported by the fact that the only outliers found in the analyses of these scales were studies exploring correctional personnel in American institutions (Saylor & Wright, 1992; Whitehead, Lindquist, & Klofas, 1987), with the remaining individuals employed in Canadian institutions. Researchers, however, are cautioned against forming strong conclusions based on this finding as only one American and two Canadian studies contributed to this analysis. Nonetheless, the preliminary implications of these findings are worthy of future examination.

External Generalizability: Fail Safe N Analyses

As discussed previously, one of the primary criticisms levied against meta-analytic reviews was labeled the file drawer problem. Since insufficient unpublished studies contributed to the present meta-analysis to adequately explore the extent of this bias, the fail safe N statistic was employed for each significant predictor category. It should be noted that this statistic was based on the unweighted effect size estimates and so these findings should be considered in consultation with the initial results (Table 8).

The results indicate that the vast majority of the significant predictor categories, especially at the composite level, are sufficiently robust that the confidence in these findings should not be undermined. Clearly, most of the relationships found within the demographics category were much more unstable and so additional caution must be taken when extrapolating from these findings. Other than this category, the remaining findings can be

Table 8	Fail Safe N (N_{fs}) Statistic for Each Significant Predictor Category	
Predictor category		**N_{fs}**
Demographic		
Ethnic minority, education level		0
Age		13
Marital status		0
Education level		3
Work attitudes		532
Participation in decision making		19
Job satisfaction		284
Commitment		82
Turnover intention		52
Correctional officer specific problems		471
Dangerousness		186
Role difficulties		57
Job characteristics		412
Security level		68
Years experience		90
Favorable correctional attitudes		120
Human services/rehabilitation		58
Counseling subscale		4
Unfavorable correctional attitudes		439
Custody		19
Punitive orientation		96
Social distance		8
Corruption		6

considered relatively stable, subject to additional research.

⧈ Conclusions

Although several recent literature reviews examined the predictors of job stress in correctional officers, the present study marked the first systematic and quantitative review of this literature to date. Accordingly, this study provided some clarification of the previous contradictory findings found in this literature and also highlighted several essential moderating variables that might have contributed to the past confusion.

Limitations of the Review

A note should be made about some of the limitations of this meta-analysis. One of the primary drawbacks of this study was its predominant reliance on published versus unpublished studies, which was popularly termed publication bias (Rosenthal, 1991). More specifically, this criticism argued that research published in the scientific literature tended to be biased in favor of significant results and thus meta-analyses predominantly based on published studies might artificially inflate the effect size estimate. Although the fail-safe Ns for the vast majority of the predictor categories were sufficiently robust not to undermine the confidence in the integrity and generalizability of the results, hopefully future meta-analytic work will be able to incorporate more unpublished studies to directly test whether publication status, in fact, differentially impacts the magnitude of the relationships reported here.

Another point to note concerns the measurement of the outcome variable of interest (job stress) in this review. More specifically, the scales used across the various studies to measure job stress relied on self-report. This was problematic as responses to these types of scales were subjective and open to bias (Cheek & Miller, 1983; Cullen, Link et al., 1985; Finn, 1998; Parker & DeCotis, 1983; Schaufeli & Peeters, 2000), which might be further complicated by previous findings that correctional officers tended to portray a "macho image" and underreport their experiences of stress (Cheek & Miller, 1983). Despite these concerns, this problem was endemic throughout this entire area of research and, as noted by Parker and DeCotis (1983), "there appears to be no fully acceptable alternative that does not compromise the precision of the concept" (p. 163).

✑ Recommendations and Directions for Future Research

Demographics

Research specifically focusing on ethnic minorities and female correctional officers is still an area in critical need of additional study (Gross et al., 1994). Although the findings from the present meta-analysis failed to reveal an important role for gender in predicting correctional officer stress levels, separating both male and female correctional officers might facilitate the discovery of differential specific predictors of work stress. For example, where appropriate, future studies should strive to adopt a similar analytical framework as employed by Triplett et al. (1999) who devised and tested several predictive models and found that there were significant between-group differences when they examined the predictors of job stress for male and female correctional officers. More attention should also be paid to the area of work-home conflicts and their spillover effects on stress (Triplett et al., 1999), as well as to explore the interaction between work and life stress (Cullen, Link et al., 1985).

Another critical area involves determining whether the curvilinear relationship between age and job stress in fact exists, and, if so, the reasons for it. In other words, if a curvilinear relationship exists, researchers must examine whether older correctional officers are less stressed due to accumulated job experience or due to the fact that more stressed out officers leave their jobs. If the latter interpretation is correct, then age is not really the important variable and the individual coping strategies are of paramount importance. Regardless of which interpretation is correct, the coping ability of individual correctional officers must be explored more systematically in the future, and their impacts on correctional officer stress more carefully controlled in more methodologically rigorous studies.

Work-Related Attitudes

Although few studies explored participation in decision-making by correctional officers, its strong relationship with job stress warrants further investigation. In particular, given the strict operational requirements of an institutional setting, research may focus on ways in which increasing correctional officer involvement in decisions may be facilitated. Furthermore, research which explores the impact of job morale on job stress, an important work-attitudinal variable, is also very much in need of an answer.

Specific Correctional Officer Problems

As originally noted by Philliber (1987) and reinforced by Schaufeli and Peeters (2000), role problems for correctional officers can take many forms, generally role conflict and role ambiguity. Future researchers should strive to more clearly operationalize their measures of role problems for correctional officers as these may indeed have differential impacts on job stress. Unfortunately, the limitations of the role difficulty measures reported in the primary studies precluded the present meta-analysis from exploring this question in more detail.

Although a recent review of the literature on correctional officer stress stated "perhaps the most important job stressors correctional officers are faced with are role problems of several kinds" (Schaufeli & Peeters, 2000, p. 34), the findings of this meta-analysis qualify this assertion. More specifically, although role problems are moderate predictors of job stress in correctional officers, several other predictors, such as perceived danger, have far greater importance in terms of their predictive utility. As recommended by Cullen, Link et al. (1985), correctional personnel may benefit from training programs aimed at recognizing potential dangers on the job without incorporating these concerns as major occupational stressors.

Other researchers suggested that the empowerment of correctional officers would increase their sense of control and consequently reduce work-related stress, a situation that could be achieved through increased participation in decision-making (Hepburn, 1987).

One important point should also be made regarding role difficulties experienced by the correctional officer. As discussed by Philliber (1987), several researchers recommended broadening the roles of both treatment and custody workers in the hopes of lessening the role conflict experienced by these two groups. She correctly pointed out that this suggestion was inherently problematic, as it could not explain why the lowest levels of correctional officer stress were found in custody-oriented settings where employees were fully aware of their role. Additional support for this argument was found in the present meta-analysis where correctional officers who espoused a human service/rehabilitation perspective within Canadian institutions experienced less job stress than those who did not. In other words, correctional officers who work in Canadian institutions have much clearer role demarcation and thus those who share a similar professional orientation are much less stressed. Therefore, as suggested by Cheek and Miller (1983), clearer articulation of job performance guidelines and improved communication from supervisory staff may be more effective in reducing job stress as opposed to role expansion.

Job Characteristics

The unstable relationship found between contact hours and stress, particularly when one considers the outlier studies, warrants more careful investigation. In particular, security level, job position, inmate composition, and perceived danger may play critical moderating roles in this relationship. Furthermore, as was the case with age, the possibility of a curvilinear relationship between years of experience and job stress might undermine the ability to find any substantive results. More detailed and specific analyses of

each of these variables are required before any firm conclusions can be drawn.

Professional Orientation

The important role of country of study origin in explaining discrepancies in job stress must be emphasized. Although these differences were hypothesized to be as a direct result of different correctional philosophies, this must be subjected to more systematic empirical examination. Nonetheless, these significant between-country differences support the recommendations of Schaufeli and Peelers (2000) who caution against generalizing the results of studies conducted in the United States to other countries such as Europe or Canada. Clearly, the dynamics operating within the correctional institutions in the United States are substantively different than those found in other countries. These findings also lend further support to Philliber (1987) who asserted that the attitudes of correctional officers might reasonably be expected to be linked to the areas where they live.

The Future of Researching Job Stress in Correctional Officers

There are several areas for future research suggested by this meta-analysis. First and foremost, considerable effort should be placed in reaching a universally accepted definition regarding the phenomenon of stress (Finn, 1998; Huckabee, 1992). In fact, attempts should be made to find a standard measurement instrument for job stress as the results from this meta-analysis suggest that certain measures may be tapping different dimensions of this multi-dimensional construct (i.e., prison social climate survey). Furthermore, future studies should strive to incorporate more objective measures of stress (Cullen, Link et al., 1985; Finn, 1998) such as staff turnover rate, sick leave, absenteeism and others (Finn, 1998). The simultaneous inclusion of such measures would allow researchers to empirically explore the impact of perceived stress on observable outcomes and quantify this relationship more explicitly.

Although Huckabee (1992) recommended that future research should compare stress across different regions of the United States, the findings from the present review strongly suggested that this be expanded to encompass international differences as well. Clearly, the environmental characteristics of institutions in the United States seem qualitatively different from those in other countries (particularly Canada) as evidenced by the frequent moderating impact of country of origin on the meta-analytic results. Researchers should examine these differences within a more controlled environment to isolate and identify the mechanisms that lead to these differential findings.

An extension of the above recommendation is for researchers to examine the robustness of the findings in this study regarding country of origin, professional orientation, and job stress. More specifically researchers should explore more systematically the interaction between institutional and professional orientation and its subsequent impact on self-reported stress levels. The findings from the present meta-analytic review suggested a strong interactive effect that might provide more insight into a previously contradictory area of research.

Future studies should also use prospective, longitudinal designs. The vast majority of the studies included in this present review utilized cross-sectional designs that preclude an appropriate examination of cause and effect relationships. As there is such a dearth of research using these more sophisticated methodological designs, the importance of this recommendation should not be overlooked (Schaufeli & Peeters, 2000).

One final recommendation takes a prominent place in this series of potential directions for future research. Although the groups of correctional officers who were explored within the primary studies came from various ethnic and gender backgrounds, researchers continue to report the predictors of work stress at the aggregate level. For example, although percent of the correctional officers in a study may be reported as male, the findings are reported for the entire sample. Thus, a tremendous opportunity for exploring different predictors across gender is denied. Researchers are strongly encouraged to adopt more specialized reporting procedures when discussing the findings of any primary research study. Not only will it provide a more fine-grained examination of the research findings and highlight critical ethnic and gender differences which, as stated previously, are sorely lacking, it will make meta-analytic reviews much more precise. Even if researchers would prefer not to present these data in the main text, attaching these to the Appendix would facilitate access by meta-analysts or other interested researchers.

⬛ References

Adwell, S. T., & Miller, L. E. (1985). Occupational burnout. *Corrections Today, 47*, 70–72.

Andrews, D. A., Zinger, I., Hoge, R. D., Bonta, J., Gendreau, P., & Cullen, F. T. (1990). Does correctional treatment work? A clinically relevant and psychologically informed meta-analysis. *Criminology, 28*, 419–429.

Bazemore, G., & Dicker, T. (1994). Explaining detention worker orientation: Individual characteristics, occupational conditions, and organizational environment. *Journal of Criminal Justice, 22*, 297–312.

Blau, J. R., Light, S. C, & Chamlin, M. (1986). Individual and contextual effects on stress and job satisfaction: A study of prison staff. *Work and Occupations, 13*, 131–156.

Bonta, J., Law, M. A., & Hanson, R. K. (1998). The prediction of criminal and violent recidivism among mentally disordered offenders: A meta-analysis. *Psychological Bulletin, 123*, 123–142.

Britton, D. M. (1997). Perceptions of the work environment among correctional officers: Do race and sex matter? *Criminology, 35*, 85–105.

Byrd, T. G., Cochran, J. K., Silverman, I. J., & Blount, W. R. (2000). Behind bars: An assessment of the effects of job satisfaction, job-related stress, and anxiety on jail employees' inclination to quit. *Journal of Criminal Justice, 23*, 69–89.

Cacciopo, J. T., Klein, D. J., Berntson, G. C., & Hatfield, E. (1993). The psychophysiology of emotion. In M. Lewis & J. M. Haviland (Eds.), *Handbook of emotions* (pp. 119–1420). New York: Guilford Press.

Cheek, F. E. (1984). *Stress management for correctional officers and their families.* College Park, MD: American Correctional Association.

Cheek, F. E., & Miller, M. (1983). The experience of stress for correction officers: A double-bind theory of correctional stress. *Journal of Criminal Justice, 11,* 105–120.

Cullen, F. T., Link, B. G., Wolfe, N. T., & Frank, J. (1985). The social dimensions of correctional officer stress. *Justice Quarterly, 2,* 505–528.

Dembo, R., & Dertke. M. (1986). Work environment correlates of staff stress in a youth detention facility. *Criminal Justice and Behavior, 13,* 328–344.

Dowden, C., & Andrews, D. A. (1999). What works for female offenders: A meta-analytic review. *Crime and Delinquency. 45,* 438–452.

Dowden, C., & Andrews, D. A. (2000). Effective correctional treatment and violent reoffending: A meta-analysis. *Canadian Journal of Criminology, 42,* 449–476.

Dowden, C., & Brown, S. L. (2001). The role of substance abuse factors in predicting recidivism: A meta-analysis. *Psychology, Crime and Law, 8,* 1–22.

Farber, B. A. (1983). Stress and burnout in the human service professions. New York: Pergamon Press.

Finn, P. (1998). Correctional officer stress: A cause for concern and additional help. *Federal Probation, 62,* 65–74.

Gendreau, P., Little, T., & Goggin, C. (1996). A meta-analysis of the predictors of adult offender recidivism: What works! *Criminology, 34,* 575–607.

Glass, G. V., McGaw, B., & Smith, M. L. (1981). *Meta-analysis of social research.* Beverly Hills, CA: Sage.

Gross, G. R., Larson, S. J., Urban, G. D., & Zupan, L. L. (1994). Gender differences in occupational stress among correctional officers. *American Journal of Criminal Justice, 18,* 219–234.

Grossi, E. L., & Berg, B. L. (1991). Stress and job satisfaction among correctional officers: An unexpected finding. *International Journal of Offender Therapy and Comparative Criminology, 20,* 73–80.

Hanson, R. K., & Bussiere, M. T. (1998). Predicting relapse: A meta-analysis of sexual offender recidivism studies. *Journal of Consulting and Clinical Psychology, 66,* 348–362.

Harenstam, A., Palm, U., & Theorell, T. (1988). Stress, health, and the working environment of Swedish prison staff. *Work and Stress, 2,* 281–290.

Hays, W. L. (1981). *Statistics* (3rd ed.). New York: Holt, Rinehart, and Winston.

Hedges, L. V., & Olkin, I. (1985). *Statistical methods for meta-analysis.* New York: Academic Press.

Hepburn, J. R. (1987). The prison control structure and its effects on work attitudes: The perceptions and attitudes of prison guards. *Journal of Criminal Justice, 15,* 49–64.

Hepburn, J. R., & Albonetti, C. (1980). Role conflict in correctional institutions: An empirical examination of the treatment-custody dilemma among correctional staff. *Criminology, 17,* 445–459.

Huckabee, R. G. (1992). Stress in corrections: An overview of the issues. *Journal of Criminal Justice, 20,* 479–486.

Hurst, T. E., & Hurst, M. M. (1997). Gender differences in mediation of severe occupational stress among correctional officers. *American Journal of Criminal Justice, 22,* 121–137.

Izzo, R. L., & Ross, R. R. (1990). Meta-analysis of rehabilitation programs for juvenile delinquents. *Criminal Justice and Behavior, 17,* 134–142.

Jacobs, J. B. (1978). What prison guards think: A profile of the Illinois force. *Crime and Delinquency, 24,* 185–196.

Jacobs, J. B., & Kraft, L. J. (1978). Integrating the keepers: A comparison of Black and White prison guards in Illinois. *Social Problems, 25,* 304–318.

Karlinsky, S. (1979). *Job satisfaction among living unit and nonliving unit staff in a Canadian penitentiary.* Unpublished doctoral dissertation, York University, Toronto, Ontario, Canada.

Kelloway, E. K., Desmarais, S., & Barling, J. (2000). *Absenteeism in the Correctional Service of Canada: Links to occupation and conditions of employment.* Unpublished research report. Correctional Service of Canada, Ottawa, Canada.

Klofas, J., & Toch, H. (1982). The guard subculture myth. *Journal of Research in Crime and Delinquency, 19,* 238–254.

Lariviere, M. (2001). *Raw data.* Unpublished doctoral dissertation. Carleton University, Ottawa, Canada.

Lasky, G. L., Gordon, B. C., & Srebalus, D. (1986). Occupational stressors among federal correctional officers working in different security levels. *Criminal Justice and Behavior, 13,* 317–327.

Launay, G., & Fielding, P. J. (1989). Stress among prison officers: Some empirical evidence based on self-report. *Howard Journal, 28,* 138–148.

Lindquist, C. A., & Whitehead, J. T. (1986a). Guards released from prison: A natural experiment in job enlargement, *Journal of Criminal Justice, 14,* 283–294.

Lindquist, C. A., & Whitehead, J. T. (1986b). Burnout, job stress, and job satisfaction among southern correctional officers: Perceptions and causal factors. *Journal of Offender Counseling, Services, and Rehabilitation, 10,* 5–26.

Lindquist, C. A., & Whitehead, J. T. (1986c). Correctional officers as parole officers: An examination of a community supervision sanction. *Criminal Justice and Behavior, 13,* 197–222.

Liou, K. T. (1995). Role stress and job stress among detention care workers. *Criminal Justice and Behavior, 22,* 425–436.

Lipsey, M. W., & Wilson, D. B. (1993). The efficacy of psychological, educational, and behavioral treatment: Confirmation from meta-analysis. *American Psychologist, 48,* 1181–1209.

Long, N., Shouksmith, G., Voges, K., & Roache, S. (1986). Stress in prison staff: An occupational study. *Criminology, 24,* 331–345.

Losel, F. (1995). The efficacy of correctional treatment: A review and synthesis of meta-evaluations. In J. McGuire (Ed.), *What works: Reducing reoffending. Guidelines from research and practice* (pp. 79–111). Chichester, England: John Wiley and Sons.

Maslach, C., & Jackson, S. E. (1981). The measurement of experienced burnout. *Journal of Occupational Behavior, 2,* 99–113.

Mowday, R., Steers, R., &. Porter, L. (1979). The measurement of organizational commitment. *Journal of Vocational Behavior, 14,* 224–247.

Parker, D. F., & DeCotis, T. A. (1983). Organizational determinants of job stress. *Organizational Behavior and Human Performance, 32,* 160–177.

Patterson, B. L. (1992). Job experience and perceived job stress among police, correctional, and *probation/parole officers. Criminal Justice and Behavior, 19,* 260–285.

Philliber, S. (1987). Thy brother's keeper: A review of the literature on correctional officers. *Justice Quarterly, 4,* 9–37.

Pollack, C., & Sigler, R. (1998). Low levels of stress among Canadian correctional officers in the northern region of Ontario. *Journal of Criminal Justice, 26,* 117–128.

Poole. E. D., & Regoli, R. M. (1980). Role stress, custody orientation, and disciplinary actions: A study of prison guards. *Criminology, 18,* 215–226.

Robinson, D., Porporino, F. J., & Simourd, L. (1997). The influence of educational attainment on the attitudes and job performance of correctional officers. *Crime and Delinquency, 43,* 60–77.

Rosenthal, R. (1991). *Meta-analytic procedures for social research* (Rev. ed.). Newbury Park, CA: Sage.

Rutter, D. R., & Fielding, P. J. (1988). Sources of occupational stress: An examination of British prison officers. *Work and Stress, 2,* 291–299.

Saylor, W. G., & Wright, K. N. (1992). Status, longevity, and perceptions of the work environment among federal prison employees. *Journal of Offender Rehabilitation, 17,* 133–160.

Schaufeli, W. B., & Peeters, M. C. W. (2000). Job stress and burnout among correctional officers: A literature review. *International Journal of Stress Management, 7,* 19–48.

Schmidt, F. L. (1996). Statistical significance testing and cumulative knowledge in psychology: Implications for training of researchers. *Psychological Methods, 1,* 115–129.

Selye, H. (1982). History and present status of the stress concept. In L. Goldberger & S. Breznitz (Eds.), *Handbook of stress* (pp. 7–17). New York: Free Press.

Shamir, B., & Drory, A. (1982), Occupational tedium among prison officers. *Criminal Justice and Behavior, 9,* 79–99.

Simourd, L. (1997). *Staff attitudes towards inmates and correctional work: An exploration of the attitude–work outcome relationship.* Unpublished doctoral dissertation, Carleton University, Ottawa, Canada.

Slate, R. N., & Vogel, R. E. (1997). Participative management and correctional personnel: A study of the perceived atmosphere for participation in correctional decision-making and its impact on employee stress and thoughts about quitting. *Journal of Criminal Justice, 25,* 397–408.

Smith, B. L., & Ward, R. A. (1983). Stress on military and civilian police personnel. *American Journal of Police, 3,* 111–126.

Steel, R. P., & Ovalle, N. K. (1984). A review and meta-analysis of research on the relationship between behavioral intentions and employee turnover. *Journal of Applied Psychology, 69,* 671–686.

Stohr. M. K., Self, R. L., & Lovrich, N. P. (1992). Staff turnover in new generation jails: An investigation of its causes and prevention. *Journal of Criminal Justice, 20,* 455–478.

Tellier, C., & Robinson, D. (1995). *Correlates of job stress among front-line correctional staff.* Paper presented at the annual convention of the Canadian Psychological Association, Charlottetown. PEI, Canada.

Triplett, R., Mullings, J. L., & Scarborough, K. E. (1999). Examining the effect of work-home conflict on work-related stress among correctional officers. *Journal of Criminal Justice, 27,* 371–385.

Van Voorhis, P., Cullen, F. T, Link, B. G., & Wolfe, N. T. (1991). The impact of race and gender on correctional officers' orientation to the integrated environment. *Journal of Research in Crime and Delinquency, 28,* 472–500.

Whitehead, J. T., & Lab, S. (1989). A meta-analysis of juvenile correctional treatment. *Journal of Research in Crime and Delinquency, 26,* 276–295.

Whitehead, J. T., & Lindquist, C. A. (1986). Correctional officer job burnout: A path model. *Journal of Research in Crime and Delinquency, 23,* 23–42.

Whitehead, J. T, Lindquist, C. A., & Klofas. J. (1987). Correctional officer professional orientation: A replication of the Klofas-Toch measure. *Criminal Justice and Behavior, 14,* 468–486.

Wolf, F. M. (1986). *Meta-analysis: Quantitative methods for research synthesis.* Newbury Park, CA: Sage.

Wright, K. N., & Saylor, W. G. (1992). A comparison of perceptions of the work environment between minority and non-minority employees of the federal prison system. *Journal of Criminal Justice, 20,* 63–71.

DISCUSSION QUESTIONS

1. Explain why work in corrections is often not regarded as "professional" vis-à-vis other commonly referenced professions.

2. Note which jobs in corrections are most sought after and why. Discuss how all correctional work might become more appealing to educated workers.

3. Discuss how the focus on maximum security institutions has served to pervert the study and understanding of corrections. What effect has this focus had on public opinion about corrections, the need to professionalize, and the potential for reform of offenders, inmates, and clients?

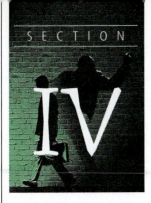

IV

LEGAL ISSUES IN CORRECTIONS

Introduction

◤ Historical Background: Prisoners as Slaves of the State

Prisons are not very nice places; they were never meant to be. This does not mean, however, that society is ever justified in treating prisoners in less than humane ways. Winston Churchill (and perhaps many others before him) once said that a civilization could be judged by the way it treated its prisoners. We do not treat them very well and never have, but many see them getting better treatment than they deserve, as summed up in the phrase "If you can't do the time, don't do the crime." Convicts are perhaps the most despised group of people in any society because they are generally viewed as evil misfits who prey on decent people. So why worry about what happens to them while they are in custody paying their debt to society?

This so-called "hands-off doctrine" has also been the general attitude of the courts throughout much of American history. The doctrine basically articulated the reluctance of the judiciary to interfere with the management and administration of prisons. It also rested on the low status of prisoners, who suffered a kind of legal and civil death upon conviction. Most states had civil death statutes, which meant that convicts lost all citizenship rights such as the right to vote, to hold public office, and even, in some jurisdictions, to marry. In essence, convicted criminals became "slaves of the state" (*Ruffin v.*

Commonwealth, 1871). In *Pervear v. Massachusetts* (1866), the Supreme Court said that prisoners did not even enjoy the protections of the Eight Amendment, which forbids "cruel and unusual punishment."

The hands-off doctrine also prevailed because the courts viewed correctional agencies as part of the executive branch of government and did not wish to violate the Constitution's separation of powers doctrine. Correctional officials were considered quite capable of administering to the needs of prisoners in a humane way without having to deal with the complicating intrusions of another branch of government. Besides, if prisoners have no rights, what is it that the courts have to monitor and protect? Convicted felons thus found themselves without any kind of constitutional protection and totally at the mercy of prison officials and of fellow prisoners.

⚑ The Beginning of Prisoners' Rights

As part of a slowly growing trend toward an overall greater respect for individual rights in the mid-twentieth century, the courts began to insinuate themselves into the uncharted area of prisoners' rights. The major issue in prisoner litigation has been the conditions of confinement, but the first significant case was *Ex parte Hull* (1941), which dealt with the denial by prison officials of a Michigan inmate's petition for an appeal of the legality of his confinement (the term *ex parte* refers to situations in which only one party appears before the court). In *Hull*, the United States Supreme Court ruled that inmates had the right to unrestricted access to federal courts to challenge the legality of their confinement. This ruling was the beginning of the end of the hands-off doctrine, although, as we shall see, there is growing evidence of its return in some respects (see the Federman article in this section).

The technical term for a challenge to the legality of confinement is a writ of habeas corpus. *Habeas corpus* is a Latin term that literally means "you have the body," and a writ of habeas corpus is basically a court order requiring that an arrested person be brought before it to determine the legality of his or her detention. Habeas corpus, a very important concept in common law, precedes even the Magna Carta of 1215, although its precise origins are unknown. It has been called the "Great Writ" and was formally codified into English common law by the Habeas Corpus Act of 1679. Indicative of the respect in which habeas corpus was held by the Founding Fathers is that it is only one of three individual rights mentioned in the United States Constitution (the other two are the prohibition of bills of attainder—imposing punishment without trial—and the prohibition of ex post facto laws—legislation making some act criminal after the fact). The other individual rights that Americans enjoy were formalized in the first 10 amendments to the constitution (the Bill of Rights) almost as an afterthought. An important point to remember as you read this section is that a writ of habeas corpus is not a direct appeal of a conviction (based on some point of law), but rather an indirect appeal of the legality of confinement (the original meaning) or the conditions of confinement (as the meaning of habeas corpus has evolved).

The writ was most famously used in *Somerset v. Stewart* (1772). This English case involved James Somerset, a slave who had traveled to England with his Virginian master. Somerset escaped, was captured again, and held aboard ship for transportation back to

Virginia. British antislavery activists filed a writ of habeas corpus on Somerset's behalf, and he was freed by the English High Court Lord Chief Justice Mansfield in a case widely (but erroneously) viewed as banning slavery throughout the British Empire. The case is prominent because a number of historians feel that it may have been instrumental in propelling some of the more reluctant southern states into participation in the American Revolution (Parrillo, 2003).

In *Jones v. Cunningham* (1963), the Supreme Court went further than it did in *Hull* and ruled that prisoners could use a writ of habeas corpus to challenge the conditions of their confinement as well as the legality of their confinement. This ruling went well beyond the original meaning of habeas corpus, which only addressed itself to the preconviction issue of the legality of a petitioner's detainment. In *Cooper v. Pate* (1964) the Court went even further and ruled that state prison inmates could sue state officials in federal courts under the Civil Rights Act of 1871, which was initially enacted to protect southern blacks from state officials and the Ku Klux Klan. This act is now codified and known as 42 USC § 1983, or simply as "section 1983," and the relevant part reads "Every person who under color of any statute, ordinance, regulation, custom, or usage of any state or territory, subjects or causes to be subject, any citizen of the United States or other person within the jurisdiction thereof to the deprivation of any rights, privileges, or immunities secured by the Constitution and laws, shall be liable to the party injured in an action at law."

What was a mere trickle of habeas petitions before *Pate* quickly became a flood that threatened to drown the federal courts with grievances ranging from the petty (a prisoner's complaint about being served partially melted ice cream) to the deadly serious, such as torture and severe deprivation. The most serious was a federal appeals judge declaring the entire prison system of Arkansas as unconstitutional and a "dark and evil world" when he placed it under federal supervision (*Holt v. Sarver*, 1969). This case gave birth to what has come to be known as a "conditions of confinement lawsuit," and, from then on, the federal courts became very much involved in the monitoring and operation of entire prison systems.

The vast majority of corpus grievances filed today are about the *conditions* of a person's confinement not the *legality* of his or her confinement. An inmate filing a petition challenging his or her confinement faces a very uphill battle because the state's defense against such a claim is based on the inmate's conviction, which is the obvious legal basis for his or her confinement! Let us now briefly examine the constitutional basis for challenging conditions of confinement. In each case, it must be understood that the rights we will examine may be restricted if doing so is deemed necessary for legitimate penological purposes.

▨ First Amendment

The First Amendment guarantees freedom of religion, speech, press, and assembly. These rights may be provided to prison inmates, although concerns of institutional safety and security must receive primary consideration. For instance, freedom of religion cannot extend to demanding alcohol or exotic foods to satisfy real or invented religious requirements. Freedom of speech or expression allows inmates to write and publish their thoughts or sell personal memorabilia, but "notoriety-for-profit" statutes enacted by the federal government and most states forbid inmates from profiting monetarily from those activities (Schmalleger, 2002). First Amendment freedoms do not extend to activities and materials that jeopardize prison safety or security. The right of assembly allows for attendance at religious services and for visitation from family and friends, but it obviously cannot be construed as allowing inmates to assemble at a tattoo conference outside the prison walls or to assemble a passel of prostitutes inside them.

One First Amendment issue addressed in Katherine Bennett's article in this section is that of cross-gender searches and visual observation of nude inmates. A Muslim inmate named Yusaf Asad Madyun filed suit in federal court claiming that the state of Illinois violated his First Amendment religious rights by subjecting him to a frisk search conducted by a female corrections officer (*Madyun v. Franzen*, 1983). The Islamic religion forbids such contact with women other than one's mother or wife. The federal appeals court found that Madyun's religious rights had not been violated by such a frisk.

Fourth Amendment

The Fourth Amendment guarantees the right to be free from unreasonable searches and seizures. What is reasonable inside prison walls is, of course, quite different from what is reasonable outside them. For all practical purposes, inmates have no Fourth Amendment protections because their prison cells are not "homes" of personal sanctuary deserving of privacy (*Hudson v. Palmer*, 1984). The one area that Fourth Amendment rights have not been completely extinguished for inmates is that involving cross-gender body searches.

As Bennett describes in her article, the courts have had to wrestle with conflicting claims on this issue. One claim is the equal employment right of female corrections officers who want to work in male institutions where promotion prospects are greater than they are in female prisons. Working in all-male prisons necessarily means that women officers will occasionally view inmates undressed or using toilet facilities, and sometimes they may be required to perform pat-downs and visual body cavity searches (physical bodily searches may only be performed by medical personnel). The inmate claim is that cross-gender searches are "unreasonable" within the meaning of the Fourth Amendment.

Bennett notes that the great majority of cross-gender search complaints are filed by males, which is not surprising since males constitute about 94% of all state prison inmates (Bohm & Haley, 2007). On the other hand, it is surprising given the complaints of female officers that some male inmates seem to take every opportunity to expose themselves and even to masturbate in their presence (Cowburn, 1998).

In *Turner v. Safley* (1987) the Supreme Court ruled that the viewing of opposite-sex inmates is constitutionally valid "if it is reasonably related to legitimate penal interests." Bennett expounds on how lower courts have interpreted "reasonableness" and concludes that while "conducting or observing strip searches [of male inmates] by female officers are tolerated primarily only in emergency situations," such "searches and body cavity searches of female inmates by male officers are deemed unreasonable." This double standard has been justified on two grounds: (1) males do not experience loss of job opportunities if they are forbidden to frisk female inmates and (2) intimate touching of a female inmate by a male officer may cause psychological trauma because many female inmates have histories of sexual abuse.

Farkas and Rand (1999) also support gender-specific standards for cross-gender searches on the basis of prior sexual abuse and state that "Cross-gender searches have the very real potential to replicate that suffering in prison," and further abuse would constitute cruel and unusual punishment (p. 53). This raises questions of the possible legal validity of complaints about same-sex body searches if the complainant can show prior sexual abuse by a same-sex person. For instance, will such a person then be in a position to demand an opposite-sex body search?

⊠ Eighth Amendment

Cruel and unusual punishment is forbidden by the Eighth Amendment. What constitutes cruel and unusual punishment is punishment applied "maliciously and sadistically for the very purpose of causing harm" (*Hudson v. McMillian*, 1992), although the onus is on the inmate to prove that the punishment was so applied. The Eighth Amendment is offended not only by prison officials doing something to inmates that they should not but also by officials failing to do something that they have a duty to do. Prison official must provide inmates with the basic amenities of life such as food and medical attention, and they must provide them protection from the physical and sexual predations of other inmates, many of whom have histories of "maliciously and sadistically" causing harm to others.

This issue is addressed in the article by Michael Vaughn and Roland Del Carmen in this section. They make clear that liability attaches to prison officials for inmate-inmate assaults if officials display *deliberate indifference* to an inmate's needs. The courts have struggled to make plain what deliberate indifference means, but basically it occurs when prison officials know of but disregard an obvious risk to an inmate's health or safety (*Wilson v. Seiter*, 1991). In other words, prison officials must not turn a blind eye to situations that obviously imperil the health or safety of inmates entrusted to their care. Purposely placing a slightly built and effeminate young male in a cell with a known aggressive sexual predator is an example of a violation of the deliberate indifference standard, and prison authorities would be liable for any injuries suffered. For inmates to prevail in suits involving deliberate indifference claims, they must prove (1) that they suffered an objectively serious deprivation or harm and (2) that prison officials were aware of the risk that caused the alleged harm and that they failed to take reasonable steps to prevent it. *Wilson* is seen as a key decision favoring correctional agencies because of these stringent proof requirements.

The medical needs of inmates in today's prisons are at least as well taken care of as those of the average person. According to a Bureau of Justice Statistics report on nationwide inmate mortality in state prisons, prisoners between 15 and 64 years of age had a mortality rate 19% lower than the U.S. residential population (Mumola, 2007). The vast majority of the difference was attributable to the rates for black males under 45, who had a mortality rate 57% lower than the rate of black citizens in general. Of course, not all or even the major part of these figures is attributable to medical care. Being incarcerated vastly lowers the probability of being murdered, being exposed to drugs, and having access to alcohol and tobacco. Finally, prison inmates are the only group of people in the United States with a constitutional right to medical care.

The issue of inmate-inmate assault is more relevant today than in the past when the worst assaults were committed by guards on inmates. Today, it is generally agreed that, while staff violence against inmates is much less common than in the past, there is more inmate-inmate violence (Bohm & Haley, 2007). Bohm and Haley suggest a variety of reasons for this turnaround. The abandoning of the hands-off doctrine by the federal courts has played a huge part in preventing staff violence against inmates, but it has also handed over a lot of the day-to-day control of some prisons to inmates. Increased inmate-inmate assaults may be attributed to the larger number of young aggressive males being incarcerated, the increase in prison gang membership, and the increase in racial gangs and tensions in prisons.

⬚ Fourteenth Amendment

The due process clause of the Fourteenth Amendment declares that no state shall deprive any person of life, liberty, or property without due process of law. This clause was first applied to inmates facing disciplinary action for infractions of prison rules in *Wolff v. McDonnell* (1974). In *Wolff* the Supreme Court declared that, although inmates are not entitled to the same due process rights as an accused but unconvicted person on the outside, they are entitled to some. These rights are (1) to receive written notice of an alleged infraction, (2) to be given sufficient time (usually 24 hours) to prepare a defense, (3) to be allowed to produce evidence and witnesses on their behalf, (4) to have the assistance of nonlegal counsel, and (5) to receive a written statement outlining the disciplinary committee's findings.

In *Sandin v. Connor* (1995) the Supreme Court clarified and trimmed back inmate rights. In *Sandin*, the Court declared that the above due process rights are only triggered by any disciplinary action that may result in the loss of "good time," which amounts to an extension of an inmate's sentence. Conner had been given 30 days punitive segregation for using foul and abusive comments to an officer while being subjected to a strip search. The court ruled that due process rights are not triggered by actions that result in temporary placement in a disciplinary segregation unit, which does not amount to an extension of sentence. The Court also concluded that disciplinary segregation is not an atypical hardship relative to the ordinary hardships of imprisonment.

⬚ The Civil Commitment of Sex Offenders

The idea that people who engage in socially disapproved behavior are sick and require treatment is an old one. As we will see in the article by Mary Ann Farkas and Amy Stichman in this section, the idea is seen as particularly applicable to sex offenders. The Supreme Court has upheld a Kansas statute that allowed for keeping sex offenders in state custody under civil commitment laws after they have served their full prison terms if they demonstrate "mental abnormality" or are said to have a "personality disorder" (*Kansas v. Hendricks*, 1997). The federal government and many other states have since passed similar involuntary confinement laws for sex offenders. Prior to *Hendricks*, civil commitments were limited to individuals suffering from mental illness, but, seemingly for the express purpose of covering sex offenders, states have loosened the mental illness criterion in favor of the "mental abnormality" criterion.

Several observers have noted that the term "mental abnormality" can be used to cover almost anything society may disapprove of and serves as a justification for imprisonment in ways "reminiscent of Soviet policies that institutionalized dissidents" (Grinfeld, 2005, p. 2). There is no doubt that Leroy Hendricks was a repeat predatory pedophile, and, although many applaud his incapacitation, what concerns civil libertarians is that the *Hendricks* decision created a special category of individuals defined as "abnormal" who may be punished indefinitely for what they *might* do if released.

Farkas and Stichman argue that what they call the "culture of fear" generated by atypically brutal sex offenses has resulted in laws that are constitutionally questionable and that have negative consequences for the treatment of sex offenders, the criminal justice system, and society in general. However, the popularity of these laws with the general public, and the extremely negative view of sex offenders that the public holds, makes it unlikely that the laws will be changed in the near future.

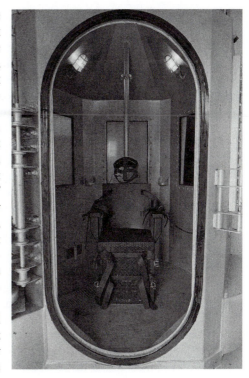

▲ **Photo 4.1** The electric chair at Parchman prison in Mississippi. Photographed in 1986.

⬛ A Return to a Modified Hands-Off Doctrine

As Cary Federman's article in this section makes clear, two acts signed into law in 1996—the Prison Litigation Reform Act (PLRA) and the Antiterrorism and Effective Death Penalty Act (AEDPA)—have severely curtailed prisoner access to the courts. Both acts were passed in part to reduce the thousands of lawsuits filed by inmates, which were clogging the federal courts. In the year of the passage of these acts, inmates filed 68,235 civil rights lawsuits in the federal courts compared to less than 2,000 in the early 1960s (Schmalleger, 2001). By 2000, the number of such lawsuits dropped to 24,519 (Seiter, 2005), a 64% decrease from the 1996 figure. Despite claims that these acts are "silencing the cells" (Vogel, 2004), the number of lawsuits filed is still more than 12 times what it was in the 1960s; even considering the increase in the prison population since that time, it still constitutes an increase of at least four times.

The primary intention of the PLRA was to free state prisons and jails from federal court supervision as well as limit prisoners' access to the federal courts. Both intentions have largely succeeded. Among the requirements of the PLRA is one that states inmates cannot bring a section 1983 lawsuit in federal court unless they first exhaust all available administrative remedies, such as filing a written grievance with the warden. The PLRA also states that inmates claiming to be unable to afford the required filing fee for the lawsuit may still have to pay a partial fee, which will be collected whenever money appears in their inmate accounts. This provision may limit the airing of genuine grievances in federal court because of financial difficulties.

As the name implies, the AEDPA is mostly about antiterrorism and the death penalty rather than an act specifically designed to limit habeas corpus proceedings. It was passed in response to the bombing of the Murrah Federal Building in Oklahoma City, with the reform of habeas corpus law being a rider to it. The AEDPA does not eliminate an inmate's right to habeas corpus, but it does restrict its availability primarily to inmates who have sought, but have been denied, state court remedies available to them. AEDPA also limits successive petitions and judicial review of evidence. The AEDPA thus takes habeas corpus partially back along the road to becoming again the preconviction remedy against unlawful imprisonment.

A review of Supreme Court decisions on habeas corpus since the AEDPA found that they have upheld the reforms largely as intended by Congress (Scheidegger, 2006). Although civil rights groups such as the American Civil Liberties Union tend to decry both statutes, the Court's attitude toward them is different for a number of reasons. First, these statutes free up the federal courts to deal with pressing inmate issues that are really repugnant to the Constitution, as well as the numerous other matters it must deal with. Second, it saves the taxpayer literally millions of dollars previously spent in frivolous legal exercises whose benefits were only to relieve the boredom of mischievous inmates (again, this does not mean that there are no substantively meaningful claims filed). Many states, aware that the courts could swing back to more active involvement, and conscious of the high cost of defending lawsuits, have established internal mechanisms to deal more effectively with inmate concerns, such as outside mediators and the creation of ombudsmen. Although PLRA and AEDPA have severely limited inmate access to the courts, in a roundabout way, they have given inmates more immediate and local ways to make their grievances known. However, given the failure of these local entities to protect the rights of correctional clients in the past, which necessitated federal involvement in the first place, there is likely room for healthy skepticism about their willingness to do so meaningfully in the future.

Legal Issues in Probation and Parole

Both probation and parole are statutory privileges granted by the state in lieu of imprisonment (in the first case) or further imprisonment (in the second case). Because of their privileged status, it was long thought that the state did not have to provide probationers and parolees any procedural due process rights either in the granting or revoking of either status. Today, probationers and parolees are granted some due process rights.

The first important case in this area was *Mempa v. Rhay* (1967). In this case, about four months after Mempa was placed on probation, he committed a burglary, which he admitted. His probation was revoked, and he was sent to prison. The issue before the Supreme Court in *Mempa* was whether probationers have a right to counsel at a deferred sentencing hearing. The Court ruled that, under the Sixth and Fourteenth Amendments, they do because Mempa was being sentenced, and the fact that sentencing took place subsequent to a probation placement does not alter the fact that sentencing is a "critical stage" in a criminal case.

A further advance in granting due process rights to offenders on conditional liberty status came in *Morrissey v. Brewer* (1972). Morrissey was a parolee who was arrested by his parole officer for a number of technical violations and returned to prison. Morrissey's petition to the Supreme Court claimed that, because he received no hearing prior to revocation, he was denied his rights under the due process clause of the Fourteenth Amendment. The Court agreed that, when a liberty interest is involved, certain processes are necessary (a liberty interest refers to government-imposed changes in someone's legal status that interfere with their constitutionally guaranteed rights to be free of such interference). However, the Court stated that, although revocation does not call for the whole array of rights due to a defendant not yet convicted, it does require an informal hearing to determine if the revocation is based on verified facts. Similar rights were granted to probationers the following year in *Gagnon v. Scarpelli* (1973).

While individuals are on probation or parole, they have limited constitutional rights, and their probation or parole officers can follow somewhat broader procedures than can police officers. Because probationers and parolees waive their Fourth Amendment search or seizure rights, probation or parole officers may conduct searches at any time without

▲ **Photo 4.2**   Justices of the U.S. Supreme Court, October 31, 2005. (First Row L-R) Justice Antonin Scalia, Justice John Paul Stevens, Chief Justice John Roberts, Justice Sandra Day O'Connor, Justice Anthony M. Kennedy, (Second Row L-R) Justice Ruth Bader Ginsburg, Justice David H. Souter, Justice Clarence Thomas, and Justice Stephen G. Breyer.

a warrant and without the probable cause needed by police officers. This "special needs" (of law enforcement) exception to the Fourth Amendment has been extended to the police under certain circumstances. The Supreme Court has stated that, if a probation order is written in such a way that provides for submission to a search "by a probation officer or any other law enforcement officer," then the police gain the same rights as probation and parole officers to conduct searches based on less than probable cause (*United States v. Knights*, 2001).

⬚ The Death Penalty: Legal Challenges to the Ultimate Sanction

All democracies except Japan and the United States have abolished the death penalty. According to Amnesty International (2006), four countries (China, Iran, Saudi Arabia, and the United States) accounted for 94% of the world's *known* executions in 2005. Given the emotional and philosophical issues surrounding it and particularly its finality, it is understandable that the death penalty has been subjected to intense legal scrutiny in recent times. This has not always been the case. Throughout much of our history the death penalty has been considered a legitimate, appropriate, and necessary form of punishment, and a clear majority of the American public still favors its retention (Radelet & Borg, 2000).

Legal challenges to the death penalty have revolved around the Eighth Amendment's prohibition of cruel and unusual punishment. Most such challenges were about the constitutionality of the method of execution, not the penalty per se, though some justices have stated their abhorrence for the punishment itself. Recently, in *Baze v. Rees* (2008), the Supreme Court, in a plurality opinion, upheld the use of lethal injection so long as its administration did not carry an "objectively intolerable risk of harm."

The first case to challenge the penalty itself was *Furman v. Georgia* (1972). Furman had shot and killed a homeowner during the course of a burglary and was sentenced to death. Furman challenged the constitutionality of his sentence, and the Supreme Court, in a 5 to 4 vote, agreed. However, the majority of the Court decided that the death penalty per se was not unconstitutional but that the arbitrary and discriminatory way in which it was imposed violated the Eighth Amendment. The Court argued that, because the death penalty is so infrequently imposed, it serves no useful purpose and that, when it is imposed, judges and juries have unbridled discretion in making life versus death decisions.

Because the way the death penalty decision was arrived at, rather than the penalty itself, was found to be unconstitutional, states began the process of changing their sentencing procedures. Some states introduced bifurcated (two-step) hearings, the first to determine guilt (the trial) and the second to impose the sentence after hearing aggravating and mitigating circumstances to determine if death were warranted. Other states removed sentencing discretion (because this discretion seemed to be the Supreme Court's problem in *Furman*) and made the death penalty mandatory for certain types of murder.

Georgia revised its statute and opted for the bifurcated hearing. Using this process, Troy Gregg was sentenced to death for two counts of murder and two counts of armed robbery. In *Gregg v. Georgia* (1976), the Supreme Court upheld the constitutionality of the bifurcated hearing and thus of Gregg's death sentence. On the same day, the Court also decided against mandatory death sentences in *Woodson v. North Carolina* (1976). The Court rejected as excessive and unduly rigid the North Carolina statute that mandated that all persons convicted of first-degree murder should receive the death penalty. (Woodson had killed a convenience store cashier and seriously wounded a customer in the course of an armed robbery.)

In *Coker v. Georgia* (1977), the Supreme Court struck down the death penalty for rape. Coker had escaped from prison where he was serving time for murder, rape, and kidnapping and promptly proceeded to commit another rape and kidnapping. Nevertheless, the Court struck down the Georgia statute authorizing death for rape under certain circumstances as "grossly disproportionate" and thus repugnant to the Eighth Amendment.

Another concern is whether the death penalty is applied in a racially discriminatory fashion. Warren McCleskey, a black man, was sentenced to death for killing a white police officer in the course of a robbery. In challenging his sentence in *McCleskey v. Kemp* (1987), McCleskey offered as evidence a statistical study demonstrating that racial disparity existed in Georgia (the state where the crime occurred) in that defendants who killed white victims were much more likely to be sentenced to death than defendants who murdered black victims. In ruling against McCleskey's claim, the Court stated that the statistical risk indicated in the study represented averages and did not establish that an individual's death sentence violates the Eighth Amendment. In other words, a study of past cases indicating average outcomes in no way serves as evidence that McCleskey was denied due process.

Further challenges to the death penalty have involved the constitutionality of imposing it on mentally retarded persons and on juveniles. In 1979, paroled rapist Johnny Penry was sentenced to death for rape and murder in Texas. Penry appealed to the Supreme Court on Eighth Amendment grounds claiming that the jury was not instructed that it could consider his low IQ (between 50 and 63) as mitigating evidence against imposing the death penalty. The Court held in *Penry v. Lynaugh* (1989) that the constitution does not prohibit the execution of a mentally retarded person.

The Court overruled itself in this regard in *Atkins v. Virginia* (2002). The Court concluded that there was a national consensus against executing the mentally retarded and that the mentally retarded, being less capable of evaluating the consequences of their crimes, are less culpable than the average offender. The Court also noted that mentally retarded individuals are more prone to confess crimes that they did not commit and therefore more prone to wrongful execution. Six of the justices concluded that the overwhelming disapproval of the world community must be considered a relevant factor in determining the imposition of capital punishment on mentally retarded individuals.

The Justices have also had to wrestle with the moral issue of imposing the death penalty on individuals who committed their crimes as juveniles. From 1973 to 2003, 22 such offenders were executed in the United States (Streib, 2003).

In 1977, 16-year-old Monty Lee Eddings and several companions stole an automobile, which was subsequently stopped by an Oklahoma Highway Patrol Officer. When the officer approached the car, Eddings shot and killed him. At Eddings's sentencing hearing, the state presented three aggravating circumstances to warrant the death penalty, but the judge only allowed Eddings's age in mitigation. In *Eddings v. Oklahoma* (1982), the Supreme Court vacated Eddings's death sentence, ruling that, in death penalty cases, the courts must consider all mitigating factors (such as that Eddings had been a victim of abusive treatment at home) when considering a death sentence.

The next juvenile case was *Thompson v. Oklahoma* (1988). Fifteen-year-old William Thompson was one of four young men charged with the murder of his former brother-in-law. All four were found guilty and sentenced to death. Thompson appealed to the Supreme Court claiming that a sentence of death for a crime committed by a 15-year-old is cruel and unusual punishment. The Court agreed, and, using the "evolving standards of decency" principle," it drew the line at age 16 under which execution was not constitutionally permissible. In *Stanford v. Kentucky* (1989) the Supreme Court underlined *Thomson* in a 5–4 decision, declaring that it was constitutionally permissible to execute individuals who were 16 or 17 at the time of their crimes.

Sixteen years later in *Roper v. Simmons* (2005), the Court redrew the age line at 18 under which it was constitutionally impermissible to execute anyone. Christopher Simmons was 17 when he and two younger accomplices broke into a home, kidnapped the owner, beat her, and threw her alive from a high bridge into the river below where she drowned. Simmons had told many of his friends before the crime that he wanted to commit a murder and bragged about it to them afterwards. His crime was a "classic" death penalty case—premeditated, deliberate, cruel, and totally unremorseful. Nevertheless, his sentence drew condemnation from around the world, with amicus curiae or "friend of the court" briefs presented to the Court arguing in support of Simmons and filed by interested parties not directly involved with the case, for example, by the European Union, the American and British Bar Associations, the American Medical and Psychological Associations, and 15 Nobel Prize winners, among others.

In the majority (5–4) opinion, Justice Kennedy noted that the United States was the only country in the world that gives official sanction to the juvenile death penalty. He also noted the growing body of evidence from neuroscience about the immaturity of the adolescent brain (the brain is undergoing numerous physical changes during adolescence: see White, 2004). The majority opinion also cited *Atkins v. Virginia* (2002). In noting that the Court in *Atkins* had ruled the execution of the mentally retarded to be cruel and unusual punishment because of the lesser degree of culpability attached to the mentally challenged; it reasoned that such logic should be applied to juveniles. The Court also pointed out that the plurality of states (30) either bar execution for juveniles or have banned the death penalty altogether, thus citing state legislation as part of the impetus behind its decision.

Figure 4.1 presents a graph of the annual number of executions in the United States from the time of the mid-1970s when states were coming to terms with *Furman* to 2006. By 2006, the number of executions declined by 56% from their peak in 1999. In 2006, there were 3,366 prisoners under a sentence of death, and 43 (1.3%) were executed (Death Penalty Information Center, 2007). Of those under sentence of death in 2006, 45% were white, 42% black, 11% Hispanic, and 2% other races. Only 1.6% of death row inmates were women. Among people sentenced to death from 1977 to 2005, 16% of whites have been executed compared to 11% of Hispanics and African Americans (Snell, 2006).

⬙ Summary

- The courts have moved through three general periods with respect to inmates' rights: the hands-off period, a short period of extending many rights to prisoners, and the current retreat to a limited hands-off policy.

Figure 4.1	Executions in the United States, 1976–2006

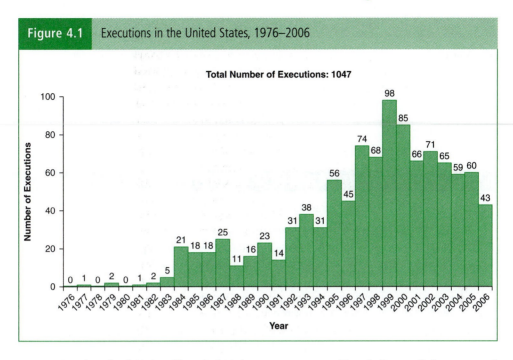

- During the hands-off period, prisoners were considered slaves of the state and had no rights at all. During the period of extending prisoner's rights, the federal courts extended a number of First, Fourth, Eighth, and Fourteenth Amendment rights to them, although these rights were obviously not as extensive as they would be outside prison walls. (However, inmates are the only group of Americans with a constitutional right to medical treatment.)
- Because of the granting of these rights, the federal courts became clogged with section 1983 suits challenging the conditions of their confinement.
- The U.S. Congress passed the PLRA in 1996 limiting prisoner access to federal courts and loosening the grip of the courts on state correctional systems. Congress also passed the AEDPA in the same year with a rider limiting inmates' habeas corpus rights.
- The courts have been active since the 1960s in providing rights to offenders under community supervision also. The Supreme Court has ruled that probationers have a right to counsel at a deferred sentencing hearing and that probationers and parolees have minimal due process rights (a fair hearing to establish cause) at revocation hearings. The Court has also extended the greater search powers of probation and parole officers to police officers under certain circumstances.
- Because the United States stands almost alone among democracies in retaining the death penalty, the issue has generated much debate and numerous cases questioning its constitutionality. Two classifications of individuals (the mentally retarded and juveniles) were removed from execution eligibility in the early 2000s based on culpability issues among others.

KEY TERMS

Amicus curiae

Antiterrorism and Effective
  Death Penalty Act

Bill of Rights

Eighth Amendment

First Amendment

Fourth Amendment

Fourteenth Amendment

Habeas corpus

Hands-off doctrine

Liberty interest

Prison Litigation Reform
  Act (1996)

Probation and Parole

Section 1983 suits

DISCUSSION QUESTIONS

1. What were the two main reasons or justifications for the hands-off doctrine?

2. Why does the concept of habeas corpus have such a revered place in common law?

3. What portions of the Bill of Rights apply to inmates and those under community supervision? What rationales are used to limit or restrict those rights?

4. Why do some scholars argue that cross-gender searchers should be different for men searching women versus women searching men? Do you think these different approaches to pat-down searchers are reasonable? Why or why not?

5. What do you think of laws permitting the civil commitment of some sex offenders? Are there any potential dangers for widespread abuse in the practice?

6. Why do you think that the United States retains the death penalty when almost all other democracies have eliminated it long ago? Should we eliminate it? Why or why not?

INTERNET SITES

American Bar Association: www.abanet.org

American Civil Liberties Union: www.aclu.org

American Correctional Association: www.aca.org

American Friends Service Committee (a Quaker organization interested in correctional reform): www.afsc.org

American Jail Association: www.aja.org

American Probation and Parole Association: www.appa-net.org

Bureau of Justice Statistics (information available on all manner of criminal justice topics): www.ojp.usdoj.gov/bjs

Death Penalty Information Center: www.deathpenaltyinfo.org

Human Rights Watch: www.hrw.org/prisons/

National Center for State Courts: www.ncsconline.org

National Criminal Justice Reference Service: www.ncjrs.gov

Office of Justice Programs, Bureau of Justice Statistics (periodic statistical reports on all manner of criminal justice topics, e.g., HIV in prisons and jails, probation and parole, and profiles of prisoners): www.ojp.usdoj.gov/bjs/periodic.htm

Vera Institute (information available on a number of corrections and other justice related topics): www.vera.org

READING

In this essay and review, Michael Vaughn and Rolando Del Carmen discuss civil liabilities imposed on prison officials for inmate-on-inmate assaults in correctional facilities. The U.S. Supreme Court decision in *Farmer v. Brennan* (1994) offers both good and bad news for prison administrators. The court's definition of "deliberate indifference" will make it difficult for prison plaintiffs to prevail in inmate-against-inmate assault litigation. Though inmates can still win their suits, the court has made it clear that prison officials who act reasonably cannot be sued successfully even if injury to an inmate occurs. A high degree of professionalism from correctional administrators and employees should help reduce assaults in their facilities and create an atmosphere in which civil liability is minimized.

Civil Liability Against Prison Officials for Inmate-on-Inmate Assault

Where Are We and Where Have We Been?

Michael S. Vaughn and Rolando V. Del Carmen

Prisons are disturbing places of "ultimate control, helplessness, and hopelessness" (Dumond, 1992, p. 141) where society institutionalizes individuals with propensities for violence. Prisons are indicative of a total institution with an intimidating environment of mistrust, fear, extortion, homosexuality, anger, and hostility. Research shows that animosity in prison frequently results in wanton intimidation, violence, brutality, and victimization. Although all inmates are susceptible to prison assault, researchers document that sex offenders, mentally ill prisoners, nonviolent property offenders, and snitches are at increased risk of victimization, leading many to request protection from potential violence.

This article examines legal liabilities imposed on prison officials for inmate-on-inmate assaults in correctional institutions. It discusses the legal issues in *Farmer v. Brennan* (1994) and highlights the Supreme Court's attempt to define "deliberate indifference," which is the standard used in many inmate-on-inmate assault cases. It also discusses federal courts of appeals cases prior to *Farmer* and how the *Farmer* case might shape the course of future lower court decisions. The article concludes with a discussion of situations in which prison officials might be liable for inmate-against-inmate assault and how correctional administrators can minimize liability risks in these types of cases.

⊠ *Farmer v. Brennan*— The Defining Case

On June 6, 1994, the U.S. Supreme Court ruled unanimously that inmates can prevail in suits against prison officials for inmate-by-inmate

SOURCE: Vaughn, M.S., & Del Carmen, R.V. (1995). Civil liability against prison officials for inmate-on-inmate assault: Where are we and where have we been? *The Prison Journal, 75*(1), 69–89. Reprinted with permission of Sage Publications, Inc.

assaults if they show that prison officials knew of a substantial risk of harm and recklessly disregarded that risk. In *Farmer v. Brennan*, plaintiff Dee Farmer was serving a lengthy federal prison term for forgery, theft, and credit card fraud. Farmer was a preoperative transsexual; although he was born a male, he possessed feminine traits and frequently wore a "T-shirt off one shoulder." Prison officials classified Farmer as a "biological male" and housed him in a male correctional institution. Farmer's condition presented problems for the Federal Bureau of Prisons (BOP) related to protection from the general inmate population. Until the time of the assault, Farmer had served most of his sentence in protective custody, which isolated him from the general inmate population. After a disciplinary infraction, however, Farmer was transferred from the Federal Correctional Institution in Oxford, Wisconsin to the U.S. Penitentiary in Terre Haute, Indiana. After the transfer to Terre Haute, prison officials first housed Farmer in administrative segregation, but later moved him to the general inmate population. About a week after that transfer to the general population, Farmer was raped and physically beaten in his cell after he spurned the sexual advances of another inmate.

After the assault, Farmer filed an action against prison officials in which he complained that placing a known transsexual with feminine traits in the general inmate population within a male prison with a history of violent inmate-against-inmate assaults amounted to "deliberate indifference" and constituted a violation of the Eighth Amendment's prohibition against cruel and unusual punishment. The U.S. District Court for the Western District of Wisconsin granted prison officials summary judgment, holding that prison administrators did not demonstrate "deliberate indifference" by failing to prevent the assault. The court said that because Farmer did not inform prison officials of concerns for his safety, officials did not possess "actual knowledge" of impending harm to Farmer and, therefore, were not reckless in a criminal law sense. On appeal, the U.S.

Court of Appeals for the Seventh Circuit affirmed without comment. The U.S. Supreme Court granted certiorari to resolve a dispute among the circuits as to the appropriate definition of "deliberate indifference" in inmate-against-inmate assault cases.

Justice Souter, writing for a unanimous Court, reaffirmed the duty of prison officials to "provide humane conditions of confinement"; maintain "adequate food, clothing, shelter, and medical care"; ensure that inmates live in a safe environment; and protect inmates from inmate-against-inmate assault. According to Justice Souter, "gratuitously allowing the beating or rape of one prisoner by another serves no 'legitimate penological objective'" and violates "'evolving standards of decency.'" In saying that the conditions of prisons may be "harsh," "restrictive," and uncomfortable, Justice Souter added that because imprisonment deprives inmates of the means to protect themselves, prison officials cannot "let the state of nature take its course" and allow assaults to occur.

The Court opined, however, that not all inmate-by-inmate assaults result in constitutional deprivation. As stated in *Wilson v. Seiter* (1991), the Court held that the prohibition against cruel and unusual punishment is violated only when two elements are proved. The so-called objective requirement is satisfied if prison officials deprive inmates of "the minimal civilized measure of life's necessities." In the context of inmate-against-inmate assaults, officials must fail to protect an inmate from a "substantial risk of serious harm." The second element pertains to the mental condition of prison officials or their "culpable state of mind." To be liable, officials must exhibit "deliberate indifference" to an actual or potential assault.

The phrase "deliberate indifference" was first defined by the Court in 1976 in *Estelle v. Gamble*, in which the Court distinguished "deliberate indifference" from "negligence." Over the years, Court cases have established that "deliberate indifference" resides on a continuum between "mere negligence . . . [and] something less than acts or omissions for the very purpose of causing harm or with knowledge that harm will result." Taking a "middle ground" on the continuum of culpability, the Court adopted the "criminal recklessness" standard, thus distinguishing "recklessness" as defined in the "civil law" from "recklessness" as defined in the "criminal law." The Court rejected the civil law definition of recklessness in which officials could be sued for failing "to act in the face of an unjustifiably high risk of harm that is either known or so obvious that it should be known," holding instead that "deliberate indifference" attached upon a "finding of recklessness only when a person disregards a risk of harm of which he is aware." This means that to hold officials liable, they "must both be aware of facts from which the inference could be drawn that a substantial risk of serious harm exists, and [they] must also draw the inference" that the risk existed. Thus, to hold officials liable for inmate-against-inmate assault, plaintiffs must show that officials "consciously" and "recklessly disregarded" a "substantial risk" of harm to an inmate.

In *Farmer*, the Court rejected plaintiff's argument that prison administrators who are unaware of "obvious" risks that "reasonable prison officials would have noticed" should be held liable, but the Court also said that prison officials cannot "ignore obvious dangers to inmates." Placing prison officials on notice that they could nonetheless be held liable for inmate-by-inmate assaults, the Court indicated that plaintiffs do not have to show that officials possessed actual knowledge of a substantial risk but failed to act. The key to finding liability is whether officials had knowledge of a substantial risk of harm and disregarded that risk. To prove that officials possessed knowledge of a substantial risk, plaintiffs can use "circumstantial evidence," including the fact that "officials knew of a substantial risk" because "the risk was obvious." The Court noted three situations in which plaintiffs might prevail: (a) if prison assaults were "long-standing, pervasive, well-documented, or expressly noted by prison officials in the past";

(b) if prison officials "refused to verify underlying facts that [they] strongly suspected to exist"; and (c) if prison officials "declined to confirm inferences of risk that [they] strongly suspected to exist." The Court emphasized that these situations might trigger a finding of "deliberate indifference," but cautioned that "it is not enough merely to find that a reasonable person would have known, or that the [prison officials] should have known."

The Court further held in *Farmer* that prison officials cannot escape liability because they did not know an inmate would be assaulted by the specific inmate who ultimately inflicted the injury. Moreover, "it does not matter whether the risk comes from a single source or multiple sources, any more than it matters whether a prisoner faces an excessive risk of attack for reasons personal to him or because all prisoners in his situation face such a risk." The Court also held that officials could be held liable if they know of a substantial risk of physical harm to a general class of inmates but no harm has yet occurred, adding that inmates do not have to wait for an assault to occur "before obtaining injunctive relief."

The *Farmer* Court suggested a number of defenses prison officials might employ to avoid liability. First, they may show that "they did not know of the underlying facts indicating a sufficiently substantial danger and that they were therefore unaware of a danger." Second, they may also show that "they knew the underlying facts but believed (albeit unsoundly) that the risk to which the facts gave rise was insubstantial or nonexistent." Third, they may show that they "knew of a substantial risk to inmate health or safety, [but] they responded reasonably to the risk, even if the harm ultimately was not averted." In sum, prison administrators are not liable if they can show that they responded reasonably to known risks.

Although the Court's decision was unanimous, it drew concurring opinions from three justices. Justice Blackmun argued for a reversal of *Wilson v. Seiter*, saying that he joined "the Court's opinion, because it create[d] no new obstacles for prison inmates to overcome, and it sen[t] a clear message to prison officials that their affirmative duty under the Constitution to provide for the safety of inmates is not to be taken lightly." Also in concurrence, Justice Stevens restated that there was no subjective intent requirement inherent in the Eighth Amendment. Justice Thomas, in a "puzzling vote," concurred in the judgment, even though he "rejected the decision's premise and . . . [argued that the Court should] reconsider [most of its] relevant [Eighth Amendment] precedents." He stated that any deprivation an inmate suffers while in custody is not protected by the Eighth Amendment. This reiterates his position that "conditions of confinement" in themselves do not violate the Eighth Amendment. He adds that because "prisons are necessarily dangerous places," the fact that inmates are physically and sexually assaulted is "unfortunate" but not unconstitutional. This interpretation is based on his reading of the original meaning of the word "punishment" in the Constitution. In the view of Justice Thomas, because the Eighth Amendment only prohibits cruel and unusual punishments and because deprivations suffered by the imprisoned "are not punishment . . . the unfortunate attack that befell [the inmate] was not part of his sentence [and therefore] it did not constitute 'punishment' under the Eighth Amendment." Pleading for juridical restraint, Justice Thomas "remain[ed] hopeful" that the Court would someday overturn *Estelle v. Gamble* and stop writing the "'National Code of Prison Regulation'" (*Farmer* at 1990).

Implications of *Farmer v. Brennan*

The implications of *Farmer* are significant because the Court's decision transcends the specific issues of the case. Two broad implications are readily identifiable. First, the Court has ruled that the Eighth Amendment is violated when inmates suffer unnecessary and wanton pain. Over the past decade, the Court has tried to clarify what standards are to be used when inmates claim their Eighth Amendment rights are violated. It has decided that whether the conduct of prison officials is wanton and unnecessary depends on the character of the Eighth Amendment claim. Two broad standards have developed: "deliberate indifference" and "malicious and sadistic." Although *Farmer v. Brennan* defines "deliberate indifference," the standard to be used in inmate-on-inmate assault cases, the implications of this case go beyond inmate assault cases.

The definition given to "deliberate indifference" will henceforth be used in cases when "deliberate indifference" governs the inquiry, such as when an inmate challenges prison medical care, conditions of confinement, smoking in prison, and inmate-on-inmate assault. The higher standard, the "malicious and sadistic" standard, governs the inquiry when an inmate alleges that prison officials used excessive deadly or nondeadly force. The difference between the two standards lies in the level of culpability attributed to prison officials.

Second, although "deliberate indifference" is used in a number of contexts in prison litigation, it is also used to hold municipalities liable. The *Farmer* Court took great pains to distinguish between the definitions of "deliberate indifference" in prison cases and "deliberate indifference" in police cases when municipalities are sued. Citing *City of Canton v. Harris* (1989), the Court restated that municipalities may be "deliberately indifferent" for failure to train police officers if the training deficiency was "'so obvious' . . . [that] 'policymakers' possessed 'actual or constructive'" knowledge that a constitutional violation would occur. The *Farmer* Court said that the "*Canton* understanding of 'deliberate indifference'" embodies an "objective" standard, which permits liability on "obviousness or constructive notice," adding that the objective test is inappropriate in prison cases because the rights protected in police liability cases differ from Eighth Amendment rights of inmates in prison cases. Thus the *Farmer* Court makes clear that liability may attach to a municipality in police civil liability cases when the customs or policies of a city cause poorly trained employees, who should have been trained in "obvious" areas, to violate citizens' constitutional rights. Conversely, prison officials are only liable if they recklessly disregard a known substantial risk of harm. The Court concluded that it would be "difficult" to "search for the subjective state of mind of a governmental entity [(i.e., city)], as distinct from that of a governmental official" (i.e., prison officer), giving more support for the Court's position that the subjective test should govern prison inquiries, but not police litigation.

▧ Classification and Incarceration of Violent Criminals

Cases on classification and incarceration of violent offenders decided in the U.S. circuit courts of appeals before *Farmer v. Brennan* show that prison officials cannot blatantly disregard substantial or pervasive risks of potential inmate danger to assault. Officials court liability when they ignore dangerous conditions or risks of assault in their facilities. Liability is more likely when a pattern of assault is pervasive and inmates are randomly assigned to the general inmate population with little regard to enemies or possible vulnerability. Officials are also susceptible to liability when inmates are victimized by violent aggressive prisoners who, although demonstrating a persistent pattern of abuse, are not segregated from the general inmate population. Specific knowledge of growing animosity and hostility between inmates that results in inmate-against-inmate assault also triggers liability if officials

did not move to protect those at risk of a pervasive risk of danger.

Supervisors have been held liable for injuries sustained by an inmate when officials fail to protect the inmate after affirmatively promising to do so. Some courts also consider the placement of an inmate in protective custody as a promise to protect; hence inmates who are assaulted by other inmates while in protective custody may also successfully sue prison officials. Similarly, when an institution issues an official protection order, prison officials have an affirmative duty to protect an inmate from harm.

Perhaps the greatest liability risk lies in failing to protect inmates in gang rivalry cases, particularly if they are known to be on a gang hit list and institutional policies or procedures fail to protect inmates from assault. Officials must protect inmates who are in a facility with prisoners against whom the others have made previously inculpatory statements. For example, inmates who testify against gang members must be segregated from all gang members, even those not testified against. Liability also attaches if officers allow an aggressive and predatory inmate into a restricted area and then watch the inmate perpetrate the assault. This occurs when officials permit aggressors to enter the cells of vulnerable inmates assigned to protective custody or inmates on gang retaliation lists. Officials may also be liable if an inmate is transferred to another facility because he "snitched" on a gang member, but officials at the new facility do not segregate him from potential gang aggressors, place him in the "toughest" part of the prison, or assign him to the general inmate population where he is murdered. In these cases, courts hold that the inmate was transferred because he was susceptible to dangerous risks of substantial harm and failure to take reasonable steps of protection at the new facility invokes liability.

Prison officials need to segregate vulnerable inmates from potential aggressors. The more vulnerable the potential victim, the greater the duty of prison officials to segregate likely aggressors. Prisons must institute classification procedures that identify weak inmates and screen them from violent cell mates. In addition, liability is more likely to occur when officials identify inmates as vulnerable and then cell vulnerable inmates with aggressive or violent offenders. Officials also invite liability when they cell inmates who have been the victims of sexual assault with aggressive inmates with criminal careers of sexual assault, particularly when the institution has taken previous steps to protect the vulnerable inmates. Moreover, officials may be successfully sued if they place inmates in administrative segregation for disciplinary reasons and the inmates are then assaulted.

Violating court orders and consent decrees also invites liability. Courts consider that court orders and consent decrees constitute proof of unconstitutional conditions within correctional facilities; hence the unconstitutional facilities are the proximate cause of the inmates' injuries sustained through inmate-by-inmate assault. Prison officials have been held liable when facilities consistently violate consent decrees with inadequate training, overcrowding, failure to segregate mentally ill inmates from the general inmate population, and general indifference by correctional staff. For example, liability has attached when no guards were present despite a consent decree order that required that "minimum custody inmates be assigned to dormitories and that at least one guard be stationed inside and one guard outside the dormitories at all times." Moreover, officials may be liable if they sit "idly by in the face of a two year violation of a consent decree and apparently [do] *absolutely nothing*" to address unconstitutional conditions (*Ryan v. Burlington County* at 1294).

⬛ Absence of Intervention by Prison Authorities in Assault Cases

Within a prison environment, correctional officials frequently know when inmates are being assaulted by other inmates. In these cases,

liability is more likely to attach when prison officials do not intervene in an appropriate way to prevent or stop an assault from occurring. For liability to attach, the case law holds that officials must either aid and abet an aggressor or fail to respond in a reasonable and appropriate manner to a known risk of assault. Because in these cases officials are highly culpable, the outcomes of the pre-*Farmer* cases in which prison officials were held liable will not significantly change under a post-*Farmer* analysis.

Courts have held officials liable for directly observing an assault and doing nothing to intervene. Liability is also likely when officials have knowledge of an ongoing altercation, fight, or dispute yet fail to respond for a significant length of time. Liability may also attach if correctional officers watch an assault unfold and inhibit a victimized inmate from escaping attack. For example, officials may be liable if they are stationed at a locked door and do not open the door to allow an inmate to escape imminent danger of assault.

Officials are likewise liable if they fail to routinely patrol areas in which assaults frequently occur. Courts have held that not patrolling an area plagued by disturbances for 45 minutes may give rise to a constitutional violation. Other courts hold officials liable for failing to monitor a holding area for several hours where predatory inmates sexually and physically assault vulnerable inmates. Neither can officials escape liability when they house nonviolent misdemeanant pretrial detainees with detainees charged with violent felonies and the violent detainees assault the nonviolent misdemeanants.

Inmates always prevail in their civil suits against prison officials when officers open cell doors of potential victims and allow aggressors to perpetrate violent assaults. Officials always lose because they know with certainty that inmates will be assaulted. By opening the victim's cell door, the correctional official gives aggressors the opportunity to perpetrate an assault. Allowing inmates to commit rapes in exchange for information about institutional security or maliciously spreading a rumor that an inmate is a "snitch" might also result in liability. Moreover, liability attaches when officials aid and abet to create a dangerous environment by motivating inmates to attack other inmates. For example, liability occurs if officers give verbal instructions to inmates to harm an individual or expose detailed information to the inmate subculture about a specific offender's criminal history, which results in assault.

⊠ Source of Weapon in Assault Cases

Courts have held that prison officials have an affirmative duty to prevent inmates from acquiring and possessing dangerous instruments that could be used to assault other inmates. Officials must monitor carefully when scrap metal is commonplace and where inmates can manufacture knives from scrap parts. Failure to control access to tools used in correctional facilities also might lead to liability. Officials are exposed to liability when their facilities have tool control policies but fail to monitor and control inmate access to potentially dangerous work tools. They are at peak risk of liability when they show continual and repeated violations of prison tool control policies and when inmates make repeated requests for protection from assault from work tools. Monitoring dangerous work tools that might be used in assaults also means that officials must not allow inmates to possess cleaning chemicals. Some cleaning chemicals, such as ammonia, bleach, or harmful acids, may be potentially used as dangerous weapons in inmate-against-inmate assaults.

⊠ Conclusion

Farmer v. Brennan is significant because it helps eliminate ambiguity as to when prison officials can be successfully sued for inmate-by-inmate assault. A viable definition of "deliberate indifference" was needed because

96 pre-*Farmer* cases in the U.S. circuit courts of appeals in inmate-on-inmate assaults indicate that the circuits used 14 different standards. The definition the Court adopted in *Farmer* appears to clarify case law by establishing a compromise between a high level of intent, "malicious and sadistic for the very purpose of causing harm," on the one hand, and "negligence," on the other. It remains to be seen, however, whether this definition can be applied more precisely than the elusive interpretation of "deliberate indifference" applied by federal courts before *Farmer*.

Only future litigation will determine the practical differences between the focus pre-*Farmer* cases place on "pervasive risk" and the focus the *Farmer* case puts on "substantial risk." In many post-*Farmer* cases, however, the final outcome of litigation may not demonstrably change from that of pre-*Farmer* cases. Officials will continue to be liable when they possess knowledge of real risks to inmates and do not act reasonably to prevent assaults from occurring. Conversely, officials will not be liable for risks they should have known about or for reasonable actions taken against known substantial risks. For these reasons, it appears that post-*Farmer* litigation will have the most significant impact in the borderline cases. But, even in the borderline cases, *Farmer* grants individual judges broad discretion to make independent case-by-case judgment calls that are plausible and therefore difficult to reverse on appeal.

The Supreme Court decision in *Farmer* offers both good news and bad news for prison administrators. The good news is that the Court's definition of "deliberate indifference" will make it difficult for prison plaintiffs to prevail in inmate-against-inmate assault litigation. The Court rejected the argument that officials are "deliberately indifferent" when they should have known of an "obvious risk" of potential harm to inmates. The bad news for prison administrators is that despite *Farmer*, inmates can, nonetheless, still win their suits. The Court said that officials cannot be ostriches and ignore "unwelcome knowledge" that "substantial risks" are present in their facilities. Prison administrators need to be aware of the Court's holding in *Farmer* and disseminate that information correspondingly to middle-management and rank-and-file employees. The key to avoiding liability is for prison personnel to act reasonably in the face of known and substantial serious risks. The Court has made it clear that prison officials who act reasonably cannot be sued successfully even if injury to an inmate occurs. A high degree of professionalism from correctional administrators and employees should help reduce assaults in their facilities and create an atmosphere in which civil liability is minimized.

◪ Reference

Dumond, R. W. (1992). The sexual assault of male inmates in incarcerated settings. *International Journal of the Sociology of Law, 20*(2), 135–157.

◪ Cases

City of Canton v. Harris, 489 U.S. 378 (1989).
Estelle v. Gamble, 429 U.S. 97 (1976).
Farmer v. Brennan, 114 S.Ct. 1970 (1994).
Ryan v. Burlington County, 674 F.Supp. 464 (D. N.J. 1987).
Wilson v. Seiter, 111 S.Ct. 2321 (1991).

DISCUSSION QUESTIONS

1. Should prison officials be civilly liable for inmate-inmate assault?

2. How does the Eighth Amendment apply (or does it) to this issue?

3. Summarize and give your opinion about what *Farmer v. Brennan* means or should mean.

READING

In this article, Katherine Bennett explores the many issues surrounding cross-gender searches and nude observations in prison settings. Litigation concerning the issues of cross-gender searches and visual observation of nude inmates in prisons and jails by opposite-sex officers involves alleged violations of several constitutional rights. Additionally, claims of two competing interest groups, prison employees and inmates, are in conflict. In 1987, the U.S. Supreme Court in *Turner v. Safley* established standards for reviewing alleged violations of inmates' constitutional rights, a ruling that clarified issues somewhat. However, the recent Ninth Circuit case of *Jordan v. Gardner* (1993) rejected Turner standards and inserted a twist in the direction courts were taking. The Religious Freedom Restoration Act of 1993 may further complicate the three issues of incidental observation of unclothed inmates by opposite-sex officers, cross-gender routine strip searches/body cavity searches, and cross-gender pat searches/clothed body searches.

Constitutional Issues in Cross-Gender Searches and Visual Observation of Nude Inmates by Opposite-Sex Officers

A Battle Between and Within the Sexes

Katherine Bennett

For years now, U.S. district courts and circuit courts of appeals have addressed the civil rights issues arising from the viewing of nude inmates by opposite-sex officers and cross-gender searches in prisons and jails. Claims of violations of constitutional rights—usually of the First, Fourth, Eighth, and Fourteenth Amendments—have been brought by both female and male correctional officers and female and male prisoners. Prior to 1980, courts held a more restricted view of cross-gender searches and observation of unclothed inmates. Illustrative of this view is *York v. Story* (1977, cited in Reisner, 1978), which indicated that forced exposure of the naked body to guards or police officers of the opposite sex is degrading and violates the privacy right. Limited work responsibilities were recommended for correctional officers of the opposite sex so as not to infringe on inmate privacy rights. Over the past decade and a half, however, courts have expanded work experiences in favor of increased employment opportunities for both female and male correctional officers. Courts have been more willing, within limits, to allow intrusion into inmate privacy.

This study shows that the majority of cases have been filed by male inmates claiming constitutional violations when searched or observed in the nude by female officers. Female inmates and both male and female correctional officers, however, have also filed cases seeking to establish the limits of their rights. In 1987, the U.S. Supreme Court in *Turner v. Safley* established standards for determining violations of inmates' constitutional rights. Several cases considering cross-gender issues have subsequently relied on *Turner*. Recent departures from *Turner* have created some controversy, however. Additionally, the Religious Freedom Restoration Act of 1993 may further exacerbate the conflict between cross-gender observation and searches, on the one side, and certain religions' tenets, on the other.

This article examines litigation that addresses three issues in a jail or prison setting: incidental observation of unclothed inmates by opposite-sex officers, cross-gender routine strip searches/body cavity searches, and cross-gender pat searches/clothed body searches. The result of *Turner* and the potential impact of the Religious Freedom Restoration Act of 1993 on these issues are also discussed.

⚑ Cases Addressing Incidental Observation of Nude Inmates by Opposite-Sex Officers

Cases filed by inmates involving incidental observation of nude inmates by opposite-sex officers generally allege violations of privacy rights, religious freedom, and claims of cruel and unusual punishment. Other cases filed predominantly by female correctional officers focus on gender discrimination in employment. In 1980, in the Second Circuit case of *Forts v. Ward*, female prisoners filed suit protesting the placement of male officers in housing and hospital areas. The female

SOURCE: Bennett, K. (1995). Constitutional issues in cross-gender searches and visual observation of nude inmates by opposite-sex officers: A battle between and within the sexes. *The Prison Journal, 75*(1), 90–112. Reprinted with permission of Sage Publications, Inc.

inmates claimed a violation of privacy rights, objecting particularly to being viewed while sleeping and/or while unclothed. The prison administrators expressed willingness to accept the remedy proposed by the state: issuing suitable sleepwear for the female inmates and allowing them to cover their cell windows for 15 minute intervals. The court added that in accepting this remedy, they "need not decide whether security interests justify impairment of inmates' privacy interests."

Cases addressing observation of nude inmates often involve the conflict between employment rights of one group and privacy rights of another. In *Griffin v. Michigan Dept. of Corrections* (1982), female correctional officers brought a class action against the Michigan Department of Corrections, alleging sex discrimination in employment because females were not allowed to work in the housing areas at all-male maximum-security institutions, thus being denied opportunities for promotion. The U.S. District Court ruled that inmates did not have any protected privacy right under the federal Constitution against being viewed naked by correctional officers of the opposite sex. It further noted that any privacy rights that inmates may have are already eroded in the interest of institutional convenience, security, and efficiency. Inmates, while nude or performing bodily functions, are constantly subject to view by other inmates and by male correctional officers. The District Court reasoned that it would constitute a "stereotypical sexual characterization" to assume that such viewing of an inmate by members of the opposite sex is "intrinsically more odious than viewing by members of one's own sex" (*Griffin* at 701).

Another case involving employment rights of officers and inmate rights was decided in *Bagley v. Watson* (1983). Female correctional officers filed suit against the Oregon state corrections division, alleging that 90% of positions for correctional officers at Oregon's prisons were designated "male only." Female officers were thus limited in the experience they might gain, available overtime, and promotion and advancement opportunities. The Oregon state constitution and the Oregon Supreme Court case of *Sterling v. Cupp* (1981) provided the state's justification for the selective work experiences.

Under the Oregon state constitution, male inmates have the right not to be treated with "unnecessary rigor" (*Bagley* at 1103). The Oregon Supreme Court had ruled in *Sterling* that frisk searches or other searches of anal-genital areas conducted routinely by female correctional officers violated this state constitutional right. Thus women were restricted from many positions for prison guards. The district court in *Bagley* stated that the U.S. Constitution does not expressly provide for a right of privacy and, further, that the U.S. Supreme Court has not judicially created a constitutional right of privacy such as the defendants were advocating. The court concluded that, in this case, male inmates had no federal constitutional rights to freedom from clothed, pat frisk searches and/or visual observations in states of undress performed by female correctional officers. The court noted, however, that the opinion in *Bagley* applied only to male inmates and female officers (*Bagley* at 1105). The question of searches of female inmates by male officers was not at issue. Additionally, the court did not address body cavity searches by female correctional officers.

Although the *Griffin* and *Bagley* cases implied that there is no constitutional privacy right not to be viewed naked, a circuit court of appeal, in *Cumbey v. Meachum* (1982), did agree that inmates have a more general right to privacy. In this case, the Tenth Circuit Court case cited the Supreme Court ruling in *Bell v. Wolfish* (1979), which stated that "although convicted prisoners are not entitled to the full protection of the Constitution, they do not forfeit all constitutional protections by reason of their conviction and confinement in prison" (*Bell* at 520). Moreover, the state may restrict a prisoner's constitutional rights only to the extent necessary to further the legitimate goals and policies of the correctional system. Although the Court of Appeals in *Forts* was spared from deciding if security interests would justify impairing inmates' privacy interests, *Cumbey* stated that the inmate's right to privacy must yield to the correctional system's need for security, although the right does not vanish completely.

The Court of Appeals in *Cumbey* ruled that a claim that the viewing of nude male inmates by female correctional officers was a violation of the right to privacy was not necessarily frivolous; what was important was the frequency of such a practice. The court did find that a single comment made by a female correctional officer regarding the male inmates' nudity could not constitute cruel and unusual punishment under the Eighth Amendment.

The frequency with which nude inmates are viewed was also a factor in *Miles v. Bell* (1985). Male inmates claimed a violation of their Eighth Amendment rights, alleging that the unannounced presence of female officers constituted cruel and unusual punishment. The District Court ruled that, because female officers saw inmates unclothed or using the shower or toilet facilities only rarely, privacy rights were not violated. Inmates would have to show that such viewing occurred on a regular basis in order to show that privacy rights had been violated.

A similar case in 1985 claimed violation of privacy rights based in five constitutional amendments. Male inmates in *Grummett v. Rushen*, a Ninth Circuit case, alleged that the prison policy and practice of allowing female correctional officers to view male inmates when partially or totally nude while dressing, showering, being strip searched, or using toilet facilities violated their rights of privacy under the First, Fourth, Eighth, Ninth, and Fourteenth Amendments (*Grummett* at 492–493). The court concluded that there was an appropriate balance among the institution's security needs, female guards' rights to equal employment opportunities, and prisoners' privacy interests. The inmates appealed; the Court of Appeals looked at the several amendments inmates were claiming had been violated.

In 1987, male inmates again claimed that their rights to privacy and religious beliefs were violated by being viewed naked by female correctional officers. In *Johnson v. Pennsylvania Bureau of Corrections*, the District Court stated that prisoners have privacy rights under the Fourth Amendment, including the right not to be viewed naked by members of the opposite sex. However, the right of privacy exists "only so far as it is not fundamentally inconsistent with prisoner status or incompatible with legitimate objectives of incarceration" (*Johnson* at 426). The court noted a reasonable attempt to balance the conflict of privacy rights and employment opportunities in that a policy was in place whereby correctional officers of opposite sex to the inmate population were not assigned to posts where they had to work in open view of unclothed inmates and where they were required to announce their presence when entering housing areas. Thus a practice of selective work responsibilities was in place. Female officers saw unclothed inmates only on a limited number of occasions, so such viewings did not violate privacy rights. Moreover, women were not assigned to showers or strip-and-search areas.

Also in 1987, the Sixth Circuit Court of Appeals in *Kent v. Johnson* heard a Michigan case, but by this time, *Turner v. Safley* (1987) had been decided and could be used as precedent. In *Turner*, the U.S. Supreme Court rejected the standard of heightened scrutiny in prison cases in favor of a rational relationship test, stating that "when a prison regulation impinges on inmates' constitutional rights, the regulation is valid if it is reasonably related to legitimate penological interests" (*Turner* at 2254). The Court enumerated four factors to be used when applying this test:

1. Existence of valid national connection between prison regulation and legitimate governmental interest put forward to justify it;

2. Existence of alternative means of exercising rights that remain open to prison inmates;

3. Impact that accommodation of asserted constitutional right will have on guards and other inmates and on allocation of prison resources generally; and

4. Absence of ready alternatives as evidence of the reasonableness of the regulation (the presence of obvious easy alternatives may point to the unreasonableness of the regulation). (*Turner* at 2262)

In *Kent,* a male inmate challenged the prison policy of allowing female correctional officers "full and unrestricted access to all areas of the prison housing unit" (*Kent* at 1221). He claimed violations of his First, Fourth, Eighth, and Fourteenth Amendment rights. Prison officials justified the practice by saying that they were required to follow nondiscriminatory staffing practices as mandated in *Griffin v. Michigan Dept. of Corrections.* The Sixth Circuit Court remanded the District Court's decision and instructed the lower court to incorporate *Turner* standards, noting in particular the Court's statement on "deferential review" (*Kent* at 1229). In *Turner,* the Supreme Court ruled that, when ascertaining whether regulations that infringe on inmates' constitutional rights "are reasonably related to legitimate penological interests," the courts are to "accord great deference to prison authorities' judgments regarding the necessity of such regulations" (*Turner* at 2259).

Despite *Turner,* by 1990 courts had not completely affirmed that a right to privacy existed. In *Timm v. Gunter* (1990), the Eighth Circuit considered an appeal from prison administrators in Nebraska. Male prisoners had originally brought a class action suit alleging that allowing female correctional officers to perform pat searches and to see the inmates nude or partially nude violated their right to privacy. Female correctional officers countered with a claim that they had a right to equal employment. In a footnote to the decision, the Court of Appeals noted: "We proceed in our analysis on the assumption that inmates possess a constitutional right to privacy. It is not necessary to decide whether this unenumerated right exists or to define its constitutional foundation or scope, and we expressly decline to do so" (*Timm* at 1097).

In summary, the above cases illustrate that, when courts are faced with alleged constitutional violations resulting from observation of nude inmates by officers of the opposite sex, the tendency has been to look at the frequency of such observations. Generally, increased employment opportunities for officers justify cross-gender observation. The Religious Freedom Restoration Act of 1993, conversely, may reverse this trend, as noted below.

⊠ Cases Addressing Cross-Gender Routine Strip Searches/Body Cavity Searches

Unlike *Forts,* which was filed by female inmates objecting to the presence of male officers in housing areas, *Edwards v. Dept. of Corrections* (1985) involved a male correctional officer claiming that the refusal to promote him in the state prison for women was because of his gender. Although this correctional officer worked as acting shift commander at a prison for women, he was told that he could not be promoted to that position because the policy restricted such positions to women. No males in the women's prison could serve as shift supervisors; the department asserted that femaleness was a bona fide occupational qualification (BFOQ) for shift commander at this prison. The court concluded that shift commanders could perform their duties without frequent patrol of dorms, restrooms, showers, or regular searches of or contact with inmates, thus femaleness was not a BFOQ for shift commanders. The court further noted that all employees were in agreement that it would be appropriate for a male correctional officer to search or subdue a female inmate in an emergency situation.

The majority of cross-gender searches and visual observation cases brought before the courts involve male inmates and female officers. However, in *Bonitz v. Fair* (1986), the First Circuit Court of Appeals considered a case involving female prisoners and body cavity searches. The court ruled that female prisoners had a "clearly established" Fourth Amendment right to be free from body cavity searches in a general security search of the prison. The court ruled that, in light of the then-prevailing case law from the U.S. Supreme Court, courts of appeals, and the local district court, body cavity searches of female inmates conducted by police officers in a nonhygienic manner in the presence of male officers and involving touching clearly established a violation of the inmates' Fourth Amendment right to be free from an unreasonable search (*Bonitz* at 173).

In 1988, the Ninth Circuit heard an appeal involving male inmates' privacy rights in strip searches and female correctional officers' equal employment opportunities. The Court of Appeals relied heavily on *Turner* for precedent. *Michenfelder v. Sumner* (1988) affirmed the District Court's ruling that the fact that female correctional officers might be able to observe routine strip searches, which included visual body cavity searches, of male prisoners housed in state prison maximum-security units did not violate prisoners' privacy rights. Female correctional officers conducted such strip searches only in severe emergencies and were not routinely present during such searches. *Michenfelder* observed that *Turner* had ruled that as long as a prisoner is presented with an opportunity to obtain contraband or a weapon while outside the cell, visual strip searches have a legitimate penological purpose. Further, the existing allocation of responsibilities among male and female employees already represented a reasonable attempt to accommodate prisoners' privacy concerns consistent with internal security needs and equal employment concerns, such as was ruled in *Grummett*.

In 1991, the First Circuit heard an appeal by New Hampshire prison officials in *Cookish v. Powell.* A male inmate had alleged that his Fourth Amendment right to be free from unreasonable searches was violated when female correctional officers supervised and/or observed him during a visual body cavity search. The defendants claimed, first, that there had been no Fourth Amendment violation and, second, that they were entitled to qualified immunity for their actions. The trend, according to the First Circuit Court and citing *Cumbey*, was that an inmate's constitutional right to privacy was violated when officers of the opposite sex regularly observed him/her engaged in personal activities, such as undressing, showering, and using the toilet. At the same time, citing both *Grummett* and *Bonitz*, the courts held that even assuming such a protected right to privacy, "occasional and inadvertent observations or casual observations restricted by distance, of inmates dressing, showering, being strip searched or using toilet facilities" did not impinge on constitutional rights (*Cookish* at 446). Further, courts acknowledged that, during times of emergency, temporary violation of inmate privacy was justified to protect the safety of both inmates and officers.

Defendants in *Cookish* had relied on the *Turner* standard and argued that the search at issue was reasonably related to legitimate penological objectives and thus was constitutional. The Court of Appeals also applied the reasonableness test and stated that, even if the search had violated plaintiffs' constitutional rights, it was objectively reasonable for the defendants to conclude that having female correctional officers supervise and/or observe this search in these circumstances was lawful, and thus they were entitled to qualified immunity.

To summarize, although courts have acknowledged that a general right to privacy does exist, *Turner* has allowed suspension of that right in prison when related to security interests. However, conducting and observing strip searches by female officers are tolerated primarily only in emergency situations. Strip searches and body cavity searches of female inmates by male officers are deemed unreasonable.

⬛ Cases Addressing Cross-Gender Pat Searches/Clothed Body Searches

Cases addressing cross-gender pat searches and clothed body searches are perhaps midspectrum in intrusiveness, but court decisions are in even bigger conflict. Courts generally uphold claims by both male and female correctional officers of gender discrimination in employment. Male inmates citing violations of privacy and free exercise of religion and claims of cruel and unusual punishment usually have been unsuccessful. On the other hand, in 1993, female inmates succeeded in substantiating a cruel and unusual punishment claim when subjected to cross-gender clothed body searches.

In *Smith v. Fairman* (1982), male inmates alleged that frisk searches by female officers constituted cruel and unusual punishment (*Smith* at 52). The court stated that such searches clearly fall short of the "shocking, barbarous treatment proscribed by the Eighth Amendment" (*Smith* at 53). The frisk searches were described as pat searches that excluded the genital area.

In 1983, the Seventh Circuit Court of Appeals heard *Madyun v. Franzen*. A male inmate from the same prison as in *Smith* appealed a U.S. District Court ruling that a frisk search by a female officer is not a violation of either First, Fourth, or Fourteenth Amendment rights. The male inmate said that submitting to frisk searches by women correctional officers violated his First Amendment rights of privacy and free exercise of his religion. His Islamic religion forbade physical contact with women other than his wife and mother. He claimed further a Fourth Amendment right to be free from unreasonable searches and a Fourteenth Amendment

right to equal protection of the laws because female inmates in Illinois were not subjected to frisk searches by male correctional officers. As to the free exercise of religion claim, the Court of Appeals acknowledged that a frisk search by a member of the opposite sex might well be incompatible with the tenets of the inmate's religion. Such an intrusion, however, is "justified by a state interest of sufficient magnitude" (*Madyun* at 957). The state interests at issue here were that of providing adequate prison security and equal employment opportunities for women. The Seventh Circuit Court further ruled that, because female officers in Illinois are instructed to exclude genital areas in a frisk search, violation of Fourth Amendment rights was not substantiated.

On the claimed Fourteenth Amendment violation, the court stated that the Fourteenth Amendment requires that any gender-based distinction drawn by the state "must serve important governmental objectives and must be substantially related to the achievement of those objectives" (*Madyun* at 962). The state of Illinois allowed female officers to frisk search male inmates, whereas male officers could not frisk search female inmates, in an effort to provide more employment opportunities for women. The court could find no indication that males had experienced the absence of opportunities to be employed as correctional officers because they were precluded from frisk searching female inmates. Thus this gender-based distinction did serve an important government interest of equalizing employment opportunities for women in corrections.

In *Grummett v. Rushen* (1985), although the female officers did not routinely conduct or observe unclothed body searches, they could conduct pat searches, including the groin area. The Ninth Circuit Court observed that, because the right of privacy is a fundamental right, such a right may be restricted only if the limitation is justified by a "compelling state interest" (*Grummett* at 491). The court concluded that routine pat searches were justified by security needs and did not violate the Fourteenth Amendment when conducted by correctional officers of the opposite sex. The court then considered whether, under the Fourth Amendment, "reasonable" searches conducted by members of the opposite sex could become unreasonable. It found that searches in this case were not unreasonable because they were done briefly and while inmates were fully clothed. Intimate contact with inmates' bodies was not involved, and searches were performed in a professional manner, with respect for inmates. Female officers very rarely observed strip searches, and those were in emergency situations. In a concurring opinion, one Court of Appeals judge cautioned that the court's ruling could increase the "further constitutionalization of prison procedures" (*Grummett* at 496-497). Although constitutionalization is necessary in cases of "gross neglect by governments of their prison systems," it can be carried too far (*Grummett* at 496–497).

In *Timm v. Gunter* (1990), an earlier ruling by a magistrate's court had stated that an inmate, when subjected to a pat search, be allowed to demand that the search of his groin area be conducted by a male officer. A subsequent District Court ruling resulted in an order that all officers be instructed to "refrain from deliberately touching genital and anal areas while conducting a pat search" (*Timm* at 1098). Additionally, the prison administration had to make physical and scheduling modifications to accommodate inmates' privacy rights, and the prison administration appealed. The Court of Appeals noted that restricting female officers to conducting pat searches but excluding the groin area completely would severely diminish the effectiveness of the search. Further, allowing female officers to pat search male prisoners on the same basis as male officers was a reasonable regulation. The court concluded that sex-neutral surveillance of prisoners and the goal of prison security are rationally connected. Thus the Eighth Circuit veered away from selective work responsibilities as have been

noted in other cases, at least in the area of pat searches of male inmates.

▨ *Jordan v. Gardner*— An Aberration

In *Jordan v. Gardner* (1993), the Ninth Circuit Court of Appeals affirmed the District Court's finding for the plaintiff, female inmates who challenged the constitutionality of cross-gender clothed body searches. The Ninth Circuit Court found that a Washington State prison policy requiring male guards to conduct random, nonemergency, suspicionless clothed body searches on female prisoners was cruel and unusual punishment in violation of the Eighth Amendment. According to expert witness testimony, these searches constituted unnecessary and wanton infliction of pain in that many female inmates had been sexually abused prior to their incarceration, hence concluding that "unwanted intimate touching by men was likely to cause psychological trauma." Additionally, such searches "did not ensure equal employment opportunities for male officers" (*Jordan* at 1521–1522).

The Court of Appeals stated that the *Turner* "reasonableness" standard, used to evaluate the constitutionality of prison officials' actions, did not apply to the Eighth Amendment. In *Jordan*, a District Court for Washington initially found the search policy to violate female prisoners' First, Fourth, and Eighth Amendment rights. A three-judge panel decision reversed the district court. A year later and sitting en banc, the 11-member Circuit Court of Appeals vacated their earlier opinion and affirmed the District Court on Eighth Amendment grounds, thus not reaching the inmates' other constitutional claims. Four judges dissented.

Interestingly, female correctional staff in *Jordan* had first filed a grievance in 1988 against the same-gender search policy. Female officers objected to being interrupted while on meal breaks to conduct searches at fixed checkpoints.

The Washington Department of Corrections denied this grievance. A new superintendent took over in 1989. To control the flow of contraband, he decided to implement a "random" search policy and to order routine cross-gender clothed body searches of inmates, recognizing that an increase in same-gender searches might lead to additional grievances and lawsuits by the female guards. The new policy was implemented, and on the same day inmates filed a civil rights action, obtaining an injunction from the cross-gender searches. Writing for the majority, Judge O'Scannlain acknowledged that privacy rights protecting inmates from cross-gender body searches have not as yet been judicially recognized. On the other hand, the Eighth Amendment right of incarcerated persons to be free from the unwarranted infliction of pain is clearly established. The court departed from precedent established in *Grummett* and *Michenfelder*, with the majority concluding that the frequency and scope of the searches in those previous cases were "significantly less invasive than the searches at issue" in *Jordan* (*Jordan* at 1524).

The court further noted that "*Turner* has been applied only where the constitutional right is one which is enjoined by all persons, but the exercise of which may necessarily be limited due to the unique circumstances of imprisonment" (*Jordan* at 1530). The majority observed that the U.S. Supreme Court had never applied *Turner* to an Eighth Amendment case, possibly because Eighth Amendment rights do not conflict with imprisonment but instead limit the actions and treatment that may be exercised on the inmate as "punishment." The decision in *Jordan* did not extend to cross-gender searches under emergency situations, cross-gender searches of men by women, or cross-gender searches at female institutions other than the institution under question.

Writing for the dissent, Judge Trott relied on the Supreme Court's ruling in *Whitley v. Albers* (1986) that to constitute cruel and unusual punishment, conduct that is *not* purported to be punishment "must involve more than ordinary lack of due care for the prisoner's interests or safety" (*Jordan* at 1545). The Supreme Court ruled that the conduct must involve "a state of mind on the part of the antagonist that is wanton" (*Jordan* at 1545). The dissent noted further that, in *Wilson v. Seiter* (1991), "it is obduracy and wantonness, not inadvertence or error in good faith, that characterize the conduct prohibited by the Cruel and Unusual Punishment Clause" (*Jordan* at 1545). As stated in *Wilson v. Seiter*, wantonness in the context of cases addressing confinement conditions is defined as "deliberate indifference."

Judge Trott, in dissent, characterized the *Jordan* opinion as a "management nightmare," stating that any individual female or male prisoner previously abused sexually would be immune from random pat searches conducted by a person of the gender of the prisoner's abuser. Thus a "male prisoner with a history of abuses as a child by a man—and our prisons are full of them—will surely be able to make a case against random pat searches by male correctional officers" (*Jordan* at 1561). A woman previously abused by a woman may be able to do the same. There could exist victimized prisoners claiming that they cannot have their genitals touched by anyone without enduring psychological damage. Judge Trott suggested that Judge O'Scannlain's opinion could create "the real specter of a special class of untouchable prisoners" (*Jordan* at 1561). Applying this ruling to cross-gender searches of felons by police officers in the community poses additional problems. In conclusion, Judge Trott quoted from Justice Holmes, saying:

> While the courts must exercise a judgment of their own, it by no means is true that every law is void which may seem to the judges who pass upon it excessive, unsuited to its ostensible end, or based upon conceptions of morality with which they disagree. Considerable latitude must be allowed for differences of view as well as for

possible peculiar conditions which this court can know but imperfectly, if at all. (*Jordan* at 1566)

Jordan was referred to by the Tenth Circuit 5 months later in *Hovater v. Robinson* (1993). The Tenth Circuit reversed a Kansas District Court that had found for a former female jail inmate who sued the county sheriff for civil rights violations by failing to protect her and failing to supervise and train jail employees adequately, allegedly resulting in a sexual assault by a detention officer. Although *Jordan* was discussed in some detail, the court noted that, unlike the policy in *Jordan*, the jail policy did not permit physical contact between the male officers and inmates. "Moreover, the record in *Jordan* was replete with evidence to support the risk of harm likely to result from the policy" (*Hovater* at 1067–1068).

⊠ Implications of *Jordan*

The *Jordan* decision led court observers to suggest that administrators of correctional institutions in the Ninth Circuit consider revising any policies they might have had that allowed male officers to pat search female inmates in other than emergency situations. A representative of the Washington State Attorney General's office noted that approximately six states and the Federal Bureau of Prisons permitted cross-gender pat searches of female inmates. Finding cross-gender pat searches of female inmates to be cruel and unusual punishment but not so for male inmates may generate more equal protection grievances and lawsuits. In an era of gender equality, such differential treatment becomes difficult to justify, absent a compelling state interest. Moreover, employment opportunities thus may be more restricted for male correctional officers than female officers in some institutions.

Jordan also weakened *Turner* by implying that *Turner* does not apply to Eighth Amendment rights. Although some have characterized the *Turner* test as a reformulation of the "hands-off" doctrine (Giles, 1993), Turner has been further modified by the passage of the Religious Freedom Restoration Act of 1993, at least in freedom of religion cases. This act may affect significantly the issue of cross-gender observation and searches of nude inmates.

⊠ The Potential Impact of the Religious Freedom Restoration Act

The Religious Freedom Restoration Act of 1993 states that "government may substantially burden a person's exercise of religion only if it . . . is in furtherance of a compelling governmental interest; and is the least restrictive means of furthering that compelling governmental interest." The test for alleged violations of religious practices is thus no longer the *Turner* test. It is instead a heightened standard compared to the *Turner* standard and is much more difficult for prison authorities to establish. Claims that cross-gender observation and searches violate an inmate's First Amendment rights may very well be evaluated in a much different light than in the recent past. As discussed previously, a male inmate cited Islamic religious tenets forbidding physical contact with the opposite sex in his objection to frisk searches by female officers in *Madyun v. Franzen*. In that case, however, the court justified the intrusion into religious practices, saying that compelling governmental interests of security and equal employment opportunities are furthered by cross-gender searches. Courts at present, on the other hand, may not see cross-gender searches as the least restrictive means to furthering compelling state interests of security and equal employment opportunities. Courts may in fact urge prison authorities to seek actively other alternatives before restricting or intruding on inmates' religious practices. Whereas *Turner* urged courts to apply "great deference" to correctional officials' judgments, the Religious Freedom Restoration Act of 1993 places a heavier burden on those officials to now justify such judgments that impinge on inmates' religious beliefs. In the case of *Johnson v. Pennsylvania Bureau of Corrections* (1987), inmates referred to the modesty requirement of the Muslim faith that forbids an individual from exposing his or her naked body to strangers, but the District Court for Pennsylvania ruled that the plaintiffs were required to show that their beliefs were sincerely held and religious in nature. It is up to the courts to establish whether such a requirement is still a "threshold requirement" under the Religious Freedom Restoration Act of 1993. The Religious Freedom Restoration Act of 1993 may yet further complicate an already difficult issue.

Conclusion

One observer notes that "cross-gender supervision issues remain troublesome" ("Allowing Male Officer," 1993). That may be an understatement. In cases involving observation of nude inmates by opposite-sex officers, the courts generally have not found constitutional violations, upholding instead equal employment opportunities for officers. No hard-and-fast rules exist, however, for when cross-gender searches and observations become constitutional violations. Of importance to the courts is the frequency of such observations. Courts tend to look at cross-gender routine strip searches and body cavity searches as much more objectionable. Cross-gender strip searches may be justified in emergency situations, but same-sex officers for such searches are definitely preferable. Pat searches and clothed body searches of male inmates by female officers are accepted, but such searches of female inmates may constitute cruel and unusual punishment, depending on the nature of the inmate population.

The complex issues involved in cross-gender searches and visual observation of inmates are far from resolved. Basic interests between two groups are in conflict—equal employment opportunities for prison personnel, on the one hand, and First, Fourth, Eighth, and Fourteenth Amendment rights for inmates, on the other. It is not always opposite genders that are at conflict. In *Jordan*, female correctional officers were the initial complainants about a same-gender search policy, yet female inmates promptly brought suit when a cross-gender search policy was implemented. Court decisions are at odds concerning inmates' constitutional right to privacy. *Turner* offers some standards for the courts but was not used in *Jordan*, which significantly departed from precedent. Due to the recent passage of the Religious Freedom Restoration Act of 1993, *Turner* standards also may not apply when the conflict is between religious practices and cross-gender observation and searches. To ensure employee morale and advancement opportunities, both female and male officers must be allowed to perform the same duties. It is unclear, however, whether equal employment opportunities merit precedence over religious freedom and privacy rights. A Supreme Court decision should go a long way toward providing guidance, but whether that is forthcoming in the near future is a matter of conjecture.

References

Allowing male officer to be alone with female inmate does not violate Eighth Amendment. (1993, December). *Correctional Law Reporter, 5*(4), 54.

Giles, C. D. (1993). *Turner v. Safley* and its progeny: A gradual retreat to the "hands-off" doctrine? *Arizona Law Review, 35,* 219–236.

Reisner, S. L. (1978). Balancing inmates' rights to privacy with equal employment for prison guards. *Women's Rights Law Reporter, 4,* 243–251.

Religious Freedom Restoration Act (RFRA) now law of the land. (1994, February). *Correctional Law Reporter, 5*(5), 65–73.

◩ Cases

Bagley v. Watson, 579 F.Supp. 1099 (D.Or. 1983).

Bell v. Wolfish, 441 U.S. 520 (1979).

Bonitz v. Fair, 804 F.2d 164 (1st Cir. 1986).

Cookish v. Powell, 945 F.2d 441 (1st Cir. 1991).

Cornwell v. Dahlberg, 963 F.2d 912 (6th Cir. 1992).

Cumbey v. Meachum, 684 F.2d 712 (10th Cir. 1982).

Edwards v. Dept. of Corrections, 615 E.Supp. 804 (D. C. Ala. 1985).

Forts v. Ward, 621 F.2d 1210 (2d Cir. 1980).

Griffin v. Michigan Dept. of Corrections, 654 F.Supp. 690 (E.D.Mich. 1982).

Grummett v. Rushen, 779 F.2d 491 (9th Cir. 1985).

Hovater v. Robinson, 1 F.3d 1063 (10th Cir. 1993).

Johnson v. Pennsylvania Bureau of Corrections, 661 F.Supp. 425 (W.D.Pa. 1987).

Jordan v. Gardner, 986 F.2d 1521 (9th Cir, 1993).

Jordan v. Gardner, 968 F.2d 984 (9th Cir. 1992).

Jordan v. Gardner, 953 F.2d 1137 (9th Cir. 1992).

Kent v. Johnson, 821 F.2d 1220 (6th Cir. 1987).

Madyun v. Franzen, 704 F.2d 954 (7th Cir. 1983).

Michenfelder v. Sumner, 860 F.2d 328 (9th Cir. 1988).

Miles v. Bell, 621 F.Supp. 51 (D.C.Conn. 1985).

Smith v. Fairman, 678 F.2d 52 (7th Cir. 1982).

Sterling v. Cupp, 290 Or. 611, 625 P.2d 123 (1981).

Timm v. Gunter, 917 F.2d 1093 (8th Cir. 1990).

Turner v. Safley, 107 S.Ct. 2254 (1987).

Whitley v. Albers, 475 U.S. 312 (1986).

Wilson v. Seiter, 111 S.Ct. 2321 (1991).

DISCUSSION QUESTIONS

1. Why do you think that it is permissible for female corrections officers to view nude male inmates and perform pat-downs and not permissible for male officers to perform the same functions with female inmates?

2. Given the sexual tensions and all the other problems created by allowing opposite-sex guards in prisons, would it be better to mandate same-sex guards only in our prisons regardless of the objections of female officers?

3. Do you think that most male inmates complaining about cross-gender body searches really believe that they are being subjected to cruel and unusual punishment, or are they merely engaging in frivolity?

READING

In this article, Cary Federman seeks to place the Antiterrorism and Effective Death Penalty Act (AEDPA) of 1996 within a political and historical framework. More specifically, he describes the effort by the Supreme Court and various interested parties to restrict prisoners' access to the federal courts by way of filing writs of habeas corpus. Of principal concern to Federman is how an act of terrorism against the United States came to provide an opportunity for the U. S. Congress to restrict death row prisoners from obtaining habeas corpus review. Along with an analysis of Supreme Court decisions, Federman examines three attempts to limit federal habeas corpus review for state prisoners from the late 1980s to the middle 1990s, all of which helped Congress to pass the AEDPA, a law that ratified the Supreme Court's most restrictive habeas corpus decisions dating back some 35 years.

Who Has the Body?

The Paths to Habeas Corpus Reform

Cary Federman

Habeas corpus is the principal means by which state prisoners attack the constitutionality of their convictions in federal courts. Variously called "the great writ of liberty," a "human right," and "a bulwark" and "palladium" of English liberties, the ancient writ of habeas corpus has achieved a status in American jurisprudence that has surpassed even those rights deemed by the U.S. Supreme Court to be preferred or fundamental, such as free speech and the right to privacy. In part because of the writ's historic association with the Magna Carta, many jurists and legal scholars consider habeas corpus as a tool of liberty in the fight against governmental oppression.

Today, however, the writ of habeas corpus stands accused of setting the guilty free. Critics charge that habeas corpus releases the convicted not on innocence grounds or even for reasons of clemency but on technical principles of law. Habeas corpus allows a solitary federal judge— so many miles removed from the crime scene and perhaps some 10 years after the initial conviction was rendered, after memories have faded and witnesses have either moved away or died— to find a due-process violation sufficient enough to overturn the judgment of numerous state judges and 12 jurors. To add insult to injury, habeas corpus interferes with the workings of what Supreme Court Justice Felix Frankfurter once called "our federalism" (Collins, 1992).

Recent decisions by the Supreme Court regarding habeas corpus, particularly death penalty cases such as *Tyler v. Cain* (2001) and *Felker v. Turpin* (1996), have done nothing more regarding the rights of prisoners who attack their convictions in federal courts than uphold restrictions passed by Congress in 1996 that prevent state prisoners from successfully attacking their convictions in federal habeas courts. And yet, the limitations on state prisoners' access to habeas corpus found within the Antiterrorism and Effective Death Penalty Act (AEDPA) of 1996 do nothing more than uphold more than 30 years of Supreme Court decisions on the subject. The purpose of this article is to explain the various paths that led to the AEDPA. There are three: congressional, interest group, and the Supreme Court's habeas decisions. Regarding interest groups, I focus on three important proposals from 1988 to 1990 that sought to modify and limit habeas appeals. I also examine the ways in which the Supreme Court's view of habeas petitioners as convicted criminals and drains on the judicial system through time hardened the attitudes of members of Congress to the point that they passed antiterrorist legislation in 1996 that was largely symbolic but which has had real effects on the lives of convicted criminals. My focus will be on the writ's relation to capital cases only, because most (if not all) of the criticisms of habeas corpus stem from concerns that the writ is responsible for the length of time it takes to go from conviction to execution.

Source: Federman, C. (2004). Who has the body? The path to Habeas Corpus reform. *The Prison Journal, 84*(3), 317–339. Reprinted with permission of Sage Publications, Inc.

⚂ From the Common Law to the Gilded Age

In Anglo-American jurisprudence, habeas corpus ("you have the body") is an institutional means to test the proposition that individuals have the right to be free from arbitrary arrests. Between the American founding and Reconstruction, the justices of the Supreme Court understood habeas corpus as it had operated in England (*Ex parte Kearney*, 1822). That is, it could not be used after conviction to contest a court's decision to incarcerate. Only executive clemency could rectify claims of a miscarriage of justice. Yet by situating the writ in Parliament and not in the king's courts, the English by the 17th century embedded habeas corpus within the language of civil liberties and legislative autonomy, thereby rejecting the idea, later adopted by the U.S. Supreme Court, that habeas corpus is a mere formal process. Both the king and Parliament saw clearly what the Americans would later learn with much difficulty: A granted writ alters jurisdictional boundaries.

Regardless of what the English knew of the great writ's jurisdictional capabilities and how they sought to contain it, it is clear that both the framers of the Constitution and the justices of the Supreme Court (both past and present) feared its effects on the states' criminal procedures. Consequently, they chose to filter habeas corpus through the federal structure of the new American state. Notably, the Judiciary Act of 1789 prohibited state prisoners from petitioning federal courts for habeas corpus. However, following a series of antebellum sectional crises in which various states arrested (or threatened to arrest) federal revenue officers (in 1815 and 1833), military personnel (1863), and a foreign national (1842), Congress expanded federal habeas corpus and removal jurisdiction to the state level. Although explicitly temporary in nature and designed to protect federal officers, not state convicts, these various removal and habeas corpus statutes created a pathway to the states that eased passage of the Habeas Corpus Act of 1867.

The 1867 act states in part that

> The several courts of the United States . . . within their respective jurisdictions, in addition to the authority already conferred by law, shall have power *to grant writs of habeas corpus in all cases where any person may be restrained of his or her liberty* [italics added] in violation of the constitution, or of any treaty or law of the United States. (Habeas Corpus Acts, 1815–1867)

The 1867 act stands as an example of Reconstruction-era state making. With a statutory command to the federal courts to *have the body* of any prisoner seeking relief, the 39th Congress ignored state sovereignty concerns regarding finality of punishment and codified the budding relationship prisoners would have with the national judiciary under the Fourteenth Amendment.

But the Radical Republican victory of the post–Civil War era was short-lived. Despite Congress's boldness in the face of a history of state control of punishment, the wording of the Habeas Act does not establish a bright-line relationship between the incarcerated and the federal courts that bypasses the state court system (*Wade v. Mayo*, 1948). The Supreme Court was free to ignore the nationalist intent of the legislation (particularly after the end of Reconstruction in 1877) and focus on the writ's common-law legacy in the United States—and it did. The act probably assumes common-law practices of using the writ as a preconviction remedy, but nothing in the language of the act prevents the writ from being used by state prisoners as a postconviction attack on a state court's judgment, because, as Marc Arkin (1995) wrote, "incarceration was not routinely imposed as a means of postconviction punishment for criminal acts until the nineteenth century" (p. 11). The problem with the 1867 habeas law is that, if used as a postconviction remedy, the determination of equitable relief falls not to Congress or to the states but to Supreme Court justices and federal court judges who presumably are free to investigate the prisoner's complaints and set him free. To be sure, the original state court decision stands because a writ operates on the judicial fiction that habeas relief does not overturn state court decisions; it releases individuals from unlawful confinement. But the decision to release a prisoner on habeas corpus can upset a delicate federal-state balance cultivated through time by congressional leaders and Supreme Court justices and embarrass the state court.

Rather than accept the burden of acting like a clemency commission or a supervisor of state court criminal justice decisions, the Supreme Court held in *Ex parte Royall* (1886) that any constitutional infraction of a defendant's rights (such as they were in the 19th century) could be dealt with by state courts in an effort to nurture federal-state comity relations following the end of Reconstruction. A policy of deference on criminal matters would, moreover, provide the federal courts with sufficient time to deal with property claims arising from the Supreme Court's decisions equating property with persons. In short, in *Royall*, the Court considered federal review of state prisoners' claims wasteful of important (and limited) judicial resources as well as constitutionally unnecessary, given the dual nature of the judicial system and the historic reliance on state courts to dispense justice to criminals. Unlike U.S. military and revenue officers before the Civil War who used the writ during confinement but before trial and claimed the writ's historic mission as a pretrial check against arbitrary executive authority, post-Reconstruction Supreme Court justices considered state habeas petitioners convicted criminals and therefore were extremely reluctant to release prisoners found guilty of crimes ranging from forgery to murder.

As the 19th century came to an end and as convicted criminals, Blacks, Asians, and indigents replaced federal officers who needed national protection from hostile state courts, the Supreme Court abjured any jurisdiction over criminal persons. The Court relegated the "great writ of liberty" to a backstop measure, an "extraordinary remedy" of the federal judiciary for overt instances of illegal criminal confinement. From the end of Reconstruction to the middle of the 20th century, purposeful congressional forbearance from civil rights violations at the state level, and pressing property cases at the national, allowed the Supreme Court to give a narrow and procedural meaning to the due process clause of the Fourteenth Amendment in criminal matters as well as to habeas corpus. The Supreme Court's extension of property rights under the Fourteenth Amendment shaped habeas' development in the 19th and early 20th centuries by squeezing out civil rights claims in the federal judiciary. The Court's propertarian understanding of due process rights created an underlying pattern of chaos within American civil rights development. Throughout the 19th century, and well into the 20th, claims of constitutionally questionable arrests, confessions, and trials went unheeded in the state appellate courts while the federal judiciary's defense of property regularized the American state.

The Road to AEDPA

On April 24, 1996, just 5 days after the first anniversary of the bombing of the Alfred P. Murrah Federal Building in Oklahoma City, Oklahoma, Congress passed the AEDPA. Senator Robert Dole had introduced the bill in the Senate, noting that it "should go a long way in preventing violent criminals from gaming the system—with more delays, more unnecessary appeals, and more grief for the victims of crimes and their families" (Sessions, 1996/1977, p. 1515). Apart from the AEDPA's stated purpose—to grant federal authorities greater powers to investigate domestic terrorist threats and provide justice for victims—the AEDPA also limits the ability of state prisoners, particularly those on death row, to get federal habeas corpus relief.

The AEDPA's four main features regarding habeas procedures include the following:

1. It imposes a 1-year limit on filing habeas petitions; previously, there was no deadline (AEDPA, 1996, § 101).

2. Habeas petitioners have only one chance for federal review, except in extraordinary circumstances; previously, there were no limits on the number of habeas filings a state prisoner could make (AEDPA, 1996, § 106).

3. It establishes an opt-in provision for states to provide counsel (AEDPA, 1996, § 106). In other words, "the opt-in provisions are a quid pro quo. If a state provides counsel, the opportunities of state prisoners for federal review are reduced, thus removing roadblocks to a state's effective and expeditious use of the death penalty" (Kappler, 2000, p. 469). If a state does not opt in, and none have, then the filing deadline for habeas review expands from 180 days after final state court affirmance to 360 days.

4. "A determination of a factual issue made by a State court shall be presumed to be correct. The applicant shall have the burden of rebutting the presumption of correctness by clear and convincing evidence" (AEDPA, 1996, § 104). As in the other instances, this is a new addition to the habeas procedural maze.

In a situation that goes back to 1886 and *Ex parte Royall*'s requirement that the petitioner must first exhaust all forms of appeal (state and certiorari to the U.S. Supreme Court) before applying for habeas corpus, the new habeas law imposes a burden on the prisoner to disprove the state's version of events by clear and convincing evidence. In passing a law directed at foreign and domestic acts of violence against U.S. citizens and buildings, Congress also codified substantial portions of the Supreme Court's habeas jurisprudence since the late 19th century that limits the rights of state prisoners to gain access to the federal courts and contest their confinements on constitutional grounds.

Following the bombing of the Murrah Federal Building, President Bill Clinton proposed the Antiterrorism Amendments Act of 1995, which was designed to increase federal powers to combat domestic and international terrorism through the Omnibus Counterterrorism Act (1995). In June 1995, the Senate passed, although the House of Representatives did not, the Comprehensive Terrorism Prevention Act. In March 1996, the House passed the Effective Death Penalty and Public Safety Act of 1996, which was a modified version of the Comprehensive Terrorism Prevention Act and which made it a crime "to commit an act of international terrorism in the United States, and if someone is killed in the act, the crime is punishable by death" (Smith, 1997, p. 266). The act also included a habeas provision that limited the amount of time prisoners could appeal their state court decisions in federal courts. The result of further House and Senate compromises was the AEDPA, which passed the Senate by a vote of 91 to 8 with 51 Republicans voting for and 1 against and Democrats voting 40 to 7 in favor of the bill. In the House of Representatives, the vote was 293 for and 133 against (Republicans: 186 to 46; Democrats: 105 to 86).

In the pages that follow, I analyze three attempts to limit federal habeas corpus for state prisoners. They are, in order of appearance, by the Reagan Administration's Department of Justice (DOJ), the Powell Committee, and the American Bar Association (ABA). Lacking both a historical sense regarding the crucial role habeas corpus has played in getting prisoners' claims heard in federal courts, and an institutional framework within which to structure a debate regarding prisoners' allegations of illegal detentions in the 20th century without harming state interests in punishment, the proposals under review purposefully ignored habeas corpus's historical importance to outcast minorities and indigent prisoners. Instead, following both recent and late 19th-century Supreme Court case law that has restricted federal access to state prisoners, and relying on the general public's concerns about criminals evading punishment through frivolous appeals (Prison Litigation Reform Act, 1996), these proposals stress the writ's detrimental effects on the states' administration of the death penalty, such as time delays in capital trials. There is, however, something to learn from studying these proposals. Together, they provide a way to understand the sequence of events leading up to the AEDPA and explain how Congress's policy choices in 1996 had been narrowed through time by Supreme Court rulings and law-and-order interest groups both within Congress and without.

✉ The Case for Limiting Federal Habeas Corpus I: The Reagan Administration

The Reagan Administration's view of habeas corpus was that "the Constitution itself and historical practice are inconsistent with the existence" of a right to a federal habeas forum (U.S. DOJ, 1988, p. 42). The Reagan Administration held that, because Congress and not the Constitution created the federal judiciary, state prisoners do not have a right to a federal habeas court, only to Supreme Court certiorari review. The Reagan DOJ refused to regard any constitutional gains made by state prisoners since the Warren Court as an advance in the administration of criminal justice procedures. If the Warren Court proved anything, according to the DOJ, it demonstrated that federal review does more harm than good to state

criminal justice interests. Content, then, with the level of protection afforded criminal defendants in the states, the DOJ's report stated that federal habeas corpus was unnecessary.

The DOJ understood habeas corpus as it had existed under English common law, that is, as a pretrial remedy to contest arbitrary executive actions, not as a postconviction remedy for state criminal trials that violate constitutional protections and, as it had existed before *Brown v. Allen* (1953), a seminal habeas case that allowed the federal courts to reconsider a state prisoner's claims of unlawful detention regardless of the state's decision and fact finding. "In terms of historical practice," the report stated, "the general federal question jurisdiction of the federal courts is a late nineteenth century development" (U.S. DOJ, 1988, p. 42). The DOJ dismissed this change in legal outcomes without investigating the reasons why it had occurred. Consequently, the DOJ held that any expansion of habeas corpus as a remedy for unconstitutional confinement claims was judicial activism and dangerous to the interests of federalism and the "dignity and independent stature of the state courts" (Yackle, 1994, pp. 24–28).

The Reagan DOJ was convinced that the best solution to the problem of habeas corpus was its abolition (U.S. DOJ, 1988, p. iv). In lieu of that, the DOJ was willing to limit habeas corpus "to the role of a backstop remedy, whose availability would be conditioned on a state judicial system's failure to provide some meaningful process for raising and deciding a federal claim" (U.S. DOJ, 1988, p. vi). The DOJ was also prepared to accept limitations on habeas appeals up to 1 year following conviction; require federal deference to state court findings of fact; limit habeas appeals arising from Fifth and Sixth Amendment claims (*Duckworth v. Eagan*, 1989), as the Court had done on Fourth Amendment cases in *Stone v. Powell* (1976); allow federal judges to dismiss unreasonably delayed petitions with greater dispatch; and would have been satisfied with establishing a uniform application of restrictive standards to claims not raised in the state courts. With the exception of the Fifth and

Sixth Amendment claims, Congress included all of these suggestions in the AEDPA (1996, §§ 101, 104, 106, 107).

The DOJ's Office of Legal Policy staff never considered habeas corpus in light of the history of discrimination against minorities in particular or prisoners in general—either at arrest, during the original trial, or on appeal from the state trial—that influenced the Supreme Court to expand habeas corpus (and due process) from the 1930s to the 1960s. In assessing habeas reform and federal judicial capacities, Reagan Administration habeas critics chose to ignore the patterning of the states' criminal justice structures—their histories of racial and ethnic bias—and instead paid special attention to the administrative problem that states would encounter in implementing relief from federal habeas courts in capital cases. Shunning the need for historical investigation into the legacy and resilience of "the patterns of discrimination, segregation, unequal justice and racial violence" (Litwack, 1998, p. xvii) that existed in the South from the 1870s through the 1960s, the Reagan DOJ concluded that federal habeas corpus for state prisoners is the result of both judicial activism (the incorporation of the Bill of Rights during the Warren Court era) and unnecessary legislative intervention (during Reconstruction, because the Habeas Corpus Act of 1867 created a second forum for appealing a state conviction). Selectively analyzing habeas's development throughout the years, the DOJ maintained that habeas corpus as a postconviction remedy is without deep roots in American constitutional law and in no way should be regarded as an unlimited right by state prisoners to use in the federal courts.

⬚ The Case for Limiting Federal Habeas Corpus II: The Powell Report

In June 1988, Chief Justice William Rehnquist appointed former Supreme Court Justice Lewis Powell, a longtime critic of federal habeas relief for state prisoners, the chair of the Ad Hoc Committee on Federal Habeas Corpus in Capital Cases. The committee included four federal district and appellate court judges from the Fifth and Eleventh Circuits. The two principal findings of the Powell Committee were (a) that the present system of federal habeas review fosters delay in the movement of habeas appeals from the state to the federal courts and (b) that there is a "pressing need for qualified counsel to represent inmates in collateral review" (Habeas Corpus Reform, 1990, p. 9). To rectify these problems, the Powell Committee proposed,

> Capital cases should be subject to one complete and fair course of collateral review in the state and federal system, free from the time pressure of impending execution, and with the assistance of competent counsel for the defendant. When this review has concluded, litigation should end. (Habeas Corpus Reform, 1990, p. 13)

Justice Powell blamed state prisoners' habeas petitions for interfering with the successful implementation of the death penalty. Indeed, the view persists today, both among the public and among judges, lawyers, and Supreme Court justices, that habeas corpus petitions undermine state court judgments in spite of the fact that state prisoners' habeas filings have not kept pace with the increase in the total number of prisoners. Since 1964, habeas filings by state prisoners have not constituted more than half of all filings in U.S. district courts by state prisoners, as they did between 1960 and 1963. "Only in the 1960s," Thomas (1988) has written, "did the increase in the proportion of suits to prisoners roughly match the increase in the number of filings, but both the number of filings and the proportion of prisoners filing them peaked in 1970" (p. 98). Since then, "Fewer prisoners have been filing habeas claims, declining from about 5 suits for

100 prisoners in 1970 to less than 2 since 1984. Hence, the increase in the number of filings may reflect little more than the increase in the prison population. This suggests that most prisoners—contrary to the claims of critics— accept the finality of their confinement, because very few prisoners challenge their conviction after incarceration" (Thomas, 1988, p. 98).

Choosing to ignore the fact that no study had revealed that habeas corpus undermines the states' interest in punishment or that it poses a real threat to federal-state comity concerns, Justice Powell (1988/1989) wrote in the *Harvard Law Review*, "A dozen years after the uniquely fair *Gregg*-type statute was approved, we have more than 2,000 convicted murderers on death row, and just over 100 executions. However this delay may be characterized, it hardly inspires public confidence in our criminal justice system" (pp. 1040–1041).

Appearing before Congress to defend the committee's report, Powell stated that "the hard fact is that the laws of thirty-seven states are not being enforced by the [federal] courts" (Habeas Corpus Reform, 1990, p. 40). The Powell Committee surveyed 50 postconviction cases in high death penalty states—Florida, Texas, Alabama, Mississippi, and Georgia (Habeas Corpus Reform, 1990, p. 15). At the time, the average death penalty proceeding, from sentencing to execution, took 7 years. It took 6 years if the state had no postconviction review of its own. From 1976, when the Supreme Court allowed the states to resume executions, to September 21,1989, when the Powell Committee report was issued, 118 executions had taken place, although two more followed by year's end. The average time elapsed from sentence to execution in 1989 was 95 months— the longest on record. "Delay of this magnitude," Powell stated before Congress, "is hardly necessary for fairness or for thorough review" (Habeas Corpus Reform, 1990, p. 40).

To reduce perceived delays that interfere with execution timetables, the committee proposed a 6-month period within which federal habeas petitions could be filed. At the time, there were no time limits on habeas appeals. The Powell Committee wanted to bar claims to the federal courts after 6 months that had not been made in the state court unless the defendant had a new claim that could prove his innocence. The committee agreed to start the 6-month period only on the appointment of counsel for the prisoner and suspend the time, called tolling, during state court proceedings. If the prisoner accepted the state's counsel, he would be prohibited from challenging his lawyer's effectiveness in court. Congress codified this proposal in the AEDPA (1996, § 107).

The Powell Committee barely mentioned the problem of state prisoners inadvertently waiving their federal rights, which increases the risk of executing those with constitutional claims not heard by a federal court. Because the Powell Committee regarded habeas corpus as a legislative creation divorced from constitutional principles, it saw no need to dictate to the states how they should proceed in this area of constitutional policy. Instead, the Powell Committee focused on the problem of time delays and unqualified death row attorneys from an administrative standpoint, thus failing to address the reason why these are problems. Time delays and unqualified attorneys working on death penalty cases are partly the result of the states' unwillingness to train death penalty lawyers and provide quality counsel at the trial stage and partly the result of complicated Supreme Court decisions from the 1970s and 1980s that have (a) allowed unqualified attorneys to defend habeas petitioners in capital cases and (b) denied state prisoners a federal hearing despite strong evidence that any procedural faults that occurred at the initial trial were the lawyer's doing, not the habeas petitioner's. In *Coleman v. Thompson* (1991), for example, the Supreme Court upheld a Virginia law that prohibited the state from accepting appeals after 30 days. Coleman's lawyer miscalculated the time from conviction to appeal, and the net effect of his inadvertent error was Coleman's execution. Problems of ineffective counsel have plagued capital defendants and habeas petitioners since the latter half of the 19th century—a fact the Powell Committee ignored.

Because of federalism concerns, the Powell Committee refused to set national right-to-counsel standards for all 50 states. It also recommended automatic stays of execution, but only for those states participating in the reforms. The purpose of the automatic stay is to prevent last-minute appeals to judges to delay execution. Rather than focusing on prisoners' claims, the committee presumed their invalidity and preferred to streamline the death penalty process by eliminating delays and repetitive petitions and, most important, reasserting the need for the complete exhaustion of one's state appellate remedies before going to a federal habeas court—a doctrine that goes back to *Ex parte Royall* in 1886 and a procedure that, in fact, increases time delays in executing criminals. Under the Powell Committee's proposal, each state could decide for itself what constituted quality counsel in capital cases. The committee feared that the imposition of national attorney standards would take power away from the states in terms of setting counsel fees, or fees for witnesses, and impose unnecessary financial and administrative burdens on the states. According to Justice Powell, because the states "would have little incentive to opt for a system that does not recognize" their interests, proposals for reform "must take care not to destroy the increased finality and order that will prompt the states to participate" (Habeas Corpus Reform, 1990, p. 47). To enforce finality, or the idea that "no one . . . shall tentatively go to jail today, but tomorrow and every day thereafter his continued incarceration shall be subject to fresh litigation" (*Mackey v. United States*, 1971, p. 691), the Powell Committee supported codifying the Burger and Rehnquist Courts' most restrictive (in regard to gaining access to the federal courts) habeas decisions to date.

The major innovations of the Powell Committee were the imposition of a 6-month cutoff period for state petitioners to file a federal habeas appeal and the noncoercive, voluntary nature of the proposed reforms for the states, particularly in regard to counsel. In 1990, Senators Strom Thurmond, Joseph Biden, and Arlen Specter each introduced habeas bills largely tracking the Powell Committee's proposals. Congress adjourned before resolving the several disputes.

▧ The Case for Limiting Federal Habeas Corpus III: The ABA

In 1990, the ABA proposed changes in habeas corpus that would have imposed burdens on both habeas petitioners and the states and used a cutoff period of 1 year for the filing of habeas petitions. The ABA Task Force was more diverse than the Powell Committee. It consisted not only of lawyers and judges from both the federal and state courts but also a law professor and the director of the Southern Prisoners' Defense Committee (Berger, 1990, p. 1684). The ABA thought that the tradeoff between requiring the states to provide quality counsel and a 1-year cutoff period would be a viable alternative to the more restrictive congressional plans proposed throughout the years and the Reagan Administration's 1988 proposal.

The ABA was more willing than the Powell Committee to employ the coercive powers of the federal government to ensure that the states were protecting criminal defendants. All of its provisions, if made into law, would bind the states. The ABA was also less disturbed than the Powell Committee with protecting federalism and finality and more concerned with ensuring that qualified attorneys represent death-sentenced individuals. Yet speeding up the habeas and death penalty appeals process was a key provision of the ABA plan. Following conviction, the ABA would allow state prisoners 1 year in which to file habeas corpus petitions. The ABA deemed the assurance of qualified counsel to be the *"sine qua non* of a just and efficient capital system" (Habeas Corpus Reform, 1990, p. 486). To that end, it recommended requiring states to train death penalty lawyers, compensate them adequately for their work, and have lawyers stay with their clients throughout the appellate process, including habeas corpus.

The ABA also differed from the Powell Committee by rejecting a number of Supreme Court decisions that had made prisoner access to federal habeas courts more difficult, such as procedural default, the exhaustion of remedies, and the presumption of correctness of state court findings of fact. Consequently, it advocated eliminating the waiver of federal claims from the cause-and-prejudice standard of *Wainwright v. Sykes* (1977). Under the Court's cause-and-prejudice test, developed more forcefully throughout the 1980s by the Rehnquist Court, six questions must first be addressed before a federal court can begin to discuss the merits of a state prisoner's claim. The result can be the execution of an innocent person without a federal hearing. Ostensibly done in the name of protecting the states from unnecessary or premature federal intervention, these tests serve to make the habeas corpus/capital punishment process burdensome, inefficient, and risky for prisoners who proceed with inexperienced lawyers or no lawyer at all.

The ABA recommended that all states with death penalties provide counsel at all stages of capital litigation. In the ABA's follow-up study on capital punishment in 1998, the association found that, "in case after case, decisions about who will die and who will live turn not on the nature of the offense the defendant is charged with committing, but rather on the nature of the legal representation the defendant receives" (*Law & Contemporary Problems*, 1998, p. 228). Factoring out California, where an inordinate number of ineffective counsel claims in death penalty cases arise, the number of claims regarding ineffective counsel is 48%, tied for second place after trial court error. Moreover, as Columbia University Law Professor James Liebman has recently demonstrated, federal habeas review has had a positive effect on state prisoners' claims of unlawful detention (Liebman, Fagan, & Valerie, 2000). A previous investigation into habeas success rates by Liebman (1990/1991, p. 541, n. 15), which the current study relies on, revealed that "the federal courts found constitutional error in 40% of the 36 capital judgments of conviction and sentence that those courts finally reviewed in habeas corpus proceedings between 1976 and mid-1991" (Liebman et al., 2000, p. 6).

Like the Powell Committee's report, the ABA failed to garner enough support in Congress. Representative Henry Hyde submitted a habeas bill that embodied much of the Powell Committee's recommendations, replacing a slightly more liberal bill that was closer to the ABA's proposal. But, when Congress passed a crime bill in 1990, habeas and death penalty provisions were not included. The technicalities of habeas corpus and the intensity of the debate about death penalty appeals wore Congress down.

⚐ The Judicial Politics of Habeas Corpus Reform

From 1954, just after the Court decided *Brown v. Allen* (1953), to 1964, 1 year after the Court decided *Fay v. Noia* (1963), a decision that fundamentally overthrew 80 years of restrictive habeas jurisprudence, Congress tried four times, unsuccessfully, to limit federal habeas corpus review for state prisoners. From 1966, when Congress last ratified certain significant Supreme Court decisions concerning habeas corpus procedures, to 1996, "the only congressional revisions of any kind to the federal habeas corpus statutes were those already approved by Supreme Court jurisprudence" (Sessions, 1996/1997, p. 1518). Indeed, going back to 1886 when the *Royall* decision first connected habeas claims to prisoners' rights, Congress has refrained from active involvement in prisoners' rights issues by preferring to validate Supreme Court decisions regarding criminal justice. "When the judicial train arrives at the station before the legislative one, there is little reason to enact a statute from a policy standpoint" (Tushnet & Yackle, 1997/1998, p. 2).

Habeas corpus has become a symbol, not of the abstract notion of the liberty of the individual as the history books portray it but of prisoner hubris in the face of the states' criminal justice administration. If habeas corpus was ever a get-out-of-jail card for prisoners, it was only so from 1963 to 1969 when the Warren Court issued its most expansive habeas decisions. Yet from the moment the Supreme Court expanded the scope of habeas corpus for state prisoners during the 1960s, opponents of an enlarged habeas jurisprudence, both in Congress and on the Supreme Court, began pushing for habeas restrictions for administrative and constitutional reasons relating to protecting the integrity of the capital punishment process.

By the 1970s, in an effort to redress a seeming imbalance in federal-state relations caused by Warren Court decisions that had applied the criminal provisions of the Bill of Rights to the states, the Burger Court began imposing jurisdictional burdens on state prisoners seeking habeas relief as a way to stem the flow of habeas petitions coming from state prisoners and give state court judgments greater protection against habeas attacks in federal courts. The Rehnquist Court has continued the practice of saddling habeas petitioners, but not the states, with legal responsibilities. The Rehnquist Court has imposed burdens not on the states regarding police tactics or on state courts or legislatures regarding the quality of capital defense attorneys but on prisoners who are mostly poor, illiterate, ignorant of the legal system, and who quite often operate without counsel or with poorly trained capital defense attorneys. To limit the writ's effectiveness, the Rehnquist Court relies on complex legal rulings (called retroactivity) stemming from the administrative problems of applying to habeas petitioners extensions of Warren-era constitutional holdings that addressed criminals' fundamental rights. In a series of cases in the late 1980s and early 1990s, the Rehnquist Court prevented habeas petitioners from asking the courts to apply certain cases to their situation (that could potentially help their cause) that had been decided after their state court convictions had become final.

In 1996, Congress reentered the habeas debate to codify, not alter, the Supreme Court's decisions from the early 1970s to the 1990s. Congress has involved itself in habeas reform only to minimize the degree of constitutional attacks on the state courts and maximize the states' legitimacy in the post–New Deal era. "The post–Civil War history of federal habeas," Joseph Hoffmann wrote in 1989, "reveals that Congress either has followed the lead of the Supreme Court in defining the scope of federal habeas, or has remained completely silent in the face of the Court's numerous decisions interpreting the Act of 1867" (p. 178). The Court views habeas corpus "as a subject almost completely within its own domain" (Hoffmann, 1989, p. 177), mostly because of the writ's common-law heritage but also because of congressional reluctance to increase legal protections for prisoners. The Supreme Court's labeling of habeas petitioners as convicted criminals and brutal murderers has also had the effect of discouraging habeas's mainstream supporters in Congress and delegitimizing sympathetic members of liberal, elite groups outside of Congress. (Indeed, more research needs to be done regarding the Court's description of criminals in murder cases and the effect it has on the law and politics.)

The new act, which Yackle (1996), a prominent critic of the Court's habeas decisions considers "not well drafted," "bears the influence of various bills that were fiercely debated for nearly forty years" (p. 381). The AEDPA, which expands the government's jurisdiction to anyone attempting to commit an act of terrorism against the United States, does not ban so-called cop-killer bullets. Nor does the act ban the use of firearms by terrorists, such as machine guns, sawed-off shotguns, and explosive devices. The act prohibits fundraising to aid terrorists but does not enhance the government's wiretapping powers of suspected terrorists. It is also without provisions to enlist the military "in cases involving biological and chemical weapons," and it failed to lengthen the "statute of limitations on firearms violations" (AEDPA, 1996, § 501). The legislative compromises forged between liberals concerned with expanding wiretapping functions and conservatives concerned about gun restrictions that produced the AEDPA notwithstanding, the AEDPA is not directed at stemming crime or terrorism. As Senator Joseph Biden said, "It's a habeas corpus bill with a little terrorism thrown in" (Idelson, 1996, p. 1046).

◣ Conclusion

The history of habeas corpus in the United States is of a limited rise (1787 to 1867) and precipitous fall (1886 to 1915) and of a substantive but patchwork rise (1923 to 1969) and a substantive and patterned fall (1970 to 1996). Periodizing the history of habeas corpus in the United States calls into question the mythic characterizations of habeas corpus as the "great writ of liberty" or as a legal protection against unlawful and arbitrary confinement. It is, in fact, clear that the Supreme Court and Congress consider the writ to pose a threat to the states' interest in capital punishment. But this threat transcends habeas's power to set free convicted murderers. It goes to the root of a state's reason for being. "A state unable to execute those it condemns to die would seem too impotent to carry out almost any policy whatsoever" (Sarat, 2001, p. 18). It behooves constitutional scholars interested in capital punishment to pay more attention to the political struggles of both prisoners and the Supreme Court that are "mediated by the institutional setting in which [they] take place" (Steinmo, Thelen, & Longstreth, 1992, p. 2). The political struggles of prisoners include not just trying to get out of prison but the federal-state relationship that channels prisoners' movements away from federal habeas review. Not all struggles are violent. Understanding

prisoners' struggles should also include using habeas corpus to attack convictions in federal court or pursuing suits against prison officials, which the Prison Litigation Reform Act (1996) limits. Suits against prison officials or postconviction appeals that challenge state court rulings provide a lens through which to view how institutions respond to challenges to their authority. Viewing the writ as a symbol and a sword, prisoners seek the protection of the writ from judicial detention. Habeas corpus exists to cut through the legal forms that protect state court decisions from being overturned on appeal. This is the reason why habeas corpus provokes backlash; this is why it remains necessary. But at its core, habeas corpus poses a challenge to the state's claim of the body of the prisoner.

Studying habeas corpus throughout the years highlights the plurality of jurisdictional powers that struggle for control of the movement of prisoners and underscores the problem of administering justice in a federal republic. Habeas corpus is a jurisdictional problem in a nation of jurisdictions. Those looking for a way to understand capital punishment jurisprudence need to pay closer attention to the Supreme Court's habeas corpus decisions and the long-term effects of Congress's activity and inactivity on prisoners' rights claims.

◣ References

Antiterrorism and Effective Death Penalty Act, Pub. L. No. 104-132, 110 Stat. 1214 (1996).

Arkin, M. (1995). The ghost at the banquet: Slavery, federalism, and habeas corpus for state prisoners. *Tulane Law Review, 70*, 1–73.

Berger, V. (1990). Justice delayed or justice denied?—A comment on recent proposals to reform death penalty habeas corpus. *Columbia Law Review, 90*, 1665–1714.

Brown v. Allen, 344 U.S. 443 (1953).

Coleman v. Thompson, 501 U.S. 722 (1991).

Collins, M. (1992). Whose federalism? *Constitutional Commentary, 9*, 75–86.

Ex parte Kearney, 20 U.S. 38 (1822).

Ex parte Royall, 117 U.S. 241 (1886).

Ex parte Watkins, 28 U.S. 193 (1830).

Fay v. Noia, 372 U.S. 391 (1963).

Felker v. Turpin, 518 U.S. 651 (1996).

Habeas Corpus Acts. (1815–1867). Act of Feb. 4, 1815, c. 31, 3 Stat. 105; Act of March 3, 1815, c. 94, 3 Stat. 231; Act of March 2, 1833, c.57, 4 Stat. 632; Act of Aug. 23, 1842, c. 188, 5 Stat. 516; Act of March 3, 1863, c. 81, 12 Stat. 755; Act of Feb. 5, 1867, c. 27, 14 Stat. 385.

Habeas Corpus Reform (1990). *Hearings before the Committee of the Judiciary, United States Senate, 101 Congress, 1st and 2nd sessions. Nov. 8, 1989; Feb 21, 1990.* Washington, DC: Government Printing Office.

Hoffmann, J. (1989). The Supreme Court's new vision of federal habeas corpus for state prisoners. *Supreme Court Review, 1989,* 165–193.

Idelson, H. (1996). Terrorism bill is headed to president's desk. *Congressional Quarterly Weekly Report, 54*(16), 1044–1046.

Kappler, B. (2000). Small favors: Chapter 154 of the Antiterrorism and Effective Death Penalty Act, the states, and the right to counsel. *Journal of Criminal Law and Criminology, 90*(2), 467–598.

Law & Contemporary Problems. (1998). Appendix: American Bar Association resolution (February 3, 1997). *61,* 219–231.

Liebman, J. (1990/1991). More than "slightly retro": The Rehnquist Court's rout of habeas corpus jurisdiction in *Teague v. Lane. New York University Review of Law and Social Change, 18,* 537–635.

Liebman, J., Fagan, J., & Valerie, W. (2000). *A broken system: Error rates in capital cases, 1973–1995. The Justice Project.* Retrieved June 30, 2004, from http://justice.policy.net/proactive/documents/printerfriendly/printerfriendly?docid=18200&PROACTIVE_ID=cececfc7c7ccc7cbc9c5cecfcfcfc5cececacec8cfcac9c8cdc5cf

Litwack, L. (1998). *Trouble in mind: Black Southerners in the age of Jim Crow.* New York: Knopf.

Mackey v. United States, 401 U.S. 667 (1971).

Omnibus Counterterrorism Act of 1995, H.R. 896, S. 390, 104th Congress (1995).

Powell, L. (1988/1989). Capital punishment. *Harvard Law Review, 102,* 1035–1046.

Prison Litigation Reform Act, Pub. L. No. 104-34, 110 Stat. 132 (1996).

Sarat, A. (2001). *When the state kills: Capital punishment and the American condition.* Princeton, NJ: Princeton University Press.

Sessions, P. (1996/1997). Swift justice? Imposing a statute of limitations on the federal habeas corpus petitions of state prisoners. *Southern California Law Review, 70,* 1513–1569.

Smith, R. (1997). America tries to come to terms with terrorism: The United States' Anti-Terrorism and Effective Death Penalty Act of 1996 v. British anti-Terrorism Law and international response. *Cardozo Journal of International and Comparative Law, 5,* 249–290.

Steinmo, S., Thelen, K., & Longstreth, F. (1992). *Structuring politics: Historical institutionalism in comparative analysis.* Cambridge, UK: Cambridge University Press.

Stone v. Powell, 428 U.S. 465 (1976).

Thomas, J. (1988). *Prisoner litigation: The paradox of the jailhouse lawyer.* Lanham, MD: Rowman & Littlefield.

Tushnet, M., & Yackle, L. (1997/1998). Symbolic statutes and real laws: The pathologies of the Antiterrorism and Effective Death Penalty Act and the Prison Litigation Reform Act. *Duke Law Journal, 47,* 1–86.

Tyler v. Cain, 121 S. Ct. 2478 (2001).

U.S. Department of Justice. (1988, May 27). *Office of legal policy. Report to the attorney general on federal habeas corpus review of state judgments.* Washington, DC: Author.

Wade v. Mayo, 334 U.S. 672 (1948).

Wainwright v. Sykes, 433 U.S. 72 (1977).

Yackle, L. (1994). *Reclaiming the federal courts.* Cambridge, MA: Harvard University Press.

Yackle, L. (1996). A primer on the new habeas statute. *Buffalo Law Review, 44,* 381–449.

DISCUSSION QUESTIONS

1. What was the original meaning of the writ of habeas corpus?

2. How has habeas corpus evolved in the United States, and what are its critics saying about it?

3. What is the difference between a preconviction and a postconviction remedy for the violation of a defendant's rights?

READING

In this article, Mary Ann Farkas and Amy Stichman assert that sex offenders are viewed in American society as a unique type of criminal, more "objectionable," less treatable, more dangerous, and more likely to recidivate than other kinds of offenders. In recent years, sex offenders have become the focus of intense legal scrutiny, primarily through the promulgation of laws specifically targeting them for indefinite confinement, registration and community notification, polygraph testing, and even chemical castration. Farkas and Stichman examine the underlying assumptions and justifications for these sex offender laws and ask whether treatment, punishment, and public safety can be reconciled as justifications for them. The authors conclude that, even though treatment is an implicit rationale in the laws' provisions, punishment, incapacitation, and public safety are the ostensive purposes of these special laws and policies. Moreover, this article questions the constitutionality and rationality of special sex offender laws and policies and their consequences for sex offenders, treatment professionals, the mental health and criminal justice systems, and society in general.

Sex Offender Laws

Can Treatment, Punishment, Incapacitation, and Public Safety Be Reconciled?

Mary Ann Farkas and Amy Stichman

The current political climate is one of increased public awareness of and outcry against sex offenders. The processing of sex offenders and their return to society is the subject for media and legislative frenzy, largely as a result of some particularly heinous crimes involving sex offenders raping and murdering children. It often appears that many of these laws fit into a "culture of fear," particularly when any violence or perceived violence threatens children. It is no coincidence that many sex offender laws are named after child victims, including the Jacob Wetterling Act and Megan's Law.

Schwartz (1988) contends that the politicization of sex offenses has resulted in revised rape and sexual assault statutes, increased reporting rates, and increased prosecution and conviction. The Bureau of Justice Statistics (BJS, 1997) reports that since 1980 the average annual growth in the number of prisoners has been about 7.6 percent; the number of prisoners sentenced for violent sexual assault has increased by an annual average of 15 percent—faster than any other category of violent crime, and faster than all other crime categories except drug trafficking. In 1994, there were approximately 906,000 offenders convicted of rape and sexual assault confined in state prisons, of whom 88,000 or 9.7 percent were violent sex offenders (BJS, 1997). It is this small percentage of violent offenders that evokes the fear and condemnation of society and drives

---

SOURCE: Farkas, M. A., & Stichman, A. (2002). Sex offender laws: Can treatment, punishment, incapacitation, and public safety be reconciled? *Criminal Justice Review, 27*(2), 256–283. Reprinted with permission of Sage Publications, Inc.

legislative and correctional efforts to incapacitate and control sex offenders.

Legislators have elevated sex offending to a major public policy priority, and the criminal justice system is responding with selective prosecution and processing of sex offenders. A flurry of sex offender–specific laws have been passed, including those mandating indeterminate confinement, sex offender registration and community notification, chemical castration, and polygraph testing. The recent spate of sex offender laws, policies, and practices reflects a "new penology" that focuses on managing or controlling high-risk categories of offenders rather than on transforming or rehabilitating the individual offender. Many of these laws were created because of the public's growing frustration with the criminal justice system in dealing with offenders and recidivists in general and sex offenders in particular. "The emotionally charged nature of the problem of sexual victimization, combined with what is often extreme pressure from interest groups and the general public to 'do something,' limits and narrows the discourse on this issue within the legislative process" (Edwards & Hensley, 2001, p. 84).

With many of these laws pertaining to sex offenders, the courts have granted states wide latitude and have been reluctant to uphold constitutional challenges. Pratt (2000, p. 143) asserts that society is more willing to remove "basic civil liberties" regarding sex offenders because of their perceived risk to the community, and, because of their repugnance to society, few citizens are willing to oppose this trend. The manifest intent of these laws is to protect society by incapacitating sex offenders in some fashion, whether it be through confinement in a mental or correctional institution or through specialized correctional mechanisms to control their behavior and movements in the community. Many of these laws suggest a treatment goal, although this "treatment" more fittingly resembles social control. This perception of "treatment" is common in corrections; however, what is often done under the guise of treatment is punitive

or incapacitative in nature. The attempt to reconcile treatment and punishment in corrections has been a dilemma for decades. Whereas there is little agreement on the purposes, effects, and constitutionality of these sex offender laws and policies, this article focuses on their design to treat or rehabilitate, to punish, to incapacitate, and to protect the public. With this in mind, several of these laws are analyzed for their intent and their implications for sex offenders, sex offender treatment providers, the justice system, and society.

⧆ Research Questions and Design

We approached this study with the following questions in mind. First, what are the assumptions inherent in these sex offender laws, policies, and practices with regard to sex offenders? Is there evidence of a reasoned, critical analysis of the theoretical and empirical research on sex offenders to provide a sound basis for these assumptions?

Next, what are the justifications underlying these sex offender laws, policies, and practices? Are these laws designed with the justification of punishment or treatment for the individual offender or for an over-inclusive category of "sex offenders"? Are these laws, policies, and practices meant to offer the public a measure of safety as well? If they are intended to fulfill punishment, treatment, and public safety objectives simultaneously, will the goal of treatment be overlooked or undermined? Finally, is there evidence of policy analysis prior to the passage of these laws? In other words, was there a consideration of the potential impact on sex offenders, treatment providers, the criminal justice and mental health systems, and society overall?

These research questions were explored through a comprehensive review of the legal, health science, and social science literature. We considered relevant court cases, legal analyses, and law reviews concerning the intent and goals of the laws, as well as their constitutional and practical limitations. We also examined the theoretical and research literature on the types, behaviors, and motivations of sex offenders and therapeutic approaches for the various kinds of sex offenders.

◪ Sexual Predator Laws and Civil Commitment Statutes

History

Different forms of laws directly pertaining to sex offenders have been around for decades. The first sexual psychopath laws were passed in the 1930s as a response to high-profile sex crimes. These statutes often had the explicit purposes of treatment and incapacitation for this group of offenders as well as protection for the community. Variously referred to as sexual psychopath laws, sexually dangerous persons acts, and mentally disordered sex offender acts, these laws commonly called for the civil commitment to a state mental hospital of those persons fitting these descriptions. The laws were premised on the ability of psychiatry to identify, predict dangerousness and risk for, isolate, and treat sexual psychopaths.

Although they were popular, these laws were also criticized. Often the adjudicated person was committed indefinitely with few procedural safeguards. Furthermore, there is some indication that these laws often targeted not just the habitual violent sex offender but also voyeurs, exhibitionists, and homosexuals. There were also disillusionment with rehabilitation and doubts about the capability of rehabilitating sexual psychopaths. By the late 1960s, many states were beginning to either repeal or ignore these statutes because of definitional, constitutional, and rehabilitational concerns, while others used these statutes sparingly.

The resurgence of interest in these statutes began in the early 1990s, again as the result of the sensationalization of certain violent sex crimes. The current laws are often labeled as sexually violent predator laws, and they allow the civil commitment of these individuals upon release from their state-mandated punishment or after they have been acquitted or found incompetent to stand trial on the basis of insanity or mental disease or defect. In place of the therapeutic appeal of the earlier sexual psychopath laws, the appeal of current sexual predator laws is in their social control function. The earlier sexual psychopath laws contained a stronger therapeutic component, coinciding with other forms of civil commitment, whereas current laws, although they contain elements of treatment, exhibit incapacitation and control as their key foci.

Civil commitment statutes for mentally ill persons were on the books in most states, but the sexual predator laws expanded the range of persons who may be committed to include those with a "mental abnormality or personality disorder" predisposing them to perpetrate another sexually violent crime (Falk, 1999, p. 129). After commitment, clinical experts periodically evaluate the committed offender medically and psychologically for signs of improvement in mental status. The offender may subsequently petition for release, return to court, and have the court again try to determine whether the offender is a danger to society. Confinement continues until such time that the psychologists and the court agree that the offender's mental abnormality or personality disorder has so changed that the individual is deemed safe to be released. Janus (2000) contends that in practice committed sex offenders are almost never discharged. "For example, in the two states with the longest contemporary commitment programs (both operating since 1990), Washington and Minnesota, no individuals have been discharged from commitment, and only a handful are in transitional placements" (Janus, 2000, p. 10).

Treatment and Punishment in Civil Commitment Statutes

Since these current laws were developed, numerous constitutional questions have arisen pertaining to their use, and the courts have varied in their rulings. Again, treatment, punishment, and public safety are the stated or implicit goals of these laws. The focus on punishment has been a key to many of the court rulings, the most notable of which is the *Kansas v. Hendricks* ruling in 1997 by the U.S. Supreme Court. In this case, Leroy Hendricks was convicted multiple times for child molestation. As his latest prison term neared expiration, Hendricks found himself the target of the new Kansas sexually violent predator statute). As a result, it was determined that he was a violent predator who would likely recidivate, and he was committed to a state mental hospital. Hendricks subsequently challenged the Kansas statute, stating that it violated due process and prohibitions against double jeopardy and ex post facto laws; the U.S. Supreme Court upheld the act and Hendricks's commitment as constitutional.

Interestingly, the Court rejected Hendricks's double jeopardy and ex post facto arguments mainly because, in their opinion, the civil commitment statute does not amount to punishment. Hendricks argued that, because he had not received any treatment, his commitment amounted to thinly disguised punishment. The Court disagreed. First, they believed that there were no additional criminal proceedings, and the act did not seek retribution or deterrence. They interpreted the goal of this statute to be that of incapacitation as Hendricks was untreatable. Second, they deferred to the legislature's intent. Because the law was housed within the state probate code and the law labeled "civil," the Court reasoned that the legislature's intent was not punitive. As the Court often submits to the findings of state and lower federal courts to reveal the intent of the legislature, it pointed to the provision in the statute that called for necessary long-term care and treatment of sexually violent predators. Thus, in *Hendricks*, the Court rejected the Kansas Supreme Court's argument that the statute was indeed punitive. Third, the lack of any treatment did not presume that the law was

punitive, because "it must be remembered that he was the first person committed under the Act" (*Kansas v. Hendricks*, 1997, p. 367). Furthermore, the Court stated that the liberty interests of these offenders must take second place to the goal of protection of the community.

In dissent, Justice Breyer observed that there were several similarities between the Kansas act and criminal law, such as the act's provision for involuntary detention, the use of criminal background as a determining factor, and the focus on incapacitation. Breyer also argued that the act was in error: There was no provision for treatment while the offender was imprisoned, yet the necessity of treating the individual was used as a partial justification for further confinement. Breyer said that, if treatment were necessary and available for this difficult-to-treat population, it should begin as soon after incarceration as possible.

In accordance with Justice Breyer, Friedland (1999) writes that the detention of this specific type of offender appears to be based on the desire to further punish and incapacitate. This intent is revealed in many ways. First, the law includes people with personality disorders or mental abnormalities that may not be treatable and those who would not benefit from involuntary hospitalization, although the law seems to imply that favorable treatment is possible. Some sex offenders' risk level is high enough that "no treatment could reasonably be expected to lower it to a level where release to the community" would be possible (Harris, Rice, &, Quinsey, 1998, p. 106), and other psychologists caution that certain types of sex offenders (e.g., pedophiles) can be extremely difficult to treat successfully.

Second, treatment is only given *after* the offender has been criminally incarcerated and appears to be a "pretext for extended confinement" (Friedland, 1999, p. 110). It is possible and even likely that detention may last long after efforts at treatment have occurred. Many of these sex offenders, committed through having a mental defect or personality disorder, may not be "authentically mentally ill"

(Friedland, 1999, p. 132), inasmuch as neither term corresponds to diagnostic categories in the American Psychiatric Association's *Diagnostic and Statistical Manual of Mental Disorders* (DSM-IV) or any other psychiatric classification. Indefinite, preventive detention underlies the pretext of offender treatment and "allows clinicians to collude with a non-clinical social agenda, with substantial likely harm to the offender through excessive false positive predictions of sexual violence" (Wettstein, 1992, p. 624).

The Court revisited the Kansas Act in *Kansas v. Crane* (2002). The Court ruled in *Crane* that, in addition to the demonstrated likelihood that the person would commit sex crimes if released, there must be a determination of a "serious difficulty in controlling the behavior," and the mental abnormality must distinguish the person from an ordinary recidivist (*Kansas v. Crane*, 2002, p. 409). This distinction is important, as the Court acknowledged the potential for the act to become too far-reaching. If a simple judgment of the existence of a mental abnormality is sufficient to commit the offender civilly, what prohibitions are there to prevent the civil commitment of other recidivists? Although many scholars would still be wary of a slippery slope (e.g., a pyromaniac or kleptomaniac could fit the identical criteria), this distinction is a movement to limit those who may be committed.

Ironically, these statutes may be seen to decrease public safety if examined on a larger scale. Finances for the mental health profession would be consumed by the increasing number of sex offenders and the need to incapacitate them. Janus (2000) questions whether these expensive commitment policies are the most effective use of scarce mental health resources. The reliance on the treatment community to house these offenders would divert necessary and already scarce money from treatment for both these offenders and other offenders who are in need of mental health services. Without this treatment, it is possible that many others who may be more amenable to treatment—not just sex offenders—may turn to crime.

In sum, the sexual predator statutes are designed to control a category of sexual offenders deemed incurable or, at best, resistive to therapy. Indefinite confinement of the sex offender is simply justification for lifetime preventive detention and continued state control. Although these laws have withstood a myriad of legal challenges, their intention and their use warrant closer examination. The laws should more clearly specify exactly who should be targeted, and treatment should begin prior to release from incarceration. Measurable goals and objectives should be used to determine progress in treatment and to decide whether further commitment is necessary. Assessment tools, standards of commitment and release, conditions of confinement, and the actual application of the law need to be scrutinized to ensure that the law is not abused or used solely for punishment objectives.

⊠ Registration and Community Notification for Sex Offenders

History

Registration of people who have been charged or convicted of various offenses has been used for decades, particularly for sex offenders. Probably the most important development in registration and community notification of sex offenders occurred in 1994 when Congress passed the Jacob Wetterling Crimes Against Children and Sexually Violent Offender Registration Act. This act required states without registration laws to create them within three years and targeted those offenders who committed a sexual act against a minor or a sexually violent offense. Registration of the sex offender is justified as necessary so that police can know the whereabouts of sex offenders in their jurisdictions. This information can then be used in subsequent investigations of sex offenses. In 1996, the law was amended to require the disclosure of information about registered sex offenders to the public. For both forms, the federal government threatened to withhold part of the states' federal law enforcement funding if they did not comply with the act. New Jersey in 1994 passed one of the most well-known notification statutes, referred to as "Megan's Law," after Megan Kanka was sexually assaulted and murdered by a neighbor who had been previously convicted of violent sex crimes.

The rationale for community notification is that protection of children will be more meaningful if the community has knowledge about the presence of a convicted sex offender in the neighborhood. Protecting the public is the crux of the argument, with punishment underlying this rhetoric. There is little attempt to integrate treatment into these laws, although offender reintegration into the community remains a focus in corrections. These laws are very popular politically and publicly; many legislatures have passed them unanimously because they give the public the perception of control and safety in their communities.

Many states that use community notification have a three-tiered system that determines how much proactive notification will occur based on the dangerousness of the offender. For offenders classified as the lowest risk (Level I) for endangering public safety, notification is limited to law enforcement personnel. Level II offenders are at medium risk for reoffending, so notification is expanded to include schools, day care centers, or other community organizations that serve primarily women and children. Level III offenders are seen as posing the highest risk; therefore, the highest degree of notification is implemented whereby the general public is notified. Some states use formal assessment instruments to systematically classify an offender's level of risk to reoffend, while others use an advisory committee to assess risk on a case-by-case basis. Registration and community notification provisions can be lengthy; more than half of the states require offenders to be registered and the community notified of

their presence for at least 10 years to a lifetime after the expiration of their sentences.

Benefits and Problems With Registration/Notification Laws

As with the civil commitment statutes, the courts are mixed in their determination of the constitutionality of these registration and notification statutes but generally tend to support them. Constitutional challenges have centered on allegations of ex post facto and cruel and unusual punishment. Courts frequently reject the constitutional challenges by reiterating that punishment is not being given and by supporting the liberty and protection of the community over that of the individual sex offender. Proponents of the laws argue that whatever punitiveness occurs is only incidental to the public protection function and is not intended to be punishment. There is some indication that community mobilization is occurring and that probation and parole officers and the community are taking a more proactive approach to supervising these offenders. For example, in California local police initiated surveillance of one child molester. They caught him driving slowly past a school yard, which was a violation of his parole, and reincarcerated him. Penalties for failing to register have also been increased to a major misdemeanor or low-level felony.

The current degree and public perception of community protection are debated, however. Zevitz and Farkas (2000a) found that increases in community protection strained probation and parole agent resources because of the requisite increases for supervision. Scholars caution that community protection is often lower than it is perceived to be. While these laws often frame the risk of victimization as that of assault by strangers, studies show that children are more likely to be abused by someone known to them, such as a family member or friend. The possibility of community notification may actually discourage reporting if the offender is a family member.

Victims may either fear being found out themselves or fear for their loved ones, even the abusers. Notification also does little to prevent the undetected sex abusers, who constitute the majority of sex offenders, from victimizing others or from venturing into nearby communities. Offenders may also threaten victims more seriously in order to lessen the chance of being reported. Furthermore, these laws are criticized as being overreaching: Even when risk instruments and knowledge of likelihood of reoffending are used, there tend to be false positives. In other words, police may err on the side of caution in targeting these offenders.

Additionally, as with many policies, what the legislature desires and what the criminal justice system can provide are divergent. Registration and notification bills were quickly introduced and passed in most states; however, implementation of these laws has its difficulties. Many jurisdictions have had to come up with innovative ways to disseminate the information, especially when little or no money was given to aid in this notification. Often community notification takes the form of community meetings, door-to-door visits by the police, Internet notices, newspaper articles and ads, and flyers placed in the area. In Wisconsin, Farkas and Zevitz (2000) noted that resources were not made available to carry out the law as intended. There were no additional funds for officer training on the notification procedures or for the creation of community meetings. Smaller communities were unable to spare enough officers to meet with citizens, and police in larger cities noted that their workload was increased by having to check up on these offenders more and to field the public's questions, complaints, and concerns. Probation and parole agents noted further difficulties. Because these cases were high-profile, agents described having to invest more time and energy into community notification cases, thereby spending less time on other cases. Both police and probation and parole agents also described lags in obtaining information to aid in determining notification level and supervision considerations.

Further difficulties are evident in any attempt to balance treatment with punishment and public safety. Not all sex offenders are incurable, and some respond well to therapy. "Community notification laws fail to discriminate between those capable of rehabilitation and those whose deviancy may be permanent" (Bedarf, 1995, p. 910). Unfortunately, this trend is common in correctional policy making and practice; practices are often "one size fits all" and offenders are treated identically regardless of wide individual differences. Rather than help sex offenders reintegrate into their neighborhoods, notification excludes, labels, and stigmatizes them. A commonly cited problem is the inability of many of these offenders to get on with their lives. Being subject to decades of registration and community notification could result in a perpetual burden on the offender's privacy interests, thereby impeding successful reintegration into the community.

Zevitz and Farkas (2000a) describe the problems that many probation and parole agents had in locating adequate housing for sex offenders and the problems that offenders experienced in finding meaningful employment. The laws assume that neighbors will act responsibly, provide adequate supervision for their children, and act within the law toward the offender. "Reintegration requires effort from both parties: the offender must abide by the rules of society, and the community must allow the offender to enter community life" (Bedarf, 1995, p. 910). In another study by Zevitz and Farkas (2000b), sex offenders described the humiliation in their daily lives and the negative effects on the lives of their family members, and they reported being ostracized by neighbors and lifetime acquaintances and being harassed or threatened by nearby residents or strangers. The evidence suggests that occasional vigilantism occurs in the community in efforts to remove offenders from neighborhoods or to further punish.

Psychologists are concerned that community reaction may increase an offender's anxiety, which may lead to poor decision making and ultimately a relapse event. These laws may also inhibit willingness to accept responsibility for one's crime. Ironically, under these laws, if an offender commits another sex crime, the blame may not rest solely on the offender. With the impetus placed on the community to track and monitor, these offenders can conceivably blame the community for not adequately supervising them. This transfer of blame is contrary to the law's purpose and is contraindicated in treatment. Many sex offender treatment programs require that the offenders admit to wrongdoing and develop appropriate self-control techniques to regulate their own behavior. Other criticisms include the potential of the laws to decrease offender self-reporting and desire for treatment and to serve as a disincentive for sex offenders to plead guilty and get treatment.

In reality, community notification and registration laws are predicated on the failure of treatment and rehabilitation programs in prison. These laws are intended to provide a long-term, protective strategy to incapacitate or control what is labeled as a permanently dangerous and contemptible class of offenders. However, in this legislative rush to enact another "get tough on sex offenders" law, a meaningful balance between the constitutional and civil rights of sex offenders and the interests of victims and the general public has been overlooked. Creating rational and thoughtful levels of risk and carefully controlled means of public disclosure will ideally lead to notification systems that lessen public fear and that do not impede the rehabilitation or reintegration of sex offenders. To accomplish this objective, states can learn from one another. Information sharing about the difficulties experienced in the development and implementation of sex offender registration and notification laws is recommended. States could share solutions as well as problems encountered. More research is also advised to better inform policy makers about the impact of such laws on corrections, law enforcement, sex offenders, and the community.

▨ Chemical Castration Laws

History and Use

Physical castration or orchiectomy has a long history dating back to biblical times. Under the "eye for an eye" *lex talionis* principle, for centuries rapists were castrated as punishment for their crimes. The eugenics movement in early nineteenth-century America endorsed both castration and sterilization for criminals and the mentally ill. Countries including Denmark, Germany, and Switzerland have used physical castration for years as an effective way of dealing with individuals who display unacceptable sexual behavior. Experimental data for physical castration indicates a low incidence of serious morbidity, an immediate reduction in sex drive, and significantly low recidivism rates, as low as 2.2 percent. At present, physical castration is still legally permissible in these countries, but it is by no means commonly used with sex offenders. Voluntary chemical castration combined with therapy has largely replaced physical castration because physicians believe that similarly effective results can be obtained through pharmacological treatment.

The use of physical castration in the United States has also not been widely embraced because of qualms about physical mutilation, invasiveness, the permanency and irreversibleness of the surgery, and the availability of medications as an alternative. Some sex offender laws do provide surgical or chemical castration as an alternative to punishment, such as Florida's law, yet chemical castration seems to be the more socially and morally acceptable solution, at least for the short term.

Although medication has been used on a case-by-case basis by clinicians for more than 30 years, renewed interest and focus on drugs as a means of controlling the behavior of sex offenders has occurred in the past decade. Studies have shown that sex offenders treated with a pharmacological agent, such as Depo-Provera, plus counseling have gained better self-regulation of sexual behavior; they were much less likely to reoffend than those who refused the medication or stopped the injections. The promising results occurred only as long as the offenders continued the injections

Chemical castration essentially involves the injection of the synthetic hormone medroxyprogesterone acetate (Depo-Provera), which lowers the blood serum testosterone levels in males. When taken on a regular basis, the result is a reduction in sexual impulses, the frequency of erotic fantasy, erections, and ejaculations. Depo-Provera does not cause impotence but produces a period of "erotic apathy" in which the feeling of sex drive is at rest. Upon cessation of the injections, erectile and ejaculatory capacity returns within 7 to 10 days. The side effects of Depo-Provera include migraines, hypertension, hyperglycemia, nausea, weight gain, insomnia, fatigue or lethargy, leg cramps, loss of body hair, and thrombosis. Most of the side effects are rare and are thought to be reversible once the treatment is discontinued; nevertheless, the long-term effects of Depo-Provera are unknown. Antipsychotic medication has also been known to reduce the frequency or intensity of sexual arousal and thoughts and is currently under experimentation.

Problems With Current Use of Chemical Castration as Treatment

Although chemical castration has been in use in the treatment of sex offenders since 1944, it was then administered with the informed consent of the individual and on an individual basis. Equally important was the desire expressed by sex offenders to be rehabilitated. The current chemical castration statutes are much different. First of all, the state statutes may stipulate forced administration of the pharmacological agent if the sex offender is a recidivist. In 1997, California became the first state to enact such a law mandating chemical castration for twice-convicted sex offenders whose victims were under the age of 13.

The sentencing judge has no discretion when dealing with these repeat sex offenders. Chemical injections are mandated prior to release on parole and are continued while the offender is under community supervision. If the offender refuses, parole is denied. Other states have followed suit, including Colorado, Florida, Georgia, Louisiana, Montana, Texas, and Wisconsin, with many of these states' statutes granting judicial discretion to impose a sentence of Depo-Provera administration. Critics contend that the judge may not be the most informed individual to make that determination and question the ability of medical staff within the Department of Corrections to make an impartial decision:

> Although no bill authorizes the judge to order physicians to prescribe the medication per se, all the bills require that the medication be started while the offender is still in the state's Department of Corrections. Correctional physicians typically have much less professional freedom than their colleagues and may be subject to considerable pressure to provide the treatment, regardless of their competency to administer it or any ethical objections that they may have. (Simon, 1998, p. 196)

Accordingly, the role of the physician as a healer and his or her code of ethics have also been challenged. Does the physician become an agent of social control with the best interests of the patient subordinate to the societal protection interest of the state? Nassi (1980) explored this issue in her study of treatment providers in California prisons for men. She found that prison psychiatrists and psychologists experienced "dissonance" or conflict between their role as a helping agent for the offender and as an employee of the state. As a helping agent, caring for the individual and his mental health and growth were central concerns. As a state employee, the goals of prisoner control and societal protection became the major foci.

A complete medical or psychiatric evaluation prior to the injections of the chemical agent is typically not required in the statutes. Without an evaluation, there is no clinical determination of whether the medication is indicated for a particular offender. Lumping all sex offenders into one broad category hinders their effective management and treatment. Not all sex offenders are suitable candidates. "The legislature, in enacting the chemical castration statute, has proceeded on the belief, or perhaps more accurately, on the hope, that administering a single drug—even involuntarily—can effectively alter abusive behavior in all categories of sexual offenders" (Spalding, 1998, p. 126). There is no differentiation between types of sex offenders suitable for chemical castration in the statutes. Injections of Depo-Provera will not have any meaningful effect on sex offenders who deny perpetrating the offense or the criminal nature of the act; those who admit the offense but blame their behavior on nonsexual or nonpersonal factors such as drugs, alcohol, and job stress; and those who are violent and appear to be motivated by nonsexual gain, anger, power, and violence. For example, chemical castration will not work for a sex offender motivated by anger, violence, or power. With this group of offenders, even if rendered impotent, they would find some other means to violate their victims.

Fundamental rights of privacy and procreation may also be contravened with the imposition of Depo-Provera injections. Privacy concerns arise when an individual's right to bodily autonomy, to the control of his own person and the right to refuse treatment, is compromised. In *Washington* v. *Harper* (1990), the U.S. Supreme Court held that the state could override a prisoner's refusal to take antipsychotic medication if the administration of the medication was in the inmate's best medical interest and the inmate was dangerous to himself or others: "Where an inmate's mental disability is the root cause of the threat he poses to the inmate population, the state's interest in decreasing the danger to others necessarily encompasses an interest in providing him with medical treatment for his illness" (p. 226). Before a prisoner could be treated with antipsychotic drugs involuntarily, the Washington Department of Corrections policy required a hearing before a committee consisting of a psychiatrist, a psychologist, and the associate superintendent of the prison. The committee must find by a majority that the prisoner suffered from a mental disability, was gravely disabled, and posed a likelihood of serious harm to himself or others. However, Justice Stevens dissented, asserting that a prisoner's liberty interest was seriously contravened by the administration of antipsychotic drugs that caused severe and often permanent side effects and that altered the will and mind of the subject. He also expressed concerns that institutional and administrative pressures would subvert the use of medication in the best medical interest of the prisoner.

Antipsychotic drugs can be used to simultaneously medically treat and control psychotic inmates; however, it is precisely this duality of purpose that makes them subject to misuse and abuse. The use of psychotropic drugs on prisoners is widespread, and this also presents a correspondingly high opportunity for state-sanctioned abuse in the administration of the drugs.

The legally mandated use of chemical castration brings to the fore the constitutionality and ethics of using medication for purposes other than treatment. "By using psychiatrists and their medications, the criminal justice system is attempting to legitimize and sanitize what are, at heart, punitive programs" (Miller, 1998, p. 199). Does chemical castration constitute cruel and unusual punishment in violation of the Eighth Amendment? In *People v. Gauntlett* (1986), Gauntlett challenged as cruel and unusual punishment a condition of probation that required him to be injected with Depo-Provera for five years. The Michigan Supreme Court found that treatment with the medication was not a lawful condition of probation because it had not found acceptance as a safe and reliable medical practice.

In summary, the goals of rehabilitation and reduction in recidivism can be met with chemical castration only when the drug can effectively treat the sexual disorder of the individual, administration of the drug is voluntary and with informed consent, and the individual is motivated to address his aberrant behavior. Thus, the most appropriate candidates for the procedure (e.g., paraphiliacs) must be identified based on research findings. The laws should also incorporate a counseling component to address the root causes of the sexual behavior.

Polygraph Testing

History and Use of Polygraph Testing for Sex Offenders

In the past decade, the polygraph has also been introduced as another tool to manage, supervise, and treat sex offenders under community supervision. Laws have been passed in several states, including Arizona, Colorado, Florida, Tennessee, Texas, and Wisconsin, requiring anyone convicted of two sexual offenses to undergo periodic mandatory polygraph supervision. Thirty states have at least some local jurisdictions using polygraphs to supervise sex offenders.

The polygraph is used in three basic ways: a disclosure polygraph administered after sentencing, a denial and specific issues exam, and a maintenance polygraph administered while in treatment. The disclosure polygraph requires sex offenders to answer questions in detail about their sex offense history and history of sexually deviant behavior. The denial exam is administered when there is a conflict between the sex offender's and the victim's versions of the crime or when the offender denies culpability, and a specific issues examination focuses on a specific crime, accusation, or suspicion. The maintenance polygraph monitors the offender's management of inappropriate thoughts and fantasies, reduces denial, and measures compliance with supervisory conditions. A diagnosis of deception is an inference based on increased physiological arousal (increased heart activity, rate and depth of breathing, and palmar sweating) in response to relevant questions

Benefits of Using Polygraphs With Sex Offenders

Sex offenders can be highly manipulative, deceitful, and evasive during questioning. They can minimize their culpability and deny any deviant sexual fantasies and preoccupations. They can be extremely reluctant to disclose their offending histories for a variety of psychosocial and legal reasons. "Some individuals are extremely adept at disguising their sexual pathology, while others exaggerate their symptoms in a histrionic fashion. In many cases, the sexual offender will simply not provide accurate information" (Maletsky, 1991, p. 37). In order for sex offenders to "get past" their denial mechanisms and deception, they need to be open and to divulge their sex offense history and their deviant fantasies and external behaviors.

The polygraph has been touted as an effective intervention in reducing denial, eliciting admissions of past and present sexual offending, improving treatment outcomes, and improving the supervision of sex offenders. Therapists and supervising agents who work with sex offenders point to the tremendous value of the polygraph, and even the threat of a polygraph, in compelling sex offenders to be truthful in therapy.

Difficulties in Mixing Punitiveness and Treatment in Polygraph Use

Treatment is assumed to be more effective if the sex offender fully discloses prior deviant acts and if the therapist is aware of the extent and nature of the deviant behavior. Amenability to treatment and successful rehabilitation are believed to hinge upon eliminating denial and replacing it with an admission of responsibility for past sexual deviancy. Previous offenses that are disclosed for treatment purposes are typically not reported because there may be insufficient information to consider formal charges.

Another problem with the use of the polygraph is that it changes the nature of the sex offender patient-therapist relationship. The social control aspect may overpower the support and advocacy role of the therapist. The polygraph has placed treatment providers in a double bind. If the sex offender discloses deviant thoughts, the therapist may feel obligated to inform the offender's probation or parole agent. The technique to motivate or effect behavioral change in therapy becomes the threat of criminal justice consequences, such as electronic monitoring, more frequent monitoring by the agent, or even incarceration. The therapist may feel tempted to contact the police if the sex offender discloses a previous offense. Yet the honesty of the sex offender may be an indication of a trust relationship building between therapist and patient and a sign of progress in treatment. The therapist may jeopardize that fragile trust in the interest of cooperating with the criminal justice system.

Finally, the therapist may become overly dependent on the polygraph in working with sex offenders. Instead of reading behavioral cues and learning additional techniques to elicit information, the therapist may come to rely solely on information from the polygraph as authority to make case management or legal decisions regarding sex offenders. The integrity of the therapeutic relationship may be compromised when an honest offender is misidentified as "deceptive" and a dishonest offender "beats" the polygraph.

In summation, polygraphs may elicit important information about sex offenders that may advance treatment goals, but the potential costs due to errors introduce genuine concerns. Erroneous conclusions with deceptive individuals can lead to a new sex offense, and faulty conclusions with truthful offenders can result in more stringent conditions of supervision and even revocation. Although polygraphs can be useful intimidation devices,

further study is recommended to establish their validity. Until that time, failed polygraph tests or refusals to take the tests should not be used as primary evidence for revocation.

▨ Conclusion

History has shown the intense focus on sex crimes and sex offenders and the rush to respond to the will of the people on the part of legislators and politicians. Public policy, especially concerning sex offenders, is often tainted by emotion and political rhetoric rather than grounded in factual information or empirical data. "The phenomenon of sexual abuse is intertwined with a strong emotionalism that exacts an almost visceral response in nearly everyone, and this emotionalism has confounded our lawmakers' collective abilities to separate legislative proposals that are functionally efficacious from those that are certainly well-intentioned but are nonetheless unsuccessful" (Edwards & Hensley, 2001, p. 84). As we have discussed in this article, this emotionalism and lack of foresight in the passage of sex offender laws has had many consequences for sex offenders, treatment providers, the criminal justice and mental health systems, and society in general.

How sex offenders are viewed by society is critical to our notion of how to treat, manage, and control these offenders. If the sex offender is portrayed as an evil monster deserving of punishment, punitive and incapacitative measures become the primary management strategy. Any sort of treatment effort may be co-opted by punishment. The past provides evidence of the cruelty of state-imposed "treatment" to repugnant groups—lobotomies for the mentally ill and eugenic sterilization of the mentally retarded and habitual criminals. Individual offender rights are abrogated by public safety concerns.

Furthermore, there is a pronounced tendency in policy making to characterize all sex offenders as being comparable, when in reality there is a wide range of differences in behavior represented by sex offenders. "Clinical experience suggests that sexual offenders can be as diverse in personality patterns and behavior habits as many other large diagnostic groups" (Maletsky, 1991, p. 141). This unitary view of sex offenders has resulted in a "one size fits all" approach to policy making. In reality, there is no single treatment modality or, for that matter, law, policy, or practice that will apply to all sex offenders. As an example, research on the use of chemical castration has found it to be an effective strategy for paraphiliacs when combined with treatment. Yet chemical castration is mandated for rapists in California's law. As was discussed earlier, rapists are not good candidates for chemical castration because their underlying motivation to rape is anger, power, and control, not sexual desire alone. Depo-Provera inhibits the sex drive but it does not impact violent tendencies. As an added note, the goal of public safety is certainly not met if Depo-Provera is legally mandated for all sex offenders without a consideration of the clinical research on the drug.

The suppositions that a sex offender (of any variety) cannot be truly rehabilitated, will continue to pose a danger to society, and warrants a long-term management and control strategy affect our policies and practices in the criminal justice and mental health systems. Law enforcement confronts a high cost in terms of time, effort, and resources for surveillance and enforcement of "dangerous" sex offenders in the community, and corrections must manage and supervise large caseloads of sex offenders in an intensive manner. These assumptions about sex offenders have never been reexamined or tested, and most of these laws, policies, and practices have been implemented without any form of analysis or evaluation of their intended outcome and of their social, psychological, economic, and legal consequences. This analysis is critical before one rushes to pass a law or initiate a policy in any area, but it is especially critical when the initiatives are already in place. Research and

evaluation are necessary regarding both the process and the outcomes of sex offender–specific laws, policies, and practices.

In conclusion, anticipatory social control is at the core of our laws and policies; sex offenders are punished and incapacitated for fear of what they might do in the future rather than for their crimes of conviction. Civil commitment laws provide a process to indefinitely confine a sex offender. Community notification, chemical castration, and polygraph statutes extend social control over the sex offender released to the community. The controls are justified as necessary to help sex offenders to change or to manage their behavior and to be reintegrated into society. The practical emphases of the laws are usually punitive and incapacitative, overshadowing or subverting any genuine attempt at treatment. "Our laws must provide treatment options for those who will respond to treatment in addition to protection for society from those who will not respond to treatment or do not want treatment" (Freeman-Longo, 1996, p. 317). Yet the courts have never held that the criminal justice system must subscribe to a rehabilitation model or even have rehabilitation as one of its goals. The inclusion of treatment terminology, however, is often used to pass constitutional scrutiny. Unfortunately, with the lack of differentiation among sex offenders and the lack of meaningful treatment or emphasis on long-term behavioral change, it is unlikely that these laws will truly provide community protection.

❧ References

American Psychiatric Association. (1994). *Diagnostic and statistical manual of mental disorders* (4th ed.). Washington, DC: Author.

Bedarf, A. (1995). Examining sex offender community notification laws. *California Law Review, 83,* 885–937.

Bureau of Justice Statistics. (1997). *Sex offenses and offenders: An analysis of data on rape and sexual assault.* Washington DC: U.S. Department of Justice.

Edwards, W., & Hensley, C. (2001). Contextualizing sex offender management legislation and policy:

Evaluating the problem of latent consequences in community notification laws. *International Journal of Offender Therapy and Comparative Criminology, 45*(1), 83–101.

Falk, A. (1999). Sex offenders, mental illness, and criminal responsibility: The constitutional boundaries of civil commitment after *Kansas v. Hendricks. American Journal of Law and Medicine, 25*(1), 117–147.

Farkas, M. A., & Zevitz, R. G. (2000). The law enforcement role in sex offender community notification: A research note. *Journal of Crime and Justice, 23*(1), 125–139.

Freeman-Longo, R. (1996). Feel good legislation: Prevention or calamity. *Child Abuse and Neglect, 20,* 95–101.

Friedland, S. (1999). On treatment, punishment, and the civil commitment of sex offenders. *University of Colorado Law Review, 70*(1), 73–154.

Harris, G. T., Rice, M. E., & Quinsey, V. L. (1998). Appraisal and management of risk in sexual aggressors: Implications for criminal justice policy. *Psychology, Public Policy, and Law, 4*(1/2), 73–115.

Janus, E. (2000). Sexual predator commitment laws: Lessons for law and the behavioral sciences. *Behavioral Science and the Law, 18,* 5–21.

Kansas v. Crane, 534 U.S. 407 (2002).

Kansas v. Hendricks, 521 U.S. 346 (1997).

Kansas v. Lumley, 977 P.2d 914 (Kan. 1999).

Maletsky, B. M. (1991). *Treating the sexual offender.* Newbury Park, CA: Sage Publications.

Miller, R. D. (1998). Forced administration of sex drive reducing medications to sex offenders: Treatment or punishment? *Psychology, Public Policy, and Law, 4*(1/2), 175–199.

Nassi, A. J. (1980). Therapy of the absurd: A study of punishment and treatment in California prisons and the roles of psychiatrists and psychologists. In H. J. Vetter & R. W. Rieber (Eds.), *The psychological foundations of criminal justice: Contemporary perspectives on forensic psychiatry and psychology* (Vol. 2, pp. 322–334). New York: John Jay Press.

People v. Gauntlett, 353 N.W. 2d 463 (Mich. 1986).

Pratt, J. (2000). Sex crimes and the new punitiveness. *Behavioral Sciences and the Law, 18,* 135–151.

Schwartz, B. (1988). *A practitioner's guide to treating the incarcerated male sex offender.* Washington, DC: National Institute of Corrections.

Simon, J. (1998). Managing the monsters: Sex offenders in the "new penology." *Psychology, Public Policy, and Law, 4*(1/2), 452–467.

Spalding, L. H. (1998). Article: Florida's 1997 chemical castration law. A return to the dark ages. *Florida State University Law Review, 25,* 117–139.

Washington v. Harper, 494 U.S. 210 (1990).

Wettstein, W. M. (1992). A psychiatric perspective of Washington's sexually violent predator statute. *University of Puget Sound Law Review, 15,* 597–633.

Zevitz, R., & Farkas, M. A. (2000a). The impact of sex-offender community notification on probation/parole in Wisconsin. *International Journal of Offender Therapy and Comparative Criminology, 44*(1), 8–21.

Zevitz, R., & Farkas, M. A. (2000b). Sex offender community notification: Managing high-risk criminals or exacting further vengeance? *Behavioral Science & the Law, 18*(2/3), 375–391.

DISCUSSION QUESTIONS

1. Given that fewer than 10% of sex offenders are violent, why do you think the public is apparently more hostile to them than to any other type of offender?

2. Why or why not is the civil commitment of some sex offenders a good idea?

3. Explain the usage of anti-androgen medications such as Depo-Provera in the treatment of sex offenders.

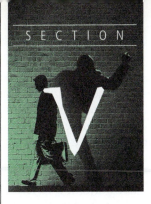

SECTION V

TREATMENT PROGRAMMING AND REHABILITATION RESEARCH

Introduction

⬙ The Rise and Fall (and Rise Again) of Rehabilitation

As we have seen, there are five primary goals of the correctional system: deterrence, incapacitation, retribution, rehabilitation, and reentry. This chapter deals with the fourth of these goals, rehabilitation. The term *rehabilitation* means to restore or return to constructive or healthy activity. Rehabilitation was the goal of the early American prison reformers, but it reached the pinnacle of its popularity from about 1950 through the 1970s when the medical model of criminal behavior was the prevailing model in corrections. The medical model viewed crime as a moral sickness that required treatment. Under the medical model, prisoners were to remain in custody under indeterminate sentences until "cured" of their criminal ways.

The rehabilitative goal was questioned among academics with the publication of Robert Martinson's article "What Works? Questions and Answers about Prison Reform" in which Martinson (1974) concluded that "with few and isolated exceptions the rehabilitation efforts that have been reported so far have no appreciable effects on recidivism"

(p. 25). Unfortunately the rhetorical question "what works?" somehow got translated into a definitive "nothing works," although, to be fair, this was the general drift of his article. Before we can decide if something does or does not work, we have to delineate thresholds for defining success. If one's criterion for success is 100 percent, then we can be sure that nothing works. A program designed to change people cannot be likened to a machine that either works or does not. Human nature being what it is, nothing will work for everybody, and nothing at all will work for anybody all of the time.

Many correctional programs Martinson surveyed did not work for a variety of reasons, such as relying on nondirective methods that were inappropriate for offenders, seeking to change behaviors unrelated to crime, or using programs that were not intensive enough or run by inadequately skilled staff. Further, few of the programs were based on the proper assessment of offender risks and needs and were often faddish "let's see what happens" programs. Such correctional quackery still exists and can include any number of little studied or validated programs (Latessa, Cullen, & Gendreau, 2002).

How have Martinson's conclusions stood up over the last 30 years? Gendreau and Ross (1987) reviewed a number of studies of treatment programs and concluded, "it is downright ridiculous to say that 'Nothing works.' This review attests that much is going on to indicate that offender rehabilitation has been, can be, and will be achieved" (p. 395). Others have stated that properly run community-based programs could result in a 30 to 50% reduction in recidivism (Van Vooris, Braswell, & Lester, 2000), although, on the basis of major meta-analytic reviews, reductions in the 10 to 20% range are more realistic (Cullen & Gendreau, 2000). The meta-analysis in the Pearson et al. article in this section, for instance, found a 55.7% success rate (measured in terms of recidivism rates) for inmates receiving cognitive-behavioral therapy versus a 42.7% success rate for a control group. This translates into a more than 20% greater success for the treatment group ($55.7 - 42.7 = 13/55.7 = 23.3$). Although there are still plenty of failures, if all treatment programs managed only half that success rate, the financial and emotional savings to society would be truly enormous.

Rehabilitation Under the Justice Model: The Cognitive-Behavioral Approach

Moving from the medical to the justice model in corrections did not mean the death of the rehabilitation goal, but terms such as "assessment" and "programming" have largely replaced terms such as "diagnosis" and "treatment." Prison officials like programming if for no other reason than it keeps inmates busy and out of trouble. Of course, the primary function of corrections is to maintain control and custody over inmates, and this function consumes about 95% of the correctional budget. Something we must keep in mind when reading the articles in this chapter, then, is that programs are typically run on a financial shoestring.

You will see that most of today's programming, regardless of its specific purpose, is run on cognitive-behavioral principles. It has been claimed that "The cognitive-behavioural approach represents the most overtly 'scientific' of all major therapy orientations" because there is "strong emphasis on measurement, assessment, and experimentation" (McLeod, 2003, p. 123). Albert Ellis (1989) claims that the great religious leaders of the past were cognitive-behavioral therapists after a fashion in that they were trying to get people to

▲ **Photo 5.1**   Texas death row inmate Carl Buntion works in the sewing factory at the Ellis Unit in Huntsville, Texas in 1997. He was convicted of killing a Houston police officer during a traffic stop. During his trial, the judge told him he was doing "God's work" in making sure Buntion was convicted and sentenced to death. Buntion's death sentence was overturned in 2006.

change their behavior from hedonism to prudence, from hatred to love, from cruelty to kindness by appealing to their rational long-term self interests. Do these things and you will not only feel good about yourself now but also go to heaven or attain nirvana in the future. This is exactly what cognitive behavioral therapy tries to do: change the antisocial and self-destructive behavior of offenders into prosocial and constructive behavior by changing the way they think.

The first lesson of cognitive-behaviorism is that criminals think differently from the rest of us. This statement is almost self-evident, but it would have been considered a radical idea in the 1970s when blaming criminal behavior on factors entirely external to those who commit it was in its heyday. Yochelson and Samenow (1976) and Samenow (1998, 1999) pioneered treatment theories based on challenging criminal *thinking errors* when they realized that modalities based on "outside circumstances" theories did not work. The task is to understand how criminals perceive and evaluate themselves and their world so that we can change them.

Criminal thinking is faulty and destructive to those around them and to the criminals themselves. It lands them in trouble with family, friends, employers, and the criminal justice system. It is thought that they perceive the world in a fatalistic fashion, believing that there is little that they can do to change the circumstances of their lives. To illustrate criminal thinking patterns, Boyd Sharp (2006) cites a Calvin and Hobbs cartoon in which Calvin says,

> I have concluded that nothing bad I do is my fault . . . being young and irresponsible I'm a helpless victim of countless bad influences. An unwholesome culture panders to my undeveloped values and it pushes me into misbehavior. I take no responsibility for my behavior. I'm an innocent pawn of society. (p. 3)

Criminals, it is believed, think like Calvin in the context of a society where many people prefer to claim victimhood rather than personal responsibility (McDonald's made me fat; Phillip Morris made me smoke, and so on). Whatever influences any outside circumstances may have on behavior, before they can affect behavior, they have to be perceived and evaluated by individuals. The strains of outside circumstances certainly influence behavior, but the important thing is not their presence per se but how we deal with them—constructively or destructively (Agnew, 2005). Some criminals are eager to jump on authoritative pronouncements that excuse their behavior, and defense lawyers are equally as quick to argue them in court. All of this reinforces the patterns of criminal denial that treatment providers find so frustrating (Sharp, 2006; Walsh, 2006).

Nevertheless, we keep on trying to rehabilitate criminals because we realize that whatever helps the offender helps the community. We are also mindful of Former Chief Justice Warren Burger's famous lines: "To put people behind walls and bars and do little or nothing to change them is to win a battle but lose a war. It is wrong. It is expensive. It is stupid" (cited in Schmalleger, 2001, p. 439). In this section, we look at various ways in which treatment personnel have been fighting the war. When reading this section, keep in mind that only between 10 and 15% of jail and prison inmates who need substance abuse treatment actually receive it (Foster, 2006, p. 301), and that surely is wrong, expensive, and stupid.

Cognitive-behavioral methods in corrections are "used to address issues such as self-control, victim awareness and relapse prevention, and to teach among other things critical reasoning and emotional control" (Vanstone, 2000, p. 172). Cognitive-behavioral therapy literally "exercises the thinking areas of the brain and thereby strengthens the [neuronal] pathways by which the thinking brain influences the emotional brain" (Restak, 2001, p. 144). A number of functional neuroimaging studies (reviewed in Linden, 2006) show that the therapy works on the brain almost exactly in the same way that drugs such as Prozac do. Unfortunately, these studies have only been conducted on patients with psychiatric problems such as depression and obsessive-compulsive disorder.

▧ Substance Abuse

Alcohol is at the same time our most popular and most deadly way of drugging ourselves. Police officers spend more than half their law enforcement time on alcohol-related offenses. One-third of all arrests (excluding drunk driving) in the United States are for alcohol-related offenses, about 75% of robberies and 80% of homicides involve a drunken offender and/or victim, and about 40% of other violent offenders in the U. S. were drinking at the time of the offense (Mustaine & Tewksbury, 2004). Illegal drug usage presents almost as big a problem, with about 67% of state and 56% of federal prisoners being regular drug users prior to their imprisonment (Seiter, 2005, p. 432). Clearly, mind altering substances, both legal and illegal, are associated with criminal behavior, and, consequently, the criminal penchant to indulge in them must be addressed by correctional agencies.

Substance abuse problems are extremely difficult to treat because individuals most at risk for becoming addicted share many of the same characteristics for engaging in chronic criminal behavior. For instance, alcoholism researchers divide alcoholics into two types: Type I and Type II. Type II alcoholics start drinking and using other drugs earlier, become more rapidly addicted, and exhibit many more character disorders, behavior problems, and criminal involvement, both prior and subsequent to their alcoholism, than do Type I alcoholics (Crabbe, 2002; DuPont, 1997). The heritability (a quantitative measure ranging between 0.0 and 1.0 used by geneticists to determine the proportion of variance in a trait attributable to genes) is about 0.90 for Type II alcoholism versus about 0.40 for Type I alcoholism (Crabbe, 2002; McGue, 1999).

Similarly, many researchers have shown that drug addiction and criminality are part of a broader propensity to engage in many forms of deviant and antisocial behavior. The National Institute of Justice attempts to monitor the extent to which drug use and crime are associated through its Arrestee Drug Abuse Monitoring (ADAM) Program, which collects urine samples from arrestees to test for the presence of drugs (Zhang, 2004). Table 5.1 shows the percentage of male and female adult arrestees in some of our largest cities who tested positive for illicit drugs. Note that the majority of both male and female arrestees tested positive for at least one drug and that the majority of males tested positive for multiple drugs (Walsh & Ellis, 2007).

Illicit drug abuse is clearly strongly *associated* with criminal behavior, but is the association a *causal* one? A large body of research indicates that drug abuse does not appear to *initiate* a criminal career, although it does increase the extent and seriousness of one (McBride & McCoy, 1993; Menard, Mihalic, & Huizinga, 2001). In other words, research seems to point to the fact that chronic drug abuse and criminality are part of a broader propensity of some individuals to engage in a variety of deviant and antisocial behaviors. Numerous studies have shown that traits characterizing antisocial individuals such as conduct disorder, impulsiveness, and psychopathy also characterize drug addicts (Fishbein, 2003; McDermott et al., 2000). The reciprocal (feedback) nature of the drugs/crime connection is explained by Menard, Mihalic, and Huizinga (2001): "Initiation of substance abuse is preceded by initiation of crime for most individuals (and therefore cannot be a cause of crime). At a later stage of involvement, however, serious illicit drug use appears to contribute to continuity in serious crime, and serious crime contributes to continuity in serious illicit drug use" (p. 295). The research indicating a genetic vulnerability to alcoholism or drug addiction helps to explain why the many millions who drink or experiment with drugs do not descend into the hell of addiction and why others are "sitting ducks" for it (Robinson & Berridge, 2003).

▧ Therapeutic Communities

A number of articles in this section address the benefits of therapeutic communities (TCs) for correctional clients with substance abuse problems and antisocial character problems. TCs are residential communities offering long-term (typically 6 to 12 months) opportunities for attitude and behavioral change and to learn constructive prosocial ways of coping with life's strains and with others (Litt & Mallon, 2003). These communities provide dynamic "mutual self-help" environments in which residents transmit and reinforce one another's acceptance of and conformity with the highly structured and stringent expectations of the TC and of the wider community. Life in a TC is extremely hard on people who have never experienced any sort of disciplined expectations from others, and, as a consequence, there are many dropouts.

The article by Mary Stohr and her colleagues in this section, however, indicates that inmates residing in residential substance abuse treatment (RSAT) communities within

| Table 5.1 | Male and Female Adult Arrestees Testing Positive for Various Drugs |

	Males					
City	**Any of 5 Drugs***	**Multiple Drugs****	**Cocaine**	**Heroin**	**Methamphetamine**	**Marijuana**
Atlanta	72.4	73.5	49.8	3.0	2.0	41.8
Chicago	86.0	86.0	50.6	24.9	1.4	53.2
Dallas	62.3	63.8	32.5	6.9	5.8	39.1
Houston	61.7	61.9	22.6	5.7	2.1	47.5
Los Angeles	68.6	68.9	23.5	2.0	28.7	47.5
New York	67.7	72.7	35.7	15.0	0.0	43.1
Philadelphia	67.0	68.8	30.3	11.5	0.6	45.8
Phoenix	74.1	76.8	23.4	4.4	38.3	45.8
San Diego	66.8	71.2	10.3	5.1	36.2	41.6
Washington	65.6	65.8	26.5	9.8	0.7	37.4

	Females*					
City	**Any of 5 Drugs***	**Multiple Drugs****	**Cocaine**	**Heroin**	**Methamphetamine**	**Marijuana**
Albany, NY	60.9	65.2	34.8	4.3	0.0	34.8
Chicago	61.1	66.7	33.3	22.2	0.0	38.9
Denver	69.1	24.9	52.5	6.1	5.0	34.3
Honolulu	74.5	27.7	8.5	6.4	57.4	29.8
Los Angeles	59.3	63.0	25.9	2.1	18.5	36.7
New York	67.7	72.7	35.7	15.0	0.0	43.1
New Orleans	58.8	17.8	37.3	13.3	0.8	30.3
Phoenix	74.6	78.5	16.8	6.1	41.6	31.6
San Diego	69.1	72.6	15.2	8.7	47.1	29.1
Washington	61.1	66.7	30.9	10.3	0.0	29.1

* The five drugs are cocaine, marijuana, methamphetamine, opiates, and phencyclidine (PCP)

** Multiple drugs are any of nine drugs which include the basic five plus barbiturates, methadone, benzodiazepines, and propoxyphene.

*** Atlanta, Dallas, Houston, and Philadelphia did not sample female arrestees; Albany, Denver, Honolulu, and New Orleans were substituted.

▲ **Photo 5.2** Inmates at the Metro-Davidson County Detention Facility participate in the LifeLine substance abuse program. Prisoners pass a stuffed animal around, each giving it to someone who helped him.

prison walls were positive about most aspects of the experience. These people are separated from the negativity and violence of the rest of the prison and provided with extensive cognitive behavioral counseling and attend Alcoholics Anonymous (AA) and Narcotics Anonymous (NA) meetings. It is interesting to note that these inmates listed cognitive self-change programs as the most positive aspect of their treatment.

An interesting program implemented in a prison setting and transitioned into the community is the Delaware Multistage Program (Mathias, 1995). In the beginning stage, offenders spend 12 months in a prison-based TC called Key; in phase 2, they spend 6 months in a pre-release TC called Crest; and, finally, in phase 3, they receive an additional 6 months counseling while on parole or in work release. Figure 5.1 compares drug use and arrest outcomes for offenders completing all phases (Key-Crest), Crest, Key, and a comparison group 18 months after release from prison. We see that 76% of Key-Crest members remained drug free and 71% remained arrest free compared with only 19% and 30%, respectively, of the control group.

Inciardi, Martin, and Butzin's article in this section provides one of the best short overviews of what a successful TC should be like. The most noteworthy finding of this study is that 48% of drug abusers who went through a residential treatment program and who received additional treatment upon release remained drug free after five years post-treatment. Of course, this means that 52% did not, but this has to be contrasted with their control group of similarly situated addicts who received no treatment. Only 23% of the control group was drug free after five years. It does show, however, how extremely difficult it is to battle addiction even after a long period of forced abstinence and extensive treatment.

Figure 5.1	Delaware Multistage Correctional Treatment Program: 18 Months After Release From Prison

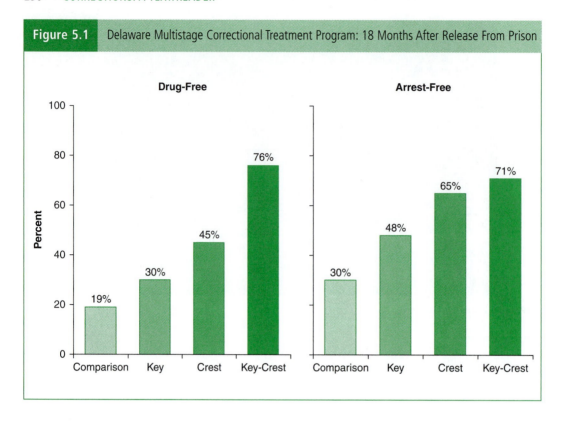

Pharmacological Treatment

This observation brings up the issue of pharmacological treatments for alcoholism and drug addiction. Addiction is a "prototypical psychobiological illness, with critical biological, behavioral, and social context elements" (Leshner, 1998, p. 5). As basically a brain chemistry problem, pharmacological treatment with drug antagonists (drugs that work by blocking the effects of other drugs) stabilizes brain chemistry and renders addicts more receptive to psychosocial counseling. Proponents of pharmacological treatment emphasize that it is no more a magic bullet than insulin is to diabetics and that it augments, not replaces, traditional treatment methods. Nevertheless, they claim that the effects of such treatment are more effective and immediate, and they wonder why (with some exceptions) the correctional system is relatively uninterested in pharmacological treatment (Kleber, 2003).

There are many drug antagonists, but we will briefly discuss only one that has claimed success in curbing both alcohol and drug addiction. This drug is naltrexone, which is sold under the trade name ReVia©. Naltrexone reduces craving (psychological craving is much tougher to beat than physical need, which is relatively quickly overcome) among alcohol and drug abstinent addicts and reduces the pleasurable effects of those who continue to use (Schmitz et al., 2004). A study of drug addicts on federal probation found that about one-third of probationers who received naltrexone plus counseling relapsed as opposed to two-thirds of those who only received counseling (Kleber, 2003). It also has been claimed that the drug reduces craving and relapse among alcoholics by 50% (DuPont, 1997).

⬛ Sex Offenders and Their Treatment

The American public harbors all sorts of very negative images of sex offenders. We lock them up under civil commitment orders after they have completed their prison terms, and all 50 states have sex offender registration laws (Talbot, et al., 2002). However, the term *sex offender* defines a very broad category of offenders ranging from relatively minor "flashers" to true sexual predators, just as the term *property offender* includes everyone from petty shoplifters to career burglars. At least 98% of all sex offenders are either in the community on probation or parole or will be some day, making the issue of sex offender treatment of the utmost importance (Carter & Morris, 2002).

Although it is part of popular lore that sex offenders are untreatable and will never stop their offending, as a category of offenders they are actually less likely to re-offend than any other category. Looking at many years of British crime statistics, it was found that burglars are the most likely of all criminals to be reconvicted (76%) within 2 years of being released from prison, with sex offenders being the least likely (19%) (Mawby, 2001, p. 182). Large-scale reviews of U.S. studies conducted by the Center for Sex Offender Management (Bynum, et al., 2006) examining recidivism among different types of sex offenders found the following recidivism rate ranges:

- Child molesters with male victims (13% to 40%)
- Child molesters with female victims (10% to 29%)
- Rapists (7% to 35%)
- Incest offenders (4% to 10%)

These rates are well below the recidivism rate for all other offender categories combined, estimated to be 67.5% by the third year after release from criminal justice supervision (Robinson, 2005, p. 222). Because recidivism rates include only those offenders who have been caught, in common with the rates of other types of offenders, the above ranges for sex offenders should be considered bare minimum figures.

State of the art treatment of sex offenders must include a thorough assessment, which includes psychosocial problem areas, deviant arousal patterns, and polygraph assessment (Marsh & Walsh, 1995). Deviant arousal patterns are assessed by a device called a penile plethysmograph (PPG), which measures blood flow in the penis (the level of tumescence) when a man is exposed to deviant sexual images. These measures are then compared with measures in response to normal (i.e., consensual adult sex) images.

Counselors are in complete agreement that effective treatment is impossible until the full extent of the offender's sex offending history is acknowledged by him or her and known to treatment personnel (Walsh, 2006). But sex offenders are notorious for hiding their sexual histories, so polygraph ("lie detector") assessment is needed to access their histories. Comparing self-reports pre- and post-polygraph testing across two decades of research, it has been found that child molesters underreport the number of sex crimes they have committed by about 500% and overreport their own childhood victimization (the "I'm a victim too" excuse) by about 250% (Hindman & Peters, 2001).

William Burdon and Catherine Gallagher examine how sex offenders are controlled through incapacitation in this section. They agree with numerous other researchers that an optimal treatment modality (following a thorough psychosocial and physiological assessment) combines the biomedical with cognitive-behavioral approaches. The bio-medical approach involves so-called "chemical castration" with a hormone called medroxyprogesterone (Depo-Provera), which is also sold as a method of female birth control. Depo-Provera works in males to reduce libido by drastically reducing the production of testosterone. Depo-Provera has been called a "limbic hypothalamic tran-quilizer" because the hypothalamus is really the controller of the sex drive, and "tran-quilizing" it "allows the offender to concentrate on his psychosocial problems without the distracting fantasies and urges accompanying androgen-driven limbic hypothalamic activity" (Marsh & Walsh, 1995, p. 87).

Following the State of California in 1997, several states now allow chemical castration ("castration" is reversible upon withdrawal from the drug), with some mandating it for repeat offenders. Not all sex offenders should be treated with this drug because there are sometimes negative side effects, and treatment can only be provided by a medical doctor. However, a number of reviews of the literature from Europe and America show that anti-androgen drugs such as Depo-Provera result in recidivism rates that are remarkably low (in the 2 to 3% range) when compared to offenders treated with only psychosocial meth-ods (reviewed in Maletzky & Field, 2003).

✉ The Responsivity Principle and Treatment

Psychosocial assessment of offenders of all types begins with the "holy trinity" of risk, needs, and responsivity principles (RNR). There is a thorough overview of RNR in the article by Andrews, Bonata, and Wormith in this section, so we will only briefly describe them here. The risk principle refers to an offender's probability of re-offending, and it suggests that those with the highest risk be targeted for the most intense treat-ment. The needs principle refers to an offender's needs, the lack of which puts him or her at risk for re-offending, and suggests that these needs receive high priority. Needless to say, in the great majority of cases, the higher the offender's level of needs, the higher his or her risk of re-offending. The responsivity principle avers that, if offenders are to respond to treatment in meaningful and lasting ways, counselors must be aware of offenders' different development stages, learning styles, and need to be treated with respect and dignity. Andrews, Bonata and Wormith's article identifies a number of major criminal risk and need factors that should be targeted for treat-ment. The RNR model fits squarely within the cognitive-behavioral approach to offender treatment.

In the final article in this section, Craig Dowden and Don Andrews find that the risks and needs of female offenders are basically the same as they are for male offenders and that the same principles of effective intervention reduce recidivism for females. Because the female inmate population rarely exceeds 6 or 7% of the adult inmate population, studies of the effects of treatment on female offenders have been neglected (Harrison & Beck, 2003). Dowden and Andrews's meta-analysis found that cognitive-behavioral counseling conducted in accordance with RNR principles significantly reduced recidivism.

▨ Mentally Ill Offenders

Mentally ill offenders under correctional supervision present a particularly difficult treatment problem. Alcoholics and drug addicts ingest substances that alter the functioning of their brains in ways that interfere with their ability to cope with everyday life, although their brains may be normal when not artificially befuddled. Mentally ill persons also have brains that limit their capacity to cope, but that limitation is intrinsic to their brains, not attributable to intoxicating substances. The World Health Organization defines mental disorders as "clinically significant conditions characterized by alterations in thinking, mood (emotions), or behaviour associated with personal distress and/or impaired functioning" . . . and adds that they "are not variations within the range of 'normal,' but are clearly abnormal or pathological phenomena" (in Brookman, 2005, p. 87).

About one in five inmates in U.S. prisons suffers from some kind of mental illness, which amounts to about three times the number of mentally ill individuals in prison than in mental hospitals in the United States (Cuellar, Snowden, & Ewing, 2007). This state of affairs results from the deinstitutionalization of all but the most seriously ill patients from mental hospitals that occurred in the 1960s. For instance, there were 559,000 persons in U.S. mental hospitals in 1955; in 2000 (with a U.S. population about 80% greater), there were only 70,000 (Gainsborough, 2002). Deinstitutionalization of the mentally ill from mental hospitals has shifted to their institutionalization in jails and prisons, which in essence has resulted in the criminalization of mental illness (Lurigio, 2000).

It has been found in studies around the world that mentally ill persons (mostly schizophrenics and manic depressives) are at least three to four times more likely to have a conviction for violent offenses than persons in general (Fisher, et al., 2006). Most mentally ill persons, however, are more likely to be victims than victimizers, and many of them make their problems worse by abusing alcohol and drugs (Walsh & Ellis, 2007). It is because of their drug taking and greater propensity for violence, in addition to mental hospital deinstitutionalization, that the mentally ill are overrepresented in the correctional system.

Mentally ill offenders in jails and prisons are often victimized by other inmates, who call them "bugs" and exploit them sexually and materially (stealing from them). Mentally ill offenders are also punished by corrections officers for behavior that, while not pleasant, is symptomatic of their illness. These behaviors include such things as excessive noise, refusing orders or medication, self-mutilation, and poor hygiene. Obviously, correctional facilities are not the ideal place for providing mental health treatment, even assuming that the staff is aware who the mentally ill are among their charges. Few officers (including probation and parole officers) have any training about mental health issues, and one nationwide survey of probation departments found that only 15% of them operated special treatment programs for the mentally ill (Lurigio, 2000). It is not that anyone expects correctional workers to become treatment providers because that's a job for psychologists and psychiatrists. However, they should be expected to recognize signs and symptoms of mental illness, to know how to deal effectively with situations involving mentally ill persons, to have a rudimentary understanding of the etiology and treatment regimens for the major mental illnesses, and to be aware of treatment facilities in the community to which they can refer their mentally ill clients.

⬚ Summary

- This chapter provides a number of contemporary articles on the subject of offender rehabilitation. Although the vast majority of the correctional budget is spent on security, and although the rehabilitation ideal has waned considerably, rehabilitation efforts have not completely ceased. When the layperson hears about the low rates of success of many of the contemporary rehabilitation programs, he or she tends to forget that these success rates are much higher than they are among similarly situated offenders who did not receive treatment.

- Most successful treatment programs today proceed by first conducting a thorough assessment of offenders' risks and needs and then addresses these issues using cognitive-behavioral techniques along with the principles of responsivity. Treatment is best accomplished for severe substance abusers in therapeutic communities, although even then there is a significant percentage of failure. Much of this failure has to do with the intense psychological craving for the substance of abuse, which is something that may be significantly alleviated by certain alcohol and drug antagonists, such as naltrexone.

- Similar observations were made about sex offenders who have difficulty refraining from acting out their sexual fantasies with inappropriate targets. Repeat sex offenders treated with anti-androgen medication combined with cognitive-behavioral counseling have very low recidivism rates compared with similar offenders treated only psychosexually.

- Female offenders respond to the same treatment modalities as men. That is, they show reduced recidivism when assessed and treated according to the principles of RNR and when cognitive-behavioral counseling methods are used.

- Mentally ill individuals are represented in the correctional system by a factor of at least three or four times their prevalence in the general population. Because of mental hospital deinstitutionalization, the criminal justice system has been the agent of social control for these people. The correctional system is not equipped to deal with mentally ill people, who are often victimized by other jail or prison inmates or disciplined by corrections officers for exhibiting behavior that is basically part of their mental disease syndromes.

KEY TERMS

Addiction

Cognitive self-change programming

Heritability

Medical model

Mental disorders

Needs principle

Rehabilitation

Responsivity principle

Risk principle

Sex offender

Therapeutic communities

DISCUSSION QUESTIONS

1. In your estimation, are the time, effort, and finances spent on rehabilitative efforts worth it given the low success rates? Would longer periods of incarceration better protect the public?

2. Cognitive-behavioral approaches stress thinking and rationality. How about emotions? Do you think that human behavior is motivated more by emotions than by rationality?

3. Given the greater genetic "loading" for Type II alcoholics, in what ways would you treat them differently than Type I alcoholics if you were a prison treatment person?

4. Should all sex offenders undergo Depo-Provera treatment? What are the ethical problems of such invasive treatment?

5. Discuss the various component parts of the responsivity principle.

INTERNET SITES

American Correctional Association: www.aca.org

American Jail Association: www.aja.org

American Probation and Parole Association: www.appa-net.org

Bureau of Justice Statistics (information available on all manner of criminal justice topics): www.ojp.usdoj.gov/bjs

Center for Sex Offender Management: www.csom.org

National Criminal Justice Reference Service: www.ncjrs.gov

National Institute of Mental Health: www.nimh.nih.gov

National Institute on Drug Abuse: www.drugabuse.gov

Office of Justice Programs, Bureau of Justice Statistics (periodic statistical reports on all manner of criminal justice topics, e.g., HIV in prisons and jails, probation and parole, and profiles of prisoners): www.ojp.usdoj.gov/bjs/periodic.htm

Pew Charitable Trust (Corrections and Safety): www.pewtrusts.org/our_work_category.aspx?id=72

Vera Institute (information available on a number of corrections and other justice related topics): www.vera.org

READING

In this article, Mary Stohr and her colleagues explore the perceptions of prison inmates in two different substance abuse programs. They note that drug and alcohol abuse by criminal offenders remains at an all-time high and that residential substance abuse and treatment (RSAT) programs were developed to address the drug and alcohol treatment needs of prison inmates. Typically, such programs range in length from 6 to 12 months, have an Alcoholics Anonymous and/or Narcotics Anonymous component, and occur in a therapeutic community environment. Some programs also include a cognitive self-change component. Inmate participation in their programming is crucial to the success of a therapeutic community treatment environment. In this research, the authors describe, compare, and contrast the perceptions of inmate clients of two RSAT programs in a northwest mountain state. They find that, overall, inmates in these programs have quite positive perceptions about them.

Comparing Inmate Perceptions of Two Residential Substance Abuse Treatment Programs

Mary K. Stohr, Craig Hemmens, Brian Shapiro, Brian Chambers, and Laura Kelley

Drug and alcohol abuse by criminal offenders remains at an all time high. According to Annual Survey of Jails data, in 1989, 67% of jail inmates had used drugs regularly or committed drug offenses prior to incarceration compared with 70% in 1998. It is not surprising that this drug use history mirrors that of prison inmates. Residential substance abuse and treatment (RSAT) programs were developed to address the drug and alcohol treatment needs of inmates in jails and prisons. Typically, such programs range in length from 6 to 12 months, have an Alcoholics Anonymous and/or Narcotics Anonymous component, and occur in a therapeutic community environment. Some programs, such as the two compared in this study, also include a cognitive self-change piece that is woven into the core of these programs.

In this research, we describe, compare, and contrast perceptions of inmate clients of two RSAT programs in a rural mountain state. Prison 1 accepts only parole violators with verifiable drug or alcohol abuse problems. Their RSAT participants usually have their parole date set after completion of the 9- to 12-month RSAT program. Program delivery at Prison 1 is conducted by a private contractor whose content and staffing is closely watched and audited by the state Department of Corrections. Prison 2 accepts regular "termers" with verifiable drug and alcohol abuse problems. These participants are usually near the completion of their sentence after they have graduated from the 9- to 12-month RSAT program. RSAT program delivery at Prison 2 is conducted by Department of Corrections personnel.

Source: Stohr, M.K., Hemmens, C., Shapiro, B., Chambers, B., & Kelly, L. (2002) Comparing inmate perceptions of two residential substance abuse treatment programs. *International Journal of Offender Therapy and Comparative Criminology*, 46(6), 699–714. Reprinted with permission of Sage Publications, Inc.

⬚ Therapeutic Communities and RSATs

Therapeutic communities have been used to treat a myriad of social and psychological disorders of clients, including substance abuse. They involve the use of a combination of intensive and secluded social and psychological therapy that is designed to alter clients' attitudes, beliefs, and behaviors. Group involvement and peer assessment is a major component of the therapeutic community process. For the program to be successful there must be engagement and searing honesty by all clients and the concomitant acceptance of responsibility. Along with the honesty in assessment of self and others, there is also an expectation of mutual assistance by all community members.

The literature on substance abuse programs and therapeutic communities indicates that successful programming can be implemented in the correctional environment. Studies of therapeutic communities with a substance abuse focus have indicated they hold some real promise for the habilation or rehabilitation of inmates.

As per the "what works" protocols emerging from the corrections treatment literature, this success is usually tempered by the strength of the treatment content, qualifications of delivery personnel, length of programming, and existence of and participation in aftercare. The most successful programs are those that combine the delivery of substantive knowledge in an environment that is suited to therapeutic change. Research also indicates that cognitive attributes, positive modeling, behavioral redirection, emotional therapy, treatment environments engendering trust and empathy, and intensive client involvement in problem solving in their own treatment are also key to attaining actual behavioral change upon release. Treatment programs directed at drug offenders also appear to achieve greater success in reducing recidivism when services were continued postrelease.

The three therapeutic communities described by Inciardi (1995) incorporated a number of the attributes that appear to be related to successful programming in correctional environments. These programs combined behavioral, cognitive, and emotional therapies as well as other techniques in treatment regimens designed to address individual needs of substance-abusing clientele. Significantly, after 6 months, clients who had participated in treatment were more likely to be drug and arrest free than was the comparison group. At 18 months, those who had completed the program were three times more likely to be drug free than were those who had not completed the program.

In the widely cited and influential review of correctional programming by Martinson (1974) and Lipton, Martinson, and Wilks (1975), it was asserted that not much has worked to reduce the recidivism rate of participants. More recent meta-analyses of correctional programming have also raised serious questions concerning the veracity of claims of success by correctional program proponents. Antonowicz and Ross (1997) reviewed the research on correctional programming published between 1970 and 1991 and found that many studies have inadequate control or comparison groups, do not report on sample size, use sample sizes that are too small to enable statistical tests, or fail to examine outcome. They determined that only 20 of 44 well-designed programs were effective. Those programs that achieved some success in theirs and other meta-analyses included a greater variety of programming options and techniques, targeted factors that were actually related to criminal involvement, and matched offender learning styles to complementary services.

It is with these caveats in mind that the two RSAT programs under study here were created. Obviously, the therapeutic community treatment programs at Prisons 1 and 2 were designed to achieve reduced recidivism of substance-abusing offenders and to collaterally decrease costs of crime and reincarceration for victims and taxpayers. The RSAT programs at Prisons 1 and 2 were also designed as therapeutic communities within the prison settings, and they employ cognitive self-change and behavioral strategies.

Whereas the efficacy of prison-based therapeutic communities has received much attention in the research literature, relatively little attention has been paid to the attitudes and perceptions of inmate clients. Lemieux (1998) examined the influence of a variety of inmate perceptions of their postrelease adjustment and found that social support was a significant factor. McCorkel, Harrison, and Inciardi (1998) compared the perceptions of program completers and dropouts and found that those who left the program viewed it more as punishment than as treatment.

In this article, we expand on this limited research on inmate client perceptions, focusing on the perceptions of the two RSAT programs among current Idaho RSAT inmates. Given that the success of therapeutic communities hinges on the meaningful involvement and decision making of their clients, the perceptions of program efficacy by those clients is a key ingredient of the program.

◤ RSAT Program Description

Social learning theories provide the framework for effective cognitive-behavioral approaches to treatment in these two prison programs. Cognitive self-change and behavioral strategies are used in such programs, as they provide inmates with both the ability to consider the thinking errors that lead to substance use and/or abuse and the means to choose alternate and less self-destructive paths. Key concepts of this treatment method include cognition and modeling.

The RSAT programs at Prisons 1 and 2 use group process, thinking reports, and journals to change cognition and behavior patterns. The first focus of the group process is to provide clients with the information to aid them in understanding the connection between thinking and behavior. The group then turns its attention to individuals' need to identify

thinking errors and proactive interventions. The group process is structured by the following five guidelines: (a) depersonalized staff authority that maintains control and adherence to rules, (b) allowing the individual offender the authority on issues related to how they think and how they should think, (c) a focus on the basic steps of cognitive change, (d) work to achieve cooperation between group members and staff, and (e) involvement of all group members in the process.

Although a cognitive approach to programming and recovery is used for treatment purposes, a behavioral change approach is employed at all times. This behavioral change model stresses individuals' accountability for their behaviors. Prosocial behaviors, which emphasize the mutual rights of others, are practiced and encouraged. Antisocial, oppositional-type behaviors, which emphasize the rights of the individual at the expense of the other "family" members, are actively discouraged.

In the therapeutic community, the rights of all family members are clearly stated and the rules are designed so that each member is supposed to be treated equally. This type of cooperative living, in which the values of responsible concern for others and their recovery are confirmed, modeled, and rewarded, serves as a behavioral model for clients to learn.

Staff members are expected to act as role models for this type of prosocial, cooperative living. This is essential because many of those in the RSAT programs have not been exposed to the cultural traditions that most prosocial individuals have experienced. Because of this lack in their socialization history, it appears that some RSAT inmates must undergo resocialization, whereas others need initial socialization or orientation to living cooperatively with others.

The Minnesota Model of Chemical Dependency (a 12-step program) is a key component of the RSAT regimen. This program includes the use of a group process with recovering alcoholics and/or addicts as counselors. Individual counseling with and facilitation by professional staff, lectures, group reading, life history work, Alcoholics Anonymous and/or Narcotics Anonymous attendance, 12-step work, and recreational and physical activity complete the program content.

Clients' involvement in their treatment is regarded as a prerequisite for successful rehabilitation programming. In the RSAT programs at Prisons 1 and 2, RSAT inmates are intimately engaged in decisions regarding their own (and each other's) treatment programming because they are involved in the selection of their own leaders or coordinators, problem solving related to their own high-risk behaviors, and maintenance of community and programmatic integrity through the use of "push-ups," "pull ups," and "haircuts," exercises designed to encourage or discourage certain behaviors by group members.

▧ Methods

The researchers created the inmate questionnaire after a review of the literature on therapeutic communities, cognitive self-change programs, and drug and alcohol treatment programs in prisons and jails. They then reworked the questions after repeated observations, over a period of months, of the RSAT program operation in Prison 1 (Stohr et al., 2002). Treatment personnel also analyzed the instrument items for face validity and offered numerous suggestions for revision. Most of those suggestions for revision were adopted. The inmate questionnaire was initially administered as part of a larger process evaluation of the RSAT program at Prison 1.

In devising the instrument, the researchers were particularly interested in how the inmate participants perceived the content of the various components of the program and the delivery of that content. They were also curious about how the inmate coordinators and staff treatment personnel were viewed, whether inmates thought the tools of a true therapeutic community were present and operating successfully, whether communication lines were open and positive, and what the perception

was of the quality of services delivered and the likely effect of those services on participants.

A 51-item Likert-type scale instrument was created to measure these perceptions of inmates of the RSAT program. Inmates were also asked to provide some demographic information, queried regarding their substance use and abuse, and given the opportunity to provide written comments about the strengths and weaknesses of the program.

▧ Findings and Analysis

The RSAT inmates at the two prisons were overwhelmingly White (86% at Prison 1 and 84% at Prison 2). This finding is reflective of the state population, which has an 89% White population. These RSAT inmates ranged in age from 19 to 50, with a mean age of 31 to 34 at Prisons 1 and 2. Almost half (45%) of Prison 1 inmates are between the ages of 20 and 29, whereas only about 20% of Prison 2 inmates are that young. In fact, 48.6% of the Prison 2 inmates are at least 35 years old. In accordance with national-level data on inmates, the RSAT inmate population is relatively undereducated. Approximately three quarters (76.2%) of the Prison 1 inmates and more than half (56.7%) of the Prison 2 inmates have no more than high school diplomas or general equivalency diplomas. Notably, none of the inmates in either program have 4-year college degrees. All in all, the Prison 2 inmates tend to be older and slightly more educated than do the Prison 1 inmates.

The RSAT program is intended to last 9 to 12 months and consists of three distinct phases. Each phase is designed to last approximately 3 to 4 months. Failure to successfully complete a phase leads to termination from the program or repetition of that phase. Regarding length of time in the program, 4 months is the mean for inmate respondents in the programs at Prisons 1 and 2. An overwhelming majority of these respondents in Prison 1 (83%) and Prison 2 (89%) are in either Phase 1 or Phase 2 of the RSAT program.

Many of the RSAT inmates in these prisons indicated they used both alcohol and drugs regularly. Frequent drug use was a particularly severe addiction for these men, requiring a fix a day for 72% of respondents. Only one inmate at each institution claimed to have never used drugs. These data suggest that drugs are the substance of choice among these populations, but most of these respondents also used alcohol. Of the respondents, 42% admitted to consuming at least one alcoholic drink per day, and 93% were sometimes or always high when they committed crimes, which would indicate that these RSAT programs were appropriately targeting the more serious substance abusers.

Our ANOVA analysis indicated that most of the respondents were quite positive in their assessments of the RSAT program operation at both prisons. Reported in Table 1 are the findings from the ANOVA analysis in which there were statistically significant differences discovered between categories.

We found that perceptions of program components varied by inmate status in the program and by their drug or alcohol usage. For instance, those in Prison 2 were more positive about their perceptions of the program content and delivery if they were at the beginning or the end of their program and if their drug use was relatively light or very heavy. Much like Prison 2, those respondents in Prison 1 who were in the second phase of their treatment had less positive perceptions of the program content and delivery.

Those who were heavy drinkers in Prison 2 were more likely to rate the treatment and involvement of program leaders higher, and the heavy drug users in Prison 2 were also quite positive about program content and delivery. Again, those inmates in the second phase of their treatment in Prison 2 were least positive about the therapeutic atmosphere of the program.

The pattern we see in these data is that people at the end of their treatment program appear to have cycled through a period in which their enthusiasm has lagged and then enjoyed a resurgence of positive perceptions of these programs. Moreover, at least for Prison 2, this RSAT program appears to be most positively perceived on a couple of subscales by those who need it most: heavy drinkers and, to some extent, heavy users of illegal drugs. This is an encouraging finding.

We next conducted a regression analysis using the findings from the ANOVA analysis as a guide. We entered the variables alcohol use, drug use, and months in RSAT (as a proxy for phases) into five regression equations, the entire 51-item scale and the four subscales. We performed a stepwise backward elimination on each equation. The results are displayed in Table 2. What was most notable about these regression findings was the dearth of differences discovered between inmates in the two programs in their regard for these RSAT programs. When effects for the independent variables were discovered, they were few and related to specific subscales. None of the independent variables was particularly discriminating for either prison when all the items are considered (Model 1).

The findings for Prison 1 indicate that those in the program longest have a negative view of the content and delivery of the program (Model 2) and of the treatment leader and involvement issues (Model 3). Conversely, those with the most serious pattern of alcohol abuse in the Prison 2 program are more likely to view the treatment leader and involvement issues (Model 3) more positively. Much like Prison 1 long-term program participants, Prison 2 inmates who have been with the program the longest are also least positive about the program on the Treatment and Involvement Issues subscale (Model 3). In other words, those in these programs longest are more likely to differ qualitatively about issues centering on the substance and modes of delivery of these programs. Inmate participants in both Prison 1 and Prison 2 were in basic agreement regarding their overall perceptions (Model 1), their perceptions of the

Table 1	Analysis of Variance (One-Way Classification) of Participants' Responses to Survey Instrument for Prisons 1 and 2				
	N	M	SD	F	P

Subscale 1: Perceptions of program content and delivery (range 13–65)

Prison 2

RSAT phase

	N	M	SD	F	P
Phase 1	13	54.9[a]	3.95		
Phase 2	14	46.5[b]	7.27		
Phase 3	6	53.7	4.23	8.183	.001

Drug use categories

Rarely or never use	9	52.8[a]	6.24		
Moderate user	6	44.8[b]	8.23		
Heavy user	18	52.4	5.68	3.640	.038

Prison 1

RSAT phase

Phase 1	13	54.9[a]	3.95		
Phase 2	14	46.5[b]	7.27		
Phase 3	6	53.7	4.23	8.183	.001

Subscale 2: Perceptions of treatment leader and involvement issues (range 8–40)

Prison 2

Alcohol use categories

Rarely or never use	24	31.5	4.81		
Moderate drinker	7	31.9	5.08		
Heavy drinker	9	35.9[c]	2.52	3.317	.048

Subscale 3: Perceptions of the therapeutic atmosphere (range 16–80)

Prison 2

RSAT phase

Phase 1	14	59.3	6.87		
Phase 2	11	51.1[b]	13.5		
Phase 3	5	63.2	6.80	3.317	.052

NOTE: Only statistically significant findings are reported here.

a. Statistically significant differences between categories 1 and 2 at the .10 level.

b. Statistically significant differences between categories 2 and 3 at the .10 level.

c. Statistically significant differences between categories 1 and 3 at the .10 level.

Table 2	Regression on the Entire Instrument and the Four Subscales by Prison and Both Prisons					
	B	SE	Beta	t	Adjusted R^2	Significance
Prison 1						
Model 2						
Constant	57.03	6.53		8.73		.000
Alcohol	−.46	1.77	−.047	−.26		.797
Drugs	−.56	1.50	−.07	−.37		.715
Months	−.92	.49	−.34	−1.88	.081	.070
Model 3						
Constant	34.50	3.88		8.89		.000
Alcohol	1.60	1.02	−.24	1.58		.124
Drugs	−.71	.87	−.13	−.81		.421
Months	−.80	.30	−.42	−2.71	.150	.010
Prison 2						
Model 3						
Constant	32.18	3.75		8.59		.000
Alcohol	2.04	1.00	.31	2.03		.050
Drugs	−.37	.88	−.07	−.42		.675
Months	−.62	.29	−.32	−2.12	.060	.042
Both prisons						
Model 3						
Constant	36.21	2.82		12.82		.000
Alcohol	1.21	.72	.20	1.68		.097
Drugs	−.80	.70	−.14	−1.13		.261
Months	−.38	.21	−.22	−1.82	.060	.073
Model 5						
Constant 1	7.99	1.94		9.28		.000
Alcohol	.66	.47	.17	1.40		.168
Drugs	.61	.48	.16	1.26		.211
Months	−.23	.14	−.20	−1.67	.026	.099

NOTE: Model 1: All perceptions model (Items 12–62 by alcohol use, drug use, and months in program); Model 2: first subscale (Perceptions of Program Content and Delivery, Items 12–26, by alcohol use, drug use, months in program); Model 3: second subscale (Perceptions of Treatment Leader and Involvement Issues, Items 27–35, by alcohol use, drug use, months in program); Model 4: third subscale (Perceptions of the Therapeutic Atmosphere, Items 36–51, by alcohol use, drug use, months in program); Model 5: fourth subscale (Perceptions of Quality of Service and Communication Issues, Items 52–62, by alcohol use, drug use, months in program).

therapeutic atmosphere (Model 4), and their perceptions of the quality of service and communication (Model 5).

Of course, the finding on alcohol is actually a positive one for the Prison 2 program, again indicating that those who perhaps need the program most are also those who value this portion of it most. However, we would note that the number of inmates who indicated they were what we would characterize as heavy drinkers (respondents who indicated they either drank 3 to 5 drinks per day or got drunk daily) totaled only 11, or about 30% of the 37 respondents in Prison 1.

When we combined the data for both prisons and regressed the same independent variables on our subscales, we again found that months in the prison had a moderately negative effect on the perceptions of the treatment leader and involvement issues (Model 3). We also found that those inmates in these programs the longest were less positive regarding quality of service and communication issues (Model 5).

In addition to answering the Likert-type scale items, inmates were asked to identify the strengths and weaknesses of the RSAT program. Responses to these two items revealed a variety of complaints and positive comments. A content analysis of the responses revealed that the most commonly listed strengths for Prison 1 included Narcotics Anonymous and Alcoholics Anonymous meetings (15 responses), counselors (11 responses), feelings of fellowship among community members (9 responses), the support system (8 responses), and the therapeutic community atmosphere (8 responses). Similarly, Prison 2 respondents indicated the strengths of the program were, in descending order, the cognitive self-change program (15 responses), staff counselors (11 responses), the unity of the program (9 responses), the therapeutic community itself (8 responses), and the Alcoholics Anonymous and/or Narcotics Anonymous programs (8 responses).

The most commonly listed weaknesses identified by Prison 1 respondents included the presence of people who retaliated against others (8 responses), the prison location of the therapeutic communities (5 responses), petty requirements and rules (5 responses), and poor cognitive self-change (CSC) instructors (5 responses). The weaknesses identified by Prison 2 respondents included retaliatory people and/or booking slips (8 responses), prison environment and/or location (7 responses), shared compound with blue shirts (non-RSAT; 5 responses), petty requirements and/or rules (5 responses), and poor CSC instructors (5 responses).

As indicated by these data, some of the strengths listed by some inmate participants at one prison are reported as weaknesses by other inmates at the same facility. This ambivalence was somewhat reflected in the ANOVA and regression analysis. Generally, respondents were positive in their assessment of the programs, but that support varied by program component (the subscales), by status in the program (phases and months in), by individual characteristics of inmates (alcohol use and drug use), and by RSAT program (Prison 1 or 2).

⬛ Conclusions

What we might reasonably glean from these findings is that inmate participants in these two RSAT programs were generally quite positive in their perceptions of most of the program components. The ANOVA and regression analysis highlighted the few discernible differences in inmate perceptions—that those who were in the program longest were less positive about a few aspects of it, whereas those who were heavy alcohol users were more positive about other aspects. We did not find that level of inmate education had any effect on perceptions when controlling for other variables in the regression equation.

Our side-by-side comparisons of these prisons and our regression equations including both sets of respondents would indicate that inmates in each are remarkably similar in their perceptions of their respective programs. This is true despite the fact that Prison 1 is an RSAT program geared toward parole violators and delivered by a private contractor and Prison 2 is an RSAT program created for inmates nearing the end of their prison sentence and delivered by public employees. What we might conclude from this analysis of perceptions of inmates is that the public or private delivery system is not as important to the viability of these RSAT programs for inmates as is the substance and delivery of the program itself.

The comments made by inmates would further substantiate the point that if the staff are trained and experienced, the facilities are adequate, the program is operated as a true therapeutic community, and the cognitive self-change component is developed and practiced, the RSAT program will be perceived positively by inmate participants regardless of the consideration as to whether it is delivered by the private or public sector. Of course, perceptions of the staff members who deliver these programs, particularly when private sector entities pay less and provide fewer benefits and opportunities for training, are another matter that merits study but that cannot be measured here.

We would note that the importance of measuring inmate perceptions of programming is sometimes discounted or ignored by those who work in and study corrections. We would argue that researchers cannot effectively evaluate the operation of a treatment program, particularly a therapeutic community-based program, without actively and independently soliciting the input of the inmate participants.

Ironically, it has become standard practice in the business and public sector to allow for the input of training or programming participants. This has become accepted practice because it is widely recognized that program participants are best able to effectively evaluate the important aspects of training or programming and its application to their circumstances. Yet, this widely accepted practice in the wider world is ignored or even regarded as naïve by some that deliver programming in the correctional setting. Their justification is that inmates will lie or mislead or may not understand the relevance of a portion of the programming. We think these possible problems in garnering inmate input can in part be overcome when the input is structured with specific questions about program components that are anchored in theory and practice. We also think that, because much programming delivered in correctional settings cannot accurately be described as wholly voluntary, it is vitally important that input from inmates be collected anonymously and by an independent entity that is not obliged to or connected with the program under review.

In sum, to not collect and analyze the input of inmate participants in programming is to ignore an important source of information regarding a program and its operation. Preventing such input also abrogates a central tenet of therapeutic communities, which is that inmates must wholeheartedly participate in their own programming to habilitate.

References

Antonowicz, D. H., & Ross, R. R. (1997). Essential components of successful rehabilitation programs for offenders. In J. W. Marquart & J. R. Sorensen (Eds.), *Correctional contexts: Contemporary and classical readings* (pp. 291–310). Los Angeles: Roxbury.

Inciardi, J. A. (1995). The therapeutic community: An effective model for corrections-based drug abuse treatment. In K. C. Haas & G. P. Alpert (Eds.), *The dilemmas of corrections: Contemporary readings* (3rd ed., pp. 406–417). Prospect Heights, IL: Waveland.

Lemieux, C. M. (1998). Determinants of expectation of treatment efficacy among incarcerated substance abusers. *International Journal of Offender Therapy and Comparative Criminology, 42,* 233–245.

Lipton, D., Martinson, R., & Wilks, J. (1975). *The effectiveness of correctional treatment: A survey of treatment evaluation studies.* New York: Praeger.

Martinson, R. (1974). What works? Questions and answers about prison reform. *Public Interest, 35,* 22–54.

McCorkel, J., Harrison, L. D., & Inciardi, J. A: (1998). How treatment is constructed among graduates and dropouts in a prison therapeutic community for women. *Journal of Offender Rehabilitation, 27,* 37–59.

Stohr, M. K., Hemmens, C., Kjaer, K., Gornik, M., Dayley, J., Noon, C., et al. (2002). Inmate perceptions of residential substance abuse treatment programming. *The Journal of Offender Rehabilitation, 35,* 1–30.

DISCUSSION QUESTIONS

1. Explain in your own words the ideas and workings of an RSAT program.

2. What are the main findings of this study?

3. Do you think that the voluntary nature of participation makes a difference to success rates; what difference would it make if inmates were made to participate?

READING

In this article, the authors concern themselves with the effectiveness of cognitive and behavioral therapy on recidivism rates among offenders. They look at a number of studies of treatment or intervention programs in prison, jail, probation, or parole settings reported from 1968 through 1996 via meta-analyses of 69 primary research studies on the effectiveness of behavioral and cognitive-behavioral treatment in reducing recidivism for offenders. They report that results on this diverse collection of studies show that this type of treatment is associated with reduced recidivism rates. However, they found that the effect is mainly due to cognitive-behavioral interventions rather than to standard behavior modification approaches. The specific types of programs shown to be effective include cognitive-behavioral social skills development programs and cognitive skills (reasoning and rehabilitation) programs.

The Effects of Behavioral/Cognitive-Behavioral
Programs on Recidivism

Frank S. Pearson, Douglas S. Lipton,
Charles M. Cleland, and Dorline S. Yee

In 1994, the National Institute on Drug Abuse (NIDA) funded the Correctional Drug Abuse Treatment Effectiveness (CDATE) project for 4 years to develop a database and conduct meta-analyses on correctional treatment evaluation studies completed between January 1, 1968

Source: Pearson, F. S., Lipton, D. S., Cleland, C. M., & Yee, D. S. (2002). The effects of behavioral/cognitive-behavioral programs on recidivism. *Crime and Delinquency, 48*(3), 438–452. Reprinted with permission of Sage Publications, Inc.

and December 31, 1996. Researchers have, over the past 20 years, begun to use meta-analysis to assess the research literature on corrections-based treatment programs.

⊠ The CDATE Project

In 1994, the NIDA funded the CDATE project for 4 years to develop a comprehensive information database of correctional treatment evaluation studies appearing in published and unpublished research reports between January 1, 1968 and December 31, 1996. CDATE coded over 2,176 research comparisons of experimental groups with comparison groups assessing the impact of the various interventions on several outcome measures, particularly drug abuse and recidivism. Of these, 69 were of programs in which the most important treatment was behavioral or cognitive behavioral.

⊠ Behavioral and Cognitive-Behavioral Therapy

We consider behavioral and cognitive-behavioral therapy as a general category comprising two subcategories of treatments: (a) behavior modification/behavior therapy and (b) cognitive-behavioral treatments. Behavior modification/behavior therapy focuses on arranging contingencies of positive reinforcement to develop and maintain appropriate patterns of behavior. In addition to standard behavior modification procedures, we included contingency contracting: A contract is written under which specific desirable behaviors by the client earn specific rewards. (In some contingency contracts there are also punishments specified for specific named undesirable behaviors by the client.) Also included in the behavior modification/behavior therapy subcategory were token economies (i.e., contingencies of reinforcements applied to groups of people—such as inmates in a prison dormitory) whereby specific desirable behaviors (e.g.,

cleaning a particular area) earn tokens that can be exchanged later for goods or privileges.

The other subcategory includes cognitive-behavioral treatments, treatments that include attention to cognitive and emotional processes that function between the stimuli received and the overt behaviors enacted. For example, social learning theory is broader than behaviorist reinforcement theory because it includes as variables cognitions, verbalizations, and social modeling to explain (and to change) behavior patterns.

As McGuire (1996) points out, "there is no single cognitive-behavioural method or theory. Work of this kind is best thought of as a 'family' or collection of methods rather than any single technique easily and clearly distinguished from others" (p. 7). McGuire includes in his discussion of cognitive-behavioral approaches: social skills training (which uses modeling, role-play practice, and feedback), social problem-solving training, rational-emotive therapy, the cognitive skills program (also known as the Reasoning and Rehabilitation program), and the relapse prevention model. He also mentions aggression replacement training for violent offenders as a program with cognitive-behavioral components.

Hypotheses

The hypotheses presented here are limited to treatments rated as the most important treatment modality in a program.

1. Behavioral/cognitive-behavioral programs (a broad category encompassing programs ranging from token economy programs up to cognitive-behavioral programs) are more effective than treatment-as-usual comparison group interventions at reducing recidivism.

   1.1 Behavioral reinforcement and incentive programs are more effective than treatment-as-usual comparison group interventions at reducing recidivism.[1]

1.2 Cognitive-behavioral programs are more effective than treatment-as-usual comparison group interventions at reducing recidivism.[2]

Measurement of Variables

Behavioral

- Standard behavior modification. (Arranging contingencies of positive reinforcement to develop and maintain appropriate patterns of behavior.)
- Contingency contracting. (System in which the offender signs a contract with the person supervising him or her in which specific behaviors by the offender are linked with particular punishments, e.g., a stricter curfew for a positive urine test, and other behaviors are linked with rewards, e.g., good time credits for satisfactory work performance.)
- Token economy. (A reinforcement system in which offenders/inmates who perform specific behaviors satisfactorily, e.g., cleaning their living area, helping other inmates, etc., are rewarded with tokens, which can later be exchanged for privileges, e.g., more time to watch television, or desired goods, e.g., snacks from the canteen.)

Cognitive Behavioral

These are counseling/training programs to develop one or more important cognitive skills that are important to avoid serious problems or deficits in society.

- Social skills development training (e.g., developing skills in communication, giving and receiving positive and negative feedback, assertiveness, conflict resolution, etc.).
- Problem solving skills training.
- Cognitive skills training such as program materials on cognitive skills and/or reasoning and rehabilitation.

- Thinking errors approach.
- Other social skills training.
- Social learning focused. (These approaches include as variables cognitions, verbalizations, and social modeling to explain—and to change—behavior patterns.)
- Cognitive behavioral. (Other counseling/training programs that teach self-reinforcement, self-instruction, self-rehearsal, role-taking, self-control, and problem solving.)
- Self-control training.
- Training in anger management. (Training in anger management or aggression management.)
- Relapse prevention. (Programs preparing the offender to deal with cravings, peer pressure, etc., to prevent relapse to the illicit behavior.)

In the analyses presented here, programs were considered to be of the relevant treatment type if that treatment was rated by CDATE coders as the most important treatment difference between the experimental and the comparison groups. Because a large percentage of studies in this field have features of the research methods used that seem likely to bias the results in favor of the experimental group, we include a rating of the overall quality of the research methods used in the study as a (potential) moderator variable. The dependent (outcome) variable in these studies was recidivism, consisting mainly of rearrest and/or reincarceration. When findings were adequately reported for both recidivism measures, the measure used in the CDATE meta-analysis was arrest because it is procedurally and temporally closer to the crime event.

▨ Results

As background for the meta-analyses, Figure 1 shows a correlation coefficient, r, effect size with recidivism as the outcome variable for all

69 studies (undifferentiated by specific treatment type). Each *r* is plotted in the middle of a vertical bar representing the 95% confidence interval for that *r*. The studies are sorted by the overall rating of the research methods used in the study (1 = *poor, barely acceptable, very low confidence, 16 studies;* 2 = *fair, a low level of confidence, 29 studies;* 3 = *good, a mid-level of confidence, 17 studies;* 4 = *excellent, a high level of confidence, 7 studies*) from left to right on the graph and, within that rating, by the favorability of the outcome. The study with the largest effect size (*r* = +.81) appears to be an outlier that does not seem to "belong" with this

collection of studies. This was a study with especially weak methods: It did not have a separate set of subjects to serve as a comparison group, so it used, for the experimental group only, arrest rates before and after the program.

There is an association between the research design used and the CDATE rating of the overall quality of the research methods: The poor studies typically used classical or quasi-experimental research designs and the good and excellent studies typically used true experimental designs. However, the preference in CDATE was not to use research design itself as the moderator variable because there are many flaws it

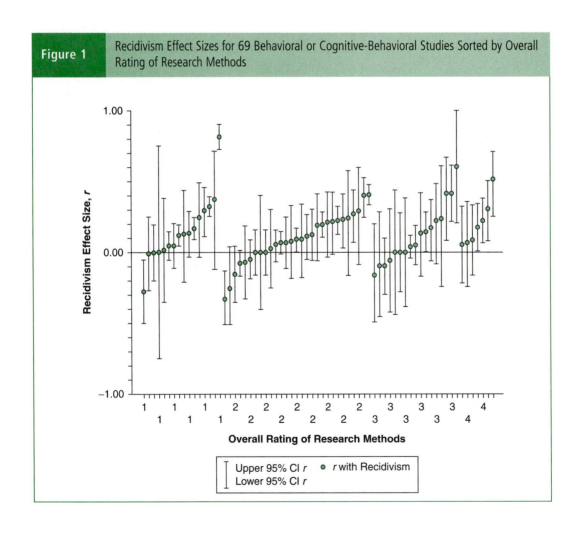

Figure 1 Recidivism Effect Sizes for 69 Behavioral or Cognitive-Behavioral Studies Sorted by Overall Rating of Research Methods

does not capture (e.g., preexisting differences between the experimental and comparison subjects that can persist in any research design, differential attrition of research subjects, biases in the statistical analyses conducted, etc.). The overall flaws (and strengths) of the research are more likely to be accurately reflected in the rated quality of research methods variable.

We considered an hypothesis to be verified when the inverse-variance-weighted mean effect size, r, was greater than or equal to .05 (with no clear evidence of research method artifact) and the t test resulted in a one-tailed probability less than .05. If r was less than .05, we considered the hypothesis to be disconfirmed. There is a conversion relationship, the Binomial Effect Size Display (BESD), which provides some indication of the practical importance of the effect size. A BESD relates a Pearson correlation coefficient, r, to a percentage differential between the experimental and comparison group, using 50% as a midpoint anchor. For

example, a correlation of $r = .05$ can be thought of as the experimental group being 5 percentage points better than the comparison group, using 50% as a midpoint anchor. Thus, the BESD would be 52.5% successes in the experimental group versus 47.5% successes in the comparison group. We set this as our minimal criterion of practical significance.

In our assessments there was also a possibility for results to fall in a borderline "gray area." One reason for this designation could be that, although the r was greater than or equal to .05, the t test did not result in a one-tailed probability less than .05. Another reason could be that there are indications of a research method artifact in which the poorer quality studies show a substantial effect whereas the better quality studies do not show any substantial effect.

Analyses relating to Hypothesis 1, the effectiveness of all behavioral/cognitive-behavioral programs combined, are presented in Table 1. The number of studies (here, $k = 68$) refers to

Table 1	Hypothesis 1: Recidivism by All Behavioral or Cognitive-Behavioral Programs			
Verified?	**Yes**			
Number of studies, k	68			
Total N, Windsorized	10,428			
Weighted mean of r	.118			
One-tailed probability	0.0000003			
Homogeneity	0			
Method rating beta	.012			
Method Rating	**k Studies**	**Mean r**	**Median r**	**SD**
Excellent	7	.207	.177	.165
Good	17	.119	.046	.209
Fair	29	.091	.092	.173
Poor	15	.110	.123	.164
Total	**68**	**.114**	**.092**	**.179**

NOTE: This test excludes the outlier study, Davidson and Robinson (1975).

the number of independent comparisons (one experimental group relative to one comparison group). The total N refers to the number of individual subjects (persons) in the experimental plus comparison groups of each study. The summary table also includes the mean inverse-variance-weighted Pearson correlation coefficient and the null-hypothesis exact probability associated with the t test used.

There are three conventionally used indicators of homogeneity: (a) at least 75% of the observed variance is accounted for by sampling error, (b) the "Q" chi-square test is not significant, and (c) the amount of residual variance is less than 25% of the estimated population effect size. The entry for homogeneity in the summary tables designates how many of these three criteria indicate homogeneity. As expected, the general category of behavioral and cognitive-behavioral studies was not homogeneous; none of the three criteria indicated homogeneity.

The "Method Rating beta" entry in the table gives the inverse-variance-weighted hierarchical linear regression beta coefficient for the overall rating of research methods. The beta coefficient of +.012 indicates that the effect sizes show a very slight (negligible) positive linear relationship with the quality of research methods used. That there is no research method artifact can also be seen in the lower section of Table 2, in which the unweighted correlation effect sizes are shown within each of the research method rating categories.

Table 2 summarizes the results of the hypothesis test for the subcategory of treatments comprising behavioral programs, including contingency contracting, token economy programs, and other standard behavior modification programs. Although the result was not quite statistically significant at the .05 level, the inverse-variance-weighted mean r of .066 does exceed our .05 criterion. The corresponding BESD is 53.3% success in the experimental groups and 46.7% success in the comparison groups. The research method rating moderator variable showed a slight

Table 2	Hypothesis 1.1: Recidivism by Behavioral Reinforcement/Incentive Programs			
Verified?	**Borderline**			
Number of studies, k	23			
Total N, Windsorized	1,935			
Weighted mean of r	.066			
One-tailed probability	.0686			
Homogeneity	0			
Method Rating beta	.042			
Method Rating	**k Studies**	**Mean r**	**Median r**	**SD**
Excellent	4	.171	.159	.117
Good	3	.181	.000	.366
Fair	11	.048	.074	.201
Poor	5	−.017	.000	.154
Total	**23**	**.073**	**.060**	**.206**

positive linear relationship with the effect sizes (beta = +.042), indicating that the better-quality studies found slightly larger effect sizes, on the average. In the detail method rating panel in the lower section of the same table, the studies rated as excellent and good have unweighted mean *r*s of .17 and .18. There are two weaknesses, however, that should be pointed out. First, post hoc exploration of the cases revealed that three of the four excellent entries are actually three independent comparisons from one overarching study. Second, the median unweighted *r* in the good category is .00. Therefore, we are unwilling to state that the effectiveness of behavioral reinforcement/incentives programs has been confirmed or that it has been disconfirmed, but, rather, we characterize it as being on the borderline of verified effectiveness.

As Table 3 summarizes, our meta-analyses showed the other subcategory (i.e., cognitive-behavioral treatments) to be effective in reducing recidivism. For the 44 studies the weighted

mean *r* was .144. The corresponding BESD is 57.2% successes in the experimental group versus 42.8% successes in the comparison group. The correlation coefficients are above .05 in both the good and excellent research method rating categories. However, as the summary table showed, these 44 studies are not statistically homogeneous, suggesting that some moderator variable or variables might further partition the studies into those with relatively stronger effect sizes and those with weaker effect sizes.

After seeing the results of the above three hypothesis tests, the following post hoc, exploratory analyses were conducted. First, because there was enough overlap in the cognitive-behavioral studies by age, this variable was explored as a possible moderator variable. However, the results did not differ by age for the cognitive-behavioral studies. Still restricting the analyses to independent comparisons, the mean *r* for juveniles (19 effect sizes) is .14 and that for adults (25 effect sizes) is also .14.

Table 3	Hypothesis 1.2: Recidivism by Cognitive-Behavioral Programs			
Verified?	**Yes**			
Number of studies, *k*	44			
Total *N*, Windsorized	8,435			
Weighted mean of *r*	.144			
One-tailed probability	0.0000002			
Homogeneity	0			
Method rating beta	−.0003			
Method Rating	***k* Studies**	**Mean *r***	**Median *r***	**SD**
Excellent	3	.254	.177	.234
Good	14	.106	.091	.179
Fair	17	.129	.093	.149
Poor	10	.173	.150	.133
Total	**44**	**.140**	**.127**	**.161**

Next, specific treatment types were explored. Within behavioral and incentives treatments, there were five or more studies in each of the three specific types of treatment (i.e., contingency contracting, token economies, and other standard behavior modification), but none of these specific behavioral treatments could be verified (post hoc) as effective in reducing recidivism.

Within the cognitive-behavioral subcategory, only three specific types of treatment included five or more studies: social skills development training, cognitive skills training, and studies coded as other cognitive-behavioral programs. Based on 14 primary research studies, social skills development programs do meet our criteria as verified effective in reducing recidivism. The inverse-variance-weighted mean r is .17 ($p = .0036$), which corresponds to a BESD of 58.5% success in the experimental group and 41.5% in the comparison group. The lone study rated as having excellent methods had an $r = .07$, and the seven studies rated as having good methods had a mean $r = .16$.

There are only seven research studies dealing with the cognitive skills programs developed by Ross and his colleagues (1988) (also known as Reasoning and Rehabilitation programs), but they still meet our criteria of verified effectiveness. The weighted mean r is .147, which corresponds to a BESD of 57.4% successes in the experimental group and 42.7% successes in the comparison group. The one study rated as having excellent methods had an $r = .52$, and the three studies rated as having good methods had a mean $r = .15$.

Last, there are 10 primary research studies of other programs CDATE researchers coded as specifically cognitive behavioral. Simply relying on our criteria for verification, these studies would be considered effective. The weighted mean r of .114 corresponds to 55.7% successes in the experimental and 44.3% successes in the comparison groups. The method rating beta for this set of studies is just about zero. However, there were no excellent quality studies and only one good study—and that study has a negative effect size ($r = -.16$). At this time, these other cognitive-behavioral programs might be considered to be in a "gray area" of verification.

Discussion

Although Garrett (1985) and Gottschalk et al. (1987) found that they could not reject the null hypothesis for behavioral/cognitive-behavioral programs and recidivism, these constituted the first wave of meta-analyses in this area that, necessarily, relied on only the 16 or so studies available to them. On the one hand, the CDATE meta-analyses drawing from a total of 69 independent comparisons are in line with the general findings of Andrews et al. (1990) in support of "behavioral/social learning/cognitive behavioral" treatment (as a general category). The CDATE findings confirm and expand Izzo and Ross's (1990) meta-analysis of cognitive-behavioral treatment (as a specific category of treatment), showing that cognitive-behavioral programs can reduce recidivism rates by significant amounts. This was found to be true for the overall collection of cognitive-behavioral studies and also for the subcategories social skills development training and cognitive skills training. On the other hand, the CDATE meta-analyses did not allow us to reject the null hypothesis for contingency contracting, token economies, and other standard behavior modification as effective in reducing recidivism (see following).

We present these results with confidence. However, it would be prudent to bear in mind two practical problems. First, some program developers who refer to their programs as "cognitive behavioral" may not be thinking of the elements used in the cognitive-behavioral models discussed in the studies cited here. (Indeed, the subcategory we referred to as other cognitive-behavioral programs was not convincingly verified because not enough good quality research has been conducted.) If programs do not incorporate the types of treatment found in the primary research studies examined here, they may not be as effective. Second, even when the models intended are the same as those discussed here, some program directors may find themselves unable to implement adequately the cognitive-behavioral model in their particular correctional program; for example, they may not be able to obtain the right kind of training in cognitive-behavioral methods for their treatment staff. Naturally, to the extent that implementation of cognitive-behavioral programming may fall short, the results may fall short as well.

These cautions lead to a call for more research on cognitive-behavioral programs to provide more specific information needed about the programming and its effects. The next wave of research should describe the details of the cognitive-behavioral programs provided to the clients, including the specifics of the treatment models and curricula being used, the training and credentials of the treatment staff, how frequent the treatment sessions are, information on supervision procedures to insure that the quality of the treatment provided is maintained, and the planned and actual time in the program for the clients. Research is also needed to investigate how effective cognitive-behavioral modules are in the context of other treatment modalities. How effective is cognitive-behavioral treatment in the context of a therapeutic community or in the context of a 12-Step program?

Before analyzing these studies, we expected not only the cognitive-behavioral programs but also the behavioral reinforcement/incentives programs to be verified effective. Based on the meta-analyses, we think that the behavioral reinforcement/incentives programs (i.e., standard behavior-modification programs, token economies, and contingency-contracting programs) should, at this time, still be viewed as neither confirmed nor disconfirmed. Some might find this surprising

because the behavioral reinforcement approach has been shown in many good laboratory studies to control subhuman and human behavior patterns. In retrospect, we think that programs that focus on contingencies of reinforcement will be verified effective in establishing and increasing desirable behaviors by the offenders, while the contingencies of reinforcement are operating (a straightforward principle of operant conditioning). However, the standard behavioral model includes the idea that, when the contingencies of reinforcement are no longer kept operating, the targeted desirable behaviors are likely to decrease in frequency ("extinguish") and old patterns of behavior reestablish themselves. Thus, if the necessary contingencies of reinforcement are not in effect after the program, the clients are likely to resume committing undesirable behaviors. (It is common in these studies for the time at risk for recidivism to extend from 6 months to 2 years after the contingencies of reinforcement program has ceased.) Several of the studies allude to this behavioral reality:

> The question remains whether concentrated training in specific behaviors can be transferred to the wide variety of situations that confront children after discharge. (Handler, 1974, p. 15)

> The immediate efficacy of its behavior-modification techniques is verified. However, follow-up results . . . indicate the failure of the program to produce desirable social outcomes for discharged youth. (Davidson & Wolfred, 1977, p. 296)

> Post-treatment persistence of appropriate social behavior does not follow naturally from effective control over institutional behavior. The meaningfulness of the predictive value of institutional adaptation is questionable. (Ross & McKay, 1976, p. 171)

The results showed difference [*sic*] during treatment favoring the Teaching-Family programs on rate of alleged criminal offenses. . . . In the post-treatment year, none of the differences between the groups was significant on any of the outcome measures. (Kirigin, Braukmann, Atwater, & Wolf, 1982, p. 1)

As noted in the presentation of findings above, we regard the behavioral reinforcement/incentive programs to be in a borderline area of verification. The three independent comparisons from one study (Davidson, Redner, & Amdur, 1990) all used experimental methods we rated as excellent. All three treatments consisted of setting up behavioral contracts for juvenile offenders and engaging in advocacy on behalf of those youth. In two of the conditions, undergraduates were trained in the techniques and implemented them. These two treatment conditions showed the highest effect sizes ($r = .23$ and $r = .31$). The third treatment condition involved teaching the juvenile's family to do behavioral contracting with, and advocacy for, the juvenile ($r = .06$). The only other contingency contracting study using excellent research methods used as its treatment setting up behavioral contracts between the youth and teachers and between the youth and parents ($r = .09$). In our opinion, most volunteers from outside the client's family, peer group, and school (or job) environments will probably greatly reduce or cease their involvement with the offender when the 6-week or 6-month program interval ends, so the effective contingencies of reinforcement will end at that time. Our opinion is that behavioral reinforcement programs will be verified effective if and only if they can develop and maintain strong contingencies of reinforcement in the natural environment of the clients, for example, maintained by parents (or spouses) and teachers (or employers).

A different focus—one that cognitive-behavioral programs tend to adopt—is to use behavioral learning techniques to change the general adaptive behaviors of the clients, that is, to have the clients return to their natural environment with new repertoires of skills, so they can obtain reinforcement in socially acceptable ways instead of illegal ways. This may be part of the reason why the cognitive-behavioral programs discussed above are effective.

Reducing recidivism has been notoriously difficult. It is a relief to know that some correctional programs can indeed work to reduce recidivism by significant amounts. The policy implication is that directors of rehabilitation programs should consider having cognitive-behavioral programming as a primary or secondary component of their treatment programming. Two examples of the kinds of programs that policy makers should review for possible adoption are the cognitive skills training program developed by Ross, Fabiano, Ewles, and colleagues (Kownacki, 1995; Ross, Fabiano, & Diemer-Ewles, 1988) and the aggression replacement training program developed by Goldstein and Glick (1994). It is up to the treatment policy makers and program directors to review whether the (purportedly) cognitive-behavioral program under consideration is evidence based, employing principles and procedures corresponding to those present in the research reviewed here, and to assess how suitable it is for their particular clients and program environment. The broader programming challenge now is to help promote the "technology transfer," so the effective program models (and related staff recruitment, training, and quality-control processes) diffuse throughout the correctional community and become well implemented. The research challenge is to expand and develop the existing body of research evidence, so the effective elements of the behavioral/cognitive-behavioral models can be specified, and then used to improve the program models still further.

◪ Notes

1. This category included contingency contracting and token economy programs but excluded aversive conditioning programs.

2. Here cognitive-behavioral programs are broadly defined to encompass a variety of approaches including social learning approaches. Such non-cognitive-behavioral programs as token economies are excluded, however.

✖ References

Andrews, D. A., Zinger, I., Hoge, R. D., Bonta, J., Gendreau, P., & Cullen, F. T. (1990). Does correctional treatment work? A clinically relevant and psychologically informed meta-analysis. *Criminology, 28*(3), 369–404.

Davidson, W. S., Redner, R., &. Amdur, R. (1990). *Alternative treatments for troubled youth: The case of diversion from the justice system.* New York: Plenum.

Davidson, W. S., & Wolfred, T. R. (1977). Evaluation of a community-based behavior modification program for prevention of delinquency: The failure of success. *Community Mental Health Journal, 13,* 296–306.

Garrett, C. J. (1985). Effects of residential treatment of adjudicated delinquents. A meta-analysis. *Journal of Research in Crime and Delinquency, 22,* 287–308.

Goldstein, A. P., & Glick, B. (1994). *The Prosocial Gang: Implementing aggression replacement training.* Thousand Oaks, CA: Sage.

Gottschalk, R., Davidson, W, S., II, Mayer, J. P., & Gensheimer, L. K. (1987). Behavioral approaches with juvenile offenders: A meta-analysis of long-term treatment efficacy. In E. K. Morris & C. J. Braukman (Eds.), *Behavioral approaches to crime and delinquency: A handbook of application, research and concepts* (pp. 399–420). New York: Plenum.

Handler, E. (1974). *The unspectacular results of three local residential treatment programs.* Urbana, IL: Jane Adams School of Social Work.

Izzo, R. L., &. Ross, R. R. (1990). Meta-analysis of rehabilitation programs for juvenile delinquents. *Criminal Justice and Behavior, 17,* 134–142.

Kirigin, K. A., Braukmann, C. J., Atwater, J. D., & Wolf, M. M. (1982). An evaluation of Teaching-Family (Achievement Place) group homes for juvenile offenders. *Journal of Applied Behavior Analysis, 15,* 1–16.

Kownacki, R. J. (1995). The effectiveness of a brief cognitive-behavioural program on the reduction of antisocial behaviour in high-risk adult probationers in a Texas community. In R. R. Ross & R. D. Ross (Eds.), *Thinking straight: The reasoning and rehabilitation program for delinquency prevention and offender rehabilitation* (pp. 249–257). Ottawa, ON: Air Training and Publications.

McGuire, J. (1996). *Cognitive behavioral approaches: An introductory course on theory and research.* Liverpool, UK: Department of Clinical Psychology, University of Liverpool.

Raudenbush, S. W. (1994). Random effects models. In H. Cooper & L. V. Hedges (Eds.), *The handbook of research synthesis* (pp. 301–321). New York: Russell Sage.

Ross, R. R., Fabiano, E. A., & Diemer-Ewles, C. (1988). Reasoning and rehabilitation. *International Journal of Offender Therapy and Comparative Criminology, 32,* 29–35.

Ross, R. R., & McKay, H. B. (1976). A study of institutional treatment programs. *International Journal of Offender Therapy and Comparative Criminology, 20,* 165–173.

DISCUSSION QUESTIONS

1. What is cognitive-behavioral therapy?

2. What were the primary conclusions of this study?

3. Explain why meta-analyses in general are more useful than single studies.

READING

The authors of this article note that, with growing numbers of drug-involved offenders, substance abuse treatment has become a critical part of corrections. A multistage therapeutic community implemented in the Delaware correctional system has as its centerpiece a residential treatment program during work release—the transition between prison and community. The authors' evaluation of this program followed 690 individuals. At 5 years, those who participated in the program were significantly more likely to be drug and arrest free. Furthermore, treatment graduates with or without aftercare had significantly greater probabilities of remaining both arrest free and drug free than did a no-treatment comparison group in regular work release. Dropouts were also significantly more likely to be drug free, although not significantly less likely to have a new arrest than those without treatment. The authors conclude that their data show that the implementation of such programs could bring about significant reductions in both drug use and drug-related crime.

Five-Year Outcomes of Therapeutic Community Treatment of Drug-Involved Offenders After Release From Prison

James A. Inciardi, Steven S. Martin, and Clifford A. Butzin

The linkages between drug abuse and crime have been well documented, and recent field-based research has provided a general understanding of various aspects of the drugs-crime connection. In extensive follow-up studies of addict careers in Baltimore, for example, researchers found high rates of criminality among heroin users during those periods when they were active users and markedly lower rates during times of nonuse. Studies conducted in New York City targeting the economics of the drug-crime relationship documented a clear correlation between the amount of drugs used and the amount of crime committed. Furthermore, Miami-based research demonstrated that the amount of crime committed by drug users is far greater than anyone had previously imagined, that drug-related crime can at times be exceedingly violent, and that the criminality of street-drug users is far beyond the control of law enforcement. The overall findings suggest that, although the use of heroin, cocaine, crack, and other illegal drugs does not necessarily initiate criminal careers, drug use does intensify and perpetuate criminal activity. That is, street drugs seem to lock users into patterns of criminality that are more acute and enduring than those of other offenders.

The presence of substance abusers in criminal justice settings has also been well documented. A concomitant of drug-related

SOURCE: Inciardi, J. A., Martin, S. S., & Butzin, C. A. (2004). Five-year outcomes of therapeutic community treatment of drug-involved offenders after release from prison. *Crime and Delinquency, 50*(1), 88–111. Reprinted with permission of Sage Publications, Inc.

criminality and the "war on drugs" of the 1980s and 1990s has been the increased numbers of drug-involved offenders coming to the attention of the criminal justice system. In fact, it has been reported that perhaps two thirds of those entering state and federal penitentiaries have histories of substance abuse. This suggests that criminal justice settings offer excellent opportunities for assessing the treatment needs of drug-involved offenders and for providing treatment services in an efficient and clinically sound manner. As a result, most prison systems in the United States have implemented substance abuse treatment programs as part of the overall correctional process. This process was greatly facilitated by the Office of Justice Programs' Residential Substance Abuse Treatment (RSAT) program, which provided significant funding for prison treatment to each state. The enabling legislation focused on therapeutic community treatment and highlighted the Delaware KEY/CREST program as models to emulate. Most of the new programs instituted have at least purported to be therapeutic communities. Although some programming is available in all states, however, programs are still unavailable in the majority of prison facilities; where they do exist, few have undergone rigorous, long-term evaluation.

For drug-involved criminal justice clients, it appears that those who remain in some type of treatment do better than those who drop out, are involuntarily discharged, or do not participate in treatment at all. Outcome research, however, in the few places where it has been attempted, has usually involved short follow-up time frames and has included only limited use of comparison groups, standardized measurement instruments, multivariate models, and appropriate control variables. Longitudinal outcome studies of recidivism or relapse are uncommon. As such, an appropriate evaluation of the effectiveness of treatment for drug-involved offenders should longitudinally examine outcomes with a large enough sample to allow multivariate analyses. This

article reports such an analysis of a multistage therapeutic community approach to the treatment of drug-involved offenders in correctional settings.

▧ Therapeutic Community Treatment

Numerous drug abuse clinicians and researchers have expressed the opinion that the *therapeutic community*, commonly referred to as the *TC*, is perhaps the most viable form of treatment for drug-involved offenders, particularly for those whose criminality has resulted in incarceration. Drug-involved offenders who come to the attention of state and federal prison systems are typically those with long arrest histories and patterns of chronic substance abuse, and the intensive nature of the TC regimen tends to be best suited for their long-term treatment needs. Moreover, the therapeutic community is especially efficacious in a correctional institution because the TC is a total treatment environment isolated from the rest of the prison population— separated from the drugs, violence, and other aspects of prison life that tend to militate against rehabilitation. The primary clinical staff members in such programs are typically former substance abusers who also underwent treatment in therapeutic communities. The treatment perspective in the TC is that drug abuse is a disorder of the whole person, that the problem is the *person* and not the drug, that addiction is a *symptom* and not the essence of the disorder, and that the primary goal is to change the negative patterns of behavior, thinking, and feeling that predispose drug use.

Research on community-based residential TCs has found them to be most effective for those who remain in treatment the longest. In fact, there is consensus throughout studies and modalities that the longer a client stays in treatment, the better the outcome in terms of declines in drug use and criminal behavior.

◩ A Multistage Therapeutic Community Treatment Continuum

Based on a wide body of literature in the fields of both treatment and corrections, combined with clinical and research experiences with correctional systems and populations, it would appear that the most effective strategy would involve three stages of therapeutic community treatment intervention. Each stage in this continuum is an adaptation to the client's changing correctional status: incarceration, work release, and parole (or another form of community supervision). This approach recognizes that "the connection between rehabilitation efforts in prison and the process of integration into society after release is probably one of the most feeble links in the criminal justice system" (Wexler & Williams, 1986, pp. 221–230).

The primary stage of treatment should consist of a prison-based therapeutic community. Segregated from the negativity of the prison culture, recovery from drug abuse, and the development of prosocial values in the prison TC involve essentially the same mechanisms seen in community-based TCs. Therapy in this stage should be an ongoing and evolving process over 12 months with the potential for the resident to remain slightly longer, if needed. Moreover, it is important that TC treatment for inmates begin *while they are still in the institution*. In a prison situation, time is one of the few resources that most inmates have in abundance. The competing demands of family, work, and neighborhood peer groups are absent. Thus, there is the time and opportunity for focused and comprehensive treatment, perhaps for the first time in a drug offender's career. In addition, there are other new opportunities presented: to interact with "recovering addict" role models, to acquire prosocial values and a positive work ethic, and to initiate a process of understanding the addiction cycle.

The secondary stage of treatment should be a *transitional* therapeutic community in a work-release setting. Since the 1970s, work release has become a widespread correctional practice for felony offenders. It is a form of partial incarceration whereby inmates who are approaching their release dates are permitted to work for pay in the free community but must spend their nonworking hours either in the institution or, more commonly, in a community-based work-release facility. Although graduated release of this sort carries the potential for easing an inmate's process of community reintegration, there is a negative side as well, especially for those whose drug involvement served as the gateway to prison in the first place. Inmates are exposed to groups and behaviors that can easily lead them back to substance abuse, criminal activities, and reincarceration. Since work-release populations mirror the institutional populations from which they came, there are still the negative values of the prison culture, but, in addition, street drugs and street norms abound. As such, the transitional work-release TC should be similar to that of the traditional therapeutic community. There should be the *family setting* removed from as many of the external negative influences of the street and inmate cultures as is possible. The clinical regimen in the work-release TC must, however, be modified to address the correctional mandate of work release. That is, in addition to intensive therapeutic community treatment, clients must prepare for and obtain employment in the free community.

In the tertiary stage (aftercare), clients will have completed work release and will be living in the community under the supervision of parole or some other supervisory program. For those individuals who entered work release after serving mandatory fixed sentences, there is no parole requirement and hence no community supervision. Treatment intervention in this stage should involve outpatient counseling and group therapy. Clients should be encouraged to return to the work-release TC for refresher/reinforcement sessions, to attend weekly groups, to call on their counselors on a regular basis, and to spend one day each month at the facility.

This multistage model has been operating in the Delaware correctional system since the mid-1990s. The treatment regimen is intensive, and the in-prison phase follows the traditional models of residential therapeutic community treatment. The treatment regimen in the work-release TC follows a 5-phase model over a 6-month period. Phase 1 is composed of entry, assessment, evaluation, and orientation, and it lasts approximately 2 weeks. New residents are introduced to the house rules and schedules by older residents. Each new resident is also assigned a primary counselor, who initiates an individual needs assessment. Participation in group therapy is limited during this initial phase so that new residents can become familiarized with the norms and procedures at the facility.

Phase 2 emphasizes involvement in the TC community, including such activities as morning meetings, group therapy, one-on-one interaction, confrontation of other residents who are not motivated toward recovery, and the nurturing of the newer people in the environment. During this phase, residents begin to address their own issues related to drug abuse and criminal activity in both group sessions and during one-on-one interactions. As well, they begin to take responsibility for their own behaviors by being held accountable for their attitudes and actions in group settings and in informal interactions with residents and staff. Residents are assigned job functions aimed at assuming responsibility and learning acceptable work habits, and they continue to meet with their primary counselors for individual sessions. The primary emphasis in Phase 2 is, however, on becoming an active community member through participating in group therapy and fulfilling job responsibilities necessary to facility operations. This phase lasts approximately 8 weeks.

Phase 3 continues the elements of Phase 2 and stresses role modeling and overseeing the working of the community on a daily basis (with the support and supervision of the clinical staff). During this phase, residents are expected to assume responsibility for themselves and to hold themselves accountable for their attitudes and behaviors. Frequently, residents in this phase will confront themselves in group settings. They assume additional job responsibilities by moving into supervisory positions, thus enabling them to serve as positive role models for newer residents. They continue to have individual counseling sessions, and, in group sessions, they are expected to help facilitate the group process. Phase 3 lasts for approximately 5 weeks.

Phase 4 initiates preparation for gainful employment, including mock interviews, seminars on job seeking, making the best appearance when seeing a potential employer, developing relationships with community agencies, and looking for ways to further educational or vocational abilities. This phase focuses on preparing for reentry to the community and lasts approximately 2 weeks. Residents continue to participate in group and individual therapy and to be responsible for their jobs in the treatment facility. Additional seminars and group sessions are, however, introduced to address the issues related to finding and maintaining employment and housing as well as returning to the community environment.

Phase 5 involves *reentry*, that is, becoming gainfully employed in the outside community while continuing to live in the work-release facility and serving as a role model for those at earlier stages of treatment. This phase focuses on balancing work and treatment. As such, both becoming employed and maintaining a job are integral aspects of the TC work-release program. During this phase, residents continue to participate in house activities such as seminars and social events. They also take part in group sessions addressing issues of employment and continue treatment after leaving the TC work-release facility. In addition, residents begin to prepare to leave the facility. They open a bank account and begin to budget for housing, food, and utilities. At the end of approximately

7 weeks, which represents a total of 26 weeks at the work-release TC, residents have completed their work-release commitment and are free to live and work in the community as program graduates.

A comprehensive research program has been established to examine the effectiveness of various components and combinations of the model. The basic hypothesis of this article is that drug-involved offenders receiving transitional treatment in a work-release therapeutic community, followed by aftercare, will have significantly lower rates of relapse and recidivism in both the short and long term than those receiving little or no treatment. Earlier studies have examined the effectiveness of the Delaware program 18 months and 42 months after release. Studies examining the 18-month follow-up data found significant effects for those who received any transitional treatment in work release, even if they did not complete the treatment. Subsequent analyses looking at 42-month follow-up data indicate, however, that significant and substantial outcome effects (reduced relapse and recidivism) really occur for those who complete the transitional treatment and particularly for those who undertake aftercare. A notable finding from both the 18- and 42-month follow-up studies is a lack of substantial long-term effects for in-prison treatment alone. This calls into question reliance on only prison treatment for criminal justice offenders. The Office of Justice Programs' funding of the RSAT initiative forced most states to focus their new treatment initiatives on prison treatment, yet it appears that long-term effects are most apparent when residential treatment is followed by aftercare. The following analyses take the consideration forward in time by examining the 5-year outcomes of the same cohort of correctional clients.

⬚ Methods

In the Delaware correctional system, those reaching eligibility for work-release status are classified based on criminal history and correctional counselor interviews. As such, work-release TC program assignments are made by treatment and correctional staff. Those classified as approved for work release with a recommendation for drug treatment between 1991 and 1997 comprise the present sample ($N = 1,077$). Because the number of those so classified exceeded the capacity of the treatment programs during that period, however, those eligible were assigned to either treatment or to regular work release—depending on the availability of a treatment opening at the time of assignment. As such, a *no treatment* group was available for comparison. An exception to this general process was that priority for entering the program was given to graduates of treatment programs within the prison and to those with direct judicial sentencing orders that required treatment participation as a condition for release. Additional comparisons of treatment graduates with and without aftercare were possible because the aftercare component was not operational until 1996, whereas the other stages of treatment had been implemented several years earlier. Once aftercare was fully established, all graduates were expected to participate.

The research protocol includes baseline and multiple follow-up interviews with all treatment and comparison clients as well as HIV and urine testing at each contact. The baseline interview is administered in prison prior to an inmate's transfer to work release. The first follow-up interview occurs 6 months hence, corresponding with graduation from the work-release TC (for the treatment groups) or completion of regular work release (for the comparison group). Subsequent interviews have been conducted 18, 42, and 60 months after baseline. Treatment dropouts are also followed. Interviews at baseline and each subsequent follow-up interview are lengthy, representing 700 variables per administration, including data on basic demographics, living situations, criminal history, drug-use history, treatment history, sexual behavior and attitudes,

HIV risks, self-esteem, sensation-seeking, and physical and mental health. Previous use of a series of illegal drugs was measured on a Likert-type scale ranging from 0 (*no use*) to 6 (*use more than once a day*) in the 6 months prior to incarceration. The data collection instruments include much of the Addiction Severity Index and the Risk Behavior Assessment developed by the National Institute on Drug Abuse. It is important to note that these instruments were administered by the researchers *after* client selection and not as part of the client recruitment process. Follow-up surveys elicit detailed event history information on the intervening periods.

The dependent variables for the analyses presented here are dichotomous measures of relapse to illegal drug use and rearrest through the 42 and 60 months of follow-ups. To be considered *drug free*, the respondent must have reported *no* illegal drug use *and* have tested negative for drugs on the urine screen at every follow-up point. Similarly, the criteria for *arrest free* included no self-reports of arrest and no official arrest records for new offenses since release from prison.

The first set of analyses reported below include all participants distinguished by intent to treat at 42 and 60 months after release from prison. This is done to verify the significant effects of treatment. Subsequent analyses then explore differences among the treated group in terms of completion of transitional treatment and participation in aftercare. Statistical analyses used multivariate logistic regression with treatment status, gender, race, age, criminal history, and history of drug use and treatment as predictors of relapse and recidivism. Use of these *covariate* predictors was designed to control for potential group differences caused by the inclusion of some participants who were not randomly selected into the treatment groups. Group means and percentages on these covariates are shown in Table 1.

▧ Results

Table 2 shows the results from the logistic regression analyses for drug-free status at 42 months after release from prison. Treatment participation was the substantially largest predictor, and age and frequency of prior drug use were additional significant predictors. Participation in the transitional treatment program more than quadrupled the odds of

Table 1	Sample Characteristics by Treatment Experience for Respondents With 42-Month Follow-Up				
	All Treatment Groups	**No Treatment Group**	**Dropouts**	**Graduates**	**Graduates With Aftercare**
N	472	218	166	138	168
African American	74.6%	66.7%	69.9%	81.8%	73.3%
Male	75.9%	82.2%	82.5%	74.5%	70.3%
Prior drug treatment	68.1%	74.0%	70.5%	72.3%	62.2%
Mean prior level of drug use, 0 (none) to 6 (several times daily)	4.8	4.2	4.8	4.7	4.9
Mean age	30.4	29.6	29.2	30.6	31.4
Mean number of times in prison	4.1	4.1	4.4	4.0	3.8

remaining drug free. Older participants were significantly more likely to have been abstinent. Each increment in the frequency of prior drug use had a negative association, reducing the odds of remaining drug free by over 10%.

Treatment participation, although a smaller effect than that for drug status, was also a significant predictor of criminal recidivism at 42 months (Table 3), with a 70% reduction in the odds of a new arrest for those assigned to treatment. Of course, the incidence of drug use represents something of an outside limit to the incidence of rearrest. The drug use monitored here is illicit behavior, and, indeed, the largest category of rearrests is for drug charges. In contrast to drug use, criminal recidivism was also related to gender, with women significantly less likely to have been rearrested. As with drug use, the probability of rearrest was significantly lower for older participants. The latter effect has been a common finding of the "maturing out" of criminal behavior. In addition, previous number of incarcerations was a significant predictor of subsequent rearrest.

Table 4 shows the results from the logistic regression analyses for drug-free status at 60 months after release from prison. The results are reasonably consistent with those at 42 months. Treatment participation remained the substantially largest predictor, and baseline frequency of drug use remained a significant predictor, although age was no longer a significant predictor. Participation in the transitional treatment program still more than tripled the odds of remaining drug free. Additionally, having no previous treatment experience was now a significant predictor of relapse to drug use.

Table 5 shows the results from the logistic regression analyses for arrest-free status at 60 months after release from prison. Treatment participation, age, and gender remained significant predictors, each with generally equivalent predictive power. Number of previous incarcerations was, however, no longer a significant predictor.

Table 2	Logistic Regression Predicting No Illicit Drug Use Through 42 Months After Release From Prison, as a Function of Participating in Treatment and of Demographic, Drug Use, and Criminal History Variables			
	Regression Coefficient	Standard Error	Odds Ratio	Significance
Treatment participant	1.50	.31	4.49	< .001
Age (years)	.04	.02	1.04	.008
Female = 1, male = 0	.03	.25	1.03	.907
African American = 1, other = 0	−.16	.22	0.85	.474
Number of times in prison	−.04	.04	0.96	.401
No prior drug treatment	.26	.23	1.30	.256
Prior level of drug use, 0 (none) to 6 (several times daily)	−.14	.05	0.87	.007
Constant	−3.37	.82	—	< .001

Nagelkerke R^2 = .115.

Table 3	Logistic Regression Predicting No New Arrest Through 42 Months After Release From Prison, as a Function of Participating in Treatment and of Demographic, Drug Use, and Criminal History Variables			
	Regression Coefficient	Standard Error	Odds Ratio	Significance
Treatment participant	.54	.18	1.71	.003
Age (years)	.07	.01	1.07	.001
Female = 1, male = 0	.43	.20	1.54	.015
African American = 1, other = 0	.17	.17	1.19	.317
Number of times in prison	−.13	.04	0.88	.001
Prior level of drug use, 0 (none) to 6 (several times daily)	−.04	.04	0.96	.325
Constant	−3.14	.58	—	< .001

Nagelkerke R^2 = .108.

Table 4	Logistic Regression Predicting No Illicit Drug Use Through 60 Months After Release From Prison, as a Function of Participating in Treatment and of Demographic, Drug Use, and Criminal History Variables			
	Regression Coefficient	Standard Error	Odds Ratio	Significance
Treatment participant	1.27	.31	3.54	< .001
Age (years)	.02	.02	1.02	.295
Female = 1, male = 0	.34	.30	1.41	.252
African American = 1, other = 0	.15	.28	1.16	.592
Number of times in prison	−.01	.06	0.99	.899
No prior drug treatment	.52	.27	1.69	.049
Prior level of drug use, 0 (none) to 6 (several times daily)	−.20	.06	0.82	.001
Constant	−3.84	.98	—	< .001

Nagelkerke R^2 = .111.

Table 5	Logistic Regression Predicting No New Arrest Through 60 Months After Release From Prison, as a Function of Participating in Treatment and of Demographic, Drug Use, and Criminal History Variables			
	Regression Coefficient	**Standard Error**	**Odds Ratio**	**Significance**
Treatment participant	.48	.20	1.61	.017
Age (years)	.05	.02	1.05	.002
Female = 1, male = 0	.54	.22	1.71	.015
African American = 1, other = 0	.27	.20	1.31	.189
Number of times in prison	−.06	.05	0.94	.197
Prior level of drug use, 0 (none) to 6 (several times daily)	−.03	.05	0.97	.541
Constant	−3.26	.67	—	< .001

Nagelkerke R^2 = .065.

Subsequent to establishing that the treatment group as a whole was significantly different than the comparison group, additional analyses examined the impact of particular components of treatment, dividing the treatment group into three stages: dropouts, graduates without aftercare, and graduates who did participate in aftercare. Because any effects of these gradations could be interpreted as reflecting that simply more time in treatment is likely to produce better outcomes, an additional variable of time in treatment was added to the regression model for these analyses.

Of particular interest are differences between those who successfully completed the program and those who failed to complete. The treatment program, in contrast to regular work release, does not allow any outside unsupervised time during the first 3 months of the 6-month program. Most of the treatment failures thus come after that time, when the opportunities for violations such as curfew infractions and positive urine tests become much more probable. Therefore, treatment dropouts have a substantial amount of treatment experience.

Figure 1 shows consistent effects throughout these time periods, reflected in identical patterns of significance for both the 42- and 60-month data. Prison releasees who completed therapeutic community treatment with or without aftercare had significantly greater probabilities of remaining both arrest free and drug free at both time points than did those without treatment. Treatment dropouts were slightly, though not significantly, less likely to be arrested on a new charge as those without treatment, but they were significantly more likely to be drug free. Any participation in treatment, whether successful or not, produced significantly beneficial effects for subsequent drug use but not on subsequent arrests. Those who completed treatment were, however, significantly more likely to have had positive outcomes, and those who completed treatment and attended aftercare were the least likely to have a new arrest or to have lapsed into drug use. Just under half of those who completed treatment and then attended aftercare would be expected to have a new arrest, compared to more than 75% of the group without treatment. When contrasted with the group with no treatment, those in the treated groups are 15–20 times more likely to be drug free. Of course, these larger effects are to be expected, both because the programs are specifically focused on drug use and because, without treatment, the probability of refraining from illicit drug use for 5 years is spectacularly low.

▧ Discussion

The typically longstanding drug and criminal careers of offenders coming to the attention of the criminal justice system are not specific to Delaware, and any endeavor to curtail these behaviors requires substance abuse treatment that is both intensive and extensive. Delaware has responded to this need by instituting its continuum of primary (in-prison), secondary (work-release), and tertiary (aftercare) TC treatment corresponding to sentence mandates. Earlier analyses of the Delaware continuum as well as the new 5-year outcome data presented here all indicate that clients who completed secondary treatment (some of whom also completed primary treatment) were significantly more likely than those with no treatment or those who dropped out of treatment to remain

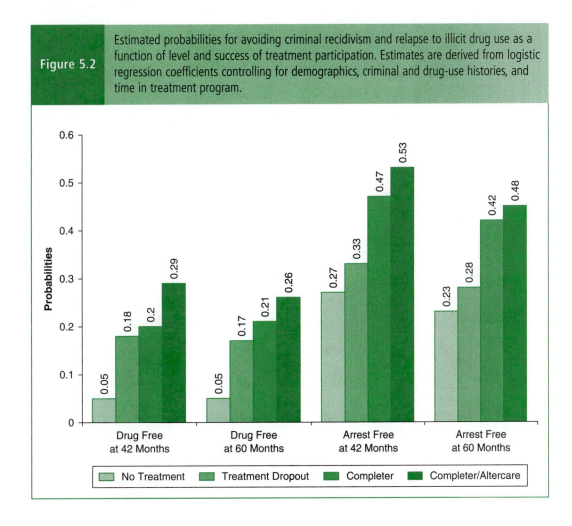

Figure 5.2 Estimated probabilities for avoiding criminal recidivism and relapse to illicit drug use as a function of level and success of treatment participation. Estimates are derived from logistic regression coefficients controlling for demographics, criminal and drug-use histories, and time in treatment program.

drug free and arrest free at 18 months, 42 months, and now 60 months after release from prison. In addition, the first analyses of 60-month data now available on Delaware clients who received tertiary treatment (the TC aftercare program implemented in 1996) suggest that treatment graduates who participate in aftercare programming surpass treatment graduates who do not receive continuing care in remaining drug free and arrest free at both 42 and 60 months. These results provide continuing support for the beneficial effects of participation in transitional and community TC treatment for drug-involved offenders.

The data presented in this article suggest that long-term treatment in correctional settings can have a major impact on the potential for relapse and recidivism among drug-involved offenders. One might still argue, however, that the effects of treatment are short-lived. After all, among the treatment graduates, 58% had been rearrested and 79% had relapsed to drug use by the time of the 60-month follow-up. Moreover, even among the treatment graduates who also had aftercare, 52% had been rearrested and 71% had relapsed. But, in counterpoint, therapeutic communities in general and corrections-based therapeutic communities in particular are dealing with the most difficult of all substance abusers: they are the most drug involved, the most criminally involved, and the most socially dysfunctional. As such, positive changes typically occur in small increments.

Additionally, when considering that substance abuse is a chronic and relapsing disease, one begins to realize how conservative the measure used in this analysis is. As noted earlier, *relapse to drug use* as used here was defined as *any* illegal drug use since release. As such, the use of cocaine or marijuana on even one day or one occasion constituted relapse according to this definition. Nevertheless, some 30% of the graduates with aftercare remained abstinent for 5 years, as measured by self-report and urine testing. This effect is quite possibly a lower limit of the estimate of treatment efficacy that would increase with less stringent criteria for relapse to drug use.

Treatment participation and completion are the focus of these analyses, but it is obvious that treatment participation comprises only part of the explanation for the phenomena of relapse and recidivism. The current analyses attempted to address some other possible explanations by modeling the effects of demographic characteristics, prior criminal activities, and drug-use history on relapse and rearrest. Nonetheless, important control or confounding variables are undoubtedly lacking. It is obvious that the models estimated, although significant, are not accounting for all of the variance in predicting relapse and recidivism. It is probable that important control variables and confounding variables for group effects have not been modeled. One large potential area that needs to be considered is the selection of clients into treatment and the suitability of the treatment that clients receive. Similarly, the differences between voluntary versus compulsory treatment have not been addressed. Both of these issues need to be operationalized and incorporated in any comprehensive model involving treatment effects predicting to relapse and recidivism.

It should also be noted that the 60-month outcome data presented here are based on modest sample sizes that are insufficient for analyses of all possible combinations of treatment participation. For example, the analyses presented here did not examine the unique effects of participating in the in-prison TC program (KEY). In fact, the long-term data actually do not find a unique effect of in-prison treatment alone, although graduates of the institutional TC are more likely to remain in treatment through work release and aftercare. Given the widespread existence in corrections of prison TCs as opposed to traditional TCs, this is an area to be examined carefully in future research.

Despite these limitations, the present data speak to the value of treatment in work-release and parole settings and the importance of retention in treatment in increasing long-term abstinence from drug use and criminal activity. More importantly, the data also support some long-held beliefs about the beneficial effects of transitional and aftercare treatment during reentry. Transitional programming appears to provide a critical bridge between institutional confinement and community reentry by providing assistance for the psychological, social, and legal obstacles that can place drug-involved offenders at risk for relapse and recidivism during work release and parole. In 1935, for example, just a year after the term *addiction* first appeared in the American Psychiatric Association's *Standard Classified Nomenclature of Disease*, the U.S. Public Health Service opened its first "narcotics farm" in Lexington, Kentucky. A second facility was opened in Fort Worth, Texas, 3 years later. They were called *farms* because clients participated in agricultural work during their treatment sojourns. In reality, however, both facilities were prison hospitals established for the treatment of addiction to narcotics while at the same time designed to alleviate prison crowding in other parts of the federal system.

It was anticipated from the outset that the treatment approach at the new federal establishments would be highly effective because, at the very least, the hospitals were designed to treat not only the physical dependence but also the mental and emotional problems thought to be related to addiction. This was an advanced conception because, until then, treatment for a narcotics problem had focused exclusively on physical dependence. And there were other innovations, including a drug-free environment and access to educational and vocational services as well as recreational and religious activities. When the patients were followed-up,

however, treatment outcomes were disappointing. In a study of 1,881 patients discharged from the Lexington facility to the New York City area, for example, some 90% had become readdicted within 1 to 4.5 years (Hunt & Odoroff, 1962). Other studies documented that 90% to 96% treated at Lexington returned to active addiction, most within 6 months of discharge. Although Dr. Victor H. Vogel, the medical officer in charge of the Lexington program, tried to put a positive spin on the outcome data by suggesting that the treatment results were better than those of such other chronic diseases as diabetes and cancer (White, 1998), the studies underscored the limited role of institutionally based treatment alone in reducing the likelihood of subsequent relapse. What the Lexington and Fort Worth experiences did suggest was the need for a continuum of treatment with community aftercare as a crucial component. The data presented in this article present some interesting parallels and comparisons with the Lexington and Fort Worth experiences, emphasizing the importance of treatment followed by aftercare for criminal justice clients.

✎ References

Hunt, G. H., & Odoroff, M. E. (1962). Follow-up study of narcotic drug addicts after hospitalization. *Public Health Reports, 77,* 41.

Wexler, H. K., & Williams, R. (1986). The "stay 'n out" therapeutic community: Prison treatment for substance abusers. *Journal of Psychoactive Drugs, 18,* 221–230.

White, W. L. (1998). *Slaying the dragon: The history of addiction treatment and recovery in America.* Bloomington, IL: Chestnut Health Systems.

DISCUSSION QUESTIONS

1. Summarize the main points of this article.

2. What do the authors have to say about the relationship between drug abuse and crime; is it causal?

3. Why is follow-up treatment after release from prison perhaps even more important than treatment inside prison walls?

READING

This article by Andrews and his colleagues is concerned with coming to understand criminal offenders so that they can be better treated using the RNR (risk-needs-responsivity) principle. The authors claim that the theoretical, empirical, and applied progress in the psychology of criminal conduct has been almost revolutionary over the past decade or so. Although not completely dismissing subjective clinical judgment, they obviously prefer structured clinical judgment. They are particularly enamored with actuarial instruments that assess offenders' risks (of re-offending) and needs (for rehabilitation), noting that the validity of such instruments greatly exceeds the validity of unstructured clinical judgments. Proper risk and needs assessment is responsive to each offender's specific requirements allowing treatment personnel to deliver effective services while at the same time protecting the public.

The Recent Past and Near Future of
Risk and/or Needs Assessment

D. A. Andrews, James Bonta, and J. Stephen Wormith

The history of risk assessment in criminal justice has been written on several occasions. Here we assess progress since Andrews, Bonta, and Hoge's (1990) statement of the human service principles of risk-need-responsivity (RNR) and professional discretion. In those articles, the corrections-based terms of *risk* and *need* were transformed into principles addressing the major clinical issues of who receives treatment (higher risk cases), what intermediate targets are set (reduce criminogenic needs), and what treatment strategies are employed (match strategies to the learning styles and motivation of cases: the principles of general and specific responsivity). General responsivity asserts the general power of behavioral, social learning, and cognitive-behavioral strategies. Specific responsivity suggests matching of service with personality, motivation, and ability and with demographics such as age, gender, and ethnicity. Nonadherence is possible for stated reasons

under the principle of professional discretion. Expanded sets of principles now include consideration of case strengths, setting of multiple criminogenic needs as targets, community-based, staff relationship and structuring skills, and a management focus on integrity through the selection, training, and clinical supervision of staff and organizational supports.

The review is conducted in the context of the advent of the fourth generation of offender assessment. Bonta (1996) earlier described three generations of risk assessment. The first generation (1G) consisted mainly of unstructured professional judgments of the probability of offending behavior. A variation of this approach is now called "structured clinical judgment." Second-generation (2G) assessments were empirically based risk instruments but atheoretical and consisting mostly of static items (e.g., the Salient Factor Score or SFS). Third-generation (3G) assessments were also empirically based but included a wider sampling of

SOURCE: Andrews, D. A., Bonta, J., & Wormith, J. S. (2006) The recent past and near future of risk and/or needs assessment. *Crime and Delinquency, 52*(1), 7–27. Reprinted with permission of Sage Publications, Inc.

dynamic risk items, or criminogenic needs, and tended to be theoretically informed (e.g., Level of Service Inventory–Revised or LSI-R). The fourth generation (4G) guides and follows service and supervision from intake through case closure. With postclosure follow-up, outcome may be linked with intake assessments of risk, strengths, need, and responsivity, with reassessments, and with service plans, service delivery, and intermediate outcomes. With systems that recognize the criminogenic-noncriminogenic distinction, the achievement of less and more relevant intermediate outcomes may be compared in relation to recidivism and with measures of well-being. The point is not only the development of management information systems but also the development of human service assessment and treatment systems. A major goal of the 4G instruments is to strengthen adherence with the principles of effective treatment and to facilitate clinical supervision devoted to enhance public protection from recidivistic crime. The best known of the 4G systems are the original and classic Wisconsin (now known as Correctional Assessment and Intervention System [CAIS]), the Correctional Offender Management Profiling for Alternative Sanctions (COMPAS), the Offender Intake Assessment (OIA) of Correctional Service Canada, and the Level of Service/Case Management Inventory (LS/CMI).

To begin, we note that theoretical, empirical, and applied progress within the psychology of criminal conduct (PCC) has been nothing less than revolutionary. This is important because the 1990 articles opened with a statement that the PCC was crucial to effective correctional treatment. Second, this article takes a brief look at clinical judgment (1G) with a nod to structured clinical judgment, notes a new energy in 2G actuarial instruments, and a renewed appreciation of the assessment of change (3G). Third, the challenge faced by forensic mental health approaches from general correctional instruments, even within mental health samples, is reviewed. Fourth, the widely known principles of effective service for offenders

are supplemented by additional principles derived from meta-analytic evidence. Finally, the article closes with a discussion of some negative evaluations of RNR and the challenges that feminist, critical criminological perspectives, and humanistic perspectives present to the future of risk and/or need assessment.

Theory

The general personality and social psychology of crime, with special attention to social learning and/or social cognition theory, is now the prominent theoretical position in criminology. This development has been traced in some detail by Andrews and Bonta (2003) and need not be repeated here. Moreover, it is clear that personality constructs such as low self-control and social learning constructs such as antisocial cognition and antisocial associates make independent contributions to the analysis of criminal behavior. Antisocial personality pattern itself appears to reflect several factors. The overtly behavioral one, reflecting early and continuing involvement in diverse antisocial conduct, may be better conceptualized as antisocial behavioral history. Others are more clearly temperamental: weak self-control (low conscientiousness) and high antagonism (low agreeableness). Psychopathy may be understood in terms of fundamental dimensions of temperament. Attitudes, associates, history, and personality, the big four in theory, are also of major empirical importance.

Empirical Understanding of Predictors

Meta-analyses of the risk and/or need factors with diverse offender groups have clarified our knowledge of major, moderate, and minor risk factors. Moreover, the Psychopathy Checklist–Revised (PCL-R) and the Violence Risk Appraisal Guide (VRAG) have lifted forensic mental health out of its dreary reliance on clinical judgment. Now, psychologists, criminologists, and mental health and justice practitioners have a common language, a shared knowledge base, and the shared technology of RNR.

One nonquantitative summary of the findings regarding the more and less powerful risk and/or need factors is provided in Table 1. We summarize the content of Table 1 by reference to the "big four" (the first four in the table) and the "central eight" (all eight of the major risk and/or need factors). Notably, the major risk and/or criminogenic need factors and the power of social learning and/or cognitive-behavioral influence strategies are readily identified within general personality and social learning perspectives on criminal behavior. Also noteworthy is that the relatively mild predictive validity of the minor risk factors, when present at all, most likely reflects contributions through the big four. For example, the predictive validity of mental disorder most likely reflects antisocial cognition, antisocial personality pattern, and substance abuse, whereas the contributions of socially disadvantaged neighborhoods reflect, in part, the lower strength and higher personal and interpersonal risk levels of residents.

Table 1	Major Risk and/or Need Factors and Promising Intermediate Targets for Reduced Recidivism	
Factor	**Risk**	**Dynamic Need**
History of antisocial behavior	Early and continuing involvement in a number and variety of antisocial acts in a variety of settings	Build noncriminal alternative behavior in risky situations
Antisocial personality pattern	Adventurous pleasure seeking, weak self-control, restlessly aggressive	Build problem-solving skills, self-management skills, anger management and coping skills
Antisocial cognition	Attitudes, values, beliefs, and rationalizations supportive of crime; cognitive emotional states of anger, resentment, and defiance; criminal versus reformed identity; criminal versus anticriminal identity	Reduce antisocial cognition, recognize risky thinking and feeling, build up alternative less risky thinking and feeling, adopt a reformed and/or anticriminal identity
Antisocial associates	Close association with criminal others and relative isolation from anticriminal others; immediate social support for crime	Reduce association with criminal others, enhance association with anticriminal others
Family and/or marital	Two key elements are nurturance and/or caring and monitoring and/or supervision	Reduce conflict, build positive relationships, enhance monitoring and supervision
School and/or work	Low levels of performance and satisfactions in school and/or work	Enhance performance, rewards, and satisfactions
Leisure and/or recreation	Low levels of involvement and satisfactions in anticriminal leisure pursuits	Enhance involvement, rewards, and satisfactions
Substance abuse	Abuse of alcohol and/or other drugs	Reduce substance abuse, reduce the personal and interpersonal supports for substance-oriented behavior, enhance alternatives to drug abuse

NOTE: The minor risk and/or need factors (and less promising intermediate targets for reduced recidivism) include the following: personal and/or emotional distress, major mental disorder, physical health issues, fear of official punishment, physical conditioning, low IQ, social class of origin, seriousness of current offense, other factors unrelated to offending.

Empirical Understanding of Effective Treatment

The meta-analyses of Andrews, Dowden, and colleagues (Andrews & Bonta, 2003) were explicitly designed to test the principles of effective correctional treatment. At the risk of making statistical purists unhappy, we summarize the enhanced understanding of effective treatment by examining the correlation of adherence to RNR (less-more) with effect size (Pearson's r) in our meta-analytic data set. It is now apparent that support for the risk principle increases from very modest to strong with increases in the precision of the risk assessment. For example, the correlation of risk with effect size varies from a mild $r = .12$ when crude aggregate risk classifications are used through $r = .54$ when risk is assessed as the recidivism rate in the control groups. It is also apparent that the validity of the need principle was underestimated in our original 1990 meta-analysis. Now, considering the full metric of the number of targeted criminogenic needs, the correlation of effect size with adherence to the need principle becomes $r = .58$ compared to the $r = .25$ value found using the more crude measure of adherence employed in the original 1990 study. Thus, multimodal has been added as a principle of effective treatment.

Adherence with general responsivity in relation to effect size remains strong but can be augmented through explicit assessment of staff relationship and staff structuring skills. With explicit additional attention to integrity through management of the setting and the selection, as well as training and clinical supervision of staff members, the correlations with effect size reach into the .60 range.

Adherence with specific responsivity and professional discretion has yet to be explored meta-analytically. Of course, many other issues are raised within this meta-analytic databank, but, with due respect for replication and extension, prevention and treatment programs aimed at reducing reoffending are well advised to attend to the RNR principles. All in all, the psychology of criminal conduct provides a base for RNR that is much more solid in 2005 than it was in 1990.

▨ Predictive Criterion Validity of 1G and 2G Assessments

Actuarial instruments (2G) that rely on a few static criminal history items with perhaps a minor sampling of dynamic domains continue to function well. The overall mean predictive validity derived from the three Bonta and Hanson reviews was .42 for general recidivism and .39 for violent recidivism and are dramatically higher than the 1G estimates. It is interesting to note, and subject to further study. Hanson and Morton-Bourgon (2004) found that mean validity of general criminality risk and/or need scales equaled or exceeded that of specialized sex-offending instruments even in the prediction of sexual violence.

The classic correctional 2G instruments did well in the prediction of general recidivism: Wisconsin Risk has a mean of .32 and SFS has a mean of .30. Their predictive success is particularly notable in that, for decades, probation and parole officers were successfully predicting criminal recidivism, while forensic mental health professionals were failing miserably. Fortunately, forensic mental health assessments have advanced with the standardization and quantification of Cleckley's clinical description of psychopathy with the PCL-R. Hare and colleagues are adamant that the PCL-R is not a risk and/or need scale but a diagnostic instrument. So be it; however, obviously the PCL-R is also a systematic survey of items tapping antisocial personality and a history of antisocial behavior (two of the big four). Mean PCL-R predictive criterion validity estimates are indeed impressive (.27 for general and for violent recidivism) and in the same range although somewhat lower than the correctional instruments.

The PCL-R is a key component of the 2G instrument VRAG. As evident in Table 2, the

| Table 2 | Mean Predictive Criterion Validity Estimates (*r*) From Meta-Analytic Studies by Generation (k) |

Study	Recidivism		
	General	Violence	Scale
First-generation unstructured clinical judgment			
Bonta, Law, & Hanson (1998)[a]	.03 (5)	.09 (3)	
Hanson & Bussière (1998)[b]	.14 (8)	.10 (10)	
Hanson & Morton-Bourgon (2004)[b]	.12 (7)	.20 (9)	
Mean	.10	.13	
Second-generation actuarial (mechanical)			
Bonta, Law, & Hanson (1998)[a]	.39 (6)	.30 (7)	General
Hanson & Bussière (1998)[b]	.42 (5)	.46 (6)	General
Hanson & Morton-Bourgon (2004)[b]	.46 (3)	.40 (3)	General
Mean general scales	.42	.39	
Gendreau, Little, & Goggin (1996)	.26 (15)		SFS
Gendreau et al. (1996) Wisconsin	.31 (14)		
Gendreau et al. (1996)	.29 (9)		PCL-R
Gendreau, Goggin, & Smith (2002)	.24 (30)	.23 (26)	PCL-R
	.26 (6)	.30 (5)	PCL-R[d]
Hemphill & Hare (2004)	.30 (7)	.28 (5)	PCL-R[d]
Mean PCL-R	.27	.27	
Rice & Harris (telephone communication, December 10, 2004)		.39	VRAG
Third-generation mechanical with dynamic items			
Gendreau et al. (1996)	.33 (28)		LSI-R
Gendreau et al. (2002)	.39 (33)	.28 (16)	LSI-R
	.40 (6)	.24 (5)	LSI-R
Hemphill & Hare (2004)	.33 (7)	.23 (5)	LSI-R[d]
Mean LSI-R	.36	.25	
Fourth-generation clinical assessment systems (from intake through closure)			
Andrews, Bonta, & Wormith (2004) (compiled from chapter 6)	.41 (8)	.29 (7)	LS/CMI

NOTE: SFS = Salient Factor Score; PCL-R = Psychopathy Checklist–Revised; VRAG = Violence Risk Appraisal Guide; LSI-R = Level of Service Inventory–Revised; LS/CMI = Level of Service/Case Management Inventory.

k = number of primary estimates.

a = Mentally disordered offenders.

b = Sex offenders.

c = General risk scales.

d = Within-sample comparisons.

VRAG is outstanding in the prediction of violence. The overall mean estimate of .39 is substantially greater than all other mean estimates in the violence column of Table 2 but for the Hanson estimates (and primary studies of VRAG contributed to the Hanson estimates). If one believes the PCL-R rescued forensic mental health assessment, then the VRAG carried the whole field of violence prediction, based in corrections and forensic mental health, to a new level. Thus, the VRAG demands special attention.

Reflecting 2G respect for multiple regression approaches to item selection and the 2G dustbowl atheoretical tradition, the VRAG (first known as the Violence Prediction Scheme) was built through careful and comprehensive coding of psychosocial history and clinical files in a maximum-security forensic psychiatric facility. The findings established, overwhelmingly so, that the major predictors of violence in that forensic sample were not mental health variables but the risk factors already well established in general corrections and the psychology of criminal conduct (recall Table 1). Much of the content of the VRAG was drawn from the central eight risk factors. However, some "minor" risk factors were selected and were scored as strengths: being schizophrenic, having a female victim of index offense, and inflicting serious injury are each scored as factors that reduce risk. The VRAG team has already shown that a short objective historical scale (the Child and Adolescent Taxon Scale–CATS) can replace the PCL-R. Perhaps simple checklists could also replace the diagnoses of "schizophrenia" and "any personality disorder" in addition to providing substitutions for inversely scored serious injury and female victim (if the latter two, in fact, are found to be required at all).

The possibility of achieving satisfactory prediction without the use of clinical items was demonstrated in a study of the Offender Group Reconviction Score. The OGRS consists of age, gender, and criminal history items selected based on their predictive validity in general offender samples. In a study of 315 offenders who were mentally disordered (Gray et al., 2004) and without sampling a single mental health item, OGRS outperformed forensic instruments. The massive superiority of the general risk assessment approach was apparent across diagnosis, and the clinical scales had no incremental predictive validity.

⧓ Predictive Criterion Validities of 3G and 4G Assessments

Only the 3G LSI-R and 4G LS/CMI are listed in Table 2 because we have been unable to discover meta-analytic summaries or accessible listings of the predictive criterion validity estimates for the risk and need elements of the Wisconsin, COMPAS, OIA and Offender Assessment System. Given the limited availability of results on other 3G and 4G systems, our discussion focuses on the LSI-R and LS/CMI.

The overall mean predictive criterion validity estimates for the LSI-R (.36) and the LS/CMI (.41; see Table 2) are quite respectable with the latter equaling or exceeding all other overall mean validity estimates in the general recidivism column of Table 2. It does appear that the LSI-R is more strongly associated with general recidivism (.36) than with violent recidivism (.25). Corresponding values for LS/CMI are .41 and .29, respectively (with the latter being substantially lower than the mean correlation of VRAG with violence).

The predictive validity of the LSI-R in regard to violence may be enhanced in the LS/CMI wherein the General Risk/Need assessment (Section 1) across the central eight domains has been strengthened by the introduction of an Antisocial Personality Pattern subcomponent. It is a behavior-based assessment of early and diverse problems. Moreover, sexual assault, violence, and diversity of antisocial behavior are now also surveyed systematically. In the first prospective validation of the LS/CMI, the correlation with violent recidivism of the enhanced assessment of personality pattern and history of aggression was $r = .42$ in the follow-up of incarcerated individuals.

Substantial improvements in the predictive criterion validity of risk assessments may reside in reassessments of dynamic risk factors. The incremental dynamic criterion validity of the LSI-R is evident from several studies. Based on the available evidence, we anticipate reassessments will double and, perhaps, triple the outcome variance explained by intake assessments. More important, with assessments of acute factors, opportunities for timely preventive action are enhanced.

The underlying RNR in the psychology of criminal conduct is intended to apply widely. The expectation is that being human means that variation on the big four of attitudes, social support, behavioral history, and temperament will account for much of the variability in antisocial behavior across a host of situational variables. Patterns of satisfaction and dissatisfaction in the behavioral settings of family, school and/or work, leisure and/or recreation, and substance abuse and a host of more distal factors such as socioeconomic indicators will be sources of variability in the big four. The expectation is that male or female, Black or White, the predictive criterion validity of assessments of major risk factors will be evident in a variety of contexts. In fact, correlations between the LS/CMI's General Risk/Need subscale and reoffending were substantial and robust in a large sample of adult female offenders in Ontario. The validity was maintained within institutional and community samples and for women suffering from psychiatric problems, severe histories of abuse, and poverty.

A great promise of meta-analysis is discovering the moderators of variation in predictive criterion validity estimates. One, of course, is assessment generation (1G vs. the later three generations), and another is specific instruments (the apparent superiority of the LSI instruments relative to others in predicting general recidivism: the apparent superiority of VRAG in predicting violence). Still, considerable variability exists within the results of studies on particular instruments. For example, single-study predictive criterion validity estimates vary from .22 to .63 from recent LS/CMI studies. Some of the sources of variation in the criterion validity estimates for particular instruments are known. The issue of general versus violent recidivism was already noted. The training, experience, and clinical supervision of users may also be important moderators of predictive criterion validity. Agencies whose staff has not been trained by certified trainers yield much smaller validity estimates than agencies with better-trained staff members. The use of intake assessments in treatment-rich agencies is also an issue; that is, the simple predictive criterion validity of original risk and/or need scores will be greatly reduced when following up cases that have been appropriately treated. In those situations, retests are more valued.

The Outcome Validity of Differential Programming: Meta-Analytic Evidence and Examples With 4G RNR Assessment

The validation of risk assessments in the RNR context requires demonstrations that adherence to the risk principle is rewarded by enhanced public protection from recidivistic crime. The meta-analytic evidence was previously noted in that estimates of the correlation with effect size of adherence with the risk principle varied with precision of the risk estimate. Back in 1990, we were able to construct a table illustrating interactions from 12 studies. After

15 years, we are able to identify only a few additional examples. The exploration of Risk × Treatment interactions has not become routine in the evaluation literature. However, the meta-analytic pattern is clear. Program service delivery to the offenders who are higher risk produces larger decreases in recidivism than it does for offenders who are lower risk.

It is expected that the Risk × Treatment interaction will only be found when the service is otherwise appropriate in regard to need and responsivity. In brief, the correlation between risk and effect size increases when treatment also adheres to general responsivity. With non-adherence to need and general responsivity, the correlation of adherence to risk and effect size actually turned negative ($r = -.28$).

Turning to need, and as already noted, our original statement of principles underestimated the power of need, and the multimodal principle has been added. The simple measure of "number of criminogenic needs targeted exceeded number of noncriminogenic needs targeted" is best supplemented by explicit consideration of the full differential between number of criminogenic needs targeted and number of noncriminogenic needs targeted. This result speaks volumes. There are solid ethical, legal, decent, and even just reasons to focus on some noncriminogenic needs; however, to do so without addressing criminogenic need is to invite increased crime and to miss the opportunity for reduced reoffending. The validity of general responsivity is overwhelming in the meta-analytic literature. Once again, of course, general responsivity is less important when service is not conforming with the risk and need principles.

Specific responsivity remains the least explored of the RNR principles. A number of specific responsivity approaches have been outlined. They include interview-based and questionnaire-based classification approaches such as Interpersonal Maturity Level and Client Management Classification (CMC). A major addition to this list is stages of change theory and motivational interviewing.

Elsewhere Van Voorhis (1994) suggested that the many categories of offenders suggested by different specific responsivity personality systems reduce to four basic types: committed criminal, character disordered, neurotic anxious, and situational. Another major development is the interest in gender-specific and culturally specific programming. The literature on gender-specific programming is large and detailed although largely nonevaluated.

With outstanding exceptions, evaluations of responsivity systems have not been conducted in the context of risk and need. A priority issue is an analysis of specific responsivity systems in terms of the extent to which the classifications do incorporate risk (and hence differential levels of service and supervision) and need (intermediate targeting) issues in addition to responsivity (differential styles, modes and strategies of intervention). CMC directs differential targeting and differential styles of intervention and has demonstrated that implementation of CMC affects revocations with moderate-risk and high-risk cases.

Advances in the assessment of psychopathy permit the conceptualization of the construct in RNR terms. Proposals have been presented of breaking down the PCL-R items into static criminal history, dynamic criminogenic needs, and responsivity items. In fact, Wong and Hare (2005) incorporated elements of RNR into their treatment prescriptions. The risk presented by people who are psychopaths is well documented; however, a discussion of criminogenic need, multimodal, general responsivity, and specific responsivity is a high-priority issue. In 1990, Andrews and colleagues proposed that the many personality-based responsivity systems could be reduced to a few differential treatment hypotheses involving several sets of offender characteristics including cognitive and/or interpersonal skills, anxiety, antisocial personality, motivation, social support, gender/race/ethnicity, and mental disorder. We now add strengths to the list as in "design a plan that builds on the strengths of the person." We await new primary studies and meta-analytic evidence in regard to specific responsivity. Research should be stimulated with the addition of a specific responsivity section to LS/CMI.

These 4G approaches have opened up opportunities for advances in service and public safety. In addition, knowledge gains will be evident in regard to many important service and theoretical issues in the near future. At least modest gains may be expected in predictive criterion validity through continuing work on the incremental value of strength ratings and expanded or refined assessments of criminal history and antisocial personality pattern. The possibility of personal and/or emotional well-being interactions with gender or other risk factors must be explored. Studies of change and acute risk factors may soon establish, on average, just how much of an improvement in predictive accuracy may be expected. Simultaneous monitoring of changes in criminogenic and noncriminogenic needs may help reshape the content of the categories of criminogenic and noncriminogenic needs. Studies that integrate the process of service planning, delivery, and intermediate outcome from intake through case closure and follow-up are most desirable. We see responsivity issues as priority ones in this process.

⧄ Does Implementation of RNR Enhance Outcomes? Agency-Level Structural Effects

The Correctional Program Assessment Inventory (CPAI 2000) was developed to assess the degree of adherence to the principles of RNR demonstrated by a program or correctional agency. It is possible to assess the strengths and weaknesses of agencies and their programs in regard to implementation of RNR and to identify areas in which improvements might be considered. In research and applied terms, it is possible to explore agency-level practice, structure, and culture not simply on the usual financial and/or staffing issues but in terms of adherence to demonstrably powerful principles of effective correctional treatment.

Edward Latessa and Christopher Lowenkamp and their colleagues and students (Lowenkamp, 2004; Lowenkamp & Latessa, 2002) have demonstrated that agency-level variation in adherence with RNR is associated with the success rates of correctional agencies. Each agency had completed a CPAI that included an assessment of the extent to which risk and/or need assessments were done. Agencies that actively employed standardized risk and need assessments had a greater impact on recidivism than agencies that did not (correlations with effect sizes of .33 and .16, respectively). Reassessment with standardized instruments was also linked with outcome (.39). Of course, consideration of general responsivity enhanced outcomes even further. The correlation between total CPAI scores and effect size was $r = .41$. It is interesting to note, Aleksandra Nesovic's (2003) meta-analytic evaluation of the CPAI yielded a correlation with effect size in the same range (.50). We can look forward to agency-level experimental investigations in which agencies are differentially exposed to and trained in RNR, and the CPAI may be used as an intermediate check on integrity of implementation.

⧄ Promise and Challenge of Alternative Models to RNR

With the possible exception of those self-consciously engaged in ongoing critical discourse, we assume that challengers of RNR are committed to the ethical, legal, efficient, decent, and just pursuit of reduced victimization (recidivism) through human service delivery. We also assume a shared commitment to implementation of the least onerous interpretation of the sanction. In one sense, feminist and critical criminological critiques of RNR are an important and valuable reminder that a commitment to rational empiricism is fundamental to the design and improvement of classification procedures. Skepticism, indeed "unsparing criticism," is a major defining component of rational empiricism. Another component is respect for evidence. Thus, the rational empiricists involved in classification research have devoted energy to the documentation of reliability and validity issues and must continue to do so.

Assertions that "nothing is a risk in itself: there is no risk, in reality" and yet "anything can be a risk" is a postmodern diversion coming from acknowledged skeptics in regard to prediction (Hannah-Moffat & Shaw, 2001, pp. 12, 18). Skepticism, indeed, but consider further the following: "risk and the enterprise of risk management appears on the surface to be moral, efficient, objective, and non-discriminatory, but they are not" (p. 12). Consider this: "the compartmentalizing of risk identities is actually a spuriously correlated constellation of traits that, in reality, hinge upon the actual predictors of socioeconomic status, ethnicity, gender and age" (Rigakos, 1999, p, 145). Actually, in our opinion, the demonstrated superior predictive criterion validity of RNR assessments relative to social location variables is being trivialized and human diversity discounted.

Once again, everyone agrees that immoral and discriminatory practices are to be avoided. Indeed, some have argued "Failure to conduct actuarial risk assessment or consider its results is irrational, unscientific, unethical, and unprofessional" (Zinger, 2004, p. 607). To sample factors unrelated to offending in a risk assessment instrument is to invite overclassification. Positively, rational empiricists are fortunate that a new subfield of critical criminology has arisen to assist us in our agreed-on goal of decent and just applications. RNR researchers and practitioners (and RNR products and processes) are a focus of the new critical discourse of understanding risk management and the risk society. This is good because criticism can be helpful and perhaps particularly so when it comes from outside the RNR professional group.

Some feminists and some clinical psychologists have been concerned with the lack of attention paid to personal well-being in the RNR approach. This concern extends to a failure to recognize the special needs of female offenders in terms of victimization, poverty, ethnicity, child care, and so on. Tony Ward and colleagues (Ward & Brown, 2004; Ward & Stewart, 2003) more generally noted that RNR is too negative (reducing too much to risk and need), does not recognize strengths, and is not devoted to enhancing human potential and achievement. We strongly endorse explorations of the issues surrounding RNR and other principles. The idea of enhancing RNR through greater attention to human motivation is very attractive. However, reduction of criminal victimization and enhancement of well-being are too important to be pursued in other than an ethically informed rational empirical manner. One should explore alternative strategies and theories but be very careful about who is treated, what is targeted, the influence strategies employed, and the quality of the direct service staff and program management.

⊠ Conclusions and Directions for the Future

Advances in the psychology of criminal conduct, on which RNR is based, have been dramatic. Meta-analyses of the prediction and treatment literature have strongly advanced knowledge, and dissemination has been impressive. General personality and social cognition perspectives are the dominant theoretical position. The predictive criterion validity of actuarial assessments of major risk and/or need factors greatly exceeds the validity of unstructured clinical judgment. New studies will address the methodological, training, supervision, and other factors that account for validity estimates that reside on the lower and upper edges for various instruments.

Evidence supports the incremental predictive criterion validity of reassessments of the major criminogenic need factors, including acute needs. The limits of this aspect of risk and/or need assessment have yet to be explored in anything but a most preliminary manner. In the absence of reassessment, it is not apparent that the predictive criterion validity of the best constructed of 3G and 4G assessments of risk and or need exceeds that of the best constructed 2G assessments of risk.

Specific responsivity assessment has a history in corrections, and there are many promising leads. However, understanding the interactions of offender and treatment characteristics remains a high-priority issue. The existing personality-based systems need to be analyzed according to the elements of risk, need, and responsivity. Advancements in the domains of self-control, psychopathy, stages of change, treatment readiness, and strengths are highly relevant here. Overall, we also continue to recommend systematic exploration of the domains of interpersonal and/or cognitive maturity, gender, and ethnicity and/or culture as responsivity issues.

The promise of 4G assessments is that linkages among assessment and programming, and of each with reassessments, and ultimate outcome will be very rewarding in theory and practice. The value of the assessments resides in planning and delivering effective service. The 4G assessment instruments promote good planning and delivery. They should greatly enhance clinical supervision of direct contact staff members. The roles for noncriminogenic need influence strategies beyond social learning and/or cognitive-behavioral interventions, and general well-being as an outcome is of additional interest.

Positive findings from the new generation of studies using instruments such as the CPAI may well have a dramatic impact on the development and successful implementation of RNR. The available evidence, and that structural evidence is brand new, suggests that agency adoption of RNR is rewarded by enhanced public protection at least when assessment and service are conducted with integrity.

✒ References

Andrews, D. A., & Bonta, J. (2003). *The psychology of criminal conduct* (3rd ed.). Cincinnati, OH: Anderson.

Andrews, D. A., Bonta, J., &. Hoge, R. D. (1990). Classification for effective rehabilitation: Rediscovering psychology. *Criminal Justice and Behavior, 17,* 19–52.

Andrews, D. A., Bonta, J., & Wormith, S. J. (2004). *The Level of Service/Case Management Inventory (LS/CMI).* Toronto, ON: Multi-Health Systems.

Andrews, D. A., & Dowden, C. (2005). Managing correctional treatment for reduced recidivism: A meta-analytic review of program integrity. *Legal and Criminological Psychology, 10*(2), 173–187.

Andrews, D. A., Zinger, I., Hoge, R. D., Bonta, J., Gendreau, P., & Cullen, F. T. (1990). Does correctional treatment work? A psychologically informed meta-analysis. *Criminology, 28,* 369–404.

Bonta, J. (1996). Risk-needs assessment and treatment. In A. T. Harland (Ed.), Choosing correctional options that work: Defining the demand and evaluating the supply (pp. 18–32). Thousand Oaks, CA: Sage.

Bonta, J., Law, M., & Hanson, R. K. (1998). The prediction of criminal and violent recidivism among mentally disordered offenders: A meta-analysis. *Psychological Bulletin, 123,* 123–142.

Gendreau, P., Goggin, C., & Smith, P. (2002). Is the PCL-R really the "unparalleled" measure of offender risk? A lesson in knowledge cumulation. *Criminal Justice and Behavior, 29,* 397–426.

Gendreau, P., Little, T., & Goggin, C. (1996). A meta-analysis of the predictors of adult offender recidivism: What works! *Criminology, 34,* 575–607.

Gray, N. S., Taylor, J., Snowden, R. J., MacCulloch, S., Phillips, H., & MacCulloch, M. J. (2004). Relative efficacy of criminological, clinical, and personality measures of future risk of offending in mentally disordered offenders: A comparative study of HCR-20, PCL: SV, and OGRS. *Journal of Consulting and Clinical Psychology, 72,* 523–531.

Hannah-Moffat, K., & Shaw, M. (2001). *Taking risks: Incorporating gender and culture into the classification and assessment of federally sentenced women in Canada* [Policy research report]. Ottawa: Status of Women in Canada.

Hanson, R. K., & Bussière, M. T. (1998). Predicting relapse: A meta-analysis of sexual offender recidivism studies. *Journal of Consulting and Clinical Psychology, 66,* 348–363.

Hanson, R. K., & Morton-Bourgon, K. (2004). *Predictors of sexual recidivism: An updated meta-analysis* (User report 2004-02). Ottawa: Public Safety and Emergency Preparedness Canada.

Hemphill, J. F., & Hare, R. D. (2004). Some misconceptions about the PCL-R and risk assessment: A reply to Gendreau, Goggin, and Smith. *Criminal Justice and Behavior, 31,* 203–243.

Lowenkamp, C. T. (2004). *Correctional program integrity and treatment effectiveness: A multisite, program-level analysis.* Unpublished doctoral dissertation, University of Cincinnati, Ohio.

Lowenkamp, C. T., & Latessa. E. J. (2002). *Evaluation of Ohio's community based correctional facilities and halfway house programs* [Technical report]. Cincinnati OH: University of Cincinnati.

Nesovic. A. (2003). *Psychometric evaluation of Correctional Program Assessment Inventory (CPAI).* Unpublished doctoral dissertation, Carleton University, Ottawa, Canada.

Rigakos, G. S. (1999). Risk society and actuarial criminology: Prospects for a critical discourse. *Canadian Journal of Criminology, 41,* 137–151.

Van Voorhis, P. (1994). *Psychological classification of the adult male prison inmate.* Albany: State University of New York Press.

Ward. T., & Brown, M. (2004). The good lives model and conceptual issues in offender rehabilitation. *Psychology, Crime, and Law, 10,* 243–257.

Ward, T., & Stewart, C. (2003). Criminogenic needs and human needs: A theoretical model. *Psychology, Crime, and Law, 9,* 125–143.

Wong, S., & Hare, R. D. (2005). *Guidelines for a psychopathy treatment program.* Toronto, ON: Multi-Health Systems.

Zinger, I. (2004). Actuarial risk assessment and human rights: A commentary. *Canadian Journal of Criminology and Criminal Justice, 46,* 607–621.

DISCUSSION QUESTIONS

1. In your own words, explain the responsivity principle.

2. Why is it important for correctional counselors to have a firm grasp of the psychology of criminal conduct?

3. Are well-designed assessment tools superior to the experienced judgment of correctional counselors? Why are they superior, or why are they not superior?

❖

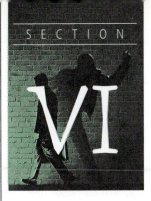

PROBATION AND COMMUNITY CORRECTIONS

Introduction

▧ The Origins of Probation

The practice of imprisoning convicted criminals is a relatively modern and expensive way of dealing with them. Up to two or three hundred years ago, they were dealt with by execution, corporal punishment, or humiliation in the stocks. All these punishments took place as community spectacles, and even with community participation in the case of individuals sentenced to time in the stocks. Assuming that a convicted person was not executed, he or she remained in the community, enduring the shame of having offended it (think of Hester Prynne's punishment in Nathaniel Hawthorne's *The Scarlet Letter*). Sex offenders are subjected to this kind of shaming today when their pictures are displayed on the Internet, and they are identified to their neighbors through community notification orders.

More enlightened ages saw punishments move away from barbaric cruelties and the emergence of the penitentiary, where offenders could contemplate the errors of their ways and perhaps redeem themselves while residing there. But, as we have seen, penitentiaries were not very nice places, and some benevolent souls in positions to do so sought ways to spare deserving or redeemable offenders from being consigned to them. This practice had its legal underpinnings in the *judicial reprieve* sometimes practiced in English courts in bygone days. A judicial reprieve was a delay in sentencing following a

▲ **Photo 6.1** Former professional baseball player Jose Canseco appears in a Miami-Dade (Florida) court during a bond hearing, June 23, 2003. Canseco was ordered to jail to await a hearing on whether he violated his probation by testing positive for steroids.

conviction, a delay that most often would become permanent contingent on the offender's behavior. In those days, there were no probation officers charged with supervising reprieved individuals; the nosey and censorious small communities typical then were more than adequate for that task.

Early American courts also used judicial reprieve whereby a judge would suspend the sentence and the defendant would be released on his or her own recognizance. However, reprieved offenders received no formal supervision or assistance to help them to mend their ways. The first real probation system in which a reprieved person was supervised and helped was developed in the United States in the 1850s by a Boston cobbler named John Augustus. Augustus would appear in court and offer to take carefully selected offenders into his own home where he would do what he could to reform them as an alternative to imprisonment. Probation soon became his full-time vocation, and he recruited other civic-minded volunteers to help him. By the time of his death in 1859, he and his volunteers had saved more than 2,000 offenders from imprisonment (Schmalleger, 2001).

In 1878, the Massachusetts legislature authorized Boston to hire salaried probation officers to do the work of Augustus's volunteers, and a number of states quickly followed suit. However, the probation idea almost died in 1916 when the United States Supreme Court ruled that judges may not indefinitely suspend a sentence (*Ex parte United States [Killits]*, 1916). In this case an embezzler was sentenced to five years imprisonment, which the judge (federal judge John Killits) suspended contingent on the embezzler's good behavior. What Killits had done was place an offender on a probation system without there having been such a system established by state law. This ruling led to the passage of the National Probation Act of 1925 allowing judges to suspend sentences and place convicted individuals on probation.

⊠ Why Do We Need Community Corrections?

Community corrections may be defined as any activity performed by agents of the state to assist offenders to establish or reestablish functional law-abiding roles in the community while at the same time monitoring their behavior for criminal activity. Monitoring and assisting offenders while they remain in the community protects society from criminal predation without the taxpayer having to shoulder the awesome financial cost of incarcerating all its villains. Even if as a society we were willing and able to bear the monetary cost of imprisoning all offenders, incarceration imposes other costs on the community. These costs can and must be borne where seriously violent and chronic criminals are concerned, but to consign each and every felony offender to a prison cell would be socially and morally counterproductive. Allowing relatively minor first- (perhaps even second-) time offenders to remain in the community under probation supervision to prove themselves (the term *probation* comes from the Latin *probatio*, meaning a period of proving or testing one's self) offers many benefits to the offender and to his or her community.

The general public has the notion that a probation sentence is "getting away with it," a notion not shared by many offenders. The probationer receives a prison sentence upon conviction, which is suspended during the period of proving that he or she is capable of living a law-abiding life. This sentence hangs over probationers' heads like the sword of Damocles, ready to drop if they fail to provide that proof. It may be for this reason that a number of studies have actually found that "experienced" offenders (i.e., offenders who have done prison time, probation time, and parole time) often prefer prison to probation, particularly day-reporting and intensive supervision probation (Crouch, 1993; May, Wood, Mooney, & Minor, 2005). Probation requires them to work, submit to treatment schedules, and comply with a number of conditions. Serving time in prison is less of a hassle for them, and many know they will end up there anyway because they will not live up to probation conditions (May et al., 2005).

Many other less criminally involved offenders obviously prefer probation. Of the over 4 million Americans on probation in 2005, 59% of them successfully completed their conditions of supervision and were released from probation (Glaze & Bonczar, 2006). In the event of a failure to live up to the conditions of probation, the prison sentence is then typically imposed. Thus, although many fail the probation period, more than half succeed, so surely providing nonviolent offenders the opportunity to try to redeem themselves while remaining in the community is sensible criminal justice policy. But what are the benefits to the community?

1. Probation costs between $700 and $1,000 per year as opposed to $20,000 to $30,000 per year for imprisonment, a saving to the taxpayer of at least $19,000 per year per nonincarcerated felon (Foster, 2006, p. 478). Moreover, many jurisdictions require probationers to pay for their own costs of supervision, which means that the taxpayer pays nothing.

2. Employed probationers stay in their communities and continue to pay taxes; offenders who were unemployed at the time of conviction may obtain training and help in finding a job. This adds further to the tax revenues of the community and, more important, allows offenders to keep or obtain the stake in conformity that employment offers. A job also allows them the wherewithal to pay fines and court costs, as well as restitution to victims.

3. In the case of married offenders, community supervision maintains the integrity of the family whereas incarceration could lead to its disruption and all the negative consequences such disruption entails.

4. Probation prevents felons from becoming further embedded in a criminal lifestyle by being exposed to chronic offenders in prison. Almost all prisoners will leave the

institution someday, and many will emerge harder, more criminally sophisticated, and more bitter than they were when they entered. Furthermore, they are now ex-cons, a label that is a heavy liability when attempting to reintegrate into free society.

5. Many more offenders get into trouble because of deficits than because of pathologies. Deficits such as the lack of education, a substance abuse problem, faulty thinking patterns, and so forth, can be assessed and addressed using the methods discussed in Section V. If we can correct these deficits to some extent, then the community benefits because it is a self-evident truth that whatever helps the offender protects the community.

We do not wish to seem Pollyannaish about this; there are people who are incapable of making it on probation. For instance, among criminals in the witness protection program for having given evidence against high-level criminals, despite being given new identities, jobs, housing, and basically a new start in life, 21% are arrested under their new identities within two years of entry (Albanese & Pursley, 1993. p. 75). These criminals are used to relatively high incomes made in a life filled with excitement and personal power and independence, and many find it extremely difficult to adjust to a mundane job with minimal financial rewards. Furthermore, economic considerations are not the primary concern of corrections; protecting the community is. Community-based corrections is the solution only for those offenders who do not pose a significant risk to public safety.

▧ The Decreasing Probation-Inmate Ratio

Although community corrections is sensible justice policy, and more people are under such supervision than ever before, there has been a decline in its use from its heyday in the rehabilitation-oriented 1970s and 1980s, relative to the use of incarceration. Figure 6.1 shows that use of probation has gone up from just over three million probationers in 1990 to just over four million in 2005 (Glaze & Bonczar, 2006). However, the ratio of proba-tioners to prison inmates was 4.02:1 in 1980 and 2.91:1 in 2004. This change is more the result of the increased reliance on incarceration than on the decreased use of probation and is, partially, a consequence of "get tough" policies such as mandatory sentences, three strikes laws, and truth in sentencing laws.

▧ The Probation or Prison Decision

In addition to complying with the "get tough" laws, some high profile cases of individu-als committing crimes while on probation have rendered many judges more careful about to whom they will grant the privilege of probation. Some offenses are nonprobationable by statute; other offenses are so serious, or the offenders so persistently antisocial, that probably no judge would consider granting probation.

To assist judges in making the "in/out" (prison/probation) decision, presentence investigation reports (PSIs) and sentencing guidelines are commonly used. PSIs are writ-ten by probation officers informing the judge about various aspects of the offense for which the defendant is being sentenced as well as about the defendant's background (edu-cational, family, and employment history), character, and criminal history. On the basis

| **Figure 6.1** | Number of Persons Under Correctional Supervision, 1990–2005, by Type of Supervision |

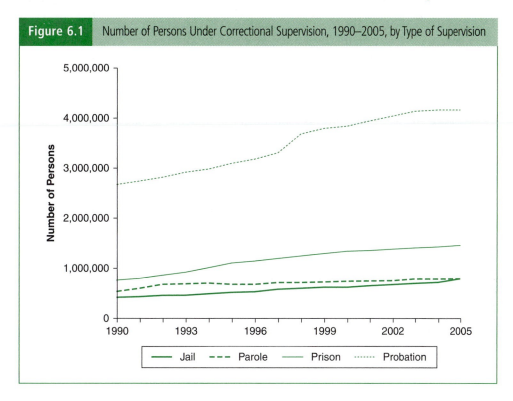

of this information, officers make recommendations to the court regarding the sentence the offender should receive. Because probation officers enjoy considerable discretionary power relating to the favorableness or unfavorableness of their reports, many scholars view them as the agents who really determine the sentences that offenders receive (Champion, 2005).

The advent of sentencing guidelines, however, has somewhat curtailed the discretionary powers of both judges and probation officers. As the name implies, sentencing guidelines were devised to guide the sentencing decisions of judges by providing them with standard criteria for tailoring sentences to the crime and the offenders before the bench. In those 16 states that use them, the typical guideline assigns numerical scores to various aspects of the crime for which an offender is being sentenced (statutory degree of seriousness, amount of harm done, if the offender was on bail, probation, or parole at the time, and so on), and his or her criminal history (prior felonies and misdemeanors, prior periods of incarceration, and so forth). These numbers are then applied to a grid at the point at which they intersect, which contains the appropriate sentence. In some jurisdictions, adherence to the guidelines is mandatory; in others, it is merely discretionary (the Supreme Court has recently concluded that the formerly mandatory federal sentencing guidelines are now only advisory). In any case, sentencing guidelines are the criminal justice system's way of attempting to put into practice Aristotle's famous definition of justice: "Justice consists of treating equals equally and unequals unequally according to relevant differences" (cited in Walsh 2006, p. 125).

According to Lubitz and Ross (2001), guidelines have achieved a number of outcomes reasonably consistent with this definition; for example, they have

1. Reduced sentencing disparity,

2. Achieved more uniform and consistent sentencing,

3. Made the sentencing process more open and understandable,

4. Decreased punishment for certain categories of offenses and offenders and increased it for others,

5. Aided in prioritizing and allocating correctional resources, and

6. Provided a rational basis for sentencing, and increased judicial accountability.

≋ Community Corrections Assessment Tools

Assuming probation placement, the information in the PSI is supplemented by a variety of assessment tools used to guide probation officers in their supervision and rehabilitative duties. Among these various assessment instruments is the Client Management Classification System (CMC), the effectiveness of which is addressed in the article by Patricia Harris, Raymond Gingerich, and Tiffany Whittaker in this section. The CMC contains offender risk and needs scales that embody the principle of responsivity discussed in this text's section on rehabilitation, and it is also used to determine the level of supervision that offenders receive.

It goes without saying that some offenders need more services and higher levels of behavioral monitoring than others and that to treat all of them as equals in this respect would be counterproductive. There is evidence that low-risk offenders (a category known as *selective intervention* in the CMC system) actually become worse if they are over-supervised and subjected to treatment modalities that they do not need (LowenKamp & Latessa, 2004). Placing such offenders in the same restrictive programs as high-risk offenders exposes them to bad influences and may disrupt the very factors (family, employment, prosocial activities and contacts) that made them low-risk in the first place. The other supervisory categories are *environmental structure* (offenders at the low end of medium risk), *casework/control* (offenders at the medium to high end of medium risk), and *limit setting* (high-risk offenders).

Harris, Gingerich, and Whittaker sound a pessimistic note on the effectiveness of the CMC. Dividing their sample into an experimental group of offenders receiving "differential supervision" as dictated by the CMC and a control group of regularly supervised clients, the researchers uncover disconcerting and surprising results that reflect poorly on the efficacy of the CMC as a means of shaping offender behavior in prosocial directions. Differentially supervised offenders were no less likely to be arrested than control offenders, were *more* likely to have technically violated their conditions of probation, but were less likely to have their probation revoked. If revocation had been the only measure of effectiveness, Harris and her colleagues could have reported findings in favor of the effectiveness of the CMC, but the inclusion of other measures of effectiveness (arrests, technical violations) led them to a different conclusion. The researchers did not attempt to discredit the CMC but rather faulted the probation officers' lack of compliance with the full range of its recommended classification and supervision practices. They provided a number of reasons for their unwelcome findings, which basically sum up to the old saying that any tool is only as good as the dedication and skills of those who use it.

▨ Probation Officer Stress

One of the reasons that probation officers in the Harris, Gingerich, and Whittaker study did not properly follow the recommendations of the CMC is that many have become cynical about their jobs and what they are trying to accomplish. In common with police officers and correctional officers, probation and parole officers are dealing with difficult and needy people on a daily basis without the tools and support needed to do the job. As the article by Risdon Slate, Terry Wells, and W. Wesley Johnson in this section points out, doing a demanding and sometimes dangerous job under less than adequate conditions can, and does, lead to stress.

Stress is essentially a physical and emotional state of tension as the body reacts to environmental challenges (stressors). Factors that provoke stress are everywhere, and the stress response (the "fight or flight" response) is an evolved adaptation that helps us to meet the challenge. A certain amount of stress is good for us because it is a motivator and activator that keeps us on our toes. However, protracted stress leads to all kinds of physical and mental problems such as headaches, stomach and chest pains, anxiety, panic, insomnia, depression, and irritability. No one can be expected to do a very good job while experiencing constant (albeit usually mild) stress, although obviously not all stressors lead to all kinds of stress reactions.

The most important job stressors identified by the officers surveyed by Slate, Wells, and Johnson were poor salaries, poor promotion opportunities, excessive paperwork, lack of resources from the community, large caseloads, and a general frustration with the inadequacies of the criminal justice system. These stressors eventually lead to psychological withdrawal from the job, which means that probationers, and thus the community, are getting shortchanged. High stress levels in the department also led to frequent absenteeism and high rates of employee turnover; thus, it is imperative that the issue of probation and parole officer stress be meaningfully addressed.

The authors emphasize that the attempt to address the problem of probation officer stress should not be one of counseling officers on how to cope with stress because the problem is organizational (inherent in the probation system) not personal. They suggest that participatory management strategies be instituted so that each person in the department participates in the decision-making process and thus feels valued and empowered. The researchers found that personnel who did participate in decision making reported fewer stress symptoms and were happier on the job. Participatory management (workplace democracy) leads to a happier and more productive workforce, even if nothing else changes—"contented cows give better milk."

▨ Engaging the Community to Prevent Recidivism

The criminological literature provides abundant support for the notions that social bonds (Hirschi, 1969) and social capital (Sampson & Laub, 1999) are powerful barriers against criminal offending. Social bonds are connections (often emotional in nature) to others and to social institutions that promote prosocial behavior and discourage antisocial behavior. Social capital refers to a store of positive relationships in social networks built on norms of reciprocity upon which the individual can draw for support. It also means, almost as a corollary, that a person with social capital has acquired an education and other solid credentials that enable him or her to lead a prosocial life. Those who have opened their social capital accounts early in life (bonding to parents, school, and other prosocial networks) may spend much of it freely during adolescence but, nevertheless, manage to salvage a sufficiently tidy nest egg by the time they reach adulthood to keep them on the straight and narrow. The idea is that they are not likely to risk losing this nest egg by engaging in criminal activity. Most correctional clients, however, lack social bonds, and, largely because of this, they lack the stake in conformity provided by a healthy stash of social capital.

If we consider the great majority of felons in terms of deficiency (good things that they *lack*), we are talking about a deficiency in social capital. The community can be seen as a bank in which social capital is stored and from which offenders can apply for a loan. That is, the community is the repository of all of those things from which social capital is derived, such as education, employment, and networks of prosocial individuals in various organizations and clubs (e.g., Alcoholics Anonymous, churches, hobby or interest centers, and so on). Time spent in involvement in steady employment and with prosocial others engaged in prosocial activities is time unavailable to spend in idleness in the company of antisocial others planning antisocial activities. The old saying that "the devil finds work for idle hands" may be trite, but it is also very true.

Thus, good case management in community corrections requires getting correctional clients involved in their communities. No community corrections agency is able to address the full range of offender needs (mental health, substance abuse, vocational training, welfare, etc.) by itself. Probation and parole officers must not only assess the needs of their charges but also be able to locate and network with the social service agencies that address those needs as their primary function. The community corrections worker is a broker who takes input about offenders from offenders themselves, police, courts, friends, neighbors, family, employees, and so on, and refers the problem out to the appropriate agency. This brokerage function can be best achieved with fewer offenders who are intensively supervised on an officer's caseload than with many who are infrequently seen and haphazardly supervised.

Intermediate Sanctions

Intermediate sanctions refer to a number of innovative alternative sentences that may be imposed in place of the traditional prison/probation dichotomy. Such sanctions are considered intermediate because they are seen as more punitive than straight probation but less punitive than prison. They are also a way of easing prison overcrowding and the financial cost of prison while providing the community with higher levels of security from victimization through higher levels of offender surveillance than is possible on regular probation. As we shall see, these supposed benefits are not always realized. We have already seen that many experienced offenders would choose prison over some of the more strict community-based alternatives. Furthermore, because offenders placed in them have recidivism rates not much different from offenders released from prison within the first and subsequent years, the costs of state incarceration are deferred rather than avoided (Marion, 2002). The first alternative we examine is intensive supervision probation (ISP).

◤ Intensive Supervision Probation

ISP involves more frequent surveillance of the probationer and is typically limited to more serious offenders than those on regular probation in the belief that there is a fighting chance that they may be rehabilitated or in an effort to save the costs of incarceration. ISP is examined in Doris Mackenzie and Robert Brame's article in this section in which they hypothesize that ISP supervision coerces offenders into prosocial activities which, in turn, leads to a lower probability of them reoffending. Intensive supervision means that probation and parole officers maintain more frequent contact with probationers or parolees and intrude into their lives more than is the normal with other probationers or parolees. Intensive supervision offenders are supervised at that level because they have the greatest probability of reoffending (they are high risk) and are the most deficient in social capital (they have high needs). Higher levels of supervision allow officers to coerce offenders into a wide variety of educational and treatment programs and other prosocial activities designed to provide offenders with social capital.

The term *coercion* has negative connotations for the more libertarian types among us ("You can lead a horse to water . . ." and all that), but the great majority of people being treated for problems such as substance abuse have very large boot prints impressed on their backsides. Sometimes this kind of coercion is merited for those probationers and parolees who will not voluntarily place themselves in the kinds of programs and activities we would like them to be in. The criminal justice system must provide that motivation via the judicious use of carrots and sticks. Reviews of the U.S. (Farabee, Pendergast, & Anglin, 1998) and U.K. (Barton, 1999) literature on coerced substance abuse treatment concluded that coerced treatment often has more positive outcomes than voluntary treatment, probably because of the threat of criminal justice sanctions.

Although Mackenzie and Brame's results are equivocal, overall they find them to be acceptably consistent with the model they proposed. That is, intensive supervision did result in offenders being coerced into more prosocial activities, and there was a slight reduction in recidivism. The issue with which they grapple is whether participating in prosocial activities in the short term enabled offenders to acquire the skills that provided them with social capital that they put to good use or whether intensive supervision per se accounted for the researchers' findings. They end by suggesting strategies for further evaluating their hypothesis, including one that we heartily endorse, namely, random assignment of offenders to different levels of supervision (and perhaps different levels of coercion into prosocial activities).

◤ Work Release

Work release programs are designed to control offenders in a secure environment while at the same time allowing them to maintain employment. Work release centers are usually situated in or adjacent to a county jail. Residents of work release centers have typically been given a suspended sentence and placed on probation with a specified time to be served in work release but they may also be parolees under certain circumstances. Surveillance of work release residents is strict; they are allowed out only for the purpose of attending their employment and are locked in the facility when they are not working. The advantage of such programs is that they allow offenders to maintain ties with their families and with employers. Such programs also save the taxpayer money because offenders pay the cost of their accommodation with their earnings.

Offenders on work release are generally the least likely of all community-based corrections offenders to be rearrested and imprisoned within one year and five years of successful completion (Marion, 2002). However, 64% of offenders successfully released and 71% unsuccessfully released had further arrests within five years (Marion, 2002). Offenders selected to partake in work release are typically chosen because, although they have committed a crime deemed too serious for regular probation, they are employed. Being employed is incompatible with a criminal lifestyle (although obviously from the above statistics, not completely), especially if the offender is a probationer rather than a parolee.

⬧ Shock Probation/Parole and Boot Camps

Shock probation was initiated in Ohio in the 1970s and was designed to literally shock offenders into desisting from crime. It was limited to first offenders who had perhaps been unimpressed with the realities of prison life until given a taste. Under this program, offenders were sentenced to prison and released after (typically) 30 days and placed on probation. In some states, a person may receive shock parole, which typically means that he or she has remained in prison longer than the shock probationer and is released under the authority of the parole commission rather than the courts. Most of the research on this kind of shock treatment was conducted in the 1970s and 1980s and concluded that shock probationers or parolees had lower recidivism rates than incarcerated offenders not released under shock conditions (Vito, Allen, & Farmer, 1981). This should not be surprising given the fact that those selected for shock probation or parole were either first offenders or repeaters who had not committed very serious crimes.

When we hear of shock incarceration today it is typically incarceration in a so-called boot camp. Boot camp placement is the subject of the article by Jean Bottcher and Michael Ezell in this section. Correctional boot camps are facilities modeled after military boot camps. Relatively young and nonviolent offenders are most typically the kinds of offenders sent to correctional boot camps for short periods (90–180 days) where they are subjected to military-style discipline and physical and educational programs. Boot camps are most unpopular with offenders. In May et al.'s (2005) analysis of "exchange rates" discussed earlier, offenders who had served time in prison would only be willing to spend an average of 4.65 months in boot camp to avoid 12 months in prison.

The idea of boot camps for young or first-time offenders was once a popular idea among the general public, as well as among a considerable number of correctional personnel and criminal justice academics. Boot camps conjured up the movie image of a surly, slouching, and scruffy misfit forced into the army who, 2 years later, proudly marched back into the old neighborhood sporting a crew cut, sparklingly clean, and properly motivated and disciplined. Yes, the drill sergeant with righteous fire and brimstone would do what the family and social work–tainted juvenile probation officers could never do—make a silk purse out of a sow's ear.

Of course, such magical transformations rarely happened in real life. The army merely provided many such scoundrels with new opportunities to offend, and they spent much of their time either avoiding the MPs or lodged in the brig while awaiting their dishonorable discharges. Bottcher and Ezell's evaluation of offenders sent to correctional boot camp revealed the same sorry outcome. Specifically, they found no significant differences between their experimental group (boot campers) and a control group of similar offenders not sent to boot camp in terms of either property or violent crime reoffending. In other words, boot camps have joined the woeful list of correctional programs that have proven ineffective.

▨ Victim-Offender Reconciliation Programs (VORPs)

VORPs are an integral component of the restorative justice philosophy. This philosophy differs from models (retributive, rehabilitative, etc.) that are offender driven (what do we do with the offender?) in that it considers the offender, the victim, and the community as partners in restoring the situation to its pre-victimization status. Restorative justice has been defined as "every action that is primarily oriented toward justice by repairing the harm that has been caused by the crime," and it "usually means a face-to-face confrontation between victim and perpetrator, where a mutually agreeable restorative solution is proposed and agreed upon" (Champion, 2005, p. 154). Restorative justice is often referred to as a balanced approach in that it gives approximately equal weight to community protection, offender accountability, and offender competency.

Many crime victims are seeking fairness, justice, and restitution *as defined by them* as opposed to revenge and punishment. Central to the VORP process is the bringing together of victim and offender in face-to-face meetings mediated by a person trained in mediation theory and practice. Meetings are voluntary for both offender and victim and are designed to iron out ways in which the offender can make amends for the hurt and damage caused to the victim.

Victims participating in VORPs gain the opportunity to make offenders aware of their feelings of personal violation and loss and to lay out their proposals of how offenders can restore the situation. Offenders are afforded the opportunity to see firsthand the pain they have caused their victims, and perhaps even to express remorse. The mediator assists the parties in developing a contract agreeable to both. The mediator monitors the terms of the contract and may schedule further face-to-face meetings.

VORPs are used most often in the juvenile system but rarely for personal violent crimes in either the juvenile or adult systems. Where they are used, about 60% of victims invited to participate actually become involved, and a high percentage (mid to high 90s) results in signed contracts (Coates, 1990). Mark Umbreit (1994) sums up the various satisfactions expressed by victims who participate in VORPs:

1. Meeting offenders helped reduce their fear of being revictimized.

2. They appreciated the opportunity to tell offenders how they felt.

3. Being personally involved in the justice process was satisfying to them.

4. They gained insight into the crime and into the offender's situation.

5. They received restitution.

VORPs do not suit all victims, especially those who feel that the wrong done to them cannot so easily be "put right" and who want the offender punished (Olson & Dzur, 2004). Additionally, the value of VORPs for the prevention of further offending has yet to be properly assessed.

⬚ Summary

- Community-based corrections is a way of attempting to control the behavior of criminal offenders while keeping them in the community. Although conditional release (judicial reprieve) was practiced in ancient times in common law, it wasn't until the mid-nineteenth century that such individuals were supervised by officers of the court, whom we call probation officers today. Although often considered too lenient, community corrections benefits the public in many ways, not the least of which are huge financial savings. Probation rather than prison also helps the offender in many ways, and what helps offenders to develop prosocially automatically helps the communities to which they belong.

- Because of many "get tough" policies, and although there are more Americans than ever under community supervision, the use of community corrections is declining relative to the use of incarceration. To make better decisions about whom to grant probation, judges use presentence investigation reports and sentencing guidelines. These devices, while far from perfect predictors of success or failure on probation, have helped to make the sentencing process more fair and rational.

- The Client Management Classification System (CMC) is another innovation designed to guide supervision and treatment strategies for offenders placed on probation. As with any tool, however, the CMC is only as good as the skill and conscientiousness with which it is used. The level of stress of the probation officers' job leads many of them to burn out and to put less than adequate care and effort into attending to the many and varied tasks they need to carry out.

- For those lucky enough to possess them, social bonds and social capital are powerful inducements to remain crime free. Those lacking them are relatively free to follow the impulse of the moment, a "freedom" that often gets them into trouble with the law. It makes sense, then, to try to provide offenders with opportunities to form social bonds with prosocial others and to earn social capital, even if they have to be coerced. Probation officers with relatively small intensively supervised caseloads have the time to spend on pushing their charges in this direction, evidently with some positive results. However, we cannot always change ingrained character problems so easily, as the dismal failure of the boot camp experiment has shown.

- Intermediate sanctions are considered to be more punitive than regular probation but less punitive than prison, although experienced correctional clients do not necessarily share that view. Some of these programs, particularly work release, show positive results although this is doubtless more a function of the kinds of offenders placed in them rather than the programs themselves. Most participants in these programs, however, tend to recidivate at rates not significantly different from parolees.

- The victim-offender rehabilitation program (VORP) is a fairly recent addition to community corrections. It considers the victim, the offender, and the community as equal partners in returning the situation to its pre-victimization status. This idea of restorative justice is mostly used with juvenile offenders and adults convicted of minor offenses.

KEY TERMS

Community corrections

Correctional boot camps

Intensive Supervision Probation

Judicial reprieve

Presentence investigations

Probationer

Restorative justice

Sentencing guidelines

Shock probation

Work release programs

DISCUSSION QUESTIONS

1. Looking at all the pros and cons of community-based corrections, do you think probation is too lenient for felony offenders? If so, what do we do with them?

2. In your opinion, what is the single biggest benefit of probation for the community and its single biggest cost?

3. Studies show that about 95% of probation officers' sentencing recommendations to the courts are followed. Why do you think this is so, and is it a good or bad thing that probation officers control the flow of information to judges?

4. If the CMC is an adequately reliable instrument for structuring supervision and treatment strategies along rational lines, why do you think probation officers are not (at least according to the Harris et al. study) adhering to them?

5. Boot camps have full and total control of offenders for up to 6 months, so why are they not able to change offenders' attitudes and behaviors?

INTERNET SITES

American Probation and Parole Association: www.appa-net.org

Bureau of Justice Statistics (information available on all manner of criminal justice topics): www.ojp.usdoj.gov/bjs

Center for Sex Offender Management: www.csom.org

National Criminal Justice Reference Service: www.ncjrs.gov

National Institute of Mental Health: www.nimh.nih.gov

National Institute on Drug Abuse: www.drugabuse.gov

Office of Justice Programs, Bureau of Justice Statistics (periodic statistical reports on all manner of criminal justice topics, e.g., HIV in prisons and jails, probation and parole, and profiles of prisoners): www.ojp.usdoj.gov/bjs/periodic.htm

Vera Institute (information available on a number of corrections and other justice related topics): www.vera.org

READING

Doris MacKenzie and Robert Brame examine the relationship between intensity of community supervision and recidivism. Prior research leads to two possible conclusions: (1) there is no consistent relationship, or (2) the relationship is indirect. In their paper, Mackenzie and Brame consider a theoretical model that is consistent with the latter possibility. They hypothesize that increased intensity of supervision leads to an increased involvement in conventional and therapeutic activities; this, in turn, should be associated with a reduction in future offending using data from four states. From their findings, they draw the following conclusions: (1) intensity of supervision is associated positively with involvement in prosocial activities, (2) intensity of supervision is not associated consistently with involvement in new criminal activities, and (3) an inverse association exists between involvement in prosocial activities and involvement in new criminal activities after adjusting for measured factors that are thought to affect both outcomes.

Community Supervision, Prosocial Activities, and Recidivism

Doris Layton MacKenzie and Robert Brame

Theorists in the control tradition (Gottfredson and Hirschi 1990; Hirschi 1969; Nye 1958; Sampson and Laub 1993) have emphasized the importance of coercion in normal socialization processes. Although many control theorists focus on the importance of coercion and direct control during childhood, this approach also may be useful for explaining behavior changes and transitions during adulthood. From this perspective, both youthful and adult offenders may be coerced into becoming involved in prosocial behavior patterns that can increase their investments in conformity and their attachments to other people, employment, school, or the community. That is, the direct control of behavior through legal coercion is predicted to increase compliance with conventional norms, rules, and prosocial activities. These activities, in turn, may, enhance the socialization process leading to increased social controls.

In this paper we hypothesize that the use of coercive interactions beyond childhood will be useful for decreasing the probability of future criminal behavior. Specifically, coercion of offenders to engage in prosocial behaviors should be associated with increased involvement in such behaviors. In addition, individuals who exhibit the greatest involvement in prosocial behaviors will exhibit reduced risk of subsequent involvement in criminal activity. Figure 1 presents a summary of the hypothesized relationships.

SOURCE: Community Supervision, Prosocial Activities, and Recidivism, Doris Layton MacKenzie and Robert Brame; *Justice Quarterly* 18(2): 429–448 (2001); Taylor & Francis, Ltd., http://www.informaworld.com, reprinted by permission of Taylor & Francis, Ltd.

Figure 1	Hypothesized Model of Relationship Between Supervision Intensity, Prosocial Activities, and Recidivisms

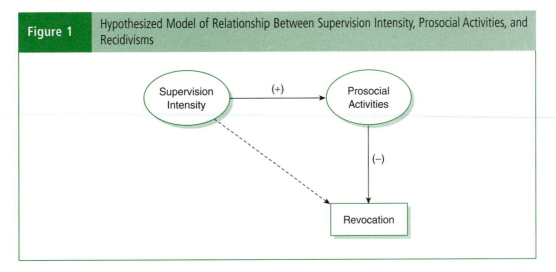

Socialization and the Internalization of Social Norms

The importance of coercion is perhaps most apparent in the literature on the implications of parenting practices for youths' future antisocial behavior. Rankin and Wells (1990:142), following Nye (1958), argue that coercion or "direct controls" on behavior take the form of "normative regulation" (i.e., the establishment of ground rules for children), "monitoring" (i.e., surveillance or direct supervision), and discipline or punishment for misbehavior. An important feature of such coercion is that it involves the "immediate application (or threat) of punishments and rewards to gain compliance with conventional norms" (Rankin and Wells 1990:142).

Coercive interactions in themselves have not traditionally been viewed as viable long-term solutions for controlling problem behavior (Gottfredson and Hirschi 1990:94–105; Nye 1958:7; Sampson and Laub 1993:65–71). The long-term value of coercive and controlling practices lies in their ability to induce socialization, for example, compliance with and (indeed) internalization of social norms (Gottfredson and Hirschi 1990:97; Sampson and Laub 1993:67). This individual internalization creates a web of disincentives for engaging in antisocial behaviors (Dishion and Patterson 1997; Nagin and Paternoster 1993, 1994).

Although some theorists assert that the tendency toward antisocial behavior is established early in life and does not change thereafter (Gottfredson and Hirschi 1990; Wilson and Herrnstein 1985), others take a different view (Nagin and Paternoster 1993, 1994; Sampson and Laub 1993). According to the latter, social bonds or "capital" that can be acquired in childhood, adolescence, or adulthood are predicted to be important barriers to offending in adulthood. From this viewpoint, individuals who have few social skills or who engage in frequent antisocial behavior are hypothesized to be relatively unlikely to acquire and maintain valuable relationships and ties to conventionality, but the possibility is not ruled out. For example, on the basis of their research on the Glueck and Glueck (1950) subjects, Sampson and Laub (1993) contend that factors such as conventional occupations and affective attachments to spouses constitute key barriers to adult offending, even among adults who offended as youths. Such attachments are associated with reduced criminal activity and are important in ending criminal careers. Yet, if social controls such as these are not strengthened, argue Sampson and Laub, offending is likely to continue.

In support of theories that identify temporally proximate causes of offending during the adult years, Horney, Osgood, and Marshall (1995) have shown how initiation and disruption of both informal and formal social controls, even if temporary or transient, can account respectively for short-term desistance from and resumption of criminal offending. Although these sources of social control may not produce meaningful increases in long-term socialization (i.e., "deep" change), Horney et al. (1995:670) speculate that these influences are likely to be important discriminators between those who exhibit meaningful change and those who do not.

Control theorists have not fully specified factors leading to the development of social controls beyond childhood. The etiology appears to be exogenous or even fortuitous (Laub and Sampson 1993:317–18; Nagin and Paternoster 1994:582–83; Sampson and Laub 1993:141–42, 250). Thus social controls may develop through chance or highly circumstantial occurrences. In short, as Nagin and Paternoster observed, the development of social control may represent a critical pathway leading away from offending, but there is little available evidence or theoretical guidance indicating how these controls might develop initially.

In this paper we consider the idea that direct controls over the behavior of adult offenders may start a process of socialization, as in childhood. That is, the imposition of direct controls may coerce offenders to participate in prosocial activities; the activities, in turn, may initiate changes in offenders that increase their ties or commitments to conventional behavior patterns; these ties, in turn, create barriers to future offending. Thus intensive supervision is important not because of its direct impact on

criminal activity, but because of its influence on the initiation of prosocial activities.

Intensive Community Supervision

Intensive supervision programs typically subject offenders to more frequent contact with supervising officers and compel them to engage in certain activities as conditions of supervision. Research on intensive community supervision has not yet found a consistent relationship between increased supervision and reduced recidivism (see Clear and Braga 1995; Erwin 1986; Land, McCall, and Williams 1990; Lurigio and Petersilia 1992; Pearson 1988; Petersilia and Turner 1993; Turner, Petersilia, and Deschenes 1992). Yet this research has demonstrated with some consistency, that offenders in intensive supervision programs are more likely to enter counseling and treatment programs and to secure employment. This finding suggests that they are coerced, in some sense, into participating in these prosocial activities (Land et al. 1990; Latessa and Vito 1988; Pearson 1988; Petersilia and Turner 1993; Turner et al. 1992).

In line with the evaluations of intensive supervision, research on releasees from boot camps, conducted by MacKenzie and her colleagues (MacKenzie, Shaw, and Souryal 1992; MacKenzie and Souryal 1994), also yielded inconsistent results when these authors examined the direct association between supervision intensity (for probationers, parolees, and boot camp releasees) and recidivism. They found evidence, however, that the intensity of supervision was associated with increased involvement in prosocial activities such as accepting responsibility for actions, achieving financial and residential stability, making satisfactory progress in treatment and education programs, and job stability.

The drug treatment literature offers further evidence that prosocial activities can be coerced. Anglin and Hser (1990), for example, reviewed a large number of drug treatment evaluation studies. Some of these evaluations focused on the effects of legally coerced drug treatment on subsequent behavior. Traditionally the major objection to coerced treatment is that therapeutic interventions will not be effective unless the treated individual is motivated to respond. Yet Anglin and Hser concluded, from their review, that clients who participated in legally coerced treatment stayed in treatment for longer periods and that the length of participation was associated with better outcomes regardless of whether treatment was coerced (also see DeLeon 1988; Gendreau, Cullen, and Bonta 1994; Wexler, Lipton, and Johnson 1988; Wish and Johnson 1986).

Overall, previous research found no consistent relationship between intensity of community supervision and recidivism. Studies of drug treatment and intermediate sanctions, however, suggest that it may be possible to legally coerce involvement in prosocial activities. We hypothesize that involvement in such activities will be associated with a reduced probability of future offending. Unlike earlier analyses, this investigation is concerned with the relationship between involvement in prosocial activities and revocation for new criminal activity after taking into account all measured factors (including supervision intensity) thought to affect both outcomes. In previous analyses, these outcomes were examined separately; here we focus on their joint distribution. Although our analysis suffers several important limitations (which we discuss below), we believe that it provides a useful initial assessment of the problem and offers a framework for future analyses of this problem.

⬚ Methods

Research Design

The data for this study were collected as part of a larger eight-state evaluation of boot camp prisons (also called shock incarceration

programs) (MacKenzie 1994; MacKenzie and Souryal 1994). In the analysis presented here we relied on data collected in four of these states (Florida, Georgia, Louisiana, and South Carolina), where data sufficient for examining the hypothesized links (see Figure 1) were available. Within each state, offenders from several comparison subsamples (boot camp completers, dropouts, prison parolees, and probationers) were followed during the first year of community supervision. Offenders were not assigned randomly to these samples, but they all met formal eligibility criteria for inclusion in the boot camp sample in their respective states. These subsamples' comparative performances have been examined elsewhere (MacKenzie 1994; MacKenzie and Souryal 1994). Here we emphasize the empirical validity of the hypothesized patterns of association between supervision intensity, involvement in prosocial activities, and recidivism (as measured by revocation for new criminal activity).

Study Variables

As shown in Table 1, the samples from the different states varied somewhat. Florida had the largest number of dropouts from the boot camp in the sample and included only prison releasees in the comparison sample. For the other three states, the samples included both probationers and prison releasees. Also, compared with the other states, the Florida sample contained more violent offenders and fewer convicted of drug-related or property offenses, and the offenders were younger. Louisiana offenders were older, and more of them had a prior record of offending. Small percentages of cases were lost because of missing data in each state: Florida, 5.2 percent; Georgia, 9.2 percent; Louisiana, 11.1 percent; and South Carolina, 5.3 percent. All information was based on offenders' official records; all subjects were males.

Variables available for analysis included subsample membership (boot camp, prison, probation), age (in years) at the beginning of

Table 1	Descriptive Statistics							
	Florida (N = 274)		Georgia (N = 238)		Louisiana (N = 247)		South Carolina (N = 230)	
Variables	Mean (SD) or %	Range	Mean (SD) or %	Range	Mean (SD) or %	Range	Mean (SD) or %	Range
Sample Categories								
Boot camp completers (%)	38.0		30.7		27.9		35.2	
Boot camp dropouts (%)	23.7				5.7			
Probationers (%)			32.4		37.7		37.8	
Prison releasees (%)	38.3		37.0		28.7		27.0	
Individual Characteristics								
Race (% nonwhite)	56.9		60.9		64.4		55.2	
Age (in years)	19.4 (1.8)	6 to 25	21.7 (2.8)	17 to 33	25.1 (5.4)	17 to 47	21.1 (2.3)	18 to 30
Current Offense								
% Violent	32.8		14.3		9.3		13.9	
% Drug-related	14.6		27.7		29.6		22.6	
% Property & other	52.6		58.0		61.1		63.5	
Prior offending record (% Yes)	27.4		31.1		83.4		58.7	
Community Supervision Data								
Supervision intensity	1.4 (.9)	0 to 4.2	1.0 (.5)	0 to 3.4	.7 (.8)	0 to 4	.9 (.5)	0 to 2.6
Prosocial activities	.0 (1.0)	−1.4 to 2.3	.0 (1.0)	−1.7 to 2.5	.0 (1.0)	−2.6 to 2.7	.0 (1.0)	−1.7 to 2.0
% revoked by end of 12 months	11.3		16.4		6.5		9.6	

community supervision, race (1 = nonwhite, 0 = white), and presence of a prior offending record (1 = yes, 0 = no). We also included indicator variables for type of offense (1 = violent, 0 = otherwise; 1 = drug-related offense, 0 = otherwise; 1 = property/other, 0 = otherwise). Although we control for all of these variables in the analyses below, we are most interested in the empirical associations between intensity of supervision, involvement in prosocial activities, and revocation for new criminal activity.

Supervision intensity is the independent variable in which we are primarily interested. In three of the four states (excluding Louisiana), supervising officers maintained monthly records of their number of contacts with offenders. Because the distributions of these contact variables exhibited a strong positive skew, we used a natural log transformation of this variable in all of the analyses. In Louisiana, supervising officers maintained a monthly index measuring their surveillance of offenders. High scores on the composite represent high levels of supervision intensity (see Appendix A). For individuals who were revoked for new criminal activities, we measured supervision by the average level of intensity until revocation. For individuals who were not revoked, we measured supervision by the average level of intensity throughout the one-year follow-up period.

Our key intervening variable, involvement in prosocial activities, was measured by an eight-item index (except in Louisiana, where we used a 13-item index) that tapped offenders' performances in an array of areas such as meeting family responsibilities, making satisfactory progress in education and treatment programs, and achieving residential and financial stability. The eight- and 13-item indexes included some (but not all) of the items used respectively by Latessa and Vito (1988) and by MacKenzie and her colleagues (MacKenzie and Brame 1995; MacKenzie et al. 1992). All of the items were binary (yes/no) indicators of whether offenders were judged by their officers to be performing well in each area. In Florida, Georgia, and South Carolina these evaluations were compiled for three-month periods of the year following release as long as offenders remained in the community. In Louisiana the evaluations were conducted every month.

For individuals who were revoked for new criminal activities, we measured involvement in prosocial activities by the average level of involvement until the revocation. For individuals who were not revoked, we measured prosocial activities by the average level of involvement throughout the one-year follow-up period. Within each state, we standardized this variable to have zero mean and unit variance.[1]

The outcome of primary interest in this analysis is a binary variable that discriminates between those whose community supervision was revoked for new criminal activities and those whose supervision was not revoked within a one-year follow-up period. Revocation for new criminal activities differs from other possible outcomes such as being arrested or experiencing a revocation for technical violations of supervision conditions. Indeed, some of the items that we describe as prosocial activities also may be requirements for avoiding revocation for technical violations. Because arrests may initiate a process that leads to revocation proceedings for technical violations, arrests and technical violations may be related tautologically to our measure of involvement in prosocial activities.

To avoid the ambiguity that such recidivism outcomes would introduce into our analysis, we measure recidivism by relying exclusively on involvement in new criminal activities. Table 1 presents descriptive statistics for each of the four states in our analysis.

▧ Results

Here we report results for each of the four states in our analysis. Although we control for a number of individual characteristics in each model, we are interested primarily in (1) whether supervision intensity is associated positively with involvement in prosocial activities; (2) whether supervision intensity exhibits any consistent pattern of association with recidivism across states; and (3) whether involvement in prosocial activities is associated inversely with recidivism after adjusting for individual characteristics.

Table 2 presents the results of this analysis across each of the four states. Our preliminary analysis suggested that the relationship between supervision intensity and prosocial activities could be modeled with a cubic polynomial regression function in Florida, Georgia, and South Carolina. Therefore, the effects of supervision intensity displayed in Table 2 include linear (supervision intensity), quadratic (supervision intensity squared), and cubic (supervision intensity cubed) coefficients. In addition, our preliminary analysis revealed that a linear coefficient adequately captured the relationship between supervision and prosocial activities in Louisiana. Even though a different functional form appears to capture the relationship between supervision intensity and involvement in prosocial activities, all of the analysis results imply a positive association between the two variables.[2]

The second interesting aspect of the results shown in Table 2 is the absence of a consistent relationship between supervision intensity and recidivism across the four states. The effect of supervision intensity on recidivism is negative and approaches statistical significance in Georgia; just the opposite is true in South Carolina. In neither Florida nor Louisiana do we find any evidence of a relationship between these two variables. Because the direct links between intensive supervision and recidivism are not clearly understood, we do not wish to read much into this result at this point. We merely note that weak and inconsistent correlations between supervision and recidivism seem to predominate in the literature; our analysis provides further support for this general finding.

The third and final noteworthy observation about the findings shown in Table 2 is the negative and statistically significant association between involvement in prosocial activities and recidivism after taking into account the effects of other covariates.[3] Net of the covariates in our models, then, individuals who exhibit high levels of prosocial activities are significantly less likely than others to be revoked for new criminal activities. This result is consistent with what we would expect to see if the model outlined in Figure 1 is faithful to the process that generated the data.

The four panels of Figure 2 present a summary of our results. These results do not vindicate the model depicted in Figure 1; they are merely consistent with that model. Any unmeasured variables (e.g., intelligence, early childhood differences in impulsivity, quality of childrearing) that affect both prosocial activities and recidivism would render the correlations shown in Table 2 at least partially spurious. Other possibilities also exist. For example, individuals who are supervised most intensively may be those who are most accessible to supervising officers. Indeed, they could be most accessible because they are involved in prosocial activities. In sum, Figure 1 depicts a plausible process by which the results in Table 2 could have been generated, but we must be aware that data involving more elaborate sets of covariates could produce different conclusions. Still, the results strike us as interesting, and the matter appears to deserve further study.

Discussion and Conclusions

Theoretically we are proposing that intensive supervision acts as a direct control on these offenders' behavior. If it has the effect we anticipate, such control will coerce individuals to participate in prosocial activities that plausibly can be linked to socialization processes such as those described by Nye (1958) and by Rankin and Wells (1990). These, in turn, may decrease future involvement in crime. To make this possibility concrete, we considered a model of the relationship between supervision intensity, prosocial activities, and recidivism. We gathered three hypotheses from the extant literature relating to these variables.

First, we hypothesized that those who were supervised more intensively would exhibit greater involvement in prosocial activities. Second, we predicted that recidivism would be associated inversely with prosocial activities. Third, we anticipated that the main effects of supervision intensity on recidivism would be weak and/or erratic and that the important effects would be indirect, through prosocial activities.

Our analysis suggests that these predictions are consistent with the data. Nonetheless,

Table 2	Parameter Estimates for Statistical Model of Prosocial Activities and Recidivism

Predictor Variables	Florida		Georgia		Louisiana		South Carolina	
	Estimate	|z|-ratio	Estimate	|z|-ratio	Estimate	|z|-ratio	Estimate	|z|-ratio
Prosocial Activities								
Intercept	−3.803	5.48	−1.447	2.40	−1.132	4.56	−1.494	2.42
Boot camp completers	.335	2.48	−.096	.59	.212	1.29	−.032	.56
Boot camp dropouts	.001	.01			.219	1.04		
Probationers			Reference		Reference		Reference	
Prison releasees	Reference		−.181	.98	.124	1.04	−.067	.44
Nonwhite	−.416	3.28	−.392	2.99	−.426	4.32	−.576	4.76
Age	.140	4.28	.014	.55	.031	3.31	.030	1.09
Violent offense	.372	2.63	.275	1.51	−.209	1.25	.131	.72
Drug-related offense	.150	.88	.439	2.96	.174	1.57	.262	1.75
Property offense	Reference		Reference		Reference		Reference	
Prior record	−.287	2.10	−.372	2.53	−.054	.42	−.288	2.34
Supervision (linear)	1.781	3.32	3.119	4.68	.729	8.69	3.295	4.12
Supervision (quadratic)	−.812	2.45	−1.854	3.54			−2.162	2.69
Supervision (cubic)	.125	2.11	.342	2.91			.419	1.82
σ_1	.871	22.45	.886	20.78	.689	21.21	.855	20.37
Revocation								
Intercept	1.003	.69	.056	.05	−.518	.55	−.652	.50
Boot camp completers	−.707	2.53	.812	2.25	.377	.71	.085	.27
Boot camp dropouts	−.625	2.00			.225	.31		
Probationers			Reference		Reference		Reference	
Prison releasees	Reference		.767	1.94	.247	.59	−.254	.69
Nonwhite	.438	1.61	.372	1.39	1.174	2.46	.192	.71
Age	−.087	1.19	−.082	1.57	−.092	2.33	−.074	1.16
Violent offense	−.640	2.03	−.733	1.86	−.443	.71	.060	.14
Drug-related offense	−.392	1.06	−.379	1.29	.080	.22	−.149	.42
Property offense	Reference		Reference		Reference		Reference	
Prior record	−.259	.86	1.106	4.18	.140	.35	.534	1.78
Supervision	−.133	.99	−.441	1.49	−.156	.57	.403	1.46
δ (Residual Association)	−.391	2.76	.283	2.04	−.444	2.11	−.431	2.84
Log-likelihood	−435.83		−390.66		−306.58		−353.18	

NOTE: |z|-ratios exceeding 1.96 are statistically significant at the 95% confidence level (two-tailed).

| Figure 2 | Summary of Estimated Relationships Between Supervision Intensity, Prosocial Activities, and Revocation |

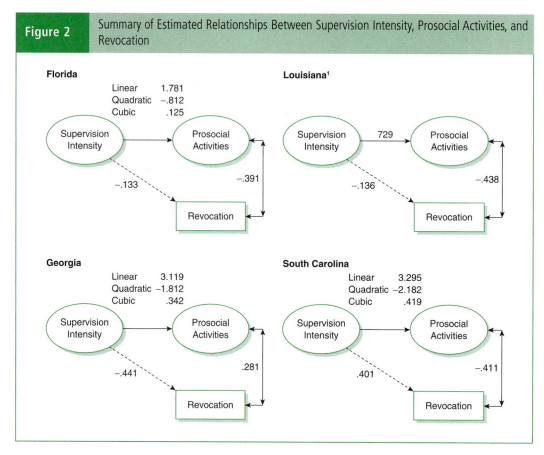

NOTE: None of the arrows from supervision intensity to revocation are statistically significant at a two-tailed 95 % confidence interval. All other arrows represent estimates that are statistically significant at that level.

1. Supervising officers' number of contacts with offenders not available for this state.

as we have suggested, a number of alternative models are also consistent with these results. Therefore, it would be premature to conclude that our findings demonstrate the validity of the model described in Figure 1.

As work in this area proceeds, we think that several issues deserve careful consideration. First, our analysis has emphasized the possibility of a link between supervision, involvement in prosocial activities, and recidivism. A key question left unanswered here and in other related work, such as that conducted by Horney and her colleagues (1995), is whether these short-term connections have more lasting implications for behavior. In other words, do individuals who experience increases in short-term prosocial behavior accumulate social capital or tools that will restrain them from future involvement in crime?

Second, the data for the current study were not collected with these hypotheses in mind. Ordinarily this would not be a major limitation, but in this study we cannot rule out the possibility that the relationship between prosocial activities and recidivism is spurious. As we have suggested, it would be difficult using data such as the information collected here, to rule out this possibility. The development of more sophisticated statistical models does not strike us as productive for future research on this question. It would be more useful, we believe, to analyze other data sets containing more extensive sets of covariates. Through such work, we should be able to see whether the introduction of more control variables alters the conclusions obtained from the present analysis.

The possibility of reciprocal effects for supervision intensity and involvement in prosocial activities is an important rival hypothesis for our results. Like other researchers who have studied supervision effects by simply assuming their exogeneity, we assume that the effects of supervision intensity on prosocial activities can be captured with a recursive model. Nevertheless, we recognize that this is not a satisfactory long-term research strategy. In the future, researchers should analyze data in which individuals are assigned randomly to varying levels of supervision intensity. Such an approach will be a strong methodological platform for examining whether analysis results depend on the exogeneity of supervision intensity (see, e.g., Petersilia and Turner 1993).

Finally, more thought is needed for the kinds of supervision that would be most likely to lead to greater levels of prosocial activity. What mechanisms under the control of community supervision personnel hold the greatest promise for increasing involvement in prosocial activities? What is the optimal set of tactics for coercing individuals to engage in constructive behaviors? Researchers not only must continue studying the plausibility of the theoretical links discussed here, but also must give more attention to how these connections work in practice.

⬕ Appendix A: Prosocial Activities Measures Used in Florida, Georgia, Louisiana, and South Carolina and Surveillance Index Used in Louisiana

Eight Item Prosocial Activities Index Used in Florida, Georgia, and South Carolina[4]

1. Employed, enrolled in school, or participating in a training program for more than 50% of the follow-up period.

2. Held any one job (or continued in educational or vocational program) for more than a three month period during the follow-up.

3. Attained vertical (upward) mobility in employment, educational, or vocational program.

4. For the last half of follow-up period, individual was self-supporting and supported any immediate family.

5. Individual shows stability in residency. Either lived in the same residence for three months or moved at suggestion or with the agreement of supervising officer.

6. Individual has avoided any critical incidents that show instability, immaturity, or inability to solve problems acceptably.

7. Attainment of financial stability. This is indicated by the individual living within his means, opening bank accounts, or meeting debt payments.

8. Participation in self-improvement programs. These could be vocational, educational, group counseling, alcohol, or drug maintenance programs.

Thirteen Item Prosocial Activity Index Used in Louisiana[5]

1. Subject was employed.

2. Employer evaluates subject favorably.

3. Subject required to attend AA and is making satisfactory progress.

4. Subject required to attend drug treatment and is making satisfactory progress.

5. No positive alcohol-sensor tests.

6. No positive drug screen.

7. Subject actively pursuing job training or attending school and is making satisfactory progress.

8. Subject is experiencing no problems in relationships with family.

9. Subject is not spending time with other offenders.

10. Subject has satisfactory attitude or appearance.

11. Subject cooperates and complies with supervising officer decisions.

12. Subject completing community service requirements satisfactorily.

13. Subject exhibits no symptoms of emotional instability.

Five Item Surveillance Index Used in Louisiana[6]

1. Employer was contacted for favorable or unfavorable evaluation of offender.

2. Alcohol sensor test administered.

3. Drug screen administered.

4. Teachers/administrators contacted about progress in school, work, or job training.

5. Community service supervisor contacted about subject's performance.

◪ Notes

1. As an anonymous reviewer of this manuscript observed, the prosocial activities indexes used in this paper do not measure the actual acquisition of social skills; instead they measure the extent to which individuals are immersed in environments where they might reasonably be expected to acquire those skills.

2. We also estimated bivariate Spearman rank-order correlation coefficients between supervision intensity and involvement in prosocial activities in each state: Florida (+.312), Georgia (+.243), Louisiana (+.678), and South Carolina (+.234). All of these coefficients were statistically significant at .05.

3. We also calculated simple t-tests for the difference in prosocial activity means between those who recidivated within 12 months and those who did not: Florida ($t = -3.384$), Georgia ($t = -3.204$), Louisiana ($t = -2.141$), and South Carolina ($t = -3.595$). Each of these t-statistics was statistically significant at .05.

4. This index is based on measures described by Latessa and Vito (1988).

5. This index is based on measure described by MacKenzie and colleagues (MacKenzie et al. 1992).

6. This index is based on measures described by MacKenzie and colleagues (MacKenzie et al. 1992).

◪ References

Anglin, M.D. and Y. Hser. (1990). "Treatment of Drug Abuse." Pp. 393–461, in *Crime and Justice: A Review of Research*, vol. 13, edited by M. Tonry and J.Q. Wilson. Chicago, IL: University of Chicago Press.

Catalano, P.J. and L.M. Ryan. (1992). "Bivariate Latent Variable Models for Clustered Discrete and Continuous Outcomes." *Journal of the American Statistical Association* 87:651–58.

Clear, T.R. and A.A. Braga. (1995). "Community Corrections." Pp. 421–44 in *Crime*, edited by J.Q. Wilson and J. Petersilia. San Francisco, CA: ICS.

Cox, D.R. and E.J. Snell. 1989. *Analysis of Binary Data.* 2nd ed. London: Chapman and Hall.

DeLeon, G. (1988). "Legal Pressure in Therapeutic Communities." Pp. 160–77 in *Compulsory Treatment of Drug Abuse: Research and Clinical Practice*, edited by C.G. Leukefeld and F.M. Tims, Washington, DC: National Institute on Drug Abuse.

Dishion, T.J. and G.R. Patterson. (1997). "The Timing and Severity of Antisocial Behavior: Three Hypotheses Within an Ecological Framework." Pp. 205–17 in *Handbook of Antisocial Behavior*, edited by D.M. Stoff, J. Breiling, and J.D. Maser, New York: Wiley.

Eliason, S.R. (1993). *Maximum Likelihood Estimation: Logic and Practice.* Newbury Park, CA: Sage.

Erwin, B.S. (1986). "Turning Up the Heat on Probationers in Georgia." *Federal Probation* 50:17–24.

Fitzmaurice, G.M. and N.M. Laird. (1995). "Regression Methods for a Bivariate Discrete and Continuous Outcome With Clustering." *Journal of the American Statistical Association* 90:845–52.

Gendreau, P., F.T. Cullen, and J. Bonta. (1994). "Intensive Rehabilitation Supervision: The Next Generation in Community Corrections?" *Federal Probation* 58:72–78.

Glueck, S. and E. Glueck. (1950). *Unraveling Juvenile Delinquency.* New York: Commonwealth Fund.

Gottfredson, M.R. and T. Hirschi. (1990). *A General Theory of Crime.* Stanford, CA: Stanford University Press.

Hirschi, T. (1969). *Causes of Delinquency.* Berkeley, CA: University of California Press.

Horney, J., D.W. Osgood, and I.H. Marshall. (1995). "Criminal Careers in the Short Term: Intra-Individual Variability in Crime and Its Relation to Local Life Circumstances." *American Sociological Review* 60:655–73.

Johnson, N.L. and S. Kotz. (1970). *Continuous Univariate Distributions—1.* New York: Wiley.

King, G. (1989). *Unifying Political Methodology: The Likelihood Theory of Statistical Inference.* New York: Cambridge University Press.

Land, K.C., P.L. McCall, and J.R. Williams. (1990). "Something That Works in Juvenile Justice: An Evaluation of the North Carolina Court Counselors' Intensive Protective Supervision Randomized Experimental Project, 1987–1989." *Evaluation Review* 14:574–606.

Latessa, E.J. and G.F. Vito. (1988). "The Effects of Intensive Supervision on Shock Probationers." *Journal of Criminal Justice* 16:319–30.

Laub, J.H. and R.J. Sampson. (1993). "Turning Points in the Life Course: Why Change Matters to the Study of Crime." *Criminology* 31:301–25.

Lurigio, A.J. and J. Petersilia. (1992). "The Emergence of Intensive Probation Supervision Programs in the United States." Pp. 3–17 in *Smart Sentencing: The Emergence of Intermediate Sanctions*, edited by J. Byrne, A. Lurigio, and J. Petersilia. Newbury Park, CA: Sage.

MacKenzie, D.L. (1994). "Results of a Multi-Site Study of Boot-Camp Prisons." *Federal Probation* 58:60–67.

MacKenzie, D.L. and R. Brame. (1995). "Shock Incarceration and Positive Adjustment During Community Supervision." *Journal of Quantitative Criminology* 11:111–42.

MacKenzie, D.L., J.W. Shaw, and C. Souryal. (1992). "Characteristics Associated With Successful Adjustment to Supervision: A Comparison of Parolees, Probationers, Shock Participants, and Shock Dropouts." *Criminal Justice and Behavior* 19:437–54.

MacKenzie, D.L. and C. Souryal. (1994). *Multisite Evaluation of Shock Incarceration.* Washington, DC: National Institute of Justice.

Maddala, G.S. (1983). *Limited-Dependent and Qualitative Variables in Econometrics.* New York: Cambridge University Press.

Nagin, D.S. and R. Paternoster. (1993). "Enduring Individual Differences and Rational Choice Theories of Crime." *Law and Society Review* 27:467–96.

——. (1994). "Personal Capital and Social Control: The Deterrence Implications of a Theory of Individual Differences in Criminal Offending." *Criminology* 32:581–606.

Nye, F.I. (1958). *Family Relationships and Delinquent Behavior.* New York: Wiley.

Pearson, F.S. (1988). "Evaluation of New Jersey's Intensive Supervision Program." *Crime and Delinquency* 34:437–48.

Petersilia, J. and S. Turner. (1993). "Intensive Probation and Parole." Pp. 281–335 in *Crime and Justice: A Review of Research*, edited by M. Tonry and N. Morris. Chicago, IL: University of Chicago Press.

Rankin, J.H. and L.E. Wells. (1990). "The Effect of Parental Attachments and Direct Controls on Delinquency." *Journal of Research in Crime and Delinquency* 27:140–65.

Sampson, R.J. and J.H. Laub. (1993). *Crime in the Making: Pathways and Turning Points Through Life.* Cambridge, MA: Harvard University Press.

Turner, S., J. Petersilia, and E.P. Deschenes. (1992). "Evaluating Intensive Supervision Probation/Parole (ISP) for Drug Offenders." *Crime and Delinquency* 38:539–56.

Wexler, H.K, D.S. Lipton, and B.D. Johnson. (1988). *A Criminal Justice System Strategy for Treating Cocaine-Heroin Abusing Offenders in Custody.* Washington, DC: National Institute of Justice.

Wilson, J.Q. and R. Herrnstein. (1985). *Crime and Human Nature.* New York: Simon and Schuster.

Wish, E.D. and B.D. Johnson. (1986). "The Impact of Substance Abuse on Criminal Careers." Pp. 52–88 in *Criminal Careers and "Career Criminals,"* edited by A. Blumstein, J. Cohen, J.A. Roth, and C.A. Visher. Washington, DC: National Academy Press.

DISCUSSION QUESTIONS

1. Explain the theoretical basis of the authors' assumption that even forced prosocial activities can have the effect of reducing recidivism.

2. What is social capital, and how does it prevent antisocial behavior?

3. Summarize the main findings of this study.

READING

In this article, Slate, Wells, and Johnson take a look at the toll of job-related stress on probation officers and their agencies. Stress can be costly to individuals and to organizations. Stress researchers have recommended participatory management as a means for reducing probation officer stress. This article examines self-report surveys of probation personnel in terms of the relationship of a number of demographic variables with employee perceptions of participation in workplace decision making, job satisfaction, and organizational and physical stress levels. Construction of a structural model revealed that employee perceptions of participation in workplace decision making was an important variable in relation to job satisfaction and its influence on both reported organizational and physical symptoms of stress. The results lend further credence to the use and development of participatory management schemas within probation organizations.

Opening the Manager's Door

State Probation Officer Stress and Perceptions of Participation in Workplace Decision Making

Risdon N. Slate, Terry L. Wells, and W. Wesley Johnson

Probation is a people business, oftentimes requiring intense, stressful confrontations with recalcitrant offenders. Today, there are more felons receiving probation than ever before (Jones & Johnson, 1994). The impact of current probation supervision conditions on probation officers has not been fully documented. Because human capital is the greatest investment an organization can make (Maggio & Terenzi, 1993), the ongoing assessment and management of employee stress effects becomes critical to maximizing the effectiveness of probation.

Although the vast majority of research on job stress in criminal justice has focused on police and correctional officers (Patterson, 1992; Simmons, Cochran, & Blount, 1997; Slate, Johnson, & Wells, 2000; Whisler, 1994), there are only a few studies that document the effects of current probation working environments on probation officers. Considering the fact that probation officers have contact with more offenders than most other justice practitioners, and probation caseloads have continued to increase at unprecedented levels, there is a need to understand more about the work of probation.

The short- and long-term effects of stress are well documented (Ganster & Schaubroeck, 1991), and there have been a number of studies linking stress to a myriad of health problems of workers. Consequently, based on National Institute for Occupational Safety and Health figures, more than 75% of trips to primary care physicians and up to 85% of workplace accidents involve stressed employees ("Stress: The Workplace Disease of the 1990s," 1995). Stress affects workers, managers, executives, significant others, and families. Stress has a direct effect on productivity, employee turnover, health care costs, disability payments, workers' compensation awards, and sick leave, with estimates that more than half of all absences are stress-related (Elkin & Rosch, 1990) amounting to more than 1 million stress-induced employee absences a day in the workplace (Dillon, 1999). DeCarlo and Gruenfeld (1989) found that approximately 40% of job turnover is a result of stress. The financial costs of stress cannot be ignored. Stress costs American organizations between $200 billion to $300 billion each year; this is more than the combined annual profits of all Fortune 500 companies (Dillon, 1999).

Stress Defined

The work of Dr. Hans Selye on job stress is well documented in the research literature, and many perceive him as a pioneer in this field. Selye (1976) described stress as being a nonspecific response of the body to any demand. He contended that stress can be produced by positive as well as negative circumstances, and prolonged, extreme stress can manifest itself in withdrawal from work, emotional exhaustion, and "burnout." Burnout is often used interchangeably with the term stress and been found to be a problem particularly among "people-oriented professions" such as probation, prison, and police work (Whitehead, 1981, 1985).

▧ Probation Officer Stress Literature

The stress levels of probation officers have been found to be higher than those of the general population (Tabor, 1987). The dangers inherent to the job and having to make recommendations that result in custodial sentences have been identified as two potential stressors for probation officers (Thomas, 1988). The linkage between stress and chronic health problems has also been discussed in the probation officer literature on stress (Brown, 1987). According to Simmons et al. (1997), as probation officers experience more stress, they are significantly more likely to be dissatisfied with their jobs and exhibit a strong inclination to quit their jobs, and minority officers have been found to be more likely to reflect this propensity for turnover. High turnover, in turn, can

SOURCE: Slate, R.N., Wells, T.L., & Johnson, W. (2003). Opening the manager's door: State probation officer stress and perceptions of participation of workplace decision making. *Crime and Delinquency, 49*(4), 519–541. Reprinted with permission of Sage Publications, Inc.

translate into increased costs for training and recruiting and can result in increased caseloads that may weaken supervision and increase the potential for revocation and recidivism (Simmons et al., 1997).

A curvilinear relationship between the amount of probation officer work experience and stress or burnout has been reported in several studies; in other words, those at the beginning of their careers or toward the end of their careers in probation work have been found to be less stressed than those situated somewhere in the middle of their careers (Patterson, 1992; Tabor, 1987; Whitehead, 1981). Thus, those at the beginning of their careers may enjoy somewhat of a honeymoon period, although those toward the end of their careers may have settled in with their eyes on the benefits of retirement. Contrary to the aforesaid findings, Thomas (1988) reported that probation officer burnout was associated with seniority and explained that this phenomenon may be linked to the perceived fairness of observed promotions within an agency over time. Seasoned officers who had witnessed promotions based on what they perceived as favoritism, politics, or simply seniority instead of qualifications, experience, and ability were more likely to show signs of burnout (Thomas, 1988).

An inverse relationship between seniority and turnover has also been found in a study by Simmons et al. (1997); as probation officers' time on the job increased, their reported propensity to quit their jobs decreased. At a certain point, individuals would become vested and perhaps feel there was too much to lose by quitting. Simmons et al. (1997) also reported that older probation officers were more apt to be satisfied with their jobs and less occupationally stressed than their younger colleagues.

Married probation officers have been found to be less occupationally stressed and exhibit more job satisfaction than their unmarried cohorts (Simmons et al., 1997; Tabor, 1987). In terms of gender, female probation officers have been found to demonstrate greater levels of stress than their male counterparts (Simmons et al., 1997), as have female probation supervisors, and male managers have exhibited more depersonalization toward probationers—with probation managers in general reflecting a less dehumanizing attitude toward probationers (Thomas, 1988). Tabor (1987) has reported a linkage between impersonal treatment of probationers and heightened stress in juvenile probation officers. Probation supervisors have also been found to be less likely than line officers to experience burnout and feel stressed (Thomas, 1988; Whitehead, 1986). As explained by Whitehead (1986), occupational level and level of job satisfaction tend to be directly correlated, as managers are given more input into workplace decision making than line officers, have more challenging and interesting work, and have less contact with probationers. Too little time to get work done was found by Thomas (1988) to be the most frequently reported cause of stress, and suspense dates on reports have been determined to be a cause of probation officer stress (Simmons et al., 1997). Religiosity has been found to be a mediating factor with burnout, as Thomas (1988) noted that the more reportedly religious probation officers were less likely to show signs of burnout.

According to Whisler (1994), too much leniency on the part of the courts has been identified as a primary stressor for probation officers. Role ambiguity and role conflict have been cited as causes of probation officer burnout (Brown, 1987; Whitehead, 1985, 1986), and responsibility for supervision of special caseloads, such as alcohol offenders, has been linked to probation officer stress (Tabor, 1987). Among the activities conducted by probation officers, Pettway and VanDine (2000) found paperwork to be the most frequently performed task reported by probation officers, even more so than offender-related activities. Inundation with paperwork has been identified as a stressor or source of burnout for probation officers in several studies (Brown, 1987; Simmons et al., 1997; Thomas, 1988; Whisler, 1994). Financial concerns (Thomas, 1988), failure to give appropriate accolades at work (Whisler, 1994), insufficient salaries, lack of promotional opportunities (Simmons et al., 1997; Whisler, 1994; Whitehead, 1986), and boredom (Whitehead, 1985) have all been specified as sources of stress and burnout for probation officers.

Although Thomas (1988) reported heightened burnout for officers who indicated their current contact with probationers was less now than when they started their career, Tabor (1987) found that probation officer stress increased as personal involvement with probationers increased and the realization set in that more and more of the probationers under their supervision would fall short of being productive citizens. In terms of prior work experience within the criminal justice system, probation officers with prior correctional experience were found to be more apt to be satisfied in their current jobs than were probation officers that had been previously employed as police officers (Simmons et al., 1997). Other interesting findings from the Simmons et al. (1997) study included the discovery that roughly 90% of their probation officer respondents indicated a dislike for their supervisors, with approximately 80% reporting that they perceived their immediate supervisor as being incompetent at his job, and almost 50% of those surveyed indicated that they often thought about quitting their jobs as probation officers.

Potential Organizational Intervention for Probation Officer Stress

As identified by Whisler (1994), the most significant stressors probation officers encounter are internal to the organization. This finding is consistent with the pioneering work in the area of Total Quality Management (TQM) by Dr. W. Edwards Deming, who has indicated that in excess of 90% of organizational problems are not the employees' fault and are endemic to the organization (Janes, 1993). As such, decentralized decision making has been advocated by Wiggins (1996) and a team

environment recommended by Siegel (1996) as means of empowering probation officers and improving job satisfaction and morale. The positive aspects of TQM have been outlined in the probation officer literature (Alston & Thompson, 1996), and Janes (1993) has offered guidance on how to introduce and implement TQM within probation organizations.

The elimination of the causes of stress within an organization is considered to be the most effective approach to undertake for the alleviation of employee stress but has received the least emphasis (Maslach, 1982; Terry, 1981, 1983). Many organizations attempting to maximize their productivity focus on individuals and not the source of the problem—the stress of the job. Managers appear to be more comfortable with trying to change people to fit organizations than modifying organizations to accommodate individuals (Ivancevich, Matteson, & Richards, 1985; Newman & Beehr, 1979).

Employee involvement in workplace decision making has been touted as a means for cost savings, increased productivity, positive affect among line workers, and reduced turnover (Tjosvold, 1998). Likewise, participation in decision making has long been recognized as a means of alleviating stress for front-line employees in the workplace (DeCarlo & Gruenfeld, 1989). Within the criminal justice research literature, employee participation in decision making has been discussed as a possible means of reducing stress in a number of forums pertaining to both law enforcement personnel (Archambeault & Weirman, 1983; Kuykendall & Unsinger, 1982; Lawrence, 1984; Melancon, 1984; Morash & Haarr, 1995; Patterson, 1992; Reiser, 1974; Rodichok, 1995; Terry, 1983) and correctional employees (Farkas, 2001; Honnold & Stinchcomb, 1985; Lasky, Gordon, & Strebalus, 1986; Lindquist & Whitehead, 1986; Patterson, 1992; Sims, 2001; Slate & Vogel, 1997; Slate, Vogel, & Johnson, 2001; Ulmer, 1992). Moreover, mechanisms for the assurance of employee participation in workplace decision making have consistently been recommended throughout the literature as a means of reducing probation officer stress and burnout (Brown, 1986, 1987; Holgate & Clegg, 1991; Simmons et al., 1997; Tabor, 1987; Whisler, 1994; Whitehead & Lindquist, 1985; Whitehead, 1981, 1986). Instead of routine adoption and implementation of such participatory management programs, employee assistance programs (Hardaway, Wence, Bingaman, & Selvik, 1996) are often unrealistically relied on to fix individual employees instead of looking to the real source of the problem—the organization. Thus, after a review of the literature, as recommended by Slate et al. (2000), it was determined that an investigation into the relationship between probation officers' perceptions of participation in workplace decision making and stress, job satisfaction, and other pertinent variables would be explored.

▨ Purpose of the Study

This research identifies the contributors to stress for probation officers and how such stressors can manifest themselves in the deterioration of the physical health of those employed in the probation field. The linkage between stress and the depletion of one's coping ability, which may lead to physical illness, injury, or psychological disorder, has been well established (e.g., Cohen & Williamson, 1991; Coyne & Downey, 1991; Pearlin, 1989). In addition, it was believed that employee perceptions of involvement in workplace decision making would prove to be a critical variable with potential consequences for the organization as well as employees. Other variables gleaned from the literature review would also prove pertinent for examination.

▨ Methodology

The Sample

The authors met with state probation administrators to gain approval for the study and to determine the particulars for dissemination and retrieval of the surveys. It was decided that all sworn probation officers would have the survey electronically mailed to them, and they would be provided with a postage-paid envelope to mail the questionnaire directly back to the researchers.

The sample in the present study consists of 636 probation officer respondents responsible for overseeing adult offender supervision in a southern state. This constitutes a 69% response rate. Of the probation officer stress studies uncovered in the literature review, our findings reflect the largest sample size and response rate for a single state. This population was selected for study due to the cooperativeness of state probation administrators. Although the response rate is fairly high for this state population, generalizations to other state probation officer populations should be cautioned.

The Questionnaire

The vast majority of questions selected for the survey instrument resulted from a review of the literature. Although none of the questions proposed by the researchers for the research instrument were modified by the state probation administrators, they were desirous of adding some additional questions to the survey that were of interest to them. Because these additional questions were rather innocuous, did not detract from the intended purpose of the survey, and anonymity was ensured, the decision to incorporate these questions into the questionnaire was made.

The questionnaire consisted of several parts that included components aimed at measuring the level of external, internal, job or task, personal, and physical stress experienced by respondents. The first four stress subscales were adapted from Whisler (1994), who utilized these subscales to measure probation officer stress. For each question that composed each subscale, a Likert-type format was used with response categories that ranged from 1 (*not stressful*) to 6 (*very stressful*). The first

14 items on the survey included rankings in terms of the stress of the ineffectiveness of the prison system, the leniency of the courts, and politics outside the agency. The first 14 items combined to form a subscale indicative of stressors external to the organization.

Items 15 through 40, which included the stressfulness of inadequate salary, lack of recognition for good work, and lack of adequate training represented stressors internal to the organization. The job- or task-related stress subscale was made up of items 41 through 54 on the questionnaire and included assessment of such stressors as difficulty supervising offenders, excessive paperwork, and an expectation to do too much in too little time. The personal stress subscale was composed of items 55 through 61 and included family demands, negative effects of the job on social life, and concern that one might make a mistake. The four subscales that emerged from the first 61 questions, though clustered by category on the survey instrument, were not labeled as external, internal, job or task, or personal on the questionnaire. These four subscales could also be collapsed into an overall stressor scale identified as the Total Stress Scale reflecting each respondent's answers to questions 1 through 61 on the survey.

The Attitudes on Participation portion of the survey has been used previously to measure attitudes of criminal justice personnel about participation in workplace decision making (Slate & Vogel, 1997; Slate et al., 2001). Twelve questions compose the scale, with each item presented in a Likert-type format, ranging from 1 (*strongly disagree*) to 5 (*strongly agree*). Two subscales compose this scale, with items number 1 through number 7 combining to form attitudes about participation and items number 8 through 12 representing the atmosphere for participation in workplace decision making. Using Cronbach's alpha, a reliability analysis of the internal consistency of these two subscales was conducted, with the atmosphere subscale yielding a moderately high reliability coefficient of .84 and the attitudes subscale

rendering a .83 reliability coefficient. The survey items that composed the atmosphere for participation subscale included

- My superiors ask me for input on decisions that affect me at work.
- I am encouraged to offer my opinions at work.
- There is opportunity for me to have a say in the running of this institution on matters that concern me.
- Management responds in a satisfactory manner to what I have to say.
- From past experience at this institution, I feel it is a waste of time and energy to tell management anything.
- I feel comfortable about offering my opinion to supervisors at work.
- Those who actually do the work are involved in the writing of the policies at this institution.

The attitudes about participation subscale included the following items:

- The quality of decisions increase as worker participation in decision making increases.
- Participation in decision making tends to make individuals feel they have a stake in running the organization.
- Participation in decision making tends to make individuals feel more a part of the team.
- Everyone should be allowed to participate in decision making in the workplace on matters that affect them.

The last scale considered as an independent variable in the study and aimed primarily at measuring job satisfaction was titled Job Opinion and consisted of six Likert-type scale items. Five of the Likert-type items had response categories that ranged from 5 (*strongly agree*) to 1 (*strongly disagree*). The other Likert-type item reflected response categories that ranged from 5 (*most of the time*) to 1 (*rarely or never*), which required reverse scoring and is listed below as the last item in the scale. The following questions composed the scale:

- ◆ I am proud of what I am doing for a living.
- ◆ Probation work with this agency is meaningful.
- ◆ If I had it to do over again, I would choose this occupation.
- ◆ I would recommend this job to others.
- ◆ I believe I will remain with this agency until I retire.
- ◆ I seriously think about quitting this job.

Using Cronbach's alpha, the job opinion scale yielded a .86 reliability coefficient.

The selection of demographic questions to be incorporated into the survey was guided by the literature review, and items selected for the questionnaire included gender, ethnicity, age, marital status, probation officer experience, prior criminal justice experience, and military experience. Other questions distinguished managers from line officers, determined the caseload size of respondents, identified those officers with special caseloads and type of caseload supervised, allowed officers to specify what function is the primary focus of their job, permitted probation officers to indicate whether they viewed their role more as police officers or social workers, asked probation personnel to report days of sick leave used in the past year, and covered other employment-related matters.

The Dependent Variable

Physical stress, the dependent variable chosen for the study, was measured via the Selye Health Scale, which has been previously utilized to measure physical stress levels of criminal justice personnel (Cheek, 1984; Cheek & Miller, 1982a; Slate & Vogel, 1997; Slate et al., 2001). The Selye Health Scale, developed by Cheek and Miller (1982b), is a 54-item Likert-type scale questionnaire that ranks severity of physical symptoms and illnesses reported by respondents. Thus, physical stress in the present study is measured by self-reports, not by the following: information from medical professionals, results of diagnostic analysis given directly to the researchers, or a review of medical records. The severity of the symptoms and illnesses increase as one progresses through the scale and is weighted and scored accordingly; the range of possible scores is 54 to 324. Utilizing the Cronbach's alpha, a reliability coefficient of .96 was obtained for the Selye Health Score in this study. Physical stress can carry dire consequences for employees as well as organizations, and its potential causes are worthy of investigation.

▧ Results

Demographics

Reflected below initially is the percentage for each demographic category that was representative of the 636 probation officer respondents. Percentages set aside in parentheses below are representative of the demographics of those respondents remaining in the sample ($n = 417$) after the restrictive structural model requirements were instituted. A comparison of the two sets of demographics, before and after structural model implementation, reveals no glaring differences in terms of the demographic makeup of the two samples.

Of the respondents, 52% were female (48%, after structural model implementation [ASMI]), and 48% were males (52%, ASMI). The majority of those responding to the survey were married (61%; 64%, ASMI), with 24% (22%, ASMI) reporting that they had never been married, 3% (4%, ASMI) separated, and 12% (10%, ASMI) divorced. The mean age was 36.2 years (36.7 years, ASMI), with a range from 21 to 70 (22 to 70, ASMI). In terms of ethnicity, 72% (76%, ASMI) of respondents reported that they were Caucasian, 25% (21%, ASMI) indicated African American, and 3% (3%, ASMI) specified other. The vast majority of the sample had no prior military experience (86%; 85%, ASMI), with an average of 2 years (1.9 years, ASMI) of work experience within the criminal justice system prior to becoming a probation officer. The majority of respondents were line personnel (73%; 70%, ASMI), and the mean amount of time for respondents with this probation agency was 8.5 years (8.05 years, ASMI).

Correlation Among Independent Variables

Pearson product-moment correlations among the dependent variable and the five independent variables are presented in Table 1. As noted previously, cases with data missing on any of the six variables were excluded from the analysis—resulting in 417 cases with useable, complete data for inclusion in the structural model. As reflected in the first column (Y1) of Table 1, the dependent variable (physical stress) was most closely associated with higher levels of total stress, perceived negative atmosphere for participation in workplace decision making, negative job opinion, seniority, and females.

The strength of the linear relationships is revealed by the correlations in the first column (Y1) of Table 1. However, if two variables are affected by the same prior variables, a high correlation may be the result of spuriousness. Due to the potential of simultaneous effects among the variables, path analysis was used to disentangle the direct, indirect, and spurious relationships.

Causal Analysis

Using a hierarchical model, the causal effects were estimated for the current study with the inclusion of the variables in Table 1. Beginning with the gender variable, the arrangement was from top to bottom as shown in Figure 1. Thus, all five independent variables were included as predictors of the dependent variable—physical stress.

The causal model depicted in Figure 1 reflects only variables and paths with significant effects (path coefficients with an absolute value of .10 and significance beyond the .05 level). The four variables with direct paths to the criterion combined to account for 36% of the variance.

Table 2 reflects the summary of the direct, indirect, and spurious effects of the variables. Together, Table 2 and Figure 1 show that, although the relationship of gender with physical stress is significant at $p < .05$, the correlation between these two variables is weak ($r = .11$). Number of years employed within the probation

Table 1	Correlations Between Predictor Variables Included in the Model and Physical Stress ($N = 417$)					
	Y1	**X1**	**X2**	**X3**	**X4**	**X5**
Y1 Physical stress	1.00					
X1 Gender	.107*	1.00				
X2 Number of years employed	.178**	−.070	1.00			
X3 Atmosphere for participation	−.309**	.054	.013	1.00		
X4 Job opinion	−.299**	.063	.080	.469**	1.00	
X5 Total stress	.535**	−.108*	.034	−.477**	−.410**	1.00

*$p < .05$; **$p < .01$ (two-tailed).

Figure 1	Structural Model of Physical Stress

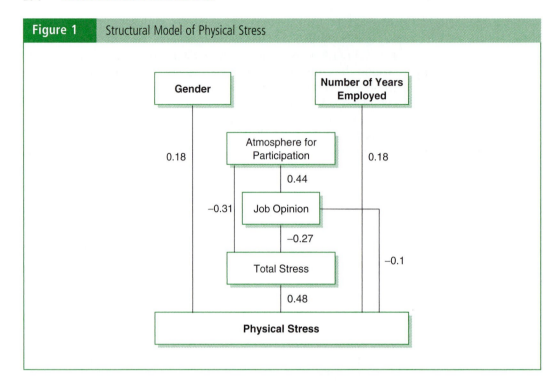

organization is weakly correlated with physical stress ($r = .18$, $p < .01$) and is explained by direct effects.

As expected, the atmosphere for participation variable proved to be a pivotal variable in the structural model. Although this variable was moderately correlated with the dependent variable, physical stress ($r = -.309$, $p < .01$), an insignificant direct path coefficient ($-.06$) to the criterion variable was found. The correlation of the atmosphere for participation variable with physical stress was largely explained by indirect effects ($-.28$), and very little explanation was offered by spurious effects ($.03$). The atmosphere for participation was a significant, direct contributor to one's opinion of

Table 2	Summary of Direct, Indirect, and Spurious Effects on Physical Stress ($N = 417$)				
Predictor	Description	Correlation	Direct	Indirect	Spurious
X1	Gender	.11*	.18***	−.07	—
X2	Number of years employed	.18**	.18***	—	—
X3	Atmosphere for participation	−.31**	−.06	−.28	.03
X4	Job opinion	−.30**	−.10*	−.12	−.08
X5	Total stress	.54**	.48***	—	.06

NOTE: Tests of significance are for correlations and direct effects only. Cumulative $R^2 = .36$. Adjusted $R^2 = .35$.

*$p < .05$; **$p < .01$; ***$p < .001$.

the job (.44) as well as to the total amount of stress experienced (–.31). This finding signifies the importance of the atmosphere for participation variable in the structural model in that one's opinion of the job and the total amount of stress experienced by an individual were determined to be significant, direct predictors of the criterion variable—physical stress.

The variable representing one's opinion of the job was determined to have a significant, yet moderate, correlation with physical stress, the dependent variable ($r = -.299$, $p < .01$). However, the variable's significant, direct effects (–.10) on physical stress were overshadowed by spurious effects (–.08) and indirect effects (–.12). The significant, direct, path coefficient from one's opinion of the job to total stress experienced (–.27) accounted for the indirect effects, reflecting the fact that the effects of one's opinion of the job on physical stress were primarily mediated indirectly by the intervention of total stress experienced.

The direct effects of total stress experienced (.48) explain the majority of the correlation ($r = .54$) with physical stress. Previously entered causal variables, such as one's opinion of the job and the atmosphere for participation, interacted to create a minimal spurious effect (.06).

Overall, several significant relationships were uncovered among the variables selected for the structural model. First, females were found to be significantly more physically stressed than males in the probation agencies surveyed ($r = .11$; direct effect = .18). Second, as the number of years employed with a probation agency increased, physical stress levels also increased ($r = .18$; direct effect = .18). Third, respondents who perceived a positive atmosphere for participation in workplace decision making were more likely to express a positive job opinion ($r = .469$; direct effect = .44), and, in an inverse relationship, those with a negative perception of the atmosphere for participation in workplace decision making were more apt to reflect higher total stress indices ($r = -.477$; direct effect = –.31). Fourth, the opinion of the job was directly predictive of each variable that followed in the structural model and resulted in two inverse relationships. Those with negative opinions of the job were more likely to exhibit higher total stress ($r = -.41$; direct effect = –.27) and higher levels of physical stress ($r = -.299$; direct effect = –.10). Finally, as total stress increased, physical stress significantly increased as well ($r = .535$; direct effect = .48).

◿ Discussion and Conclusion

Contrary to previous research findings, a curvilinear relationship between probation officer stress and years of work experience (Patterson, 1992; Tabor, 1987; Whitehead, 1985) was not detected in this study. This might be explained by the fact that these previous studies did not operationally define stress in terms of its manifestation into physical maladies as done in the present study via inclusion of the Selye Health Score in the research design. In this study, as one's probation work experience increased, the propensity for indicating physical health symptoms also increased. As noted previously, support for increased seniority as a possible cause of probation officer stress has been documented (Thomas, 1988).

Female probation officers, as noted earlier, have been found to exhibit greater levels of occupational stress than their similarly situated male counterparts (Simmons et al., 1997; Thomas, 1988), and female probation officers were found to significantly report greater physical symptoms of stress than their male counterparts in the current study. These findings jibe with results ascertained in other research pertaining to corrections work (Cullen, Link, Wolfe, & Frank, 1985; Lovrich & Stohr, 1993; Stohr, Lovrich, & Mays, 1997; Voorhis, Cullen, Link, & Wolfe, 1991; Wright & Saylor, 1991). Although several reasons are posited as possible explanations for our findings regarding stress levels and gender, the relationship between gender and stress is a complex one; the rationales offered to explain our results are speculative in nature and should not be considered conclusive.

The explanation for our findings regarding stress levels and gender may be related to the male-dominated work environment that often permeates criminal justice organizations (Cullen et al., 1985) and can breed an atmosphere conducive to the promulgation of gender and sexual harassment (Dantzker & Kubin, 1998; Morash & Haarr, 1995; Stohr et al., 1997). An internal assessment document that was provided to the researchers by the probation agency administration revealed that there had been a history of racism and/or sexism in some field offices, though not agencywide. Although an initiative for investigating and eradicating this problem was specified by the agency, perhaps, as represented by the results of this study, the organizational directive has not been fully realized.

Although females tend to experience the same stressors as males in the workplace, some stressors may be unique to females (Morash & Haarr, 1995). As noted by Hendrix, Spencer, and Gibson (1994) and DeCarlo (1987), in addition to responsibilities in the workplace, females are typically more apt than males to be inundated with operation of a household and serving as primary caregiver for a family.

Furthermore, in terms of physical stress, although females generally tend to have lower mortality rates than males, it has been predicted that as females continue to enter the workforce, their overall health will decline; although females seem less susceptible to serious illness than males, they have been found to be more prone to suffer from mild psychological distress and to exhibit more acute symptoms (DeCarlo & Gruenfeld, 1989). Being more sensitive to the early signs of stress, women perhaps are more likely than men to take steps to try to alleviate it (such signs are much of what composes the Selye Health Score); this, coupled with the so-called John Wayne Syndrome, which has been found within male-dominated criminal justice occupations

such as policing and is characterized by male officers who maintain a stiff upper lip and a lack of interpersonal communication, may serve to further explain this phenomenon (Bartol & Bartol, 1994; Reiser, 1974; Wrightsman, Nietzel, & Fortune, 1994). In other words, males may be less apt to talk about their maladies as opposed to females. The observation that women are more likely to express or communicate such symptoms might serve to explain why female probation officers reported higher physical stress levels than males in the present study. Furthermore, according to DeCarlo and Gruenfeld (1989), females have been found more likely than males to use sickness as a coping strategy. This finding might explain why female probation officers in the present study purportedly averaged nine sick days from work per year whereas their male cohorts reported a yearly average of four sick days. Thus, female probation personnel reportedly averaged more than twice as many sick days a year as male probation officers. Of course, though beyond the scope of this study, it could be that females were succumbing to not only the stressors their male counterparts were experiencing, but their greater propensity to use sick days could be the result of the condition of female probation officers being exacerbated by stressors that are unique to females as previously discussed.

As noted previously, Likert-type response categories ranged from 1 (*not stressful*) to 6 (*very stressful*) for each of the 61 items composing the Total Stress Scale. The 11 items rated as most stressful by probation officer respondents from the Total Stress Scale are listed with their rank and mean score in our study, followed by, where possible, rankings and mean scores from a study by Whisler (1994) of probation officers in the state of Florida. The most influential stressors that composed the Total Stress Scale were as follows:

1. Inadequate salary, mean = 4.93 (Whisler [2] mean = 4.87).

2. Courts being too lenient on offenders, mean = 4.53 (Whisler [4] mean = 4.47).

3. Lack of promotional opportunities, mean = 4.34 (Whisler [3] mean = 4.73).

4. Frustration with the criminal justice system, mean = 4.19 (Whisler [9] mean = 4.19).

5. Excessive paperwork, mean = 4.17 (Whisler [1] mean = 4.89).

6. Ineffectiveness of the judicial system, mean = 4.16 (Whisler [12] mean = 3.96).

7. Expectations to do too much in too little time, mean = 4.12.

8. Lack of recognition for good work, mean = 4.06 (Whisler [8] mean = 4.26).

9. Ineffectiveness of the correctional system, mean = 3.80 (Whisler [10] mean = 4.11).

10. Inadequate support from the agency, mean = 3.62 (Whisler [8] mean = 4.21).

11. Lack of adequate community resources, mean = 3.62 (Whisler [20] mean = 3.44).

Although not all of these stressors are within the control of probation administrators, several of the stressors could be combated through organizational means.

A comparison of the Total Stress Score category means, with a possible range of 1 to 6, revealed an overall mean score of 3.05 for this analysis compared to an average score of 3.29 in the Whisler (1994) study. Furthermore, although we determined a mean Total Stress Scale score, from a possible range of 61 to 366, to be 180.10 from this study, unfortunately no determinations of an average Total Stress Scale score could be gleaned from Whisler (1994) for the sake of comparison.

Beyond comparison of our results to stress levels of other probation officers, continuity in stress research has been recommended so that comparisons across occupations can be made to determine, for example, if criminal justice practitioners are actually among the most highly stressed types of workers (Cullen et al., 1985; Triplett, Mullings, & Scarborough, 1996). The Selye Health Scale utilized in our study is one such instrument that can be used to measure physical stress across occupations and has been used to assess physical stress levels of correctional personnel. In this study, from a possible range of 54 to 324, the mean Selye Health Scale score was 231. This compares to average Selye Health Scale scores of 223 in 1990 and 258 in 1997 for employees from a private correctional institution (Slate et al., 2001), 240 of correctional employees in South Carolina and Kentucky (Slate, 1993), and 278 in a study of correctional personnel in Illinois, New Jersey, New York, Pennsylvania, and Washington (Cheek & Miller, 1982a). Although, for the most part, the average stress levels of probation officers in our study are lower than those of employees in previous research, this examination is aimed at identifying those factors associated with those respondents reporting the greatest levels of stress in the present study.

As cited previously, participatory management within the probation organizational environment has been recommended by a number of researchers as a means of reducing probation officer stress and burnout (Brown, 1986, 1987; Holgate & Clegg, 1991; Simmons et al., 1997; Tabor, 1987; Whisler, 1994; Whitehead, 1981, 1986; Whitehead & Lindquist, 1985). Likewise, employee perceptions of participation in decision making proved to be a pivotal variable in the current study, as the perceived atmosphere for participation in workplace decision making significantly influenced one's opinion of his/her job and the Total Stress Scale. In other words, those who did not perceive a positive atmosphere for participation in decisions that affect them in the workplace were significantly more likely to have a negative opinion of their job and to be the most stressed. Those who scored high on the Total Stress Scale were significantly more likely to exhibit physical symptoms of stress, and those physically stressed were significantly more likely to be female and/or have seniority on the job.

Ideally, especially in view of the identification of participatory management as a pivotal variable concerning probation officer stress, comparison of probation agencies that have implemented participatory management with similarly situated organizations that have not done so could prove enlightening. Also, if possible, a study that examined probation officer stress levels prior to implementation of participatory management and then looked at such stress levels after participatory management was implemented would be beneficial as well. Ultimately, longitudinal studies of probation officer stress could prove informative.

Although perceptions of employees have been found to significantly impact organizational outcomes and behavior, such as employee job satisfaction and well-being (Griffin, 1999), employee perceptions also played an integral role in respondents' answers to survey questions in this study. However, examination of employees' blood pressure, as has been done in correctional research (Wright & Sweeney, 1990), could add another dimension to future investigations. Even those reticent to reveal their physical maladies would not be able to disguise their blood pressure. Of course, researchers can only rely on information that is available to them.

Federal probation officers have been severely neglected by researchers focusing on occupational stress. More attention needs to be paid to this area, with the hope that, eventually, sufficient numbers of respondents can be garnered to produce meaningful results, and comparisons with state probation officers and other occupations can be made. There is also something to be said for the standardization of research instruments so that studies can more readily be replicated and comparisons across studies and even occupations can be made.

According to Brown (1987), government usually lags behind the private sector in a number of ways, and the use of participatory management styles in the probation setting proves to be no exception. However, as noted by Taylor and Card (1985), empowering employees by giving them a voice in the running of the organization lets them know that they are valued and serves to instill a sense of worth, a phenomenon that will lend support to those decisions that are made together as well as those that must be made alone and will result in trust and dedication on the part of employees. To paraphrase the old Carnation Milk commercial, "Contented cows give better milk; likewise, contented employees give better performances" (Slate, 1993, p. 54). The results from this analysis indicate that employees who perceive that they have input into workplace decision making are more likely to express higher opinions of their job and are less likely to report physical symptoms of stress, which can translate to greater productivity and morale, with less absenteeism, health care costs, and employee turnover. Thus, participatory management strategies are critical to maximizing the functioning of human capital. Furthermore, as probation agencies continue to be pushed beyond their designed "hull speed" and asked to do more with less, public safety demands that probation managers and policy makers stay in tune and remain responsive to the well-being of those on the front lines of community control. The results of this analysis suggest that participatory management can be a critical factor in this process.

◁ References

Alston, K. H., & Thompson, M. J. (1996). Adaptability: The hallmark of a good probation/pretrial services supervisor in the 1990s. *Federal Probation, 60*(1), 83–85.

Archambeault, W. J., & Weirman, C. L. (1983). Critically assessing the utility of police bureaucracies in the 1980s: Implications of management theory z. *Journal of Police Science and Administration, 11*, 420–429.

Bartol, C. R., & Bartol, A. M. (1994). *Psychology and law* (2nd ed.). Pacific Grove, CA: Brooks/Cole.

Brandt, L. P. A., &. Nielsen, C. V. (1992). Job stress and adverse outcome pregnancy: A causal link or recall bias. *American Journal of Epidemiology, 135,* 302–311.

Brown, P. W. (1986). Probation officer burnout: An organizational disease/an organizational cure. *Federal Probation, 50*(1), 4–7.

Brown, P. W. (1987). Probation officer burnout: An organizational disease/an organizational cure, part II. *Federal Probation, 51*(3), 17–21.

Cheek, F. E. (1984). *Stress management for correctional officers and their families,* College Park, MD: American Correctional Association.

Cheek, F. E., & Miller, M. D. (1982a). *Prisoners of Life.* Washington, DC: American Federation of State, County and Municipal Employees.

Cheek, F. E., & Miller, M. D. (1982b). *Managerial stress in correctional facilities handbook.* Trenton, NJ: New Jersey Department of Corrections.

Cohen, S., & Williamson, G. M (1991). Stress and infectious disease in humans. *Psychological Bulletin, 109*(1), 5–24.

Cooper, C. L., & Watson, M. (1991). Cancer and stress: Psychological, biological and coping studies. New York: John Wiley.

Coyne, J. C., & Downey, G. (1991). Social factors and psychopathology: Stress, social support, and coping processes. *Annual Review of Psychology, 42,* 401–425.

Cullen, F. T., Link, B. G., Wolfe, N. T., & Frank, J. (1985). The social dimensions of correctional stress. *Justice Quarterly, 2*(4), 505–533.

Dantzker, M. L., & Kubin, B. (1998). Job satisfaction: The gender perspective among police officers. *American Journal of Criminal Justice, 23*(1), 19–31.

DeCarlo, D. T. (1987). *Workplace stress: Trends, outlook and perspectives.* New York: American Insurance Association.

DeCarlo, D. T., & Gruenfeld, D. H. (1989). *Stress in the American workplace: Alternatives for the working wounded.* Fort Washington, PA: LRP.

Dillon, P. (1999). The cost of stress is mind-numbing. *Orlando Business Journal, 16,* 36–45.

Does Stress Kill? (1995). *Consumer Reports on Health, 7*(7), 73–77.

Elkin, A. J., & Rosch, P. J. (1990). Promoting mental health at the workplace: The prevention side of stress management. *Occupational Medicine: State of the Art Review, 5*(4), 739–754.

Falk, A., Hanson, B. S., Isacsson, S., & Ostergren, P. (1992). Job strain and mortality in elderly men: Social network, support, and influence as buffers. *American Journal of Public Health, 82,* 1136–1139.

Farkas, M. A. (2001). Correctional officers: What factors influence work attitudes? *Corrections Management Quarterly, 5*(2), 20–26.

Ganster, D. C., & Schaubroeck, J. (1991). Work stress and employment health. *Journal of Management, 17*(2), 235–271.

Griffin, M. L. (1999). The influence of organizational climate on detention officers' readiness to use force in a county jail. *Criminal Justice Review, 24*(1), 1–26.

Hagan, F. E. (1989). *Research methods in criminal justice and criminology* (2nd ed.). New York: Macmillan.

Hardaway, E., Wence, F., Bingaman, D., & Selvik, R. (1996). The employee assistance program: Help for court managers and employees in the changing workplace. *Federal Probation, 60*(1), 60–66.

Hendrix, W. H., Spencer, B. A., & Gibson, G. S. (1994). Organizational and extraorganizational factors affecting stress, employee well-being, and absenteeism for males and females. *Journal of Business and Psychology, 9*(2), 103–128.

Holgate, A.M., & Clegg, I.J. (1991). The path to probation officer burnout: New dogs, old tricks. *Journal of Criminal Justice, 19,* 325–337.

Homer, C. J., Sherman, J. A., & Siegel, E. (1990). Work-related psychosocial stress and risk of preterm, low birthweight delivery. *American Journal of Public Health, 80,* 173–177.

Honnold, J. A., & Stinchcomb, J. B. (1985). Officer stress: Costs, causes and cures. *Corrections Today, 47,* 46–51.

Ivancevich, J. M., Matteson, M.T., & Richards, E. P. (1985). Special report: Who's liable for stress on the job? *Harvard Business Review, 70,* pp. 60–62, 66, 70, 72.

Janes, R. W. (1993). Total Quality Management: Can it work in federal probation? *Federal Probation, 57*(4), 28–33.

Johnson, J. V., Hall, E.M., & Theorell, T. (1989). Combined effects of job strain and social isolation on cardiovascular disease mortality in a random sample of Swedish male working population. *Scandinavian Journal of Work Environment and Health, 15,* 271–279.

Johnson, J. V., & Johansson, G. (Eds.) (1991). *The psychosocial work environment: Work organization, democratization and health.* New York: Baywood.

Jones, M., & Johnson, W. (1994). The increased felonization of probation and its impact on the function of probation: A descriptive look at county level data from the 1980s and the 1990s. *Perspectives, 18*(4), 42–46.

Karasek, R. A., &. Theorell, T. (1990). *Healthy work: Stress, productivity and the reconstruction of working life.* New York: John Wiley.

Kuykendall, J., & Unsinger, P. C. (1982). The leadership styles of police managers. *Journal of Criminal Justice, 10,* 311–321.

Lasky, G. L., Gordon, B., & Strebalus, D. J. (1986). Occupational stressors among federal correctional officers working in different security levels. *Criminal Justice and Behavior, 13*(3), 317–327.

Lawrence, R. A. (1984). Police stress and personality factors: A conceptual model. *Journal of Criminal Justice, 12,* 247–263.

Lindquist, C. A., & Whitehead, J. T. (1986). Burnout, job stress, and job satisfaction among southern correctional officers: Perceptions and causal factors. *Journal of Offender Counseling, Services and Rehabilitation, 10*(4), 5–26.

Lovrich, N. P., & Stohr, M. K. (1993). Gender and jail work: Correctional policy implications of perceptual diversity in the work force. *Policy Studies Review, 12*(1/2), 66–84.

Maggio, M., & Terenzi, E (1993). The impact of critical incident stress: Is your office prepared to respond? *Federal Probation, 57*(4), 10–16.

Maslach, C. (1982). *Burnout: The cost of caring.* Englewood Cliffs, NJ: Prentice Hall.

Melancon, D. D. (1984). Quality circles: The shape of things to come? *The Police Chief, 51*(11), 54–55.

Morash, M., & Haarr, R. N. (1995). Gender, workplace problems, and stress in policing. *Justice Quarterly, 12*(1), 113–140.

Muntaner, C., Tien, A., Eaton, W. W., & Garrison, R. (1991). Occupational characteristics and the occurrence of psychotic disorders. *Social Psychiatry and Psychiatric Epidemiology, 26,* 273–280.

Newman, J. E., & Beehr, T. A. (1979). Personal and organizational strategies for handling job stress: A review of research and opinion. *Personnel Psychology, 32,* 1–43.

Palmer, S. (1989, August). Occupational stress. *Health and Safety Practitioner*, pp. 16–18.

Patterson, B. L. (1992). Job experience and perceived job stress among police, correctional, and probation/parole officers. *Criminal Justice and Behavior, 19*(3), 260–285.

Pearlin, L. I. (1989). The sociological study of stress. *Journal of Health and Social Behavior, 30,* 241–256.

Pettway, C., & VanDine, S. (2000, November). *One moment in time: Using pagers to measure how parole/probation officers spend their time.* Paper presented at the American Society of Criminology, San Francisco, CA.

Reiser, M. (1974). Some organizational stresses on policemen. *Journal of Police Science and Administration, 2,* 156–159.

Rodichok, G. J. (1995). *A quantitative and qualitative survey of job stress among African-American police officers.* Unpublished doctoral dissertation, Temple University, Philadelphia, PA.

Schnall, P. L. (1990). Heartbreaking work: How job strain may harm the ticker. *Prevention, 11,* 14.

Schnall, P. L., Pieper, C., Schwartz, J. E., Karasek, R. A., Schlussel, Y., Devereux, R. B., et al. (1990). The relationship between job strain, workplace diastolic blood pressure, and left ventricular mass index: Results of a case-control study. *Journal of the American Medical Association, 263,* 1933–1934.

Selye, H. (1976). *The stress of life.* New York: McGraw-Hill.

Siegel, M. E. (1996). Reinventing management in the public sector. *Federal Probation, 60*(1), 30–35.

Simmons, C., Cochran, J. K., & Blount, W. R. (1997). The effects of job-related stress and job satisfaction on probation officers' inclinations to quit. *American Journal of Criminal Justice, 21*(2), 213–229.

Sims, B. (2001). Surveying the correctional environment: A review of the literature. *Corrections Management Quarterly, 5*(2), 1–12.

Slate, R. N. (1993). *Stress levels and thoughts of quitting of correctional personnel: Do perceptions of participatory management make a difference?* Unpublished doctoral dissertation, The Claremont Graduate School, Claremont, CA.

Slate, R. N., Johnson, W. W., & Wells, T. (2000). Probation officer stress: Is there an organizational solution? *Federal Probation, 64*(1), 56–60.

Slate, R. N., & Vogel, R. E. (1997). Participative management and correctional personnel: A study of the perceived atmosphere for participation in correctional decision making and its impact on employee stress and thoughts about quitting. *Journal of Criminal justice, 25,* 397–408.

Slate, R. N., Vogel, R. E., & Johnson, W. W. (2001). To quit or not to quit: Perceptions of participation in correctional decision making and the impact of organizational stress. *Corrections Management Quarterly, 5*(2), 68–78.

Stohr, M. K., Lovrich, N. P., & Mays, G. L. (1997). Service v. security focus in training assessments: Testing gender differences among women's jail correctional officers. *Women and Criminal Justice, 9*(1), 65–85.

Stress: The workplace disease of the 1990s. (1995). *San Antonio Business Journal, 9*(13), 28–37.

Tabor, R. W. (1987). *A comparison study of occupational stress among juvenile and adult probation officers.* Unpublished manuscript, Virginia Polytechnical Institute and State University, Blacksburg, VA.

Taylor, R. G., Jr., & Card, R. H. (1985). *Power sown: Power reaped.* Augusta, ME: Felicity.

Terry, W. C., III. (1981). Police stress: The empirical evidence. *Journal of Police Science and Administration, 9,* 61–75.

Terry, W. C., III. (1983). Police stress as an individual and administrative problem: Some conceptual and theoretical difficulties. *Journal of Police Science and Administration, 11,* 97–106.

Thomas, R. L. (1988). Stress perception among select federal probation and pretrial services officers and their supervisors. *Federal Probation, 52*(3), 48–58.

Tjosvold, D. (1998). Making employee involvement work: Cooperative goals and controversy to reduce costs. *Human Relations, 51*(2), 201–214.

Triplett, R., Mullings, J. L., & Scarborough, K. E. (1996). Work-related stress and coping among correctional officers: Implications from organizational literature. *Journal of Criminal Justice, 24*(4), 291–308.

Ulmer, J. T. (1992). Occupational socialization and cynicism toward prison administration. *Social Science Journal, 29*(4), 423–443.

Voorhis, P. V., Cullen, F. T., Link, B. G., & Wolfe, N. T. (1991). The impact of race and gender on correctional officers' orientation to the integrated environment. *Journal of Research in Crime and Delinquency, 28*(4), 472–500.

Whisler, P. M. (1994). *A study of stress perception by selected state probation officers.* Unpublished master's thesis, University of South Florida, Tampa, FL.

Whitehead, J. T. (1981). The management of job stress in probation and parole. *Journal of Probation and Parole, 13,* 29–32.

Whitehead, J. T. (1985). Job burnout in probation and parole: Its extent and intervention implications. *Criminal Justice and Behavior, 12*(1), 91–110.

Whitehead, J. T. (1986). Job burnout and job satisfaction among probation managers. *Journal of Criminal Justice, 14,* 25–35.

Whitehead, J. T., & Lindquist, C. A. (1985). Job stress and burnout among probation/parole officers: Perceptions and causal factors. *International Journal of Offender Therapy and Comparative Criminology, 29*(2), 109–119.

Wiggins, R. R. (1996). Ten ideas for effective managers. *Federal Probation, 60*(1), 43–49.

Wright, K. N., & Saylor, W. G. (1991). Male and female employees' perceptions of prison work: Is there a difference? *Justice Quarterly, 8*(4), 505–524.

Wright, T. A., & Sweeney, D. (1990). Correctional institution workers' coping strategies and their effect on diastolic blood pressure. *Journal of Criminal Justice, 18,* 161–169.

Wrightsman, L. S., Nietzel, M., & Fortune, W. (1994). *Psychology and the legal system* (3rd ed.). Pacific Grove, CA: Brooks/Cole.

DISCUSSION QUESTIONS

1. Why is it important to reduce job stress in any occupation?

2. What do the authors recommend as a strategy to reduce probation officer stress.

3. What was the biggest factor contributing to probation officer stress?

READING

This article by Patricia Harris, Raymond Gingerich, and Tiffany Whittaker is an evaluation of a method for assessment and differential supervision of offenders that embodies the principle of responsivity known as the Client Management Classification System (CMC). They found that, as in prior evaluations of the CMC, probationers whose officers were trained in CMC techniques experienced lower rates of revocation compared with regularly supervised subjects. However, the experimental group incurred similar or higher rates of rules violations and arrests. Of particular interest, the study found that supervision of experimental subjects did not conform to recommended CMC strategies. In combination, these results suggest the possibility that training in CMC successfully heightened officers' understanding of offender motivations and needs, leading them to view probationer misconduct in a more lenient and flexible context—and thereby producing the appearance of favorable outcomes. The findings have implications for the design of evaluations of efforts to implement principles of effective offender treatment in community corrections agencies.

The "Effectiveness" of Differential Supervision

Patricia M. Harris, Raymond Gingerich, and Tiffany A. Whittaker

This article reports the findings of an evaluation of the Client Management Classification System (CMC), a method of offender management that incorporates the principle of responsivity into correctional case supervision. The first section, a backdrop to the current effort, summarizes overlooked areas in prior research on the use of classification by community corrections agencies. The second section provides an overview of the CMC and its application to community supervision. The third section synthesizes findings and limitations of prior studies of the CMC and underscores the need for its more rigorous evaluation. Two subsequent sections describe the method and findings of a comprehensive evaluation of the CMC. A final section presents implications for the design and interpretation of future efforts

to understand the relation between community supervision practices and effective correctional treatment.

✑ Community Corrections and the Practice of Classification

Meta-analytic research on offender rehabilitation (Andrews et al., 1990; Dowden & Andrews, 1999) establishes three requirements of successful treatment. To be effective, programs must target high-risk offenders (the risk principle), address offenders' criminogenic needs (the needs principle), and match program characteristics to offenders' learning styles and cognitive abilities (the responsivity principle).

As various studies demonstrate (see, e.g., Bonta; Wallace-Capretta, & Rooney, 2000;

SOURCE: Harris, P.M., Gingerich, R., & Whittaker, T.A. (2004). The "effectiveness" of differential supervision. *Crime and Delinquency, 50*(2), 235–260. Reprinted with permission of Sage Publications, Inc.

Polizzi, MacKenzie, & Hickman, 1999; Wilson, Gallagher, & MacKenzie, 2000), community supervision agencies are particularly well suited to carry out the principles of effective treatment. In observance of the risk principle, agencies can adopt reliable and valid risk assessment tools that incorporate both static and dynamic predictors of recidivism, ensure that officers have sufficient training and incentives to correctly administer the instruments, and refrain from providing treatment to other than high-risk offenders. In observance of the principle of criminogenic needs, agencies can contract with service providers willing to target offender deficits whose remediation is associated with greatest reductions in reoffending. In observance of the principle of responsivity, community supervision officers can offer programs or impose treatment referrals that address their clients' learning styles, level of motivation, and intellectual capabilities.

Appropriate selection and correct application of classification instruments are essential to implementation of effective correctional treatments. To date, much research on classification in community corrections has helped to establish the predictive efficacy of various risk assessment instruments (see, e.g., Bonta & Motiuk, 1992; Hoffman, 1994; Rice & Harris, 1995; Silver, Smith, & Banks, 2000). Research has also confirmed the potency of criminogenic needs to the assessment of offender risk (Gendreau, Little, & Goggin, 1996). In contrast, however, little is known about how community corrections agencies use classification to enhance responsivity. This is partly because few responsivity factors have been studied (Andrews & Bonta, 1998), and a scarcity of standardized measures is available for their assessment (Kennedy, 1999).

A second weakness in research on classification in community corrections has been the failure to measure the integrity of classification practices. According to Gendreau, Goggin, and Smith (1999, p. 180), "Implementation issues . . . have been virtually ignored in the corrections literature." To alleviate this shortcoming,

scholars have begun to promote an "evidence-based approach" to correctional treatment (Andrews, 1999). One example of this approach is the Correctional Program Assessment Inventory (CPAI), which assesses the extent to which particular treatment programs operationalize the principles of effective correctional interventions (Andrews, 1999; Gendreau & Andrews, 1994). High scores on the CPAI are equated with effective correctional programs (Latessa & Holsinger, 1998).

Although there are instances of treatment programs earning high scores on the CPAI (see, e.g., Bonta et al., 2000), in the aggregate, evaluations using the CPAI offer a pessimistic outlook on the status of treatment providers' compliance with the principles of effective treatment. In a survey of 170 substance abuse programs by Gendreau and Goggin (1997), only 10% received a satisfactory rating, although the agency responsible for administering the programs was considered a leader in offender treatment. Latessa and Hoisinger's (1998) survey of 51 treatment programs also produced disappointing results: The authors noted that programs rarely matched staff, offender, and program characteristics on criteria relevant to the principle of responsivity.

Regrettably, research on implementation generally skirts the issue of how well community corrections agencies, in particular, have executed classification. Mainly, prior studies describe the extent to which probation and parole agencies have embraced emerging case classification practices or classification objectives.

Descriptive accounts from the early 1980s reported the misuse of classification instruments. Clear and Gallagher (1983) observed that officers mistook assessment scores for actual probabilities of reoffending, that cutoff scores delineating high risk cases often failed to take available supervision resources or agency mission into account, and that agencies administered different supervision standards to cases with similar overall risk scores. In a later effort, they created a typology of probation agencies to summarize the diversity of approaches to classification that prevailed in community supervision agencies during the 1980s (Clear & Gallagher, 1985). "Level I" agencies, which were without formal classification systems, entrusted supervision officers with the broadest discretion. "Level II" agencies integrated classification into supervision decision making but chose instruments arbitrarily, often without regard for the validity of their application to the offender population in question. Level II agencies' inadequate classification tools undermined their officers' confidence in assessment results and caused them to deviate from classification results, rendering the adopted systems useless. "Level III" agencies, in contrast, either developed their own assessment tools using analyses of their offender populations or validated those that were adopted.

Community supervision officers' ambivalence toward quantitative classification tools persisted into the 1990s. According to a statewide survey of Oklahoma probation officers by Schneider, Ervin, and Snyder-Joy (1996), only about one third of officers supported the use of risk and needs assessment tools in shaping supervision decisions. However, three quarters of those surveyed felt that officer discretion should play a greater role in shaping supervision decisions than was permitted by the classification system rules. As many as 85% of officers believed that officer judgment yielded more accurate choices regarding offender supervision than did formal classification tools. A somewhat more positive view of objective risk and needs classification instruments was reported by Jones, Johnson, Latessa, and Travis (1999), who surveyed U.S. probation and parole agencies. Four fifths of respondents reported using standardized, objective classification tools, and nearly as many rated standardized case classification as at least "very important."

Two studies examined the relation between community supervision officers' receptivity to the treatment objective of offender classification and officers' role performance. Clear and Latessa (1993) administered two questionnaires to 31 probation officers in one control-oriented and one treatment-oriented organization. The first questionnaire assessed officers' acceptance of the law enforcement and treatment functions of community supervision, respectively. The second captured officers' ratings of the appropriateness of 60 supervision tasks in response to descriptions of five cases. An analysis that included a control for site revealed a stronger relation between attitudes supportive of law enforcer roles and preference for control approaches than between attitudes supportive of treatment roles and assistance approaches— though preferences for authoritative roles did not suppress selection of assistance-related tasks, nor did preferences for treatment roles suppress selection of control-related tasks. Organizational philosophy supportive of treatment approaches was a better predictor of officers' selection of assistance-related activities than were officers' attitudes themselves.

Fulton, Stichman, Travis, and Latessa (1997) examined the attitudes of 11 intensive supervision probation (ISP) and 61 regular probation officers. Survey items consisted of 33 semantic differentials reflecting disparate perspectives on officer roles, strategies, and supervision goals. The ISP officers, who had undergone training on the principles of effective intervention, exhibited stronger support for rehabilitation and strategies favoring behavioral change than did the untrained regular supervision officers. The study did not include a pretest of the trained officers, however.

As vehicles for understanding the extent to which community corrections agencies have implemented classification to facilitate effective correctional treatment, these studies have two main limitations. The first is the reliance on officers' attitudes in place of measures of actual case supervision decision making. The second is the exclusion of indicators of offender behaviors as dependent variables. The failure to explore the relation between various officer role orientations and recidivism and, more important, classification-related decision making and recidivism, is a substantial deficit of research to date.

In summary, renewed credibility of rehabilitation as a correctional goal coupled with growing awareness that treatment efforts often fail to embody the principles of risk, needs, or responsivity prompts closer attention to the role played by community corrections agencies, a major conduit for offender rehabilitation, in implementing classification for effective correctional treatment. Especially lacking are studies of how community corrections agencies use classification to achieve the principle of responsivity.

▧ Client Management Classification System: An Overview

The CMC, a component of the Wisconsin classification model, is used by roughly one quarter of the probation and parole agencies nationwide that practice case classification (Jones et al., 1999). The CMC is a structured, interview-based assessment leading to the classification of offenders into one of five "strategy groups," the basis for their differential supervision (Lerner, Arling, & Baird, 1986). These include the selective intervention–situational (SI-S), selective intervention–treatment (SI-T), casework/control (CC), environmental structure (ES), and limit setting (LS) classifications. The offender's criminogenic needs, officer's manner of interaction with the offender, and types of programs to which the offender should be referred all differ depending upon the strategy group into which he or she has been classified. When completed, the CMC differentiates offenders with respect to their stability, motivation, number and type of needs, suggestibility, and antisocial values.

The interview includes 45 questions, accompanied by a scoring guide to facilitate interrater reliability (Lerner et al., 1986). The 45 items query subjects on their attitudes regarding their offense, criminal history, family background, interpersonal relationships, current problems, and prospects for the future. Also included are 11 objective measures of criminal history and client background, 8 measures assessing the client's behavior during the interview, and 7 items that capture the officer's impressions of the relation of the client's needs to his or her criminal involvement. Items have either positive or negative scores attached, depending on whether each is indicative or contraindicative of a particular strategy group. The higher the score attached to an item, the greater its interrater reliability.

Upon completion and scoring of the interview, officers complete a force-field analysis that facilitates their identification of offenders' dynamic (versus static, or unalterable) needs, which requires them to consider the salience of those needs to offenders' ongoing involvement in criminal activity. Officers then rank the needs, giving highest priority to those needs that are alterable, pose the strongest relation to the offender's ongoing involvement in crime, give rise to other needs, and can be addressed the most quickly. During the final assessment stage, each officer completes a supervision plan for the highest ranked need or needs. The completed plan identifies (a) a behavioral, time-oriented objective for case supervision, which, if attained, would eliminate the offender's involvement in crime; (b) referrals and strategies the officer will use to help the offender bring about the change; and (c) specific steps the offender will take to meet the supervision objective. The officer then reviews the plan with the offender, eliciting the latter's agreement to its provisions.

Agencies that incorporate the CMC into supervision planning can implement principles of effective correctional intervention. Though the CMC is not a risk assessment tool per se, its application may be reserved for cases designated high risk by other means. (In some locales, this aspect of the CMC may have less to do with regard for the principle of risk than concern for the extra time—one hour—involved in conducting and scoring each interview.) Presuming treatment referrals are withheld from lower risk cases, the selective application of the CMC facilitates the dedication of treatment resources to highest risk offenders.

The CMC is particularly useful in helping officers to identify offenders' criminogenic needs. After scoring the completed CMC, the officer uses the information developed from the interview to develop a case supervision plan. Officers begin supervision planning by compiling a force-field analysis to identify supervision priorities. For each of 11 needs factors, officers record both offender strengths and weaknesses. Officers then prioritize the problem areas, eventually targeting only those that are a direct cause of the offender's criminal activity and for which intervention is likely to bring about a decrease in recidivism. In this respect, the CMC not only heightens officers' awareness of criminogenic needs, it enhances their appreciation for those that are dynamic as well.

Finally, the CMC takes the principle of responsivity into account, and it is one of only a few contemporary classification systems that do (Van Voorhis, 1997). Responsive interventions match offender characteristics with therapist characteristics, and they match therapist skills with program type (Gendreau, 1996; Serin & Kennedy, 1997; Latessa & Holsinger, 1998). Relevant offender characteristics include "specific competencies, interests, and learning styles that a client must possess in order to benefit from particular types of programs" (Andrews & Bonta, 1998, p. 89). These include level of social skills and cognitive style, and they can include gender and race considerations, inasmuch as responsive treatment may require cultural sensitivity (Bonta, 1995). Other important offender characteristics are motivation and treatment readiness (Kennedy & Serin, 1997).

The CMC acknowledges the importance of building offender motivation to participate in treatment. Raising the offender's awareness of his or her respective problem behavior and its consequences is the recommended starting point for supervising SI-T, CC, ES, and LS offenders. For example, officers are trained to help raise the ES client's awareness of the negative consequences attached to manipulation by antisocial companions. The CMC calls attention to the tendency of the CC offender to relapse, and to the need of the supervising officer to restore the CC client's sustained program involvement. The CMC cautions officers to question the SI-T client's denial and the LS client's superficial conformity to supervision or program rules (National Institute of Corrections, 1991).

The CMC also helps officers to recognize for which offender types cognitive interventions would be helpful, and for which they would not. Officers who comply with CMC guidelines help CC and LS offenders to recognize and overcome thinking errors that, if left uncorrected, would result in further antisocial or illegal conduct. By assisting them to recognize the intellectual and social deficits of the ES offender, the CMC helps officers to favor behavioral over inappropriate cognitive strategies and to avoid reliance on abstract concepts when conveying information to members of this group. Finally, the CMC forewarns officers of measures that would be counterproductive for each respective strategy group (National Institute of Corrections, 1991).

Data and Method

Description of the Sample

The sample consists of 1,017 adult offenders who entered felony probation supervision in a large urban south-central county of the United States between March 15, 1991 and February 18, 1992. Subjects were those members of two cohorts of 1,292 felony probationers ordered to report to two regional offices of the probation department, and for whom follow-up data were available. Group assignment was made on the basis of probation referral location. Treatment group subjects were composed of the cohort of offenders entering the north regional office of the probation department; control group subjects consisted of offenders entering the central regional office. The cohorts exclude probationers who never reported for supervision and those who could not be assessed by their officers within the first several months of supervision. Probationers' case files were the source of most of the data collected for this study. Records of new arrests were compiled from case files and the state-administered criminal history information system.

Of the 1,017 offenders for whom follow-up data were available, 581 subjects composed the treatment group cohort and 436 subjects composed the control group cohort. Each of the 581 treatment group subjects was classified and differentially supervised by officers trained in the proper use of CMC assessment and supervision strategies. The remaining 436 control subjects were classified within 2 months of entry but not subjected to differential supervision. A confederate of the researchers, not an employee of the probation department, interviewed and classified control group subjects. Classification results for the group of 436 controls were subsequently withheld from the subjects' supervising officers, who received no training in CMC assessment and supervision techniques.

By the end of the assessment data collection period, the sample consisted of 349 offenders classified as SI-S (34.3% of the sample); 310 classified as SI-T (30.5%); 166 classified as CC (16.3%); 122 classified as LS (12.0%); and 70 classified as ES (6.9%).

Dependent Variables

The dependent variable in all analyses is case outcome or recidivism. The study considered the following three outcome measures: (a) whether the subject experienced a write-up for a technical violation, (b) whether the subject experienced revocation, and (c) whether the subject experienced a new arrest while

under supervision. Of the 1,017 offenders in the study, 44% engaged in a technical rule violation, 23.2% were revoked, and 31.1% experienced an arrest while under supervision. Officers enjoyed considerable discretion in making recommendations regarding the disposition of violations and motions to revoke. Department policies required officers to document all violations, but whereas policy stipulated various *reporting* requirements, it did not mandate particular *outcomes*. That is, with respect to most violations, officers were able to use discretion to determine whether to file on probationers who violated the terms of their supervision, as well as to recommend outcomes other than revocation in response to offenders who were ultimately filed on.

For example, although policy directed officers to follow up positive urinalyses with a documented meeting with the offender and to request additional specimens, it did not require officers to file on the violator. Where subjects failed to complete community service as ordered, policy required officers only to notify the court of the subject's noncompliance. Officers whose subjects failed to complete programs were required to advise the court that subjects had not completed classes by the required due date. When probationers failed to pay fees for 3 consecutive months, policy required officers to review the case with a senior officer. Officers were, however, required to file motions to revoke in response to probationers who failed to report, provided certain criteria were met. Because it was not always possible to determine the precise duration of subjects' unauthorized absences (or more generally, when officers were required to file versus when they merely exercised discretion to file), offenders who failed to report were excluded from analyses where revocation was the dependent variable.

Independent Variables

Independent variables used in the logistic regression analyses include generally accepted predictors of recidivism risk (Gendreau et al., 1996). Risk predictors consisted of measures of subjects' criminal history, substance abuse, criminal companions, education deficits, employment history, age, and gender.

Measures of criminal history included the variables prior felony arrest, prior period of probation or incarceration, and prior assaultive behavior. Prior incarcerations were combined with prior periods of probation inasmuch as the former category comprised only 3.6% of the sample. Subjects whose committing offense was a repeat DWI, whose criminal histories included offenses committed under the influence of alcohol or drugs, or whose response to depression included alcohol or drug use were coded as substance abusers. Subjects who did not graduate from high school, who experienced remedial education, or who experienced difficulty performing school work were coded as having educational deficits. Subjects who were not self-supporting, whose officer had identified lack of job skills as a primary work problem, or who were employed less than full-time for most of their working life were coded as having employment problems. The multivariate analyses also included controls for race and ethnicity, using the dummy variables Black and Hispanic. White non-Hispanic subjects served as the reference category for each of the latter variables.

The analysis also controlled for various conditions of probation, in light of prior research linking community supervision outcomes with variations in intensity of supervision. Offenders who are observed more closely, either through more frequent officer contacts, electronic surveillance, or some other form of intensive supervision, experience a greater probability of being caught breaking rules than do peers who are subjected to less scrutiny (Byrne, Lurigio, & Baird, 1989; Petersilia, 1998).

Although most subjects experienced only regular supervision throughout the course of the study, 213 probationers (21% of the sample) experienced periods of closer supervision during the follow-up period. Of the 213,

roughly one third experienced a period of ISP or day reporting as an initial condition of probation. For all others, transfer occurred following an arrest or technical rules violation. Whether or not a subject was coded as having experienced intensive supervision (a category that includes either ISP or day reporting) depended on the timing of the transfer with respect to the outcome in question. That is, some individuals violated prior to (or absent any) transfer to more intensive supervision, and some violated while on intensive supervision. As revocation is a terminal event, all 213 individuals who had experienced any periods of intensive supervision over the course of their probation were coded as experiencing intensive supervision in the analysis that used revocation as an outcome measure. However, with respect to the outcomes rearrest or technical violation, subjects were coded as having experienced intensive supervision only if they had not already engaged in the misconduct in question while on regular supervision. Thus, subjects who transferred to intensive supervision following an arrest but not before would not be coded as having been on intensive supervision in the analysis that used arrest as an outcome measure because the subject had experienced the negative outcome while on regular, not intensive, supervision. Similarly, subjects who engaged in technical violations prior to a transfer to intensive supervision were coded as regular supervision cases because their misconduct had occurred under regular and not intensive supervision. With respect to either outcome, the group of subjects coded as regular supervision includes only cases for which no intensive supervision had been imposed at any time in the current probation, up to and including the date of misconduct in question. In all, there were 164 intensive supervision subjects (16.2% of the sample) in the analysis of technical violations and 120 intensive supervision subjects (11.8% of the sample) in the analysis of rearrest.

Finally, the analysis controlled for number of helpful probation conditions, including referrals to treatment, educational, and employment programs, for the same reason it controlled for punitive ones. Offenders who are referred to several programs at once are at higher risk of failing to attend them than are counterparts who are referred to fewer (or no) programs, if only because they are exposed to more frequent opportunities to miss meetings. In addition, program referrals provide opportunities for collateral contacts (such as treatment providers) to catch offenders in the act of violating other conditions of probation (e.g., the prohibition against substance abuse).

All categorical variables were coded 1 if the subject had the characteristic in question and 0 if he or she did not. Age, an interval variable, was left as is. Subjects were followed up for the duration of their probation supervision, or for 4 years following assessment, whichever was shorter. In the aggregate, subjects were tracked for an average of 43.5 months, with a standard deviation of 10.6 months. There was no statistically significant difference between the control and differentially supervised groups with respect to average months of follow-up.

⊠ Sample Characteristics

Table 1 summarizes the distribution and coding schemes for all independent variables used in the regression analyses, disaggregated by study group (control or differential supervision). The table reveals various statistically significant differences between the groups, underscoring the need for and importance of a multivariate analysis of outcomes. Control group subjects were significantly more likely to be Hispanic ($p = .003$), associate with criminal companions ($p < .001$), have employment problems ($p < .001$), and exhibit prior assaultive behavior ($p = .012$) compared with differential supervision subjects. On the other hand, differential supervision subjects were more likely to have experienced prior probation supervision or incarceration ($p < .001$)

than were control group subjects. In addition, they were more likely to have experienced ISP at any time over the course of their supervision than were control group subjects ($p = .001$). No significant differences between control and differentially supervised offenders emerged with respect to gender, substance abuse, education, prior felony arrests, number of program referrals, age, and race (i.e., percentage African American).

⊠ Measures of Program Integrity

Independent variables used in the second phase of the analysis facilitate an examination of the extent to which the subject's case supervision met the expectations of differential supervision called for by the CMC. Focus was limited to features of differential supervision routinely documented by officers in their clients' case files and subject to routine audit by the state's probation oversight agency. Measures of case supervision of interest to this effort included the following nine variables:

Quality of the problem statement. Problem statements summarize the officer's understanding of reasons for the offender's involvement in crime. In measuring the quality of the officer's problem statement, raters assessed whether the officer had correctly identified (a) the offender's criminogenic need, (b) the offender's behavior in response to the criminogenic need, and (c) the consequences of the offender's behavior. The problem statement was rated as having demonstrated full compliance if it correctly identified all three characteristics, and some compliance if it identified just two. Problem statements that met one or none of the three criteria were considered seriously deficient.

Quality of the force-field analysis. The force-field analysis provides the officer with a framework for summarizing the salient

Table 1	Distribution of Independent Variables Used in the Logistic Regressions (in percentages)	
Variable	Control (n = 436)	Differential Supervision (n = 581)
Hispanic**	19.0	12.2
Black	45.2	45.1
Male	78.7	76.8
Has criminal companions***	70.3	52.6
Has employment problems***	67.4	55.9
Substance abuse indicated	25.9	21.3
Educational deficits indicated	69.5	67.3
Has prior felony arrests	28.4	26.9
Has prior probation or incarceration***	23.9	35.9
Prior assaultive behavior indicated*	33.0	25.8
Experienced intensive supervision (for analysis of revocations)***	16.1	24.6
Experienced intensive supervision (for analysis of rearrest)***a	12.4	18.9
Experienced intensive supervision (for analysis of technical violations)***a	10.3	12.9
Number of treatment referrals		
None (reference category)	28.9	28.2
One	27.1	24.4
Two	18.8	19.3
Three or more	25.2	28.1
Age (interval)	$M = 29.9$	$M = 28.9$
	$SD = 11.3$	$SD = 9.8$

a. Subjects who experienced the negative outcome without having a prior placement of intensive supervision were coded as regular supervision cases.

$*p \leq .05; **p \leq .01; ***p \leq .001.$

features of an offender's background. The analysis facilitates the officer's ability to prioritize problem areas and to distinguish between offender problems that likely result in criminal behavior and those that likely do not.

Appropriateness of supervision priority. The supervision priority was judged to be fully compliant if the problem identified as the top priority was the proximate cause of the offender's criminal behavior and if addressing this problem was consistent with the strategy group (classification) to which the offender was assigned.

Validity of case classification. Cases were coded as correctly classified, and in full compliance, if the auditor agreed with the officer's selection of strategy group for the offender in question.

Adequacy of supervision objectives. Cases were rated as fully compliant with the expectations if the supervision objectives met the following characteristics: The objectives (a) addressed the offender's criminogenic need, (b) were consistent with CMC goals for the strategy group in question, (c) were positively stated, (d) stated an outcome, and (e) were behavioral.

Adequacy of the offender action plan. The offender action plan outlines the nature and scope of the offender's contributions to meeting supervision objectives. An offender action plan was rated as fully compliant if it possessed the following characteristics: The plan (a) reflected a CMC strategy for the offender, (b) stated how the offender will achieve a change in behavior, and (c) established how the officer would verify the offender's change.

Adequacy of the officer action plan. The officer action plan identifies how the officer will assist the offender in meeting supervision objectives. Fully compliant officer action plans met all of the following criteria: (a) The plan reflected a CMC strategy for the officer, (b) the plan established how the officer would assist the offender in achieving change, and (c) the plan described how the officer would verify the offender's behavioral change.

Timeliness of supervision plan. Supervision plans exhibited full compliance if they were completed within 45 days of the date the offender entered probation and some

compliance if completed within 60 days but not sooner than 46 days. Supervision plans completed at later dates were regarded as seriously deficient.

Compliance with CMC strategy. Case supervision was coded as fully compliant with the CMC strategy if the file documentation indicated that the officer used referrals and interactions appropriate to the classification group in question. A rating of some compliance was attached to those cases where the officer mixed appropriate and inappropriate referrals and interactions. Cases rated seriously deficient were those involving only incorrect referrals and interactions.

▧ Findings

Analysis of Program Outcomes

Table 2 summarizes contingency analyses and chi-square tests of differences in case outcome for regularly and differentially supervised cases. Results are presented for the sample as a whole and for the sample controlling for classification (i.e., strategy group) type. When all classifications were considered together, differentially supervised cases did not perform significantly better or worse than their regularly supervised counterparts with respect to new substance abuse, absconding, technical violations as a whole, or new arrests during supervision. Differentially supervised cases did incur significantly higher rates of failure to comply with other program conditions (i.e., failure to attend program[s]; failure to pay fees, fines, or restitution; and failure to work) compared with controls. Given their record of misconduct relative to control subjects, it is interesting that differentially supervised probationers were only two thirds as likely to experience revocation compared with the controls ($p = .001$).

Strategy group–specific comparisons produced results that, in most instances, were

Table 2	Summary of Contingency and Chi-Square Analyses of the Relation Between Group Membership and Various Study Outcomes[a]

	Group				
	Control		Differential Supervision		
Outcome	%	n	%	n	χ^2
All classifications ($N = 1,017$)					
New substance abuse	20.4	`89	18.9	110	.347
Absconded	14.4	63	16.5	96	.812
Violation of other technical conditions**[b]	9.4	41	19.1	111	18.441
Any technical violation	41.1	179	46.1	268	2.601
Revoked***[c]	25.2	94	16.3	79	10.404
Arrested during probation supervision	30.7	134	31.3	182	.041
SI-S classification only ($n = 349$)					
New substance abuse	16.7	23	10.9	23	2.424
Absconded	11.6	16	18.5	39	2.983
Violation of other technical conditions**[b]	7.2	10	18.5	39	8.729
Any technical violation	34.1	47	41.2	87	1.816
Revoked[c]	16.4	20	10.5	18	2.229
Arrested during probation supervision	23.2	32	26.5	56	.497
SI-T classification only ($n = 310$)					
New substance abuse	18.3	19	22.3	46	.688
Absconded	11.5	12	17.0	35	1.597
Violation of other technical conditions[b]	9.6	10	14.1	29	1.251
Any technical violation	35.6	37	46.1	95	3.140
Revoked[c]	19.6	18	15.8	27	.601
Arrested during probation supervision	26.0	27	28.6	59	.247
CC classification only ($n = 166$)					
New substance abuse	23.6	21	27.3	21	.295
Absconded	19.1	17	18.2	14	.023
Violation of other technical conditions*[b]	11.2	10	24.7	19	5.172
Any technical violation	48.3	43	58.4	45	1.700
Revoked [c]	31.9	23	30.2	19	.050
Arrested during probation supervision	38.2	34	46.8	36	1.238

ES classification only ($n = 70$)					
New substance abuse*	9.7	3	28.2	11	3.706
Absconded	16.1	5	12.8	5	.154
Violation of other technical conditions*[b,d]	6.5	2	28.2	11	5.405
Any technical violation	32.3	10	51.3	20	2.522
Revoked[c]	34.6	9	17.6	6	2.262
Arrested during probation supervision	35.5	11	35.9	14	.001
LS classification only ($n = 122$)					
New substance abuse	31.1	23	18.8	9	2.288
Absconded	17.6	13	6.3	3	3.273
Violation of other technical conditions*[b]	12.2	9	27.1	13	4.385
Any technical violation	56.8	42	43.8	21	1.972
Revoked*[c]	39.3	24	20.0	9	4.520
Arrested during probation supervision	40.5	30	35.4	17	.323

NOTE: SI-S = selective intervention–situational; SI-T = selective intervention–treatment; CC = casework/control; ES = environmental structure; LS = limit setting.

a. Table presents percentages only for subjects experiencing the negative outcome.

b. Includes failure to attend required programs; failure to pay fees, fines, or restitution; and failure to work.

c. Excludes probationers who were revoked for failing to report.

d. Reported probability is from Fisher's Exact Test, due to expected cell counts of less than 5.

*$p \le .05$; **$p \le .01$; ***$p \le .001$

consistent with the findings from the aggregate analysis, in that differentially supervised subjects fared worse than controls with respect to most outcomes measured, while all five strategy groups experienced lower rates of revocation. Relations in the disaggregated analyses, however, mainly did not achieve statistical significance (though with markedly reduced Ns, statistical significance would be more difficult to achieve).

Differentially supervised SI-S subjects were more likely to abscond supervision, violate other technical rules, engage in any technical violation, and experience rearrest than SI-S subjects who were not differentially supervised. The experimental group fared better only with respect to new substance abuse. Nonetheless, the differential supervision group was only two thirds as likely to encounter revocation compared with the control group. Differentially supervised SI-T clients fared worse than controls on all outcome measures except for revocation. Differentially supervised CC offenders fared worse on all measures of violations, except for absconding, where they were only slightly less likely to violate. The difference in revocation rates (30.2% for differentially supervised offenders and 31.9% for controls) is negligible.

Analyses of ES cases produce roughly comparable findings, though the number of cases available for study—70—is modest.

Differentially supervised ES offenders were somewhat less likely to abscond but far more likely to engage in new substance abuse, other technical violations, and any technical violation than were control group ES offenders. The groups were roughly as likely to experience a new arrest (35.9% for differentially supervised versus 35.5% for controls), but differentially supervised ES subjects were only half as likely to experience revocation.

Findings for the LS group differ from the other four strategy groups in that differentially supervised subjects fared noticeably better than control group subjects on all but one outcome measure (violation of other technical conditions). However, they were only slightly less likely to experience rearrest compared with the controls (35.4% versus 40.5%). Differentially supervised LS offenders were half as likely to be revoked compared with their regularly supervised counterparts.

Table 3 presents the results of logistic regression analyses for each of three outcomes: revocation, arrest while under supervision, and any technical violation. In all three analyses, risk-related variables were entered in an initial block, and group membership (i.e., differential versus regular supervision) was entered in the second block. This analytic sequence permits an assessment of the contribution of supervision type to case outcome, after controlling for demographic characteristics and factors typically associated with risk of offender failure.

As was expected, offender risk factors and number of program referrals contribute to an understanding of supervision outcomes but to varying extents, depending upon the outcome in question. Risk factors make the greatest contribution in the analysis of technical violations. Probationers who keep criminal companions and those who have employment problems experience nearly one and one half times the odds of engaging in technical violations than probationers who do not share these characteristics. Substance abusers have greatly reduced odds of engaging in technical violations than those who are not abusers, though they have increased odds of being revoked. As expected, the number of program referrals heightens the odds of incurring any negative supervision outcome, though the relation between referrals and outcome is strongest for technical violations.

Also interesting is a comparison of the effect of intensive supervision on outcome. For both arrest and technical violations, placement under intensive supervision has an apparently deterrent effect: B coefficients are negative and statistically significant, and intensively supervised cases have markedly lower odds of arrest or rules violations than cases under regular supervision. However, in the revocation analysis, the opposite is true. Intensively supervised cases encounter roughly 1.3 times the odds of being revoked than regularly supervised cases, though the relation is not statistically significant. This finding suggests a greater willingness to impose revocation as a punitive alternative in cases where probationers who violate conditions of supervision have already been exposed to less severe intermediate sanctions than in cases where they have not. Thus, taking into account the results of all three analyses, although intensively supervised cases are less likely to be arrested or observed violating rules than are probationers who are regularly supervised, intensively supervised cases who do violate can expect a higher likelihood of revocation.

Turning now to the variable of greatest interest to this study, a perusal of the goodness-of-fit statistics for the three models reveals that, although all three are significant, group membership (differential supervision or control) makes a statistically significant contribution to only the model predicting revocation. Group membership fails to achieve statistical significance in either of the models predicting arrest or technical violations. A comparison of odds ratios for the three models produces an unexpected contrast:

| Table 3 | Results of Logistic Regression Analyses for Three Study Outcomes |

	Study Outcome								
	Revocation (N = 843)[a]			Arrest During - Supervision (N = 1,000)[a]			Any Technical Violation (N = 1,001)[a]		
Variable	B	SE	Odds Ratio	B	SE	Odds Ratio	B	SE	Odds Ratio
Age at assessment	−.027**	.010	0.974	−.035***	.008	0.966	−.011	.007	0.989
Prior assault	−.043	.202	0.958	.254	.158	1.290	.031	.160	1.031
Black	.870***	.215	2.387	.045	.165	1.046	.355*	.160	1.427
Criminal companions	.315	.213	1.370	−.174	.163	0.841	.353*	.158	1.423
Employment problems	.326	.204	1.385	−.029	.156	0.971	.233	.154	1.262
Hispanic	.136	.296	1.146	−.166	.216	0.847	.216	.212	1.241
Education deficits	.279	.215	1.321	.087	.169	1.091	−.041	.160	0.960
Intensive supervision	.274	.218	1.315	−.572**	.208	0.565	−.535*	.223	0.585
Prior felony	.082	.208	1.086	.280	.166	1.323	.015	.168	1.015
Prior probation/ incarceration	.271	.219	1.312	.189	.171	1.209	−.087	.171	0.917
Number of program referrals									
None	—	—	—	—	—	—	—	—	—
One	.499	.273	1.647	.166	.207	1.180	.763***	.194	2.145
Two	.623*	.285	1.865	.546*	.215	1.727	.851***	.210	2.343
Three or more	.453	.274	1.572	.775***	.206	2.171	2.178***	.212	8.830
Sex	.459	.250	1.582	.654**	.200	1.924	.197	.180	1.218
Substance abuse	.481*	.232	1.617	.260	.185	1.297	−.405*	.190	0.667
Group (control/ differential supervision)	−.549**	.192	0.578	−.014	.152	0.986	.273	.149	1.313
−2LL likelihood	769.98	1,159.06	1,197.99						
Model χ^2	74.80	82.32	175.03						
df	16	16	16						
p (full model)	< .001	< .001	< .001						
Block χ^{2b}	8.25	0.008	3.35						
p (block only)	.004	.928	.067						

a. N of cases represents total remaining following listwise exclusion of missing cases. Revocation N also excludes subjects who failed to report.

b. Change in model χ^2 due to group type (control or differential supervision) alone.

*$p \leq .05$; **$p \leq .01$; ***$p \leq .001$.

After taking other independent variables into account, differential supervision cases encounter roughly similar odds of rearrest, somewhat higher odds of new technical violations, *but less than two thirds the odds of revocation* than their control counterparts.

The outcomes displayed in Table 4 with respect to the role played by differential supervision are clearly inconsistent. On one hand, controlling for a variety of factors ordinarily associated with program effects, differentially supervised cases encountered higher success rates relative to control group subjects when revocation is used to represent failure. On the other hand, they fared as poorly or worse than controls when failure is represented by rearrest and technical rules violations, respectively. Can this inconsistency be reconciled?

Revocation differs from either of the other two outcomes in that it can occur following an accumulation of misconduct. Whereas the measures of rearrest and technical violation indicate whether each subject ever engaged in even one of these behaviors, revocation can take place as a response to aggregates of misconduct by individual subjects. One might speculate, then, that the lower rate of revocation noted among differentially supervised cases may be due to a lower average number of violations per differentially supervised probationer, or lesser severity of violations, compared with controls.

Table 4	Use of Revocation as a Response to Various Probation Violations and Combinations of Violations, by Study Group	
	Percentage Committing This Violation Who Were Revoked	
Violation Type	**Differential Supervision**	**Control Group**
New law violation ($n = 317$)**	36.3	52.6
New substance abuse ($n = 199$)*	35.5	51.7
Failure to work ($n = 16$)[a]	45.5	60.0
Failure to attend program ($n = 85$)	12.5	19.0
Other violations ($n = 57$)[a,b]	10.0	11.8
One violation total ($n = 275$)***[c]	22.4	42.3
Two violations total ($n = 146$)[c]	31.9	47.3
Three violations total ($n = 71$)*[c]	33.3	57.7
Only violation is a law violation ($n = 146$)	35.5	50.0
Two law violations ($n = 68$)	35.4	60.0
Three law violations ($n = 13$)[a]	50.0	80.0

a. Reported probability is from Fisher's Exact Test, due to expected cell courts of less than 5.

b. Includes failure to pay fees, fines, or restitution.

c. Excludes failures to report. Categories of two and three violations, respectively, encompass any combination of law and technical violations.

*$p \leq .05$; **$p \leq .01$; ***$p \leq .001$.

Table 5 explores this possibility by reporting the likelihood of revocation as a response to particular misconduct, as well as to overall number and combination of violations. As the table indicates, neither number nor type nor combination of violations can account for the lower rate of revocations among the differentially supervised cases. The differential supervision group encounters a lower rate of revocation than the controls, no matter the type of violation (law, substance abuse, failure to work, failure to attend programs, and failure to comply with financial conditions), the number of violations (total of one, two, or three violations), or the number of rearrests (total of one, two, or three rearrests). That is, compared with control counterparts with similar violation records and no matter the extent or severity of their records of misconduct, differentially supervised cases are simply less likely to be revoked. In most of the categories depicted in Table 5, the rate of revocation for differentially supervised cases is substantially lower than for the controls. Moreover, with respect to four categories (any new law violation,

any new substance abuse, total of one violation, and total of three violations), the relations are statistically significant.

Analysis of Implementation of Differential Supervision

The second stage of the analysis examines the extent to which officers carried out differential supervision as trained, and explores the impact of variation in integrity of differential supervision on case outcome. Table 6 depicts the distribution of nine characteristics of case supervision. The table reveals that officers' compliance with the requirements of differential supervision frequently fell short of the desired standards. Judging from the "full compliance" column of Table 6, offenders' CMC classifications were correctly identified in roughly 7 of every 10 cases, but indication that the appropriate supervision strategy was followed occurred in roughly one quarter of cases. Officers developed satisfactory force-field analyses approximately 30% of the time and correctly prioritized offender problems only once in every 5 cases.

Table 5	Distribution of Nine Measures of Integrity of Case Supervision					
	Non-compliance		Some Compliance		Full Compliance	
Characteristic of Supervision	%	n	%	n	%	n
Problem statement correct	44.7	84	36.2	68	19.1	36
Force field is adequate	21.1	41	49.0	95	29.9	58
Priorities set correctly	57.7	113	21.9	43	20.4	40
Classified into correct strategy	11.9	25	16.2	34	71.9	151
Objective correct	57.2	107	34.2	64	8.6	16
Offender action plan correct	46.8	88	38.8	73	14.4	27
Officer action plan correct	42.6	80	43.1	81	14.4	27
Plan completed within 45 days	32.6	62	12.1	23	55.3	105
Strategy followed	38.0	79	36.1	75	26.0	54

Table 6 examines the strength of the relation between the nine indicators of the integrity of differential supervision and case outcome. Due to the small numbers of cases exhibiting full compliance with each of the indicators in question, this analysis groups cases in which there was either some or full compliance into one category, referred to as "at least some compliance."

The table contains several interesting findings. First, although probationers who did experience at least some differential supervision were less likely to exhibit negative case outcomes than those who received none, for most supervision factors studied, differences were typically slight and nonsignificant. Second, where significant differences emerge, there is a negative relation between the officer's recognition of an offender's criminogenic need and case outcome (for outcomes of revocation and arrest under supervision). Fewer than 15% of subjects whose supervision plans included

Table 6	Relation Between Characteristics of Case Supervision and Various Outcomes[a]

		At Least Some Compliance With Feature of Supervision								
		% Revoked[b]			% Arrested[c]			% Any Technical Violation[d]		
Characteristic	n	No	Yes	φ	No	Yes	φ	No	Yes	φ
Problem statement correct	188	34.5	14.4	−.24***	40.5	25.0	−.17*	58.3	59.6	.01
Force field is adequate	194	31.7	19.6	−.12	48.8	27.5	−.19**	61.0	60.1	−.01
Priorities set correctly	196	25.7	19.3	−.08	37.2	25.3	−.13	59.3	61.4	−.02
Classified into correct strategy	210	24.0	23.2	−.01	40.0	32.4	−.05	68.0	59.5	−.06
Objective correct	187	24.3	21.3	−.04	35.5	27.5	−.09	57.9	60.0	.02
Offender action plan correct	188	27.3	20.0	−.09	35.2	29.0	−.07	58.0	60.0	.02
Officer action plan correct	188	23.8	23.1	−.01	38.8	26.9	−.13	56.3	61.1	.05
Plan completed within 45 days	190	32.3	19.5	−.14*	38.7	29.7	−.09	62.9	57.0	−.06
Strategy followed	208	24.1	22.5	−.02	41.8	27.9	−.14*	57.0	60.5	.04

a. Tables present percentages only for subjects experiencing the negative outcome.

b. Percentage of subjects revoked, by characteristics of case supervision.

c. Percentage of subjects arrested during supervision, by characteristics of case supervision.

d. Percentage of subjects engaged in a technical violation, by characteristics of case supervision.

*$p \le .05$; **$p \le .01$; ***$p \le .001$.

an adequate problem statement were revoked, versus 34.5% of subjects whose supervision plans contained inaccurate or missing problem statements ($p < .001$; $\varphi = -.24$). Similarly, just one quarter of subjects whose supervision plans included an adequate problem statement experienced a new arrest while under supervision, compared with 40.5% of subjects whose supervision plans contained inaccurate or missing problem statements ($p < .05$; $\varphi = -.17$). In other words, the greater the officer's understanding of the problems driving the offender's behavior, the lower the offender's likelihood of revocation or rearrest.

That there is a significant relation between the officer's understanding of the offender's underlying problem and outcome for both of the most serious outcomes studied, as opposed to just revocation, suggests the possibility that some officers may have successfully employed some methods to reduce the incidence of new offending that were not considered part of the recommended CMC strategy for the subjects in question. However, the values of even the largest phi statistics are still fairly modest, and associations between all measures of strength of supervision integrity and recidivism are negligible and nonsignificant when failure is defined as involvement in technical violations.

⊠ Discussion and Conclusions

The outcome evaluation portion of this research finds that, for this sample, the apparent effectiveness of the CMC depends on how recidivism is defined. Use of revocation, typical in studies of the CMC, yields results that strongly favor differential supervision as an effective means for shaping offender behavior. Alternately, use of measures of recidivism that are less dependent on officer decision making, and more closely indicative of the actual behaviors of probationers (i.e., technical violations and rearrest), produces results unfavorable

to the CMC. With respect to rearrest, differentially supervised cases in the aggregate fared roughly the same as subjects who were not exposed to differential supervision; with respect to most measures of technical violations, they fared worse. These results remain even when the analysis includes controls for offender risk and other factors typically associated with supervision failure.

The process evaluation portion of this research, on the other hand, indicates that, for large proportions of offenders, differential supervision was not delivered as intended. Further, the integrity of case supervision was at best weakly correlated with each of the case outcomes studied. When outcome and process evaluation findings are combined, one sees that officers trained in differential supervision techniques who failed either to implement those techniques or to use them as directed resorted less frequently to revocation in response to probationers' technical violations and rearrests than did their nontrained counterparts.

At least two explanations compete for the results observed here. The first is that officers trained in differential supervision displayed the so-called Hawthorne effect—i.e., the artificially positive response brought about in subjects who are aware that their behavior is being measured. The second is that training in the need for and techniques of differential supervision heightened officers' understanding of offender behavior, such that officers viewed offender misconduct in a more lenient and flexible context.

The second explanation is more compelling than the first. Given the low level of enthusiasm expressed by officers for assessment activities, the authors find it inconceivable that officers who participated in the study would alter responses to violations so as to artificially inflate the effectiveness of the CMC—and consequently, the probability that the CMC would be required for additional cases. Assessment-related tasks suffer from very low popularity among community supervision officers because of the perception of heightened workload associated with them. In addition, a theory that officers simply manipulated outcomes in favor of the CMC is inconsistent with the higher rates of recorded technical violations observed among the cases targeted for differential supervision. If officers were intentionally manipulating study outcomes, one would expect greater uniformity in their decisions to suppress records of specific forms of misbehaviors than is exhibited across the disaggregated analyses reported in Table 3. More likely, implementation of the CMC altered behavior in officers by increasing their sensitivity to offender motivations and needs, thereby raising their thresholds for revocable behaviors.

This study should not be construed as proof that the CMC is ineffective, given that the CMC was never implemented as intended. The audit of differentially supervised cases revealed both a high rate of classification error and officers' low level of compliance with recommended supervision practices. Nor does the study invalidate the positive findings of prior evaluations of the CMC, which link differential supervision with reductions in revocation.

What this research does is lend credence and urgency to calls for additional study about the implementation of correctional treatment generally and the role played by community corrections agents in the delivery of correctional services specifically. The lack of compliance with recommended classification and supervision practices highlights the importance of incorporating measures of the integrity of treatment program implementation into evaluations of correctional treatment. As the current research demonstrates, determining only that a generally accepted classification system has been adopted, or that staff have been trained in the system (versus measuring whether staff actually supervise or treat as they have been trained) can produce a misleading picture of the quality of program implementation.

Audits of the caliber of individual case supervision are especially important where interventions aim to increase officers' responsivity to differences in offender learning styles and motivations for engaging in criminal behavior. The current study questions conclusions of research that affirms responsivity as a principle of effective correctional treatment, where revocation is employed as the principal measure of program outcome. Development of case supervision audit instruments composed of measures with demonstrated interrater reliability would produce useful accompaniments to the CMC or other classification systems that aim to alter officers' responsivity to offenders. Whereas the current study limited the period being audited to the first year of supervision, repeated measurements of supervision integrity over the course of a probationer's sentence would be more valuable and informative.

More generally, the study highlights the perils associated with reliance on single measures of program failure. The question of which justice system outcome measure (rule violations, new arrests, convictions, or incarceration) best represents program failure has long been controversial (Blumstein, Cohen, Roth, & Visher, 1986; Vasoli, 1967; Waldo & Griswald, 1979). However, as Gendreau et al. (1996, p. 586) note, "the issue is rarely, if ever, addressed in the literature." There is likely to be little disagreement that any one of the traditional outcome measures is wanting. Even the use of reconviction—apparently superior as an outcome measure due to the stringent standard of proof required for a finding of guilt—is unfeasible in jurisdictions whose decision makers routinely refrain from formal prosecutions when revocations can more easily be obtained.

Possibly, risk of misinterpretation of study outcomes is highest in evaluations that attempt to measure the impact of efforts to increase responsivity by community supervision officers or treatment specialists, versus manipulation of other features of correctional treatment. Consideration of multiple outcome measures safeguards researchers against misinterpretation of results. Alternately, scales of offender misbehavior that take a variety of traditional outcome measures into account may provide more credible and reliable indicators of program failure.

Although the research failed to document lower rates of technical violations and rearrests in differentially supervised offenders, it is not without positive findings. Training in the use of the CMC appeared to increase officers' receptivity to less drastic means than revocation (and eventual incarceration) for addressing misconduct in offenders. The recognition that classification can bring about change in officer behavior is consistent with growing appreciation of the need to expand the focus and beneficiaries of classification on constituents other than merely the offender (Dooley, 1999). Less frequent use of revocation reduces demand for more expensive prison resources.

✑ References

Andrews, D. A. (1999). Assessing program elements for risk reduction: The Correctional Program Assessment Inventory. In P. M. Harris (Ed.), *Research to results: Effective community corrections* (pp. 151–190). Lanham, MD: American Correctional Association.

Andrews, D. A., & Bonta, J. (1998). *The psychology of criminal conduct* (2nd ed.). Cincinnati, OH: Anderson.

Andrews, D. A., Zinger, I., Hoge, R. D., Bonta, J., Gendreau, P., & Cullen, F. T. (1990). Does correctional treatment work? A clinically relevant and psychologically informed meta-analysis. *Criminology, 28*, 369–404.

Baird, C., & Neuenfeldt, D. (1990). The Client Management Classification System. *NCCD Focus, 1990*, 1–7.

Blumstein, A., Cohen, J., Roth, J. A., & Visher, C. A. (1986). *Criminal careers and career criminals* (Vol. 1). Washington, DC: National Academy Press.

Bonta, J. (1995). The responsivity principle and offender rehabilitation. *Forum on Corrections Research, 7*, 34–37.

Bonta, J., & Motiuk, L. L. (1992). Inmate classification. *Journal of Criminal Justice, 20*, 343–353.

Bonta, J., Wallace-Capretta, S., & Rooney, J. (2000). A quasi-evaluation of an intensive rehabilitation supervision program. *Criminal Justice and Behavior, 27*, 312–329.

Byrne, J. M., Lurigio, A. J., & Baird, C. (1989). The effectiveness of new intensive supervision programs. *Research in Corrections, 2*, 1–48.

Clear, T. R., & Gallagher, K. W. (1983). Screening devices in probation and parole: Management problems. *Evaluation Review, 7*, 217–234.

Clear, T. R., & Gallagher, K. W. (1985). Probation and parole supervision: A review of current practices. *Crime & Delinquency, 31*, 423–443.

Clear, T. R., & Latessa, E. J. (1993). Probation officers' roles in intensive supervision: Surveillance versus treatment. *Justice Quarterly, 10*, 441–462.

Dooley, M. (1999). Classification and restorative justice: Is there a relationship? In *Topics in community corrections. Annual issue 1999: Classification and risk assessment* (pp. 38–42). Washington, DC: National Institute of Corrections.

Dowden, C., & Andrews, D. A. (1999). What works for female offenders: A meta-analytic review. *Crime & Delinquency, 45*, 438–452.

Eisenberg, M., & Markley, G. (1987). Something works in community supervision. *Federal Probation, 51*, 28–32.

Fulton, B., Stichman, A., Travis, L., & Latessa, E. (1997). Moderating probation and parole officer attitudes to achieve desired outcomes. *Prison Journal, 77*, 295–312.

Gendreau, P. (1996). The principle of effective intervention with offenders. In A. T. Harland (Ed.), *Choosing correctional options that work: Defining the demand and evaluating the supply* (pp. 117–130). Thousand Oaks, CA: Sage.

Gendreau, P., & Andrews, D. A. (1994). *The Correctional Program Assessment Inventory* (4th ed.). St. John, New Brunswick, Canada: University of New Brunswick.

Gendreau, P., & Goggin, C. (1997). Correctional treatment: Accomplishment and realities. In P. Van Voorhis, M. Braswell, & D. Lester (Eds.), *Correctional counseling and rehabilitation* (pp. 271–280). Cincinnati: Anderson.

Gendreau, P., Goggin, C., & Smith, P. (1999). The forgotten issue in effective correctional treatment: Program implementation. *International Journal of Offender Therapy and Comparative Criminology, 43*, 180–187.

Gendreau, P., Little, T., & Goggin, C. (1996). A meta-analysis of the predictors of adult offender recidivism: What works! *Criminology, 34*, 575–607.

Harris, P. M. (1994). Client Management Classification and prediction of probation outcome. *Crime & Delinquency, 40,* 154–174.

Hart, T. C., & Reaves, B. A. (1999). *Felony defendants in large urban counties, 1996.* Washington, DC: Bureau of Justice Statistics.

Hoffman, P. B. (1994). Twenty years of operational use of a risk prediction instrument: The United States parole commission's salient factor score. *Journal of Criminal Justice, 22,* 477–494.

Jones, D. A., Johnson, S., Latessa, E., & Travis, L. F. (1999). Case classification in community corrections: Preliminary findings from a national survey. In *Topics in community corrections. Annual issue 1999: Classification and risk assessment* (pp. 4–10). Washington, DC: National Institute of Corrections.

Kennedy, S. (1999). Responsivity: The other classification principle. *Corrections Today, 61,* 48–51.

Kennedy, S., & Serin, R. (1997). Treatment responsivity: Contributing to effective correctional programming. *The ICCA Journal on Community Corrections, 7,* 46–52.

Latessa, E. J., & Holsinger, A. (1998). The importance of evaluating correctional programs: Assessing outcome and quality. *Corrections Management Quarterly, 2,* 22–29.

Leininger, K. (1998). *Effectiveness of client management classification.* Unpublished report prepared for the Florida Department of Corrections.

Lerner, K., Arling, G., & Baird, S. C. (1986). Client management classification strategies for case supervision. *Crime & Delinquency, 32,* 254–271.

McManus, R. F., Stagg, D. I., & McDuffie, R. C. (1988). CMC as an effective supervision tool: The South Carolina perspective, *Perspectives, 1988,* 30–34.

National Institute of Corrections. (1991). *National Academy of Corrections lesson plans for strategies for case supervision.* Unpublished manuscript.

National Institute of Corrections. (2000). *Promoting public safety using effective interventions with offenders.* Retrieved from http://www.ncic.org/resources/curriculum/016296.htm

Petersilia, J. (1998). Experience with intermediate sanctions: Rationale and program effectiveness. In J. Petersilia (Ed.), *Community corrections: Probation, parole, and intermediate sanctions* (pp. 68–70). New York: Oxford University Press.

Polizzi, D. M., MacKenzie, D. L., & Hickman, L. J. (1999). What works in adult sex offender treatment? A review of prison- and non-prison-based treatment programs. *International Journal of Offender Therapy and Comparative Criminology, 43,* 357–374.

Rice, M. E., & Harris, G. T. (1995). Violent recidivism: Assessing predictive validity. *Journal of Consulting and Clinical Psychology, 63,* 737–748.

Schneider, A. L., Ervin, L., & Snyder-Joy, Z. (1996). Further exploration of the flight from discretion: The role of risk/need instruments in probation supervision decisions. *Journal of Criminal Justice, 24,* 109–121.

Serin, R., & Kennedy, S. (1997). *Treatment readiness and responsivity: Contributing to effective correctional programming* (Research Report R-54). Ottawa, Ontario: Correctional Service of Canada.

Silver, E., Smith, W. R., & Banks, S. (2000). Constructing actuarial devices for predicting recidivism: A comparison of methods. *Criminal Justice and Behavior, 27,* 733–764.

Van Voorhis, P. (1997). Correctional classification and the "responsivity principle." *Forum on Corrections Research, 9,* 46–50.

Vasoli, R. H. (1967). Some reflections on measuring probation outcome. *Federal Probation, 31,* 24–32.

Waldo, G., & Griswald, D. (1979). Issues in the measurement of recidivism. In L. Sechrest, S. O. White, & E. D. Brown (Eds.), *The rehabilitation of criminal offenders: Problems and prospects* (pp. 225–250). Washington, DC: National Academy of Sciences.

Wilson, D. B., Gallagher, C. A., & MacKenzie, D. L. (2000). A meta-analysis of corrections-based education, vocation, and work programs for adult offenders. *Journal of Research in Crime and Delinquency, 37,* 347–368.

DISCUSSION QUESTIONS

1. Why is differential supervision largely ineffective according to the authors?

2. What is the purpose of the CMC, and how did it perform in practice in this study?

3. What can be done to improve the effectiveness of the CMC/differential supervision connection?

READING

Jean Bottcher and Michael Ezell note that, in the early 1960s, the boot camp model became a correctional panacea for juvenile offenders, promising the best of both worlds—less recidivism and lower operating costs. Studies examining these assumptions have shown mixed results and have mostly relied on nonrandomized comparison groups. The California Youth Authority's (CYA's) experimental study of its juvenile boot camp and intensive parole program (called LEAD)—versus standard custody and parole—was an important exception, but its in-house evaluation was prepared before complete outcome data were available. The present study capitalizes on full and relatively long-term follow-up arrest data for the LEAD evaluation provided by the California Department of justice in August. Using both survival models and negative binomial regression models, the results indicate that there were no significant differences between groups in terms of time to first arrest or average arrest frequency. Thus, as shown by many other studies of lesser quality, boot camps do not result in lower recidivism rates.

Examining the Effectiveness of Boot Camps

A Randomized Experiment With a Long-Term Follow Up

Jean Bottcher and Michael E. Ezell

Despite tragic, highly publicized consequences (Clines, 1999; Selcraig, 2000) and disappointing evaluative research results (MacKenzie et al., 2001), correctional boot camps are still supported in some areas of the country (Buckley, 2000; Walker, 2002). Documented instances of extreme abuse have led to the closure of some camps, for example, in Arizona, Georgia, and Maryland (Schnurer and Lyons, 2000), yet camps in other states, such as Florida, Illinois, Oregon (personal survey, May 12–14, 2003), and Pennsylvania (Kempinen and Kurlychek, 2003) remain active. Most of the evaluative research is flawed by poor comparative data or program-implementation problems. For example, MacKenzie (2000) located only four evaluations based on experimentally derived comparison groups. Three of these evaluated camps (Peters, 1996a, 1996b; Thomas and Peters, 1996) apparently experienced relatively serious problems with staff turnover and an unhealthy balance between military discipline and treatment (Bourque et al., 1996). The fourth, a legislatively mandated study of a California juvenile boot camp, came due before complete outcome data were available (California Youth Authority [CYA], 1997).

The present study capitalizes on a full and relatively long-term set of arrest outcome data for that fourth experimental evaluation. Designed as an alternative placement for the CYA's least serious male offenders, the program (called LEAD for expected participant

SOURCE: Bottcher, J., & Ezell, M. E. (2005) Examining the effectiveness of boot camps: A randomized experiment with a long-term follow up. *Journal of Research in Crime and Delinquency, 42*(3), 309–332. Reprinted with permission of Sage Publications, Inc.

outcomes—leadership, esteem, ability, and discipline) was typical of other juvenile boot camps around the country in targeting cost savings and lower rates of recidivism as major goals and in incorporating treatment components. Although the military model was politically initiated (by Governor Pete Wilson's administration), LEAD'S enabling legislation was crafted by CYA administrative staff based also on their sense of "good program elements" (Gary Maurer, personal communication, February 18, 2003). The four-month institution phase opened in September 1992, and the six-month aftercare phase with the first release of graduates in January 1993; LEAD was quietly phased out during the summer of 1997.

Despite an increasing focus upon tight security and a concomitant declining focus upon correctional treatment, the CYA developed LEAD with interest, even enthusiasm. Enriched line staffing was an important element of design. Eligibility criteria included a nonserious, nonviolent juvenile court commitment; an age of at least 16 (later modified to 14); a history or risk of substance abuse; informed consent; medical clearance; and Youthful Offender Parole Board (YOPB) approval. Additional criteria (established jointly by the CYA and YOPB) included ineligibility for special, mental-health programs, lack of recent violent behavior, and citizenship or legal presence in the United States. The selection process isolated an eligible population representing about 14 percent of the entire male juvenile court intake pool.

CYA management planned and administered the program. In response to potential problems of ward abuse, a California National Guard (CNG) consultant suggested using an officer training (or leadership) model with a critical focus on mentoring. The two camp sites were run primarily by 12 youth counselors (rather than the standard 7), newly titled TAC officers. TAC stands for teach, advise, and counsel—key elements of the officer mentoring role. The envisioned program was not notably theoretical. Implicitly, it seemed based

on assumptions that program diversity, along with a little individualized treatment, would reach more wards; that a military environment would rub off as self-discipline; and that newly developed skills and positive attitudes would "produce" less criminal behavior. An exception was the TAC mentoring role, which was explicitly and theoretically related to the manner by which LEAD might reduce recidivism. Such an effect could occur through the "referent power" of the TACs, the possibility that cadets would identify with TACs and emulate their good qualities. The program and its experimental evaluation were implemented as specified by the enabling legislation (Bottcher and Isorena, 1994; Bottcher et al. 1995; Isorena and Lara, 1995). Thus, this study with its experimental design and long-term follow-up likely represents one of the most rigorous evaluations of a correctional boot-camp program in the United States.

This article begins with a review of the literature on boot camps in corrections and a summary of the CYA's (1997) in-house evaluation findings on LEAD. It proceeds with a section on the data and methods and the results of this analysis. A concluding discussion places this study's findings in the context of contemporary corrections.

⚑ Literature Review

Correctional use of quasi-military regimentation may be traced to the "perfected" American prisons of the 1820s and 1830s (Rothman, 1995), as well as to the earliest American reform schools for juveniles (Schlossman, 1995). As described by Rothman (1995), these early prisons were designed for reform and organized toward that end around silence, discipline, and hard work. A military model fleshed out the disciplinary milieu—routines to the sound of bells, marching in lockstep, uniforms for guards and inmates, requisite deference by prisoner to guard, even the symmetry and regularity of architectural design.

The contemporaneous inventions of prison and factory and their marked similarities prompted the notion that the former was designed to support the latter. Prisons, some historians suggested, were developed to support the nascent industrial order. Rothman (1990) prefers a different interpretation—the resemblance of prisons and factories was a product of the same historically specific assumptions about how people should be controlled. Invented at a time of enormous national growth and unregulated social change, prisons, based on Rothman's historical analyses, were envisioned as models of order in exciting but uncertain, even frightening times.

The contemporary correctional boot camp is a relatively new (but declining) phenomenon often analyzed in conjunction with other recent innovative sanctions (like drug courts and day reporting centers) called "intermediate sanctions" (Petersilia, 1998). Tonry (1998) attributes the development of intermediate sanctions to political and ideological trends regarding crime control since 1980—a declining belief in rehabilitation, an increasing commitment to the "just deserts" rationale, and a receptivity to harsher penalties. Precursors of the boot camps appeared earlier, though, as the tumultuous social changes of the 1960s and beyond were beginning to play out. Austin, Jones, and Bolyard (1993) and Flowers, Carr, and Ruback (1991) trace the concept to "shock probation" (brief incarceration that first appeared in the 1960s) and somewhat later "scared straight" programs. Over time, increasing reliance on deterrence and harsher sanctions helped produce an enormous increase in prison populations during the 1980s. New penalties were developed that toughened probation or justified less incarceration. Boot camps formed a popular but relatively less common version of these intermediate sanctions.

At one level, then, contemporary correctional boot camps appear a useful midrange sanction for selected offenders judged ready for just that amount of cost-effective deterrence

or reform. Initially, though, the combination of their deliberate harshness and rigid format, popular appeal and bipartisan political support, brevity, and thin reformative veneer suggest another level of interpretation—a search for order amidst turbulent social change and an unmapped future, conditions comparable to Rothman's (1995) perceptions of pre–Civil War America. With more historical distance, we may come to see the unusual correctional boot-camp movement explained in ways comparable to Rothman's explanation of our earliest prisons.

Despite popular support, correctional boot camps elicited criticism from the beginning (Sechrest, 1989), and they remain controversial (Lutze and Brody, 1999). The primary argument surrounds the appropriateness of harsh confrontational tactics in corrections (MacKenzie et al., 2001). A vast literature (Andrews et al., 1990; Cowles, Castellano, and Gransky, 1995; Gendreau and Goggin, 2000; MacKenzie, 2000), as well as professional expertise (see, for example, Chamberlain, 1998), suggests that effective correctional treatment includes state-of-the art theoretical grounding, qualified treatment providers, prosocial modeling and reinforcement, consistent discipline, individualization, and interpersonally warm, supportive staff. Fear tactics and verbal confrontation find no support in the literature on correctional treatment. In addition, critics contend that camps incorporate conflicting goals, may expand the correctional population (net-widening), pave the way for inmate abuse, and promote sexist attitudes (Dieterich, Boyles, and Colling, 1999; Morash and Rucker, 1990; Parent, Snyder, and Blaisdell, 1999).

Although the evaluative literature suggests that some boot camps provide more positive environments than some contemporary correctional institutions (Lutze, 1998; MacKenzie and Souryal, 1995), many of the early critics' worst fears have been realized. Criminal charges and lawsuits based on physical brutality in juvenile correctional boot camps have arisen in at least eight states in recent years; and at least six deaths have been attributed to boot-camp negligence and abuse (Schnurer and Lyons, 2000; Selcraig, 2000).

Based on an extensive search of the literature and a subsequent meta-analysis of 29 evaluations that included 44 samples with a "reasonable" comparison group and postprogram measure of recidivism, MacKenzie et al. (2001) found no overall differences in recidivism between boot camp and comparison groups. A close examination of the 9 comparison samples (from 5 studies) that yielded a statistically significant difference in favor of the boot-camp group revealed that none were randomized experiments and, furthermore, that all were based on rough comparison groups (Flowers, Carr, and Ruback, 1991; Farrington et al., 2000; Jones, 1999; MacKenzie and Souryal, 1994; Marcus-Mendoza, 1995). The MacKenzie et al. (2001) meta-analysis did not attempt to incorporate any rating of the quality of the boot-camp programs (and data to accomplish that would be rough in any event). However, as noted above, three of the most rigorously evaluated programs were not model boot camps (Bourque et al., 1996). In contrast, evaluation of New York State's highly touted and elaborately refined Youth Leadership Academy (YLA; MacKenzie et al., 1997) was based on a retrospectively (albeit very carefully) generated comparison group of youths locked up in similarly secure facilities during the same time frame (early 1993 through early 1996) but still, for some reason, not selected for the program. According to the study authors, though, the director did not consider YLA an "efficient model" until 1996 and by then—based on his descriptions and video illustrations (Cornick, 1996; Office of Juvenile Justice and Delinquency Prevention [OJJDP], 1996)—YLA had become a largely demilitarized, treatment-oriented (and ultimately unevaluated) boot camp. Granted its limitations, then, evaluative research to date provides no methodologically rigorous support for the contention that boot camps lower recidivism. Although beyond the scope of this study, there is evidence that boot camps may lower costs if designed with that purpose in mind (MacKenzie and Piquero, 1994; Parent et al., 1999).

Lead In-House Evaluation

Legislation called for initial process evaluations at each camp site and an impact study with a rigorous experimental design. Random selection procedures (described in the following Data and Methods section of this study) were designed around the required intake process for LEAD. In addition to observation and interview data, evaluators located or developed other measures of program delivery and performance, including documentation of aftercare services and parolee performance via monthly phone calls with parole agents.

The California Department of Justice (CDOJ) provided the primary source of outcome data, statewide arrests by law enforcement agencies. Because parole agents can arrest and detain parolees in the same manner as law enforcement officers (and potentially with the same effect), the CYA evaluators' monthly parole agent follow-up contacts provided these additional outcome data. An arrest was defined as any charge (technical or legal, by law enforcement or parole agent) that resulted in a law-enforcement citation or in any custody. Sources of data were coded so that CDOJ-verified arrests could be distinguished from arrests based only on CYA parole-agent contacts. Overall program attrition rates were 25 percent at the first site and 31 percent at the second, but all LEAD dropouts were retained in the CYA evaluation, as well as the present study.

Boot-camp sites generated lively, lengthy daily schedules of physical training, military drill and ceremony exercises, school classes, group counseling sessions, substance abuse treatment groups, and various unit maintenance routines. The program developed creatively, with varied additional elements by site, such as a bereavement-therapy group at one site in response to the many cadets who had experienced tragic losses. Both sites demonstrated similar positive characteristics, including a relatively safe and healthy environment. Comparative survey data, for example, indicated that LEAD wards, compared to control wards, felt less fear of being hurt by each other and more physically fit. LEAD wards were generally enthusiastic about the military milieu. Interview data revealed, for example, their clear awareness of the positive effects of leadership rotation, daily shifts in ward leadership roles that seemed to dampen gang conflicts considerably. Wards most liked the physical training, 12-step drug treatment, and discipline of LEAD. On average, LEAD wards were incarcerated 4.6 months less than control wards. Evaluators noted that LEAD attracted many dedicated staff and provided a location where some treatment efforts could be generated. Monthly graduation ceremonies celebrated LEAD, a CYA showpiece in the mid-1990s.

Although the six-month intensive parole phase (as opposed to the standard two-month intensive parole) was initially envisioned around selected agents with caseloads of 15, the realities of ward-program recruitment forced a less auspicious design. The parole phase was rather hastily developed after the first camp opened. Nonetheless, LEAD parolees, compared to control parolees, received more face-to-face contacts and more drug tests per month during their first six months on parole, clear indicators of a higher level of supervision.

Prominent among the limitations that seemed to plague LEAD were its lack of an underlying treatment philosophy that clearly explained how the program was expected to change its participants for the better, unresolved conflicts between cost savings and rehabilitation goals, and the need for more cohesiveness between institution and parole phases, as well as the need for continued development of the parole phase. Very few institution staff touted TAC mentoring as developed in the leadership training model, which failed to play a central role in the treatment orientation. Relatively high camp attrition rates based largely on disciplinary problems reflected staff judgments that many cadets were misplaced in an early release program. The military milieu required vigilance against abusive and demeaning confrontation tactics from line staff.

When the final evaluation report was prepared (CYA, 1997), 12-month follow-up arrest and disposition data were available on only 90 percent of the LEAD group and 86 percent of the control group (because some wards had been released only a short time or were still incarcerated). Analyses showed that LEAD wards, compared to control wards, were more likely to be arrested for any offense (technical or law violation), but that neither group was more likely to be arrested for a CDOJ-verified law violation, to be arrested with a weapon, to cause injury during an arrest event, or to be arrested more times in 12 months. An analysis of disposition data revealed that LEAD wards, compared to control wards, were somewhat more likely to be returned to CYA custody following their first arrest. Furthermore, analyses by source of arrest data (CYA parole agent contact only or CDOJ verified) showed that the LEAD group, compared to the control group, received more arrests for law violations that were not verified by CDOJ rap sheets and were more likely to be arrested initially by a parole agent. The evaluation clearly indicated that, on average, LEAD parolees, compared to control parolees, were more tightly supervised and subjected to more arrests (and more detention) by parole agents. In sum, though, the final evaluation concluded that LEAD did not reduce recidivism.

⊠ Data and Method

This analysis relies on the experimentally derived comparison group data generated by CYA researchers for their impact evaluation of LEAD (CYA, 1997) and on a relatively long-term set of official follow-up arrest data provided by the CDOJ in August 2002. Thus, in contrast to the CYA final evaluation summarized above, the present study relies only on CDOJ arrest data (which does not provide full information on final dispositions or subsequent incarceration). In this section we describe randomization procedures, sources of data, variables, and data analysis plan.

Selection of Experimental and Control Group Members

Recall that program eligibility criteria were fairly stringent such that only 14 percent of the male juvenile court intake pool was found eligible for LEAD. Following the screening and YOPB approval process, CYA reception center staff called in the names of LEAD-approved wards to the research office during each monthly cycle. Groups of eligible wards were then stratified by ethnic categories (and parole-violator status, if possible) and selections were made using a table of random numbers. During the first year, monthly random selection procedures worked almost flawlessly except that reception center crowding sometimes forced periodic monthly selection groups, which were not always evenly numbered. Thus, the probability of being selected for LEAD was sometimes not .50. During the next year, after the second boot-camp site came online, each of the two reception centers was expected to generate enough wards to sustain one LEAD site (15 wards for each incoming platoon), as well as a control group, single-handedly. However, the reception centers were rarely able to come up with 30 eligible wards for a 50-50 split each month. Randomization was then always put off until the end of each monthly cycle and, if there were only enough wards to fill a standard platoon on a given month, random procedures were suspended and all eligible wards were sent to LEAD. However, nonrandomly placed LEAD wards were never included in the experimental study. Even when there were more

than enough eligible wards to fill a platoon, though, the eligible wards rarely numbered 30, and the probability of being selected for LEAD was virtually never .50.

The final CYA evaluation study file was formed of all eligible wards randomly placed during the first two years of LEAD operation. Based on data presented in their final report (CYA, 1997), study group attrition following random assignment was impressively small. Overall, 10 (or 2 percent) of the randomly selected wards were lost, 9 from the experimental (or LEAD) group (representing a 3 percent loss), leaving a total of 632 wards (348 in the LEAD group and 284 in the control group).

Source of Arrest Data

Follow-up arrest data were retrieved from the CDOJ. These data are known as the California Information and Identification (or CII rap sheet) information. When an individual is committed to the CYA, he or she is assigned a CII identification number and a computerized CII rap-sheet file is initiated and maintained by the CDOJ. When an adult is arrested in California, the arrest is reported by the arresting law-enforcement agency to the CDOJ. Thus, any time one of the wards in our samples was arrested as an adult, the arrest record, including the date of arrest and information on the arrest charges was forwarded to CDOJ. If wards are released by the CYA while still minors (under age 18), the CYA reports any subsequent criminal arrests to the CDOJ until they become adults.

The files of the CDOJ were searched in late August 2002. We permitted eight months of "lag time" for any arrests to be entered by CDOJ into the case's "rap-sheet" file. Thus, the arrests were "censored" as of December 31, 2001, and any arrests occurring between that date and August 2002 were not included in the analyses for this study. Postrelease follow-up (or exposure) periods for the sample averaged just over 7.5 years. The minimum follow-up time was just over 2 years and the maximum just over 9 years. Of the 632 wards in the CYA study file, 11 (4 LEAD and 7 control) cases could not be located through CDOJ. This left us with 621 (or 344 LEAD and 277 control) subjects for the analyses presented in this article.

⬚ Results

Bivariate Comparisons of Experimental and Control Groups

We begin the presentation of results (in Table 1) with a comparison of ward characteristics by group. Although probabilistically speaking, the randomization procedures should by themselves ensure comparability, these analyses are particularly important because the randomization procedures were not always 50/50. The chi-square statistic (from a two-way tabular analysis) was used for the categorical variables, and the t-statistic (from an independent samples t test) was employed for the continuous variables.

These analyses indicate that the two groups were composed of comparable youths. For example, upon admission for the current stay, roughly 83 percent of each group were "first commitments" (that is, this was the first time they had been committed to the CYA), and roughly 17 percent were parole violators. The bulk of each group (about 73 percent of the LEAD group and 66 percent of the control group) were initially committed for property offenses and, at program admission, wards in each group averaged 17.5 years of age. None of the variations in subject characteristics was statistically significant.

Descriptive Comparisons of Arrest Charge Outcomes

Table 2 presents arrest-charge outcomes for the two groups at various follow-up interval lengths and disaggregated by different offense-type categories. We include results here for four different follow-up time periods and for four different offense categorizations, but focus our discussion and subsequent statistical analyses (using the negative binomial regression model) on the total number of arrest charges. It is important to keep in mind that, because we did not have access to incarceration data in the postrelease period, we do not have a measure of "street time" for the members of each group. Thus, if any differences in this description (or in the analyses that follow) are due to differences in amount of street time, we will not be able to verify this. Recall, however, that, on average, the LEAD group was released to a longer period of intensive parole than the control group, and, according to the CYA's final evaluation, LEAD members were significantly more likely to be taken off the street for parole technicalities than control wards (CYA, 1997).

In the first year after release, 44 percent of the boot-camp wards and 50 percent of the control wards were arrested for new criminal offenses. The average ward in each group accumulated about 1 new arrest charge during that year. Looking at serious offenses only, we find that 30 percent of the LEAD group had been arrested for at least 1 serious charge, whereas 37 percent of the control group had been so arrested. The LEAD group averaged 0.54 serious arrest charges, whereas the control group averaged 0.60. Based on the accumulated data on total offenses and serious offenses for the two-and three-year periods, the small differences between the two groups seem to widen a bit. For example, after two years, about 60 percent of the LEAD group, compared to about 69 percent of the control group, had been arrested for at least one criminal offense. However, using all of the follow-up data, we find a considerable degree of similarity between the two groups. About 92 percent of each group had been arrested at least once, and roughly 80 percent of each

| Table 1 | Characteristics by Experimental Group | | |

Characteristics	Experimental Group		Chi-square/ *t* Test *p* Value
	LEAD (n = 344)	Control (n = 277)	
Ethnicity (%)			
White	24.4	25.6	.903
Latino	43.9	42.2	
African American	24.1	25.6	
Other	7.6	6.5	
Prior local confinements			
Mean	1.6	1.7	.328
SD	1.4	1.4	
Age at initial CYA commitment			
Mean	17.1	17.0	.631
SD	0.9	0.9	
County of initial commitment (%)			
San Francisco Bay area	16.3	14.4	.922
Other Northern California	44.5	46.6	
Los Angeles	27.9	27.8	
Other Southern California	11.3	11.2	
Initial commitment offense (%)			
Drug, minor	5.2	8.7	.125
Property	72.7	66.3	
Person	22.1	25.0	
Age at program admission			
Mean	17.5	17.5	.697
SD	1.3	1.3	
Program admission status (%)			
First commitment	83.4	83.0	.895
Parole violator	16.6	17.0	

NOTE: All subjects in this study were males. The modal response was substituted for one subject with missing information on initial commitment offense and the mean number was substituted for 10 subjects with missing information on prior local confinements.

Table 2	Summary Arrest Charge information by Experimental Group Status, Offense Type, and Length of Follow-up							
	Offense Type							
	All-Offense Charges		**Serious-Offense Charges[a]**		**Violent-Offense Charges**		**Property-Offense Charges**	
Length of follow-up	**LEAD**	**Control**	**LEAD**	**Control**	**LEAD**	**Control**	**LEAD**	**Control**
One year								
Mean number	0.97	1.04	0.54	0.60	0.32	0.31	0.30	0.26
% with any	43.6	50.18	30.23	37.18	19.77	19.86	18.6	18.77
Two years								
Mean number	1.63	2.06	0.85	1.09	0.55	0.69	0.46	0.51
% with any	59.59	68.59	44.77	53.07	29.65	35.02	26.45	29.96
Three years								
Mean number	2.59	3.02	1.31	1.52	0.79	1.04	0.72	0.69
% with any	75.29	77.26	55.81	60.65	39.24	45.85	34.30	36.82
All available data								
Mean number	6.68	7.25	3.17	3.18	2.15	2.48	1.61	1.49
% with any	91.57	91.70	82.27	80.87	68.31	71.12	54.65	53.07

Note: Analyses of total arrest events produced comparable findings. Mean numbers of arrest events for one year were 0.66 (LEAD) and 0.68 (Control), for two years, 1.10 (LEAD) and 1.33 (Control), for three years, 1.71 (LEAD) and 2.00 (Control), and for all years, 4.28 (LEAD) and 4.59 (Control).

a. Serious offense charges included homicide, forcible rape, robbery, aggravated assault, kidnap/extortion, child molestation, sodomy/forced oral copulation, weapon discharge, burglary, auto theft, arson, drug sales/trafficking, and drug possession/possession for sale.

group had been arrested at least once for a serious offense.

Analyses of Time to First Arrest

We now turn our focus to the lengths of time that, on average, wards from each of the two groups managed to "survive" without being arrested. Figure 1 presents the survival curves for time to first criminal arrest. The survival curves represent the fraction of each group still arrest free at given time points (represented by days in the figure). As seen in the curves of both groups, the survival curves drop quite steadily during the first one thousand days after release. For example, at the 200th day, 30 percent of the boot camp wards and 33 percent of the control group wards had already been arrested for a new criminal offense. At the end of the first year, only 50 percent of the boot camp wards and 46 percent of the control wards remained free from a new criminal offense arrest. At the end of this study (using all available data), the estimated survivor function indicates that just 8 percent of each group survived arrest free. A graph of the survival curves for time to first serious criminal arrest generated survival curves for the groups that

Figure 1	Survivor Function for Any Offense

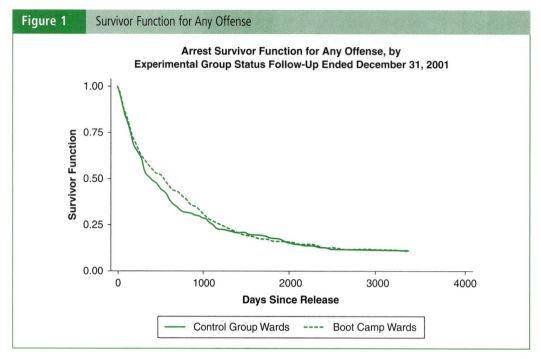

NOTE: Log-rank for equality of survivor functions: $\chi^2 = 0.57$, p value = .4144.

were substantively comparable to those presented in Figure 1, except that the curves dropped somewhat more slowly and ended at 0.18 (Control) and 0.19 (LEAD). (This graph is available upon request.)

Our next set of analyses examines the possible effect of LEAD on the time until a first arrest occurs. Table 3 contains the results: Model 1 with just the boot-camp variable and model 2 with the boot-camp variable and other available independent variables. The hazard ratios substantively indicate how the hazard rates (or instantaneous rate of event occurrence) either varies between two groups (categorical variables) or changes with increasing values of a variable (continuous variables; Allison, 1995). By the nature of the Cox model specification, the hazard ratios are calculated independent of time and assumed to be proportional over the entire follow-up period.

Model 1 indicates that the LEAD wards had hazard rates that were 7 percent [(1– 0.933)*100] lower than control wards.

However, this estimate was not statistically significant (z value = – 0.81; p value = .418). Model 2 estimates the LEAD effect while holding constant the effects of other variables (which is important in this study because perfect 50/50 randomization was not always possible). Examining these results, we still find that the hazard rates of the boot camp and control groups were not significantly different from one another (z value = 0.11; p value = .909).

Hazard ratios for the remaining variables in model 2 indicate that several predicted time to first arrest. Although white wards had hazard rates that were not significantly different from those of Latino wards (the reference group), African American wards had rates that were significantly higher (by 27 percent) and wards in the "Other" ethnic group had rates that were significantly lower (by 30 percent). Wards from Los Angeles County, an area previously associated with more subsequent ward arrests than other areas (CYA, 1997), had hazard

| Table 3 | Estimates From Cox Proportional Hazards Model: Time to First Criminal Arrest | | | | | |

	Model 1			Model 2		
Variable	Hazard Ratio	Robust *SE*	*p* Value	Hazard Ratio	Robust *SE*	*p* Value
LEAD	0.933	.079	.418	1.010	0.089	.909
Ethnicity						
White				0.934	0.105	.544
African American				1.272	0.143	.033
Other				0.701	0.119	.036
Prior local confinements				1.107	0.029	.000
County of initial commitment						
Los Angeles				1.133	0.126	.262
Initial commitment offense						
Drug, minor				1.382	0.272	.101
Property				0.983	0.111	.883
Admission status						
Parole violator				0.779	0.113	.085
Age at release				1.092	0.039	.013

NOTE: The modal response was substituted for one subject with missing information on initial commitment offense and the mean number was substituted for 10 subjects with missing information on prior local confinements.

rates that were no different from the wards committed from other California counties combined (the reference group). Compared to wards committed for person offenses (the reference group), wards committed for drug or minor offenses had elevated hazard rates that were marginally significant (*p* value = .101). Because only 5 percent of the sample were committed for drug or minor offenses, statistical power may be responsible for the lack of significance (given the size of the estimated effect). Wards admitted for parole violations had hazard rates that were lower (based only on marginal significance) compared to those committed to the CYA for the first time (the reference group). Surprisingly, older age at release was significantly related to shorter time to first arrest, with each additional year increasing the hazard rate by about 9 percent.

Although we cannot interpret this finding definitively, it is likely that this variable is merely picking up unobserved heterogeneity in the criminal propensity of these wards such that those who were older at release (usually because of behavioral problems that delayed their release) were likely to reoffend faster. Finally, and typically, the number of prior local confinements (the best available measure of prior record) was significantly related to higher hazard rates, with each additional local confinement leading to roughly an 11 percent increase in the hazard of a first criminal offense arrest.

Analyses of Counts of Arrest Charges

The final statistical analyses examine differences in average numbers of all follow-up arrest charges between the two groups during four time periods. Recall that these data were presented descriptively in Table 2. Table 4 presents the results of eight negative binomial models: four with just the experimental group variable and four full-specification models.

The results of model 1 reveal no significant difference in average arrest charges between LEAD and control-group wards during the first year of release. The parameter estimate indicates that boot-camp wards, compared to control group wards, had a 7 percent reduction in the expected number of arrest charges [i.e., $(100^*((\exp(-.071)) -1) = -.07]$, but the associated p value (.543) indicates that this difference was not statistically significant. Adding the other available independent variables into the model specification only confirms the findings from model 1. Substantively, the parameter estimate in model 2 indicates only about a 2 percent difference in expected arrest counts by group and, again, this estimate is not significantly different from zero (p value = .831).

The findings for the analyses of the cumulated two-year arrest charges were slightly different. The parameter estimate in model 3 indicates that LEAD wards had an expected arrest-charge count that was about 21 percent less than the control wards, and the associated p value (.021) shows that this difference was statistically significant. In the full model (model 4), the difference still seems notable in size (about 15 percent) but it is only marginally significant (p value = .094). Note, too, that this difference seems visually confirmed in Figure 1. Because there is no reason to expect such a delayed but positive effect on subsequent criminal activity from the boot camp experience, these (albeit modest) findings are virtually impossible to interpret definitively. Recall, however, that LEAD graduates (who comprised about 72 percent of the boot-camp group) were referred to a lengthier period of

intensive parole and were subjected to more arrests (using a more encompassing definition than the present study, which included technical violations) and to more detention by parole agents than other study wards (CYA, 1997). Thus, the most likely possibility is that those higher rates of custody among boot-camp wards, compared to control wards, slightly dampened the cumulative arrest charge totals during the first couple of years following release.

Expanding the length of follow-up to three years of data, models 5 and 6 of Table 4 present the results of the bivariate and multivariate specifications of the negative binomial regression model. In model 5, the parameter estimate for the boot-camp variable is marginally significant (with a p value of .084) and substantively indicates that the boot-camp wards had expected arrest charges that were about 14 percent lower than the control-group wards. However, after we control for the effects of the other variables, the boot-camp coefficient is no longer significant. Considering all available data (in models 7 and 8), differences between the experimental groups are not statistically significant. Furthermore, in the full specification model 8, age at release is no longer a significant predictor and, except for the relatively small "Other" category, ethnicity is no longer a significant variable. Prior record (measured by prior local county confinements) still remains a predictor of arrest charge counts, and wards committed for drug and minor offenses and for property offenses still had expected arrest event counts that were significantly greater than wards committed for person offenses.

The substantive conclusions of the models discussed above were replicated when we used (1) the count number of serious arrest charges and (2) the count number of total arrest events ("number of arrests") as the dependent variables.

| | Arrest Charges in First Year | | | | | | | | Arrest Charges in First Two Years | | | | | |
| | Model 1 Robust | | | Model 2 Robust | | | Model 3 Robust | | | Model 4 Robust | | |
Variable	Coefficient	SE	p Value	Coefficient	SE	p Value	Coefficient	SE	p Value	Coefficient	SE	p Value
LEAD	−0.071	.118	.543	−0.024	.113	.831	−0.234	.101	.021	−0.159	.095	.094
Ethnicity												
White				0.101	.148	.496				0.081	.119	.496
African American				0.425	.140	.002				0.329	.118	.005
Other				−0.254	.267	.342				−0.219	.257	.395
Prior local confinements				0.090	.036	.013				0.050	.034	.143
County of initial commitment												
Los Angeles				0.026	.123	.833				−0.120	.106	.258
Initial commitment offense												
Drug, minor				0.948	.211	.000				0.662	.167	.000
Property				0.429	.151	.004				0.280	.122	.022
Admission status												
Parole violator				−0.057	.181	.753				−0.109	.152	.473
Age at release				0.214	.045	.000				0.249	.041	.000
Constant	0.039	.080	.625	−4.671	.876	.000	0.723	.073	.000	−4.379	.760	.000

(Continued)

	Arrest Charges in First Three Years						Arrest Charges in All Available Years					
	Model 5 Robust			Model 6 Robust			Model 7 Robust			Model 8 Robust		
Variable	Coefficient	SE	p Value	Coefficient	SE	p Value	Coefficient	SE	p Value	Coefficient	SE	p Value
LEAD	-0.153	.088	.084	-0.081	.083	.332	-0.082	.069	.239	-0.070	.069	.309
Ethnicity												
White				0.206	.106	.052				0.048	.085	.570
African American				0.265	.103	.010				0.109	.086	.204
Other				-0.318	.209	.130				-0.460	.160	.004
Prior local confinements				0.057	.028	.040				0.066	.023	.005
County of initial commitment												
Los Angeles				-0.033	.094	.728				-0.092	.075	.222
Initial commitment offense												
Drug, minor				0.587	.144	.000				0.403	.124	.001
Property				0.246	.107	.021				0.283	.081	.000
Admission status												
Parole violator				-0.034	.124	.783				0.131	.100	.190
Age at release				0.229	.036	.000				-0.017	.028	.546
Constant	1.105	.064	.000	-3.642	.657	.000	1.980	.053	.000	1.930	.519	.000

Note: The modal response was substituted for one subject with missing information on initial commitment offense and the mean number was substituted for 10 subjects with missing information on prior local confinements. The same analyses as those presented in this table using only serious arrest charges (defined in the note on Table 2) and using number of arrests as dependent variables produced substantively similar findings. These additional analyses are available upon request.

⊠ Discussion and Conclusion

This study capitalized on a long-term set of outcome arrest data for a previously incomplete but rigorously designed experimental evaluation of a relatively well-developed and implemented juvenile boot camp and intensive aftercare program (called LEAD). In sum, it found no significant differences between boot camp and control youths in average time to first arrest or in average overall arrest charges during the first year, during the first three years, and during all available years following release to parole. An anomalous difference in the two-year follow-up period, which favored the LEAD group and held up with marginal significance controlling for available independent variables, cannot be explained but was likely due to the dampening effects of tighter parole supervision, including more time in custody, for LEAD wards versus control wards during the first year following release to parole. We conclude that the LEAD boot camp (which incorporated a shorter period of incarceration that averaged 4.6 months) and its intensive aftercare program neither reduced crime nor placed the public at any greater risk of crime.

The bulk of the evidence from previous studies supports the conclusion that boot camps are ineffective as correctional treatment (MacKenzie et al., 2001). In contrast to most prior studies though, this study comprised three notably strong features—the experimental design, virtually complete long-term follow-up data, and a relatively impressive focal boot camp. It was also limited, though, in significant ways—most notably by the lack of consistent 50/50 randomization procedures, as well as the lack of subsequent incarceration data (for street-time information) and the reliance on only official arrest outcome data. Nonetheless, this study's strengths markedly increase our confidence in previous research findings.

Why did LEAD fail to reduce recidivism? Recall the elements of effective correctional treatment (discussed in the literature review above): theoretical grounding in state-of-the-art treatment modalities, trained treatment staff, prosocial role modeling and reinforcement, avoidance of confrontational tactics, consistent discipline, individualization, and interpersonally warm, supportive staff. Although many LEAD staff were good role models and clearly cared about their cadets, the program itself was not *specifically* designed to incorporate any of these important dimensions of effective treatment. In particular, LEAD was not theoretically grounded in the best contemporary treatment methods; and CYA youth counselors were not trained in state-of-the-art treatment techniques. Furthermore, the officer-mentoring model did not take hold in the program, confrontational tactics were commonly employed, and most program activities were focused on group performance.

Once noted for its progressive, experimental treatment programs, the CYA had become, by the 1990s, a politically driven, less professional, and increasingly punitive agency (Broder, 2004; Palmer and Petrosino, 2003). Faced with political pressure from the governor, the CYA administration was unable to sustain its initial decision not to pursue a boot-camp program. Furthermore, having largely abandoned its mission of rehabilitation, the agency did not have many professionally trained treatment staff to develop the program. Continuously refined in an ad hoc but often creative manner, LEAD's boot-camp phase was still fundamentally a militarized quick fix and its aftercare a hastily designed and unevenly implemented, albeit longer term and overall somewhat more diversified, supervision service. As many staff repeatedly complained, as well, the two major goals of the program really did conflict. In short, although the boot-camp's regimentation, impressive array of daily activities, and enriched staffing generally improved the institution environment and its intensive aftercare clearly provided more surveillance, LEAD did not focus much on individual needs or provide much by way of treatment services.

Thus, in our opinion, the program was, at the outset, unlikely to reduce rates of recidivism among its participants.

References

Allison, Paul D. (1995). *Survival Analysis Using the SAS System: A Practical Guide.* Cary, NC: SAS Institute.

Andrews, D. A., Ivan Zinger, Robert D. Hoge, James Bonta, Paul Gendreau, and Francis T. Cullen. (1990). "Does Correctional Treatment Really Work? A Clinically Relevant and Psychologically Informed Meta-analysis." *Criminology* 28:369–404.

Austin, James, Michael Jones and Melissa Bolyard. (1993). *Assessing the impact of a County Operated Boot Camp: Evaluation of the Los Angeles County Regimented Inmate Diversion Program.* San Francisco: National Council on Crime and Delinquency.

Bottcher, Jean and Teresa Isorena. (1994). *LEAD: A Boot Camp and Intensive Parole Program: An Implementation and Process Evaluation of the First Year.* Sacramento, CA: Department of the Youth Authority.

Bottcher, Jean, Teresa Isorena, and Marietta Belnas. (1996). *LEAD: A Boot Camp and Intensive Parole Program: An Impact Evaluation: Second Year Findings.* Sacramento, CA: Department of the Youth Authority.

Bottcher, Jean, Teresa Isorena, Jeff Lara and Marietta Belnas. (1995). *LEAD: A Boot Camp and Intensive Parole Program: An Impact Evaluation: Preliminary Findings.* Sacramento, CA: Department of the Youth Authority.

Bourque, Blair B., Roberta C. Cronin, Frank R. Pearson, Daniel B. Felker, Mei Han, and Sarah M. Hill. (1996). *Boot Camps for Juvenile Offenders: An Implementation Evaluation of Three Demonstration Programs.* Washington, DC: Department of Justice, National Institute of Justice.

Broder, John M. (2004). "Dismal California Prisons Hold Juvenile Offenders." *The New York Times.* February 15, p. 12.

Buckley, Frank. (2000). New Jersey bucks trend of correctional boot camps' demise. *CNN.com.* March 11. Retrieved January 24, 2003, from www.cnn.com/2000/US/03/11/n.j.boot.camp

California Youth Authority. (1997). *LEAD: A Boot Camp and Intensive Parole Program: The Final Impact Evaluation.* Sacramento, CA: Author.

Chamberlain, Patricia. (1998). *Family Connections: A Treatment Foster Care Model for Adolescents with Delinquency.* Eugene, OR: Northwest Media, Inc.

Clines, Francis X. (1999). "Maryland Is Latest of States to Rethink Youth 'Boot Camps.'" *The New York Times.* December 19. Retrieved April 22, 2003, from web.lexis-nexis.com/universe/document?_m= 8c6924ded

Cornick, Thomas H. (1996). "The Youth Leadership Academy Boot Camp: An Examination of a Military Model." Pp. 119–32 in *Juvenile and Adult Boot Camps* edited by Alice Fins. Lanham, MD: American Correctional Association.

Cowles, Ernest L., Thomas C. Castellano, and Laura A. Gransky. (1995). "Boot Camp" *Drug Treatment and Aftercare Interventions: An Evaluation Review* (Research in Brief, NIJ 155062). Washington, DC: Department of Justice, National Institute of Justice.

Dieterich, William, Cecilia E. Boyles, and Susan Colling. (1999). *Colorado Regimented Juvenile Training Program Evaluation Report.* Golden: Colorado Department of Human Services, Division of Youth Corrections.

Ezell, Michael E. and Lawrence E. Cohen. (2005). *Desisting from Crime: Continuity and Change in Long-Term Crime Patterns of Serious Chronic Offenders.* Oxford, UK: Oxford University Press.

Ezell, Michael E., Kenneth C. Land, and Lawrence E. Cohen. (2003). "Modeling Multiple Failure Time Data: A Survey of Variance-Corrected Proportional Hazards Models with Empirical Applications to Arrest Data." *Sociological Methodology* 33:111–67.

Farrington, David, Gareth Hancock, Mark Livingston, Kate Painter, and Graham Towl. (2000). *Evaluation of Intensive Regimes for Youth Offenders* (Research Findings No. 121). London, UK: Home Office Research, Development and Statistics Directorate.

Flowers, Gerald T., Timothy S. Carr, and R. Barry Ruback. (1991). *Special Alternative Incarceration Evaluation.* Atlanta: Georgia Department of Corrections.

Geerken, Michael R, (1994). "Rap Sheets in Criminological Research: Considerations and Caveats." *Journal of Quantitative Criminology* 10: 3–21.

Gendreau, Paul and Claire Goggin. (2000). "Correctional Treatment: Accomplishments and Realities." Pp. 289–98 in *Correctional Counseling & Rehabilitation,* edited by Patricia Van Voorhis, Michael Braswell, and David Lester. Cincinnati, OH: Anderson.

Isorena, Teresa and Jeff Lara. (1995). *LEAD: A Boot Camp and Intensive Parole Program: An Implementation* and *Process Evaluation of the First Year at the Fred C. Nelles School*. Sacramento, CA: Department of the Youth Authority.

Jones, Robert J. (1999). *1998 Annual Report to the Governor and the General Assembly: Impact Incarceration Program*. Springfield: Illinois Department of Corrections.

Kempinen, Cynthia A. and Megan C. Kurlychek, (2003). "An Outcome Evaluation of Pennsylvania's Boot Camp: Does Rehabilitative Programming Within a Disciplinary Setting Reduce Recidivism?" *Crime & Delinquency* 49:581–602.

King, Gary. (1988). "Statistical Models for Political Science Event Counts: Bias in Conventional Procedures and Evidence for the Exponential Poisson Regression Model." *American Journal of Political Science* 32:838–63.

Land, Kenneth C., Patricia L. McCall, and Daniel S. Nagin. (1996). "A Comparison of Poisson, Negative Binomial, and Semiparametric Mixed Poisson Regression Models." *Sociological Methods & Research* 24:387–442

Long, J. Scott. (1997). *Regression Models for Categorical and Limited Dependent Variables*. Thousand Oaks, CA: Sage.

Lutze, Faith E. (1998). "Are Shock Incarceration Programs More Rehabilitative Than Traditional Prisons? A Survey of Inmates." *Justice Quarterly* 15:547–66.

Lutze, Faith E. and David C. Brody. (1999). "Mental Abuse As Cruel and Unusual Punishment: Do Boot Camp Prisons Violate the Eighth Amendment?" *Crime & Delinquency* 45:242–55.

MacKenzie, Doris Layton. (2000). "Evidence-Based Corrections: Identifying What Works." *Crime & Delinquency* 46:457–71.

MacKenzie, Doris Layton and Alex Piquero. (1994). "The Impact of Shock Incarceration Programs on Prison Crowding." *Crime & Delinquency* 40: 222–49.

MacKenzie, Doris Layton and Claire Souryal. (1994). *Multisite Evaluation of Shock Incarceration* (NCJ 142462). Washington, DC: Department of Justice, National Institute of Justice.

MacKenzie, Doris Layton and Claire Souryal. (1995). "Inmates' Attitude Change during Incarceration: A Comparison of Boot Camp with Traditional Prison." *Justice Quarterly* 12:325–54.

MacKenzie, Doris Layton, Claire Souryal, Miriam Sealock, and Mohammed Bin Kashem. (1997).

Outcome Study of the Sergeant Henry Johnson Youth Leadership Academy (YLA). College Park, MD: Evaluation Research Group, Department of Criminology and Criminal Justice.

MacKenzie, Doris Layton, David B. Wilson, and Suzanne B. Kider, (2001). "Effects of Correctional Boot Camps on Offending." *The Annals of the American Academy of Political and Social Science* 578:126–43.

Marcus-Mendoza, Susan T. (1995). "A Preliminary Investigation of Oklahoma's Shock Incarceration Program." *Oklahoma Criminal Justice Research Consortium Journal*. Retrieved February 11, 2002, from www.doc.state.ok.us/DOCS/OCJRC/OCJRC 95/950725d.htm

Martínez, Rubén. (2001). *Crossing Over: A Mexican Family on the Migrant Trail*. New York: Henry Holt.

Morash, Merry and Lila Rucker. (1990). "A Critical Look at the Idea of Boot Camp as a Correctional Reform." *Crime & Delinquency* 36:204–22.

Office of Juvenile Justice and Delinquency Prevention. (1996). *Juvenile Boot Camps: OJJDP Teleconference*, February 14. Washington, DC: U.S. Department of Justice.

Palmer, Ted and Anthony Petrosino. (2003). "The 'Experimenting Agency.'" *Evaluation Review* 27:228–66.

Parent, Dale G., R. Bradley Snyder, and Bonnie Blaisdell. (1999). *Final Report: Boot Camps' Impact on Confinement Bed Space Requirements* (NCJ 189788). Washington, DC: Department of Justice, National Institute of Justice.

Peters, Michael. (1996a). *Evaluation of the Impact of Boot Camps for Juvenile Offenders: Mobile Interim Report*. Washington, DC: Department of Justice, Office of Juvenile Justice and Delinquency Prevention.

Peters, Michael. (1996b). *Evaluation of the Impact of Boot Camps for Juvenile Offenders: Denver Interim Report*. Washington, DC: Department of Justice, Office of Juvenile Justice and Delinquency Prevention.

Petersilia, Joan. (1998). "Probation and Parole." Pp. 563–88 in *The Handbook of Crime and Punishment*, edited by Michael Tonry. New York: Oxford University Press.

Rothman, David J. (1990). *The Discovery of the Asylum: Social Order and Disorder in the New Republic*. Rev. Ed. New York: Aldine de Gruyter.

Rothman, David J. (1995). "Perfecting the Prison." Pp. 110–29 in *The Oxford History of the Prison*, edited by Norval Morris and David J. Rothman. New York: Oxford University Press.

Schlossman, Steven. (1995). "Delinquent Children: The Juvenile Reform School." Pp. 362–89 in *The Oxford History of the Prison*, edited by Norval Morris and David J. Rothman. New York: Oxford University Press.

Schnurer, Eric B. and Charles R. Lyons. (2000). "Juvenile Boot Camps: Experiment in Trouble." *CNP Stateline*. Retrieved January 24, 2003, from www.cnponline.org/Issue%20Briefs/Statelines/ statelin0200.htm

Sechrest, Dale K. (1989). "Prison 'Boot Camps' Do Not Measure Up." *Federal Probation* 53 (No. 3): 15–20.

Selcraig, Bruce. (2000). "Why Do So Many States Still Insist That Humiliation and Abuse Will Straighten Out Troubled Kids?" *Mother Jones*, November/ December, 66–71.

Thomas, David and Michael Peters. (1996). *Evaluation of the Impact of Boot Camps for Juvenile Offenders: Cleveland Interim Report*. Washington, DC: Department of Justice, Office of Juvenile Justice and Delinquency Prevention.

Tonry, Michael. (1998). "Intermediate Sanctions." Pp. 683–711 in *The Handbook of Crime & Punishment*, edited by Michael Tonry. New York: Oxford University Press.

Walker, Ronald Boyce. (2002). "Second Chances." *The Houston Chronicle*, July 11, p. 1, Retrieved April 22, 2003, from web.lexisnexis.com/universe/document ?_m=2fd201c

DISCUSSION QUESTIONS

1. In your own words, explain the idea of randomized experiments and why they are the best ways of studying a phenomenon.

2. What is the basic idea behind boot camps?

3. What were the conclusions of this study?

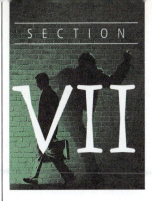
PAROLE AND PRISONER REENTRY

Introduction

⊠ What Is Parole?

The term *parole* comes from the French phrase *parole d'honneur*, which literally means "word of honor." In times when a person's word really meant something, parole was used by European armies to release captured enemy soldiers on condition (their word of honor) that they would take no further part in the present hostilities (Seiter, 2005). Modern parole refers to the release of convicted criminals from prison before they have completed their full sentences. Parole is different from probation in two basic ways. First, parole is an administration function practiced by a parole board, which is part of the executive branch of government, while probation is a judicial function. Second, parolees have spent time in prison before being released into the community whereas probationers typically have not. In many states, both parolees and probationers are supervised by state probation and parole officers or agents while, in others, they are supervised by separate probation or parole agencies.

⊠ Brief History of Parole

The philosophical foundation of parole, as applied to convicted criminals, was laid by the superintendent of the Norfolk Island penal colony off the coast of Australia in the 1830s. The superintendent was a retired British naval officer and geography professor named Alexander Maconochie, who had experienced imprisonment himself as a captive of the

French during the Napoleonic Wars. The horrendous conditions in the colony offended the compassionate and deeply religious Maconochie, who was a firm believer in the primacy of human dignity. As prison superintendent, he operated on three basic principles: (1) cruel and vindictive punishment debases both the criminal and the society that allows it; (2) the purpose of punishment should only be the reformation of the convict; and (3) criminal sentences should not be seen in terms of time to be served but rather in terms of tasks to be performed. To implement his programs, Maconochie required indefinite (open-ended) prison terms rather than determinate (fixed) terms so that convicts would have an incentive to work toward release.

With respect to the third principle, Maconochie devised a *mark system* involving credits earned for the speedy and efficient performance of these tasks, as well as for overall good behavior. When a convict had accumulated enough credits, he could apply for a *ticket of leave* (TOL), which was a document granting him freedom to work and live outside the prison before the expiration of his full sentence. TOL convicts were free to work, acquire property, and marry, but they had to appear before a magistrate when required and church attendance was mandatory. Maconochie's system appeared to have worked very well. It has been determined that only 20 out of 900 of Maconochie's TOL convicts were convicted of new felonies; a recidivism rate of 2.2% that modern penologists are scarcely able to comprehend (Hughes, 1987). Nevertheless, and perhaps predictably, when Maconochie returned home to England and tried to institute his reforms there, he was accused of coddling criminals and relieved of his duties (Petersilia, 2000).

Nevertheless, the TOL system was adapted to the differing conditions in Britain under Walter Crofton, who devised the so-called *Irish system*. This system involved four stages beginning with a period of solitary confinement, followed by a period in which convicts could earn marks through labor and good behavior, then movement to an open prerelease prison when enough marks had been accumulated, and, finally, a TOL. TOL convicts were supervised in the community by either police officers or civilian volunteers (forerunners of the modern parole officer) who paid visits to their homes and attempted to secure employment for them (Foster, 2006). Of the 557 men released on TOL under the Irish system in the 1850s, only 17 (3.05%) were revoked for new offenses (Seiter, 2005).

Elements from both Maconochie's and Crofton's systems were brought into practice in the United States in the 1870s by Zebulon Brockway, superintendent of the Elmira Reformatory in New York. Brockway's system required indeterminate sentencing so that "good time" earned through good conduct and labor could be used to reduce inmates' sentences. However, there were no provisions for the supervision of offenders who obtained early release until 1930 when the U.S. Congress established the United States Board of Parole. Eventually, parole came to be seen as a way of maintaining order in prisons by holding out the prospect of early release, as a method to reintegrate offenders back into the community, and as a partial solution to the problem of prison overcrowding. In any case, parole became an essential and much valued part of the American correctional system.

◼ The Modern Parole System

It was not to remain "much valued" for long, however. The skyrocketing crime rates in the 1970s and 1980s, the perception that rehabilitation did not work, and the general conservative turn that the country took led to the "tough on crime" approaches to punishment discussed in the last section. The heinous kidnapping, rape, and murders of 12-year-old Polly Klaas by parolee Richard Davis and 7-year-old Megan Kanka by parolee Jesse Timmendequas led to understandable calls for the abolition of parole. A number of states and the federal government heeded this call, and we had the return of fixed determinate or mandatory sentencing in many jurisdictions, practices that Maconochie so disliked.

This all sounds very tough and goes over quite well politically, but the reality is something different. Prisoners are still released early for reasons of overcrowding and budgetary concerns, but there is much less control now over who is released. It is discretionary parole that has really been abolished in favor of mandatory parole. *Discretionary parole* is parole granted at the discretion of a parole board for selected inmates who have earned it; *mandatory parole* is automatic parole for almost all inmates (Petersilia, 2000). Tragically, both Davis and Timmendequas were granted mandatory parole determined by mathematical norms generated by a computer based solely on time served. Had their cases gone before a parole board, where board members can peruse parole applicants' criminal history and target violent and dangerous criminals for longer incarceration, odds are that neither man would have been released (Petersilia, 2000).

As we see from the graph in Figure 7.1, discretionary parole releases have been dropping significantly since 1980, while mandatory parole releases have gone up significantly. Both discretionary and mandatory release parolees are supervised post-release; those 17 to 18% of inmates who are released at the expiration of their sentences (they have completed every day of their sentence in prison or "maxed out") are not supervised (Hughes, Wilson, & Beck, 2001). The graph in Figure 7.2 supports those who favor a

▲ **Photo 7.1** Kenneth Foster, Sr. walks out of "the Walls" prison in Huntsville, Texas after Governor Perry commuted his son's death sentence. His son, Kenneth Foster, Jr., had been scheduled to be executed that evening, but the Texas Board of Pardons and Paroles, in a vote of 6–1 recommended that the Governor spare the life of Kenneth Foster, Jr.

return to discretionary parole and the elimination of mandatory parole. Note that in 1999 about 52% of discretionary parolees successfully completed parole while only about 32% of mandatory parolees did.

◤ What Goes in Must Come Out: Prisoner Reentry

Pundits of all persuasions consume an incredible amount of ink and paper writing about America's imprisonment binge. However, rarely outside of the academic literature do we find much concern about the natural corollary of the binge: What goes in must come out. Understanding the process of prisoner reentry and reintegration into the community is a very pressing issue in corrections today. In 2004, 503,200 adult correctional clients entered American prisons and 483,000 left them (Glaze & Palla, 2005). Except for those few who leave prison in a pine box or who make their own clandestine arrangements to abscond, all prisoners are eventually released back into the community. Unfortunately, among this huge number, one in five will leave with no post-release supervision, rendering "parole more a legal status than a systematic process of reintegrating returning prisoners" (Travis, 2000, p. 1). Pushing inmates out the prison door with $50, a bus ticket to the nearest town, and a fond farewell is a strategy almost guaranteeing that the majority of them will return. With the exception of those who max out, then, prisoners will be released under the supervision of a parole officer charged with monitoring offenders' behavior and helping them to readjust to the free world.

Because parolees have been in prison and are thus, on average, more strongly immersed in a criminal lifestyle, we should expect them to be more difficult to supervise than probationers, and they are, and to have lower rates of success, and they do. While Glaze and Palla (2005) indicate a success rate of 60% for probationers in 2004, the same

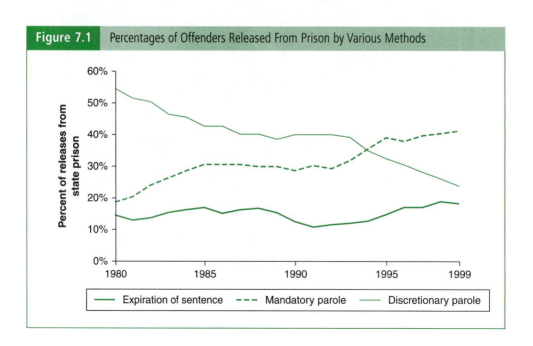

Figure 7.1   Percentages of Offenders Released From Prison by Various Methods

| Figure 7.2 | Percentages of Parolees Successfully Completing Parole by Release Type |

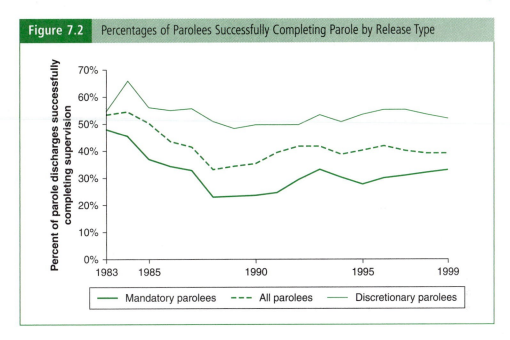

figure for successful completion of parole was only 46% (although note the difference between the success rates of discretionary versus mandatory release parolees above).

The longer people remain in prison, the more difficult it is for them to readjust to the outside world. Inmates spend a considerable amount of time in prison living by a code that defines as "right" almost everything that is "wrong" on the outside. Adherence to that code brings them acceptance by fellow inmates as "good cons." Over time, this code becomes etched into an inmate's self-concept as the prison experience becomes his or her comfort zone. When they return to the streets, they do not fit in, they feel out of their comfort zone, and their much sought-after reputations as good cons become liabilities rather than advantages. As prison movie buffs are aware, these readjustment problems were dramatically presented in the suicide of Brooks in *The Shawshank Redemption* and in the final crazy hurrahs of Harry and Archie in *Tough Guys.*

Brooks, Harry, and Archie were all old men who had served very long periods of incarceration and who had thoroughly assimilated the prison subculture by the time of their release into an alien and less than welcoming world. Thus, one recommendation might be to reduce the length of prison sentences so that those unfortunate enough to be in them do not have time to become "prisonized." Such a recommendation gains support from statistics showing that the shorter the time spent in prison the greater the chance of success on parole (Travis & Lawrence, 2002), but, if we are looking for something causal in those statistics, we are surely sniffing around the wrong tree. Shorter sentences typically go to those committing the least serious crimes and who have the shortest arrest sheets; such people are already less likely to commit further crimes than those who commit the more serious offenses and have long rap sheets. It is not for nothing that former U. S. Attorney General Janet Reno called prisoner reentry "one of the most pressing problems we face as a nation" (Petersilia, 2001, p. 370).

✎ The Impact of Imprisonment and Reentry on Communities

Because crime is highly concentrated in certain neighborhoods, a disproportionate number of prison inmates come from, and return to, those same neighborhoods. There are those who believe that, though high incarceration rates may reduce crime in the short run, the strategy provides only a temporary reprieve and will eventually lead to increased crime rates by weakening families and communities and reducing the supervision of children (DeFina & Arvanites, 2002). According to this body of literature, the loss and return of individuals concentrated in certain communities reduces community organization and cohesion, disrupts families economically and socially, and adds a plethora of other problems that will eventually lead to more crime than would have occurred if offenders had been allowed to remain in the community.

The article by Todd Clear, Dina Rose, and Judith Ryder in this section is in this tradition, with their working hypothesis being that, when public control (incarceration) occurs at high levels, private control (informal normative control) functions at low levels and ultimately results in more crime. The authors' interviews with members of high incarceration rate neighborhoods identified a number of problems faced by the neighborhood and its residents because of the recycling of many of its members through the criminal justice system.

However, the authors do not deny that the neighborhood and its residents were better off and safer when the bad apples were pulled from the shelf and shipped off to prison. This implies the opposite of their hypothesis, i.e., when private control functions at low levels, public control occurs at high levels. When you read of the negative impact to the community in the article, especially as it pertains to the moral and financial impact on families, think of the findings from a Bureau of Justice Statistics report (Mumola, 2000) indicating that 48% of imprisoned parents were never married, 28% of those who were ever married were divorced or separated, and very few men lived with their children prior to imprisonment. Moreover, the majority of parents had been convicted of violent or drug crimes and 85% had drug problems. A longitudinal study of 1,116 British families showed that the presence of a criminal father in the household predicted antisocial behavior of his children and that the harmful effects increased the longer he spent with the family (Moffitt, 2005). All this makes it clear that, when correctional clients reenter communities, it is paramount that every effort is made to ensure the programming is available to make that reentry successful for that former inmate's sake, his or her family's sake, and the sake of the larger community.

✎ What Is Reentry, and What Makes for a Successful One?

Reentry is the process of reintegrating offenders back into the community after release from jail or prison. Part of that process is preparing offenders through the use of various programs targeting their risks and needs so that they will have a fighting chance of remaining in the community. The United States Department of Justice (Solomon,

Dedel Johnson, Travis, McBride, 2004) lists three phases believed necessary for successful reentry:

Phase 1—Protect and Prepare: The use of institution-based programs designed to prepare offenders to reenter society. These services include education, treatment for mental health and substance abuse issues, job training, mentoring, and a complete diagnostic and risk assessment.

Phase 2—Control and Restore: The use of community-based transition programs to work with offenders prior to and immediately following their release from jail or prison. Services provided in this phase include education, monitoring, mentoring, life-skills training, assessment, job-skills development, and mental health and substance abuse treatment.

Phase 3—Sustain and Support: This phase uses community-based long-term support programs designed to connect offenders no longer under the supervision of the justice system with a network of social services agencies and community-based organizations to provide ongoing services and mentoring relationships.

The above programming strategy is an ideal rather than a reality most of the time, but many offenders released from incarceration do have access to some parts of each of these phases, and some manage to complete parole successfully. To be successful in phase I, offenders have to be willing to apply themselves to the training and programming offered to them. They have to make a conscious decision to set long-term goals and convince themselves that a criminal career is not for them. Obviously, these are decisions that are typically only made by individuals not fully immersed in the criminal lifestyle. In phase II, they have to implement those decisions in the real world where it counts and not be sidetracked by either the frustration generated by the monitoring they are subjected to by their parole officers or by the criminal opportunities that may present themselves. Phase III may be the most difficult of all for some offenders who rely on authoritative figures to give their lives direction. This is why it is so important for community corrections officers to plug their clients into noncriminal justice agencies prior to termination of supervision.

Perhaps the most important tool for the successful reintegration of offenders who want to go straight is employment. Unfortunately, the typical correctional client is not prepared for much other than a low-skill manufacturing job, the kind of jobs that the United States has been losing in truly staggering numbers due to technological advances and companies moving operations overseas. Job prospects are thus fairly limited unless parolees can improve themselves educationally. Amy Solomon and her colleagues (2004) report that 53% of Hispanic inmates, 44% of black inmates, and 27% of white inmates have not completed high school or obtained a GED as opposed to 18% of the general population. Add offenders' criminal record to their general lack of preparedness, and you will understand why employers are reluctant to hire ex-inmates. Perhaps as a result of this lack of preparedness and employer reluctance to hire ex-offenders, economists find that incarceration reduces employment opportunities by about 40%, wages by about 15%, and wage growth by about 33% (Western, 2003).

Determining Parole "Success"

Although one would not think so, defining parole success is difficult. Does it mean (1) a completed crime-free/technical violation-free period of parole, or does it mean (2) that the offender was released from parole without being returned to prison despite his or her behavior while on parole? It obviously means vastly different things in different states because parole "success" rates in 1999 ranged from 19% in Utah to 83% in Massachusetts (Travis & Lawrence, 2002). Does this mean that the Mormons of Utah are over four times more resistant to taking the straight and narrow road than the Catholics of Massachusetts? Of course not; it most likely means that conservative Utah follows our first definition of success while liberal Massachusetts follows our second definition. Given the national average rate of 42% that year, it is plain that many parolees are forgiven many technical violations, or perhaps even a petty arrest or two, in most states. Thus, "success" has as much or more to do with the behavior of parole authorities than it does with the behavior of parolees. When we speak of "success," then, we are generally speaking in middle-of-the-road terms in which certain parolee peccadilloes are forgiven occasionally in the interest of maintaining him or her on a trajectory that is at least somewhat positive.

In this section, Richard Seiter and Karen Kadela identify programs that work, that do not, and that are promising in helping prisoners in the long process of successfully reentering the community. They look at transitional community programs such as halfway houses and work release programs and programs that initiated treatment for inmate deficits (drug dependency, low education, poor life skills, etc.) while they were in prison and continued in the community after release. Not surprisingly, the programs that worked best were concrete programs that provided offenders with skills to compete in the workforce and with intensive drug programs. Programs that were located in the community, such as halfway houses, were more effective than prison-based programs. We have explored many treatment and programmatic aspects in previous sections so will limit ourselves to discussing halfway houses and electronic monitoring in this introduction.

Halfway Houses

As the name implies, halfway houses are transitional places of residence for correctional clients that are, in terms of strictness of supervision, "halfway" between the constant supervision of prison and the much looser supervision in the community. In addition to being a transition between prison and the community, such places (also referred to as *community residential centers*) may serve as an intermediate sanction for offenders not sent to prison but needing greater supervision than straight probation. The rationale behind halfway houses is that individuals with multiple problems, such as substance abuse and lack of education, may have a better chance to tackle these problems positively and to comply with court orders if they are placed in residential centers where they will be strictly monitored while, at the same time, being provided with support services to address the problems that got them there. Halfway houses may be operated by corrections personnel, but they are more likely to be operated by faith-based organizations such as the Salvation Army and Volunteers of America. Nevertheless, residents are still under the control of probation or parole authorities and may be removed and sent to prison if they violate the conditions of their probation or parole.

In times of rising correctional expenses and prison overcrowding, cost-conscious legislators tend to view community-based alternatives to prisons like diet-conscious drinkers view calorie-reduced beer—as "prison lite." Community-based residential programs supposedly provide public safety at a fraction of the cost while allowing offenders to remain in the community and at work earning their own keep, and, best of all, these programs are assumed to reduce recidivism. Halfway houses are also a valuable resource in that they provide offenders released from prison, who would otherwise be homeless, with an address for employment purposes.

Nancy Marion's article in this section questions some of these assumptions and provides us with an excellent overall view of residential community-based alternatives to prison. Her case study finds that some of the assumptions are true some of the time, but it generally paints a disappointing picture for those who believe that keeping offenders out of prison aids in rehabilitation. However, Lowenkamp and Latessa's (2002) examination of 38 such facilities in Ohio found that, although all were not effective, most were. They found that community-based programs were of no use for low-risk offenders (indicating that they didn't need them in the first place) but that the majority of these programs were effective in substantially reducing recidivism among medium- and high-risk offenders.

Halfway houses should not be viewed as another way of coddling criminals, a claim that is supported by the "exchange rate" for halfway house placement (that is, how much time in this alternative sanction an offender is willing to serve to avoid 12 months in prison). May and his colleagues (2005) determined that the halfway house "exchange rate" was an average of 12.77 months for offenders who had served time in prison (14.42 months for all offenders). "Experienced" correctional clients may therefore see the halfway house as almost as punitive as prison. Much of this has to do with the level of responsibility expected of the parolee, which likely includes programming, working or looking for work, being subjected to frequent and random testing for drug and alcohol intake, and, possibly, electronic monitoring (Shilton, 2003).

▧ House Arrest, Electronic Monitoring, and Global Positioning Systems

House arrest is a program used by probation and parole agencies that requires offenders to remain in their homes at all times except for approved periods to travel to work or school and, occasionally, to other approved destinations. As a system of social control, house arrest is typically used primarily as an initial phase of intensive probation or parole supervision, but it can also be used as an alternative to pretrial detention or a jail sentence. As is the case with so many other criminal justice practices, house arrest was designed primarily to reduce financial costs to the state by reducing institutional confinement.

House arrest did not initially gain widespread acceptance in the criminal justice community because there was no way of assuring offender compliance with the order short of having officers constantly monitoring the residence. However, house arrest gained in popularity with the advent of *electronic monitoring* (EM). EM is a system by which offenders under house arrest can be monitored for compliance using computerized technology. In modern EM systems, an electronic device worn around the offender's ankle sends a continuous signal to a receiver attached to the offender's house phone. If the offender moves beyond 500 feet from his or her house, the transmitter records it and relays the information to a centralized computer. A probation or parole officer is then dispatched to the offender's home to investigate whether he or she has absconded or removed or tampered with the device. As of 2004, almost 13,000 offenders were under house arrest, with 90% of them being electronically monitored (Bohm & Haley, 2007).

An even more sophisticated method of tracking offenders uses *global positioning systems* (GPS). GPS requires offenders to wear a removable tracking unit that constantly communicates with a non-removable ankle cuff. If communication is lost, the loss is noted by a Department of Defense satellite, which records the time and location of the loss in its database. This information is then forwarded to criminal justice authorities so that they can take action to determine why communication was lost. Unlike EM systems, GPS can be used for surveillance purposes as well as detention purposes. For instance, it can let authorities know if a sex offender goes within a certain distance of a schoolyard or if a violent offender is approaching his or her victim's place of residence or work (Black & Smith, 2003).

Brian Payne and Randy Gainey's article on EM in this section indicates that offenders released from jails or prisons and placed in EM programs are generally positive about the experience (not that they enjoyed it, but rather that it was better than the jail or prison alternative). Their findings mirror findings from a larger sample of correctional clients on EM programs in New Zealand (Gibbs & King, 2003). Many see it as jail or prison time simply served in a less restrictive and less violent environment, although the average experienced offender would exchange 11.35 months on EM for 12 months in prison.

Although the authors are positive about the alleged rehabilitative promise of allowing offenders to serve time at home and thus to maintain their links to family, and although successful completion rates are high, recidivism rates, which are the litmus tests for any corrections program aimed at rehabilitation, were not any better than those for probationers or parolees not on EM programs, when offenders were matched for offender risk in several Canadian provinces (Bonta, Wallace-Capretta,

▲ **Photo 7.2**   Charles Manson reads a statement at his parole hearing in San Quentin in 1986. He was turned down for parole and remains in prison to this day.

& Rooney, 2000). This finding may be viewed positively, however, as a function of the greater ability to detect noncompliance with release conditions among those under EM supervision.

An additional problem with EM is that, because its low cost relative to incarceration is alluring to politicians, it may be (and is) used without sufficient care being taken about who should be eligible for it. While offenders can be monitored and more readily arrested if they commit a crime while on EM, EM does not prevent them from committing further crimes. Several high-profile cases including rapes and murders have been committed by offenders who succeed in removing their electronic bracelets (Reid, 2006). Such cases make it understandable that the public thinks EM is too lenient for offenders, especially violent offenders, and would rather see iron balls and chains attached to those types of offenders rather than plastic bracelets.

▨ Concluding Remarks on Reentry and Recidivism

We have learned that reentry into the community, whether an ex-inmate is on supervised parole or not, is an extremely difficult process. Everything appears to be working against the offenders' successful reintegration. Work is difficult to find; old addictions, crime partners, and the criminal lifestyle might tempt people to reoffend (Rengert & Wasilchick, 2001; Walsh & Ellis, 2007; Wright & Decker, 1994). Nevertheless, correctional work is premised on the assumption that people can change. Although some correctional clients are particularly problematic in this regard, many others can be reformed.

When we find the best reentry programs (programs that have repeatedly been shown empirically to reduce recidivism significantly) and implement them, what kind of success can we expect? Joan Petersilia (2004, pp. 7–8), the preeminent reentry researcher today, sums up the combined Canadian and American reentry "what works" literature and states that effective programs took place mostly in the community (as opposed to institutional settings), were intensive (at least six months long), focused on high-risk individuals (with risk level determined by classification instruments rather than clinical judgments), used cognitive-behavioral treatment techniques, and matched therapist and program to the specific learning styles and characteristics of individual offenders. As the individual changed his or her thinking patterns, he or she would be provided with vocational training and other job-enhancing opportunities. Positive reinforcers would outweigh negative reinforcers in all program components. Every program begun in jail or prison would have an intensive and mandatory aftercare component.

Petersilia also suggests that, if we could design programs that combined all these things, we might be able to reduce recidivism by about 30%. So with all the putatively best methods currently available, caring and knowledgeable counselors, and a budget sufficient to meet the needs of all the identified programs, the best we can hope for is a 30% reduction in recidivism. As much as we would all love to find a way to turn offenders into respectable citizens, this reminds us that human beings are not lumps of clay to be molded to someone else's specifications. Although correctional workers might regret that they cannot mold their charges' minds as they might wish to, in the long run, the fact that they cannot is a vindication of human dignity.

⌧ Summary

- ◆ Parole is the legal status of a person who has been released from prison prior to completing his or her full term. The concept can be traced in the nineteenth century to Maconochie's ticket of leave system in Australia and Walter Crofton's Irish system. Parts of these systems were brought to the United States in the 1870s by Zebulon Brockway, superintendent of the Elmira Reformatory in New York. Brockway's system required indeterminate sentencing so that "good time" earned through good conduct and labor could be used to reduce inmates' sentences.

- ◆ In the modern United States, we have two systems of parole—discretionary and mandatory. Discretionary parole is parole granted by a parole board based on its members' perceptions of the inmate's readiness to be released; mandatory parole is based simply on a mathematical formula of time served. Discretionary parolees are significantly more likely to complete parole successfully than mandatory parolees.

- ◆ The reentry of prisoners into the community is a very difficult process. The ex-con stigma makes getting employment problematic, and the period of absence make it tough to reestablish relationships. Successful reentry depends on several factors, not the least of which is how success is defined by various parole authorities, as the huge gap between the Massachusetts and Utah "success" rates indicates. Nevertheless, providing parolees with concrete help such as job skills and drug rehabilitation programs can go a long way in helping them to remain crime free. This effort may be particularly fruitful if it is made in some form of a community-based residential program.

- ◆ Electronic monitoring and global positioning systems technology is increasingly used in corrections. It helps by increasing the level of offender monitoring and apprehension, but it cannot altogether prevent additional crimes while offenders are on the program, which is why candidates for this type of supervision must be chosen carefully. In sum, few programs can be said to work for most offenders if we define "working" unrealistically. Human nature is complicated, often ornery, and resistant to change. Even "ideal" programs such as those defined by Petersilia could only be expected (according to her) to reduce recidivism by about 30%.

KEY TERMS

Discretionary parole	Mandatory parole
Electronic monitoring	Marks system
Halfway houses	Parole
House arrest	Reentry

DISCUSSION QUESTIONS

1. Compare the recidivism rates claimed for the TOL parolees with the modern American recidivism rates of parolees. What do you think may account for the huge differences?

2. Why do we still continue to utilize mandatory parole in the face of evidence that discretionary parole is a safer bet?

3. What do you think may be the single most difficult problem for a parolee to overcome, one who has just spent five years in prison but who is interested in staying out of trouble?

4. Would it be a good thing to have a number of community-based residential facilities located in high-crime communities so that some of the problems noted by Clear, Rose, and Ryder might be avoided?

5. Given that expert opinion says that a 30% reduction in the recidivism rate is about the best we can accomplish, do you think that trying to rehabilitate criminals is a waste of time and that the money would be better spent in some other way?

INTERNET SITES

American Probation and Parole Association: www.appa-net.org

Bureau of Justice Statistics (information available on all manner of criminal justice topics): www.ojp.usdoj.gov/bjs

Center for Sex Offender Management: www.csom.org

National Criminal Justice Reference Service: www.ncjrs.gov

National Institute of Mental Health: www.nimh.nih

National Institute on Drug Abuse: www.drugabuse.gov

Office of Justice Programs, Bureau of Justice Statistics (periodic statistical reports on all manner of criminal justice topics, e.g., HIV in prisons and jails, probation and parole, and profiles of prisoners): www.ojp.usdoj.gov/bjs/periodic.htm

Vera Institute (information available on a number of corrections and other justice related topics): www.vera.org

READING

The article by Seiter and Kadela is an excellent companion to the Petersilia article given the renewed interest in prisoner reentry in corrections. This renewed interest is due to a change in many of the factors surrounding the release of prisoners and their reentry into the community. These changes include a modification of sentencing from the use of parole to determinate release with fewer ex-offenders having supervision in the community, an increased emphasis on surveillance rather than assistance for those under supervision, less community stability and the decreasing availability of community social service support, and dramatically larger numbers returning to the community. More releasees are being violated and returned to the community than ever before. Therefore, it is important to identify prisoner reentry programs that work. Seiter and Kadela define reentry, categorize reentry programs, and use the Maryland Scale of Scientific Method to determine the effectiveness of program categories and to conclude that many such categories are effective in aiding reentry and reducing recidivism.

Prisoner Reentry

What Works, What Does Not, and What Is Promising

Richard P. Seiter and Karen R. Kadela

The United States has had prisons as a sanction for those who violate criminal laws since William Penn and the Quakers of Pennsylvania created a wing of the Walnut Street Jail to house sentenced offenders in 1790. During the next 200 years, there have been many changes in how prisons were operated, what correctional goals were emphasized, and what programs were offered. Throughout this period, the pendulum has repeatedly swung from harsh discipline and tight security to a focus on individual prisoner rights and rehabilitative treatment.

Correctional institutions have been the holders of prisoners sent to their authority under many different sentencing structures. During the 1800s, prisoners served a set amount of time in very crowded prisons, with little emphasis on rehabilitation or preparation for release. During much of the 1900s, sentences were indeterminate; therefore prisons accentuated the provision of rehabilitation, and parole board experts made the decision about when prisoners would be released based on their readiness for returning to the community. During the past 20 years, there has been a return to set, determinate sentences. With determinate sentences, offenders have often been limited in the amount of good time they can earn from their sentences, as many truth-in-sentencing laws have been passed, requiring completion of 85% of the sentence before prisoners are eligible for release.

Prisons have also experienced a changing makeup of offenders. Currently, prison populations are increasingly diverse regarding race and ethnicity, age, gender, type of crime, and affiliation with organized crime or organized gangs. Prisons have therefore become increasingly sophisticated in classifying and separating populations by security level, medical problems, special program needs, and even work programs. What has remained constant is that almost every inmate is still released from prison. Prisoners have historically returned to the communities from which they were sentenced, generally to live with family members, attempt to find a job, and successfully avoid future criminality. The world to which they return is drastically different from the one they left regarding availability of jobs, family support, community resources, and willingness to assist ex-offenders.

The current status of prisoner reentry is very different from that of only a few decades ago. There are many more offenders released from prisons than in the past. Many are released after serving a determinate sentence (without a parole board), and some have no supervision requirements after release. Overall, prisoners are serving significantly longer prison terms, and only a small percentage is receiving the benefit of extensive rehabilitation or prerelease programs. The communities are more disorganized, their families are less likely to be supportive, and the releasees find fewer social services available to them in the community. Most distressing is that a large number of releasees are returned to prison, either for committing new crimes or

SOURCE: Seiter, R. P., & Kadela, K. R. (2003). Prison reentry: What works, what does not, and what is promising. *Crime and Delinquency, 49*(3), 360–390. Reprinted with permission of Sage Publications, Inc.

for violating the technical conditions of their parole or release supervision.

The goal of this article is to provide an overview and background of prisoner reentry and to examine the current evaluations of reentry programs to determine what works. Prisoner reentry has changed in many ways, including an altered sentencing structure for many states, an increase in the number of inmates and releasees, a more diverse offender population, and a changing community to which offenders return. These changes create many issues that were not critical or even considered until recently. Although many evaluations of prison and community correctional programs exist, few are labeled specifically as prisoner reentry programs. Therefore, we have created a definition of prisoner reentry for purposes of identifying which evaluations to include in the examination of what works, what does not work, and what is promising.

The Changing Realm of Prisoner Reentry

As noted, many things have changed when considering the current status of prisoner reentry. These changes are the result of many forces, including a tough-on-crime attitude, reduced funding for prison programs and community social services, a weakening of the traditional support structures within communities and neighborhoods, and less (sometimes zero) tolerance for lapses by prison releasees under official supervision. The issue of prison reentry is one that covers a broad base of social and governmental networks. Contributing to the current status of reentry are the types of sentences and release mechanisms, the types of programs provided by the departments of corrections, the types and intensity of supervision provided by the parole or release agency, the family support available to the offender, community funding of social services, and the economic status and availability of jobs. The changing nature of prisoner reentry has made successful transitions from prison to community more difficult. Although it is not suggested that prisoner reentry was successful in the past—or that it was without problems—there is no question that the current system of incarceration and reentry creates unique challenges for ex-offenders.

The Traditional Approach to Reentry

For much of the 20th century, preparation for release was considered an important part of the prison experience, and correctional systems were organized to provide programs to prepare inmates for the community transition. During the mid-1900s, all states used indeterminate sentences with release by parole boards (Clear & Cole, 1997). By 1977, release on parole reached its peak, as 72% of all prisoners were released on parole (Bureau of Justice Statistics, 1977). For almost 20 years preceding this high-water mark, the medical model, with a focus on rehabilitation, was embraced, and prisons created programs to prepare inmates for release. Education and vocational programs, substance abuse and other counseling programs, therapeutic communities and other residential programs, and prison industry work programs were important parts of prison operations. Many of these programs were mandatory, and when they were voluntary, inmates still participated at high rates to impress the parole board and improve their chances of a favorable parole decision.

Once decisions to release prisoners were made, there were usually extensive efforts to ensure the prisoners were prepared for reentry. The parole boards closely reviewed inmates' release plans in consideration of parole.

Community parole officers investigated the plans and reported on their acceptability to the parole board. When plans were less than solid, inmates were usually released to a halfway house, with the express purpose of assisting in transitional areas, such as housing, employment, family relationships, and mental health or substance abuse counseling. Correctional officials recognized the difficulty in the prison-to-community transition, and reintegrative programs were expanded and developed to ease the transition. There was experimentation with specialized caseloads, the use of volunteers in parole, and even ex-offenders as parole officer aides.

The pattern during this era, emphasizing rehabilitation and reintegration, was clear and consistent. Prisons diagnosed inmate problems and provided rehabilitative programs to reduce these problems. Parole boards considered inmates' prison program participation and attitude in determining preparation for release and weighed the acceptability of the inmates' release plans in the parole decision-making process. The inmates' return to the community was intensely supervised. If the resources and community ties were not strong, inmates were placed in halfway houses. In addition, for the first year or two, parole officers (whose primary responsibility was to guide the offender to programs and services) supervised offenders. From the 1950s through the 1970s, there was significant attention focused on prisoner preparation and the transition to the community.

Two Decades of Change

Since the early 1980s, the traditional pattern just noted has begun to deteriorate. The demise of the medical model, the tough-on-crime attitude by the public and elected officials, the belief that rehabilitation did not work (as a result of the Martinson study and "nothing works" conclusion), the reduced funding for prison and transitional programs, and the change in parole supervision from a casework (helping) to a surveillance (policing) model had an effect on changing the traditional approach that was accepted prior to the 1980s.[1]

Although these changes did not transpire overnight, the current model of prison operations and prisoner reentry does not focus on inmate rehabilitation and preparation for release, but on punishment, deterrence, and incapacitation to prevent future crimes. Many offenders currently serve a determinate sentence that is much longer than in the past, in hopes of producing a proper deterrent value. Inmates are not seen as sick, as they were under the medical model, but as making a conscious decision to commit crimes. Prison programs are seen as valuable to keep inmates busy and maintain order, more than for release preparation. Without parole boards in many states, there is no gatekeeper to review the inmate's preparation and release plans. After release, if the offender is under supervision, there is zero tolerance for drug use, technical violations, and minor criminal behavior. If a violation occurs, the offender is returned to prison. The following presents some of the evolutionary changes that have had an effect on prisoner reentry.

Changing Sentencing and Supervision Policies

Indeterminate sentencing was the dominant model used across the United States for most of the 20th century. Under this structure, parole served many positive functions. First, extremely dangerous inmates were often maintained in prisons longer than they would have been under a determinate sentence structure. Determinate sentences are usually shorter than indeterminate sentences, and parole boards regularly require dangerous, high-risk inmates to serve the maximum sentence. The state of Colorado abolished parole as a release mechanism in 1979 but reinstated it after finding out that the length of prison sentences served was decreasing, particularly for high-risk offenders.

Second, parole boards do act as a gatekeeper to ensure inmates have solid release plans when they return to the community.

Parole boards always ask inmates questions, such as "Where will you live when you get out of prison?" and "What job opportunities are available to you?" The boards also had reports available to them from parole officers who had investigated the inmates' release plans. It is true that prison staff—working with inmates as they near release—can ask some of the same questions. However, with a firm release date looming, there is less incentive for staff and inmates to try to improve a weak plan, and there is usually no way to delay a release due to an insufficient plan.

Third, the existence of parole and parole consideration is an incentive for good behavior by inmates and for program participation that can be beneficial, even if not truly voluntary. Some of the criticism of parole during the 1970s had to do with the involuntary nature of program participation. Opponents to this system suggested that programs would be more effective if there was no coercion regarding participation. If rehabilitative programming were to be fully effective, it was argued that it had to be carried out in a noncoercive fashion. Even though many correctional programs were considered voluntary, parole board decisions considered the efforts toward rehabilitation put forth by offenders, judged primarily by the number of programs that they completed. Release from parole supervision also considered offenders' efforts toward rehabilitation.

Psychiatrist Seymour Halleck argued that it was almost impossible to distinguish between fully voluntary and coercive treatment participation, especially in a correctional setting in which decisions affecting offenders (parole) considered such participation (Halleck, 1971). Norval Morris (1974) convincingly asserted that although rehabilitation is valuable as a correctional goal, it could not be effective if coercive in the eyes of offenders, or if they saw it as an element of the punishment they were receiving for their criminal offenses. In addition, David Fogel (1975) argued for fully voluntary prison programs in his justice model.

However, there was no evidence that nonvoluntary program participation was less effective than participation with some coercion. With more recent data indicating the benefit of a variety of prison program participation (cognitive skills training, drug treatment programs, education and work programs, and treatment of sex offenders) on reducing recidivism (Gaes, Flanagan, Motiuk, & Stewart, 1999), there is a renewed interest in encouraging inmates to become involved in prison programs. As such, many states more recently have made participation in programs, such as basic literacy and substance abuse treatment, mandatory.

Finally, parole consideration sets the framework for supervision and treatment needs following release. Parole boards represent a group of experienced professionals considering the inmates' level of risk and the chance for success. To respond to both of these, parole boards create conditions under which parolees must be supervised and attend treatment programs from which they would benefit. Without parole, many states do have a form of mandatory supervision following determinate-sentencing release. However, this supervision is less individualized and based on risk rather than need, setting supervision levels based primarily on offenders' history of criminal behavior.

Currently, many states have opted to abolish parole, and 15 states and the federal government have now ended the use of indeterminate sentencing with release decisions made by a parole board.[2] As well, 20 states have severely limited the parole eligible population. Only 15 states still have full discretionary parole for inmates. As noted previously, in 1977 more than 70% of prisoners were released on discretionary parole. However, by 1997, this had reduced to 28% (Bureau of Justice Statistics, 1997). Twenty-seven states have adopted truth-in-sentencing statutes, under which inmates must serve 85% of their determinate sentence before release. The U.S. Congress encouraged truth in sentencing

(TIS), providing that only states enacting such laws may qualify for federal funds to aid in prison construction. TIS statutes not only eliminate parole but also dramatically reduce the amount of good time that prison officials may grant inmates as incentives for good behavior or program participation (Ditton & Wilson, 1999). Because of these changes, the prison population grew more rapidly than at any other period of time since prisons were first established (Blumstein, Cohen, & Farrington, 1988). From 1980 to 1996, the number of prisoners in state and federal prisons went from 330,000 to 1,054,000, an increase more than threefold (Furniss, 1996), and reached 1.32 million on January 1, 1999 (Camp & Camp, 1999).

For most of the 1990s, community supervision (probation and parole) underwent a transition from helping and counseling offenders to one of risk management and surveillance (Feeley & Simon, 1992). The focus on risk management is accompanied by new allocations of resources toward incarceration, rather than probation and parole, and management of internal system processes. This perspective is referred to as the "new penology" (Feeley & Simon, 1992). Rhine (1997) described this perspective as one in which

> crime is viewed as a systemic phenomenon. Offenders are addressed not as individuals but as aggregate populations. The traditional corrections objectives of rehabilitation and the reduction of offender recidivism give way to the rational and efficient deployment of control strategies for managing (and confining) high-risk criminal populations. Though the new penology refers to any agency within the criminal justice system that has the power to punish, the framework it provides has significant analytic value to probation and parole administrators. (p. 73)

Issues Regarding Prisoner Reentry

As has been well established, there has been a tremendous growth in the prison population in the United States. Almost all the attention is on the number of offenders in prison. Receiving little attention is the fact that the large number of prisoners becomes a large number of releasees. Camp and Camp (1998) reported that 626,973 prison inmates were released from prison during 1998. In New York City alone, the New York State Department of Correctional Services releases approximately 25,000 people a year to the city, and the New York City jails release almost 100,000 (Nelson, Deess, & Allen, 1999). In the state of California, there were 124,697 prisoners leaving prisons after completing their sentences, almost 10 times the number of releases only 20 years earlier (Petersilia, 2000).

When there were only a few hundred thousand prisoners, and a few thousand releasees per year, the number did not seem significant, and the issues surrounding the release of offenders were not overly challenging for communities. However, with the high number of offenders now returning to their communities—many without parole and some with no supervision—there has been a call for academics and correctional administrators to identify the effect of this phenomena on the offenders, their families, and their communities (Petersilia, 1999).

A study by the Vera Institute of Justice in New York City identified many issues that confront inmates released from prison (Nelson et al., 1999). The study included 88 randomly selected inmates released from state prisons in July 1999. Of those selected, 49 (56%) completed the study by allowing interviews to determine their progress and successful transition from prison to the community. Several issues were identified, including finding housing, creating ties with family and friends, finding a job, alcohol and drug abuse, continued involvement in crime, and the effect of parole supervision. It is interesting to note that, even at the point of release, the process had an ominous beginning. The study found that 50 out of the 66 who were interviewed on release reentered the community alone, with no one to meet them as they exited prison or got off the bus in New York City (Nelson et al., 1999).

Most offenders end up living with family or friends until they find a job, can accumulate some money, and then find their own residence. Finding a job is often the most serious concern among ex-inmates, who have few job skills and little work history. Their age at release, their lack of employment at time of arrest, and their history of substance abuse problems make it difficult to find a good job. Many released inmates quickly return to substance abuse. Release is a stressful time, making it even more difficult to avoid a relapse to drug or alcohol abuse. These issues make it difficult for ex-inmates to avoid a return to crime, and it is critical that prisons have programs to prepare inmates for what they will face on release and return to the communities.

Another issue is the effect on social cohesion and community stability by the return of so many ex-inmates. Anderson (1990) identified how the attitudes and behaviors of ex-inmates are transmitted to those in the community on release, concluding that, as issues such as poverty and unemployment persist, the community becomes vulnerable to problems of crime, drugs, family disorganization, and generalized demoralization (Anderson, 1990). In reviewing the effects of imprisonment and the removal of an offender from a Tallahassee, Florida neighborhood, Rose, Clear, and Scully (1999) found an increase in crime precipitating a questioning on the deterrent and rehabilitative effect of prison. They further suggested that returning a large number of parolees released from prison back to the community destabilizes the communities' ability to exert informal control over its members, as there is little opportunity for integration, often resulting in increased isolation, anonymity, and, ultimately, higher crime.

As much of a concern as these issues are in practical, social, and economic terms, there is another dire result. Whether it is a result of tougher parole and release supervision with no tolerance for mistakes or the failure of the system to prepare inmates for release, there are an increasing number of inmates being returned as parole and release violators. During 1998, there were 170,253 reported parole violators from the states, representing more than 23% of new prison admissions (Beck & Mumola, 1999). Even more alarming is that 76.9% of all parole violations were for a technical violation only, without commission of a new felony (Camp & Camp, 1998). There is a trend to violate releasees for minor technical violations, as administrators and parole boards do not want to risk keeping offenders in the community. If these minor violators later commit a serious crime, those deciding to allow them to continue in the community after demonstrating less-than-responsible behavior could face criticism or even legal action. This risk-free approach represents an "invisible policy" not passed by legislatures or formally adopted by correctional agencies. However, these actions have a tremendous effect on prison populations, cost, and community stability.

⊠ A Definition of Prisoner Reentry Programs

This article reports on a review of evaluations of prisoner reentry programs. To analyze evaluations of correctional programs that address prisoner reentry, it was first necessary to develop a definition of prisoner reentry. It can be argued that every prison and even every community correctional program contributes to prisoner reentry and that prisoner reentry begins at the point of admittance to a prison. Reentry should be the focus of classification decisions, prison program participation, and assignment to prison-community transition programs. As well, postrelease community supervision should have a goal of successful reentry, meaning in most cases the offender leads a productive and crime-free life.

However, it would be an inaccurate assessment of prisoner reentry to evaluate every aspect of correctional operations and programs and suggest that the evaluations describe prisoner reentry programs. Therefore, for purposes of this analysis, we created the following two-part definition of prisoner reentry programs as

1. correctional programs (United States and Canada) that focus on the transition from prison to community (prerelease, work release, halfway houses, or specific reentry programs) and

2. programs that have initiated treatment (substance abuse, life skills, education, cognitive/behavioral, sex/violent offender) in a prison setting and have linked with a community program to provide continuity of care.

This definition is appropriate for a review of prisoner reentry for many reasons. First, prisoner reentry programs historically have addressed the difficult transition from prison to community life. Although every program reasonably contributes to the successful return of inmates to society, for purposes of developing policy to improve the reentry process, it should be limited to the prison-community transition. Second, there are some very specific prison programs near the end of a sentence that are designed to aid in the transition to the community. Almost every state and the federal prison system have prerelease programs. Many are only a few hours of orientation by parole or mandatory release supervision officers about supervision conditions and how to make the initial report to the offender's officer on release. Others, however, are very thorough and are excellent

preparation for the challenges that face offenders in the community.

Wilkinson (2001) described one example of prerelease programming that began in 1985 in Ohio. Inmates within the last 6 months of their sentence were transferred to a prerelease center and received extensive programming on basic community skills, such as how to prepare a resume, search for a job, and respond to a job interview. The program also included how to open a bank account and apply for credit and how to find a place to live. Center staff also conducted counseling regarding reuniting with family and friends and what to expect in these tenuous relationships. However, there have been no empirical data available that suggests the program has had an effect on recidivism. Therefore, Ohio is redesigning the centers to ensure there is value added by requiring individual reentry plans be developed for each offender released from prison.

Third, there are community supervision programs that target successful reentry by emphasizing new approaches to individualizing offender management to deal with their risks and needs. Lehman (2001) described Washington State's implementation of the Offender Accountability Act (E2SSB 5421). Washington uses the Level of Service Inventory-Revised (LSI-R), as developed by Andrews and Bonta (1995), to predict chance of recidivism based on offenders' risk and need. Washington also assesses the individual in terms of the nature of potential harm, the effect of relationships (particularly with victims), community risk, and public safety. In this regard, the state not only supervises offenders based on the likelihood to reoffend and the nature of harm but also includes the community (victims, police, and citizens) as partners in managing and mitigating risk.

Finally, there are many programs focused on dealing with a specific issue, such as substance abuse or sex offender treatment. Some of these begin in prison and continue the treatment into the community. An example is the

Federal Prison System Residential Drug Treatment Program. Rhodes et al. (2001) described how this program begins with residential treatment within the prison, and after completion, continues with a 6-month placement in a community halfway house and further follow-up in the community. Programs with a link from prison to community have therefore been included within the definition of prisoner reentry, as they specifically address reentry with the linkage from prison to community, even though the program content does not specifically target reentry.

Using the earlier definition of prisoner reentry, we identified and analyzed several evaluations of correctional programs to identify "what works" in prisoner reentry. The evaluations include published studies from programs in the United States and Canada that evaluate such interventions with adult offender populations. Although there are studies from outside North America and others having to do with juvenile offenders, we limited the definition as noted.

▧ Research Study Method

After determining which studies fall within the reentry definition, a criterion had to be developed to determine if they work or not. Deciding what works for prisoner reentry programs required applying rigorous means for determining which programs have had a demonstrated effect on the recidivism rates of ex-offenders, as well as increased job placement, academic achievement, and remaining drug free. One important criterion was to identify evaluations that provided evidence on the effect of programs on outcome measures. Many evaluations are process evaluations that describe what was done but do not include the effect that the program had on the target population.

Scientific evaluations of program effectiveness have limitations and strengths. The major limitation is that scientific knowledge is provisional because the accuracy of generalizations to all programs drawn from one or even several tests of specific programs is uncertain. The major strength of scientific evaluations is that the rules of science provide a consistent and reasonably objective way to draw conclusions about cause and effect.

Rating Prisoner Reentry Studies

Research methods. To determine whether a program was successful, we used the Maryland Scale of Scientific Methods (MSSM) developed by Sherman et al. (1998) for the National Institute of Justice to identify crime prevention programs that work. This scale ranks each study from 1 (*weakest*) to 5 (*strongest*) on overall internal validity. This scale would not work for secondary reviews or meta-analyses, but an overall study rating based on the following three factors would be sufficient:

- ◆ control of other variables in the analysis that might have been the true causes of any observed connection between the program and an outcome measure,
- ◆ measurement error from such things as participants lost over time or low interview response rates, and
- ◆ statistical power to detect program effects (including sample size, base rate of crime, and other factors affecting the likelihood of the study detecting a true difference not due to chance).

Generally, the MSSM applies across all settings and includes these core criteria, which define the five levels of the MSSM. The following list represents the levels used by the MSSM to categorize evaluative studies by the rigor or their scientific method. There is an assumption of employing univariate and multivariate statistics when considering Level 2 through Level 5 categories.

- ◆ Level 1: correlation between a type or level of reentry program (intervention, i.e., substance abuse treatment, violent or sex offender treatment, vocational training, work release, life skills) and an outcome measure at a single point in time (recidivism, return to custody, employment rate, drug use, academic achievement).

♦ Level 2: temporal sequence between the program (intervention) and outcome measure clearly observed or the presence of a comparison group without demonstrated comparability to the treatment group.

♦ Level 3: comparison between two or more comparable units of analysis, one with and one without the program.

♦ Level 4: comparison between multiple units with and without the program, controlling for other factors, or using comparison units that evidence only minor differences.

♦ Level 5: random assignment and analysis of comparable units to program and comparison groups.

Threats to internal validity. Sherman et al. (1998) identified the rigor of the evaluation by examining the research design and the threats to internal validity. Sherman et al. (1998) stated,

Each higher level of the scale from weakest to strongest removes more of these threats to validity, with the highest level on the scale generally controlling all four of them and the lowest level suffering all four. The progressive removal of such threats to demonstrating the casual link between the program effect and recidivism is the logical bias for the increasing confidence scientists put into studies with fewer threats to internal validity. (p. 5)

Description of Studies

There were 32 studies identified that fit the definition of prisoner reentry. Each study was placed into an MSSM level, and evaluations of similar programs were grouped into (a) vocational training and work, (b) drug rehabilitation, (c) educational programs, (d) sex/violent offender programs, (e) halfway house programs, and (f) prison prerelease programs. The Appendices A through F present the studies and the MSSM level assigned.

◙ Results of the Review

The next critical question after a review of the studies is to decide what works. For guidance, we again used the framework used by Sherman et al. (1998) in their evaluation of whether crime prevention programs effectively reduced crime. These authors asked the question, "How high should the threshold of scientific evidence be for answering the congressional question about program effectiveness?" They developed the following criteria to determine whether a crime prevention program was effective or ineffective.

What Works

For a program to be considered "working," there must be at least two Level 3 evaluations with significance tests indicating that the intervention was effective, and the preponderance of the remaining evidence must support that conclusion.

What Does Not Work

For a program to be coded as "not working," there must be at least two Level 3 evaluations with statistical significance indicating the ineffectiveness of the program, and the preponderance of the remaining evidence must support the same conclusion.

What Is Promising

These are programs for which the level of certainty from available evidence is too low to support generalizable conclusions. However, there is some empirical basis for predicting that further research could support such conclusions, such as programs are found effective in at least one Level 3 evaluation, and the preponderance of the remaining evidence supports that conclusion.

What Is Unknown

Any program not classified in one of the three previous categories is defined as having unknown effects.

⊠ Results of the Analysis

The following represents a summary of the findings of prisoner reentry studies identified within the various reentry categories, using the Sherman methodology and the MSSM criteria to determine effectiveness. For at least all Level 4 and 5 studies, a brief description of each program's design and outcome measures of effectiveness are included. In some instances, Level 3 studies are described when no Level 4 or 5 studies are available.

Vocational and Work Programs

Seven programs were evaluated in this area that included two Level 4 studies (Saylor & Gaes, 1992, 1997) and one Level 5 study (Turner & Petersilia, 1996). The Turner and Petersilia (1996) experiment implemented random assignment to treatment and control groups that allows for greater confidence in asserting that observed differences result from participating in work release rather than from preexisting background differences. The study compared recidivism of 218 offenders in Seattle, Washington, one half of

whom participated in a work release program and one half of whom completed their sentences in prison. Generally, the program achieved its primary goal of preparing inmates for final release and facilitating their adjustment to the community. The offenders who participated in work release were somewhat less likely to be rearrested; however the results were not statistically significant.

The Saylor and Gaes (1992, 1997) studies evaluated the Post-Release Employment Project during a 4-year period. Data were collected on more than 7,000 federal offenders, comparing those participating in training and work programs with similar offenders who did not take part and with a baseline group of all other inmates. The longitudinal results demonstrated significant and substantive training effects on both in-prison (misconduct reports) and postprison (employment and arrest rates) outcome measures.

We can conclude from the results of the three previous studies that vocational training and/or work release programs are effective in reducing recidivism as well as in improving job readiness skills for ex-offenders. There were also three Level 2 studies (Finn, 1999) and one Level 1 study (Finn, 1999) that could have added increased promise for vocational work programs if a predesign and postdesign and comparison control groups were implemented.

Drug Rehabilitation

Twelve programs were evaluated in this area. There is one Level 5 study (Rhodes et al. 2001) and eight Level 4 studies. Three studies evaluated the same prison-based treatment assessment (PTA) program over time (Hiller, Knight, & Simpson, 1999; Knight, Simpson, Chatham, & Camacho, 1997; Knight, Simpson, & Hiller, 1999), and four other studies evaluated the Key-Crest program over time (Butzin, Scarpetti, Nielsen, Martin, & Inciardi, 1999; Inciardi, Martin, Butzin, Hooper, & Harrison,

1997; Martin, Butzin, & Inciardi, 1995; Martin, Butzin, Saum, & Inciardi, 1999). The other Level 4 program, Stay'n Out, was evaluated by Wexler, Falkin, and Lipton (1990).

The Rhodes et al. (2001) study examined 2,315 federal inmates: 1,193 treatment individuals, 592 comparison participants, and 530 control participants. A quasi-experimental design was implemented to test for treatment effectiveness; however, three different statistical approaches were used to minimize selection bias as an explanation for treatment outcomes. The two outcome variables measured were recidivism rates and rates of relapse to drug use. In general, for recidivism and relapse to drug use, drug treatment is statistically significant in reducing both outcomes for men but not for women.

The in-prison therapeutic communities (TCs) evaluated by Knight and colleagues (1997, 1999) show effectiveness of intensive treatment when integrated with aftercare, with benefits most apparent for offenders with serious crime and drug-related problems. The earliest study demonstrates that 80% of the 222 offenders who took part in the TC graduated and had marked reductions in their criminal and drug use activity from the 6 months before entering prison to the 6 months after leaving prison. Those who completed the first phase of their aftercare program had lower relapse and recidivism rates than did the parolees in the comparison sample (Knight et al., 1997). A 3-year follow-up study, based on 291 follow-up eligible parolees, showed that those who completed the TC program and aftercare are the least likely to be reincarcerated (25%), as compared to 64% of aftercare dropouts and 42% of untreated comparison groups (Knight et al., 1999). Another study of 293 treated inmates and 103 untreated inmates showed that in-prison TC programs—especially when followed by residential aftercare—reduce the likelihood of postrelease rearrest by 12% (Hiller et al., 1999).

The Key-Crest in-prison TC and work release program evaluated by Inciardi and colleagues (1997) demonstrated marked success in its 6-month and 3-year follow-ups. The Key is a prison-based TC whereas the Crest Outreach Center is a 6-month residential, community-based, work release treatment and aftercare program located in Delaware. Together these two programs formed Key-Crest, which allowed for three stages of treatment for seriously drug-involved offenders: prison, work release, and parole or other form of community supervision.

In the first evaluation (Martin et al., 1995), baseline data at release from prison and outcome data 6 months after release were analyzed for 457 offenders. Four different groups of offenders were evaluated. The first group consisted of offenders who participated in neither of the TCs and was compared to groups that either participated in the TC in prison only, the transitional TC only, or both the TCs. The latter two groups had significantly lower rates of drug relapse and criminal recidivism when adjusted for other risk factors. Eighteen-month follow-up data also indicated that the participants in the two- and three-stage models had significantly lower rates of drug relapse and criminal recidivism (Inciardi et al., 1997).

A third evaluation compared participants in only the Crest Outreach Center ($n = 334$) to a group of drug-involved inmates who entered a traditional work release program ($n = 250$) (Butzin et al., 1999). Results showed that compared with the noncompleters ($n = 122$), completers ($n = 212$) are less likely to be incarcerated at 18 months and more likely to be employed. When comparing completers to those not exposed to the program, not only are completers less likely to be incarcerated and more likely to be employed but also those completers who are unemployed used fewer drugs, less frequently than the unemployed comparison group. This suggests that exposure to a TC work release environment can moderate expected negative effects (drug use) of unemployment.

The final evaluation examined the success of the TC outcomes when the time at risk was moved to 3 years after release (Martin et al.,

1999), program effects declined; however, effects remained significant when program participation, completion, and aftercare were taken into account. Clients who completed secondary treatment ($n = 101$) did better than those with no treatment ($n = 210$) and those who dropped out ($n = 109$). Clients who received aftercare ($n = 69$) did even better in remaining both drug free and arrest free. The authors concluded that the TC continuum has value in work release and parole settings and that retention in treatment is important in predicting long-term success in reducing the likelihood of recidivism.

Wexler et al. (1990) performed an evaluation of New York City's Stay'n Out TC that is based on more than 1,500 participants. The quasi-experimental design compares the program participants ($n = 682$) with inmates who volunteered for the program but never participated ($n = 197$) and inmates who participated in other types of in-prison drug abuse treatment programs in different prisons ($n = 947$). Results showed that after 3 years at risk, those who completed the TC program had a significantly lower arrest rate (26.9%) than those who had different drug treatment (34.6%, 39.8%) and those who received no treatment (40.9%). In general, the TC was effective in reducing recidivism, and this positive effect increased as time in program increased but tapered off after 12 months. This can be explained by the fact that when 12 months have passed and the offender is repeatedly denied parole, the client is frustrated and slowly reduces his involvement in the TC. Accordingly, 9 to 12 months is the optimal treatment duration for success in the TC program.

In addition, there was one Level 3 study (Hartmann, Wolk, Johnston, & Coyler, 1997) and two Level 2 studies (Field, 1985; Knight & Hiller, 1997) that contributed to the success of drug treatment programs. However, potential selection bias with respect to program completion and participation in aftercare cannot be completely ruled out. Overall, drug rehabilitation programs represent the strongest area of quasi-experimental and experimental design for prisoner reentry programs. In most of the evaluations, threats to internal validity were controlled for as a function either of the design or with statistical methods. From the evidence presented here, it can be concluded that drug treatment programs do work in easing the transition from prison to the community.

Education Programs

Only two education programs identified were within the definition of prisoner reentry, yet both implemented a quasi-experimental design to help control for threats to internal validity. The evaluations measured rearrest and return to custody rates, increases in academic achievement after program graduation, and time the offender was exposed to educational services.

There are mixed results in this area of reentry programs. Vito and Tewksbury (1999) evaluated the Learning, Instruction, and Training = Employment (LITE) program in Kentucky, which is aimed at increasing the literacy levels of state and local inmates and reducing recidivism. Out of 662 inmates who were tested for program entry, 105 inmates participated in and completed the program. The results showed that during a 6-week time period, graduates increased their reading and math competencies up to three levels: However, the educational component did not seem to have an effect on their recidivism rates when compared to nongraduates (Vito & Tewksbury, 1999). Recidivism was measured 12 to 15 months after program involvement. It should be noted that the employment component of the program was never fully implemented and may have had an effect on recidivism.

Adams et al. (1994) studied prison behavior and postrelease recidivism of more than 14,000 Texas inmates who were released between March 1991 and December 1992. Some of the inmates participated in prison education programs (treatment group), whereas others did not participate (control group). The cohort was assessed on release and followed-up after 14 to 36 months, depending on their release date. The results of the study showed increases in academic achievement, but recidivism rates were only affected if the offender participated in 200 or more hours of educational programs (Adams et al., 1994). The baseline level of academic achievement of the offender affects this outcome, in that only the offenders with the lowest levels of academic achievement have a decreased likelihood of recidivism with 200 or more hours of educational programs.

The programs were evaluated as a Level 4 and a Level 3, respectively, with selection bias and comparability of groups as the threats to internal validity. From the evidence presented here, we can state that education programs increase educational achievement scores but do not decrease recidivism. Educational reentry programs that link prison programs to community-based resources after release are needed, and the programs that do exist are promising at best.

Sex Offenders and Violent Offenders

Five programs were evaluated in this area—one with a Level 4 rating, one with a Level 3 rating, and three with a Level 2 rating. These studies measured recidivism, level of risk of recidivism, and time at risk of recidivism. Each of these studies present alternative findings. The Level 4 study by Robinson (1996) randomly assigned 2,125 offenders either to a cognitive skills training program or to a control group. All offenders were subject to at least 12 months follow-up after release. The study indicated that the completion of cognitive behavioral therapy does reduce the offenders' return-to-custody rate by 11%, as compared to offenders who did not complete the therapy. This study also noted that therapy is most effective for offenders with a moderate level of risk of recidivism, as compared to a high level (Robinson, 1996).

The Level 3 study by Barbaree, Seto, and Maric (1996) assessed violent sex offenders' risk of recidivism and suggested treatment alternatives. Of the original 250 offenders, 193 completed treatment and were tracked on release. In general, the results of the program do not indicate that a significant difference exists between recidivism rates of offenders who completed treatment (18%) and those who refused treatment (20%). Yet this study indicated that the treatment refusers were at risk of recidivism significantly less time than the treatment completers; therefore, the refusers had a higher failure rate (38.9%) than the treatment completers (22.2%) when a comparable follow-up period was used (Barbaree et al., 1996).

There were three Level 2 studies that simply measured recidivism rates of violent and sexual offenders. However, because the integrity of the internal validity of these studies is weak, it is difficult to make a decision as to whether this treatment is effective. This area is one of the fastest growing in-prison reentry programs, and additional Level 3 and Level 4 evaluations need to be performed.

Halfway House Programs

Four halfway house programs met our criteria for prisoner reentry. There was one Level 4 study, one Level 3 study, and two Level 2 studies. The Level 4 study was an evaluation of Ohio halfway houses, comparing 236 house clients to a 404-parolee comparison group, with statistical controls for selection bias (Seiter, 1975). The study examined outcome in terms of the frequency and severity of criminal offenses by both groups but also using a score of relative adjustment, which was a measure of positive activities, such as finding and holding a job, being self-supporting, and participating in self-improvement programs. The halfway house group performed better on the positive activities than the comparison group but not at a statistically significant level. However, the halfway house group did commit fewer and less severe offenses (a statistically significant level) during a 1-year outcome analysis than the comparison group.

The Level 3 study was an evaluation of a California halfway house for women. Results indicated that the average number of crimes in the treatment group ($n = 60$) was one half that of the control group ($n = 134$) (Dowell, Klein, & Krichmar, 1985). In addition, the severity of the crimes committed by the treatment group was less than two thirds of the control group. The Level 2 studies look at success rates of participants living in Ohio and Colorado halfway houses (Donnelly & Forschner, 1984; Department of Criminal Justice, 2001). The results from the two Level 2 studies were consistent with the findings from the other two halfway house evaluations. From the evidence presented here, it can be concluded that halfway house programs do work in easing the transition from prison to the community.

Prison Prerelease Programs

We were only able to find two prerelease programs that met the evaluation criteria. The PreStart program in Illinois was labeled as a Level 3. This statewide program was very inclusive in its efforts to prepare ex-offenders for life in the community through a two-phase system: prerelease education and postrelease assistance. The rearrest rates within 1 year of release were 40%, as compared to 48% of the comparison group (Castellano et al., 1994). Return-to-prison rates showed an even greater success of 12% for the treatment group and 32% for the comparison group. Of course, there are some limitations to these findings, because randomization was not possible, and the comparison group was a sample of inmates released from similar facilities 2 years earlier. Because selection bias and chance factors pose threats to the internal validity of the results, this program only shows promise as a model for other states to use for prerelease programs.

In a Level 4 study by LeClair and Guarino-Ghezzi (1991), the researchers drew five separate study samples, one that consisted of all men released from Massachusetts Department of Corrections (DOC) facilities in 1974 ($n = 840$) to test the effect of prerelease participation on recidivism rates. The subsample consisted of 212 inmates who completed the prerelease program in 1974 and were tracked for 12 months from the date of each individual's release. Recidivism rates were compared to those of other releasees who had not participated in the program ($n = 629$). The researchers used a predictive attribute analysis to calculate base expectancy prediction tables to test for any nonrandom selection effects. Results showed that the expected recidivism rate for the 212 inmates who participated in prerelease programs was 21.1%. However, the postdischarge behavior only showed that 11.8% of the offenders recidivated. This difference is not significant, but it does indicate an intervention effect. When compared to recidivism rates of offenders who did not participate in prerelease programs (29%), there is support that the prerelease intervention is effective. In combination, these programs demonstrate that prerelease centers and programs can be effective in reducing recidivism rates of ex-offenders

⊠ Conclusions and Recommendations for Further Research

It is encouraging to note the positive results of many prisoner reentry programs as identified in this review. Results indicate a positive result for vocational training and/or work release programs (found to be effective in reducing recidivism rates as well as in improving job readiness skills for ex-offenders), for drug rehabilitation (graduates of treatment programs were less likely than other parolees and noncompleters to have been arrested, commit a drug-related offense, continue drug use, or have a parole violation), to some extent for education programs (only to increase educational achievement scores, but not to decrease recidivism), for halfway house programs (found effective in reducing the frequency and severity of future crimes), and for prerelease programs (effective in reducing recidivism rates of ex-offenders). In addition, there are promising results for sex- and violent-offender programs. One general point that needs to be made regarding prison reentry programs is that to fully determine what types of programs work to assist in the success of offenders in the community, there is a need to evaluate additional reentry programs currently in operation.

Prisoner reentry is a problem for many reasons. First, the number and makeup of prisoners released has increased and changed considerably during the past 2 decades. Second, the communities to which offenders return are less stable and less able to provide social services and support to these large numbers of returning prisoners. Third, there is less availability of prison rehabilitative programs to meet inmate needs. Fourth, the focus on supervision and monitoring rather than casework and support by parole and release officers of prisoners reentering society has confounded the problem of lack of programs. Last, there are large numbers of released prisoners failing in the community and being returned to prison, with more than three fourths of those returned for technical violations rather than the commission of new crimes.

Even with the problems noted, this analysis of prisoner reentry programs has identified several categories of programs in which there is evidence of success. Correctional administrators should take note of these programs; implement or expand the use of vocational training and/or work release programs, drug rehabilitation programs, education programs, halfway house programs, and prerelease programs that have proven success; and expand the use of sex- and violent-offender programs that show promise. These programs can be expanded significantly with only a small portion of funding that is currently used for imprisoning offenders. These programs should be further examined as they are expanded.

Research should also be conducted regarding the supervision styles of parole officers (surveillance vs. casework) to determine the effect on failure in the community. There should be an examination of the role of the community and the community's ability to respond to the number and needs of returning ex-offenders. There should be an examination of the causes of the increasing number of ex-inmates returned to prison for minor crimes or only technical violations of their release conditions. Last, there should be additional evaluations of those programs that show promise yet cannot at this time be concluded to improve the likelihood of the success of reentering prisoners.

The nation has invested billions of dollars into locking up offenders. The policies around reentry have become increasingly an avoidance of risk. As a result, we have created a revolving door of offenders who will be committed to prison time and again as they fail in the community. This is not only a failure of the inmate, it is a failure of our release and reentry policies. As this analysis pointed out, we do know that certain programs can improve prisoner reentry and reduce the revolving-door syndrome. With billions of dollars focused on imprisonment, it is only fitting that a few million more be focused on prisoners' return to the community.

Notes

1. For a review of the "nothing works" conclusion see Lipton, Martinson, and Wilks (1975).

2. These states abolishing discretionary parole release include Arizona (1994), Delaware (1990), Illinois (1978), Indiana (1977), Kansas (1993), Maine (1975), Minnesota (1980), Mississippi (1995), New Mexico (1979), North Carolina (1994), Ohio (1996), Oregon (1989), Virginia (1995), Washington (1984), and Wisconsin (1999).

References

Adams, K., Bennet, K. J., Flanagan, T. J., Marquart, J. W., Cuvelier, S. J., Fritsch, E., Gerber, J., Longmire, D. R., & Burton, V. S., Jr. (1994, December). Large-scale multidimensional test of the effect of prison education programs on offenders' behavior. *The Prison Journal, 74*(4), 433–449.

Anderson, E. (1990). *Streetwise: Race, class, and change in an urban community.* Chicago: University of Chicago Press.

Andrews, D. A., & Bonta, J., (1995). *The level of service inventory-revised (LSI-R) manual.* Toronto, Ontario: Multi-Health Systems.

Barbaree, H. E., Seto, M. T., & Maric, A. (1996). Effective sex offender treatment: The Warkworth Sexual Behavior Clinic. *Forum on Corrections Research, 8*(3), 13–15.

Beck, A., & Mumola, C. (1999). *Prisoners in 1998.* Washington, DC: U.S. Department of Justice, Bureau of Justice Statistics.

Blumstein, A., Cohen, J., & Farrington, D. P. (1988, February). Longitudinal and criminal career research: Further clarification. *Criminology, 26*(1), 57–74.

Bureau of Justice Statistics. (1977). *National prisoner statistics.* Washington, DC: U.S. Department of Justice.

Bureau of Justice Statistics. (1997). *National prisoner statistics.* Washington, DC: U.S. Department of Justice.

Butzin, C. A., Scarpetti, F. R., Nielsen, A. L., Martin, S. S., & Inciardi, J. A. (1999). Measuring the impact of drug treatment: Beyond relapse and recidivism. *Corrections Management Quarterly, 3*(4), 1–7.

Camp, C. G., & Camp, G. M. (1998). *The corrections yearbook, 1998.* Middletown, CT: Criminal Justice Institute.

Camp, C. G., & Camp, G. M. (1999). *The corrections yearbook, 1999: Adult corrections.* Middletown, CT: The Criminal Justice Institute.

Castellano, T. C., Cowles, E. L. McDermott, J. M., Cowles, E. B., Espie, N., Ringel, C., et al. (1994). *Implementation and impact of Illinois' Prestart Program: A final report.* Carbondale: Southern Illinois University, Center for the Study of Crime, Delinquency, and Corrections.

Clear, T., & Cole, G. (1997). *American corrections.* Belmont, CA: Wadsworth.

Department of Criminal Justice, Office of Research and Statistics. (2001). *2000 community corrections study.* Manuscript in preparation.

Ditton, P., & Wilson, D. J. (1999). *Truth in sentencing in state prisons.* Washington, DC: U.S. Department of Justice, Bureau of Justice Statistics.

Donnelly, P. G., & Forschner, B. (1984). Client success or failure in a halfway house. *Federal Probation, 48,* 38–44.

Dowell, D. A., Klein, C., & Krichmar, C. (1985). Evaluation of a halfway house. *Journal of Criminal Justice, 13,* 217–226.

Feeley, M. M., & Simon, J. (1992). The new penology: Notes on the emerging strategy of corrections and its implications. *Criminology, 30*(4), 449–479.

Field, G. (1985). The Cornerstone Program: A client outcome study. *Federal Probation, 49,* 51–56.

Finn, P. (1999, July). Job placement for offenders: A promising approach to reducing recidivism and correctional costs. *National Institute of Justice Journal,* pp.1–35.

Fogel, D. (1975). *We are the living proof: The justice model for corrections.* Cincinnati, OH: Anderson.

Furniss, J. (1996). The population boom. *Corrections Today, 58*(1), 38–43.

Gaes, G. G., Flanagan, T. J., Motiuk, L. L., & Stewart, L. (1999). Adult correctional treatment. In M. Tonry & J. Petersilia (Eds.), *Prisons* (pp. 361–426). Chicago: University of Chicago Press.

Gordon, A., & Nicholaichuk, T. (1996). Applying the risk principal to sex offender treatment. *Forum on Corrections Research, 8*(2), 36–38.

Halleck, S. (1971). *The politics of therapy.* New York: Science House.

Hartmann, D. J., Wolk, J. L., Johnston, J. S., & Coyler, C. J. (1997). Recidivism and substance abuse outcomes in a prison-based therapeutic community. *Federal Probation, 61*(4), 19–25.

Hiller, M. L., Knight, K., & Simpson, D. D. (1999). Prison-based substance abuse treatment, residential aftercare and recidivism. *Addiction, 94*(6), 833–842.

Inciardi, J. A., Martin, S. S., Butzin, C. A., Hooper, R. M., & Harrison, L. D. (1997). An effective model of prison-based treatment for drug-involved offenders. *Journal of Drug Issues, 27,* 261–278.

Knight, K., & Hiller, M. (1997). Community-based substance abuse treatment: A 1-year outcome evaluation of the Dallas County Judicial Treatment Center. *Federal Probation, 61*(2), 61–68.

Knight, K., Simpson, D. D., Chatham, L. R., & Camacho, L. M. (1997). An assessment of prison-based drug treatment: Texas' in-prison therapeutic community program. *Journal of Offender Rehabilitation, 24*(3/4), 75–100.

Knight, K., Simpson, D. D., & Hiller, M. L. (1999). Three-year reincarceration outcomes for in-prison therapeutic community treatment in Texas. *The Prison Journal, 79*(2), 337–351.

LeClair, D. P., & Guarino-Ghezzi, S. (1991). Does incapacitation guarantee public safety? Lessons from Massachusetts furlough and prerelease programs. *Justice Quarterly, 8*(1), 1–40.

Lehman, J. (2001). Reinvesting community corrections in Washington state. *Correction Management Quarterly, 5*(3), 41–45.

Lipton, D., Martinson, R., & Wilks, J. (1975). *The effectiveness of correctional treatment.* New York: Praeger.

Martin, S.S., Butzin, C. A., & Inciardi, J. A. (1995). The assessment of multi-stage therapeutic community for drug-involved offenders. *Journal of Psychoactive Drugs, 27,* 109–116.

Martin, S. S., Butzin, C. A., Saum, C. A., & Inciardi, J. A. (1999). Three-year outcomes of therapeutic community treatment for drug-involved offenders in Delaware: From prison to work release to aftercare. *The Prison Journal, 79*(3), 294–320.

Morris, N. (1974). *The future of imprisonment.* Chicago: University of Chicago Press.

Motiuk, L., Smiley, C., & Blanchette, K. (1996). Intensive programming for violent offenders: A comparative investigation. *Forum on Corrections Research, 8*(3), 10–12.

Nelson, M., Deess, P., & Allen, C. (1999). *The first month out: Post-incarceration experiences in New York City.* Unpublished monograph. New York: The Vera Institute.

Petersilia, J. R. (1999). Parole and prisoner reentry in the United States. In M. Tonry & J. Petersilia, (Eds.), *Prisons* (pp. 479–529). Chicago: University of Chicago Press.

Petersilia, J. R. (2000). The collateral consequences of prisoner reentry in California: Effects on children, public health, and community. Unpublished monograph. Irvine: University of California.

Rhine, E. E. (1997). Probation and parole supervision: In need of a new narrative. *Corrections Quarterly, 1*(2), 71–75.

Rhodes, W., Pelisser, B., Gaes, G., Saylor, W., Camp, S., & Wallace, S. (2001). Alternative solutions to the problem of selection bias in an analysis of federal residential drug treatment programs. *Evaluation Review, 25,* 19–45.

Robinson, D. (1996). Factors influencing the effectiveness of cognitive skills training. *Forum on Corrections Research, 8*(3), 6–9.

Rose, D. R., Clear, T., & Scully, K. (1999, November 8). *Coercive mobility and crime: Incarceration and social disorganization.* Paper presented at the American Society of Criminology meetings, Toronto, Ontario.

Saylor, W. G., & Gaes, G. G. (1992). The post-release employment project: Prison work has measurable effects on post-release success. *Federal Prisons Journal, 2*(4), 33–36.

Saylor, W. G., & Gaes, G. G. (1997). Training inmates through industrial work participation and vocational apprenticeship instruction. *Corrections Management Quarterly, 1*(2), 32–43.

Seiter, R. P. (1975). *Evaluation research as a feedback mechanism for criminal justice policy making: A critical analysis.* Unpublished dissertation. Columbus: Ohio State University.

Sherman, L. W., Gottfredson, D. C., MacKenzie, D. L., Eck, J., Reuter, P., & Bushway, S. D. (1998). *Preventing crime: What works, what doesn't, what's promising* [Monograph]. Washington, DC: U.S. Department of Justice, National Institute of Justice.

Studer, L. H., Reddon, J. R., Roper, V., & Estrada, L. (1996). Phoenix: An in-hospital treatment program for sex offenders. *Journal of Offender Rehabilitation, 23*(1/2), 91–97.

Turner, S., & Petersilia, J. (1996). Work release in Washington: Effects on recidivism and corrections costs. *The Prison Journal, 76*(2), 138–164.

Vito, G. F., & Tewksbury, R. (1999). Improving the educational skills of inmates: The results of an impact evaluation. *Corrections Compendium, 24*(10), 46–51.

Wexler, H. K., Falkin, G. P., & Lipton, D. S. (1990). Outcome evaluation of a prison therapeutic community for substance abuser treatment. *Criminal Justice and Behavior, 17,* 71–92.

Wilkinson, R. A. (2001). Offender reentry: A storm overdue. *Corrections Management Quarterly, 5*(3), 46–51.

DISCUSSION QUESTIONS

1. What are the primary obstacles facing a prisoner returning to his or her community?

2. Should all people released from prison have a period of supervision in the community regardless of whether or not they have "maxed out"?

3. Summarize the reentry strategies that have worked, have not worked, and are promising.

READING

Nancy Marion's article seeks to explore the truth of certain claims for the superiority of community-based over institutional-based corrections. Community-based alternatives to prison claim to be more effective in reducing recidivism than are traditional prisons, to be cheaper than prisons, and to reduce overcrowding in prisons and jails. Marion's study uses a case study of a community-based program in the U.S. Midwest to determine if those community corrections alternatives achieve the results claimed for them. The findings show that the recidivism rates of community corrections are lower than those of the prison inmates only in some cases and that the costs are cheaper only in some cases. The findings also show that community corrections serves as a true alternative to prison in some instances but more often only widens the net and increases the state's control over criminal offenders.

Effectiveness of Community-Based Correctional Programs

A Case Study

Nancy Marion

Alternatives to traditional prison and jail confinement have become popular in recent years as officials attempt to deal with overcrowded correctional facilities and the ever-increasing costs of imprisoning offenders. Alternative sanctions have also been lauded for having lower recidivism rates than traditional prison settings. However, some research has shown that the presumed benefits of community corrections may not exist. For example, Jackson, de Keijser, and Michon (1995) noted that alternatives to custody "may not be so cheap, may not be so effective in reducing recidivism, and they may not always constitute real alternatives to prison" (p. 44). In writing that, Jackson et al. recognized three primary goals underlying the majority of community-based correctional programs: to

protect the public's interest (i.e., reduce recidivism and thereby protect the public), to save fiscal resources of the community, and to reduce overcrowding in jails and prisons. This analysis is an attempt to determine if community corrections achieve these goals, using a detailed case study of five community corrections programs.

◼ Effectiveness (Reducing Recidivism)

One question frequently asked of community sanctions is, Do they work? It is a difficult question to answer. Most community corrections are based on the proposition that the programs can be more effective than traditional

SOURCE: Marion, N. (2002). Effectiveness of community-based correctional programs: A case study. *The Prison Journal, 82*(4), 478–497. Reprinted with permission of Sage Publications, Inc.

corrections settings because they help the offender reintegrate into society and establish a legitimate role in the community (Lawrence, 1991). But missing from the literature is compelling evidence that sanctions decrease recidivism (Jackson et al., 1995). Many researchers find that alternative sanctions are not necessarily more effective than traditional prisons. In 1977, Pease, Billingham, and Earnshaw found that 44% of a community service group were reconvicted within 1 year of imposition of the sentence, whereas only 35% of a custodial group were reconvicted. Jones (1991) found that "more community corrections than prison clients reoffended during the follow-up period" (p. 61). Wright (1994) also found that prison was the more effective method for reducing recidivism.

At the same time, however, other research shows that alternative sanctions are more effective than a prison setting. Bol and Overwater compared persons sentenced to a community service group versus those sentenced to a prison term and found that 42% of the community service group were reconvicted, whereas 54% of the prison group were reconvicted (cited in Jackson et al., 1995). Langworthy and Latessa (1993, 1996), in both their original study of a drunk-driving program and a 4-year follow-up study, found that one program geared toward reducing recidivist behavior of alcohol-related offenses was successful in reducing rates of recidivism to below those who were sentenced to prison. Vito and Tewksbury (1998) showed that the recidivism rates for the graduates of a drug court program are remarkably lower than a comparison group who were not part of a drug court program. They concluded that drug treatment programs can effectively reduce recidivism rates. Other research supports the idea that alternative sanctions are more effective than prisons in reducing future criminal behavior by offenders (see Gendreau & Ross, 1979; Izzo & Ross, 1990; Palmer, 1992).

Yet other research findings are mixed. A study by Latessa, Travis, Holsing, Turner, and Hartman (1997) showed that rearrest rates for

community programs ranged from 34% to 63.5%, whereas the rearrest rate for prison inmates was 59%. In addition, the reincarceration rates for the community facilities ranged from 22.5% to 37%, whereas the reincarceration rate for prison inmates was 27%. Langan (1998) compared recidivism rates of offenders who received different sanction modes to those who did not receive the sanction. He found that during a 3-year follow-up period, the recidivism rates were lower for clients in the day-reporting and community-service programs, whereas the intensive-supervision clients had increased arrest rates.

Still other research concluded that there was no difference between the alternative settings and the prisons. Martinson, Lipton, and Wilks (1975); Kuehlhorn (1979); and Whitehead and Lab (1989) each found that correctional treatment generally, and probation supervision in particular, has no impact on recidivism rates. Still others found that the two methods were equivalent (Palmer, 1991). Jones (1991) found that reoffending rates were similar for both prison and community corrections clients and that placement in community corrections rather than prison had little impact on overall propensity to reoffend.

On the whole, these studies provide contradictory evidence about the effectiveness of community sanctions in reducing recidivism of inmates when compared to those of traditional prison settings. The previous work only serves to raise more questions about the effectiveness of community options. The current study will attempt to address this issue.

Cost

The second issue surrounding community corrections that needs to be addressed is cost. For many years, community corrections has claimed to be a cheaper version of the prison cell (Barajas, 1993; Larivee, 1993). But are community corrections actually cheaper than a traditional prison? Some research shows that a community can achieve crime reduction through community corrections programs at considerably less cost (Lawrence, 1991). One early study of the cost benefits of community corrections examined probation in two counties in California. The report showed that probation supervised two thirds of all correctional clients yet received only about one fourth of the financial resources allocated to corrections (Petersilia, Turner, Kahan, & Peterson, 1985). Another study completed by the International Association of Residential and Community Alternatives (IARCA) found that, across the nation, $7.4 million was saved by using community corrections programs and diverting offenders from prisons (Huskey, 1992). Finally, Buddress (1997) analyzed federal probation and pretrial services and found them to be cost-effective alternatives to incarceration. She cited a report by the Administrative Office of the U.S. Courts in which supervision costs for an offender under probation were reported to be $2,344 a year, compared to the costs of incarcerating an individual in a correctional facility, which was $21,352 per year (plus an additional $3,431 for health care expenses).

Cost savings can be measured in other ways as well. Effective anticrime programs, such as drug treatment, can benefit the state by reducing crime. Petersilia (1995) found that a drug treatment program saved money by preventing crime (p. 76). In addition, costs can be reduced through a lessened demand for prison construction (Jones, 1991; Petersilia, 1998). Another benefit of many community corrections programs is that the client is required to pay either the entire cost of a treatment program, partial costs, or even fees, which lessens the financial burden on the state (Lucken, 1997).

According to these studies, community corrections are a cost-effective way to provide alternatives to incarceration. But the intensive treatment programs and tightened control over the inmates found in alternative placements tend to increase the costs of community corrections when compared to traditional institutional placements. If the custodial sanctions being used by communities across the country are not cost effective, then they should be brought into question. If community corrections can be shown to be at least as effective as prisons but cheaper, then such strategies should be adopted (Huskey, 1992; Jones, 1991). As Wasson (1993, p. 110) stated, "All else being equal, we should pursue choices that are least expensive for the citizen." All in all, the cost issue is a concern that remains viable. This study will address the issue of costs and attempt to determine if community corrections are a cheaper alternative to the prison.

✎ Overcrowding

Overcrowding in correctional institutions has been a persistent problem across the nation for many years. There has been a quadrupling in the number of state and federal prisoners between 1975 (240,593) and midyear 1994 (1,012,851) (Tonry, 1998). In many places, community corrections were developed with the primary motive of relieving a state's prison overcrowding problem by offering a way to reduce commitments to state prisons, particularly for first-time, nonviolent offenders (Jones, 1991; Lawrence, 1991).

By using community corrections, those inmates who do not need the high level of control found in a prison are still under the supervision of the state. However, community corrections have also been accused of widening the net and being a substitute for prison. In other words, offenders are still incarcerated but in an environment other than a penal institution. Those offenders who would previously not have been given a sentence of incarceration are brought under the control of the state. Although not everyone agrees with the arguments surrounding the concept that community corrections programs widen the net of control (McMahon, 1990), the argument persists that community corrections serve to increase the number of inmates under state control. Do community facilities actually help to remove offenders from jails and prisons, thereby reducing the overcrowding problem? This issue will be addressed in this study as well.

⬚ Current Study

Despite all of the previous work that examined the issues described above, there are many remaining questions concerning community corrections. Renzema (1992) noted that research on some prison alternatives is "uninterpretable" because of weak research designs. In addition, Tonry (1998) wrote that there have been no substantial evaluations of selected community corrections programs. Furthermore, Palmer (1994) noted that there is a lack of "good to excellent" studies on prison alternatives. He noted that studies that analyze the effectiveness (e.g., the recidivism) on clients during a given postprogram follow-up are needed. The current study is an attempt to begin to fill this void in the literature by examining alternative correctional settings across a longitudinal time span. The study is a survey of some of the community-based correctional alternatives used by one correctional agency to house and provide treatment for inmates (or clients, as they are called) in lieu of a prison setting.

The study examines three issues. The first pertains to the effectiveness issue. Do the community-based programs have better recidivism rates than those associated with prison inmates? A related question is, What community-based programs have better recidivism rates than others? The study will also briefly examine the question of cost: Are these programs effective in reducing the costs associated with confining inmates? Last, the study will consider whether these programs help the problem of overcrowded prisons and jails.

The current study uses data from a not-for-profit community corrections agency located in a midwestern U.S. state that currently operates 22 separate programs for offenders. These include residential and non-residential programs for males, females, and juveniles. Unfortunately, information is not available for all 22 programs. Some of the programs are relatively new, such as the drug court, which was implemented in 1995. Obviously, the program has not been in operation for a long enough period for a recidivism study to be complete. Other information is missing because the agency simply lacks the manpower to complete the studies on each of the programs on a timely basis. Below is information that is available at the current time on the larger programs. This includes an analysis of the community-based correctional facility, the halfway house, work release, day reporting, and home incarceration program.

⬚ Methodology

One accepted definition of recidivism is a relapse into prior criminal habits, especially after punishment (Jackson et al., 1995, p. 47). Generally, recidivism is used most frequently as an indicator of effectiveness (Pease et al., 1977). Despite the fact that there is little debate over what recidivism is, there are many conceptual and practical problems associated with measuring it (Jones, 1991). Some studies use self-reported delinquency (Davidson, Redner, Blakely, Mitchell, & Emshoff, 1987), reconvictions (Walker, 1981), court appearances (Stewart, Vockell, & Ray, 1986), and rearrests (Langworthy & Latessa, 1993, 1996; Latessa et al., 1997; Menard & Covey, 1983). Jones (1991) used both rearrests and reconvictions to measure recidivism.

The recidivism reports completed by the correctional agency in this study include the following three measures of recidivism: rearrest, conviction, and imprisonment (Jones, 1991). Recidivism data for each program reported here were gathered from internal reports compiled by researchers at the correctional agency. The correctional agency staff compiles recidivism reports on a 1-year, 2-year, and 5-year basis and verifies information for the in-between years as well. This means that recidivism reports are initially computed 1 year after a program ends, then after 2 years, and then again after 5 years.

The client groups are split into whether or not they successfully complete the program. Those who finish program requirements without committing any other criminal acts are considered to have successfully left the program. Those who commit another offense (i.e., are rearrested, convicted, and/or imprisoned) are removed from the program. These, obviously, are considered unsuccessful terminations.

The recidivism rates for rearrests, convictions, and imprisonments noted below are then compared to the recidivism rate for traditional correctional institutions in the same state. Unfortunately, the state Department of Corrections only completes a 2-year follow-up for its inmates serving time in state prison facilities and only reports on reimprisonment rates. It does not report on rearrest or reconviction rates. The reimprisonment rate, which was provided by the head of the Research Section of the Department of Corrections, hovers right around 30%. Thus, a comparison of the effectiveness of community-based programs versus institutionalization can be made only on reimprisonment rather than rearrest or reconviction.[1]

☒ Findings

Below is an analysis of each of the programs included in the study. The essential elements of each program are first described, then information is provided that indicates the percentages of clients who were arrested, convicted, or imprisoned for additional offenses.

CBCF

This data in Table 1 shows that, in the CBCF, there is a high recidivism rate for both the clients who successfully completed the program and those who did not. When recidivism is measured by an arrest during the 5-year period after starting the program, we see that the offenders who began the program in 1993 (called the 1993 intake group) and were unsuccessful in completing the program had a 100% recidivism rate. For those clients who successfully completed the program, the recidivism rate was reduced to 77%. Of the clients who did not successfully complete the program, 97% were convicted of committing another crime within 5 years of starting the program, and 69% of the successful clients were convicted of another crime within 5 years of starting the program. If recidivism is measured by imprisonment, we see that almost 83% of the unsuccessful and 34% of the successful clients were imprisoned.

Although the information is not as complete for the later years, it also shows a relatively high recidivism rate for the CBCF clients. For example, 70% of the successful clients released from the 1994 intake class were rearrested within 4 years, and 66% of the successful clients from the 1995 intake group were rearrested within 4 years. In fact, the rearrest rate for all of the intake classes was more than 90% within 3 years after a successful release. As expected, the unsuccessful clients had much higher rates of rearrest, conviction, and imprisonment for all years.

The rearrest rates for the successfully released clients over time remain similar. For each of the intake groups, there is a 60% to 70% range of rearrest, around 60% for convictions, and around 30% for imprisonment. For the unsuccessful groups, the rearrest rate is somewhere around 90% to 95%, the conviction rate is upper 80% to lower 90%, and the imprisonment rate is in the 80% range.

These recidivism rates show that the CBCF clients who successfully completed the program return to prison at about the same rate as inmates from a traditional correctional setting (34% for the CBCF clients and 30% for prison inmates). This means that the CBCF is not doing any better or any worse than a traditional prison environment in terms of

Table 1	CBCF Recidivism Rates (percentages)						
		Arrests		Convictions		Imprisonment	
Intake	Year	Successful Release	Unsuccessful Release	Successful Release	Unsuccessful Release	Successful Release	Unsuccessful Release
1993	1	39	84.4	31	78.1	14	64.1
1993	2	51	87.5	46	82.8	19	71.9
1993	3	69	95.3	62	92.2	32	78.1
1993	4	76	96.9	69	96.8	34	82.8
1993	5	77	100.0	69	96.9	34	82.8
1994	1	38.3	69.9	34.2	61.6	NA	NA
1994	2	60	91.8	53.3	86.3	NA	NA
1994	3	68.3	94.5	60.8	87.7	NA	NA
1994	4	70	95.9	NA	NA	NA	NA
1994	5	NA	NA	NA	NA	NA	NA
1995	1	37.7	77.7	28.7	73.3	19.6	75.5
1995	2	62.2	84.4	51.7	80.0	33.6	80.0
1995	3	66.4	93.3	55.9	88.9	35.7	88.9
1995	4	66.4	93.3	56.6	88.9	35.7	88.9
1995	5	NA	NA	NA	NA	NA	NA
1996	1	40.3	77.1	32.5	75.5	18.6	71.4
1996	2	60.5	90.0	49.6	87.1	29.5	80.0
1996	3	62.8	91.3	50.4	88.5	29.5	82.8
1996	4	NA	NA	NA	NA	NA	NA
1996	5	NA	NA	NA	NA	NA	NA
1997	1	40.5	88.7	92.2	95.2	73.2	91.7
1997	2	NA	NA	NA	NA	NA	NA
1997	3	NA	NA	NA	NA	NA	NA
1997	4	NA	NA	NA	NA	NA	NA
1997	5	NA	NA	NA	NA	NA	NA

NOTE: CBCF = community-based correctional facility.

inmates' returning to crime. In other words, the recidivism rates for the CBCF and the prison are fairly similar.

Halfway House

Table 2 shows that, within 5 years after completing the halfway house program, the recidivism rate for the clients was high. Of the successful releases, more than 60% were rearrested or convicted, and 27% were imprisoned. Of the unsuccessful releases, more than 90% were arrested or convicted, with more than half imprisoned within 5 years of starting the program. Of those inmates who began the program in 1994, 70% of the successful releases

Table 2		Halfway House Recidivism Rates (percentages)					
		Arrests		**Convictions**		**Imprisonment**	
Intake	**Year**	**Successful Release**	**Unsuccessful Release**	**Successful Release**	**Unsuccessful Release**	**Successful Release**	**Unsuccessful Release**
1993	1	35.4	55.4	25.8	53.0	12.9	33.7
1993	2	54.8	78.3	48.4	76.0	24.7	44.6
1993	3	60.2	89.2	55.9	88.0	25.8	49.4
1993	4	65.6	90.4	61.3	90.4	26.8	50.6
1993	5	67.7	90.4	62.4	90.4	26.8	50.6
1994	1	22.0	43.2	17.1	27.0	6.1	10.8
1994	2	34.1	62.2	28.0	45.1	7.3	16.2
1994	3	42.7	70.3	36.6	64.9	11.0	21.6
1994	4	NA	NA	NA	NA	NA	NA
1994	5	NA	NA	NA	NA	NA	NA
1995	1	62.6	82.4	51.5	78.4	22.2	59.5
1995	2	NA	NA	NA	NA	NA	NA
1995	3	NA	NA	NA	NA	NA	NA
1995	4	NA	NA	NA	NA	NA	NA
1995	5	NA	NA	NA	NA	NA	NA

were rearrested within 3 years, 65% were convicted of another offense, and 22% were imprisoned. Of those not successfully completing the program, 43% were rearrested, 37% convicted of another offense, and 22% were imprisoned for an additional crime.

This information shows that, 5 years after release, the recidivism rates for the clients in the halfway house were lower than those associated with the community-based correctional facility. Nonetheless, they remain higher than the recidivism rates of inmates in traditional prison settings.

Because the information for the halfway house clients is not as complete as that for the community-based correctional facility, it is difficult to look at trends over time. It appears as though the recidivism rates for the 1994 intake class are better overall than the 1993 client group, but because the information is not complete, it is difficult to determine if that is a significant trend.

In summary, the reimprisonment rates for the successful halfway house clients, which were 27% and 22% in the different years, are again very similar to [those of] the prison inmates who return to prison (30%). This means that the rate of return for those clients successfully released from the program is about the same as for those from a correctional facility, or possibly a little lower. However, it must also be noted that more than 90% of the unsuccessful clients were rearrested or convicted, and more than half returned to prison.

▧ Work Release

Table 3 indicates that the recidivism rates for the clients involved in the work release program

| Table 3 | Work Release Recidivism Rates (percentages) |

		Arrests		Convictions		Imprisonment	
Intake	Year	Successful Release	Unsuccessful Release	Successful Release	Unsuccessful Release	Successful Release	Unsuccessful Release
1993	1	18.5	28.9	12.7	21.0	1.5	7.8
1993	2	36.5	55.3	26.3	47.3	4.3	21
1993	3	50.2	65.7	37.1	52.6	4.8	23.7
1993	4	58.5	65.7	45.9	55.2	6.8	23.7
1993	5	63.9	71.0	50.2	60.5	7.3	26.3
1994	1	15.8	21.7	12.3	8.7	4.8	0
1994	2	30.1	34.8	23.3	30.4	6.2	4.3
1994	3	41.8	39.1	34.8	35.6	8.9	4.3
1994	4	NA	NA	NA	NA	NA	NA
1994	5	NA	NA	NA	NA	NA	NA
1995	1	24.6	27.9	14.8	23.3	4.5	9.3
1995	2	40.9	55.8	28.6	51.1	7.1	20.9
1995	3	56.4	62.7	42.2	53.4	10.5	20.9
1995	4	59.0	67.4	42.8	55.8	10.5	20.9
1995	5	NA	NA	NA	NA	NA	NA
1996	1	22.6	42.8	19.8	38.8	3.4	16.3
1996	2	43.6	79.6	35.5	73.5	6.9	47.5
1996	3	45.9	83.7	37.8	81.6	7.6	50.0
1996	4	NA	NA	NA	NA	NA	NA
1996	5	NA	NA	NA	NA	NA	NA

are lower than those for the clients in other programs. It shows that only a little more than 7% of the inmates who began the program in 1993 and successfully completed it were returned to prison. At the same time, however, 64% of the successful clients were rearrested and about half were convicted of another offense. When compared to the 1994 client group, we see that more of the inmates who began the work release program in 1994 were returned to prison than those who started in 1993, but the other measures of recidivism (i.e., arrests and convictions) were lower for the 1994 group. The inmates who started in 1995 had an even higher recidivism rate as measured by imprisonment, convictions, or arrests.

It appears as though the clients in the work release program who successfully complete the program have a consistent rearrest rate of approximately 60% within 5 years after completion of the program. For those who do not complete the program successfully, their rearrest rate hovers around 70%. The successful clients have around a 50% conviction rate and a 10% imprisonment rate. The unsuccessful clients have a conviction rate in the 60% range and an imprisonment rate around 25%.

The clients who are successfully released from the work release program seem to have a lower recidivism rate for reimprisonment, on the whole, than that for inmates in a correctional institution. Of course, this could be because the clients in the work release program have probably committed offenses that are less serious in nature than those clients in the prison. In addition, the clients in the work release program, by nature, more than likely have employment and other commitments than do correctional inmates.

◼ Day Reporting

More than 70% of the clients who completed the day reporting program were rearrested within 5 years, as Table 4 indicates. More than 60% were convicted, but only 21% were imprisoned. Unfortunately, the data are not as complete for the later years, but it seems that the clients admitted to the day reporting program in 1994 have about the same rates of arrest, conviction, and imprisonment for both the successful and unsuccessful clients. The lack of complete information on the day reporting program makes it difficult to look at long-term trends. For the successful clients, the rearrest rate seems to hover around 70%, the conviction rate around 60%, and the imprisonment rate around 15%. For the unsuccessful clients, the rearrest rate is around 90%, the conviction rate around 80%, and the imprisonment rate around 50%.

Although it is hard to determine much from the little data that is provided about the success of the day reporting program when compared to a traditional prison, it seems as though the reimprisonment rates for the

Table 4		Day Reporting Recidivism Rates (percentages)					
		Arrests		Convictions		Imprisonment	
Intake	Year	Successful Release	Unsuccessful Release	Successful Release	Unsuccessful Release	Successful Release	Unsuccessful Release
1993	1	21.2	62.2	13.6	50.9	8.3	32.1
1993	2	43.9	83.0	32.5	74.5	14.4	49.0
1993	3	59.0	86.8	50.0	83.0	19.7	50.0
1993	4	66.6	89.6	58.3	86.8	21.2	50.0
1993	5	70.4	90.6	60.6	86.8	21.2	50.0
1994	1	19.3	52.0	11.4	46.7	4.4	28.0
1994	2	44.7	80.0	36.0	72.0	10.5	38.7
1994	3	57.7	86.7	47.7	78.7	11.7	44.0
1994	4	NA	NA	NA	NA	NA	NA
1994	5	NA	NA	NA	NA	NA	NA
1995	1	64.5	75.8	54.8	62.6	28.0	46.1
1995	2	NA	NA	NA	NA	NA	NA
1995	3	NA	NA	NA	NA	NA	NA
1995	4	NA	NA	NA	NA	NA	NA
1995	5	NA	NA	NA	NA	NA	NA

successful releases from the day reporting program are slightly lower than for the correctional programs (21% and 30%, respectively). In this case, the day reporting is doing better at preventing future criminal acts. Once again, however, it might be expected that a day reporting program would have lower recidivism based on the type of offender in the programs.

Home Incarceration

We can see from Table 5 that only 0.5% of the clients who were enrolled in the home incarceration program in 1993 and successfully completed it were imprisoned within 5 years. This does not hold true for the 1994 intake group, however, where the successful clients have a higher imprisonment rate, even after just 3 years. Of the successful clients from the 1993 intake group, 48% were convicted of another offense and 61.5% were rearrested. This is compared to the unsuccessful releases, in which 8.6% of the clients were imprisoned, 67.3% were convicted of another crime, and 84.8% were rearrested.

The clients successfully released from the home incarceration program seem to have a lower recidivism rate than inmates released from prison. This conclusion is tentative, however, because of the lack of data available at this time. Once again, it might be argued that clients from home incarceration might be expected to have lower recidivism because the type of offender in the programs is typically less dangerous and therefore at a lower risk of committing further criminal offenses.

Drug Court

Table 6 shows some preliminary analysis of the recidivism rates of the clients in the drug court program. None of the clients who successfully left the program were imprisoned. Only 13.7% of the successful clients were convicted of another offense, and 23.5% were rearrested. As expected, the unsuccessful releases had higher rates of imprisonment, convictions, and arrests. These percentages are 4.4%, 37.8%, and 42.2%, respectively.

Table 5		Home Incarceration Recidivism Rates (percentages)					
		Arrests		Convictions		Imprisonment	
Intake	Year	Successful Release	Unsuccessful Release	Successful Release	Unsuccessful Release	Successful Release	Unsuccessful Release
1993	1	13.2	19.5	6.8	15.2	0.0	4.3
1993	2	35.6	56.5	25.9	36.9	0.0	6.5
1993	3	49.4	82.6	39.1	63.0	0.5	8.6
1993	4	55.7	82.6	44.8	67.3	0.5	8.6
1993	5	61.5	84.8	48.8	67.3	0.5	8.6
1994	1	14.3	44.7	9.5	36.2	1.9	19.1
1994	2	30.5	76.6	19.5	74.5	3.8	29.8
1994	3	41.0	87.2	28.1	83.0	6.2	31.9
1994	4	NA	NA	NA	NA	NA	NA
1994	5	NA	NA	NA	NA	NA	NA

Table 6	Drug Court Recidivism Rates at 1996 Intake (percentages)					
	Arrests		**Convictions**		**Imprisonment**	
Year	**Successful Release**	**Unsuccessful Release**	**Successful Release**	**Unsuccessful Release**	**Successful Release**	**Unsuccessful Release**
1	23.5	42.2	13.7	37.8	0.0	4.4
2	NA	NA	NA	NA	NA	NA
3	NA	NA	NA	NA	NA	NA
4	NA	NA	NA	NA	NA	NA
5	NA	NA	NA	NA	NA	NA

These initial results seem to show that the drug court is successful in reducing recidivism by the clients. It appears from this initial data as if the drug court clients have a far lower recidivism than prison inmates. However, more analysis is required.

⬛ Cost

To examine the cost effectiveness of alternatives to prison, one can simply look at the costs of the various programs. It costs the state $70.56 per day to house an inmate in the community-based correctional facility. The average length of stay is 120 days, so the total cost per inmate is $8,467.20. The halfway house costs are lower, with a cost of $48.47 per day per client, with an average stay of 97 days. This comes to $4,217.00. If taken on a per-year basis, the halfway house costs $17,692. This is compared to a prison setting, which costs $40.47 per inmate per day, with an average stay of 243 days. This comes to $11,960.46. On a yearly basis, it costs the state $17,968.00 to house an inmate. These figures are detailed in Table 7.

It is estimated that the number of offenders in the state-funded community alternatives rose from 2,914 to 9,506. This means the state paid $126.5 million for community sanctions programs, but it would have cost the state $313.6 million if those offenders were sentenced to prison. This means that the state saved approximately $187.1 million by using community corrections programs. This is an indication that prison alternatives are cheaper than traditional prison programs.

These costs, however, may not be typical of all community-based correctional

Table 7	Community Punishment/Prison Costs			
Program	**Cost per Day in $**	**Average Stay**	**Cost per Stay in $**	**Cost per Year in $**
CBCF	70.56	120 days	8,467.20	25,754.40
Halfway house	48.47	97 days	4,217.00	17,692.00
Prison	40.47	243 days	11,960.46	17,968.00

NOTE: CBCF = community-based correctional facility.

programs. A study by Parent (1990) showed that the costs of day reporting programs vary widely, from less than $10 to more than $100 per offender per day. The average cost per offender per day was $35.04. In addition, publicly run day reporting centers cost less per offender per day than private programs. Obviously, programs with more intense surveillance were more costly than programs with less intensive surveillance.

These figures indicate that the community corrections programs provide the state with some savings initially. However, the savings are really from the shorter stays in the community facility compared to the prison. The average sentence in a community setting is much lower than the average prison term. When one considers the high recidivism rates of some of these programs, particularly the CBCF, it is clear that the community-based programs are not necessarily saving money for the state but simply delaying the costs of long-term imprisonment in a correctional facility. In fact, the state often ends up paying two times—once for the community setting and once for the prison term.

◪ Overcrowding

It is very difficult to determine whether the community sanctions available in this case are true alternatives to prison or if they serve only to widen the net of the state's supervision over offenders. With the data available here, it is only possible to surmise the effect of community sanctions on prison populations. The offenders in the CBCF are, as described earlier, typically convicted of felony drug or property offenses. Based simply on that description, it can be assumed that these offenders, without community alternatives, would probably be sentenced to either a prison or jail term. In other cases, offenders may be released early from an overcrowded prison facility and placed in a halfway house, allowing the state to maintain some control over the offenders but at the same time opening up space in a prison. In these cases,

then, the availability of community sanctions may indeed be an alternative to prison and would tend to ease prison overcrowding.

However, many halfway house clients have already served a sentence in a prison and are being eased back into society with the assistance of the halfway house. Without the halfway house program, these offenders would probably be released into society without being under an additional term of state supervision. The argument can be made, then, that this service is indeed a way that the state can maintain control over offenders who, without this program, would no longer be supervised by the state. The same is true of the work release and day reporting programs. The clients in these programs are low-risk offenders who, more than likely, would not have been given a term of supervision (especially in a correctional facility) if it were not for the community alternatives. Because of this, it is probably not the case that the halfway house program, the work release program, or the day reporting program are reducing prison overcrowding to any significant extent.

The same is true of the home incarceration program, which serves adult felony, misdemeanor, or traffic clients. These offenders may have received a short jail term but probably not a term in prison for their offenses. Thus, the home incarceration program is not reducing prison overcrowding significantly.

Thus, based solely on assumptions made from the agency's description of the clients in each of the programs, a good argument can be made that the community alternatives do not help to address the prison overcrowding issues in the state. Instead, they only serve to increase the state's control over inmates, or to widen the net.

◪ Conclusion

The first goal of community-based programs is to reduce recidivism and to protect the public from further harm by criminal acts. It has long been assumed that the clients from community corrections programs have lower recidivism rates that those of prison inmates. From the data presented above, it is not clear as to whether this is happening, at least in this one agency. The recidivism rates for the two programs for which data are readily available (CBCF and halfway house) show that the offenders who are released after successfully completing the requirements for the community-based programs are reimprisoned at about the same rate as offenders released from correctional facilities around the state. Both rates hover somewhere around 30%. This means that the community-based programs offer no more in terms of community safety than do prisons.

In the other programs (including the work release program, day reporting, and home incarceration), the clients who successfully completed the programs have lower rates of reimprisonment than offenders released from prisons. This may be an indication that these programs are more successful in treating offenders so they do not return to crime on release. But it may also simply be an indication that those inmates who are sentenced to these community programs committed less serious offenses and are less likely to commit further criminal acts. It is also unclear as to whether this pattern of lower recidivism rates in these programs holds true over time because of the lack of complete information available.

Are community-based programs successful? Palmer (1994) stated that "we could define success or promise as a recidivism reduction of at least 15% relative to a comparison or control" (p. 106). This was not met here. In short, this means that the community correctional facilities are not "successful" at reducing recidivism, but they are not doing any worse than prisons in terms of protecting society from future harm. These findings support Jones (1991) and others who found similar reoffending rates for prison releases and those released from community corrections.

The second goal of community corrections, saving money, seems to be met in this case. The per-day cost of prison versus the CBCF was fairly significant: $40.47 versus $70.56. However, the halfway house daily cost was only $48.47, which is not too much different from the costs for prison. The real savings come from the fact that inmates spend fewer days in a community setting rather than a correctional program. For those inmates who serve a shorter time in a community corrections sanction and commit no further offense, the state may save money by using a community-based program. Because recidivism rates for the CBCF and the halfway house are similar to those of the prisons, a good argument could be made for relying on community corrections rather than the prison.

However, considering a recidivism rate of 100% for the unsuccessful clients in the CBCF and 90% in the halfway house, the cost to the state is, for many offenders, only delayed by involvement in a community-based program. In fact, the state has to pay two times for those offenders: once in the community setting and then again in the correctional institution. In short, the second goal of saving money is not necessarily met by community corrections programs.

The third goal of community corrections is to reduce overcrowding in traditional penal institutions. From the descriptions given of each of the programs and the clients, it is obvious that many of the clients in the community-based settings in this particular program would probably not have been sentenced to prison for their offenses. Therefore, it is hard to call these programs true alternatives to prison. Rather, this community-based facility has increased the state's control over these offenders and effectively widened the net of people over whom it maintains control. As stated earlier by Vass (1990), prison alternatives may not always constitute real alternatives.

It is also the case that many of the offenders in this program would probably be sentenced to jails rather than prison for misdemeanor offenses. If community programs were not available, serious jail overcrowding would be the end result. So the community-based alternatives are reaching the goal of reducing overcrowding at least in the jails by opening up new methods of control.

Of course, it must be noted that the data presented here make up only a single case study, and the findings are not necessarily representative or generalizable to all community corrections facilities across the nation. It is also the case that the goals of the programs and the clients are not the same, so comparability is further weakened. In addition, the lack of complete statistics is a serious concern. But this does not mean that a case study such as this one is not valuable.

Rather, further research needs to be completed to provide a more complete understanding of what works in the correctional process. There is no doubt that questions surrounding what works to reduce recidivism by offenders is a difficult question to answer. It is virtually impossible to have a perfectly designed study with randomization and perfect matching of offenders and programs. Nonetheless, similar research to that done here is needed with more complete data to answer these questions more completely.

Furthermore, future research needs to look at not only recidivism but other measures of effectiveness as well. Jackson et al. (1995) noted that "recidivism must therefore only be considered as one of a range of possible outcome measures that should be considered, and it should be used in combination with others to evaluate the effectiveness of a particular sanction" (p. 50). More research in these areas will help policy makers understand the importance of community corrections as a correctional tool and will help administrators of the community corrections facilities be more effective in reaching their goals. Furthermore, Jackson et al. wrote,

> Unquestionably, the evaluation of sanctions is an important research task, particularly as the array of sentencing options continues to widen.

Moreover, straightforward comparisons of different sanctions will continue to be of interest to policy makers and these kind[s] of evaluation studies will still be carried out in the future. (p. 49)

✄ Note

1. The state Department of Corrections wrote that a 3-year follow-up study showed that the weighted average for returning to prison was 37.5%, and a 2-year follow-up study showed that the rate of return to prison was 30.2%. Reconviction and rearrest rates are unknown to them.

✄ References

Barajas, E., Jr. (1993). Defining the role of community corrections. *Corrections Today, 55,* 28–32,

Buddress, L. A. N. (1997). Federal probation and pretrial services: A cost effective and successful community corrections system. *Federal Probation, 61,* 5–12.

Davidson, W. S., II, Redner, R., Blakely, C. H., Mitchell, C.M., & Emshoff, J. G. (1987). Diversion of juvenile offenders: An experimental comparison. *Journal of Consulting and Clinical Psychology, 55,* 68–75.

Gendreau, P., & Ross, B. (1979). Effective correctional treatment: Bibliotherapy for cynics. *Crime and Delinquency,* pp. 463–489.

Huskey. B. L. (1992). The expanding use of CRCs. *Corrections Today, 54,* 70–73.

Izzo, R., & Ross, R. (1990). Meta-analysis of rehabilitation programs for juvenile delinquents. *Criminal Justice and Behavior, 17,* 134–142.

Jackson, J. L., de Keijser, J. W., & Michon, J. A. (1995). A critical look at research on alternatives to custody. *Federal Probation, 59,* 43–51.

Jones, P. R. (1991). The risk of recidivism: Evaluating the public safety implications of a community corrections program. *Journal of Crime and Justice, 19,* 49–66.

Kuehlhorn, E. (1979). *Non-institutional treatment and rehabilitation.* Stockholm: National Swedish Council for Crime Prevention.

Langan, P. A. (1998). Between prison and probation: Intermediate sanctions. In J. Petersilia (Ed.), *Community corrections.* New York: Oxford University Press.

Langworthy, R., & Latessa, E. J. (1993). Treatment of chronic drunk drivers: The turning point project. *Journal of Criminal Justice, 21,* 265–276.

Langworthy, R., & Latessa, E. J. (1996). Treatment of chronic drunk drivers: A four-year follow-up of the Turning Point Project. *Journal of Criminal Justice, 24,* 273–281.

Larivee, J. J. (1993). Community programs: A risky business. *Corrections Today, 55,* 20–26.

Latessa, E. J., Travis, L. F., Holsing, A., Turner, M., & Hartman, J. (1997). *Evaluation of Ohio's community-based correctional facilities.* Cincinnati. OH: University of Cincinnati.

Lawrence, R. (1991). Reexamining community corrections models. *Crime and Delinquency, 37,* 449–464.

Lucken, K. (1997). "Rehabilitating" treatment in community corrections. *Crime and Delinquency, 43,* 243–259.

Martinson, R., Lipton, D., & Wilks, J. (1975). *The effectiveness of correctional treatment: A survey of treatment evaluation studies.* New York: Praeger.

McMahon, M. (1990). "Net widening": Vagaries in the use of a concept. *British Journal of Criminology, 30,* 121–149.

Menard, S., & Covey, H. (1983). Community alternatives and rearrest in Colorado. *Criminal Justice and Behavior, 10,* 93–108.

Palmer, T. (1991). The effectiveness of intervention: Recent trends and current issues. *Crime and Delinquency, 37,* 330–346.

Palmer, T. (1992). *The reemergence of correctional intervention.* Newbury Park, CA: Sage.

Palmer, T. (1994). *A profile of correctional effectiveness and new directions for research.* Albany: State University of New York Press.

Parent, D. (1990). *Day reporting centers for criminal offenders: A descriptive analysis of existing programs.* Washington, DC: National Institute of Justice.

Pease, K., Billingham, S., & Earnshaw, I. (1977). Community service assessed in 1976. *Home Office Research Study,* p. 39.

Petersilia, J. (1995). A crime control rationale for reinvesting in community corrections. *Spectrum, 68,* 16–26.

Petersilia, J. (1998). *Community corrections.* New York: Oxford University Press.

Petersilia, J., Turner S., Kahan, J., & Peterson, J. (1985). *Granting felons probation.* Santa Monica. CA: RAND.

Renzema, M. (1992). Home confinement programs: Development, implementation, and impact. In J. M. Byrne. A. J. Lurigio, & J. Petersilia (Eds.), *Smart sentencing: The emergence of intermediate sanctions.* Newbury Park, CA: Sage.

Stewart, M. J., Vockell, E. L., & Ray, R. E. (1986). Decreasing court appearances of juvenile status offenders. *Social Casework: The Journal of Contemporary Social Work, 67,* 74–79.

Tonry, M. (1998). Evaluating intermediate sanction programs. In J. Petersilia (Ed.), *Community corrections.* New York: Oxford University Press.

Vass, A. A. (1990). *Alternatives to prison: Punishment, custody and the community.* London: Sage.

Vito, G. F., & Tewksbury, R. A. (1998). The impact of treatment: The Jefferson County (Kentucky) Drug Court Program. *Federal Probation, 62,* 46–51.

Walker, N. (1981). Reconviction rates of adult males after different sentences. *British Journal of Criminology, 21,* 357–360.

Wasson, B. F. (1993). A solid union: Community corrections department and jail form winning combination. *Corrections Today, 55,* 108–110.

Whitehead, J. T., & Lab, S. P. (1989). A meta-analysis of juvenile correctional treatment. *Journal of Research in Crime and Delinquency, 26,* 276–295.

Wright, R. (1994). *In defense of prisons.* Westport, CT: Greenwood.

DISCUSSION QUESTIONS

1. Why do individuals sentenced to community corrections have a lower recidivism rate in general than individuals sent to prison?

2. What were the study's finding with regard to halfway houses and work release?

3. What did the author conclude about the effectiveness of community-based programs?

READING

In this article, Brian Payne and Randy Gainey present a study of a neglected area of research—electronic monitoring and its consequences. A number of concerns have surfaced about the use of electronic monitoring as a sanction since its inception in 1984. Research into these concerns has examined the sanction's breadth, pitfalls, and successes. This research focuses on the way electronically monitored offenders define various issues about the sanction. Results suggest that offenders do not necessarily see the sanction in ways that are consistent with the portrayal of the sanction in the literature and the media. Implications are provided.

The Electronic Monitoring of Offenders Released From Jail or Prison

Safety, Control, and Comparisons to the Incarceration Experience

Brian K. Payne and Randy R. Gainey

House arrest with electronic monitoring can be used during various phases of the justice process as an alternative to incarceration. In some jurisdictions, electronic monitoring is used during the pretrial phase to ensure that the offender will appear for trial. More often, however, the sanction is used as a method to supervise, control, and punish offenders who have already been convicted. When used in this manner, electronic monitoring is generally applied in one of two ways. First, it may be a sanction in and of itself, which judges use for some offenders. Or, it can be used in conjunction with other sanctions wherein offenders receive a prison or jail sanction and then are placed on electronic monitoring when they are released back into the community. Focusing primarily on those who are monitored after being released from prison or jail, the current study considers the way offenders experience house arrest with electronic monitoring.

◫ Review of Literature

A handful of criminologists have devoted a great deal of attention to trying to understand the role of electronic monitoring as an alternative to incarceration. Four bodies of scientific debate and research characterize the body of literature that has developed since the sanction was initially created. These four areas include: (a) debate about the controversial issues surrounding the sanction; (b) concern about the applicability of the sanction for various kinds of offenders; (c) evaluations of the success of the sentencing alternative; and (d) examinations of the experiences of offenders on electronic monitoring.

SOURCE: Payne, B. K., & Gainey, R. R. (2004) The electronic monitoring of offenders released from jail or prison: Safety, control, and comparisons to the incarceration experience. *The Prison Journal, 84*(4), 413–435. Reprinted with permission of Sage Publications, Inc.

Controversial Issues Surrounding Electronic Monitoring

The first body of academic literature on electronic monitoring critically examines the controversial issues surrounding the sanction (see Del Carmen & Vaughn, 1986; Grace, 1990; Houk, 1984; Lilly & Ball, 1987; Muncie, 1990; Petersilia, 1986; von Hirsch, 1990). To those who opposed electronic monitoring, it was perhaps no coincidence that the sanction surfaced in 1984, as the Orwellian nature of the sanction was particularly offensive. Lilly and Ball (1987) note, "Concerns expressed over possible invasions of privacy, either by the government or by private agencies, had come about because of the enormously increased power of technology to penetrate the private realm" (p. 371). Using technological advancements to control and punish offenders has been resisted by those seeing such innovations as intrusive and barbaric. Alternately, supporters of electronic monitoring point out that new methods of punishment most always evolve with broader societal changes (Lilly & Ball, 1987).

Critics also claimed that the sanction was not really an alternative to incarceration but simply a new sentencing alternative. The belief was that this new sanction would simply widen the net of criminal justice control. In effect, some believe that offenders sentenced to electronic monitoring are actually offenders who in the past would have been informally diverted from the justice system altogether (Bonta, Wallace-Capretta, & Rooney, 2000a; Mainprize, 1992). Supporters of the sanction point out that its versatility ensures that overcrowding can in fact be reduced if it is used appropriately. When used as a form of pretrial detention, supporters argue that individuals who once would have remained in jail can now return to their homes. When used as a form of release from jail or prison, offenders who truly otherwise would spend time incarcerated can be returned to the community in a structured and rehabilitative yet retributive manner (Gainey, Payne, & O'Toole, 2000).

Furthermore, because most electronically monitored offenders must cover the costs of their incarceration, electronic monitoring is seen as more economical than incarceration (Payne & Gainey, 1999).

Other critics were less concerned with the net-widening potential and more concerned with the belief that the sanction was unsafe because offenders could easily escape into the community. Comments in the media suggest that electronic monitoring sanctions pose a great risk to the community (Payne & Gainey, 2000). Supporters of the sanction point out that generally only low-risk offenders are placed on the sanction and that, even if they did escape, they have little or no propensity toward violence (Jolin & Stipak, 1992; Loconte, 1998).

Critics of the sanction also claim that electronic monitoring with home confinement turns the home into a prison (Grace, 1990; Houk, 1984). When punishment is administered in the home, and when the offender loses contact with the outside world, it is indeed possible that sanctions do turn the home into a prison. However, two points counter this claim. First, corrections scholars note that it is nothing new to apply sanctions in offenders' homes (Lilly & Ball, 1987). Second, it is reasonable to argue, as the current research does, that the best way to determine whether the sanction turns the home into a prison is to ask offenders whether their homes were prisonized during their sanction. Philosophically speaking, the sanction may turn the home into a prison, but does this happen in reality?

Electronic Monitoring as a Versatile Sanction

The next area of literature on electronic monitoring focuses on the versatility of the sentencing alternative. Research has examined whether, and how, the sanction could be used for various offenders including juveniles and drug offenders (Courtright, Berg, & Mutchnick, 2000; Jolin & Stipak, 1992; Roy, 1997). In general, this research suggests that the sanction can be used for different types of offenders in meaningful ways.

Research has also examined how the sanction could be used during different phases of the justice process (Maxfield & Baumer, 1990). Specifically, electronic monitoring can be used in at least three parts of the justice process, including prior to trial, immediately after conviction, and postincarceration. When used prior to trial, electronic monitoring has been shown to be an effective strategy for pretrial detention (Altman, Murray, & Wooten, 1997; Cadigan, 1993; Cooprider, 1992; Cooprider & Kerby, 1990). With pretrial detention occurring in the home rather than in jail, suspects are able to avoid the criminogenic environment found in many jails. They also have more access to their attorneys, thus allowing them to assist in preparing their defenses. Moreover, it is cost effective to provide home supervision.

Electronic monitoring can also be used immediately after conviction as a form of punishment in and of itself. When used in this manner, research suggests that the sanction potentially fulfills many goals of the justice process (Payne & Gainey, 2000). It is punitive but rehabilitative. As well, the controlling nature of the sanction protects society. Moreover, studies show that electronically monitored offenders are less likely than are comparable offenders to commit new offenses (Bonta, Wallace-Capretta, & Rooney, 2000b; Courtright, Berg, & Mutchnick, 1997, 2000).

Research has also considered the use of electronic monitoring after one has served a prison or jail sanction. When used in this part of the justice process, the sanction is seen as helping to reintegrate offenders into the community. In addition, because incarceration is experienced as a shameful event, house arrest with electronic monitoring following incarceration helps to show offenders that society is placing trust back into them (Gainey et al., 2000). According to Gainey et al. (2000), "Jail incarceration followed by electronic monitoring affords offenders respect by trusting them with early release into the community" (p. 748).

Defining and Measuring the Success of Electronic Monitoring

The third body of electronic monitoring literature entails the examination of ways to define and measure the success of the sanction. This success literature defines success in three different ways. First, some researchers have defined and measured success by focusing on whether offenders violated their electronic monitoring conditions and failed to finish the electronic monitoring sanction (Baumer, Maxfield, & Mendelsohn, 1993; Lilly, Ball, Curry, & Smith, 1992). Defining success in terms of violations, most have come to agree that the longer one is on electronic monitoring, the more likely one will be to violate the terms of one's probation or parole.

Second, some have defined success in terms of whether offenders committed, or were convicted of, new offenses (O'Toole, 1999; Roy, 1997). Research examining subsequent convictions of electronically monitored offenders shows that the sanction is more effective than comparable sanctions in deterring certain types of offenses (e.g., traffic offenses and drunk driving; Courtright et al., 1997; Gainey et al., 2000). The ability of electronic monitoring to deter future misconduct varies among programs. Those that are believed to have the most deterrent effect are programs that have a strong treatment component as a core of the punishment experience (Bonta et al., 2000b).

Third, some research has defined success in terms of the public's support for electronic monitoring (Brown & Elrod, 1995). Research defining success as public support has shown that the public is generally supportive of the sanction for nondangerous offenders. These nondangerous offenders generally include younger offenders with no prior record who have been convicted of mundane offenses such as drug possession, traffic offenses, and so on. Recent research also shows that some groups may not adequately understand the sanction (Gainey & Payne, 2000).

The Electronic Monitoring Experience

The fourth area of electronic monitoring research has focused on the way that offenders experience the electronic monitoring sanction (Gainey & Payne, 2000; Payne & Gainey, 1998). This research shows that most electronically monitored offenders prefer house arrest to jail, but they still experience the sanction as punitive. Based on interviews with 29 electronically monitored offenders, Payne and Gainey (1998) report that those experiencing this sanction will experience pains of imprisonment similar to those considered by Sykes (1958; e.g., deprivation of liberty, deprivation of autonomy, etc.) along with some pains that are unique to the electronic monitoring experience. For example, electronically monitored offenders must pay a fee to be on the sanction, have to watch others do things they themselves are unable to do, may experience shame from wearing the bracelet, and may experience family problems due to the fact that they are always at home. This body of research fits in with recent research suggesting that some offenders see alternative sanctions as equally punitive as incarceration (Petersilia & Deschenes, 1994; Spelman, 1995; Wood & Grasmick, 1999).

The past literature has generally considered these areas (e.g., the controversial issues, applicability of the sanction, the sanction's success, and offender's experiences) separately. The current study pulls together the areas by considering how one group of offenders (e.g., those who have been released from jail) experience the sanction and by focusing on offenders' experiences regarding the safety, control, and effectiveness of the sanction. Some will likely claim that offenders' perceptions are not accurate indicators in terms of the sanction's safety, controlling nature, or effectiveness. We believe, however, that offenders' perceptions are important to understand for at least three reasons.

First, many of the criticisms about electronic monitoring are based on issues that would be best understood through an assessment of the offenders' perceptions of, or experiences with, the sanction. Recall that some academics claim that the sanction turns the home into a prison. Perhaps it does in theory, but does this happen in reality? The only way to find out would be to ask offenders if they think their home was turned into a prison while they were being monitored. Likewise, some claim that the sanction is too lenient. Again, lenience is a relative concept that would be best appreciated only by those who have actually experienced the sanction.

Second, as far as the safety of the sanction is concerned, asking offenders about their perceptions of the safety of the sanction seems like the most obvious starting point. To be sure, the safety of the sanction cannot be determined entirely from offenders' perceptions about safety, but it is important to consider their perceptions nonetheless. If offenders who have been on electronic monitoring tell us that they think the sanction is unsafe to the public, then it would seem that the sanction truly is a threat to public safety. Alternately, if offenders explain what keeps them from committing offenses while they are being monitored, then understanding about what is needed to maintain public safety is provided. Third, electronically monitored offenders are often ignored when punishment experiences are assessed. Lilly and Jenkins (1989) note that although many studies have been done on electronic monitoring and other sentencing alternatives, "hardly anyone asks the criminals . . . what they say" (p. 23). It seems to us that electronically monitored offenders would have much to say about the safety, effectiveness, and controlling nature of the sanction.

Because electronically monitored offenders have so much to offer insofar as our understanding of the sanction is concerned, the current research examines the following questions. First, how do offenders experience and perceive the sanction? Second, what factors influence the way offenders experience and perceive electronic monitoring? Third, what do offenders say about the protective function served by the sanction? Fourth, how do offenders respond to the controlling nature of the sanction? Last, how do offenders compare their monitoring experience with their incarceration experience? The questions were addressed by focusing primarily on offenders who had spent at least some time incarcerated.

▨ Method

To see how offenders experienced electronic monitoring after being incarcerated, the authors surveyed 49 offenders who were on electronic monitoring in a jurisdiction where the sanction was generally used in conjunction with jail sentences and work release programs. Specifically, offenders in the program, in theory, would spend one third of their sentence in jail, one third of their sentence on work release, and the remaining one third of their sentence on house arrest with electronic monitoring. Due to constraints in contacting the offenders, four survey techniques were used to administer the survey: face to face interviews in a separate room at the electronic monitoring supervisor's office at the sheriff's department ($n = 12$); phone interviews ($n = 3$); onsite completion of the survey in a separate room at the electronic monitoring supervisor's office ($n = 29$); and mail return surveys ($n = 5$).

The survey included four sections: (a) a demographic section; (b) a close-ended section asking offenders about problems they experienced on electronic monitoring; (c) a close-ended section asking offenders their perceptions of the sanction; and (d) an open-ended section asking offenders about various aspects of the sanction. Scales have been developed from items in the second and third sections and have been subjected to a number of different tests.

For instance, the second section is a 24-item instrument assessing various pains and costs that monitored offenders would experience during the course of their sanction. Four subscales from this instrument were examined in a prior study: (a) a controlling conditions subscale assessing the conditions designed to control offenders; (b) a technological conditions subscale assessing the way technology conditions influence the punishment experience; (c) a controlling restriction subscale assessing the way that restrictions control behavior; (d) a technological restriction subscale assessing limitations on behavior as a result of the technology associated with the sanction.

The third section of the instrument is a 38-item instrument assessing how offenders perceived the sanction. This instrument included subscales designed to measure how well the sanction was perceived as meeting different goals of the justice process (e.g., deterrence, rehabilitation, retribution, etc.). The current research uses items from the second and third sections to build on the previous research by examining whether specific items vary among various offenders and by combining these bivariate analyses with an examination of offenders' responses to the close-ended questions.

The close-ended questions assessing experiences of electronic monitoring included a series of statements about possible problems offenders would confront on the sanction. Respondents were asked to indicate whether each problem was no problem, a little problem, a moderate problem, or a major problem. The close-ended questions assessing perceptions about the sanction included a series of statements about controversial aspects of the sanction asking respondents their degree of agreement or disagreement.

Roughly three fourths of the sample was male, employed, and unmarried, and just under half was White. The age of respondents ranged from a low of 21 years old to a high of 63 years old, and the average age of respondents was 37 years. Also, on average, 3.7 individuals lived in the household where the offender was monitored, and one third of the offenders lived in an apartment and the rest lived in a house. At the recommendation of the program director, we did not ask offenders about their offense types. From the program director, however, we learned that when all monitored offenders in the program are considered, about half of them were traffic offenders, one third were felony offenders, and about 17% were convicted of misdemeanors. Most of the traffic offenders had drunk driving offenses as part of their records. In terms of whether these offenders represent other offenders on electronic monitoring in this particular jurisdiction, it is important to note that the characteristics of the offenders included in this study are similar to a broader study that examined 215 electronically monitored offenders (Gainey et al., 2000).

The analyses presented here are exploratory and include primarily qualitative strategies. Content analysis was performed on the open-ended questions. Four themes were examined: (a) the safety of the sanction; (b) the controlling nature of the sanction; (c) the electronic monitoring experience; and (d) comparisons to other sanctions. For the closed-ended questions, descriptive statistics are reported to provide a general understanding of the problems confronted on the sanction and to illustrate how offenders perceive the sanction. Integrating the descriptive statistics with the content analysis is a useful strategy to illustrate important patterns in the data (Berg, 2000).

▧ Findings

General Experiences and Perceptions

Table 1 includes the univariate statistics from the close-ended questions asking about the offenders' experiences with the sanction. From the close-ended questions, it becomes relatively clear that offenders do not experience the sanction as overly punitive or as overly lenient. As far as experiences are concerned, certain aspects of the sanction were rarely seen as a problem. Those experiences that were rarely seen as a problem included having to provide urine for drug tests, having

to avoid alcohol, having to worry about friends getting the offender in trouble, having to keep the house in order, having one's leisure time interrupted, having one's family know one's whereabouts at all times, not being able to have call waiting, and not being able to turn the ringer off on the phone or ignore the answering machine. The vast majority of the sample indicated that these experiences were simply not problematic aspects of the electronic monitoring sanction.

Some experiences were cited as being at least a little bit of a problem, if not a moderate or major problem, relatively frequently. Those experiences that were somewhat frequently cited as problems included shameful aspects of the sanctions (e.g., having to wear a visible monitor and the embarrassment of having to tell friends about the sanction) and those that limit one's interactions or possible interactions (e.g., not having weekends free and having to limit the length of conversations on the phone). Taken together, the shameful experiences and limiting of interactions seem to capture problems that offenders would face in regards to their social needs.

A handful of other experiences were more frequently cited as problems offenders faced on the sanction. These other experiences included not being able to go to the store when one wants, not being able to go out to eat when one wants, and not being able to go for a walk or run when one wants. These three experiences can be seen as problems meeting one's physiological needs (e.g., the need to eat, to exercise, or to buy food or other necessities).

Table 2 includes the univariate statistics from the questions asking offenders about their perceptions of electronic monitoring. In general, offenders saw the sanction as punitive, though a sizeable minority (nearly one in five) agreed that it may be too lenient. Offenders were split with regard to the belief that the sanction turns the home into a prison. Slightly over half agreed with the statement, and slightly under half disagreed. Interestingly,

Table 1	Problems Confronted When on Electronic Monitoring			
	No Problem (%)	Little Problem (%)	Moderate Problem (%)	Very Big Problem (%)
Not being able to go for walk or run when you want.	9 (18.4)	13 (26.5)	14 (28.6)	13 (26.5)
Not being able to go to the store when you want.	3 (6.1)	17 (34.7)	14 (28.6)	15 (30.6)
Not being able to stay late at work.	20 (40.8)	10 (20.4)	8 (16.3)	10 (20.4)
Not being able to meet friends after work.	26 (53.1)	10 (20.4)	6 (12.2)	7 (14.3)
Not being able to turn the ringer off on your phone.	33 (67.3)	7 (14.3)	3 (6.1)	6 (12.2)
Not being able to ignore the answering machine.	32 (65.3)	8 (16.3)	3 (6.1)	6 (12.2)
Not being able to have call waiting.	33 (67.3)	7 (14.3)	4 (8.2)	4 (8.2)
Having to limit the length of conversations on phone.	16 (32.7)	13 (26.5)	13 (26.5)	7 (14.3)
Not being able to go out to eat when you want.	6 (12.2)	18 (36.7)	10 (20.4)	15 (30.6)
Not being able to drink alcohol.	33 (67.3)	9 (18.4)	3 (6.1)	4 (8.2)
Having to provide urine for drug and alcohol tests.	41 (83.7)	6 (12.2)	0 (0.0)	1 (2.0)
Having to worry about friends showing up with alcohol or drugs and getting you in trouble.	41 (83.7)	4 (8.2)	2 (4.1)	2 (4.1)
Having your family or friends know where you are at every moment.	35 (71.4)	10 (20.4)	3 (6.1)	1 (2.0)
Embarrassment of having to tell that you can't go out.	22 (44.9)	13 (26.5)	9 (18.4)	5 (10.2)
Having to keep your house in order in case a staff person checks in on you.	39 (79.6)	8 (16.3)	2 (4.1)	0 (0.0)
Embarrassment of having to tell your friends or family that you're constrained to the house.	25 (51.0)	10 (20.4)	6 (12.2)	8 (16.3)
Having to wear a visible monitor.	15 (30.6)	10 (20.4)	10 (20.4)	14 (28.6)
Having a box on your phone that people might ask about.	27 (55.1)	13 (26.5)	5 (10.2)	4 (8.2)
Having your work interrupted by law enforcement calls.	24 (49.0)	11 (22.4)	6 (12.2)	7 (14.3)
Having your leisure time interrupted by calls from a staff person.	33 (67.3)	8 (16.3)	7 (14.3)	1 (2.0)
Having to worry about technical problems that you might get blamed for.	14 (28.6)	11 (22.4)	13 (26.5)	10 (20.4)
Not having weekends free.	11 (22.4)	12 (24.5)	10 (20.4)	15 (30.6)
Having your sleep interrupted by calls to check on you.	24 (49.0)	12 (24.5)	6 (12.2)	6 (12.2)
Not being able to get away from family or roommates when you want.	21 (42.9)	15 (30.6)	6 (12.2)	7 (14.3)

Table 2	Monitored Offenders' Perceptions About Electronic Monitoring			
I think that electronic monitoring. . .	**Strongly Disagree (%)**	**Disagree (%)**	**Agree (%)**	**Strongly Agree (%)**
as a form of punishment, may be too lenient.	13 (26.5)	23 (46.9)	9 (18.4)	4 (8.2)
is a severe punishment because it keeps the offender from going anywhere.	8 (16.3)	10 (20.4)	22 (44.9)	8 (16.3)
ensures that the offender is punished.	3 (6.1)	6 (12.2)	32 (65.3)	8 (16.3)
really isn't a form of punishment for many people.	8 (16.3)	26 (53.1)	10 (20.4)	3 (6.1)
is an effective method of controlling offenders.	1 (2.0)	3 (6.1)	29 (59.2)	15 (30.6)
is dangerous because it's too easy for the offender to escape.	16 (32.7)	27 (55.1)	5 (10.2)	1 (2.0)
turns the home into a prison.	3 (6.1)	23 (46.9)	21 (42.9)	2 (4.1)
may help rehabilitate some offenders.	1 (2.0)	2 (4.1)	35 (71.4)	10 (20.4)
is effective because it doesn't require convicts to associate with other criminals, like in jail.	3 (6.1)	5 (10.2)	21 (42.9)	17 (34.7)
helps in treating offenders by maintaining close supervision over them.	0 (0.0)	3 (6.1)	33 (67.3)	10 (20.4)
may be effective because the offender can maintain close contact with his or her family.	0 (0.0)	2 (4.1)	29 (59.2)	16 (32.7)
may be effective because the offender can continue to work.	0 (0.0)	1 (2.0)	24 (49.0)	21 (42.9)
may be effective because the offender can help with household duties.	0 (0.0)	3 (6.1)	25 (51.0)	19 (38.8)
turns the home into a prison.	3 (6.1)	23 (46.9)	21 (42.9)	2 (4.1)

very few of the offenders strongly agreed or strongly disagreed with the statement.

There was more agreement about the rehabilitative appeal of the sanction, with nearly 95% of the sample agreeing or strongly agreeing that the sanction helps in treating offenders by maintaining close supervision, may be effective because the offender can still work, may be effective because the offender can maintain contact with his or her family, and may be effective because the offender can help with his or her household duties.

◼ Perceptions About the Safety of the Sanction

In asking offenders about whether it would be easy to escape, most agreed that it may be easy to temporarily be free, but certain factors kept them from even contemplating escape. Offenders cited four factors that kept them from escaping, including threat of punishment, monitoring potential, conventional ties, and offender characteristics. With regard to threat of punishment, the offenders in this study seemed to understand that they could get into a significant amount of trouble should they tamper with their monitoring equipment. Here is how a few of them put it:

It's very easy to escape. But as soon as you go past that door seal, you're in trouble.

I wouldn't escape. That's another charge, more time.

If they left, they'd be worse off than when they started. I just do this and get it over with.

Well anyone that's on the monitoring is stupid if they try to escape because they are the one that's going to suffer consequences.

It will catch up to you [if you try to escape].

This relates to a second factor that seemed to keep offenders from considering escape: the monitoring potential of the electronic technology. One offender commented that escape was not an option "because this has a range and they will know pretty quick." In a similar vein, another offender indicated, "Wherever you go, they are ahead of you." And another said, "They'd know if the bracelet were off." More specific in terms of the monitor's strength, a fourth offender commented, "I don't see it as easy to escape. My monitor picks me up within five feet of the door. I'm out at 7:30 in the morning and in at 7:30 at night. If I'm late, she knows."

Conventional ties were also cited as factors that kept offenders in line while they were being monitored. Specifically, the fear of losing something or someone of value as a result of any tampering with the equipment seemed to keep offenders from contemplating escape. Those conventional ties that seemed to be most important to offenders included their families and their jobs. Consider the following comments from three different offenders:

It depends on a person's consequences. I love my family. It's the most important part of my life. I do not want to see another man caring for my daughter or sleeping with my woman. To someone who has nothing to lose, it's a joke; but for someone who has positive things to offer himself and family, it is everything.

I'm just glad I'm able to work.

My job is far more important. It would hurt me. It would hurt my family. It doesn't fit. I can't comprehend how you can escape, or why you'd want to escape. It is not reasonable.

Each of these comments shows that the offenders see their conventional ties with their families and their jobs as being too important to risk.

Offender characteristics were also cited as factors that would keep offenders from trying to escape. In particular, these offenders tended to define themselves as less serious offenders who are really not a societal threat. Said one offender, "People who are a real threat are not on it." Another offender voiced a similar belief, stating, "People on house arrest are generally there because of a lesser charge." As less serious offenders, the electronically monitored offenders in this study seemed to see themselves as privileged in that they were given the option of leaving jail earlier than other offenders. Implying this privilege, one offender commented that electronic monitoring "shouldn't be given to certain offenders."

Keep in mind that the responses to the close-ended questions showed that the offenders did not see this sanction as overly punitive. Given that it is not overly punitive, offenders do not want to jeopardize doing things that would result in a stiffer penalty. Tying these ideas together, these are less serious offenders who have too much to lose (e.g., conventional ties and a stiffer punishment) by trying to escape from a sanction that is virtually omnipresent. The omnipresence of the sanction relates to the offender's experiences with the controlling nature of electronic monitoring.

⬚ Electronic Monitoring and Control

Offenders generally agreed that the sanction does in fact control their lives much the same way that incarceration controls inmates' lives. When talking about the controlling nature of the sanction, offenders' comments tended to fit into two categories: concerns about freedom and retributive experiences. Certainly, these two categories cannot be entirely separated in that any time an offender loses his or her freedom, punishment has occurred. Even so, the nature of the comments made by the offenders suggests that the loss of freedom is something that is perhaps unique to certain types of community-based sanctions such as electronic monitoring, halfway houses, and other nonpenal custodial facilities.

When comments about the controlling nature of the sanction focused on losses of freedom, electronically monitored offenders pointed out that the sanction made them think about freedom and everyday activities that many nonoffenders may take for granted. Consider the following comments made by monitored offenders:

> Freedom is something you don't miss 'till you don't have it.

> It made me realize about life and being free.

> I can't go anywhere. . . . I am a very active person, and this is almost like being jailed. It takes time to go downtown and get checked. . . . I can't take my granddaughter anywhere, like the park or to friends with other children. I can't go to the grocery store.

Two points about the loss of freedom experienced on electronic monitoring are noteworthy. First, some offenders seemed to see this aspect of the sanction as a learning experience, something that would change them in the future. In the words of one of them,

> Electronic monitoring has taught me a valuable lesson of what it is like to have your freedom taken away from you. Also, not to take anything for granted as so many of us do. You don't realize what you've lost until it's gone.

Second, the loss of freedom is a relative experience, and the way offenders would experience this loss would vary from offender to offender. As one monitored offender said, whether or not electronic monitoring is seen as a mechanism of control "depends on how much someone values their freedom."

Although some offenders described the controlling nature of the sanction as a loss of freedom, others characterized this aspect of the sanction from a retributive framework. In short, the control the sanction elicited over the offender was the source of the punishment experience for some offenders. It was not the monitor or the situation that punished offenders; rather, it was the control. According to one offender, "You're confined at your own expense. Sometimes I hate these things, like on my daughter's birthday when I have no control." Another offender commented, "To me, it's punishment because you have to answer to someone every time you want to go somewhere, even to work." Other offenders were more succinct in describing the way the controlling nature of the sanction was punitive. Consider the following comments:

> It controls but punishes too! You pay for it.

> It's punishment. I work and go home. That's it.

> I am still confined. I can't go out.

> It is a form of control. No question, no doubt.

Clearly, the sanction is experienced both as a loss of freedom and as a punishment. Although offenders complained about certain aspects of the sanction, when asked to compare the sanction to their time in jail, most offenders saw electronic monitoring in a positive light.

✎ Electronic Monitoring and Jail Comparisons

The majority of offenders surveyed had spent at least some time in jail, and all of them indicated that they preferred electronic monitoring to jail. Here's a sampling of their comments:

> Electronic monitoring is heaven compared to jail.

> It's like night and day. Jail is terrible.

> But if I go to jail, it would be worse. It's in the eye of the beholder.

> It's not like jail.

> I learned a very valuable lesson but house arrest is better than jail. . . . It was . . . not as bad as the shame of jail.

When prodded about why jail was worse than house arrest with electronic monitoring, offenders generally pointed to four different areas: differences between the amount of control, the maintenance of family ties, the ability to maintain employment, and time for reflection.

Although offenders commented that electronic monitoring was a controlling sanction, many saw it as less controlling and less invasive than jail. This aspect of electronic monitoring made the sanction better than incarceration for many offenders. For example, one offender said, "My favorite thing is that I have control of the television." Another offender commented, "In jail they wake you up at anytime to eat. 3:00 in the morning, they wake you up. They have more control over you in jail. Here I can eat whenever." Echoing these previous comments, a third offender said, "It is not as bad as being in the city jail because you are allowed to go to work and eat what you want. You watch television when you feel like it." Another

offender balked at the suggestion that house arrest could even be compared to jail. He said, "You have more control on electronic monitoring than you do in jail. In jail, if you run out of toilet paper, what do you do? In jail . . . they have complete control over you."

The ability to maintain family ties was also cited as something that offenders liked about electronic monitoring as compared to jail. In jail, offenders have limited contact with family members. While on house arrest, offenders will spend virtually every moment of their free time with family members living in the same residence. The following comments illustrate the importance of these family ties for monitored offenders:

> I feel fortunate to be back in my family.

> It gives me the opportunity to be with my family each and every day.

> Mom is happy I'm home. My brother has cerebral palsy and he is glad I'm home.

> You still have physical contact with your family and friends.

> I've been able to help my mother tremendously with household duties and yard work.

> My relationship with my family has improved.

Though many offenders enjoyed the opportunity to be with their families, some commented that family problems arose as a result of the sanction. After all, electronically monitored offenders in the program we studied generally went from never being in the household to always being in the household. One offender commented that this made the sanction "stressful and nerve-racking." This same offender indicated that several arguments ensued with family members because of the offender's house arrest status. Another offender indirectly cited a possible source of arguments, stating, "You become dependent on others when you're on [electronic monitoring]."

Despite these possible relationship problems, offenders also appreciated the ability to work while on electronic monitoring. As an example, one offender said, "I am able to work because I'm not incarcerated." Though he did not cite why working was so important to him, other offenders suggested that the ability to maintain one's wealth was significant. In one offender's words, "[Electronic monitoring] gives me a chance to have half of a natural life without giving up everything I own." Another offender agreed, pointing out that electronic monitoring made it "easier to keep credit, my truck, boat, and so on."

A fourth way that offenders saw electronic monitoring favorably, at least compared to jail, was the way that the sanction gave them time to think, reflect, and plan a new future. As one offender put it, "Electronic monitoring gives you a chance to think about what you have done." Other offenders seemed just as pensive. One, for instance, said the sanction "confined me and made me think twice about doing it again." In a similar vein, another offender commented, "I got on house arrest and am able to stop and think about things. I'm able to know the consequences."

Each of these comments suggests that the offenders changed as a result of having time to think about their past misconduct. What this means is that the offenders felt that they were rehabilitated by the sanction. A couple of offenders actually directly commented that the sanction rehabilitated them or at least has the potential for rehabilitation. One offender simply suggested that electronic monitoring is "more likely to rehabilitate than jail," whereas another said that the sanction was "beneficial to rehabilitation because it gave me the time to stop and think about things."

This time to reflect led some offenders to conclude that they had been given a new lease on life, or in their words, "a second chance." Said one, "I have been given another chance and I will make the best of it, using what I know and have experienced." Keeping in mind that most of the respondents had spent some time

incarcerated, the reasons they saw electronic monitoring with house arrest as a second chance were because the sanction is less controlling than jail, allows contact with family members, allows offenders to maintain employment, and gives offenders time to think. In short, the sanction was not simply experienced as a punishment but as a guide to life. Punishment by itself for offenders who have already been incarcerated has the potential to be counterproductive. One offender we interviewed paraphrased a popular quote about the dangers of punishment: "If you have a bed wetter and keep beating the bed wetter, you end up with a bed wetter with a sore butt." The majority of these offenders did not see the sanction as leading to "a bed wetter with a sore butt."

▨ Discussion

The current study finds that offenders who spent part of their time in jail and then on electronic monitoring experienced the sanction, for the most part, in a way that can be characterized as controlling and rehabilitative. They faced a few problems on the sanction and certainly experienced it as a controlling mechanism, but they generally preferred electronic monitoring to jail. These results have important implications for policy, theory, and research.

Three general policy recommendations evolve from our findings. First, electronically monitored offenders, especially those who are placed on the sanction after they have been incarcerated, should be told what to expect before they are placed on electronic monitoring to minimize potential problems. At a minimum, offenders must be told that the sanction is controlling, that it may take a toll on their family members, and that they should use their time to reflect on their past and to think about their future. These are three areas that seemed to surface consistently as concerns among the monitored offenders we surveyed. It is likely that many of the offenders did not expect to confront these experiences. In leaving jail, they may have assumed that they would be gaining their entire freedom. Electronic monitoring, however, still restricts offenders. Also, because offenders go from never being around their family members while they were incarcerated to always being around their family members while they are monitored, the potential for family discord is ripe if a proactive approach is not taken in alerting offenders to the possibility of discord. If offenders are warned about these possible problems and shown how to effectively use their time to think about their futures, we believe that the usefulness of the sanction will be enhanced.

Second, and on a related point, other groups should also be educated about the sanction. Policy makers, the media, and citizens tend to misconceive the sanction. Policy makers tend to favor strict approaches to handle criminals, and electronic monitoring is often seen as lenient. Electronic monitoring, however, is far from lenient, and it potentially leads to offenders remaining in jail or prison far longer than they should. Some research shows that longer time in jail results in higher levels of recidivism, whereas longer periods on electronic monitoring lead to less recidivism (Gainey et al., 2000). Moreover, with the media feeding the public information about cases when violent offenders are placed on the sanction or when offenders escape from their house arrest, citizens come to see the sanction as unsafe. Consequently, it becomes more difficult to garner public support for the sanction. Judging from the offenders in this study, however, the reality is that the sanction is seen as better than jail, and offenders know that they would reap what they sow should they violate their monitoring conditions. In essence, as long as less serious, nonviolent offenders are being placed on the sanction, there should be little concern about offenders escaping into the community, and the public needs to know this.

Finally, we believe that our study shows that community-based sanctions can be effectively used in conjunction with other traditional sanctions (see also Gainey et al., 2000; Jones & Sims,

1997; Thistlewaite, Woolredge, & Gibbs, 1998). The combination of sanctions meets the demands of citizens who generally want the justice system to "incarcerate first, then rehabilitate" (McCorkle, 1993, p. 251). When applied after incarceration, electronic monitoring does just this—offenders are incarcerated and then given the control and guidance needed to think about their misdeeds so that they will be less likely to reoffend in the future. Further, many offenders tend to see the sanction as a second chance. This second chance affords them the wherewithal to become reintegrated into the community as citizens as opposed to criminals. Moreover, given that offenders are granted early release into the community, jail overcrowding is reduced.

Our findings have two implications for theory. First, comments from our respondents lend credence to assumptions underlying social control theory. Social control theory suggests that crime occurs when individuals' bonds to society are weakened (Hirschi, 1969). The key to responding to crime, then, is to make sure that offenders feel connected to society in such a way that they would feel that they have too much to lose should they violate the law. Strong ties result in a lower likelihood of crime.

Offenders in this study seemed to recognize that they had too much to lose from violating their electronic monitoring conditions, and they appreciated the opportunity to maintain their family and employment bonds. Furthermore, some indicated that the controlling nature of the sanction helped to keep them in line. Others have also suggested that house arrest with electronic monitoring is based on control theory assumptions. O'Toole (1999) writes,

> Electronic monitoring might have positive effects on offender behavior by enforcing the kind of structured lifestyle that many offenders lack. This builds attachment to family and positive peer groups. The offenders value

the good opinion of their family. . . . Offenders do not want to jeopardize their chances of success, for they risk losing their place in society by getting into trouble with the law and going [back] to jail or prison. (p. 13)

Basically, when applied after a jail or prison sanction, house arrest with electronic monitoring shows offenders that society is beginning to place trust back into the offender, theoretically strengthening the bond that the offender will have with society (Gainey et al., 2000). This stronger bond the offender has with his or her society, family, and job reduces the likelihood of future misconduct on the part of the offender.

Second, and on a related point, although we did not test Gottfredson and Hirschi's (1990) self-control theory, many of the offenders implied both directly and tacitly that the sanction had taught them self-control. That self-control can change is in direct opposition to Gottfredson and Hirschi's underlying assumption of their general theory of crime. However, other research has questioned the stability of self-control over time (see Arneklev, Cochran, & Gainey, 1998). If offenders' perceptions about the ability to change and resist temptation are any indication whatsoever, findings from the current research also question the stability hypothesis and lend credence to the possibility that punishment may actually work. Of course, the punishment considered here is less restrictive than are other sanctions and is often misinterpreted as a slap on the wrist, but electronic monitoring is still punitive and potentially rehabilitative.

Our findings also have important implications for future research. First, we were not able to fully determine whether offenders believed that this particular sanction turned the home into a prison. About half agreed that it does, and about half disagreed. On the surface, this suggests that the sanction affects offenders and their families differently. Future research should explore this question in more detail.

Second, future research should broaden the way success is operationalized. All too often researchers tend to define success solely in terms of whether offenders reoffend. Research on reoffending is certainly needed. However, researchers must realize that future reoffending is not the only way to measure the success of alternative sanctions. Instead, among other ways, success can be defined by any combination of the following questions: Is the sanction humane? Does the sanction restore trust between the offender and the community? Is the sanction cost-effective? Does the community support the sanction? Does the sanction meet the community's need for retribution? Do offenders experience the sanction in positive ways? How does the sanction affect others (e.g., family members, the public, etc.)?

Of course, the goal of deterrence cannot be dismissed. A humane, cost-effective sanction that restores trust and has community support is useless if it has no deterrent power whatsoever. At the same time, we must recognize that deterrence is not the only goal, especially when alternative sanctions are concerned. Broadly defining success will help to determine the role that alternative sanctions have in society.

Third, our research shows that offenders can serve as a source of information about the usefulness of various sanctions. One respondent, who self-administered the survey, wrote at the end of the instrument, "And I also appreciate you asking my concerns and opinions." This same feeling was demonstrated by many of the other interviewees who wanted to talk far longer than needed about the sanction or who simply seemed excited to be able to talk about their situation. They appreciated the opportunity to talk. Those who were convicted of felonies have lost their right to vote; they have not, however, lost their right or their ability to inform. We encourage others to continue to explore how offenders in the community experience their sanctions.

◩ References

Altman, R. N., Murray, R. E., & Wooten, E. B. (1997). Home confinement: A '90s approach to community supervision. *Federal Probation, 61,* 30–32.

Arneklev, B. J., Cochran, J., & Gainey, R. R. (1998). Testing Gottfredson and Hirschi's low self-control stability hypothesis: An exploratory study. *American Journal of Criminal Justice, 23,* 107–127.

Baumer, T. L., Maxfield, M. G., & Mendelsohn, R. J. (1993). A comparative analysis of three electronically monitored home detention programs. *Justice Quarterly, 10,* 121–142.

Berg, B. L. (2000). *Qualitative research methods for the social sciences.* Boston: Allyn & Bacon.

Bonta, J., Wallace-Capretta, S., & Rooney, J. (2000a). Can electronic monitoring make a difference? *Crime and Delinquency, 46,* 61–75.

Bonta, J., Wallace-Capretta, S., & Rooney, J. (2000b). A quasi-experimental evaluation of an intensive rehabilitation supervision program. *Criminal justice and Behavior, 27,* 312–330.

Brown, M., & Elrod, P. (1995). Electronic house arrest: An examination of citizen attitudes. *Crime and Delinquency, 41,* 332–346.

Cadigan, T. (1993). Technology and pretrial services. *Federal Probation, 57,* 48–53.

Clemmer, D. (1940/1958). *The prison community.* New York: Holt, Rinehart & Winston.

Cooprider, K. W. (1992). Pretrial bond supervision: An empirical analysis with policy implications. *Federal Probation, 56,* 41–49.

Cooprider, K. W., & Kerby, J. (1990). A practical application of electronic monitoring at the pretrial stage. *Federal Probation, 54,* 28–35.

Courtright, K., Berg, B. L., & Mutchnick, R. (1997). The cost effectiveness of using house arrest with electronic monitoring for drunk drivers. *Federal Probation, 61,* 19–22.

Courtright, K., Berg, B. L., & Mutchnick, R. (2000). Rehabilitation in the new machine? *International Journal of Offender Therapy and Comparative Criminology, 44,* 293–311.

Del Carmen, R., & Vaughn, J. (1986). Legal issues in the use of electronic surveillance in probation. *Federal Probation, 50,* 60–69.

Gainey, R. R., & Payne, B. K. (2000). A qualitative and quantitative consideration of offenders' experiences on electronic monitoring. *International Journal of Offender Therapy and Comparative Criminology, 44,* 84–96.

Gainey, R. R., Payne, B. K, & O'Toole, M. (2000). Time in jail, time on electronic monitoring, and recidivism: An event history analysis. *Justice Quarterly, 17,* 733–752.

Gottfredson, M., & Hirschi, T. (1990). *A general theory of crime.* Palo Alto, CA: Stanford University Press.

Grace, A. M. (1990). Home incarceration under electronic monitoring. *New York Law School Journal on Human Rights, 7,* 285–314.

Hirschi, T. (1969). *Causes of delinquency.* Berkley, CA: University of California Press.

Houk, J. M. (1984). Electronic monitoring of probationers, *Golden Gate University Law Review, 14,* 431–446.

Jolin, A., & Stipak, B. (1992). Drug treatment and electronically monitored home confinement *Crime and Delinquency, 38,* 158–170.

Jones, M., & Sims, B. (1997). Recidivism of offenders released from prison in North Carolina: A gender comparison *Prison Journal, 77*(3), 335–358.

Lilly, J. R., & Ball, R. (1987). A brief history of house arrest. *Northern Kentucky Law Review, 20,* 505–530.

Lilly, J. R., Ball, R., Curry, D., & Smith, R. (1992). The Pride, Inc. Program: An evaluation of five years of electronic monitoring. *Federal Probation, 56,* 42–53.

Lilly, J. R., & Jenkins, D. (1989). Life with a tag. *New Statesman, 2,* 20–24.

Loconte, J. (1998). Making criminals pay: A New York county's bold experiment in biblical justice. *Policy Review, 87,* 26–32.

Mainprize, S. (1992). Electronic monitoring in corrections. *Canadian Journal of Criminology, 34,* 161–180.

Maxfield, M. J., & Baumer, T. (1990). Home detention with electronic monitoring. *Crime and Delinquency, 36,* 521–536.

McCorkle, R. (1993). Research note: Punish and rehabilitate? Public attitudes toward six common crimes. *Crime and Delinquency, 39,* 240–252.

Muncie, J. (1990). A prisoner in my own home: The politics and practice of electronic monitoring. *Probation Journal, 37,* 72–77.

O'Toole, M. (1999). *Factors that affect recidivism of offenders on electronic monitoring in Norfolk, VA.* Unpublished Master's thesis, Old Dominion University, Norfolk, VA.

Payne, B. K., & Gainey, R. R. (1998). A qualitative assessment of the pains experienced on electronic monitoring. *International Journal of Offender Therapy and Comparative Criminology, 49,* 49–63.

Payne, B. K., & Gainey, R. R. (1999). Attitudes toward electronic monitoring among monitored offenders and criminal justice students. *Journal of Offender Rehabilitation, 29,* 195–208.

Payne, B. K., & Gainey, R. R. (2000). Electronic monitoring: Philosophical, systemic, and political problems. *Journal of Offender Rehabilitation, 31,* 93–112.

Petersilia, J. (1986). Exploring the option of house arrest. *Federal Probation, 50,* 50–59.

Petersilia, J., & Deschenes, E. (1994). Perceptions of punishment. *Prison Journal, 74,* 306–328.

Roy, S. (1997). Five years of electronic monitoring of adults in Lake County, Indiana. *Journal of Crime and Justice, 20,* 141–160.

Spelman, W. (1995). The severity of intermediate sanctions. *Journal of Research in Crime and Delinquency, 32,* 107–135.

Sykes, G. (1958). *A society of captives.* Princeton, NJ: Princeton University Press.

Thistlewaite, A., Woolredge, J., & Gibbs, D. (1998). Severity of dispositions and domestic violence recidivism. *Crime and Delinquency, 44,* 388–399.

Von Hirsch, A. (1990). The ethics of community based sanctions. *Crime and Delinquency, 36,* 162–173.

Wood, P. B., & Grasmick, H. G. (1999). Toward the development of punishment equivalencies. *Justice Quarterly, 16,* 19–50.

DISCUSSION QUESTIONS

1. What are some of the controversies surrounding electronic monitoring?

2. How effective is the electronic monitoring of probationers compared with ordinary supervision?

3. What were the main findings of this study?

❖

READING

In this article, Clear, Rose, and Ryder examine the impact on neighborhoods wrought by large rates of incarceration among neighborhood residents. Prior research has established that the characteristics of neighborhoods are an important aspect of public safety and local quality of life. Growth in the rates of incarceration since 1973, combined with social disparity in the experience of imprisonment among certain groups, has meant that some communities experience concentrated levels of incarceration. This article examines the spatial impact of incarceration and explores the problems associated with removing and returning offenders to communities that experience high rates of incarceration. The study analyzes data from a series of individual and group interviews designed to reveal the experiences and perspectives of a sample of 39 Tallahassee, Florida residents (including ex-offenders) who live in two high-incarceration neighborhoods. The authors then provide a series of policy recommendations to offset some of the unintended consequences of incarceration.

Incarceration and the Community

The Problem of Removing and Returning Offenders

Todd R. Clear, Dina R. Rose, and Judith A. Ryder

Criminologists have long been interested in uncovering the dynamics associated with the spatial distribution of crime in an effort to understand this phenomenon and how community context impacts the lives of people living in those neighborhoods. One vein of research has drawn on social disorganization theory, focusing on ecological characteristics such as rates of poverty, residential mobility, and single parent families (see, among others, Sampson, 1988; Shaw & McKay, 1942). Another closely related vein has examined the structural and cultural impact of entrenched poverty (Wilson, 1987), whereas others have focused on opportunities for crime provided by structural changes in lifestyles and labor force participation (Felson, 1987). Recently, Bursik and Grasmick (1993) merged social disorganization and systemic theories to specify how the three levels of social control (private, parochial, and public) mediate between deleterious environmental characteristics and crime.

One of the prima facie assumptions of all these approaches is that public control operates as a response to crime and, as such, it need not be considered one of the determinants of crime. Yet, research has underscored the fact that at least one form of public control, incarceration, affects some groups of Americans much more than others (Mauer, 2000) and, given the residential segregation realities of living in America today (South & Deane, 1993), public control is spatially concentrated, too. Recently, Rose and Clear (1998a) built on Bursik and Grasmick's (1993) work by specifying the theoretical links between public control and community instability. They theorized that the aggregate impact of high levels of incarceration would be damage to networks of private and parochial social control and decreases in the legitimacy of formal social control. Thus, in their model, when public control occurs at high levels, informal (private and parochial) controls function less effectively. The result is more crime.

To test this theory, Rose, Clear, Waring, and Scully (2000) compiled two data sets on Tallahassee, Florida neighborhoods. The first combined measures of crime (1996 and 1997) and criminal justice (1996) with social disorganization data from the 1990 census. The other data set was a 1998 telephone survey, assessing attitudes toward formal and informal social control. Their analyses to date have investigated two general propositions regarding the impact of incarceration: first, whether high levels of incarceration in a given neighborhood were associated with later increases in crime in that neighborhood and, second, whether experience with the criminal justice system, through knowing someone who has been incarcerated, influences attitudes toward social control.

One of these analyses (Rose, Clear, Waring et al., 2000) found support for the proposition that spatial concentrations of incarceration promote higher than expected rates of crime. Using neighborhood as the level of analysis, the crime rate in 1997 was regressed on 1996 rates of removal for incarceration and reentry from incarceration, controlling for social

SOURCE: Clear, T. R., Rose, D. R., & Ryder, J. A. (2001) Incarceration and the community: The problem of removing and returning offenders. *Crime and Delinquency, 47*(3), 335–351. Reprinted with permission of Sage Publications, Inc.

disorganization measures in 1990 (poverty, mobility, and racial heterogeneity) and a dummy variable for "high crime rate" crime in 1996. This analysis found a linear, positive relationship between the rates of neighborhood reentry from prison in 1996 and crime in 1997, but a curvilinear relationship was found between the rate of neighborhood admissions to prison in 1996 and crime in 1997. Low rates of prison admissions were associated with no drop in crime the following year, moderate rates of admissions were associated with moderate drops in crime, but higher rates of admission—after a "tipping point" was reached of about 1.5% of the neighborhood's total population—had a strong, positive relationship to crime in the following year. This result supports the idea that high rates of admitting people to prison can destabilize informal networks of social control and lead to increases in crime.

Another analysis (Clear & Rose, 1999) found that knowing someone personally who has been incarcerated changes the relationship between people's attitudes toward formal and informal social control. In this study, people who had a low assessment of formal social control also had a low assessment of informal social control, but only if they had been exposed to incarceration. Alternatively, people who had not been exposed to incarceration and had a low assessment of formal control were more likely not to have a low assessment of informal social control. Moreover, the effects of race on attitudes toward social control disappear when controlling for exposure to incarceration. This means that Whites who know someone who has been to prison or jail have attitudes toward social control that resemble the notably negative attitudes of Blacks (Henderson, Cullen, Cao, Browning, & Kopache, 1997).

The current article builds on this previous work by exploring the ways in which incarceration affects residents living in high-incarceration neighborhoods. We summarize the experiences and perspectives of a sample of Tallahassee, Florida residents (including ex-offenders) who live in two high-incarceration neighborhoods to learn how the processes of incarceration (removal and reentry) might both strengthen and weaken neighborhoods. We then use these data as a springboard for a series of policy recommendations designed to offset some of the unintended consequences of incarceration. We conclude by identifying research priorities for further study. Readers interested in a more developed description of the current study are referred to Rose, Clear, and Ryder (2000).

⊠ The Spatial Concentration of Incarceration

It is indisputable that incarceration affects minority males more than others. Today, for example, men are 8 times more likely to go to prison than women. Furthermore, the lifetime probabilities of spending time in prison are 28.5 per 100 for African American males and 16 per 100 for Hispanic males, about 6 and 3 times higher, respectively, than for White males (Bonczar & Beck, 1997). Recent advances in the spatial analyses of crime and justice have enabled researchers to estimate the selection rates for incarceration in different neighborhoods (rather than simply in the population at large). Because poor men of color live in concentrations in neighborhoods that are racially and economically homogeneous, some of the places where these men live are particularly hard hit by incarceration. Depending on the size of the neighborhood and the method of counting, studies have estimated that up to 25% of the adult male residents in particular neighborhoods are locked up on any given day (Lynch & Sabol, 1992; Mauer, 2000), up to 13% of adult males enter prison or jail in a given year (Center for Alternative Sentencing and Employment Services [CASES], 2000), and up to 2% of all residents enter prison in a given year (Rose, Clear, Waring et al., 2000). Although a growing body of research has established the concentration of incarceration in these places and among these groups, little is known about the socio-political implications of this phenomenon on community life. Public attitudes toward incarceration are badly studied (Doble & Green, 1999; Perry & Gorczyk, 1997), consisting mostly of random telephone surveys that seek to measure the desire for punitiveness. These methods always seem to find a powerful reservoir of support for tougher penal sanctions. When nonsurvey methods are used—focus groups, for example—more subtle opinions are unearthed (Doble & Green, 1999). Moreover, nobody has tried to understand the attitudes of citizens who are heavily affected by incarceration, although one study found that exposure to incarceration changes a person's attitudes toward it and other aspects of social control (Rose & Clear, 1998b). Thus, it is plausible that an investigation of the beliefs and experiences of residents of high-incarceration locations might uncover opinions at variance with those found through the usual telephone survey method.

The Ecology of Social Disorganization

The argument that incarceration improves neighborhood life is appealingly simple: By removing people who are engaged in crime, the remaining residents are free of victimization (Zedlewski, 1987). Fear of crime discourages public cooperation, increases residential isolation, and contributes to crime by encouraging a disengagement from collective public life (Skogan, 1990). Alternatively, the argument that incarceration, at high rates, can destabilize neighborhoods can be seen as a natural conclusion derived from the ecological perspective on crime (for a full discussion, see Rose & Clear, 1998a). This approach situates individual criminal events in their social contexts, underscoring the significance of "place" and investigating how people interact with places in ways that tend to contribute to or constrain crime (see Bursik & Grasmick, 1993; Hawkins, Laub, Lauritsen, & Cothern, 2000).

The current study explores the way people living in high-incarceration neighborhoods understand and experience the impact of incarceration on themselves, their families, and their communities. We are particularly interested in the processes of removing and returning offenders, rather than the experience of incarceration itself, because we want to better understand how incarceration affects the social networks (and subsequently the social capital) necessary for residents to positively affect community-level, informal social control. Thus, our approach was designed to identify factors associated with the incarceration process that either promote or reduce community stability and, as a result, either promote or reduce crime.

Data and Method

The data for this article were drawn from interviews in Tallahassee, Florida, conducted in June 2000 in conjunction with a larger study of two high-incarceration neighborhoods in Tallahassee, Florida (Frenchtown and South City). We used snowball sampling methods to identify potential neighborhood participants. After pilot tests and nearly 100 screening interviews, we then conducted individual and group interviews and four focus groups with people who either live or have businesses (or services) operating in the two neighborhoods.

We did not ask respondents directly about the impact of incarceration on social networks or public safety. Instead, our approach was to ask for general commentary about the impact of incarceration on the participants and their families and communities and then to explore responses to these opening probes. The focus groups were led by a group facilitator and conducted at sites in each of the neighborhoods. In an effort to minimize social distances along racial, ethnic, educational, and professional lines, each group was hosted by a representative of the local neighborhood association. The facilitator was assisted by members of a local justice advocacy organization, one of whom had previously been incarcerated. Ex-offenders were interviewed separately to maximize everyone's comfort in talking about sensitive issues of removal and reentry.

Twenty-six people participated in the resident focus groups, and 5 people participated in the ex-offenders focus groups (we augmented the ex-offender data with interviews with an additional 8 ex-offenders in individual and group settings). The resident focus group members were overwhelmingly female (71%–75%) and ranged in age from 21 to 73. Ex-offender interviewees were all males in their 20s, 30s, and 40s, All but 1 of the participants were African American, reflecting the demographic composition of the neighborhoods.

⊠ Results: Four Domains of Community Impacts

It is important to emphasize that respondents report a complicated picture of the effects of incarceration on the neighborhood. Some consequences they describe are positive in the sense that the neighborhood and its residents are better off with some residents incarcerated. Not surprisingly, respondents feel that justice is done when wrongdoers are apprehended, prosecuted, and sanctioned. Respondents did not hesitate to say that removing convicted offenders makes the streets safer and life better. At the same time, partly as a consequence of the ubiquity of incarceration as an experience in their everyday lives, they see its negative impact as well. In the focus groups and individual interviews, respondents devoted more time to describing the negative effect of incarceration on their communities than the positive, and their passions seemed to be more readily engaged by these issues than the traditional matters of public safety. Ex-offenders raised the same kinds of themes but also emphasized the pressure they feel in reentry from almost every source: the criminal justice system, society in general, neighbors, and families.

Our analysis shows that the insights our respondents describe about the impact of incarceration on the community can be categorized into four domains: the problem of stigma, financial impacts, issues regarding identity, and the maintenance of interpersonal relationships. Elsewhere (Rose, Clear, & Ryder, 2000), we describe in detail the way our respondents describe these domains and explain their perspectives on them. Here their points are summarized.

Incarceration and the problem of stigma Being involved in the criminal justice system is a negative social status. The consequences of stigma are far-reaching, as stigma damages both human and social capital. In these communities, even though the experience of incarceration is widespread, it is still stigmatizing, and incarceration is not discussed openly. Residents noted that "offender" becomes a master status. Ex-offenders find it difficult to get good jobs, and their noncriminal neighbors and associates are often suspicious and loathe to interact with them. Integrating into the community and networks that provide informal social control is made problematic for ex-offenders when many community members react with distrust. The stigma also is frequently transferred to the ex-offender's family and community. Families often feel public shame for the actions of their loved ones. Businesses that might provide jobs are reluctant to locate in a neighborhood with a "bad reputation," and residents feel that living elsewhere would provide a better situation for family members. The net result of stigma is the loss of the community's reputation as a good place to live and do business. These locations become defined as bad places, and, consequently, positive social organization becomes an uphill battle.

Financial impact of incarceration One of the most significant points our respondents repeatedly make is that incarceration has adverse effects on the financial capacity of the neighborhood. Although some families may experience financial relief from the drain caused by an offender, many families lose a breadwinner and, as a result, struggle to compensate for this loss. In addition, families often support the offender both during incarceration and following release. At the community level, local businesses experience financial losses when ex-offenders fail as employees, and they may lose customers when idle residents (frequently ex-offenders unable to find employment) hang out on the street.

The financial implications are complicated, because incarceration can bring relief from the drain of an active offender but also may impose new costs associated with caring for a family member in confinement. The implications for informal social control are mostly indirect, because family networks that suffer a depletion of financial resources struggle to build human capital and take advantage of potential sources of family strength. A clearer understanding of the financial issues emerges when one considers that these neighborhoods are already confronted with a high degree of poverty and economic deprivation, and most who live there find it difficult to absorb additional deficits.

Incarceration and the problem of identity

The people who live in communities that have high concentrations of residents flowing in and out of prison know they live in "problem" places. A consequence is that residents struggle with issues of personal identity. In particular, residents report that community members (particularly children) suffer a loss of self-worth and self-esteem. There is a loss of positive role models for children, and, overall, the community experiences a sense of hopelessness and apathy.

An important part of human and social capital is the shared sense among a social collective that life prospects may be affected through collective action. This is the essence of collective efficacy—that people believe their lives can be enhanced by shared social action. How respondents describe the impact of a high concentration of incarceration in their neighborhoods stems from the sense of powerlessness that incarceration brings out in their lives. People who feel low self-esteem may be less likely to set high personal goals and less likely to engage in goal-directed collective social activity, and this may be particularly problematic for the aspirations and activities of young people whose adult role models are incarcerated. The ability of a person to thrive socially is partly a result of that person's sense of identity, and an identity associated with the cycle of incarceration is a difficult foundation on which to build life choices.

Incarceration and the dynamics of community relationships

Interpersonal networks are disrupted by high rates of incarceration. The aggregate effect of these disruptions is a reduction in the capacity of social supports for all concerned. Whereas the removal of a problem family member can improve relationships among remaining family members, frequently both spousal and parent-child relationships are strained or severed as a result of incarceration. Families often suffer an overwhelming sense of loss, and the care and supervision of children suffers when a parent is gone. Relationships with neighbors also can become strained, especially if families isolate from their neighbors and the larger community. When ex-offenders return, neighbors are welcoming but cautious, suspicious, and frequently fearful. Neighborhood networks can be severed if families move to a different environment or if ex-offenders move to a different environment and move in with extended family members. Finally, our respondents reported restricted public social interaction because of police surveillance.

Our understanding of the social importance of place stems from a recognition of the powerful importance of local relationships, especially in low-income locations where people lack access to supports outside the local area. A high concentration of incarceration disrupts or impairs the formation and development of those relationships at the neighborhood level, and weakened relationships provide a diminished foundation for informal social control.

In sum, when asked to discuss the impact of incarceration on community life, residents of high-incarceration neighborhoods readily identify a number of ways incarceration affects their lives. Some of these effects act on the networks of offenders, that is, families and other close associations. Others are broader, having importance for abstractions such as community identity and for tangible community assets such as economic activity. Because of the adverse effects of high incarceration rates on community life, it is understandable that networks, which are the foundation of informal social control, suffer when rates of incarceration are high.

Consistent with our quantitative results reported elsewhere (Rose, Clear, Waring et al., 2000), the qualitative data do not indicate that the effects of incarceration are wholly negative. Respondents describe ways in which incarceration strengthens households by removing problem family members. They also provide strong support for law enforcement methods that interrupt criminal activities and reinforce legal norms. Residents do not ask for less criminal justice. But they express concern and regret about the way the high levels of incarceration lead to secondary community problems, and they seek ways to address those problems.

▧ Programmatic Recommendations

The perspectives of residents, including ex-offenders, can be seen as a call for change in the way justice services are provided in high-incarceration communities. We can envision a comprehensive programmatic response to the problems that arise from high rates of incarceration concentrated in certain communities.

In making these recommendations, we recognize that there are public safety issues facing the criminal justice system that call for supervision, surveillance, and enforcement, and we do not wish to undermine that fact. Not all offenders want to change. Some will resume their old lives upon reentry. Likewise, not all families are well suited to receive ex-felons upon their reentry. Nonetheless, we believe these recommendations could minimize the disruptions incarceration causes in the lives of residents and, subsequently, improve the quality of life in these neighborhoods. Our recommendations are meant to offset problems families experience when offenders are incarcerated and for the case in which an offender returning to the community wants to succeed but faces significant obstacles in doing so. They are for the families that want to be a support system but lack the capacity for doing so as fully as might be possible with services; this circumstance applies to the majority of situations involving reentry to high-incarceration neighborhoods. We see the following recommendations as potentially useful to high-incarceration locations generally, but as particularly useful to the neighborhoods of Frenchtown and South City. A detailed description of the rationale for these recommendations is provided elsewhere (Rose, Clear, & Ryder, 2000). Here we summarize them.

1. Target families of incarcerated offenders for an array of services. Appropriate services would alleviate many of the problems and lower the level of disorganization incurred immediately by many families when a member is incarcerated. These services might include

a. short-term financial assistance for food, clothing, and housing;

b. short-term crisis-oriented mental health assistance to deal with anger, depression, and self-esteem issues, particularly for children;

c. parenting classes;

d. dental and physical health assistance;

e. supervisory and recreational services for children; and

f. adult mentors for children.

2. Facilitate contact between families and incarcerated family members. Assistance would promote the family bonds that are essential for successful reintegration into community life, and it would also help individuals maintain ties with their children while incarcerated. Maintenance of family bonds, especially with children, often is an incentive for an inmate's "good behavior" while incarcerated. Assistance might include

    a. low-cost telephone service between inmates and their families and

    b. assistance with transportation to prisons.

3. Provide services to children of prisoners to help stabilize their living situation. Many children lose one or more of their parents to incarceration, and many are raised by a caretaker relative—grandmother, aunt, or sister, for example—or are placed in foster care. These children and their caretakers could benefit from the following services:

    a. counseling for common problems, such as depression, anger, shame, and low self-esteem;

    b. counseling for caretakers about how to talk with the children about the situation;

    c. intervention regarding acting-out problems; and

    d. assistance in maintaining meaningful contact with the incarcerated parent, including family-oriented programs in prison.

4. Implement comprehensive prerelease transition plans that address family needs. These plans would maximize the health of the family, optimize successful reentry, and reduce recidivism by anticipating the problems incurred when an ex-offender is released. Transition plans might

    a. determine whether incarcerated individuals should return to their families upon release;

    b. determine whether released individuals should return to their communities or move to new neighborhoods;

    c. determine whether families and released ex-offenders should move to new neighborhoods together;

    d. identify employment and housing possibilities for families and returning offenders who choose to move to new neighborhoods;

    e. link inmates to the exact services they need upon release and begin the service delivery process prior to release;

    f. address typical inmate fears, such as concern about partner faithfulness, community attitudes, and so on; and

    g. provide family-focused interventions to cope with the strains of reintroducing the ex-offender into the family.

5. Provide transitional housing for ex-offenders. This would alleviate the immediate need ex-offenders have for a place to stay and prevent people from heading to the streets or the shelters. It also would relieve the burden families sometimes experience when they house ex-offenders. Such housing, with a house monitor to assist ex-offenders in reintegrating, could function as a service center, facilitating the process of obtaining identification papers, clothing, employment, and so on.

6. Modify rules that disallow individuals with a felony record to acquire a lease. The inability of many ex-offenders to acquire a lease often forces them into transient living conditions and, in effect, undermines their acceptance of responsibility. It can also rupture marital and parental relationships when, for example, a man's wife is allowed a lease but must "sneak" him in for visits. Such an arrangement is also detrimental to the offender's self-esteem and presents a poor model of fatherhood to children.

7. Assist ex-offenders in obtaining and retaining employment. Such assistance would alleviate the financial strain ex-offenders experience and the financial burden often absorbed by families, and it would also reduce the stigma associated with incarceration and unemployment. Assistance might include

a. programs to help ex-offenders become self-employed,

b. employer-education programs to promote the hiring of ex-offenders,

c. encouraging employers to hire ex-offenders through a program of government "bonding" to reduce the risk assumed by potential employers, and

d. encouraging employers to provide full-time employment (40 hours per week) and benefits.

8. Make training, education, and legal assistance available to ex-offenders. Training and education are the foundation of quality employment. Ex-offenders who have trouble getting good jobs should be able to obtain job training. In addition, ex-offenders need basic information about legal issues and assistance in solving legal problems. Ex-offenders also need help in restoring their civil rights and closing out any pending criminal cases and legal obligations. Affordable legal help is not typically available, but internships for students from local law schools could be instituted to assist with the legal needs of ex-offenders and their families.

9. Reduce initial financial pressures faced by ex-offenders immediately upon release. This can be accomplished by reducing the unnecessary burdens imposed by the criminal justice system, such as supervision fees, and providing short-term financial assistance to pay for such needs as security deposits and the first month's rent, initiating utilities, and obtaining toiletries and other basic necessities. Such financial assistance would reduce the incentive to participate in illegal activities for quick money.

10. Increase the availability of low-cost drug treatment programs for ex-offenders and their families. Currently available programming is insufficient to meet needs or, because it is not locally based, is not easily accessible to residents of these neighborhoods.

11. Form self-help support groups for ex-offenders. These groups would help model successful reintegration into the community by allowing ex-offenders to talk to each other about the pressures and temptations they face, the frustrations of trying to make it, and the discouragements of everyday life. They can also help head off relapse and recidivism by reducing anger and bolstering self-esteem.

12. Match ex-offenders to community mentors. Mentors would serve as advisors, contacts, and support for returning offenders. They can help ex-offenders with very basic life skills, such as how to open a checking account and other mundane requirements. Mentors can also be part of the transition planning process and serve as advocates for the ex-offenders' needs and interests in reentry. The mentor system can apply to families as well, with families "adopting" other families for support.

13. Involve ex-offenders in neighborhood projects. Ex-offenders can play a role in a wide range of positive neighborhood activities, from organized sports programs to neighborhood reclamation projects. This would put ex-offenders in productive contact with fellow residents in neighborhood activities that lead to the overall improvement of the community. It also would reduce the stigma and isolation associated with incarceration. These projects might include

a. work programs that improve public space in the community,

b. renovations of housing and other building stock, and

c. recreational sports programs.

14. Develop awareness programs to reduce the stigma of incarceration for ex-offenders who attempt a clean start. De-stigmatizing individuals and communities should help reduce the pressures experienced by ex-offenders who are attempting to make a new start in the community. A broader understanding of the needs and obstacles facing ex-offenders will also enhance the quality of community life by countering some of the unintended consequences. Programs might target

a. police, to help alleviate difficult community tensions;

b. probation officers, to assist in the reintegration process;

c. employers, who may disdain or be fearful of hiring ex-offenders;

d. educators, who can talk about the problem of reentry with greater sensitivity; and

e. the community at large, to encourage tolerance for returning felons.

15. Provide services at a neighborhood-based center. Such a center would

a. promote access to services for families and returning offenders,

b. enable services to be tailored to the specific needs of the community,

c. promote integration and informal networks by locating multiple services in one place,

d. involve neighborhood groups such as neighborhood associations in the design and delivery of services, and

e. transfer resources from society at large to the community by adding a local service entity to the neighborhood and by being a site through which financial resources can be funneled into the neighborhood.

16. Provide services through coalitions and partnerships of public and private sources. Human service organizations, both public and private non-profits, can organize coalitions to develop and concentrate their work in high-incarceration communities. Private, for-profit organizations can contribute to the costs of public services financially and programmatically. This would leverage the resources of both public and private interests and direct them toward community-based strategies, which might include

a. police partnerships with resident groups to engage in problem-solving strategies and to provide families with support when they need it,

b. social service provider-neighborhood partnerships to coordinate and intensify local service delivery,

c. public-private partnerships to create new jobs for residents, and

d. expert-citizen group partnerships that help resident groups develop grant proposals and new projects.

▧ Research Recommendations

Our data add to our knowledge about the way removal and reentry affect community life, and the recommendations point to a potentially more effective way of dealing with concentrated levels of incarceration in certain communities. Yet this issue remains poorly studied, and our research points to a number of questions that deserve further inquiry. Below, we note some of the more prominent issues raised by our research.

1. Conduct replications and extensions of this research. Further studies are needed of the neighborhood-level significance of concentrations of incarceration. Ideally, such studies would apply to dense urban areas as well as high-incarceration rural areas and would continue to investigate the impact of incarceration on social networks and collective efficacy.

2. Investigate whether offenders do better upon reentry going to their old neighborhoods or to new ones.

Some anecdotal data suggest that ex-offenders who are released to new locations do better, because they are able to avoid the temptations of old patterns and acquaintances. Yet there are good empirical and anecdotal reasons to believe that support systems from the neighborhood can play a critical role in successful adjustment after incarceration. These are obviously two very different strategies of reentry. Studies that compare offender experiences with these two strategies would help to identify the circumstances under which one strategy might be preferred over the other.

3. Investigate whether offenders who have meaningful contact with their neighbors do better than those who do not.

Our respondents, residents and ex-offenders alike, suggest that there is an important role for neighbors in supporting an ex-offender's reentry. On the other hand, neighborhood notification laws have made neighbor relationships potentially more problematic. Respondents report a tendency of ex-offenders to be isolated from normal neighbor relations in ways that might impede successful adjustment. Little is known about the various ways nonfamily neighbors relate to ex-offenders at the community level. In theory, a positive relationship with a neighbor might be an important aid to adjustment, and the opposite might lead to adjustment problems. Studies that document the interaction between offenders in reentry and their nonfamily neighbors and the impact of those interactions will help formulate successful policies regarding neighbor notification and reentry support.

4. Improve our understanding of the processes that foster the intergenerational incarceration cycle.

It has been well established that having a parent or older sibling in prison is a significant risk indicator of incarceration. Little is known about why this is so, and even less is known about the circumstances that enable children living under these risk situations to avoid incarceration. A better understanding of the reasons why parental/sibling incarceration creates such a risk, and how some families successfully avoid or navigate that risk, will provide information about the necessary focus of services to families of men and women in prison.

5. Improve our understanding of the relationship between ex-offenders' ties to parochial social controls (e.g., religious institutions) and reentry adjustment.

One of the most common comments from our sample is the importance of faith and faith organizations in successful adjustment. Despite widespread agreement about the importance of such parochial social controls, very few studies have been conducted on the nature and level of impact of these associations on adjustment. A better understanding of the role of these organizations in ex-offender adjustment would inform the development of public-private partnerships for high-incarceration locations.

6. Investigate the impact of differences between types of neighborhood support structures (e.g., job markets, family structures, housing patterns) and reentry adjustment.

Whereas offenders reenter certain neighborhoods in concentrated numbers, the demographic nature of these communities varies. Little is known, however, about which neighborhood characteristics affect reentry patterns, and how. A better understanding of how certain characteristics affect reentry processes would enable us to more reliably identify the local areas requiring more comprehensive support based on these attributes.

7. Evaluate the impact of targeted services to families of incarcerated offenders. Many evaluations of services find that they have limited impact on the problems they are designed to alleviate. There is no reason to assume that services targeted to families of offenders would be any different in this common outcome. It will not be enough merely to offer services; rather, we should study the impact of those services on the problems they target and the broader public safety aims of criminal justice.

8. Evaluate the impact of best-practice transition planning methods. It is a truism that better transition planning is needed, but no reliable studies exist on how high-quality transition planning impacts reentry success. Furthermore, there is no empirical basis to say which aspects of transition planning are most important to increase successful reentry. Studies are needed that tell us more about how well transition planning works, which aspects are most important, and why.

9. Investigate the impact of reentry financial requirements, such as supervision fees, on the probability of return to prison. It is plausible that the net effect of the financial requirements on offenders is to increase their chances of failure. If so, it may follow that these financial policies cost more taxpayer resources to pay for the consequences of reentry failure (new crimes, return to prison) than is justified by the funds they generate to defray taxpayer costs of correctional and other services. On the other hand, financial requirements might impose a discipline on ex-offenders that reinforces the kind of responsible conduct required for successful reentry and thereby not only provide additional revenues but actually increase the chances of success. Currently, we do not know the impact of the financial aspects of reentry.

References

Bonczar, T. P., & Beck, A. J. (1997). *Lifetime likelihood of going to state or federal prison.* Washington, DC: U.S. Department of Justice, Bureau of Justice Statistics.

Bursik, R. J., Jr., & Grasmick, H. G. (1993). *Neighborhoods and crime: The dimensions of effective community control.* New York: Lexington.

Center for Alternative Sentencing and Employment Services. (2000). *The community justice project.* New York: Author.

Clear, T. R., & Rose, D. R. (1999). When neighbors go to jail: Impact on attitudes about formal and informal social control. In *National Institute of Justice Research in Brief.* Washington, DC: National Institute of Justice.

Doble, J., & Green, J. (1999). *Public opinion and criminal justice in Vermont.* (Final Rep. to the National Institute of Justice). Washington, DC: Dable and Associates.

Felson, M. (1987). Routine activities and crime prevention in the developing metropolis. *Criminology, 25*(4), 911–931.

Hawkins, D. F., Laub, J. H., Lauritsen, J. L., & Cothern, L. (2000, June). Race, ethnicity and serious violent juvenile offending. *Juvenile Justice Bulletin.* Washington, DC: Office of Juvenile Justice and Delinquency Prevention.

Henderson, M., Cullen, F. T., Cao, L., Browning, S. L., & Kopache, R. (1997). The impact of race on perceptions of criminal injustice. *Journal of Criminal Justice, 25,* 447–462.

Lofland, J., & Lofland, L. H. (1995). *Analyzing social settings: A guide to qualitative observations and analysis.* Belmont, CA: Wadsworth.

Lynch, J. P., & Sabol, W. J. (1992, November). *Macrosocial changes and their implications for prison reform: The underclass and the composition of prison populations.* Paper presented to the American Society of Criminology, New Orleans, LA.

Mauer, M. (2000). *Race to incarcerate.* Washington, DC: The Prison Project.

Perry, J. G., & Gorczyk, J. F. (1997). Restructuring corrections: Using market research in Vermont. *Corrections Management Quarterly, 3,* 26–35.

Rose, D. R., & Clear, T. R. (1998a). Incarceration, social capital and crime: Examining the unintended consequences of incarceration. *Criminology, 36*(3), 441–479.

Rose, D. R., & Clear, T. R. (1998b, November). *Who doesn't know someone in jail? The impact of exposure to prison on attitudes of formal and informal control.* Paper presented to the American Society of Criminology, Toronto, Canada.

Rose, D. R., Clear, T. R., & Ryder, J. A. (2000, September). *Drugs, incarceration and neighborhood life: The impact of reintegrating offenders into the community* (Final Rep. to the National Institute of Justice). New York: John Jay College.

Rose, D. R., Clear, T. R., Waring, E., & Scully, K. (2000). *Coercive mobility and crime: Incarceration and social disorganization.* Unpublished manuscript.

Sampson, R. J. (1988). Local friendship ties and community attachment in mass society: A multilevel systemic model. *American Sociological Review, 53,* 766–779.

Shaw, C. R., & McKay, H. D. (1942). *Juvenile delinquency and urban areas.* Chicago: University of Chicago Press.

Skogan, W. (1990). *Disorder and decline: Crime and the spiral of decay in American neighborhoods.* New York: Free Press.

South, S. J., & Deane, G. D. (1993). Race and residential mobility: Individual determinants and structural constraints. *Social Forces, 72,* 147–167.

Wilson, W. J. (1987). *Truly disadvantaged.* Chicago: University of Chicago Press.

Zedlewski, E. (1987). *Making confinement decisions: The economics of disincarceration.* Washington, DC: U.S. Department of Justice.

DISCUSSION QUESTIONS

1. According to the authors, what is the effect on the community of large numbers of its inhabitants being removed from it?

2. In your own opinion, is it a good or bad thing for the community when its "bad apples" are removed?

3. What is the solution to the community problems caused by removing and returning offenders to the community?

❖

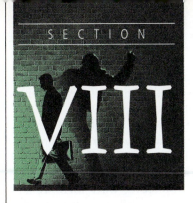
JAILS

Introduction

The American jail is a derivative of various modes of holding people for trial that have existed in Western countries for centuries (Zupan, 1991). Whether fashioned from caves or mines or old houses or as separate buildings, jails were developed originally as a primary means of holding the accused for trial, for execution, or in lieu of a fine (Kerle, 2003; Welch, 2004). They have been in existence much longer than prisons and, as mentioned in Section II, their mission is much more diverse. They hold people who are presumed innocent before trial, convicted people before they are sentenced, more minor offenders who are sentenced for terms that are usually less than a year, juveniles (usually in their own jails or separated from adults), women (usually separated from men and sometimes in their own jails), and people for the state or federal authorities. Also, depending on the particular jail population being served and the capacity of any given facility, they serve to incapacitate, deter, rehabilitate, punish, and reintegrate. Jails are multifaceted, usually local, facilities with a complex mission.

Described as correctional afterthoughts by scholars, and despite their complicated mission, jails have often received short shrift in terms of monetary support and professional regard (Kerle, 1991, 2003; Thompson & Mays, 1991; Zupan, 1991). Most jails are operated by county sheriffs whose primary focus has been law enforcement rather than corrections. As a result, jail facilities have often been neglected, resulting in dilapidated structures and jail staffs with less training and pay than correctional staffs working at the state or federal level in prisons or than deputy sheriffs working in their own organizations. The late comic Rodney Dangerfield's perennial lament "I don't get no respect" surely applies to jails more than perhaps any other social institution.

Jails, much like prisons, have felt the pressures of surging inmate populations as the drug war and other political initiatives have taken hold across the country. Because of a building binge that has led to a phenomenal increase in beds, jails are now at about 94% of capacity, rather than the 100-plus percent they were at for much of the 1980s and 1990s (Beck, 2002; Bureau of Justice Statistics, 2008b; Gilliard & Beck, 1997).

But they remain virtually full because of the drug war and the lack of mental health facilities in communities (Beck, 2002). As the secure and long-term mental health facilities closed their doors, beginning in the 1950s and 1960s, jails have held people who are only nominally criminal, because there is no other social institution to hold them (Kalinich, Embert, & Senese, 1991; Severson 2004). Moreover, when there are no or not enough beds for the homeless in any given community, jails are often the dumping grounds for those poor and dispossessed people (Irwin, 1985).

But there has been some good news about jails in the late twentieth and early twenty-first centuries too. Beginning in the 1980s, and continuing today, there has been a major shift in the paradigm that has guided jail development and operations, at least for some large city jails (Kerle, 2003; Zupan, 1991). They began to adopt the podular–direct supervision (or new generation) jail model (discussed in the following) that entailed not just a different physical design but a management philosophy that is more open than that within traditional jails (Gettinger, 1984; Nelson & Davis, 1995; Zupan & Menke, 1988). At about this same time, managers of some larger city jails (and, these days, medium-sized jails across the country) recognized that jail staff must be trained and compensated at the same level as deputy sheriffs if professional-level knowledge and behavior were to be expected of them (Kerle, 2003; Zupan, 1991).

In addition, the central community role that jails play is now being recognized, understood, and positively exploited by correctional practitioners. This has led to the notion that, as people are admitted to jail with numerous physical, mental, and substance abuse problems—not to mention educational deficits, the programming and treatment that they received on the outside should continue to follow them in the jail and upon

▲ **Photo 8.1**   A jail inmate awaits transportation to his court hearing.

release. Part and parcel of this belief is the emerging interest in community reentry programs (discussed in the last section on parole), as a means of preventing crime and addressing the multifaceted needs of jail ex-mates. In this section of the book, these emerging trends will also be explored, but first we address the all too common problems that jails in this country face.

⊠ Overcrowding

As indicated in other sections of this book, jails have to deal with the same kinds of overcrowding issues that have afflicted prisons. Overcrowding occurs when the number of inmates exceeds the physical capacity (the beds) available. Each year, and over the last several decades, the number of jail beds needed by jurisdictions has increased, though they have been filled almost as soon as they have been built. As of June 2006 (the latest data available), on average, jails were operating at 94% of their capacity (Bureau of Justice Statistics, 2008b). This percentage use of capacity is actually better than in past years when jails of the 1980s and 1990s were operating at well over their rated capacity (Cox & Osterhoff, 1991; Gilliard & Beck, 1997; Klofas, 1991). Also, and notably, even an average of 94% means that half of the jails in this country are operating at over that average, while half are operating at under it.

Overcrowding limits the ability of the jail to fulfill its multifaceted mission: Less programming can be provided, health and maintenance systems are overtaxed, and staff are stressed by the increased demands on their time and the inability to meet all inmate needs. From the inmate's perspective, their health, security, and privacy are more likely to be threatened when the numbers of inmates in their living units increase and the amount of space and possibly the number of staff do not. The jail staff members also lose their ability to effectively classify and sometimes control inmates; they may be unable to keep the serious convicted offenders away from the presumed innocent unconvicted or the more minor offending inmates. Judges and jail managers will struggle over how to keep the jail population down to acceptable limits, and, as a result, even serious offenders may be let loose in communities as a means of reducing the crowding. Therefore, though the "get tough" laws in many states were passed with the explicit intent of incarcerating more people for longer, their actual unintended effect in some jails may be to incarcerate serious offenders less (as there is no room) and all offenders in less safe and secure facilities.

Though suits by jail inmates are usually not successful, some are. Welsh (1995) found in his study of lawsuits involving California jails that the issue that courts gave greatest credence to was overcrowding. Perhaps this is because overcrowding is clearly quantifiable (the rated capacity is clear, and the inmate count is obvious), but it is likely that it was regarded as so important by courts because it can lead to a number of other seemingly intractable problems, such as those just mentioned.

⊠ Jail Inmates

Gender, Juveniles, Race, and Ethnicity

As indicated from the data supplied in Tables 8.1, 8.2, and 8.3, most jail inmates are adult minority males, though whites represent the largest racial grouping of the men.

Women comprise over 13% of jail inmates, and the number of women in jail has increased faster than the number of males. The reason often cited for the overall increases in incarceration in jails and prisons and the increases for women and minorities in jails and prisons, in particular, has been the prosecution of the drug war since the 1980s and 1990s. The "get tough" policies have led to longer periods of incarceration in prisons and also to a greater propensity to catch and keep low-level drug offenders in jails (Irwin, 2005; Owen, 2005; Welch, 2005; Whitman, 2003). The focus of arrests in the drug war has often been on the low-level sellers rather than the buyers, and that has netted more minorities and women into the system. Mandatory sentences, juvenile waivers, and sentence enhancements for certain offenses have collectively led to longer sentences for most offenders and backed up numbers of offenders in some jails, either awaiting transfer to state or federal prisons or doing their time there rather than in the overcrowded prisons.

As a consequence, and as is evident from the figures presented in Table 8.1, the number of adult males in jail from 1990 to 2006 almost doubled, while the numbers of adult females and juveniles almost tripled. Percentage increases for women and juveniles are also large: in 1990, women represented only about 9% of jail populations and juveniles about .6% whereas, by 2006, women comprised 13% and juveniles .8%.

Across the two largest racial groupings (whites and African Americans) and the largest ethnic grouping (Hispanics), there have been significant increases in jail incarceration. As indicated in Tables 8.2 and 8.3, the raw number of whites has increased from

Table 8.1	Jail Populations by Age and Gender, 1990–2006		
	Number of Jail Inmates (One-day count)		
Year	**Adult Males**	**Adult Females**	**Juveniles**
1990	365,821	37,198	2,301
1991	384,628	39,501	2,350
1992	401,106	40,674	2,804
1993	411,500	44,100	4,300
1994	431,300	48,500	6,700
1995	448,000	51,300	7,800
1996	454,700	55,700	8,100
1997	498,678	59,296	9,105
1998	520,581	63,791	8,090
1999	528,998	67,487	9,458
2000	543,120	70,414	7,615
2001	551,007	72,621	7,613
2002	581,411	76,817	7,248
2003	602,781	81,650	6,869
2004	619,908	86,999	7,083
2005	646,807	93,963	6,759
2006	661,329	98,577	6,104

1990 (when there were fewer whites incarcerated in jails than African Americans) to the point where, in 2006, they comprise the largest racial grouping. Proportionate to their representation in the population, however, African Americans are much more likely to be incarcerated in American jails than are whites or Hispanics. As reported by the Bureau of Justice Statistics (2008b), in 2006, "Blacks were almost three times more likely than Hispanics and five times more likely than whites to be in jail" (p. 2). Again, this higher proportional rate of incarceration for African Americans in particular can likely be attributed to their greater concentration in impoverished neighborhoods and to the focus of the drug war, which has tended to target such living areas.

The Poor and the Mentally Ill

John Irwin (1985) once referred to the types of people who are managed in jails as the "rabble," by which he meant "[d]isorganized and disorderly, the lowest class of people" (p. 2). These were not just the undereducated, the under- or unemployed, or even the poor and mentally ill. He meant to include all those descriptors as they related to their state of being disorganized and disorderly and as those designations might lead to permanent residence in a lower class; but he also meant that jail inmates tend to be "detached" and of "disrepute" in the sense that they offend others by committing mostly minor crimes in public places.

Table 8.2	Jail Populations by Race and Ethnicity, 1990–2006		
	Number of Jail Inmates (One-day count)		
Year	White, Non-Hispanic	Black, Non-Hispanic	Hispanic of Any Race
1990	169,400	172,300	58,000
1991	175,300	185,100	60,600
1992	178,300	196,300	64,500
1993	180,700	203,200	69,400
1994	191,800	215,300	75,500
1995	206,600	224,100	75,700
1996	215,700	213,100	80,900
1997	230,300	237,900	88,900
1998	244,900	244,000	91,800
1999	249,900	251,800	93,800
2000	260,500	256,300	94,100
2001	271,700	256,200	93,000
2002	291,800	264,900	98,000
2003	301,200	271,000	106,600
2004	317,400	275,400	108,300
2005	331,000	290,500	111,900
2006	336,600	296,000	119,200

Table 8.3	Jail Incarceration Rates by Race and Ethnicity, 1990–2006		
	Number of Jail Inmates per 100,000 U.S. Residents		
Year	**White, Non-Hispanic**	**Black, Non-Hispanic**	**Hispanic of Any Race**
1990	89	560	245
1991	92	594	247
1992	93	618	251
1993	94	633	262
1994	98	656	274
1995	104	670	263
1996	111	640	276
1997	117	706	293
1998	125	716	292
1999	127	730	288
2000	132	736	280
2001	138	703	263
2002	147	740	256
2003	151	748	269
2004	160	765	262
2005	166	800	268
2006	170	815	283

NOTE: U.S. resident population estimates for race and Hispanic origin were made using a U.S. Census Bureau Internet release with adjustments for census undercount. Estimates for 2000–2006 are based on the 2000 Census and then estimated for July 1 each year.

Certainly, the fact that one is homeless puts that person at a greater risk for negative contact with the police; for one who lacks a home, private matters are more likely to be subject to public viewing in public spaces. Those who are mentally ill are more likely to be homeless, as they are unable to manage the daily challenges that work and keeping a roof over one's head and food in one's mouth require (McNiel, Binder, & Robinson, 2005; Severson, 2004).

Jails in this country are full of the mentally ill and the poor. The latest data from the Bureau of Justice Statistics (based on interviews of local jail inmates in 2002) indicate that about 64% of jail inmates (75% of females and 63% of males) have a mental health problem (as compared to 56% of state prisoners and 45% of federal prisoners) (James & Glaze, 2006, p.1). In contrast, about 10.6% of the United States population has symptoms of mental illness. Moreover, for virtually every manifestation of mental illness, more jail inmates than state or federal prisoners were likely to exhibit symptoms, including displaying 50% more delusions and twice as many hallucinations (James & Glaze, 2006, p. 2). Jail inmates with a mental health problem had specific diagnoses that included mania (54%), major depression (30%), and a psychotic disorder (24%). The specific identification of a mental illness for each inmate by the BJS research team was based on a recent clinical diagnosis or on the prisoner's demonstration of symptoms that fit the criteria of the *Diagnostic and Statistical Manual of Mental Disorders* (DSM-IV).

A whole host of problems has been found to be associated with mental illness including homelessness, greater criminal engagement, prior abuse and substance use (McNiel et al., 2005). Among the findings from this BJS study of jails was that those with a mental illness were almost two times as likely to be homeless as those jail inmates without a mental illness designation (17% as opposed to 9%) (James & Glaze, 2006, pp. 1–2). More inmates with a mental health problem had prior incarcerations than those without such a problem (one quarter as opposed to one fifth). About three times as many jail inmates with a mental health problem had a history of physical or sexual abuse than those without such a problem (24% as opposed to 8%). Almost three quarters of the inmates with a mental health problem were dependent on, or abused, alcohol or illegal substances (74% as opposed to 53% of those without a mental health problem). In short, mental illness and poverty were entangled in a whole array of societal issues for jail inmates.

Further evidence for this supposition was found by McNiel, Binder, and Robinson (2005) in their study of San Francisco County. They found that mental illness, substance abuse, and jail incarcerations were inextricably connected as life events. Those who were mentally ill and homeless were also more likely to have a substance abuse problem; or those who had a substance abuse problem and were homeless were also likely to be mentally ill. And, in both sets of instances, it is likely that jail incarcerations have been part of their existence as well.

Medical Problems

One of those social issues that is particularly problematic for jail inmates, and the people who manage them, is the relatively poor health of people incarcerated in jails (Williams, 2007). According to that same 2002 study of jail inmates by the Bureau of Justice Statistics, more than a third of jail inmates, or 229,000 people, reported a medical problem more serious than a cold or the flu (Maruschak, 2006, p. 1). Most of these medical maladies preceded placement in jail and included (in order of prevalence): arthritis, hypertension, asthma, heart problems, cancer, paralysis, stroke, diabetes, kidney problems, liver problems, hepatitis, sexually transmitted diseases, tuberculosis, and HIV. A small percentage of inmates (2%) were so medically impaired that they needed to use a cane or a walker or a wheelchair.

As one might expect, the elderly are much more prone to some of these medical maladies than would be younger inmates. In the BJS study, 61% of those over 45 reported a medical problem (Maruschak, 2006, p. 1). With the exception of asthma and HIV, which tended to be more prevalent among younger inmates, the older inmates were much more likely to have the other medical problems tallied in this report, which means that older inmates are more costly to manage in jails because they are more likely to need medical care.

Like the older inmates, women were much more likely to report medical problems to the BJS researchers (53% for women as opposed to 35% for men) (Maruschak, 2006, p. 2). They reported a rate of overall cancers that was almost eight times that of men (831 per 10,000 inmates compared to 108 per 10,000 inmates), with the most common cancer being cervical for women and skin for men. In fact, for every medical problem documented in the study, the women reported more prevalence than the men, with the exception of paralysis, where they were even with men, and tuberculosis, where a slightly greater percentage of men reported more (4.3% for men as opposed to 4.0% for women) (Maruschak, 2006, p. 2).

Incarcerated youth have their own potentially debilitating health problems that also present an immediate health risk to communities. In a recent study of adolescents in a juvenile detention center in Chicago, about 5% of the teens had contracted gonorrhea and almost 15% chlamydia (Broussard, Leichliter, Evans, Kee, Vallury, & McFarlane, 2002, p. 8). Girls were over three times more likely to have both diseases than were boys in this study.

According to the 1976 Supreme Court case *Estelle v. Gamble*, inmates have a constitutional right to reasonable medical care. The court held that to be deliberately indifferent to the medical needs of inmates would violate the 8th Amendment prohibition against cruel and unusual punishment. Needless to say, treating such problems requires that a jail of any size have budgetary coverage for the salaries of nurses; a contract with a local doctor, mental health provider, and dentist; and an arrangement with local hospitals. Moreover, regular staff need basic training in CPR and other medical knowledge (e.g., to know when someone is exhibiting the symptoms of a heart attack or stroke or the symptoms of mental illness), so that, when a problem arises, they recognize how serious it might be and know how to address it or who to call (Kerle, 2003; Rigby, 2007).

Some jails are addressing these issues by contracting with private companies to provide medical services or by using telemedicine as a means of delivering some services. The National Commission on Correctional Health Care recommends that, should jails go the route of private provision of services, they make sure that such programs are properly accredited so that the services provided meet national standards (Kerle, 2003). When such matters as obtaining and maintaining quality care are not attended to, as sometimes they are not in jails and prisons (Vaughn & Carroll, 1998; Vaughn & Smith, 1999), jail inmates are likely to suffer the consequences in continued poor health (Sturgess & Macher, 2005). In addition, jails may be sued for failure to provide care, and communities might be exposed to contagious diseases, along with the legal bills (Clark, 1991; Macher, 2007; Rigby, 2007). Clearly the provision of decent health care to incarcerated persons is important not just because the Supreme Court mandates it or because it is the moral thing to do for people who are not free to access health care on their own but also because the vast majority of jail inmates return to the community, most within a week or two (Kerle, 2003). Therefore, to prevent the spread of diseases and to save lives both inside and outside of jails, some medical care would appear to be called for. Some jails are clearly expending energy to address this area of incarceration as four in ten of the inmates in that 2002 BJS study reported that they had received a medical exam since their admission (Maruschak, 2006, p. 1).

Substance Abuse and Treatment in Jails

It is one of those oft-cited beliefs that people in prisons and jails have substance abuse problems, and this is one area of social commentary that actually fits social reality. According to the BJS 2002 study of jail inmates, fully 68% of jail inmates reported substance abuse or dependence problems (Karberg & James, 2005, p. 1). In fact, half of convicted inmates reported being under the influence at the time they committed their offense, and 16% said they committed the crime to get money for drugs. Women and white inmates were more likely to report usage at the time of the offense (Karberg & James, 2005, p. 5). For convicted offenders who used at the time of offense, alcohol was more likely to be in their system than drugs (33.3% for alcohol as opposed to 28.8% for drugs). The drugs of choice for abusers and users varied and included, by prevalence of use, marijuana, cocaine or crack, hallucinogens, stimulants (including methampheta-mines), and inhalants (Karberg & James, 2005, p. 6). Not surprisingly, those who reported a substance abuse or use problem were also more likely to have a criminal record and to have been homeless before incarceration.

Violent offenders were more likely to use alcohol than other substances at the time of the offense. But then violent offenders were also least likely, with the exception of public order offenders, to report being on drugs or alcohol at the time of the offense (Karberg & James, 2005, p. 6).

Fully 63% of those with a substance abuse or use problem had been in a treatment program before (Karberg & James, 2005, p. 1). Most such programs were of the self-help variety including Alcoholics Anonymous or Narcotics Anonymous. However, 44% of these people had actually been in a residential treatment program or a detoxification program or had received professional counseling or had been put on a maintenance drug (Karberg & James, 2005, p. 8). Treatment for convicted offenders in jails, as of 2002, was at 6%. Notably, provision of treatment in jails is difficult because most inmates who are out of the facility within a week and half are typically unconvicted, so, as people who are "presumed innocent," they cannot be coerced into getting treatment. Therefore, treatment programs are usually focused on jail inmates who meet all of the following criteria: they have a substance abuse problem, they are convicted, and they are longer-term inmates. Even having said this, the amount of treatment programming in jails does not fit the obvious need (Kerle, 2003).

▧ Suicides and Sexual Violence in Jails

Suicides

As indicated from the data presented in the above, those incarcerated in jails often enter them at some level of intoxication, many will have a mental disability and are impoverished, and this may be their first experience with incarceration and being booked itself may represent both the mental and physical lowest point of their lives. Such a combination of conditions may predispose some jail inmates to not just contemplate but attempt suicide (Winfree & Wooldredge, 1991; Winter, 2003). Data obtained by the Bureau of Justice Statistics from a 2-year study (2000–2002) of deaths while in custody would indicate that other factors such as age, gender, and race, along with jail size, may also narrow the predictors of a propensity to commit suicide (Mumola, 2005). Other data published by the BJS on sexual violence in incarceration would indicate that an inmate in a jail lives far from a carefree existence (Beck, Harrison, & Adams, 2007).

One thing is clear, as indicated from the data presented in Table 8.4, jails have three times the rate of suicides as prisons do, though their homicide rates are comparable. The good news is, however, that jail and prison deaths due to suicide (and homicide) have declined precipitously from 1983 to 2002, with the rate of prison suicides declining by half during this time and jail suicides by almost two-thirds (Mumola, 2005, p. 2). In 1983, jail suicide was the major cause of death for inmates, but, by 2002, illness replaced suicide as the primary reason for death.

Suicide rates in jails varied by type of inmate and size of jail. White males under 18 and over 35 and those inmates with a more violent commitment history were more likely to commit suicide than African American males or those in other age groups or those who were not incarcerated for a violent offense (Mumola, 2005; Winter, 2003). Winter

(2003) found in her study of 10 years of suicide data from jails in one Midwestern state that those who committed suicide tended to be younger, were arrested for a violent offense, had no history of mental or physical illness, did not necessarily "exhibit suicidal tendencies," and were more likely to be intoxicated with alcohol when admitted (p. 138).

Moreover, according to the authors of the BJS study, the suicide rate at large primarily urban jails, which tend to hold fewer whites, was about half that of the smaller jails, which tend to hold more whites (Mumola, 2005). Similarly, Tartaro and Ruddell (2006) found that smaller jails (with less than a 100-bed capacity) had a two to five times greater prevalence of attempted and completed suicides than larger jails did. According to this study, crowded jails and those with "special-needs and long-term inmates" were also more likely to have a higher number of suicide attempts (Tartaro & Ruddell, 2006, p. 81).

The shock of incarceration may be one explanation for jail suicide rates, although why this shock might be greater for those in their midthirties and older or for those placed in smaller jails is not entirely clear. These data do indicate that about half of the suicides occur within the first 9 days—for women it was 4 days—and in the cell of the person committing the suicide (Mumola, 2005).

Larger jails, with their greater resources and enhanced training for staff, may be better equipped than their smaller counterparts to monitor and prevent suicides in their facilities. For instance, if the younger inmates are fearful of being housed with, and possibly abused by, adults, some less crowded and perhaps larger jails may have the luxury of segregating young men from older men, thereby lessening the fear that might precipitate some suicides. We do know that the rate of suicide in jails, despite its marked decrease over the last 20 years, is still twice as high as would be true for a comparable group of free citizens (Mumola, 2005).

Sexual Violence

The Prison Rape Elimination Act of 2003 mandated that the BJS collect data on sexual assault in adult and juvenile jails and prisons (notably, the data on incidence in juvenile facilities was not available at the time of this writing). The data, summarized here, were collected in different but, in many instances, comparable ways for 2004, 2005, and 2006 (Beck, Harrison, & Adams, 2007). Data were obtained from administrative records, surveys and interviews with current and former inmates, and from all state departments of corrections, the Federal Bureau of Prisons, and a sample of jails.

Based on just these 3 years of data, which is really not enough to establish a trend, the number of allegations of sexual violence in all adult correctional institutions increased to almost three (2.91) per 1,000 inmates in 2006 from 2.46 per 1,000 inmates in 2004 (Beck, Harrison, & Adams, 2007, p. 3). Most of these alleged incidents occurred at night in the victim's cell and involved the use or threat of force. But notably, most of these allegations were not substantiated or investigated or found to be supported by evidence, according to prison or jail officials. Having said this, however, we should recognize that, in most such instances of sexual violence, it would be very difficult to find evidence, as it is in the free world, particularly if the one perpetrating the victimization was a staff member.

Prison inmates were more likely to report sexual violence allegations than were jail inmates (Beck, Harrison, & Adams, 2007). Moreover, about 49% of the time, inmates alleged that staff perpetrated the sexual violence against inmates (36%) or sexually harassed the inmate (13%) (Beck, Harrison, & Adams, 2007, p. 4). Conversely, it was alleged that inmates were involved in the sexual victimization of other inmates roughly 51% of the time.

Table 8.4	Suicide and Homicide Rates in State Prisons and Jails

	Suicide Rate		Homicide Rate	
Year	State Prisons	Jails	State Prisons	Jails
1980	34		54	
1981	26		36	
1982	26		28	
1983	27	129	21	5
1984	25		30	
1985	26		24	
1986	19		20	
1987	18		15	
1988	20	85	12	3
1989	19		11	
1990	16		8	
1991	13		8	
1992	14		9	
1993	17	54	10	4
1994	17		7	
1995	16		9	
1996	15		6	
1997	15		7	
1998	16		5	
1999	15	54	4	5
2000	16	47	5	3
2001	14	50	3	3
2002	14	47	4	3
2003	16	43	4	2

NOTE: Local jail inmate mortality rates are based on average daily population for each year. Data on deaths for 1983–1999 are from the *Census of Jails;* data from 2000–2003 are from the Deaths in Custody Reporting Program (DCRP). State prisoner mortality rates for 1980–2000 are based on death counts of sentenced prisoners and the December 31 jurisdiction population as collected in the National Prisoner Statistics (NPS) program. Rates for 2001–2003 are based on prisoner death counts from the Deaths in Custody Reporting Program and the NPS June 30 custody population count.

Only 967 incidents of sexual violence in all correctional institutions were substantiated in 2006, whereas 885 were in 2005 (Beck, Harrison, & Adams, 2007, p. 4). Rates of substantiated incidents were lowest in federal and privately operated prisons but higher in state level prisons and in both public and private jails. This finding might mean that there was less victimization in the federal and private prisons, or it

▲ **Photo 8.1**   A typical jail cell.

might mean that they were less vigorous in investigating and thus substantiating it. Female staff were more likely perpetrators in prisons, though we know from research on Texas prisons that, when the offense was actual sexual battery, the staff offender was more likely to be male (Marquart, Barnhill, & Balshaw-Biddle, 2001). Male staff members were more likely the perpetrators of sexual violence in jails. For instance, in a recent (2007) case involving the jail in Yuma County, Arizona, three male officers were charged with unlawful sexual conduct with three female inmates (Reuffer, 2007).

Notably, based on the BJS data, in 57% of the substantiated incidents in 2006, the sexual relationship "appeared to be willing" (Beck, Harrison, & Adams, 2007, p. 6). Some states and localities maintain, however, that, if there is sexual involvement between staff and inmates in a correctional institution, it is by definition "unwilling" as incarcerated people are powerless and vulnerable, as compared to their keepers, and so do not have the ability to say "no" or to give consent. The BJS researchers (Beck, Harrison, & Adams, 2007) report the following findings regarding victims and perpetrators of sexual violence in correctional institutions:

- In state and federal prisons, 65% of inmate victims of staff sexual misconduct and harassment were male, while 58% of staff perpetrators were female.
- In local jails, 80% of victims were female, while 79% of perpetrators were male.
- 49% of staff perpetrators in prisons were age 40 or older, while 65% of victims were under age 35.
- 56% of staff perpetrators in jails were age 40 or older, while 86% of victims were under age 35.
- Among staff perpetrators in prisons and jails, 71% were white; 20%, black; and 7%, Hispanic. Among inmate victims, 66% were white; 23%, black; and 8%, Hispanic.
- A correctional officer was identified as the perpetrator in 54% of incidents in prisons and in 98% of incidents in jails.
- "Three-quarters of staff perpetrators in 2006 lost their jobs; 56% were arrested or referred for prosecution."
- Half of inmates involved in staff sexual misconduct were transferred or placed in segregation
- In most incidents of staff sexual misconduct or harassment (76%), victims received no medical follow-up, counseling, or mental health treatment. (pp. 7–8)

▧ Innovations in Jails: New Generation or Podular Direct-Supervision Jails, Community Jails, Coequal Staffing, Reentry Programs for Jails

New Generation or Podular Direct-Supervision Jails

In the 1980s, a new kind of jail was under construction in this country. Its two key components included a rounded or "podular" architecture for living units and the "direct," as opposed to indirect or intermittent, supervision of inmates by staff; in other words, staff were to be in the living units full time (Applegate & Paoline, 2007; Farbstein & Wener, 1989; Gettinger, 1984; Zupan, 1991). It was believed that the architecture would complement the staff's ability to supervise, and the presence of staff in the living unit would negate the ability of inmates to control those units. Other important facets of these jails are the provision of more goods and services in the living unit (e.g., access to telephones, visiting booths, recreation, library books) and the more enriched leadership and communication roles for staff.

Not surprisingly, several scholars recognized that the role for the correctional officer in a podular direct-supervision jail would have to change. Zupan (1991), building on the work of Gettinger (1984), identified seven critical dimensions of new generation jail officer behavior: proactive leadership and conflict resolution skills; building a respectful relationship with inmates; uniform, and predictable, enforcement of all rules; active observation of all inmate doings and occurrences in the living unit; attending to inmate requests with respect and dignity; disciplining inmates in a fair and consistent manner; and being organized and in the open with the supervisory style. Whether officers in podular direct-supervision jails are always adequately selected and trained to fit these dimensions of their role is, as yet, an open research question (Applegate & Paoline, 2007; Nelson & Davis, 1995; Wener 2006).

New generation jails, though hardly "new" anymore, became wildly popular in this country by the late 1980s and through the 1990s (Kerle, 2003; Wener, 2005). Reportedly, in the twenty-first century, about one fifth of medium and larger jails are said to be new generation facilities (Tartaro, 2002). The architecture of new generation jails, though not all their features and not always direct supervision, can be seen in most new jails and prisons built these days.

It is widely acknowledged by correctional scholars and practitioners that, though podular direct-supervision jails or prisons are not necessarily a panacea for all that ails corrections today (i.e., crowding, few resources, etc.), they often do represent a significant improvement over more traditional jails (Kerle, 2003; Perroncello, 2002; Zupan, 1991). If operated correctly and inclusive of all of the most important elements, they are believed to be less costly in the long run (due to fewer lawsuits) and safer for both staff and inmates, to provide a more developed and enriched role for staff, and to include more amenities for inmates. This is a big *if*, however, as some research has called into question these claims of a better environment for inmates and staff and a more enriched role for staff, as the implementation of the new generation model has sometimes faltered or been incomplete in many facilities (Applegate & Paoline, 2007; Stohr, Lovrich, & Wilson 1994; Tartaro, 2002, 2006). Clearly, more research on new generation jails is called for to determine their success (or failure) in revolutionizing the jail environment for staff and inmates.

Community Jails

Another promising innovation in jails has been the development of "community jails" (Barlow, Hight, & Hight, 2006; Kerle, 2003; Lightfoot, Zupan, & Stohr, 1991). Community jails are devised so that programming provided on the outside does not end at the jailhouse door, as the needs such programming was addressing have not gone away and will still be there when the inmate transitions back into the community. Therefore, in a community jail, those engaged in educational, drug or alcohol counseling, or mental health programming will seamlessly receive such services while incarcerated and again as they transition out of the facility (Barlow et al., 2006; Bookman, Lightfoot, & Scott, 2005; National Institute of Corrections, 2008;). So whether one is in and out of the facility within a few days or a few months, needs are met and services provided so that the reintegration into the community is more predictable.

Managers of community jails also recognize that they cannot staff or resource the jail sufficiently to address every need of their inmates. Rather, community experts who are regularly engaged in the provision of such services are the appropriate persons to provide such services whether the inmate is in a jail or free in the community; in both instances, it is argued, he or she is a community member and entitled to such services (Barlow, et al., 2006; Lightfoot et al., 1991).

Obviously, the development of community jails requires that some resources (particularly space) be devoted to the accommodation of community experts who provide for inmates' needs. Unfortunately, it is the rare jail that has the luxury of excess space for allocation to such programming. Therefore, the solution may lie in inclusion of such space in jail architectural plans, though this certainly is not optimal given the immediacy of inmate needs discussed in the foregoing.

The second problem that faces jail managers interested in creating "community jails" is convincing local service providers, and lawmakers if need be, that people in jails have a right and a continued need for services and that the continued provision of such services, by community experts, benefits both those inmates and the larger community. Needless to say, making this case, as reasonable as it might sound, can be a "hard sell" when addressing those social service agencies that already have scarce resources and those policy makers who are concerned that more tax dollars might be required to fund such resource provision in jails. For these reasons, larger jails and communities, with their economies of scale and a greater proportion of their populations in need of social services, might be better situated to operate community jails and thus achieve their purported benefits of less crime because of the continuous provision of services in jails (Kerle, 2003; Lightfoot, Zupan, & Stohr, 1991).

Coequal Staffing

Another promising innovation in jails that has occurred in the last couple of decades, in some sheriffs' departments, has been the development of staffing programs that provide comparable pay and benefits to those who work in the jail and those who work on the streets as law enforcement in sheriffs' departments (Kerle, 2003). Historically, jails have not been just a dumping ground (to use Irwin's [1985] terminology) for inmates but for staff as well. If a sheriff deemed that a staff person could not "make it" on the streets as law enforcement, he or she was given a job in the jail where apparently a lack of skill and ability was not seen as a problem. Moreover, jail staff members were (and often still are) paid less and received less training than their counterparts working on the streets (Stohr & Collins, 2008). As a result, jails have and do find it difficult to attract and keep the best personnel; or, even if they can attract the more talented applicants, jail jobs were and are used as "stepping stones" to better paying and better status jobs on the law enforcement side of sheriffs' agencies (Kerle, 2003).

Since the 1980s, however, many sheriffs' departments, though far from a majority, have recognized the problems created by according this second-tier status to those who work in jails (Kerle, 2003). Consequently, they have instituted programs whereby staff who work in the jails, who often are given deputy status, are trained and paid similarly to those who work in the free communities. Some anecdotal evidence from sheriffs' departments indicates that this change has had a phenomenal effect on the professional operation of jails (as they are better staffed) and on the morale of those who labor in them (Kerle, 2003).

Reentry Programs for Jails

Perhaps the newest "thing" in jails these days (and in prisons too) is a rethinking about how to keep people out of them! Rather than focusing on deterrence or incapacitation so much (that is *so* '80s and '90s), jail practitioners are studying how to make the transition from jails to the community smoother and more successful so that people do not commit more crime and return (Bookman, Lightfoot, & Scott, 2005; Freudenberg, 2006; McLean, Robarge, & Sherman, 2006; Osher, 2007). As is indicated by the discussion in the foregoing of all the medical, criminal, and social deficits that many inmates of jails have, this transition back into the community is likely to be fraught with difficulties. Consequently, any successful reentry program must include a recognition of the problems individual inmates may have (i.e., mental illness, physical illness, joblessness, and homelessness) and address them systematically in collaboration with the client and the community (Freudenberg, 2006; McLean et al., 2006). In a study by Freudenberg, Moseley, Labriola, Daniels, and Murrill (2007) conducted in New York City jails, the researchers asked hundreds of inmates what their top three priority reentry needs were: adult women needed housing, substance abuse treatment, and financial assistance; adult men needed employment, education, and housing; adolescent males needed employment, education, and financial assistance (Freudenberg, 2006, p. 15).

Effective interventions to improve reentry, in the New York study, included everything from referral to counseling to drug treatment to post-release supervision, depending on the needs of the inmate, their unique reentry situation, and the services available in the community. Because, obviously, reentry is a complex process for people with multiple problems, it requires that jail personnel prioritize the needs they will target and the interventions they will apply and then network with community agencies to provide the package of services most likely to further the goal of a successful reentry (Freudenberg, 2006; McLean et al., 2006). In fact, Bookman and her colleagues (2005) would argue that jail personnel should expect to engage in collaborative arrangements with community agencies (sounds a bit like community jails, doesn't it?), if they hope to help their inmates succeed in the reentry process.

⊠ Jail Research

The research articles featured in this section of the book cover the gamut of jail topics from gang interventions in jails to transitioning from homeless shelters to jails (and back), from assaults in Texas jails to implementation issues for new generation jails. The authors of the article on gang interventions surveyed jail administrators about their sense of the prevalence and problems that gang members bring to the jail setting (Ruddell, Decker, & Egley, 2006). Interestingly, they thought that inmates with a severe mental illness were more disruptive than gang members in the jail environment but that the latter inmates were more violent.

In the article by Metraux and Culhane (2006, p. 504) the authors explore the incarceration and shelter histories of over 7,000 persons in New York City. Not surprisingly, given what we read in the foregoing, the authors found that almost one-fourth of the shelter population had been incarcerated within the last two years.

In a study of inmate assaults in Texas county jails, Kellar and Wang (2005) analyze the nexus between importation and managerial models as they might explain inmate violence. Their findings would indicate that there was some support for the explanatory power of the importation model, though not much for the managerial.

Tartaro (2006) conducted a national survey of new generation jails and found that the ideal does not always appear in reality. It is often the devil in the details of implementation that sets many a noble project back (Pressman & Wildavsky, 1979; Rothman, 1980).

⊠ Summary

- ◆ Jails in the United States are faced with any number of seemingly intractable problems. They are either overcrowded or close to it, and they house some of the most debilitated and vulnerable persons in our communities. They house the accused, the guilty and the sentenced, the low-level offender, and the serious and violent ones. As with prisons, their mission is to incapacitate (even the untried), to deter, to punish, and even to rehabilitate. The degree to which they accomplish any of these goals is in large part determined by the political and social climate that the jail is nested in. Since the 1980s, that conservative political climate and the resultant "harsh justice" meted out by policy makers and the actors in the criminal justice system have led to the unrelenting business of filling and building prisons and jails across this country (Cullen, 2006; Irwin, 1985, 2005; Whitman, 2003).
- ◆ Jails have also served as a dumping ground for those who are marginally criminal and are unable, or unwilling, to access social services. Too often, the needs of such

persons go unaddressed in communities, and, as a result, these unresolved needs either contribute to the incarceration of individuals (in the case of substance abuse and mental illness) or make it likely (e.g., homelessness) that they will enter and reenter the revolving jailhouse door.

♦ Violence in jails remains problematic. It is likely true that the rate of violence between inmates and inmates or between staff and inmates has gone down in recent years. However, increased monitoring of this phenomenon is certainly called for and may serve to further reduce violence through the implementation of violence reduction techniques and training for staff. To that end, the implementation of the Prison Rape Elimination Act of 2003 with its reporting requirement for correctional institutions represents a positive move in that direction.

♦ Thankfully, there have been some other hopeful developments on the correctional horizon. Some states and localities are reconsidering the mandatory and harsh practices that have led to burgeoning populations in jails and prisons. Moreover, jails in a position to do so have expanded their medical and treatment options to address the needs of inmates. Architectural and managerial solutions have been applied to jails in the form of new generation jails and coequal pay for staff in sheriffs' departments, and some few jails have even experimented with community engagement to ensure that the needs of people in communities are not neglected when such folks enter jails or reenter communities.

KEY TERMS

Coequal staffing	Overcrowding
Community jails	New generation or podular direct supervision jails
Jails	Reentry

INTERNET SITES

American Correctional Association: www.aca.org

American Jail Association: www.aja.org

Bureau of Justice Statistics (information available on all manner of criminal justice topics): www.ojp.usdoj.gov/bjs

National Criminal Justice Reference Service: www.ncjrs.gov

National Institute of Corrections: www.nicic.org

Vera Institute (information available on a number of corrections and other justice related topics): www.vera.org

READING

In this article Rick Ruddell, Scott H. Decker, and Arlen Egley present a national-level study of the perceptions of 134 jail administrators in 39 states about the prevalence of gang members in their facilities. Consistent with previous empirical work, they found that approximately 13% of jail populations are thought to be gang involved. Although there are no regional differences in these estimates, small jails report having fewer gang-involved inmates. When asked about the problems that these inmates cause in their facilities, respondents report that gang members are less disruptive than inmates with severe mental illnesses but are more likely to assault other inmates. The use and efficacy of 10 programmatic responses to gangs are evaluated, with respondents rating the gathering and dissemination of gang intelligence as the most effective intervention. Implications for practitioners and gang research are outlined.

Gang Intervention in Jails

A National Analysis

Rick Ruddell, Scott H. Decker, and Arlen Egley, Jr.

Most of our knowledge about gangs in correctional facilities is based on research conducted in state or federal prison systems (Camp & Camp, 1985; Gaes, Wallace, Gilman, Klein-Saffran, & Suppa, 2002; Ralph & Marquart, 1991; Stastny & Tyrnauer, 1983). Typically, these studies have found that the proliferation of gangs and the number of gang members in prison settings have increased substantially since the 1980s (Decker, 2003). Understanding the extent of the gang problem is an important issue for prison administrators because gang-involved inmates contribute to higher rates of prison violence (Camp & Camp, 1985), increase racial tensions within prisons (Anti-Defamation League of B'nai B'rith, 2002; Ross & Richards, 2002), challenge rehabilitative programming by supporting criminogenic values (Decker, 2003; Fortune, 2003), engage in criminal enterprises within prisons (Ingraham & Wellford, 1987), and contribute to failure in community re-integration if these parolees return to gang activities on release (Adams & Olson, 2002; Fleisher & Decker, 2001b; Olson, Dooley, & Kane, 2004).

Fischer's (2001) study of Arizona prisons, for instance, reported that "members of certified prison gangs (security threat groups [STGs]), uncertified prison gangs, and street gangs commit serious disciplinary violations at rates two to three times higher than do non-gang inmates housed in units of the same security level" (p. ii). Thus, by better understanding the scope of the problem and the efficacy of different types of interventions, jail professionals can

SOURCE: Ruddell, R., Decker, S., & Egley, A., Jr. (2006). Gang Intervention in Jails: A National Analysis. *Criminal Justice Review,* *31*(1), 33–46. Reprinted with permission of Sage Publications, Inc.

work to reduce the influence that gangs have in their facilities. In addition, other stakeholders also need to better understand the extent of this social problem. Esbensen, Winfree, He, and Taylor (2001) observe that research about gangs is also important for researchers and theorists:

> For researchers, it is important to refine measurement: to assess the validity and reliability of the measures being used. For theorists, it is important to better understand factors associated with gang membership and associated behaviors, whether testing or constructing theory. (p. 122)

These scholars also acknowledge the importance of information sharing and collaboration between academics and policy makers, although we also suggest that jail practitioners ought to be involved in the research process as they have intimate knowledge of the success and failure of different interventions within their institutions.

Comparatively little is known about the extent of adult gangs in jails.[1] Increasing our knowledge about gangs and tactics designed to respond specifically to these groups is an important issue given the size of the jail population—approximately 714,000 inmates are housed in local correctional facilities (Harrison & Beck, 2005b). Unlike prisons, jails are intended primarily for inmates awaiting court processes and incarceration for periods of less than 1 year. Operated by counties and local governments, there are more than 3,300 American jails that range in size from four or five beds to the Los Angeles County jail system, which held an average of 18,629 inmates in the third quarter of 2005 (Corrections Standards Authority, 2005). Parallel with federal and state prison systems, however, local jails have also experienced dramatic growth during the past two decades (Cunniff, 2002; Harrison & Beck, 2005b; Stephan, 2001). This growth has stretched county budgets (Davis, Applegate, Otto, Surette, & McCarthy, 2004), increased staff turnover (Kerle, 1998), and may contribute to higher rates of inmate violence (Tartaro, 2002). Altogether, these changes produce less predictable conditions within America's jails (Mays & Ruddell, 2004).

It is plausible that some of the problems in the day-to-day operations that jails confront are a result of expanding gang populations. As a result, an important first step is to examine the extent of the problem. Wells, Minor, Angel, Carter, and Cox (2002) surveyed jail administrators and found that approximately 16% of all jail inmates were members of STGs, whereas 13% of prison inmates were STG members. There are a number of reasons why jail populations have rates of gang involvement that closely correspond with prison systems. First, although jail populations tend to be more heterogeneous than prisons (a wide variety of persons from different demographic and socioeconomic backgrounds are admitted), this diversity is a function of the short-term nature of jail incarceration. Most inmates are held for a day or two until they make bail, but serious or persistent offenders, such as gang members, may have more difficulty securing release and may wait months or even years for the conclusion of their trials. James (2004, p. 4) found that 11.3% of jail inmates were held more than 6 months and that an additional 6.5% were held more than 1 year.

A second reason why rates of gang membership in jails parallel prison rates is that jails act as an entry point for prisons. Virtually everyone who is admitted to prison will first spend time in jails, either awaiting court dates or pending transfer to prison. Approximately 10% of all jail inmates have already been sentenced to a term within the state prison system, but overcrowding keeps these inmates in city or county facilities awaiting transfer (Harrison & Beck, 2005b). Finally, jails hold persons sentenced to periods of incarceration up to 1 year, and some may serve much longer periods of time in a local jail (James, 2004). Although early scholarly work reported that jail inmates were primarily members of the underclass held on relatively minor offenses (Goldfarb, 1975; Irwin, 1985), current research indicates that offenders in many urban jails are held on more serious crimes. In fact, Rainville and Reaves (2003, p. 33) found that nearly 75% of all persons sentenced to jail incarceration were felony offenders.

Organizational characteristics of jails might also contribute to gang membership. Jails hold a diverse mix of short- and long-term inmates, which can contribute to unpredictability. Although prison populations are fairly stable through periods of years, the population in a jail unit may change completely in a single week. Jails with high levels of population turnover are less predictable for jail officers and inmates alike (Richards, 2003). The short-term nature of jail confinement relative to prisons also makes classification and programming more challenging. Wright and Goring (1989) observed that "prisoners come in directly from the street as unknown quantities, often with alcohol, drug or psychiatric problems" (p. ii). A combination of this instability and high populations of gang-involved inmates recently lead to riots involving thousands of inmates at a Southern California jail (Cable News Network, 2006).

Although the unpredictability of jail incarceration is bearable for a few days, it may contribute to gang affiliation as months stretch into a year. As Ross and Richards (2002) remark, "in some prisons you absolutely need to affiliate with a group that will protect you. The loners, the people without social skills or friends, are vulnerable to being physically attacked or preyed upon" (p. 133). Lhotsky (2000) describes the pressures to join a gang in a large urban jail: "Everything here is gang politics and you have to be involved, one way or another . . . and you better participate or you're gonna get beat bloody" (p. 213). Consequently, the characteristics of an individual jail or the types of inmates incarcerated

within a facility (both the demographic and offense-related characteristics) may also contribute to gang membership (Santos, 2004).

Despite the problems that gang-affiliated inmates cause, there is some evidence to suggest that jails have been slow to adopt gang intervention programs, especially compared with state prison systems (see Wells et al., 2002). Moreover, there has been very little empirical attention devoted to the issue. This national-level study examines the prevalence of gang members in jails, based on information about the methods that jails use to classify gang involvement; perceptions about the types of problems that gang-affiliated inmates cause; and the efficacy of strategies intended to reduce the harm that gangs create in jails.

⬡ Interventions With Incarcerated Gang Members

Jail officers and administrators have developed a number of gang intervention strategies during the past few decades. The types and methods of intervention are based on the prevalence of gang-affiliated populations within a jail, the levels of inmate involvement in gangs, resources available to the jail, and the location of the facility. Gangs have been observed to exhibit considerable diversity across the United States in terms of organization and structure (Klein & Maxson, 1996) and criminal involvement (Howell, Egley, & Gleason, 2002). Correctional settings are not immune to this variation. A county jail located in rural Texas, for instance, is likely to have an entirely different gang problem—and response—than a state prison that draws its population from Los Angeles or Chicago.

Previous empirical work has reported that there are regional as well as urban-rural differences in the type of gang problem that a jurisdiction is likely to encounter (Egley, Howell, & Major, 2004; Weisheit & Wells, 2004). Demographic and offense-related characteristics of the gang population, for instance, may influence the involvement of gang members in disruptive activities within an institution. Younger gang members, as well as those sentenced on violent offenses, have a higher likelihood of involvement of gang members in disciplinary infractions (Fischer, 2001). In addition, some prison gangs were established generations ago, and the types of problems that teenaged "wannabe" gang members from small rural gangs cause are fundamentally different from those caused by members of the Aryan Brotherhood or Mexican Mafia, as these established gangs have established networks that transcend the jail or prison.

Decker (2003) outlined a number of tactics that correctional facilities have used to control gangs, including

> use of inmate informants, the use of segregation units for prison gang members, the isolation of prison gang leaders, the lockdown of entire institutions, the vigorous prosecution of criminal acts committed by prison gang members, the interruption of prison gang members' internal and external communications, and the case-by-case examination of prison gang offenses. (p. 58)

The types of interventions that a jail or prison develops are likely to depend on a number of organizational characteristics, such as the size of the facility, the internal and external resources that the institution can draw on, and the nature of the gang problem. Decker (2003) notes that the efficacy of the approaches outlined above have typically not been evaluated—a common limitation of many criminal justice interventions (Sherman, Gottfredson, MacKenzie, Eck, Reuter, & Bushway, 1997).

Wells et al. (2002, p. 14) solicited information about the effectiveness of different interventions to control STGs in their survey of correctional administrators and reported that 76% of prisons had established such

interventions, compared to only 44% of jails. Prisons also deployed a greater range of gang interventions. Wells et al. also found that many jails reported supervising inmates through a central monitoring unit, although this is a common characteristic in direct-supervision facilities. Contrasted to prisons, these researchers found that jails were more likely to use segregation and protective custody. Respondents from prisons, however, reported that they relied on monitoring inmate communication, collecting information from searches, and compiling this intelligence.

A number of "containment" strategies have been used to control gangs in correctional facilities. Long-term isolation or the transfer of gang leaders, for instance, has the goal of reducing a leader's ability to recruit or influence other members, and this tactic has been successfully used in some jails (Decker, 2003). Another approach, which has been used extensively in the federal prison system, is bus or "diesel" therapy. Inmates perceived to be disruptive, including gang members, are transferred from facility to facility for periods of months (Knox & Tromanhauser, 1991; see Richards, 2003, p. 134, for a discussion). Yet such transfers are beyond the ability or resources of a single county jail.

Jail interventions also include information collection from informants as well as intelligence gathered by law enforcement agents or correctional officers (Norris, 2001). Intelligence gathering may include the development of gang profile reports, classification of inmate affiliation, and information about gang involvement in offenses—either internal or external to the jail (Nadel, 1997). In some cases, aggressive prosecution of unsolved cases or current criminal activity (such as the sale or distribution of narcotics) may result in additional prison terms for the gang-involved inmates and their associates.

Containment approaches vary by jurisdiction. The Texas Department of Criminal Justice, for example, attempted to consolidate all hard-core gang members in several designated prisons, which reduced their ability to influence or recruit others (United States Department of Justice, 1992). A similar approach was undertaken in Arizona prison systems, and Fischer (2001) reported that isolating (or incapacitating) these inmates resulted in declines of "rates of assault, drug violations, threats, fighting . . . by over 50%" (p. ii). Furthermore, Rivera, Cowles, and Dorman (2003) outlined how one state sent all nonaffiliated inmates to "gang free" prisons, a strategy that has also been used in Illinois (Olson et al., 2004). Ultimately, this approach attempts to reduce the influence of gangs and provide a safer environment (which, in turn, may reduce the likelihood of nonaffiliated inmates joining gangs to increase their feelings of safety).

Data and Method

In June 2004, 418 surveys were sent to jails throughout the nation soliciting information from jail administrators about their experiences with "special needs" jail populations, including persons with mental illness, gang members, repeat offenders, and long-term inmates. With the exception of six states that had integrated state jail systems, all states were included in the sample. A random sample of jails was completed, choosing facilities listed in the American Jail Association's (2003) *Who's Who in Jail Management.* The exception to the random selection was an oversampling of large jails. Harrison and Beck (2005a) found that the largest 50 jails (mostly located in urban areas) held approximately 30% of all jail inmates, so all jails with a rated capacity of more than 1,500 inmates were surveyed.

To enhance the response rate, each facility was contacted by phone, and survey team members spoke with administrators and encouraged their participation. Survey instruments were either mailed or faxed to the administrators, although, in some cases, the surveys were directed to mental health professionals or classification officers at the request of the official who was contacted. Responses to faxed surveys were somewhat better than those that were mailed and tended to be promptly returned. The survey instrument solicited responses from jail administrators about their experiences with gang members, including asking how gang affiliation or membership was defined, estimating the prevalence of these populations within their jail, and analyzing strategies that worked or that were not effective in responding to gangs. Although most respondents returned the survey within 2 or 3 weeks, we continued to receive responses to the survey for several months afterward. Altogether, 134 surveys from 39 states were returned, a response rate of approximately 32%.

There were several limitations with the survey results. Jails in the northeastern states, for example, were underrepresented in the surveys that were returned, as were returns from small jails. This underrepresentation rests on the sampling strategy used in this study. Integrated jail-prison systems in Alaska, Connecticut, Delaware, Hawaii, Rhode Island, and Vermont were not sent surveys. Fewer surveys were also sent to smaller institutions, and the smallest jail that returned a survey had 28 beds. More than 10% of the sample were smaller jails, but the responses from these facilities were disappointing. Of one group of 25 facilities of 35 beds or less, for instance, only 1 was returned. It is plausible that larger jails are more likely to have classification experts, administrators, and mental health specialists who have more time to respond to such requests for information. Thus, the generalizability of the findings in this study is limited somewhat by the facilities that did not respond to the survey or were not included in the sampling strategy. Although the response rate was somewhat less than the Wells et al. (2002) study, the estimates in this research are based on a sample more than 3 times as large.[2]

Table 1	Jail Characteristics: National Jail Sample
Participating jails	134
States	39
Rated capacity (beds)	941.8 ($SD = 1,279.4$)
Average daily population	898.7 ($SD = 1,261.2$)
Percentage rated capacity	93.8 ($SD = 26.2$)
Daily cost	55.4 ($SD = 19.1$)
Turnover (annual admits/average daily population)	30.8 ($SD = 64.4$)
Total rated capacity (beds)	125,259
Region	
Northeast	7
Midwest	36
South	45
West	44

Table 1 reveals the organizational characteristics of the facilities that were represented in this study. With an average rated capacity of 941.8 inmates ($mdn = 512$ inmates), jails from large urban areas were overrepresented. The standard deviation of 1,279.37 inmates, however, indicates considerable skewness. Altogether, the facilities represented in the survey had a total rated capacity of 125,259 beds, or 19% of all jail inmates nationwide. The jails examined in this study were also smaller than those in the research reported by Wells et al., as the mean size of their sample was 5,638 beds.

Respondents reported that their facilities operated near capacity: The average jail operated at 94% of its rated capacity ($mdn = 91\%$, $SD = 26\%$). Based on the average daily population and admissions data, we extrapolated several statistics, including an average inmate turnover of approximately 31 inmates per year and an average stay of approximately 12 days.[3]

As most jail inmates only remain in custody for a day or two, this finding suggests that there are a large number of inmates who serve lengthy periods in county jails, either awaiting their court dates, serving a sentence of less than 1 year, or some combination of these two factors. Consequently, one question on the survey asked respondents to estimate the percentage of inmates that had been in the jail for periods in excess of 1 year, and the mean was 12.7% ($mdn = 5.0\%$, $SD = 16.8\%$).

▨ Results

Jail administrators were asked to select methods of classifying gang affiliation in their facilities. Table 2 outlines the five different options for classifying or defining gang membership that were provided. Respondents overwhelmingly reported that they defined gang membership on the basis of tattoos, clothing (gang colors), or hand signs, although designation of gang membership by another law enforcement agency was commonly used to define gang membership. Eighty-one percent of respondents reported that an individual's self-declaration as a gang member was used as a method of classification. Correspondingly, law enforcement agencies report frequent use of this technique as well (Egley & Major, 2004). It is important to report that there is considerable research that supports the validity of this measurement approach (see Esbensen et al., 2001). Furthermore, almost three quarters of respondents based definitions of membership on the basis of the inmate's associates. Respondents were least likely, however, to base definitions of gang membership on claims by informants. These results indicate that self-report remains a viable and frequent method of identification among criminal justice practitioners. In addition, there is considerable overlap between police and jail methods of classification, extending the validity of the measurement approach to yet another group.

Using the methods of designating gang affiliation reported above, respondents were asked to estimate the prevalence of gang members in their jail. Fong and Buentello (1991) outline how correctional administrators have historically been reluctant to provide information to researchers about internal problems, such as gangs or gang violence. Anecdotal information from administrators suggests that gang members may not be forthcoming about their gang affiliation with classification or intake officers. As a result, the true rate of gang membership is likely to be undercounted in many places,

Table 2	Jail Administrator Definitions of Gang Membership			
		%	SD	Range
Inmate has been designated a gang member by another law enforcement agency		83.3		
Inmate has been identified as a gang member by a reliable informant		65.3		
Inmate claims to be a gang member		81.7		
Inmate displays symbols of membership: clothing, "colors," hand signs, or tattoos		86.8		
The inmate is known to associate with and/or has been arrested with known gang members		71.4		
Average reported gang membership (84 jails reporting estimates)		13.2	15.5	0 to 70
Jails reporting no gang membership (11 jails)[a]		13.1		
Jails reporting 1% to 10% gang membership (42 jails)		50.0		
Jails reporting 11% to 25% gang membership (19 jails)		21.4		
Jails reporting more than 25% gang membership (12 jails)		14.3		
Admissions of gang members have increased during past 5 years		45.0		
Gang members younger than other jail inmates		55.0		

NOTE: All figures are percentages.

a. Totals might not add up to 100 % because of rounding.

especially if jails do not collect such data on admission or do not have gang intelligence officers that track these populations.

The mean (unweighted) estimate of gang membership among jail inmates was 13.2%, which closely approximates the national-level estimate of 16% reported by Wells et al. (2002). Estimates varied greatly, from 11 respondents who reported having no gang members in their facilities to 1 California jail administrator who reported that 70% of inmates in their facility were gang involved. In fact, by subtracting the facilities that reported having no gang members in their facility, the mean increased to 15.2%. Overall, half of the respondents estimated that the prevalence of gang members in their facilities ranged from 1% to 10%.

A number of statistical tests were completed on these data to determine the relationships between jail characteristics and the prevalence of gang members, and these results are presented in Table 3. First, a bivariate correlation between average daily population and percentages of gang membership was estimated, and there was a significant positive association between the rated capacity of the jail and the percentage of gang members. Analysis of variance (ANOVA) models were also estimated to compare the means between gang membership and jail characteristics, including the size of the facility, the daily cost to house an inmate, and the region where the jail was located (using the four U.S. Census Bureau regions). The only variable that had a statistically significant association with the percentage

of gang members was the average daily population of these facilities. Bigger jails were likely to house a greater number of persons thought to be gang members. Although gangs are more apt to be found in urban areas, another plausible reason for this finding is that smaller jails have been known to underestimate the true rate of "special needs" populations, such as persons with mental illness (see McClearen & Ryba, 2003). This suggests the need for greater attention to classification issues in such facilities, perhaps not only for gangs.

Forty-five percent of jail administrators reported that the gang problem had increased during the past 5 years, and an additional 40% said it had stayed the same. The finding that a majority of respondents report a decreasing or stabilizing gang problem corresponds to law enforcement reports of local gang problems during the same time period. From 1996 through 2002, declining or stabilizing gang problems and populations were increasingly reported among police and sheriff's agencies, especially among those serving the less-populated, rural areas (Egley et al., 2004). These parallel findings suggest a dynamic and divergent gang problem both within the general population and within correctional settings.

Table 4 presents the perceptions of jail administrators regarding the likelihood of problem behaviors in special needs populations. These problem behaviors included suicide, incidents of self-harm, victimization, assault (either inmates or staff), disruptive behavior within the facility, escapes (or attempts), or other criminal

Table 3	Tests of Association Between Jail Characteristics and Gang Membership			
Jail Characteristic	*r*		**Test**	**F**
Average daily population	.240*			
Average daily population			ANOVA	2.349**
Daily cost			ANOVA	.689
Region			ANOVA	.478

* $p < .05$; ** $p < .01$.

conduct. Gang members were compared to three other special needs populations: inmates with severe mental illness, "frequent flyers" (repeat offenders with more than 20 admissions), and long-term inmates (prisoners who had served more than 1 year of jail incarceration). A disruption index was calculated where 1 point was counted each time a respondent checked that a special needs group was likely to be involved in one of the disruptive behaviors listed above. Of a possible total value of 1,072 (if inmates were coded as being disruptive in all categories by all respondents), inmates with mental illness had the highest disruption score of 610.[4] Gang members followed with a value of 367, whereas the remaining groups had much lower scores. This finding reinforces the findings of previous research that found gang members represent a significant challenge to correctional operations (Fischer, 2001).

In addition to estimating the overall potential for disruption, perceptions of involvement in violent behavior were also collected. Jail administrators reported that gang members were more likely to assault other inmates than any other group of special needs inmates. Fischer's (2001) surveys of 463 Arizona prisoners found that

> inmates believe that inmates who are not members of a gang are safer than

those who are: 69% said that inmates who are not members of a gang are very safe, safe, or somewhat safe, while only 57% said inmates who were members of a gang were very safe, safe, or somewhat safe. (p. 172)

Furthermore, although inmates with mental illness were perceived as the most likely special needs group to assault officers and staff, gang members were also considered to pose a physical threat to jail staff.

Administrators were also asked about the efficacy of 10 different jail-based interventions to respond to gangs. Of the tactics presented in Table 5, the three that were considered most effective were segregation or separation of gang members, intelligence gathering, and sharing this information with other agencies. Thus, these results are consistent with the findings reported in the earlier national study of jail interventions (Wells et al., 2002) as well as Decker's (2003) research. It is interesting to note that, for every category, the "not effective" category was the lowest for every intervention strategy. Stated differently, jail administrators were highly inclined to view any intervention strategy as effective.

The strategies perceived as least successful in reducing the influence of jail gangs were placing restrictions on outside visitors, transfers

Table 4	Perceptions of Jail Administrators on the Involvement of Special Needs Populations in Problem Behaviors		
Population	Disruption[a]	Assault Inmates	Assault Staff
Gang members	367	90	65
Frequent flyers	233	44	30
Inmates with mental illness	610	78	81
Long-term inmates	201	35	21

a. Disruption Index = sum of the following categories: likelihood of suicide, self-harm, victimization, assault (other inmates or staff), disruptive behavior, escapes (or attempts), and other criminal conduct (highest possible value = 1,072).

Table 5	Perceptions of Jail Administrators About the Efficacy of Gang Interventions (Percentage of Respondents)			
Programs or Responses	**Very Effective**	**Somewhat Effective**	**Not Effective**	**Not Applicable**
Segregation or separation	36.2	30.8	1.5	31.5
Restrict outside visitors	10.9	15.5	5.4	68.2
Transfer (e.g., other jail)	5.5	20.5	4.7	69.3
Facility sanctions for gang behavior	27.5	36.6	6.1	29.8
Legal sanctions for gang behavior	11.6	20.9	9.3	58.1
Loss of "good time" credits	19.8	22.9	9.2	48.1
Information sharing (other agencies)	36.6	35.1	3.8	24.4
Intelligence gathering	33.1	39.2	3.1	24.6
Written policies or procedures	26.8	40.9	2.4	29.9
Limit program participation	15.4	26.9	11.5	46.2

Jails that report using none of these tactics = 15

Jails that report using two or fewer of these tactics = 34

of gang members to other facilities, or legal sanctions for criminal behavior. Few administrators reported that they used these tactics, and those facilities that used these approaches did not deem them very effective. Transferring gang members is not, for example, a feasible approach in most local jails. Respondents also indicated that some approaches were ineffective, including the loss of "good time" credits and limiting program participation for gang-involved jail inmates.

The finding that 15 jail administrators reported their facilities used none of the interventions outlined above was surprising. Chi-square analyses were used to evaluate the relationships between the number of gang interventions used by a given jail (split at the median) and the following variables split at their median values: the daily cost to house an inmate, the estimated gang membership, and the rated capacity of the facility. Consistent with expectations, the only variable that had a significant association with high levels of gang interventions was the size of facility; larger facilities used a greater variety of interventions.

▨ Discussion and Conclusion

This research examined the perceptions of jail administrators about the problems that gang members cause in their facilities, the prevalence of these populations, methods of classifying gang membership, and approaches that may reduce the disruption or violence associated with these groups. The findings reported above suggest that the estimate of gang populations vary greatly by location, and, although there were no statistically significant regional differences, smaller jails were less likely to report the presence of gangs. Although 11 facilities reported that they had no gang members in their populations, a number of jails reported that more than half of their populations were gang involved.

Estimating the true population of gang members is problematic, especially considering that there are different degrees of membership. Silverman (2001, p. 284) outlined seven different classifications of membership from hard core members to "sympathizers and wannabes." The United States Department of Justice (1992) estimated that hard core members represent only 15% to 20% of the total gang membership. It is important to distinguish between these different categories of gang membership, although there is little evidence to suggest this is regularly performed within prisons and jails. There is an intuitive conceptual appeal to the notion that strategies for long-term hard core members need to be fundamentally different than [for] those who have less commitment to the gang.

Depending on the categories and methods of classifying gang members, estimates of the prevalence of gang members within jails are likely to vary greatly. The definition of gang membership used in this study was broadly inclusive in that it asked respondents to base their estimates on the five categories outlined in the survey and did not distinguish between "streetgangs," "prison-gangs," "unaffiliated gangs," and "STGs." Still, the finding that 13.2% of all jail inmates in this national sample were gang involved suggests that accurately estimating the gang populations is an important step for jail administrators. Extrapolating this estimate to the 2003 national jail population, for example, would result in 90,000 gang members held in American jails on any given day.

A second relevant research question is to evaluate the commitment to the gang of these 90,000 inmates: Are the estimates of 15% to 20% of hard core members accurate today? Is gang membership, for instance, a situational or temporary condition for most inmates? Furthermore, at what point of a person's incarceration does that person become gang affiliated? Knowing the answers to such questions may enable us to prevent gang affiliation in the first place or create more effective interventions to discourage jail inmates from joining such groups. Such questions can only be answered through further empirical work, including interviews with gang members.

Eleven jails reported having no gang members. Of these facilities, the mean rated capacity was 271 inmates (*mdn* = 119 beds), which is approximately one third the size of the average jail in this study. Yet one of the jails that reported having no gang members had a capacity of nearly 1,500 inmates, which does not seem plausible. Although rural communities are less likely than other localities to develop gang problems, a number of scholars have found that these areas are not immune to gangs (Egley & Major, 2004; Weisheit & Wells, 2004). It is likely that some administrators are not aware of the scope of the problem, are in denial, deliberately underreport the percentages of gang-involved inmates, or do not regard gangs as problematic (Fong & Buentello, 1991), concerns that apply elsewhere to law enforcement officials (Huff, 1990).

Most administrators reported that gang members challenged the operations of their facilities through illegal or disruptive behaviors and a greater involvement in violence. In response to these problems, jails have adopted a number of strategies to reduce the prevalence or harm that these special needs inmates pose. The foremost of these strategies was gathering intelligence and disseminating this information within the facility and to other law enforcement agencies. Also effective was the segregation or separation of gang-involved jail inmates, although this strategy may not be feasible in all locations or in some jails. An examination of the American Jail Association's (2003) national inventory of jails reveals that there are some 650 facilities of 25 beds or less, and these institutions are unlikely to have the ability to separate or segregate any inmate.

Despite the fact that most inmates serve short terms of temporary incarceration, there are long-term jail populations that may be vulnerable to gang recruitment. In some cases, inmates may serve years in local jails (James, 2004) and serve part (or all) of their state prison sentence there (see Ruddell, 2005). Moreover, of the estimated 13% of gang members in this sample, some may have weak ties with the gang. If jail-based interventions can prevent gang recruitment or prevent those "wannabes" from becoming full-fledged gang members, the benefits may be felt throughout justice systems. First, by reducing gang populations, jails will be safer. Safe jails are important not only from a human rights perspective, but high levels of violence might also contribute to increased membership as inmates affiliate themselves with gangs in search of safety (Lhotsky, 2000; Ross & Richards, 2002).

The second advantage of jail-based gang interventions is that jails serve as the entry point for prison populations. Reducing gang membership in local facilities and information sharing with state correctional systems may enhance the effectiveness of prison-based gang interventions. Furthermore, high-visibility interventions may deter nonaffiliated inmates from joining gangs by increasing the "costs" of membership. Some correctional systems, for instance, provide inmates with "guidelines" of the lost opportunities that occur when they affiliate with a gang (see Connecticut Department of Corrections, 1995).

One issue that requires careful consideration is whether society can provide realistic alternatives to the safety and status that gangs offer. Returning vulnerable young people from jails to the community with little hope of meaningful opportunities makes joining a gang more likely. Fleisher and Decker (2001a, pp. 69–70) outline the many barriers to successful reintegration of gang members into the community. Thus, although correctional interventions can attempt to reduce the prevalence of gangs and control their criminal behaviors within these institutions, these jail or prison-based programs need to be supplemented with a corresponding increase in community-based gang-intervention programs to support prisoner reentry.

⊠ Notes

1. Jails are operated by city or county governments, and some jurisdictions may have a jail system where the population resides in several facilities. Further, some states, such as West Virginia, are moving toward regional jails, where operations are consolidated between several counties. Last, jails in "Indian

Country" or detention centers and Metropolitan Correctional Centers operated by the Federal Bureau of Prisons also house jail populations. We use the term *jail* to describe these different variations.

2. Within the sample, the response rate for jails larger than 1,500 beds was 42.6%, whereas the response rate for jails smaller than 1,500 beds was 29.8%.

3. Average length of stay in jail depends on a number of factors, including the number of sentenced and detained inmates in a particular facility. The California Board of Corrections (2004) notes that an increasing number of detainees will reduce overall length of stay.

4. In some cases, the respondents did not complete all categories, so the total of 1,072 represents the highest possible total of all participants who provided a response. It is likely that these categories are not mutually exclusive, as there is some possibility of overlapping categories (e.g., a long-term inmate could also be a gang member).

⊠ References

Adams, S., & Olson, D. E. (2002). *An analysis of gang members and non-gang members discharged from probation.* Springfield: Illinois Criminal Justice Information Authority.

American Jail Association. (2003). *Who's who in jail management.* Hagerstown, MD: Author.

Anti-Defamation League of B'nai B'rith. (2002). *Dangerous convictions: An introduction to extremist activities in prisons.* New York: Author.

Cable News Network. (2006). *One dead in California jail riot.* Retrieved February 8, 2006, from http://www.cnn.com/2006/US/02/04/prison.riot.ap/?section=cnn_mostpopular

California Board of Corrections. (2004). *Jail profile survey 2004, 1st quarter results.* Sacramento, CA: Author.

Camp, G. M., & Camp, C. G. (1985). *Prison gangs: Their extent, nature, and impact on prisons.* Washington, DC: U.S. Government Printing Office.

Connecticut Department of Corrections. (1995). *Gang membership.* Wethersfield, CT: Author.

Corrections Standards Authority. (2005). *Jail profile survey 2005, 3rd quarter results.* Sacramento, CA: Author.

Cunniff, M. A. (2002). *Jail crowding: Understanding jail population dynamics.* Washington, DC: National Institute of Corrections.

Davis, R. K., Applegate, B. K., Otto, C. W., Surette, R., & McCarthy, B. J. (2004). Roles and responsibilities: Analyzing local leaders' views on jail crowding from a systems perspective. *Crime & Delinquency, 50,* 458–482.

Decker, S. H. (2003). *Understanding gangs and gang processes.* Richmond: Eastern Kentucky University.

Egley, A. H., Howell, J. C., & Major, A. K. (2004). Recent patterns of gang problems in the United States: Results from the 1996–2002 National Youth Gang Survey. In F. Esbensen, L. Gaines, & S. G. Tibbetts (Eds.), *American youth gangs at the millennium* (pp. 90–108). Long Grove, IL: Waveland.

Egley, A. H., & Major, A. K. (2004). *Highlights of the 2002 youth gangs survey.* Washington, DC: Office of Juvenile Justice and Delinquency Prevention.

Esbensen, F., Winfree, L. T., He, N., & Taylor, T. J. (2001). Youth gangs and definitional issues: When is a gang a gang, and why does it matter? *Crime & Delinquency, 47,* 105–130.

Fischer, D. R. (2001). *Arizona Department of Corrections: Security threat group (STG) program evaluation, final report.* Retrieved September 25, 2004, from http://www.ncjrs.org/pdffiles1/nij/grants/197045.pdf

Fleisher, M. S., & Decker, S. H. (2001a). Going home, staying home: Integrating prison gang members into the community. *Corrections Management Quarterly, 5,* 65–77.

Fleisher, M. S., & Decker, S. H. (2001b). Overview of the challenge of prison gangs. *Corrections Management Quarterly, 5,* 1–9.

Fong, R. S., & Buentello, S. (1991). The detection of prison gang development: An empirical assessment. *Federal Probation, 55,* 66–69.

Fortune, S. H. (2003). *Inmate and prison gang leadership.* Unpublished doctoral dissertation, East Tennessee State University, Johnson City.

Gaes, G. G., Wallace, S., Gilman, E., Klein-Saffran, J., & Suppa, S. (2002). Influence of prison gang affiliation on violence and other prison misconduct. *The Prison Journal, 82,* 359–385.

Goldfarb, R. (1975). *Jails: The ultimate ghetto of the criminal justice system.* Garden City, NY: Anchor.

Harrison, P. M., & Beck, A. J. (2005a). *Prison and jail inmates at midyear 2004.* Washington, DC: Bureau of Justice Statistics.

Harrison, P. M., & Beck, A. J. (2005b). *Prisoners in 2004.* Washington, DC: Bureau of Justice Statistics.

Howell, J. C., Egley, A. H., & Gleason, D. K. (2002). *Modern day youth gangs.* Washington, DC: Office of Juvenile Justice and Delinquency Prevention.

Huff, C. R. (1990). Denial, overreaction and misidentification: A postscript on public policy. In C. R. Huff (Ed.), *Gangs in America* (pp. 310–317). Newbury Park, CA: Sage.

Ingraham, B. L., & Wellford, C. F. (1987). The totality of conditions test in eighth-amendment litigation. In S. D. Gottfredson & S. McConville (Eds.), *America's correctional crisis: Prisons populations and public policy* (pp. 13–36). New York: Greenwood.

Irwin, J. (1985). *The jail: Managing the underclass in American society.* Berkeley: University of California Press.

James, D. J. (2004). *Profile of jail inmates, 2002.* Washington, DC: Bureau of Justice Statistics.

Kerle, K. E. (1998). *American jails: Looking to the future.* Boston: Butterworth-Heinemann.

Klein, M. W., & Maxson, C. L. (1996). *Gang structures, crime patterns, and police responses.* Los Angeles: Social Science Research Institute.

Knox, G. W., & Tromanhauser, E. D. (1991). Gangs and their control in adult correctional institutions. *The Prison Journal, 71,* 15–22.

Lhotsky, N. (2000). The L.A. county jail. In R. Johnson & H. Toch (Eds.), *Crime and punishment: Inside views* (pp. 211–213). Los Angeles: Roxbury.

Mays, G. L., & Ruddell, R. (2004, November). *Frequent flyers, gang-bangers, and old-timers: Understanding the population characteristics of jail populations.* Paper presented at the annual meetings of the American Society of Criminology, Nashville, TN.

McLearen, A. M., & Ryba, N. L. (2003). Identifying severely mentally ill inmates: Can small jails comply with detection standards? *Journal of Offender Rehabilitation, 37,* 25–40.

Nadel, B. A. (1997). Slashing gang violence, not victims: New York City Department of Corrections reduces violent jail incidents through computerized gang tracking data base. *Corrections Compendium, 22,* 20–22.

Norris, T. (2001). Importance of gang-related information sharing. *Corrections Today, 63,* 96–99.

Olson, D. E., Dooley, B., & Kane, C. M. (2004). *The relationship between gang membership and inmate recidivism.* Springfield: Illinois Criminal Justice Authority.

Rainville, G., & Reaves, B. A. (2003). *Felony defendants in large urban counties.* Washington, DC: U.S. Department of Justice.

Ralph, P. H., & Marquart, J. W. (1991). Gang violence in Texas prisons. *The Prison Journal, 71,* 38–49.

Richards, S. C. (2003). My journey through the Federal Bureau of Prisons. In S. C. Richards & J. I. Ross

(Eds.), *Convict criminology* (pp. 120–149). Belmont, CA: Thompson/Wadsworth.

Rivera, B. D., Cowles, E. L., & Dorman, L. G. (2003). Exploratory study of institutional change: Personal control and environmental satisfaction in a gang-free prison. *The Prison Journal, 83,* 149–170.

Ross, J. I., & Richards, S. C. (2002). *Behind bars: Surviving prison.* Indianapolis, IN: Alpha Books.

Ruddell, R. (2005). Long-term jail populations: A national assessment. *American Jails, 19,* 22–27.

Santos, M. G. (2004). *About prison.* Belmont, CA: Thompson/Wadsworth.

Sherman, L. W., Gottfredson, D., MacKenzie, D., Eck, J., Reuter, P., & Bushway, S. (1997). *Preventing crime: What works, what doesn't, what's promising.* Washington, DC: U.S. Department of Justice.

Silverman, I. (2001). *Corrections: A comprehensive review.* Belmont, CA: Wadsworth.

Stastny, C., & Tyrnauer, G. (1983). *Who rules the joint? The changing political culture of maximum-security prisons in America.* Lanham, MD: Lexington Books.

Stephan, J. J. (2001). *Census of jails, 1999.* Washington, DC: Bureau of Justice Statistics.

Tartaro, C. (2002). The impact of density on jail violence. *Journal of Criminal Justice, 30,* 499–510.

U.S. Department of Justice. (1992). *Management strategies in disturbances and with gangs/disruptive groups.* Washington, DC: Author.

Weisheit, R. A., & Wells, L. E. (2004). Youth gangs in rural America. *National Institute of Justice Journal, 251,* 2–7.

Wells, J. B., Minor, K. I., Angel, E., Carter, L., & Cox, M. (2002). *A study of gangs and security threat groups in America's adult prisons and jails.* Indianapolis, IN: National Major Gang Task Force.

Wright, C., & Goring, S. (1989). Litigation can stop unnecessary jail building. *National Prison Project Journal, 18,* i–vii.

DISCUSSION QUESTIONS

1. Why should criminal justice researchers and policy makers care about gangs in jails?

2. What sort of interventions have jail personnel employed to deal with adult gangs in jails?

3. How might gang intervention programs in jails lead to a reduction of the gang problem in prisons and in the free world?

❖

READING

In this article, Stephen Metraux and Dennis Culhane examine incarceration histories and shelter use patterns of 7,022 homeless persons staying in public shelters in New York City. Through matching administrative shelter records with data on releases from New York State prisons and New York City jails, 23.1% of a point-prevalent shelter population was identified as having had an incarceration within the previous 2-year period. The authors found that persons entering shelter following a jail episode (17.0%) exhibited different shelter stay patterns than did those having exited a prison episode (7.7%), leading them to the conclusion that different dynamics predominate and that different interventions are called for in preventing homelessness among persons released from jail and from prison.

Recent Incarceration History
Among a Sheltered Homeless Population

Stephen Metraux and Dennis P. Culhane

It is widely assumed that there are increased rates of incarceration among the homeless population (Fischer, 1992; Snow, Baker, & Anderson, 1989). Although research has offered explanations for this relationship, there is little in the research literature that outlines its empirical dimensions. This study addresses this gap as it examines incarceration histories of persons staying in the public, single adult shelter system in New York City and the associations between incarceration histories and shelter use patterns.

Demographics alone would suggest there to be a substantial overlap among the sheltered and incarcerated populations. Compared to the overall U.S. adult population, both the homeless and the incarcerated populations are disproportionately male, young, and Black (Burt, Aron, Lee, & Valente, 2001; Culhane & Metraux, 1999; Langan & Levin, 2002; Mauer, 1999). Poverty and unemployment are endemic to both populations (Burt et al., 2001; Lichtenstein & Kroll, 1996; Western & Beckett, 1999). High rates of mental illness and substance abuse have been widely documented in research on both populations (Burt et al., 2001; Conklin, Lincoln, & Tuthill, 2000; Freudenberg, 2001; Lamb & Weinberger, 1998; Peters, Greenbaum, Edens, Carter, & Ortiz, 1998). And the convergence of characteristics also manifests itself spatially, as both incarceration and homelessness disproportionately affect persons in low-income, urban Black neighborhoods (Correctional Association of New York, 1990; Culhane, Lee, & Wachter, 1996; Wacquant, 2000).

Prior research presents a broad range of findings on rates of incarceration among homeless population samples. Schlay and Rossi (1992) summarized 60 studies on the characteristics and compositions of the homeless population from 1981 to 1988. Among these studies, 26 reported findings on incarceration history among the homeless population. Depending on the study, between 8% and 82% of the homeless populations studied reported having been previously incarcerated, with a mean across the studies of 41%. A later review by Eberle, Kraus, Pomeroy, and Hulchanski (2000) reported that surveys showed prior "rates of arrest and incarceration among the homeless, ranging from 20% to 67%" (p. 35). Burt et al. (2001), drawing on results from a nationally representative sample of the homeless population and a comparison group of nonhomeless soup kitchen users, reported that 49% disclosed ever having spent time in a jail and 18% reported spending time in a state or federal prison and that history of incarceration was associated with a significantly higher likelihood of being homeless.

Although these findings provide support for the salience of the link between homelessness and increased criminal activity, they provide little detail beyond general, self-reported prevalence rates of persons who have spent time in jails and/or prisons and who have records of previous arrests or convictions. Yet despite the vagueness of these findings, researchers point to high rates of criminal activity as evidence of a criminalization of homelessness where homeless persons, because of their marginal economic and social status and the public nature of their existence, are more prone to arrests and incarceration for

SOURCE: Metraux, S., & Culhane, D. P. (2006). Recent Incarceration History Among a Sheltered Homeless Population. *Crime and Delinquency, 52*(3), 504–517. Reprinted with permission of Sage Publications, Inc.

misdemeanors and a range of minor crimes (Barak & Bohm, 1989; Snow et al., 1989). The argument that arrests and incarcerations serve as a mechanism of social control over the homeless population has a long history (e.g., Bittner, 1967; Spradley, 1970) and is consistent with Irwin's (1985) description of "rabble management." Fischer (1992) also points out that, through these incarcerations, the criminal justice system functions as a provider of services such as housing, substance abuse treatment, and mental health care that are ordinarily received from other systems. Finally, shelters, jails, and prisons may be part of a larger "institutional circuit" that includes sequential stints in a series of institutions in place of a stable living situation (Hopper, Jost, Hay, Welber, & Haugland, 1997).

An alternative viewpoint is that homelessness may be one result of more general readjustment problems that follow release from incarceration. Shelter use among persons released from incarceration is seen here as one outcome related to a problematic community reentry process (Petersilia, 2001; Travis, Solomon, & Waul, 2001). Metraux and Culhane (2004) found that 11.9% of persons released from New York State Prison to New York City experienced a shelter stay in 2 years following release, a rate that is comparable to shelter rates among persons released from public psychiatric hospitals (Kuno, Rothbard, Averyt, & Culhane, 2000). Furthermore, of these released prisoners who stay in shelters, 54.4% enter within 30 days of their release from prison (Metraux & Culhane, 2004).

This study outlines the prevalence of incarceration history among a point-prevalent sheltered homeless population by matching records from the municipal shelter system in New York City to records of persons released from both New York State prisons and New York City jails. In doing so, it adds to the scant knowledge about the extent of the intersection of homelessness after incarceration. Furthermore, this study examines whether there are associations between these incarceration histories and basic shelter use dynamics and whether or not these associations can provide support for the criminalization and reentry explanations.

▧ Data and Method

The data used in this study came from three administrative databases: records of users and utilization of single adult shelters administered by the New York City Department of Homeless Services (DHS); records of all jail discharges (related to convictions) from the New York City Department of Corrections (DOC); and all releases from prison to New York City from New York State Department of Correctional Services (NYSDOCS). DHS administers the largest shelter network of any American city and covers approximately 85% of all New York City shelter beds (Culhane, Dejowski, Ibanez, Needham, & Macchia, 1994; New York City Department of Human Services, 2003), whereas DOC and NYSDOCS operate the second largest municipal jail and third largest prison systems in the United States, respectively.

This study selected all persons who were in a DHS single adult shelter on December 1, 1997 (i.e., the index date), and matched these records with records of jail and prison discharges for the 2-year period preceding this date. Matches of NYSDOCS observations to observations from the DHS data were based on common name, date of birth, sex, and social security number. The same identifiers, except for social security number, which was unavailable, were used to match DOC data. When a match with either jail or prison was determined, the matching record was appended onto the corresponding DHS record. In the event of matches with multiple incarceration records, the most recent jail and prison record was retained. Jail episodes that led to transfers to prison were considered part of the prison episode.

Descriptive and multivariate regression techniques were used to assess (a) the extent to which persons in the DHS single adult shelter system on a specific night had recent histories of incarceration, (b) how incarceration histories intersected with shelter use patterns, and (c) whether there are differences in these areas between persons who have been jailed and persons who have been imprisoned. The multiple regression analyses focused on four dependent measures: (a) the number of shelter stays prior to the instant stay, (b) the length of instant stay subsequent to the index date, (c) the occurrence of a subsequent shelter stay, and (d) the time between release from incarceration and shelter admission.

Three different regression techniques were applied to model these outcomes. For the previous shelter stays regression model, a Poisson distribution was fitted to accommodate the discrete, highly skewed nature of count variables such as this (Allison, 1999).[1] Ordinary least squares (OLS) regression was used to examine, for the entire shelter population, the length of the instant stay from the index date onward and the incarceration to shelter gap for those among the shelter population with an incarceration history.[2] Finally, a Cox proportional hazards regression model was fitted to assess the association of various factors on the hazard of incurring another shelter stay subsequent to exiting the instant stay, given that the majority of persons in the study group will be "censored" (i.e., not experience a subsequent shelter stay; Allison, 1995). All data management, matching, and analyses were performed using SAS statistical software, version 8.02.

⊠ Results

Table 1 presents descriptive demographic and shelter utilization results for the overall point-prevalent shelter population and the subgroups in which jail and prison releases occurred up to 2 years prior to the index date. Altogether, 23.1%, just less than one fourth, of the 7,022 persons staying in the single adult shelter system that night had a record of an incarceration. This included 17.0% with a jail

release and 7.7% with a prison release. These two groups were not discrete, as 113 persons or 1.6% of the overall population (21% of the previously imprisoned population and 9.5% of the previously jailed population) had been incarcerated in both jail and prison.

Shelter utilization is represented by three measures: the number of DHS shelter stays (prior and instant), the prospective length of the instant stay (i.e., the duration of the shelter stay after the index date), and whether or not a repeat stay occurred within 1 year from exiting the instant stay. Summarizing these measures, when compared to the overall group, the prison subgroup had about the same number of stays, but their stays were shorter, whereas the jail subgroup also had shorter stays but stayed in shelters more frequently both before and after the instant stay.

There were also significant demographic differences among the subgroups and the general shelter population. Among a predominantly Black and Hispanic shelter population, the prison subgroup featured a higher proportion of persons of Hispanic ethnicity, whereas the jail subgroup contained a higher proportion of persons of (non-Hispanic) Black race. The single adult shelter population was 81.5% male, and both the prison and jail subgroups had even higher proportions of males. Both subgroups were significantly younger than the general shelter population.

These descriptive characteristics were fitted into three multivariate models to estimate the associations of jail or prison release on three measures of shelter utilization, controlling for demographic and shelter utilization measures. The first set of results was from a Poisson model regressing on the number of previous shelter stays experienced by each person in the study group. Although being incarcerated during this period, especially in prison, reduced the opportunity for persons to accrue shelter stays, having a history of jail release showed a highly significant association with a greater number of past shelter stays, whereas history of prison release had a nonsignificant

Table 1	Persons in New York City (NYC) Municipal Single Adult Shelters on December 1, 1997: Incarceration, Shelter Use, and Demographic Characteristics

	Overall Sheltered Population %[a]	Sheltered Population With Prison History %[b]	Sheltered Population With Jail History %[c]
Incarceration			
Any history[d]	23.1	100.0	100.0
Jail[c]	17.0	21.0	100.0
Prison	7.7	100.0	9.4
Shelter stay history			
1st stay	40.4	37.9	26.1
1–5 stays	42.6	45.5	46.3
6–10 stays	11.4	12.2	17.6
More than 10 stays	5.6	4.4	10.0
Days in instant shelter stay (after December 1)[e,f,g]			
1–7 days	5.4	5.7	6.9
8–30 days	9.7	12.8	11.4
31–180 days	41.9	48.1	41.7
181–365 days	20.5	19.6	22.2
365+ days	22.5	13.9	17.8
Subsequent shelter stay[f,h]	26.9	25.2	35.4
Race/ethnicity[f,g]			
Black (non-Hispanic)	60.4	56.2	65.1
White (non-Hispanic)	13.6	8.9	9.0
Hispanic	20.6	31.9	22.5
Other or Unknown	5.4	3.0	3.3
Male[f,g]	81.5	92.6	87.8
Age[f,g]			
18–25	5.8	3.2	5.8
26–35	23.1	35.8	31.4
36–45	35.8	41.6	41.6
46–55	23.0	15.6	17.7
56+	12.2	3.9	3.5

a. $n = 7,022$.

b. $n = 539$.

c. $n = 1,196$. Does not include episodes where persons were transferred directly from jail to prison.

d. Incarceration (prison and/or jail) histories are limited to releases from New York State prisons and NYC jails within the 2-year period prior to December 1, 1997.

e. For all persons in the study group, shelter stays are truncated in this measure to begin on December 1, 1997, because of prison and jail subgroups having less opportunity to accrue pre-December 1 shelter days given their incarceration histories.

f. Appropriate tests of significance (chi-square and t test) indicate significant differences ($p < .001$) between the prison subgroup and the rest of the study group.

g. Appropriate tests of significance (chi-square and t test) indicate significant differences ($p < .001$) between the jail subgroup and the rest of the study group.

h. Subsequent shelter stay occurred either within 1 year after instant stay exit or, if this exit occurred in 2001, before December 31, 2001.

association and no negative effect. The more days accrued during the part of one's current stay that occurred prior to the index date, the further opportunity to accrue stays is reduced, and this was borne out by a significant, negative association between this measure and number of past shelter stays.

In the second model, which used OLS regression on the number of days in the instant shelter stay that occurred after index date, a prison stay was significantly associated with a

shorter shelter stay, whereas a jail stay had a nonsignificant association. The number of past shelter stays was significantly associated with a reduced length of shelter stay, and accruing more shelter days prior to the index date was associated with a longer stay after this date.

The final model in Table 2 was a Cox regression model estimating the association of the covariates with the hazard of returning for a subsequent shelter stay in the year following exit from the instant shelter stay. Here both jail and

| Table 2 | Regression Results From Three Models on Shelter Utilization Measures for Persons Staying in New York City Municipal Single Adult Homeless Shelters on December 1, 1997 |

	Past Shelter Stays (Poisson)		Partial Length of Instant Shelter Stay—Post-December 1 (Ordinary Least Squares)		Hazard for Repeat Shelter Stay (Cox Regression)	
	Coefficient Estimate	CI	Coefficient Estimate	CI	Hazard Ratio	CI
Incarceration						
Prison release	−0.10	−0.23, 0.01	−39.6	−68.6, −10.6**	0.80	0.67, 0.95*
Jail release	0.47	0.40, 0.55***	−5.9	−28.7, 15.0	1.15	1.02, 1.28*
Shelter utilization						
Number of prior stays	not in model		−3.0	−5.5, −0.4*	1.12	1.10, 1.13***
Length of stay (total stay)	not in model		not in model		0.9999	−0.99, 1.00*
Length of stay (pre-December 1)	−0.0003	−0.01, 0.00***	0.21	0.20, 0.23***	not in model	
Race/ethnicity						
Black (non-Hispanic)	reference category		reference category		reference category	
White (non-Hispanic)	−0.57	−0.68, −0.46***	0.7	−22.6, 24.0	1.00	0.87, 1.15
Hispanic	−0.31	−0.39, −0.22***	3.7	−15.9, 23.4	0.98	0.87, 1.10
Other or Unknown	−1.26	−1.52, −1.01***	−13.6	−48.4, 21.4	0.74	0.57, 0.95*
Male	0.46	0.36, 0.56***	−25.1	−45.1, −5.0*	1.54	1.34, 1.77***
Age	0.02	0.01, 0.02***	3.5	2.8, 4.2***	0.993	0.99, 1.00**
Intercept	−0.13	−0.29, 0.03	90.7	55.5, 125.8		
Scale	2.018					

NOTE: CI indicates 95% confidence interval.

$*p < .05$; $**p < .01$; $***p < .001$.

prison history had significant ($p < .05$) associations with the dependent variable, but whereas a jail stay history was associated with an increased hazard (by 15%) of a repeat shelter stay, having had a prison stay history was associated with a 20% decrease in the hazard of experiencing a repeat shelter stay. The higher the number of past shelter stays, the greater the hazard for experiencing a subsequent stay, whereas the number of days in the instant stay (total stay length) had a significant but small incremental association with a decreased risk of a subsequent shelter stay (0.01% reduction in hazard per shelter day).

Tables 3 and 4 focused on the incarceration episodes of the 1,622 persons in the study group and demonstrate further differences related to jail and prison histories.[3] Table 3 shows that unsurprisingly, prison incarcerations on average lasted considerably longer than jail incarcerations. But, in looking at the gap between the end of incarceration and the start of the index shelter stay, the prison to shelter gap, on average, was considerably shorter than the jail to shelter gap. More than half (54.3%) of the former lasted 1 week or less, compared to 32.9% of the latter. The

Table 3	Length of Most Recent Incarceration Episodes and Length of Time Between the End of the Incarceration Episodes and the Start of the Corresponding Shelter Stays Associated With Persons in the New York City Single Adult Shelter System on December 1, 1997, With an Incarceration Record in the 2-Year Period Prior to This Date		
	All Incarcerations %[a]	**Prison Episodes %[b]**	**Jail Episodes %[c]**
Length of incarceration			
1 day	6.3	0.0	8.9
2–7 days	30.2	0.0	42.4
8–30 days	14.9	1.1	20.5
31–365 days	29.1	33.5	27.4
More than 366 days	19.5	65.5	0.9
Length of incarceration release to shelter entry (gap)			
0–1 day	26.9	37.5	22.6
2–7 days	12.2	16.8	10.3
8–30 days	9.7	7.5	10.6
31–180 days	22.9	17.9	25.0
181–365 days	16.2	11.7	18.0
366–730 days	12.0	8.5	13.4

NOTE: Where both jail and prison histories preceded one person's shelter stay ($n = 113$), only the incarceration episode that was closest to shelter stay was included.

a. $n = 1,622$.

b. $n = 469$.

c. $n = 1,153$.

Table 4	Regression Model for Assessing the Effects of Incarceration Type on the Incarceration to Shelter Gap Length for Persons in New York City Single Adult Shelter System With an Incarceration Record in 2-Year Period Prior to December 1, 1997

	Coefficient Estimate	CI
Days incarcerated	−0.02	−0.04, 0.0001
Incarcerated in prison	−31.0	−54.0, −8.1***
Intercept	178.0	134.6, 221.5***

NOTE: CI indicates 95% confidence interval. *Incarcerated in prison* is as compared to *incarcerated in jail* as the reference group. These results control for demographic variables (race/ethnicity, age, sex), whose results are not included here and are all nonsignificant.

***$p < .001$.

median gap length (not shown in the table) for the jail gap (64 days) was also considerably longer than that for the prison gap (5 days).

Table 4 presents the results of an OLS regression model that assesses whether the association between prison stay and shorter incarceration-shelter gap remained after controlling for the differences in length of the incarceration episode (and for race/ethnicity, age, and sex). After controlling for these covariates, prison stay was still associated with a considerably shorter gap length compared to jail stay, whereas length of incarceration falls just outside of being significant at the .05 level.

⬛ Discussion and Conclusion

This study, which matched prison and jail records to records of individuals staying in municipal homeless shelters in New York City on December 1, 1997, found that 23.1%, or nearly one fourth of the study population, had been incarcerated in a New York State prison or a New York City jail within the previous 2 years. This overall rate, when broken down by incarceration type, has 17.0% experiencing a jail episode and 7.7% experiencing a prison episode. These rates are almost certainly

understated because of limitations related to the relatively short period studied, the lack of data on incarcerations outside of NYSDOCS and New York City DOC, and the undetermined number of missed matches because of inconsistent identifying information being collected by the different systems. Nonetheless, the findings indicate that incarceration affects a substantial minority of the single adult sheltered population and that criminal justice issues, whether recognized or not, figure prominently among the homeless milieu.

The extent to which findings such as this are generalizable is always a matter of concern. As the largest city in the United States, New York City also has the largest shelter system. However, when taken as a proportion of its population, New York City's shelter population falls into the middle of a range of other different-sized urban jurisdictions (Metraux et al., 2001). With respect to its jail population, New York City ranks second in overall size to Los Angeles and, when viewed as a proportion of its overall population, ranks behind numerous other cities (Harrison & Karberg, 2003). Similarly, although New York State has one of the largest inmate populations in the United States, its rate of incarceration ranks it among the middle of the states (U.S. Department of

Justice, 2002). Other dynamics specific to individual cities are more difficult to quantify, but there is no indication that factors particular to New York City would preclude these findings from being considered more generally.

The distinct patterns of shelter use associated with prison releases and jail releases each have different implications for developing effective interventions to ameliorate homelessness on release from incarceration. Among many of the 7.7% of the study group who had a prior prison stay, shelter use appears to have been related to reentry issues. History of a prison release in the 2 years prior to the index date was associated with a shorter instant shelter stay, a reduced hazard of experiencing a subsequent stay, and, compared to those released from jails, a shorter gap between incarceration exit and shelter entry. The finding that 61.8% of those in the study population who were released from prison commenced their instant shelter stay within 30 days of release is consistent with findings that these 30 days represent a critical period when released prisoners are most vulnerable to a variety of negative outcomes (Nelson, Deess, & Allen, 1999; Travis et al., 2001).

Thus, shelter stays among persons released from prison appear more likely to be of a transitional nature rather than part of a long-term pattern of homelessness. However, it is unclear whether the long-term outcomes following this transition are more likely to include eventual economic and residential stability or less desirable outcomes such as reincarceration. Other research using these data shows shelter use, among a cohort of released prisoners, to be associated with a modest increase in the hazard of returning to prison (Metraux & Culhane, 2004). Conversely, the short period between prison release and most subsequent shelter use suggests that housing assessments prior to release could identify many of those who will be at risk for homelessness. Housing, if made available on prison release on either a transitional or a permanent basis, might preclude the need for homeless services among

persons released from prison and facilitate the more general community reentry process (Osher, Steadman, & Barr, 2003).

Among the 17.0% in the study group entering the shelter system from a recent jail stay, a different shelter use pattern emerges. Compared to the overall study group, this subgroup tended to have a more extensive history of prior shelter stays and a greater hazard for experiencing a subsequent shelter stay. Not only did shelter stays follow a more prolonged, episodic pattern, but the incarceration stay was typically of a relatively brief duration, with 71.8% staying in jail for 30 days or less. This sequential pattern of shelter and jail use points to a more prolonged pattern of residential instability.

This pattern offers support, albeit tentative, for other broader paradigms describing the similar functions that jails and shelters play among extremely poor populations. In this context, these serial jail and shelter stays alternately represent pieces of an institutional circuit that acts as a surrogate for stable housing (Hopper et al., 1997), a means for rabble management in which jails and shelters exercise social control over an undesirable population (Irwin, 1985), and a process of socialization into a long-term, deviant lifestyle described as "a life sentence on the installment plan" (Spradley, 1970, p. 252; see also Grunberg & Eagle, 1990). Kuhn and Culhane (1998) have found that homeless persons with such episodic patterns of shelter use tend to be younger and have higher rates of mental illness and substance abuse when compared to the overall population of single adult shelter users. Interventions suited for this group would require a more structured residential treatment format, although supported housing programs have also reported success with persons who have such institutional stay patterns (Tsemberis, 1999).

To summarize, jail and prison releases were each associated with different shelter stay patterns, and each type of incarceration calls for a different intervention approach. In making these conclusions, this study has emphasized the dynamics among shelters and jails and prisons instead of the individual characteristics of the persons in the study group, who are usually the focus of such studies (Snow, Anderson, & Koegel, 1994). Indeed, the subgroups with jail and prison records are likely to have overlapping constituencies who share similar individual characteristics, and the extent to which this is so further highlights the different effects of jail and prison on homelessness following release. Instead, an institutional focus underscores the roles that carceral institutions play in subsequent patterns of homelessness and their potential roles as intervention points.

Data limitations preclude a more in-depth look at these dynamics and create an agenda for future research. The interaction of shelters, jails, and prisons with other institutional dynamics is one such area. Mental health and substance abuse services, as well as income support and other poverty amelioration services, have all figured prominently in proposed interventions for sheltered and formerly incarcerated populations and may provide additional insight into understanding and intervening in these different shelter use patterns. Furthermore, data on already existing community supervision services that the criminal justice system provides, and particularly probation and parole, could show how they play a role in either preventing or facilitating postincarceration shelter use and ways to render these services more effective. In the meantime, the need for different approaches to preventing homelessness on prison release and jail release is apparent, as is the potential for such interventions to substantially reduce the demand for shelter among single adults.

Notes

1. A shelter *stay* is here considered to be a span of shelter utilization that both followed and preceded a 30-day absence from a shelter (Culhane & Kuhn, 1998; Piliavin, Wright, Mare, & Westerfelt, 1996; Wong, Culhane, & Kuhn, 1997). By using this 30-day exit criterion, a stay hereby precedes an extended period away from shelters and assumes that, after an exit, alternate living arrangements have supplanted, not just provided temporary relief from, shelter use. However, leaving a shelter may not mean leaving homelessness as, depending on the living situation and the definition of homelessness used (Cordray & Pion, 1991), a person exiting a shelter stay may still, by virtue of subsequently living on the streets or in doubled up situations with other households, be considered homeless.

2. The length of the instant shelter stay is measured prospectively from the index date (December 1, 1997) to reduce the extent to which the group differences are an artifact of incarceration history. As was already explained, to be considered to have an incarceration (prison or jail) history, a person must have experienced an incarceration within the 2-year period prior to the index date. Depending on the release date and the length of incarceration, time spent in jail or prison reduces the opportunity to accrue an extended shelter stay prior to the index date (as one cannot simultaneously be incarcerated and sheltered). To avoid confounding, the stay length measure only includes time accrued after the index date.

3. The 113 instances where both jail and prison histories preceded a shelter stay were grouped by whichever incarceration episode was closest to shelter entry.

References

Allison, P. D. (1995). *Survival analysis using the SAS system: A practical guide.* Cary, NC: SAS Institute.

Allison, P. D. (1999). *Logistic regression using the SAS system.* Cary, NC: SAS Institute.

Barak, G., & Bohm, R. M. (1989). The crimes of the homeless or the crime of homelessness: On the dialectics of criminalization, decriminalization, and victimization. *Contemporary Crises, 13,* 275–288.

Bittner, E. (1967). Police on skid row: Study of peacekeeping. *American Sociological Review, 32,* 699–715.

Burt, M. R., Aron, L. Y., Lee, E., & Valente, J. (2001). *Helping America's homeless: Emergency shelter or affordable housing?* Washington, DC: Urban Institute.

Conklin, T. J., Lincoln, T., & Tuthill, R. W. (2000). Self-reported health and prior health behaviors of newly admitted correctional inmates. *American Journal of Public Health, 90,* 1939–1941.

Cordray, D. S., & Pion, G. M. (1991). What's behind the numbers? Definitional issues in counting the homeless. *Housing Policy Debate, 2,* 587–616.

Correctional Association of New York. (1990). *Young men under criminal justice custody in New York State.* New York: Author.

Culhane, D. P., Dejowski, E., Ibanez, J., Needham, E., & Macchia, I. (1994). Public shelter admission rates in Philadelphia and New York City: The implications of turnover for sheltered population counts. *Housing Policy Debate, 5*(2), 107–140.

Culhane, D. P., & Kuhn, R. S. (1998). Patterns and determinants of shelter utilization among single adults in New York City and Philadelphia: A longitudinal analysis of homelessness. *Journal of Policy and Management, 17*, 23–43.

Culhane, D. P., Lee, C. M., & Wachter, S. (1996). Where the homeless come from: A study of the prior address distribution of families admitted to public shelters in New York City and Philadelphia. *Housing Policy Debate, 7*, 327–360.

Culhane, D. P., & Metraux, S. (1999). Assessing relative risk for homeless shelter usage in New York City and Philadelphia. *Population Research and Policy Review, 18*(3), 219–236.

Eberle, M., Kraus, D., Pomeroy, S., & Hulchanski, D. (2000). *Homelessness—Causes and effects: A review of the literature, Vol. 1.* Vancouver, Canada: British Columbia Ministry of Social Development and Economic Security.

Fischer, P. J. (1992). The criminalization of homelessness. In M. J. Robertson (Ed.), *Homelessness: A national perspective* (pp. 57–64). New York: Plenum.

Freudenberg, N. (2001). Jails, prisons, and the health of urban populations: A review of the impact of the correctional system on community health. *Journal of Urban Health, 78*, 214–235.

Grunberg, J., & Eagle, P. F. (1990). Shelterization: How the homeless adapt to shelter living. *Journal of Hospital and Community Psychiatry, 41*, 521–525.

Harrison, P. M., & Karberg, J. C. (2003). *Prison and jail inmates at midyear, 2002.* Washington, DC: U.S. Department of Justice, Bureau of Justice Statistics.

Hopper, K., Jost, J., Hay, T., Welber, S., & Haugland, G. (1997). Homelessness, severe mental illness, and the institutional circuit. *Psychiatric Services, 48*, 659–665.

Irwin, J. (1985). *The jail: Managing the underclass in American society.* Berkeley: University of California Press.

Kuhn, R. S., & Culhane, D. P. (1998). Applying cluster analysis to test a typology of homelessness by pattern of shelter utilization: Results from the analysis of administrative data. *American Journal of Community Psychology, 26*, 207–232.

Kuno, E., Rothbard, A. B., Averyt, J. M., & Culhane, D. P. (2000). Homelessness among persons with severe mental illness in an enhanced community-based mental health system. *Psychiatric Services, 51*, 1012–1016.

Lamb, H. R., & Weinberger, L. E. (1998). Persons with severe mental illness in jails and prisons: A review. *Psychiatric Services, 49*, 483–492.

Langan, P. A., & Levin, D. J. (2002). *Recidivism of prisoners released in 1994.* Washington, DC: U.S. Department of Justice, Bureau of Justice Statistics.

Lichtenstein, A. C., & Kroll, M. (1996). The fortress economy: The economic role of the U.S. prison system. In E. Rosenblatt (Ed.), *Criminal injustice: Confronting the prison crisis* (pp. 16–39). Boston: South End.

Mauer, M. (1999). *Race to incarcerate.* New York: New Press.

Metraux, S., & Culhane, D. P. (2004). Homeless shelter use and reincarceration following prison release: Assessing the risk. *Criminology & Public Policy, 3*, 201–222.

Metraux, S., Culhane, D. P., Raphael, S., White, M., Pearson, C., Hirsch, E., et al. (2001). Assessing homeless population size through the use of emergency and transitional shelter services in 1998: Results from the analysis of administrative data in nine US jurisdictions. *Public Health Reports, 116*, 344–352.

Nelson, M., Deess, P., & Allen, C. (1999). *The first month out, post-incarceration experiences in New York City.* New York: Vera Institute of Justice.

New York City Department of Human Services. (2003). *Homeless outreach population survey results.* New York: Author.

Osher, F., Steadman, H. J., & Barr, H. (2003). A best practice approach to community reentry from jails for inmates with co-occuring disorders: The APIC model. *Crime & Delinquency, 49*, 79–96.

Peters, R. H., Greenbaum, P. E., Edens, J. F., Carter, C. R., & Ortiz, M. M. (1998). Prevalence of *DSM-IV* substance abuse and dependence disorders among prison inmates. *American Journal of Drug and Alcohol Abuse, 24*, 573–587.

Petersilia, J. (2001). Prisoner re-entry: Public safety and reintegration challenges. *The Prison Journal, 81*, 360–375.

Piliavin, I., Wright, B. R., Mare, R. D., & Westerfelt, A. H. (1996). Exits and returns to homelessness. *Social Service Review, 70*(1), 33–57.

Schlay, A. B., & Rossi, P. H. (1992). Social science research and contemporary studies of homelessness. *Annual Review of Sociology, 18*, 129–160.

Snow, D. A., Anderson, L., & Koegel, P. (1994). Distorting tendencies in research on the homeless. *American Behavioral Scientist, 37*, 461–475.

Snow, D. A., Baker, S. G., & Anderson, L. (1989). Criminality and homeless men: An empirical assessment. *Social Problems, 36*, 532–549.

Spradley, J. P. (1970). *You owe yourself a drunk: An ethnography of urban nomads.* Boston: Little, Brown.

Travis, J., Solomon, A. L., & Waul, M. (2001). *The dimensions and consequences of prisoner reentry.* Washington, DC: Urban Institute.

Tsemberis, S. (1999). From streets to homes: An innovative approach to supported housing for homeless adults with psychiatric disabilities. *Journal of Community Psychology, 27*, 225–241.

U.S. Department of Justice. (2002). *Prisoners in 2001.* Washington, DC: Author.

Wacquant, L. (2000). Deadly symbiosis: When ghetto and prison meet and mesh. *Punishment and Society, 3*(1), 95–134.

Western, B., & Beckett, K. (1999). How unregulated is the US labor market? The penal system as a labor market institution. *American Journal of Sociology, 104*, 1030–1060.

Wong, Y. L., Culhane, D. P., & Kuhn, R. S. (1997). Predictors of exit and reentry among family shelter users in New York City. *Social Service Review, 71*, 441–462.

DISCUSSION QUESTIONS

1. Why are jails the "dumping" ground for so many people in our communities? What are the consequences of this social policy?

2. What are the relative advantages and disadvantages of community jails?

3. How might reentry programs prevent recidivism?

❖

READING

In this article Mark Kellar and Hsiao-Ming Wang examine factors surrounding inmate assaults in local jail. The authors analyzed survey responses and existing documentation from 138 county jail administrators in Texas to determine relationships between select input variables and the incidence of inmate assaults on staff and other inmates. Logistic regression was used to examine the effects of importation model and managerial model variables on inmate assaults. The logistic model of inmate-on-inmate assaults was significant, whereas that of inmate-on-staff was not. The importation model approach was supported by a strong relationship between proportion of maximum security inmates and inmate assaults and by the relationship between inmates in metropolitan as opposed to nonmetropolitan jails and inmate assaults. Findings indicated a weaker relationship between two managerial model variables. No discernible relationships were found between the degree of rehabilitative philosophy expressed by the administrator, the ethnic breakdown of staff, or the type of facility structure and inmate assaults.

Inmate Assaults in Texas County Jails

Mark Kellar and Hsiao-Ming Wang

▧ Background: Prisoner Assaults

Prisoner assaults on both staff and inmates in correctional settings have been a topic of fundamental interest to practitioners and theorists alike since the inception of incarceration as a sanctions strategy. Essential individual safety is a prerequisite condition for any justice institution. To the extent that inmates may continuously and brutally assault one another or staff at will, the system will inevitably become dysfunctional and fail to accomplish its fundamental missions. Liabilities associated with assaults further underscore the proposition that internal violence must be clearly understood and, when possible, controlled. Therefore, scholars have conducted numerous studies and developed elaborate theories over the decades that contribute to the collective understanding of prisoner violence in correctional settings.

The Prison Environment

Early researchers (Clemmer, 1958; Sykes, 1958) addressed the issue from the focus of behaviors developed by inmates within the prison subculture. Others such as Irwin and Cressey (1962) rebutted this position by noting that many behaviors, including propensities toward violence, are acquired in social and cultural environments where the inmates live and then imported into the prison where the behaviors are institutionalized. More recently, DiIulio (1987) has argued that prisoner behavior is a function of managerial constraints (or a lack thereof) that should be controlled by prison management. Each of these orientations

has led to a prescribed model and a corresponding theory complete with cases in point, explanations for instances where deviation seems to occur, and a wealth of replication studies. Most are grounded on incident-based reporting, and many rely heavily on inmate trait-based reports. Most important, as it relates to this study, all are based on experiences in state or federal prisons. It is interesting that almost no research has focused on inmate assaults in local jails.

The question arises, Can the knowledge relative to assaults gained from prison studies be readily generalized to local jails, or is concurrent construct validity automatically transferable from one institutional environment to another? This study is presented in an attempt to examine the relationships that may exist in local jails between the occurrence of inmate assaults and various institutional and managerial factors. One underlying assumption of this approach is that behaviors such as prisoner assaults occur in the full context of an institutional setting made up of complex social, cultural, organizational, and political factors that may or may not be reflective of the state prison environment. Because few projects have concentrated on the jail culture, specifically, much of the work will necessarily be exploratory in nature. Obviously, any results will reflect a preliminary or "first pass" finding and should be enhanced, expanded, or replicated through future efforts.

A comprehensive survey was developed to poll Texas county jail administrators in an attempt to address a number of managerial issues such as general corrections philosophy, human resource and staffing demographics, agency organization, and various operational

SOURCE: Kellar, M., & Wang, H.-M. (2005). Inmate Assaults in Texas County Jails. *The Prison Journal, 85*(4), 515–534. Reprinted with permission of Sage Publications, Inc.

topics (Kellar, Jaris, & Manboah-Roxin, 2001). The specific issue of inmate assaults was considered by asking for the number of serious assaults in a calendar year committed by inmates on other inmates and by inmates on staff within each institution. To fully appreciate the contextual ramifications of these inquiries, one must consider the nature of local jails as justice institutions in addition to those factors related to violent or aggressive behaviors by prisoners.

Local Jails: An Alternative Research Environment

Local jails are, perhaps, the most misunderstood institutions in the criminal justice system. They are an important component of the justice complex because they represent a kind of systemic hub where law enforcement, the courts, and corrections interface in a complex series of processes. Mays and Winfree (2002) refer to jails as "the gateway to the criminal justice system" (p. 90). Among the institutions and programs of the corrections system, jail is the one most neglected by scholars and least known to the public (Clear & Cole, 1997, p. 143).

Jails are generally viewed as extensions of a broader correctional system that is firmly grounded, as far as most academic researchers are concerned, in the study of state prisons (O'Toole, 1999). Because both prisons and jails are penal institutions, they each comply with those principles related to corrections theory that have been meticulously constructed by scholars trained in the study of prison systems. However, certain assumptions associated with state prisons may be invalid, or at least misleading, when applied to county or local jails.

A negative image is often associated with local jails (Zupan, 2002, p. 38). This may be attributable to the lack of a firm and positive institutional identity for the agency. Most county sheriffs, the individuals charged with the administration of some 70% of local jails (Kerle & Ford, 1982), consider law enforcement as their primary duty. The responsibility for jail management may be maintained with reluctance or

relegated to marginal subordinates. The typical county sheriff and his or her deputies may have little interest in corrections or jails (Clear & Cole, 1997; Moynahan & Stewart, 1980). In systems where the jail staff is made up of law enforcement officers, assignment to jail duty is often considered boring, less glamorous than law enforcement billets, or even demeaning (Peak, 2001). Few, if any, organized lobbies or influential interest groups advocate increased spending for jails. Perhaps, from a political perspective, this is because jails have no "sex appeal." Zupan (2002) attributes Huey Long with the adage, "There ain't no votes in prison!" (p. 48). With the exception of reforms initiated by judicial intervention during the 1970s, local jails often reflect practices and structures dating back to the 19th century. This lack of administrative priority and workplace status inevitably leads to a negative institutional environment (Gaines, Kaune, & Miller, 2000). Administrative structures for local detention operations vary greatly from one jurisdiction to another, and this makes for confusing or even invalid comparisons. As mentioned earlier, elected county sheriffs manage most jails, but some fall under the authority of a local police chief. Some jails are managed as a department within a county government and are headed by an appointed director. A few states maintain local jails as a division of the state department of corrections (Mays & Winfree, 2002).

Local jails house a variety of inmate classification types. Virtually all convicted felons spend some time in county jails before they are transferred to state prisons. The judicial process may be extended for long periods of time, thereby assuring that some jail inmates are housed for lengthy stays. Further, jail inmates represent a virtual "hodge-podge" of classification categories in addition to those serving misdemeanor sentences and those awaiting trial. Jail inmates include felons bench-warranted as witnesses in other cases, parolees, probationers, absconders, individuals placed in protective custody, prisoners held for other state or federal jurisdictions, persons awaiting transfer to mental institutions, and even civil detainees (Mays & Winfree, 2002, p. 90).

Prisoner Assaults: A Review of Literature

Historically, researchers and scholars have constructed and defined a number of criminological theories and paradigms to explain violent behavior among incarcerated populations as well as individual propensities for aggression. At least three conceptual models have been widely studied and analyzed as a means of classifying aggressive behaviors exhibited by prison inmates. Almost all of these studies were conducted in prison rather than jail settings.

The deprivation model is based on Donald Clemmer's (1958) process of "prisonization" as described in his classic study, *The Prison Community*, and advocated by such notables as Gresham Sykes (1958). According to this view, incarcerated individuals adopt various elements of the prison culture through time and evolve a unique set of behavioral codes including a preference for violent behavior as a means of settling disputes. The prison experience itself dehumanizes the inmate and deprives the individual of self-esteem, personal value, and traditional cultural traits. This leads to frustration, anxiety, and ultimately those behaviors condoned by the subculture, including assault and other aggressions. Some researchers have conducted studies that lend support for the deprivation model. Zingraff (1980) concluded that deprivation variables including length of time incarcerated, alienation from organized social units, and alienation from general society were important predictors of inmate behaviors, especially among male delinquents. Paterline and Peterson (1999) found that deprivation model variables were better predictors of prisonization in maximum-security federal prisoners than were others. Petersen (1997) found that deprivation factors such as length of sentence affected ethnic inmate groups differently. Other researchers have concluded that,

although deprivation variables are not necessarily the best predictors of inmate behaviors, some linkage between prison-initiated cultural factors and inmate behavioral outcomes does exist (Alpert, 1979; Brown, 1990; Sorensen, Wrinkle, & Gutierrez, 1998).

A second approach to explaining prisoner violence is found in the importation model. This construct was originally proposed by Irwin and Cressey (1962), who theorized that various environmental and cultural behaviors were brought into the prison setting by the criminals themselves. The importation model defines the source of inmate violence as those factors external to the institution that caused the individuals to resort to violence as a means of coping in a hostile environment. Generally, importation theorists maintain that the inmate subculture reflects a lower class mentality influenced by poverty, lack of education, drug abuse, and gang life. Such variables as inmate race, age, and type of conviction may therefore be used to predict violent behavior or, in this case, prisoner assaults.

A large body of research supports the importation model. A decade ago, Fry and Frese (1992) noted that the importation model had gained ascendancy over a strict deprivation model in studies at the time. Brown (1990) found that age of inmate and proportions of inmates convicted of violent offenses were the best predictors of violence in California prisons. Wolf, Freinek, and Shaffer (1996) determined that both the frequency and severity of disciplinary infractions were negatively correlated with age in a group of youthful male offenders. A study by Cao, Zhao, and Van Dine (1997) supported the importation model. It concluded that inmate race and gender were the best predictors of violence in Ohio prisons. In a study of Washington state inmates, Alpert (1979) found that race and criminal history had the greatest effect on prisoner violence. A study of federal prisoners found that African American prisoners had higher rates of violence than did other groups (Harer & Steffensmeier, 1996).

A third approach to understanding prison or jail violence is labeled the *managerial* or *institutional model*. DiIulio (1987) maintains that applying appropriate managerial standards in what might otherwise be a cauldron of deprivation and importation variables can control inmate conduct. He rejects the notion that "anything that disrupts the inmates must also disrupt the prison" (see McCorkle, Miethe, & Drass, 1995). To the extent that the prison or jail authority can indeed manipulate prisoner behavior, the managerial model can offset the effects of both the deprivation and the importation models. DiIulio argues that the total prison experience does not necessarily equate with the negative behaviors associated with both the deprivation and importation models. Proponents of this orientation view levels of inmate violence as a function of those rules and regulations imposed by the managerial authority to control acts of assault. In addition, the managerial model has an innate appeal to practitioners because of the obvious policy implications contained in the orientation. Even if one accepts the validity of the other models, they offer no practical assistance to the administrator because the prison authority is obligated to accept inmates as they are. The manager can do nothing about those sociological factors that may have led an individual to a lifestyle of violence or those deep-seeded prison norms that reflect an entire counterculture behind bars.

From an institutional perspective, however, a great deal can be done to control various managerial variables. For DiIulio, the outbreak of prisoner violence is the result of poor prison management. Factors such as appropriate inmate classification, proper security procedures, staff professionalism, training and positive inmate-oriented programming all contribute to the reduction of prison violence, a legitimate goal of the correctional institution.

Of all research conducted relative to prisoner violence, the managerial model yields the most impressive results. McCorkle et al. (1995) reviewed cases involving incidents of individual and collective violence from 371 state prisons and found that higher African American to White officer ratios were related to lower levels of inmate-on-inmate and inmate-on-staff violence. Further, institutions in which large proportions of inmates were involved in formal programs were less likely to have high rates of inmate violence. Larger facilities were more likely to experience higher rates of inmate violence. A comparative study of juvenile correctional institutions by Poole and Regoli (1983) measured the effects of several variables on offender violence and found that, even though specific factors were related to outcome aggression, variations in institutional context mediated the effect of all independent variables. Memory, Guo, and Parker (1999) evaluated North Carolina's Structured Sentencing Law for prison inmates who violated institutional rules and concluded that the imposition of this law resulted in reduced rates of inmate violence, especially violence that was carried out in a calculated manner. Atlas (1983) reviewed architectural designs to determine if one style of prison was associated with lower rates of inmate violence. Although no design mode was found superior, he did conclude that lines of sight for corrections officers should be unobstructed, that dormitories were associated with higher rates of inmate violence than "single-bunking," and that direct supervision was related to a reduced rate of inmate violence.

Other dimensions of managerial variables have been subjected to rigorous analysis including a variety of sophisticated classification systems (Kennedy, 1986; Leeke & Mohn, 1986; Sechrest, 1991), a number of jail and prison programs (Wilkinson et al., 1994), and even inmate telephone systems (LaVigne, 1994). Many managerial factors seem related to inmate violence.

It should be noted that the so-called models delineated above may not be mutually exclusive in all cases. The researcher must apply a level of practical discretion to each circumstance so that generalized tenants are properly analyzed and evaluated. For example, McCorkle et al. (1995) classified prison size as a "prison management variable" because it is assumed that decision makers at the state level can consciously decide to build or close facilities or transfer prisoners from one unit to another in order to maintain proper housing levels. When one considers the local jail, the size of the institution becomes more correctly an "importation" variable because the inmate population reflects the demographics of a geographically defined area. Densely populated counties such as Harris County (Houston) or Dallas County (Dallas) will obviously have jails with large inmate populations, whereas sparsely populated counties will have jails with few inmates regardless of managerial preference. Management variables must be defined in the context of their occurrence. For the purpose of this study, emphasis is placed on the relationship of factors rather than on some theoretical model concocted from the manipulation of variables.

⬚ Method

An extensive survey including 116 topics was sent to a total of 241 Texas jail administrators (Kellar et al., 2001). Responses were returned from 145 or 60.2% of those institutions polled. This return rate equaled 66.7% (4 of 6) of institutions with more than 1,000 inmates, 40.0% (6 of 15) of jails with 500 to 999 inmates, 75.0% (12 of 16) of those with 250 to 499 inmates, 53.0% (35 of 66) of those with 50 to 249 inmates, and 63.8% (88 of 138) of those with 1 to 49 inmates. Although reasonably representative of the state at large, because of the large proportion of small jails in Texas, responses reflect a strong representation of those institutions with relatively few inmates. Among those 145 returned surveys, some respondents did not answer certain key questions; therefore, those responses were dropped and the total sample size was adjusted downward to 138. The number of inmates in each facility was retrieved from official records and monthly population reports maintained by the Texas Commission on Jail Standards (2001) and

arranged in a database format to correspond with the responses reported by local managers.

The effects of several variables on inmate-on-staff and inmate-on-inmate assaults were investigated. Entries from the survey were compiled into two dependent and nine independent variables. Logistic regressions were used to analyze the relationships among input variables and jail assaults. To present a manageable research design, the authors concentrated on three formal hypotheses and treated variables "years in operation," "jail structural type," "male jail staff," and "White jail staff" as controls.

Dependent Variables

Inmate-on-staff assault in Texas jails (question 99a in Kellar et al., 2001) and inmate-on-inmate assault in Texas jails (question 100a) were treated as dependent variables. These responses estimated the number of serious assaults that occurred from January 1, 2000, through December 31, 2000. The authors purposely requested "serious" assaults so that administrators would refrain from reporting minor incidents or mere arguments among inmates or inmates and staff. The responses were dummy coded (1 = yes and 0 = no).

Independent Variables

Two variables were used to measure the effects of the managerial model. The first was the Correction Philosophy Index from four entries designed to measure the institution's prevailing corrections philosophy (Kellar et al., 2001). These entries included a rating for each of the following statements:

1. Jail should be a punishing experience for inmates.

2. Jail should keep dangerous criminals off the street.

3. Jail should encourage inmates to obey the law.

4. Jail should be used to rehabilitate inmates.

In the survey, respondents were asked to choose answers from a 5-point Likert-type scale in each of the entries from *strongly agree* to *strongly disagree*. Researchers summed responses in the first two questions from 1 to 5 and for questions 3 and 4, from 5 to 1. Recorded values were then added for each item to obtain an index ranging from 4 to 20. The higher the index, the more likely the institution reflected a rehabilitative philosophy.

Some research suggests that, when inmates are involved in rehabilitative programs, the rate of violence against both staff and other inmates will be reduced (Burlew et al., 1994; Gaes & McGuire, 1985; McCorkle et al., 1995). By structuring the rehabilitative index, it was possible to evaluate the relationship between an institution's commitment to inmate rehabilitative programs and inmate violence.

The second managerial model variable considered was the average monthly pay of corrections officers. Labor unions and employee advocacy groups have long maintained that underpaid officers are less professional and less effective than those who receive a more liberal compensation. Historically, jail officers have been grossly underpaid (Clear & Cole, 1997; National Sheriffs' Association, 1982; Zupan, 2002) and the resulting deficit in professionalism may be related to inmate violence. The average monthly pay of an experienced corrections officer was chosen as an input variable. The question was designed to test whether there was a discernible relationship between officer pay and inmate violence. Maximum monthly pay before deductions made up an ordinal level of measurement.

Three variables were developed to test the effects of the importation model. The first was the proportion of inmates in a facility classed as maximum security. Unlike many state prison facilities, the typical local jail houses a variety of inmates from risk categories including minimum, medium, and maximum custody. The survey asked administrators to estimate the percentage of maximum security inmates housed in each facility. A number of

studies (Alpert, 1979; American Correctional Association, 1993; Brown, 1990) concluded that an individual's security classification and the severity of offense for which an individual is incarcerated are important determinants of whether a prisoner will demonstrate assaultive behavior. It therefore stands to reason that jails with more maximum security inmates are more likely to report inmate assaults than those with lower proportions of maximum security inmates. To test this hypothesis, researchers created an interval variable that indicates the percentage of maximum security inmates reported by each administrator.

A second measure of the importation model was the total inmate population of the institution. The study used inmate population within each jail as of April 1, 2001 (Texas Commission on Jail Standards, 2001), as a variable to estimate such effects. Jail populations constantly fluctuate, but the recorded population on a specific date represents a reasonable estimate of the actual number of inmates housed at any time during the calendar year. This factor may be treated as an importation variable because it represents a demographic density associated with a particular county as opposed to a managerial or policy result that can be adjusted administratively (as may be the case in state prison populations). Studies by McCorkle et al. (1995), California Finance Department (1975), and Camp and Camp (2000) suggest that relationships do exist between size of an institution and rates of inmate violence. If one considers the local jail population as a representation of community urbanization, it is reasonable to hypothesize that city criminals are more likely to engage in violence than are rural criminals. A sociological basis for this hypothesis, in one form or another, can be traced to the "Chicago School" almost a century ago (Park, 1915; Park, Burgess, & McKenzie, 1925; Wirth, 1925). Recent Bureau of Justice Statistics (2000) crime reports indicate that the average urban violent crime rate from 1993 to 1998 was 74% higher than the rural violent crime rate and 34% higher than the suburban violent crime rate. One would not be surprised to find that inmates in urban jails are more violent than those in rural jails. Population values were entered as an ordinal variable.

A final importation variable was made up of the jail region within the state. The State of Texas was divided into six geographic zones to determine if trends in one part of the state were consistent with those in other locations, thereby creating a cultural variable. The regions included 44 counties in mostly rural East Texas; the Dallas/Fort Worth "Metroplex," the area including the 10 suburban counties contiguous to and including the two largest cities in the zone; the Greater Houston/Galveston/Beaumont area including Harris, Galveston, and Jefferson counties and those suburban counties contiguous to them; the 44 counties in the geographic center of the state or Central Texas; South Texas, including the 52 counties with large Hispanic populations and traditions along the Rio Grande River from Brownsville to El Paso; and West Texas, including the 92 counties of the Texas "Panhandle" and "High Plains" regions. The geographic regions were coded as a binary variable. The Greater Houston/Galveston/Beaumont and the Dallas/Fort Worth areas were coded as "metropolitan," whereas the remaining regions were coded as "non-metropolitan." It was therefore possible to divide jails into two types of regions: metropolitan and non-metropolitan.

Control Variables

To further clarify the relationship between the independent and dependent variables, four control variables were used. These variables included jail structural type, age of facility, percentage of male jail staff, and percentage of White staff.

Jail structural type During the late 1970s, architects began designing jails to encourage more interaction between guards and inmates

to reduce inmate violence and create a safer environment (Wener, Frazier, & Farbstein, 1987). This led to an "open concept" that placed the corrections officer alongside inmates in the designated living areas in what became known as "direct supervision" jails. Unlike traditional "linear" jails, direct supervision facilities enable the officer to view the inmates without obstructions, and unlike the "podular remote model," the officer has direct physical contact with the inmates (Nelson, 1993). Some researchers conclude that direct supervision jails have been shown to substantially reduce inmate assaults (Farbstein, Liebert, & Sigurdson, 1996; O'Toole, 1982).

Texas jail managers were asked to classify their facility as "linear," "podular remote," "direct supervision," or "some combination of the preceding" (Kellar et al., 2001). An inmate supervision type variable was thereby established to separate assaults by building design. Linear building design was coded 1, whereas all nonlinear building design was coded 0.

Age of facility Historically, many local jails are made up of structures that have been in continuous use for decades. By examining the age of the facility as it corresponds to inmate assaults, it was possible to evaluate the overall effectiveness of the facility as it affects violence including enhanced staff communication, visual surveillance, individual comfort, and efficiency of operations.

The percentage of male staff Some authorities believe that women corrections officers are as well suited, if not better suited, than men to supervise male inmates because they often use alternatives to the physical control of inmates (Tewksbury, 1999; Zimmer, 1986). Females tend to "talk through" controversies that might elicit force to resolve controversy (Grana, 2002). Respondents reported number of employees by position and gender so it was possible to ascertain the relative number of female corrections officers for each jail. Officer gender cannot easily be categorized as a management variable even though hiring practices are traditionally viewed as managerial prerogatives. Case law as well as cultural norms influence the available employment pool from which all corrections officers are drawn. It is interesting that smaller jails tend to employ a greater proportion of females than do larger jails (Kellar et al., 2001, p. 41). The proportion of female corrections officers was entered as an ordinal value.

The percentage of White staff The race of corrections officers has been suggested as a factor in explaining inmate violence (Irwin, 1977). According to this logic, officers of the same ethnicity as many of the inmates tend to have a clearer cultural understanding of the injustices inherent in prisons and provide role models for the prisoner, which in turn reduces violence. A review of literature for this particular variable yields mixed results. Fisher-Giorlando and Jiang (2000) found no significant differences in the volume of disciplinary reports written by Black or White officers, but McCorkle et al. (1995) concluded that higher White to Black staffing was associated with higher rates of inmate violence.

Hypotheses

Three research hypotheses were developed to define the relationship of independent variables to inmate assaults as discussed above. To test each hypothesis, zero-order correlations were calculated to examine possible occurrences of multicollinearity. Next, multiple logistic regression was applied to predict the likelihood of assaults, both inmate-on-staff and inmate-on-inmate. Research hypotheses of the study included the following:

1. that the incidence of inmate assaults would be less likely in those jurisdictions with a higher rehabilitative index;

2. that metropolitan areas would have more assaults than non-metropolitan areas; and

3. that assaults would be more prevalent in those jurisdictions reporting a higher proportion of maximum security inmates.

Findings

Sample Characteristics

Data in Table 1 represent the statewide sample ($N = 138$). Note that 15% of Texas jails reported serious inmate-on-staff incidents in the calendar year 2000, whereas 30% of Texas jails reported serious inmate-on-inmate assaults. The mean rehabilitative index (11.67) suggests that, on average, jail administrators lean toward the rehabilitative position. The daily population for each Texas jail varies from 0 to 7,097 inmates. On average, each Texas jail accommodates about 237 inmates on a given day. About 22% of Texas jail inmates are classed as maximum security. The average jail facility has been in operation for 27.4 years. Most jails (91%) are located in non-metropolitan regions. Approximately 57% of jail employees are male and 71% are White. More than half (51.4%) of the Texas jail facilities are linear in design.

Zero-Order Correlation

Zero-order correlation is a method for diagnosing multicollinearity, an important relationship in multivariate analysis. Coefficients may range from +1.0 (perfect positive correlation) to −1.0 (perfect negative correlation). Although there is no statistical rule to establish a "hard and fast" criterion for multicollinearity, convention suggests that correlations above .5 present potential instances wherein strong relationships between independent variables may lead to misrepresentations of relationships with dependent variables. Tables 2 and 3 represent the zero-correlation matrix of variables associated with inmate-on-staff and inmate-on-inmate assault, respectively. Note that jail region and monthly salary

Table 1	Variable and Measures (N = 138)			
Concept	**Variable**	**Level of Measurement**	**Mean**	**SD**
Dependent variables				
	inmate-on-staff assault	0 = no .15	.36	
		1 = yes		
	inmate-on-inmate	0 = no .30	.46	
	assault	1 = yes		
Independent variables Managerial model				
	Correction Philosophy Index	ordinal (6–16)	11.67	1.76
	Monthly salary	interval, in dollars	1830	512.66
Importation model				
	jail size	interval, in # of inmates	237	852
	jail region	0 = non-metropolitan	.09	.28
		1 = metropolitan		
	max. security inmates	interval, in %	21.77	17.51
Control variables				
	jail structural type	0 = nonlinear	.51	.50
		1 = linear		
	years in operation	interval, in year	27.40	25.26
	male jail staff	interval, in %	57.49	21.59
	White jail staff	interval, in %	71.26	26.06

are exceptionally correlated (.514). The authors have interpreted these data to infer that monthly salary is logically a function of jail region rather than inmate assaults being a function of officer pay. To suggest that paying officers less will result in reduced inmate violence cannot be logically defended; however, jails located in metropolitan regions are more likely to both pay higher salaries and to have a greater likelihood of inmate-on-inmate violence. Thus, in the logistic models, the variable Monthly Salary was excluded. Further, the variable Jail Size was not included because it also was highly correlated with Jail Region (.526).

Logistic Model: Inmate-on-Staff Assaults

Table 4 presents the logistic model of inmate-on-staff assaults in jails. Logistic regression is appropriate with dichotomous dependent variables because it uses the maximum likelihood method to estimate the parameters in the sample population. In this model, the chi-square statistic (7.466; $df = 7$) is not significant at the .05 level. This finding suggests that the model does not contribute to an understanding of the issue, inmate-on-staff assaults. However, the Wald statistic indicates that the percentage of maximum security

Table 2	Zero-Order Correlation Coefficients of Inmate-on-Staff Assault Model									
	Y1	**X1**	**X2**	**X3**	**X4**	**X5**	**X6**	**X7**	**X8**	**X9**
Y1 Inmate-on-staff	1.000									
X1 Rehabilitation index	−.059	1.000								
X2 Monthly salary	.144	−.092	1.000							
X3 Jail structural type	−.002	−.043	−.009	1.000						
X4 Jail size	.081	−.083	.498**	.116	1.000					
X5 Max. security inmates	.163	.002	.123	.105	−.079	1.000				
X6 Years in operation	−.033	.158	−.175*	.019	.070	−.125	1.000			
X7 Jail region	.012	−.162	.514**	−.062	.526**	−.080	−.070	1.000		
X8 Male jail staff	.134	−.094	.281**	−.033	.117	.027	−.200*	.064	1.000	
X9 White jail staff	−.102	.165	−.072	−.135	−.154	.047	.078	−.043	−.164	1.000

*Significant at .05 level. **Significant at .01 level.

Table 3	Zero-Order Correlation Coefficients of Inmate-on-Inmate Assault Model									
	Y1	**X1**	**X2**	**X3**	**X4**	**X5**	**X6**	**X7**	**X8**	**X9**
Y1 Inmate-on-inmate	1.000									
X1 Rehabilitation index	.033	1.000								
X2 Monthly salary	.467**	−.092	1.000							
X3 Jail structural type	.077	−.043	−.009	1.000						
X4 Jail size	.175*	−.083	.498**	.116	1.000					
X5 Max. security inmates	.174*	.002	.123	.105	−.079	1.000				
X6 Years in operation	−.205*	.158	−.175*	.019	.070	−.125	1.000			
X7 Jail region	.243**	−.162	.514**	−.062	.526**	−.080	−.070	1.000		
X8 Male jail staff	.196*	−.094	.281**	−.033	.117	.027	−.200*	.064	1.000	
X9 White jail staff	−.121	.165	−.072	−.135	−.154	.047	.078	−.043	−.164	1.000

*Significant at .05 level. **Significant at .01 level.

inmates in a specific jail is significantly related to inmate-on-staff assaults at the .05 level. Exp(b) presents the odds ratio, which indicates the odds change when a particular independent variable increases by one unit. Findings suggest that a higher proportion of maximum security inmates are more likely to be associated with inmate-on-staff assaults.

Table 4	Logistic Model of Inmate-on-Staff Assault		
Variable	B	Wald	Exp(b)
Correction Philosophy Index	−.053	.134	.948
Jail structural type	−.094	.130	.910
Max. security inmates	.026	3.908*	1.026
Years in operation	.003	.075	1.003
Jail region	.193	.052	1.213
Male jail staff	.017	1.754	1.017
White jail staff	−.011	1.237	.989
Constant	−1.868	.674	

NOTE: Chi-square = 7.466; df = 7; N = 138.

*α ≤ .05.

Logistic Model: Inmate-on-Inmate Assaults

Table 5 presents the logistic model of inmate-on-inmate assaults in the jail. In this model, the chi-square statistic (26.658; df = 7) is significant at the critical level of .01. This finding suggests that the model contributes significantly to an understanding of the issue of inmate-on-inmate assaults in jails. As the data in Table 5 show, two independent variables—percentage of maximum security inmates and jail region—are significant; the first at the .05 level and the latter at the .01 level, respectively. The exp(b) presents the odds ratio, which indicates the odds change when a particular independent variable increases by one unit. Exp(b) findings indicate that jails located in urban areas and those with a higher percentage of maximum security inmates are more likely to experience assaults by inmates on other inmates.

Table 5	Logistic Model of Inmate-on-Inmate Assault		
Variable	B	Wald	Exp(b)
Correction Philosophy Index	.214	2.794	1.239
Jail structural type	.250	1.273	1.284
Max. security inmates	.023	4.107*	1.023
Years in operation	−.021	3.709	.979
Jail region	2.077	8.576**	7.980
Male jail staff	.021	3.564	1.021
White jail staff	−.010	1.501	.990
Constant	−4.880	6.125	

NOTE: Chi-square = 26.658**; df = 7; N = 138.

*α ≤ .05.

**α ≤ .01.

⬚ Conclusions and Discussion

Hypothesis 1, "that the incidence of inmate assaults would be less likely in those jurisdictions with a higher rehabilitative index," was rejected. Both inmate-on-staff and inmate-on-inmate models indicated little if any relationship between general rehabilitative philosophy and inmate assaults. If a more positive attitude toward inmate rehabilitation is more conducive to non-assaultive behavior, such influence may be counterbalanced by increased opportunities for inmates to engage in assaultive behavior when more programs are available. It is also possible that those jail administrations espousing a more punitive viewpoint also employ more restrictive managerial controls and thereby reduce assaultive acts.

Hypothesis 2, "that metropolitan areas would have more assaults than non-metropolitan areas," produced a strong relationship with inmate-on-inmate assaults but no discernible relationship with inmate-on-staff assaults. It is noteworthy that the zero-order correlation coefficient between jail region and inmate-on-inmate assaults was statistically significant at the .01 level. This suggests that there is a relatively strong relationship between urban jails and increased assaults on other inmates. However, the relationships did not hold for inmate-on-staff assaults. This may be because larger jails typically exact severe punishments on those inmates who assault staff. If this postulate is true, the managerial variable of prevention may offset the importation factor of otherwise increased assaults.

Hypothesis 3, "that assaults would be more prevalent in those jurisdictions reporting a higher proportion of maximum security inmates," was confirmed by comparisons of the proportionality of maximum security inmates with inmate-on-staff assaults and with inmate-on-inmate assaults. Maximum security inmates are more likely to be involved in assaultive behavior than lower risk inmates.

These findings reinforce the necessity of viable classification systems in order to provide safety for staff and inmates alike. Inmate classification is especially difficult in local jails given the transient nature of the population and the time necessary to properly complete the complex process.

Like most multivariate studies of human behavior, this attempt to investigate can be characterized as yielding a certain degree of mixed results. One view supports the importation model in that one variable, the proportion of inmates classified as maximum security, was statistically significant at the .05 confidence interval as it related to inmate-on-staff assaults. A similar relationship was found between the proportion of maximum security inmates and inmate-on-inmate assaults ($\alpha \leq$ .05). The study indicated that inmate-on-inmate violence was more likely in metropolitan areas than in non-metropolitan jails. Likewise, this finding seemed to confirm the importation model approach but the evidence was not overwhelming.

In addition, two control variables—years in operation and male jail staff—were not significant in the logistic models, but zero-order correlations indicated that both were correlated to inmate-on-inmate assaults at the .05 level. The correlation between age of a facility and inmate-on-inmate assaults was negative. That is, the newer the facility, the more likely inmate-on-inmate assaults occur. This may be explained by noting that newer facilities often require complete changes in operations, whereas older facilities are more stable relative to the administration of routine operations. Therefore, older facilities seem to provide an atmosphere more conducive to tranquility and overall institutional stability accompanied by a resulting reduction in assaultive behaviors.

The correlation between the percentage of male jail staff and inmate-on-inmate assaults was positive. That is, the higher the percentage of male staff in a unit, the more likely inmate-on-inmate assaults became. On its surface, this finding seems to support the proposition that female officer supervisors are more nurturing and less likely to elicit violent responses from their charges (Grana, 2002; Tewksbury, 1999). Note, however, that the relationship did not result in a finding of statistical significance in the full logistic model.

Other variables such as racial breakdown of staff, type of facility design structure, and rehabilitative philosophy seemed to have only marginal, if any, effects on inmate assaults. Further research is recommended concerning these issues. It may well be that these and other forces interact to shape assaultive trends in jails, but more precise and sensitive analytical modeling techniques and additional research should be applied.

In conclusion, the study seems to support the importation model approach, but some factors suggest influence by managerial model variables. The deprivation model could not be adequately examined given the nature of the variables. Conceptually, the absence of deprivation variables may be explained by the fact that inmates spend shorter times in county jails and those variables associated with deprivation theory do not have time to emerge.

Local jails represent a unique challenge. Although this study is in no way a definitive work, it emphasizes the complexity posed by this important component of the justice system. It is hoped that future endeavors will address the questions initiated by this effort.

⊠ References

Alpert, G. P. (1979). Patterns of change in prisonization: A longitudinal analysis. *Criminal Justice and Behavior, 6*(2), 159–173.

American Correctional Association. (1993). *Gangs in correctional facilities: A national assessment.* Laurel, MD: Author.

Atlas, R. (1983). Crime site selection for assaults in four Florida prisons. *The Prison Journal, 58*(1), 59–72.

Brown, G. C. (1990). *Violence in California prisons: A test of the importation and deprivation models.* Doctoral dissertation, University of California at Irvine.

Bureau of Justice Statistics. (2000). *National crime victimization survey special report: Urban, suburban and rural victimization, 1993–1998.* Washington, DC: Author.

Burlew, K., Dinitz, S., Griffin, B., et al. (1994). *Final report of the (Ohio) governor's select committee on corrections.* Columbus: State of Ohio.

California Finance Department. (1975). *Prison violence in California: Issues & alternatives.* Sacramento, CA: Author.

Camp, C., & Camp, G. (2000). *Corrections yearbook 2000: Jails.* Middletown, CT: Criminal Justice Institute.

Cao, L., Zhao, J., & Van Dine, S. (1997). Prison disciplinary tickets: A test of the deprivation and importation models. *Journal of Criminal Justice, 25*(2), 103–113.

Clear, T., & Cole, G. (1997). *American corrections.* Belmont, CA: Wadsworth.

Clemmer, D. (1958). *The prison community.* New York: Holt, Rinehart and Winston.

DiIulio, J. (1987). *Governing prisons.* New York: Free Press.

Farbstein, J., Liebert, D., & Sigurdson, H. (1996). *Audits of podular direct supervision jails.* Longmont, CO: U.S. Department of Justice, National Institute of Corrections, Jails Division.

Fisher-Giorlando, M., & Jiang, S. (2000). Race and disciplinary reports: An empirical study of correctional officers. *Sociological Spectrum, 20*(2), 169.

Fry, L. J., & Frese, W. (1992). Bringing the convict back in: An ecological approach to inmate adaptations. *Journal of Criminal Justice, 20*(4), 355.

Gaes, G. G., & McGuire, W. J. (1985). Prison violence: The contribution of crowding versus other determinants of prison assault rates. *Journal of Research in Crime and Delinquency, 22*, 41.

Gaines, L., Kaune, M., & Miller, L. (2000). *Criminal justice in action.* Belmont, CA: Wadsworth.

Grana, S. D. (2002). *Women in justice.* Boston: Allyn and Bacon.

Harer, M. D., & Steffensmeier, D. J. (1996). Race and prison violence. *Criminology, 34*(3), 323.

Irwin, J. (1977). The changing structure of men's prisons. In D. Greenberg (Ed.), *Corrections and punishment* (pp. 21–40). Beverly Hills, CA: Sage.

Irwin, J., & Cressey, D. (1962). Thieves, convicts and the inmate culture. *Social Problems, 10*, 142–155.

Kellar, M., Jaris, M., & Manboah-Roxin, J. (2001). *Texas jail survey 2001: A status report.* Austin: Texas Commission on Jail Standards.

Kennedy, T. D. (1986). Trends in inmate classification: A status report of two computerized psychometric approaches. *Criminal Justice and Behavior, 13*(2), 165.

Kerle, K. E., & Ford, F. R. (1982). *The state of our nation's jails.* Washington, DC: National Sheriff's Association.

LaVigne, N. (1994). Rational choice and inmate disputes over phone use on Riker's Island. In R. V. Clark (Ed.), *Crime prevention studies* (Vol. 3, pp. 109–126). Monsey, NY: Criminal Justice Press.

Leeke, W., & Mohn, H. (1986). Violent offenders: AIMS and unit management maintain control. *Corrections Today, 48*(3), 22.

Mays, G. L., & Winfree, L. T. (2002). *Contemporary corrections.* Belmont, CA: Wadsworth.

McCorkle, R. C., Miethe, T. D., & Drass, K. A. (1995). The roots of prison violence: A test of the deprivation, management and "not so total" institution models. *Crime and Delinquency, 41*(3), 317–331.

Memory, J. M., Guo, G., & Parker, K. (1999). Comparing disciplinary infraction rates of North Carolina fair sentencing and structured sentencing inmates: A natural experiment. *The Prison Journal, 79*(1), 45–71.

Moynahan, J., & Stewart, E. (1980). *The American jail: Its development and growth.* Chicago: Nelson-Hall.

National Sheriffs' Association. (1982). *The state of our nation's jails.* Washington, DC: Author.

Nelson, W. R. (1993). *New generation jails, podular direct supervision jails.* Longmont, CO: U.S. Department of Justice, National Institute of Corrections, Jails Division.

O'Toole, M. (1982). *New generation jail survey: Comparative data from 1981 & 1982 on assaults & escapes.* Boulder, CO: U.S. Department of Justice, National Institute of Corrections, Jails Division.

O'Toole, M. (1999). *Jails and prisons: The numbers say they are more different than generally assumed.* Retrieved from http://www.corrections.com/aja/

Park, R. (1915). The city: Suggestions for the investigation of behavior in the city environment. *American Journal of Sociology, 20,* 579.

Park, R., Burgess, E., & McKenzie, R. (1925). *The city.* Chicago: University of Chicago Press.

Paterline, B. A., & Peterson, D. M. (1999). Structural and social psychological determinants of prisonization. *Journal of Criminal Justice, 27*(5), 427–441.

Peak, K. J. (2001). *Justice administration: Police, courts, and corrections management* (3rd ed.). Upper Saddle River, NJ: Prentice Hall.

Petersen, R. D. (1997). *Inmate subcultures of female youth: An examination of social systems and gang behavior* (Criminal Justice Abstracts, No. 11742). Ann Arbor: University of Michigan.

Poole, E. D., & Regoli, R. M. (1983). Violence in juvenile institutions: A comparative study. *Criminology, 21*(2), 213.

Sechrest, D. K. (1991). The effects of density on jail assaults. *Journal of Criminal Justice, 19*(3), 211.

Sorensen, J., Wrinkle, R., & Gutierrez, A. (1998). Patterns of rule-violating behaviors and adjustment to incarceration among murderers. *The Prison Journal, 78*(3), 222–231.

Sykes, G. (1958). *The society of captives.* Princeton, NJ: Princeton University Press.

Tewksbury, R. (1999). Should female corrections officers be used in male institutions? In C. B. Fields (Ed.), *Controversial issues in corrections* (pp. 187–193). Boston: Allyn & Bacon.

Texas Commission on Jail Standards. (2001). *Monthly inmate population report for April, 2001.* Austin: Author.

Wener, R., Frazier, F. W., & Farbstein, J. (1987). Direct supervision of correctional institutions. In *Podular direct supervision jails, 1993.* Longmont, CO: U.S. Department of Labor, National Institute of Corrections, Jails Division.

Wilkinson, R. A., Austin, C. P., Baugh, S., et al. (1994). Stemming the violence. *Corrections Today, 56*(5), 64.

Wirth, L. (1925). *The ghetto.* Chicago: University of Chicago Press.

Wolf, S., Freinek, W. R., & Shaffer, J. W. (1996). Frequency and severity of rule infractions as criteria of prison maladjustment. *Journal of Clinical Psychology, 22,* 244–247.

Zimmer, L. E. (1986). *Women guarding men.* Chicago: University of Chicago Press.

Zingraff, M. T. (1980). Inmate assimilation: A comparison of male and female delinquents. *Criminal Justice and Behavior, 7*(3), 275–292.

Zupan, L. (2002). The persistent problems plaguing modern jails. In T. Gray (Ed.), *Exploring corrections* (pp. 37–63). Boston: Allyn and Bacon.

DISCUSSION QUESTIONS

1. According to the literature, what are the causes of inmate assaults on correctional staff?

2. What is the relationship between high levels of male staff and inmate assaults on staff? According to the authors, how might this situation be dealt with?

3. How do inmate assaults on jail staff differ from inmate assaults on prison staff?

READING

Christine Tartaro examines jail characteristics within Gettinger's (1984) framework of the psychology of new generation jails. The podular direct-supervision jail design and supervision strategy have gained popularity among criminal justice researchers and administrators during the past 20 years. These "new generation" jails are a radical departure from the traditional linear intermittent design that dominated jail architecture for two centuries. Tartaro's study involves an examination of characteristics of 76 jails identified by their administrators as podular direct-supervision facilities. The purpose is to determine the extent to which the jails adhere to what researchers and administrators have identified as elements vital to the success of this type of institution. Results indicate that few jails are strictly adhering to the training and design techniques that have been recommended for the successful operation of these institutions.

Examining Implementation Issues
With New Generation Jails

Christine Tartaro

Jails in the United States retained the same general design and supervision plan from the birth of the country until the 1970s. Institutional design features have emphasized the isolation of inmates from society and from most staff members. Few changes have been made to the interior of these facilities, even though the purpose of these institutions has changed. Whereas inmates residing in jails 200 years ago were isolated from the outside world and provided no services, modern-day corrections systems are expected to respond to

SOURCE: Tartaro, C. (2002). Examining Implementation Issues With New Generation Jails. *Criminal Justice Policy Review, 13*(3), 219–237. Reprinted with permission of Sage Publications, Inc.

the inmates' needs. Sechrest and Price (1985) recommended that the institutional mission, security needs, and management issues should be considered when planning for the design of a correctional facility.

During the early 1970s, the Federal Bureau of Prisons (BOP) commissioned architects to present new designs for their Metropolitan Correctional Centers (MCCs). The BOP directed the architects to include four characteristics: individual rooms for inmates, living units that house fewer than 50 inmates, direct supervision of inmates by officers, and restricted movement within the facility (Gettinger, 1984). The resulting facility design is known as podular direct supervision. These "new generation" facilities are a radical departure from the traditional linear intermittent design that dominated jail architecture for two centuries.

⊠ History

The podular direct-supervision model originated at three MCCs operated by the BOP (Nelson & Davis, 1995). A presidential directive in 1969 ordered the BOP to improve the corrections system so that it would serve as a model for state and local systems (Nelson, 1988). As part of this effort, the bureau began work on a new facility design (Gettinger, 1984). The BOP intended the new institutions to incorporate the unit management approach, which involved the reduction of a facility into smaller, manageable units that allow for concentration of staff members and services to particular sections of the facility (Farbstein, 1986; Levinson, 1999). Finally, the jails were to have a "normalized" living environment, meaning that the interior of the facilities would have a less institutional feel to them. According to Nelson and Davis (1995), the BOP adopted the philosophy that "if you can't rehabilitate, at least do no harm" (p. 2). Architects were commissioned to design three MCCs in New York, Chicago, and San Diego.

The newly designed facilities in San Diego and New York opened in 1974, and the institution in Chicago opened in 1975 (Nelson & Davis, 1995). Evaluations of the new facilities produced favorable findings. Violent incidents in the three facilities dropped 30% to 90% (Wener, Frazier, & Farbstein, 1993).[1] Although these institutions experienced sharp reductions in assaults, suicides, graffiti, and vandalism, local jail administrators were apprehensive about adopting this design. Nelson and Davis (1995) explained that local corrections officials initially attributed the success of the MCCs to the inmate composition instead of environmental and supervision factors. There was some merit to this argument because federal institutions typically held less violent offenders prior to the passage of legislation to broaden the category of federal offenses to include drug trafficking in the late 1980s (Welch, 1996).

Despite the concerns about the potential of placing local jail populations in podular direct-supervision facilities, decision makers in Contra Costa, California, were impressed by the success of the Chicago MCC (Wener et al., 1993). The Contra Costa podular direct-supervision jail opened in 1981 (Zupan, 1991). The facility featured an open booking area with a large lounge, television sets, coffee machines, and comfortable furniture. Intoxicated inmates and others who were not able to adapt to the normalized booking or living area were placed in a more secure setting, and inmates under the influence of alcohol were monitored by volunteers from Alcoholics Anonymous (Gettinger, 1984; Zupan, 1991). Contra Costa County's experience with this model was as positive as that of the Metropolitan Correctional Centers, with assault rates 95% lower than in the old facility. Critics were quick to assume that, like the MCCs, the Contra Costa County Jail housed "softer" inmates. As Gettinger (1984) observed, Contra Costa's jail population consisted of few misdemeanants, with most inmates charged with burglary, narcotics offenses, armed robbery, murder, or escape.

The inmate population was also racially mixed with Blacks, Whites, and Hispanics.

As the MCCs and the Contra Costa County facility continued to be successful, other counties gradually followed their example. The dramatic change in a New York City jail, called "the Tombs" because of its terrible conditions prior to a shift to podular direct supervision, provided additional evidence of this model's success. The reductions in vandalism and violence in an institution that clearly held "hard" criminals helped to bolster support for new generation jails (Wener et al., 1993). The National Institute of Corrections encouraged the development of more direct-supervision facilities by assisting local jurisdictions, and the American Jail Association, the American Correctional Association, and the American Institute of Architects officially announced their support of the direct-supervision model (Nelson, 1988). By 1995, 147 podular direct-supervision jails housing more than 70,000 inmates were operational (Kerle, 1995).

▧ Philosophy

Inmates in podular direct-supervision jails are supposed to be housed in mostly single- (some double-) occupancy cells located around a common dayroom. These triangle- or podular-shaped units improve officer sight lines, thereby making it more difficult for inmates to commit infractions without being detected. In addition to improving the observation of inmates, the podular direct-supervision model aims to enhance interaction between inmates and staff members (Bayens, Williams, & Smykla, 1997a). The goal of these jails is to reduce destructive and violent behavior while maximizing control of the inmate population. This is to be accomplished through the reporting of minor rule infractions or diffusion of problems before they escalate (Bayens et al., 1997a; Nelson & Davis, 1995). The two main

components of this model are a normalized physical environment and a management strategy focusing primarily on corrections officers' interpersonal communication skills. Ray Nelson, the first warden of the Chicago MCC, defined a normalized environment as "inmate living areas outfitted with normal commercial fixtures, finishes, and furnishings" (p. 10). The supervision strategy involves direct supervision, which Bayens and associates (1997a) defined as a

> method of correctional supervision in which one or more jail officers are stationed inside the living area and are in direct physical interaction with those housed within the pod throughout the day with the ultimate goal of keeping negative behavior in check. (p. 54)

Gettinger (1984) discussed problems experienced in traditional jails and explained how they can be alleviated in a new generation facility. Specifically, Gettinger highlighted five problems—(a) the fear-hate syndrome, (b) the power vacuum, (c) privacy, (d) positive expectations, and (e) isolation—as serious issues with which jail administrators must contend.

The first two problems—the fear-hate syndrome and the power vacuum—will be dealt with simultaneously. Inmates' and officers' fear of other inmates creates an environment that is conducive to the creation of weapons, formation of gangs, and increased conflict among inmates and between inmates and officers. Past research has indicated that officers in traditional facilities have at times relinquished power to inmates in exchange for their cooperation (Sykes, 1972). The officers' lack of control of inmates results in a power vacuum that fosters the creation of gangs and increased violence to compensate for the absence of leadership. Staff control of the institution will reduce the need for inmates to band together for protection, and inmates will learn that defying authority is disadvantageous. New generation jails essentially aim to establish control over every part of the institution (Zupan & Stohr-Gilmore, 1988) by using corrections officers' interpersonal skills in the direct-supervision approach.

In contrast to the reactive approach officers take in traditional facilities, officers in new generation jails are expected to prevent negative behavior before it occurs (Nelson, 1983; Wener, et al., 1993). Officers with extensive training in interpersonal relations and conflict management remain in the pods with the inmates at all times (Zupan & Stohr-Gilmore, 1988). Because officers are located in the living areas with inmates, it is expected that they will detect signs of tension between inmates and defuse the problem by either talking to them or by restricting privileges. In addition, new generation jails typically have an interior design that provides a much more comfortable living environment for inmates. Officers are better able to use the threat of removal of privileges or expulsion from the pod as incentives for good behavior. The increased officer control and improved living conditions should result in a more stable living environment in which inmates and officers alike are less fearful.

Zupan and Stohr-Gilmore's (1988) study of podular direct-supervision facilities illustrates the importance of training in the functioning of these jails. One of the four institutions evaluated by the researchers differed in training procedures. Although other facilities held 6-to 8-week training sessions for its officers, one jail provided only 2 weeks. The jails that required 6 to 8 weeks of training covered communication skills, but the facility offering the 2-week session focused on work found in traditional facilities, with emphasis on the physical control of inmates. When the facility opened, inmates were occasionally locked into cells without supervision, contrary to the direct-supervision philosophy. Inmates living in this institution had more negative evaluations of the facility and staff than those living in a true podular direct-supervision jail.

Houston, Gibbons, and Jones (1988) found similar results in another facility. The researchers examined a county jail system as it moved into a new facility. The new jail contained officer workstations within inmate living quarters, a common characteristic of direct-supervision jails. Despite the change in architecture, the county provided no training for officers as they made the transition to the new facility. The researchers found that the new environment had little effect on inmate and staff attitudes about their living and work environment, respectively.

The third issue that Gettinger (1984) presented is privacy. Humans need privacy that cannot be found in double-bunked cells or dormitories. Single cells or rooms can give inmates the protectable space that they need. The rooms provide inmates a place to which they may retreat if they want to be alone. This type of living arrangement gives inmates their own private space, and the ability to enter and exit the rooms as they wish gives them a feeling of control.

Fourth, if inmates are given positive expectations for their behavior, many will refrain from functioning in an antisocial manner (Gettinger, 1984). Traditional jails, with their vandal-resistant furnishings and bars on the doors, convey the message that inmates are expected to misbehave. Expensive steel or concrete beds and steel toilet seats and sinks are used because they are more difficult to damage than wood furniture or porcelain fixtures. Heavy metal doors and bars are also used to prevent escapes (Zupan, 1991). Resser (1989) explained that "vandal-proof" fixtures express to an inmate that "you are a vandal" and "you cannot damage this fitting." Resser reasoned that inmates are likely to respond negatively to this message and participate in destructive behavior. Wener and associates (1993) reported that this type of setting sends a message that animal-like behavior is expected of the inmates because they are placed in cages while staff maintain a safe distance on the other side of the bars. Environmental stressors in jails, such as lack of privacy, boredom, excessive heat, noise, unpredictability, and crowding, all of which are beyond the control of inmates, may influence inmates' physical and mental health as well as social behavior (Zupan, 1991; Zupan & Menke, 1991). Zupan (1991) summarized the problem with the traditional jail environment: "Architectural and operational features used to prevent violence and destruction may foster misbehavior by communicating and reinforcing negative expectations to inmates" (p. 87).

Accordingly, if inmates are given the impression that destructive behavior will not be tolerated and they are given incentives to behave, proponents of the new generation jail agree that inmates will be more likely to conform. Nelson (1988) stated that "in essence, an environment designed to be indestructible evokes destructive behavior, while an environment designed for normal usage evokes normal behavior" (p. 10). Living areas where inmates can socialize in a normal environment are essential. Wener, Frazier, and Farbstein (1985) described new generation environments as including comfortable furniture and a tile or carpet floor, ample telephones, television sets, and recreation areas. The absence of bars and vandal-resistant settings reduce the high noise levels often found in correctional facilities. Individual rooms or cells should also have noninstitutional furniture. The comfortable furnishings serve to reduce stress by making the unit look and sound "noninstitutional" (Wener et al., 1985).

Because the environment within the pod is "normalized," inmates have much to lose by being removed (Farbstein & Wener, 1989). Officers are expected to handle minor incidents within the pods, usually by temporarily locking down an inmate in his cell, but inmates are also given the opportunity to regain privileges by improving their behavior (Farbstein & Wener, 1989; Senese, Wilson, Evans, Aguirre, & Kalinich, 1992). Removal of an inmate from the pod and placement in a more secure, less enjoyable portion of the facility is also an option. Unlike a traditional jail setting where inmates have little to lose, the residents of these jails have an incentive to behave because failure to conform with rules could result in movement to a less desirable setting. Nelson (1988) discussed the impact that this had on inmates in the Chicago MCC. He noted that

> inmates, in effect, had been given the concrete choice of behaving appropriately and, as a result, being housed in a relatively desirable general population housing unit or behaving noncompliantly with the certainty of being placed in administrative segregation. Apparently, this was not a difficult decision for most inmates. (p. 10)

Fifth, inmates' separation from their home environment and isolation from friends or relatives may serve to loosen external controls, resulting in an "anything goes" attitude. Gettinger (1984) recommended that reduction of a sense of isolation from the outside world and increased social controls could be accomplished "through its design, furnishings, visiting rules, telephone privileges and contact with staff members" (p. 15). Specific design and furnishing issues have been included in earlier sections of this article. Contact with those from home may reduce isolation, so the use of contact visits, regular telephone access, and packages from home may make inmates' incarceration periods less difficult. As such, it was decided to include these variables to determine the extent to which podular direct-supervision facilities are using them. With regard to contact with staff, the purpose of the unit management approach is to increase the frequency and intensity of contact between inmates and staff (Levinson, 1999). Part of this plan involves the concentration of services, so inmates have easier access to programs and recreation.

The new generation jail approach is intended to address some of the most difficult problems that jail administrators face. Through proper operation of this type of facility, the inmate-officer power struggles, the negative expectations that are conveyed to inmates, and the inmates' sense of isolation have the potential to be diminished. This is not to say that the jail will be converted into a luxurious setting, but if architects and administrators work to develop a normalized physical and social environment, the potential for the reduction of inmate and officer stress, violence, and property damage exists (Bayens et al., 1997a; Bayens, Williams, & Smykla, 1997b; Farbstein & Wener, 1989; Nelson, 1983; Nelson & Davis, 1995; Senese et al., 1992; Williams, Rodeheaver, & Huggins, 1999; Zupan & Menke, 1988; Zupan & Stohr-Gilmore, 1988).

A common problem in criminal justice administration is the partial implementation of a program. A treatment program, policing initiative, or, in this case, a new type of jail will receive positive evaluations. Shortly thereafter, decision makers nationwide will support the development of similar programs. The problem, however, is that criminal justice agencies will frequently adopt only parts of the successful program without realizing that it is a combination of factors on which its success hinges. Simply erecting a new jail and placing officers into the pods without communication skills training or normalizing the physical environment without placing officers in the pods has the potential to backfire. As Houston et al.'s (1988) and Zupan and Stohr-Gilmore's (1988) research has indicated, building a jail and then neglecting to strictly adhere to every aspect of the new generation jail philosophy will likely result in a jail with similar stress and violence levels for inmates and officers as in the traditional facilities.

The current study involves an examination of characteristics of 76 jails identified by their administrators as podular direct-supervision facilities. The purpose of this study is to determine the extent to which the jails adhere to what researchers and administrators have identified as vital elements to the success of this type of institution. Specifically, jail characteristics will be presented within Gettinger's (1984) framework of the psychology of new generation jails (fear-hate syndrome, power vacuum, privacy, positive expectations, and isolation).

Data

The original purpose of this study was to compare characteristics of different types of jails (podular direct, podular indirect, and linear intermittent). As part of this effort, surveys were sent to all jails in the American Jail Association's directory with a rated capacity of 51 or higher. Of the 666 completed surveys that were returned, administrators from 76 of the responding jails identified their facility's architectural design as podular and their supervision strategy as direct. Direct supervision was defined on the instrument as "officers are stationed inside the living areas and have direct physical interaction with the inmates."

Results

Fear-Hate Syndrome and Power Vacuums

The work of officers in traditional facilities has been described by Zupan and Menke (1988) as "fragmented, routinized and menial—in a word, impoverished" (p. 615). Work in new generation jails, however, involves effective communication with inmates. Given the greater responsibilities that officers have inside these jails, it was expected that they would be provided with extensive training, particularly training to improve the officers' ability to interact with difficult populations. The current survey included a question about the weeks of training new corrections officers receive during their 1st year of employment. Eleven percent (8) of the responding jails provided a maximum of 3 weeks of general training for incoming officers, and 30% (23) provided 4 to 6 weeks. Twenty-seven percent (20) offered 7 to 12 weeks, whereas 33% (25) offered 13 or more weeks of training to incoming officers. The results are displayed in Table 1.

One responding podular direct-supervision jail reported that it did not provide any communication skills training to its officers. Thirty-six percent (27) of the responding facilities reported providing a maximum of 1 day of this type of training (1 to 8 hours), and 34% (26) offered 9 to 16 hours of training. Eleven percent (18) provided between 17 and 25 hours of communications skills training, 13% (10) provided 25 to 40 hours, and one jail offered more than 1 work week (40 hours) of such training.

With regard to education requirements for incoming corrections officers, most jails required only a high school or equivalent degree (84% [64]). Seven jails (9%) required 2 years of college, and one jail required a 4-year college degree. Two of the responding jails (3%) had no education requirements for new officers.

Table 1	Corrections Officer Characteristics		
		Frequency	**Percentage**
Weeks of training during first year			
0 to 3 weeks		8	10.5
4 to 6 weeks		23	30.3
7 to 9 weeks		15	19.7
10 to 12 weeks		5	6.6
13 weeks or more		25	32.9
Total		**76**	**100.0**
Hours of communication skills training			
0 hours		1	1.3
1 to 8 hours		27	35.5
9 to 16 hours		26	34.2
17 to 24 hours		8	10.5
25 to 40 hours		10	13.2
41 or more hours		1	1.3
Nonresponses		3	3.9
Total		**76**	**100.0**
Minimum education requirements for corrections officers			
None		2	2.6
High school/GED		64	84.2
2 years of college		7	9.2
4 years of college		1	1.3
Nonresponses		2	2.6
Total		**76**	**100.0**
Percentage with shift spent circulating throughout living areas ($M = 80.43$, $SD = 21.84$)		74	

A fundamental component of the direct-supervision inmate-management strategy is the placement of officers in the living areas with inmates. Administrators were asked about the percentage of officers' shifts spent circulating throughout the living areas. The mean response to this question was 80% (median = 80, SD = 21.84). Ten of the jails (14%), however, reported that officers spend 50% or less of their shifts in the pods.

Privacy

When asked about the use of single-occupancy cells in their jails, 17% (13) of the responding jails reported housing 75% to 100% of their inmates in single cells (see Table 2). Almost half (49% [37]) indicated that they house less than one quarter of their inmates in single cells. More than three quarters of the participating jails (78% [59]) reported allowing inmates to enter and exit their cells throughout most of the day.

Positive Expectations

Of the podular direct-supervision jails that responded to this survey, 65% (49) used vandal-resistant metal and concrete or plastic fixtures and furnishings in the inmate living areas (see Table 3). Eighteen percent (14) of the jails used wood or fabric furniture, and 15% (11) reported using a combination of the vandal-resistant and the fabric furniture. Two jails (3%) indicated that they used fixtures and furnishings other than the options of metal, plastic, concrete, fabric, or wood that were listed on the survey. Because bolting furniture to the floor is a common feature in vandal-resistant settings, the instrument included a question about the percentage of furniture bolted to the floor. Forty-three percent (33) of jails reported bolting less than one quarter of their furniture to the ground.

The survey included questions about other factors that may affect inmates' comfort inside the jails. Administrators' responses indicated that nearly all of the jails participating in this survey were air conditioned (94% [71]). Inmates in half of the responding institutions (38) were able to control the lighting in their cells. Sixty-three percent (48) of responding facilities indicated that bumpers on doors and other tools were used to reduce noise and echoes inside the jail. Thirty-four percent (26) of the jails permitted inmates to possess electronic devices, 29% (22) provided inmates with access to weight-lifting equipment, and

Table 2	Privacy		
		Frequency	Percentage
Percentage with inmates housed alone in single cells			
0–24		37	48.7
25–49		6	7.9
50–74		15	19.7
75–100		13	17.1
Nonresponses		5	6.6
Total		**76**	**100.0**
Inmates can enter and exit cells during most of the day (n = 76)		59	77.6

Table 3	Physical Characteristics of Responding Jails		
		Frequency	Percentage
Type of furnishing			
Metal, concrete, or plastic		49	64.5
Wood and fabric		14	18.4
Combination		11	14.5
Other		2	2.6
Total		**76**	**100.0**
Percentage of furniture bolted to the floor			
0–24		33	43.4
25–49		0	0.0
50–74		5	6.6
75–100		25	32.9
Nonresponses		13	17.1
Total		**76**	**100.0**

92% (70) provided inmates with access to televisions throughout the day. Ninety-three percent (71) of the jails allowed inmates access to showers throughout most of the day.

Isolation

Administrators were asked a series of questions about privileges that would allow inmates to maintain a closer connection to their homes and their loved ones (see Table 4). When asked if inmates were permitted to receive packages from home, 26% (20) of participating administrators responded affirmatively. Ninety-one percent (69) of the jails permitted continuous telephone access during waking hours, and 28% (21) allowed contact visits. Because availability of services in each unit or pod has been identified as a promising technique for facilitating stronger inmate-staff contact (Levinson, 1999), several questions were asked about the availability and location of programs. The survey included questions about whether education, drug counseling,

religious services, recreation, and nondrug therapy are offered on the pod level. With the exception of recreation, few jails offered these services within the pods (see Table 5). Fifty-nine percent (41) of responding facilities provided recreation services in this manner.

⊠ Discussion

Sechrest and Price (1985) studied the planning, design, and construction of correctional facilities, and they recommended the following for functional institutions:

A greater sense of safety for staff and inmates, less destructive (or "normalized") inmate behavior, minimal staff turnover through greater worker satisfaction, and a more humane and positive environment based on the implementation of design concepts consistent with the standards of good practice of the field. (p. 7)

Table 4

	n	Frequency	Percentage
Additional characteristics of responding jails			
Air-conditioned living areas	75	71	93.8
Inmates can control lighting	76	38	50.0
Bumpers to reduce noise	76	48	63.2
Possess electronic devices	75	26	34.2
Access to weight-lifting equipment	76	22	28.9
Access to TV in living areas throughout most of the day	76	70	92.1
Access to showers throughout most of the day	76	71	93.4
Isolation-reduction techniques			
Receive packages	75	20	26.3
Continuous phone access during waking hours	76	69	90.8
Have physical contact with visitors	76	21	27.6

Table 5   Concentration of Services on the Pod Level

Program	n	Jails Offering the Program	Jails Offering the Program on the Pod Level (%)
Education	68	66	9 (11.8)
Drug counseling	69	66	16 (21.1)
Therapy	67	61	11 (14.5)
Religious services	76	76	15 (19.7)
Recreation	76	76	41 (59.4)

Evaluations of individual jails that have moved from a linear intermittent to a podular direct-supervision design as well as studies examining small numbers of jails of different designs have been encouraging. Researchers have reported positive results with each of Sechrest and Price's characteristics of functional institutions. Farbstein (1996) and Williams et al. (1999) surveyed inmates who reported feeling safer in a new generation jail than a traditional facility. Nelson (1986) reported finding fewer morale problems among staff members in direct-supervision facilities. Although Williams and associates (1999) did not find differences in staff members' perceptions of the safety in the different jails, they did report that both staff members and inmates were more positive about their physical surroundings in the new generation jails than in traditional facilities. Zupan and Stohr-Gilmore (1988) also found that inmates were significantly more positive in their evaluations of the climate and safety issues in direct-supervision jails. Zupan and Menke (1988) found that officers reported a higher level of job satisfaction in the new generation jails.

Most of the new generation jails included in these positive evaluations, though, strictly adhered to the recommendations for changes to the physical design of the facility as well as adjustments to the training curriculum to allow for thorough communication skills training. The results of these studies indicate that, if built and operated properly, the jails can reduce violence and increase perceptions of safety among incarcerated populations. Jails that only partially implemented these recommendations were not found to be as successful in terms of staff member and inmate feelings of safety.

A surprising finding of the current study involves the little training that officers in these facilities receive. Jail administrators and researchers who are familiar with new generation jails have discussed the importance in training officers so they can best protect themselves and the inmates while working in the living areas. In these facilities, a single officer is often placed in a pod with anywhere from 12 to 46 inmates (Bayens et al., 1997b; Zupan & Stohr-Gilmore, 1988). The corrections officers are expected to maintain a peaceful environment in a setting consisting entirely of convicted and alleged offenders. Regardless, nearly one third of the responding jails provided only 6 weeks or less of general officer training. More than two thirds of the participating jails required a maximum of 2 workdays of communication skills training. Only one of the responding jails dedicated more than 1 week to effective communication training.

It is interesting to note that the national average for local police officer training is 1,100 hours. Put in terms of a 40-hour workweek, that is 27.5 weeks (U.S. Bureau of Justice Statistics, 2001). The findings of this study indicate that, although most of the responding jails are placing their officers in the pods, the officers are not receiving enough training to be adequately prepared for the challenges they will face in a direct-supervision jail. Although there is no guarantee that college education will be an effective component of preparation for jail work or other criminal justice jobs, education requirements are typically associated with efforts at professionalizing a workforce. Several studies concerning education and police performance have revealed that college-educated police perceive themselves as better performers on the job (Suman, 1998), were rated by supervisors as performing better than their non-college-educated counterparts (Smith & Aamodt, 1997; Truxillo, Bennett, & Collins, 1998), and expressed greater concern with ethical conduct (Shernock, 1992). Corrections systems should consider exploring the impact that college education requirements might have on their staff performance.

Crowding is a problem that has affected state and local corrections systems for years. The use of one cell per inmate remains problematic at a time when incarceration has become a tool of first resort. Nearly half of the responding jails indicated that less than one quarter of their inmates are actually housed alone in single cells. Without the use of single cells, inmates do not have the privacy that the BOP described as being an important factor in stress reduction and order maintenance. An encouraging finding was that more than three quarters of the responding jails allow inmates to enter and exit their cells throughout most of the day. The lack of privacy that inmates apparently have in the cells, however, has the potential to reduce the benefit derived from the design's intent. Although the use of single cells would be more in line with a normalized environment, it seems impractical to expect many counties to be able to use this housing option as long as incarceration continues to be used as widely as it is today.

A normal living environment inside these jails is expected to communicate to the inmates that they are expected to behave. Researchers concerned with physical aspects of buildings such as public housing projects and jails have noted the negative message that vandal-resistant fixtures sends to residents (Newman, 1972; Resser, 1989; Wener et al., 1993). Despite the BOP and some counties' understanding of the impact of the physical environment on residents, this has mostly been ignored by decision makers involved in construction of podular direct-supervision jails. Only 18% of the responding jails reported using wood or fabric furniture in the living areas, and an additional 15% used this in conjunction with traditional vandal-resistant furnishings. The majority of these facilities are continuing to use metal, concrete, or plastic materials for their furniture and fixtures. Forty percent of the jails reported bolting at least half of their furniture to the floor in the pods. Fortunately, more promising findings regarding physical characteristics of the institution were found. About 60% of the participating jails reported using bumpers or other material to reduce echoes, and nearly all provide air-conditioned living spaces for the inmates. Half of the jails allowed inmates to control the lighting in their cells. Although this is encouraging, those responsible for constructing these institutions need to understand the role that a normalized living environment plays in new generation jails.

Television appears to be a widely accepted form of entertainment for the inmates, with more than 90% of the jails providing inmates with access to televisions throughout most of the day. Possession of electronic devices and access to weight-lifting equipment were allowed in about one third of the institutions. Inmates were given a certain degree of freedom in most of the responding jails, with the majority of administrators reporting that inmates have access to showers and telephones throughout most of the day.

With regard to some of the variables assessing isolation of inmates from their loved ones, only one quarter of the jails permitted inmates to receive packages and have physical contact with their visitors. It is understood that there are security concerns associated with contact visits and packages, but these privileges may allow inmates to maintain a better connection with loved ones. One of the management strategies that the BOP wished to incorporate into the podular direct-supervision design was the unit management approach. This is expected to increase the quality of inmate and staff contact as well as have actual services closer to the inmates. The results of the current study suggest that most responding jails have declined to concentrate their services. Although more research is necessary to study the effectiveness of this technique, Farbstein (1986) noted that it has the potential of being beneficial to both staff members and inmates. The availability of activities on the pod reduces management problems associated with intrafacility movement of inmates.

⊠ Conclusion

After the positive results found in the federal MCCs and the success in a few county jails were publicized, decision makers chose increasingly to build podular direct-supervision facilities. The survey results presented here indicate that, although several counties in the United States are constructing jails that are podular in design and are built for direct supervision, other characteristics of the new generation jail are not being included. Simply changing the physical design from a linear model to a podular, triangle-shaped design does improve officer sight lines, as does the placement of the officers inside the pods. As the data indicate, administrators reported that officers spend most of their time inside the living areas. What seems to have become lost, though, is the importance of other characteristics that were included in the MCCs and the Contra Costa County Jail.

Some improvements to living areas have been made, such as ample access to televisions, air conditioning, bumpers to reduce the noise of slamming doors, and continuous telephone access. Nevertheless, there still is need for improvement in several areas. Positive expectations are generated in part by an environment that is noninstitutional in appearance. Unfortunately, such a setting is not being offered in most of the responding jails. Similar to traditional facilities, the majority of podular direct-supervision jails responding to this survey use vandal-resistant furnishings. Concentration of services on the pod level should also be considered. Although it is expensive and difficult to house one inmate per cell during this get-tough-on-crime period, this jail model was designed to operate with single cells.

The work of officers in podular direct-supervision jails is complex and difficult. It is a job that requires extensive training, but counties are not providing this. It is unlikely that recruits will be able to fully understand and properly utilize the intercommunication skills information after only a day or two of training. Jail and prison officers should be provided with a similar number of hours of training as law enforcement officers. Mandating higher education for incoming officers might also present jails with candidates who are better prepared to communicate with inmates and staff members alike.

Decision makers in counties need to realize that it is a combination of factors, including proactive officer work, a normalized environment, privileges for inmates, and efforts to reduce their feelings of isolation, that make this model work. Inmates who are under direct supervision and are housed in a barren, institutionalized setting will have little incentive to behave. County officials should be informed of the philosophy and important components to the new generation jails, with an emphasis on

how the components come together to make the model a success.

Limitations

The instrument used to generate these findings did include questions about many of the training issues and physical characteristics that have been identified as important components to the new generation design. Surveys, however, do need to be relatively short to facilitate higher response rates. Ideally, additional and more specific questions about the exact types of general training offered to staff members would have been beneficial. Perhaps the most suitable methodology for this study would have involved site visits to every facility. Site visits clearly would have generated more information and provided a better understanding of each jail's operations and the philosophy of each administration. Conducting this type of study on such a large number of jails nationwide, however, was financially prohibitive. Every effort was made to capture the essence of the included jails, but it is understood that this was not the ideal methodological approach for the issue of jail operations.

Note

1. It should be mentioned that the conditions of confinement in the New York Metropolitan Correctional Center was the subject of a lawsuit (*Bell v. Wolfish*, 1979). The complaint cited overcrowded conditions, specifically double bunking, inadequate recreational facilities, and lack of educational facilities. The court found that the conditions did not constitute punishment, which is prohibited prior to adjudication of guilt (del Carmen et al., 1998).

References

Bayens, G. J., Williams, J. J., & Smykla, J. O. (1997a). Jail type and inmate behavior: A longitudinal analysis. *Federal Probation*, 61(3), 54–62.

Bayens, G. J., Williams, J. J., & Smykla, J. O. (1997b). Jail type makes a difference: Evaluating the transition from a traditional to a podular, direct supervision jail across ten years. *American Jails, 11*(2), 32–39.

Bell v. Wolfish 441 U.S. 520 (1979).

del Carmen, R. V., Ritter, S. E., & Witt, B. A. (1998). *Briefs in leading cases in corrections* (2nd ed.). Cincinnati, OH: Anderson.

Farbstein, J. (1986). *Correctional facility planning and design.* New York: Van Nostrand Reinhold.

Farbstein, J., & Wener, R. (1989). *A comparison of "direct" and "indirect" supervision correctional facilities.* Washington, DC: National Institute of Corrections.

Gettinger, S. H. (1984). *New generation jails: An innovative approach to an age-old problem.* Longmont, CO: NIC Jails Division.

Houston, J. G., Gibbons, D. C., & Jones, J. F. (1988). Physical environment and jail social climate. *Crime and Delinquency, 34,* 449–466.

Kerle, K. (1995). Direct supervision: The need for evaluation. *American Jails, 9*(3), 5.

Levinson, R. B. (1999). *Unit management in prisons and jails.* Lanham, MD: American Correctional Association.

Nelson, W. R. (1983, April). New generation jails. *Corrections Today,* pp. 108–112.

Nelson, W. R. (1986). Can cost savings be achieved by designing jails for direct supervision inmate management? In J. Farbstein & R. Wener (Eds.), *Proceedings of the First Annual Symposium on Direct Supervision Jails.* Boulder, CO: National Institute of Corrections.

Nelson, W. R. (1988). The origins of the podular direct supervision concept: An eyewitness account. *American Jails, 2*(1), 8–16.

Nelson, W. R., & Davis, R. M. (1995). Podular direct supervision: The first twenty years. *American Jails, 9*(3), 11–22.

Newman, O. (1972). *Defensible space.* New York: Collier.

Resser, J. P. (1989). *The design of safe and humane police cells.* Royal Commission Into Aboriginal Deaths in Custody.

Sechrest, D. K., & Price, S. J. (1985). *Correctional facility design and construction management.* Washington, DC: U.S. National Institute of Justice.

Senese, J. D., Wilson, J., Evans, A. O., Aguirre, R., & Kalinich, D. B. (1992). Evaluating jail reform: Inmate infractions and disciplinary response in a traditional and a podular/direct supervision jail. *American Jails, 6*(4), 14–23.

Shernock, S. (1992). The effects of college education on professional attitudes among police. *Journal of Criminal Justice Education, 3*(1), 71–92.

Smith, S. M., & Aamodt, M. G. (1997). The relationship between education, experience and police performance. *Journal of Police and Criminal Psychology, 12*(2), 7–14.

Suman, K. (1998). Self-evaluations of police performance: An analysis of the relationship between police officers' education level and job performance. *Policing—An International Journal of Police Strategies and Management, 21,* 632–647.

Sykes, G. (1972). *Society of captives: A study of a maximum security prison.* Princeton, NJ: Princeton University Press.

Truxillo, D. M., Bennett, S. R., & Collins, M. L. (1998). College education and police job performance: A ten-year study. *Public Personnel Management, 27,* 269–272.

U.S. Bureau of Justice Statistics. (2001). *State and local law enforcement statistics.* Washington, DC: U.S. Department of Justice. Also available from http://www.ojp.usdoj.gov/bjs/sandlle/htm

Welch, M. (1996). *Corrections: A critical approach.* New York: McGraw-Hill.

Wener, R., Frazier, W., & Farbstein, J. (1985). Three generations of evaluation and design of correctional facilities. *Environment and Behavior, 17,* 71–95.

Wener, R., Frazier, F. W., & Farbstein, J. (1993). Direct supervision of correctional institutions. In National Institute of Corrections (Eds.), *Podular, direct supervision jails* (pp. 18). Longmont, CO: NIC Jails Division.

Williams, J. L., Rodeheaver, D. G., & Huggins, D. W. (1999). A comparative evaluation of a new generation jail. *American Journal of Criminal Justice, 23,* 223–246.

Zupan, L. L. (1991). *Jails: Reform and the new generation philosophy.* Cincinnati, OH: Anderson.

Zupan, L. L., & Menke, B. A. (1988). Implementing organizational change: From traditional to new generation jail operations. *Policy Studies Review, 7,* 615–625.

Zupan, L. L., & Menke, B. A. (1991). The new generation jail: An overview. In J. A. Thompson & G. L. Mays (Eds.), *American jails* (pp. 180–194). Chicago: Nelson-Hall.

Zupan, L. L., & Stohr-Gilmore, M. K. (1988). Doing time in the new generation jail: Inmate perceptions of gains and losses. *Policy Studies Review, 7,* 626–640.

DISCUSSION QUESTIONS

1. What is the best use for a jail? What factors might make it difficult to operate jails so that they are able to focus on this best use?

2. What factors are likely to compromise the ability of new generation jails to achieve their promise?

3. Why are jail personnel in most facilities and sheriffs' departments still paid less than those on patrol? What argument can be made for the same, and even higher, pay for jail staff?

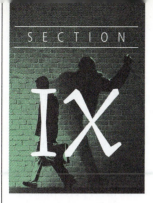

IX

GENDER

Introduction

As far back as anyone can remember, there have been fewer women and girls incarcerated or under correctional supervision than there have been men or boys. Of course, there have been exceptions for some crimes (e.g., prostitution for adults and juveniles or status offending for girls), but correctional populations have always been overwhelmingly male. Though the percentage of women and girls in those populations has increased in recent years, it is still true that males form the vast majority of those under correctional supervision in this country.

What this numerical "minority" status for girls and women has meant is that institutions and programming are, and have been, typically, geared toward boys and men. Our very history of institution building illustrates this fact. As Young (1994) found from her research on the construction of juvenile facilities in the southern portion of the United States, males, and particularly white males, were much more likely to have juvenile prisons constructed specifically for them than were females. Women and girls accused (in the case of jails) or sentenced (in the case of jails or prisons) were much more likely to "do their time" in male facilities, initially as part of those facilities (e.g., in bridewells and other poor houses) and later in separate sections of jails and prisons or completely distinct facilities (Baunach, 1992; Belknap, 2001; Chesney-Lind & Shelden, 1998; Kerle, 2003; Pollock, 2002; Rafter, 1985). We will explore this history of the female correctional experience a bit more in this section.

More recently and by any measure, however, all observers note that the number of women and girls as inmates or supervisees in corrections has grown exponentially over the last several years. In 2000, women comprised 11.4% of jail populations, but, by 2006, that figure was 12.9% (Sabol, Minton, & Harrison, 2007). In 2000, women comprised 6.3% of prison populations, and, by 2006, that figure was 6.9% (Sabol, Couture, &

Harrison, 2007). Girls confined in residential facilities increased from 13.6% of all juveniles in 1997 to 15.1% in 2003 (Sickmund, Sladky, & Kang 2005). But the largest growth, as far as correctional populations are concerned, has come in probation and parole, and particularly probation. As the cost of incarceration continues to escalate and the propensity to incarcerate has not abated, the number of women and girls on probation has risen a commensurate amount. Women constituted 22% of probation and 12% of parole caseloads in 2000; by 2006, that percentage had increased to 24% for probation and remained the same for parole (Glaze & Bonczar, 2007). Though some of these percentage increases seem small, they represent increases, in the case of probation, of thousands of women on probation, as total caseloads in 2006 included 5 million people. For girls on probation, the increases do not even seem small. In 1985 girls constituted 19.3% of juveniles on probation; by 2004, that percentage had grown to 27.2 (Puzzanchera & Kang, 2007).

The trajectory of employment of female correctional officers has not been as steep or as steady as it has been for those women and girls under correctional supervision. Women were employed as matrons, to a limited degree, to work with females in some of the earliest prisons and jails (Pollock, 2002; Stohr, 2006; Zupan, 1992). But they really did not make significant inroads into the correctional profession until the 1964 Civil Rights Act was amended in 1972 and they began suing to gain employment in both female and male correctional institutions. According to the *Sourcebook of Criminal Justice Statistics* (Maguire & Pastore, 2002), women occupied 40% of correctional officer positions in jails, but their employment in prisons can no longer be determined from national statistics—as these statistics are no longer available in publications by the federal government—though it is thought to be much less than in jails (Stohr, 2006). Nor are staff demographic statistics regarding probation and parole officers for adults or children readily available. We do know that the Federal Bureau of Prisons claims to have a staff that is 28.3% female, though they do not indicate how many of these women are correctional officers (Bureau of Justice Statistics, 2003). About 22% of correctional officers in Canadian minimum- and medium-security federal prisons are women (Correctional Service of Canada, 2003), which is probably close to the American employment level. Though these numbers are not particularly impressive, they do represent an increase since 1988 when Zupan (1992) reported that women constituted only 15% of correctional officers nationally and 10% in Canadian federal prisons in 1984 (Correctional Service of Canada, 2003).

In a number of other sections in this book, we have integrated a discussion of the experience of women and girls in corrections. For this reason, we will focus in this section of the book on a few unaddressed areas as they relate to that experience. We will explore the history and the nature of the female experience in corrections, whether that is as a staff member or as an inmate or supervisee. Certainly, that history and the current status of women and girls in corrections would not be the same without the collective efforts of thousands of dedicated women and men, who often practiced—and espoused—feminist beliefs about the true worth, needs, and abilities of the female of the species.

☒ Feminism

Women staff would not be employed at the level they are and female inmates would not have the attention and programming they do, albeit much less than men and boys, if not for the sustained efforts of feminist scholars and practitioners agitating for their rights and their needs (Pollock, 2002; Smykla & Williams, 1996; Rafter, 1985; Stohr, 2006; Young, 1994; Zimmer, 1986; 1989; Zupan, 1992). As indicated in the article by Rafter (included at the beginning of this book), the proponents of change in female corrections in the last half of the 1800s and first half of the 1900s tended to be of two minds. There were those moralists (who were sometimes social feminists, as Rafter terms them) who believed that women and girls involved in the criminal justice system were, in effect, morally impaired and therefore in need of religious and social remedies (prayers, efforts to keep them chaste, etc.). There were others, those who espoused a liberal feminist perspective, who believed that the problem lay more with the social structure around these women and girls (e.g., poverty and lack of sufficient schooling or training, along with patriarchal beliefs) and that the solution lay in preparing them for an alternate existence; sometimes this involved "traditional women's work," so they would not turn to crime (Rafter). Some of these early feminists believed, as liberal feminists do today, that men and women are inherently equal and, as such, women and girls are entitled to the same rights, liberties, and considerations (e.g., in corrections, this would be programming, quality of institutions, and equal employment as staff) as men and boys (Belknap, 2001; Daly & Chesney-Lind, 1988).

The moralists triumphed, though not completely, in the argument over what lay at the heart of female criminality. As a consequence, we have had over a century of correctional operation that has tended to be overly concerned with the sexuality of females. Another consequence of this triumph was that reform efforts were directed at training female inmates to be proper wives and mothers while forgetting that, as members of the lower classes, they would need to make a living for themselves and their children once they reentered their communities. Despite this morals-of-the-fallen-woman focus, the soiled dove if you will, feminist women and men were able to agitate for and get separate facilities for women and some other services (e.g., educational and job training) that was geared toward helping women and girls become independent and self-supporting in the free world (Hawkes, 1998; Yates, 2002).

One societal obstacle to achieving equal treatment has been *patriarchy*. Patriarchy involves the attitudes, beliefs, and behaviors that value men and boys over women and girls (Daly & Chesney-Lind, 1988). Members of patriarchal societies tend to believe that men and boys are worth more than women and girls. They also believe that women and girls, as well as men and boys, should have certain restricted roles to play and that those of the former are less important than those of the latter. Therefore, education, work training that helps one make a living, and better pay are more important to secure for men and boys than for women and girls who are best suited for more feminine and, by definition in a patriarchal society, less worthy professions. Feminist scholars have determined that many cultures hold such beliefs and engage in the practices that derive from them (Stohr, 2008).

In the United States, much effort has been expended over the centuries, by male and female feminists, to address the patriarchal belief system, and there has been some success in this regard (Dworkin, 1993; Martin & Jurik, 1996; Morash, 2006; Spohn, 1990; Whittick, 1979). As regards corrections, feminists have been instrumental in pushing for more and better programming for incarcerated and supervised women and girls, for the reduction in the incarceration of girls for status offenses, for the attention to the sexual abuse of women and girls while incarcerated, and for the greater employment of women in adult male and female correctional institutions.

⊠ Females in Corrections: Growth, History, Needs, Programming, and Abuse

Growth and History

As mentioned in the foregoing, the number of incarcerated and supervised women under the correctional umbrella has never been larger, but it was not always so. In the past, the number of women inmates and supervisees was proportionately smaller. For instance, we know from United States Census reports that women constituted three or 4% of state and federal prison populations from 1910 through the 1970s (Cahalan, 1986, p. 65). In 1980, that percentage had risen to 5% and has only increased since (see the earlier discussion in this section about current percentages). If you add in reformatories, women and girls accounted for, on average, about 5% of those incarcerated in correctional facilities, from 1910 through 1959 (Cahalan, 1986, p. 66). Jail inmates were anywhere from 9 to 5% female from 1910 to 1983 (Cahalan, 1986, p. 91). Juvenile institutions averaged about 21.2% female residents (aged 15–19) from 1880 to 1980 (Cahalan, 1986, p. 130). Unfortunately, a gender breakdown of parolees and probationers is not available, but given the overall increase in the percentage of incarcerated women generally and women

▲ **Photo 9.1**   Jail inmates share a laugh in squalid "Tent City," where over 1,000 inmates serve their sentence. The tents are worn ex-army tents, purchased by Sheriff Joe Arpaio from the U.S. government. Razor wire and a corrugated fence separate the women from the men. These inmates have not committed violent crimes. Their crimes range from petty theft to drug use to prostitution.

and girls on probation or parole currently, it is likely that, historically, they were not as subject to the criminal justice system as they are today either.

The best explanation for the lack of female offenders in the criminal justice system has been the fact that they commit fewer street crimes that would garner this distinction. Most murders, robberies, rapes, burglaries, and even larcenies are committed by men and boys (Federal Bureau of Investigation [FBI], 2007). Even among corporate/white collar and environmental crimes, the more likely offender is male—if for no other reason than more males are in a position to commit such crimes than females. As mentioned in other sections, the drug war of the last 20 to 30 years has brought more female offenders into the system, and this fact has resulted in their greater proportional growth among correctional populations, but, even with this sort of offense, they are in the clear minority (FBI, 2007).

Needs and Programming

As a practical matter, then, if not just because women and girls have historically been valued less by this society (patriarchy) but perhaps more because crime has generally been the purview of men and boys, correctional facilities and correctional practices have tended to focus on men. This focus led to disparate treatment that disadvantaged women and girls from the beginning and resulted in little concern for their needs (Muraskin, 2003).

Yet, as was discussed in previous sections, women and girls in custody are more likely to have mental and physical health problems than incarcerated men and boys. They are also more likely to have substance abuse problems than their male counterparts. Moreover, they have the same kinds of educational and job training deficits and needs as men and boys (DeKeseredy 2000; Gray, Mays, & Stohr, 1995; Morash, Haarr, & Rucker, 1994; Owen & Bloom, 1995; Pollock, 2002). Their need for gainful employment is likely as great as, if not greater than, men's and boys', as they most often have to support themselves and their children, whereas fewer men had custody of their children before they were incarcerated. Moreover, a greater percentage of them, perhaps as high as 50%, were the victims of sexual abuse before incarceration, and this is likely to shape their self-concept and their relations with others negatively, thus necessitating more programming (Belknap, 2001; Comack, 2006; Morash, 2006, Pollock, 2002).

Given all of these needs, and assuming that policy makers would not want women and girls to reenter the correctional system, one would think that all of their needs would be met with adequate programming. There should be a particularly urgent incentive to develop these programs if for no other reason than that females cost much more than males to incarcerate because of their multiple needs and a reduction in economies of scale: separate female institutions house fewer inmates but require almost the same number of administrative staff as much larger male institutions. Unfortunately, programming for incarcerated girls and women has been far from adequate in most jurisdictions. Though there has been some recognition by the federal government of the need to develop programming that fits the needs of women and girls, it is unclear how much this view has spread to state and local facilities (Morash, 2006).

Though most of these needs are far from met in correctional environments, and far less than a majority of women are involved in meaningful programming, Pollock (2002) notes that some states have made renewed efforts to address the needs of women and girls, especially in terms of tackling educational, vocational, parenting, and substance abuse issues and their histories of past victimization.

Some states and localities, recognizing the special needs of women and girls in corrections, have devised programming that suits them. For instance, many states have parenting programs, substance abuse programming, vocational programming, and counseling to deal with past abuse. However, the numbers of these programs and their quality (very few are rigorously evaluated in terms of desired outcomes) leave much to be desired (Pollock, 2002). The sad truth is that most women and girls who need programming in corrections are not able to access it, or, if they are, it is sometimes of dubious worth.

Abuse

One of the primary reasons for removing women and girls from male facilities and hiring female staff to supervise them was that they were targets of sexual abuse by correctional staff and male inmates (Henriques & Gilbert, 2003). Though the separation from male inmates has reduced the abuse of females in custody, the sexual abuse by male staff, though likely much less prevalent than it once was, in part because of the inclusion of more female staff, has not been eliminated.

One of the authors (Stohr) had occasion to serve as an expert witness for the plaintiffs in a civil suit in 2004 against a city in New Mexico whose judge and a few correctional officers for the local jail were involved in the sexual abuse of female jail inmates (*Salazar v. City of Espanola,* 2004). The male judge and a few male correctional staff had an arrangement whereby female offenders whom the judge found attractive would be placed in the jail (whether their offense merited it or not), and then the judge would have access to them when they were sent over to "clean" his chambers. Inevitably, he would make passes at them, using the threat of more jail time, denied privileges, or a lengthened sentence as a way to coerce them into sexual activity with him. Meanwhile, a few of the correctional staff were harassing the female inmates, for example, by watching and commenting on their bodies as they showered, making sexual advances toward them, and touching them inappropriately. Two male officers were even involved in removing a few females from their cells and having sex with them in the control room at night when no one else was around. After this kind of activity occurred for a period of time, and because of the concerted efforts of several ex-inmates and their attorneys, the judge was convicted of rape, and the judge, correctional staff, and city lost a million-dollar lawsuit (*Salazar v. City of Espanola,* 2004).

Such abuse is particularly damaging when one considers that about half of incarcerated women and girls have experienced some form of sexual abuse in the past (Gray, Mays, & Stohr, 1995; Henriques & Gilbert, 2003). Recognizing this fact, the Ninth Circuit put some restrictions on the body searches of female inmates by male staff, noting that such searches may serve to revictimize the women with sexual abuse histories (*Jordan v. Gardner,* 1993—see the discussion of this practice in another section).

Efforts to reduce sexual abuse in correctional institutions have centered on ensuring that staff have the proper training and are supervised sufficiently to prevent abuse. Moreover, the value of disciplinary measures to reinforce appropriate practices cannot be overstated. Staff members who violate the rights of their charges, in a way that is as serious as sexual abuse, should be fired and prosecuted.

The hiring of more women officers to cover living units is another way that correctional agencies have worked to keep sexual predators from gaining access to relatively powerless female victims. There is no question, however, that lawsuits have been successful in spurring some of these needed changes in correctional practice. But the problem

▲ **Photo 9.2**   Inmate in her cell during "lock down" in Estrella Jail in Phoenix, Arizona. Estrella Jail is home to America's only female chain gang. Inmates stay in their tiny 8 x 12 foot cells 23 hours per day unless they are out on assigned chain gang duty. No television or organized recreation is provided. Inmates must memorize 10 rules of conduct, addressing grooming, behavior, and attitude. Chain gang and other privileged duties can be suspended for infractions such as swearing.

with lawsuits is that their application is hit and miss at best, and the success of plaintiffs is always iffy. Therefore, the best preventative measures are those that focus on hiring competent people, training them to behave professionally, rewarding them when they do, and punishing them (up to and including firing them) when they do not.

⊠ Female Correctional Officers

Overcoming Employment Obstacles

As with the accused and convicted in the system, women have always constituted a minority in terms of correctional staff (as discussed at the beginning of this section). Though one would expect that women might comprise a greater percentage of staff, given their representation in the larger community, the current figures actually represent a significant improvement over 30 or 40 years ago. Then, women, with exceptions allowed for matrons in women's and girl's facilities and at lower pay than men working in male facilities, were prohibited by practice, tradition, or law from working in the more numerous men's and boys correctional institutions or in probation and parole. As was mentioned earlier, it was not until the Civil Rights Act of 1964 was passed and amended in 1972 that women were given the legal weapon to sue for the right to work and be promoted in all prisons, jails, detention centers, halfway houses and as staff in community corrections. Many women did, in fact, sue; they had to if they wanted the same kinds of jobs and promotional opportunities then available only to men in both corrections and policing (Harrington 2002; Hawkes, 1998; Stohr, 2006; Yates, 2002). As a result of this agitation and advocacy, and slowly, jobs and promotional opportunities became open to these pioneering feminist women, resulting in the more diverse correctional workforce we see today.

Current Status: Equal Employment Versus Privacy Interests of Inmates, Qualifications for the Job, and Sexual and Gender Harassment

As the number of women employed in corrections has increased, three issues have been particularly problematic for them in the workplace: (1) whether women's rights to equal employment in male correctional facilities is more important than male inmates' rights to privacy in those same facilities (see the brief mention of this topic in an earlier section of this book), (2) whether women are physically and mentally suited to do correctional work with men, and (3) how to deal with the sexual and gender harassment—primarily from other staff—that female corrections officers encounter while on the job.

As mentioned before in this section, women achieved the legal right to equal employment in corrections through law and lawsuit. Most of the jobs in institutional corrections are in dealing with male inmates. On the other hand, one can certainly understand the male inmate perspective that they would like some privacy when engaged in intimate bodily functions, such as using the toilet or showering. The courts in this matter, however, have tended to side with the female employees' or prospective employees' right to equal protection over the male inmates (Maschke, 1996). Their reasoning was likely as much influenced by the fact that inmates in this country have very limited rights while incarcerated, with no real "right to privacy," as by the idea that correctional staff should be respectful and professional, no matter their gender, in their dealings with inmates.

The issue of whether women were physically and mentally qualified to work with male inmates has, for the most part, become a settled matter in most institutions and states: They are, and they can. But when they were first making inroads into the correctional workplace, there were plenty of doubts about the ability of women to handle the work (Jurik, 1985, 1988; Jurik & Halemba, 1984; Zimmer, 1986, 1989). Moreover, the Supreme Court has left open the possibility that, if there is a bona fide job qualification that a woman could not do in a male prison, they can be excluded from that work (Bennett, 1995; Maschke, 1996).

Clearly, because of their biology, most women are not as physically strong as most men, and sometimes strength is called for in dealing with an unruly inmate. However, the use of brute force is rather rare in most correctional institutions (note the discussion of violence in corrections in an earlier section). Second, there are defensive (and offensive tactics) that give a trained and armed woman some advantage in a physical altercation with a male inmate. Third, there is some evidence that female staff may have a calming effect on male inmate aggression because they are more inclined to use their interpersonal and communication skills and are less likely to be seen as a threat (Jurik, 1988; Jurik & Halemba, 1984; Zimmer, 1986).

More recent research has indicated that both male and female correctional officers value a service over a security orientation to their work so that work styles and preferences may be more similar than dissimilar (Farkas, 1999; Hemmens, Stohr, Schoeler, & Miller, 2002; Stohr, Lovrich, & Mays, 1997; Stohr, Lovrich, & Wood, 1996). In fact, in research by Kim and colleagues (2003) on the attitudes of male and female wardens, they found that 90 of the female wardens (out of a total of 641 wardens surveyed) were more inclined to value programming and amenities in their prisons that promoted the health, education, and rehabilitation of inmates than were their male colleagues. However, these researchers also found that there were many more similarities than differences between the two genders as they viewed and appreciated their work. Taken in total, what all of this research on women's ability to do the work and on the differences and similarities in work styles between men and women indicate is that men and women mostly view and do correctional work similarly and that some women, perhaps more than some men, can calm an agitated inmate and that some men, perhaps more than some women, are better at physically containing an agitated inmate.

Unfortunately, the problem with gender and sexual harassment of female staff by male staff has not become a settled matter. This is not to say, of course, that male staff members are not harassed by female staff. This does happen, and it can be as debilitating for the male employee as it is for the female. But a number of studies have shown, over a period of years, that females are much more likely to be the victims of male harassment in the workplace and that, when men are victims, they are as likely to be harassed by other men as by women (Firestone & Harris, 1994, 1999; Gruber, 1997, 1998; Mueller, De Coster, & Estes, 2001; O'Donohue, Downs, & Yeater, 1998; Pryor & Stoller, 1994). Male institutions in particular, with their smaller percentages of women employees and managers and their traditions aligned with male power in the workplace, are more susceptible to this kind of behavior than are other correctional workplaces (Lawrence & Mahan, 1998; Lutze & Murphy, 1999; Pogrebin & Poole, 1997). The harassment that occurs can be of the quid pro quo type (something for something, as in, you give me sexual favors and you get to keep your job) or the less serious, but still work stultifying, hostile environment (when the workplace is sexualized with jokes, pictures, or in other ways that are offensive to one gender).

Thankfully, there are remedies, imperfect and cumbersome though they might be, that can be employed to stop or at least significantly reduce such harassment. Again, women initially had to sue to stop the harassment. In one mid-1990s case, one of the authors of this book (Stohr) served as an expert witness for the plaintiff against the State of California's San Quenton Prison. The female victim won over a million dollars for enduring harassment by several male staff and one inmate, harassment that started in the 1970s, ended when she quit in frustration in the 1990s, and was never stopped by the prison administration (*Pulido v. State of California*, 1994). As successful as this case was, there was incontrovertible evidence of the harassment (as provided by memos and diaries and staffing logs and witnesses), evidence which is usually not available to support most victims' stories. Moreover, the female victim lost her job and had to endure almost two years of an uncertain legal battle before the case was tried and the judge ruled. Even then, the State of California appealed, and it took another year before the matter was finally settled in the plaintiff's favor.

What this story illustrates is that there are few true "winners" when sexual and gender harassment cases go to trial. Most such cases fail as there is not sufficient evidence of the abuse, beyond a he said/she said scenario. Victims of such abuse suffer untold harm, in terms of their psychological and physical well-being, both during the abuse and as they relive it during the legal process. Even when cases are successful at trial, taxpayers (not just the instigators of the abuse who often do not have the "deep pockets" of their employers) have to pay for the illegal practices of their own governmental entities and actors.

In other words, there has got to be a better way, and there is! Researchers and correctional practitioners have agreed that there are proactive steps that managers and other employees can take to prevent or stop sexual and gender harassment in the correctional workplace. Such steps would involve hiring, training, firing, and promoting based on respectful treatment of other staff and clients. Training, in particular, can reinforce the message of a "no tolerance" policy as regards harassment. But, to be effective, employees need to see that people are rewarded when they do, or punished when they do not, adhere to the policy.

As a discussion of the current status of female staff working in corrections would indicate, women have made some significant advances in these workplaces. They have not just made gains in employment, but female supervisors and managers are no longer anomalies in most states. Though nowhere near matching the numbers of men as staff and management in corrections, and while still grappling with the pernicious problem of sexual and gender harassment, women, nonetheless have come a long way since the days when they worked as matrons for lower pay than their male counterparts, a position that typified their work for most of the history of corrections.

◙ The Research

Research on women and girls in corrections has often been geared toward highlighting the differences between how men and women experience corrections. For correctional staff, researchers have explored the idea that women might have more of a human services perspective than their male colleagues do. Though there is some evidence of such a difference, it is clear that men and women are more similar than dissimilar in their supervision styles and attitudes about their work.

For those women and girls who experience corrections as supervisees or inmates, the historical research was inordinately focused on their sexuality both in and outside of correctional control (Giallombardo, 1966; Hefferman, 1972; Owen, 1998; Rafter, 1985). Today, the researchers, such as those whose work is featured in this section, find that women and girls have programmatic needs and styles that determine whether some rehabilitative approaches are more effective than others (e.g., see the Wright, Salisbury, and Van Voorhis article or the article by Staton-Tindall, Garner, Morey, Leukefeld, Krietemeyer, Saum, and Oser featured in this section). One type of programming with particular relevance for women, given that most had physical custody of their children prior to incarceration, is parenting (e.g., see the Loper and Tuerk article in this section).

Taken in tandem, the research presented in this section should shift our perspective from the much more frequent and normative study of males to the study of females in correctional settings. By shifting our gaze to the feminine side, we are as likely to see the sameness of the genders as the contrasts that distinguish them (Rodriguez, 2007; Smith & Smith, 2005). We might also see that the life course of a woman or girl entangled in the criminal justice system (e.g., see the Brown article) forms a predictable pattern that might be fruitfully addressed if we only had the will do so.

◙ Summary

The study of women and girls in corrections was not always a priority for scholars (Flavin & Desautels, 2006; Goodstein, 2006). Patriarchal perceptions and beliefs, along with the status of women and girls as numerical minorities, have served to shape organizational and scholarly priorities in a way that favors men and boys. Since the 1970s, however, there has been more scholarly focus on the reality of women and girls who work, live, or are supervised under the correctional umbrella.

◆ Part of this shift in focus has occurred as the result of feminist work to equalize the work of women in corrections and the living and supervision arrangements for women and girls under correctional supervision.

◆ Recent research on women and girls under correctional supervision has highlighted the outstanding needs they have for educational, substance abuse, work training, parenting, and surviving abuse programming. Unfortunately, it has also shown that little programming is provided in either jails or prisons or in the communities to meet these needs.

◆ Female correctional officers have also faced a number of legal and institutional barriers to their full and equal employment in corrections.

◆ For the most part, many of these formal barriers have been removed as female officers have demonstrated their competencies in handling correctional work.

◆ Some researchers have even found a feminine style of, or approach to, correctional officer work that employs the successful use of interpersonal communication skills to address inmate needs.

◆ However, there is still evidence that sexual and gender harassment and the sexual abuse of female inmates continues in some correctional environments. Though organizational remedies exist to "deal" with such abuse, they are not always employed by managers.

KEY TERMS

Hostile environment sexual harassment

Liberal feminist

Patriarchy

Quid pro quo sexual harassment

INTERNET SITES

American Correctional Association: www.aca.org

American Jail Association: http://www.aja.org

American Probation and Parole Association: www.appa-net.org

Bureau of Justice Statistics (information available on all manner of criminal justice topics): www.ojp.usdoj.gov/bjs

National Criminal Justice Reference Service: www.ncjrs.gov

National Institute of Corrections www.nicic.org

Office of Justice Programs, Bureau of Justice Statistics (periodic statistical reports on all manner of criminal justice topics, e.g., HIV in prisons and jails, probation and parole, and profiles of prisoners): www.ojp.usdoj.gov/bjs/periodic.htm

Vera Institute (information available on a number of corrections and other justice related topics): www.vera.org

READING

In this article, Wright, Salisbury, and Van Voorhis examine the responsivity needs of females under the assumption that the needs of women offenders may be qualitatively different than the needs of male offenders. The "pathways" and "gender-responsive" perspectives of female offending have recently garnered attention in both practitioner and scholarly arenas. The pathways perspective focuses attention on the co-occurrence and effects of trauma, substance abuse, dysfunctional relationships, and mental illness on female offending, while the gender-responsive perspective also suggests that problems related to parenting, child care, and self-concept issues are important needs of women offenders. The authors examine whether or not these are risk factors for poor prison adjustment with a sample of 272 incarcerated women offenders in Missouri. They look at how gender-responsive need is related to 6- and 12-month prison misconducts, and whether the inclusion of such needs to traditional static custody classification items increases the predictive validity of such tools. Results suggest that women offenders do, in fact, display gender-responsive risk factors in prison.

Predicting the Prison Misconducts of Women Offenders

The Importance of Gender-Responsive Needs

Emily M. Wright, Emily J. Salisbury, and Patricia Van Voorhis

Institutional custody classification tools have been adopted by correctional agencies throughout the United States (Van Voorhis & Presser, 2001) and are used to inform offender placement into community, minimum-, medium-, and maximum-security custody levels. For prisons, placement into an appropriate custody level facilitates safety, housing, privileges, movement, and programming (Brennan, 1998; Van Voorhis & Presser, 2001). Because male offenders make up the majority of prisoners in the United States, it is not surprising that custody classification systems were developed from male samples and designed with male offenders in mind (Salisbury, Van Voorhis, & Spiropoulos, in press).

Until recently these classification systems were applied to women offenders with little regard to their applicability and appropriateness. However, the increasing number of women offenders being sentenced to prison and the increasing attention granted to the "gender-responsive" needs of females has amplified scrutiny over the usefulness of such systems for women offenders. Gender-responsive scholars suggest that institutional classification systems that were designed for male offenders are less useful for women offenders and in many cases are invalid. They contend that females are very different from male offenders, as evidenced by their unique paths into criminal behavior, the offenses in which

Source: Wright, E. M., Salisbury, E. J., & Van Voorhis, P. (2007). Predicting the Prison Misconducts of Women Offenders: The Importance of Gender-Responsive Needs. *Journal of Contemporary Criminal Justice, 23*(4), 310–340. Reprinted with permission of Sage Publications, Inc.

they engage, their decreased threat of violence across criminal justice settings, and their unique needs relating to victimization, substance abuse, mental health, self-concept, child care, and relationship issues (Bloom, Owen, & Covington, 2003; Covington, 2000). Furthermore, these scholars criticize current systems for ignoring women's needs and failing to adequately inform their treatment and programming.

These criticisms are not without merit; a growing body of empirical research reports that women offenders are more likely than male offenders to be victims of sexual and physical abuse, exhibit mental health problems, engage in substance abuse, encounter parenting and child care problems, be affected by relationship issues, and have problems with self-concept (Bloom et al., 2003; Koons, Burrow, Morash, & Bynum, 1997; Lindquist & Lindquist, 1997; Sheridan, 1996). Moreover, current evidence indicates that prison classification systems do work better for male offenders than for female offenders (Hardyman & Van Voorhis, 2004). Custody classification systems that are used today tend to overclassify women into higher risk categories than is warranted by their behavior, thus increasing the limitations placed on women's freedoms and access to programming (Brennan, 1998; Van Voorhis & Presser, 2001). Although gaining increased attention in practitioner and scholarly debate, these gender-responsive needs have been understudied with regard to women offenders and institutional outcomes.

In a recent pilot study to the current research, however, Salisbury et al. (in press) noted that some of these needs were more relevant to prison adjustment than the criminal history variables typically used to predict serious prison misconducts. Specifically, substance abuse, mental health, child abuse, self-concept, and relationship issues significantly affected women's chances of becoming involved in serious prison misconducts.[1]

Given that current prison classification systems may be doing more harm than good for women offenders, scholars have begun to question whether changes to the systems are needed to increase their predictive accuracy. Of particular concern is whether the inclusion of important gender-responsive needs would increase the validity of these assessment tools for women offenders and more appropriately inform their treatment and programming. Of course, the importance of this research extends beyond the classification instruments themselves to the very issue of the mission of women's prisons. The discovery that troubled inmates make poorer adjustment to prison than those currently classified as high custody through offense-related variables may question many of the current policies for managing women's prisons. Prison safety from this needs-based perspective emanates not solely from the practice of holding women with serious offenses at higher custody levels but rather from sound plans to accommodate, program for, and promote well-being. In this sense, gender-responsive classification systems are intended to serve as tools to more accurately guide gender-responsive placement, programming, and correctional policies in settings that place high emphasis on treatment, case management, and effective community transition.[2]

To this end, the current study examines the role that gender-responsive needs relating to trauma and abuse, mental health, parenting, relationships, and self-concept play in women's adjustment to prison. We consider as well those needs that are currently identified by gender-neutral, risk/needs assessments (e.g., employment, education, substance abuse, antisocial attitudes, and antisocial associates; see Andrews & Bonta, 1995; Brennan, Dieterich, & Oliver, 2006).[3] The current study expands on the pilot study reported by Salisbury et al. (in press) but utilizes a larger sample of women offenders and tests more gender-responsive needs.[4] Specifically, this article examines whether gender-responsive needs function as risk factors to women offenders' institutional misconducts. Finally, we examine whether the inclusion of gender-responsive needs increases the predictive validity of custody classification for women offenders.

The gender-responsive risk factors of interest to the current study are drawn from the "pathways" perspective (see Chesney-Lind & Pasko, 2004; Daly, 1992, 1994; Owen, 1998; Reisig, Holtfreter, & Morash, 2006) and recent gender-responsive work (see Bloom et al., 2003; Chesney-Lind, 2000b; Covington, 1998; Green, Miranda, Daroowalla, & Siddique, 2005). According to the pathways perspective, the confluence of trauma, substance abuse, and mental health puts women on "pathways" to crime that are inherently different from the pathways into crime that males take. Chesney-Lind (2000a), for example, noted that early victimization, trauma, and exploitation of females by family members or close friends provide the incentive for girls to run away, increasing their chances of later engaging in crime. Daly's (1992, 1994) groundbreaking research provided a framework for understanding several women's criminal pathways that were organized by their life experiences, offending contexts, and social location. Four of the five pathways found by Daly (1992, 1994) can be considered "gendered" pathways reflecting offending contexts not typically seen with men. Other researchers have suggested that the early victimization of girls leads to depression and low self-concept that then may promote drug use, subsequent victimization, and crime in adult years; such a trajectory may not be comparable to that of male offenders (McClellan, Farabee, & Crouch, 1997). Scholars asserted that gender-specific theories of female offending cannot discard the important roles of trauma, substance abuse, relationships, and mental health in female offending (Covington, 1998). Thus, at the least, a gender-specific perspective on female misbehavior entails that women are more likely than men to experience childhood and adult victimization, substance abuse, and diagnoses of mental illness. Additional needs related to parenting, child care, and self-concept have also been suggested as influencing women's criminal behavior (Bloom et al., 2003).

Victimization and Abuse

Data from incarcerated women offenders support the assertion that female offenders are more likely to experience abuse or victimization. As many as 47% and 39% of women in corrections report experiencing some sort of physical or sexual abuse, respectively, during their lifetimes (Bureau of Justice Statistics [BJS], 1999; McClellan et al., 1997). The estimates of male abuse are much lower—just up to 13% and 6% report experiencing physical or sexual abuse, respectively (BJS, 1999). Estimates of such rates can be widely variable, however; some researchers have reported rates of physical abuse among women offenders as high as 75% (e.g., Browne, Miller, & Maguin, 1999; Greene, Haney, & Hurtado, 2000; Owen & Bloom, 1995) and sexual abuse as high as 65% (e.g., Browne et al., 1999; Islam-Zwart & Vik, 2004).

Although research on prevalence rates indicates that women offenders often experience abuse as children as well as adults (Browne et al., 1999), conclusions are mixed concerning the importance of childhood abuse versus adult abuse per se, as well as their importance to community versus institutional outcomes. Whereas some researchers have found no association between adult victimization and community recidivism (Bonta, Pang, & Wallace-Capretta, 1995; Loucks & Zamble, 1999; Rettinger, 1998), others have found negative relationships (Blanchette, 1996; Bonta et al., 1995) once these variables were entered into multivariate models; still other researchers have found positive associations between adult victimization and recidivism (Salisbury et al., in press).[5] Results regarding the effect of childhood abuse on community outcomes are also mixed—some researchers suggest that childhood abuse is a significant predictor of community recidivism (Law, Sullivan, & Goggin, 2006), whereas other researchers have found that childhood abuse is not significantly related to community outcomes (Salisbury et al., in press).

Moreover, whether these relationships are stable across community and institutional settings is understudied. Although Law et al. (2006) found that childhood abuse was predictive of recidivism in the community, this relationship did not hold when assessing institutional adjustment. However, Salisbury et al. (in press) found that, though adult victimization was predictive of community recidivism and childhood victimization was not, these relationships flipped when assessing institutional misconducts—child abuse became a significant predictor of institutional misconducts while adult emotional victimization was the only type of adult victimization which remained significant.[6] Islam-Zwart and Vik (2004) also assessed childhood and adult physical and sexual abuse on women's adjustment to prison. These researchers found that female inmates who were sexually victimized during adulthood reported more external adjustment problems such as fighting and arguing, while childhood sexual abuse was also associated with internal adjustment problems such as having anger toward others (Islam-Zwart & Vik, 2004). These studies demonstrate mixed results regarding the importance of adult and childhood victimization, especially when assessing their impact on institutional misbehavior.

Mental Health and Related Personal Distress

Mental illness, alone as well as in interaction with other factors, is a major hindrance to prison adjustment among women offenders and has been found to be predictive of such problems (Law et al., 2006; Salisbury et al., in press; Warren, Hurt, Booker Loper, & Chauhan, 2004). It is a well-established observation that incarcerated women experience high levels of distress on many mental health indices (Center for Substance Abuse Treatment, 1999; Jordan, Schlenger, Fairbank, & Caddell, 1996; Singer, Bussey, Song, & Lunghofer, 1995; Teplin, Abram, & McClelland, 1996) and that the prevalence of mental health problems is greater among incarcerated women than among incarcerated men (Lindquist & Lindquist, 1997; Sheridan, 1996).

Within prediction assessments, mental health needs have often been considered "personal distress" factors; these factors have been found to exert only weak to moderate relationships with criminal justice—related outcomes among male and female offenders (Andrews & Bonta, 2003; Gendreau, Little, & Goggin, 1996; Simourd & Andrews, 1994). Although mental health problems are considered gender-neutral risk/need factors in this context, the mental health needs of female offenders may differ substantially from those of male offenders. Depression, anxiety, and self-injurious behavior are more prevalent among female than male populations (Belknap & Holsinger, 2006; Bloom et al., 2003; McClellan et al., 1997; Peters, Strozier, Murrin, & Kearns, 1997), and women often suffer from several co-occurring mental health needs such as depression and substance abuse (Bloom et al., 2003; Holtfreter & Morash, 2003; Owen & Bloom, 1995) at higher rates than men (Blume, 1990).

There may be two potential problems concerning the measurement of "personal distress" in gender-neutral assessments that may have masked the true importance of mental health among female offenders. First, some forms of mental illness may be overlooked in current risk assessment instruments. For example, women who suffer from major mood disorders may be ignored, especially if they have not been previously diagnosed and recorded. As such, the mental health problems of stress, depression, fearfulness, and suicidal thoughts or attempts have shown to be strong predictors of women's recidivism (Benda, 2005; Blanchette & Motiuk, 1995; Brown & Motiuk, 2005), though not for men's recidivism (Benda, 2005).

Second, prediction studies frequently aggregate mental illness indicators into broad mental health domains that could potentially confound relevant associations. For example, a recent meta-analysis by Law et al. (2006) suggested that women offenders' mental health is significantly related to institutional and community outcomes. Although the mean effect sizes reported from that study are relatively weak in strength (Mz^+[mean effect size] = .07, and .09 for institutional and community outcomes, respectively), the study's mental health domains reflected a mixture of heterogeneous indicators of mental illness. This method of aggregation could mask important relationships between specific types of mental illness and criminal behavior. To address these potential problems, we examine specific, symptom-based measures and general measures of mental illness in the current study.

Substance Abuse and Addiction

Substance abuse and addiction are related to male and female offending (McClellan et al., 1997) and are currently assessed in gender-neutral needs and risk/needs assessments. However, some scholars have suggested that substance abuse has unique effects on females, given its high co-occurrence with mental illness, relational problems, and histories of victimization (Covington & Bloom, 2007). There is some evidence to support this argument. McClellan et al. (1997) found that overall illicit drug use was higher for female inmates than male inmates, and the severity of substance abuse was more predictive of property crime for women than for men. In addition, a recent meta-analysis showed that substance abuse was a significant criminogenic need in predicting women's general and violent recidivism (Law et al., 2006; see also Salisbury et al., in press), and women who reported problems with substance abuse have been shown to incur more prison misconducts than women without such problems (see Salisbury et al., in press).

The prevalence of substance abuse among female offenders is high. Among state prisoners, over 60% of women met the Diagnostic and Statistical Manual of Mental Disorders (DSM-IV; American Psychiatric Association, 1994) criteria for having a drug dependence or abuse problem during the year prior to their incarceration, and 59% reported having abused substances in the month prior to their offense (BJS, 2006). In addition, mandatory drug sentences may have affected women offenders more than male offenders (Austin, Bruce, Carroll, McCall, & Richards, 2001); in 1998, more than a quarter of a million female drug arrests were reported, accounting for 18% of all female arrests for drug law violations (BJS, 1999). Given that a substantial proportion of females being sentenced to prison are characterized by substance abuse (Austin et al., 2001), it is important to determine whether this need also acts as a risk factor to prison adjustment and misconduct.

Relationships With Significant Others

Proponents of "gender-responsive" approaches also focus on needs that do not fall under the rubric of physical or mental health. With calls for holistic and comprehensive approaches to the treatment of offenders, additional needs relating to relationships, self-concept, parenting, and child-rearing warrant consideration (Bloom et al, 2003). For example, deeply rooted in feminist scholarship is the notion that most aspects of women's lives are contextualized according to their relationships with others (Gilligan, 1982; Miller, 1976). According to relational theory, a woman's identity, self-worth, and sense of empowerment are said to be defined by the quality of relationship she has with others (Gilligan, 1982; Kaplan, 1984; Miller, 1976; Miller & Stiver, 1998).

Research indicates that women are more relational than men and tend to place great emphasis on the importance of developing and maintaining healthy and supportive relationships with others in their lives (Bloom et al., 2003). Female offenders are no different. However, because of the high rates of abuse and trauma experienced by female offenders, their ability to achieve healthy relationships may be severely limited (Covington, 1998). Relationships characterized by high levels of conflict and dysfunction between partners and low levels of support may influence women's criminality prior to, during, or after incarceration.

In fact, Salisbury et al. (in press) found that women whose relationships were characterized by high codependency incurred more misconducts while incarcerated, whereas relationships characterized by low codependency decreased the likelihood that a woman would have problems adjusting to prison. Many women offenders may engage in relationships that facilitate their criminal behavior (Koons et al., 1997; Richie, 1996), may be involved in abusive relationships (Bloom et al., 2003; BJS, 1999), or may turn to substance abuse as a result of problems with their inmate relationship (Langan & Pelissier, 2001; Peters et al., 1997). All of these factors have been hypothesized to relate to women offenders' criminal behavior.

Institutional misbehavior can also be influenced by the nature of women's relationships with significant others on the outside. Support from family members may be important in this regard. Emotional support, warmth, contact, and encouragement from family members may alleviate some of the strife that incarceration may bring on women offenders; however, limited support from or high conflict with family members may also make adjustment more difficult.

Parenting

Relationships with children may also affect women's behavior while institutionalized. This is an important issue to consider given that the prevalence of women offenders with children is so high; female offenders in the criminal justice system are more likely than male offenders to be the primary caregiver for dependent children prior to and immediately after their experience with the criminal justice system (Bloom et al., 2003; Mumola, 2000). In fact, more than 70% of women under supervision in the criminal justice system are mothers to minor and dependent children, whereas more than 40% of those women are single and often experience no help from intimate others in raising those children (Bloom et al., 2003).

Thus, concern for children may loom as a major source of anxiety among incarcerated women. Women offenders with dependent children may feel overwhelmed and worry about their ability to ensure the safety and security of their children while incarcerated (Greene et al., 2000). Furthermore, they may worry about their ability to manage their children and provide for their needs on release. Whether or not such problems affect institutional misconduct is still being investigated; however, much research indicates that access to children and family are focal concerns for women (Fogel & Martin, 1992; Koons et al., 1997; Warren et al., 2004). Despite such evidence, Salisbury et al. (in press) found no significant relationship between parental stress and institutional misconduct.[7]

Self-Esteem and Self-Efficacy

A significant amount of research has addressed whether self-esteem is a dynamic risk factor. Most results from these studies have shown that low self-esteem, often aggregated into the category of personal distress, was not a risk factor for recidivism and that programs targeting self-esteem were not promising (Andrews & Bonta, 2003). In fact, some programs actually increased the likelihood of recidivism (Andrews, 1983; Andrews, Bonta, & Hoge, 1990; Gendreau et al., 1996; Wormith, 1984).

Again, the majority of these studies focused on male offenders. The gender-responsive literature emphasizes the importance of self-esteem and self-efficacy in that high levels of each aids women in taking control of their lives and circumstances (Task Force on Federally Sentenced Women, 1990). Such needs are often cited by correctional treatment staff, researchers, and women offenders themselves as critical to their desistance (Carp & Schade, 1992; Case & Fasenfest, 2004; Chandler & Kassebaum, 1994; Koons et al., 1997; Morash, Bynum, & Koons, 1998; Prendergast, Wellisch, & Falkin, 1995; Schram

& Morash, 2002; Task Force on Federally Sentenced Women, 1990).

Gender-responsive scholars contend that trauma, victimization, and abusive relationships may contribute to lower self-concept, self-esteem, and feelings of self-efficacy and self-worth (Bloom et al., 2003). In support, the psychological literature puts forward a large body of knowledge showing negative associations between women's abusive experiences and self-esteem among women in the general population (Aguilar & Nightingale, 1994; Cascardi & O'Leary, 1992; Clements, Ogle, & Sabourin, 2005; Clements, Sabourin, & Spiby, 2004; Orava, McLeod, & Sharpe, 1996; Resick, 1993; Williams & Mickelson, 2004; Zlotnick, Johnson, & Kohn, 2006). However, whether women's self-esteem, in turn, is related to their institutional misconduct is understudied.

Likewise, little is known about the importance of self-efficacy to institutionalized women offenders, although it has been suggested as playing a major role (Rumgay, 2004). Self-efficacy reflects a person's confidence in achieving her or his specific goals. Although high self-efficacy may function as a protective factor in the community (e.g., by increasing the likelihood of goal attainment), it may operate as a risk factor for prison misbehavior. This is because self-efficacious women may be more likely to question institutional authority, thereby instigating citations from staff who have difficulty managing female inmates. Indeed, Salisbury et al. (in press) found support for self-efficacy increasing the likelihood of prison misconducts but decreasing the likelihood of community recidivism.

Current Institutional Classification Systems: Problems for Women Offenders

Increased attention to the gender-responsive needs has brought gender disparity to the forefront of research and practice, especially with respect to women's prisons. Scholars suggested that institutional classification systems that are not gender sensitive overclassify female offenders and do not adequately identify or treat their needs. As such, current evidence indicates that prison classification systems work better for male offenders than for female offenders (Bloom et al., 2003; Hardyman & Van Voorhis, 2004). That this situation has not been corrected is largely attributable to the fact that most states have not validated their classification systems on women offenders (Van Voorhis & Presser, 2001).

Institutional custody classification systems currently focus on factors relating to prior record, seriousness of the current offense, history of violent offenses, and age to assess risk (Brennan & Austin, 1997). Within prison settings, *risk* refers to the degree to which an offender poses a threat to himself or herself, other offenders, prison workers, or the secure management of a correctional facility. Custody classification assessments based on risk inform custody-level placement, which allows prison administrators to allocate resources properly, determine eligibility for and access to programs, determine appropriate housing and cellmate assignments, and maintain safety and security within prison by protecting prisoners against self-inflicted violence and victimization from other prisoners (Warren et al., 2004).

Overclassification occurs when women are placed into higher risk/custody categories than is warranted by their behavior. Overclassification sometimes occurs when the same cutpoints for differentiating custody levels are applied to men and women. Women's scores typically have to be higher than men's before a given custody level (e.g., maximum) shows similar rates of misconduct for men and women. Of course, if the custody assessment is not valid to begin with, the problem cannot be corrected simply by changing cutoff scores. Overclassification can be detrimental for females because their inflated custody score may lead to excessive and inappropriate custody measures, such as limited movement, more restraints, inappropriate housing, and inappropriate programming (Brennan, 1998). Overclassification is evident by staff overrides of custody scores; in a recent survey of state classification systems, Van Voorhis and Presser (2001) found that 20% of state correctional agencies used overrides between 18% and 70% of the time when classifying their female offenders.

Integrating needs into the institutional custody assessment practices is a prospect that stands in stark contrast to the current custody classification process. However, doing so appears to improve the prediction of women's prison misconducts (Salisbury et al., in press). Including the assessment of needs in risk-based classification systems is not a new idea in the assessment and classification literature; however, it is something that has not been widely considered in prison assessments. Early assessments used in community and institutional settings distinguished between the assessment of risk and the assessment of needs (Van Voorhis, 2004). Early risk assessments included static variables linked to criminal history and current offense behavior (Bonta, 1996). In this sense, community risk assessments looked much like current custody classification systems. A second distinct assessment was used to measure needs so that offenders could be referred to programs related to educational, employment, substance abuse, mental health, or family problems (Lerner, Arling, & Baird, 1986). More recently, researchers have found that certain needs are also predictive of recidivism (Andrews et al., 1990). The most recent generation of risk assessment instruments, known as dynamic risk/need assessments, include the assessment of static risk factors (e.g., measures of prior criminal history and the seriousness of the current offense) as well as criminogenic needs (e.g., education difficulties and substance abuse) to predict an offender's likelihood of future criminal behavior.

Again, though community agencies have largely integrated dynamic risk/needs assessment

in their operations, prisons have been slower to include measures of dynamic needs in their assessments, preferring to rely on static measures of criminal history (Hardyman & Van Voorhis, 2004). However, with prisoner reentry initiatives (Petersila, 2003; Travis, 2005) and the notion that offender's needs affect one's risk of reoffending on release, a number of states are beginning to use dynamic risk/needs assessments in prisons (Salisbury et al., in press). The most commonly used instruments of this type are the Northpointe COMPAS (Brennan et al., 2006) and the Level of Service Inventory—Revised (LSI-R; Andrews & Bonta, 1995). These assessments do not incorporate gender-responsive needs, however.

Given these considerations, including an assessment of needs in prison classification systems might increase the validity of institutional classification systems for women offenders. If cut-points are set appropriately, such assessments may also reduce overclassification and simultaneously inform women's treatment and programming.[8] We examine gender-responsive as well as gender-neutral needs in the current study.

▧ The Current Study

There have been no comprehensive, large-scale, and ongoing empirical investigations into the specific risk factors for women, their unique needs and adjustment to incarceration, or their institutional misconduct rates after lengthy follow-up periods. What evidence does exist in this area of research suggests that gender-responsive needs are prevalent among women offenders. The limited research conducted to date indicates that gender-responsive needs are predictive of prison misconducts, and assessment of these needs improves the prediction of such behavior (Salisbury et al., in press). However, further research is clearly warranted to provide more conclusive statements regarding the importance of needs for women offenders.

To this end, the current study expands on the study conducted by Salisbury et al. (in press) and seeks additional understanding of the role that needs play in women's adjustment to prison. Two research questions are posed: First, do gender-responsive needs function as risk factors to women's institutional misconducts? Second, does the inclusion of gender-responsive needs increase the predictive validity of custody classification among women offenders?

Method

Data collection and analyses were funded by the National Institute of Corrections (NIC), as part of a larger research agenda to improve classification, assessments, and programs for women. The sample consisted of 272 newly admitted women offenders to the Missouri Department of Corrections. All women admitted between February 11, 2004 and July 28, 2004 were asked to participate: of 322 women, 84.5% consented to the research under recruitment and consent procedures approved by the University of Cincinnati's Institutional Review Board. Follow-up data describing the incidence and prevalence of prison misconducts were obtained between August 2004 and July 2005.

Participants

Table 1 describes the demographic characteristics, criminal histories, offense characteristics, and institutional misbehavior for the 272 institutionalized women who participated in the current study. On average, the participants were age 33 years, with the majority being White, followed by African American (79.6% and 19.5%, respectively). Consistent with previous findings regarding the female correctional population (see Bloom et al., 2003), most of the women in this sample had children younger than age 18 years (74.6%), although only 27% were married. Also in line with previous research (see Austin et al. 2001;

BJS, 1999, 2006), 44% of the participants were convicted of drug offenses, with forgery or fraud cited second most frequently (20.6%). Only 10% of incarcerated women committed a violent offense against a person. Of the 272 women offenders, roughly 56% had been convicted of a prior felony, 25% had been previously incarcerated, and 6% had previously engaged in a prior violent offense. Table 1 demonstrates that around 47% of the incarcerated women incurred a serious misconduct 6 months into their prison term, and that increased to almost 52% after 12 months.

Assessment Instruments

Scales derived from one of the two sources were included in the analyses as potential risk factors (predictors) for misconducts. These sources included (a) the Missouri Women's Risk Assessment interview created by the Missouri Women's Issues Committee in conjunction with the University of Cincinnati and National Institute of Corrections and (b) the Trailer, a self-report, paper-and-pencil instrument created by the University of Cincinnati staff. A more detailed description of each assessment follows.

Missouri Women's Risk Assessment The Missouri Women's Risk Assessment is an intake interview that was created by the Missouri Women's Issues Committee as a way to integrate gender-specific questions into Missouri's custody classification system. Twelve subscales make up the Women's Risk Assessment; these subscales assess areas regarding women's criminal history, family lives, relationships, parenting issues, substance use or abuse, economic issues, mental health issues, friends outside of prison, anger, educational and employment attainments, adult and childhood victimization, and criminal attitudes. This interview incorporated gender-responsive questions and gender-neutral items that are used in various other assessment tools (e.g., the LSI-R or Northpointe COMPAS).

Table 1	Sample Descriptive Statistics, Missouri Prison ($N = 272$)	

Characteristic	n	Percentage
Participant age		
18–20 years old	12	4.5
21–30 years old	86	32.2
31–40 years old	110	41.2
41–50 years old	55	20.6
51 years and older	4	1.5
$M = 33.8$ years ($SD = 8.3$)		
Participant race		
White	211	79.6
African American	53	19.5
Asian	1	0.4
Indian	1	0.4
Participant currently married		
Yes	74	27.2
Participant has children younger than age 18 years		
Yes	203	74.6
Participant employment		
Employed full- or part-time	228	84.8
Unemployed	41	15.2
Participant holds high school diploma		
Yes	155	57.0
Current offense		
Drug-related offense	121	44.5
Forgery/fraud offense	56	20.6
Property offense	30	11.0
Violent offense	28	10.3
DUI/DWI/motor vehicle offense	23	8.5
Prior felonies		
Yes	145	55.6
None	116	44.4
1–2	111	42.5
3–5	30	11.5
6 or more	4	1.5

(Continued)

Table 1	(Continued)	
Characteristic	**n**	**Percentage**
M = 2.0 felonies (SD = 1.6)		
Prior incarcerations		
Yes	69	25.4
M = 1.4 terms (SD = 1.0)		
Prior violent offense		
Yes	15	5.5
6-month misconducts		
Yes	129	47.4
M = 1.00 misconducts (SD = 1.43)		
12-month misconducts		
Yes	141	51.8
M = 1.39 misconducts (SD = 1.98)		

NOTE: DUI = driving under the influence; DWI = driving while intoxicated.

Gender-Responsive "Trailer" The "Trailer" is a self-report survey that was created by University of Cincinnati research staff to measure gender-responsive needs of women offenders. The survey comprises multiple subscales; each asks several questions to tap an underlying domain. These domains pertain to self-esteem, self-efficacy, parenting and relationship problems, and childhood and adult victimization.

Measures

Dependent variables All outcome variables used in the analyses are described in Table 1. The dependent variables were intended to tap institutional adjustment as measured by serious prison misconducts. In this case, serious misconducts excluded minor rule violations such as being in unauthorized areas. These measures were collected 6 and 12 months after intake and are reported as incidence (frequency) and prevalence (presence/absence) measures.

Gender-neutral independent variables The mean, standard deviation, and ranges for the

scales tested in this study are provided in Table 2. For the ease of presentation, the subscales have been designated as either gender-neutral scales or gender-responsive scales. The gender-neutral scales reflect domains in offenders' lives that are often incorporated in risk and needs assessment tools, such as the LSI-R, and have been shown to be predictive of criminal behavior among males and females (e.g., Andrews & Bonta, 2003; Gendreau et al., 1996; Simourd & Andrews, 1994). The gender-responsive scales were designed to reflect those areas in women's lives that may be particularly important to their criminal behavior and institutional misconduct, such as self-concept, trauma or victimization, relationships, and mental health problems.

The scales presented here were identified through factor analyses using principle component extraction with varimax rotation. Final scales were created through principle component analysis of the selected items. Scales are coded so that higher scores reflect the presence of a risk factor; to accommodate differences in ranges among the scales, all individual measures

Table 2	Descriptive Statistics for Assessment Scales, Missouri: Prison Sample ($N = 272$)		
Scale Item	**M**	**SD**	**Range**
Gender-Neutral Scales			
Antisocial attitudes	1.49	2.04	0–7
Antisocial friends	2.18	1.61	0–5
Low education	3.34	1.62	0–5
Employment/financial difficulties	3.34	1.88	0–8
High family conflict	0.80	0.87	0–4
Low family support	2.13	1.83	0–6
Static substance abuse	5.90	3.00	0–10
Dynamic substance abuse	2.27	1.51	0–5
History of mental illness	2.43	1.91	0–6
Anger control	1.54	1.55	0–7
Gender-Responsive Scales			
Low self-esteem	1.36	1.16	0–3
Low self-efficacy	1.45	1.11	0–3
Childhood abuse	1.29	1.12	0–3
Adult emotional abuse	1.33	1.07	0–3
Adult physical abuse	1.47	1.14	0–3
Adult harassment	1.36	1.11	0–3
Low relationship support	5.86	3.32	0–10
High relationship conflict	0.89	1.18	0–5
High relationship dysfunction	2.72	2.52	0–10
Parental stress ($N = 203$)	1.40	1.06	0–3
Current depression/anxiety	2.00	1.99	0–6
Current psychosis	0.08	0.32	0–2
Risk Scale			
Institutional risk	1.02	1.01	0–5
Needs Scales			
Gender-neutral needs	11.73	5.46	1–30
Gender-responsive needs	9.22	4.36	1–19
Modified Risk/Needs Scales			
Gender-neutral risk/needs	12.80	5.70	1–31
Gender-responsive risk/needs	10.28	4.56	1–22
Final Scale			
Gender-neutral and gender-responsive risk/needs	22.09	8.35	3–46

with ranges higher than 0 to 10 were divided into quartiles.

The Antisocial Attitudes scale was designed to assess the degree to which an offender had internalized criminal values or denied responsibility for her actions. Seven items pertaining to attitudes such as harm minimization, denial of responsibility, and blaming others were included in this scale. The summed items resulted in a scale with an eigenvalue of 3.91 and an alpha reliability of .87.

Antisocial Friends scale included six items (eigenvalue = 2.29, alpha = .70) to assess whether the offender associated with friends who engaged in criminal behavior. Questions relating to whether the participant had friends outside of prison who had been incarcerated or been in trouble with the law made up this scale.

Educational issues were tapped by a four-item scale incorporating questions about whether the offender had difficulty reading and writing, had learning disabilities, or never graduated from high school or received her General Equivalency Diploma (GED). The scale produced an eigenvalue of 2.12 and an alpha reliability of .66, which was marginal.

Employment and financial difficulties were measured with eight items (eigenvalue = 2.16, alpha = .61). This scale comprised questions relating to whether participants had difficulty finding or keeping a job, paying their bills, and supporting themselves.

Family problems were measured with the Family Conflict and Family Support scales. The Family Conflict scale consisted of three items indicating that there was much conflict, criminality of other family members, and the family's refusal to communicate with the inmate. Factor loadings for these items were high (eigenvalue = 1.28); however, the alpha for the scale was unacceptably low (.29). The items did, however, form a Guttman scale with a coefficient of reproducibility equal to .83, so the scale was retained for further analysis. The Family Support scale included five items (eigenvalue = 2.50, alpha = .73) that measured how supportive an offender's family members

had been during incarceration; questions regarding whether family members had visited or helped the woman while incarcerated and were willing to help after the prison term were included in this scale.

Data reduction analyses produced two substance abuse factors, a 10-item History of Substance Abuse scale measuring past substance use or abuse (eigenvalue = 4.63, alpha = .86), and a 5-item Dynamic Substance Abuse scale (eigenvalue = 2.14, alpha = .66). The History of Substance Abuse scale comprised items pertaining to prior substance-related offenses, prior drug treatment, and whether the use of drugs affected daily life. The Dynamic Substance Abuse scale assessed the degree to which substance use presented a problem for an offender within 6 months prior to her incarceration and incorporated questions relating to whether the offender associated with other substance users, missed treatment programs, or was violated for using substances.

Mental illness has also been incorporated in gender-neutral needs assessments and is often denoted as a personal distress variable. The 6-item History of Mental Illness scale used in the current study was designed to evaluate whether an offender had ever experienced delusions, attempted suicide, been hospitalized, received medication, or been diagnosed with a mental illness (eigenvalue = 3.02, alpha = 0.80).

Anger Control scale (eigenvalue = 2.25, alpha = 0.62) measured the degree to which women reported difficulties managing their anger. The scale consisted of seven questions related to whether the participants felt they had strong tempers or engaged in physical violence toward others when upset or angry, and whether such behaviors ever resulted in law enforcement involvement.

Gender-Responsive Independent Variables

The Self-Esteem scale was based on the Rosenberg Self-Esteem Scale (Rosenberg, 1979) and consisted of 10 items tapping the degree to which participants feel positive feelings about themselves, such as self-respect, self-worth, and self-satisfaction (eigenvalue = 5.29, alpha = .90). The purpose of the Self-Efficacy scale was to measure the degree to which participants felt that they were capable of achieving their goals and dealing with problems in their lives. This 17-item scale was based on the Sherer Self-Efficacy Scale (Sherer et al., 1982) (eigenvalue = 7.01, alpha = .91).

Abuse and victimization were measured with the Childhood Abuse, Adult Emotional Abuse, Adult Physical Abuse, and Adult Harassment scales. These scales were informed by the writings of Crowley and Dill (1992), Fischer, Spann, and Crawford (1991), and Roehling and Gaumond (1996). The 19-item Childhood Abuse scale (eigenvalue = 10.92, alpha = .95) was designed to assess the degree to which a participant experienced physical and emotional abuse as a child. Questions included whether the participant had been pushed, kicked, beaten, dragged, choked, and burned, as well as forced to do something embarrassing, or insulted or ridiculed, among other things during childhood. The 17-item Adult Emotional Abuse scale (eigenvalue = 11.37, alpha = .97) measured the degree to which participants had been controlled, insulted, humiliated, disrespected, and harassed by others during adulthood. The purpose of the Adult Physical Abuse scale was to determine the degree of physical abuse experienced by the participant as an adult. Fifteen items made up this scale; questions relating to physical violence such as being kicked, beaten, dragged, scratched, and choked, as well as being threatened with weapons were used (eigenvalue = 10.26, alpha = .96). Finally, the Adult Harassment scale tapped participants' experience of harassment, such as being stalked or followed, as well as having a restraining order violated and having their home broken into. Eleven items made up this scale (eigenvalue = 6.71, alpha = .93).

Several scales were created to measure relationships with intimate partners, including two from the interview and one from the self-report Trailer. The Relationship Support scale (eigenvalue = 5.07, alpha = .86) consisted of seven items relating to whether participants' significant other was encouraging of treatment, as well as their expected level of support and help on release. High scores on this scale reflect little support in relationships. The Relationship Conflict scale was designed to tap the amount of conflict and control within the relationship. This five-item scale produced an eigenvalue of 2.21 and an alpha reliability of .66. Finally, the six-item Relationship Dysfunction scale (eigenvalue = 2.94, alpha = .77) measured notions of codependency and loss of power while in relationships. Its development was informed by Crowley and Dill (1992), Fischer et al. (1991), and Roehling and Gaumond (1996).

The Parental Stress scale was based on the scale developed by Avison, Turner, and Noh (1986). Modifications were made to the scale to include 12 items that measured the degree that women felt that their lives were out of control, their children were unmanageable, and they received little to no support from family members or significant others (eigenvalue = 4.31, alpha = .82).

Variables measuring mental illness in gender-neutral needs and risk/needs assessment have potentially masked the effect of specific mental illnesses such as depression and psychosis among women offenders. We examined depression, anxiety, and psychosis as gender-responsive needs to assess whether these specific measures of mental illness were more important to women offenders than less specific measures of mental illness. The six-item Current Depression/Anxiety scale (eigenvalue = 3.13, alpha = .82) measured the degree to which participants were currently experiencing symptoms of depression and anxiety. Questions pertaining to loss of appetite and worry interfering with daily functioning were incorporated in this scale. The purpose of the two-item Current Psychosis scale ($r = .36$, $p < .001$) was used to assess whether participants were presently experiencing delusions or having thoughts that others are out to harm them.

Institutional risk scale The Institutional Risk Assessment scale was designed to reflect custody assessment tools that are used in many prison institutions throughout the United States. This scale summed six items pertaining to the severity of the current offense, history of violence, history of escapes, multiple prior felonies, prior violent offenses, prior incarcerations, and forms of noncompliance during prior terms of correctional supervision (eigenvalue = 2.37, alpha = .63). The low alpha improved to .70 when items pertaining to current and prior assaults were removed from the scale. However, that decision would not be acceptable to correctional managers charged with supervising high-stakes offenders.

Needs scales The current study examines whether institutional classification systems used today benefit from the assessment of needs. Although we are primarily interested in the importance of gender-responsive needs among female offenders, we include an examination of the significance of gender-neutral needs among women offenders as well. The Gender-Neutral Needs scale is the composite of the significant gender-neutral needs. Only items that reached significance at the $p < .05$ level when correlated with the institutional misconduct measures at the bivariate level were included in this scale. The total Gender-Neutral Needs assessment scale summed the totals of six gender-neutral scales, including the Antisocial Attitudes, Employment/ Financial Difficulties, Family Conflict, Family Support, Mental Illness, and Anger Control scales. The Gender-Responsive Needs scale is the composite of the gender-responsive needs that were significantly correlated at the $p < .05$ level with any of the institutional misconduct measures. Thus, this scale summed the totals of four of the gender-responsive scales, including the Childhood Abuse, Relationship Support, Depression/Anxiety, and Psychosis scales.

Modified risk/needs scales The Modified Risk/Needs Scales incorporated risk and need factors to predict institutional misconducts. The Gender-Neutral Risk/Needs scale was designed to measure an offender's criminal risk level and gender-neutral needs. The Institutional Risk and the Gender-Neutral Needs scales were combined to create this measure. This scale summed the totals of the severity of the current offense, history of violence, history of escapes, multiple prior felonies, prior violent offenses, prior incarcerations, antisocial attitudes, employment/ financial difficulties, family conflict, family support, mental illness, and anger control.

The Gender-Responsive Risk/Needs scale This scale was designed to measure an offender's criminal risk level and gender-responsive needs. The Institutional Risk and Gender-Responsive Needs scales were combined to create this measure. This scale summed the totals of the severity of the current offense, history of violence, history of escapes, multiple prior felonies, prior violent offenses, prior incarcerations, childhood abuse, low relationship support, current depression or anxiety, and current psychosis scales.

Final scale The Gender-Neutral and Gender-Responsive Risk/Needs scale was designed to measure the degree to which the inclusion of gender-responsive and gender-neutral needs with institutional risk factors increased the predictive validity of such tools. Therefore, the Institutional Risk scale was combined with the gender-neutral needs assessment and the gender-responsive needs assessment. This scale summed the totals of the severity of the current offense, history of violence, history of escapes, multiple prior felonies, prior violent offenses, prior incarcerations, antisocial attitudes, employment/financial difficulties, family conflict, family support, mental illness, anger control, childhood abuse, low relationship support, current depression or anxiety, and current psychosis scales.

◪ Results

Results of this study are shown in Tables 3 and 4. The first goal of this research was to determine whether certain gender-responsive needs function as risk factors to institutional adjustment. Table 3 presents the bivariate relationships between gender-neutral and gender-responsive needs and institutional misconducts. As can be seen, many gender-neutral and gender-responsive needs are highly correlated with 6- and 12-month institutional misconducts. Gender-responsive needs such as *experiencing childhood abuse, depression or anxiety, psychosis,* and *involvement in unsupportive relationships* were highly correlated with all measures (e.g., prevalence and incidence) of institutional misconducts. Experiencing childhood abuse increased the likelihood of women engaging in institutional misconduct within 6 and 12 months of incarceration (correlation coefficients ranging from $r = .20$ to $r = .25$, all significant at $p < .01$), as does having an unsupportive significant other on the outside ($r = .10$ to $r = .16$, significance at all levels). Currently experiencing depression, anxiety, or psychosis also dramatically increased the likelihood of institutional misconducts (correlation coefficients ranging from $r = .13$, $p < .05$ to $r = .23$, $p < .01$ for depression and anxiety, and $r = .16$ to $r = .31$, all significant at $p < .01$ for psychosis). Parental stress was marginally correlated with 6-month institutional misconducts ($r = .09$ for the number of 6-month misconducts; $r = .10$ for the occurrence of any 6-month misconducts, both significant at $p < .10$), but not with 12-month misconducts. Likewise, experiencing harassment by others as an adult was significantly correlated with the number of 6-month misconducts ($r = .08$, $p < .10$), and dysfunctional relationships were significantly correlated with the prevalence of 6-month misconducts ($r = .09$, $p < .10$); however, these were relatively weak relationships and did not hold with any other outcomes. In general, the coefficients for the gender-responsive needs

were as strong as or stronger than the coefficients among the gender-neutral needs. In their relationships with institutional misconducts, gender-neutral need correlation coefficients ranged from $r = .09$ to $r = .20$, whereas the gender-responsive need coefficients ranged from $r = .09$ to $r = .31$.

This is not to imply that gender-neutral needs were not predictive of institutional outcomes. Indeed, they were; gender-neutral need factors pertaining to antisocial attitudes, employment and financial difficulties, conflict with family members, limited family support, a history of mental illness, and limited anger control were highly predictive of institutional misconducts during 6- and 12-month periods. Having antisocial attitudes while incarcerated increased the likelihood that women would engage in institutional misconduct (correlation coefficients ranging from $r = .14$, $p < .05$ to $r = .18$, $p < .01$). Employment and financial difficulties prior to incarceration increased the incidents of 6- and 12-month misconducts ($r = .10$, $p < .05$ and $r = .09$, $p < .10$ for the number of 6- and 12-month misconducts, respectively). High family conflict and little to no family support also increased the chances that a woman would incur institutional misconducts (correlation coefficients ranging from $r = .12$ to $r = .19$, significant at $p < .05$ and $p < .01$, respectively, for high family conflict, and $r = .12$ to $r = .20$, significant at $p < .05$ and $p < .01$, respectively, for low family support). Having experienced previous indicators of mental illnesses was also predictive of institutional misbehavior (correlation coefficients ranging from $r = .11$ to $r = .19$, significant at $p < .05$ and $p < .01$, respectively). Anger control was predictive of the number of misconducts ($r = .12$ and $r = .13$, $p < .05$, for 6- and 12-month misconducts, respectively) and the prevalence of six-month misconducts ($r = .09$, $p < .10$).

The second objective of the current study was to determine whether the inclusion of gender-responsive needs increased the predictive validity of institutional classification

Table 3	Relationships Between Gender-Neutral Assessment Scales, Gender-Responsive Assessment Scales, and Prison Misconducts, Missouri Prison Sample (Pearson r, one-tailed)

	6-Month Outcomes		12-Month Outcomes	
Assessments and Subscales	# Misconducts	Any Misconducts	# Misconducts	Any Misconducts
Gender-neutral scales				
Antisocial attitudes	.16***	.18***	.14**	.15***
Antisocial friends	—	—	—	—
Low education	—	—	—	—
Employment/financial difficulties	.10**	—	.09*	—
High family conflict	.18***	.14***	.19***	.12**
Low family support	.19***	.15***	.20***	.12**
Static substance abuse	—	—	—	—
Dynamic substance abuse	—	—	—	—
History of mental illness	.12**	.11**	.19***	.13**
Low anger control	.12**	.09*	.13**	—
Gender-responsive scales				
Low self-esteem	—	—	—	—
Low self-efficacy	—	—	—	—
Childhood abuse	.25***	.22***	.22***	.20***
Adult emotional abuse	—	—	—	—
Adult physical abuse	—	—	—	—
Adult harassment	.08*	—	—	—
Low relationship support	.10*	.16***	.13**	.16***
High relationship conflict	−.09*	—	−.16***	−.09*
High relationship dysfunction	—	.09*	—	—
Parental stress ($N = 203$)	.09*	.10*	—	—
Current depression/anxiety	.20***	.14**	.23***	.13**
Current psychosis	.26***	.19***	.31***	.16***

*$p < .10$; **$p < .05$; ***$p < .01$.

systems that are often used today. This was accomplished through a three-step process. First, total risk and needs scales were created. These scales, the Institutional Risk scale, Gender-Neutral Needs Scale, and Gender-Responsive Needs Scale, were described in the Method section. Each scale was correlated with 6- and 12-month prevalence and incidence measures of institutional misconducts. Second, Modified Risk/Needs Scales were created. The Institutional Risk scale was combined with the Gender-Neutral Needs scale to create the Gender-Neutral Risk/Needs scale; this scale determined the relative importance that gender-neutral needs play in predicting misconducts. The Gender-Responsive Needs

	6-Month Outcomes		12-Month Outcomes	
Assessments and Subscales	**# Misconducts**	**Any Misconducts**	**# Misconducts**	**Any Misconducts**
Risk Scale				
Institutional Risk Scale[a]	.11**	.16***	.23***	.17***
Needs Scales				
Gender-Neutral Needs Scale[b]	.26***	.22***	.28***	.19***
Gender-Responsive Needs Scale[c]	.25***	.25***	.28***	.25***
Modified Risk/Needs Scales				
Gender-Neutral Risk/Needs Scale[d]	.29***	.26***	.33***	.23***
Gender-Responsive Risk/Needs Scale[e]	.27***	.28***	.34***	.27***
Final Scale				
Gender-Neutral and Gender-Responsive Risk/Needs Scale[f]	.33***	.31***	.38***	.28***

Table 4 Comparison of Risk, Need, and Gender-Responsive Assessment Scales, Missouri Prison Sample (Pearson *r*, one-tailed)

a. Scale includes factors pertaining to severity of the current offense, history of violence, prior escapes, prior felonies, prior violent offenses, and prior incarcerations.

b. Scale includes gender-neutral needs pertaining to antisocial attitudes, employment/financial difficulties, high family conflict, low family support, mental illness, and low anger control.

c. Scale includes gender-responsive needs pertaining to childhood abuse, low relationship support, depression/anxiety, and psychosis.

d. Scale includes all factors in the Institutional Risk Scale plus the gender-neutral needs included in the Gender-Neutral Needs Scale.

e. Scale includes all factors in the Institutional Risk Scale plus the gender-responsive needs included in the Gender-Responsive Needs Scale.

f. Scale includes all factors in the Institutional Risk Scale and the Gender-Neutral Needs Scale, plus gender-responsive needs included in the Gender-Responsive Needs Scale.

$^*p < .10$; $^{**}p < .05$; $^{***}p < .01$.

scale was combined with the Institutional Risk scale to determine the importance of gender-responsive needs in predicting institutional misconducts; this scale is denoted as the Gender-Responsive Risk/Needs scale. Last, a final scale assessing gender-neutral needs, gender-responsive needs, and risk factors was created. The Institutional Risk scale was combined with the Gender-Responsive Needs scale and the Gender-Neutral Needs scale to determine the importance that gender-responsive needs play in addition to gender-neutral risk and need factors in predicting institutional

misconducts. This scale is denoted as the Gender-Neutral and Gender-Responsive Risk/Needs scale.

Table 4 illustrates the results of the above analyses. There are five important results evident in this table. First, the traditionally used institutional assessment was a comparatively weak predictor of institutional misconduct among women offenders (correlations ranging from $r = .11$ to $r = .23$, significant at $p < .05$ and $p < .01$, respectively).

Second, Gender-Neutral and Gender-Responsive Needs were more important than the Institutional Risk scale in predicting institutional misconduct. That is, by themselves, needs assessments were somewhat stronger predictors of institutional misconducts than risk assessments currently being used by many correctional agencies.

A third finding evident in Table 4 is that the predictive power of institutional misconducts was greatly increased when needs were added to the assessment of risk. For instance, when needs were added to the Institutional Risk scale, the predictive power of the new scales (i.e., the Gender-Neutral Risk/Needs Scale and the Gender-Responsive Risk/Needs scale) increased. This increase went beyond the assessment of gender-neutral or gender-responsive needs only. The strengths of these relationships were quite strong; institutional misconduct and gender-neutral risk/needs and gender-responsive risk/needs were strongly related (correlations ranging from $r = .23$ to $r = .33$ for the Gender-Neutral Risk/Needs scale and $r = .27$ to $r = .34$ for Gender-Responsive Risk/Needs scale). Thus, the assessment of gender-responsive needs in addition to traditional risk factors seems quite promising, given the results provided here.

A fourth noteworthy finding from the current study is that gender-responsive needs were important to consider when predicting institutional misconducts. Although gender-responsive needs and gender-neutral needs performed at similar levels, the correlations between gender-responsive needs and institutional misconducts appear to be more consistent than correlations between gender-neutral needs and prison misconducts (correlation coefficients ranging from $r = .19$ to $r = .28$ for the Gender-Neutral Needs scale compared to correlations ranging between $r = .25$ to $r = .28$ for the Gender-Responsive Needs scale), and they increase the predictive power of risk assessments slightly more than gender-neutral needs do ($r = .27$ to $r = .34$ for the Gender-Responsive Risk/Needs scale compared to $r = .23$ to $r = .33$ for the Gender-Neutral Risk/Needs scale). Thus, it appears that gender-responsive

needs are, in fact, important factors to consider when predicting institutional misconducts.

Finally, the inclusion of gender-responsive needs in risk assessments with gender-neutral needs yielded the strongest relationship with institutional misconducts and increased the predictive power of such behavior beyond the assessment of risk, needs, and gender-neutral or gender-responsive risk/needs alone. Certainly, the prediction of all types of institutional misconducts was increased when risk and gender-neutral as well as gender-responsive needs were included; relationships between the Gender-Neutral and Gender-Responsive Risk/Needs scale and institutional outcomes were stronger than all other relationships presented in Table 4 (correlations ranging from $r = .28$ to $r = .38$, all significant at $p < .01$). This presents convincing evidence that needs, gender-neutral and gender-responsive, are important to consider when predicting institutional misconducts.

⊠ Discussion

Results from the current study indicate that gender-responsive needs are indeed predictive of institutional misconducts. Furthermore, these gender-responsive needs performed as well as and, in some instances, slightly better than gender-neutral needs when predicting institutional misbehavior. In particular, childhood abuse, unsupportive relationships, experiencing anxiety or depression, and psychosis were highly related to the likelihood that a woman might incur institutional misconducts within 6 and 12 months of incarceration. Other research has also found support for child abuse as a risk factor. Salisbury et al. (in press) found that childhood abuse was predictive of institutional misconducts ($r = .16$, $p < .05$), though it was not so with community recidivism outcomes. Thus, a pattern appears to be emerging with regard to the effect of child abuse on women offenders, particularly in institutional settings. Women who experienced abuse as children may be at risk for prison misconducts because they are acutely sensitive to the traumatizing aspects of prison life. These results highlight the importance of implementing trauma-informed protocols and services in women's prisons.[9]

Lack of support from significant others outside of prison also appeared to be quite critical in identifying women who have difficulty adapting to the institutional environment. It is important that women have a satisfying relationship with their partner, as well as the expectation of continued support on their release. This is consistent with relational theory and pathways research that emphasize the significant impact of relationships in women's lives (Gilligan, 1982; Miller, 1976). Such findings may also translate into the need for supportive relationships inside the institution, from staff and other inmates.

It is interesting to note that women who reported high relationship conflict at intake actually incurred fewer misconducts than women with lower levels of relationship conflict. At first glance, this appears to contradict our findings related to supportive intimate relationships. However, one explanation for this unexpected finding may be that women who experienced conflict-ridden relationships (characterized by power and control and resulting in physical violence) actually felt more behavioral stability, and perhaps even safety, once admitted to prison as a result of being removed from their current relational situation. Such relationships may be more pertinent as a risk factor for women in the community, or it may be an important element in establishing a pathway toward offending (Koons et al., 1997; Richie, 1996), perhaps exhibiting an indirect relationship with crime through other risk factors (Salisbury & Van Voorhis, 2007).

Results regarding women's current mental health, specifically *depression, anxiety,* and *psychosis,* were consistent with previous research under the pathways perspective (Bloom et al., 2003; Covington, 1998; McClellan et al., 1997). Women's mental health needs cannot be

overlooked as risk factors for prison adjustment; adequate treatment for women's mental illnesses is essential.

Parental stress, dysfunctional relationships, and experiencing adult harassment were also predictive of 6-month institutional misconducts; however, the importance of these gender-responsive needs to prison adjustment appears to be marginal. Because these needs were measured during the intake process, it may be that they are more critical to prison adjustment only for an initial short-term period. Recall that parental stress measured the degree that women felt that their lives were out of control, their children were unmanageable, and they received little to no support from family members or significant others. Such stressful aspects of parenting understandably might diminish for women once they become stabilized in prison. It is important to note that our measure of parental stress was not focused on potential child custody stressors, which may be strong predictors of institutional misconducts.

Similarly, the deleterious effects of dysfunctional relationships and harassment appeared to be important only in the short term. Similar results come from Salisbury et al. (in press), who found that relationship dysfunction was predictive of women's misconducts after 6 months. However, the same study indicated that adult harassment was not predictive of institutional misconducts but was predictive of rearrest once released (Salisbury et al., in press).

The findings from the current study also suggest that self-esteem, self-efficacy, adult emotional abuse, and adult physical abuse are not significantly related to institutional misbehavior and thus do not function as risk factors for misconduct. Once again, this is in partial support of the findings reported by Salisbury et al. (in press), who also found that self-esteem and adult physical abuse were not significantly related to institutional misconduct.

Gender-neutral needs relating to antisocial attitudes, employment and financial difficulties, family problems, mental illness, and anger were predictive of institutional misbehavior and thus functioned as risk factors to women offenders' institutional misconduct. Thus, these needs cannot be dismissed as irrelevant to women offenders' risk. On the other hand, several gender-neutral needs were not predictive of women's misconducts after either 6 or 12 months, including antisocial friends, low education, and substance abuse. This suggests that researchers cannot assume that all risk factors pertinent to men are applicable to women.

Findings with respect to substance abuse are not consistent with other studies of women offenders. They may implicate the assessment scales themselves, except for the fact that the scales predicted in other samples are yet to be published. The findings may also be an artifact of the fairly good control Missouri officials had over in-prison substance abuse–related misconducts.

Finally, our analyses of risk and needs assessment scales indicated that risk assessments based primarily on static criminal history measures were relatively weak predictors of institutional misconducts among women offenders. It is unfortunate that this static, offense-based risk assessment is the common classification system in place today for women inmates across the United States (Van Voorhis & Presser, 2001). If most systems are only able to predict female misbehavior marginally well, there is an ethical obligation to attempt to improve them, particularly because classification affects not only custody level but also a variety of additional privileges, including movement around the facility, access to programs, work release, and prerelease/parole decisions (Brennan, 1998; Van Voorhis & Presser, 2001).

Our results demonstrated that needs, gender-neutral and gender-responsive, were more predictive of women's institutional adjustment than offense-based items. It is important to note that we found that the most predictive power was achieved when static, offense-related risk factors were combined with gender-neutral and gender-responsive need factors. The utility of a needs-based institutional classification system for women lies not only in its predictive power but also in its ability to (a) identify women's treatment needs, (b) triage women into appropriate treatment programs, and (c) serve as a seamless tool across supervision settings. We recognize that implementing such a model would require careful policy discussions surrounding the translation of needs into risks. Our intent is certainly not to punish women for having a multitude of needs, nor should it be the intent of any correctional agency. For a needs-based approach to work effectively, institutional settings must (a) be treatment intensive, (b) have competent case management, and (c) strive for wrap-around and reentry services.

Last, results from the current study provide evidence that women's risks and needs do not necessarily emerge from solely "gender-neutral" or "gender-responsive" domains. Factors from both perspectives are relevant, and thus neither perspective should be dismissed as irrelevant to women. Undoubtedly, we still have much more to learn about women's complex lives and the factors that contribute to their success in institutional as well as community settings.

It will be for policy makers and practitioners to sort out the implications of findings such as these. We maintain, however, that gender-responsive risk assessment instruments are best used in treatment-intensive settings, including regional community-based correctional centers focused on wrap-around services and facilities where inmate transition is a priority. These assessments could facilitate continuity of care concerns and efforts to plan prison transition even at the point of prison intake. States that reserve some facilities for intensive programming and others for more limited approaches to low-risk offenders might also benefit from these systems because the assessments also differentiate between high-need inmates and low-need inmates. It would be unacceptable, however, to elevate custody beyond a medium level according to issues pertinent to trauma, mental health, and other needs identified by this research. In addition to rather obvious ethical issues, the need for maximum custody placement of women is being reevaluated by prison scholars and correctional practitioners alike because aggression among women inmates is dramatically lower than rates for male inmates (Hardyman & Van Voorhis, 2004). Finally, the nature of the risk factors observed in this research may also be suggesting that prisons may need to reevaluate policies and conditions that aggravate inmates. Trauma-informed policies, family reunification, improved mental health services, and enhanced staff skills for managing women offenders all appear to be warranted.

≥ Notes

1. Mental health predicted aggressive prison misconducts but did not predict nonaggressive misconducts.

2. Using the tools solely for the purpose of elevating custody according to one's problems would clearly be a misuse of the gender-responsive systems.

3. *Risk/needs classification* refers to the emerging dynamic risk assessment systems where offender outcomes are predicted by needs and criminal history characteristics. Most custody classification systems reach a risk score through the consideration of static criminal history items. As is explained later in this article, the risk/needs assessments are used primarily in community corrections but are valid for institutional corrections as well.

4. The pilot study compared existing custody variables to the variables identified by the Level of Service Inventory-Revised (Andrews & Bonta, 1995) and to the seven gender-responsive variables (mental health, self-esteem, self-efficacy, loss of power in relationships, parental stress, child abuse, and adult victimization). As will be seen, the current study considers a larger array of gender-responsive factors.

5. No multivariate model was conducted by Salisbury, Van Voorhis, and Spiropoulos (in press).

6. Salisbury et al. (in press) found that a composite scale of adult victimization, as well as emotional victimization and harassment, were associated with community rearrests.

7. The variable did, however, correlate with recidivism on release.

8. Valid assessment does not fully resolve the issue of overclassification, however. Researchers must also take special care to set cut-points that effectively differentiate the different risk classifications.

9. For a detailed discussion of trauma-informed services, please see Elliott, Bjelajac, Fallot, Markoff, and Reed (2005).

≥ References

Aguilar, R. J., & Nightingale, N. N. (1994). The impact of specific battering experiences on the self-esteem of abused women. *Journal of Family Violence, 9,* 35–45.

American Psychiatric Association. (1994). *Diagnostic and statistical manual of mental disorders* (4th ed.). Washington, DC: Author.

Andrews, D. A. (1983). Assessment of outcome in correctional samples. In M. J. Lambert, E. R. Christensen, & S. S. DeJulio (Eds.), *Assessment of psychotherapy outcome* (pp. 160–201). New York: John Wiley.

Andrews, D. A., & Bonta, J. (1995). *Level of Service Inventory—Revised.* North Tonawanda, NY: Multi-Health Systems.

Andrews, D. A., & Bonta, J. (2003). *The psychology of criminal conduct* (3rd ed.). Cincinnati, OH: Anderson.

Andrews, D. A., Bonta, J., & Hoge, R. D. (1990). Classification for effective rehabilitation: Rediscovering psychology. *Criminal Justice and Behavior, 17,* 19–52.

Austin, J., Bruce, M. A., Carroll, L., McCall, P. L., & Richards, S. C. (2001). The use of incarceration in the United States: American Society of Criminology National Policy Committee. *Critical Criminology: An International Journal, 10,* 17–41.

Avison, W., Turner, R., & Noh, S. (1986). Screening for problem parenting: Preliminary evidence on a promising instrument. *Child Abuse & Neglect, 10,* 157–170.

Belknap, J., & Holsinger, K. (2006). The gendered nature of risk factors for delinquency. *Feminist Criminology, 1,* 48–71.

Benda, B. B. (2005). Gender differences in life-course theory of recidivism: A survival analysis. *International Journal of Offender Therapy and Comparative Criminology, 49,* 325–342.

Blanchette, K. (1996). *The relationship between criminal history, mental disorder, and recidivism among federally sentenced female offenders.* Unpublished master's thesis, Carleton University, Ottawa, Canada.

Blanchette, K., & Motiuk, L. L. (1995, June). *Female offender risk assessment: The case management strategies approach.* Poster session presented at the Annual Convention of the Canadian Psychological Association, Charlottetown, Prince Edward Island.

Bloom, B., Owen, B., & Covington, S. (2003). *Gender responsive strategies: Research, practice, and guiding principles for women offenders.* Washington, DC: U.S. Department of Justice, National Institute of Corrections.

Blume, S. B. (1990). Chemical dependency in women: Important issues. *American Journal of Drug and Alcohol Abuse, 16,* 297–307.

Bonta, J. (1996). Risk-needs assessment and treatment. In A. T. Harland (Ed.), *Choosing correctional options that work: Defining the demand and evaluating the supply* (pp. 18–32). Thousand Oaks, CA: Sage.

Bonta, J., Pang, B., & Wallace-Capretta, S. (1995). Predictors of recidivism among incarcerated female offenders. *The Prison Journal, 75,* 277–294.

Brennan, T. (1998). Institutional classification of females: Problems and some proposals for reform. In R. T. Zaplin (Ed.), *Female offenders: Critical perspectives and effective interventions* (pp. 179–204). Gaithersburg, MD: Aspen.

Brennan, T., & Austin, J. (1997). *Women in jail: Classification issues.* Washington, DC: U.S. Department of Justice, National Institute of Corrections.

Brennan, T., Dieterich, W., & Oliver, W. (2006). *COMPAS: Technical manual and psychometric report Version 5.0.* Traverse City, MI: Northpointe Institute.

Brown, S. L., & Motiuk, L. L. (2005). *The Dynamic Factor Identification and Analysis (DFIA) component of the Offender Intake Assessment (OIA) process: A meta-analytic, psychometric, and consultative review* (Research Report R-164). Ottawa: Correctional Service Canada.

Browne, A., Miller, B., & Maguin, E. (1999). Prevalence and severity of lifetime physical and sexual victimization among incarcerated women. *International Journal of Law and Psychiatry, 22,* 301–322.

Bureau of Justice Statistics. (1999). *Special report: Women offenders.* Washington, DC: U.S. Department of Justice.

Bureau of Justice Statistics. (2006). *Special report: Drug use and dependence, state and federal prisoners, 2004.* Washington, DC: U.S. Department of Justice, Bureau of Justice Statistics.

Carp, S. V., & Schade, L. (1992, August). Tailoring facility programming to suit female offender's needs. *Corrections Today,* 154–158.

Cascardi, M., & O'Leary, K. D. (1992). Depressive symptomatology, self-esteem, and self-blame in battered women. *Journal of Family Violence, 7,* 249–259.

Case, P., & Fasenfest, D. (2004). Expectations for opportunities following prison education: A discussion of race and gender. *Journal of Correctional Education, 55,* 24–39.

Center for Substance Abuse Treatment. (1999). *Substance abuse treatment for women offenders: Guide to promising practices.* Rockville, MD: U.S. Department of Health and Human Services.

Chandler, S. M., & Kassebaum, G. (1994). Drug-alcohol dependence of women prisoners in Hawaii. *Affilia, 9,* 157–170.

Chesney-Lind, M. (2000a). What to do about girls? Thinking about programs for young women. In M. McMahon (Ed.), *Assessment to assistance: Programs for women in community corrections* (pp. 139–170). Lanham, MD: American Correctional Association.

Chesney-Lind, M. (2000b). Women in the criminal justice system: Gender matters. In P. Modley (Ed.), *Annual review of topics in community corrections: Responding to women offenders in the community* (pp. 7–10). Washington, DC: U.S. Department of Justice, National Institute of Corrections.

Chesney-Lind, M., & Pasko, L. (2004). *The female offender: Girls, women, and crime* (2nd ed.). Thousand Oaks, CA: Sage.

Clements, C. M., Ogle, R., & Sabourin, C. M. (2005). Perceived control and emotional status in abusive college student relationships: An exploration of gender differences. *Journal of Interpersonal Violence, 20,* 1058–1077.

Clements, C. M., Sabourin, C. M., & Spiby, L. (2004). Dysphoria and hopelessness following battering: The role of perceived control, coping, and self-esteem. *Journal of Family Violence, 19,* 25–36.

Covington, S. (1998). The relational theory of women's psychological development: Implications for the criminal justice system. In R. T. Zaplin (Ed.), *Female offenders: Critical perspectives and effective interventions* (pp. 113–128). Gaithersburg, MD: Aspen.

Covington, S. (2000). Helping women to recover: Creating gender-specific treatment for substance-abusing women and girls in community corrections. In M. McMahon (Ed.), *Assessment to assistance: Programs for women in community corrections* (pp. 171–233). Lanham, MD: American Correctional Association.

Covington, S. S., & Bloom, B. E. (2007). Gender-responsive treatment and services in correctional settings. In E. Leeder (Ed.), *Inside and out: Women, prison, and therapy* (pp. 9–34). Binghamton, NY: Haworth.

Crowley, J. D., & Dill, D. (1992). The Silencing the Self Scale. *Psychology of Women Quarterly, 16*, 97–106.

Daly, K. (1992). Women's pathways to felony court: Feminist theories of lawbreaking and problems of representation. *Southern California Review of Law and Women's Studies, 2*, 11–52.

Daly, K. (1994). *Gender, crime, and punishment.* New Haven, CT: Yale University Press.

Elliott, D. E., Bjelajac, P., Fallot, R. D., Markoff, L. S., & Reed, B. G. (2005). Trauma-informed or trauma denied: Principles and implementation of trauma-informed services for women. *Journal of Community Psychology, 33*, 461–477.

Fischer, J., Spann, L., & Crawford, D. (1991). Measuring codependency. *Alcoholism Treatment Quarterly, 8*, 87–99.

Fogel, C. I., & Martin, S. L. (1992). The mental health of incarcerated women. *Western Journal of Nursing Research, 14*, 30–47.

Gendreau, P., Little, T., & Goggin, C. (1996). A meta-analysis of the predictors of adult offender recidivism: What works! *Criminology, 34*, 575–607.

Gilligan, C. (1982). *In a different voice: Psychological theory and women's development.* Cambridge, MA: Harvard University Press.

Green, B., Miranda, J., Daroowalla, A., & Siddique, J. (2005). Trauma exposure, mental health functioning, and program needs of women in jail. *Crime & Delinquency, 51*, 133–151.

Greene, S., Haney, C., & Hurtado, A. (2000). Cycles of pain: Risk factors in the lives of incarcerated mothers and their children. *The Prison Journal, 80*, 3–23.

Hardyman, P. L., & Van Voorhis, P. (2004). *Developing gender-specific classification systems for women offenders.* Washington, DC: U.S. Department of Justice, National Institute of Corrections.

Holtfreter, K., & Morash, M. (2003). The needs of women offenders: Implications for correctional programming. *Women & Criminal Justice, 14*, 137–160.

Islam-Zwart, K. A., & Vik, P. W. (2004). Female adjustment to incarceration as influenced by sexual assault history. *Criminal Justice and Behavior, 31*, 521–541.

Jordan, B. K., Schlenger, W. E., Fairbank, J. A., & Caddell, J. M. (1996). Prevalence of psychiatric disorders among incarcerated women II. Convicted women felons entering prison. *Archives of General Psychiatry, 53*, 513–519.

Kaplan, A. G. (1984). *The "self in relation": Implications for depression in women* (Publication No. 14). Wellesley, MA: Stone Center.

Koons, B. A., Burrow, J. D., Morash, M., & Bynum, T. (1997). Expert and offender perceptions of program elements linked to successful outcomes for incarcerated women. *Crime & Delinquency, 43*, 512–532.

Langan, N. P., & Pelissier, B. (2001). Gender differences among prisoners in drug treatment. *Journal of Substance Abuse, 13*, 291–301.

Law, M. A., Sullivan, S. M., & Goggin, C. (2006). *Security classification measures for female offenders and predictors of female criminal conduct: A literature review.* Ottawa: Correctional Service Canada.

Lerner, K., Arling, G., & Baird, C. (1986). Client management classification: Strategies for case supervision. *Crime & Delinquency, 32*, 254–271.

Lindquist, C. H., & Lindquist, C.A. (1997). Gender differences in distress: Mental health consequences of environmental stress among jail inmates. *Behavioral Sciences and the Law, 15*, 503–523.

Loucks, A., & Zamble, E. (1999). Predictors of recidivism in serious female offenders: Canada searches for predictors common to both men and women. *Corrections Today, 61*, 26–31.

McClellan, D. S., Farabee, D., & Crouch, B. M. (1997). Early victimization, drug use, and criminality: A comparison of male and female prisoners. *Criminal Justice and Behavior, 24*, 455–476.

Miller, J. B. (1976). *Toward a new psychology of women.* Boston: Beacon.

Miller, J. B., & Stiver, I. P. (1998). *The healing connection: How women form relationships in therapy and life.* Wellesley, MA: Wellesley Centers for Women.

Morash, M., Bynum, T. S., & Koons, B. A. (1998). *Women offenders: Programming needs and promising approaches.* Washington, DC: U.S. Department of Justice, National Institute of Justice.

Mumola, C. (2000). *Incarcerated parents and their children* (No. 182335). Washington, DC: U.S. Department of Justice, Office of Justice Programs.

Orava, T. A., McLeod, P. J., & Sharpe, D. (1996). Perceptions of control, depressive symptomatology, and self-esteem of women in transition from abusive relationships. *Journal of Family Violence, 11*, 167–186.

Owen, B. (1998). *In the mix: Struggle and survival in a woman's prison.* Albany: State University of New York Press.

Owen, B., & Bloom, B. (1995). Profiling women prisoners: Findings from national surveys and a California sample. *The Prison Journal, 75,* 165–185.

Peters, R. H., Strozier, A. L., Murrin, M. R., & Kearns, W. D. (1997). Treatment of substance abusing jail inmates: Examination of gender differences. *Journal of Substance Abuse Treatment, 14,* 339–349.

Petersilia, J. (2003). *When prisoners come home: Parole and prisoner reentry.* New York: Oxford University Press.

Prendergast, M., Wellisch, J., & Falkin, G. (1995). Assessment of services for substance-abusing women offenders and correctional settings. *The Prison Journal, 75,* 240–256.

Reisig, M. D., Holtfreter, K., & Morash, M. (2006). Assessing recidivism risk across female pathways to crime. *Justice Quarterly, 23,* 384–405.

Resick, P. A. (1993). The psychological impact of rape. *Journal of Interpersonal Violence, 8,* 223–255.

Rettinger, L. J. (1998). *A recidivism follow-up study investigating risk and need within a sample of provincially sentenced women.* Unpublished doctoral dissertation, Carleton University, Ottawa, Canada.

Richie, B. E. (1996). *Compelled to crime: The gender entrapment of Black battered women.* New York: Routledge.

Roehling, P., & Gaumond, E. (1996). Reliability and validity of the Codependent Questionnaire. *Alcoholism Treatment Quarterly, 14,* 85–95.

Rosenberg, M. (1979). *The concept of self.* New York: Basic Books.

Rumgay, J. (2004). Scripts for safer survival: Pathways out of female crime. *Howard Journal of Criminal Justice, 43,* 405–419.

Salisbury, E. J., & Van Voorhis, P. (2007, March). *Gendered pathways: An investigation of women offenders' unique paths to crime.* Paper presented at the Annual Meeting of the Academy of Criminal Justice Sciences, Seattle, WA.

Salisbury, E. J., Van Voorhis, P., & Spiropoulos, G. V. (in press). The predictive validity of a gender-responsive needs assessment: An exploratory study. *Crime & Delinquency.*

Schram, P. J., & Morash, M. (2002). Evaluation of a life skills program for women inmates in Michigan. *Journal of Offender Rehabilitation, 34,* 47–70.

Sherer, M., Maddus, J., Mercandante, B., Prentice-Dunn, S., Jacobs, B., & Rogers, R. (1982). The Self-Efficacy Scale: Construction and validation. *Psychological Reports, 51,* 663–671.

Sheridan, J. J. (1996). Inmates may be parents, too. *Corrections Today, 58,* 100–103.

Simourd, L., & Andrews, D. A. (1994). Correlates of delinquency: A look at gender differences. *Forum on Corrections Research, 6,* 26–31.

Singer, M. I., Bussey, J., Song, L. Y., & Lunghofer, L. (1995). The psychosocial issues of women serving time in jail. *Social Work, 40,* 103–113.

Task Force on Federally Sentenced Women. (1990). *Creating choices: The report of the Task Force on Federally Sentenced Women.* Ottawa: Ministry of the Solicitor General Canada.

Teplin, L. A., Abram, K. M., & McClelland, G. M. (1996). Prevalence of psychiatric disorders among incarcerated women. *Archives of General Psychiatry, 53,* 505–512.

Travis, J. (2005). *But they all come back: Facing the challenges of prisoner reentry.* Washington, DC: Urban Institute Press.

Van Voorhis, P. (2004). An overview of offender classification systems. In P. Van Voorhis, M. Braswell, & D. Lester (Eds.), *Correctional counseling and rehabilitation* (5th ed., pp. 133–160). Cincinnati, OH: Anderson.

Van Voorhis, P., & Presser, L. (2001). *Classification of women offenders: A national assessment of current practices.* Washington, DC: U.S. Department of Justice, National Institute of Corrections.

Warren, J., Hurt, S., Booker Loper, A., & Chauhan, P. (2004). Exploring prison adjustment among female inmates: Issues of measurement and prediction. *Criminal Justice and Behavior, 31,* 624–645.

Williams, S. L., & Mickelson, K. D. (2004). The nexus of domestic violence and poverty. *Violence Against Women, 10,* 283–293.

Wormith, J. S. (1984). Attitude and behavior change of correctional clientele: A three year follow-up. *Criminology, 23,* 326–348.

Zlotnick, C., Johnson, D. M., & Kohn, R. (2006). Intimate partner violence and long-term psychosocial functioning in a national sample of American women. *Journal of Interpersonal Violence, 21,* 262–275.

READING

Ann Booker Loper and Elena Tuerk introduce us to the world of programming for incarcerated parents. Large increases in the number of incarcerated parents have led to the implementation of a variety of parent training programs in prisons. This article examines the current state of research on parenting interventions, including the types of programs available, the outcomes measured in each study, and the overall effectiveness of parent training. Variables that may affect program effectiveness, such as sentence length, educational level, and parent gender, are considered. The authors conclude by emphasizing the importance of primary prevention through parent training, and they include implications for social welfare and for further scientific work in this area of inquiry.

Parenting Programs for Incarcerated Parents

Current Research and Future Directions

Ann Booker Loper and Elena Hontoria Tuerk

During the past decade, a growing number of Americans have been incarcerated. In 2002, more than 2 million inmates were held in American jails or prisons (P. M. Harrison & Beck, 2003), representing a 28% increase since 1995. Parallel to this trend has been an increase in the number of children with parents in prison. Between 1991 and 1999, the number of minor children with a parent in a state or federal prison increased by 60%, representing a total of approximately 1.5 million children with at least one parent in prison during 1999 (Mumola, 2000).

Source: Loper, A. B., & Tuerk, E. H. (2006). Parenting Programs for Incarcerated Parents: Current Research and Future Directions. *Criminal Justice Policy Review, 14*(4), 407–427. Reprinted with permission of Sage Publications, Inc.

Women, in particular, have shown a steep increase in incarcerations and consequent separation from minor children. During 2002 alone, the number of female prisoners increased by nearly 5%, double the rate of increase for male offenders (P. M. Harrison & Beck, 2003). As mothers are very likely to have been the primary—often sole—caregiver for their children, the impact of incarceration on children is particularly acute. In a recent survey of incarcerated parents, nearly all fathers (89.6%) reported that at least one of their minor children was living with the other parent, in contrast to only 28.0% of incarcerated mothers, who were more likely to report their children as living with grandparents or other relatives (Mumola, 2000). Mumola also noted that approximately 10.0% of the female inmates in state prisons reported having children in foster care, in contrast to 1.8% of the male inmates.

To help improve relationships between this growing population of incarcerated parents and their children, prisons have implemented education programs designed to teach inmates how to promote healthy interactions with their children. These efforts have included parenting classes, programs allowing children and parents to live together at prison while receiving support and consultation, relationship-building visitation activities, parent counseling, and postrelease assistance. The Family and Corrections Network (www.fcnetwork.org) lists numerous organizations that provide extensive support for families affected by incarceration.

Recent research supports the potential benefits to incarcerated parents and their families of focusing on inmates' parental roles. Several investigations have documented the negative impact of parental separation on children (Luke, 2002; Meyers, Smarsh, Amlund-Hagen, & Kennon, 1999; Trice & Brewster, 2004). Trice and Brewster (2004) found that adolescents with mothers in prison were 4 times more likely to drop out of school than a cohort of friends without incarcerated mothers. They also noted an association between reduced mother-child contact and the increased likelihood of school suspension.

Children of incarcerated parents are themselves at increased risk for future incarceration. In an investigation of incarcerated juveniles in Virginia between 1993 and 1998, McGarvey and Waite (1998) found that 21% of boys and 17% of girls had a history of paternal incarceration and that approximately 8% of both boys and girls had a history of maternal incarceration. Reed and Reed (1997) point to the impact of this intergenerational trend on communities of color, where high incarceration rates among families can lead to a normalization of the prison experience and expectations among youth that time in prison is a typical milestone.

The constellation of problems seen in children of incarcerated parents may make them particularly resistant to intervention. Dalley (1997) surveyed mothers at three Montana correctional facilities and the attorneys representing their children's legal rights. Results indicated high levels of problems that often mirrored those of their inmate mothers. Many of the children had experienced separation from their mothers before the current imprisonment, often because of maternal substance abuse. A large number of the children experienced mental and physical health problems and exhibited a range of problems such as aggressive behavior, tantrums, and frequent crying. Approximately one third of the associated children had school problems including inattention and learning disabilities. Most of the problems of inmate children were evident before incarceration, which undermines a conclusion that prison separation directly causes negative child outcomes. More likely, the significant number of problems among mothers before incarceration, including drug abuse, personality dysregulation, and mental illness, can lead to poor outcomes for children. Many of these children, already at risk prior to separation, experience new stress with maternal incarceration. The cumulative risks and stressors increase the likelihood of continued emotional and behavioral problems that may lead to the child's own eventual incarceration.

Stress related to parenting in prison is also associated with poor inmate adjustment. Houck and Loper (2002) queried 362 inmate mothers on the degree to which they experienced parenting stress, as measured by a modification of the Abidin (1995) Parenting Stress Index. Results indicated that inmate mothers with higher levels of parenting stress were more likely to report higher levels of depression and anxiety. In addition, higher levels of reported parenting stress were associated with increased frequency of institutional infractions, suggesting that mothers experiencing parenting stress had greater difficulty adjusting to the rules and constraints of prison life. Several studies have conducted in-depth interviews with mothers concerning their experiences in prison. For example, on interviewing 52 women in a minimum-security, prerelease correctional facility, Coll, Surrey, Buccio-Notaro, and Molla (1998) documented the significant amount of stress and pain that can be directly attributed to child separation experiences. They noted difficulties that many women experienced with the loss of daily contact with their children, given the centrality of motherhood to the inmates' identities. Similar recurrent themes of stress and loss have been reported in other studies that used in-depth prisoner interviews (Enos, 2001; Ferraro & Moe, 2003; Forsyth, 2003).

Given these trends, a large number of correctional facilities have become interested in providing interventions for incarcerated parents. Responses to inmate parenting needs, however, have been characterized by tremendous variability among prison parent-training programs. For example, some programs differ by targeting specific parent populations within the institution (i.e., parents of young children). Providers of the parenting programs may also vary and include educational staff, mental health staff, volunteers from religious or community groups, or even inmates themselves. In addition, programs may differ in their level of financial and institutional support. The prison's security level may also limit the range of training and inmate contact opportunities, with fewer options available at high-security institutions. In a questionnaire mailed to state facilities in the United States, Clement (1993) collected information concerning various ongoing programs. Results illustrated that, although parenting programs were frequently offered at various institutions, there was no obvious consistency among programs in terms of length, depth, or content. Furthermore, most classes tended to be taught by volunteers with varying levels of expertise in parent training.

With the increase in inmate population and number of children and parents separated, there is a growing need for effective services. The purpose of this article is to examine peer-reviewed articles that describe specific parent-training interventions for prison inmates. Selected articles include qualitative or quantitative information about the effectiveness of the intervention. A literature search for potentially appropriate programs was conducted in scientific journal databases, which resulted in the selection of 17 studies that met the above criteria. Our goals are to describe interventions, identify commonalities among programs, discern which practices are most effective, and highlight unanswered questions related to optimal prison parent-training interventions.

We do not claim that this list is exhaustive and assume that there may be appropriate but unpublished interventions that have not been included. However, we believe that the included studies adequately represent the available range of options for parent training.

▨ Parent Training Interventions for Prisoners

Most prison programming for parents has the overarching goal of improving outcomes for inmates and their children, both during and after incarceration. Despite having similar goals, parenting programs differ significantly in design, execution, and method of assessment. Typical parenting curricula include education regarding effective parenting techniques and child development; other components may include enhanced visiting, parental rights training, nursery programs, or support groups. Table 1 provides a description of parenting programs and their methods for self-assessment.

Evaluation of parenting programs provides researchers with information regarding the effectiveness of their curricula. Many studies, however, do not have sufficiently large sample sizes, use random assignment of participants, use control groups for comparison, or use pre- and posttests to examine the effects of the intervention. As a result of their non- or quasiexperimental designs, most of the studies cannot support conclusions regarding program effectiveness.

In the majority of studies, the criterion for effectiveness is a measure that is only indirectly related to parenting. This is largely because of the pragmatic difficulties of assessing inmate-child relationships. Parents have limited opportunities to implement knowledge gained from parenting programs, necessitating the use of measures that examine theoretical constructs instead of direct parenting behaviors. There may be significant obstacles for researchers who seek access to children of inmates (e.g., difficulty obtaining consent from a parent, caregiver, or custodial agency, distance to child's residence, and institutional concerns about research activities during visitation). As a result, most measures of program efficacy are based on constructs related to inmates' adjustment as incarcerated parents. Typically, researchers have examined one or more of the following constructs: self-esteem, parenting attitudes, and institutional adjustment.

Self-Esteem

Several researchers have used positive changes in inmate self-esteem as a criterion measure for program effectiveness (D. H. Browne,

1989; Harm, Thompson, & Chambers, 1998; K. Harrison, 1997; Kennon, 2003; Moore & Clement, 1998; Thompson & Harm, 2000). The most frequently used measure is the 25-item Index of Self-Esteem (Hudson, 1982). Although some studies have shown an increase in self-esteem at posttest (D. H. Browne, 1989; Kennon, 2003), others have found no significant changes (K. Harrison, 1997; Moore & Clement, 1998). Still other studies have found that additional variables such as substance abuse history, prior victimization, and frequency of inmate-child contact mediate changes in self-esteem (Harm et al., 1998; Thompson & Harm, 2000). Those interventions that showed positive changes in self-esteem (D. H. Browne, 1989; Kennon, 2003; Thompson & Harm, 2000) did not have control groups for comparison. The two studies that did not see changes used experimental and control groups (K. Harrison, 1997; Moore & Clement, 1998) but did not use random group assignment and had relatively small sample sizes. Overall, there is limited support for a connection between participation in a parenting program and increased self-esteem. Experimental designs with larger sample sizes are needed to further assess this possibility. If self-esteem is to be a typical outcome measure for parenting interventions, it bears noting that studies have not yet examined whether an increase in inmate self-esteem produces improved parenting or improved confidence in one's parenting, whether good or bad.

Parenting Attitudes

Inmates' parenting attitudes are important areas of intervention for parenting curricula. Specifically, beliefs regarding affection, discipline, and family roles can have repercussions for children's psychological development. Furthermore, primary prevention through the alteration of parenting attitudes is a politically viable method of intervention because of its potential cost-saving benefits. For this reason,

Table 1	Examinations of Prisoner Parenting Programs			
Authors	**Participants**[a]	**Program Description**	**Research Design**	**Results**
Block and Potthast (1998)	16 female inmates and daughters	Girl Scouts Beyond Bars developed ongoing troops for inmates and daughters. Program included enhanced-visit "troop meetings" at the prison. Program provided transportation for daughters to these visits.	Hudson (1982) Parent-Child Contentment Scale at pre- and posttest	Qualitative results indicated increased visiting and improvements in parent-child relationship after participation. No significant differences in means of Hudson Parent-Child Contentment Scale over time.
D. H. Browne (1989)	29 females residing in structured community treatment program used as alternative to incarceration	Class met twice a week for 6 months. Focused on children's needs, emotional involvement, development of individual personalities within a family setting, and self-esteem.	Pre- and posttest using AAPI[b] and SEI[c]	Significant differences on SEI self-esteem construct.
Boudin (1998)	10 female inmates	Inmates met 5 days a week for 3 months. Focused on mothers' own traumatic experiences, shame and guilt about choices, and grief at separation from children.	No measures included	Qualitative results concerning parenting difficulties.
Carlson (2001)	37 female inmates residing in prison nursery	Inmates participated in parenting classes while caring for child in nursery program during 18 months.	Institutional records, follow-up survey	Reduced misconduct and recidivism by participants. Positive attitude endorsed on survey.
Eddy, Powell, Szubka, McCool, and Kuntz (2001)	152 males at city detention center, children, and caregivers	Class met 4 times in 1-hour sessions. Experimental group received parenting manual and training in parenting skills, communication, and child development. Control group received parenting manual.	Interviews	No quantitative results presented. Described difficulties in implementing intervention.

(Continued)

Table 1	(Continued)			
Authors	**Participants**[a]	**Program Description**	**Research Design**	**Results**
Gat (2000)	16 female inmates in parenting class, 5 in nursery and parenting class, 4 in control group, record review of 117 recidivist mothers	Inmates participated in Mother/Offspring Life Development (MOLD) parenting program.	IPPA,[d] IRI,[e] PROM-R,[f] recidivism records	No significant differences among inmates on measures or recidivism rates.
Harm, Thompson, and Chambers (1998)	104 female inmates	Inmates participated in 15-week program based on Bavolek and Comstock's (1985) Nurturing Parent curriculum.	AAPI and ISE[g] at pre- and posttest interviews	Women with frequent use of drugs or alcohol showed significant improvement on ISE at posttest. Women with histories of victimization showed significantly lower ISE scores at both pre- and posttest than did nonvictimized participants.
K. Harrison (1997)	30 male inmates, divided into experimental and control groups	Experimental group met 3 times a week for 6 weeks. Received training in child development, behavior management, family relationships, and communication. Control group watched tapes and participated in discussions. Children were mailed a questionnaire.	Fathers: AAPI and ISE at pre- and posttest; children: Self-Perception Profile for Children	Improvement in fathers' attitudes on the AAPI. No significant differences on ISE or Self-Perception Profile for Children. (Note: control group received alternate parenting intervention.)
Kennon (2003)	66 female inmates	Class met for 12 sessions. Targeted self-esteem, communication, legal knowledge, and parenting attitudes.	SES,[h] PARQ,[i] Incarcerated Parent Legal questionnaire, communication questionnaire, program satisfaction survey	Increases in positive parenting attitudes, self-esteem, and legal knowledge. No increase in amount of communication between mothers and children.
Landreth and Lobaugh (1998)	16 male inmates in	Inmates received 10-week	Inmates: PPAS,[j] PSI,[k] FPC;[l]	Fathers in experimental

Authors	Participants[a]	Program Description	Research Design	Results
	experimental group, 16 in control group	training in filial therapy (play therapy). Play therapy sessions with inmates and children.	children: Joseph Preschool and Primary Self Concept Scale	group scored significantly higher than did controls on PPAS at posttest. Experimental group showed significantly lower parenting stress on the PSI than did controls at posttest. Children in experimental group showed a significant increase in self-concept at posttest.
Marsh (1983)	Three families of incarcerated fathers (10 children total)	Both parents attended training on communication and child management.	Pre- and posttest 1-hour home observations of family, behavior reports, communication checklist, Adjective Checklist	Qualitative results indicating better father-child communication skills at posttest and improved child behavior.
McKeown (1993)	40 female inmates in experimental group, 49 in control group	Extended visits for participants and children during a 4-month period.	Postvisit questionnaire, institutional records	Positive qualitative reactions. Decrease in requests for medical treatment.
Moore and Clement (1998)	20 female inmates in experimental group, 20 in waitlist control group	Mothers Inside Loving Kids (MILK) parenting group and enhanced visiting.	ISE, AAPI, Nurturing Quiz	No differences between groups.
Showers (1993)	203 female inmates in experimental group, 275 in control group	Class met 1.5 hours per week for 10 weeks. Used modified STEP program.	Pre- and posttest behavior management questionnaire	Significant increase on behavior management questionnaire for experimental group at posttest.
Snyder-Joy and Carlo (1998)	31 female inmates in visitation program, 27 in waitlist control group	Visitation experiences enhanced with volunteers providing special activities and child-friendly visitation area.	Interviews	Treatment group reported more letters received from children and more frequent child-focused

(Continued)

Table 1	(Continued)			
Authors	**Participants**[a]	**Program Description**	**Research Design**	**Results**
				communication with family and fellow inmates. No significant differences in perceptions of child's well-being or expressed concerns about child.
Thompson and Harm (2000)	104 female inmates	Inmates participated in 15-week program based on Bavolek and Comstock's (1985) Nurturing Parent curriculum.	AAPI and ISE at pre- and posttest	ISE for mothers who had at least some visits from children or frequent letters improved significantly. Significant improvements for the total group on the AAPI were found in the areas of expectations, belief in corporal punishment, and parent-child roles.
Wilczak and Markstrom (1999)	42 male inmates, divided into experimental and control groups	Class met for 8 sessions in 3 weeks. Used modified STEP program.	Content test, Parent Locus of Control, CGPSS	Experimental group showed significant improvement in knowledge, content test, Parent Performance subscale of the CGPSS, and Locus of Control.

a. Unless otherwise noted, all participants are prison inmates.

b. Adult-Adolescent Parenting Inventory (Bavolek, 1984).

c. Self-Evaluation Inventory (Schaefer, Edgerton, & Hunter, 1984).

d. Inventory of Parent and Peer Attachment (Armsden & Greenberg, 1987).

e. Interpersonal Reactivity Index (Davis, 1980).

f. Prosocial Moral Reasoning-Revised (Carlo, Eisenberg, & Knight, 1992).

g. Index of Self-Esteem (Hudson, 1982).

h. Self-Esteem Scale (Rosenberg, 1965).

i. Parental Acceptance Rejection Questionnaire (Rohner, 1990).

j. Parent-Peer Attachment Scale (Porter, 1954).

k. Parenting Stress Index (Abidin, 1995).

l. Filial Problem Checklist (Horner, 1974).

interventions that produce research related to parenting attitudes may be more relevant to policy makers than programs that mainly target inmate self-esteem.

Many studies of parenting programs include measures to examine changes in parenting attitudes before and after participation in a program (D. H. Browne, 1989; Gat, 2000; Harm et al., 1998; K. Harrison, 1997; Kennon, 2003; Landreth & Lobaugh, 1998; Moore & Clement, 1998; Showers, 1993; Thompson & Harm, 2000; Wilczak & Markstrom, 1999). Several scales have been used to measure changes in parenting attitudes, most frequently the Adult-Adolescent Parenting Inventory (Bavolek, 1984). Other scales include the Porter Parental Acceptance Scale (Porter, 1954), the Parenting Stress Index (Abidin, 1995), the Child Behavior Management Survey (Showers, 1993), and the Cleminshaw-Guidubaldi Parent Satisfaction Scale (Guidubaldi & Cleminshaw, 1985). Some parenting programs have developed their own content measures to evaluate inmate retention of psychoeducational material.

Some studies using measures of parenting attitudes have found positive results, with participants showing significant improvement at posttest (K. Harrison, 1997; Kennon, 2003; Landreth & Lobaugh, 1998; Thompson & Harm, 2000). Others (Landreth & Lobaugh, 1998; Showers, 1993; Wilczak & Markstrom, 1999) have shown positive and significant differences between experimental and control group members; however, these groups were not randomly assigned. Those studies that did not show changes at posttest or between groups (D. H. Browne, 1989; Gat, 2000; Moore & Clement, 1998) had relatively small sample sizes. Given the group differences seen in Showers's (1993) large sample and in smaller samples where effects would be harder to detect (Landreth & Lobaugh, 1998; Wilczak & Markstrom, 1999), current evidence suggests that parenting attitudes do improve with intervention. Experimental designs in future studies can further illuminate this possibility.

Institutional Adjustment

Poorer institutional adjustment has been shown to be related to parenting stress (Houck & Loper, 2002), suggesting that inmate behavior may improve if parenting stress decreases. Using institutional records, Carlson (2001) found reduced misconduct and recidivism of parenting program participants. In a study that also used institutional records, however, Gat (2000) did not find any significant differences in recidivism between participating inmates and controls. McKeown (1993) observed that mothers who participated in an extended visit intervention were less likely to seek medical treatment in institution facilities. These studies add important information that is often easily obtainable and of value when assessing the institutional merits of parenting interventions. Parents who are less stressed regarding their children may be better citizens of the institutions and thus more amenable to rehabilitative efforts. Future studies may wish to assess how other informal indicators of incarceration adjustment may be affected by parenting stress and intervention. These areas of adjustment may include relationships with other inmates, coping strategies (e.g., smoking, meditation), undetected rule breaking, and vocational or educational functioning.

Qualitative and Relational Measures

Many studies of parenting programs have described qualitative improvements in participants' relationships with their children (Boudin, 1998; Eddy, Powell, Szubka, McCool, & Kuntz, 2001; Marsh, 1983; McKeown, 1993; Snyder-Joy & Carlo, 1998). Several of these studies offer rich descriptions of inmates' experiences as parents in prison that can be used to further develop parenting interventions. Block and Potthast's (1998) study uses pretests, posttests, and delayed posttests to assess relationship satisfaction and visitation patterns. Their work suggests that women who participate in structured programming with their daughters receive more visits than matched nonparticipant mothers, but conventional statistical support for this finding is not provided. Block and Potthast's study would be further enhanced by a larger sample size, random selection of eligible participants, and comparisons of relationship satisfaction with a control group.

Because one goal of parenting interventions is to improve family relations, it is important to assess how interventions may affect children of parents who have received such training. Landreth and Lobaugh (1998) provide encouraging initial evidence. After their 10-week filial therapy parent training with incarcerated fathers, they observed an improvement in the self-concepts of the participants' children. Block and Potthast (1998) describe qualitative evidence that daughters' self-perceptions and grades improved after participation in a visitation program. In contrast, K. Harrison (1997) observed changes in parenting attitudes that were not accompanied by measured changes in child self-perceptions. It is possible that the relatively short time frame for K. Harrison's intervention (6 weeks) was insufficient for the detection of generalized effects on children. Clearly, more emphasis on measuring the short- and long-term indirect effects of parenting training on the children of inmates is needed.

⬚ Where Do We Go From Here?

The most remarkable feature of the literature on optimal parenting training for prisoners is its paucity. Despite the fact that there are more than 700,000 parents in prison (Mumola, 2000), there are only a handful of published studies of parent training, and there is very little uniformity among methods. Some of the studies provide rich qualitative information (Block & Potthast, 1998; Boudin, 1998; Eddy et al., 2001; Marsh, 1983; McKeown, 1993), but complementary follow-up quantitative studies

are rare. The variables chosen to measure effects most frequently focus on changes among parents rather than children. Although this focus is an essential initial achievement for such programs, the impact of improved parenting skills on family relationships ultimately needs to be evaluated. Interventions that include child measures (K. Harrison, 1997; Landreth & Lobaugh, 1998) provide important information about these effects.

Several studies detected an improvement in inmates' self-esteem (D. H. Browne, 1989; Harm et al., 1998; Thompson & Harm, 2000). With the exception of D. H. Browne's study (1989), parent training was also associated with improved attitudes toward parenting (Harm et al., 1998; K. Harrison, 1997; Landreth & Lobaugh, 1998; Showers, 1993; Thompson & Harm, 2000; Wilczak & Markstrom, 1999). Given the generally poor mental health and familial experiences of inmates (Beck & Maruschak, 2001; Ditton, 1999; Warren et al., 2002), these findings are important. However, more focus on the ultimate goal of such training (i.e., improved family relationships) is an essential next step.

In part because of the scarcity of studies of parenting interventions with prisoners, there are many unanswered questions concerning the optimal nature of training for incarcerated parents. Broadly, these questions include: (a) In what ways should parenting-training programs be distinguished from those designed for nonincarcerated populations? (b) Can a uniform prison-specific intervention meet the diverse needs of an incarcerated population? and (c) Are there conditions where parenting training is contraindicated?

What, If Any, Should Be the Unique Features in Prison Parenting Training?

Several of the reviewed interventions emphasize the need for parents to understand the developmental needs of children, form meaningful attachments to their children, and employ appropriate child-management techniques (e.g., K. Harrison, 1997; Kennon, 2003; Showers, 1993). These emphases are not surprising in a parenting curriculum and are also significant in parenting interventions for nonincarcerated parents (Abidin, 1982; Schaeffer & Briesmeister, 1989; Serketich & Dumas, 1996).

Given the unique demands, stressors, and contexts for maintaining a parent relationship from prison, however, parenting training for inmates also needs to attend to the unique features of prison parenting. Several features present in one or more of the programs examined appear to be promising elements.

Peer support Training programs can benefit from the establishment of a positive peer culture among inmate parents and the development of support structures that inmates can access outside the training milieu. For example, Moore and Clement (1998) describe a program developed in Virginia, Mothers Inside Loving Kids, which includes an array of support services offered to a selected group of women who meet regularly for mutual support and education. This program, which is still in operation, continues to rely on an established support network so that inmates can collaboratively plan activities for upcoming visits, take advantage of offered parent training, and counsel one another through difficult times (S. Dunn, personal communication, May 8, 2006).

Communication. Unlike typical communication patterns among nonincarcerated parents, communication between inmate parents and their children is generally limited to visits, letter writing, and phone calls. Each of these forms of communication is different in terms of the amount of time spent on each activity and the resulting qualitative experience of parenting. In an examination of the relationship between parenting stress and amount of contact during incarceration, Tuerk and Loper (2006) documented a relationship between lowered parent stress, as measured by scales from the Abidin (1995) Parenting Stress Index,

and the amount of mother-child contact. In particular, the amount of letter writing was a stronger predictor of reduced parenting stress than were either the amount of visitation or the number of phone calls. The authors interpreted the results as because of the greater degree of control mothers experienced with letter writing relative to the other two forms of communication. Parenting programs designed for nonincarcerated populations do not target long-distance relationships and do not cover topics such as phone and letter communication. Incarceration-specific programs need to emphasize explicit training in the best methods for communicating via mail and telephone, with advice on what to say, how much emotional content to convey, the best ways of voicing support, and other strategies.

Emotional well-being As a group, inmates experience high levels of mental illness and emotional distress that may undermine their ability to cope with the stresses of parenting from prison (Beck & Maruschak, 2001; Ditton, 1999; Warren et al., 2002). Particularly among female inmates, there are likely to be relatively high levels of sexual and physical abuse history (A. Browne, Miller, & Maguin, 1999; Maeve, 2000), which will likely affect relationships with children. Because of their mental health histories, some inmates may lack the emotional reserves to cope with the stressful nature of visitation, particularly when problems arise. Boudin (1998) countered this by placing emphasis on helping mothers cope with their own emotional issues, particularly their trauma histories, guilt about past choices, and grief at separation from their family. A sadly familiar story in women's prisons is that of an inmate who eagerly looks forward to a visit, only to have an angry meltdown during the visit when she learns disturbing news. For most inmate parents, visits with children are rare events, with approximately half of the inmate mothers and fathers never receiving any visits (Mumola, 2000). Because of the rarity of these visits, emotionally charged or explosive events during visitation can be particularly toxic, leading children and caretakers to resist returning. Inmate parents need to cope with their own strong feelings about themselves as parents so that they can better handle the stresses of separation. They benefit from strategies to monitor and reevaluate their own automatic thoughts and feelings prior to reacting. These strategies need to be explicitly taught, however, and related to the context of parenting.

Provision for lower educational levels

Inmates in the United States tend to be more poorly educated than the general public (Harlow, 2003). Materials adapted from training materials designed for nonincarcerated populations may be too difficult for some inmates to understand. Eddy et al. (2001) responded to this challenge by designing training materials that emphasized the use of visuals and simple language. Recognition of the varied educational and social backgrounds of incarcerated parents presents a special challenge for working with this population.

Liaison with child caretakers

In any training program for parents of incarcerated children, concern about the caretaker who is responsible for the day-to-day rearing of the child is often the "elephant in the room." Virtually every aspect of inmates' relationships with their children relates to the quality of the relationship between the inmate and caregiver. Caregivers are usually the major factor in determining when and how often children visit. They can facilitate or impede phone calls or letters between children and inmates, and they bear the greatest responsibility for handling crises. Caretakers typically experience significant stress while raising a child who has at least one parent in prison (Burton, 1992; Minkler & Roe, 1993; Ruiz, 2002). Grandparents most frequently raise the children when their daughter is incarcerated (Mumola, 2000) and are faced with the emotional and physical burdens that parenthood brings. In a series of in-depth qualitative analyses of interviews and observations

of Black grandparents rearing children of drug-addicted parents, Burton (1992) documented a pervasive sense of fatigue and concern about the long-term nature of the grandparents' responsibilities. Several lamented the loss of freedom that they had expected to enjoy during their senior years. Grandparents also voiced concerns about having the cognitive and physical energy to keep up with children's school needs.

Parenting training for inmates needs to include issues related to the caretaker. Incarcerated parents may need guidance in understanding caretaker stressors so that they can better empathize and establish realistic expectations about their own parenting role. Learning how to "partner parent" with another caretaker requires smooth communication and relationship building, skills that need to be explicitly taught and emphasized in training.

Dealing with institutional constraints

Unlike parenting training designed for the general population, there are a number of institutional factors that can alter the content and process of training. For example, Schram and Morash (2002) lamented the intrusion of loudspeaker announcements that frequently interrupt classes and a poor understanding of the value of parenting programming by some institutional staff. Eddy et al. (2001) also noted initial difficulties in gaining the cooperation of skeptical officers. Routine institutional security features also affect programming; problems with completing daily inmate counts, for example, can delay or cancel the return of inmates to training. Light switches that can only be operated by certain staff members may make it difficult for inmates to view overheads, videos, or other projections. Inmates may be restricted in the types of handouts and materials they are allowed to use. Successful implementation of parenting interventions in prisons requires the instructor to gain the cooperation of institutional officials as to the value of programming and the need for an adequate instructional environment.

Unanswered Question: Does One Size Fit All?

The context for parenting training in prison can vary considerably. The parent's gender, ethnic background, length of sentence, time served, level of institutional security, and other factors may influence the optimal type of training intervention. At this stage of research into parenting training programs, however, very little information is available about possible interaction effects between an inmate's particular situation and the most effective training content.

Gender of parent Incarcerated mothers and fathers have unique needs and types of relationships with their children. Because more incarcerated mothers than fathers were single parents prior to incarceration, parental imprisonment generally causes more disruption for the children of incarcerated mothers. Compared to children with a father in prison, a greater proportion of children with a mother in prison move into the home of a relative or foster parent (Mumola, 2000). In many cases, this transition necessitates moving to a new town or city and developing new friendships, which compounds adjustment to the mother's absence. Kazura (2001) used a self-report measure to assess the perceived needs of incarcerated mothers and fathers regarding themselves and their family members. Women were relatively more interested than men in the effects of separation and how to talk with their children. Women were also more concerned than men with finding transportation for children to the facility. These perceived needs make sense considering the greater likelihood that the mothers' children would be placed with a nonparent caretaker, which makes developing new types of relationships with their children a high priority. The greater frequency of a nonparent guardian also necessitates that incarcerated mothers learn skills for negotiating and collaborating with caretakers to remain in contact with their children.

Men may also have gender-specific stressors that require conceptual and practical alterations to parent training (Palm, 2001). For example, men may have more difficulty relating to their children when their previous role as the family's financial support is lost. Magaletta and Herbst (2001) noted the helplessness of men who view their paternal role as one of activity or doing concrete things for their children. When constrained from contact by prison, many could not name any ways in which they could continue to provide fatherly support. Men who have not enjoyed emotionally close relationships with their children may have more difficulty finding methods of expressing affection and caring while in prison. Developing affective skills was a feature of Landreth and Lobaugh's (1998) intervention, which specifically taught fathers how to attend to and empathize with their children's emotions. Children of fathers who received this training demonstrated a concomitant increase in self-concept at posttest.

Men who were not the primary caretaker prior to incarceration may have a more tangential relationship with their children and may experience difficulty connecting with their children because of their ambiguous parental status. For example, Eddy et al. (2001) reported that approximately one fifth of the fathers participating in their training program had no legal or biological connection to the children but felt a connection because of their relationship with the mother. This tangential status of some inmate fathers can affect their understanding of the impact of separation. Martin (2000) interviewed jailed fathers and observed that men with less preincarceration day-to-day contact with their children were poorly attuned to their children's needs. These fathers minimized the impact of their separation on their family. In contrast, men with frequent preincarceration contact were extremely upset by the separation. These fathers were more likely to agonize over the separation, even to the point of sometimes refusing visitation efforts because of a perception that these visits would be too emotionally difficult for both themselves and their children. The emotional impact of parental separation may affect men and women differently because incarcerated men typically have less preincarceration contact.

Sentence conditions Parenting interventions may likewise need to be tailored according to sentence length. Inmates with short sentences will likely be reunited with minor children within a relatively short period. These parents may therefore wish or need to maintain a more direct role in decisions regarding their children. They may need training in how to maintain their own status as the primary parent while collaborating with the caretaker. Long-term inmates, however, may need more instruction on how to maintain an affective bond in the context of a lengthy separation.

Similar to the questions of whether treatments for long- and short-term offenders should be different is the question of the importance of distinguishing between inmates who will be released at or near the end of training from those whose incarceration will continue. For parents who will soon resume day-to-day care of their children, there may be a greater need to learn behavior-management techniques or other strategies for coping with full-time contact. In these situations, interventions designed for nonincarcerated populations may be more transferable. For example, Wilczak and Markstrom (1999) designed an intervention for fathers in a minimum-security facility who had relatively short amounts of time to serve prior to reunification. Their intervention incorporated a widely used parenting training intervention designed for the general population (STEP; Dinkmeyer & McKay, 1989). In contrast, parents who face a longer separation may need support to maintain healthy relationships with children and long-term caregivers.

Unanswered Question: Is Parenting Training Always a Good Idea?

An explicit goal in several of the programs reviewed is to enhance the connection between an incarcerated parent and his or her child. Although there is evidence that this goal is achievable and can have beneficial effects, it is not clear whether there are contexts in which children might be better served if the relationship were not encouraged. For example, an effort to reconnect an estranged parent with his or her child might have negative effects on the child if the parent is unreliable and loses interest after initiating the reunion. In addition, inmates with severe personality disorders, including psychopathy, may use the relationship for manipulative or self-gratifying purposes that are unhealthy for the child. In a series of interviews with 24 female inmates, Forsyth (2003) noted that the discourse of 2 of the inmates was characterized by little real feeling for their children and the assumption that by behaving in a way that appeared motherly, they would make a better impression on their parole board. Although there is available research that evaluates the overall benefits of specific programs, it would be helpful to understand the personality characteristics or other features of inmates that distinguish parent-child relationships that may benefit from parent training from those that either will not benefit or that will be negatively affected.

✑ Conclusion

As the number of incarcerated Americans continues to rise, it is likely that correctional facilities will strive to meet family needs by providing parenting services. Although research on effective parenting training in prison is still in its infancy, existing studies offer promise and provide support for the value of helping incarcerated mothers and fathers become better parents. Interventions may improve attitudes toward parenting, self-confidence, institutional behavior, and, in some cases, child outcomes. Information on best practices for these programs, however, is limited.

Although there is some evidence of effective programming, as demonstrated in this review, there are few efforts to replicate or improve specific interventions. Little is known about how to best modify existing curricula to meet prison-specific needs. Many existing programs, run by well-intentioned volunteers, are rarely evaluated in a systematic way. Better dissemination of information concerning effective interventions is greatly needed in this area.

The available body of literature and practice provides guideposts to develop effective interventions. Primary among these is the need to be aware of the importance of placing parenting training in a prison-specific context. Parenting from prison is substantially different from parenting on the outside. An intervention that helps a mother effectively handle tantrums at the mall is of little value to a mother who only sees her daughter for a few hours every month in a confined and typically inhospitable setting. Parents in prison need instruction on how to maintain healthy bonds, communicate in truthful and developmentally sensitive ways, and collaborate with caretakers within the context of incarceration.

The model parenting program provides personal support for inmate parents. This can include peer support from other inmates, personal support from facilitators, and sensitivity from institutional staff. For many inmates, discussing their children's needs can provoke intense feelings of guilt, sadness, and anger. Optimal training provides a forum for inmate parents to process these feelings in a safe and constructive way.

Similarly, a model program provides a strategy for dealing with problems in a flexible and creative way. Correctional settings are, by nature, structured and rule bound. They embody the black-and-white thinking that categorizes actions as either acceptable or unacceptable. Although this cognitive template has a longstanding history in corrections, it is not particularly useful for parenting. The ideal parenting program in prison works with inmates to develop flexible plans for dealing with the problems of parenting. For example, visits often do not turn out as parents may wish. Inmate parents need to know how to stay focused on their goals and be ready to think up creative, on-the-spot solutions.

Parenting instruction for inmates needs to be done in conjunction with actual child contact. It is difficult to practice newly developed skills without opportunities to use the training. This does not mean, however, that the sole focus should be either visitation or upcoming reunification. Parenting programs should explore the full range of communication methods that are available to the inmate, such as letter writing, phone calls, and collaboration with child caregivers. Many inmates do not receive regular visits, and, for those who do, such visits can be limited by logistical and emotional concerns. Effective parenting training shows inmates how to make the best of all their contact opportunities.

It is important to recognize that, among parenting programs for inmates, there will not be a single program that meets all needs; there is room for many different types. Rather than seeking the gold standard for inmate parenting training, there is a need for well-researched programs, each defining a particular target audience, goals, and method. Furthermore, policy makers and prison administrators are becoming increasingly knowledgeable about the financial benefits of using empirically based treatments in programming; the move toward evidence-based practice in prisons requires that we, as researchers of parenting interventions in prison, improve the scientific rigor of program assessments to maintain legitimacy. As we refine our methods of intervention and assessment, our work will be more effective for the population we hope to serve. The rewards of improved parent-child relationships and better adjusted inmates will be well worth the effort of refining this important body of work.

◪ References

Abidin, R. (1982). *Parenting skills workbook* (2nd ed.). New York: Human Sciences Press.

Abidin, R. R. (1995). *Parenting Stress Index: Professional manual* (3rd ed.). Odessa, FL: Psychological Assessment Resources.

Armsden, G. C., & Greenberg, M. T. (1987). The Inventory of Parent and Peer Attachment: Individual differences and their relationship to psychological well-being in adolescence. *Journal of Youth and Adolescence, 16,* 427–454.

Bavolek, S. (1984). *Adult-Adolescent Parenting Inventory (AAPI).* Park City, UT: Family Development Resources.

Bavolek, S. J., & Comstock, C. (1985). *The nurturing program.* Eau Claire, WI: Family Development Resources.

Beck, A. J., & Maruschak, L. M. (2001). *Bureau of Justice Statistics special report: Mental health treatment in stat prisons, 2000.* Washington, DC: U.S. Department of Justice, Office of Justice Programs.

Block, K. J., & Potthast, M. J. (1998). Girl Scouts beyond bars: Facilitating parent-child contact in correctional settings. *Child Welfare, 77,* 561–578.

Boudin, K. (1998). Lessons from a mother's program in prison: A psychosocial approach supports women and their children. *Women & Therapy, 21*(1), 103–125.

Browne, A., Miller, B., & Maguin, E. (1999). Prevalence and severity of lifetime physical and sexual victimization among incarcerated women. *International Journal of Law & Psychiatry, 22,* 301–322.

Browne, D. H. (1989). Incarcerated mothers and parenting. *Journal of Family Violence, 4,* 211–221.

Burton, L. M. (1992). Black grandparents rearing children of drug-addicted parents: Stressors, outcomes, and social service needs. *Gerontologist, 32,* 744–751.

Carlo, G., Eisenberg, N., & Knight, G. P. (1992). An objective measure of adolescents' prosocial moral reasoning. *Journal of Research on Adolescence, 2,* 331–349.

Carlson, J. R. (2001). Prison nursery 2000: A five-year review of the prison nursery at the Nebraska Correctional Center for Women. *Journal of Offender Rehabilitation, 33,* 75–97.

Clement, M. J. (1993). Parenting in prison: A national survey of programs for incarcerated women. *Journal of Offender Rehabilitation, 19*(1–2), 89–100.

Coll, C. G., Surrey, J. L., Buccio-Notaro, P., & Molla, B. (1998). Incarcerated mothers: Crimes and punishments. In C. G. Coll, J. L. Surrey, & K. Weingarten (Eds.), *Mothering against the odds: Diverse voices of contemporary mothers* (pp. 255–274). New York: Guilford.

Dalley, L. P. (1997). *Montana imprisoned mothers and their children: A case study on separation, reunification, and legal issues.* Unpublished doctoral dissertation, Indiana University of Pennsylvania, Indiana, PA.

Davis, M. H. (1980). A multidimensional approach to individual differences in empathy. *JSAS Catalog of Selected Documents in Psychology, 10,* 85–105.

Dinkmeyer, C., & McKay, G. D. (1989). *The parent's handbook: STEP systematic training for effective parenting* (3rd ed.). Circle Pines, MN: American Guidance Services.

Ditton, P. (1999). *Bureau of Justice Statistics special report: Mental health and treatment of inmates and probationers.* Washington, DC: U.S. Department of Justice, Office of Justice Programs.

Eddy, B. A., Powell, M. J., Szubka, M. H., McCool, M. L., & Kuntz, S. (2001). Challenges in research with incarcerated parents and importance in violence prevention. *American Journal of Preventive Medicine, 20*(Suppl. 1), 56–62.

Enos, S. (2001). *Mothering from the inside: Parenting in a women's prison.* Albany: State University of New York Press.

Ferraro, K. J., & Moe, A. M. (2003). Mothering, crime, and incarceration. *Journal of Contemporary Ethnography, 32*(1), 9–40.

Forsyth, C. J. (2003). Pondering the discourse of prison mamas: A research note. *Deviant Behavior, 24,* 269–280.

Gat, I. (2000). *Incarcerated mothers: Effects of the Mother/Offspring Life Development Program (MOLD) on recidivism, prosocial moral development, empathy, hope, and parent-child attachment.* Lincoln: University of Nebraska.

Guidubaldi, J., & Cleminshaw, H. K. (1985). The development of the Cleminshaw-Guidubaldi Parent Satisfaction Scale. *Journal of Clinical Child Psychology, 14,* 293–298.

Harlow, C. W. (2003). *Bureau of Justice Statistics special report: Education and correctional populations.* Washington, DC: U.S. Department of Justice, Office of Justice Programs.

Harm, N. J., Thompson, P. J., & Chambers, H. (1998). The effectiveness of parent education for substance abusing women offenders. *Alcoholism Treatment Quarterly, 16*(3), 63–77.

Harrison, K. (1997). Parental training for incarcerated fathers: Effects on attitudes, self-esteem, and children's self perceptions. *Journal of Social Psychology, 137,* 588–593.

Harrison, P. M., & Beck, A. J. (2003). *Bureau of Justice Statistics bulletin: Prisoners in 2002.* Washington, DC: U.S. Department of Justice, Office of Justice Programs.

Horner, P. (1974). *Dimensions of child behavior as described by parents: A monotonicity analysis.* Unpublished doctoral dissertation, Pennsylvania State University, College Park.

Houck, K. D., & Loper, A. B. (2002). The relationship of parenting stress to adjustment among mothers in prison. *American Journal of Orthopsychiatry, 72,* 548–558.

Hudson, W. (1982). *The clinical measurement package: A field manual.* Chicago: Dorsey.

Kazura, K. (2001). Family programming for incarcerated parents: A needs assessment among inmates. *Journal of Offender Rehabilitation, 32*(4), 67–83.

Kennon, S. (2003). *Developing the parenting skills of incarcerated parents: A program evaluation.* Richmond: Virginia Commonwealth University.

Landreth, G. L., & Lobaugh, A. F. (1998). Filial therapy with incarcerated fathers: Effects on parental acceptance of child, parental stress, and child adjustment. *Journal of Counseling & Development, 76*(2), 157–165.

Luke, K. P. (2002). Mitigating the ill effects of maternal incarceration on women in prison and their children. *Child Welfare, 81,* 929–948.

Maeve, M. (2000). Speaking unavoidable truths: Understanding early childhood sexual and physical violence among women in prison. *Issues in Mental Health Nursing, 21,* 473–498.

Magaletta, P. R., & Herbst, D. P. (2001). Fathering from prison: Common struggles and successful solutions. *Psychotherapy: Theory, Research, Practice, Training, 38*(1), 88–96.

Marsh, R. L. (1983). Services for families: A model project to provide services for families of prisoners. *International Journal of Offender Therapy & Comparative Criminology, 27,* 156–162.

Martin, J. S. (2000). *Inside looking out: Perceptions of jailed fathers regarding separation from children.* Unpublished doctoral dissertation, Indiana University of Pennsylvania, Indiana, PA.

McGarvey, E., & Waite, D. (1998). *Profiles of incarcerated adolescents in Virginia correctional facilities: Fiscal years 1993–1998.* Richmond: Commonwealth of Virginia, Department of Juvenile Justice.

McKeown, M. (1993). An evaluative study of the extended visits scheme. *Issues in Criminological & Legal Psychology, 20,* 32–40.

Minkler, M., & Roe, K. M. (1993). *Grandmothers as caregivers: Raising children of the crack cocaine epidemic.* Thousand Oaks, CA: Sage.

Moore, A. R., & Clement, M. J. (1998). Effects of parenting training for incarcerated mothers. *Journal of Offender Rehabilitation, 27*(1/2), 57–72.

Mumola, C. (2000). *Incarcerated parents and their children* (No. 182335). Washington DC: U.S. Department of Justice, Office of Justice Programs.

Myers, B. J., Smarsh, B. S., Amlund-Hagen, K. A., & Kennon, B. S. (1999). Children of incarcerated mothers. *Journal of child and family studies, 8,* 11–25.

Palm, G. (2001). Parent education for incarcerated fathers. In J. Fagan & A. J. Hawkins (Eds.), *Clinical and educational interventions with fathers* (pp. 117–141). Binghamton, NY: Haworth Clinical Practice Press.

Porter, B. (1954). Measurement of parental acceptance of children. *Journal of Home Economics, 46,* 176–182.

Reed, D. F., & Reed, E. L. (1997). Children of incarcerated parents. *Social Justice, 24*(3), 152–170.

Rohner, R. P. (1990). *Handbook for the study of parental acceptance and rejection* (3rd ed.). Storrs, CT: Rohner Research.

Rosenberg, M. (1965). *Society and the adolescent self-image.* Princeton, NJ: Princeton University Press.

Ruiz, D. S. (2002). The increase in incarcerations among women and its impact on the grandmother caregiver: Some racial considerations. *Journal of Sociology & Social Welfare, 29*(3), 179–197.

Schaefer, E., Edgerton, M., & Hunter, S. (1984). *Background of the Self-Evaluation Inventory.* Unpublished manuscript.

Schaeffer, C. E., & Briesmeister, J. M. (1989). *Handbook of parent training: Parents as co-therapists for children's behavior problems.* New York: John Wiley.

Schram, P. J., & Morash, M. (2002). Evaluation of a Life Skills Program for women inmates in Michigan. *Journal of Offender Rehabilitation, 34*(4), 47–70.

Serketich, W. J., & Dumas, J. E. (1996). The effectiveness of behavioral parent training to modify antisocial behavior in children: A meta-analysis. *Behavior Therapy, 27*(2), 171–186.

Showers, J. (1993). Assessing and remedying parenting knowledge among women inmates. *Journal of Offender Rehabilitation, 20*(1-2), 35–46.

Snyder-Joy, Z. K., & Carlo, T. A. (1998). Parenting through prison walls: Incarcerated mothers and children's visitation programs. In S. L. Miller (Ed.), *Crime control and women: Feminist implications of criminal justice policy* (pp. 130–150). Thousand Oaks, CA: Sage.

Thompson, P. J., & Harm, N. (2000). Parenting from prison: Helping children and mothers. *Issues in comprehensive pediatric nursing, 23*, 61–81.

Trice, A. D., & Brewster, J. (2004). The effects of maternal incarceration on adolescent children. *Journal of Police and Criminal Psychology, 19*(1), 27–35.

Tuerk, E. H., & Loper, A. B. (2006). Contact between incarcerated mothers and their children: Assessing parenting stress. *Journal of Offender Rehabilitation, 43*, 23–43.

Warren, J. I., Hurt, S., Loper, A., Bale, R., Friend, R., & Chauhan, P. (2002). Psychiatric symptoms, history of victimization, and violent behavior among incarcerated female felons: An American perspective. *International Journal of Law and Psychiatry, 25*, 129–149.

Wilczak, G. L., & Markstrom, C. A. (1999). The effects of parent education on parental locus of control and satisfaction of incarcerated fathers. *International Journal of Offender Therapy & Comparative Criminology, 43*(1), 90–102.

DISCUSSION QUESTIONS

1. Why have prisons implemented parenting programs in recent years?

2. According to the authors, what types of parenting programs are most effective? Which are least effective? Why?

3. How do male and female inmates vary in their need for and receptiveness to parenting programs? How do the authors explain this difference?

❖

READING

This article by Stanton-Tindall and her colleagues examines gender differences in treatment engagement, psychosocial variables, and criminal thinking among a sample of male and female substance abusers. The sample was drawn from inmates enrolled in 20 prison-based treatment programs in five different states as part of the Criminal Justice Drug Abuse Treatment Studies cooperative agreement funded by the National Institute on Drug Abuse. The authors state that inmates in female treatment programs report more psychosocial dysfunction, less criminal thinking, and higher engagement than do inmates in male facilities, and there is a more negative relationship between psychosocial variables and treatment engagement among those in female as compared to male programs. They also find that criminal thinking had a significant gender interaction, with males showing a significantly stronger relationship between cold-heartedness and low treatment engagement.

Gender Differences in Treatment Engagement Among a Sample of Incarcerated Substance Abusers

Michele Staton-Tindall, Bryan R. Garner, Janis T. Morey,
Carl Leukefeld, Jennifer Krietemeyer, Christine A. Saum, and Carrie B. Oser

Prior research has identified differences between men and women receiving community-based substance abuse treatment. For example, women in community treatment are often less likely to report an employment history, less likely to have an arrest record, and more likely to report histories of sexual and physical abuse than men (Bartholomew, Rowan-Szal, Chatham, Nucatola, & Simpson, 2002; Grella & Joshi, 1999; Jainchill, Hawke, & Yagelka, 2000; McCance-Katz, Carroll, & Rounsaville, 1999; Robinson, Brower, & Gomberg, 2001; Wallen, 1992; Westermeyer & Boedicker, 2000). In addition, studies have shown that men entering substance abuse treatment are more likely to be married, but women are more likely to be divorced; and women are more likely to have children living with them, more likely to have a low income, and less likely to be able to meet their financial needs (Chatham, Hiller, Rowan-Szal, Joe, & Simpson, 1999; Robinson et al., 2001; Rowan-Szal, Chatham, Joe, & Simpson, 2000).

Differences in drug-use characteristics of men and women also have been noted in the literature. For example, it has been suggested that women may become addicted to drugs and/or alcohol more easily than men because of biological or metabolic differences (Blume, 1999; Frezza, Di Padova, & Pozzato, 1990). Women also may engage in drug use in a manner that is closely linked to their relationships with significant others, particularly males (Boyd, 1993; Hutchins, 1995). With regard to frequency and type of use, women in community treatments are more likely to report crack cocaine as their drug of choice (Arfken, Klein, di Menza, & Schuster, 2001; Wechsberg, Craddock, & Hubbard, 1998) as well as to report more days of cocaine use in the month before treatment entry compared to men (Robinson et al., 2001). Men drink alcohol more frequently and in larger amounts than women do, and this finding has been noted across multiple samples of substance users (Minugh, Rice, & Young, 1998; Nunes-Dinis & Weisner, 1997; Robinson et al., 2001; Staton et al., 1999a, 1999b). In addition, men entering community treatment are more likely to be diagnosed with alcohol abuse or dependence than are women (Robinson et al., 2001).

These findings suggest that much of what is known about gender differences among substance abusers is derived from community-treatment samples. However, there has been an emerging body of literature in the past few years on gender differences among samples of incarcerated men and women. Similar to community samples, these studies suggest that, compared to incarcerated men, incarcerated women are more likely to be polysubstance abusers, more likely to have experienced sexual and/or physical abuse, more likely to have a lifetime episode of serious depression, more likely to report poor health, and more likely to have been involved in problem relationships (Blitz, Wolff, Pan, & Pogorzelski, 2005; Langan & Pelissier, 2001; Messina, Burdon, Hagopian, & Prendergast, 2006; Pelissier, Camp, Gaes, Saylor, & Rhodes, 2003; Pelissier & Jones, 2005; Peters, Strozier, Murrin, & Kearns, 1997; Sheridan, 1996). Men are more likely to report being employed prior to incarceration, more likely to have an antisocial personality diagnosis, and more likely to have a prior criminal commitment (Blitz et al., 2005; Messina et al., 2006; Pelissier et al., 2003).

Gender differences likewise have been noted in factors associated with substance-abuse-treatment entry and treatment outcomes among incarcerated samples. Specifically, one study found that severity of the current offense, a longer sentence, and having a history of violence were associated with not completing treatment for males, whereas having a depression diagnosis was negatively associated with treatment entry for incarcerated women (Pelissier, 2004). In addition, factors associated with postrelease relapse for incarcerated males include higher severity of pre-incarceration drug use and not living with a spouse, compared to mental health issues and mental health treatment for females (Messina et al., 2006; Pelissier et al., 2003).

Source: Staton-Tindall, M., Garner, B. R., Morey, J. T., Leukefeld, C., Krietemeyer, J., Saum, C. A., & Oser, C. B. (2007). Gender Differences in Treatment Engagement Among a Sample of Incarcerated Substance Abusers. *Criminal Justice and Behavior 34*(9), 1143–1156. Reprinted with permission of Sage Publications, Inc.

These studies suggest that there are differences between men and women that include demographic characteristics, differences in patterns and intensity of drug use and related problems, and differences in factors associated with treatment entry, retention, and completion. These differences are consistently reported across community and correctional samples. The research literature has not, however, focused clearly on factors that may differentially influence treatment engagement among males and females. Treatment engagement has been described as "early engagement," meaning "the extent to which new admissions show up and actively engage in their role as patient" (Simpson, 2004, p. 106). Treatment engagement also has been described as "therapeutic engagement," meaning "cognitive appraisals of commitment to the treatment episode and recovery" (Hiller, Knight, Leukefeld, & Simpson, 2002, p. 64). In general, it has been shown that measures of treatment engagement—including treatment participation (i.e., session attendance and psychological engagement), counselor rapport (i.e., the therapeutic relationship, therapeutic alliance, and client/counselor bonding), and client confidence in treatment (i.e., satisfaction with services and access to counselors)—contribute to positive behavioral changes and psychological functioning in treatment (Simpson, 2004; Simpson & Knight, 2001). The present study built on findings related to treatment engagement by examining factors that might differentially influence treatment engagement within male and female prison-based treatment facilities.

One factor that may be differentially related to treatment engagement for males and females is psychosocial functioning. Studies have consistently indicated that substance-abusing women tend to have more psychological distress, particularly depression and anxiety, when compared to men (i.e., Brady, Grice, Dustan, & Randall, 1993; Gray & Saum, 2005; Rowan-Szal et al., 2000; Sacks, 2004). For example, female substance users typically score higher on measures of anxiety when compared to males in numerous treatment settings, including inpatient and outpatient substance abuse treatment (McCance-Katz et al., 1999), therapeutic communities (Jainchill et al., 2000), methadone treatment (Chatham et al., 1999; Rowan-Szal et al., 2000), and multiple community-treatment modalities (Compton et al., 2000). In addition, among participants admitted to community drug treatment, women have reported higher rates of mood disorders, phobias, panic disorder, and obsessive-compulsive disorder than men (Compton et al., 2000). Co-occurring disorders of substance abuse and mental health also are common among females in the criminal justice system (Sacks, 2004), and the noted findings among community-treatment populations are consistent for females in correctional-based treatment programs (Messina et al., 2006; Pelissier et al., 2003; Peters et al., 1997; Sheridan, 1996), especially when considering psychological issues such as depression and anxiety.

In addition to these commonly noted issues, other studies have found that women in substance abuse treatment scored significantly higher on measures of hostility when compared to men (Petry & Bickel, 2000; Robinson et al., 2001). Although this finding may be counterintuitive, one study found that hostility scores actually predicted early treatment termination for female clients (Petry & Bickel, 2000). Females also reported lower self-esteem, lower confidence in decision making, and lower scores on social conformity when compared to males (Rowan-Szal et al., 2000), but they demonstrated more improvements in these indicators following treatment when compared to males (Chatham et al., 1999). These indicators of psychosocial functioning for women—particularly hostility, risk taking, and depression—can be more problematic if women have a history of sexual abuse and victimization (Bartholomew et al., 2002). Consequently, psychosocial problems may be more common among females than males admitted to treatment, which might differentially influence treatment engagement for women.

Another factor that may be differentially related to treatment engagement for male and female offenders is criminal thinking. Walters (1995) conceptualized criminal thinking as the cognitive thought processes that are believed to precede criminal behavior. This is typically more commonly associated with males because of traditional masculine attitudes (Walters, 2001) and is closely aligned with thought patterns associated with antisocial personality disorder, which likewise tends to be more prevalent among males (Compton et al., 2000; Jainchill et al., 2000). For male offenders, criminal thinking can have a powerful effect on drug use, criminal acts, and increasing rates of recidivism (Walters, 1998). Walters developed a scale (Psychological Inventory of Criminal Thinking; Walters, 1995, 2002), which measures domains of criminal thinking. The scale has been normed on male (Walters, 1995) and female (Walters, Elliott, & Miscoll, 1998) inmate populations. Although studies on gender differences in criminal thinking are limited in the research literature, one study found that females scored higher than males on only one dimension of criminal thinking that included problem avoidance (i.e., suggesting more avoidance of problems by eliminating them, finding a shortcut, or distraction; Walters, 2001). Thus, these findings suggest that criminal thinking is more readily recognized among male offenders. Although it appears that criminal thinking is associated with criminal behavior and subsequent risk for recidivism (Knight, Garner, Simpson, Morey, & Flynn, 2006), limited attention has been given to criminal thinking as a potential influence on treatment engagement.

Most of the gender research in the treatment literature focuses on individual differences. However, research examining gender differences across treatment programs has been limited. In general, male and female offenders receive substance abuse treatment in gender-specific prison-based treatment programs, which have been found to vary in both content and structure (Bouffard & Taxman, 2000). These programmatic differences may be problematic for traditional individual-level analytic approaches because program-level differences may result in offenders from the same program being systematically more similar to one another than to offenders from other programs. Therefore, the current study explored gender differences at the program level, which took into account gender-specific clustering in treatment programs as well as potential treatment programming differences.

Few studies have been found that focus on treatment engagement in relation to gender-related factors, particularly as those factors may vary across gender-specific treatment programs. Because previous studies identified gender differences on psychosocial factors and criminal thinking, it was expected that there would be differences in the relationships between these factors and treatment engagement for male and female offenders. The study had three primary objectives: (a) to examine gender differences in treatment engagement, psychosocial functioning, and criminal thinking; (b) to examine gender as a moderator of the relationship between treatment engagement and psychosocial functioning; and (c) to examine gender as a moderator of the relationship between treatment engagement and criminal thinking. It was expected that there would be a stronger relationship between high treatment engagement and high psychosocial functioning in female programs than in male programs. It was also expected that there would be a stronger relationship between measures of high treatment engagement and low scores on the criminal thinking measures for male programs when compared to female programs.

▧ Method

Participants

As part of the Performance Indicators for Corrections study, in the National Institute on

Drug Abuse–funded Criminal Justice Drug Abuse Treatment Studies (CJ-DATS) project, participants were recruited from five collaborating research centers, including Texas Christian University (TCU), University of Kentucky, National Development and Research Institutes, University of Delaware, and University of California at Los Angeles. Participants included a total sample of 3,266 male and female offenders enrolled in 26 prison-based treatment programs (17 male programs, 6 female programs, and 3 mixed-gender programs) drawn from six states. All clients who were currently enrolled in treatment at these facilities were eligible to participate in the survey. However, the sample for the present study included all participants with available data for age, race or ethnicity, and time in treatment, which were used as control variables, as well as those who were in all-male or all-female programs for gender comparisons. This reduced sample included 1,950 male offenders from 15 programs and 824 female offenders from 5 programs ($N = 2,774$ offenders from 20 treatment programs in five states). Participants were mostly non-White (56%), with a mean age of 34 years and a mean length of time in treatment of 141 days. These demographics did not differ significantly by gender.

Measures

The CEST (Client Evaluation of Self and Treatment) was originally developed at TCU (Joe, Broome, Rowan-Szal, & Simpson, 2002), and recent revisions (Garner, Knight, Flynn, Morey, & Simpson, 2007) were made to the assessment to make it appropriate for use in criminal justice correctional settings. The CJ CEST uses a 5-point, Likert-type scale from 1 (*disagree strongly*) to 5 (*agree strongly*) and includes scales that represent conceptually distinct key factors as delineated in the TCU treatment process model (Simpson, 2004; Simpson & Knight, 2001). For these analyses, subscales measuring psychosocial functioning, criminal thinking, and treatment engagement were examined. The psychometric properties,

including reliability and validity of each of the scales used in this study, are discussed in Garner et al. (2007).

Psychosocial functioning Scores on six psychosocial subscales were examined in this study: Anxiety, Depression, Self-Esteem, Decision Making, Risk Taking, and Hostility. Higher scores on Anxiety, Depression, Hostility, and Risk Taking indicate poorer functioning, whereas higher scores on Self-Esteem and Decision Making indicate better functioning. Anxiety has seven items, including measures of excessive nervousness, increased arousal and/or apprehension, restlessness, difficulty concentrating, irritability, muscle tension, disturbed sleep, and fear. The Depression scale (six items) measures the degree by which the participant is experiencing pervasive feelings of sadness, fatigue, worry, loneliness, low self-esteem, and hopelessness. The Self-Esteem scale (six items) is an indicator of self-respect that is assessed by feelings of self-satisfaction, pride, and the sense of one's own value or worth as a person. Decision Making (nine items) provides information about the act or process of deciding, future planning, problem solving, risk assessment, and the impact that one's actions will have on others. The Hostility scale, with its eight items, is a measure of the participant's level of self-reported angry feelings and aggression and behaviors associated with that anger.

Risk Taking (seven items) assesses the preferences of the participant for taking chances, reckless living, and choosing "wild" friends.

Criminal thinking The addition of the Criminal Thinking scales to the CJ CEST included rewording and modifying versions of scales developed by Walters (1995; Walters & Geyer, 2005) and used in the Federal Bureau of Prisons (BOP) Survey of Program Participants (available from the BOP Office of Research and Evaluation: http://www.bop.gov). For each of the Criminal Thinking subscales, a higher score indicates greater negativity in the criminal thinking. The subscales include Cold-Heartedness, Entitlement, Personal Irresponsibility, Criminal Rationalization, Justification, and Power Orientation. Cold-Heartedness (eight items) portrays the depth (or lack) of emotional involvement that the offender has in his or her relationships with others. Entitlement (seven items) is indicative of the extent to which an individual feels ownership of privileges or benefits that are automatic and unrelated to societal restrictions (i.e., the world "owes them," they deserve special consideration, and they perceive themselves to be above the law). Personal Irresponsibility (eight items) involves a lack of accountability and a general unwillingness to accept ownership for actions and for choices and is also reflective of an unwillingness to accept responsibility, including a readiness to cast blame on others. Criminal Rationalization (eight items) measures a generally negative attitude toward law and authority figures. Offenders who score high on this scale view their behaviors as being no different than the criminal acts that they believe are committed every day by authority figures. Justification (six items) reflects a thinking pattern characterized by minimizing the seriousness of antisocial acts and justifying actions based on external circumstances. Power Orientation (nine items) measures the need for power and control. Offenders who score high on this scale typically show an outward display of aggression in an attempt to control their external environment and try to achieve a sense of power by manipulating others.

Treatment engagement Two subscales from the treatment engagement domain—Counselor Rapport and Treatment Participation—were included in this analysis. These two subscales were included because they have been established in other studies as core measures of treatment engagement (Simpson & Joe, 2004). Higher scores indicate more positive response to treatment generally or to a specific element of treatment. Treatment Participation (12 items)

measures willingness to participate in group sessions and to receive and provide input from counselors and peers. Counselor Rapport (13 items) focuses on the relationship between the participant and his primary counselor, providing information about the extent to which the counselor is perceived as helpful (e.g., encourages, understands, motivates, respects, prepares).

Procedures

Prior to data collection, project approval was obtained from each of the participating CJ-DATS Research Center's Institutional Review Board and each state's correctional department. Written consents and protected health-information authorizations were obtained from individuals who volunteered to participate after a full explanation of the project and a question-and-answer period. Participants had been in substance abuse treatment longer than 2 weeks, although they were at various points in the treatment process at the time of assessment.

During survey administration, participants were directed to follow along as designated staff read the directions and each item aloud for approximately 25 respondents per group. Reading items aloud helped to accommodate those with poorer reading and language skills. Also, whenever necessary, bilingual participants were paired with non-English-speaking participants to translate the items being read aloud. Although participants were encouraged to seek clarification for items that they did not understand, they were also instructed to leave items blank that they did not feel comfortable answering or that continued to be unclear after an explanation was given. The assessment took approximately 1 hour to administer.

Analytic Overview

In the present study, because offenders were clustered within different treatment programs, a multilevel analytic approach using hierarchical linear modeling (HLM; Raudenbush & Bryk, 2002) was deemed necessary. For example, because offenders in different treatment programs were included, it is likely that those from the same program shared more similarities with one another than with offenders from other programs. This systematic similarity among offenders within the same program (e.g., because of factors such as regional variations, prison cultures, placement practices, etc.) violates the ordinary least squares regression assumption of independent error terms and may lead to misestimated standard errors. HLM has the capacity to overcome this limitation by specifying relationships among offender measures within programs and allowing these relationships to vary from one program to another. In addition, because treatment programs in the current study were either all-male or all-female, gender (as a variable) did not vary within any specific treatment program. Consequently, gender is a characteristic of the program and any differences by gender should be analyzed at the program level (i.e., program gender).

A series of ANCOVA models with random effects was conducted first to examine the degree to which male and female programs differed in ratings for each of the scales (controlling for background differences such as age, race, and length of time in treatment). To meet the second study objective, HLM was used to examine gender as a moderator of the relationship between offenders' engagement in treatment and their ratings of psychosocial functioning, controlling for age, race, and length of time in treatment. The third study objective was addressed by conducting an HLM analysis to examine gender as a moderator of the relationship between offenders' engagement in treatment and their scores on criminal thinking measures, controlling for age, race, and length of time in treatment.

⬛ Results

Gender Differences

The first set of analyses examined the degree to which male and female programs differed in their ratings for each of the measures, controlling for age, race or ethnicity, and length of time in treatment (see Table 1). For measures of psychosocial functioning, there were significant differences on the Anxiety scale at $p < .05$, with the female programs reporting greater anxiety. Although not statistically significant when compared to males, female programs had higher ratings for depression and lower ratings for self-esteem, hostility, and risk taking. With respect to criminal thinking, the male programs reported significantly higher scores than the female programs for each of the subscales. Female programs reported significantly higher scores on each of the treatment engagement measures, including counselor rapport and treatment participation.

Treatment Engagement and Psychosocial Functioning

The second set of analyses examined gender as a moderator of the relationship between offenders' engagement in treatment and ratings of psychosocial functioning, controlling for age, race, and length of time in treatment (see Table 2). Overall, significant relationships were found between the psychosocial variables and engagement, with the exception of the relationship between risk taking and counseling rapport ($p = .264$). Better psychosocial functioning (e.g., lower depression, anxiety, hostility, risk taking; higher self-esteem and decision making) was significantly related to better treatment engagement, as indicated by higher ratings of both counselor rapport and treatment participation.

A number of significant cross-level interactions were found, suggesting that gender of the program was a moderator of the relationship between psychosocial functioning and treatment engagement. Specifically, female programs had a significantly stronger negative relationship between counselor rapport and depression ($p = .016$), anxiety ($p = .014$), and decision making ($p = .006$) compared to male programs. Similarly, female programs also had

Table 1	Average Scores on Psychosocial Functioning, Criminal Thinking, and Treatment Engagement for Male and Female Programs (controlling for age, race, and length of time in treatment)				
Subscale	Male Programs	Female Programs	t Ratio	p	Effect Size
Psychosocial Functioning					
Depression	23.52	25.32	−1.99	.061	0.94
Anxiety	26.51	29.17	−2.94	.009	1.39
Self-Esteem	37.27	36.21	1.23	.236	0.58
Decision Making	37.45	37.58	−0.22	.828	0.10
Hostility	25.82	23.86	1.79	.091	0.84
Risk Taking	29.68	28.94	0.80	.435	0.38
Criminal Thinking					
Entitlement	20.01	17.44	4.44	<.000	2.09
Justification	21.63	20.16	2.31	.033	1.09
Power Orientation	26.69	23.94	3.29	.004	1.55
Cold-Heartedness	23.30	20.40	3.77	.002	1.78
Criminal Rationalization	32.60	28.88	3.85	.001	1.81
Personal Irresponsibility	21.99	18.55	4.30	<.000	2.03
Treatment Engagement					
Counselor Rapport	35.50	41.84	−2.92	.010	1.38
Treatment Participation	40.62	43.39	−2.64	.017	1.24

Note: Based on 20 treatment programs and 2,774 offenders.

stronger relationships between treatment participation and depression ($p = .029$), anxiety ($p = .010$), poor decision making ($p < .000$), and hostility ($p = .044$) compared to the male programs. Last, male programs had significantly stronger relationships between low risk taking and increased counselor rapport ($p = .021$).

Treatment Engagement and Criminal Thinking

The third set of analyses examined gender as a moderator of the relationship between offenders' engagement in treatment and criminal thinking scores, controlling for age, race, and length of time in treatment (see Table 3). Results revealed a significant association between criminal thinking and treatment engagement, with the exception of power orientation and counselor rapport ($p = .187$). Generally, higher criminal thinking scores were significantly related to poorer treatment engagement, as indicated across ratings of both counselor rapport and treatment participation. However, only the relationship between cold-heartedness and treatment engagement (both counselor rapport and treatment participation) was significantly different by program gender. Male programs reported stronger relationships between high cold-heartedness and low treatment engagement. Results suggest that, although higher scores on the Criminal Thinking scales are associated with lower scores on treatment engagement overall, this relationship did not differ significantly by program gender, despite the earlier finding that male programs report significantly higher levels of criminal thinking.

| Table 2 | Gender of Program as a Moderator of the Relationship Between Psychosocial Functioning and Engagement (controlling for age, race, and length of time in treatment) | | | | | | | |

	Counselor Rapport				Treatment Participation			
Scale	*B*	*SE*	*t* Ratio	*p*	*B*	*SE*	*t* Ratio	*p*
Depression (DP)								
DP slope	−.24	.04	−6.62	<.000	−.18	.03	−6.84	<.000
Male Program × DP Slope	.11	.04	2.42	.016	.07	.03	2.19	.029
Anxiety (AX)								
AX slope	−.17	.03	−5.28	<.000	−.12	.02	−4.90	<.000
Male Program × AX Slope	.10	.04	2.46	.014	.07	.03	2.58	.010
Self-Esteem (SE)								
SE slope	.16	.03	4.76	<.000	.14	.02	5.78	<.000
Male Program × SE Slope	−.05	.04	−1.28	.200	.02	.03	0.79	.430
Decision Making (DM)								
DM slope	.16	.05	3.16	.002	.32	.03	9.82	<.000
Male Program × DM Slope	.16	.06	2.77	.006	.20	.04	5.38	<.000
Hostility (HS)								
HS slope	−.12	.03	−3.83	<.000	−.13	.02	−5.92	<.000
Male Program × HS Slope	.02	.04	0.59	.558	.05	.03	2.02	.044
Risk Taking (RT)								
RT slope	−.04	.04	−1.12	.264	−.10	.03	−4.11	<.000
Male Program × RT Slope	−.10	.04	−2.32	.021	−.03	.03	−1.04	.300

NOTE: Based on 20 treatment programs and 2,774 offenders. *B* values should be interpreted with females as the referent category.

✑ Discussion

Differences between male and female substance users have been identified in the literature, including patterns and intensity of drug use and related problems, psychological functioning, criminal thinking, and factors associated with treatment entry and treatment completion. Much of this research has targeted individual differences in male and female clients and has been conducted in single community- and corrections-based substance-abuse-treatment programs. In addition, there

has been a limited amount of research on treatment engagement and gender-related factors that might be associated with measures of treatment engagement, particularly as factors vary across gender-specific substance-abuse-treatment programs. This gap in the literature was the focus of the current study.

The first objective of this study was to examine differences between male and female programs in treatment engagement, psychosocial functioning, and criminal thinking. With regard to psychosocial functioning, female programs reported significantly higher levels

Table 3	Gender of Program as a Moderator of the Relationship Between Criminal Thinking and Engagement (controlling for age, race, and time in treatment)

	Counselor Rapport				Treatment Participation			
Scale	*B*	*SE*	*t* Ratio	*p*	*B*	*SE*	*t* Ratio	*p*
Entitlement (EN)								
EN slope	−.23	.05	−4.56	<.000	−.26	.03	−7.55	<.000
Male Program × EN Slope	.02	.06	0.28	.781	.02	.04	0.56	.577
Justification (JU)								
JU slope	−.08	.04	−2.06	.039	−.12	.03	−4.48	<.000
Male Program × JU Slope	−.01	.05	−0.11	.911	−.02	.03	−0.65	.513
Power Orientation (PO)								
PO slope	−.05	.03	−1.57	.116	−.11	.02	−4.45	<.000
Male Program × PO Slope	−.05	.04	−1.32	.187	.03	.03	1.04	.301
Cold-Heartedness (CH)								
CH slope	−.14	.04	−3.19	.002	−.21	.03	−7.16	<.000
Male Program × CH Slope	−.26	.05	−5.07	<.000	−.24	.03	−7.03	<.000
Criminal Rationalization (CN)								
CN slope	−.23	.04	−6.55	<.000	−.11	.03	−4.39	<.000
Male Program × CN Slope	.06	.04	1.48	.138	.03	.03	1.15	.249
Personal Irresponsibility (PI)								
PI slope	−.22	.04	−5.05	<.000	−.22	.03	−7.23	<.000
Male Program × PI Slope	.02	.05	0.42	.678	.04	.04	1.04	.299

NOTE: Based on 20 treatment programs and 2,774 offenders. *B* values should be interpreted with females as the referent category.

of anxiety than did male programs. Although not statistically significant, consistent trends were also observed, with female programs reporting higher ratings for depression and lower ratings for self-esteem, hostility, and risk taking (compared to male programs). For criminal thinking, significant gender differences also were found in the expected direction, with male programs reporting higher scores overall. On each of the treatment engagement measures, including counselor rapport and treatment participation, female programs reported significantly higher scores.

The finding that female programs reported higher anxiety scores is consistent with previous research because female substance users have been consistently shown to score higher, when compared to males, in a number of treatment settings (Chatham et al., 1999; Compton et al., 2000; Jainchill et al., 2000; McCance-Katz et al., 1999; Peters et al., 1997; Rowan-Szal et al., 2000; Sacks, 2004). Despite trends in the expected direction, however, other psychosocial measures did not reach statistical significance when compared by program gender. One possible explanation for the lack of significance is the lower statistical power associated with examining differences at the program level ($N = 20$).

The second study objective was to examine program-based gender as a moderator of the relationship between treatment engagement and psychosocial functioning. It was expected that there would be stronger relationships between psychosocial functioning and treatment engagement in female programs than in male programs. Evidence supported the hypothesis that the relationship between treatment engagement and psychosocial functioning was moderated by gender. Specifically, depression, anxiety, and decision making were more negatively associated with measures of treatment engagement for female programs than male programs. In addition, measures of hostility were more negatively associated with the treatment participation measure within female programs.

Because the literature suggests that psychosocial problems are more common among female treatment clients, it was suspected that treatment engagement in female programs would be more influenced by these problems. The analytic approach used provided more than simple group comparisons to help understand relationships that may be unique within male and female programs. Findings suggest that, in female programs, the degree of severity of depression and anxiety was associated with lower scores on measures of counselor rapport and treatment participation. In addition, female programs whose inmates reported less hostility and more confidence in future planning, problem solving, and assessing risks also had higher treatment participation.

The third study objective was to examine gender as a moderator of the relationship between treatment engagement and criminal thinking. It was expected that there would be a stronger relationship between high treatment engagement and reduced criminal thinking in male programs than in female programs. Overall, higher scores on the Criminal Thinking scales were significantly related to

lower treatment engagement. This relationship, however, did not differ consistently by gender. The one exception was the finding that cold-heartedness and lower treatment engagement were more strongly related in male programs (compared to female programs).

Studies on gender differences in criminal thinking are limited in the research literature. Although research on criminal thinking has been conducted more often with male samples, findings from this study are consistent with those reported by Walters (2001), who did not find significant gender differences on most of the criminal thinking factors examined. These findings are therefore notable in that, although male programs scored significantly higher on each of the Criminal Thinking scales (compared to female programs), the relationship between criminal thinking and treatment engagement was generally similar for both types of programs. The only exception involved Cold-Heartedness, and it is possible that this subscale taps a construct that is more closely related to antisocial personality characteristics that are more commonly associated with males. The influence of the Criminal Thinking subscales on measures of treatment process seems to warrant further research, particularly to understand distinctions in these measures among male and female offenders.

This study has some limitations. Participants were not randomly selected, but all clients enrolled in one of the targeted drug treatment programs were eligible for the study (and participation rates were very high). Nevertheless, generalizability to the larger population of male and female clients in prison-based treatment programs may be limited based on potential selectivity of inmate attributes. Another limitation is that data were self-reported, and participant honesty in responding to the questions is a possible concern. However, Garner et al. (2007) report solid reliability and validity measures for the scales and data used in this study. Another possible limitation is that the influence of only two primary factors on treatment engagement was examined. There are a variety of other factors that might also influence a client's engagement in treatment, particularly when surveyed at one point in time. Additional factors, such as individual differences and motivation for treatment, are therefore being considered for future analyses.

Despite these limitations, findings from this study provide notable contributions to the literature. For example, research has shown that attention to the unique differences between males and females, particularly with regard to their response to supervision, can promote better outcomes for offenders (Bloom, Owen, & Covington, 2005). Overall, females reported a higher level of treatment engagement (both counselor rapport and treatment participation) than did males. However, these findings may suggest that, at an individual level, a woman's level of depression and anxiety when in treatment may negatively affect her ability to build counselor rapport and to actively participate in treatment. In addition, women's ability to engage in treatment also may be affected by confidence in decision making and hostile feelings. Assessing for these factors early in treatment and incorporating them into treatment planning could enhance treatment engagement for women.

These implications are consistent with a recent report that proposed guiding principles for gender-responsive strategies for corrections policy and programming (Bloom et al., 2005). This plan suggests that it is important to incorporate treatment for mental health issues into corrections-based programming for women and to begin to understand and focus on factors that contribute to depression, anxiety, and other issues such as a history of victimization and violence. It is particularly important to incorporate these services in prison settings because a large number of women experiencing depression, anxiety, and other mental health issues do not receive services on the street (Staton, Leukefeld, & Webster, 2003).

Findings from this study also have implications for male prison-based treatment programs. Male programs consistently scored higher on each of the Criminal Thinking subscales. In addition, the Cold-Heartedness subscale (described as a general lack of emotional involvement with others) was negatively related to treatment engagement among male program participants. Identifying and addressing this thinking pattern early in the treatment process could help treatment providers increase treatment engagement among male substance-using offenders. Although presently there are no customized "best practice" guidelines for working with male offenders in treatment, there is evidence to suggest a link between criminal thinking, the incidence of criminal behavior, and subsequent risk for recidivism for male offenders (Gendreau, Little, & Goggin, 1996; Knight et al., 2006). Therefore, targeted programming that focuses on criminal thought processes, along with treating substance abuse issues, might enhance treatment engagement and subsequent treatment outcomes for male offenders.

Examining factors that influence treatment engagement is important for future research because clients who do not engage in treatment are less likely to complete treatment and less likely to have positive treatment outcomes (Rowan-Szal, Flynn, & Simpson, 2002). By assessing factors that may influence treatment engagement—such as psychosocial issues and criminal thinking—early in the treatment process, programs may be able to target treatment interventions designed to address these problems in an effort to enhance the retention rates, treatment experience, and treatment outcomes for their clients. These targeted interventions should consider gender-specific features and include a focus on psychosocial issues among female offenders and criminal thinking for male offenders. Targeted interventions also should be grounded in research examining additional factors that may differentially influence treatment engagement among male and female offenders.

⊠ References

Arfken, C. L., Klein, C., di Menza, S., & Schuster, C. R. (2001). Gender differences in problem severity at assessment and treatment retention. *Journal of Substance Abuse Treatment, 20,* 53–57.

Bartholomew, N. G., Rowan-Szal, G. A., Chatham, L. R., Nucatola, D. C., & Simpson, D. D. (2002). Sexual abuse among women entering methadone treatment. *Journal of Psychoactive Drugs, 24,* 347–354.

Blitz, C. L., Wolff, N., Pan, K., & Pogorzelski, W. (2005). Gender-specific behavioral health and community release patterns among New Jersey prison inmates: Implications for treatment and community reentry. *American Journal of Public Health, 95,* 1741–1746.

Bloom, B., Owen, B., & Covington, S. (2005). *A summary of research, practice, and guiding principles for women offenders* (No. 020418). Washington, DC: U.S. Department of Justice, National Institute of Corrections.

Blume, S. B. (1999). Addiction in women. In M. Galanter & H. D. Kleber (Eds.), *Textbook of substance abuse treatment* (2nd ed., pp. 485–490). Washington, DC: American Psychiatric Press.

Bouffard, J. A., & Taxman, F. S. (2000). Client gender and the implementation of jail-based therapeutic community programs. *Journal of Drug Issues, 30,* 881–901.

Boyd, C. J. (1993). The antecedents of women's crack cocaine abuse: Family substance abuse, sexual abuse, depression and illicit drug use. *Journal of Substance Abuse Treatment, 10,* 433–438.

Brady, K. T., Grice, D. E., Dustan, L., & Randall, C. (1993). Gender differences in substance use disorders. *American Journal of Psychiatry, 150,* 1707–1711.

Chatham, L. R., Hiller, M. L., Rowan-Szal, G. A., Joe, G. W., & Simpson, D. D. (1999). Gender differences at admission and follow-up in a sample of methadone maintenance clients. *Substance Use & Misuse, 34,* 1137–1165.

Compton, W. M., Cottler, L. B., Abdallah, A. B., Phelps, D. L., Spitznagel, E. L., & Horton, J. C. (2000). Substance dependence and other psychiatric disorders among drug dependent subjects: Race and gender correlates. *The American Journal on Addictions, 9,* 113–125.

Frezza, M., Di Padova, C., & Pozzato, G. (1990). High blood alcohol levels in women: The role of decreased gastric alcohol dehydrogenase activity and first-pass metabolism. *New England Journal of Medicine, 322,* 95–99.

Garner, B. R., Knight, K., Flynn, P. M., Morey, J. T., & Simpson, D. D. (2007). Measuring offender attributes and engagement in treatment using the Client Evaluation of Self and Treatment. *Criminal Justice and Behavior, 34,* 1113–1130.

Gendreau, P., Little, T., & Goggin, C. (1996). A meta-analysis of the predictors of adult offender recidivism: What works? *Criminology, 34,* 575–607.

Gray, A. R., & Saum, C. A. (2005). Mental health, gender, and drug court completion. *American Journal of Criminal Justice, 30,* 55–69.

Grella, C. E., & Joshi, V. (1999). Gender differences in drug treatment careers among clients in the National Drug Abuse Treatment Outcome Study. *American Journal of Drug and Alcohol Abuse, 25,* 385–406.

Hiller, M. L., Knight, K., Leukefeld, C., & Simpson, D. D. (2002). Motivation as a predictor of therapeutic engagement in mandated residential substance abuse treatment. *Criminal Justice and Behavior, 29,* 56–75.

Hutchins, E. (1995). *Psychosocial risk factors associated with drug use during pregnancy.* Unpublished doctoral dissertation, Johns Hopkins University, Baltimore, MD.

Jainchill, N., Hawke, J., & Yagelka, J. (2000). Gender, psychopathology, and patterns of homelessness among clients in shelter-based TCs. *American Journal of Drug and Alcohol Abuse, 26,* 553–567.

Joe, G. W., Broome, K. M., Rowan-Szal, G. A., & Simpson, D. D. (2002). Measuring patient attributes and engagement in treatment. *Journal of Substance Abuse Treatment, 22,* 183–196.

Knight, K., Garner, B. R., Simpson, D. D., Morey, J. T., & Flynn, P. M. (2006). An assessment for criminal thinking. *Crime & Delinquency, 52,* 159–177.

Langan, N. P., & Pelissier, B.M.M. (2001). Gender differences among prisoners in drug treatment. *Journal of Substance Abuse Treatment, 13,* 291–301.

McCance-Katz, E. F., Carroll, K. M., & Rounsaville, B. J. (1999). Gender differences in treatment-seeking cocaine abusers: Implications for treatment and prognosis. *The American Journal on Addictions, 8,* 300–311.

Messina, N., Burdon, W., Hagopian, G., & Prendergast, M. (2006). Predictors of prison-based treatment outcomes: A comparison and men and women participants. *American Journal of Drug and Alcohol Abuse, 32,* 7–28.

Minugh, P. A., Rice, C., & Young, L. (1998). Gender, health beliefs, health behaviors, and alcohol consumption. *American Journal of Drug & Alcohol Abuse, 24,* 483–497.

Nunes-Dinis, M. C., & Weisner, C. (1997). Gender differences in the relationship of alcohol and drug use to criminal behavior in a sample of arrestees. *American Journal of Drug and Alcohol Abuse, 23,* 129–141.

Pelissier, B. (2004). Gender differences in substance use treatment entry and retention among prisoners with substance use histories. *American Journal of Public Health, 94,* 1418–1424.

Pelissier, B.M.M., Camp, S. D., Gaes, G. G., Saylor, W. G., & Rhodes, W. (2003). Gender differences in outcomes from prison-based residential treatment. *Journal of Substance Abuse Treatment, 24,* 149–160.

Pelissier, B., & Jones, N. (2005). A review of gender differences among substance abusers. *Crime & Delinquency, 51,* 343–372.

Peters, R. H., Strozier, A. L., Murrin, M. R., & Kearns, W. D. (1997). Treatment of substance-abusing jail inmates. *Journal of Substance Abuse Treatment, 14,* 339–349.

Petry, N. M., & Bickel, W. K. (2000). Gender differences in hostility of opioid-dependent outpatients: Role in early treatment termination. *Drug & Alcohol Dependence, 58,* 27–33.

Raudenbush, S. W., & Bryk, A. S. (2002). *Hierarchical linear models: Applications and data analysis methods* (2nd ed.). Thousand Oaks, CA: Sage.

Robinson, E. A., Brower, K. J., & Gomberg, E. S. (2001). Explaining unexpected gender differences in hostility among persons seeking treatment for substance use disorders. *Journal of Studies on Alcohol, 62,* 667–675.

Rowan-Szal, G. A., Chatham, L. R., Joe, G. W., & Simpson, D. D. (2000). Services provided during methadone treatment: A gender comparison. *Journal of Substance Abuse Treatment, 19,* 7–14.

Rowan-Szal, G. A., Flynn, P. M., & Simpson, D. D. (2002, June). *Investigation of gender and treatment modality differences using the TCU client evaluation of self and treatment.* Poster presented at the annual meeting of the College of Problems on Drug Dependence, Quebec City, Canada.

Sacks, J. Y. (2004). Women with co-occurring substance use and mental disorders (COD) in the criminal justice system: A research review. *Behavioral Sciences & the Law, 22,* 449–466.

Sheridan, M. J. (1996). Comparison of the life experiences and personal functioning of men and women in prison. *Families in Society: The Journal of Contemporary Human Services, 77,* 423–434.

Simpson, D. D. (2004). A conceptual framework for drug treatment process and outcomes. *Journal of Substance Abuse Treatment, 27,* 99–121.

Simpson, D. D., & Joe, G. W. (2004). A longitudinal evaluation of treatment engagement and recovery stages. *Journal of Substance Abuse Treatment, 27,* 89–97.

Simpson, D. D., & Knight, K. (2001). The TCU model of treatment process and outcomes in correctional settings. *Offender Substance Abuse Report, 1,* 51–58.

Simpson, D. D., & Knight, K. (2007). Offender needs and functioning assessments from a national cooperative research program. *Criminal Justice and Behavior, 34,* 1105–1112.

Staton, M., Leukefeld, C., Logan, T., Zimmerman, R., Lynam, D., Milich, R., et al. (1999a). Gender differences in substance use and initiation of sexual activity. *Population Research and Policy Review, 18,* 89–100.

Staton, M., Leukefeld, C., Logan, T., Zimmerman, R., Lynam, D., Milich, R., et al. (1999b). Risky sex behavior and substance use among young adults. *Health and Social Work, 24,* 147–154.

Staton, M., Leukefeld, C., & Webster, J. M. (2003). Substance use, health, mental health: Problems and service utilization among incarcerated women. *International Journal of Offender Therapy and Comparative Criminology, 47,* 224–239.

Wallen, J. (1992). A comparison of male and female clients in substance abuse treatment. *Journal of Substance Abuse Treatment, 9,* 243–248.

Walters, G. D. (1995). The Psychological Inventory of Criminal Thinking Styles, Part I: Reliability and preliminary validity. *Criminal Justice and Behavior, 22,* 307–325.

Walters, G. D. (1998). *Changing lives of crime and drugs: Intervening with substance abusing offenders.* New York: John Wiley.

Walters, G. D. (2001). Relationship between masculinity, femininity, and criminal thinking in male and female offenders. *Sex Roles, 45,* 677–698.

Walters, G. D. (2002). The Psychological Inventory of Criminal Thinking Styles (PICTS). *Assessment, 9,* 278–291.

Walters, G. D., Elliott, W. N., & Miscoll, D. (1998). Use of the Psychological Inventory of Criminal Thinking Styles in a group of female offenders. *Criminal Justice and Behavior, 25,* 125–134.

Walters, G. D., & Geyer, M. D. (2005). Construct validity of the Psychological Inventory of Criminal Thinking Styles in relationship to the PAI, disciplinary adjustment and program completion. *Journal of Personality Assessment, 84,* 252–260.

Wechsberg, W. M., Craddock, S. G., & Hubbard, R. L. (1998). How are women who enter substance abuse treatment different than men? A gender comparison from the Drug Abuse Treatment Outcome Study (DATOS). *Drugs & Society, 13*, 97–115.

Westermeyer, J., & Boedicker, A. E. (2000). Course, severity, and treatment of substance abuse among women versus men. *American Journal of Drug and Alcohol Abuse, 26*, 523–539.

DISCUSSION QUESTIONS

1. According to the literature, what differences exist between men and women receiving community-based substance abuse treatment?

2. What does this study add to the literature on treatment engagement?

3. How do the findings from this study support other research that indicates there is a need for gender-responsive correctional treatment strategies?

❖

READING

Marilyn Brown's article develops what some researchers call the "pathways" approach to understanding women's criminality. This perspective argues that women's offending is an outgrowth of histories of violence, trauma, and addiction—conditioned by race, culture, gender inequality, and class. Brown expands this perspective on crime across the life course for females, providing a more nuanced analysis of the nature of intimate relationships and developmental turning points for women. Brown states that whereas men's assumption of adult responsibilities such as marriage and childrearing may be turning points away from delinquency and crime, the matter is far more complex and may even be the inverse for some women. The major findings of this paper are that women of native Hawaiian ancestry have more negative experiences with education and employment and poorer outcomes on parole compared to women without Hawaiian ancestry.

Gender, Ethnicity, and Offending Over the Life Course

Women's Pathways to Prison in the Aloha State

Marilyn Brown

Life Course Development and the Pathways Perspective

Versions of the life course development perspective highlight the links across stages of youthful development, delinquency, and adult crime and their relation to social bonds. Sampson and Laub's *Crime in the Making: Pathways and Turning Points Through Life* (1993) examines the interplay between structural and individual level variables over the life course, wherein life events mark processes of continuity and change in patterns of delinquency and offending. This genre examines the effects of social bonds that arise from affiliation with the institutions of family, school, and work and their relationship to human development over the life course. Sampson and Laub (1990) see social bonds like marriage or employment as important to explanations of changes in offending behavior. But these events are not separate from the larger influences of culture, political economy, and history. Feminist scholars have adopted concepts from life course development theories to talk about women's pathways to crime and incarceration. Owen and Belknap (2002), Belknap (2001), and Gaarder and Belknap (2002) have begun to pull together a pathways framework for understanding the gendered life course events that propel women into delinquency, crime, and ultimate involvement with the criminal processing system.

This paper is an account of women on parole in the state of Hawai'i,[1] examined in the context of change and continuity in a unique penal setting. The state is undergoing a painful transition from a mainly rural plantation-based economy to one based on its identity as a sought-after destination in a globalized visitor market. This transformation has resulted in massive economic, cultural, and social dislocations, and accompanying uncertainties which differentially affect segments of Hawai'i's population. The impact of these changes has been hardest felt by Native Hawaiian and newer immigrant groups from Asia and elsewhere in the Pacific. Indeed, the colonial transformations that ultimately alienated the indigenous people of Hawai'i from their land have been ongoing for over 200 years. This paper argues that women's offending is grounded in the larger concerns of culture, society, and history which condition their life course development, a perspective that merits more attention in theories about patterns of women's offending.

The pathways perspective, as used in recent feminist research on offending, balances attention to macro-level issues such as race, gender, and class with elements of personal action (Daly 1994; Widom 2000; Belknap 2001; Bloom et al. 2002). This framework pays significant attention to relations of power, arguing that women experience multiple forms of marginality related to delinquency and crime (Bloom 1996; Owen 1998). Women's vulnerabilities, that is, those factors that make them likely to become involved with the law, are produced by the intertwined effects of race, gender, and class (Bloom 1996) and the effects of personal agency. Studies of incarcerated women have demonstrated the ubiquitous nature of abuse and trauma (especially sexual exploitation and assault), chaotic family lives, personal losses (of family and children), involvement in violent relationships, and drug use in their lives. All of these conditions, highly structured by gender, are central to women's involvement in crime. Male offenders clearly experience many of these problems, but these factors are less clearly related to criminality and offending among males than females. Men's lives are far less likely to be defined by sexual abuse, exploitation, and violent victimization by a loved one. Nor are men's major life events marked to the same extent as women's by pervasive sexism and patriarchal oppression (Owen 1998: 3). In the pathways perspective, gendered experiences are considered crucial to explaining the different patterns of offending behavior and the reasons underlying them.

SOURCE: Brown, M. (2006). Gender, Ethnicity, and Offending Over the Life Course: Women's Pathways to Prison in the Aloha State. *Critical Criminology, 14*, 137–158. © American Society of Criminology.

The pathways approach shows that the life events and the social controls shaping outcomes for women are highly gendered, a perspective missing in Sampson and Laub's early work. Maeve, for example, describes how women's sexual victimization and experience of violence "emerge as life-shaping events" in the lives of incarcerated women (2000: 474). In connection with these traumas, Maeve continues, sexual and other forms of self-destructive, risk-taking behaviors follow as sequelae. Chief among these reactions to trauma is using drugs and alcohol to suppress or medicate feelings of anxiety, tension, and sadness.

The family setting and the dynamics of intimate relationships have a special and sometimes paradoxical meaning in the lives of female offenders. Theories that call attention to social control and women's offending have been supported by studies of how conventional families constrain girls' freedom and limit opportunities for delinquency (Bottcher 1995). Chesney-Lind and Shelden (1998) argue that a troubled home life plays a more crucial role in producing girls' delinquency than for boys. Family cuts both ways in the lives of girls and women who become involved as offenders in the criminal processing system. Families and intimate relationships can also be crucibles for the formation of addiction, delinquency, and crime. So, rather than affording protection, families may well be the source of victimization and abuse that give rise to offending among women. And, as stories told by the women in my study show, women are often introduced to drug use and related criminal activities through the significant others in their lives—often the very people upon whom they are economically or otherwise dependent. The following discussion centers on aspects of social structure and culture that are pertinent to this study of women on parole in Hawai'i.

⧊ Gender, Culture, and Offending: The Hawai'i Context

Hawai'i is making the difficult transition from a largely agrarian economy dominated by industrial agriculture to a modern service economy. Hawai'i's cultural diversity and residual stratification is largely the legacy of its plantation past and successive waves of immigration from East Asia, Pacific Islands, Europe, and the Americas. There is also the indigenous population, which faces marginalization in its own homeland under the impact of Western (and especially American) colonialism. Therefore, I adopt a more complex view of gender and structure, more in keeping with Elder's work (1985) concerning the role that historical and cultural transitions play in shaping life events of individuals. This perspective, this convergence of social action and structure, resonates with depictions of the social transitions described in accounts of the deindustrialization of urban centers and the consequent dislocations of African American families by Wilson (1987). The sorts of social, cultural, and economic transformations that have occurred in Hawai'i have conditioned the life events of women in this study.

It is important to appreciate the role of historical and cultural change in shaping who comes under the gaze of the law and why they do. Adjudicated populations (i.e., criminal offenders) come to the attention of the law not solely by virtue of their vices, but as a result of a range of personal, social, economic, and political vulnerabilities. The transition of Hawai'i from traditional rule to Western law and jurisprudence, with its early missionary-inspired obsession with sexual morality, brought large numbers of Hawaiian women and men under carceral control during the 19th century (Merry 2000). Women's reproductive and productive roles changed as a result of the very same transformations that brought so many Hawaiian women into 19th-century

Hawaiian courts. Later, the numbers of women in the courts and prisons were substantially reduced as society and the family adapted to the capitalist formations of plantation agriculture and modernist governance, formations that were less focused on sexual morality than on maintaining an orderly workforce (Brown 2003).

At the end of the 20th century, Hawai'i, along with the rest of the nation, saw explosive growth in women's incarceration. Such trends were not initially analyzed to reveal the vulnerabilities of women offenders. The emancipation hypotheses linking the emerging female crime problem with the women's movement were simplistic, ignoring the fact that women were still committing the same types of crimes they had always committed (see Anderson and Bennett 1996: 36, for discussion). The beneficiaries of the expanded field of social and economic opportunities won by second wave feminism tended to be white women, not the women of color who make up most inmate populations here in Hawai'i and the nation. As my data show, the individual biographies of the female carceral population in Hawai'i, 84% of whom are non-white, do not in the least reflect the experiences of the emancipated woman. Rather, the expansion of social, cultural, and economic capital that accompanied the growing equality of opportunity for some women as a result of more expanded educational, economic, and social rights, missed this group almost completely. Women offenders in Hawai'i were far more affected by the increasingly punitive criminal processing policies and cutbacks in social welfare of the 1980s and 1990s than by legislation that expanded rights and opportunities to other women.

⬚ Patterns of Offenses: Current and Previous Offenses

To an extent greater than for men, women's substance abuse is linked to their involvement in the criminal processing system (Chesney-Lind and Sheldon 1998; Owen and Bloom 1995). In looking at the patterns of offending within the aggregate universe of conviction offenses,[2] it is clear that the majority of women in this study were incarcerated for crimes involving property (48%) and drug offenses (44.5%). Very few crimes involved violence or crimes against persons (5.6%), and crimes having to do with public order and the like (1.8%) were uncommon. A clear majority (78.5%) of all offenses fall into Hawai'i's Felony C category: those offenses resulting in a maximum sentence of five years.[3] In terms of serious crime, very few conviction offenses (3.5%) were Class A Felonies (involving 20-year sentences) while only about 18% involved Class B Felonies. As in other jurisdictions (Maher and Daly 1996), Hawai'i women offenders involved in drug offenses tend to be low-level users and distributors indeed none of their drug-related crimes fell into the Class A Felony category. And only about 8% of such offenses were categorized as intermediate Class B felons.

Half of the women in this study had been convicted of an offense (felonies or misdemeanors) prior to their current commitment. The largest single source of previous involvement with the law is in the area of misdemeanor property crimes (35.1%) and in other sorts of misdemeanors (22.9%) like prostitution and public order offenses. Very few women (10.4%) had records for crimes against persons, and most of these were non-serious misdemeanors (7.6%) for issues like making threats (referred to as second degree terroristic threatening in Hawai'i statutes) and minor assaults. Slightly over one-third of the women in the study (36%) had a juvenile record. Although juvenile records are not available for analysis, this pattern of youthful offenses is significant for girls and young women because it often signals traumatic and chaotic home lives, followed by more serious offending as girls attempt to survive on the streets (Chesney-Lind 1997).

⌧ Intersections of Family Life and Offending

As in virtually all other jurisdictions, the majority of the women in this study (85%) are mothers. Yet this ubiquitous finding has yet to be integrated into our understandings of offending and the life course. The conventional life course development perspective, which is grounded mainly in the experience of males, suggests that the assumption of adult responsibilities would foster turning points away from crime. Moreover, early attachment to family structures theoretically strengthens self-control. But for many girls in working class and poor families, responsibilities begin early and do not always protect young women from risks. Women in my study often tell about childhoods that included domestic chores and the care of siblings—routines that continued into adulthood with their own children—that are merged in these narratives with drug use and status offenses. Behavior outside the law, and drug use in particular, do not preclude engagement with conventional gender roles like domestic chores and child-care, calling into question the supposed protective role of attachment to the family.

On average, women in this sample who had a previous offense came into the criminal processing system at age 23. Of parenting women in the sample (85%), 71% had already had at least one child at the time they were initially sentenced. These women had a relatively late entry into the criminal processing system, and many had already taken on adult responsibilities for children. These are events that life course theory suggests might spell a cessation to delinquency and offending for males. But, for women who continue on the trajectory to prison, childbearing and domestic ties do not seem to preclude continuing drug use or other types of offending.

This is a group for whom motherhood is often a troubled condition. Official records say little about the history and quality of maternal relationships with children, but some serious indications are obvious. The case files contain records of official actions or contacts with child protective services. Nearly 24% (48) of the 203 mothers in this study had been involved with Child Protective Services. In other words, there were references to such investigations in their case files. In addition, Hawai'i or some other state jurisdiction terminated the parental rights of nearly 17% of mothers. These investigative or termination proceedings involved one or more children.

My interviews with women on parole suggest two versions of lifestyle that are important to unraveling these apparent paradoxes. One is that, for some women, drug use begins in adolescence and continues on into adulthood. This pattern converges with most life course theory research, especially the control versions (see Gottfredson and Hirschi 1990) of continuity in offending. It should be noted that these women often ceased their drug use and drinking during pregnancy and during their children's infancy, sometimes resuming the lifestyle as children got older. However, a second group of women explained that their drug abuse began in the context of marriage or relationship with a significant other—often with the father of their children—as adult women. Momi, who had the first of her children when she was 16, tells a story that is typical of the second pattern:

> I was in my 20s, my youngest one wasn't in school yet. I think maybe she was three or four when I started [doing drugs]. But, that's the thing, because of him. But it's not like he pushed me to do it. It was because I wanted to lose weight and to try to please him by losing weight. But it really made no difference because the person he was married to was six feet and over 200 pounds, so it really don't matter how they look. But the way he used to treat me and the way he still treats me is that I cannot do anything [that's] . . . enough to please him.

This pattern of later onset of serious drug use, methamphetamine in this case, is unusual given that the onset of street drug use tends to precede the adoption of adult roles. But clearly, this later incidence is an important reality for women offenders in my study. Such findings suggest that the pathways approach to women's offending should consider the role of intimate relationships as potential risk factors in the evolution of drug problems and crime for women and that marriage and relationships with intimates may be turning points in the lives of these women toward breaking the law rather than desistance. This may be especially problematic for women who are economically dependent or dependent in other ways and for whom autonomy is a very complex issue. Therefore, the pathways perspective also needs to consider the degree of autonomy which women can exercise within relationships that are often marked by dependence, abuse, and victimization.[4]

⧖ Education and Employment

Studies of women involved in the criminal processing system routinely find that their subjects are poorly educated and lacking in basic job skills and experience. Hawai'i's female offenders share these characteristics, and, in some respects, educational achievement is even worse for these women than in some national studies. Not only do these conditions bode ill for economic survival and family stability, but deficits in these areas may shape the length of time women spend on parole. Continuous full-time employment is a basic requirement for anyone hoping to gain an early discharge from parole in Hawai'i.[5] Adequate employment and income are basic to establishing households where women can reunite with their children. As O'Brien notes, inadequate income within a comprehensive parole plan sets a woman up for failure upon leaving prison. Attaining what O'Brien refers to as "environmental sufficiencies"

(2001: 27) in an environment of diminished social welfare support for women means that ex-inmates will have to depend on themselves for financial resources or upon family members who are themselves strapped by the high cost of living in Hawai'i.

Women parolees in Hawai'i fare less well than women in state prisons as a group when it comes to educational attainment. While a handful of women in my study had some college prior to incarceration, only two had college degrees. One woman had four years of college and a second had an advanced college degree, an MBA, prior to her incarceration. Nearly 54% of the study group had not graduated from high school; nor had they obtained their GEDs prior to their incarceration. In contrast, the Bureau of Justice Statistics reported in 2001 that 44% of women in state prisons lacked a high school diploma, while 39% had a high school diploma or equivalent. About 17% of women in state prisons had at least one year of college or more (see Greenfeld and Snell 1999).

Owen and Bloom's study (1995) of California's female inmates found that 24% of their sample had less than a high school education, while 26% graduated from high school or had a GED and nearly 26% had some post-high school education (college or vocational school) (Owen and Bloom 1995). Hawai'i's basic educational statistics for women offenders (in terms of finishing high school or getting a GED) are inferior to both California's and the nation's, especially when it comes to post-secondary education. Fewer than 10% of Hawai'i women in this study had any post-secondary education compared to 17% across the nation and nearly 26% in California.

In addition, I found that educational attainment has a strong relationship to ethnicity among these parolees. Women of Hawaiian descent were less likely to have graduated from high school compared to other women in this study. In looking at overall educational attainment for women, Hawai'i compares very favorably with other states, ranking 11th in the nation in 1990 (United States Census 1990). Thus, the educational statistics for women in this sample reflect a departure from this economic indicator for women in our state. The pathways perspective suggests looking at this pattern of early separation from school as another of the turning points in the trajectory of women who get into trouble with the law. Moreover, women of Hawaiian ancestry are even more marginalized in this respect compared to other women.

Given these dismal educational statistics, the data collected by the Hawai'i Paroling Authority as part of its assessment of employment-related skills offer few surprises. In the year before sentencing for their offenses, few women had any substantial employment. The assessments also showed that fewer than 50% had adequate educational or vocational skills. Making a living on parole is crucial to setting up a household and stable living situation. Those of us with secure housing probably take for granted the extent to which our residences, our homes, structure our lives and make access to other resources possible. Our intimate relationships, safe or dangerous, take place in the context of the home. Stable housing is important to children's attending school and qualifying for needed services. O'Brien (2001: 25) notes that home has both metaphorical and concrete aspects for women offenders leaving prison and that the quality of home in both these regards is crucial to successful reentry into society. Clearly, these subjective and objective factors overlap. Financial instability precludes attaining secure housing, and women's economic dependence increases the potential for living in an unsuitable and even dangerous familial environment. Economic dependency is a common reason that women remain in violent households or in households plagued with family troubles and conficts. Problems with housing are a primary concern for women leaving prison, but these difficulties frequently existed prior to incarceration for over a third of the study group, according to Hawai'i Paroling Authority assessment data.

Economic dependence shapes relationships with intimates. Several months after I interviewed Donna, the paroled mother of a 15-year-old son, I ran into her in downtown Honolulu. She had had a fight with her boyfriend, who accused her of using drugs because she returned home late the night before. Despite her denials, he was threatening to put her and her teenaged son out on the streets. Distraught at this possibility, Donna wasn't sure where to turn next. Her concerns were both emotional and economic— and she has a son who is dependent upon her for his support. Donna told me at our first meeting that she wanted to attend cosmetology school after leaving prison, but her parole officer had quashed the idea, telling her instead that she was expected to work. Policies like these leave women like Donna less skilled and less self-sufficient economically than they might otherwise be—and dependent on the good graces of intimates. The fates of women on parole and their children are shaped extensively by such links to social and economic capital, links that often rest on the few tenuous social ties women are able to build with family, friends, and lovers.

⌧ Substance Abuse

For the majority of the women in this study, drug and alcohol problems have conditioned their entry into the criminal processing system. The Hawai'i Paroling Authority assesses[6] inmates for substance abuse problems, using a scale that differentiates alcohol and drug usage. Data for this population of female parolees indicate that the majority of this group has experienced significant alcohol and drug problems and that many require substance abuse treatment. Slightly more than two-thirds of this population (for whom assessments were available) experienced some disruption or severe disruption of their lives as a result of alcohol use. Over a third of these women had problems so severe that they required alcohol dependence treatment.

In conversations with women in this study, I came to an understanding of their alcohol and drug problems as overlapping, with polydrug patterns of substance use being more typical than not. Pharmaceuticals prescribed by physicians played a role in these accounts as well. Kassebaum and Chandler's (1994) study found that polydrug use was the norm for newly sentenced felons in Hawai'i prisons. They found nearly universal alcohol and marijuana use, along with extensive use of cocaine, crack, heroin, crystal methamphetamine (ice), and other drugs. Heavy polydrug use was more common among females than the males in their sample. These findings resonate with other studies that find that the link between addiction and property crime is stronger among female than male inmates (Owen and Bloom 1995; Chesney-Lind and Sheldon 1998).

Not surprisingly, serious drug use can be said to be typical of the women in my study, with 80% of those for whom data were available experiencing serious disruption of their lives due to their addiction. While data are not collected by the Hawai'i Paroling Authority on the specific drugs used by parolees, the women I interviewed had patterns of drug use that were similar to those found by Kassebaum and Chandler, with methamphetamine (ice), being especially common.

The stories told to me by women about their drug use show a good deal of diversity, with some repeating refrains. Two stories represent addiction and drug use as matters of late onset in the context of family life. Anna's drug use had a relatively late onset; she was in her early 30s. Her husband was, she said, "a local boy" and wanted her to stay home when her kids were young. He was very controlling, and, when she took a job as a waitress, he constantly complained about her working. Her husband began selling drugs, and they both began using, starting with him. Their marital troubles became so bad that she left the family and began running the streets. She picked up a prostitution charge this way. Darleen, on the other hand, held off using coke during her pregnancy and, after the birth of her child, allowed her husband to introduce her to smoking crack. She didn't know about addiction—her family had problems with alcohol but drugs weren't part of their lives.

By contrast, Betty's childhood was rendered chaotic by her parents' alcoholism and drug use. She was raised in a home where she could smoke marijuana and drink from the age of 12. "It was okay, it was the normal thing," she explained. She characterized her home as broken and dysfunctional. She said that she "grew up with this mentality" and that drug use and drinking were condoned, even for a child of 12. Thus while families can serve as sources of support, they may also serve as means by which addiction and other problems are transmitted.

⧖ Violence, Abuse, and Trauma

Angela lowered her hand from her mouth to respond to my question about the violence in her life before prison. She told me that she'd been abused "plenty," adding, "it's part of the deal, right?" When asked who abused her, she replied, "who didn't?" Angela told me she'd been beaten by lovers, friends, strangers, dates, and New Orleans cops. Sometimes she sought legal protection from attackers, and sometimes she didn't. She called the police one time here in Hawai'i when an acquaintance held a knife to her throat. He ended up going to jail. Later on, she was raped and reported it, but she didn't follow through with the case. Angela avoids smiling and often shields her mouth when talking to conceal her missing front teeth. The other marks of her abuse are less apparent. She receives therapy now for the effects of her many traumatic experiences living on the streets; this is the first time that she has received any counseling.

Gender violence, especially trauma due to childhood sexual assault, has now been established as a precursor to adult drug use and offending among women (Gilfus 1992, 1999). Victimization is being recognized as an event in women's lives that accelerates women's trajectories into offending (Gaarder and Belknap 2002). Macmillan (2001) notes that violent victimization has profoundly negative effects over the life course. The social and economic status of women places them at risk for violent victimization throughout the life course.

Often, the mediating link between histories of abuse and adult offending appear to be drug use and involvement with prostitution. Indeed, as Stein (2001) points out in her study of Hawai'i women in an inmate work furlough program, street life and prostitution are associated with ubiquitous violence, "part of the deal," as Angela told me. Women who break the law themselves occupy an especially marginalized position that affords them little protection against violence and sexual assaults. Nor do they have much recourse when violence occurs.

Laura, the mother of six children, told me that her husband was a very abusive man who beat her. But her calls to the police did not result in help. She said that she didn't know her rights and that the police never told her about temporary restraining orders. She commented that, because she and her husband were known to be involved with drugs, the police didn't feel that she was worthy of help. They discounted her calls for help because of the addict label she carried.

It is not uncommon for young women from disruptive family backgrounds to enter into adult relationships that are fraught with violence. Betty, whose troubled family of origin has already been described, experienced such a transition from adolescence to adulthood. This mother of four reports that her future husband began hitting her during her first pregnancy. She told me that lots of mental, emotional, and physical abuse followed. There were many trips to the hospital as a result. Her husband threatened her repeatedly, sometimes with a gun. She described what she felt were a number of near-death experiences from this violence. She felt that the continual abuse made her "weaker and weaker." She talked about living with continual violence and the impact of the abuse on her spirit:

In order to bring my strength back from being broken from just everything—it's hard. To build yourself up again is hard. You start believing all of the things he says.

Living at the margins, as Laura's story illustrates, often precludes getting effective help from the police or other mainstream helping institutions. Indeed, the law often fails to protect women who are leading conventional lives. And, it is not uncommon for the perpetrators of the violence to be, themselves, the police.[7]

Trauma related to sexual abuse and violence during childhood and adolescence has begun to be well documented among women offenders. And its link with drug addiction continues to be validated by biographical and life course research on women offenders and addicts (Gaarder and Belknap 2002; Widom 1995, 2000). But trauma continues well into adulthood for many of these women, and sometimes it arises as a result of their gendered responsibilities for children. Sometimes, a woman's first brush with the authorities arises when she comes under the gaze of child protective services and loses a child through court actions—a common problem for incarcerated women. Termination of parental rights, now even more common since the passage of the 1997 Adoption and Safe Families Act, is a singular source of grief for women. Motherhood is extraordinarily powerful as a motivating force in the lives of women, and women offenders are no exception. Their maternal lives are, however, rendered much more complicated by their marginalized statuses, unlawful life styles, and incarceration. They often cannot depend on families for help, nor are they particularly successful at securing help from agencies like substance abuse treatment facilities. Of course, there are few facilities that offer treatment services to pregnant or parenting drug addicts along with their young children.[8]

Compared to what is now known about the nexus of abuse, trauma, and addiction, very little is known about how the loss of children through the termination of parental rights affects women's trajectories through life. Angela, whose son was born in 1993, viewed her pregnancy as a turning point. She got off the streets and stopped taking drugs for a time. She attributes her descent into addiction to the loss of her child to her mother who coerced her into giving him up. This loss, she said, "really caused me to bottom out." That event prompted a return to drugs and a cascade of losses culminating in her arrest. She told me that she used to think that she probably wouldn't have become an addict had it not been for this loss. "Well, at the time," she added, "I didn't understand the disease [of addiction]." Nevertheless, the birth of Angela's child was an important motivator in Angela getting off the street. The loss of the child became an enduring source of pain, one that was linked for her with a return to and exacerbation of her drug problem.

⧈ Ethnicity: Being Hawaiian

The life course perspective emphasizes individual life events and turning points in its causal explanation of delinquency and crime. But there is, in the work of Elder (1985) and a number of feminist writers looking at the pathways approach, an understanding of how individuals' life events are positioned by larger structural factors that affect the likelihood of coming before criminal courts. The "triple jeopardy" (Bloom 1996) of race, gender, and class are, of course, cultural and historical outcomes that have an important bearing on women's trajectories into crime—and their fates once they enter the criminal processing system. Contemporary Hawai'i still bears the legacy of its ethnically stratified plantation past, a system that dominated 19th century society. As a result of its colonial history, *kanaka maoli* (as persons of Native Hawaiian ancestry refer to themselves) and newer immigrant groups from Asia and the Pacific Islands dominate the ranks of the poorer classes—and populate correctional institutions.

Statistical data on cultural, racial, and ethnic identity in Hawai'i are limited as to how well they can describe the lived experience of people and the dynamics of identity in this place. Identities are always layered, shifting from context to context, and one runs the risk of reifying human identity when attempting to describe these realities statistically. Ethnic and cultural identities are inherently multidimensional; these characteristics of experience cannot be effectively reduced to numerical assessments.[9] However, there are aspects of Hawai'i's diversity that are salient to this population of women offenders and their lives. And, numerical data are a way to begin talking about these realities. In my study, I have used what have become standard conventions in official data for describing race and ethnicity in Hawai'i. Distinguishing Hawaiian ancestry is particularly important in this context because the state's indigenous population is overrepresented in the criminal processing system.

In the 2000 census, about 20% of the population in Hawai'i claimed some Native Hawaiian ancestry (United States Census 2000), but carceral populations routinely reflect the group as present at more than double this rate. Nearly 53% of women parolees in this study are Hawaiian or Part-Hawaiian. Caucasians, on the other hand, at 24% of the overall population, are underrepresented in this group at only 16%. The overrepresentation of Native Hawaiians in the carceral population is consistent from study to study: Stein (2001) reported that women of Hawaiian ancestry made up 61% of her sample of inmates at a work-furlough program in Honolulu. Kassebaum and Davidson-Coronado's (2001) study of both male and female parolees in Hawai'i found that 45% were of Hawaiian ancestry. In July of 2000, 33% of Hawai'i's 2319 female probationers were of Native Hawaiian ancestry as were 40.2% of Hawai'i's 542 adult female inmates.[10] Persons of Native Hawaiian ancestry not only make up the largest category of offenders in this study but are overrepresented in terms of poverty indices, major life-style diseases, and a range of other problem categories across the spectrum of social distress (Blaisdell 1993). There is considerable consensus among scholars of Hawaiiana that colonialism, racism, and extensive alienation from aboriginal culture and lands have severely damaged the life chances of the indigenous population (Blaisdell 1993, 1996; Trask 1993; Merry 2000).

The difficulties of women of native ancestry were themes in a number of the interviews that I conducted. Malia, one of the very few women incarcerated for a violent crime, wants to establish a claim for land from the Hawaiian Homelands program, a long-standing federal and state program whose beneficiaries are those individuals with more than 50% Hawaiian ancestry.[11] Documenting this blood quantum issue is a long and arduous process, requiring genealogical research at the state archives among records going back for more than a century. With her 11th grade education and lack of relevant skills, Malia feels overwhelmed at the prospect and wishes she had some help to establish her claim. Other women of Native Hawaiian ancestry, having no basis for land claims, will have to rely on meager work experience, family resources, or marginal state assistance for their survival.

Like the women in Bloom's study of incarcerated women in California (Bloom 1996), Hawai'i's female offender population is marginalized by race, gender, and class. For women of Hawaiian ancestry, another layer is added to this "triple jeopardy" of vulnerability and marginalization: that of being a colonized indigenous people. Native peoples are present in larger than expected numbers in carceral populations in Canada, the United States, Australia, and other nations where indigenous ties to ancestral lands and ways of life have been interrupted by capitalism and urbanism. Hawaiian cultural capital, with its metaphors of aloha, reciprocity, and respect, was frankly overpowered in a social field dominated by Western models that prioritize economic values and materialist ideologies (Kame'eleihiwa 1992: 317–318; Trask 1993). The resources that culture provides for individual identity are often difficult to articulate. Kepuhi's story illustrates a particular feeling about cultural identity that can be described as a way of knowing, one that was lost with adulthood:

> But spiritually I've lost that too. It's kind of hard to explain. Say like, I know something or I would feel something or I would sense something and I've lost that. Or see something, nobody else can see it now. You know what I mean?

Kepuhi described two instances of knowing things about people and events that, as a child, she should have had no means of knowing. These intuitive feelings were essential to her Hawaiian identity, giving her a sense of connection to the past and to a reality that Western empiricism denies. She describes these as spiritual gifts linked to being Hawaiian. Accounts of such experiences are met by Westerners with derision, but to native peoples, they are crucial cultural resources that link them to their past and to one another. Looked at in this way, Hawaiian ethnicity and culture are not risk factors for crime. Rather, in the post-colonial context, alienation from indigenous culture and the loss of resources it affords are important and often unanalyzed factors in the overrepresentation of native peoples in corrections populations (Merry 2002). The loss of traditional means of subsistence, relationships, spirituality, and identity in modern contexts of poverty and institutional alienation may well be a better explanation for problems like addiction, violence, and the other significant problems Native Hawaiians face.

Women of Native Hawaiian ancestry are not only more likely to be overrepresented in correctional populations, but their criminal involvement begins earlier. This risk factor for recidivism was found significantly more often for women of Native Hawaiian ancestry in the study compared to others. Additionally, Native

Hawaiian women tended to have their first convictions at earlier ages (19 and younger). Table 1 illustrates criminal convictions in adult court for Hawaiian and Non-Hawaiian women. Additionally, as Table 2 illustrates, women of Native Hawaiian ancestry were also significantly more likely to have a juvenile arrest record.

Early trouble with the law is strongly related to ethnicity, illustrating that the trajectory to prison begins earlier for these *kanaka maoli* women.

Finally, women of Hawaiian ancestry, while experiencing problems on parole such as using drugs, domestic violence, and homelessness at rates similar to those of non-Hawaiians, were more likely to be taken into custody or have a warrant issued for their arrest (although this relationship was not statistically significant). Thus, while the burden of parole and continuing personal difficulties affect both groups at equivalent rates, Native Hawaiian women are significantly more likely to have legal repercussions.

≋ Conclusion

The rate of female incarceration has increased exponentially during the latter part of the 20th century. Offending for women is more likely to involve drug and property offenses, rather than crimes against persons (see Government Accounting Office 1999). Women's patterns of offending, as my Hawai'i data indicate, are deeply embedded in patterns of drug and alcohol use, accompanied by property crimes. The trajectories propelling women into breaking the law are all too often linked to victimization within intimate relationships.

The conventional life course perspective, while powerful, shares an androcentric weakness with other criminological theories in that it fails to see crime as gendered social action. Nor, regarding female offenders, is sufficient notice taken of unequal power relationships that produce unexpected patterns of criminality among women. Bottcher (2001), for example, notes that events throughout the life course are informed

Table 1	Age at First Conviction (Adult) for Women of Hawaiian and Non-Hawaiian Ancestry		
Ancestry	**First Conviction at Age 20 or Older**	**First Conviction at Age 19 and Below**	**Total**
Hawaiian ancestry	70	43	113
Other ancestry	80	17	97
Total	**150**	**60**	**210**

$\chi^2 (1, N = 210) = 10.78, p < 0.05$.

Table 2	Juvenile Arrest Record		
Ancestry	**No Juvenile Record**	**Juvenile Record**	**Total**
Hawaiian ancestry	70	53	113
Other ancestry	74	32	97
Total	**144**	**85**	**210**

$\chi^2 (1, N = 210) = 7.50, p < 0.05$.

and structured by gendered divisions of labor. She notes that girls take on child raising responsibilities earlier than boys, and usually age out of delinquency earlier than their male peers (2001: 918–922). The accounts that Hawai'i women give about the onset of their troubles with the law, however, show that the influence of adult relationships, such as the drug addiction of a spouse, can precipitate a late onset into drug use and offending. Traumatic events throughout the life course may propel women into drug use and related types of offending. The loss of a child, especially the termination of parental rights, has been largely unexplored as part of the cascade of losses these women experience as adults. Given that contemporary adoption policies put many incarcerated women in danger of losing their children, the implications for women who are punished a second time (by both imprisonment and the termination of their parental rights) are dire in terms of staying out of prison after release.

The women in my study were far more likely to be imprisoned for low-level property crimes and drug offenses than any other cause. Their major thefts and other property crimes were often in service to sustaining their drug addiction. And, their criminality often took place in the confines of relationships where they played highly dependent emotional and economic roles under the influence of abusive, drug-addicted spouses. Unfortunately for the women in this study and others like them, the dismantling of social welfare policies, with the erosion of entitlement programs that benefit women and children, has been accompanied by more punitive moves in penalty. Mandatory minimums for drug offenses, three strikes laws, and reduced judicial discretion have come to dominate late-modern crime control (Garland 2001). These trends, coupled with changes in the labor market and other economic sectors, increased the exposure of poor women to criminalization, making them more likely to be incarcerated.

The structural elements that theorists such as Elder (1985) find important to shaping life events are embodied in feminist versions of the pathways perspective, but perhaps need to be made even more explicit. I have argued that structural factors, shaped by historical and cultural change, have made it more likely that women of Hawaiian ancestry will come under the gaze of the law. Hawai'i's female inmates, more than others in the nation, are likely to have serious educational deficits and to be returned to prison for violations of parole. Merely returning women to their pre-incarceration state will do little to alter their lives in any significant way. The assumptions underlying rehabilitative ideologies need to be unpacked and assessed against the real lives of women, starting with an understanding of the life events and processes that placed them on their trajectory to prison—as well as how their biographies intersect with various forms of marginality.

Notes

1. This paper examines data from 25 interviews and 240 case files of female parolees in Hawai'i taken from a larger study of gender and penality in the state (see Brown 2003).

2. More often than not, women are serving concurrent sentences on a number of charges. My data reflect aggregate crimes. That is, the 240 women in my sample were convicted for a total of 762 crimes.

3. Hawai'i's criminal statutes assign indeterminate maximum sentences by felony level, designated A through C in order of most to least serious. In general, commission of the most serious crimes, "A level felonies," results in a maximum sentence of 20 years. Convictions of B and C level felonies result in maximum sentences of 10 and 5 years respectively. A feature of Hawai'i's indeterminate sentencing system, time served is usually a proportion of the maximum sentence as determined by the Hawai'i Paroling Authority as a function of the minimum-sentence setting process. Exceptions have entered into Hawai'i's sentencing system through mandatory minimum requirements which been attached to certain of these statutes—mostly offenses involving crystal methamphetamine. Hawai'i Law §712—1241 promoting a dangerous drug in the first degree, for example, requires mandatory minimum sentences for the possession, distribution, or distribution to a minor of any amount, or manufacture of methamphetamine (or its derivatives).

4. It should be noted that these patterns exist in same-sex as well as heterosexual relationships. These cases, about which little is known, may be especially problematic because of the extent to which lesbian relationships are stigmatized and their problems concealed.

5. Standard terms of Hawai'i parole include actively seeking and retaining employment. To qualify for early discharge from parole, individuals must demonstrate a record of "prosocial attitudes and employment" for a period of time based on their total sentence. For example, a parolee with a 10-year sentence must adhere to employment and attitude guidelines for at least three consecutive years. In practice, this means full-time employment. In addition, parolees must pay any outstanding restitution orders from the court (Hawai'i Administrative Rules, Chapter 700, Title 23, January, 1991). According to the "letter" of the regulations, education is not considered a substitute for full-time employment in securing early release from parole.

6. These assessments are not as specific as more clinical assessment instruments used by substance abuse treatment specialists.

7. Two high-profile murders of wives by their police-officer spouses occurred in Hawai'i over the past decade.

8. Hawai'i has one residential treatment center that accepts pregnant women or mothers with children. In fact, they can only accommodate one child under the age of two years for each mother accepted into the program.

9. Racial and ethnic identity in Hawai'i in daily speech is referred to as "nationality"—an interesting acknowledgment of the plantation heritage of most Hawai'i residents. Local people routinely list a number of ancestries in response to the everyday question (which has nothing to do with citizenship): "What nationality are you?"

10. Female offender population statistics are from the Hawai'i Adult Probation Division Circuit Court and the Hawai'i Department of Public Safety (see Hawai'i Symposium, 2000).

11. Congress established the Hawaiian Homeland program in 1920, setting aside 200,000 acres of land to lease to full or nearly full-blooded Native Hawaiians. (The program establishes a minimum qualification, known as the "50% blood quantum.") Over the past 75 years, the Territory of Hawai'i, followed by the State, has failed to settle more than a handful of Hawaiians on the undeveloped lands, selling and sometimes leasing the lands to non-Native Hawaiians.

≋ References

Anderson, T.L. and Bennett, R.R. (1996). Development, gender, and crime: The scope of the routine activities approach. *Justice Quarterly*, 13, 31–56.

Brown, M. (2003). *Motherhood on the Margins: Rehabilitation and Subjectivity among Female Parolees in Hawaii.* Dissertation, Sociology, University of Hawaii at Manoa.

Belknap, J. (2001). *The Invisible Woman: Gender, Crime, and Justice.* Belmont, CA: Wadsworth.

Blaisdell, K. (1993). Historical and cultural aspects of Native Hawaiian health. In P. Manicas (ed.), *Social Process in Hawaii: A Reader.* New York: McGraw-Hill, Inc, pp. 37–57.

Blaisdell, K. (1996). Kanaka Maoli: Indigenous Hawaiians. In R.H. Mast and A.B. Mast (eds.), *Autobiography of Protest in Hawai'i.* Honolulu: University of Hawai'i Press, pp. 363–373.

Bloom, B. (1996). *Triple Jeopardy: Race, Class, and Gender as Factors in Women's Imprisonment.* Dissertation Thesis, Sociology Department, University of California, Riverside.

Bloom, B., Owen, B. and Covington, S. (2002). *Gender-Responsive Strategies: Research, Practice, and Guiding Principles for Women Offenders.* National Institute of Corrections, U.S. Department of Justice.

Bottcher, J. (1995). Gender as social control: A qualitative study of incarcerated youths and their siblings in greater Sacramento. *Justice Quarterly* 12, 33–58.

Bottcher, J. (2001). Social practices of gender: How gender relates to delinquency in the everyday lives of high-risk youths. *Criminology* 39, 893–932.

Chesney-Lind, M. (1997). *The Female Offender: Girls, Women, and Crime.* Thousand Oaks: Sage.

Chesney-Lind, M. and Sheldon, R.G. (1998). *Girls, Delinquency, and Juvenile Justice.* Thousand Oaks, CA: Sage.

Daly, K. (1994). *Gender, Crime and Punishment.* New Haven: Yale University Press.

Elder, G.H. (1985). *Life Course Dynamic: Trajectories and Transitions 1968–1980.* New York: Cornell University Press.

Gaarder, E. and Belknap, J. (2002). Tenuous borders: girls transferred to adult court. *Criminology* 40, 481–517.

Garland, D. (2001). *The Culture of Control: Crime and Social Order in Contemporary Society.* Chicago: University of Chicago Press.

Gilfus, M. (1992). From victims to survivors to offenders: Women's routes of entry into street crime. *Women and Criminal Justice* 4, 63–89.

Gilfus, M. (1999) The price of the ticket: A survivor-centered appraisal of trauma theory. *Violence Against Women* 11, 1238–1257.

Gottfredson, M. and Hirschi, T. (1990). *A General Theory of Crime.* Palo Alto, CA: Stanford University Press.

Government Accounting Office. (1999). *Women in Prison: Issues and Challenges Confronting U.S. Correctional Systems.* Washington, D.C: United State General Accounting Office.

Greenfeld, L.A. and Snell, T.L. (1999). *Special Report: Women Offenders.* Washington, D.C: Bureau of Justice Statistics, U.S. Department of Justice.

Hawai'i Symposium on Women Offenders. (2000). *Treat the Women: Save the Children.* Conference Proceedings, December 1–3, 2000. Kauai, Hawai'i.

Kame'eleihiwa, L. (1992). *Native Land and Foreign Desires.* Honolulu: Bishop Museum Press.

Kassebaum, G. and Chandler, S.M. (1994). Polydrug use and self control among men and women in prisons. *Journal of Drug Education* 24, 333–350.

Kassebaum, G. and Davidson-Coronado, J. (2001). *Parole Decision Making in Hawai'i: Setting Minimum Terms Approving Release, Deciding on Revocation, and Predicting Success and Failure on Parole.* Honolulu: Social Science Research Institute, University of Hawai'i at Manoa.

Macmillan, R. (2001). Violence and the life course: The consequences of victimization for personal and social development. *Annual Review of Sociology* 17, 1–22.

Maeve, M.K. (2000). Speaking unavoidable truths: Understanding early childhood sexual and physical violence among women in prison. *Issues in Mental Health Nursing* 2, 473–498.

Maher, L. and Daly, K. (1996). Women in the street level drug economy: Continuity or change? *Criminology,* 34, 465–491.

Merry, S.E. (2000). *Colonizing Hawai'i: The Cultural Power of Law.* Princeton: Princeton University Press.

Merry, S.E. (2002). Crime and criminality: historical differences in Hawai'i. *Contemporary Pacific* 14, 412–423.

O'Brien, P. (2001). *Making It in the "Free World": Women in Transition from Prison.* Albany: State University of New York Press.

Owen, B. (1998). *In the Mix: Struggle and Survival in a Women's Prison.* Albany: State University of New York Press.

Owen, B. and Bloom, B. (1995). Profiling women prisoners: Findings from national surveys and a California sample. *The Prison Journal* 75, 165–185.

Owen, B. and Belknap, J. (2002). *Elements of the pathways perspective.* San Diego: Paper presented at the meetings of the Western Society of Criminology.

Sampson, R.J. and Laub, J.H. (1990). Crime and deviance over the life course: The salience of adult social bonds. *American Sociological Review* 55, 609–627.

Sampson, R.J. and Laub, J.H. (1993). *Crime in the Making: Pathways and Turning Points Through Life.* Cambridge, MA: Harvard University Press.

Stein, L.Y. (2001). The psychosocial needs of Hawaiian women incarcerated for drug-related crimes. *Journal of Social Work Practice in the Addictions* 1, 47–69.

Trask, H. (1993). Hawaiians, American colonization, and the quest for independence. In P. Manicas (ed.), *Social Process in Hawai'i: A Reader.* New York: McGraw-Hill, Inc, pp. 1–36.

United States Census. (1990). *State Ranking: Educational Attainment by Sex.* Washington, DC: U.S. Census Bureau.

United States Census. (2000). *Ranking of Races for the State of Hawai'i, 2000.* Washington, DC: U.S. Census Bureau.

Widom, C.S. (1995). *Victims of Childhood Sexual Abuse—Later Criminal Consequences.* Washington, DC: National Institute of Justice. Research in Brief. U.S. Department of Justice, Office of Justice Program.

Widom, C.S. (2000). Childhood victimization, early adversity, later psychopathology. *National Institute of Justice Journal,* 242, 2–9.

Wilson, W.J. (1987). *The Truly Disadvantaged.* Chicago: University of Chicago Press.

DISCUSSION QUESTIONS

1. What is feminism, and what do feminists advocate? How have they had an effect on the work of women in corrections and on the female experience with incarceration?

2. What is patriarchy? What kind of effect did and does patriarchy have on corrections for women and girls? How might it negatively affect the experience of men and boys?

3. In what ways have women and girls occupied a minority status in corrections? How has that status affected how they are treated in the system?

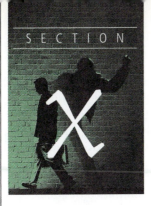

THE JUVENILE JUSTICE/ CORRECTIONS SYSTEM

Introduction

▧ Juvenile Delinquency

The juvenile justice system falls under the broad umbrella of the civil law rather than criminal law. This placement emphasizes the distinction that the law makes between adults and juveniles who commit the same illegal acts. When a juvenile commits an act that would be considered criminal if it were committed by an adult, the offender is called a *delinquent* rather than a criminal, conveying the idea that the juvenile has *not done something right* (behaved lawfully) rather than that he or she has *done something wrong* (behaved unlawfully). This difference is a subtle one and, according to some, no more than a word game, but it does reflect the rehabilitative rather than retributive philosophy of American juvenile justice.

Juveniles are subject to laws that make certain actions that are legal for adults, such as smoking, drinking, not obeying parents, staying out at night to all hours, and not going to school, illegal for them. These acts are called *status offenses* because they apply only to individuals having the status of a juvenile, and they exist because the law assumes that juveniles lack the maturity to appreciate the long-term consequences of their behavior. Many of these acts can jeopardize juveniles' future acquisition of suitable social roles because they may lead to defiance of all authority, inadequate education, addiction, and teenage parenthood (Binder, Gies, & Bruce, 2001). If parents are unwilling or unable to shield their children from harm, the juvenile justice system becomes a substitute parent. Status offenses constitute the vast majority of juvenile offenses and consume an inordinate amount of juvenile court time and resources (Bynum & Thompson, 1999). Consequently, some states have relinquished court jurisdiction over status offenses to other social service agencies where terms such as "child in need of supervision" (CHINS) or "person in need of supervision" (PINS) are used to differentiate status offenders from juveniles who have committed acts that are crimes when committed by adults.

✎ The Extent of Delinquency

Figure 10.1 shows the juvenile proportion of all reported arrests in the FBI's (2006) 2005 Uniform Crime Reports. Juveniles ("youths under 18") accounted for 16% of all violent crime arrests and 26% of all property crime arrests. According to the United States Census Bureau (2004), the percentage of the population between 10 and 17 years of age (defined legally as *juvenile* and within the age of responsibility) averaged across all states was about 11.5% in 2005. Juveniles are thus overrepresented in most of the crime categories presented in Figure 10.1.

Figures such as these are troubling, but we should realize that antisocial behavior is normative (although certainly not welcome or excusable) for juveniles; juveniles who do *not* engage in it are statistically abnormal (Moffitt & Walsh, 2003). Adolescence is a time when youths are "feeling their oats" and temporarily fracturing parental bonds in their own personal declaration of independence. Looking at data from 12 different countries, Junger-Tas (1996) concluded that delinquent behavior is a part of growing up and that the peak ages for different types of crimes were similar across all countries (16–17 for property crimes and 18–20 for violent crimes). Biologists tell us that adolescent rebellion is an evolutionary design feature of all social primates. Fighting with parents and seeking out age peers with whom to affiliate "all help the adolescent away from the home territory" (Powell, 2006, p. 867). As Caspi and Moffitt (1995) put it, "every curfew broken, car stolen, joint smoked, or baby conceived is a statement of independence" (p. 500). The juvenile courts are thus dealing with individuals at a time in their lives when they are most susceptible to antisocial behavior.

Neuroscience research over the past 15–20 years has thrown much light on why there is a sharp rise in antisocial behavior during adolescence, across time and cultures, and some very important court decisions in juvenile justice have been influenced by this research (Garland & Frankel, 2006). What has emerged from this research is that the immaturity of adolescent behavior is matched by the immaturity of the adolescent brain (White, 2004). The onset of puberty brings with it a 10- to 20-fold increase in testosterone in males, a hormone linked to aggression and dominance seeking (Ellis, 2003). Additionally, brain chemicals that excite behavior increase in adolescence while chemicals that inhibit it decrease (Collins, 2004; Walker, 2002). Many other events are reshaping the adolescent's body and brain during this period, leading to the conclusion that there are *physical* reasons why adolescents often fail to exercise rational judgment and why they tend to attribute erroneous intentions to others. When the brain reaches its adult state, a more adult-like personality emerges, e.g., one with greater conscientiousness and self-control (McCrea et al., 2000).

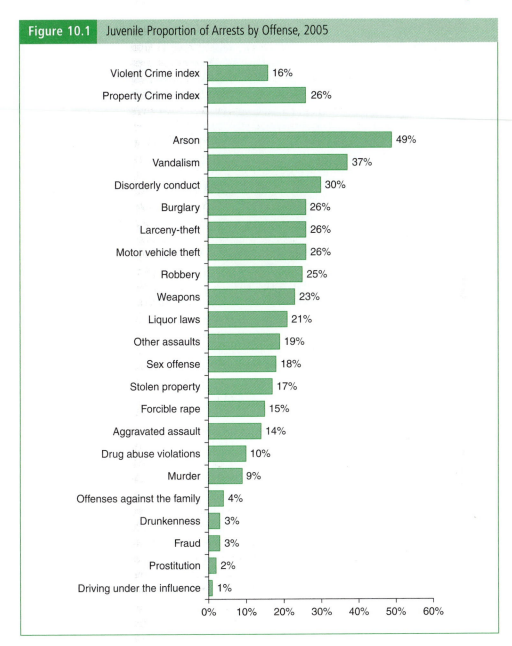

Figure 10.1 Juvenile Proportion of Arrests by Offense, 2005

⬚ History and Philosophy of Juvenile Justice

Up until about 300 years ago, the concept of childhood was not recognized, children were considered as being not much different from property, and no special allowances for children were acknowledged in matters of determining culpability and punishment. The minimum legally defined age of criminal responsibility was defined in early English

▲ **Photo 10.1** Students are escorted through the corridors of the Shelby Training Center, a private juvenile detention facility in Memphis owned by Corrections Corporation of America. While walking through the halls, students are required to keep their hands clasped behind their backs. Freed from the need to navigate through the state for funds, privately operated prisons have the ability to be more efficient and cost-effective. Opponents contend that privatization removes the levels of responsibility and accountability that government run facilities offer.

common law as 7 (in the modern United States it ranges from 6 in North Carolina to 10 in Arkansas, Colorado, Kansas, Pennsylvania, and Wisconsin, as it is in modern England) (Snyder, et al., 2003, p. 2). Historical records show that children as young as 6 were often housed in jails with adults and were even executed for relatively minor offenses (Schmalleger, 2003).

Under the increasing influence of Christianity, the English courts in the Middle Ages began to accept the religious doctrine that exempted children below the age of 7 from criminal responsibility, and children between the ages of 7 and 14 could only be held criminally responsible if it could be shown that they were fully aware of the consequences of their actions. Fourteen was the cutoff age between childhood and adulthood for the purpose of assigning criminal responsibility because individuals were considered rational and responsible enough at this age to marry (Springer, 1987).

Although parents were responsible for the behavior of their children, ever since the formation of the English chancery courts in the 13th century, there was movement toward greater state involvement. Chancery courts adopted the doctrine of *parens patriae*, which essentially means "state as parent." *Parens patriae* gave the state the right to intercede and act in the best interest of the child or any other legally incapacitated persons, such as the mentally ill. This doctrine meant that the state and not the parents had ultimate authority over children and that children could be removed from their families and placed in the custody of the state if they were delinquent (Schmalleger, 2001).

Despite *parens patriae*, the family was still considered the optimal setting for rearing children, and, consequently, orphans or children with inadequate parents were assigned to foster families through a system known as *binding out*. Children whose parents could not control them or who were too poor to provide for them were apprenticed to richer families who used them for domestic or farm labor. This period saw the establishment of the first laws directed specifically at children, including laws that condemned begging and vagrancy (Sharp & Hancock, 1995). The concern over vagrancy led to the creation of workhouses in which "habits of industry" were to be instilled. The first one, called Bridewell, was opened in 1555 and was considered so successful that in 1576 the English Parliament passed a law establishing "bridewells," or workhouses, in every English county (Whitehead & Lab, 1996). The idea behind these institutions was that, if vagrant youths were removed from the negative influences of street life, they could be reformed by discipline, hard work, and religious instruction.

⚐ Childhood in the United States

American notions of childhood and how to deal with childhood misconduct were imported whole from England. Based on the Bridewell model, the New York House of Refuge was established in 1825 to house orphans, beggars, vagrants, and juvenile offenders. Several other cities, counties, and states soon established their own homes for "the perishing and dangerous classes," as they were deemed (Binder, Geis, & Bruce, 2001, p. 202). Children in houses of refuge lived highly disciplined lives and were required to work at jobs that brought income to the institution. The indeterminate nature of children's residence allowed the institutions a great deal of latitude in their treatment. Children were required to work long hours, often received little or no training, and were frequently mistreated (Whitehead & Lab, 1996).

It was a frequent practice for poor parents to place their children in residence for idle and disorderly behavior, making it clear that the courts would have to create standards for admission. The courts did this in *Ex parte Crouse* (1838). (The term *ex parte* denotes a hearing in the presence of only one of the parties to a case.) The subject of the case was a child named Mary Ann Crouse who was placed in the Pennsylvania House of Refuge by her mother against the wishes of her father. Mary's father argued that it was unconstitutional to incarcerate a child without a jury trial, but the Pennsylvania supreme court ruled that parental rights are superseded by the *parens patriae* doctrine. This landmark decision established *parens patriae* as settled law in American juvenile jurisprudence (del Carmen, Parker, & Reddington, 1998).

As the 19th century drew to a close, an increasingly influential middle class called for reforms in the way juveniles were treated. Although many reformers believed that poverty was a sign of personal defects such as laziness and feeblemindedness, they thought that these defects could be rectified with discipline and training. A group of well-funded and highly educated liberal reformers known as the "child savers" began attacking the operations of the houses of refuge, which they saw as punitive rather than rehabilitative. The child savers argued that children should be deinstitutionalized and placed in settings that provided a more family-like atmosphere where they could be taught the value of hard work and other solid middle-class values. Toward this end, they urged placement with farm families (often idealized as "God's reformatories" [Mennel, 1973]) in the western United States, where it was assumed that wholesome surroundings would cure children of their delinquent ways.

⊠ The Beginning of the Juvenile Courts

The child savers created an impetus for change in the way juvenile offenders were handled by the courts, as it became increasingly obvious that adult criminal courts were not equipped to apply the spirit of the *parens patriae* doctrine. In 1899, Cook County, Illinois, enacted legislation providing for a separate court system for juveniles, and, by 1945, every state in the union had established juvenile court systems (Hemmens, Steiner, & Mueller, 2003). These courts combined the authority of social control with the sympathy of social welfare in a single institution and afforded judges a great deal of latitude in determining how "the best interests of the child" could be realized.

The creation of a separate system of justice for juveniles brought with it a set of terms (euphemisms, if you like) describing the processing of children accused of committing delinquent acts, which differentiated it from the adult system. These terms reflect the protective and rehabilitative nature of the juvenile system in contrast with the punitive nature of the adult system. Table 10.1 lists the terms used to describe the procedure or event from the initial to the last contact with authorities in both the adult and juvenile justice systems today.

⊠ Juvenile Waiver to Criminal Court

Juveniles can sometimes be *waived* (transferred) to adult criminal court where they lose their status as minors and become legally culpable for their alleged crime and subject to criminal prosecution and punishments. A transfer to adult court is called a waiver because the juvenile court waives (relinquishes) its jurisdiction over the child and passes this jurisdiction to the adult system. Waivers are designed to allow the juvenile courts to transfer to a more punitive system juveniles over a certain age who have committed particularly serious crimes or who have exhausted the juvenile system's resources for rehabilitating chronic offenders. Juveniles become increasingly more likely to be waived if they are chronic offenders approaching the upper age limit of their state's juvenile court's jurisdiction. It should be noted that only about 1.5% of juvenile cases nationwide are waived to the criminal courts (Schmalleger, 2003). There are three primary (non-mutually exclusive) ways in which juveniles can be waived to criminal court.

1. *Judicial waiver:* A judicial waiver involves a juvenile court judge deciding after a "full inquiry" that the juvenile should be waived (48 states at present use this judicial discretionary model). In some states there are mandatory waivers for some offenses, but the judges are involved in determining if the criteria for a mandatory waiver are met. Twelve states use a system of presumptive waivers in which the burden of proof is on juveniles to prove that they are amenable to treatment and therefore should not be waived, not on the prosecutor to prove they should.

Table 10.1	Comparing Procedural/Event Terminology in Adult and Juvenile Court Systems	
Procedure or Event	**Adult System**	**Juvenile System**
Police take custody of offender	Placed under arrest	Taken into custody
Official who makes initial decisions about entry into the court system	Intake Officer	Magistrate
Place accused may be held pending further processing	Jail	Detention
Document charging the accused with specific act	Indictment or information	Petition
Person charged with illegal act	Defendant	Respondent
Accused appears to respond to charge(s)	Arraignment	Hearing
Accused verbally responds	Enters a plea of guilty, not guilty, or no contest	Admits or denies
Court proceeding to determine if accused committed the offense	Public jury trial	Adjudicatory hearing: No jury; not public
Decision of the court as to whether accused committed offense	Verdict of jury	Adjudication by judge
Standard of proof required	Beyond a reasonable doubt	Beyond a reasonable doubt
Court proceeding to determine what to do with person found to have committed offense	Sentencing hearing	Dispositional hearing
Institutional Confinement	Prison	Juvenile correctional facility
Community supervision	Probation; parole if had been imprisoned	Probation; aftercare if had been confined to uvenilej correctional facility

2. *Prosecutorial discretion:* This model allows prosecutors to file some cases in either adult or juvenile court. In such cases (usually limited by age and seriousness of the offense), the prosecutor can file the case directly with the adult court and bypass the juvenile court altogether. Fourteen states and the District of Columbia allow prosecutorial discretion waivers.

3. *Statutory exclusion:* These are waivers in cases in which state legislatures have statutorily excluded certain serious offenses from the juvenile courts for juveniles over a certain age, which varies from state to state. These automatic waivers were found in 31 states.

Studies have shown that juveniles waived to adult courts are more likely to recidivate than youths adjudicated for similar crimes in juvenile court, although remember that only the most delinquent-prone youths are waived (Butts & Mitchell, 2000). Neither does a waiver necessarily guarantee a more punitive disposition. Waived juveniles who commit violent crimes are likely to be incarcerated, but juveniles waived for property and drug offenses often receive more lenient sentences than they would have received in juvenile courts (Butts & Mitchell, 2000).

In the article by Benjamin Steiner and Andrew Giacomazzi in this section, the authors examine recidivism among juveniles waived to adult court and placed into a boot camp program, compared with a control group of juveniles who were also waived to adult court but placed on probation rather than placed in the boot camp. The researchers found no difference between the boot camp and control groups on rates of recidivism, but boot camp juveniles were significantly less likely to be reconvicted.

⊠ Extending Due Process to Juveniles

Contrary to the "best interests of the child" philosophy, juvenile courts often punished children in arbitrary ways that would not be tolerated in the adult system. Critics argued that the *parens patriae* doctrine allowed too much latitude for courts to restrict the rights of juveniles and that, because the courts could remove juvenile rights to liberty, juveniles should be afforded the same due process protections as adults. Supporters of *parens patriae* countered that it was suitable and proper for the treatment of children and that any problems concerning juvenile court operation were problems of implementation, not philosophy (Whitehead & Lab, 1996).

The United States Supreme Court itself had maintained a "hands off" policy with regard to the operation of the juvenile courts until 1966 when it agreed to hear *Kent v. United States*. In 1961, when he was 16 years old, Morris Kent broke into a woman's apartment, raped her, and stole her wallet. Because of Kent's chronic delinquency and the seriousness of the current offense, the juvenile judge waived his case to adult court. The adult court found Kent guilty of six counts of housebreaking and robbery, for which the judge sentenced him to 30 to 90 years in prison. Had Kent remained in juvenile court, he could have been sentenced to a maximum of 5 years (the remainder of his minority). Kent appealed, arguing that the waiver process had not included a "full investigation," no findings were made, no reasons were stated for the waiver, and counsel was denied access to the files upon which the juvenile judge presumably relied to make its waiver determination.

The Supreme Court remanded Kent's case back to district court, with Justice Abe Fortas commenting that "there is no place in our system of law for reaching a result of such tremendous consequences without ceremony—without hearing, without effective assistance of counsel, without a statement of reasons. . . . the admonition to function in a 'parental' relationship is not an invitation to procedural arbitrariness." Justice Fortas also noted that, under the *parens patriae* philosophy, the child receives the worst of both worlds: "he gets neither the protections accorded to adults nor the solicitous care and regenerative treatment postulated for children." The Kent decision determined that juveniles must be afforded certain constitutional rights, and thus began the process of formalizing the juvenile system into something akin to the adult system (Hemmens, Steiner, & Mueller, 2003).

The Supreme Court heard a second case concerning the civil rights of juveniles one year later in *In re Gault* (1967). (*In re* literally means "in the matter of"; it is used in non-adversarial proceedings.) In 1964, 15-year-old Gerald Gault was adjudicated delinquent for making obscene phone calls, and he was sentenced to 6 years in the State

Industrial School. An adult convicted of the same offense would have faced a $5 to $50 fine and a maximum of 60 days in jail. The Supreme Court used this case to establish five basic constitutional due process rights for juveniles: (1) the right to proper notification of charges, (2) the right to legal counsel, (3) the right to confront witnesses, (4) the right to privilege against self-incrimination, and (5) the right to appellate review. All of these rights had been denied to Gault.

A third significant juvenile case is *In re Winship* (1970). In 1967, 12-year-old Samuel Winship was accused of stealing $112 from a woman's purse taken from a locker. Winship was adjudicated delinquent based on the civil law's preponderance of evidence standard of proof and was sent to a state training school. Upon appeal, the Supreme Court ruled that, when the possibility of commitment to a secure facility is a possibility, the "beyond a reasonable doubt" standard of proof must extend to juvenile adjudication hearings.

In another case, *McKeiver v. Pennsylvania* (1971), the sole issue before the Court was "Do juveniles have the right to a jury trial during adjudication hearings?" The Supreme Court ruled that they do not. The Court did not rule that the states cannot provide juveniles with this due process right, only that the states are not constitutionally required to do so.

In yet another case (*Breed v. Jones, 1975*), the Supreme Court ruled that the prohibition against double jeopardy applied to juveniles once they have had an adjudicatory hearing (which is nominally a civil process and not technically a trial). Breed had an adjudicatory hearing and was subsequently waived to adult court. The Court ruled that he had been subjected to the burden of two trials for the same offense and therefore the double jeopardy clause of the Fifth Amendment had been violated.

In the case of *Schall v. Martin* (1977), the issue before the Supreme Court was whether the preventative detention of a juvenile charged with a delinquent act is constitutional. The Court ruled that it was permissible because it serves a legitimate state interest in protecting both society and the juvenile from the risk of further crimes committed by the person being detained while awaiting a hearing. This ruling established that juveniles do not enjoy the right to bail consideration and reasserted the *parens patriae* interests of the state.

The issue of the juvenile death penalty was addressed in Section 4 and will not be repeated here except as part of Table 10.2, which presents a summary of important Supreme Court decisions regarding juveniles from *Kent* (1966) to *Roper* (2005). Taken as a whole, what these cases essentially mean is an erosion of the distinction between juvenile and criminal courts. On the positive side, these rulings have helped to create a juvenile court system that more closely reflects the procedural guidelines established in adult criminal courts. On the negative side, they have, in effect, criminalized juvenile courts. In order to gain the due process rights enjoyed by adults, juveniles have surrendered some benefits such as the informality of the solicitous treatment they nominally enjoyed previously. Only time will tell if this convergence of systems results in more just outcomes for juveniles than they received under unmodified *parens patriae*.

The article by Peter Benekos and Alida Merlo in this section examines the issue of the increasingly punitive juvenile justice policies of the 1990s. Their basic argument is that the data on the neurological immaturity of the juvenile brain had a large impact on the *Roper* decision declaring the juvenile death penalty to be unconstitutional and may have the overall effect of softening the punitive practices of the 1990s. They see a fairly large decrease in juvenile waivers as indicative of more enlightened policies toward juveniles. They also examine the effects of juveniles being incarcerated in adult prisons and offer a critique of life without parole for juvenile offenders.

Table 10.2	Supreme Court Cases Altering the Nature of Juvenile Court Proceedings, 1966–2005
Case	**Result**
Kent v. United States (1966)	Courts must provide essentials of due process when waiving juveniles to adult system.
In re Gault (1967)	In hearings that could result in institutional commitment, juveniles have four basic constitutional rights.
In re Winship (1970)	The state must prove guilt beyond a reasonable doubt in delinquency matters.
McKeiver v. Pennsylvania (1971)	Jury trials are not required in juvenile court hearings.
Eddings v. Oklahoma (1982)	All mitigating factors should be considered in deciding to apply the death sentence to juveniles.
Stanford v. Kentucky (1989)	It is constitutionally permissible to impose the death penalty on 16- and 17-year-olds.
Schall v. Martin (1984)	Pretrial preventive detention of juveniles is permissible under certain circumstances.
Roper v. Simmons (2005)	The death penalty for juveniles is unconstitutional.

Juvenile Community Corrections

As seen from Figure 10.2, juvenile corrections mirrors the adult system in that the majority of adjudicated delinquents are placed into some form of community-based corrections and less than a quarter are sent to residential facilities. Juvenile community corrections offers a wide variety of options, all ostensibly designed to implement the three-pronged goal of the juvenile justice system: (1) to protect the community, (2) to hold delinquent youths accountable, and (3) to provide treatment and positive role models for youths. This is known as the *balanced approach* to corrections (Carter, 2006).

When juveniles are taken into custody, a complicated process is initiated of determining how to best deal with them in light of the above goals. Juveniles may be released to their parents or detained in a detention center until this determination is made. The most lenient disposition of a case is known as *deferred adjudication*. Depending on the jurisdiction, a deferred adjudication decision can be made by the police, the prosecutor, the juvenile's probation officer, or the juvenile court's magistrate or judge. A deferred adjudication means an agreement reached between the youth and his or her probation officer, and without any formal court appearance, that the youth will follow certain probation conditions. This form of disposition is only used for status offenses or minor property offenses, and, as long as the juvenile has no further charges, he or she is discharged from probation within a short time. No formal record of the proceedings of the case is made in deferred adjudications.

Other juveniles may be placed on formal probation after adjudication in juvenile court by a judge. In such cases, there are records of the proceedings and probationers are more strictly monitored. As in the adult courts, in juvenile courts, judges typically make their dispositional decision based on recommendations made by probation officers. Juvenile system probation officers write *predisposition reports* (analogous to the adult presentence

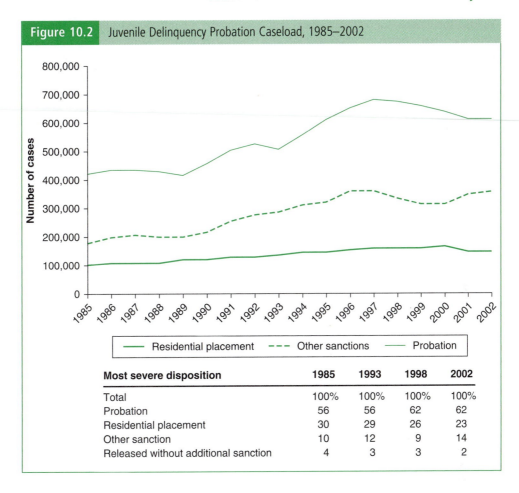

Figure 10.2 Juvenile Delinquency Probation Caseload, 1985–2002

Most severe disposition	1985	1993	1998	2002
Total	100%	100%	100%	100%
Probation	56	56	62	62
Residential placement	30	29	26	23
Other sanction	10	12	9	14
Released without additional sanction	4	3	3	2

investigation report) and will have a variety of classification instruments very similar to those used in the adult system to help them to formulate their recommendations.

Once the youth is adjudicated and placed on probation, under the doctrine of *parens patriae*, the probation officer becomes a surrogate parent to the youth. In reality, probation officers are saddled with such large caseloads (a nationwide average of around 42 [Taylor, Fritsch, & Caeti, 2007]) that they can do very little parenting. Probation officers may see their charges only once a month for perhaps 30 minutes whereas the juveniles' natural parents see them (or should) every day. Juvenile probation officers therefore insist on parental support in working with their children, because parental involvement in the rehabilitative effort of juveniles is considered a "must" (Balazs, 2006). It is a must because, while probation serves the positive goals of keeping youths in the community and letting them avoid the stigma of institutionalization and the exposure to other seriously delinquent youths, the potential danger is that the probationer may view the disposition as a slap on the wrist and return with more confidence to the old ways that led to his or her adjudication.

Of course, if the child comes from an antisocial family rife with substance abuse and criminality, officers are not likely to get any sort of positive support. Even if juvenile probationers come from prosocial families, there is often resistance by parents to juvenile authorities "poking their noses" into family affairs and "picking on" their children, who of course are victims of "bad company" (Walsh, 2006, p. 410). If parents are anxious to help their children, however, there are some excellent parent effectiveness training programs out there. The relatively short-term *Prosocial Family Therapy System* described by Bleckman and Vryan (2000) is a good comprehensive system with some very encouraging results reported.

⬛ Intensive Probation

There will always be some juveniles who require more extensive supervision and treatment than others. To meet their needs and the needs of community protection, a variety of methods have been devised. One such method is *intensive supervision probation* (ISP). ISP is usually imposed on youths as a last chance before incarceration. Juvenile probation officers with an ISP caseload typically supervise only 15–20 juveniles and may carry a gun (Taylor, Fritsch, & Caeti, 2007). Officers may make daily contact with their charges, visiting them at home, school, and work to monitor their behavior and progress in these settings. Officers will also enlist the help of other agencies (both public and private) that can provide probationers with more specialized and concrete help of the kind outside the purview of the juvenile court. These agencies will include mental health clinics, substance abuse centers, educational and vocational guidance centers, and welfare agencies (to help juveniles' families). ISP officers know that they cannot possibly provide for all the needs of their probationers themselves and that efficient case management consists of them delivering services by using networks of collaborative providers. Delany, Fletcher, and Shields (2003) point out the importance of collaborative efforts to assist youths with multiple problems: "Without some level of collaboration among agencies, the odds of relapse and recidivism, which often leads to repeated institutionalization, are high" (p. 66).

Other forms of more intense supervision include electronic monitoring and/or house arrest. These sanctions were discussed in Section 6, and since they operate for juveniles exactly as they do for adults, they will not be discussed again here.

Youths who commit property crimes are frequently made to pay restitution to their victims to compensate for the victims' losses. This both compensates the victim and holds the youth accountable for his or her actions. To compensate the community as a whole, adjudicated delinquents may receive a *community service* order. A community service order is part of a disposition requiring the probationer to work a certain number of hours doing some tasks to help their communities. This work can range from cleaning graffiti from walls to picking up trash along highways or in parks. Restitution and community service orders can go a long way toward helping juveniles develop a sense of responsibility and the ability to accept the consequences of their actions without rancor. For these reasons, community service and restitution have been called "integral components of the restorative justice philosophy" (Walsh, 2006, p. 408).

The article by James Wells and his colleagues in this section offers a quasi-experimental evaluation of juveniles in a shock incarceration (boot camp) and an aftercare program compared with a matched control group released from more traditional residential placements. They found a lower recidivism rate among the boot camp group during a 4-month initial aftercare period but no difference during their 8- and 12-month follow-up periods. They conclude by calling for longer periods of aftercare to complement shock incarceration sanctions.

Restorative justice may be defined as "every action that is primarily oriented toward justice by repairing the harm that has been caused by the act" and "usually means

▲ **Photo 10.2**   Students line up near their bunks in a dormitory at the West Texas State School. This is a juvenile correctional facility.

face-to-face confrontation between victim and perpetrator, where a mutually agreeable restorative solution is proposed and agreed upon" (Champion, 2005, p. 154). Restorative justice defines delinquency as an offense committed by one person against another rather than against the state, and, by doing so, it personalizes justice by engaging the victim, the offender, and the community in a process of *restoring* the situation to its pre-offense status. Thus, restorative justice gives equal weight to the needs of offenders, victims, and the community (Bazemore & Umbreit, 1994). An integral part of restorative justice is victim-offender reconciliation programs (VORPS).

▨ Institutional Corrections

A commitment to a juvenile institutional corrections facility is a serious matter and is typically the disposition reserved for juveniles who have committed violent offenses or for chronic repeat offenders. There are two broad categories of institutional correctional facilities: long term and short term. *Short-term* facilities include reception and detention centers (the equivalent of adult jails), where children may be held before being released to parents or youth shelters or while awaiting court adjudication. *Long-term* facilities are facilities used for housing juveniles after adjudication. They include secure detention centers or training schools (the equivalent of an adult prison) and boot camps, as well as less secure youth centers, ranches, and adventure forestry camps.

Juveniles sent to long-term secure correctional facilities tend to have committed very serious delinquent acts or to be chronic offenders. A study of juveniles sent to long-term secure facilities found that 35% were committed for violent offenses and the remaining 65% for property, drug, or status offenses (Sickmund & Wan, 2003). Minority youths are even more overrepresented in secure juvenile correctional facilities than minority adults are in adult prisons. Martin (2005) reports that, whereas there are about 204 white juveniles per 100,000 in secure facilities, there are 1,018 per 100,000 African American juveniles and that about 70% held in custody for violent offenses are minorities (p. 247).

Other important differences between juvenile and adult facilities are that juvenile facilities are almost always much smaller (rarely more than 250 juveniles), that the costs associated with incarceration are considerably higher, and that much more money is spent on programming relative to security (Taylor, Fritsch, & Caeti, 2007). For instance, the California Youth Authority spends 52% of its budget on academic and vocational training, case planning, counseling, and skills training as opposed to only 13% on custody and security (Taylor, Fritsch, & Caeti, 2007). Nevertheless, many of the same problems seen in adult prisons are also seen in juvenile facilities, especially in the larger institutions with a large staff to resident ratio. In juvenile facilities as in adult prisons, gangs exist formed along racial, ethnic, and neighborhood lines, and violence and sexual assault are always dangers for the unaffiliated (Martin, 2005).

Caeti, Hemmens, Cullen, and Burton demonstrate that the monolithic view of corrections is not supported by the beliefs of the directors of juvenile correctional facilities—that these beliefs differ from those of adult prison managers on issues such as support for rehabilitation and the ancillary effects that this support has on job satisfaction. Notably, other predictors of job satisfaction for juvenile facility directors included years on the job, their stress level, and pay.

⬛ Summary

- The juvenile justice system is based on civil law and deals with status offenses (applicable only to juveniles) and delinquency (crimes if committed by adults). Until recently, proof standards and courtroom procedures were much the same. However, recent Supreme Court decisions have created a juvenile justice system that more closely reflects the adult criminal justice system, especially in cases involving serious delinquent behavior.

- Juveniles commit a disproportionate number of both property and violent crimes, and this has always been true across time and cultures. Recent scientific evidence relates this situation to the hormonal surges of puberty juxtaposed with an adolescent brain undergoing numerous changes. Although most adolescents commit antisocial acts, only a small proportion continue to commit antisocial acts after brain maturation is completed.

- The history of juvenile justice has three distinct periods. Originally, Western culture relied heavily on parents to control children. As society has changed, so have the expectations regarding juvenile delinquency. Institutional control of wayward youth was the model from the mid 1500s until the emergence of the juvenile courts in the United States in the late 1800s and early 1900s. The juvenile court follows the doctrine of *parens patriae*, but, recently, there has been a movement away from the broad discretion formerly accorded to juvenile courts and toward a model more closely reflecting the constitutional protections afforded adult offenders. Much of this change has issued from the increased waivers of juveniles to adult courts and from the often arbitrary control juvenile justice authorities have exercised over juveniles.

- Much of what constitutes juvenile corrections mirrors what we have written about in other sections in this book; thus, we have only briefly highlighted differences between the juvenile and adult systems. Major differences are a greater emphasis on rehabilitation, as exemplified by the ratio of programming to security expenditures in juvenile correctional facilities and the lesser likelihood of juveniles being sent to secure facilities relative to adults.

KEY TERMS

Balanced approach

Bridewells

Child savers

Community service order

Delinquents

Juvenile waiver

Parens patriae

Restorative justice

Status offenders

Status offenses

DISCUSSION QUESTIONS

1. Discuss the development of the concept of childhood in western culture.

2. Discuss the doctrine of *parens patriae* in relation to the development of the juvenile court system in the United States.

3. Why did the Supreme Court give juveniles more due process rights? What rights do they have and don't they have?

4. Given the science about the brain that indicates adolescents may have difficulty understanding right from wrong some of the time, how do you think the juvenile justice system should adjust?

5. Does waiver to adult court always result in a harsher sentence for juveniles? What does the research indicate?

6. Do you think that restorative justice is workable? In what circumstances would and would it not be?

INTERNET SITES

American Correctional Association: www.aca.org

American Friends Service Committee (a Quaker organization interested in correctional reform): www.afsc.org

American Jail Association: www.aja.org

American Probation and Parole Association: www.appa-net.org

Bureau of Justice Statistics (information available on all manner of criminal justice topics): www.ojp.usdoj.gov/bjs

National Criminal Justice Reference Service: www.ncjrs.gov

Office of Justice Programs, Bureau of Justice Statistics (periodic statistical reports on all manner of criminal justice topics, e.g., HIV in prisons and jails, probation and parole, and profiles of prisoners): www.ojp.usdoj.gov/bjs/periodic.htm

Office of Juvenile Justice and Delinquency Prevention: ojjdp.ncjrs.org

Vera Institute (information available on a number of corrections and other justice related topics): www.vera.org

READING

In this article, Peter Benekos and Alida Merlo discuss what they consider to be the legacy of a punitive policy toward juveniles in the United States. They state that, despite declining juvenile crime rates, the juvenile justice system in the United States continues to include punitive sanctions. The authors also claim that attitudes toward offenders are ambivalent, but they see evidence that legislators and the public are reluctant to abandon the punitive policies (for instance, waivers to adult courts) of the 1990s. At the same time, the authors see indications of more enlightened approaches to juvenile justice, and, in this context, they review the state of juvenile justice policy and outline trends in waiver and sentencing as they exist in the early twenty-first century.

Juvenile Justice

The Legacy of Punitive Policy

Peter J. Benekos and Alida V. Merlo

Although the future of juvenile justice policy is uncertain, the impact of policies from the 1990s is clear: Despite declining juvenile crime rates, the adultification of youth continues to include punitive and exclusionary sanctions. Attitudes toward offenders are ambivalent, but there is evidence that legislators and the public are reluctant to abandon the punitive policies of the 1990s. Simultaneously, there are indications of more enlightened approaches to juvenile justice. In this context, the authors review the state of juvenile justice policy and review trends in waiver and sentencing.

This article examines the current state of juvenile justice policy and considers whether there has been a softening in public attitudes toward youthful offenders and how these attitudes have affected existing policies. Specifically, the authors review recent data on juvenile waiver, the incarceration of juveniles with adults, and juveniles who are serving life sentences without the possibility of parole. If there is evidence of a softening in attitudes toward juvenile offenders, it would indicate a retreat from the "get tough" philosophy that has characterized juvenile justice policy in the last 20 years. However, as Hutchinson (2005) observed, "Lawmakers are loath to do anything that can make them appear soft on crime. That is still considered the kiss of death for political careers" (para 6). This would suggest reluctance to change the course of policy.

⬚ Lessons of *Roper v. Simmons*

In March 2005, the U.S. Supreme Court determined that executing juveniles under the age of 18 constituted cruel and unusual punishment in violation of the Eighth Amendment (*Roper v. Simmons*, 2005). Death row inmates who had been sentenced as juveniles typically received life

SOURCE: Benekos, P., & Merlo, A. (2008). Juvenile Justice: The Legacy of Punitive Policy. *Youth Violence and Juvenile Justice, 6,* 28–46. Reprinted with permission of Sage Publications, Inc.

sentences after *Roper*. For example, Governor Rick Perry commuted the death sentences of 28 offenders to life in prison. In capital murder cases, Texas juries previously could sentence an offender to death by lethal injection or to life in prison, which meant that after 40 years the offender could be considered for parole. However, after commuting the death sentences of the 28 offenders to life, the Governor also signed into law a new bill that prohibits parole in life sentences, but it is not to be applied retroactively. For those offenders sentenced to life after September 1, 2005, life in prison in Texas is life without possibility of parole.

Even before the Court's decision in *Roper*, support for the death penalty, as evidenced by the number of juveniles sentenced to be executed, had been receding. For example, in 1999, 14 juveniles were sentenced to death. By contrast, 2 were sentenced to death in 2004. These data suggest that juries and judges appeared reluctant to impose the death sentence on offenders who were under 18 at the time of the crime. The reduction in the number of death sentences could be interpreted as an indication of a softening in attitudes toward youthful offenders. In *Roper*, the justices concluded that juveniles, compared to adults, were perceived as less culpable. In particular, they noted that juveniles are less blameworthy than adults because they are more immature and less responsible than adults, more likely to be influenced by external pressure including peer pressure, and more vulnerable, in part, because they have less control over the environment than adults (Benekos & Merlo, 2005). Finally, the justices noted the differences in character between juveniles and adults: "The personality traits of the juveniles are more transitory, less fixed" (*Roper v. Simmons*, 2005, p. 16).

Furthermore, Justice Kennedy, in writing for the majority, referred to Article 37 of the United Nations Convention on the Rights of the Child. Currently, the United States and Somalia are the only countries that have not ratified the Convention, which includes a prohibition on capital punishment for juveniles who commit crimes under the age of 18. Including information regarding international perceptions of sanctions in the United States suggests that, although the opinions of other countries did not determine the decision, the current international policies informed the decision by providing "respected and significant information for our own conclusions" (*Roper v. Simmons*, 2005, p. 24).

Perceptions of juvenile sanctions in the United States, both international and domestic, are an important dimension of policy. In this article, the authors consider whether that attitudinal shift regarding the execution of juvenile offenders affects other aspects of criminal justice policy toward youthful offenders. We begin with an examination of how the United States has fared with the issue of juvenile waiver or transfer to adult criminal court.

⬚ Criminalizing Juvenile Delinquents

Even though juvenile crime, as measured by arrest, has continued to decline since the mid-1990s, the get-tough legislation enacted during that decade, which targeted youthful offenders, resulted in adultification policies that increased the number of youth in criminal court and the number of youth incarcerated in adult prisons.

Regarding juvenile crime, Snyder (2005) determined that juvenile arrests for violent crime (murder, forcible rape, robbery, and aggravated assault) were the lowest since 1987 and represent "about one-third of 1% of all juveniles ages 10–17 living in the U.S." (p. 4). In 2003, there were an estimated 92,300 juvenile arrests for violent crime; in 2004, the number of arrests was 91,000, a 2% drop (*OJJDP Statistical Briefing Book*, 2005). Of the "2.2 million arrests of persons under age 18," in 2003, about "71% were referred to juvenile court and 7% were referred directly to criminal court" (2005, p. 5).

In describing the mechanisms for referring youth to criminal court, Griffin (2003) focused on three primary mechanisms for jurisdictional transfer: judicial waiver, statutory exclusion, and direct file. Based on his assessment of state transfer laws "through the 2002 legislative sessions" (p.3),

- 46 states have judicial waiver
- 29 states have statutory exclusion
- 15 states have direct file

In addition, 25 states have "reverse waiver" and 34 have "once adult/always adult" transfer provisions (Griffin, 2003, p.3). The criteria for discretionary judicial waiver generally emphasize "the best interests of the child and the public" but identify age, offense, and prior record as "threshold" considerations in determining jurisdictional transfer (Griffin, 2003, p. 4).

The threshold criteria also determine which youth qualify for statutory exclusion from juvenile court and therefore begin their judicial process in criminal court. With direct file, offense seriousness generally "triggers" appearance in criminal court (Griffin, 2003, p. 10).

In enacting tougher waiver policies in the 1990s, legislatures accomplished the following (Urbina, 2005, p. 148):

- Increased the number of crimes eligible for judicial waiver
- Lowered the threshold age for waiver
- Designated certain crimes for automatic waiver
- Specified certain crimes for presumptive waiver
- Expanded prosecutorial authority to review cases.

As a result of these legislative changes, "approximately 210,000 adolescents nationwide are now prosecuted in adult courts each year." The National Campaign for Youth Justice (n.d.) also reports that about 250,000 youth under 18 are "tried and sentenced in adult courts each year" (p. 1). Mattingly (2006)

reports that this adultification policy occurs "despite the fact that research shows that trying and sentencing youth as if they were adults does not increase public safety or reduce crime" (p. 11). Nonetheless, as Sontheimer and Volenik (2004) observed, "As a society, we have decided that people who break the law as children should pay heavier and longer lasting consequences for that behavior than we exacted from them in the past" (p.1).

Judicial Waiver

Based on data accessible from the National Center for Juvenile Justice (2006), the use of judicial waiver continued to decline in the early 2000s. In 2003, of 928,849 formally handled cases, 6,735 (0.7%) were judicially waived. This is a 49% drop from 1994, when the number of judicially waived cases peaked at 13,089, to 6,735 cases in 2003. From 1986 to 2003, 1.1% (163,094) of all formally handled cases (14,698,959) were waived to criminal court (Stahl, Finnegan, & Kang, 2006).

From 1986 to 2003, of all the person offenses (5,788,243), 1.1% (63,214) were waived to criminal court. Of the person offenses formally handled (3,356,466), 1.9% (63,214) were judicially waived (National Center of Juvenile Justice, 2006; see Figure 1). Of the total number of formally handled property offenses (6,800,062), about 1.0% (65,673) were waived. And of the formally handled drug offenses (1,474,241), 1.5% (21,376) were judicially waived.

Of all the judicially waived cases from 1986 to 2003 (163,094), about 39% (63,214) were person offenses, 40% (65,673) were property offenses, and 13% (2/3:6) were drug offenses.

The trend in judicial waivers for all offense categories in the late 1990s decreased. In addition, the age pattern for waiver did not appear to have changed, and 17-year-old youth were most likely to be waived (52%), followed by 16-year-olds (29%). Fifteen-year-old youth, however, were more likely to be waived in the 1990s than in the 1980s. For example, in 1986,

6.2% of all waivers were of 15-year-old youth compared to 8.5% in 1994, 10.6% in 1998, 12.1% in 1995, and 8.5% in 2003 (National Center for Juvenile Justice, 2006).

Although the percentage of older-than-17 remained fairly constant at about 7.0%, the 17-year-olds were less likely to be judicially waived: 60% in 1986, compared to 47% in 2003. These national data on waiver indicate that a small percentage of youth are judicially waived (1.1%).

☒ Youth Incarcerated in Adult Institutions

The incarceration of juveniles with adults has a long history in the United States. The deleterious effects associated with housing children with adults were cited by reformers in Cook County, Illinois, to support the creation of a separate juvenile court in 1899. Beginning in the 1990s, there is evidence that the United States reverted to this approach with little consideration of the long-term and short-term consequences.

The exact number of juveniles in jail is unknown, but there are statistics that provide estimates of the number of youth who are younger than 18 and who are incarcerated. For example, Snyder and Sickmund (2006) reported that in June 2004, there were 7,083 youth (under 18) in jails (p. 236). These youth comprised approximately 1% of the jail population (p. 236). By contrast, in 1998, Austin, Dedel Johnson, and Gregoriou (2000, p. x) reported that there were 9,100 youth under 18 who were incarcerated in local jails in 1998. These recent data suggest that the number of youth in jails has decreased since the 1990s.

Using data derived from states that reported the number of youth under age 18 in state prisons, Snyder and Sickmund (2006) found that there were approximately 4,100 youth who comprised new court commitments in 2002, and they represented 1.1% of all new prison commitments. Most of these

youth (79%) were 17 years old when they were admitted (Snyder & Sickmund, 2006, pp. 237–238). Robbery was the primary offense for which these youth were admitted to prison, and it accounted for 4.3% of all new court commitments to prison in 2002 (Snyder & Sickmund, 2006, p. 237). There were 5,400 juveniles incarcerated in adult prisons in 1998. Woolard, Odgers, Lanza-Kaduce, and Daglis (2005, p.1) estimate that there are more than 10,000 juvenile offenders in adult correctional settings. Snyder and Sickmund found that the steady increase in the number of new admissions to state prisons between 1986 and 1995 has been followed by a considerable decrease in the number of youth under 18 who have been admitted to prison between 1996 and 2002.

Recent prison data suggest a slowing in the incarceration of youth in adult prisons, but it has occurred along with an overall decline in violent offending as demonstrated by the number of arrests for violent offenses. Rather than discontinuing the processing and sentencing of juveniles like adults, these data suggest that juveniles continue to be adversely affected by legislative initiatives that were implemented in the 1990s. In 1985, 18 youth for every 1,000 arrests were incarcerated in an adult prison. By 1997, there were 33 youth incarcerated for every 1,000 juveniles arrested. Similarly, in Florida, Greene and Dougherty (2001a) reported that one in 13 Florida inmates was doing time for a crime committed as a juvenile.

In Pennsylvania, the Department of Corrections reported that the number of youth under 18, who were received in state prisons in the early 2000's, was decreasing (Hartman, 2006). In 2003, however, the number more than doubled from 32 in 2002 to 66 in 2003. This represents less than 1% (0.75%) of all the new commitments received by the Pennsylvania Department of Corrections in 2003 ($N = 8,760$; Hartman, 2006, p. 15).

Furthermore, commitment to an adult institution does not necessarily signal that the juvenile has exhausted the remedies available in juvenile corrections. For example, Annino (2001) found that, of the approximately 1,000 youths sentenced to adult prisons in Florida, more than 40% were never previously committed in the juvenile court (Annino, 2001, p. 477). In short, jail and prison sentences are sometimes used as the first rather than the last disposition in a case.

Demographically, youth in prison tend to be overwhelmingly male. In 2002, males comprised 96% of the new court commitments. In addition, new prison commitments of youth under 18 were disproportionately Black. For example, when Black and White inmate admissions were compared, Blacks outnumbered Whites by 2 to 1 in 2002 (Snyder & Sickmund, 2006, p. 238).

Although it is generally assumed that youth in prison are primarily sentenced as adults only after conviction for a violent offense, this is not always the case. In fact, there is evidence that juveniles may be sentenced more harshly than adults for similar kinds of criminal activity. According to data from 344 counties, juveniles who were transferred to adult court and convicted of larceny, burglary, or weapons offenses in 1996 faced a greater likelihood of incarceration in prison than adult offenders who were convicted of similar crimes. In addition, juveniles who were convicted of murder and weapons offenses were also more likely to be sentenced to longer terms of incarceration in prison than their adult counterparts.

These disparate sentencing practices are not inconsequential. Kurlychek and Johnson (2004) examined the sentences of juveniles (under 18) and young adults (18–24) in Pennsylvania. They found that, during a 3-year period from 1997 to 1999, juveniles were sentenced more harshly than young adults (p. 500). "Overall, juveniles appear to be more likely than young adults to be incarcerated for lesser offenses and they tend to receive considerably longer sentence lengths for more serious offenses" (Kurlychek & Johnson, 2004, p. 502). They contend that it is possible that judges

may view youth who are transferred to adult court as more culpable and dangerous than young adult offenders.

Effects of Incarcerating Youth in Adult Institutions

One of the consequences of sentencing juveniles to adult jails and prisons is the increased risk of suicide. This risk occurs for youth under 18, who are incarcerated in local jails as well as in state prisons. For example, inmates in local jails who were under 18 had the highest rate of suicide between 2000 and 2002; their rate was 101 for every 100,000 inmates (Mumola, 2005, p. 5). These data contradict the overall trends in jail suicide rates. Typically, it is the oldest inmates, aged 55 or older, who have the highest rate of suicide. However, their rate was 58 suicides for every 100,000 jail inmates in the 2000–2002 data.

The situation is even worse for youth under the age of 18 incarcerated in state prisons. Although the suicide rate of state inmates ranged from 13 to 14 suicides for every 100,000 inmates for all age groups over 18, "the suicide rate of State prisoners under 18 was 4 times higher (52 per 100,000), but this group accounted for less than 0.3% of State prisoners and had 3 suicides nationwide over 2 years" (Mumola, 2005, p. 6). In short, the suicide rate for offenders under 18 incarcerated in jail or prison is high, but the actual number of suicides is low compared to the other age groups.

Juveniles incarcerated in adult prisons face greater risks of being physically and sexually abused than adults. In their research, Austin et al. (2000, p. 8) and Schiraldi and Zeidenberg (1997) report that the incidence of sexual attack or rape, being "beaten up" by staff, and the likelihood of being attacked with a weapon were much higher among juveniles in adult prisons than juveniles in juvenile institutions. Similarly, Greene and Dougherty (2001b, para 8) reported that juveniles in Florida who are incarcerated in adult male prisons were "four times as likely as adults to report being assaulted in DOC facilities." Between 1995 and 1999, there were 362 assault complaints where a juvenile was the victim, which constituted one for every two juvenile offenders incarcerated with adults. By contrast, for adults, the rate was one complaint for every seven adult offenders. When juveniles housed with adults were compared to juveniles in juvenile facilities, the findings were even more striking. Youth in the adult system were almost "21 times as likely to be assaulted or injured as teens in Department of Juvenile Justice facilities" (Green & Dougherty, 2001b, para 12). As Woolard et al. (2005) note, the victimization of youth in adult institutions is widely known, yet "there are few safeguards in place to prevent such incidents" (p. 9).

When a juvenile enters prison, he/she frequently lacks the coping skills (both mental and physical) that older offenders employ to sustain their self-respect and their mental health. These teenagers are ill equipped to deal with the prison milieu, "and it is also an unlikely place for them to gain the life experiences and education necessary for healthy mental and physical development" (Amnesty International and Human Rights Watch, 2005, p. 52). In short, prison is not a rehabilitative environment designed for youth. Prisons are designed to incapacitate offenders rather than treat them, and youth are particularly disadvantaged in this setting.

Rose (1999) contends that a safe and secure environment is a critical component of treatment programs for young offenders. It is particularly important that the staff and professionals establish a positive relationship and use a consistent approach with the offenders (Rose, 1999, p. 17). In their interviews with 44 young offenders in Florida who had been sentenced to adult incarceration, Lane, Lanza-Kaduce, Frazier, and Bishop (2002) found that more than 60% of the youth perceived their experience negatively. According to Lane et al. (2002), "They [the respondents] felt staff took their hope from them and were generally too mean or apathetic, that the environment was always unsafe, and that they learned too much about how to be better criminals" (p. 448). Youth who experience victimization in adult prisons by inmates, guards, visitors, and other juveniles may not only fear for their safety but also fail to develop positive relationships with adult mentors and role models.

Classification and screening processes for juveniles are not available in all adult facilities. In addition, juveniles in adult prisons are also less likely to have rehabilitation programs, medical, mental health, and academic programs that are appropriate for their age and level of development. For example, based on information from the *Survey of Inmates in Local Jails, 2002*, Karberg and James (2005) report that about 61% of jail inmates who were 24 years of age or younger "had the highest rate of drug dependence or abuse" (p. 6). In their interviews of girls who had been transferred to adult court and sentenced to adult prisons, Gaarder and Belknap (2002) also found that drug and alcohol dependence was prevalent in their sample. Unfortunately, prison treatment programs designed to help young offenders deal with these problems are sorely lacking. Woolard et al. (2005, p. 8) contend that it is not just fitting existing adult offender programs and policies for juvenile offenders but rather a requirement that correctional administrators make qualitative changes in approach. Whether it is the juvenile offenders' special housing needs, educational needs, or developmental differences, their treatment plans are significantly different from those of their adult counterparts (Woolard et al., 2005, p. 9).

Juveniles are also affected by the lack of appropriate medical services in prisons. These youth require education programs that address their physical and sexual development. They have nutrition needs that are related to their physical development, as well as vision and dental care concerns that typically change in adolescence.

The long-term consequences of juvenile incarceration in adult institutions are not fully understood. The socialization of young offenders in prison may affect their ability to successfully adjust outside the prison when they are eventually paroled or released. As Singer (2003) contends, "The absence of familial and noncriminal attachments may predict the extent to which juvenile offenders are unable to adjust to life outside of prison" (p. 125). In deciding to transfer juveniles into adult courts and sanction them as adults, the sustaining effects of the prison social environment have been largely ignored. Clearly, this is an area worthy of further research.

In addition to their suicide risk, victimization by other inmates and staff, lack of specialized services, and socialization in the prison environment, youth are also disadvantaged by their cognitive functioning. Rather than assume that juvenile and adult brains function identically, Gur (2005) reviewed research conducted on youth and adult brains and contends that the brain does not reach full maturity until the early-to-mid-20s. Furthermore, with respect to moral culpability, those parts of the brain that deal with judgment, impulsive behavior, and foresight develop in the 20s rather than in the teen years. Evidence obtained from magnetic resonance imaging (MRI) data has consistently found that children do not have the same physiological means of controlling themselves that adults have (Amnesty International and Human Rights Watch, 2005, p. 47). Despite these data, legislatures have authorized and judges continue to sentence youth to life in prison without the possibility of parole. It is this dimension of juvenile justice policy that we now examine.

⬚ Juvenile Life Without Parole (JLWOP) Sentences

Another consequence of the get-tough, punitive legislation that characterizes adultification policies is life sentences for offenders who commit their crimes before age 18. In a comprehensive study of "child offenders" sentenced to life without parole (JLWOP), Amnesty International and Human Rights Watch (2005) identified that 2,225 prisoners in the United States "have been sentenced to spend the rest of their lives in prison for the crimes they committed as children" (p. 1). Because there is "no national depository of these data" this report is a "first-ever" attempt to collect data from state departments of correction (p. 1). The data indicate that "59 percent received the sentence (i.e., LWOP) for their first-ever criminal conviction" and "16 percent were between 13 and 15 years old at the time they committed the crimes" (p. 1). On average, 98 youth under 18 have been admitted to prison with a sentence of life without possibility of parole in each year from 1990 to 2003 (Hartney, 2006, p. 3).

One report concluded that the number of offenders serving LWOP in Pennsylvania is actually higher than that indicated by the data published by Amnesty International Human Rights Watch in 2004. According to Levin (2007), there were 440 such offenders incarcerated in Pennsylvania prisons in February 2007. Although the number of life without parole sentences imposed on "children" peaked in 1996 (50 in 1989, 152 in 1996, 54 in 2002), the use of life sentences without parole increased during the 1990s (Amnesty International and Human Rights Watch, 2005, p. 2):

> For example, in 1990 there were 2,234 youths convicted of murder in the United States, 2.9 percent of whom were sentenced to life without parole. Ten years later, in 2000, the number of youth murderers had dropped to 1,006, but 9.1 percent were sentenced to life without parole.

In 2003, even though 54 life without parole sentences were imposed, the rate is "three times higher today than it was fifteen years ago" (p. 2). And as previously noted, 59% of young offenders serving life without parole were sentenced on their "first-ever criminal conviction" (Amnesty International and Human Rights Watch, 2005, p. 1).

Table 1	Total Number of Youth Serving Life Without Parole by State		
State	**Youth LWOP Total**	**State**	**Youth LWOP Total**
Alabama	15	Montana	1
Arizona	30	Nebraska	21
Arkansas	46	Nevada	16
California	180	New Hampshire	3
Colorado	46	New Jersey	0
Connecticut	10	North Carolina	44
Delaware	7	North Dakota	1
Federal	1	Ohio	1
Florida	273	Oklahoma	49
Georgia	8	Pennsylvania	332
Hawaii	4	Rhode Island	2
Idaho	Data missing	South Carolina	26
Illinois	103	South Dakota	9
Indiana	2	Tennessee	4
Iowa	67	Utah	0
Louisiana	317	Vermont	0
Maryland	13	Virginia	48
Massachusetts	60	Washington	23
Michigan	306	Wisconsin	16
Minnesota	2	Wyoming	6
Mississippi	17		
Missouri	116	Nationwide	2225

SOURCE: Data provided by 38 state correctional departments and additional sources for the states of Alabama and Virginia.

⊠ Critique of Life Without Parole for Youthful Offenders

As discussed above, although the legal issue of the death penalty for young offenders has ended, the policy issue of harsh penalties has not. Get-tough policies are reflected in jurisdictional waiver, adult sentences, and life without parole. As reported by Amnesty International and Human Rights Watch (2005), "Although it has never ruled on the constitutionality of life without parole for children, the U.S. Supreme Court has often highlighted the inherent differences between youth and adults in the criminal law context" (p. 86). In *Roper* (2005), the Court recognized the immaturity, irresponsibility, and diminished culpability of youth; in response, Amnesty International and Human Rights Watch concluded that punishment for children "should acknowledge that substantial difference" (p. 45).

As noted in this article, post-*Roper* attention has shifted focus to such harsh sentences as life without parole imposed on juveniles. In Mississippi, the Court of Appeals upheld the life sentence of Tyler Edmonds who was 13 when he killed his half-sister's husband (*Edmonds v. Mississippi*, 2006). One of the challenges in *Edmonds* was the trial court's failure to inform the jury that conviction would result in a life sentence. The Appeals Court found that no error occurred "in refusing to inform a jury of the mandatory life sentence that the defendant would receive if convicted" (*Edmonds v. Mississippi*, 2006, p. 48). Essentially, the Mississippi court did not rely on *Roper* in determining that youth and culpability did not offset the harsh mandatory sentence.

A case that specifically challenges the constitutionality of life without parole for juveniles (JLWOP) was filed in the Superior Court of Pennsylvania. The case involves Aaron Phillips, who was found guilty of second degree murder in 1988 for a crime he committed when he was 17. The *Amicus Curiae* filed on June 5, 2006 by the Defender Association of Philadelphia and the Juvenile Law Center, cites the language and rationale of *Roper* in arguing that the imposition of "life imprisonment without the possibility of parole as here, is unconstitutional" (*Brief of Amicus Curiae*, 2006, p. 7).

From another perspective, in her critique of JLWOP sentences, Massey (2006) concluded that the sentences were "grossly disproportionate" (p. 1083), and, based on the Court's reasoning in *Roper*, sentences of life without parole for juveniles violate the Constitution when the punishment exceeds the seriousness of the offense (p. 3).

Acknowledging that life without parole may be excessive punishment, legislators are still reluctant to deviate from the get-tough policies adopted in the 1990s. One state senator in Pennsylvania, who opposes life without parole for juveniles, Vincent Fumo, D-Philadelphia, stated that it would be "political suicide" to propose legislative changes (DiFilippo, 2006, para 45). "The minute you do, you're 'soft on crime,' and then your opponents use it against you, and in today's society, that's all they need" (DiFilippo, 2006, para 46).

The emphasis on harsh punishments such as life without parole—characterized as "grossly disproportionate" (Logan, 1998, p. 681)—suggests that "commitment to a juvenile justice system and the youth rehabilitation principles embedded in it" have been abandoned (Amnesty International and Human Rights Watch, 2005, p. 2). As Kurlychek and Johnson (2004) observed, tougher sentencing reflects an emphasis on public safety, a discounting of rehabilitation, and disregard for diminished culpability or blameworthiness of youthful offenders.

This review of trends in waiver, adult incarceration, and life without parole for young offenders indicates that these punitive legislative reforms of the last decade are well entrenched in juvenile justice policy. Urbina (2005) concluded that "young criminals today are being punished for the behavior of their counterparts who committed serious offenses 15 to 20 years ago" (p. 148). Although severe, punitive, and lengthy sanctions are inconsistent with the rationale used by the Court in *Roper* (2005), it is too soon to measure what, if any, impact *Roper* will have on juvenile justice. Some developments, however, may portend a "softening" or "balancing" in policy.

In Michigan, which confines the third largest number of young life without parole offenders, a recent survey found that 95 percent of Michigan citizens opposed juveniles being sentenced to life without parole (Charney, 2005). Also in Michigan, legislation has been introduced to abolish life-without-parole sentences for young offenders" (DiFilippo, 2006, para 54). Senate Bill 944 stipulates that courts "Shall not sentence an individual who was less than 18 years of age when the crime was committed to imprisonment for life without parole eligibility" (Senate Bill 944, 2006, p. 9). The bill was referred to the Senate Judiciary Committee in January 2006. In January 2007, the Michigan legislature again introduced legislation (Senate Bills 6, 9, 28, and 40) to remove this sanction from the penal code. The bills were referred to the Senate Judiciary Committee (Michigan State Senate, 2007). Although it did not pass the Colorado legislature, a bill was introduced in 2005 to eliminate life without parole and "other particularly long sentences for youth offenders, giving judges the ability to periodically re-examine a youth offender's progress in prison" (Amnesty International and Human Rights Watch, 2005, p. 89). Also in 2005, Florida legislators considered Senate Bill 446 to "ensure parole for some children sixteen years old and younger sentenced to life" (Amnesty International and Human Rights Watch, 2005, p. 89). The legislation did not pass.

In Pennsylvania, which confines the largest number of life without parole youth and is "one of the 15 states that have no age minimum on JLWOP sentences" (Rubin, 2006, p. 15), the guiding mission of juvenile justice is balanced and restorative justice (Juvenile Court Judges' Commission, 2005). This offers a commitment to "redemption" for youthful offenders while protecting the community and holding juveniles accountable for their offending. Apparently, the state has not abandoned the spirit of juvenile justice even though harsh sentences persist.

Also in response to tough adult sentences for juveniles, Colorado is being scrutinized for the broad authority prosecutors have to charge youth as adults (Moffeit & Simpson, 2006). In examining cases of youth sentenced as adults, disparities and injustices are striking "a nerve with judges, jurors, lawyers and legislators who believe the adult system has mishandled some juveniles' cases" (Moffeit & Simpson, 2006, para 9). Because prosecutors are authorized to direct file to adult court, waiver hearings have been eliminated, and youth do not have an opportunity for judicial review of waiver criteria or circumstances of the case.

Similarly, *The New York Times* featured a series of articles in October 2005 that reported on prisoners serving life sentences and included a story on youth serving life without parole (Liptak, 2005). Liptak noted that some youth were 15 years or younger when they committed their crimes and could spend their entire lives incarcerated.

Whether other states will follow Michigan, Colorado, and Florida and begin reassessing their juvenile justice policies and move away from adultification remains to be seen. The rationale of *Roper* may provide legislators with language to justify reform of policies enacted in the 1990s when fear, frustration, and anger pervaded the discourse on youth crime.

▧ Discussion

It may be premature to assess what effect, if any, the *Roper* decision has had on juvenile policy regarding waiver to adult court, incarceration in adult prisons, and sentences to life imprisonment without the possibility of parole. In his review of challenges to JLWOP sentences, Rubin (2006) concluded that "JLWOP relaxation is not an easy redirection, and current efforts are a long way from becoming a movement" (p. 1). Because the Supreme Court only recently determined that the death penalty was unconstitutional for juvenile offenders, it may be unrealistic to anticipate significant changes this soon. Whether the Pennsylvania court applies the *Roper* rationale in finding JLWOP unconstitutional remains to be seen.

Prior to *Roper*, however, there was evidence that there were (a) fewer admissions to prison for both White and Black youth in 1999 compared to 1995, (b) fewer cases judicially waived in 2002 compared to the mid-1990s, and (c) fewer youth being sentenced to life without parole in 2002 compared to 1996. Ironically, as Hutchinson (2005) observed, without the death penalty, legislatures may be reluctant to remove the life without parole sanction. In addition, LWOP is seen as a "far more humane" sentence than the death penalty (para 9).

Considering the relatively low rate of juvenile crime, these data may signify a more lenient or softer approach in handling juvenile offenders. As previously discussed, restorative justice is an alternative that holds some promise in dealing with youthful offenders. It is also conceivable that more states will emulate Michigan and consider rescinding life without possibility of parole sentences for juvenile offenders. Conversely, more states, like Texas, might move to amend life sentence statutes that previously permitted consideration for parole after a specified number of years and now preclude any eligibility for parole. In terms of economic considerations, the exorbitant costs and uncertain benefits associated with life without parole sentences may also affect willingness to reconsider legislation. For example, in Michigan, it is estimated that keeping a youth in prison for life "will cost the state at least one million dollars, and the value of keeping them in prison will never be re-evaluated" (LaBelle, Phillips, & Horton, 2004, p. 24).

From a pragmatic perspective, the effectiveness of harsh penalties on juvenile crime is also questioned. As reported by Amnesty International and Human Rights Watch (2005), in 1989, youth represented 11% of the offenders arrested for homicide. In 1999, they comprised 10% of the homicide offenders. "If harsh sentencing were the answer to deterring serious and violent juvenile crime, the United States should be among the countries with the lowest percentages of youth murderers" (p. 109).

As Rubin (2006) noted, "efforts to reduce criminalization of juveniles are taking place" (p. 16). He cites Connecticut's legislative review for raising the "current maximum juvenile age of 16 years to 18 years" (p. 16). In addition, in Delaware, automatic transfer for robbery offenders has been reversed, and, in Wisconsin and New Hampshire, policies for returning eligibility for juvenile court to a youth's 18th birthday are under review (Rubin, 2006, p. 16).

These conflicting approaches to young offenders illustrate the ambiguity in juvenile justice policy and reflect Bernard's (1992) thesis that policy is cyclic and shifts between severe and soft handling of youth. Although Applegate and Davis (2006) found some evidence among Florida respondents of a softening in public attitudes toward sentencing of juvenile offenders, and Nagin, Piquero, Scott, and Steinberg (2006) found Pennsylvania respondents more willing to pay for rehabilitation versus incarceration, there is also a punitive *Zeitgeist* that prevails.

This review of waiver, incarceration, and JLWOP demonstrates the punitive aspects and legacy of juvenile justice policy: Although the Court's reasoning in *Roper* provides rationale for more tolerant and therapeutic responses to youthful offending, the legislative reforms at the end of the 20th century portend a dualistic model for juvenile justice: a system that engages in prevention and intervention for some youth and punishment and exclusionary sanctions for others.

In 2004, Sontheimer and Volenik asked, "Is the Juvenile Justice System Still Relevant?" (p. 1). Although they recognized the "blurred" line between juvenile and criminal court and the emphasis on public safety rather than treatment, based on their review of recent trends and research, they concluded that the juvenile justice system is capable of providing public safety while also using a wider range of dispositions and interventions than does the adult criminal justice system (2004, p. 4).

Although the prevailing rhetoric of juvenile justice continues to emphasize criminalization and punishment of youth, it does not supplant the juvenile court's original mission to intervene on behalf of youthful offenders. The Court's ruling in *Roper* reaffirms the assumptions of juvenile justice that the diminished capacity and culpability of youth require a degree of benevolence in the accountability and punishment of juvenile offenders. Whether the *Roper* reaffirmation that youth are different from adults will mitigate harsh punishments such as JLWOP will be determined by emergent legislative reforms and appellate reviews. Conflicting images of youth and the dual but overlapping juvenile and criminal justice systems ensure that the sanctioning of juvenile offenders will continue to be a salient policy issue.

⊠ References

Amnesty International and Human Rights Watch. (2005). *The rest of their lives: Life without parole for children offenders in the United States.* New York: Human Rights Watch.

Annino, P. G. (2001). Children in Florida adult prisons: A call for a moratorium. *Florida State University Law Review, 28,* 471–490.

Applegate, B. K., & Davis, R. K. (2006). Public views on sentencing juvenile murderers: The impact of offender, offense, and perceived maturity. *Youth Violence and Juvenile Justice, 4,* 55–74.

Austin, J., Dedel Johnson, K., & Gregoriou, M. (2000). *Juveniles in adult prisons and jails: A national assessment.* Bureau of Justice Assistance. Washington, DC: U.S. Department of Justice.

Benekos, P.J., & Merlo, A.V. (2005). Juvenile offenders and the death penalty: How far have standards of decency evolved? *Youth Violence and Juvenile Justice, 3,* 316–333.

Bernard, T. J. (1992). *The cycle of juvenile justice.* New York: Oxford University Press.

Brief of Amicus Curiae, Commonwealth of Pennsylvania, Appellee v. Aaron Phillips, Appellant, No. 2729, Superior Court of Pennsylvania, Eastern District, EDA 2005.

Charney, J. (2005). *WSU: Michiganians oppose state law on sentencing youths as adults.* Retrieved September 5, 2007, from http://life.wayne.edu/article.php?id=1412

DiFilippo, D. (2006). No future: Pa leads nation in juveniles serving life sentences. *Philadelphia Daily News.* Available from http://www.philly.com

Edmonds v. Mississippi, No. 2004-KA-02081-COA (Miss. App. January 31, 2006).

Gaarder, E., & Belknap, J. (2002). Tenuous borders: Girls transferred to adult court. *Criminology, 40,* 481–517.

Greene, R., & Dougherty, G. (2001a, March 18). Kids in prison tired as adults, they find trouble instead of help and rehabilitation. *The Miami Herald,* Retrieved September 6, 2007, from http://vachss.com/help_text/archive/kids_prison.html

Greene, R., & Dougherty, G. (2001b, March 19). Kids in prison: Young inmates report highest rate of assault scalding water, handmade knives, locks among weapons used to attack. *The Miami Herald,* Retrieved September 6, 2007, from http://www.vachss.com/help_text/archive/kids_prison.html

Griffin, P. (2003). *Trying and sentencing juveniles as adults: An analysis of state transfer and blended sentencing laws.* Special Project Bulletin. Pittsburgh, PA: National Center for Juvenile Justice and Office of Juvenile Justice and Delinquency Prevention.

Gur, R. C. (2005, January/February). Brain maturation and the execution of juveniles. *The Pennsylvania Gazette, 103.* Retrieved October 23, 2006, from http://www.upenn.edu/gazette/0105/index.html

Hartman, M. (2006). *Annual statistical report 2003.* Office of planning, research, statistics and grants. Pennsylvania Department of Corrections.

Hartney, C. (2006). *Youth under age 18 in the adult criminal justice system.* San Francisco, CA: National Council on Crime and Delinquency.

Hutchinson, E. O. (2005, November 16). *No-parole sentences hurt black teens.* AlterNet. Retrieved September 5, 2007, from http://www.alternet.org/story/28376/

Juvenile Court Judges' Commission. (2001–2006). *Pennsylvania juvenile court dispositions* (Annual Editions). Shippensburg, PA: Center for Juvenile Justice Training and Research.

Karberg, J. C., & James, D. J. (2005). *Substance dependence, abuse, and treatment of jail inmates, 2002.* Bureau of Justice Statistics, Special Report. Washington, DC: U.S. Department of Justice.

Kurlychek, M. C., & Johnson, B. D. (2004). The juvenile penalty: A comparison of juvenile and young adult sentencing outcomes in criminal court. *Criminology, 42,* 485–515.

LaBelle, D., Phillips, A., & Horton, L. (2004). *Second chances: Juveniles serving life without parole in Michigan prisons.* Detroit, MI: American Civil Liberties Union.

Lane, J., Lanza-Kaduce, L., Frazier, C. E., & Bishop, D. M. (2002). Adult versus juvenile sanctions: Voices of incarcerated youth. *Crime & Delinquency, 48,* 431–455.

Levin, S. (2007, February 18). How teens end up put away for life. *Pittsburgh Post-Gazette,* pp. A1, A12.

Liptak, A. (2005, October 3). Jailed for life after crimes as teenagers. *The New York Times.* Available from http://www.nytimes.com

Logan, W. A. (1998). Proportionality and punishment: Imposing life without parole on juveniles. *Wake Forest Law Review, 33,* 681–725.

Massey, H. J. (2006). Disposing of children: The Eighth Amendment and juvenile life without parole after *Roper. Boston College Law Review, 47,* 1083. Retrieved September 5, 2007, from http://ssrn .com/abstract=926758

Mattingly, M. (2006). New organization aims its focus in trying juveniles as adults. *Juvenile Justice Update, 11*(6), 11.

Moffeit, M., & Simpson, K. (2006, February 19). Teen crime, adult time. *Denver Post.* Available from http://www.denverpost.com

Michigan State Senate. (2007). *Senate fiscal agency.* "Senate Bill Analysis." Retrieved September 5, 2007, from http://www.senate.michigan.gov/sfa

Mumola, C. J. (2005). *Suicide and homicide in state prisons and local jails.* Bureau of Justice Statistics. Washington, DC: Office of Justice Programs.

Nagin, D.S., Piquero, A. R., Scott, E., & Steinberg, L. (2006). Public preferences for rehabilitation versus incarceration of juvenile offenders: Evidence from a contingent valuation survey. *Criminology and Public Policy, 5,*627–652.

National Campaign for Youth Justice. (n.d.). *Facts.* Available from http:/www.campaign4youthjus tice.org

National Center of Juvenile Justice (producer). (2006). *National Juvenile Court Data Archive: Juvenile Court Case Records* 1985–2003 [machine-readable data files]. Pittsburgh, PA: Author.

OJJDP statistical briefing book. (2005). Retrieved February 28, 2005, from http://ojjdp.ncjrs.gov/ ojstatbb/ crime/qa0510l.asp?qaDate=20050228

Roper v. Simmons, 543 U.S. 551, 578 (2005).

Rose, J. (1999). Young lives—long sentences. *Prison Service Journal, 122,* 16–19.

Rubin, H. T. (2006). Challenges to juvenile life without parole sentences and adult punishments: New beginnings. *Juvenile Justice Update, 12*(5), 1–2, 14–16.

Schiraldi, V., & Zeidenberg, J. (1997). *The risks juveniles face when they are incarcerated with adults.* Retrieved February 15, 2006, from the website of the Center on Juvenile and Criminal Justice: http://www.cjcj.org/pubs/risks/risks.html

Senate Bill No. 944 (2006). *Ban life without parole sentences for juveniles.* Retrieved September 8, 2007, from http://www.senate.mo.gov/00info/pdf-bill/ intro/SB944.pdf

Singer, S. (2003). Incarcerating juveniles into adulthood. *Youth Violence and Juvenile Justice, 1,* 115–127.

Snyder, H.N. (2005). *Juvenile arrest 2003.* Juvenile Justice Bulletin. Washington, DC: U.S. Department of Justice, Office of Juvenile Justice and Delinquency Prevention.

Snyder, H. N., & Sickmund, M. (2006). *Juvenile offenders and victims: 2006 national report.* Washington, DC: U.S. Department of Justice, Office of Juvenile Justice and Delinquency Prevention.

Sontheimer, H., & Volenik, A. (2004). Is the juvenile justice system still relevant? *Juvenile Justice Update, 9*(6), 1–4.

Stahl, A., Finnegan, T., & Kang, W. (2006). *Easy access to juvenile court statistics: 1985–2003.* Retrieved September 8, 2007, from http://ojjdp.ncjrs.org/ ojstatbb/ezajcs/

Urbina, M. G. (2005). Transferring juveniles to adult court in Wisconsin: Practitioners voice their views. *Criminal Justice Studies, 18,* 147–172.

Woolard, J. L., Odgers, C., Lanza-Kaduce, L., & Daglis, H. (2005). Juveniles within adult correctional settings: Legal pathways and developmental considerations. *International Journal of Forensic Mental Health, 4,* 1–18.

DISCUSSION QUESTIONS

1. What do the authors mean by the "adultification" of juvenile justice?

2. Do you have any personal objections to juveniles being waived to adult court if they have committed a violent crime or are chronic property offenders?

3. What is the opinion of Amnesty International and the Human Rights Watch about sentencing juveniles to life without the possibility of parole, and do you agree with these organizations?

READING

In this article, Benjamin Steiner and Andrew Giacomazzi explore the effects of juvenile waivers and boot camp on recidivism. They note that waivers of juveniles to adult criminal court have increased in recent years and that they transfer young offenders out of the juvenile system and into the adult criminal justice system where the range of sanctions is presumably greater. Boot camps are one such sanction that is typically designed for youthful, first-time offenders. Because boot camp placement is an intermediary sentence, waived youths are likely candidates for placement there. The authors examine the effectiveness of a boot camp program in terms of recidivism for juveniles waived to criminal court in a northwestern state. They compare juveniles in the boot camp program to juveniles waived to criminal court and sentenced to probation using a 2-year follow-up period.

Juvenile Waiver, Boot Camp, and Recidivism in a Northwestern State

Benjamin Steiner and Andrew L. Giacomazzi

Serious and violent juvenile offenders pose a significant challenge to the juvenile justice system. They are responsible for a disproportionate amount of all crime and yet are still developing, which suggests they may be highly amenable to change. The decision regarding whether a youth is dangerous or amenable to rehabilitative treatment is one of the most difficult issues that confronts both juvenile courts and juvenile justice policy makers.

Throughout the 1980s and 1990s, policy makers in most states punitively responded to serious and violent juvenile offending by increasing juvenile disposition options and adding or expanding juvenile waiver statutes allowing for easier transfer of certain youthful offenders to adult criminal court for prosecution. Correspondingly, the number of juveniles waived to adult criminal court considerably increased during this period.

On the other hand, waived juveniles still represent a small percentage of the convicted felons sentenced in adult criminal court. Criminal court judges have little experience dealing with this special population of offenders. Furthermore, they are rarely provided any information regarding the youths' juvenile court histories or their prospects for rehabilitation. Yet criminal court judges are typically forced to decide between placing waived juveniles on probation or sentencing them to prison. Rarely do they have other sentencing options at their disposal.

Although states are being encouraged to develop specialized programs for youthful offenders, most states have not yet done so. Austin, Johnson, and Gregoriou (2000) revealed that 44 states place transferred juveniles who are sentenced to prison into the general population, typically without any

Source: Steiner, B., & Giacomazzi, A. L. (2007). Juvenile Waiver, Boot Camp, and Recidivism in a Northwestern State. *The Prison Journal, 87*(2), 227–240. Reprinted with permission of Sage Publications, Inc.

specialized services. Although a few states have experimented with alternatives to prison, such as blended sentencing, housing waived youth in juvenile facilities until they reach 18, or segregating them within adult prisons, only a couple of states have developed sentencing options that the judiciary could impose for youthful offenders whom they determine are not suitable for probation or prison. Although the effectiveness of sentences to prison and probation, in terms of recidivism, has not been encouraging for waived juveniles when compared to similar youth retained in the juvenile system, the effects of alternative dispositions for this population are still unclear.

In this study, we evaluate the effectiveness of one northwestern state's alternative disposition, the rider program, for juveniles waived to adult criminal court. The rider program, which is subsequently described, is not restricted to transferred juvenile offenders, but it is typically reserved for first-time offenders. In addition, we reveal that the rider program was the most frequent sentence imposed for juveniles waived to criminal court in this rural northwestern state.

Bootcamps for Juveniles Waived to Criminal Court?

Boot camps are often designed for younger offenders who do not have a prior criminal record (Correia, 1997; MacKenzie, 1994). In sentencing a waived juvenile to an adult boot camp facility, the court would seemingly satisfy the harsher punishment goal of the waiver while still offering the juvenile an opportunity to better himself or herself in an environment that is not perceived as being as harsh as the prison milieu. In theory, this could be an effective manner of dealing with this dispositional challenge for the courts.

In some cases, juvenile-specific boot camps have generated significant differences in recidivism or time to recidivism when compared to other dispositions. Other studies have revealed that juveniles have not responded well

to the boot camp environment and have been equally as likely to recidivate on release as comparison groups.

Findings with regard to adult facilities have also been mixed. Some studies have found no significant differences in recidivism rates between boot camp participants and comparison groups. Others have discovered that results differ by site, whereas some have found lower recidivism rates for boot camps offenders. A few studies have revealed higher recidivism rates for boot camp groups. In their recent meta-analysis, Wilson and MacKenzie (2005) found no differences in the overall effect of boot camps relative to comparison groups in reducing recidivism for either juvenile or adult offenders. However, they noted that, when an aftercare treatment component was combined with the boot camp program, some encouraging results emerged.

Much of literature suggests that the typical offender sent to boot camp facilities is a young, male, first-time offender, which could include juveniles transferred to adult criminal court. Indeed, the samples assessed in several studies contained offenders aged 16 and 17. Although several of these studies controlled for age, no study of boot camps to date has examined their effects on juveniles waived to adult criminal court. In this study, we evaluate the effects of a boot camp program on this special population of offenders.

Rider Program Description

The state department of corrections operates the boot camp, known as the rider program, under investigation here. The rider program provides a sentencing alternative for those offenders who may, after a period of programming and evaluation, be candidates for probation rather than incarceration. During the period the offenders assessed in this study were confined, the rider program was primarily a discipline-oriented boot camp. The typical sentence length for an individual sent to the program was 180 days; however, a few offenders were sentenced to l20-day commitments. When the offenders assessed here were in its custody, the rider program provided offenders basic education services but offered little else in the way of rehabilitative treatment. Completion of the rider program was intended to foster confidence and self-esteem in graduates. If an offender violated the rules of the facility, he or she was typically transferred to prison. Offenders who successfully completed the program were released on probation.

▧ Method

The study described here was designed to evaluate the effects of the rider program, in terms of recidivism, for juveniles waived to criminal court in a rural northwestern state. The target population for the study includes all juveniles waived to criminal court between 1995 and 1999 who were sentenced to the rider program or probation.

Sample

It was determined that 102 juveniles 17 and younger were waived to criminal court between 1995 and 1999, 20 of whom were sent to prison and not examined in this study because not all of them had been released. Of the remaining 82 juveniles waived to adult criminal court, 49 were sentenced to the rider program, whereas 33 were placed on probation. Those juveniles sentenced to probation were used as a comparison group for the purposes of this study. The rider group and the probation group were not significantly different on measures of age, race, offense type, or type of county (urban or rural) from which they were waived. The two groups were significantly different on whether they were previously committed to a state juvenile institution. Those sentenced to the rider program were more likely to have been committed to a state juvenile institution. This difference between

the two groups constitutes a limitation of the study for which we adjust in part by including a control variable tapping whether a juvenile had a prior commitment.

Dependent Variables

We examined both the rider group and the probation group for a 2-year period of street time. Two measures of recidivism were used to evaluate the rider program. *Reincarcerated* is defined as whether an offender was convicted of either a new offense or a technical violation of probation. *Reoffended* is measured as a new felony conviction.

Independent Variables

To examine what other factors improved prediction of a juvenile's likelihood of recidivating, we also examined the effects of several other variables. The sample and all the measures considered in this study are described in Table 1. A dummy variable, rider program, was created to assess the effects of the two different sentences on the dependent variables.

As noted above, the rider and probation groups differed as to whether the juveniles composing each of them had been previously committed to a state juvenile institution. To control for this, a measure, prior commitment, was created and defined as whether the juvenile had ever been committed to the state department of juvenile corrections or the state department of health and welfare for a delinquent offense.

Age has been a variable that has been examined in much of the prior waiver and boot camp research. In this state, juveniles can be waived to adult criminal court once they reach the age of 14. As such, age for this sample can only range from 14 to 17 years of age. In addition to age, the type of offense the juvenile was transferred for is another variable that has been examined. For the purposes of this study, offense type was coded as a dichotomous variable, nonviolent offense. Violent

Table 1	Sample-Specific Means and Standard Deviations	
Measure	**M**	**SD**
Dependent variables		
Reincarcerated[a]	0.34	0.48
Reoffended[a]	0.26	0.44
Independent variables		
Age	16.63	0.53
Nonviolent[a]	0.39	0.49
Non-White[a]	0.37	0.49
Prior-commitment[a]	0.35	0.48
Sentenced to rider[a]	0.60	0.49

NOTE: $N = 82$

a. Dummy coded (0 = *no*, 1 = *yes*).

offenses were defined as those that could have resulted in injury to a person. Race has been examined by others who have explored these topics. The state under study here is homogenous, where the 2000 census revealed it to be 93% White. Consequently, race was defined as a dichotomous variable, non-White.

⊠ Findings

Before discussing the effectiveness of the rider program for juveniles waived to criminal court, we think it is noteworthy that nearly all (96%) of the juveniles sentenced to the rider program completed it. This finding is similar to those derived from evaluations of boot camps in some states, but it is an improvement over others. The study achieved results from the logistic model assessing the effects of the rider program while controlling for prior commitment. Although it approached significance in the model predicting whether a juvenile was reincarcerated, being sentenced to the rider program did not have an influence on either measure of recidivism.

Table 2 reports the results from the models predicting recidivism outcomes. As can be seen, when controlling for age, offense type, race, and whether the juvenile had been previously committed to a juvenile facility, juveniles sentenced to the rider program were less likely to be reincarcerated on their release. However, when we assessed whether the juveniles reoffended, no relationship was observed.

Juveniles sentenced for a nonviolent offense were also less likely to be reincarcerated than were those who were waived for a violent offense, but offense type was not important in explaining whether a juvenile reoffended. Interestingly, non-White race and whether a juvenile reoffended were related, suggesting minority offenders were more likely to reoffend than White offenders regardless of age, offense type, sentence imposed, or prior history with the state juvenile corrections system.

Given the potential policy relevance of the findings for the state in which the rider program is located and the findings regarding the rider program and offense type in the model predicting whether a juvenile was reincarcerated, we assessed an interaction term capturing the effect of the rider program for nonviolent offenders. Although we did not reveal a relationship between the interaction term and whether a juvenile was reincarcerated,

Table 2	Predicting Recidivism for Juveniles Waived to Criminal Court			
	Reincarcerated		Reoffended	
	b	SE	b	SE
Constant	0.73		−7.52	
Rider program	−0.87*	0.51	0.03	0.55
Age	−0.02	0.49	0.35	0.55
Nonviolent	−1.11*	0.58	−0.17	0.60
Non-White	−0.26	0.53	0.91*	0.55
Prior commitment	−0.23	0.54	−0.18	0.56
Model χ^2 (5 df)	7.78		3.54	
Percentage classified correctly with model	70.7		74.4	

*$p < .10$

a correlation was observed in the model predicting whether a juvenile reoffended ($b = -2.38$, $SE = 1.22$). This finding suggests that, while controlling for age, race, prior commitment, and the main effects of offense type and sentence, nonviolent offenders who were sentenced to the rider program were less likely to reoffend.

⬚ Discussion

The extant research suggests that juveniles waived to criminal court are typically sentenced to prison or probation. Boot camps represent an intermediate sanction that is typically reserved for young, first-time offenders, making waived youth probable candidates for a boot camp sentence. Boot camps, whether juvenile or adult specific, have been shown to be effective in reducing recidivism in some areas but to be ineffective in others. Yet their effect on juveniles waived to adult criminal court has not been previously evaluated.

The findings from this study suggest that when controlling for age, race, offense type, and criminal history, waived juveniles sentenced to a boot camp facility known as the rider program were not less likely to reoffend

than those offenders sentenced to probation. The rider group was, however, reconvicted at a lower rate. From this, it could be inferred that the rider program graduates were merely less likely to violate probation than the juveniles sentenced to probation. It may be that the rider program instilled discipline in these youthful offenders, which is arguably necessary to be successful on probation. Then again, it is likely that reductions in new offenses, and not probation success, is what the judiciary had in mind when sentencing these offenders to the rider program.

From a different perspective, it could be that the judiciary in this northwestern state may be using the rider program as a graduated sanction for previously committed youth. We observed that juveniles sentenced to the rider program were significantly more likely to have been previously committed to a state juvenile corrections facility than those juveniles sentenced to probation. Seemingly, the rider program would be a tougher sentence than what is typically imposed in the juvenile system. On the other hand, offenders are sentenced to the rider program for either 120 or 180 days. The average commitment to this state's department of juvenile corrections was 378 days in 1997 and 416 days in 1998 (Idaho Department of

Juvenile Corrections, 2001). Thus, it seems that those offenders sent to the rider program generally served less time than they would have had they been retained in this state's juvenile corrections system. Although we acknowledge the tenuous nature of this comparison, it does appear the rider program is not a more stringent punishment, at least with respect to time served, than what is typically doled out in the juvenile system.

An additional possibility is that the intent of the boot camp evaluated here is to reduce prison overcrowding. Indeed, some evidence suggests boot camps, if correctly applied, are effective in doing so. In an effort to explore this possibility, we conducted a subsequent analysis to determine where the waived juveniles who recidivated were sent on reconviction. The data revealed that more than 90% of those sentenced to the rider program who were reconvicted (nearly 74%) were subsequently sentenced to prison. Given these discouraging findings, it does not appear the boot camp program is reducing overcrowding or saving correctional costs in the long term with respect to this population. On the other hand, if prison were the likely sentence for the offenders who were sent to the rider program, the costs of the rider program for the juveniles who were reconvicted would need to be balanced against the projected costs of a prison sentence for those offenders (roughly 26%) who successfully completed probation after they were released from the rider program and those offenders (10%) who were reconvicted and not sent to prison. Determining these costs would provide a more complete picture of whether the rider program is reducing this state's prison population in a cost-effective manner.

In view of the differences across states in boot camp programs such as the one evaluated here and the relatively small amount of juveniles waived to adult criminal court in this state, we are not comfortable generalizing the findings from this study so as to inform policy in other states. The findings here do allow us to feel comfortable in making the claim that the rider program is generally *not* effective in reducing reoffenses for juveniles waived to criminal court in this northwestern state. On the other hand, it did not aggravate recidivism either. In fact, we found that, for nonviolent offenders, the rider program was effective in lowering a juvenile's likelihood of committing a new offense. Accordingly, this state may want to experiment with sending certain types of juveniles waived to adult criminal court, such as those waived for nonviolent offenses, to the rider program.

We also suggest that our finding in the full multivariate model that boot camp program graduates were generally reincarcerated less often than the probation group, but that they reoffended at nearly the same frequency, is important. In the past, evidence has shown that boot camps with an aftercare component are effective in reducing recidivism. This particular northwestern state may want to examine the possibility of expanding the rider program to include aftercare services. Aftercare services could be an effective way to take advantage of the compliance on probation that the rider program seems to foster in these youthful offenders.

At the time of the study, the rider program did not contain any type of treatment component beyond basic education. Evidence suggests that boot camps that contain an effective treatment component have produced some reductions in recidivism. Including a treatment component that adheres to certain principles may be a way for this state to experience more encouraging outcomes with this challenging offender group.

All told, the findings from this study may inform policy only in one state. Although we are cautious generalizing these results to other states, we do think that these findings support exploring the use of alternative sanctions for juveniles waived to criminal court. Although this study's findings overall are not supportive of a boot camp yielding reductions in recidivism for this population, it did not appear to make them worse either. Incarceration lengths in boot camps are typically shorter than prison sentences, which may ameliorate correctional system crowding and, possibly through the structure they provide, reduce young offenders' likelihood for the victimization they often suffer when placed in general population adult prisons (Redding, 2003). Boot camps could also provide an environment that may be more conducive to treatment. Although this may seem counterintuitive at first, evidence has suggested that juvenile offenders found boot camps to be safer and more therapeutic than traditional facilities, which was then associated with improvements in inmate adjustment. Taken together with the encouraging findings for boot camps that combine treatment, it may be that using boot camps that contain effective treatment and aftercare could be a way for states to attend to this dispositional challenge. In any event, future empirical inquiry is certainly warranted.

✒ References

Austin, J., Johnson, K., & Gregoriou, M. (2000). *Juveniles in adult prisons and jails: A national assessment.* Washington, DC: U.S. Department of Justice, Office of Justice Programs, Bureau of Justice Statistics.

Correia, M. (1997). Boot camps, exercise, and delinquency. *Journal of Contemporary Criminal Justice, 13,* 94–113.

Idaho Department of Juvenile Corrections (2001). *The first five years: Idaho Department of Juvenile Corrections.* Boise: Author.

MacKenzie, D. (1994). Results of a multi-site study of boot camp prisons. *Federal Probation, 28,* 60–67.

Redding, R. (2003). The effects of adjudication and sentencing juvenile as adults: Research and policy implications. *Youth Violence and Juvenile Justice, 1,* 128–155.

Thomas, D., & Peters, M. (1996). *Evaluation of the impact of boot camps for juvenile offenders: Cleveland interim report.* Washington, DC: Office of Juvenile Justice and Delinquency Prevention.

Wilson, D. B., & MacKenzie, D. L. (2005). Boot camps. In B. Welsh & D. Farrington (Eds.), *Preventing crime: What works for children, offenders, victims, and places* (pp. 73–86). Belmont, CA: Wadsworth.

DISCUSSION QUESTIONS

1. Explain the rider program alternative for juveniles waived to adult court.

2. What is the stated purpose of having such a thing as a rider program?

3. What were the main findings of this study?

READING

In this article, James Wells and his colleagues explore the consequences of juvenile shock incarceration (boot camp) programs and aftercare programs. The authors extend the literature on these subjects by comparing the recidivism of juveniles who completed a shock incarceration program that included a systematic aftercare phase with recidivism among a matched control group of juveniles released from more traditional residential placements. Findings were mixed as regards recidivism at 4-, 8-, and 12-month follow-ups. The authors found no differences in reconvictions at 8- or 12-month follow-ups and no differences in re-offense seriousness across time frames. However, they did find a significantly lower pro-portion of the boot camp group recidivated during the initial 4-month aftercare phase, and, at 12 months, a lower proportion had been recommitted to residential placements. Older juveniles had significantly higher recidivism scores than did younger ones. The authors conclude that their findings demonstrate the importance of combining shock incarceration with quality aftercare.

A Quasi-Experimental Evaluation of a Shock Incarceration and Aftercare Program for Juvenile Offenders

James B. Wells, Kevin I. Minor,
Earl Angel, and Kelli D. Stearman

Shock incarceration programs, or boot camps, have generated both appeal and controversies that are by now well known across the fields of corrections and juvenile justice. Stinchcomb and Terry (2001) state that for some people, the appeal of boot camps "stems from the intu-itive belief that military discipline promotes law-abiding behavior. For . . . others, it is the result of desperation fueled by a deficit of more worthwhile alternatives" (p. 221). Despite the appeal of these programs, controversies have arisen over such interrelated issues as their appropriateness for particular groups of offenders (e.g., juveniles and women), pro-gram conditions and practices, and, perhaps above all, effectiveness in achieving objectives.

Although the use of quasi-militaristic pro-gramming with incarcerated populations dates to at least the reformatory movement of the latter 19th century, the contemporary era of adult boot camps dates to the early to mid-1980s. Programs for juveniles proliferated about a decade later amid more encompassing efforts to crack down on juvenile crime and promote accountability; a good deal of the impetus came from the Office of Juvenile Justice and Delinquency Prevention (Howell, 2003). Whether designed for adults or juve-niles, however, these programs are meant to control three interrelated phenomena, as iden-tified by Reid-MacNevin (1997): (a) institu-tional crowding, (b) correctional costs, and (c) offender recidivism. The purpose of this article is to examine the latter. Specifically, we compare the recidivism of juveniles who completed a shock incarceration program that included a systematic aftercare phase to the recidivism of juveniles receiving more tradi-tional residential placements.

⧉ Previous Research

The majority of research on correctional boot camps has focused on adults. Indeed, this literature is quite voluminous addressing heterogeneous

SOURCE: Wells, J., Minor, K., Angel, E., & Stearman, K. (2006). A Quasi-experimental Evaluation of a Shock Incarceration and Aftercare Program for Juvenile Offenders. *Youth Violence and Juvenile Justice, 4*, 219–233. Reprinted with permission of Sage Publications, Inc.

programs, employing diverse research methods, and producing some inconsistent findings. On balance, though, the studies show that, when compared to incarceration in more traditional facilities, shock incarceration is often associated with more positive attitudes and with similar proxies for institutional adjustment and personal improvement, such as educational gains. However, such positive effects can be short lived and diminish quickly on return to the community. In terms of recidivism control, boot camp programs generally do not seem to be an improvement over alternative types of incarceration or sanctions; one possible exception to this is when the boot camp incorporates therapeutic interventions, especially interventions that have been effective in other settings and are appropriately matched with offenders' risks and criminogenic needs.

There is no reason to suppose that research on adult shock incarceration programs is generalizable to juvenile offenders. Juveniles are at earlier stages of physical and psychosocial development. There are obvious philosophical, legal, and operational differences between juvenile justice and adult criminal justice. Juvenile programs need to be examined in their own right. In fact, much concern has been aired that shock incarceration programs may be physically or mentally harmful, thus exacerbating the likelihood of future problems among young offenders.

Although some boot camp research has focused on juveniles, much of this work has examined attitudes, perceptions, values, institutional adjustment, educational gains, and facility characteristics, as opposed to recidivistic behavior. For instance, Styve et al. (2000) found that in comparison to juveniles in traditional facilities, those in boot camps perceived the environment as more controlled and structured, safer, and more conducive to transition to the community. Similarly, Steele (1997) reported that boot camp participation was associated with improved self-esteem and expectations for self-efficacy at avoiding future problems.

Yet one would be hard pressed to argue that recidivism control is not ultimately a major goal of nearly all shock incarceration programs and indeed a goal of most correctional programs generally. A concern with recidivism is often implicit in studies that focus on other outcomes, such as attitudes, and obviously crowding in other facilities and financial costs are more likely to be contained by shock incarceration programs when these programs are associated with lower recidivism. Furthermore, some studies that have measured recidivism have failed to use experimental or quasi-experimental designs capable of increasing the confidence with which cause-effect conclusions can be drawn. This article employs a quasi-experimental, matched-groups design to study recidivism.

Previous Recidivism Research on Juvenile Boot Camps

Peters, Thomas, and Zamberlan (1997) conducted one of the earliest recidivism evaluations of juvenile boot camps. Studying demonstration programs located in Alabama, Colorado, and Ohio, Peters et al. reported that the reconviction rates of boot camp participants were higher than, or comparable to, those found among members of control groups. In addition, boot camp participants displayed shorter times to failure. However, in Ohio, the researchers reported some evidence that reoffending dropped sharply among those youths who were exposed to an enhanced aftercare component that was implemented around the middle of the project.

Similar findings have been reported by the Florida Department of Juvenile Justice (1997) and the California Department of the Youth Authority (1997). The Florida study found that almost two thirds of boot camp graduates were rearrested within a year of graduation and that nearly half were reconvicted, results comparable to those for a matched comparison group. In California, boot camp graduates experienced more arrests (especially for technical violations) than did controls, and offense severity was comparable for the groups. Extending on this earlier experimental study, Bottcher and Ezell (2005) concluded that the program in California "neither reduced crime nor placed the public at any greater risk of crime" (p. 328).

More recently, investigators have studied boot camp programs located in school settings (Trulson, Triplett, & Snell, 2001). Although the researchers reported favorable attitudinal findings from parents, teachers, and program participants, recidivism findings were not encouraging. More than half of boot camp participants were rearrested, compared with 36% of a control group of juveniles assigned to intensive supervision probation. At the adult level, Jones and Ross (1997) also found that boot camp participants had higher recidivism rates than did offenders placed on probation. Thus, although the recidivism associated with juvenile shock incarceration programs has been understudied relative to that associated with adult programs, the evidence that has emerged is less than encouraging.

Aftercare Research

The pessimism about recidivism control that pervades the empirical research on both adult and juvenile boot camps is not as applicable to the literature on juvenile aftercare (sometimes called reintegration or reentry) programs. Although aftercare research is not abundant, on the whole the data show a more mixed pattern of results, with some investigators reporting that aftercare programming can have rather impressive effects on recidivism. Programming stressing a punitive or pure surveillance orientation is unlikely to be as effective as that which incorporates developmentally appropriate interventions and services meant to help youth overcome problems and address needs that have contributed to delinquent behavior.

For example, Josi and Sechrest (1999) evaluated a California-based aftercare program designed to build life skills and help juveniles overcome barriers to successful reentry through better decisions. During a limited follow-up period of 3 months, program participants were significantly less likely than were members of a control group to experience rearrest. In general, positive effects were sustained after the 90-day intervention ceased, up to a year. An earlier experimental study by Fagan (1990) demonstrated that, across four urban sites, 2-year recidivism was significantly lower among serious juvenile offenders who received intensively implemented services than among equivalent groups of juveniles who received no services. Evaluation of a Philadelphia program also yielded some positive findings.

But the findings of other studies have been less positive. Three of the evaluation studies reviewed by Altschuler et al. (1999) did not demonstrate significant findings favoring intensive aftercare over control conditions. Likewise, Wiebush et al.'s (2005) evaluation of the Intensive Aftercare Program (IAP) in three states found very few significant differences between IAP youth and youth in control groups on multiple measures of recidivism.

Based on the evidence, juvenile boot camps seem less likely to positively affect recidivism when these programs are not accompanied by quality aftercare. As described above, the study by Peters et al. (1997) uncovered little evidence that juvenile boot camps positively affected recidivism, except for the finding in Ohio that the introduction of an enhanced aftercare component was associated with a decline in new offenses. Even boot camp programs that incorporate interventions to address criminogenic needs may show higher postrelease recidivism if no effort is made to address those and similar needs in the community. Peters et al. point to the importance of achieving continuity between the residential and aftercare phases of boot camp programs. Furthermore, even programs followed by quality aftercare may be at risk of having higher recidivism rates if aftercare intervention is too short term and characterized by dissipating effects. The present study extends on past work to examine recidivism among juveniles completing a shock incarceration program with a systematic aftercare component.

Program Overview

The Kentucky Department of Juvenile Justice (DJJ) established the Cadet Leadership Education Program (CLEP) in 1999 as the only boot camp program in the state for juvenile offenders. Integral to the stated mission of the program is the reduction of further delinquency following program completion. Program goals include promoting discipline through physical conditioning and teamwork, instilling responsibility and prosocial values, and increasing educational functioning.

CLEP is a 40-bed facility for males aged 14 to 18 years who have been committed to DJJ by the juvenile court for residential placement. Juveniles are identified as possible placements for CLEP by staff working at the state's centralized assessment center. The main factors assessment center staff consider in screening for CLEP placements include lack of history of (a) substance abuse, (b) mental health problems, and (c) sexual offending. The state has other programs designed for juveniles with these characteristics. In addition, juveniles are screened to evaluate their mental and physical suitability for CLEF. Youth displaying the following characteristics are not considered good candidates: (a) history of runaway or being absent without leave from residential placement, (b) history of physical or emotional abuse that may culminate in serious outbursts of anger and negative reactions to a boot camp environment, and (c) physical conditions that may preclude participation, such as asthma or heart problems. Although Kentucky has a separate, highly secure facility for youth with the most serious backgrounds of and potential for violence, an absence of

past violent behavior is not among the criteria for placement in CLEP.

On being identified by assessment staff as possible candidates for CLEP, juveniles are interviewed at the assessment center by CLEP staff. CLEP officials describe the program and its expectations and seek to identify those youth whom officials believe could derive genuine benefits from CLEP. These officials make the final decision irrespective of whether a given youth desires placement in the program.

The residential stay is 4 months in duration. Typical of adult and juvenile shock incarceration programs, CLEP consists of a quasi-military regime emphasizing physical exercise and work. It also emphasizes both academic and vocational education, Junior ROTC instruction, and community service. In addition, group and individual counseling is provided to help prepare youth for success in the community. Counseling services target criminogenic needs and behaviors in an effort to (a) change attitudes, orientations, values favorable to law violations and association with anticriminal role models; (b) reduce problems associated with alcohol and drug use; (c) reduce anger and hostility toward the self and others; (d) increase self-control, self-management, and problem-solving skills; (e) encourage constructive use of leisure time; (f) improve skills in interpersonal conflict resolution; and (g) promote more positive attitudes toward education and improved school performance.

Juveniles enter CLEP in platoons of approximately 10 persons. In turn, each platoon is assigned three staff members—two group workers and a counselor. The counselor (but not the group workers) continues to work with the youth in his or her platoon during the intensive aftercare phase. The aftercare phase is also 4 months in duration and begins on graduation from the residential phase.

Aftercare programming focuses on (a) preparing youth for, and providing them with, progressively increased responsibilities and freedom; (b) teaching youth to become involved and interact successfully with the community; (c) working with targeted community support systems, such as the family and school, to establish constructive interaction patterns and promote successful adjustment; (d) developing new supports and resources where these are identified as lacking; and (e) continually monitoring the interaction between the youth and the surrounding community to ensure that the chances of prosocial adjustment are maximized. During the aftercare phase, there are mandatory weekly sessions between the counselor and the youth in the home setting. Counselors frequently spend additional time with youth by taking them on outings in the community.

◼ Method

Study Design and Participants

A quasi-experimental, matched-groups design was employed in this research to compare the recidivism of CLEP graduates with that of juveniles receiving other residential placements administered by the Kentucky DJJ. The CLEP group ($n = 68$) consisted of the first 7 platoons to graduate from the program.

The control group ($n = 68$) was not exposed to a correctional boot camp. Instead, control group members had been released during a comparable time frame from either group homes ($n = 11$) or youth development centers (YDCs; $n = 57$) administered by DJJ. Members of the control group received varying levels of aftercare depending on the circumstances surrounding their cases and the discretion of their aftercare workers. These aftercare services were not structured around the focal points of CLEP aftercare, as described above. As such, these services did not represent a systematic extension of a program of residential intervention. In addition, aftercare in the control group generally involved higher staff caseloads. More detailed information about aftercare services received by individuals

in the control group was only available in hardcopy community service worker case files located at dozens of local DJJ offices throughout the state. Unfortunately, it was not feasible to visit each of these offices and collect individual-level data on control group aftercare.

CLEP and control youth were matched individually (vs. aggregately) based on age, race, prior offenses, and release date. Gender was a constant because CLEP is a male facility. The control group was drawn from a pool of 275 cases supplied by DJJ and was selected to parallel the CLEP group as closely as possible on the matching variables.

It was possible to achieve identical matches on the prior offense variable (i.e., prior felony matched to prior felony and prior misdemeanor matched to prior misdemeanor) for only 24 of the 68 matched pairs. That is, for 24 pairs both youth had identical numbers of prior misdemeanor convictions and identical numbers of prior felony convictions. For the other matches, we tried to ensure that members of the CLEP group had more pronounced prior records so as to produce a greater probability of recidivism in the CLEP group and thus render the test of CLEP more conservative. However, this was not always feasible. For 31 matches, CLEP youth had more prior misdemeanors, but for 5 matches, control youth had more such convictions. The average discrepancy in number of convictions across these 31 pairs was 2 (i.e., CLEP youth averaged 2 or more misdemeanors), whereas the average discrepancy across the 5 pairs was 1.8. Likewise, in 21 matches, CLEP youth had more prior felony convictions (average discrepancy = 1.71 convictions), whereas in 5 matches, control youth had more felonies (average discrepancy = 1.4 convictions). There were no pairs in which a member of the control group had both more misdemeanor and more felony convictions, but there were 11 pairs wherein CLEP youth had more of both than did the control counterparts (average discrepancy = 3.1 convictions).

⊠ Results

CLEP and Control Group Characteristics

The mean age of CLEP graduates was 16.9 years (SD = 0.84), compared with a mean of 17.0 years for the control group (SD = 0.92). Both groups consisted of 77.9% Whites and 22.1% African Americans. Just more than 55% of the CLEP graduates and just more than 58% of the control group had prior misdemeanor convictions. Similarly, about 27% of both groups had previous felony convictions; 17.7% of the CLEP group and 14.3% of the control group had past convictions for status, code, and/or probation violations.

Recidivism

Chi-square tests were computed to study the association between group membership and recidivism at 4-, 8-, and 12-month intervals. The proportions are summarized in Table 1.

Although 23.5% of the control group had been convicted of some type of new offense at 4 months following release from a facility, the same was true for only 10.3% of the CLEP group, χ^2 (1) = 4.239, p = .040. Chi-squares could not be computed for the other offense categories because of low frequencies. Matched-samples t tests were used to compare the recidivism scores of the control and CLEP groups across the three time frames. Recall that these scores were calculated based both on the number and seriousness of offenses applying the weights presented above.

Four 2-way ANOVAs were performed to examine the 12-month recidivism scores of the CLEP and control groups by the age and race variables and by prior offense and instant offense scores. The 2 × 2 ANOVAs incorporating group membership by race (White vs. Other), prior offense scores, and instant

	% Convicted Within 4 Months		% Convicted Within 8 Months		% Convicted Within 12 Months	
Group or Offense Category	**Yes**	**No**	**Yes**	**No**	**Yes**	**No**
Control group						
Any offense or violation	23.5	76.5	35.3	64.7	48.5	51.5
Status	1.5	98.5	1.5	98.5	1.5	98.5
Code violations	8.8	91.2	10.3	89.7	11.8	88.2
Probation violation	0.0	100.0	0.0	100.0	1.5	98.5
Misdemeanor	19.1	80.9	33.8	66.2	42.6	57.4
Felony	1.5	98.5	4.4	95.6	10.3	89.7
Cadet Leadership Education Program group						
Any offense or violation	10.3	89.7	22.1	77.9	39.7	60.3
Status	0.0	100.0	0.0	100.0	1.5	98.5
Code	0.0	100.0	4.4	95.6	8.8	91.2
Probation violations	0.0	100.0	0.0	100.0	0.0	100.0
Misdemeanor	7.4	92.6	17.6	82.4	32.4	67.6
Felony	7.4	92.6	7.4	92.6	11.8	88.2

Table 1   Proportion of Groups Recidivating at Various Time Intervals by Offense Category

offense scores, respectively, produced neither significant main nor significant interaction effects. However, the 2 x 2 ANOVA based on median divided age groups (i.e., 14.00–17.03 years vs. 17.04–19.00 years) yielded a significant main effect for age, $F(3, 132) = 4.26$, $p = .04$. Specifically, youth in the older age group had significantly higher recidivism scores ($M = 10.1$, $SD = 22.48$) than did those in the younger group ($M = 4.1$, $SD = 6.87$). This analysis yielded no other significant main or interaction effects. Thus, older youth recidivated more, irrespective of group standing.

As a supplement to the data presented above, we examined data on an alternative operationalization of recidivism—return to residential placement in either a juvenile or adult facility. This differs from operationalizing recidivism as new adjudications because a youth may have been adjudicated for a new offense but not recommitted to a facility or, alternatively, may have been recommitted without having been adjudicated for new offending. Although 33.8% of the control group was recommitted during the entire follow-up period, only 14.7% of the CLEP group was recommitted. The difference in proportions was statistically significant, $\chi^2(1) = 6.76$, $p = .009$. Further analysis revealed that, although comparable proportions of the control (13.2%) and CLEP (10.3%) groups were recommitted to facilities for new criminal charges, a significantly greater proportion of controls (20.6%) than CLEP graduates (4.4%) were recommitted for technical violations, $\chi^2(1) = 9.01$, $p = .011$.

◼ Discussion

The matching procedure used in this study resulted in reasonably close equivalence between the groups on the matching variables. The exception to this is that, compared to the control group, CLEP graduates had significantly higher prior and instant offense scores. In addition to implying a more conservative

test of CLEP, this finding is evidence against a net-widening conceptualization of the shock incarceration program, as that program was evaluated in this research. Whenever a new program such as CLEP is touted as an alternative to more traditional residential placements and is in fact associated with net-widening, the new program would be expected to receive less serious cases than traditional facilities; the program should receive offenders who, were it not for the program's existence, would have been retained in the community, most likely on some form of probation. Although our research was not specifically designed to examine potential net-widening effects of CLEP, it seems most likely that, in the absence of CLEP as a dispositional option, the youths in this study would have received more traditional residential placements.

The recidivism findings of this research are mixed. On the negative side, although lower proportions of the CLEP group were reconvicted at all three follow-up periods, the differences were not significantly different at the 8- and 12-month follow-up periods. Furthermore, between-group recidivism scores, which take both the volume and seriousness of new charges into account, were not significantly different; in fact, the recidivism scores of CLEP graduates were somewhat higher than were those of the control group at both 8 and 12 months. Older youth had higher recidivism scores regardless of CLEP versus control group standing. These findings are largely consistent with the results of previous recidivism research on juvenile boot camps.

However, the findings of this research were not entirely negative. A significantly lower proportion of shock incarceration graduates than members of the control group had been reconvicted at 4 months, the same time frame during which the former CLEP residents were exposed to systematic aftercare programming in the community.

Because youth were not randomly assigned to comparison groups, and because the control group did receive varying levels of aftercare services, the design of our study does not allow unambiguous causal inference about the efficacy of aftercare. But our evidence is clearly consistent with the presence of positive CLEP aftercare effects on the likelihood of reconviction. A related positive finding was that a significantly lower proportion of the CLEP group (14.7%) than the control group (33.8%) was recommitted to residential placements during the year follow-up period. Of those juveniles from both groups who were recommitted, comparable numbers were recommitted for new criminal charges. However, a significantly greater proportion of youth in the control group was placed for technical violations.

Because recidivism scores (the best proxy for offense seriousness in this study) did not significantly differ between the groups, one possible interpretation is that officials were simply more reluctant to react to instances of recidivism associated with the novel shock incarceration program by reincarcerating youth, particularly for technical violations. In any event, there is certainly no evidence in any of our data that the systematic aftercare received by the CLEP group as a whole was associated with increased detection of technical violations. This seems consistent with the fact that CLEP aftercare was more focused around intervention and community adjustment than around surveillance per se. It is possible that much of the aftercare directed toward the control group was more oriented toward simple surveillance, but we have no direct data on this question.

Based on the results of this study in the context of the literature, future research in this area should carefully attend to measuring the recidivism effects of juvenile boot camp placements with and without aftercare components. This should be done using experimental research designs or designs that approximate experiments as closely as is feasible. Likewise, when possible, recidivism should be operationalized in various ways and measured at varying time intervals, and the duration of

aftercare should be tracked so that any short-term, positive effects can be detected. To help overcome one of the limitations of this study, an effort should be made to gather systematic data on any aftercare directed toward members of a control group.

If future research can establish that, during the period community-based services are being provided, shock incarceration followed by systematic aftercare is associated with lower recidivism than is shock incarceration alone, there will be little justification (other than containment of short-term costs) for not including aftercare as a routine aspect of juvenile boot camp placement. Furthermore, if future research shows that positive outcomes are short lived when aftercare is of short duration, there will be little rationale for not extending the duration of aftercare. If other residential placements lacking systematic aftercare are ultimately associated with significantly higher rates of reincarceration, as was true in this study, and/or if boot camp graduates who do not receive systematic aftercare are recommitted at higher rates than are those who do, then aftercare services could save revenue during the long term—even when the aftercare is of considerably longer duration than 4 months. The intervention taking place following residential placement is likely to be as, or more, salient than what takes place during, and long-term resources expended on the latter can become much greater when due consideration is not given to the former.

▨ References

Altschuler, D. M., Armstrong, T. L., & MacKenzie, D. L. (1999). *Reintegration, supervised release, and intensive aftercare.* Washington, DC: U.S. Department of Justice, Office of Juvenile Justice and Delinquency Prevention.

Bottcher, J., & Ezell, M. E. (2005). Examining the effectiveness of boot camps: A randomized experiment with a long-term follow up. *Journal of Research in Crime and Delinquency, 42,* 309–332.

California Department of the Youth Authority. (1997). *LEAD: A boot camp and intensive parole program:*

The final impact evaluation. Sacramento: California Department of the Youth Authority.

Fagan, J. A. (1990). Treatment and reintegration of violent juvenile offenders: Experimental results. *Justice Quarterly, 7,* 233–263.

Florida Department of Juvenile Justice. (1997). *Bay County Sheriff's Office Juvenile Boot Camp: A follow-up study of the first seven platoons.* Tallahassee: Florida Department of Juvenile Justice.

Howell, J. C. (2003). *Preventing & reducing juvenile delinquency: A comprehensive framework.* Thousand Oaks, CA: Sage.

Jones, M. A., & Ross, D. L. (1997). Is less better? Boot camp, regular probation and rearrest in North Carolina. *American Journal of Criminal Justice, 21,* 147–161.

Josi, D., & Sechrest, D. K. (1999). A pragmatic approach to parole aftercare: Evaluation of a community reintegration program for high-risk youthful offenders. *Justice Quarterly, 16,* 51–80.

Peters, M., Thomas, D., & Zamberlan, C. (1997). *Boot camps for juvenile offenders.* Washington, DC: U.S. Department of Justice, Office of Juvenile Justice and Delinquency Prevention.

Reid-MacNevin, S. A. (1997). Boot camps for young offenders: A politically acceptable punishment. *Journal of Contemporary Criminal Justice, 13,* 155–171.

Steele, W. R. (1997). Changes in self-efficacy expectancy, outcome expectancy, and self-esteem in incarcerated juvenile boot camp participants. *Dissertation Abstracts International, 58,* 5-B. (Item No. 1997-95022-382)

Stinchcomb, J., & Terry, W. C., III. (2001). Predicting the likelihood of rearrest among shock incarceration graduates: Moving beyond another nail in the boot camp coffin. *Crime & Delinquency, 47,* 221–242.

Styve, G. J., MacKenzie, D. L., Gover, A. R., & Mitchell, O. (2000). Perceived conditions of confinement: A national evaluation of juvenile boot camps and traditional facilities. *Law and Human Behavior, 24,* 297–308.

Trulson, C. R., Triplett, R., & Snell, C. (2001). Social control in a school setting: Evaluating a school-based boot camp. *Crime & Delinquency, 47,* 573–609.

Wiebush, R. G., Wagner, D., McNulty, B., Wang, Y., & Le, T. (2005). *Implementation and outcome evaluation of the Intensive Aftercare Program: Final report.* San Francisco: National Council on Crime and Delinquency.

DISCUSSION QUESTION

1. Explain what a quasi-experiment is.

2. Explain the program setup of CLEP.

3. What are the primarily findings of this study?

❖

READING

This article by Caeti, Hemmens, Cullen, and Burton examines what they consider to be the increasing punitiveness of the juvenile justice system in the United States. Noting that there has been a variety of research assessing the attitudes of adult correctional system administrators and personnel and that relatively few studies have examined the attitudes of their juvenile system counterparts, the authors conducted a national survey of juvenile correctional facility directors. The authors wanted to determine the attitudes of these directors on several issues in juvenile corrections and their managerial problems and issues. Comparisons are made with adult prison wardens on several dimensions including demographics, job satisfaction, correctional orientation, and correctional programming emphasis and operation. Results show striking differences between perceptions of juvenile facility directors and those of directors of adult facilities. Several other managerial issues such as job-related stress, confidence in staff role conflict, and attitudes toward juveniles and juvenile corrections are also discussed.

Management of Juvenile Correctional Facilities

Tory J. Caeti, Craig Hemmens, Francis T. Cullen, and Velmer S. Burton, Jr.

The hundred-year history of the juvenile justice system in the United States is marked by at least three major philosophical shifts in thinking about how to deal with juvenile offenders. The most recent shift is toward a more punitive juvenile justice. State after state has radically altered the form, structure, and function of the juvenile justice system during the past 20 years. Indeed, the stated purpose of the juvenile system has been altered in several states. It is clear that juvenile justice is becoming more punitive across the United States.

Public support for more punitive policies toward young offenders in particular is increasing. In addition, several high-profile cases involving juvenile offenders garnered significant media attention. The result was a belief that rehabilitation had failed and that the goals of the entire justice system should be redirected. Although conservatives complained that the system was "soft" on juvenile offenders, liberals complained that juvenile court judges and prosecutors often abused their discretion. The odd coupling of liberal and conservative interests led to what became known as the "justice model," which resulted in increased determinate sentencing, mandatory institutionalization, and the widening use of the adult court for handling serious juvenile offenders.

Research indicates that the general public and many system actors hold very punitive ideologies. In spite of the increased emphasis on punishment, most politicians, system actors, and the general public still want juveniles rehabilitated. In short, it is clear that the public wants both punishment and rehabilitation.

The juvenile correctional institution's mission is to incarcerate, care for, and rehabilitate juvenile delinquents. The youths' fate is affected by the officers and administrators who are responsible for their care. The facility director is like the manager of a business, and the business is determined by the type of manager. DiIulio (1987) proposes that the administrator's managerial style is the most important determinant of an institution's health and organizational philosophy. The directors of juvenile facilities are in a unique position to affect the goals and objectives of the institution. Their correctional orientation can directly affect whether the institution manifests a punitive or rehabilitative atmosphere. Previous research indicates that detention professionals make a significant impact on the facility and

Source: Caeti, T. J., Hemmens, C., Cullen, F. T., and Burton, V. S. (2003). Management of Juvenile Correctional Facilities. *The Prison Journal, 83*(4), 383–405. Reprinted with permission of Sage Publications, Inc.

programs, and, in turn, the staff, facility, and programs make an impact on the inmates or detainees. The director can affect the operation of the institution in several ways: the allocation of funds, the types of training provided for staff, evaluation criteria, the rewards associated with good performance by staff, and ultimately, the structure and management of the program. Although they are key actors, juvenile facility directors have rarely been studied.

Through a national survey of juvenile facility directors, we examine several dimensions of managerial practices and attitudes. This research is unique in that it focuses on upper level administrators as opposed to line staff, and its scope is national. Questions were posed to juvenile facility directors regarding their correctional orientation, their attitudes toward the juveniles and staff in their facilities, and several other issues regarding the management and operation of their facilities. The results are compared with similar research conducted on prison wardens.

Research on Correctional Orientation

Research dealing with job-related attitudes of correctional officers has grown exponentially. Previous literature has examined the correctional orientation of correctional staff and probation and parole officers. The findings are varied. Cullen et al. (1989) found that over 75% of prison guards feel that prison is too soft; however, they were also very supportive of rehabilitation. Similar findings have been recorded in other studies.

A survey of the general public, lawyers, circuit court judges, correctional administrators, and members of the Illinois legislature assessed support for juvenile rehabilitation. The findings indicate that a high proportion (81.6%) of the sample felt that it would be "irresponsible" to stop trying to rehabilitate delinquents (Cullen et al., 1983). "This study indicates that the future of juvenile justice will

likely involve a mixture of these same conflicting philosophies (rehabilitation and punishment), which creates a special set of complications" (p. 11).

Some researchers studying the rehabilitation/ punishment dichotomy have characterized it as a fundamental problem and confusion of ideals. Others, while noting the shift in thinking, have characterized it as commonplace among many people, especially those actors who work in correctional systems. The varying ideologies expressed in the popular media, legislative initiatives, and academic literature send mixed messages to correctional personnel.

Correlates of Correctional Personnel Beliefs and Orientation

Two general lines of reasoning dominate thinking concerning the variables that determine correctional beliefs and attitudes. The importation model holds that attitudes about work are affected by the experiences brought to the job by the employee. In contrast, the work role-prisonization model holds that attitudes are shaped predominately by organizational conditions and work policies.

Cullen et al. (1993) argue that correctional attitudes are shaped by more than the policy changes that have happened over time. Correctional orientation is also influenced by other factors such as the occupational conditions under which people work. Bazemore and Dicker (1994) demonstrate that improvements in organizational climate result in a less punitive orientation toward detainees. Whitehead and Lindquist (1989) note that occupational climate has a substantial influence on the orientation of correctional professionals. The mere fact of working with a population of young offenders has been found to increase punitive attitudes. Many facility directors are subject to external scrutiny that negatively affects their discretion and autonomy. Perhaps a combination of both individual attributes and organizational variables influence attitudes toward inmates.

Cullen et al. (1993) argue that, even though research shows that correctional orientation is affected by occupational climate, it is crucial to understand what causes some facility directors to endorse rehabilitation and others to endorse punishment.

Farmer (1977) discovered high levels of cynicism among correctional officers and concluded that cynicism is a defense mechanism to the conflicting goals of punishment and rehabilitation. The levels of cynicism were particularly high at institutions emphasizing treatment. Some research shows that the support for rehabilitation declines with age. Yet officers who enter the field at a later age are more likely to support rehabilitation. Other research shows no relationship between age and attitude. Education has been found to be related to greater support for the rehabilitative ideal, more professional orientation, and more positive attitudes toward inmates.

Several studies examined the impact of race and gender on correctional personnel attitudes. African Americans are more likely to support rehabilitation and hold more favorable attitudes toward offenders. Whitehead and Linguist (1989) found that Black correctional officers expressed less support for harsher punishment than their White counterparts, and minorities also were found to have lower job-related stress. Still other studies found no significant relationships between race and attitudes.

Several studies have found women to be more supportive of a human services orientation. Jurik (1985) shows that 55% of female respondents took the job to work in rehabilitation, whereas only 23% of the male respondents did. Crouch and Alpert (1982) report that, during the first 6 months on the job, men are more punitive and women are less punitive.

Other studies determined no significant impact of gender on correctional attitudes. The job itself influences attitudes, specifically, variables such as salary, overtime, and organizational culture. Burton et al. (1991) found

that higher salary yields higher support for rehabilitation. There is a theory that tenure at a facility may diminish support for rehabilitation and heighten custodial views. However, Bazemore and Dicker (1994) uncovered a weak but significant relationship between tenure and support for treatment. Shamir and Drory (1981) discovered that the higher the rank, the less one believes in the rehabilitative potential of the prison.

However, rank is unrelated to attitudes concerning the detainees' ability to be rehabilitated. Philliber (1987) notes that research indicates that length of service affects orientation, usually in the direction of heightening a custodial orientation or negative attitudes toward inmates. For Arthur (1994), job satisfaction had the strongest correlation with rehabilitation. Job security is negatively related to support for treatment, and job involvement is positively correlated with rehabilitation. This suggests that the more involved employees are in the job, the more service oriented they are toward detainees.

Philliber (1987) observes that research finds correctional officers to be "alienated, cynical, burned out, stressed but unable to admit it, suffering from role conflict of every kind, and frustrated beyond imagining" (p. 9). Philliber found that education is one characteristic of corrections that seems to defy clear findings. She also determines that there are conflicting and inconsistent findings on the relationship of race to punitiveness and other attitudes toward inmates. In addition, the role of gender has proven to be quite complex in both affecting attitudes and stress among correctional personnel. In short, there is no clear indication of which variables are or are not predictive of correctional orientation or other job-related attitudes.

▧ Role Conflict

The definitions of role conflict vary in the studies that have either measured the concept directly or have used it as an independent predictor. An early definition and explanation by Grusky (1959) used in an analysis of role conflict in prisons remains sufficiently descriptive and succinct:

> The official goals of an organization determine in large part the types of role expectations associated with the positions that make up the social structure of the system. If an organization is assigned a new major goal, and if this goal is in conflict with what formerly was the only primary goal of the system, then we would expect that conflict between the goals would create new stresses for many members of the organization. These two or more sets of conflicting role expectations, defined by the organization as legitimate by the fact that they are derived from an official goal, create role conflict. (pp. 452–453)

Previous research has focused on the variables that predict or cause role conflict. For example, role conflict is higher in minimum-security facilities, higher in institutions with mixed treatment and custody goals (Zald, 1962), higher among treatment staff, and higher among those who report a sense of calling to the field. Other studies report that the frequency of inmate contact is not related to role conflict and that commitment to a professional ideology reduces role conflict. Another line of research examines the effects that role conflict has on several other dimensions of attitudes. Relationships between role conflict and stress, punitiveness, custody orientation, and burnout have all been documented. Role conflict was found to be negatively related to support for rehabilitation and related to hostile or negative attitudes toward prisoners. Hepburn and Albonetti (1980) note that job satisfaction is negatively related to role conflict; furthermore, role conflict has been found to affect worker alienation and levels of cynicism among juvenile probation officers.

The body of knowledge regarding role conflict is mixed at best. Indeed, Toch and Klofas (1982) found no evidence of role conflict among officers. Philliber (1987) notes that the literature on role conflict is unsystematic and disorganized. Furthermore, there are problems in both the measurement and operationalization of the concept. Finally, she echoes the conclusions of several role conflict researchers in observing that the conceptual ordering of role conflict with correctional orientation is unclear. Does role conflict *cause* attitudes in correctional staff or vice versa?

The bulk of the literature on role conflict has focused on adult correctional personnel. Logically, juvenile correctional personnel should have attitudes that support rehabilitation more than their counterparts in the prison system. In addition, how a juvenile facility is operated should be different from how an adult facility is. We would expect that greater attention is paid to education and rehabilitation in a juvenile facility and more attention is paid to security and retribution in an adult facility. The exploration of juvenile correctional management practices and administrator attitudes has not been systematically investigated. Therefore, this research should shed light on whether the punitive ideology has been manifested in the juvenile correctional setting.

Rehabilitative Ideal Scale

A six-item scale was used to measure the respondent's personal belief in the rehabilitative ideal. The scale was scored so that a high score indicates greater support for the rehabilitative ideal. Possible scores on the scale could range from 1 (indicating low support for rehabilitation) to 7 (indicating high support for rehabilitation). The average scaled score for the facility directors was 5.79 ($\alpha = .87$), and scores ranged from a low of 2.67 to 7.2. The directors responded to such items as (a) The best way to stop juveniles from engaging in crime is to rehabilitate them, not punish them; and (b) It would be irresponsible for us to stop trying to rehabilitate juveniles and thus save them from a life of crime. Most directors had high support for the rehabilitative ideal and scored high on the items making up the scale as well.

Punitive Ideal Scale

A four-item scale was used to measure respondents' belief in a punitive ideal. The scale was scored so that a high score indicates greater support for a punitive ideal. Possible scores on the scale could range from 1 (indicating low support for punishment) to 7 (indicating high support for punishment). The average score for the Punitive Ideal Scale was 3.63 ($\alpha = 1.09$), and scores ranged from 1 to 7. The directors responded to such items as (a) Most juveniles who commit crimes know full well what they are doing and thus deserve to be punished for their offenses; and (b) All juveniles who commit violent crimes should be tried as adults and given adult penalties. Most directors indicated little or no support for punitive items. Analysis revealed that those who strongly supported rehabilitation were least likely to support punishment. Scores on the rehabilitation scale were inversely predictive of scores on the punitive scale and vice versa.

Custody Orientation Scale

An eight-item scale was used to measure the degree of emphasis placed on and the success of the respondent's facility regarding custody and security issues. Directors ranked such items as (a) degree of emphasis on creating conditions that prevent juvenile escapes; (b) success of your institution in preventing the flow of contraband into the facility; and (c) success of your institution in preventing flow of contraband within the facility. This scale was scored so that a higher score represents a higher degree of emphasis on security and custody concerns. Possible scores could range from 1 (indicating low emphasis on security issues) to 10 (indicating high emphasis on

security issues). The average score for the scale was 8.09 ($\alpha = 1.14$), and scores ranged between 5.12 and 10.

Job Satisfaction Scale

Respondents were asked to indicate their level of job satisfaction with their position as a juvenile facility director. The scale included items such as (a) All in all, how satisfied would you say you are with your job? and (b) Knowing what you know now, if you had to decide all over again whether to take the job you now have, what would you decide? The scale was scored so that a high score indicates a high level of job satisfaction (scores could range from 1 to 10). The average score on the scale was 8.48 ($\alpha = 1.5$), and scores ranged from 3.6 to 10.5.

Job-Related Stress Scale

An 11-item scale was constructed to measure the level of job-related stress a juvenile facility director experiences. The scale was scored so that a high score was indicative of a high level of job-related stress. Directors responded to such items as (a) I often feel that the control of my institution is slipping out of my hands; (b) No matter how explicit I make my directives, staff always find a way to get around them; and (c) I can generally trust my staff to handle matters when I am away from the institution. Possible scores on the scale could range from 1 (indicating a low level of job-related stress) to 7 (indicating a high level of job-related stress). The average score for this scale was 3.9 ($\alpha = .89$), with scores ranging from 1.82 to 7.

Job Autonomy Scale

A four-item rating scale was used to measure the amount of pressure placed on juvenile directors by actors outside the facility. The directors were asked to indicate the degree of influence exerted by the courts, their state office, the parents of the children, and the general public on the day-to-day operations of their facility. The scale was scored so that a high score indicated a high level of outside influence on the facilitiy's operation. Possible scores on the scale could range from 1 (indicating high job autonomy) to 10 (indicating low job autonomy). The mean score for this scale was 5.05 (α = 1.67). The mean of this scale masks the fact that scores in this scale ranged from 1.5 to 10; thus, there was wide variation in the sample concerning this particular measure.

Results

The juvenile facility directors are comparable to their counterparts in the adult prisons in terms of age and years of education. Twenty percent of the juvenile facility directors were women and 80.2% were White. The adult prison wardens had fewer minorities, and data on gender were not collected. There were some striking differences between the two groups in terms of their background and current working conditions. Juvenile facility directors tended to have less experience in the military, were less likely to have worked as a correctional officer, and more likely to have worked in a treatment position. The juvenile directors had worked for longer periods of time in their current institution but for less time overall in corrections. Perhaps the most striking difference was in the numbers of inmates or juveniles that each group supervised: an average of 118.6 in the juvenile facility versus an average of 862.1 in the adult prisons. There was wide variation in the size of the juvenile correctional institutions ranging from a low of 4 to a high of 1,500.

Directors and prison wardens were asked to indicate the emphasis they placed on the day-to-day activities of their facilities. In general, the juvenile facility directors tended to report greater emphasis on rehabilitation and less emphasis on custody and institutional-order items. However, the differences in rankings were not as great as might be expected.

Thus, it appears that juvenile facility directors and prison wardens place comparable emphasis on rehabilitation and security issues.

Although the differences between directors and wardens in their emphasis on programming was not greatly different, several other questions showed impressive differences. We asked directors to rank order the four primary goals of corrections (rehabilitation, deterrence, incapacitation, and retribution) in order of importance. Most directors ranked rehabilitation the number one goal of juvenile corrections followed by deterrence, incapacitation, and retribution. Of note, only 1.9% ranked rehabilitation fourth and only 0.4% ranked retribution first. The primary differences between directors and wardens can be seen in several attitudinal areas and dimensions.

Directors and wardens were both asked to rank order the four goals of corrections in order of importance. Prison wardens ranked retribution first, followed in order by deterrence, rehabilitation, and incapacitation. When asked about the percentage of their clients that would be rehabilitated, the directors estimated nearly 50%, whereas the wardens estimated a little greater than 25%. In addition, differences can be noted in several general attitudinal items concerning the treatment of the clients and the nature of their institutions. Juvenile facility directors tended to be much more positive on items concerning rehabilitation and less positive on punitive items when compared with prison wardens. Finally, each group was asked to assign 100 points to reflect the importance of several activities they would engage in their ideal facility. Both groups reported a similar importance in both maintaining security and keeping clients busy by having them work. However, the facility directors, compared to the wardens, placed a stronger importance on rehabilitation and much less importance on preventing escapes and punishing clients.

We asked the facility directors to assess the juveniles they had in their care on a variety of items. The results were consistent with a more rehabilitative attitude; however, the results also showed wide variation on several items. For the most part, the variation was not attributable to differences in the demographics of the facility directors or the facilities in which they worked. Interestingly, the number of juveniles estimated to be rehabilitated was almost equal to the number estimated to recidivate. The directors estimated low numbers of dangerously violent juveniles in juvenile corrections.

In addition to asking about their attitudes toward corrections and the clients they serve, we asked the directors a variety of questions regarding the management and operation of their facilities. Based on the data regarding their attitudes toward corrections and rehabilitation, we found that juvenile facility directors expressed a great deal of role conflict. The directors were asked to rank their perceptions of their influence and the influence of other stakeholders in juvenile corrections.

In terms of the influence the directors feel they exert, as well as the influence of several other groups who might have influence over the day-to-day operations of the facility, directors report the greatest influence from those groups actually inside their facility, that is, themselves, their staff, and the juveniles. The state office and top administrators were judged to have strong influence, but the courts, parents, and general public did not. Therefore, if facility directors were to report role conflict, the pressures would have to be coming from those inside their facilities or inside their state departments of juvenile corrections. The vast majority of facility directors indicated strong support for rehabilitation and little or no support for punitive ideals. Thus, it is reasonable to conclude that little or no role conflict exists in juvenile correctional facilities.

In terms of their perceptions of their staff and the operation of their facility, the directors strongly agreed with the statement that they could generally trust staff to handle matters while they were away from the facility. However, the other items concerning the staff were not as strong. The directors tended to agree with or be neutral in their perception of staff. They also tended to disagree with the statement that the control of the institution was slipping out of their hands.

A standardized job satisfaction index was included in the survey that has been used in the past to assess prison wardens, corrections officers, and the general public. Overall, juvenile facility directors were a little less satisfied than prison wardens, more satisfied than correctional officers, and much more satisfied than the general public. However, for the most part facility directors appear to be satisfied with their jobs. We were interested in what distinguishes facility directors reporting high job satisfaction from those reporting lower job satisfaction. To this end, a multivariate procedure was conducted to determine which variables predicted higher job satisfaction.

▧ Conclusion

The importance of research into the attitudes and beliefs of juvenile facility directors is clear. Their ability to affect the lives of juveniles in the system is arguably greater than any legislature. The ability of criminal justice personnel to frustrate policy change is obvious. The director of a juvenile institution decides what form the correctional process takes at the most important level. In fact, the new punitive philosophy being embraced by states across the nation may even come to help those who manage juvenile corrections.

Violent and repeat juvenile offenders are frequently the target of new legislation that seeks to exclude them from the juvenile system altogether. As more and more juveniles are waived to adult court, fewer "hard-core" juveniles will be sent to juvenile facilities.

Juvenile facility directors are in a unique position such that they can affect the shape and direction of juvenile corrections probably more so than outside attempts to "toughen" things up. This research shows that, in the near future at least, most juvenile facility directors are unwilling to abandon the rehabilitative ideal. Furthermore, the level of role conflict may in fact be rising, but the nefarious effects measured in adult systems appear not to be the case in juvenile facilities, at least at the upper administrative level. In previous research on adult correctional personnel, a variety of contradictory findings muddled any substantive conclusions regarding which variables predict correctional orientation or job satisfaction. Nonetheless, this research found that belief in rehabilitation was a strong predictor of job satisfaction. In addition, variables that would support the idea of work-role factors influencing attitudes were significant in the job satisfaction model. Years in juvenile corrections, job-related stress, and salary were all significant predictors.

The results of this research show that juvenile facility directors are more rehabilitative in outlook than their adult counterparts. In addition, juvenile facility directors report relatively low levels of influence from those outside their facility. The fact that the directors overwhelmingly support rehabilitation over punishment suggests that these individuals suffer from relatively low levels of role conflict. Outside pressures to become more punitive do not appear to be affecting the attitudes of juvenile facility directors.

This research also showed substantial differences in the attitudes and operational philosophies of juvenile facility directors compared to their adult prison warden counterparts. Future research should examine the effect of the drive for punishment on the attitudes and operation of juvenile facilities over time. Although this research shows that the punitive ideal has not found its way into the operation of the juvenile facility, will the same hold true for the future? This question takes on more salience as more and more punitive juvenile justice legislation takes hold.

References

Arthur, B. A. (1994). Correctional ideology of Black correctional officers. *Federal Probation, 58,* 57–66.

Bazemore, G., & Dicker, T. J. (1994). Explaining detention worker orientation: Individual characteristics, occupational conditions, and organizational environment. *Journal of Criminal Justice, 22,* 297–312.

Burton, V. S., Ju, X., Dunaway, R., & Wolfe, N. (1991). The correctional orientation of Bermuda prison guards: An assessment of attitudes toward punishment and rehabilitation. *International Journal of Comparative and Applied Criminal Justice, 15,* 71–80.

Crouch, B., & Alpert, Q. (1982). Sex and occupational socialization among prison guards: A longitudinal study. *Criminal Justice and Behavior, 9,* 159–176.

Cullen, F. T., Golden, K. M., & Cullen, J. B. (1983). Is child saving dead? Attitudes toward rehabilitation in Illinois. *Journal of Criminal Justice, 1,* 1–13.

Cullen, F. T., Lutze, F., Link, B., & Wolfe, N. (1989). The correctional orientation of prison guards: Do officers support rehabilitation? *Federal Probation, 53,* 33–42.

Cullen, F. T., Latessa, E., Burton, V. S., & Lombardo, L. (1993). The correctional orientation of prison wardens: Is the rehabilitative ideal supported? *Criminology, 31,* 69–92.

DiIulio, J. J., Jr. (1987). *Governing prisons: A comparative study of correctional management.* New York: Free Press.

Farmer, R. (1977). Cynicism: A factor in corrections work. *Journal of Criminal Justice, 5,* 237–246.

Grusky, O. (1959). Role conflict in an organization: A study of prison camp officials. *Administration Science Quarterly, 3,* 452–472.

Hepburn, J., & Albonetti, C. (1980). Role conflict in correctional institutions: An empirical examination of the treatment-custody dilemma among correctional staff. *Criminology, 17,* 445–459.

Jurik, N. (1985). Individual and organizational determinants of correctional officer attitudes toward inmates. *Criminology, 23,* 523–539.

Philliber, S. (1987). Thy brother's keeper: A review of the literature on correctional officers. *Justice Quarterly, 4,* 9–37.

Shamir, B., & Drory, A. (1981). Some correlates of prison guards' beliefs. *Criminal Justice and Behavior, 8,* 233–249.

Toch, F. L., & Klofas, J. (1982). Alienation and desire for job enrichment among correction officers. *Federal Probation, 46,* 35–44.

Whitehead. J., & Lindquist, C. (1989). Determinants of correctional officers' professional orientation. *Justice Quarterly, 6,* 69–87.

Whitehead, J. & Lindquist, C. (1992). Determinants of probation and parole officer orientation. *Journal of Criminal Justice, 20,* 13–24.

Zald, M. (1962). Power balance and staff conflict in correctional institutions. *Administrative Science Quarterly, 6,* 22–49.

DISCUSSION QUESTIONS

1. How do the perceptions of juvenile facility directors differ from those of adult facility directors? What explanations do the authors provide for these differences?

2. Should juvenile facility directors have a different approach to their job than directors of adult facilities? Why or why not?

3. Why do the perceptions of juvenile facility directors matter to criminal and juvenile justice policy makers?

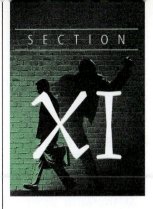

CORRECTIONS IN THE 21ST CENTURY

Introduction: Learning From the Past So That We Have Hope for the Future

Americans have a tendency to revisit old themes, efforts, and programming every generation or so, even when such endeavors were clear failures and rejected by generations past. Perhaps it is because we are a relatively new nation and are seemingly remade as new generations of immigrants flood our shores and bring other histories and cultures that do not include memories of past correctional efforts made in this country. Perhaps we keep retrying old endeavors because of the media influence that reduces very complex problems to a brief and simplistic message, and we do not under-stand that, despite the new packaging and marketing, we have been there and done that before. Or maybe it is a hand and glove collusion by the media and politicians in this reductionism of complex topics and collective memory loss. Whatever the reason, we do not seem to learn much from the experience of those who have come before us, at least as that is related to correctional practice. Or more accurately, we certainly could learn more from our past than we have! It is an oft-cited truism, courtesy of the twentieth-century philosopher Santayana and already mentioned in this book, that those who do not know their history are likely to repeat it. This adage bears reiterating, as it clearly applies to the history and future of correctional programs, operation, and practice.

◪ Punitive Policies Yield Overuse of Corrections

To illustrate this point, all we need do is consider the corrections efforts in vogue over the last two decades, which are now declining in popularity: namely, the drug war, mandatory sentencing, super-max prisons, and the abandonment of treatment programming. Spurred by punitive sentiments that swept the political, social, and economic systems, an emphasis on retribution as opposed to rehabilitation worked its way into the statutes, declarations, and practices that derived from these efforts and profoundly changed corrections as Americans experienced it.

First, this emphasis vastly increased the use of all forms of corrections in this country. Our imprisonment rate (just prisons, not including jails) was stabilized at about 125 persons per 100,000 residents for 50 years (1920 to 1970) until the drug war, mandatory sentences, and other punitive policies increased it (Ruddell, 2004). At the end of 2004, this number had risen to 486 persons per 100,000 residents, or approximately four times the imprisonment rate of that 50-year period (Harrison & Beck, 2005). In raw numbers, the offenders sentenced to prison more than quadrupled for an increase by 445%, from 319,598 in 1980 to 1,421,911 in 2004 (Bureau of Justice Statistics, 2005a).

A similarly steady and swift increase in the use of jails has occurred since punitive policies have been put in place: In 1986, the incarceration rate for jails in this country was 108 persons per 100,000 residents; by 2005, it was 252 or almost a 250% increase in the use of jails in more recent years (Bureau of Justice Statistics, 2005b). Put another way, the numbers of persons incarcerated in America's jails more than tripled, or increased by 388%, from 183,988 in 1980 to 713,990 in 2004 (Bureau of Justice Statistics, 2005a).

Similarly astounding increases can be found in the use of probation and parole because of these efforts. From 1980 to 2004, the number of people on probation more than tripled, with an increase of 351% (1,118,097 in 1980 to 4,151,125 in 2004) (Bureau of Justice Statistics, 2005a). Likewise, the numbers of persons on parole more than tripled during this period (from 220,438 to 765,355), even though 16 states and the federal government eliminated all or most of their parole programs (Bureau of Justice Statistics, 2005a).

Yet, as best we can tell, we do not have proportionately more crime these days than at any other period in our history (see Figure 11.1). In fact, based on victimization and police reports, it appears that, by the early 2000s, we had the lowest rate of victimization since the National Victimization Survey had begun in 1973. Violent crime did increase in the 1970s and through the early 1990s, but it has since dropped precipitously, though clearly the use of incarceration has not tracked this decrease (Ruddell, 2004).

Nor is our use of corrections in sync with what other countries are doing. We have similar crime rates, for instance (Farrington, Langan, & Tonry, 2004), yet our incarceration rate is more than 14 times that of Japan; close to seven times that of Canada, France, Germany, or Italy; and over five times that of the United Kingdom (Ruddell, 2004). Russia is the only other developed nation that gets close in its incarceration rate, and we still outpace them by almost 100 more people per 100,000 residents. As of 2001, almost 2.7% of Americans served time in prison, a figure that is twice that of 1974 (1.3%) (Bureau of Justice Statistics, 2006a).

| Figure 11.1 | Violent Crime Rates: "Since 1994, violent crime rates have declined, reaching the lowest level ever in 2005." |

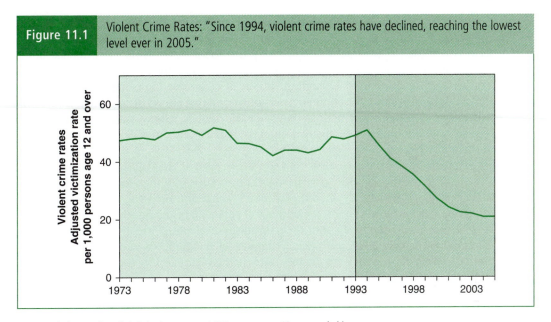

NOTE: Graph shows adjusted victimization rate per 1,000 persons age 12 years and older.

Second, interestingly enough, these punitive policies have not had the effect of increasing sentencing length. The average sentence to prison in 1992 was 6½ years (though offenders could expect to serve only about a third of that time), as compared to 4½ years in 2000 (of which offenders could expect to serve about a half) (Langan & Cohen, 1996). Why, when more punitive policies are in effect, has sentencing length decreased, though actual time spent in prison has remained virtually the same? Clearly, decreased sentencing length might be one of those unintended consequences of the overuse of incarceration. The capacity of prisons and jails, along with probation and parole caseloads, has been vastly increased over the last 20 years, but it may not have increased enough to accommodate the numbers of processed felons in the courts. What this means is that courts are forced to adjust their sentences to the lower relative capacity of prisons, and parole boards are pressured to release inmates as prisons and jails fill up.

Third, punitive policies have led to the explosion in the number of women and minority group members who were incarcerated or under some form of correctional supervision (Bureau of Justice Statistics, 2006a; Irwin, 2005; Irwin & Austin, 2001; Pollock, 2004; Zimring, Hawkins, & Kamin, 2001). Until the current version of the drug war was resurrected—yes there were others in American history (Abadinsky, 1993)—the proportion of women to men or of racial and ethnic minorities to whites in prisons and jails or on parole and probation was somewhat stable (Bureau of Justice Statistics, 2006a).

Fourth, such policies have favored the use of more isolation, "punishment," and ware-housing to deal with both bulging correctional populations and recalcitrant inmates. The number of super-max facilities has exploded, as has the number of super-max inmates, as the management of the number of inmates turned to punishment and warehousing over treatment (Irwin, 2005).

As a result of such policies, a fifth outcome has been the abandonment of a core principle of some correctional institutions and practices (e.g., probation, parole, minimum security institutions, work releases)—namely, treatment—on insufficient evidence. Some correctional programs and institutions were formulated on the premise that treatment is a major goal of corrections. Though the public has continued to believe this (Applegate, Cullen, & Fisher, 2001), for all intents and purposes, real efforts at treatment in prisons, jails, and in community corrections (beyond basic AA/NA, religious, and GED programs) received little funding and virtually disappeared for 20 years in many places (mid-1970s to the mid-1990s).

Though some of these endeavors and efforts, such as the drug war and mandatory sentencing, continue and even grow in some communities, for the most part, scholars and some policy makers have deemed the former a failure and the latter a spectacular waste of money. Furthermore, though the number of super-maxes has grown in the recent past, there is not nearly the hype about their promise for eliciting inmate reform (Pizarro & Narag, 2008). Finally, all indications are that the belief in, and embrace of, treatment programming, albeit ones that can demonstrate their worth, is on the upswing both in correctional institutions and in the communities (see Section V). And though all of these changes in attitudes and perceptions are positive, it is frustrating to know that we knew, or we should have known from our own past, that drug wars, mandatory sentencing, isolation, and pure punishment in the form of warehousing are not likely to reduce crime in this country, let alone reform those under correctional control. The long and the short of it is that we should have known better, because it had all been done before.

⬛ Professionalization

As we look to the future, however, there are a number of concerns that should preoccupy those of us interested in effective correctional practice. One issue that affects almost all areas of practice is that of professionalism. As indicated in Section III, the effort to professionalize corrections has not yet yielded consistent fruit around the country. Some correctional institutions and programs have moved to enforce professional standards for their new hires, such as a required college level educational background, sufficient training, and pay that is commensurate with job requirements. However, most correctional organizations, perhaps primarily because of a lack of resources, have failed to move in a similar direction.

Yet hiring and keeping a professional staff is key to moving correctional institutions into the twenty-first century. When correctional practitioners do not have the kind of education that acquaints them with the history, background, concepts, and research on corrections, then the correctional organization is simply ill prepared to meet the challenges it faces. Moreover, when turnover is high, because training and pay are insufficient, the organization becomes less stable and less equipped to problem solve regarding pressing concerns. Therefore, if we ever hope to move beyond the past and failed correctional endeavors and perspectives, the ranks of correctional practitioners need to be professionalized.

⬛ Corrections Is a Relationship Business

The correctional experience for clients, offenders, inmates, and staff and the success of treatment and probation or parole programming all hinge on the relationships between the people in these organizations. It is often said that the greatest expense for any public service organization is its staff. A collateral expense for correctional institutions and programs is the care of their inmates and clients. Notably, these expenses wax and wane to some degree based on the relationship between actors. If that relationship is characterized by respect and concern among staff members and respect and care (coupled with a healthy degree of control) vis-à-vis staff and clients or offenders, then costly lawsuits, staff turnover, riots, and just general stress that produces discord in the workplace are less likely.

In his groundbreaking work on less explored and identified types of intelligence—emotional and social—Goleman (1995, 2006, p. 4) argues that scientific research on the brain indicates that we are "wired to connect" to others, which means that, every time we engage with other human beings, we affect and are affected by their thoughts and consequently their behavior. Those relationships that are the most prolonged and intense in our lifetime are the most likely to affect us, not just socially or emotionally but biologically.

> To a surprising extent, then, our relationships mold not just our experience but our biology. The brain-to-brain link allows our strongest relationships to shape us on matters as benign as whether we laugh at the same jokes or as profound as which genes are (or are not) activated in T-cells, the immune system's foot soldiers in the constant battle against invading bacteria and viruses. (Goleman, 2006, p. 5)

Goleman (2006) identifies a "double-edged sword" in relationships because those that are positive are healthful, but those that are negative can lead to stress, fear, frustration, anger, and despair, all emotions that can manifest themselves in physical ailments (p. 5). Of course, correctional environments are chock full of stressed, fearful, frustrated, angry, and despairing people, and we are not just referring to the inmates here! So this means that, unless correctional environments can foster some positive relationships between and amongst staff and clients or inmates, both will suffer psychologically and physically.

Recognition of the need to provide opportunities for inmates to "maturely cope" while under correctional supervision (see Section II) would appear to be an acknowledgment that something positive can come out of the decent incapacitation of offenders. Moves to democratize workplaces and give people a voice and choice in their work (as discussed in Section III) may serve to reduce some of the negative emotions associated with working in corrections. More recent attempts to "treat" rather than just "warehouse" inmates in institutions and offenders on probation and parole also represent a move to more positive relationships and so to a future for corrections. As indicated by the findings from the research presented in Section V, VI, and VII, there is reason to believe that some treatment and supervision tactics can work to help offenders as they endeavor to deal with their substance and other abuse issues.

▨ Current Directions: Privatization, Correctional Programming, Religion, Health Care, Older Inmates

It is a stunning realization that much of the future looks like the past, but it is true, in a way. Current trends in corrections mimic those themes we laid out in the introduction and Section I of this book. However, as has been demonstrated by the research presented throughout this book, there is also great progress in refining how we handle correctional practice and programming.

There is little doubt that most correctional experiences for clients and inmates are not tinged with violence or brutality. The vast majority of correctional staff, whether they work in communities or institutions, act professionally, whether the attributes of their work fit that designation or not. Basic health care, clean housing, and nutritious food are provided to most incarcerated persons in this country. Probation and parole officers do provide referrals to their clients when time permits and programs are available. Despite crowded caseloads, these officers usually make every effort to watch the most dangerous of their charges carefully. Jails, though often overcrowded or at least overused, are generally helpful at ensuring the safety of suicide-prone or mentally disturbed inmates or of people detoxing off drug or alcohol induced highs; jails may not represent the best places for such people, but they are usually safer than the street and do provide a minimum of much needed services to them. There is much more programming for those incarcerated in prisons, and even jails, and available to probationers and parolees than there was even 10 years ago. In short, though we tend to repeat our past mistakes, there has been some learning from them as well, and that is manifested in improved correctional practice.

Continuing that trend, and with a wary eye to the future, we present in this section research on the issues related to privatization (Ogle), the value of correctional programming (Latessa), health care (Marquart, Merianos, Cuvelier, & Carroll), and older inmates (Lemieux, Dyeson, & Castiglione). Collectively, the research presented in this section, and that from the other sections, reviews the challenges, options, and promise of corrections in the future. As the political winds shift away from purely punishment-oriented corrections, it will be interesting to see how correctional organizations, programs, and their actors will adjust.

KEY TERM

Professional standards

DISCUSSION QUESTIONS

1. Review the attributes of a professional and why and how the presence of those characteristics would serve to "improve" correctional operation.

2. What is the connection between biology and environment in correctional operation? How do positive and negative environments affect the "biology" of those who work in corrections and those who are clients of and inmates in them?

INTERNET SITES

American Correctional Association: www.aca.org/

American Jail Association: www.aja.org/

American Probation and Parole Association: www.appa-net.org

Bureau of Justice Statistics (information available on all manner of criminal justice topics): www.ojp.usdoj.gov/bjs

National Criminal Justice Reference Service: www.ncjrs.gov

Office of Justice Programs, Bureau of Justice Statistics (periodic statistical reports on all manner of criminal justice topics, e.g., HIV in prisons and jails, probation and parole, and profiles of prisoners): www.ojp.usdoj.gov/bjs/periodic.htm

Pew Charitable Trust: www.pewtrusts.org

Vera Institute (information available on a number of corrections and other justice related topics): www.vera.org

READING

In this article, Robbin Ogle examines the various legal and philosophical arguments relating to the viability of prison privatization. In contrast to these kinds of arguments, Ogle takes an organizational behavior perspective. This is a perspective in which organizations are viewed as active entities, interactively engaged with their environment. Correctional organizations exist in an institutional environment; they were created to serve a set of socially embedded and legally supported beliefs. Ogle points out that private prison firms must interact with two environments: the institutional environment of corrections and the competitive market environment of business, a situation that creates significant conflict for such companies and raises questions about how they will respond to ensure their survival. As private correctional organizations are forced to comply with the survival requirements in both of these environments, they will face an "environmental catch-22": Actions that they take to promote survival in one environment will violate requirements of the other. Thus, Ogle believes that this catch-22 situation reduces the likelihood of the long-term survival of private prisons.

Prison Privatization

An Environmental Catch-22

Robbin S. Ogle

Much of the literature on prison privatization has been devoted to arguing the practical, legal, and moral merits of public versus private corrections. In this article I do not argue about whether public or private corrections are "better" or "worse"; certainly both of these positions have been documented extensively by others. In this theoretical examination I attempt to go beyond that research by offering an organizational behavior perspective. Such a perspective is warranted because it provides a distinctive theoretical basis for understanding and predicting the actions of private prison firms as aggressive

SOURCE: Prison Privatization: An Environmental Catch-22, Robbin S. Ogle; *Justice Quarterly* 16(3): 579-600 (1999); Taylor & Francis, Ltd., http://www.informaworld.com, reprinted by permission of Taylor & Francis, Ltd.

entities acting to ensure their long-term survival under conflicting environmental conditions. I offer a theoretical examination of the conflict that occurs when organizations must operate in both the institutional and the competitive market environment. This discussion provides the basis for examining how such a conflict requires organizations to respond in predictable ways in order to ensure their long-term survival.

In this paper, *prison privatization* refers specifically to private companies owning and operating prisons for profit. Many have hailed these privatization efforts as a particularly positive solution for the problems plaguing government prisons. Proponents believe that privatization offers significant benefits:

Private companies will be able to build prisons more cheaply and more quickly to alleviate overcrowding and save money through cost-effectiveness;

Private entities will be able to operate those prisons with more efficiency and effectiveness, doing so at a lower cost than government while also making a profit;

Since private companies must function in the competitive market environment, they will have the necessary incentives to keep costs low and to provide higher-quality services in order to survive.

Additionally, there seems to be an assumption that these are things that correctional organizations should do in our society.

Opponents offer several practical and philosophical arguments against these assumptions:

The necessity for detailed contract development, monitoring, and regulation will be so significant that it will eat up any savings achieved through privatization;

The government will retain its legal liability and therefore will be liable for actions

of contractors over which it has only limited control;

Privatization violates the moral and ideological underpinnings of corrections.

In this paper I seek to address this third argument in greater detail because although it seems vague, it actually refers to the institutional environment of correctional organizations. DiIulio (1988) discussed this point facetiously but clearly in his discussion of prison privatization as a form of denial of group integrity and moral duty:

Imagine that a private consulting firm consisting of the nation's wisest, most seasoned, and most widely respected statesman could do a superb job of selecting our next president, at a tiny fraction of what it now costs to stage primaries and a general election. The very thought sends a moral chill up my spine because I believe, as most Americans no doubt do, that the public, democratic nature of the process of presidential selection has a value independent of the outcome. . . . I would not hesitate to choose the public process even though it was far more expensive. . . . [S]uppose that CCA [Corrections Corporation of America] has made it really big. They have proven that they can do everything the privatizers have promised and more. The corporation decides to branch out. The company changes its name to CJCA—Criminal Justice Corporation of America. It provides a full range of criminal justice services: cops, courts, and corrections. In an unguarded moment, a CJCA official boasts that "our firm can arrest 'em, try 'em, lock 'em up, and, if need be, fry 'em for less." Is there anything wrong with CJCA? (pp. 80–81)

DiIulio and other scholars (for example, Ericson, McMahon, and Evans 1987; Robbins 1986; Shichor 1998) have examined these moral issues at length. They seem to conclude that some government tasks simply should not be privatized because of their ideological significance.

Schiflett and Zey (1990) reached a similar conclusion based on their organizational comparison of private product producing companies and public service organizations. Public service organizations are environmentally constrained by many different legal, social, and political means, as well as some secondary economic factors. Private organizations generally face fewer environmental constraints. These are primarily economic, and are evaluated only by each company's board of directors, the investors, and (indirectly) by their consumers (Pfeffer and Salancik 1978; Schiflett and Zey 1990; Scott 1995).

Private companies control their own revenues through loans and profits, in contrast to the complex and conflict-ridden tax allocation process faced by public organizations. They face very little legal regulation of domain, technology, goals, structure, or revenues, unlike the multiple, varied, conflicting political, legal, and social constraints on public organizations. Private companies have the freedom to adjust their goals and technologies as necessary to ensure efficiency and profitability, whereas most public organizations were created to serve specific existing institutionalized goals and technologies (Ogle 1998). Private companies' revenues are not generally subject to severe fluctuations based solely on ideological variation or political whim (Meyer and Scott 1983; Schiflett and Zey 1990; Wamsley and Zald 1973). Conversely, public organizations are subject to significant and varying legal and political mandates on their domain, goals, technology, and structure. In fact, they can be and have been eliminated for political or ideological reasons regardless of their economic efficiency or effectiveness (Meyer and Rowan 1977; Pfeffer and Salancik 1978; Schiflett and Zey 1990).

Consequently, private companies have more direct control of their survival, which is based on profitability, because they can focus primarily on efficiency and effectiveness; these are evaluated only by the board of directors, the investors, and demand. Ultimately they must be committed to serving their investors' interests to survive in the competitive market; such service involves the rational pursuit of profit. Because of these constraints, private for-profit organizations generally have centralized decision making at the top of the organization and standardized procedures that allow relatively strict control of technological efficiency. These features should result in effectiveness leading to increased profits and ultimately to survival (Meyer and Scott 1983; Scott 1995).

In contrast, decentralized decision making by various line-level professionals and nonstandardized technological applications are necessary in the institutional environment. In this environment, efficiency and effectiveness become undefinable and unmeasurable; thus organizational survival is more dependent on legitimacy than direct outcomes (Garland 1990; Hasenfeld and English 1974; Ogle 1998; Schiflett and Zey 1990). Ultimately these types of environmental differences were the major variables leading Schiflett and Zey (1990) to conclude that some public service organizations do not lend themselves well to privatization.[1]

Private correctional service organizations face all of these environmental constraints. They must attend both to the economic environment, which affects survival through profitability, and to the institutional environment (social, legal, and political), which influences survival through legitimacy. Although their first allegiance, because of the competitive market, must be to their investors and to profit, they cannot simply ignore the institutional environment, which is the essence of corrections.

⬚ The Institutional Environment of Corrections

Historically, efforts at prison privatization seemed to be based on the same restrictive assumptions as pervaded early organizational theories (i.e., closed system theories). These perspectives assume that organizations are closed off and protected from their environment so that they can pursue technical efficiency within, regardless of what happens outside. Early prison privatizers acted to maximize profit with internal efficiency, without considering how this would be received externally or how it might affect their operations over time. In effect, they ignored the institutional environment of corrections; this oversight resulted in their demise.

Today it is an accepted tenet in organizational studies that organizations are actually open systems which interact with their external environment so as to promote their survival. In addition, some early scholars noted that certain organizations, particularly public service agencies, seem to exist in an environment steeped in ideological concerns and symbolic requirements quite different from those of the typical organization in the competitive market (Selznick 1957; Wamsley and Zald 1973). Subsequently, organizational studies began to reflect the recognition that public service organizations often exist and must interact in an environment quite different from the technical market.

In recent years, organizational scholars have identified and differentiated this institutional environment from the technical efficiency-oriented environment of the market (DiMaggio and Powell 1991; Meyer and Rowan 1977; Meyer and Scott 1983; Scott 1995). Scott (1992) most clearly defines the difference between technical and institutional environments:

Technical environments are those in which organizations produce a product or service that is exchanged in a market such that they are rewarded for effective and efficient performance. These are environments that foster the development of rationalized structures that efficiently coordinate technical work. . . . By contrast, institutional environments are characterized by the elaboration of rules and requirements to which individual organizations must conform in order to receive legitimacy and support. In institutional environments organizations are rewarded for utilizing correct structures and processes, not for the quantity and quality of their outputs. (p. 132)

Scott (1992) also notes that both environments require interaction and responsiveness by organizations. The type of environmental contingencies faced and the organizational response required are different, however. This difference indicates a problem for private organizations attempting to adapt to the contingencies of both types of environments.

Institutionalized environments are characterized by rules and rituals arising from the societal belief system, which pertain to issues in the world for which we have no scientific basis. Meyer and Rowan (1977) called these beliefs "myths": Once institutionalized in society, they become what everyone "knows" about the issue at hand. Such myths actually become the foundation for the creation of organizations and for the rules and rituals forming the function of such organizations. To maintain legitimacy and support in the larger society, these organizations develop their structures and behave so as to reflect this institutionalized nature. Legitimacy—rather than technical efficiency—becomes the basis for survival for institutionalized organizations because there exists no scientific (cause/effect) basis for the organization's goals or technologies. These goals and technologies—these social myths— and their political, social, and legal supports become the institutionalized environment in which such organizations must operate and survive.

The identification of correctional organizations as existing in an institutional environment is nothing new. Many scholars have used this perspective to analyze various aspects of corrections (Crank 1996; Garland 1990; Hagan, Hewitt, and Alwin 1979; Ogle 1998; Rosecrance 1988). Their research has established the institutional nature of correctional organizations by examining the function and necessity of various organizational structures in the institutional environment.

The correctional organization was established to serve a set of intangible (scientifically undefinable and unmeasurable) goals and technologies which, over time, have come to represent and maintain our belief system about criminal behavior and its correction (Garland 1990; Ogle 1998).[2] The organization is "infused with value beyond the technical requirement of the task at hand" (Selznick 1957: 17). In our society, the process of corrections is far more important than the evaluation of the results. The political, social, and legal requirements and expectations for humaneness and fairness in that process are quite high regardless of the technical outcomes. Thus legitimacy is more important for organizational survival than is any cost-benefit analysis (Crank 1996; Garland 1990; Hagan et al. 1979; Ogle 1998; Rosecrance 1988).

Legitimacy, in this sense, refers to "the property of a situation or behavior that is defined by a set of social norms as correct or appropriate" (Scott 1992: 305). Thus the organization continually appears to be doing what we expect it to do (pursuing our vague preset goals) as we expect it to be done (using the preset technologies) regardless of the technical results. By obtaining and maintaining legitimacy over time, as well as resolving legitimacy crises as they arise, correctional organizations have achieved long-term stability and survival, whether or not offenders were changed or crime was reduced.

Concomitantly, maintenance of legitimacy for survival precludes correctional organizations from certain actions, regardless of the technical efficiency of those actions. For example, correctional organizations do not attempt to change their goals or technologies for the sake of technical efficiency or cost-effectiveness.[3] In fact, they often create costly and technically inefficient structures whose only function is to represent and dramatically enact our institutional beliefs and values (Ritti and Silver 1986). This is particularly true of beliefs and values depicted in moral or legal requirements—that is, lengthy and detailed procedures to acknowledge rights and lengthy presentence investigations evaluating individual assets and liabilities, even though most offenders are sentenced on the basis of prior record and the seriousness of their offense.

From the viewpoint of technical efficiency, creation of such structures might appear to be the rational pursuit of technical failure. Correctional organizations, however, recognize that these institutionalized beliefs and values, although intangible, are the social basis for their existence. Their survival depends on the organization's consistently being perceived as the legitimate embodiment of those ideals, rather than displaying technical efficiency in correcting crime or criminals.

The existence of corrections in an institutional environment creates a set of very different constraints for the organization, which are not measurable in the instrumental or technical sense. They do not readily lend themselves to the cost-benefit analysis to which we are accustomed in the competitive market environment. These constraints often conflict with technical efficiency, and such conflicts are not easily resolved. In fact, history teaches us that most prison privatizers have been unable or unwilling to resolve such conflicts, and that organizational failure has been the result.

⬛ The Environmental Catch-22

The heart of this argument concerns the existence of an "environmental catch-22" that plagues private prison companies. The term is intended to imply an external environmental conflict that will place private prison companies in a situation where they are "damned if they do and damned if they don't": financially damned if they conform to the institutional environment, and ideologically damned—socially, legally, and politically—if they do not conform.

Private contracting for secondary or fairly limited support functions in government (e.g., waste collection or public vehicle maintenance) has been reasonably successful. Private contracting in major government operations, however, particularly those embedded in strong institutional environments, has not fared so well; these operations include the defense industry and health care. In fact, many opponents have framed the ideological problems in privatizing the operations of the criminal justice system in terms of a loss of legitimacy for government. Such ideological problems are based on moral questions concerning whose interests are being served (DiIulio 1988; Ericson et al. 1987; Robbins 1986, 1989; Shichor 1995, 1998). These issues arise from the institutional environment of corrections.

Most private companies contracting to provide government services such as waste collection and vehicle maintenance do not enter an environment as highly institutionalized as that of corrections. In this environment, the key to survival is legitimacy rather than technical efficiency and cost-effectiveness. Here, technical efficiency and cost-effectiveness are

undefinable and relatively unmeasurable—based on beliefs rather than on science—and therefore become secondary to legitimacy with reference to the organization's survival.

Thus it is inappropriate to compare prison privatization with privatization of services such as waste disposal. Waste collection and disposal is monitored and measured relatively easily: Either the garbage is removed or it is not. Americans do not hold strong ideological expectations about the type of garbage trucks used, the qualifications of waste collection employees, or the manner in which such work is done. In contrast, for other activities, particularly those based primarily on our institutionalized belief system, merely changing the type of charter does not reduce the environmental constraints facing the organization. In fact, it adds a whole new set of environmental constraints associated with economics and the competitive market.

Corrections, and criminal justice in general, illustrate precisely this type of institutional environment. As stated earlier, this environment is highly institutionalized—as opposed to an environment based on technical efficiency—in that correctional organizations represent the enactment of powerful beliefs and values that have been institutionalized in the larger society (DiIulio 1988; Garland 1990; Ogle 1998). Such an environment differs significantly from the environment of private companies in the competitive market. The organization must create structures and direct its behavior outward into the institutionalized environment, and must dramatically enact these beliefs and values regardless of the technical outcomes (Ritti and Silver 1986). Private prison companies, however, also must exist and operate in the competitive market, where they are rewarded for technical efficiency and cost-effectiveness. These qualities are essential for survival in the competitive market, and are among the primary reasons proffered by supporters of privatization as the basis for its likely success.

This environmental conflict has important implications for organizational survival. I propose here that these private companies face an "environmental catch-22" which public correctional organizations do not face, because they employ methods designed to create technical efficiency for market survival and because those methods conflict with the institutionalized environment of corrections.[4]

✑ Environmental Constraints and Organizational Action

In this highly institutionalized environment, private prison companies are subject to evaluation both by their boards and investors interested in profit and by the complex social, legal, and political mechanisms of the institutional environment: government regulators, the legal system, professional organizations, the public, and prisoners. All of these evaluators will be as interested in ideology as in costs, and it is quite likely that all of those interests will vary and conflict. Consequently the task of managing these conflicting environmental constraints on private prison companies will involve both growth/profit and legitimacy. Profit is fairly straightforward: legitimacy is much more complex to manage, but is also essential to organizational survival.

Organizational survival is the primary focus of this analysis because it is the most basic need of any organization. It requires "the maintenance of the integrity and continuity of the system itself" (Selznick 1948: 29). To examine organizational survival, Selznick (1948) recommends focusing on the organization's adaptations to its environment over time, rather than on its tasks and structure. These adaptations will reflect the decisions and actions taken by the organization in response to its environment, to ensure its long-term survival. Contrary to much of the literature on prison privatization, this approach implies that organizations are active entities in an

open system rather than "passive pawns" in their environment (Feeley 1991; Meyer and Scott 1983; Oliver 1991; Pfeffer and Salancik 1978; Scott 1995). Consequently it seems unlikely that the competitive market (such as it exists in corrections) and sophisticated contract regulation will automatically result in acquiescent compliance by private prison companies. Assertive interaction with the environment, intended to manage constraints that threaten growth and profit and/or legitimacy, is far more likely.

Oliver (1991: 152) states that organizations can interact with their environment in five ways to manage their conflicting environmental constraints and enhance their likelihood of survival: (1) acquiescence, (2) compromise, (3) avoidance, (4) defiance, and (5) manipulation. Acquiescence is the only one of these adaptations that could be considered passive; the others involve active efforts by the organization to negotiate, placate, or change elements in their environment. In fact, Oliver (1991) notes that the active adaptations are far more likely in environments where the organization faces highly conflicting and diverse demands. The history of private corrections reflects the existence and the importance of such aggressive adaptations to generate growth and profit for survival (Feeley 1991).

Historical Adaptations

Incarceration began in this country as a private enterprise. Early jails often were privately owned and operated for profit. Jails charged the government, the prisoners' families, or the prisoners for their keep (Hirsch 1992; Morris and Rothman 1995; Rothman 1971). This opportunity for profit led to such widespread prisoner abuse and corruption, which violated our institutionalized beliefs about the humane treatment of citizens, that government was forced to heed public outcry. Accordingly it established a system that could be accepted as legitimate, representing and enacting our institutionalized beliefs and values (Garland 1990; Morris and Rothman 1995). Privatization also has occurred at various other periods in American history; each time it failed either because it lost environmental legitimacy due to corruption and abuse or because the profits were inadequate to sustain the private institutions.

Historically, private correctional organizations did not manage these conflicting environmental constraints. They aggressively defied, manipulated, or avoided them. According to Carleton (1971), for example, the Louisiana prison system underwent several periods of privatization; the process failed each time, even with political support. In Louisiana in the 1860s, the government leased its prison to a private concern that proceeded to work prisoners to death for profit. The private owner disregarded all government and legal efforts to intervene until the legislature passed a law preventing prison leases (Carleton 1971; Durham 1994). These private contractors virtually ignored the institutional environment. They used avoidance methods—dragging out legal actions for long periods, disregarding legal mandates, ignoring contractual requirements, and exploiting offenders to maximize profits—before they were trampled by the institutional environment.

Ethridge and Marquart (1993) note that Texas had a similar experience in the 1870s. The state leased a prison to several different private contractors for about 12 years, but finally abolished prison leases because of the abuse of prisoners for profit. McAfee and Shichor (1990) point to similar problems concerning defiant and grossly negligent avoidance by prison privatizers in California during the 1850s. These are only a few of the examples in the literature concerning the abusive and aggressively defiant history of prison privatization in the United States.

During that privatization era, all of those state governments changed their leases and selected new contractors several times in an effort to obtain adequate contract compliance with what are primarily institutional requirements. None of the states succeeded in their endeavor; each one finally succumbed to public outcry reflecting institutionalized beliefs and values, which forced legislation banning prison leases. Prison privatization in the nineteenth century might have been pragmatically workable, but the privatizers underestimated the power of the institutional environment over cost analyses, profit generation, and political fervor.

The history of private prisons in the United States reflects a misconception that the competitive market would create compliance and competitiveness without disturbing the institutional environment. This approach led to corruption, abuse, political and contractual manipulation, and ultimately failure because these organizations were not passive entities. Rather, they were aggressive actors trying to adapt to the conflicts created by operating in both the competitive market and institutional environments. This dismal history is most often tied to the private contractors' neglect of the institutional environment of corrections for the sake of the profit and growth essential to their survival in the market. These actions resulted in organizational loss of legitimacy, increased regulation, and ultimately bankruptcy or contractual failure.

Environmental Changes in the Interim

According to Ethridge and Marquart (1993), the reasons for prison privatization are the same today as in the 1860s: increasing prison populations, cost containment, and economic difficulties for governments. The goals of privatization are the same in both periods; so are the problems. For example, private companies fail to abide by their overenthusiastic contracts, and the quality and quantity of services decrease steadily over time. In addition, as in the past, today's private contractors are attempting to manipulate politics and our institutionalized beliefs and values in order to justify their contractual inadequacies and to survive in the competitive market. Yet we observe one important difference between the two periods: Governments now fully acknowledge that they are ultimately responsible for the care, custody, and treatment of inmates (Ethridge and Marquart 1993). In addition, government has become increasingly constitutionally and legally bound to that responsibility (DiIulio 1988; Robbins 1986, 1989).

These constitutional and legal developments contribute to and solidify the institutionalized environment of corrections. Our society has significant political, social, and legal expectations for the care and custody of prisoners. Care and custody are monitored today, as in the past, by outside philanthropic groups, and also by prisoners and their families, who continue to file lawsuits in ever-increasing numbers (Taylor 1997). Often the only defenses that government correctional organizations can mount against these demands are ideological defenses invoking "the service of a significant state interest" or "good faith." Such defenses are based on our institutionalized beliefs and values; they are supported by the government's legitimacy, which originates in the society's general belief that government acts in the public's best interests. Privatization may even increase the government's liability because it may reduce the government's legitimacy. Legitimacy crises may reduce the potential use of a "good faith" defense as a result of the government's apparent willingness to ignore our beliefs and values and to allow the profit motive or a cost/benefit analysis to guide decision making about citizens. Because private companies must make a profit and therefore must act in their own and their investors' best interest, they will find it difficult and costly to overcome these ideological demands concerning the public interest and maintenance of legitimacy.

This legitimacy problem becomes especially difficult in today's environment, where information on operations, lawsuits, and profits is immediately available to every interested citizen. Such difficulty is exemplified by the private

prison company that attempted to obtain a prison contract in Pennsylvania, with the intention of building the prison on a toxic waste site that had been purchased for one dollar (Robbins 1986). This certainly would have been a shrewd business move because it would have decreased the capital outlay and increased the likelihood of profits for investors. This private prison firm, however, failed to heed the importance of the institutional environment. The government and the people of Pennsylvania were so outraged by its blatant cost/benefit approach and its disregard for humanity that a statutory moratorium on private prisons was established in that state (Robbins 1986).

Some observers have noted that outright abuse of prisoners for profit is unlikely in present-day private prisons because prisoners now have access to powerful external controls in the institutional environment. For example, they have ready access to the media, the courts, and prisoners' rights advocates, who constantly monitor and question compliance with legal mandates and general standards of human decency. In addition, these organizations are subject to sophisticated contracts, government auditing and regulation, court orders, and the requirements of the competitive market, which force them to comply and innovate in order to maintain their contracts.

Although outright abuse for profit may be unlikely, it is also true that private prison companies will not survive in the competitive market if they do not make a profit and grow (Feeley 1991). Most of the institutional demands in corrections will conflict directly with that effort; therefore these organizations must take some action to manage this conflict and ensure their survival. How will these private prison companies handle such conflicting demands? How will those actions affect long-term survival?

Modern Adaptations

The literature is filled with examples of present-day private prison organizations attempting to use the five adaptations cited by Oliver (1991). It appears that efforts to meet requirements in the competitive market environment often create or intensify adaptive problems in the institutional environment. Rather than acquiescing, most private prison companies have chosen more aggressive adaptations.

Acquiescence Adaptation through acquiescence would involve simply complying with all of the conflicting environmental demands. Private prison companies claim that they can do this: providing higher-quality and -quantity services at a lower cost while making a profit. These companies generally claim that this is possible through the use of creative private management techniques that are not encumbered by government requirements, such as bidding for peripheral goods and services, or the public financing process.

Acquiescence, however, is an unlikely adaptation for any organization, especially if it faces conflicting environmental demands (Oliver 1991; Pfeffer and Salancik 1978; Scott 1995). These companies are much more likely to manage such conflicts actively than to passively accept the status quo. As a result of the highly institutionalized environment of corrections, private prison companies must face diversity and confront many conflicting issues concerning ideology versus profit and growth. Profit and growth will not be sufficient to sustain survival in this dual environment because efficiency and effectiveness are undefinable and unmeasurable in the institutional environment. Continuous compliance with ever-increasing and varying ideological demands, however, will escalate costs and decrease profits.

In such a situation, we would expect to see a company innovate, especially by adjusting its goals, technology, and/or structures, to achieve greater efficiency and effectiveness. For instance, some companies might choose to abandon or ignore the goals of treatment, rehabilitation, or reintegration and to pursue only custody or warehousing because it is cheaper and easier to measure that achievement. Other private contractors might decide to reduce costs by eliminating human security requiring training, salaries, and benefits. They might pursue this avenue by creating an electronically controlled prison with little regard for humane treatment of citizens or preparation for eventual release back into the community. Their institutions might resemble the underground prisons depicted in science fiction horror movies. Private prison companies, however, will not be able to do this unless they can change the institutionalized beliefs and values that make up this environment. This will be particularly difficult because, as a result of the past and present strength of these beliefs and values, private contractors are contractually required to adopt and maintain the same goals, technology, and core work structure as do government prisons. Consequently they face the same undefinable and unmeasurable task as government prisons, but with additional constraints: the necessity, imposed by the competitive market, to grow and make a profit.

This conflict will become even more salient over time as these companies face aging facilities, ever-increasing lawsuits, capricious political changes, increasingly precise contracts, and expanded monitoring and regulation as a result of their attempts to adapt. In this situation, private prison companies will be unlikely to simply acquiesce. They are much more likely to act aggressively, both internally and externally, to manage the institutional demands so that they have a minimal impact on the organization's continued growth, profit, and survival.

As noted earlier, most scholars believe that these organizations are much more likely to choose more active, more aggressive methods for managing such conflicts. In fact, the literature already contains many examples that illustrate the private prison firms' "active" efforts to ensure their survival by attempting to manage this "environmental catch-22."

Compromise and Avoidance Techniques
Compromise and avoidance seem likely in such an environment, at least initially, because

the firms must appear to be seriously pursuing achievement of institutional expectations. For example, many of these private prison companies initially agree to utilize government employees and to provide the same or better pay, training, and benefits. Over time, however, this situation changes, at least in part because employees are a major expense for any correctional organization. Therefore they are one of the most likely targets for cutbacks to save money, increase profits, and reduce the ideological impact of professionals. Consequently companies negotiate compromises with the state, which permit them to make economic allowances in this area; usually these are accompanied by placations such as American Correctional Association accreditation or other symbolic certifications (Bowditch and Everett 1987; Neilsen 1986). Yet as other scholars have observed, these offerings are purely symbolic and have little relation to the quantity or quality of the facility, the staff, or the work (DiIulio 1988; Garland 1990; Ogle 1998).

Other companies have been even more aggressive, requiring contractual power, either in the initial contract or at renewal, to make their own decisions about personnel and their salaries, training, and benefits. Such power, of course, allows them to hire less adequate employees and to provide substandard training, salaries, and benefits, which will increase profits and will allow more company control of growth and profit potential over time. Such compromised control, however, has already generated many instances of inmate abuse and increased state liability reminiscent of the period in government corrections before professionalization. In 1995, for example, the Immigration and Naturalization Service was forced to shut down one of its private detention facilities in New Jersey after a highly publicized riot of inmates due to misbehavior by untrained, unqualified, and inadequately supervised staff members (Sullivan 1996). In 1996 the State of Missouri was forced to remove some 800 inmates from several private facilities in Texas. The Missouri Department of Corrections had obtained a copy of a videotape depicting Missouri inmates being "dragged, kicked, and beaten, as well as bitten by a snarling dog" ("More Missouri Inmates" 1996: A1).

When government contract regulators stand firm and such compromise measures fail, we often see avoidance efforts aimed at the same goal. In avoidance, organizations often meet requirements in one area to the detriment of others, or drag out compliance indefinitely. For example, organizations might be required to have certified psychologists on staff because this is part of the institutionalized belief system concerning the handling of inmates. Psychologists, however, are more expensive than the average counselor, and the organization might prolong the hiring process for two years of a three-year contract. They might reach compliance only during auditing or contract renewal negotiations. This measure is often employed by other private for-profit organizations working in other areas of corrections such as private treatment centers and halfway houses. In these instances, the hard-nosed private manager attempts to creatively ignore or work around the institutionalized beliefs and values in order to minimize expenses, to the organization's long-term peril.

Texas faced similar problems in 1989, when private prison contractors failed to meet court-ordered requirements. To meet those requirements without increasing costs and losing profit potential, the companies simply cut back in other areas such as medical care, offender programs, self-monitoring, and staff training. When Texas contract auditors returned to check compliance with the court orders, they discovered that those changes had been achieved to the detriment of other services, and documented an array of new violations (Ethridge and Marquart 1993). In fact, compromise and avoidance measures have become so commonplace, and contract rebidding so expensive, as to generate discussions about the use of positive and negative performance incentives—other than contract termination—to gain private prison companies' "near" compliance (Tomz 1996).

Defiance and Manipulation Techniques

Defiance and manipulation are more aggressive measures, aimed at controlling and even changing the environment. In the past, private prison contractors simply chose to defy the institutional environment. To maximize profits, they ignored contractual requirements and public disapproval. As a result, they consistently failed to survive regardless of costs and benefits.

Although complete defiance seems unlikely today, we would expect to see some forms of defiance, particularly in areas where the institutional expectations are not legally or socially strong, or where corruption is achievable. Corruption certainly seems to be a possibility in private corrections. In many instances these companies recruit their elite managers, supporters, and sometimes investors from among current government employees and from academics with many connections to the government network (Ethridge and Marquart 1993; Press 1990; Shichor 1995, 1998).[5]

Manipulation would include activities such as lobbying for legislative changes involving increased use of incarceration, changes in offenders' rights or ideological expectations, greater control of the contract and auditing process, and decreases in regulation and legal mandates. Calabrese (1992), for example, reports that the Texas Department of Criminal Justice allowed a "third party contract monitor" to take over because of the stalemate between the state and private firms regarding performance. As another example of this manipulative activity, the Corrections Corporation of America in Tennessee backed a bill that would change a state law restricting privatization. This would have allowed the corrections commissioner to privatize more of the system without legislative approval (Shichor 1998; Taylor 1997). The benefits for the private company in this situation are clear: more control over its environment because of a decrease in political, legal, and ideological restrictions that would lessen the potential for growth and profits. Such a decrease would increase the organization's chances for survival.

Shichor (1998) reports these manipulative behaviors in Tennessee and California, particularly lobbying for legal change to benefit the company. He even notes the involvement of some former and current state legislators in these private companies, and observes their lobbying efforts on behalf of their companies. Ethridge and Marquart (1993) discuss similar problems in the Texas system. When private prison organizations begin to use these techniques to ensure their survival in the competitive market, they slowly but consistently violate the requirements in the institutional environment, thus creating the "environmental catch-22."

⊠ Discussion

This situation has only two possible outcomes. First, these violations of the institutional environment could go unchecked and eventually could change the institutionalized beliefs and values. Second, and alternatively, the violations could result in questions of legitimacy, provoking further scrutiny and eventual elimination due to financial or ideological failure. These two potential outcomes warrant further discussion.

First, over time, private correctional organizations successfully change the institutional environment by changing our institutionalized beliefs and values and altering their legal, political, and social supports. Thus private correctional organizations could select new goals, technologies, and other structures that are more conducive to growth and profits for correctional corporations. The result would probably be very different from what we currently call corrections in America.

This outcome seems unlikely for several reasons. Although the institutional environment is a web of interwoven and often conflicting ideals, it has remained relatively stable for hundreds of years, and is generally consistent with the institutional environment of other Western industrialized countries. In addition, at this point in our history, the institutional environment is supported strongly by law, constitutions, and international human rights advocacy and declarations.[6] Consequently, private correctional organizations are unlikely to achieve great success in such an endeavor. Deteriorating or inhumane prisons struggling to meet expectations are one thing; adopting such a position as our vision is something else entirely.

The second outcome seems much more likely. In many ways, it represents a repetition of history. In this instance, private correctional organizations would attempt to manipulate or change the institutional environment, but with little success. This limited success eventually would cause questions about legitimacy, which would give rise to increased monitoring and regulation. Increased scrutiny would hinder expansion of growth and profits in the competitive market and would force these private prison organizations to become more blatant and more aggressive in their efforts to change, or at least reduce, the influence of the institutional environment. Eventually these political and legal battles would accumulate, reflecting a general disregard for our institutionalized beliefs and values, and would force governments to act to protect their own legitimacy.

This reassertion of the institutional environment is likely to trigger even more intensive scrutiny of private organizations and to precipitate continual legitimacy crises. Such crises cannot be addressed with figures on technical efficiency or cost-effectiveness. They demand that the organization behave in ways that dramatically enact those beliefs and values in the institutional environment. Creating structures designed to enact institutional values is costly, especially when they do not contribute to the technical efficiency of the organization.

If private correctional organizations choose to create institutional structures that address their legitimacy problems, they will face increasing costs. This action will further reduce growth opportunities and profits, and is likely to lead to financial difficulties in the market environment. If these organizations choose not to develop structures to address legitimacy, they will face ever-increasing monitoring and regulation, lawsuits, and legitimacy problems, which may well lead to elimination for ideological reasons. This growth in regulation, lawsuits, and scrutiny increases the organization's costs as well as raising the cost to government for creating such regulation and managing any liability it may retain.

Government corrections face the same kind of problem when funding is cut and when organizational activities must be tailored to the money available without violating our beliefs about what correctional organizations should do and how they should do it. The major difference is that government corrections operate only in the institutional environment; they can simply comply with that environment and ignore growth and profit, which are not essential to their survival. In addition, government retains its use of a "good faith" or "significant state interest" defense against any violations of institutional expectations. Private corrections companies lack this option. They must comply with both legitimacy and profit, or face failure in one environment or the other. Failure in either realm will decrease their chances of survival.

⬚ Summary

Correctional organizations exist in an institutional environment composed of powerful, deeply embedded social beliefs and values, which receive strong support through the legal, political, and social systems of the larger society. These beliefs and values have no scientific or empirical grounding, but over time they are institutionalized in society and come to be what we all "know" about corrections. Sometimes they vary and conflict, but at any given time they all exist and influence what correctional organizations can and cannot do.

Long-term survival in the institutional environment requires legitimacy, which often conflicts with the survival requirements of growth and profit in the competitive market environment. The result is an "environmental catch-22" for organizations that attempt to operate in both spheres simultaneously. Private prison organizations are likely to respond with aggressive actions designed to ensure the growth and profit essential to their survival in the competitive market. They have no other choice. When they respond in this way, however, they violate the requirements of legitimacy in the institutional environment. Such violations will prompt questions about whose interests are being served; they are likely to result in a destructive cycle of regulation, violations, and increased regulation, increasing the costs to both the company and the government. This is a self-perpetuating series of legitimacy crises for both the company and government; in addition, the taxpayers bear the expense of assisting such companies, continuously increasing the monitoring and regulation, rebidding the contracts, or returning to government operation.

This point addresses the theoretical reasonableness of many of the fears expressed by opponents of prison privatization. Private prison companies, acting aggressively to ensure their survival in the market, are likely to violate institutional legitimacy. This process may take some time to become clear today, because the actions are likely to be more subtle than in the past and to occur slowly over time, rather than being immediate, blatant acts of defiance by private prison contractors. Unless this process precipitates a major change in the institutional environment, it will set off a cycle that drastically increases costs to the companies and to the government. Under the scrutiny of both domestic and international observers, if government does not address these issues, it risks its own legitimacy and that of the entire criminal justice system. In such an instance, it is likely that government will choose to end the conflict and rising costs by reasserting its control of correctional operations.

◼ Notes

1. Certainly one could examine the problems of viewing correctional organizations as product-producing companies. In such a view, correctional organizations are regarded as offering a definable and measurable product that can be evaluated in relation to the costs of processing in some market, scientific technologies that perform this production process, and the existence of a competitive market where the quality of output is evaluated. Such an examination, however, is beyond the scope of this paper. For further discussion, see Ogle (1998), where the author compares and contrasts the values of these two organizational views (institutional and technical/ rational) for understanding and evaluating the correctional organization. She maintains that our failure to understand this complex organization and its function in our society is one of the reasons that, historically, we have made less than adequate choices for reform or change in corrections. She uses privatization as an example of a fix-all based on an organizational perspective that is inadequate for understanding corrections.

2. At any given time, both rehabilitation and retribution—which appear to be conflicting positions within correctional ideology—exist and are in use, and in fact are crucial to correctional legitimacy. They are both necessary because neither is scientific; both represent reasonable and well-accepted beliefs that allow the organization to appear to be "all things to all people." Sometimes one position—such as retribution at the present—may appear to dominate, but the other positions never disappear. Consequently one state may build new prisons while its neighbor expands community corrections. In addition, we may see the development of programs such as electronic monitoring, which purport to meet the expectations of both the rehabilitative and the retributive supporters by utilizing increased surveillance and also by requiring offenders to leave their home prison for work and treatment. Even within prisons, regardless of their retributive quality, prisoners receive at least minimal opportunities for rehabilitative activities such as spiritual counseling and groups, some kind of recreation, and reading materials. Some facilities still provide education and vocational training. In fact, correctional ideology is always present and does not change much. Even though one position within the ideology may be dominant at a particular time, the others are always present as well.

3. Efforts to change the technologies of corrections, for instance, have met extreme resistance. Such

change efforts include a return to corporal punishment or the use of medical technologies such as psychosurgery, psychochemical treatment, and aversion therapy. These technologies might be more cost-effective and even more efficient for crime control. They are not acceptable for use in corrections, however, because they violate our beliefs regarding humane treatment, people's ability to change, and the right to dignity and control of one's body and mind. These are institutional beliefs and values rather than scientific imperatives.

4. This is not an argument about whether public or private charters are preferable for correctional organizations. Both charter types present significant problems, and each has its virtues. This examination is a purely theoretical discussion on the process of organizational adaptation to environmental conflict, which has been identified in the literature and in history. Here I use well-established organizational theories to identify the two separate environments, the conflicts presented by this situation, and the impact of these conflicts on the choices of behavior for private correctional organizations.

5. For example, Clear and Cole (1994) discuss a former vice president and board member of CCA who was the commissioner of Arkansas corrections when the U.S. Supreme Court found that state's correctional system to be in violation of the Eighth Amendment's "cruel and unusual punishment" clause. Becker and Stanley (1985) discuss a small private correctional company's hiring of a former warden in the federal prison system, who had been found guilty of abusing inmates. These are only a few examples of the collusion and corruption already occurring; see Shichor (1998) for a more fully detailed discussion.

6. Some scholars argue that the ideology or institutional environment of corrections is already changing because of the conservative political movement and the efforts of private prison companies. As examples they cite changes such as the removal of television sets and weight-lifting equipment from prisons. I would argue that these actions actually are small managerial adjustments designed to symbolize the conservative political climate of the moment without creating any real change in the beliefs and values that are the foundation of corrections. For example, the social expectations and legal requirements that inmates must receive adequate physical exercise and recreation still exist intact. Whether prison officials use television and weights or some other means to meet these institutional requirements, the larger institutionalized beliefs and values do not change. The requirements still exist and still receive significant social, political, and legal support.

▨ References

Becker, C. and A. D. Stanley. (1985). "Incarceration Inc.: The Downside of Private Prisons." *The Nation*, June 15, pp. 28–30.

Bowditch, C. and R. S. Everett. (1987). "Private Prison: Problems with Solutions." *Justice Quarterly* 4: 441–153.

Calabrese, W. H. (1992). "Low Cost, High Quality, Good Fit: Why Not Privatization?" Pp. 175–192 in *Privatizing Correctional Institutions*, edited by Gary W. Bowman, Simon Hakim, & Paul Seidenstat. New Brunswick, NJ: Transaction.

Carleton, M. T. (1971). *Politics and Punishment: The History of the Louisiana State Penal System*. Baton Rouge: Louisiana State University Press.

Clear, T.R. and G. F. Cole. (1994). *American Corrections* (3rd ed.). Belmont, CA: Wadsworth.

Crank, J. P. (1996). "The Construction of Meaning during Training for Probation and Parole." *Justice Quarterly* 13: 265–190.

DiIulio, J. J., Jr.)1988). "What's Wrong with Private Prisons." *The Public Interest* 92: 66–83.

DiMaggio, P. J. and W. W. Powell. (1991). "The Iron Cage Revisited: Institutional Isomorphism and Collective Rationality in Organizational Fields." Pp. 63–82 in *The New Institutionalism in Organizational Analysis*, edited by W. W. Powell and P. J. DiMaggio. Chicago: University of Chicago Press.

Durham, A. M., III. (1994). *Crisis and Reform: Current Issues in American Punishment*. Boston: Little, Brown.

Ericson, R. V., M. W. McMahon, and D. G. Evans. (1987). "Punishing for Profit: Reflections on the Revival of Privatization in Corrections." *Canadian Journal of Criminology* 29: 355–387.

Ethridge, P. A. and J. W. Marquart. (1993). "Private Prisons in Texas: The New Penology for Profit." *Justice Quarterly* 10: 29–48.

Feeley, M. M. (1991). "The Privatization of Prisons in Historical Perspective." *Criminal Justice Research Bulletin* 6(2): 1–10.

Garland, D. (1990). *Punishment in Modern Society*. New York: Oxford University Press.

Hagan, J., J. D. Hewitt, and D.F. Alwin. (1979). "Ceremonial Justice: Crime and Punishment in a

Loosely Coupled System." *Social Forces* 58: 506–527.

Hasenfeld, Y. and R. A. English, eds. (1974). *Human Service Organizations*. Ann Arbor: University of Michigan Press.

Hirsch, A. J. (1992). *The Rise of the Penitentiary*. New Haven: Yale University Press.

McAfee, W. M. and D. Shichor. (1990). "A Historical-Sociological Analysis of California's Private Prison Experience in the 1850s: Some Modern Implications." *Criminal Justice History* 11: 89–103.

Meyer, J. W. and B. Rowan. (1977). "Institutionalized Organizations: Formal Structure as Myth and Ceremony." *American Journal of Sociology* 83: 340–363.

Meyer, J. W. and W. R. Scott. (1983). *Organizational Environments: Ritual and Rationality*. Beverly Hills: Sage.

"More Missouri Inmates Withdrawn from Texas." (1996). *Kansas City Star*, August 29, p. A1.

Morris, N. and D. J. Rothman. (1995). *The Oxford History of the Prison*. New York: Oxford University Press.

Neilsen, E. (1986). *Improved Program Delivery: A Study Team Report to the Task Force on Programming Review*. Ottawa: Department of Justice.

Ogle, R. S. (1998). "Theoretical Perspectives on Correctional Structure, Evaluation, and Change." *Criminal Justice Policy Review* 9(1): 43–71.

Oliver, C. (1991). "Strategic Responses to Institutional Processes." *Academy of Management Review* 16: 145–179.

Pfeffer, J. and G. Salancik. (1978). *The External Control of Organizations*. New York: Harper & Row.

Press, A. (1990). "The Good, the Bad, and the Ugly: Private Prisons in the 1980s." Pp. 19–41 in *Private Prisons and the Public Interest*, edited by D.C. McDonald. New Brunswick, NJ: Rutgers University Press.

Ritti, R. R. and J. H. Silver. (1986). "Early Processes of Institutionalization: The Dramaturgy of Exchange in Interorganizational Relations." *Administrative Science Quarterly* 31: 25–42.

Robbins, I. P. (1986). "Privatization in Corrections: Defining the Issues." *Federal Probation* 50(3): 24–30.

———. (1989). "The Legal Dimensions of Private Incarceration." *American University Law Review* 38: 531–854.

Rosecrance, J. (1988). "Maintaining the Myth of Individual Justice: Probation Presentence Reports." *Justice Quarterly* 5: 235–256.

Rothman, D. J. (1971). *The Discovery of the Asylum: Social Order and Disorder in the New Republic*. Boston: Little, Brown.

Schiflett, K. L. and M. Zey. (1990). "Comparison of Characteristics of Private Product Producing Organizations and Public Service Organizations." *Sociological Quarterly* 31: 569–583.

Scott, W. R. (1992). *Organizations: Rational, Natural, and Open Systems* (2nd ed.). Englewood Cliffs, NJ: Prentice-Hall.

———. (1995). *Institutions and Organizations*. Thousand Oaks, CA; Sage.

Selznick, P. (1948). "Foundations of the Theory of Organization." *American Sociological Review* 13: 25–35.

———. (1957). *Leadership in Administration*. New York: Harper & Row.

Shichor, D. (1995). *Punishment for Profit: Private/Prisons Public Concerns*. Thousand Oaks, CA: Sage.

———. (1998). "Private Prisons in Perspective: Some Conceptual Issues." *Howard Journal of Criminal Justice* 37(l): 82–100.

Sullivan, J. (1996). "Talks Are Held on Reopening of Alien Unit: New Contractor Sought for a Detention Center." *New York Times*, January 19, p. B1.

Taylor, V., ed. (1997). "Tennessee Privatization Bill Stalls over Control Issue." *Corrections Journal* 2(9): 1–8.

Tomz, J. E. (1996). "Prison Privatization in the U.S." *Overcrowded Times* 7(2): 1–20.

Wamsley, G.L. and M. N. Zald. (1973). "The Political Economy of Public Organizations." *Public Administration Review* 33(1): 62–73.

DISCUSSION QUESTIONS

1. According to the literature, what are the moral and legal issues associated with the privatization of prisons?

2. According to the author, what are the benefits of privatization? What are the limitations?

3. How do correctional organizations develop and establish their legitimacy? How does privatization impact correctional legitimacy?

READING

This article by Edward Latessa is a speech given on his acceptance of the prestigious Vollmer Award in criminology and criminal justice. In his essay, he shares some observations about the challenge of change in corrections (framed in terms of alternatives to long-term imprisonment), at least as this challenge applies to implementing evidenced-based programs. He informs us of how difficult it is to make changes given the reluctance of institutions to do so. He emphasizes that, if institutions (or individuals) are to make a move to change, they must be given a reason to change, part of which involves presenting them with empirical evidence. Furthermore, this evidence must be presented in such a way that the layperson can understand it. Despite all the difficulties of presenting and implementing plans for change, Latessa has faith that, with the right evidence presented in understandable fashion, change for the better will occur.

The Challenge of Change

Correctional Programs and Evidence-Based Practices

Edward J. Latessa

Over the past ten years or so, much of my effort has been devoted to working with correctional agencies around the country. During this time, I have conducted hundreds of workshops and training sessions on evidence-based practices and programs for offenders. My current and former doctoral students and I have assessed over 360 correctional programs of all shapes and sizes, including those serving adults as well as juveniles, those in prisons, and those conducted in the community. During this time, I have seen some of the worst that corrections has to offer, as well as some of the best. I have seen programs that I would not refer anyone to and programs that have had a demonstrated effect on offender behavior and recidivism. I have also had the opportunity to work with many dedicated and committed people, and I am indebted to those who have not given up the quest for new knowledge and advancement in correctional research. Among those whom I have had the pleasure to work with and learn from include some of the giants of correctional rehabilitation: Paul Gendreau, Don Andrews, and Frank Cullen. There are many others, but space does not permit me to thank them all individually. I am forever in their debt. This brief prelude brings me to the purpose of this essay: to discuss the challenge of change in corrections.

Let me also say that my comments are going to be narrowly limited to a discussion of correctional programs and treatment efforts, which is not to say that the debate over the use of incarceration, sentencing practices, and other policy-related issues is not closely linked

SOURCE: Latessa, E. J. (2004). The challenge of change: Correctional programs and evidence-based practices. *Criminology & Public Policy, 3*(4), 547–560. © American Society of Criminology.

to rehabilitative efforts. They obviously are; it's just that my experience and research have been more focused on studying and, hopefully, improving correctional programs, rather than on more policy-related concerns. It's not that I have no interest in changing systems; it's just that I have set my sights much lower, and I am blissfully content if I can improve a small program. It is within this context that I want to present some observations I can make from venturing outside the ivory tower into the field of corrections. I will address several specific topics. First, I begin by discussing some lessons for correctional change: that change is difficult and that there is political context that needs to be addressed. In this section, I also discuss organizational readiness, the importance of leadership, and the challenge of overcoming the armchair quarterbacks that pervade the field. In the second section, I will speak to some of the reasons that researchers and academics share the blame and offer some suggestions for remedying this problem. Finally, I will end by offering some examples of states that have been trying to use evidence-based knowledge to improve correctional programs.

Lessons for Correctional Change

Change Is Hard Even for Us

First, we have to accept the premise that change is difficult. I often tell correctional staff who complain about the difficulty of changing offender behavior to try to change something about themselves. For those who have tried to lose weight, quit smoking, eat fewer sweets, or exercise more, you know it is not easy, and heaven knows many of us try. Why would we think that it would be easy for a correctional system or an agency to change, especially given the relative comfort that exists in maintaining the status quo? It is important to remember that corrections often operates under the modus operandi of "if nothing bad happened yesterday,

do the same thing today." As with people, I suspect that organizations often view change reluctantly, and ultimately ask, "why should we change?" Researchers, of course, like to point to the data and evidence as the logical reason to change; however, when you consider that 80% is the median rate for nonadherence with advised health care practices by health care professionals, why would we expect any different behavior from correctional professionals? Given the general lack of motivation that exists among people it should not be surprising that correctional organizations resist change.

Understanding the Politics of Change

Given this premise, it is important that those with an agenda for change understand the political context that exists at all levels of a correctional organization, be it large or small. Although there are exceptions, first and foremost, politicians and those they appoint, including correctional policy makers, are committed to survival, which usually translates into an aversion to what they perceive as risk taking. It is not that policy makers are necessarily opposed to evidence-based programs and practices—in fact I have found just the opposite to be true—but it is just that they have to be helped to understand the "up-side," if you will, as to why change can be beneficial. For example, when I ask policy makers to give me an estimate of the percentage of the public that they believe view the primary purpose of prisons to be punishment, they greatly overestimate the punitiveness of the public. As numerous studies have demonstrated, the public's support for effective programs and rehabilitation is still strong.[1] Helping them understand that the public is not monolithically punitive and that the research shows that a large percentage support rehabilitative efforts can be a powerful way to begin breaking down some of the resistance they may have to developing effective correctional programs and alternatives to incarceration. Of course, this is only one aspect of making change politically palatable. As public protection is often seen as a fundamental goal of correctional officials, another important step is to demonstrate to them that using research to improve correctional programs can actually increase public protection, whereas, conversely, using approaches that have not been found effective can have the opposite effect. I have found very few policy makers unwilling to at least listen to the empirical research when you frame it within the context of public protection.

Of course, getting policy makers to listen is only the first step. For decision makers, it may involve helping them understand what evidenced-based practices are, that developing more effective alternatives and programs based on research is not necessarily going to put them at odds with public opinion, and that they may in fact be more consistent with what the public wants than they know. Telling them that is not enough, however. It may also mean getting them to understand that effective treatment and incarceration are not always mutually exclusive. The fact is, we are always going to operate prisons and incarcerate a significant number of offenders, but that does not mean we should not be designing effective correctional rehabilitation programs. Similarly, the public will support community alternatives, but they also appear to want something "done" with offenders, and they want those interventions to be effective in reducing recidivism.

If meaningful change is going to occur, it is also important to communicate with correctional staff so that they can begin to understand some of the benefits for them. For example, if you are trying to implement evidenced-based programs in a prison, do you really think that telling the correctional officers that the program may help to reduce recidivism is going to convince them to lend support to your efforts? For these staff, the value of evidence-based programs is the potential reduction of critical incidents and safety issues. For the bean counters, the critical issue may be the cost savings that can accrue. The point is that the change process will require support at all levels of an organization, which will require an understanding of the concerns and issues that confront various staff. This point brings me to my next one, the need for organizational responsivity.

▧ Moving Beyond Providing Data: Organizational Responsivity

Why would we think that simply providing someone with facts and data would lead them to change their behavior? It certainly does not work with offenders (or most of us, for that matter), so it should come as no surprise to us that organizations are unlikely to change simply because we give them some information or research findings.

One of the characteristics of effective correctional treatment programs is that they assess an offender's readiness to change before they begin the actual program. Identifying barriers (and strengths) that a person may have should be part of the assessment process. This information can then be factored into a case plan so that the program can better match the offender to treatment or, as is more often the case, better prepare the offender for the program. The process for an organization should be similar: Assessing an organization's readiness to change and then developing a strategy or action plan can be an important step in the process. Preparing an organization for change can significantly increase the chances that the changes will be effectively implemented and supported.

▧ Leadership Is Important at All Levels

I have conducted workshops for correctional staff, and when I finish, they sometimes say, "you are talking to the wrong group, you need to talk to the policy makers, since they are the ones that make the decisions." Then when I address the policy makers, I am often told, "our staff need to hear this, since they are the ones who will be responsible for changes." Of course, this is not a problem for me, because I will talk with whomever will listen, but what I have learned is that if you are going to be successful in moving an organization forward, strong leadership for change needs to be in place at every level. For example, I do not find effective correctional programs in jurisdictions or organizations unless some high-level leader is willing to take charge and make things happen. Without a "hero" or two, it is very difficult for effective programs to emerge, let alone sustain over time. However, although vision and commitment from policy makers is necessary, it is not sufficient. Strong involved leaders and role models are also needed at the program level. Those persons who are responsible for the day-to-day operation of correctional programs need to be directly involved in designing, implementing, and operating their programs. In other words, they need to be "on the shop floor" so to speak. Whether it means cofacilitating groups, carrying a small caseload, or conducting some offender assessments, it is important for program directors to be involved in some aspect of service delivery. I remember one residential facility I assessed a number of years ago. When I asked the program administrator if he was involved in working with offenders, he replied that he always carried a caseload of the five highest risk offenders in the facility. Similarly, I have worked with a large urban probation department that talked for a long time about implementing evidence-based programs. They were having a tough time getting staff to change until the chief probation officer began cofacilitating cognitive behavioral groups and having his supervisors do the same. These leaders not only talked about supporting evidence-based programming, also they practiced it.

These examples illustrate good role modeling and the type of leadership that can have a significant impact on staff and the organizational culture. To further illustrate my point, I would like to use the example of the person who becomes a dean, and the first thing he/she does is quit teaching. The second thing he/she does is tell faculty how important teaching is, and how we have to do a better job. This person is not my idea of a good role model.

Everyone Is an Expert

One of the problems with crime is that everyone is an expert. I would dare say that, if I studied quantum physics, few people would offer their opinions about how I should go about my business, but because I study criminal behavior and corrections, everyone offers me advice. For example, once I was on a flight and was seated next to an older woman. She asked me what I did for a living, and I made the mistake of telling her I was a criminologist. For the next four hours she told me how to solve the crime problem. Now I just tell folks I am a proctologist and they leave me alone.

Although this story may be humorous, the problem of everyone believing they are an expert about crime affects correctional practice at many levels. From the politician to the case worker, everyone thinks they know how to deal with offenders and what we need to do to "straighten" them out. As an aside, I often ask correctional staff who work with offenders day in and day out what they think are the major risk factors associated with criminal conduct. They are often all over the map, and, needless to say, I am often amazed with the list they come up with. Interestingly, several years ago, my youngest daughter asked me to come to her fourth-grade class and talk about what I did for a living. When I asked the class to tell me why they thought some people got into trouble, they named many of the risk factors supported by the research: anti-social attitudes, hanging around with the wrong friends, personality traits, familial care and supervision, and lack of social support. As the old saying goes, "out of the mouth of babes." When you combine the "everyone is an expert" problem with the lack of credentialing (both individual and agency), little if any adequate staff training on the skills needed to change offender behavior, and an abundance of ignorance about what the research tells us, is it any wonder that we have so much correctional quackery being practiced?[2]

⬛ We Share Some of the Blame

As I have gone out into the field, I have also been struck by the failure of researchers and scholars to bridge the gap among theory, research, and practice. I am certainly not the first to voice concern about the relevancy of criminology or to suggest ways to increase it.[3] I do, however, believe that research has made a difference, at least in the area of correctional rehabilitation, and I will give some examples later. For now, let me offer some suggestions for how researchers and academics can do a better job of promoting change.

1. *Leave the Office.* It is almost a given that most practitioners and policy makers do not read the literature or published research (and this certainly helps explain why we have so many correctional programs based on half-baked theories). If we want our research to have an impact on the field, we need to recognize that it may be necessary to leave the office. We have to be willing to attend and present at nonacademic conferences, conduct workshops for local professionals, testify at legislative hearings, and in general be willing to lend our expertise and knowledge when asked to do so. Let me also add that I realize that it is a lot easier to sit in our offices and classrooms and pontificate to our students, but if we expect to have an impact, we need to be willing to get our hands a little dirty. It is not always easy to face a skeptical, if not sometimes hostile, crowd, and although it is unpleasant to be challenged and questioned, it may also be necessary if we are going to win over converts to evidence-based practices.

2. *Make research understandable.* I have gone to numerous practitioner-oriented conferences where researchers and academics spend all of their allotted time talking about the methodology or statistical techniques they used rather than focusing on their findings and its relevance. Talking about the log ratios and beta weights may be important to other researchers, but it will not do much to transfer knowledge to the field. I find that too often scholars write and talk in such a way as to make their work undecipherable by those who can most benefit from it. If we want to have an impact, it is incumbent on us to translate research and findings into understandable concepts and terms and then to present them in a way that helps practitioners understand the value of the research to what they do.

3. *Include measures of program integrity and quality in our research.* Too often, correctional researchers have focused on the offenders and their characteristics to explain the results from a study of a correctional program or intervention. Relatively few outcome studies include measures of program integrity or fidelity. We often lump together in outcome studies programs of various quality and implementation efficacy. Not only can the "failures" cancel out the "successes," but also, by failing to measure program characteristics and fidelity, we are often unable to explain some of the programmatic reasons why differences in outcome measures may occur. For example, in a recent study of similar types of residential correctional programs in Ohio, we found a wide range of effects based on the quality of the program.[4] Although most of the programs were found to have a positive effect on recidivism rates, a handful actually produced increased rates of recidivism. Fortunately, we included a number of measures of program integrity that helped explain these findings. Another example of the importance of program fidelity can be seen in the recent study by the Washington State Institute for Public Policy. These researchers found that several evidenced-based interventions had a significant

effect on recidivism, provided they were competently delivered. If not, they actually increased failure rates.[5]

4. *Do a better job of preparing our students.* The doctoral students who study with me are well versed in working with practitioners, and I hope that they have learned from it. However, I also recognize that they are not likely to be working in the trenches of corrections. It is the undergraduates that are produced that will occupy most of the positions in the field, and, to be blunt, we need to do a better job of preparing them to work in corrections. In my opinion, three important ingredients improve the preparation of undergraduate students to work in corrections: (1) teach them the knowledge base, (2) provide them better skills and competencies, and (3) expose them to other relevant disciplines. Let me touch briefly on these points.

Perhaps I can best illustrate the first point by giving an example. Most of us believe that cigarette smoking is harmful to our health. Why? Well, one might answer, "because of the research that has been conducted." Yet, I would guess that if only one or two studies were conducted on the health risks of tobacco, we might not be so sure. The reason most of us believe smoking is harmful to our health is that research has been conducted for decades all over the world by independent researchers who have concluded that, if you smoke, it can lead to cancer, heart disease, emphysema, and other health problems. In other words, a body of knowledge exists about the effects of smoking. I would argue that a significant body of knowledge exists in corrections as well. Research on risk factors and correctional treatment has been ongoing for decades, and numerous scholars have reviewed and summarized this research. This is not to say that more research is not needed, or that we have all of the answers (we clearly do not), but we do have a considerable amount of knowledge about criminal

conduct and effective (and ineffective) ways to rehabilitate offenders. What we need to do is a better job of teaching our students what we do know and its relevance to the field they have chosen to study, and perhaps someday work in.

The second and third points are related. That is, those students who are going to work in corrections with offenders need to have some core skills, many of which may have to come from related fields, such as psychology, social work, counseling, addictions, and other helping professions. Although much of the responsibility for specific training lies with the agencies that hire our students, I do believe that there are some areas in which we can better prepare students, such as giving them a working knowledge about the most effective ways to effect change and shape behavior, an understanding of risk and need factors related to criminal conduct, and an understanding of other competencies that will be important in a correctional setting.

Change Is Possible: Examples From the States

Now that I have gotten that off my chest, I would like to devote the remainder of this essay to providing some examples of states that are attempting to use correctional research to implement change and improve programs for offenders.

Oklahoma

I first visited Oklahoma in 1997. At the time, I was part of a team funded by the National Institute of Corrections that was conducting workshops on evidenced-based practices in corrections. Although it was not uncommon for a state or jurisdiction to seek subsequent technical assistance, Oklahoma's Department of Corrections wanted to begin to assess the programs it offered offenders, including those operated by the state as well as those under contract with private providers. In

1999, a team of researchers from the University of Cincinnati began reviewing programs throughout the State of Oklahoma. Initially, a total of 29 programs were selected for assessment, including those operated in prisons as well as community based. We used the Correctional Program Assessment Inventory (CPAI) as the evaluative instrument.[6] The CPAI assesses program integrity and the degree to which a correctional program meets the principle of effective intervention.[7] During our initial review, only 9% of the programs scored "satisfactory." The remaining 91% scored "needs improvement" or "unsatisfactory." Although some states might have sent us packing at this point, Oklahoma officials made a decision to not only continue the process but also institutionalize it as part of their efforts to improve the programs and services they offered offenders. Those programs that were not satisfactory were required to develop action plans on how they were going to address deficiencies and improve their programming. Programs were given specified time periods to correct deficiencies, and reassessments were scheduled. Subsequent program assessments indicated that Oklahoma was able to dramatically improve the quality of its correctional programs, at least as measured through the CPAI. The most recent results indicate that 79% of the programs in Oklahoma are now rated as "satisfactory" or higher, and none of the programs are "unsatisfactory."

Change certainly was not always easy for many of these programs, but there are several reasons it occurred. First, strong committed leadership was in Oklahoma. Second, specific and clear direction was provided, based on research, and, third, training and technical assistance was provided programs throughout the process. Granted, Oklahoma still has a very high incarceration rate, and its sentences are among the longest in the country, but at least it now offers offenders correctional programs that are based on evidence and the principles of effective intervention.

Oregon

The second example I would like to highlight is the State of Oregon. Over the past few years, the Oregon Department of Corrections under the leadership of Dr. Ben deHaan sought to develop and implement research-driven practices and programs. In the summer of 2003, I was asked to testify to a joint Oregon Judiciary Committee hearing on the use of evidenced-based programs for offenders. The Oregon legislature subsequently passed SB 267, which requires prevention, treatment, or intervention programs that are intended to reduce future criminal behavior in adults and juveniles or to reduce the need for emergency mental health services to be evidence-based. Furthermore, by 2005, 25% of funds spent by the Oregon Department of Corrections, Youth Authority, the Department of Human Services, the Criminal Justice Commission, and the Commission on Children and Families have to be allocated to evidenced-based programs. By 2007, the amount increases to 50%, and by 2009, 75%. While this state is the first state I know of to statutorily require evidenced-based programs for offenders, I suspect it will not be the last, especially as states continue to wrestle with budget deficits.

Ohio

In recent years, Ohio, like many states, has been experiencing significant budget shortfalls, and most state agencies, including corrections, have been hard hit. Ohio has also made a significant investment in residential programming for offenders, spending over $89 million this fiscal year for halfway houses and community-based correctional facilities.[8] As the budget for these programs has grown, so has the demand by the legislature to justify these expenditures by determining the effectiveness of these programs in reducing recidivism. As fate may have it, I happen to be in the research business, and I received the contract to conduct a study of all of the residential correctional programs funded by the state. In the summer of 2002, we completed the largest study ever conducted of residential correctional programs. Over 13,000 offenders were included in our study of 38 halfway houses and 15 CBCFs. Results from the study showed that treatment effects were strongest for higher risk offenders and that, for all but a handful of programs, the recidivism rates for low-risk offenders actually increased as a result of the programming (Lowenkamp and Latessa, 2002). Although the findings from our study mirrored what one would expect given the "what works" research that exists, the point of my example is to show how research can be used to change correctional practice. As a result of this study, Ohio has enacted a number of policy changes. These include:

- All programs must administer an assessment tool within five days of intake to measure risk level, determine case planning strategies, and identify special needs, such as mental health and sex offender.
- All programs need to develop a service delivery model based on individualized risk and needs assessment results. The high-risk offender should receive more intensive and additional services; conversely, the low-risk offender will receive minimal services.
- A cognitive behavioral modality should be adopted, or minimally cognitive programming skills should be implemented within other modalities.
- Criminogenic targets should be addressed in programming.
- Audit standards shall assess both processes and program outcomes. Standards will be based on a performance-based model wherever possible.
- Program evaluations will be conducted every three years.
- Programs shall conduct a Correctional Program Assessment Inventory, or similar instrument, every three years to ensure program fidelity.

Although some critics may see these policies as basic, they constitute major changes in the operation and monitoring of community-based correctional programs in Ohio, all brought about through research.

Other Examples

Many other examples exist of jurisdictions that are paying attention to the research on evidence-based practices in correctional treatment. These examples include Washington, which through the work of Steve Aos and his colleagues at the Washington State Institute for Public Policy have promoted and studied the effects of these efforts. Maine and Illinois were recently awarded demonstration project grants by the National Institute of Corrections to implement and promote evidence-based practices throughout these states. Likewise, the Florida Division of Juvenile Justice has recently begun an extensive effort to promote evidence-based programming throughout the system. Other states that have been paying attention to the research and have been attempting to implement evidence-based programs include Idaho, Iowa, Wisconsin, Indiana, Pennsylvania, Utah, Colorado, and Minnesota. Of course I would be remiss not to mention the Correctional Services of Canada, which has made evidenced-based programming the hallmark of its correctional system. Undoubtedly, other examples exist.

▧ Conclusions

In the above essay, I have shared some observations that I can make about the challenge of change in corrections, at least as it applies to implementing evidenced-based programs. These can be briefly summarized as follows:

- Change is difficult, and many of the reasons that organizations resist change are the same reasons we as people do.

◆ Giving reasons to change is part of the process, but sometimes it takes some planning and persistence to remove barriers, as well as someone to motivate.

◆ Having strong leadership from top to bottom is a necessary ingredient for the change process to occur and be sustained over time.

◆ Everyone may think he or she is an expert about corrections and criminal behavior, but few are. Rely on the empirical evidence rather than your neighbor.

◆ Researchers and academics who are interested in having their work used by corrections need to leave the office, make research findings more understandable, and expand the measures we use to evaluate corrections to include program characteristics. We also need to do a better job of teaching our students the body of knowledge in corrections.

◆ Despite the challenges, change is occurring, and several examples were offered, including Oklahoma, which has improved correctional programs through assessment and evaluation; Oregon, which has legislated evidenced-based programs throughout the correctional system; and Ohio, which has used research to change policy and auditing processes for programs.

I end with a leap of faith. I am not naive enough to believe that there is a magic wand that can be waved to dramatically change the policies and practices that have lead to over 2 million people being incarcerated in America, or that we can somehow have a dramatic effect on policies that are embedded in politics, tradition, custom, and imitation. I do, however, believe that we can make a difference, and that change, although difficult, is possible.

✎ Notes

1. Cullen has written extensively about public support for correctional rehabilitation. For an excellent summary of this research, see Cullen et al. (2002).

2. For a better explanation of correctional quackery and some of the theories that are used in treating offenders see Latessa et al. (2002).

3. For a discussion of these issues, see Petersilia (2000) and Austin (2003).

4. See Lowenkamp and Latessa (2002).

5. See Barnoski (2004).

6. The CPAI was developed by Paul Gendreau and Don Andrews to assess the integrity of correctional programs. For a discussion of the instrument, see Latessa and Holsinger (1998).

7. Several studies have found significant correlations between scores on the CPAI and recidivism rates. See Holsinger (1999) and Lowenkamp (2004).

8. Figures provided by the Ohio Department of Rehabilitation and Correction.

✎ References

Austin, James. 2003. Why criminology is irrelevant. *Criminology and Public Policy* 2:557–563.

Barnoski, Robert. 2004. *Outcome Evaluation of Washington State's Research-Based Programs for Juvenile Offenders.* Olympia, WA: Washington State Institute for Public Policy.

Cullen, Francis T., Jennifer A. Pealer, Bonnie S. Fisher, Brandon K. Applegate, and Shannon A. Santana. 2002. Public support for correctional rehabilitation: Change or consistency. In Julian V. Roberts and Michael Hough (eds.), *Changing Attitudes to Punishment: Public Opinion, Crime and Justice* (pp. 128–147). Devon, UK: Willan Publishing.

Holsinger, Alexander. 1999. *Opening the "Black Box": Assessing the Relationship Between Program Integrity and Recidivism.* Doctoral Dissertation, University of Cincinnati, Ohio.

Latessa, Edward J. and Alexander Holsinger. 1998. The importance of evaluating correctional programs: Assessing outcome and quality. *Correctional Management Quarterly* 2: 22–29.

Latessa, Edward J., Francis T. Cullen, and Paul Gendreau. 2002. Beyond correctional quackery—professionalism and the possibility of effective treatment. *Federal Probation* (September): 43–49.

Lowenkamp, Christopher. 2004. *Correctional Program Integrity and Treatment Effectiveness: A Multi-site, Program-level Analysis.* Doctoral Dissertation, University of Cincinnati, Ohio.

Lowenkamp, Christopher and Edward J. Latessa. 2002. *Halfway House and CBCF Findings: Program Characteristics.* Available online: http://www.uc.edu/criminaljustice.

Lowenkamp, Christopher and Edward J. Latessa. 2002. *Evaluation of Ohio's Community Based Correctional Facilities and Halfway House Programs.* Available online: http://www.uc.edu/criminal justice.

Petersilia, Joan. 2000. Policy relevance and the future of criminology. In Barry W. Hancock and Paul M. Sharp (eds.), *Public Policy, Crime, and Criminal Justice* (pp. 383–395). Upper Saddle River, NJ: Prentice Hall.

DISCUSSION QUESTIONS

1. Discuss the evidence that indicates our correctional practices do not fit the amount of crime in this country. Note how we compare with other countries in terms of the use of incarceration.

2. What problems do "get tough" policies create for correctional operation? What benefits, if any, do they provide?

3. In your opinion, what current initiatives in corrections offer the most promise for the future? Support this opinion with research and from the readings provided in this section or the rest of the text.

READING

Catherine Lemieux, Timothy Dyeson, and Brandi Castiglione's article critically reviews the scholarly literature on older U.S. prison inmates and examines correctional responses to this sub-population of incarcerated offenders. Like many other such articles, this review shows that the number of incarcerated men and women who are older is increasing, as can be expected given higher life expectancies and better health. The article finds that older offenders are primarily incarcerated for violent offenses, and many report one or more chronic health conditions. Health care management is the most prevalent theme in the current literature on older prison inmates, and there is tremendous variation in how jurisdictions accommodate older inmates. This review of the literature reveals that there are major gaps in the scientific understanding of older inmates, and concludes with recommendations for policy- and practice-relevant research.

Revisiting the Literature on Prisoners Who Are Older

Are We Wiser?

Catherine M. Lemieux, Timothy B. Dyeson, and Brandi Castiglione

The number of older inmates in U.S. facilities has increased by more than 50% since 1996 (American Correctional Association, 1999).

This unprecedented growth has mobilized correctional administrators and other criminal justice professionals to examine policies that balance a number of equally compelling issues: public safety, economic costs, institutional management, and humanitarian concerns (W. E. Adams, 1995; Faiver, 1998; Hunsberger, 2000; Wall, 1998). The literature on older prisoners was last reviewed by Rubenstein in 1984, who at that time described the research as "sparse and sporadic" (p. 153) and "fraught with methodological limitations" (p. 164). The purpose of this article, therefore, is to take stock of the scholarly literature published since the late 1970s that describes older inmates and examines the biopsychosocial characteristics that are unique to this subpopulation of incarcerated offenders.

Few scholars expressed any intellectual interest in older prisoners prior to the 1970s. Instead, the earliest literature explored theoretical paradigms about criminal behavior among older persons. In the 1970s and 1980s, researchers were concerned about the prevalence and types of crimes committed by older persons. As the population of older prisoners steadily increased during this time period, scholars, policy makers, and the media focused on issues such as living conditions, institutional adjustment, recreation and socialization needs, and health care and rehabilitation. The most recent literature on older inmates is preoccupied with their health care needs and with the costs of providing specialized care.

The goal of this review is to provide a comprehensive and critical review of what is known, to facilitate additional empirical analysis. We accomplish this by providing an overview and profile of older inmates (e.g., incidence, demographics, types) and by describing their biopsychosocial characteristics (e.g., adjustment to prison, mental and physical health status). We also examine the extent to which policy and practice recommendations are supported by data. This review

concludes with recommendations for further research.

⊠ Methods and Results of Search

To determine the type and quantity of available literature, we conducted searches of Internet resources and relevant electronic databases for the years 1977 through 2001, using the parameters "older inmates" and "older offenders." Our search uncovered numerous publications about older offenders in general (i.e., theories about deviance, crime patterns, and arrest rates). However, the current review is restricted to 49 scholarly publications written since 1977 that specifically focus on older U.S. prisoners, 24 of which are empirically based descriptive investigations (see Table 1). We also included criminal justice population data sources that are routinely disseminated by national organizations. Two nationwide studies (Aday, 1994b; Goetting, 1984b) surveyed state prison systems about existing facilities and programs for older inmates. Twenty-two empirical investigations examined inmate characteristics: of these, 2 were at the national level and 20 were at the state or local level. The research methodology used for these latter 22 studies included surveys ($n = 10$, 45.4%), secondary analysis ($n = 6$, 27.2%), or a combination of both ($n = 3$, 13.7%). Three (13.7%) used a case study approach (Aday, 1994a; Golden, 1984; Reed & Glamser, 1979). Among the 15 state-level studies that incorporated secondary analysis or survey research methods, the sample size of older inmates ranged from 40 to 469, with a median sample size of 92 inmates. As seen in Table 1, general descriptions of older inmates compose the largest proportion of subject matter for both types of publications.

SOURCE: Lemieux, C. M., Dyeson, T. B., & Castiglione, B. (2002). Revisiting the literature on prisoners who are older: Are we wiser? *The Prison Journal, 82*(4), 440–458. Reprinted with permission of Sage Publications, Inc.

Table 1	Primary Subject Focus of Scholarly Publications ($N = 49$) on Older U.S. Inmates	
Subject	**n**	**%**
Empirically based publications ($n = 24$)		
Overview/descriptive profile	10	41.7
Institutional adjustment/behavior	7	29.2
Health-related issues	4	16.7
Program descriptions	2	8.3
Program evaluation	1	4.1
Nonempirical and impressionistic publications ($n = 25$)		
Overview/descriptive profile	10	40.0
Health-related issues	8	32.0
Program descriptions	3	12.0
Housing issues	2	8.0
Correctional staff training issues	2	8.0

NOTE: Scholarly publications include published books and book chapters and articles published in refereed journals since 1977.

⊠ Overview and Profile of Older Inmates

Incidence

A major and persistent shortcoming is the lack of agreement among scholars on an age criterion by which older prisoners are designated (W. E. Adams, 1995; Aday, 1994b). Among the empirical reports listed in Table 1, for example, 7 used an age criterion of 55 and older, 9 used a criterion of 50, and 1 used 60. Thus, the estimated proportion of older inmates reported in the literature varies with both the age criterion and the sources of data that were used (e.g., Goetting, 1985; Rubenstein, 1984).

Using an age criterion of 55, data indicate that older inmates currently compose about 4% of the total inmate population. In 1999, the American Correctional Association reported that 42,930 inmates over the age of 55 were incarcerated in state ($n = 35,686$) and federal ($n = 7,244$) facilities. This represents 3.7% of

the total adult inmate population ($N = 1,168,713$). Of the 37 states that provided 1998 data, 20 account for almost three fourths ($n = 1,662, 73.8\%$) of the inmate population over 55 years of age. As seen in Table 2, eight states reported more than 1,000 older inmates each and account for more than half of the total older inmate population. The greatest number of older inmates was reported by Texas, followed by California, and Florida (see Table 2).

Older Inmate Demographics

Available, albeit dated, evidence suggests that older inmates tend to be unmarried White men who were employed prior to incarceration but who never graduated from high school. The most representative demographic data are available from two nationwide studies. Goetting (1984a) surveyed 11,397 randomly selected state prisoners, 248 of whom were 55 years of age or older. Falter (1999b) conducted a secondary analysis of Federal Bureau of Prisons medical data for 1,051 randomly

Table 2	Frequency and Proportion of Older Inmates in State Prison Systems With More Than 1,000 Older Inmates			
State	**n**	**Cumulative n**	**%**	**Cumulative %**
Texas	8,436	8,436	19.7	19.7
California	4,385	12,821	10.2	29.9
Florida	2,396	15,217	5.6	35.5
New York	1,963	17,180	4.6	40.1
Ohio	1,871	19,051	4.4	44.5
Pennsylvania	1,545	20,596	3.6	48.1
Georgia	1,298	21,894	3.0	51.1
Arizona	1,015	22,909	2.4	53.5

NOTE: An older inmate is 55 years of age or older. Reported percentages refer to proportions of the total U.S. older inmate population ($N = 42,930$).

selected inmates over the age of 50. Older inmates in Goetting's (1984a) and Falter's (1999b) studies were predominantly male (at 96% and 93%, respectively). In terms of race, approximately half of the inmates were Black (49%) and half were White (48%) in Goetting's (1984a) study. Falter (1999b) reported a far greater proportion of White (78%) than Black (20%) older inmates in the federal system. Descriptive statewide studies (e.g., Aday & Webster, 1979; Fry, 1987; Kratcoski & Babb, 1990) also report a greater proportion of White than Black older inmates.

Approximately one third (34%) of Goetting's (1984a) respondents were married. They reported two children, on average. Falter (1999b) did not gather data on these latter variables. A far greater proportion of unmarried than married participants were reported in state-level studies (e.g., Colsher, Wallace, Loeffelholz, & Sales, 1992; Fry, 1987; Moore, 1989). In Goetting's (1984a) study, approximately three fourths (74%) were employed prior to incarceration, and the mean completed education level was seventh grade. Wilson and Vito (1986) also found that more than two thirds of the inmates in their sample (from one Kentucky institution) reported less than a high school education. Conversely, there was a greater proportion of high school graduates in studies conducted by Fry (1987) (at one California facility) and by Kratcoski and Babb (1990). This latter study included older inmates from 8 Federal Bureau of Prisons, 1 Pennsylvania, 3 Florida, and 3 Ohio facilities.

Older Inmate Types

Older male inmates appear to compose different categories of offender types. Goetting (1984a) delineated four inmate types from her representative survey of 248 inmates: career criminals (prison recidivists), old offenders (inmates whose first incarceration occurred at age 55 or older), first offenders (who were first incarcerated before the age of 55), and old-timers (inmates who were growing old in prison). Aday and Webster (1979) distinguished between first and chronic older offenders among 95 inmates from Arkansas and Oklahoma facilities. Fry (1987) also distinguished between first and chronic older offenders in a profile of 62 California inmates, and he identified four patterns among first offenders: those who are violent, white-collar, or drug or alcohol involved. Fry (1987) characterized chronic offenders either as drug or alcohol abusers or as property offenders.

Offenses Committed by Incarcerated Older Offenders

Since the 1970s, criminologists have expressed considerable concern about escalating crime rates among older persons as this segment of the population continues to grow (for a review, see Forsyth & Gramling, 1988). Gewerth (1988) found that older persons are most frequently arrested for fraud, embezzlement, larceny-theft, alcohol-related offenses, assault, and sexual offenses. Shichor and Kobrin (1978) examined changes in Uniform Crime Reports arrest data for older offenders (age 55 or older) from 1964 through 1974. These authors found that older persons were most likely to be arrested either for aggravated assault (among violent crimes) or for larceny-theft (among property crimes). Sapp (1989) replicated this study using Uniform Crime Reports data from 1972 through 1981, and he obtained similar findings. These data, however, provide no information about convictions or sentencing decisions.

There is some evidence to suggest that criminal justice agencies exercise lenience with older offenders (B. R. McCarthy & Langworthy, 1987; Shichor & Kobrin, 1978). Champion (1988) examined sentencing decisions ($N = 2,365$) from randomly selected federal judges, and he found a negative correlation between sentencing severity and the offender's age. Although violent crimes were more severely punished compared with property offenses across all ages, Champion (1988) found that probation was used in lieu of incarceration more than 4 times as much for older offenders than for their younger counterparts. B. R. McCarthy and Langworthy (1987) similarly found that older offenders were more likely to be under community supervision for violent crimes than younger offenders. These authors' data suggested that older nonviolent offenders are more rigorously screened out by the criminal justice system than are their younger counterparts.

Male offenders who commit violent crimes such as aggravated assault, sex offenses, and homicide are more prevalent than nonviolent offenders in samples of incarcerated older inmates. Aday's (1994a) case study data suggested that first-time older offenders had committed violent crimes against family members. Kratcoski and Walker's (1988) research indicated that homicides committed by older persons are most likely to be perpetrated by male offenders against their spouses in the home. However, Kratcoski and Walker did not specify whether their sample was drawn from incarcerated offenders. More than two thirds (70%) of the 248 older prisoners in Goetting's (1984a) nationwide study were serving sentences for violent offenses, whereas approximately one fourth (26%) were incarcerated for property offenses.

Unlike the state prisons, Kratcoski and Pownall (1989) found that only 13% ($n = 707$) of Federal Bureau of Prisons inmates over the age of 50 ($N = 5,522$) were incarcerated for violent offenses. One third (32%) of Goetting's (1984a) respondents reported a prior probation experience, and half (51%) experienced at least one incarceration. Fewer than half of Federal Bureau of Prisons inmates ($n = 2,494$, 45%) reported a prior commitment (Kratcoski & Pownall, 1989). Older inmates with criminal justice histories were also prevalent in state-level descriptive studies (Faiver, 1997; Teller & Howell, 1981; Walsh, 1989; Wilson & Vito, 1986).

There is no available research that specifically investigates whether older offenders are more likely to commit violent acts or if older offenders are more likely to be incarcerated if they are violent. Moreover, systematic investigation is hampered by the numerous limitations of Uniform Crime Reports data (Sapp, 1989), as well as the broad discretion granted to law enforcement and criminal justice agents with whom older offenders interact.

⊠ Biopsychosocial Characteristics of Older Inmates

Adjustment to Prison

The literature suggests that numerous interacting individual and institutional variables contribute to an older inmate's adjustment to incarceration. Older male inmates have been portrayed both as loners (Kratcoski & Babb, 1990; Vega & Silverman, 1986) and as being involved in friendship networks (Sabath & Cowles, 1988). In one of the few empirical studies that included women, Kratcoski and Babb (1990) found that older women lacked a support system from both inside and outside of the institution, and they experienced a greater sense of isolation than men. Wilson and Vito (1986) found that all of the older male inmates in their study ($N = 87$) reported difficulties with loneliness and with missing their families.

Sabath and Cowles (1988) found that being involved in work and social interactions had positive effects on adjustment to prison life. Rubenstein (1984) and Goetting (1983) concluded from their respective literature reviews that inmate adjustment is related to various individual (e.g., educational level, health status, and so forth) and institutional characteristics (e.g., security level, degree of mainstreaming). These authors criticized earlier researchers (viz., Reed & Glamser, 1979; Teller & Howell, 1981) for failing to properly account for these variations.

Mental Health Issues

Although Gewerth (1988) cited evidence that the offenses of older persons can often be attributed to age-related disturbances (e.g., organic brain syndrome), the actual prevalence of mental illness among older inmates cannot be ascertained from the literature. The limited research to date suggests that some older inmates experience symptoms of depression and anxiety. Colsher et al. (1992) found little evidence of either cognitive impairment or psychotic symptoms among older male inmates in their sample ($N = 119$). However, some participants reported symptoms of depression, loneliness, and anxiety (at 15.4%, 7.1%, and 8.1%, respectively). Aday (1994a) also found evidence of depression, guilt, and psychological stress among 25 older male inmates who were first-time offenders.

Correctional health professionals express conflicting opinions based on their personal experiences and impressions. Kelsey (1986) believed that, except for the long-term deterioration that occurs with aging, older inmates demonstrate few mental health problems. Conversely, Booth (1989) and Dugger (1988) stated that older inmates are at high risk for developing depression, especially those who are experiencing ill health and age-related changes and losses. Chaiklin (1998) speculated that older inmates with mental illness are overlooked because of limited resources, cramped environments, and improperly trained staff. Chaiklin further attributed low detection rates of mental illness to staff suspicions about malingering.

Physical Health Status of Older Inmates

The health status of older inmates has been measured through either self-report or secondary analysis of prison medical records. In a nationwide study of Federal Bureau of Prisons health care data, Falter (1999b) examined the relationship between medical encounters and the prevalence of five chronic health care conditions (viz., noninsulin-dependent diabetes, insulin-dependent diabetes, arteriosclerotic heart disease, chronic obstructive pulmonary disease, hypertension). The author found that hypertension and heart disease explained almost 25% of the variance in total number of medical encounters. Few state-level, health-related surveys have been conducted with older inmates. Colsher et al. (1992) found that 65% of the older male inmates in their health survey of Iowa state prisoners ($N = 119$) self-reported good health; however, almost half believed that their health had worsened since incarceration. The most commonly reported chronic illnesses were arthritis (45%), hypertension (40%), venereal disease (22%), ulcers (21%), prostate problems (20%), myocardial infarction (19%), and emphysema (19%). Incontinence, sensory and flexibility impairment, and limitations in gross physical functioning were also prevalent (Colsher et al., 1992). Most inmates smoked cigarettes and had a history of alcohol consumption. Wilson and Vito (1986) similarly observed high rates of alcohol abuse among older inmates. Colsher et al. (1992) recommended modifying prison environments and implementing specialized health promotion and prevention programs to meet the needs of older inmates.

Moore (1989) examined the characteristics of 41 older male inmates in Michigan and found that 83% had at least one chronic health problem, and 49% had three or more. Inmates most frequently self-reported cardiac, vision, respiratory, and gastrointestinal health problems (at 25.9%, 17.0%, 14.8%, and 9.0%, respectively). In one of the few studies that included women, Kratcoski and Babb (1990) found that twice as many older women than men reported heart, respiratory, and degenerative illnesses. Thus, the findings of Colsher et al. (1992), Kratcoski and Babb (1990), and Moore (1989) parallel those of Falter (1999b).

Moore (1989) examined the impact of prison environments on the use of health care, and he found that male inmates who were segregated made more demands for medical care than those who were not. In fact, older inmates who were relocated to a segregated unit demonstrated a 100% increase in medical demands. Wilson and Vito (1986) surveyed both correctional staff and inmates to assess medical needs among 87 older male prisoners who were housed in the general population ($n = 62$) and in a segregated geriatrics unit ($n = 25$) of a Kentucky institution. Many inmates felt that the prison environment exacerbated their respiratory and arthritis problems. Other medical complaints were related to their beliefs that services and medications for serious illnesses were not available to them in prison. These concerns, as well as the fear of dying in prison, were more frequently expressed among inmates in the geriatrics unit (Wilson & Vito, 1986). Staff believed that inmates' medical problems were aggravated by the stress of incarceration and prison conditions. Staff also believed that older inmates were preoccupied with medical concerns because of high levels of lethargy and inactivity, especially among those in the geriatrics unit (Wilson & Vito, 1986).

In sum, empirical evidence suggests that older inmates are more likely to demand health care services than are their younger counterparts. The prison environment, housing considerations, and the availability of specialized programming also influence the extent to which older inmates consume services. Very little is known about the health status of females and non-White older inmates because these populations are underrepresented in the research. The available evidence suggests, however, that considerably more older women than men report debilitating chronic and degenerative illnesses.

Social Status

The empirical literature provides a conflicting picture of the social status of older inmates within the prison hierarchy. Goetting (1983, 1985) and Rubenstein (1984) cited evidence that younger inmates respect older inmates for their knowledge of prison life, which enables older inmates to establish behavioral norms and occupy leadership roles. According to this perspective, incarcerated older inmates are accorded prestige and respect by their peers. Conversely, other researchers (e.g., Kratcoski & Babb, 1990; Wilson & Vito, 1986) found that older inmates are vulnerable to and fearful of predatory younger inmates.

Security Issues

Available evidence suggests that older male prisoners are less likely to commit disciplinary infractions than are their younger counterparts. Wiegand and Burger (1979) described older inmates as loners who "often live quietly, not bothering anybody and nobody bothering with them" (p. 51). Goetting (1984a) found that older prisoners were substantially less likely than their younger counterparts to be guilty of rule breaking within the prison. Wilson and Vito (1986) also found that older inmates violated prison rules less frequently than younger inmates.

McShane and Williams (1990) compared disciplined with nondisciplined older inmates, and these authors identified a subtype of older inmate whose most common offense was refusing to carry out an order. Disciplined older inmates had longer prison sentences and a history of incarceration. They were relatively younger and healthier and had fewer outside contacts than the nondisciplined group (McShane & Williams, 1990). This latter finding is consistent with Sabath and Cowles's (1988) research indicating that family contact has a positive impact on the morale and conduct of older inmates. Although insubordination was the most common infraction for disciplined inmates, McShane and Williams (1990) did not raise the issue of whether staff training might impact these behaviors. Kratcoski and Babb (1990) found that older inmates appear to be violating rules when, in fact, they may have neither heard nor understood them. Gewirtz (1984), Goetting (1983, 1984a), and Morton (1993, 1994) emphasized the importance of training correctional staff to properly understand and effectively respond to the behaviors and heightened needs exhibited by older inmates.

⊠ Correctional Responses to Older Inmates

Formal Provisions for Older Inmates

Available data suggest that there is tremendous variability in the extent to which correctional systems supply older and other special-needs inmates with formal provisions (i.e., housing accommodations and specialized educational, vocational, medical, and recreational programs). We found numerous policy recommendations (e.g., Aday, 1994b; Duffee, 1984; Goetting, 1985; Vito & Wilson, 1985) for specialized programming, and several program descriptions (e.g., Aday, 1977; Anderson, 1994; Hunsberger, 2000; Wilson & Vito, 1986); however, we were unable to locate one controlled outcome-oriented evaluation of any type of corrections-based geriatric program.

According to Kratcoski and Pownall (1989), the Federal Bureau of Prisons has designated specific institutions to provide health care to inmates who are chronically ill or handicapped. These facilities either house Comprehensive Medical Units or they provide access to specialized medical care in the community. Thus, attention is given to health status along with security needs and regional considerations when determining how older offenders are distributed throughout Federal Bureau of Prisons institutions.

Geriatric Programs and Services

One uncontrolled evaluation of a therapeutic program for older inmates was located in the literature (viz., Aday, 1977). The Geriatrics Unit Program offered specialized activities (e.g., visitation, toy repair, arts, and crafts) to 55 male inmates over the age of 55. Aday (1977) found that measures of life satisfaction among inmate participants were comparable to those of older respondents in the community. Specialized programs for older prisoners should prevent deterioration and should promote alertness and physical activity

(Rosefield, 1993; Vito & Wilson, 1985). Recommendations for such programs include the following: (a) appropriate work assignments (Goetting, 1984b); (b) arts and crafts (Aday, 1977); (c) wellness, exercise, and passive recreational activities (Aday, 1994b; Falter, 1999a); (d) leisure-time activities (Rosefield, 1993); and (e) opportunities for visitation and socializing (Neeley, Addison, & Craig-Moreland, 1997). Several authors (viz., Moore, 1989; Wilson & Vito, 1986) believe that gardening is of particular interest and utility to older inmates.

✉ Summary

The incarceration of older inmates has been deemed problematic for social justice and humanitarian reasons, because of practical or logistical issues, and, most recently, because of the costs associated with caring for inmates with numerous health-related conditions. Thus, the problem of older inmates has proven to be a slippery one. Despite a limited amount of current research on older inmates, several findings of this review merit highlighting:

1. The proportion of state and federal inmates 55 years of age and older is steadily increasing. The number of inmates older than 75 will continue to increase in the future if current sentencing practices remain in place. Proportional increases appear more dramatic than numerical increases for this latter subgroup of older offenders and for females in particular.

2. The most reliable and representative descriptive data on older inmates were last gathered in 1979 (Goetting, 1984a). At that time, older state inmates were most likely to be unmarried White men with children who reported less than a 12th-grade education.

3. The population of older inmates is likely composed of heterogeneous subtypes, with the needs of first-time offenders being different from those of more chronic offenders and old-timers.

4. Older offenders are most likely to be incarcerated because of violent crimes such as aggravated assault, homicide, and sex offenses. These crimes of violence are often perpetrated against family members in the home.

5. Older inmates are likely to report one or more chronic health conditions. Some may also report symptoms of depression and anxiety. Gender differences exist with respect to older inmates' health and mental health status, with greater reported health concerns among older women. A history of alcohol use appears prevalent among older male prisoners.

6. Most states and the Federal Bureau of Prisons have implemented limited provisions to accommodate older inmates and those with special needs.

Our review indicates that scholarly concern about the needs and experiences of older inmates has dissipated over the past three decades. There is widespread support for specialized programming, staff training, physical environment improvements, health promotion programs, and alternative sentencing. These recommendations are based not on reliable and representative data that detail the health condition of older inmates but on their presumed deteriorated health status.

As of 1990, most states did not have written policies that specifically addressed older inmates, one third offered no special provisions, and another third had no plans to implement specialized programming (Aday, 1994b).

American Correctional Association survey data from 1998 indicate that, among eight states with the greatest number of older inmates (see Table 1), only two (Texas, Ohio) provided specialized geriatric programs. These same data (American Correctional Association, 1998) also indicate that approximately one third of state correctional administrators see no need to provide older inmates with any formal specialized services. Thus, the care and management of older inmates appears to be more of a problem for some states than for others.

▨ Conclusions and Recommendations

Empirical research on older inmates is believed to lag behind because of the stereotypes associated with aging (E. H. Johnson, 1988) and because of their low priority status relative to younger inmates (H. W. Johnson, 1989). In the early 1980s, Goetting (1983) and Rubenstein (1984) each reviewed approximately 20 early studies (most of which overlapped) about older inmates. Both authors found conflicting results and large gaps in the knowledge base, and they similarly concluded that research designs did not properly control for influential individual and institutional characteristics. At that time, Goetting (1983) and Rubenstein (1984) criticized investigators for not complying with acceptable standards of research, for relying on invalid and nonrepresentative data, and for failing to develop consistent operational definitions. Correctional researchers since then have done little to correct these problems: Impressions are uncritically accepted, primary data sources are misused, sample sizes are small, and measures are unreliable. On the other hand, it should be noted that much of the available literature is exploratory in nature, and it addresses the specific need to generate information to develop responsive policies and

programs, not scientific findings. This policy-research focus is not unlike the knowledge-building process that accompanied the dramatic increases in women's imprisonment that occurred in the early 1990s.

It is recommended that researchers employ more rigorous designs to control for the relative contributions of influential variables. Based on our review, the following relevant inmate characteristics emerge: (a) level of preincarceration functioning, (b) criminal history, (c) health and mental health status, (d) institutional behavior, (e) adjustment to the aging process, and (f) social support. The most influential institutional variables include (a) housing arrangements, (b) access to medical care, (c) qualifications and training of staff, and (d) availability of specialized programming.

Intellectual interest in issues affecting older inmates has been supplanted by more practical concerns about how to manage them in institutions geared toward younger, healthier, and more predatory inmates. Information needs since the 1980s have been altered by a highly influential political agenda and by pressing economic priorities, and the research on older inmates has not kept pace with these changes. We would argue that, unless gross violations of human rights are discovered, the problem of older inmates will continue to be one of bed management and cost containment.

We have identified two research priorities as a result of our review: an empirically derived typology of older inmates and a representative survey of older inmates' health status. Investigations should be conducted with older inmates in both the federal and state systems.

A national-level descriptive study of inmates over the age of 55 is long overdue. The number and type of older prisoners who should be considered special-needs inmates is unclear. Replication of Goetting's (1984a) study could lead to the development of an empirically based typology of older inmates. For example, correctional researchers (viz.,

Lightfoot & Hodgins, 1993) have devised an empirically derived typology of substance-abusing offenders to inform the development of more comprehensive and cost-effective treatment programs. A typology of older offenders would enable correctional administrators to allocate treatment resources in an informed and responsible manner. It is further advisable to measure how the number and characteristics of older offenders change over time, with particular emphasis on racial and gender differences. An empirically based typology would also inform future research. If, for example, sufficient evidence indicates that first-time offenders are more prevalent than other types of older inmates, then it would behoove researchers to systematically examine socioeconomic contributors to criminality among older persons.

State-level surveys should be administered to document the prevalence and types of certain health conditions among older offenders. Such conditions include the following: (a) chronic illness (e.g., diabetes, chronic obstructive pulmonary disease, and hypertension), (b) acute illness (e.g., stroke and myocardial infarction), (c) neurological disorders (e.g., Parkinson's and Alzheimer's diseases), (d) terminal diseases (e.g., cancer), and (e) mental illness (e.g., depression and anxiety). Research designs should allow for an examination of gender differences and, if necessary, statistical controls should be used to separate older from other special-needs inmates.

Our review suggests that the relationship between health status and health care utilization may be moderated by certain institutional variables. Thus, utilization data should not be used as a proxy measure for health status.

Rather, these data should be gathered in concert, along with descriptive information about relevant aspects of the institutional environment (e.g., housing arrangements, availability of care, access to community-based resources) that could prove influential. This level of health-related information will enable administrators to develop the most cost-effective and comprehensive health care arrangements for their particular institutions. Such an approach is especially important when investigating the health status of older female inmates who are more likely to be the recipients of differential treatment because of their segregation in a dual prison system (Goetting, 1985).

Contrary to Newman and Newman's (1984) belief that elderly criminals are not dangerous, our review indicates that sex offenders and perpetrators of family violence are overrepresented among older state inmates, relative to younger inmates. A history of alcohol use is also prevalent. However, we did not locate any descriptions of treatment programs that specifically addressed these problems. It is unknown whether their needs are overlooked, or if older inmates simply participate in programs offered to their younger counterparts who have similar histories. These are critical areas for program development and testing. We further caution correctional administrators about over-reliance on community supervision. Although Hoffman and Beck (1984) found that recidivism rates decline with age, other researchers (e.g., Ellsworth & Helle, 1994; B. R. McCarthy & Langworthy, 1987) believe that this decline has more to do with negligent and discriminatory parole and probation practices than with actual changes in offenders' behaviors. Given the lack of proper supervision for violent and alcohol-involved older offenders, shifting the burden of care may place offenders' family members and the larger community at risk.

The most important conclusion that can be drawn from this review is that we would be wise to recognize that existing research on older inmates has been conducted primarily to stimulate the development of needed correctional policies and programs. Policy researchers are currently challenged to find and disseminate facts that will illuminate both the most compelling issues affecting older inmates and the provisions designed to accommodate them. Such knowledge will serve to benefit diverse prison systems that vary in their institutional populations, problems, and resources.

📚 References

Adams, M. E., & Vedder, C. (1961). Age and crime: Medical and sociological characteristics of prisoners over age 50. *Journal of Geriatrics, 16,* 177–180.

Adams, W. E., Jr. (1995). The incarceration of older criminals: Balancing safety, cost, and humanitarian concerns. *Nova Law Review, 19,* 465–486.

Aday, R. H. (1977). Toward the development of a therapeutic program for older prisoners. *Journal of Offender Counseling, Services, & Rehabilitation, 1*(4), 343–348.

Aday, R. H. (1994a). Aging in prison: A case study of new elderly offenders. *International Journal of Offender Therapy and Comparative Criminology, 38*(1), 79–91.

Aday, R. H. (1994b). Golden years behind bars: Special programs and facilities for elderly inmates. *Federal Probation, 58*(2), 47–54.

Aday, R. H., & Webster, E. L. (1979). Aging in prison: The development of a preliminary model. *Offender Rehabilitation, 3*(3), 271–282.

Alston, L. T. (1986). *Crime and older Americans.* Springfield IL: Charles C Thomas.

American Correctional Association. (1998). Inmate health care, Part II. *Corrections Compendium, 23*(11), 11–25.

American Correctional Association. (1999). *1999 Directory: Juvenile and adult correctional departments, institutions, agencies, and paroling authorities.* Lanham, MD: Author.

Anderson, J. (1994). Incarceration alternatives: A special unit for elderly offenders and offenders with disabilities. *Forum on Corrections Research, 6*(2), 35–36.

Baier, G. F. (1961). The aged inmate. *American Journal of Corrections, 23,* 4–6, 30, 34.

Bintz, M. T. (1974). Recreation for the older population in correctional institutions. *Geriatrics, 28,* 149–157.

Booth, D. E. (1989). Health status of the incarcerated elderly: Issues and concerns. *Journal of Offender Counseling, Services, & Rehabilitation, 13*(2), 193–212.

Chaiklin, H. (1998). The elderly disturbed prisoner. *Clinical Gerontologist, 20*(1), 47–62.

Champion, D. J. (1988). The severity of sentencing: Do federal judges really go easier on elderly felons in plea-bargaining negotiations compared with their younger counterparts? In B. McCarthy & R. Langworthy (Eds.), *Older offenders: Perspectives in criminology and criminal justice* (pp. 143–156). New York: Praeger.

Colsher, P. L., Wallace, R. B., Loeffelholz, P. L., & Sales, M. (1992). Health status of older male prisoners: A comprehensive survey. *American Journal of Public Health, 82*(6), 881–884.

Cullen, F. T., Wozniak, J. F., & Frank, J. (1985). The rise of the elderly offender: Will a "new" criminal be invented? *Crime and Social Justice, 15,* 151–165.

Duffee, D. E. (1984). A research agenda concerning crime by the elderly. In E. S. Newman, D. J. Newman, M. L. Gewirtz, & Associates (Eds.), *Elderly criminals* (pp. 211–223). Cambridge, MA: Oelgeschlager, Gunn & Hain.

Dugger, R. (1988). The graying of America's prisons: Special care considerations. *Corrections Today, 50,* 26, 28–31, 34.

Ellsworth. T., & Helle, K. A. (1994). Older offenders on probation. *Federal Probation, 58*(4), 43–50.

Faiver, K. L. (1997). Perspective from the field: Golden years and iron gates. In A. Walsh (Ed.), *Correctional assessment, casework, and counseling* (2nd ed., pp. 351–354). Lanham, MD: American Correctional Association.

Faiver. K. L. (1998). *Health care management issues in corrections.* Lanham, MD: American Correctional Association.

Falter, R. G. (1999a). Special geriatric housing. In L. R. Witke (Ed.), *Planning and design guide for secure adult and juvenile facilities* (pp. 137–144). Lanham. MD: American Correctional Association.

Falter, R. G. (1999b). Selected predictors of health service needs of inmates over age 50. *Journal of Correctional Health Care, 6*(2), 149–175.

Forsyth, C. J., & Gramling, R. (1988). Elderly crime: Fact and artifact. In B. McCarthy & R. Langworthy (Eds.), *Older offenders: Perspectives in criminology and criminal justice* (pp. 3–13). New York: Praeger.

Freedman, H. L. (1948). Rehabilitation of the older prisoner. *Journal of Clinical Psychopathology, 9,* 226–232.

Fry, L. J. (1987). The older prison inmate: A profile. *The Justice Professional, 2*(1), 1–12.

Fry, L. J. (1988). The concerns of older inmates in a minimum prison setting. In B. McCarthy & R. Langworthy (Eds.), *Older offenders: Perspectives in criminology and criminal justice* (pp. 164–177). New York: Praeger.

Gewerth, K. E. (1988). Elderly offenders: A review of previous research. In B. McCarthy & R. Langworthy (Eds.), *Older offenders: Perspectives in criminology and criminal justice* (pp. 14–31). New York: Praeger.

Gewirtz, M. L. (1984). Social work practice with elderly offenders. In E. S. Newman, D. J. Newman, M. L. Gewirtz, et al. (Eds.), *Elderly criminals* (pp. 193–208). Cambridge, MA: Oelgeschlager, Gunn & Hain.

Gillespie, M. W., & Galliher, J. R. (1972). Age anomie and the inmate's definition of aging in prison: An exploratory study. In D. F. Kent, R. Kostenbaum, & S. Sherwood (Eds.), *Research planning and action for the elderly* (pp. 465–483). New York: Behavioral Publications.

Goetting, A. (1983). The elderly in prison: Issues and perspectives. *Journal of Research in Crime and Delinquency, 20,* 291–309.

Goetting, A. (1984a). The elderly in prison: A profile. *Criminal Justice Review, 9*(4), 14–24.

Goetting, A. (1984b). Prison programs and facilities for elderly inmates. In E. S. Newman, D. J. Newman, M. L. Gewirtz, et al. (Eds.), *Elderly criminals* (pp. 169–175). Cambridge, MA: Oelgeschlager, Gunn & Hain.

Goetting, A. (1985). Racism, sexism, and ageism in the prison community. *Federal Probation, 49,* 10–22.

Golden, D. (1984). Elderly offenders in jail. In E. S. Newman, D. I. Newman, M. L. Gewirtz, et al. (Eds.), *Elderly criminals* (pp. 143–152). Cambridge, MA: Oelgeschlager, Gunn & Hain.

Ham, J. N. (1980, July/August). Aged and infirm male prison inmates. *Aging,* pp. 24–31.

Hoffman, P. B., & Beck, J. L. (1984). Burnout: Age at release from prison and recidivism. *Journal of Criminal Justice, 12,* 617–623.

Hunsberger, M. (2000). A prison with compassion. *Corrections Today, 62*(7), 90–92.

Johnson, E. H. (1988). Care for elderly inmates: Conflicting concerns and purposes in prisons. In B. McCarthy & R. Langworthy (Eds.), *Older offenders: Perspectives in criminology and criminal justice* (pp. 157–163). New York: Praeger.

Johnson, H. W. (1989). If only: The experience of elderly ex-convicts. *Journal of Gerontological Social Work, 14*(1/2), 191–208.

Keller, O., & Vedder, C. (1968). Crimes that old persons commit. *Gerontologist, 8,* 43.

Kelsey, O. W. (1986). Elderly inmates: Provide safe and humane care. *Corrections Today, 48,* 56, 58.

Kratcoski, P. C., & Babb, S. (1990). Adjustment of older inmates: An analysis of institutional structure and gender. *Journal of Contemporary Criminal Justice, 6*(4), 264–281.

Kratcoski, P. C., & Pownall, G. A. (1989). Federal Bureau of Prisons programming for older inmates. *Federal Probation, 53*(2), 28–35.

Kratcoski, P. C., & Walker, D. B. (1988). Homicide among the elderly: Analysis of the victim/assailant relationship. In B. McCarthy & R. Langworthy (Eds.), *Older offenders: Perspectives in criminology and criminal justice* (pp. 62–75). New York: Praeger.

Lightfoot, L. O., & Hodgins, D. (1993). Characteristics of substance-abusing offenders: Implications for treatment programming. *International Journal of Offender Therapy and Comparative Criminology, 37*(3), 239–250.

Malinchak, A. A. (1980). *Crime and gerontology.* Englewood Cliffs, NJ: Prentice-Hall.

McCarthy, B. R., & Langworthy, R. H. (1987). Older offenders on probation and parole. *Journal of Offender Counseling, Services, & Rehabilitation, 12*(1), 7–25.

McCarthy, M. (1983). The health status of elderly inmates. *Corrections Today, 45,* 64–65.

McShane, M. D., & Williams, F. P. I. (1990). Old and ornery: The disciplinary experiences of older prisoners. *International Journal of Offender Therapy and Comparative Criminology, 34*(3), 197–212.

Moberg, D. (1953). Old age and crime. *Journal of Criminal Law, Criminology, and Police Science, 43,* 773–775.

Moore, E. O. (1989). Prison environments and their impact on older citizens. *Journal of Offender Counseling, Services, & Rehabilitation, 2,* 175–191.

Morton, J. B. (1993). Training staff to work with elderly and disabled inmates. *Corrections Today, 55*(1), 42, 44–47.

Morton, J. B. (1994). Training staff to work with special needs offenders. *Forum on Corrections Research, 6*(2), 32–34.

Neeley, C. L., Addison, L., & Craig-Moreland, D. (1997) Addressing the needs of elderly offenders. *Corrections Today, 59*(5), 120–123.

Newman, E. S., & Newman, D. J. (1984). Public policy implications of elderly crime. In E. S. Newman, D. J. Newman, M. L. Gewirtz, et al. (Eds.), *Elderly criminals* (pp. 225–242). Cambridge, MA: Oelgeschlager, Gunn & Hain.

Reed, M. B., & Glamser. F. D. (1979). Aging in a total institution: The case of older prisoners. *The Gerontologist, 19*(4), 354–360.

Rosefield, H. A. (1993). The older inmate: Where do we go from here? *Journal of Prison and Jail Health, 12*(1), 51–58.

Rubenstein, D. (1984). The elderly in prison: A review of the literature. In E. S. Newman, D. J. Newman, M. L. Gewirtz, et al. (Eds.), *Elderly criminals* (pp. 153–168). Cambridge, MA: Oelgeschlager, Gunn & Hain.

Sabath, M. J., & Cowles, E. L. (1988). Factors affecting the adjustment of elderly inmates to prison. In B. McCarthy & R. Langworthy (Eds.), *Older offenders: Perspectives in criminology and criminal justice* (pp. 178–195). New York: Praeger.

Sapp, A. D. (1989). Arrests for major crimes: Trends and patterns for elderly offenders. *Journal of Offender Counseling, Services, & Rehabilitation, 13*(2), 19–44.

Shichor, D., & Kobrin, S. (1978). Note: Criminal behavior among the elderly. *The Gerontologist, 18*, 213–218.

Teller, F. E., & Howell, R. J. (1981). The older prisoner. *Criminology, 18*(4), 549–555.

van Wormer, K. (1981). To be old and in prison. In S. T. Letman, L. French, H. Scott, Jr., & D. Weichman (Eds.), *Contemporary issues in corrections* (pp. 86–101). Jonesboro, TN: Pilgrimage.

Vega, M., & Silverman, M. (1986). Stress and the elderly convict. *International Journal of Offender Therapy and Comparative Criminology, 32*(2), 153–162.

Vito, G. F., & Wilson. D. G. (1985). Forgotten people: Elderly inmates. *Federal Probation, 49*, 18–24.

Wall, J. (1998). Elder care: Louisiana initiates program to meet needs of aging inmate population. *Corrections Today, 60*(2), 136–138, 195.

Walsh, C. E. (1989). The older and long term inmates growing old in the New Jersey prison system. *Journal of Offender Counseling, Services, & Rehabilitation, 13*(2), 215–248.

Whisken, F. (1968). Delinquency and the aged. *Journal of Geriatric Psychiatry, 1*, 242–262.

Wiegand. N. D., & Burger, J. C. (1979). The elder offender and parole. *Prison Journal, 59*(2), 48–57.

Wilson, D. G., & Vito, G. F. (1986). Imprisoned elders: The experience of one institution. *Criminal Justice Policy Review, 1*(4), 399–421.

DISCUSSION QUESTIONS

1. What do we know about older inmates? Why should we focus our attention on them?

2. Why does it matter to corrections administrators that inmates are getting older? Should it matter to legislators and policy makers?

3. How do older inmates differ from younger inmates, and what implications do these differences have for corrections administrators?

❖

READING

In this article, James Marquart, Dorothy Merianos, Steven Cuvelier, and Leo Carroll explore how health conditions within lower socioeconomic segments of the population influence the health characteristics of prisoner admissions (what they bring with them with respect to health status). Thus, they demonstrate how health conditions within the wider society have major implications for prisoner health care systems. Prison organizations are not institutions isolated from the broader society; thus social and economic change in the wider society affects the internal dynamics of prisons. The article also examines the effects of recent conservative crime control ideologies on institutional health care programs and concludes with the development of a research agenda on prisoner health care issues.

Thinking About the Relationship Between Health Dynamics in the Free Community and the Prison

James W. Marquart, Dorothy E. Merianos,
Steven J. Cuvelier, and Leo Carroll

⊠ Introduction

Nearly two decades ago, prison scholar James B. Jacobs (1977) made the following observation: "Prisons do not exist in a vacuum: they are part of a political, social, economic, and moral order" (p. 89). This insight suggests that prison walls are permeable and that many characteristics of, and changes in, the wider society find their way into prison settings. A good example of this interplay between the larger society and prisons is the recent revolution in prisoners' rights. *Brown v. Board of Education* (1954) ushered in a new era of civil rights in which many disadvantaged persons were afforded the same constitutional protections held by citizens in the wider community (Jacobs 1980).

The Supreme Court, buttressed by *Brown* and the civil rights movement, held in *Cooper v. Pate* (1964) that prisoners could file suit against correctional administrators. *Cooper* opened a floodgate of prisoner suits and judicial intervention that led to sweeping changes in prison management and eliminated the notion that prisoners were "slaves of the state" (DiIulio 1990; Carroll 1974; Crouch and Marquart 1989; Feeley and Hanson 1990; Martin and Ekland-Olson 1987). The prisoners' rights movement clearly illustrates the linkages between broader macrostructural changes and the internal dynamics of prison organizations. In short, much research has demonstrated that prison organizations, like other formal organizations, are affected by their environment (Irwin 1970; Haas and Drabek 1973).

Legal, moral, and political issues are not, however, the only social forces that impact prison organizations. One would expect, for example, that health trends in the wider society would affect prisoner populations. The most obvious example in this regard is HIV. In 1992, the HIV incidence rate among the U.S. population was 18 cases per 100,000, while the rate among prisoners was estimated to be 362 per 100,000 (McDonald 1995). This huge differential can be attributed in part to the recent "War on Drugs." Large numbers of injection drug users, who are at high risk of developing AIDS, were incarcerated (Blumberg 1990). In the New York, Florida, California, and Texas prison systems, AIDS is the leading cause of death among prisoners and an increasingly expensive focus of the prison health care delivery system (Hopper 1995; Lyons, Griefinger, and Flanery 1994; Weiner and Anno 1992). Death in prison today is far more likely to result from AIDS-related complications than from other forms of death—for example, the stereotypical knifing in the shower or executions.

SOURCE: Marquart, J. W., Merianos, D. E., Cuvelier, S. J., & Carroll, L. (1996). Thinking about the relationship between health dynamics in the free community and the prison. *Crime and Delinquency, 42*(3), 331–360. Reprinted with permission of Sage Publications, Inc.

HIV is one example of how "health matters" for prison organizations. On a broader level, correctional personnel in three state prison systems have claimed that there has been a major decline over the past decade in the overall health of inmates entering their prison systems (Belbot, Cuvelier, and Marquart 1995). Although specific empirical data are not readily available to test this assertion, these officials' perceptions are indirectly supported by data from the National Health Interview Survey that show "American men's and women's health has worsened since the late 1950s, especially since the 1970s" (Verbrugge 1985, p. 156). The central question is, how, and in what ways, do health characteristics and trends in the wider society affect the prisoner population? Additionally, what effect might crime control policies implemented today have on correctional health care systems? We begin to answer these questions by describing general health patterns in American society.

Prisoners are not randomly drawn from the larger population (Jacobs 1977); instead, they come from the lower socioeconomic strata of our society. This fact suggests that health conditions within the lower socioeconomic segments of the noninstitutionalized population shape and influence population morbidity and mortality within prison settings. In short, to account for the health condition of the institutionalized population, one must examine health conditions within the wider population, especially among those segments of the population that contribute a disproportionate number of individuals to the prison—the urban poor (Wilson 1993).

This article extends previous work on prison organizations in two primary ways: (a) it examines the relationship between health conditions in the wider society and prisoner populations, and (b) it explores the implications of this relationship for prisoner management and policy making. We review the relevant research on the complex relationship between health patterns and various individual and societal dynamics. Research evidence

suggests that health conditions, like wealth, status, power, and prestige, are socially stratified and that the circumstances of poverty result in poor health conditions, which in turn lead to great variability in levels of illness and death. This evidence suggests that continued declines in the health status of the American population in general, and the poor specifically, coupled with punitive sentencing policies, will strain correctional health care delivery systems.

◼ The American Prisoner: Background, Health Condition in the Free World, and Health Condition at Admission to Prison

Social scientists, policy analysts, and government researchers have produced an immense body of data on the demographic characteristics of prisoners. *The Sourcebook of Criminal Justice Statistics* and the Bureau of Justice Statistics publication, *Survey of State Prison Inmates,* have been instrumental in providing a demographic portrait of the "typical" state prisoner admittee.

From reviewing the information in these publications, we know that most offenders admitted to American prisons are male, non-White, between the ages of 17 and 30, and from metropolitan areas (Bureau of Justice Statistics 1993). Research on prisoners has also revealed strong links between offending and low socioeconomic status (SES) (Braithwaite 1989; Duster 1987; Blau and Blau 1962; Good and Pirog-Good 1987; Rieman 1990; Krisberg 1975), poor education (in 1991, 65% had not completed high school), prior history of illicit drug use (in 1991, 62% regularly used drugs), lack of attachment (in 1991, 82% were unmarried), and low income (less than $15,000 annually prior to incarceration) (Bureau of Justice Statistics 1993).

This profile would not be complete without a discussion of regional variations. In

1992, the South accounted for 44% of all new prison admissions (Perkins 1994). In the same year, that region had the largest prisoner population (324,454) compared to the Northeast (138,156) or West (174,385). Further, in 1992, the national incarceration rate was 329 prisoners per 100,000 residents; 19 states exceeded that rate, and 12 of those were located in the South (Bureau of Justice Statistics 1993).

Free World Health Conditions

Assessing the health condition of "potential prisoners" prior to confinement is difficult because relevant data are not readily available, but we can make inferences about the free society health condition of most prisoners prior to confinement by reviewing research on health conditions in the wider society as they relate to the aforementioned prisoner characteristics.

Socioeconomic Status and Health

Prisoners are disproportionately from "underclass" areas typified by limited job opportunities and social isolation in impoverished neighborhoods (Wilson 1993, p. 20). Research shows marked variations in health status and access to health care by socioeconomic status (Feinstein 1993). The poor exist in "triple jeopardy," for they are typically uninsured, generally live in medically underserved areas, have difficulty obtaining needed health care services, and continue to have higher mortality rates than higher income persons (Rice 1989; Davis, Gold, and Makuc 1981, p. 179; Davis and Rowland 1983).

Lower income individuals with illnesses are more likely to obtain diagnosis and hospitalization at later points in the course of illnesses than are upper social class individuals (Farley and Flannery 1989; Walker, Neal, Ausman, Whipple, and Doherty 1989), to be sicker upon admission to emergency rooms (Latour, Lopez, Rodriguez, Nolasco, and Alvarez-Dardet 1991), and to make infrequent

visits to dentists (Davis et al. 1981). The poor are also less likely to be insured (Davis and Rowland 1983) and have lower survival probabilities following diagnosis and treatment (Cella, Orav, and Kornblith 1991). After reviewing the research on the relationship between health and income, Feinstein (1993) concluded,

> The evidence is convincing that individuals of lower socioeconomic status do less well in the health care system. It is reasonably clear that both materialist and behavioral factors contribute to inequalities in health care. Thus, lower income and lower status employment (or unemployment) restrict the choice of physician, health care plan, and treatment option; lessened educational attainment, reduces awareness, and attenuates decision-making skills; and cultural idiosyncrasies may make it more difficult to communicate with health-care workers, trust physicians, and play the system. (p. 314)

The advantages of wealth, education, power, and prestige enable higher SES individuals to avoid health risks/hazards and to better mobilize health-protective factors across the life course (House et al. 1990, p. 406).

Race/Ethnicity and Health

Three quarters of all state prison admissions in 1992 were African Americans and Hispanics. Studies of race/ethnicity and health condition consistently report that life expectancy among African Americans lags behind that for Whites and that the gap is expanding; Black men have a higher risk of cancer than non-African American men, diabetes is 33% more common among African Americans than Whites, the rate of AIDS among African Americans is three times that of Whites, African Americans have higher levels of hypertension and higher risks of stroke, and the incidence of primary and secondary syphilis has increased among African Americans despite a decrease in the general population (Department of Health and Human Services 1990).

Manton, Patrick, and Johnson (1987) have reported that African Americans have higher cirrhosis mortality suggestive of greater alcohol consumption in adulthood, twice the risk of diabetes related to obesity, and higher levels of arthritis and chronic nonlethal mental and nervous disorders. African Americans make fewer visits annually to physicians than do Whites, are more likely to use hospital emergency rooms and clinics as their source of medical care (Lewin-Epstein 1991), and are more likely to be limited in their everyday activity due to chronic conditions (Davis et al. 1981). Manton et al. (1987) reviewed the research on African American and White health differentials and concluded, "Though both blacks and whites exhibited increases in life expectancy and health improvements, blacks remain significantly disadvantaged on a broad range of health measures" (p. 192).

Among Hispanics, the leading causes of death are heart disease and cancer, at rates higher than for non-Hispanic Americans. Compared to non-Hispanics, Hispanics also have higher rates of chronic liver disease, cirrhosis, and AIDS (Department of Health and Human Services 1990). Selik, Castro, Papaionnou, and Ruehler (1989) noted that Hispanics' rates of AIDS are almost three times higher than for non-Hispanic Whites, and the incidence of AIDS among Hispanic women is nearly eight times higher than among non-Hispanic women. Hispanics, like African Americans, receive less preventive health care (Lewin-Epstein 1991).

Gender and Health

Differences exist between males and females based on their genes or reproductive physiology, which confers differential risks of morbidity (Verbrugge 1985, p. 164). Health differentials between the sexes is a well-researched topic (Verbrugge 1976, 1980, 1984; Ortmeyer 1979; Waldron 1983; Nathanson 1977, 1978, 1984). The most ubiquitous health difference between the sexes concerns mortality. In 1990, life expectancy at birth for females was 78.8 years compared to 71.8 years for males. The death rate was 928.9 for males and 819.9 for females (per 100,000). Males also have higher rates of heart disease, cancer, malignant neoplasms, and HIV (National Center for Health Statistics 1992; Waldron 1983; Wingard 1982), while, on the other hand, females have higher rates of cerebrovascular diseases (National Center for Health Statistics 1992). The risk of death is much higher for males than for females at all ages for all the leading causes of death in our society (Kitagawa and Hauser 1973).

Lifestyle and Health

Over the life course, an individual's health habits bear directly on his or her risk of illness and injury. Job-related injuries, household tasks, nutrition levels, leisure activities, management and reaction to stress, and social bonds *across the life course* affect an individual's health condition. Acquired risks also include the use of such harmful substances as cigarettes, alcohol, PCP, cocaine, marijuana, heroin, and harmful vapors/inhalants, as well as such high-risk physical characteristics as obesity, high levels of serum cholesterol, and elevated blood pressure. These latter risk factors, alone or in combination, are associated with the five major causes of death (cancer, heart disease, stroke, injury, and chronic lung disease) in the United States (Department of Health and Human Services 1990).

Acquired health risks are unevenly distributed in American society. Men are more likely than women to engage in activities that pose major health risks (e.g., use tobacco and illicit drugs, consume alcohol). Although tobacco use has declined in the general population, it has remained constant among persons with less than a high school education and for those employed in blue-collar jobs. Health risks and socioeconomic status are strongly related, with poverty being a predisposing factor in obesity and high blood pressure, which in turn are major risk factors for heart disease and stroke. Low-income persons are more likely than those with high incomes to acquire and be exposed to health risks across their life course (Department of Health and Human Services 1990).

Region and Health

The prisoner profile presented earlier took note of the large numbers of prisoner admissions from the South. Research indicates geographical differentials in mortality in the United States. Makuc, Feldman, Kleinman, and Pierre (1990) examined this relationship and found that

> Among white men the highest death rates were concentrated in the Southeast and lowest were found west of the Mississippi. Among white women the highest rates were east of the Mississippi, primarily in the Middle Atlantic states. Almost all areas with the lowest death rates were west of the Mississippi. For black persons the lowest death rates were found in the West and highest in the Southeast. Among white persons death rates were similar for metropolitan and nonmetropolitan residents. Among black persons death rates for cardiovascular diseases were lower in metropolitan areas. (p. 156)

The South, when compared to other areas of the country, has a higher concentration of poor and minority persons (who are more likely to be uninsured), and these disparities help explain the higher rates of mortality.

Aging and Health Condition

The majority of prisoners at admission in 1991 were under 30 years old (Bureau of Justice Statistics 1994). However, the prisoner population is "graying" (Dugger 1988; Enter 1995; Goetting 1983).

Research has consistently demonstrated that aging is accompanied by (a) a decrease in immunological defenses of humans; (b) a nonspecific, genetically determined "wearing out" with age (Lilienfeld 1976, pp. 92–93); and (c) an interaction effect between aging and wearing out that leads to greater disability, more chronic conditions, and the need for accommodation (Rice and Feldman 1983). Kart and Dunkle (1989) found a strong relationship between aging and the need for assistance with activities of daily living (ADLs, which include such routine activities as bathing/showering, dressing, eating, getting in and out of bed, and getting to and using a toilet). These findings illustrate that major advances in life expectancy do not automatically translate into improved health (Verbrugge 1984).

The natural processes of aging and the increase in coping with chronic conditions and illnesses have major economic and social implications. Rice (1989) found that

> In 1986, elderly people comprised 11.6 percent of the population but accounted for 17.6 percent of all physician contacts, 27.2 percent of hospital discharges, and 35.3 percent of the hospital days of care. Eleven percent of the population aged 65 and over consumed 29 percent of the total health expenditures in 1980; by 2040, the elderly are projected to comprise 21 percent of the population and almost half of the expenditures will be made in their behalf. (pp. 344–45)

These figures suggest that the natural process of aging creates additional complications, especially economic ones, because the elderly are extremely heavy users of health services (Wolinsky, Mosely, and Coe 1986). Most important, aging and socioeconomic status are highly correlated with health condition, race/ethnicity, education, lifestyle, and education, which affect the incidence and prevalence of morbidity and mortality within various population subgroups. House et al. (1990) found that "the lowest socioeconomic stratum manifests a prevalence of chronic conditions at ages 35–44 that is not seen in the highest socioeconomic stratum until after age 75" (p. 398). The lowest socioeconomic groups manifest the highest levels of chronic conditions and accompanying limitations of functional capacity not evident in the highest socioeconomic groups until after age 75.

Summary of Health Factors

In the preceding section, we have reviewed the most fundamental research findings on the relationship of socioeconomic status, gender, aging, race/ethnicity, lifestyle, and region and the health condition of the noninstitutionalized population. This research indicated that those with the poorest health are typically poor non-White males, who are uneducated, primarily from the South, and typically underemployed. The data also suggested that many of these individuals also lead high-risk lifestyles, often involving excessive use of alcohol, tobacco products, and illicit drugs. Coincidentally, this "at risk" health profile closely parallels the characteristics of felons admitted to penal institutions on any given day in the United States.

Health Condition of Prisoners at Admission

In the first systematic prison health study, Frank Rector (1929) surveyed the medical conditions in American penitentiaries. Since his book appeared, researchers have examined the standards of health care in prisons (Dubler

1986), the health care demands of women prisoners (McGaha 1987), mortality among prisoners (King and Whitman 1981), elderly prisoners (Dugger 1988), the health risks of imprisonment (Jones 1976), prison health care professionals (Maier, Bernstein, and Musholt 1989), the management of infectious diseases in correctional settings (Blumberg 1990), and legal/ethical issues concerning prison medicine (Jamieson 1985). Despite the diversity of this research, scholars have not systematically analyzed prisoner health characteristics at admission or across confinement (Novick and Al-Abrahim 1977), What we know comes primarily from studies of jail inmates.

Raba and Obis (1983) conducted one of the first systematic studies on the health conditions of male admissions ($n = 987$) at the Cook County Jail in Chicago. They collected data, through physical exams, on inmates' medical condition, psychiatric referrals, surgical history, blood pressure, genitourinary abnormalities, drug history, and tuberculin tests. After analyzing the data, they concluded that

> The admissions were young and largely members of minority groups. In spite of the young overall age, over 17% were on chronic medication for asthma, hypertension, psychosis, diabetes mellitus, or seizure disorder. The group was found to have a high incidence of tobacco, alcohol, and drug use. Psychoses, venereal diseases, seizure disorders, asthma, hypertension, tuberculosis contacts, and posttrauma sequelae were detected in higher incidence than found nationally. (p. 21)

Fitzgerald, D'Atri, Kasl, and Ostfeld (1984) examined the health conditions of 366 male prisoners in a Massachusetts house of corrections, where most inmates were confined for 12 months or less. Data were collected from inmate self-reports on current and past medical complaints. These researchers concluded that

> Psychological problems, trauma, oral diseases, and the use of alcohol, tobacco, and illicit drugs were the most frequent health conditions observed or reported among these male prisoners at intake. . . . [The self-report data] support the contention that these young male prisoners suffered disproportionately from poor health. . . . [Most] were predominantly of low socioeconomic status. . . . Lifestyle factors also are involved. The frequent use of alcohol and illicit drugs among these men, for instance, may explain their relatively high prevalences of hepatitis and seizure disorders. (p. 63)

Finally, the inmates' self-report data were compared to similar medical conditions in the U.S. population, which revealed that the inmates had higher prevalence rates of TB, tobacco use, epilepsy, heroin use, diabetes, asthma, barbiturate use, high blood pressure, amphetamine use, ulcers, alcoholism, and arthritis (Fitzgerald et al. 1984, p. 69). Support for these findings can also be found in the work of Whalen and Lyons (1962), Jones (1976), King and Whitman (1981), Derro (1978), Litt and Cohen (1974), Baird (1977), Twaddle (1976), and Novick, Della Penna, Schwartz, Remmlinger, and Loewenstein (1977). Fogel (1988) reported similar findings among women prisoners in a southern U.S. state prison.

◼ Correctional Health Care Standards and Utilization of Prison Health Services

Health data consistently show evidence of health inequalities in our society. Individuals from lower socioeconomic groups are more likely to have been, or be, in the poorest health condition. By inference, those admitted to prison—because of lifestyle factors, poor education, and limited resources—possess the poorest health condition of perhaps any segment of the American population. They experience health problems far in excess of those of the noninstitutionalized population. Once institutionalized, what level of care must the staff provide to inmates? Are there medical standards available to serve as guideposts for action?

Judicial Principles and Standards for Prison Health Care

Prisoners have the constitutional right to medical care and services. Prisoners' right to medical care became a reality through litigation and judicial intervention in the late 1960s and early 1970s (Jacobs 1980). Early court rulings on prison medical care, based on an interpretation of the Eighth Amendment that prohibited torture and other barbarous treatment, centered on personal hygiene issues and deprivations that endangered life and limb (*Jackson v. Bishop* 1978; Murton and Hyams 1969).

Judicial decisions often cited "conduct that shocks the conscience" or "barbarous acts" or "obvious neglect or intentional mistreatment" or "conduct so grossly incompetent, inadequate, or excessive as to shock the general conscience or to be intolerable to fundamental fairness" (Neisser 1977, pp. 923–24; *Holt v. Sarver* 1971; *Newman v. Alabama* 1972). The "barbarous act" standard became inadequate. Many inmate suits not involving life-threatening situations cited inadequate services and the standard neglected misconduct by prison medical personnel (Neisser 1977).

In 1976, the Supreme Court in *Estelle v. Gamble* set forth a deliberate indifference standard as to whether or not a state breached its constitutional duties regarding medical care. The court held that "[d]eliberate indifference by prison personnel to a prisoner's serious illness or injury constitutes cruel and unusual punishment contravening the Eighth Amendment." The court in *Gamble* also articulated a governmental obligation to provide medical care to those it punishes by incarceration. The deliberate indifference standard, like the barbarous act standard it replaced, was very broad and provided no clear standards or principles to guide prison officials in implementing medical service programs consistent with evolving standards of decency.

In response to mounting litigation and to helping clarify court decisions, the American Correctional Association (ACA) developed plans for prison accreditation in 1970 and implemented these plans in 1974 (Sechrest 1976). At the same time, the American Medical Association (AMA) examined and found many deficiencies in the level of health care in American prisons (Briscoe and Kuhrt 1992). To assist prison organizations to establish minimum requirements for health services, the AMA published in 1979 a set of standards for prison health services that later were revised by the National Commission on Correctional Health Care (NCCHC). Although the AMA and ACA were already examining health care, *Gamble* provided the catalyst to make accreditation a reality.

NCCHC is a nonprofit organization that uses standards to evaluate correctional institutions seeking voluntary accreditation, which "serves as recognition that an organization and its staff are performing at a level which experts have determined to be acceptable" (National Commission on Correctional Health Care [NCCHC] 1992, p. 161). According to NCCHC, accreditation provides additional justification for budgetary requests, a sense of accomplishment and achievement, and maintenance of appropriate levels—as determined by national standards—of health care (Fasano and Anno 1988). NCCHC also maintains that accreditation helps reduce "liability premiums and protect[s] facilities from lawsuits related to health care" (NCCHC 1995, pp. 2–3): In 1994, NCCHC reported that 280 correctional facilities had earned accreditation status; 13 more were accredited in 1995 ("National Commission Accredits" 1994; "Thirteen Facilities" 1995).

The NCCHC accreditation manual lists 68 standards, of which 34 are identified as "essential." To be accredited, an institution must comply with all "essential" standards and 85% of the "important" ones. Essential standards include the development/maintenance of written policy and procedures, health service personnel qualifications, emergency services, and the management of pharmaceuticals. Further, all inmates must receive a health/dental assessment within 7 days of admission, and all of them must have access to health services. One essential NCCHC standard concerns sick call and related staffing patterns. Sick call must be conducted by physicians or qualified health personnel in "prisons with fewer than 200 inmates, a minimum of three days a week; in prisons with from 200 to 500 inmates, a minimum of four days a week; and in prisons with 500 or more inmates, a minimum of five days a week" (NCCHC 1992, p. 47). Inmates who are about to be placed in disciplinary segregation must also be evaluated prior to their placement and on a daily basis thereafter.

The ACA established standards for voluntary accreditation of prisons in 1978, and since that time more than 1,500 correctional facilities—including 363 adult correctional institutions and programs—have been accredited (ACA 1995). The ACA accreditation manual lists over 350 standards for accreditation; agencies must comply with all of the mandatory standards and 90% of the nonmandatory ones. There are 53 health care standards; 11 are specified as mandatory.

According to the ACA, the purposes of accreditation are to offer agencies the "opportunity to evaluate their operations against national standards, to remedy deficiencies, and to upgrade the quality of correctional programs and services" (ACA n.d., p. 3). ACA promotes accreditation as a "defense against lawsuits through documentation and the demonstration of a 'good faith' effort to improve conditions of confinement" (ACA n.d., p. 3).

Table 1 depicts the mandatory standards for the two accrediting bodies. ACA mandatory health standards are broadly written and involve requirements for written policy, procedures, and personnel qualifications. Finally, ACA-accredited institutions must provide "unimpeded access to health care" and a system to process grievances against the facility health care system. Inmates who present themselves at the infirmary through legitimate channels must be seen by qualified medical personnel. Custodial staff cannot screen or restrict inmates from attending sick call.

In summary, accreditation was launched as a mechanism to set minimum standards for inmate health care and to help prevent or minimize litigation. Accreditation of a service delivery program, however, does not guarantee immunity from inmate lawsuits, and the number of jurisdictions using accreditation (either with the NCCHC or ACA) as a defense is not available. However, ACA researchers have found, in a recent review of court decisions, that judges consult the ACA standards as a reference point but do not automatically adopt the standards as a judicial remedy (Miller 1992). Both accrediting organizations expect growth in the number of correctional facilities seeking accreditation.

Issues of accreditation aside, utilization of prison medical services is affected by access, and accredited facilities are required to maintain an "open door" policy. More admissions result in increased demand on prison medical services. It is reasonable to assume that most offenders admitted to state prisons are in poor health, thus the question arises, how often do inmates utilize prison health services? Relatedly, what impact does a prison term have on offenders' health conditions?

Utilization of Prison Health Services and Health Condition While in Prison

Data to address these questions are rarely available. Few studies have systematically assessed utilization of health services in prison settings. From the health-related literature presented previously, we know that the health demands of a 25-year-old male felony offender sentenced to 20 years in prison, for example, will change over the life course.

Prison health utilization studies, excluding dental visits, report usage rates ranging from 18.5 visits per inmate per year (Fitzgerald et al. 1984), to 17.8 (Twaddle 1976), to 7.7 (Derro 1978). In these studies, the researchers did not control for age, sex, or institutional context. Research has indicated that African American prisoners have the highest utilization rates. In spite of these problems with the data, it can be inferred that inmates utilize prison medical services at greater rates than does the noninstitutionalized population (Suls, Gaes, and Philo 1991; Fogel 1988). Sheps, Schechter, and Prefontaine (1987) found that Canadian prisoners averaged 5.2 health encounters per prisoner over 12 months, or "2.4 times higher than the mean annual physician visit rate for noninstitutionalized men in Canada" (p. 4).

Many factors influence utilization. Prisoners may use health services because they are "free," because they have greater access to these services than in the free world, because they have greater health needs, or because a visit to the clinic may represent a break in the institutional routine. Institutions engender dependency among inmates, and the numerous rules and regulations may contribute to overutilization of prison health services (Neisser 1977; Twaddle 1976). For example, diabetic inmates must present themselves to health officials for insulin injections. Inmates typically cannot possess pharmaceuticals and must therefore visit the infirmary to receive medication. Only trained personnel can dispense medication. Although these contacts seem innocuous and fleeting, they take time, resources, and personnel. Addition of more inmates to prisons will increase the numbers of these contacts.

Service utilization is a complex issue affected by human and organizational variables.

Table 1	Mandatory Health Standards

NCCHC Standards	**ACA Standards**
Responsible health authority	Medical autonomy
Medical autonomy	Unimpeded access to care
Administrative meetings and reports	Facilities and equipment
Policies and procedures manual	Personnel
Comprehensive quality improvement program	Administration of treatment
Disaster plan	Pharmaceutical services
Infection control program	Health screenings and examinations
Environmental health and safety	Transfers
Credentialing	Emergency care
Continuing education for qualified health services personnel	Training for correctional and other personnel
Training for correctional officers	
	Prohibition against inmate participation in medical, pharmaceutical, or cosmetic research
Medication administration training	
Written policy prohibits inmates being used as health care workers	
Pharmaceutical services	
Receiving screening	
Access to health care services	
Health assessment	
Mental health evaluation	
Dental care	
Daily handling of nonemergency medical requests	
Sick call	
Direct orders	
Emergency services	
Health evaluation of inmates in disciplinary segregation	
Special needs treatment plans	
Intoxication and withdrawal	
Infirmary care	
Suicide prevention	
Prenatal care	
Health record format and contents	
Confidentiality of health records	
Confidentiality of health information	
Medical restraints and therapeutic seclusion	
Policies and procedures to guide the use of forced psychotropic medication	

Another important institutional issue concerns an inmate's health status across the confinement period. Wallace, Klein-Saffran, Gaes, and Moritsugu (1991) examined a sample of Federal Bureau of Prison inmates at admission and at release to assess changes in their overall health condition. They found that almost 90% of the inmates had experienced no change in their health status during incarceration, while 4% showed improvement, and 7% worsened. Due to long histories of illicit drug use, Hispanic inmates were more likely to exhibit a *decline* in health status during confinement. General population inmates classified in high-security statuses were *more likely* to experience a worsening in health status during confinement (Wallace et al. 1991, p. 148).

Preliminary evidence suggests that most inmates experience no change in their health status during incarceration. Although many prisoners have access to and often receive better health care in prison than they would have outside, the health condition of many of them is so poor at admission that the influence of the preprison factors and the discovery of new conditions may frustrate or counteract the benefits of prison health care. The persistent effects of lower socioeconomic status chart a health course that may be permanent. House et al. (1990) reported that "the impact of socioeconomic status on health may be like a powerful river. If you identify its present course and alter or block that course, it may simply find a new route to its destination" (p. 406). If this is the case, it is important to examine the impact of criminal justice policy on the delivery of correctional health care.

⚎ Criminal Justice Policy and Correctional Health Care

Criminal justice policy making today is dominated by a conservative crime control ideology (Packer 1968). For example, California recently enacted an indeterminate term of 25 years to life for offenders with two or more prior convictions for violent offenses. Between March 1994, when the law took effect, and August 1994, 713 men and women were admitted to California prisons as a result of this legislation (Levinson 1995). Georgia is one of five states with a "two strikes" law, Maryland has a "four strikes" law, Oregon now has a "one strike" law covering 16 specific felonies that applies to all offenders 15 years old or over, Mississippi requires inmates to serve at least 85% of their sentences, and Nebraska recently implemented a 25-year minimum sentence for repeat violent offenders ("States Leaping" 1995, pp. 1–2).

This list of states recently enacting tough sentencing laws is by no means complete, as many other jurisdictions are weighing similar punitive options. These sentencing "reforms" will result in more inmates spending more time in prison (*Corrections Yearbook* 1995). In addition to the usual programming and staffing needs associated with prison expansion, correctional managers will have to provide medical care to an increasingly unhealthy population that will have high rates of service usage for decades, especially in view of three rapidly growing inmate population groups.

Three Growing Long-Term Prisoner Groups

Three offender populations present a burgeoning demand on prison health services: natural lifers (confined to prison forever, with no chance of release), life sentence inmates (sentenced to a term in excess of 60 years or "life," but the sentence can be "served" by accumulating good time), and nonlifers with sentences in excess of 20 years.

Table 2 documents the growth of these three populations between 1986 and 1995. The rise in these three populations is best seen in the last column in the table. In 1986, the three groups totaled 17% of all prisoners, but, by the beginning of 1995, these groups accounted for one quarter of all prisoners. The composition of these three inmate groups will for the most part be similar to the prisoner profile discussed earlier. These prisoners will be primarily young non-White males, with little education, of low socioeconomic status, from inner-city areas, with a probable history of drug usage, and will be confined in southern states. These offenders, although in the free society, will probably have had low health care utilization rates, high acquired health risks, little preventive care, few health care opportunities (e.g., immunization, visits to a doctor's office, routine medical screening), and will have entered prison in poor health condition (Prout and Ross 1988).

It can be anticipated that these long termers will have a major impact on the institutional health delivery system throughout their confinement. Their service demands will be felt in the areas of health programming, personnel, and budgets. Yet we know little about the relationship between sentence length and service demands. To address this issue, we must return to the research on the noninstitutionalized population in order to make inferences about *prisoners.*

Wolinsky et al. (1986) examined the use of health services by elderly persons in several age cohorts. Table 3 shows the mean number of physician visits per person in the preceding 12 months by age cohorts in three different years.

Given the generally poor health condition of prisoners at arrival, the sequencing of their service demands over time and with age will increase. It has been suggested that, by the year 2000, there will be 125,000 inmates over 50 years of age, 50,000 of whom will be over the age of 65, in America's prisons (*Criminal Justice Newsletter* 1989). In all likelihood, health services for older prisoners will constitute a disproportionate amount of the overall correctional health care resources and budget—a pattern already under way among the elderly in the wider society (Anno 1990).

Table 4 presents recent data on the three long-term populations and those over the age of 50 by jurisdiction. Close inspection of those data uncovered an interesting pattern. Eighteen states had 25% (or more) of the total prisoner population consisting of natural lifers, lifers, and those sentenced to 20 or more years in prison. Specifically, Louisiana reported that 34% of its prisoner population consisted of natural lifers and those sentenced to 20 or more years in prison. The Louisiana prison system, listed as accredited by the ACA, also

Table 2	State Prison Sentences by Year and Type				
Year	Total Prisoner Population	Natural Life Sentence	Life Sentence	20 Years or More	Percentage of Three Groups
1986	485,321	7,399	26,178	51,256	17
1987	522,744	5,469	37,191	78,042	23
1988	554,626	8,569	38,874	71,848	21
1989	597,603	10,370	41,005	88,343	23
1990	673,559	11,246	43,961	96,921	23
1991	732,236	11,759	44,451	105,881	22
1992	776,059	13,937	52,054	125,996	24
1993	824,901	17,071	55,856	127,915	24
1994	892,270	17,446	53,650	148,026	24
1995	980,513	17,853	64,686	163,811	25

Table 3	Mean Physician Visits by Cohort by Survey Year		
Cohort Age	**1972**	**1976**	**1980**
56–59	3.48 (12.5)		
60–63	3.65 (13.1)	3.62 (13.6)	
64–67		3.92 (15.8)	3.71 (15.2)
68–71			3.96 (16.4)

NOTE: Figures in parentheses represent the percentage of individuals with one or more hospital episodes in the preceding 12 months.

spent an average of $4.59 per day per inmate on medical care while the national average was $6.07 per day. Whether or not $4.59 per day will be enough to satisfy the health-related demands of a large group of prisoners expected to be incarcerated for decades is an unresolved question (Dennis 1994). What happens when, and if, the medical costs for inmates exceed the average costs per day to incarcerate them? Finally, 11 of these 18 states (61%) were located in the South—the region of the nation with the fastest growing prisoner population.

Policy Implications

"Getting tough on crime" policies differentially affect lower socioeconomic groups, the same groups that researchers have consistently shown to have the poorest health in the free society. Most prisoners come from low-income backgrounds. Current discussions of the dilemmas of enhancing criminal penalties and building more prisons consistently center on construction and personnel costs, but "sentencing reforms" aimed at lengthening time served under the policy goal of "protection of the public" must include frank discussion of the anticipated health costs over time. To date, few policy makers have considered the long-term implications of the medical costs related to incarceration. For example, ACA-accredited facilities must medically examine all intrasystem transfers. It is not unusual for prison organizations to use "bus therapy" as a management

device for unruly prisoners. The cost in terms of time, medical resources/personnel, and money of this "therapy" may eventually exceed the benefits for accredited agencies or those contemplating accreditation.

Policy makers must also address the objectives of a prison health care delivery system. Should prison health care meet only *minimum* standards? Or, should prisons deliver health services *above* what their clientele would have access to in the free society? Research to date suggests that most prisoners during confinement experience neither an improvement nor slippage in their health condition (Wallace et al. 1991). Should prisoners "get less" because it may be economically convenient? Should they "get more" because conscience demands efforts to help or improve the conditions of the less fortunate? The type, goals, and quality of prisoner health care services remain an issue to be settled in the political and legal sectors.

Maintaining programs with higher standards, however, poses two problems. First, high-standard services, while sounding good, cost money. Second, the principle of less eligibility means that prisoners, for example, should not be treated better than the poorest noncriminal individuals in our society (Mannheim 1939). Prison programs meeting a high standard run the risk of creating a medical services imbalance between low socioeconomic status offenders and noncriminals.

The last column in Table 4 reports the average cost per day per inmate for medical

Table 4	Comparison of Selected Long-Term Populations and Health Costs per Day							

	In Prison	Over Age 50	Percentage Over Age 50	Natural Lifers	Lifers	20 Years or More	Percentage of All Inmates	Daily Health Cost ($)
Alabama	17,039	1,179	6.5	858	2,178	4,461	41.6	3.00
Alaska	2,791	282	8.5	0	178	204	11.5	8.50
Arizona	19,582	13,633	6.9	5	864	1,498	12.2	5.83
Arkansas	8,806	434	4.9	888	0	2,872	42.7	5.03
California	125,605	5,541	4.4	1,545	11,634	4,338	13.1	8.55
Colorado	8,037	606	6.2	99	509	1,550	22.0	0
Connecticut	14,246	441	2.7	49	112	572	4.6	8.06
Delaware	4,388	202	4.6	149	273	0	9.6	5.44
District of Columbia	10,621	426	4.0	28	935	1,098	19.1	6.39
Florida	57,139	3,051	5.3	3,191	237	9,438	26.5	9.13
Georgia	33,383	1,718	5.1	19	4,169	3,891	23.8	7.62
Hawaii	2,905	191	5.7	23	145	518	20.6	7.54
Idaho	2,253	234	9.7	54	242	297	24.6	5.00
Illinois	36,531	1,209	3.3	697	13	7,541	22.6	6.45
Indiana	14,111	1,078	7.6	0	377	3,191	25.3	7.80
Iowa	5,437	300	5.5	433	0	351	14.3	3.45
Kansas	6,299	397	6.2	0	571	2,500	47.7	6.61
Kentucky	9,097	708	6.8	8	600	1,672	21.9	0
Louisiana	15,623	1,349	8.4	2,482	0	4,169	41.5	4.59
Maine	1,386	43	3.0	29	12	238	19.3	4.66
Maryland	20,256	879	4.2	0	1,642	3,800	26.1	3.73
Massachusetts	10,591	533	5.0	495	664	1,915	28.8	6.8
Michigan	40,352	0	0	2,139	1,395	2,429	14.8	5.52
Minnesota	4,488	259	5.6	6	203	293	10.8	8.31
Mississippi	9,746	0	0	169	1,287	0	14.9	2.94
Missouri	17,334	904	5.0	446	1,070	3,070	25.6	3.62
Montana	1,700	165	9.5	12	19	576	34.8	2.81
Nebraska	2,686	159	5.8	138	68	10	7.9	6.86
Nevada	6,909	565	7.8	280	841	254	19.0	0
New Hampshire	2,066	97	4.4	36	28	98	7.3	6.35
New Jersey	19,241	1,062	5.0	0	863	2,959	18.0	8.43
New Mexico	3,868	256	6.5	0	197	275	12.0	8.21
New York	66,758	2,988	4.5	0	10,349	5,726	24.1	6.00
North Carolina	22,653	1,104	4.6	0	2,828	6,246	38.0	7.56

(Continued)

Table 4	(Continued)							
North Dakota	526	35	6.4	0	17	37	9.9	3.15
Ohio	41,718	2,454	5.8	0	3,586	14,932	44.0	4.07
Oklahoma	13,398	1,156	6.8	168	1,139	3,736	29.7	2.73
Oregon	6,915	498	7.1	30	477	917	20.2	5.75
Pennsylvania	27,522	1,846	6.5	2,795	0	5,480	29.2	7.40
Rhode Island	2,938	121	3.8	8	116	0	3.9	0
South Carolina	17,359	912	4.8	33	1,484	3,699	27.4	4.54
South Dakota	1,661	147	8.5	112	0	210	18.6	5.60
Tennessee	12,421	889	7.1	14	1,209	4,048	42.0	9.12
Texas	97,650	4,662	4.8	0	5,165	32,567	38.6	0
Utah	3,239	264	7.5	4	45	792	24.0	8.60
Vermont	976	0	0	0	25	102	9.6	7.67
Virginia	20,893	944	4.5	70	1,516	9,220	51.6	3.34
Washington	10,847	761	7.0	147	533	837	13.9	8.15
West Virginia	1,865	200	10.4	180	130	65	19.5	5.00
Wisconsin	10,020	516	5.0	0	665	1,521	21.2	2.78
Wyoming	1,065	98	8.1	14	84	103	16.5	7.97
FBOP	85,573	10,055	10.6	0	1,692	7,495	9.7	8.77

NOTE: 0 = not reported.

care and reveals great disparity between agencies in the amount of inmate health care expenditures, but masks the fact that prison health service costs, like those in the free world, have exploded in recent years. Anno (1990) examined inmate care expenditures between fiscal years 1982 and 1989. She found that in 1989 "the percent of the total DOC's expenditures devoted to health ranged from a low of 2.8 in South Dakota to a high of 18.9 in Texas. The mean percentage expended on health was 9.5 percent whereas the median was 8 percent" (p. 107). The total mean expenditure in 1989 was $25 million per state. Trend data for the 1982–1989 period reveal that, on average, the agencies increased their per inmate annual health expenditure by 104%. Given the large numbers of prison commitments in recent years and their poor health at admission, we fully expect that prison health care costs will continue to rise.

Much of our previous discussion has centered on long-term prisoners. One prisoner population group destined to grow and, consequently, consume high levels of health care resources will be prisoners with HIV. Recent statistics show that less than 3% of state and federal prisoners were infected with HIV (Brien and Harlow 1995). However, there has been a 52% increase in infected inmates between 1991 and 1993. A joint survey conducted by the National Institute of Justice and the U.S. Centers for Disease Control found that correctional agencies reported nearly "3,500 deaths attributed to AIDS" (Hammett, Harrold, Gross, and Epstein 1994, p. xi).

African American and Hispanic inmates are overrepresented among AIDS cases in prison. The infection rate is also higher among female than male inmates (Brien and Harlow 1995). The cost of providing care and services to HIV-infected inmates will most certainly grow in the years to come. Punitive crime control policies aimed at incarcerating drug users will contribute to the growth and spread of HIV in prison settings and the wider society.

In 1994, 47,756 women were admitted to state and federal prisons (*Corrections Yearbook* 1995). Although this number pales in comparison to that for men (458,841), the female prisoner population rapidly increased throughout the 1980s and shows no signs of abating. The health care demands of women are also more complex because of their more complicated reproductive system. According to the NCCHC (1994),

> Research regarding the provision of gynecological services for women in correctional settings has been limited, but it consistently has indicated that such services are inadequate. Annual gynecological exams are not done routinely in either jails or prisons, nor are they regularly performed upon admission. Appropriate initial screening questions about a woman's gynecologic history may not be asked, and, in many correctional facilities, there are no physicians who are trained in obstetrics and gynecology, leading to inadequate and inappropriate gynecologic care. As a result, women in jails and prisons are at risk for the lack of detection of some diseases such as breast cancer, ovarian cancer, and abnormal Pap smears. In addition, because of past medical histories of many incarcerated women, their pregnancies tend to be more complicated. Further, many women enter jails and prisons while pregnant and they need adequate prenatal care and reproductive counseling such as family planning and birth control. In addition, postpartum screening for physical and psychiatric complications is needed. (p. 2)

Research has shown that women prisoners utilize health care services more often than do men (NCCHC 1994). The full impact of the growth in the female prisoner population and concomitant medical costs has yet to be fully explored or considered. Incarcerating more women, coupled with their unique health demands, will be a costly crime control policy. Other special needs populations that will place significant demands on health resources include death row prisoners, those with mental and physical impairments (in conjunction with ADA categories), and inmates with drug and alcohol impairments.

Options

The medical costs associated with lengthy determinate sentences may well drive future policy makers to enact legislation to release large numbers of long-term prisoners. Correctional policy analysts may also conduct research on the "health trajectories" of older prisoners (or other high-risk, high-cost prisoners such as those with HIV) to determine the point at which an *individual's* health care costs exceed ordinary confinement costs and then seek to have these "costly" prisoners released to the community. Once released, the care of such citizens will be borne by community hospitals and other health services.

Another policy option available to contain prisoner medical costs will be to use a copayment system. Currently, most correctional agencies provide inmates with health services at no cost. However, nine correctional agencies (Arizona, Colorado, Florida, Kansas, Maine, Maryland, Nevada, Oklahoma, and Oregon) charge inmates a minimum of $3 for routine medical care (*Corrections Yearbook* 1995, p. 53). Copayment plans have been found to reduce health costs ("National Commission Accredits" 1994, p. 3). In Virginia, a copayment program was recently implemented and raised $32,000 in the first month of operation and "reduced inmate sick calls by about 35 percent" ("VA Copayment Plan" 1995, p. 3).

The trend will certainly be toward copayment programs. However, a no pay/no care system would violate the deliberate indifference standard set forth in *Gamble* and might actually discourage inmates with legitimate problems from reporting them. Case law on this issue is sparse and represents a new area for judicial scrutiny. In the end, other methods will be used to trim medical costs, including obtaining physicians' services at reduced costs, enacting price controls, operating hospitals within prison systems instead of purchasing services from community hospitals, and aggressive negotiating for better prices for pharmaceuticals and other services (McDonald 1995).

Prison systems will also seek to cut costs through privatizing prisoner health services. Recently, the state of Texas created, through enabling legislation, a managed health care system in which the prison system contracts with two public state medical schools—the University of Texas Medical Branch and the Texas Tech University Health Sciences Center—to provide medical services to inmates (Caruso. 1995). Like all health maintenance organizations in the public sector, the prison health providers will seek to contain costs. To this end, they have implemented telemedicine and the capitation of prison health care services as strategies to cut $100 million from the Texas Department of Criminal Justice budget over the next 2 years (Mooney and Mendelsohn 1995, pp. 4–5). How and to what extent private vendors can balance the constitutional obligation to provide health care and at the same time contain costs at a profit is a question that deserves serious attention.

Maintaining an accredited facility or prison system may well be too costly and force some agencies to relinquish accreditation. Prisoner health care costs, like those in the free community, will increase in the future. Increasing health care costs coupled with the costs of complying with standards may force some agencies to rethink the overall worth of accreditation. Ironically, accreditation may also be locking present prison managers into providing health services that surpass free world care and may be beyond current fiscal and programmatic realities. Maintaining compliance with accepted practices is good policy. We do not suggest that agencies give up accreditation, but the unintended consequences of accreditation may create fiscal problems in the long term.

Finally, one of the main concerns in the current debate on health care in our society involves health services for the uninsured. Nearly three quarters of those persons in poverty lack health insurance and typically have the highest levels of morbidity and mortality (Zaldivar 1995). There is also considerable evidence showing the overlap between the health condition of individuals in poverty and those in the incarcerated population. Federal policies designed to improve the health conditions of the poor must include prisoners (King and Whitman 1981). Improvements in the health condition of prisoners during confinement will be felt in the urban society at large (Raba and Obis 1983; Schilling et al. 1994). Perhaps prison infirmaries could be dynamically linked with the community health care system to assist in the detection and prevention of many chronic, acute, and communicable diseases, including HIV.

⬚ Directions for Future Research

Further research on prison health services must go in four directions. First, researchers should examine the consequences of confinement on an inmate's health condition. To do this, they need to reconceptualize incarceration as an aspect of a criminal career within the broader context of life course research (Hagan 1989). The life course metaphor has direct applicability to prisoner health studies because it links health-related factors in the free world to health conditions in the prison and beyond. Longitudinal studies, involving panels of inmates with different offenses and custody levels, must be conducted to assess health status and medical utilization patterns at different points in time. Of special interest would be "prisoner life course" research on the conditions of extreme confinement (i.e., death row, administrative segregation, and protective custody).

Second, given the current interest in enacting long sentences, researchers could examine population mortality within the prison. For example, the overall life expectancy in the wider society has increased to approximately 75 years. Future inquiry might explore life expectancy in prison and the psychosocial processes of death and dying in prison. How long do prisoners live? Epidemiological research conducted on the prisoner population will contribute to our understanding of the dynamic relationship between health conditions in the wider society and in prisoner populations. Policy makers could then utilize this information in formulating crime control and correctional health care policies.

Third, understanding who utilizes correctional health care services is linked to health conditions and patterns in the wider society. Researchers in general and correctional health care providers specifically might monitor health and economic trends in the wider society (particularly in the South), especially among those in the lower socioeconomic levels, to anticipate the health condition of prisoners at admission and to assist in the planning and resource allocation process.

Fourth, accreditation and compliance with essential or mandatory standards may become too expensive, and some facilities may give up their accredited status. Researchers need to examine the relationship between accreditation, health care costs, and litigation.

✒ Conclusion

The prison is not immune from or isolated from health inequalities and trends in the wider society. Most prisoners come from lower socioeconomic groups; thus the inverse relationship between health and income has important consequences for criminal justice organizations, and especially prisons. Research over the last several decades on the prison population consistently portrays its members to be poor, non-White, uneducated, inner-city dwellers. Offenders also enter prison with health problems far in excess of those reported in the noninstitutionalized population.

Currently, criminal justice policy making is dominated by conservative thought. As a policy goal, protection of the public is thought to be achieved by enacting long punitive sentences and the reduction or elimination of good time to increase time served in prison. To house the growing numbers of prison admissions, many states have embarked on massive prison construction programs. Whether or not policy makers have given sufficient thought to prison health services and resources remains to be seen. One major fact of life is that as more offenders (who bring to the prison a significant number of health risks and conditions) are sentenced to prison for long periods of time, their demands on health services will grow and so too will the costs of such services. Health care costs are increasing throughout society, and prison organizations can expect the same trend.

Should we provide prisoners with health programs to simply maintain their health, or should we provide them with health services to improve their health condition while in prison *above* what they could expect on the streets? Should prisoners pay for their medical services? Deliberate indifference to the health needs of prisoners violates the constitution. If we classify some inmates as high security risks and confine them for decades, and their health condition worsens, who is responsible? Inmates may open a new round of litigation to obtain answers to these questions.

Who now reads Clemmer? Nearly 60 years ago Donald Clemmer (1940) suggested that the prison was a community impacted by social trends and processes far beyond the walls. His notion of the prison community is important for understanding the link between health inequalities in the free world and the parallels of such inequalities within the prison community. Scholars and policy makers can no longer ignore the relationship between health status and the administration of justice, because this relationship has powerful implications for the way we treat criminal offenders. We hope that future research on prisoner health issues can be used to answer perennial questions about the goals of incarceration and expand our knowledge about the consequences of confinement.

⬚ References

American Correctional Association. (ACA). (1995). *Accredited Facilities and Programs.* Laurel, MD: American Correctional Association, Commission for Accreditation for Corrections. (tel: 301-206-5044)

——. Undated. *Accreditation Guidelines* (BluePrint for Corrections). Laurel, MD: American Correctional Association, Commission for Accreditation for Corrections. (tel: 301-206-5044)

Anno, B. Jaye. (1990). "The Cost of Correctional Health Care: Results of a National Survey." *Journal of Jail and Prison Health* 9: 105–127.

Baird, J. (1977). "Health Care in Correctional Facilities." *Journal of the Florida Medical Association* 64: 813–818.

Belbot, Barbara, Steve J. Cuvelier, and James W. Marquart. (1995). "Assessing Current Prisoner Classification Systems, Legal Environments, and Technological Developments." Unpublished final report, National Institute of Justice, Washington, DC.

Blau, Judith and Peter Blau. (1962). "The Cost of Inequality: Metropolitan Structure and Violent Crime." *American Sociological Review* 47: 114–129.

Blumberg, Mark. (1990). "Issues and Controversies With Respect to the Management of AIDS in Corrections." Pp. 195–210 in *AIDS: The Impact of the Criminal Justice System,* edited by M. Blumberg. Washington, DC: National Institute of Justice.

Braithwaite, John. (1989). *Crime, Shame, and Reintegration.* Cambridge: Cambridge University Press.

Brien, Peter and Caroline Harlow. (1995). "HIV in Prisons and Jails, 1993." *Bureau of Justice Statistics Bulletin,* Washington, DC: U.S. Department of Justice.

Briscoe, Kathy and Joyce Kuhrt. (1992). "How Accreditation Has Improved Correctional Health Care." *American Jails* (September/October): 48–52.

Brown v. Board of Education, 347 U.S. 483 (1954).

Bureau of Justice Statistics. (1993). *Survey of State Prison Inmates, (1992).* Washington, DC: U.S. Department of Justice.

Bureau of Justice Statistics. (1994). *National Corrections Reporting Program, 1991.* Washington, DC: U.S. Department of Justice, Office of Justice Programs, Bureau of Justice Statistics.

Carroll, Leo. (1974). *Hacks, Blacks, and Cons.* Lexington, MA: Lexington Books.

Caruso, Kleanthe. (1995). "Correctional Managed Health Care in the State of Texas." *CorrectCare* 9: 6.

Cella, David E., John Orav, and Alice Kornblith. (1991). "Socioeconomic Status and Cancer Survival." *Journal of Clinical Oncology* 9: 1500–1509.

Clemmer, Donald. (1940). *The Prison Community.* Boston: Christopher.

Cooper v. Pate, 388 U.S. 546 (1964).

Corrections Yearbook. (1986–1995). South Salem, NY: Criminal Justice Institute.

Criminal Justice Newsletter. (1989). November, p. 15.

Crouch, Ben M. and James W. Marquart. (1989). *An Appeal to Justice: Litigated Reform of Texas Prisons.* Austin: University of Texas Press.

Davis, Karen, Marsha Gold, and Diane Makuc. (1981). "Access to Health Care for the Poor: Does the Gap Remain?" *Annual Review of Public Health* 2: 159–182.

Davis, Karen and Diane Rowland. (1983). "Uninsured and Underinsured: Inequities in Health Care in the United States." *The Milbank Quarterly* 61: 149–176.

Dennis, Douglas. (1994). "The Living Dead." *The Angolite* 19(September/October): 20–57.

Department of Health and Human Services. (1990). *Healthy People 2000.* Washington, DC: U.S. Government Printing Office.

Derro, Robert. (1978). "Admission Health Evaluation of Inmates of a City-County Workhouse." *Minnesota Medicine* 61: 333–337.

DiIulio, John. (1990). *Courts, Corrections, and the Constitution: The Impact of Judicial Intervention of Prisons and Jails.* New York: Oxford University Press.

Dubler, Nancy. (1986). *Standards for Health Services in Correctional Institutions.* Washington, DC: American Public Health Association, Prisons and Jails Task Force.

Dugger, Richard. (1988). "The Graying of America's Prisons." *Corrections Today* (June): 26–34.

Duster, Troy. (1987). "Crime, Youth Unemployment and the Underclass." *Crime & Delinquency* 33: 300–316.

Enter, Jack. (1995). "Aging Populations in a Correctional Facility." *Corrections Managers' Report* (June/July):11, 15.

Estelle v. Gamble, 429 U.S. 97 (1976).

Farley, Thomas and John Flannery. (1989). "Late-Stage Diagnosis of Breast Cancer in Women of Lower Socioeconomic Status: Public Health Implications." *American Journal of Public Health* 79: 508–512.

Fasano, Charles and B. Jaye Anno. (1988). "Health Care Accreditation—Is It Worth It?" *American Jails* (Spring): 24–28.

Feeley, Malcolm and Roger Hanson. (1990). "The Impact of Judicial Intervention on Prisons and Jails: A Framework for Analysis and a Review of the Literature." Pp. 12–46 in *Courts, Corrections, and the Constitution: The Impact of Judicial Intervention of Prisons and Jails,* edited by J. DiIulio. New York: Oxford University Press.

Feinstein, Jonathan. (1993). "The Relationship Between Socioeconomic Status and Health: A Review of the Literature." *The Milbank Quarterly* 71: 279–322.

Fitzgerald, Edward, David D'Atri, Stanislav Kasl, and Adrian Ostfeld. (1984). "Health Problems in a Cohort of Male Prisoners at Intake and During Incarceration." *Journal of Jail and Prison Health* 4: 61–76.

Fogel, Catherine. (1988). "Expecting in Prison: Preparing for Birth Under Conditions of Stress." *Journal of Obstetrics and Gynecology and Neonatal Nursing* 15: 454–458.

Goetting, Ann. (1983). "The Elderly in Prison: Issues and Perspectives." *Journal of Research in Crime and Delinquency* 2: 291–309.

Good, David and Maureen Pirog-Good. (1987). "A Simultaneous Probit Model of Crime and Employment for Black and White Teenage Males." *Review of Black Political Economy* 16: 109–127.

Haas, J. Eugene and Thomas Drabek. (1973). *Complex Organizations.* New York: Macmillan.

Hagan, John. (1989). *Structural Criminology.* New Brunswick, NJ: Rutgers University Press.

Hammett, Theodore, Lynne Harrold, Michael Gross, and Joel Epstein. (1994). *1992 Update: HIV/AIDS in Correctional Facilities.* Washington, DC: National Institute of Justice.

Holt v. Sarver, 442 F. Supp. 83 (1971).

Hopper, Leigh. (1995). "Biggest Killer in Texas Prisons: AIDS." *Austin American-Statesman,* September 16.

House, James, Ronald Kessler, A. Regula Herzog, Richard Mero, Ann Kinney, and Martha Breslow. (1990). "Age, Socioeconomic Status, and Health." *The Milbank Quarterly* 68: 383–411.

Irwin, John. (1970). *The Felon.* Englewood Cliffs, NJ: Prentice Hall.

Jackson v. Bishop, 404 F.2d. 571 (8th Cir. 1978).

Jacobs, James B. (1977). "Macrosociology and Imprisonment." Pp. 89–107 in *Corrections and Punishment,* edited by D. Greenberg. Beverly Hills, CA: Sage.

——. (1980). "The Prisoners' Rights Movement and Its Impacts, 1960–1980." Pp. 429–470 in *Crime and Justice: An Annual Review of Research,* edited by N. Morris and M. Tonry. Chicago: University of Chicago Press.

Jamieson, Scott. (1985). "The Effect of Incarceration on the Right to Die." *New England Journal of Criminal and Civil Confinement* 11: 395–419.

Jones, David. (1976). *Health Risks of Imprisonment.* Lexington, MA: Lexington Books.

Kart, Cary and Ruth Dunkle. (1989). "Assessing Capacity for Self-Care Among the Aged." *Journal of Aging and Health* 1: 430–450.

King, Lambert and Steven Whitman. (1981). "Morbidity and Mortality Among Prisoners: An Epidemiologic Review." *Journal of Prison Health* 1: 7–29.

Kitagawa, Evelyn and Philip Hauser. (1973). *Differential Mortality in the United States: A Study in Socioeconomic Epidemiology.* Cambridge: Harvard University Press.

Krisberg, Barry. (1975). *Crime and Privilege.* Englewood Cliffs, NJ: Prentice Hall.

Latour, Jaime, Vicent Lopez, Manuel Rodriguez, Andreu Nolasco, and Carlos Alvarez-Dardet. (1991). "Inequalities in Health in Intensive Care Patients." *Journal of Clinical Epidemiology* 44: 889–894.

Levinson, Arlene. (1995). "State Three Strike Laws Swing, Miss at Deterring Criminals." *Houston Chronicle,* October 1, p. 1.

Lewin-Epstein, Noah, (1991). "Determinants of Regular Source of Health Care in Black, Mexican, Puerto Rican, and Non-Hispanic White Populations." *Medical Care* 29: 543–557.

Lilienfeld, Abraham. (1976). *Foundations of Epidemiology.* New York: Oxford University Press.

Litt, Iris and Michael Cohen. (1974). "Prisons, Adolescents, and the Right to Quality Medical Care." *American Journal of Public Health* 64: 894-97.

Lyons, James, Robert Griefinger, and Terrence Flanery. (1994). *Deaths of New York State Inmates 1978–1992.* Albany: N. Y. S. Department of Correctional Services.

Maier, Gary, Michael Bernstein, and Edmund Musholt. (1989). "Personal Coping Mechanisms for Prison Clinicians: Towards Transformation." *Journal of Jail and Prison Health* 8: 29–40.

Makuc, Diane, Jacob Feldman, Joel Kleinman, and Mitchell Pierre. (1990). "Sociodemographic Differentials in Mortality." Pp. 155–171 in *Health Status and Well-being of the Elderly,* edited by J. Cornoni-Huntley, R. Huntley, and J. Feldman. New York: Oxford University Press.

Mannheim, Hermann. (1939). *The Dilemma of Penal Reform.* London: George Allen and Unwin Ltd.

Manton, Kenneth, Clifford Patrick, and Katrina Johnson. (1987). "Health Differentials Between Blacks and Whites: Recent Trends in Mortality and Morbidity." *The Milbank Quarterly* 65: 129–199.

Martin, Steve and Sheldon Ekland-Olson. (1987). *Texas Prisons: The Walls Came Tumbling Down.* Austin: Texas Monthly Press.

McDonald, Douglas. (1995). *Managing Health Care and Costs.* Washington, DC: U.S. Department of Justice, National Institute of Justice.

McGaha, Glenda. (1987). "Health Care Issues of Incarcerated Women." *Journal of Offender Counseling Services and Rehabilitation* 12: 53–59.

Miller, Rod. (1992). "Standards and the Courts: An Evolving Relationship." *Corrections Today* 54 (3): 58–60.

Mooney, Sarah and Bruce Mendelsohn. (1995). "TX Conducts Health Care by Headcount." *Corrections Alert* 2(11): 4–5.

Murton, Tom and Joe Hyams. (1969). *Accomplices to the Crime.* New York: Grove.

Nathanson, Constance. (1977). "Sex Roles as Variables in Preventive Health Behavior." *Journal of Community Health* 3: 142–155.

——. (1978). "Sex Roles as Variables in the Interpretation of Morbidity Data." *International Journal of Epidemiology* 7: 253–262.

——. (1984). "Sex Differences in Mortality." Pp. 191–213 in *Annual Review of Sociology,* Vol. 10, edited by R. Turner and J. Short. Palo Alto, CA: Annual Reviews Inc.

National Center for Health Statistics. (1992). *Health, United States, and Healthy People 2000 Review.* Hyattsville, MD: Public Health Service.

"National Commission Accredits Forty Facilities." (1994). *CorrectCare* 8(4): 3–4, 10.

National Commission on Correctional Health Care (NCCHC). (1992). *Standards of Health Services in Prison, 1992.* Chicago: NCCHC.

——. (1994). *Women's Health Care in Correctional Settings.* Chicago: NCCHC.

——. (1995). *Health Services Accreditation.* Chicago: NCCHC.

Neisser, Eric. (1977). "Is There a Doctor in the Joint? The Search for Constitutional Standards for Prison Health Care." *Virginia Law Review* 63: 921–973.

Newman v. Alabama, 349 F. Supp. 278 (1972).

Novick, Lloyd and Mohamed Al-Abrahim. (1977). *Health Problems in the Prison Setting.* Springfield, IL: Charles C Thomas.

Novick, Lloyd, Richard Della Penna, Melvin Schwartz, Elaine Remmlinger, and Regina Loewenstein. (1977). "Health Status of the New York City Prison Population." *Medical Care* 15: 205–216.

Ortmeyer, Linda. (1979). "Females Natural Advantage?" *Women and Health* 42: 121–133.

Packer, Herbert. (1968). *The Limits of Criminal Sanction.* Stanford, CA: Stanford University Press.

Perkins, Craig. (1994). *National Corrections Reporting Program, 1992.* Washington, DC: U.S. Department of Justice, Bureau of Justice Statistics.

Prout, Curtis and Robert Ross. (1988). *Care and Punishment: The Dilemma of Prison Medicine.* Pittsburgh: University of Pittsburgh Press.

Raba, John and Clare Obis. (1983). "The Health Status of Incarcerated Urban Males: Results of Admission Screening." *Journal of Jail and Prison Health* 3: 6–24.

Rector, Frank. (1929). *Health and Medical Service in American Prisons and Reformatories.* New York: National Society of Penal Information.

Rice, Dorothy and Jacob Feldman. (1983). "Living Longer in the United States: Demographic Changes and Health Needs of the Elderly." *The Milbank Quarterly/Health and Society* 61: 362–396.

——. (1989). "Health and Long-Term Care for the Aged." *American Economic Review* 79: 343–348.

Rieman, Jeffery. (1990). *The Rich Get Richer and the Poor Get Prison.* New York: Macmillan.

Schilling, Robert, Nabila El-Bassel, Andre Ivanoff, Louisa Gilbert, Kuo-Hsien Su, and Steven Safyer. (1994), "Sexual Risk Behavior of Incarcerated, Drug-Using Women. 1992." *Public Health Reports* 109: 539–546.

Sechrest, Dale. (1976). "The Accreditation Movement in Corrections." *Federal Probation* 40(4): 15–19.

Selik, Richard, Kenneth Castro, Marguerite Papaionnou, and James Ruehler. (1989). "Birthplace and the Risk of AIDS Among Hispanics in the United States." *American Journal of Public Health* 79(7): 836–839.

Sheps, Samuel, Martin Schechter, and Real Prefontaine. (1987). "Prison Health Services: A Utilization Study." *Journal of Community Health* 12: 4–22.

"States Leaping on Anti-Crime Bandwagon." (1995). *Corrections Alert* 2: 1.

Suls, Jerry, Gerald Gaes, and Virginia Philo. (1991). "Stress and Illness Behavior in Prison: Effects of Life Events, Self-Care Attitudes, and Race." *Journal of Prison and Jail Health* 10: 117–132.

"Thirteen Facilities Awarded Initial NCCHC Accreditation." (1995). *CorrectCare* 9(3): 3.

Twaddle, Andrew. (1976). "Utilization of Medical Services by a Captive Population." *Journal of Health and Social Behavior* 17: 236–248.

"VA Copayment Plan Yields Impressive Results." (1995). *Corrections Alert* 2: 3.

Verbrugge, Lois. (1976). "Sex Differentials in Morbidity and Mortality in the United States." *Social Biology* 23: 275–296.

——. (1980). "Recent Trends in Sex Mortality Differentials in the United States." *Women and Health* 5: 17–37.

——. (1984). "Longer Life but Worsening Health? Trends in Health and Mortality of Middle-Aged and Older Persons." *The Milbank Quarterly* 62: 475–519.

——. (1985). "Gender and Health: An Update on Hypotheses and Evidence." *Journal of Health and Social Behavior* 26: 156–182.

Waldron, Ingrid. (1983). "Sex Differences in Human Mortality: The Role of Genetic Factors." *Social Science and Medicine* 17: 321–333.

Walker, Alonzo, Loni Neal, Robert Ausman, Julianne Whipple, and Barbara Doherty. (1989). "Per Capita Income in Breast Cancer Patients." *Journal of the National Medical Association* 81: 1065–1068.

Wallace, Susan, Jody Klein-Saffran, Gerald Gaes, and Kenneth Moritsugu. (1991). "Health Status of Federal Inmates: A Comparison of Admission and Release Medical Records." *Journal of Jail and Prison Health* 10: 133–151.

Weiner, Janet and B. Jaye Anno. (1992). "The Crisis in Correctional Health Care: The Impact of the National Drug Control Strategy on Correctional Health Services." *Annals of Internal Medicine* 117: 71–77.

Whalen, Robert and John Lyons. (1962). "Medical Problems of 500 Prisoners on Admissions to a County Jail." *Public Health Reports* 77: 497–502.

Wilson, William Julius. (1993). *The Ghetto Underclass.* Newbury Park, CA: Sage.

Wingard, Deborah. (1982). "The Sex Differential in Mortality Rates." *American Journal of Epidemiology* 115: 205–216.

Wolinsky, Fredric, Ray Mosely, and Rodney Coe. (1986). "A Cohort Analysis of the Use of Health Services by Elderly Americans." *Journal of Health and Social Behavior* 27: 209–219.

Zaldivar, R. A. (1995). "Poverty Rate Falls, but Median Income Remains Stagnant." *Houston Chronicle,* October 6.

DISCUSSION QUESTIONS

1. According to the authors, what is the relationship between health conditions in the free world and in the prison?

2. What impact will the decline in the health status of the American population in general have on prison populations?

3. Why are inmates generally in poorer health than people in the free world?

❖

Glossary

Addiction: A "prototypical psychobiological illness, with critical biological, behavioral, and social context elements" (Leshner, 1998, p. 5).

Amicus curiae: "Friend of the court" briefs presented to the court arguing in support of one side or the other by interested parties not directly involved with the case.

Antiterrorism and Effective Death Penalty Act: Mostly about antiterrorism and the death penalty rather than an act specifically designed to limit habeas corpus proceedings. It was passed in response to the bombing of the Murrah Federal Building in Oklahoma City, with the reform of habeas corpus law being a rider to it. The AEDPA does not eliminate an inmate's right to habeas corpus but does restrict its availability.

Balanced approach: A three-pronged goal of the juvenile justice system: (1) to protect the community, (2) to hold delinquent youths accountable, and (3) to provide treatment and positive role models for youths (Carter, 2006).

Bill of Rights: The first 10 Amendments to the United States Constitution.

Bridewells: Institutions also called workhouses that were established in every English county in 1576. The idea behind these institutions was that, if vagrant youths (or adults) were removed from the negative influences of street life, they could be reformed by discipline, hard work, and religious instruction.

Child savers: A group of well-funded and highly educated liberal reformers. They were social reformers of the late 1800s and early 1900s who attacked the operations of the houses of refuge, which they saw as punitive rather than rehabilitative. The child savers

argued that children should be deinstitutionalized and placed in settings that provided a more family-like atmosphere, where they could be taught the value of hard work and other solid middle-class values.

Classical explanation of criminal behavior: Assumed that humans are hedonistic, rational, and endowed with free will. This was essentially the Enlightenment concept of human nature.

Coequal staffing: Personnel processes that provide comparable pay and benefits to those who work in the jail and those who work on the streets as law enforcement in a sheriff's department.

Cognitive self-change programming: Attempts to change the antisocial and self-destructive behavior of offenders into prosocial and constructive behavior by getting them to identify and end the thinking errors associated with criminal engagement.

Community corrections: Any activity performed by agents of the state to assist offenders to establish or reestablish functional law-abiding roles in the community while at the same time monitoring their behavior for criminal activity.

Community jails: Facilities organized so those inmates engaged in education, drug or alcohol counseling, or mental health programming will seamlessly receive such services while incarcerated (primarily by community experts in those areas) and again as they transition out of the facility.

Community service order: Part of a disposition requiring the probationer to work a certain number of hours doing some tasks to help their communities. This work can range from cleaning graffiti from walls to picking up trash along highways or in parks.

Congregate, but silent labor systems: Included in early prisons (such as Auburn and Sing Sing in New York state), these enabled prisons to offset the cost of incarceration by allowing inmates to work together and hence produce more. When inmates were allowed out of their cells, they could also maintain the prison, which also reduced the cost of incarceration. Usually inmates were required to be silent.

Convict lease system: Inmates' labor was sold by the prison to farmers or other contractors. As the supply of prisoners, mostly in the form of ex-slaves, was plentiful during the post–Civil War period, their lives were treated cheaply, and they were often not fed or clothed or sheltered adequately.

Correctional boot camps: Facilities modeled after military boot camps. Relatively young and nonviolent offenders are most typically the kinds of offenders sent to correctional boot camps for short periods (90–180 days) where they are subjected to military-style discipline and physical and educational programs.

Corrections: A generic term covering a wide variety of functions carried out by government (and, increasingly, private) agencies having to do with the punishment, treatment, supervision, and management of individuals who have been accused or convicted of criminal offenses. These functions are implemented in prisons, jails, and other secure institutions, as well as in community-based correctional agencies such as probation and parole departments.

Decent prison: According to Johnson (2002), not a facility that necessarily has more programming, staffing, or amenities than the norm (though he thinks more of these might be helpful) but rather an institution or program that is relatively free of violence and includes some opportunities so that inmates might find a *niche* to be involved in. In order for inmates to find this niche, however, decent prisons need to include some opportunities for inmates to act autonomously.

Delinquents: Juveniles who commit acts that would be considered criminal if committed by adults.

Deterrence: The prevention of crime by the threat of punishment and a more complex justification for punishment than retribution. The principle that people respond to incentives and are deterred from crime by the threat of punishment is the philosophical foundation behind all systems of criminal law. Deterrence may be either specific or general.

Discretion: The ability to make choices and to act or not act on them.

Discretionary parole: Parole granted at the discretion of a parole board for selected inmates who have earned it.

Double deviants: Women and girls who violate the criminal law. They have been deviant criminally (like men), but they also have deviated from societal expectations for their gender.

Eighth Amendment: Forbids cruel and unusual punishment. What constitutes cruel and unusual punishment is punishment applied "maliciously and sadistically for the very purpose of causing harm" (*Hudson v. McMillian*, 1992), although the onus is on the correctional client to prove that the punishment was so applied.

Electronic monitoring: A system by which offenders under house arrest can be monitored for compliance using computerized technology. In modern EM systems, an electronic device worn around the offender's ankle sends a continuous signal to a receiver attached to the offender's house phone. If the offender moves beyond 500 feet from his or her house, the transmitter records it and relays the information to a centralized computer. A probation or parole officer is then dispatched to the offender's home to investigate whether he or she has absconded or removed or tampered with the device.

Enlightenment: A period in human history when a major shift in the way people began to view the world and their place in it occurred. This new worldview questioned traditional religious and political values and began to embrace humanism, rationalism, and a belief in the primacy of the natural world. The Enlightenment also ushered in the beginnings of a belief in the dignity and worth of all individuals, a view that would eventually find expression in the law and in the treatment of criminal offenders.

Factory prisons: Inmates labored together to produce goods, sometimes for the state and sometime for private contractors.

First Amendment: Guarantees freedom of religion, speech, press, and assembly. These rights may be provided to prison inmates, although concerns of institutional safety and security must receive primary consideration.

Fourteenth Amendment: Contains the due process clause, which declares that no state shall deprive any person of life, liberty, or property without due process of law. This clause was used by the courts to incorporate (or apply) the Bill of Rights to citizens in state courts.

Fourth Amendment: Guarantees the right to be free from unreasonable searches and seizures. What is reasonable inside prison walls is, of course, quite different from what is reasonable outside them. For all practical purposes, inmates and all other correctional clients have no Fourth Amendment protections.

General deterrence: Refers to the preventive effect of the threat of punishment on the general population and is thus aimed at potential offenders. Punishing offenders serves as an example to the rest of us of what might happen if we violate the law.

Habeas corpus: Latin term that literally means "you have the body" and that is basically a court order requiring that an arrested person be brought before it to determine the legality of his or her detention.

Hack: A correctional officer in a prison who is a violent, cynical, and alienated keeper of inmates (Johnson, 2002).

Halfway houses: Transitional places of residence for correctional clients that are, in terms of strictness of supervision, "halfway" between the constant supervision of prison and the much looser supervision of a probationer or parolee in the community.

Hands-off doctrine: A court articulated belief that the judiciary should not interfere with the management and administration of prisons (or corrections in general). The doctrine rested primarily on the status of prisoners, who suffered a kind of legal and civil death upon conviction.

Hedonism: A doctrine that maintains that all life goals are desirable only as means to the end of achieving pleasure or avoiding pain. Pleasure is intrinsically desirable and pain is intrinsically undesirable, and we all seek to maximize the former and minimize the latter.

Hedonistic calculus: A method by which individuals are assumed to weigh logically the anticipated benefits of a given course of action against its possible costs.

Heritability: A quantitative measure, ranging between 0.0 and 1.0, used by geneticists to determine the proportion of variance in a trait attributable to genes.

Hostile environment sexual harassment: Behavior resulting in a workplace sexualized with jokes, pictures, or in other ways that are offensive to one gender.

House arrest: A program used by probation and parole agencies that requires offenders to remain in their homes, at all times, except for approved periods to travel to work or school and, occasionally, to other approved destinations.

Human service officer: A correctional officer who provides "goods and services"; serves as an "advocate" for inmates, when appropriate; and assists them with their "adjustment" through the use of "helping networks" (Johnson, 2002, p. 242–259).

Incapacitation: Refers to the inability of criminals, while they are locked up, to victimize people outside prison walls.

Intensive supervision probation: Involves more frequent surveillance of the probationer and is typically limited to more serious offenders than those on regular probation in the belief that there is a fighting chance that they may be rehabilitated or in an effort to save the costs of incarceration. Higher levels of supervision allow officers to coerce offenders into a wide variety of educational and treatment programs and other prosocial activities designed to provide offenders with social capital.

Jails: First type of correctional institution created. When some sort of legal representative, either a sheriff or a designee, holds people so that they will appear at trial, then wherever these detainees are placed is, by definition, a jail. Jails developed originally as a primary means of holding the accused for trial or the convicted for execution or in lieu of their paying a fine, and the form jails took varied by the sophistication and size of the community and its relative remoteness (Kerle, 2003; Welch, 2004). They have been in existence much longer than prisons, and their mission is much more diverse. They hold people who are presumed innocent before trial, convicted people before they are sentenced, more minor offenders who are sentenced for terms that are usually less than a year, juveniles, women, and people at the behest of state or federal authorities. Jails are multifaceted, usually local facilities with a complex mission.

Judicial reprieve: A delay in sentencing following a conviction that often became permanent contingent on the offender's behavior. Early American courts also used judicial reprieve, whereby a judge would suspend the sentence, and the defendant would be released on his or her own recognizance.

Juvenile waiver: The transfer of juveniles to adult criminal court where they lose their status as minors and become legally culpable for their alleged crime and subject to criminal prosecution and punishments.

Liberal feminist: One who believes that the problem for girls and women involved in crime lies more within the social structure around these women and girls (e.g., poverty and lack of sufficient schooling or training, along with society's patriarchal beliefs) and that the solution lies in preparing women for an alternate existence so that they will not turn to crime.

Liberty interest: Refers to government-imposed changes in someone's legal status that interfere with his or her constitutionally guaranteed rights to be free of such interference.

Mandatory parole: Automatic parole for almost all inmates (Petersilia, 2000).

Marks system: Involves credits earned for the speedy and efficient performance of mandated tasks, as well as for overall good behavior.

Mature coping: "Means, in essence, dealing with life's problems like a responsive and responsible human being, one who seeks autonomy without violating the rights of others, security without resort to deception or violence, and relatedness to others as the finest and fullest expression of human identity" (Johnson, 2002, p. 83).

Medical model: A paradigm defining crime as a moral sickness that required treatment. Under the medical model, prisoners were to remain in custody with indeterminate sentences until "cured" of their criminal ways.

Mental disorders: Defined by the World Health Organization as "clinically significant conditions characterized by alterations in thinking, mood (emotions), or behavior associated with personal distress and/or impaired functioning"; additionally, they "are not variations within the range of 'normal,' but are clearly abnormal or pathological phenomena" (in Brookman, 2005, p. 87).

Mortification: Occurs as inmates enter the prison and suffer from the loss of the many roles they occupied in the wider world (Goffman, 1961; Sykes, 1958). Instead, only the role of "inmate" is available and that role is formally powerless and dependent.

Needs principle: A principle of effective offender rehabilitation that refers to an offender's needs, the lack of which puts him or her at risk for re-offending, and suggests that these needs receive high priority.

New generation or podular direct-supervision jails: These kinds of jails have two key components that include a rounded or "podular" architecture for living units and the "direct," as opposed to indirect or intermittent, supervision of inmates by staff; in other words, staff are to be in the living units full time. Other important facets of these jails are the provision of more goods and services in the living unit (e.g., access to telephones, visiting booths, recreation, library books, etc.) and the more enriched leadership and communication roles for staff.

Overcrowding: Occurs when the number of inmates exceeds the physical capacity (the beds) available.

Pains of imprisonment: Gresham Sykes (1958) described the pains of imprisonment as the "deprivation of liberty, the deprivation of goods and services, the deprivation of heterosexual relationships, the deprivation of autonomy, and the deprivation of security" (pp. 63–83).

Parens patriae: A phrase meaning, essentially, "state as parent." *Parens patriae* gives the state the right to intercede and act in the best interest of the child or of any other legally incapacitated persons, such as the mentally ill. This meant that the state and not the parents had the ultimate authority over children and that children who were delinquent could be removed from their families and placed in the custody of the state.

Parole: Refers to the release of convicted criminals from prison before they have completed their full sentences. Parole is different from probation in two basic ways. First, parole is an administration function practiced by a parole board, which is part of the executive branch of government, while probation is a judicial function. Second, parolees have spent time in prison before being released into the community whereas probationers typically have not.

Patriarchy: Involves the attitudes, beliefs, and behaviors that value men and boys over women and girls (Daly & Chesney-Lind, 1988). Members of patriarchal societies tend to believe that men and boys are worth more than women and girls. They also believe that women and girls, as well as men and boys, should have certain restricted roles to play and that those of the former are less important than those of the latter. Therefore, education, work training that helps one make a living, and better pay are more important to secure for men and boys than for women and girls, who are best suited for more feminine and, by definition in a patriarchal society, less worthy professions.

Penitentiary: A term coined by John Howard, who considered the penitentiary a place of penitence and contemplation for convicted criminals.

Plantation prisons: In the post–Civil War period, these acted as proxies for slavery by keeping poor ex-slaves essentially enslaved.

Positivism: The belief that human actions have causes and that these causes are to be found in the uniformities that typically precede those actions; such causes might also be discovered through the use of the scientific method. The search for causes of human behavior led positivists to dismiss the classical notion that humans are free agents who are alone responsible for their behavior.

Power: Dahl (1957) defines power as, essentially, the ability to get others to do what they otherwise wouldn't.

Presentence investigations: Reports written by probation officers informing the judge about various aspects of the offense for which the defendant is being sentenced as well as about the defendant's background (educational, family, and employment history), character, and criminal history.

Prison Litigation Reform Act (1996): The primary intention of the PLRA was to free state prisons and jails from federal court supervision as well as to limit prisoners' access to the federal courts.

Prison with a philosophy of penitence: Hence penitentiary. A grand reform, it represented, in theory at least, a major improvement over the brutality of punishment that characterized early English law and practice (Orland, 1995).

Prisons: Correctional institution used for long-term and convicted offenders who are to be simultaneously punished, deterred, and reformed while being isolated from the community and incapacitated in regards to perpetrating further crimes (Goldfarb, 1975).

Probation and Parole: Statutory privileges granted by the state in lieu of imprisonment (in the first case) or further imprisonment (in the second case). Because of their privileged status, it was long thought that the state did not have to provide probationers and parolees any procedural due process rights either in the granting or revoking of either status. Today, probationers and parolees are granted some due process rights.

Probationer: A person who, upon conviction, receives a prison sentence that is suspended during the period of proving that he or she is capable of living a law-abiding life.

Profession: An occupation typified by four things: prior educational attainment, formal training on the job, pay and benefits that are commensurate with the work, and the ability to exercise discretion.

Professional standards: For their new hires, these include required college level educational background, sufficient training, and pay that is commensurate with job requirements.

Punishment: Referred to as *moralistic* or *retaliatory aggression* aimed at discouraging cheats. Punishment is observed in every species of social animal, leading evolutionary biologists to conclude that it is an evolutionarily stable strategy designed by natural selection both for the emergence and the maintenance of cooperative behavior.

Quid pro quo sexual harassment: Behavior, especially by someone in authority, that involves exchanging something for something, as in you give me sexual favors, and you get to keep your job.

Reentry: The process of reintegrating offenders or detainees back into the community after release from prison or jail. Part of that process is preparing people through the use of various programs targeting their risks and needs, so they can make the transition from jails or prisons to the community smoother and more successful, and so they do not commit more crime and return to a corrections facility. Successful reentry programs must include a recognition of the problems individual inmates may have (i.e., mental illness, physical illness, joblessness, and homelessness) and address them systematically in collaboration with the client and the community.

Reformatories: Created in the last half of the nineteenth century and into the twentieth, these were originally geared primarily toward juveniles, young adult male offenders, and women. They tended to reflect a distinctly rehabilitative purpose. Developing in tandem with the reformatory era and the reformatory were the indeterminate sentence, probation parole, and education or vocational training programs (Rotman, 1990; Sullivan, 1990). The reformatory style, with its more open and college-like campuses and rehabilitative focus, eventually came to influence adult male institutions as well, and reformatories remain in widespread use in some form in many states today.

Rehabilitation: To restore or return to constructive or healthy activity. Whereas deterrence and incapacitation are primarily justified philosophically on classical grounds, rehabilitation is primarily a positivist concept. The rehabilitative goal is based on a medical model that views criminal behavior as a moral sickness requiring treatment.

Reintegration: Strategy aiming to use the time criminals are under correctional supervision, either in institutions or in the community, to prepare them to reenter the free community as well equipped to do so as possible. This goal is also known as reentry or restoration.

Repressive justice: Principle of moral rightness driven by the natural passion for punitive revenge that "ceases only when exhausted . . . only after it has destroyed" (Durkheim, 1893/1964, p. 86).

Responsivity principle: A principle of effective offender rehabilitation averring that, if offenders are to respond to treatment in meaningful and lasting ways, counselors must be aware of offenders' different development stages, learning styles, and need to be treated with respect and dignity.

Restitutive justice: Principle of moral rightness driven by simple deterrence and the need of reparation for wrongs done. Restitutive justice is more humanistic and tolerant than repressive justice, although it is still "at least in part, a work of vengeance" because it is still "an expiation" (Durkheim, 1893/1964, p. 88–89).

Restorative justice: Defined by Champion (2005) as "every action that is primarily oriented toward justice by repairing the harm that has been caused by the crime." It "usually means a face-to-face confrontation between victim and perpetrator, where a mutually agreeable restorative solution is proposed and agreed upon" (p. 154). Restorative justice is often referred to as a balanced approach in that it gives approximately equal weight to community protection, offender accountability, and offender competency.

Retribution: The justification for punishment underlined by the concept of *lex talionis*. It is a "just deserts" model that demands that punishments match the degree of harm criminals have inflicted on their victims, i.e., what they justly deserve.

Risk principle: A principle of effective offender rehabilitation that refers to an offender's probability of re-offending and suggests that those with the highest risk be targeted for the most intense treatment.

Section 1983 suits: Allowed by the Supreme Court as a mechanism for state prison inmates to sue state officials in federal court regarding their confinement and their conditions of confinement. Section 1983 is part of the Civil Rights Act of 1871, which was initially enacted to protect southern blacks from state officials and the Ku Klux Klan. This act is now codified and known as 42 USC § 1983 or, simply, as "section 1983."

Sentencing guidelines: Devised to guide the sentencing decisions of judges by providing them with standard criteria for tailoring sentences to the crime and the offenders before the bench.

Sex offender: A very broad category of offenders ranging from relatively minor "flashers" to true sexual predators.

Shock probation: Programs that involve placing sentenced offenders in prison and releasing them after (typically) 30 days and placing them on probation. The idea is to shock them with the reality of prison so that they will be deterred from further criminal engagement.

Silent and penitent systems: Practiced in early prisons (such as the Eastern and Western Pennsylvania prisons), these isolated inmates in their cells, restricted their contact with others, and reinforced the need for penitence. When labor was allowed at all, it was a solitary affair in one's cell.

Specific deterrence: Refers to the effect of punishment on the future behavior of persons who experience the punishment.

Status offenders: Juveniles who commit certain actions that are legal for adults but not for children, such as smoking, drinking, not obeying parents, staying out at night to all hours, and truancy.

Status offenses: These apply only to individuals having the status of a juvenile, and they exist because the law assumes that juveniles lack the maturity to appreciate the long-term consequences of their behavior.

Subcultures: Also called environments with their own norms, values, beliefs, and even language, these tend to solidify when people are isolated from the larger culture and when members have continual contact with each other for an extended time.

Therapeutic communities: Residential communities offering long-term opportunities (typically 6 to 12 months) for attitude and behavioral change and to learn constructive prosocial ways of coping with life's strains and with others. These communities provide dynamic "mutual self-help" environments in which residents transmit and reinforce one another's acceptance of and conformity with the highly structured and stringent expectations of the TC and of the wider community.

Total institution: According to Goffman, a "place of residence and work where a large number of like-situated individuals, cut off from the wider society for an appreciable period of time, together lead an enclosed, formally administered round of life" (p. xiii). Another key component of this social world is that there are clear social strata in such institutions dividing the "inmates" and the "staff" (Goffman, 1961, p. 7). There are formal prohibitions against even minor social interactions between these two groups, and all of the formal power resides with one group (the staff) over the other group (the inmates).

Work release programs: Designed to control offenders in a secure environment while at the same time allowing them to maintain employment.

Credits and Sources

Introduction to the Book

Photo I.1: © Getty Images.

Photo I.2: © Corbis.

Figure I.1: FBI (2006). Crime in the United States, 2005. Washington, DC. Government Printing Office.

Figure I.2: The Sentencing Project (2005). Reproduced with permission.

Section I

Photo 1.1: © Getty Images.

Photo 1.2: © Getty Images.

Table 1.1: Information retrieved from the Georgia Department of Corrections (2007) website on February 28, 2007: www.dcor.state.ga.us/AboutGDC/CorrectionsCosts.

Section III

Photo 3.1: © Getty Images.

Photo 3.2: © Getty Images.

Figure 3.1: Bureau of Justice Statistics Correctional Surveys (The Annual Probation Survey, National Prisoner Statistics, Survey of Jails, and The Annual Parole Survey) as presented in Correctional Populations in the United States, Annual, Prisoners in 2004 and Probation and Parole in the United States, 2004.

Section IV

Photo 4.1: © Corbis.

Photo 4.2: © Corbis.

Figure 4.1: Death Penalty Information Center; Facts about the death penalty (2007).

Section V

Photo 5.1: Getty Images.

Photo 5.2: Corbis.

Table 5.1: In Walsh, A., and Ellis, L. (2007). *Criminology: An Interdisciplinary Approach.* Thousand Oaks, CA: Sage. Adapted from the Arrestee Drug Abuse Monitoring Program (Zhang, 2004).

Figure 5.1: Mathias, R. (1995, July/August). Correctional treatment helps offenders stay drug and arrest free. *NIDA Notes, 10*(4). Retrieved June 29, 2008, from http://www.drugabuse.gov/Nida_Notes/NNVol10N4/Prison.html

Section VI

Photo 6.1: Corbis.

Figure 6.1: Glaze, L. & S. Bonczar (2006). Probation and parole in the United States, 2005. *Bureau of Justice Statistics Bulletin.*

Section VII

Photo 7.1: Corbis.

Photo 7.2: Corbis.

Figure 7.1: Hughes, Wilson, & Beck (2001). Trends in state parole, 1990-2000. Washington, D.C.: Bureau of Justice Statistics.

Figure 7.2: Hughes, Wilson, & Beck (2001). Trends in state parole, 1990-2000. Washington, D.C.: Bureau of Justice Statistics.

Section VIII

Photo 8.1: Getty Images.

Photo 8.2: Corbis.

Table 8.1: Bureau of Justice Statistics Correctional Surveys (The Annual Survey of Jails and Census of Jail Inmates) as presented in Correctional Populations in the United States, 1997, and Prison and Jail Inmates at Midyear series, 1998-2006.

Table 8.2: Bureau of Justice Statistics Correctional Surveys (The Annual Survey of Jails and Census of Jail Inmates) as presented in Correctional Populations in the United States, 1997, and Prison and Jail Inmates at Midyear series, 1998-2006.

Table 8.3: Bureau of Justice Statistics Correctional Surveys (The Annual Survey of Jails and Census of Jail Inmates) as presented in Correctional Populations in the United States, 1997, and Prison and Jail Inmates at Midyear series, 1998-2006.

Table 8.4: Mumola, Christopher J. 2005. *Suicide and Homicide in State Prisons and Local Jails.* U.S. Department of Justice, Office of Justice Programs. www.ojp.usdoj.gov/bjs/pub/pdf/shplj.pdf

Section IX

Photo 9.1: Corbis.

Photo 9.2: Corbis.

Section X

Photo 10.1: Corbis.

Photo 10.2: Corbis.

Figure 10.1: Snyder, H. Juvenile Arrests 2005. [2007]. Washington, D.C.: Office of Juvenile Justice and Delinquency Prevention.

Figure 10.2: Sarah Livsey (2006) Juvenile delinquency probation caseload, 1985-2002. Office of Juvenile Justice and Delinquency Prevention: U.S. Department of Justice.

Table 1: Amnesty International and Human Rights Watch (2005). *The rest of their lives: Life without parole for children offenders in the United States* (p. 35). New York: Human Rights Watch.

Section XI

Figure 11.1: Bureau of Justice Statistics (2005c), U.S. Department of Justice, Office of Justice Programs.

Table 2: Adapted from 1999 Directory: Juvenile and Adult Correctional Departments, Institutions, Agencies, and Paroling Authorities, by the American Correctional Association, 1999 (p. xxxii).

2nd Table 2: Corrections Yearbook (1986-1995).

Table 4: Corrections Yearbook (1995).

References

Abadinsky, H. (1993). *Drug abuse: An introduction.* Chicago: Nelson-Hall Publishers.

Agnew, R. (2005). *Why do criminals offend?* Los Angeles: Roxbury.

Albanese, J., & Pursley, R. (1993). *Crime in America: Some existing and emerging issues.* Englewood Cliffs, NJ: Regents/Prentice-Hall.

Alcock, J. (1998). *Animal behavior: An evolutionary approach* (6th ed.). Sunderland, MA: Sinauer Associates.

American Correctional Association. (1983). *The American prison: From the beginning . . . A pictorial history.* Lanham, MD: American Correctional Association.

Amnesty International. (2005). *Death penalty: 3,977 executed in 2004.* Amnesty International: London.

Amnesty International. (2006). *Death sentences and executions in 2005.* Retrieved July 14, 2008, from http://web.amnesty.org

Anderson, T. L. (2006). Issues facing women prisoners in the early twenty-first century. In C. M. Renzetti, L. Goodstein, & S. L. Miller (Eds.), *Rethinking gender, crime and justice* (pp. 200–212). Los Angeles: Roxbury Publishing Company.

Andrews, D. A., Zinger, I., Hoge, R. D., Bonta, J., Gendreau, P., & Cullen, F. T. (2001). Does correctional treatment work? A clinically relevant and psychologically informed meta-analysis. In E. J. Latessa, A. Holsinger, J. W. Marquart, & J. R. Sorensen (Eds.), *Correctional contexts: Contemporary and classical readings* (2nd ed., pp. 291–310). Los Angeles, CA: Roxbury Publishing Company.

Applegate, B. K., Cullen, F. T., & Fisher B. S. (2001). Public support for correctional treatment: The continuing appeal of the rehabilitative ideal. In E. J. Latessa, A. Holsinger, J. W. Marquart, & J. R. Sorensen (Eds.), *Correctional contexts: Contemporary and classical readings* (2nd, pp. 506–519). Los Angeles, CA: Roxbury Publishing Company.

Applegate, B. K., & Paoline, E. A., III. (2007). Jail officers' perceptions of the work environment in traditional versus new generation facilities. *American Journal of Criminal Justice, 31,* 64–80.

Arax, M., & Gladstone, M. (1998, July 5). State thwarted brutality probe in Corcoran Prison, investigators say. *Los Angeles Times,* p. 1.

Augustus, J. (1972) *John Augustus: First probation officer* (Patterson Smith Reprint No. 130). Montclair, New Jersey: Patterson Smith. (Original work published 1852 in Boston by Wright & Hasty Printers under the title *A report of the labors of John Augustus* and reprinted in 1939 by the National Probation Association)

Austin, J., & Irwin, J. (2001). *It's about time: America's imprisonment binge.* Belmont, CA: Wadsworth.

Balazs, G. (2006). The workaday world of a juvenile probation officer. In A. Walsh *Correctional assessment, casework, and counseling* (4th ed., pp. 414–416). Alexandria, VA: American Correctional Association.

Barker, V. (2006). The politics of punishing: Building a state governance theory of American imprisonment variation. *Punishment & Society, 8,* 5–32.

Barlow, L. W., Hight, S., & Hight, M. (2006). Jails and their communities: Piedmont regional jail as a community model. *American Jails, 20,* 38–45.

Barton, A. (1999). Breaking the crime/drugs cycle: The birth of a new approach? *The Howard Journal, 38,* 144–157.

Baunach, P. J. (1992). Critical problems of women in prison. In I. L. Moyer (Ed.), *The changing role of women in the criminal justice system* (pp. 99–112). Prospect Heights, IL: Waveland Press.

Bazemore, G., & Umbreit, M. (1994). *Balanced and restorative justice.* Washington, DC: Office of Juvenile Justice and Delinquency Prevention, U.S. Department of Justice.

Beaumont, G. de., & Tocqueville, A. de. (1964). *On the penitentiary system in the United States and its application in France.* Carbondale, IL: Southern Illinois University Press. (Original work published 1833)

Beccaria, C. (1963). *On crimes and punishment* (H. Paulucci, Trans.). Indianapolis: Bobbs-Merrill. (Original work published 1764)

Beck, A. J. (2002). Jail population growth: National trends and predictors of future growth. *American Jails, 16,* 9–14.

Beck, A. J., Harrison, P. M., & Adams, D. B. (2007). *Sexual violence reported by correctional authorities, 2006.* Washington, DC: U.S. Department of Justice, Office of Justice Programs. Retrieved January 28, 2008, from www.ojp.usdoj.gov/bjs/pub/pdf/svrca06.pdf

Becker, G. (1997). The economics of crime. In M. Fisch (Ed.), *Criminology 97/98* (pp. 15–20). Guilford, CT: Dusskin.

Belknap, J. (2001). *The invisible woman: Gender, crime, and justice* (2nd ed.) Belmont, CA: Wadsworth.

Bennett, K. (1995). Constitutional issues in cross-gender searches and visual observation of nude inmates by opposite-sex officers: A battle between and within the sexes. *The Prison Journal, 75,* 90–112.

Bentham, J. (1930). *The rationale of punishment.* London: Robert Heward. (Original work published 1811)

Bentham, J. (1948). *A fragment on government and an introduction to the principles of morals and legislation* (W. Harrison, Ed.). Oxford: Basil Blackwell. (Original work published 1789)

Bergner, D. (1998). *God of the rodeo: The quest for redemption in Louisiana's Angola Prison.* New York: Ballantine Books.

Binder, A., Geis, G., & Bruce, D. (2001). *Juvenile delinquency: Historical, cultural, and legal perspectives.* Cincinnati: Anderson.

Black, M., & Smith, R. (2003). *Electronic monitoring in the criminal justice system* (Trends and Issues in Crime and Criminal Justice No. 254). Canberra, Australia: Australian Institute of Criminology.

Blalock, H. M. (1967). *Toward a theory of minority group relations.* New York: John Wiley & Sons, Inc.

Bleckman, E., & Vryan, K. (2000). Prosocial family therapy: A manualized preventative intervention for juveniles. *Aggression and Violent Behavior, 5,* 343–378.

Bohm, R., & Haley, K. (2007). *Introduction to criminal justice.* New York: McGraw-Hill.

Bonta, J., Wallace-Capretta, S., & Rooney, J. (2000). Can electronic monitoring make a difference? *Crime & Delinquency, 46,* 61–75.

Bookman, C. R., Lightfoot, C. A., & Scott, D. L. (2005). Pathways for change: An offender reintegration collaboration. *American Jails, 19,* 9–14.

Britton, D. M. (2003). *At work in the iron cage: The prison as gendered organization.* New York: New York University Press.

Brookman, F. (2005). *Understanding homicide.* Thousand Oaks, CA: Sage.

Broussard, D., Leichliter, J. S., Evans, A., Kee, R., Vallury, V., & McFarlane, M. M. (2002). Screening adolescents in a juvenile detention center for gonorrhea and chlamydia: Prevalence and reinfection rates. *The Prison Journal, 82,* 8–18.

Bureau of Justice Statistics (1997). *Correctional populations in the United States, 1997.* Washington, DC: Office of Justice Programs, U.S. Department of Justice. Retrieved on March 29, 2007, from http://www.ojp.usdoj.gov/bjs/.

Bureau of Justice Statistics. (2001). *Trends in state parole, 1990–2000* (NCJ 184735). Washington, DC: Author. Retrieved July 27, 2008, from http://www.ojp.usdoj.gov/bjs/abstract/tsp00.htm

Bureau of Justice Statistics. (2002). *Sourcebook of criminal justice statistics* (NCJ 203301). Washington, DC: Author. Retrieved July 17, 2008, from http://www.albany.edu/sourcebook/

Bureau of Justice Statistics. (2003). *Census of state and federal correctional facilities 2000.* Washington, DC: U.S. Department of Justice, Office of Justice Programs.

Bureau of Justice Statistics. (2005a). *Key facts at a glance: Correctional populations.* Washington, DC: U.S. Department of Justice, Office of Justice Programs. Retrieved November 3, 2006, from http://www.ojp.usdoj.gov/bjs/

Bureau of Justice Statistics. (2005b). *Sourcebook of criminal justice statistics online.* Washington DC: U.S. Department of Justice, Office of Justice Programs.

Bureau of Justice Statistics. (2005c). *Since 1994, violent crime rates have declined, reaching the lowest level ever in 2005.* Washington, DC: U.S. Department of Justice, Office of Justice Programs. Retrieved November 3, 2006, from http://www.ojp.usdoj.gov/bjs/

Bureau of Justice Statistics. (2006a). *Prevalence of imprisonment in the United States.* Washington, DC: U.S. Department of Justice, Office of Justice Programs. Retrieved November 5, 2006, from http://www.ojp.usdoj.gov/bjs/

Bureau of Justice Statistics. (2006b). *Prisoners in 2005.* Washington, DC: Office of Justice Programs, U.S. Department of Justice. Retrieved March 29, 2007, from http://www.ojp.usdoj.gov/bjs/

Bureau of Justice Statistics. (2008a). *Demographic trends in jail populations.* Washington, DC: Author. Retrieved July 14, 2008, from http://ojp.usdoj.gov/bjs/gcorpop.htm#DemJailPop

Bureau of Justice Statistics. (2008b). *Jail statistics: summary of findings.* Washington, DC: U.S. Department of Justice, Office of Justice Programs. Retrieved January 9, 2008, from www.ojp.usdoj.gov/bjs/jails

Butts, J., & Mitchell, O. (2000). Brick by brick: Dismantling the border between juvenile and adult justice. In C. M. Friel (Vol. Ed.), *Criminal justice 2000: Vol. 2 Boundary changes in criminal justice organizations* (pp. 167–213). Washington, DC: National Institute of Justice.

Bynum, J., & Thompson, W. (1999). *Juvenile delinquency: A sociological approach.* Boston: Allyn & Bacon.

Bynum, T., Carter, M., Matson, S., & Onley, C. (2006). Recidivism of sex offenders. In E. Latessa & A. Holsinger (Eds.), *Correctional contexts* (pp. 277–296). Los Angeles: Roxbury.

Cahalan, M. W. (1986). *Historical corrections statistics in the United States, 1850–1984.* Washington, DC: U.S. Department of Justice, Bureau of Justice Statistics.

Carroll, L. (1974). *Hacks, blacks, and cons: Race relations in a maximum security prison.* Lexington, MA: Lexington Books.

Carter, K. (2006). Restorative justice and the balanced approach. In A. Walsh, *Correctional assessment, casework, and counselling* (4th ed., pp. 8–11). Lanham, MD: American Correctional Association.

Carter, M., & Morris, L. (2002). *Managing sex offenders in the community.* Washington, DC: Center for Sex Offender Management.

Caspi, A., & Moffitt, T. (1995). The continuity of maladaptive behavior: From description to understanding in the study of antisocial behavior. In D. Ciccheti & D. Cohen (Eds.), *Manual of developmental psychology* (pp. 472–511). New York: Wiley.

Champion, D. (2005). *Probation, parole, and community corrections* (5th ed.). Upper Saddle River, NJ: Prentice Hall.

Chesney-Lind, M. (2001). Patriarchy, prisons, and jails: A critical look at trends in women's incarceration. In M. McShane, & F. P. Williams, III (Eds.), *Criminal justice: Contemporary literature in theory and practice* (pp.71–87). New York: Garland Publishing.

Chesney-Lind, M. (2002). The forgotten offender: Women in prison. In T. Gray (Ed.), *Exploring corrections: A book of readings.* Boston: Allyn & Bacon.

Chesney-Lind, M., & Shelden, R. G. (1992). *Girls: Delinquency and juvenile justice.* Pacific Grove, CA: Brooks/Cole Publishing Company.

Chesney-Lind, M., & Shelden, R. G. (1998). *Girls: Delinquency and juvenile justice* (2nd ed.). Belmont, CA: West/Wadsworth Company.

Clark, J. (1991). Correctional health care issues in the nineties: Forecast and recommendations. *American Jails Magazine, 5,* 22–23.

Clear, T. R. (1994). *Harm in American penology: Offenders, victims, and their communities.* New York: State University of New York Press.

Clemmer, D. (2001). The prison community. In E. J. Latessa, A. Holsinger, J. W. Marquart, & J. R. Sorensen (Eds.), *Correctional contexts: Contemporary and classical reading* (pp. 83–87). Los Angeles, CA: Roxbury Publishing Company.

Clutton-Brock, T., & Parker, G. (1996). Punishment in animal societies. *Nature, 373,* 209–216.

Coates, R. (1990). Victim-offender reconciliation programs in North America: An assessment. In B. Galaway & J. Hudson (Eds.), *Criminal justice, restitution, and reconciliation* (pp. 125–134). Monsey, NY: Criminal Justice Press.

Collins, B. (2004). Personal email correspondence with the Executive Director of the Juvenile Justice Trainers Association. Accessed at www.jta.org.

Collins, R. (2004). Onset and desistence in criminal careers: Neurobiology and the age-crime relationship. *Journal of Offender Rehabilitation, 39,* 1–19.

Comack, E. (2006). Coping, resisting, and surviving: Connecting women's law violations to their history of abuse. In L. F. Alarid, & P. Cromwell (Eds.), *In her own words* (pp. 33–44). Los Angeles, CA: Roxbury Publishing Company.

Conover, T. (2001). *Newjack: Guarding Sing Sing.* New York: Vintage Books.

Correctional Service of Canada (2003). *Women correctional officers in male institutions, 1978.* Retrieved April 29, 2004, from: http//www.csc.scc.gc.ca.

Corrections Compendium. 2003. Correctional officer education and training. *Corrections Compendium, 28*(2), 11–12.

Cowburn, M. (1998). A man's world: Gender issues in working with male sex offenders in prison. *The Howard Journal, 37,* 234–251.

Cox, N. R., & Osterhoff, W. E. (1991). Managing the crisis in local corrections: A public-private partnership approach. In J. Thompson & G. L. Mays (Eds.), *American jails: Public policy issues* (pp. 227–239). Chicago: Nelson-Hall Publishers.

Crabbe, J. (2002). Genetic contributions to addiction. *Annual Review of Psychology, 53*, 435–462.

Crouch, B. M. (1993). Is incarceration really worse? Analysis of offenders' preferences for prison over probation. *Justice Quarterly, 10*, 67–88.

Cuellar, A., Snowden, L., & Ewing, T. (2007). Criminal records of persons served in the public mental health system. *Psychiatric Services, 58*, 114–120.

Cullen, F. T. (2006). Assessing the penal harm movement. In E. J. Latessa & A. M. Holsinger (Eds.), *Correctional contexts: Contemporary and classical readings* (pp. 61–74). Los Angeles, CA: Roxbury Publishing Co.

Cullen, F. T., & Gendreau, P. (2000). Assessing correctional rehabilitation: Policy, practice, and prospects. In J. Horney (Ed.), *NIJ Criminal Justice 2000. Vol. 3: Policies, processes, and decisions of the criminal justice system* (pp. 109–175). Washington, DC: National Institute of Justice.

Cullen, F. T., & Gilbert, K. (1982). *Reaffirming rehabilitation.* Cincinnati, Ohio: Anderson Publishing Co.

Currie, E. (1999). Reflections on crime and criminology at the millennium. *Western Criminology Review, 2*(1). Retrieved July 14, 2008, from http://wcr.somoma.edu/v2n1/currie.html

Dahl, R. (1957). The concept of power. *Behavioral Science, 2*(3), 201–215.

Dahl, R. (1961). *Who governs: Democracy and power in an American city.* New Haven: Yale University Press.

Daly, K., & Chesney-Lind, M. (1988). Feminism and criminology. *Justice quarterly, 5*, 497–535.

Davis, R. K., Applegate, B. K., & Otto, C. W. (2004). Roles and responsibilities: Analyzing local leaders' views on jail crowding from a systems perspective. *Crime & Delinquency, 50*(3), 458–482.

De Quervain, D., Fischbacher, U., Valerie, T., Schellhammer, M., Schnyder, U., Buch, A., & Fehr, E. (2004). The neural basis of altruistic punishment. *Science, 305*, 1254–1259.

Death Penalty Information Center. (2007). *Facts about the death penalty.* Retrieved July 14, 2007, from www.deathpenaltyinfo.org

DeFina, R., & Arvanites, T. (2002). The weak effect of imprisonment on crime: 1971–1998. *Social Science Quarterly, 83*, 635–653.

Del Carmen, R., Parker, V., & Reddington, F. (1998). *Briefs of leading cases in juvenile justice.* Cincinnati, OH: Anderson Publishing.

Delany, P., Fletcher, B., & Shields, J. (2003). Reorganizing care for the substance using offender—the case for collaboration. *Federal Probation, 67*, 64–69.

DeLisi, M. (2005). *Career criminals in society.* Thousand Oaks, CA: Sage.

DeKeseredy, W.S. (2000). *Women, crime and the Canadian criminal justice system.* Cincinnati, OH: Anderson Publishing Co.

DuPont, R. (1997). *The selfish brain: Learning from addiction.* Washington, DC: American Psychiatric Press.

Durant, W., & Durant, A. (1967). *Rousseau and revolution.* New York: Simon and Schuster.

Durkheim, E. (1964). *The division of labor in society.* New York: Free Press. (Original work published 1893)

Dworkin, A. (1993). Against the male flood: Censorship, pornography, and equality. In P. Smith (Ed.), *Feminist jurisprudence.* Oxford: Oxford University Press.

Ellis, A. (1989). The history of psychotherapy. In A. Freeman, K. Simon, L. Beutler, & H. Arkowitz (Eds.), *Comprehensive handbook of cognitive therapy.* New York: Plenum.

Ellis, L. (2003). Genes, criminality, and the evolutionary neuroandrogenic theory. In A. Walsh & L. Ellis (Eds.), *Biosocial criminology: Challenging environmentalism's supremacy* (pp. 13–34). Hauppauge, NY: Nova Science.

Farabee, D., Pendergast, M., & Anglin, M. (1998). The effectiveness of coerced treatment for drug abusing offenders. *Federal Probation, 109*, 3–10.

Farbstein, J., & Wener, R. (1989). *A comparison of "direct" and "indirect" supervision of correctional facilities: Final report.* Washington, DC: National Institute of Corrections.

Farkas, M. A. (1999). Inmate supervisory style: Does gender make a difference? *Women and Criminal Justice, 10*, 25–45.

Farkas, M. A. (2001). Correctional officers: What factors influence work attitudes. *Corrections Management Quarterly, 5*(2), 20–26.

Farkas, M. A., & Rand, K. (1999). Sex matters: A gender-specific standard for cross-gender searches of inmates. *Women and Criminal Justice, 10*, 31–55.

Farrington, D. P., Langan, P. A., & Tonry, M. (2004). *Cross national studies in crime and justice.* Washington, DC: Bureau of Justice Statistics, U.S. Department of Justice, Office of Justice Programs.

Federal Bureau of Investigation. (2005). *Crime in the United States, 2004: Uniform crime reports.* Washington, DC: U.S. Government Printing Office.

Federal Bureau of Investigation. (2006). *Crime in the United States, 2005: Uniform crime reports.* Washington, DC: U.S. Government Printing Office.

Federal Bureau of Investigation. (2007). *2006: Crime in the United States.* Retrieved February 29, 2008, from http://www.fbi.gov/ucr/cius2006/arrests/index .html/

Feeley, M. M. (1991). The privatization of prisons in historical perspective. *Criminal Justice Research Bulletin, 6*(2).

Fehr, E., & Gachter, S. (2002). Altruistic punishment in humans. *Nature, 415,* 137–140.

Ferri, E. (1917). *Criminal sociology.* Boston: Little, Brown. (Original work published 1897)

Fielding, H. (1967). *Inquiry into the causes of the late increase of robbers.* Oxford: Oxford University Press. (Original work published 1751)

Firestone, J. M., & Harris, R. J. (1994). Sexual harassment in the military: Environmental and individual contexts. *Armed Forces and Society, 21,* 25–43.

Firestone, J. M., & Harris, R. J. (1999). Changes in patterns of sexual harassment in the U.S. military: A comparison of the 1988 and 1995 DOD surveys. *Armed Forces and Society, 25,* 613–632.

Fishbein, D. (2003). Neuropsychological and emotional regulatory processes in antisocial behavior. In A. Walsh & L. Ellis (Eds.), *Biosocial criminology: Challenging environmentalism's supremacy* (pp. 185–208). Hauppauge, NY: Nova Science.

Fisher, W., Roy-Bujnowski, K., Grudzinkas, A., Clayfield, J., Banks, S., & Wolff, N. (2006). Patterns and prevalence of arrest in a statewide cohort of mental care consumers. *Psychiatric Services, 57,* 1623–1628.

Fishman, J. (1969). *Crucibles of crime: The shocking story of the American jail.* Englewood Cliffs, NJ: Prentice-Hall, Inc. (Original work published 1923)

Flavin, J., & Desautels, A. (2006). Feminism and crime. In C. M. Renzetti, L. Goodstein, & S. L. Miller (Eds.), *Rethinking gender, crime, and justice* (pp. 11–28). Los Angeles, CA: Roxbury Publishing Company.

Foster, B. (2006). *Corrections: The fundamentals.* Upper Saddle River, NJ: Prentice Hall.

Foucault, M. (1979). *Discipline and punish: The birth of the prison.* New York: Vintage.

Foucault, M. (1991). *Discipline and punish.* Harmondsworth, UK: Penguin.

Freudenberg, N. (2006). Coming home from jail: A review of health and social problems facing U.S. jail populations and of opportunities for reentry interventions. *American Jails, 20,* 9–24.

Freudenberg, N., Moseley, J., Labriola, M., Daniels, J., & Murrill, C. (2007). Comparison of health and social characteristics of people leaving New York City jails by age, gender, and race/ethnicity: Implications for public health interventions. *Public Health Reports, 122*(6), 733–743.

Gainsborough, J. (2002). *Mentally ill offenders in the criminal justice system: An analysis and prescription.* Washington, DC: The Sentencing Project.

Garland, B., & Frankel, M. (2006). Considering convergence: A policy dialogue about behavioral genetics, neuroscience, and law. *Law and Contemporary Problems, 69,* 101–113.

Garofalo, R. (1968). *Criminology.* Montclair, NJ: Patterson Smith. (Original work published 1885; original English translation published 1914)

Gendreau, P., & Ross, R. R. (1987). Revivification of rehabilitation: Evidence from the 1980s. *Justice Quarterly, 4,* 349–407.

Georgia Department of Corrections. (2007). *FY 2006 costs of adult sanctions.* Retrieved July 17, 2008, from http://www.dcor.state.ga.us/AboutGDC/

Gettinger, S. H. (1984). *New generation jails: An innovative approach to an age-old problem.* Washington, DC: National Institute of Corrections, U.S. Department of Justice.

Giallombardo, R. (1966). *Society of women: A study of a women's prison.* New York: John Wiley.

Gibbons, J. J., & Katzenbach, B. (2006). *Confronting confinement: A report of the commission on safety and abuse in America's prisons.* New York, NY: Vera Institute of Justice.

Gibbs, A., & King, D. (2003). The electronic ball and chain? The operation and impact of home detention with electronic monitoring in New Zealand. *Australian and New Zealand Journal of Criminology, 36,* 1–17.

Gilliard, D. K., & Beck, A. J. (1997). *Prison and inmates at midyear 1996.* Washington, DC: U.S. Department of Justice, Office of Justice Programs. Retrieved February 4, 2008, from www.ojp.usdoj .gov/bjs/pub/pdf/mhppji.pdf

Glaze, L. E., & Bonczar, T.P. (2006). *Probation and parole in the United States, 2005.* Washington, DC: Office of Justice Programs, U.S. Department of Justice. Retrieved March 29, 2007, from http:// www.ojp.usdoj.gov/bjs/pub/pdf/ppus05.pdf.

Glaze, L. E., & Bonczar, T.P. (2007). *Probation and parole in the United States, 2006.* Washington, DC: Bureau of Justice Statistics, U.S. Department of Justice, Office of Justice Programs. Retrieved February 27, 2008, from http://www.ojp.usdoj .gov/bjs/

Glaze, L. E., & Palla, S. (2005). *Probation and parole in the United States, 2004.* Washington DC: U.S. Department of Justice.

Glueck, S. (Ed.). (1939). Introduction. In J. Augustus, *A report of the labors of John Augustus, for the last ten years, in aid of the unfortunate.* New York: National Probation Association. (Original work published 1852)

Goffman, E. (1961). *Asylums: Essays on the social situation of mental patients and other inmates.* Garden City, NY: Anchor Books.

Goldfarb, R. (1975). *Jails: The ultimate ghetto.* Garden City, New York: Anchor Press.

Goleman, D. (1995). *Emotional intelligence:* New York: Bantam Books.

Goleman, D. (2006). *Social intelligence: The new science of human relationships.* New York: Bantam Books.

Goodstein, L. (2006). Introduction: Gender, crime, and criminal justice. In C. M. Renzetti, L. Goodstein, & S. L. Miller (Eds.), *Rethinking gender, crime, and justice* (pp. 1–10). Los Angeles, CA: Roxbury Publishing Company.

Gray, T., Mays, L. G., & Stohr, M. K. (1995). Inmate needs and programming in exclusively women's jails. *The Prison Journal, 75*(2), 186–202.

Grinfeld, M. (2005). Sexual predator ruling raises ethical, moral dilemmas. *Psychiatric Times, 16,* 1–4.

Gruber, J. E. (1997). An epidemiology of sexual harassment: Evidence from North America and Europe. In W. O'Donohue (Ed.), *Sexual harassment: Theory, research, and treatment* (pp. 84–98). Boston: Allyn and Bacon.

Gruber, J. E. (1998). The impact of male work environments and organizational policies on women's experiences of sexual harassment. *Gender and Society 12,* 301–320.

Haney, C., Banks, C., & Zimbardo, P. (1981). Interpersonal dynamics in a simulated prison. In R. R. Ross (Ed.), *Prison guard/Correctional officer* (pp. 137–168). Toronto, Canada: Butterworth. (Original work published 1973 in the *International Journal of Criminology and Penology*)

Harrington, P. E. (2002). Advice to women beginning a career in policing. *Women & Criminal Justice, 9,* 1–21.

Harrison, P. M., & Beck, A. J. (2003). *Prisoners in 2002.* Washington, DC: Bureau of Justice Statistics, Office of Justice Programs, U.S. Department of Justice.

Harrison, P. M., & Beck, A.J. (2005). *Prisoners in 2004.* Washington, DC: Bureau of Justice Statistics, Office of Justice Programs, U.S. Department of Justice.

Hassine, V. (1996). *Life without parole: Living in prison today.* Los Angeles, CA: Roxbury Publishing Company.

Hawkes, M. Q. (1998). Edna Mahan: Sustaining the reformatory tradition. *Women & Criminal Justice, 9,* 1–21.

Hefferman, E. (1972). *Making it in prison: The square, the cool, and the life.* New York: Wiley Interscience.

Hemmens, C., Steiner, B., Mueller, D. (2003). *Significant cases in juvenile justice.* Los Angeles: Roxbury

Hemmens, C., Stohr, M. K., Schoeler, M., & Miller, B. (2002). One step up, two steps back: The progression of perceptions of women's work in prisons and jails. *Journal of Criminal Justice, 30,* 473–489.

Henriques, Z.W., & Gilbert, E. (2003). Sexual abuse and sexual assault of women in prison. In R. Muraskin (Ed.), *It's a crime: Women and justice* (pp. 258–272). Upper Saddle River, NJ: Prentice Hall.

Hensley, C., Tewksbury, R., & Wright, J. (2006). Exploring the dynamics of masturbation and consensual same-sex activity within a male maximum security prison. In R. Tewksbury (Ed.), *Behind bars: readings on prison culture* (pp. 97–109). Upper Saddle River, NJ: Pearson Prentice Hall.

Hickman, M. J., & Reeves, B. A. (2006). *Local police departments, 2003.* Washington, DC: Bureau of Justice Statistics, U.S. Department of Justice. Retrieved August 30, 2006, http://www.usdoj.gov/

Hindman, J., & Peters, J. (2001). Polygraph testing leads to better understanding adult and juvenile sex offenders. *Federal Probation, 65,* 1–15.

Hirsch, A. J. (1992). *The rise of the penitentiary.* New Haven: Yale University Press.

Hirschi, T. (1969). *The causes of delinquency.* Berkeley: University of California Press.

Howard League for Penal Reform. (2006). *Information pamphlet.* Retrieved July 17, 2007, from http://www.howardleague.org/index.php?id=fact1

Hughes, K. (2006). *Justice expenditure and employment in the United States, 2003.* Washington, DC: Bureau of Justice Statistics, Office of Justice Programs, U.S. Department of Justice.

Hughes, R. (1987). *The fatal shore.* New York: Vintage books.

Hughes, T., Wilson, D., & Beck, A. (2001). *Trends in state parole, 1990–2000.* Washington, DC: U.S. Department of Justice, Bureau of Justice Statistics.

Ireland, C. A., & Ireland, J. (2006). Descriptive analysis of the nature and extent of bullying behavior in a maximum-security prison. In R. Tewksbury (Ed.), *Behind bars: readings on prison culture* (pp. 87–96). Upper Saddle River, NJ: Pearson Prentice Hall.

Irwin, J. (1985). *The jail: Managing the underclass in American society.* Berkeley, CA: University of California Press.

Irwin, J. (2005). *The warehouse prison: Disposal of the new dangerous class.* Los Angeles, CA: Roxbury Publishing Company.

Irwin, J., & Austin, J. (1994). *It's about time: America's imprisonment binge.* Belmont: Wadsworth.

Irwin, J., & Austin, J. (2001). *It's about time: America's imprisonment binge* (3rd ed.). Belmont, CA: Wadsworth Publishing Company.

Irwin, J., & Cressey, D. R. (1962). Thieves, convicts and the inmate culture. In M. K. Stohr & C. Hemmens (Eds.), *The inmate prison experience* (pp. 3–16). New Jersey: Prentice Hall.

Jacobs, J. B. (1977). *Stateville: The penitentiary in mass society.* Chicago: The University of Chicago Press.

Jacoby, S. (1983). *Wild justice: The evolution of revenge.* New York: Harper & Row.

Jafee, S., Moffitt, T., Caspi, A., & Taylor, A. (2003). Life with (or without) father: The benefits of living with two biological parents depend on the father's anti-social behavior. *Child Development, 74,* 109–126.

James, D. J., & Glaze, L. E. (2006). *Mental health problems of prison and jail inmates.* Washington, DC: U.S. Department of Justice, Office of Justice Programs. Retrieved January 10, 2008, from www.ojp.usdoj.gov/bjs/pub/pdf/mhppji.pdf

JFA Institute. (2007). *Public safety, public spending: Forecasting America's prison population, 2007–2011.* Washington, DC: Public Safety Performance Project, Part of the Pew Charitable Trusts. Retrieved July 28, 2008, from http://www.pewcenteronthestates.org/report_detail.aspx?id=32076

Johnson, R. (2002). *Hardtime: Understanding and reforming the prison* (3rd ed.). Belmont, CA: Wadsworth/Thomson Learning.

Johnston, N. (2004). The world's most influential prison: Success or failure? *The Prison Journal, 84*(4), 20S–40S.

Joseph, J., & Taylor, D. (Eds.). (2003). *With justice for all: Minorities and women in criminal justice.* Upper Saddle River, NJ: Prentice-Hall.

Joseph, J., Henriques, Z. W., & Richards-Ekeh, K. (2003). Get tough policies and the incarceration of African Americans. In J. Joseph & D. Taylor (Eds.), *With justice for all: Minorities and women in criminal justice* (pp.105–120). Upper Saddle River, NJ: Prentice-Hall.

Junger-Tas, J. (1996). Delinquency similar in Western countries. *Overcrowded Times, 7,*10–13.

Jurik, N. C. (1985). An officer and a lady: Organizational barriers to women working as correctional officers in men's prisons. *Social Problems, 33,* 375–388.

Jurik, N. C. (1988). Striking a balance: Female correctional officers, gender-role stereotypes, and male prisons. *Sociological Inquiry, 58,* 291–305.

Jurik, N. C., & Halemba, G. J. (1984). Gender, working conditions, and the job satisfaction of women in a non-traditional occupation: Female correctional officers in a men's prison. *Sociological Quarterly, 25,* 551–566.

Kalinich, D., Embert, P., & Senese, J. (1991). Mental health services for jail inmates: Imprecise standards, traditional philosophies, and the need for change. In J. A. Thompson & G. L. Mays (Eds.), *American jails: Public policy issues* (pp. 79–99). Chicago: Nelson-Hall.

Karberg, J. C., & James, D. J. (2005). *Substance dependence, abuse, and treatment of jail inmates, 2002.* Washington, DC: U. S. Department of Justice, Bureau of Justice Statistics.

Kellar, M., & Wang, H-M. (2005). Inmate assaults in Texas county jails. *The Prison Journal, 85,* 515–534.

Kerle, K. E. (1991). Introduction. In J. Thompson & G. L. Mays (Ed.), *American jails: Public policy issues* (pp. 1–3). Chicago: Nelson-Hall Publishers.

Kerle, K. E. (1998). *American jails: Looking to the future.* Boston: Butterworth-Heinemann.

Kerle, K. E. (2003). *Exploring jail operations.* Hagerstown, MD: American Jail Association.

Kim, A-S., DeValve, M., DeValve, E. Q., & Jonson, W. W. (2003). Female wardens: Results from a national survey of state correctional executives. *The Prison Journal, 83,* 406–425.

Kitano, H. H. L. (1997). *Race relations* (5th ed.). Upper Saddle River, NJ: Prentice Hall.

Kleber, H. (2003). Pharmacological treatments for heroin and cocaine dependence. *The American Journal on Addictions, 12,* S5–S18.

Klofas, J. M. (1991). Disaggregating jail use: Variety and change in local corrections over a ten-year period. In J. Thompson & G. L. Mays (Eds.), *American jails: Public policy issues* (pp. 40–58). Chicago: Nelson-Hall Publishers.

Kluger, J. (2007). The paradox of supermax. *Time, 169*(6), 52–53.

Knight, K., Simpson, D. D., & Hiller, M. L. (2004). Three-year reincarceration outcomes for in-person therapeutic community treatment in Texas.

In M. K. Stohr & C. Hemmens (Eds.), *The inmate prison experience* (pp. 316–328). Upper Saddle River, New Jersey: Pearson/Prentice Hall.

Krebs, A. (1978). John Howard's influence on the prison system of Europe with special reference to Germany. In J. C. Freeman (Ed.), *Prisons past and future* (pp. 35–52). London, UK: Heinemann.

Langan, P. A., & Cohen, R.L. (1996). *State court sentencing of convicted felons, 1992*. Washington, DC: U.S. Bureau of Justice Statistics.

Langan, P. A., & Farrington, D. (1998). *Crime and justice in the United States and England and Wales, 1981–1996*. Washington, DC: Bureau of Justice Statistics.

Latessa, W., Cullen, F., & Gendreau, P. (2002). Beyond correctional quackery—Professionalism and the possibility of effective treatment. *Federal Probation, 66*, 43–50.

Lawrence, R., & Mahan, S. (1998). Women corrections officers in men's prisons: Acceptance and perceived job performance. *Women & Criminal Justice, 0*, 63–86.

Leshner, A. (1998). Addiction is a brain disease—and it matters. *National Institute of Justice Journal, 237*, 2–6.

Lewis, O. F. (1922). *The development of American prisons and prison customs: 1776 to 1845*. New York: Prison Association of New York.

Lightfoot, C., Zupan, L. L., & Stohr, M. K. (1991). Jails and the community: Modeling the future in local detention facilities. *American Jails, 5*, 50–52.

Linden, D. (2006). How psychotherapy changes the brain—the contribution of functional neuroimaging. *Molecular Psychiatry, 11*, 528–538.

Litt, M., & Mallon, S. (2003). The design of social support networks for offenders in outpatient drug treatment. *Federal Probation, 67*, 15–22.

Livsey, S. (2006, November). Juvenile delinquency probation caseload, 1985–2002. *OJJDP Fact Sheet, 4*, p.1. Retrieved July 24, 2008, from http://ojjdp.ncjrs.org/

Logan, C., & Gaes, G. (1993). Meta-analysis and the rehabilitation of punishment. *Justice Quarterly, 10*, 245–263.

Lombardo, L. X. (2001). Guards imprisoned: Correctional officers at work. In E. J. Latessa, A. Holsinger, J. W. Marquart, & J. R. Sorensen (Eds.), *Correctional contexts: Contemporary and classical readings* (2nd ed., pp. 153–167). Los Angeles, CA: Roxbury Publishing Company.

Lowenkamp, C., & Latessa, E. (2002). *Evaluation of Ohio's community based corrections facilities and halfway house programs: Final report*. Cincinnati, OH: Center for Criminal Justice Research, University of Cincinnati.

Lowenkamp, C., & Latessa, E. (2004). Understanding the risk principle: How and why correctional interventions can harm low risk offenders. In D. Faust (Ed.), *Assessment issues for managers* (pp. 3–7). Washington, DC: U.S. Department of Justice, National Institute of Corrections.

Lubitz, R., & Ross, T. (2001). *Sentencing guidelines: Reflections on the future* (Sentencing & Corrections: Issues for the 21st Century, No. 10). Washington, DC: U.S. Department of Justice, National Institute of Justice.

Lurigio, A. (2000). Persons with serious mental illness in the criminal justice system: Background, prevalence, and principles of care. *Criminal Justice Policy Review, 11*, 312–328.

Lutze, F. E., & Murphy, D. W. (1999). Ultra-masculine prison environments and inmates' adjustment: It's time to move beyond the 'boys will be boys' paradigm. *Justice Quarterly, 16*, 709–734.

Macher, A. M. (2007). Issues in correctional HIV care: Neurological manifestations of patients with primary HIV infection. *American Jails, 21*, 49–55.

MacKenzie, D. L. (1987). Age and adjustment to prison interactions with attitudes and anxiety. *Criminal Justice and Behavior, 14*, 427–447.

Maguire, K., & Pastore, A. L. (Eds.). (2002). *Sourcebook of criminal justice statistics* (NCJ 203301). Washington, DC: Bureau of Justice Statistics. Retrieved July 17, 2008, from http://www.albany.edu/sourcebook/

Maitland, A. S., & Sluder, R. D. (1996). Victimization in prisons: A study of factors related to the general well-being of youthful inmates. *Federal Probation, 60*, 24–31.

Maletzky, B., & Field, G. (2003). The biological treatment of dangerous sexual offenders, a review and preliminary report of the Oregon pilot Depo-Provera program. *Aggression and Violent Behavior, 8*, 391–412.

Marion, N. (2002). Effectiveness of community-based programs: A case study. *The Prison Journal, 82*, 478–497.

Marquart, J. W., Barnhill, M. B., & Balshaw-Biddle, K. (2001). Fatal attraction: An analysis of employee boundary violations in a southern prison system, 1995–1998. *Justice Quarterly, 18*, 877–910.

Marquart, J. W., & Roebuck, J. B. (1985). Prison guards and snitches: Deviance within a total institution. *The British Journal of Criminology, 25*(3), 217–233.

Marsh, R., & Walsh, A. (1995). Physiological and psychosocial assessment and treatment of sex offenders: A comprehensive victimoriented program. *Journal of Offender Rehabilitation, 22,* 77–96.

Martin, G. (2005). *Juvenile justice: Process and systems.* Thousand Oaks, CA: Sage.

Martin, S., & Jurik, N. (1996). *Doing justice, doing gender.* Thousand Oaks, CA: Sage.

Martinson, R. (1974). What works? Questions and answers about prison reform. *The Public Interest, 35,* 22–54.

Maruschak, L. M. (2006). *Medical problems of jail inmates.* Washington, DC: U.S. Department of Justice, Office of Justice Programs. Retrieved January 10, 2008, from www.ojp.usdoj.gov/bjs/pub/pdf/mpji.pdf

Maschke, K. J. (1996). Gender in the prison setting: The privacy-equal employment dilemma. *Women & Criminal Justice, 7,* 23–42.

Maslow, A. H. (1998). *Maslow on management.* New York: John Wiley & Sons.

Mathias, R. (1995, July/August). Correctional treatment helps offenders stay drug and arrest free. *NIDA Notes, 10*(4). Retrieved June 29, 2008, from http://www.drugabuse.gov/Nida_Notes/NNVol10N4/Prison.html

Mauer, M. (2005). *Comparative international rates of incarceration: An examination of causes and trends.* Washington, DC: The Sentencing Project.

Mawby, R. (2001). *Burglary.* Colompton, Devon: Willan Publishing.

May, D., Wood, P., Mooney, J., & Minor, K. (2005). Predicting offender-generated exchange rates: Implications for a theory of sentence severity. *Crime & Delinquency, 51,* 373–399.

McBride, D., & McCoy, C. (1993). The drugs-crime relationship: An analytical framework. *The Prison Journal. 73,* 257–278.

McCrae, R., Costa, P., Ostendorf, F., Angleitner, A., Hrebickova, M., Avia, M., Sanz, J., Sanchez-Bernardos, M., Kusdil, M., Woodfield, R., Saunders, P., & Smith, P. (2000). Nature over nurture: Temperament, personality, and life span development. *Journal of Personality and Social Psychology, 78,* 173–186.

McDermott, P. A., Alterman, A. I., Cacciola, J. S., Rutherford, M. J., Newman, J. P., & Mulholland, E. M. (2000). Generality of psychopathy checklist-revised factors over prisoners and substance-dependent patients. *Journal of Consulting and Clinical Psychology, 68*(1), 181–6.

McGue, M. (1999). The behavioral genetics of alcoholism. *Current Directions in Psychological Science, 8,* 109–115.

McLean, R. L., Robarge, J., & Sherman, S. G. (2006). Release from jail: Moment of crisis or window of opportunity for female detainees. *Journal of Urban Health, 83,* 382–393.

McLeod, J. (2003). *An introduction to counseling.* Buckingham, UK: Open University Press.

McNiel, D. E., Binder, R. L., & Robinson, J. C. (2005). Incarceration associated with homelessness, mental disorder, and co-occurring substance abuse. *Psychiatric Services, 56,* 840–846.

Menard, S., Mihalic, S., & Huizinga, D. (2001). Drugs and crime revisited. *Justice Quarterly, 18,* 269–299.

Mennel, R. (1973). *Thorns and thistles.* Hanover, NH: The University of Hanover Press.

Merlo, A.V., & Benekos, P. J. (2000). *What's wrong with the criminal justice system.* Cincinnati, OH: Anderson Publishing Company.

Metraux, S., & Culhane, D. P. (2006). Recent incarceration history among a sheltered homeless population. *Crime & Delinquency, 52,* 504–517.

Miller, D., & Vidmar, N. (1981). The social psychology of punishment reactions. In M. Lerner & S. Lerner (Eds.), *The justice motive in social behavior* (pp. 145–172). New York: Plenum.

Moffitt, T. (2005). The new look of behavioral genetics in developmental psychopathology: Gene-environment interplay in antisocial behavior. *Psychological Bulletin, 131,* 533–554.

Moffitt, T., & Walsh, A. (2003). The adolescence-limited/Life-course persistent theory of antisocial behavior: What have we learned? In A. Walsh & L. Ellis (Eds.), *Biosocial criminology: Challenging environmentalism's supremacy* (pp. 125–144). Hauppauge, NY: Nova Science.

Moore, J., & Pachon, H. (1985). *Hispanics in the United States.* Englewood Cliffs, NJ: Prentice-Hall, Inc.

Morash, M. (2006). *Understanding gender, crime and justice.* Thousand Oaks, CA: Sage.

Morash, M., Haarr, R., & Rucker, L. (1994). A comparison of programming for women and men in U.S. prisons in the 1980s. *Crime & Delinquency, 40,* 197–221.

Morris, N. (2002). *Maconochie's gentlemen: The story of Norfolk Island and the roots of modern prison reform.* Oxford: Oxford University Press.

Mueller, C. W., De Coster, S., & Estes, S. (2001). Sexual harassment in the workplace. *Work and Occupations, 28,* 411–446.

Mumola, C. J. (2000). *Incarcerated parents and their children.* Washington, DC: Bureau of Justice Statistics.

Mumola, C. J. (2005). *Suicide and homicide in state prisons and local jails.* Washington, DC: U.S.

Department of Justice, Office of Justice Programs. Retrieved January 21, 2008, from http://www.ojp.usdoj.gov/bjs/pub/pdf/shplj.pdf

Mumola, C. J. (2007). *Medical causes of death in state prisons, 2001–2004* (Bureau of Justice Statistics Data Brief). Washington, DC: U.S. Department of Justice.

Muraskin, R. (2003). Disparate treatment in correctional facilities. In R. Muraskin (Ed.), *It's a crime: Women and justice* (3rd ed., pp. 220–231). Upper Saddle River, NJ: Prentice Hall.

Murton, T. O., & Hyams, J. (1976). *Accomplices to crime: The Arkansas prison scandal.* New York: Grove Press.

Mustaine, E. E., & Tewksbury, R. (2004). Alcohol and violence. In S. Holmes & R. Holmes (Eds.), *Violence: A contemporary reader* (pp. 9–25). Upper Saddle River, NJ: Prentice-Hall.

Nagin, D. (1998). Criminal deterrence research at the onset of the twenty-first century. In M. Tony (Ed.), *Crime and justice: A review of research* (pp. 1–42). Chicago: University of Chicago Press.

National Center For Policy Analysis. (1998a, August 17). Does punishment deter? *NCPA Policy Backgrounder, 148.*

National Center for Policy Analysis. (1998b). *Falsified crime data.* Retrieved July 28, 2008, http://www.ncpa.org/pi/crime/aug98a.html.

National Institute of Corrections. (2008). *Today's jails: Are you looking for a way to coordinate jail and community health services?* Washington, DC: Jails Division, National Institute of Corrections. Retrieved on February 4, 2008, from http://nicic.org/JailsDivision.

Nelson, W. R., & Davis, R.M. (1995). Podular direct supervision: The first twenty years. *American Jails, 9,* 11–22.

O'Donohue, W., Downs, K., & Yeater, E. (1998). Sexual harassment: A review of the literature. *Aggression and Behavior 3,* 111–128.

O'Sullivan, S. (2006). Representations of prison in nineties Hollywood cinema: From *Con Air* to *The Shawshank Redemption.* In R. Tewksbury (Ed.), *Behind bars: readings on prison culture* (pp. 483–498). Upper Saddle River, NJ: Pearson PrenticeHall.

Olson, S. M., & Dzur, A. W. (2004). Revisiting informal justice: Restorative justice and democratic professionalism. *Law & Society, 38,* 139–176.

Orland, L. (1995). Prisons as punishment: An historical overview. In K. C. Haas & G. P. Alpert (Eds.), *The dilemmas of corrections: Contemporary readings* (3rd ed.). Prospect Heights, IL: Waveland Press, Inc.

Osher, F. C. (2007). Short-term strategies to improve reentry of jail populations: Expanding and implementing the APIC model. *American Jails, 20,* 9–18.

Oshinsky, D. M. (1996). *Worse than slavery: Parchman farm and the ordeal of Jim Crow justice.* New York: The Free Press.

Owen, B. (1998). *In the mix: Struggle and survival in a women's prison.* Albany, State University of New York Press.

Owen, B. (2005). Afterword: The case of the women. In J. Irwin (Ed.), *The warehouse prison: Disposal of the new dangerous class* (pp. 261–289). Los Angeles, CA: Roxbury Publishing Company.

Owen, B., & Bloom, B. (1995). Profiling women prisoners: Findings from national surveys and a California sample. *The Prison Journal, 75,* 165–185.

Palmer, T. (1983). The 'effectiveness' issue today: An overview. *Federal Probation, 47,* 3–10.

Parrillo, V. (2003). *Strangers to these shores: Race and ethnic relations in the United States.* Boston: Allyn and Bacon.

Perroncello, P. (2002). Direct supervision: A 2001 odyssey. *American Jails, 15,* 25–32.

Petersilia, J. (2000, Summer). Parole and prisoner reentry in the United States. *Perspectives, American Probation and Parole Association,* pp. 32–46.

Petersilia, J. (2001). Prisoner reentry in the United States. In J. Lynch & W. Sabol (Eds.), *Crime policy report: Vol. 3. Prisoner reentry in perspective.* Washington, DC: Urban Institute.

Petersilia, J. (2004). What works in prison reentry? Reviewing and questioning the evidence. *Federal Probation, 68,* 1–8.

Phelps, R. (1996). *Newgate of Connecticut: Its origin and early history.* Camden, ME: Picton Press.

Pizarro, J. M., & Narag, R. E. (2008). Supermax prisons: What we know, what we do not know, and where we are going. *The Prison Journal, 88,* 23–42.

Pogrebin, M. R., & Poole, E. D. (1997). The sexualized work environment: A look at women jail officers. *The Prison Journal, 77,* 41–57.

Pollock, J. M. (2002). *Women, prison, and crime* (2nd ed.). Belmont, CA: Wadsworth Thomson Learning.

Pollock, J. M. (2004). *Prisons and prison life: Costs and consequences.* Los Angeles, CA: Roxbury Publishing.

Powell, K. (2006). How does the teenage brain work? *Nature, 442,* 865–867.

Pressman, J. L., & Wildavsky, A. (1979). *Implementation.* Berkeley, CA: University of California Press.

Pressman, J. L., & Wildavsky, A. (1984). *Implementation: How great expectations in Washington are dashed in Oakland—Or, why it's amazing that federal programs work at all this being a saga of the economic development administration* (3rd ed.). Berkeley, CA: University of California Press.

Pryor, J., & Stoller, L. M. (1994). Sexual cognition processes in men high in the likelihood to sexually harass. *Personality and Social Psychology Bulletin, 20,* 163–169.

Puzzanchera, C., & Kang, W. (2007). *Juvenile court statistics databook.* Retrieved February 27, 2008, from http://ojjdp.ncjrs.org/ojstatbb/jcsdb/

Radelet, M., & Borg, M. (2000). The changing nature of death penalty debates. *Annual Review of Sociology, 26,* 43–61.

Radzinowicz, L. (1978). John Howard. In J. C. Freeman (Ed.), *Prisons past and future* (pp. 7–14). London: Heinemann.

Radzinowicz, L., & King, J. (1979). *The growth of crime: The international experience.* Middlesex, UK: Penguin Books.

Rafter, N. H. (1985). *Partial justice: Women in state prisons, 1800–1935.* Boston: Northeastern University Press.

Reid, S. (2006). *Criminal justice* (7th ed.). Cincinnati, OH: Atomic Dog Publishers.

Reiman, J. (1998). *The rich get richer and the poor get prison: Ideology, class and criminal justice.* Boston: Allyn and Bacon.

Rengert, G., & Wasilchick, J. (2001). *Suburban burglary: A tale of two suburbs.* Springfield, IL: Charles C Thomas.

Restak, R. (2001). *The secret life of the brain.* Washington, DC: Joseph Henry Press.

Reuffer, D. M. (2007). Arizona jail sex results in charges for guards, prisoner. *Prison Legal News, 18,* 31.

Rigby, M. (2007). Dallas, Texas, jail pays $950,000 for neglecting mentally ill prisoners. *Prison Legal News, 18,* 22.

Roberts, J. W. (1997). *Reform and retribution: An illustrated history of American prisons.* Lanham, MD: American Correctional Association.

Robinson, M. (2005). *Justice blind: Ideals and realities of American criminal justice.* Upper Saddle River, NJ: Prentice Hall.

Robinson, T., & Berridge, K. (2003). Addiction. *Annual Review of Psychology, 54,* 25–53.

Rodney, E., & Mupier, R. (1999). Behavioral differences between African American male adolescents with biological fathers and those without biological fathers in the home. *Journal of Black Studies, 30,* 45–61.

Rodriguez, N. (2007). Restorative justice at work: Examining the impact of restorative justice resolutions on juvenile recidivism. *Crime & Delinquency, 53,* 355–379.

Rosenfeld, R. (2000). Patterns in adult homicide. In A. Blumstein & J. Wallman (Eds.), *The crime drop in America* (pp. 130–163). Cambridge: Cambridge University Press.

Rothman, D. J. (1980). *Conscience and convenience: The asylum and its alternatives in progressive America.* Glenview, Illinois: Scott, Foresman and Company.

Rothman, S., & Powers, S. (1994). Execution by quota? *The Public Interest, 116,* 3–17.

Rotman, E. (1990). *Beyond punishment: A new view on the rehabilitation of criminal offenders.* New York: Greenwood Press.

Ruddell, R. (2004). *America behind bars: Trends in imprisonment, 1950 to 2000.* New York: LFB Scholarly Publishing LLC.

Ruddell, R., Decker, S. H., & Egley, A. (2006). Gang interventions in jails: A national analysis. *Criminal Justice Review, 31,* 33–46.

Ruden, R. (1997). *The craving brain: The biobalance approach to controlling addictions.* New York: Harper/Collins.

Sabol, W. J., Couture, H., & Harrison, P. M. (2007). *Prisoners in 2006.* Washington, DC: Bureau of Justice Statistics, U.S. Department of Justice, Office of Justice Programs. Retrieved February 27, 2008, from http://www.ojp.usdoj.gov/bjs/.

Sabol, W. J., Minton, T. D., & Harrison, P. M. (2007). *Bureau of Justice Statistics, prison and jail inmates at midyear 2006* (No. NCJ217675). Washington, DC: U.S. Department of Justice.

Samenow, S. (1998). *Straight talk about criminals.* Livingston, New Jersey: Jason Aronson.

Samenow, S. (1999). *Before it's too late: Why some kids get into trouble and what parents can do about it.* New York: Times Books.

Sampson, R., & Laub, J. (1999). Crime and deviance over the lifecourse: The salience of adult social bonds. In F. Scarpitti & A. Nielsen (Eds.), *Crime and criminals: Contemporary and classical readings in criminology* (pp. 238–246). Los Angeles: Roxbury Press.

Santayana, G. (1905). *The life of reason: Reason in common sense.* New York: Dover Publications, Inc.

Scheidegger, K (2006). *Supreme Court decisions on habeas corpus.* Washington, DC: The Federalist Society. Retrieved July 17, 2007, from http://www.fed-soc.org/Publications/practicegroupsnewsletters/habeas.htm.

Schmalleger, F. (2001). *Criminal justice today* (6th ed.). Upper Saddle River, NJ: Prentice Hall.

Schmalleger, F. (2002). *Criminal law today.* Upper Saddle River, NJ: Prentice Hall.

Schmalleger, F. (2003). *Criminal justice today* (7th ed.) Englewood Cliffs, NJ: Prentice Hall.

Schmitz, J., Stotts, A., Sayre, S., DeLaune, K., & Grabowski, J. (2004). Treatment of cocaine-alcohol dependence with naltrexone and relapse prevention therapy. *The American Journal on Addictions, 13,* 333–341.

Seiter, R. (2002). *Correctional administration: Integrating theory and practice.* Saddle River, NJ: Prentice Hall.

Seiter, R. (2005). *Corrections: An introduction.* Upper Saddle River, NJ: Prentice Hall.

Severson, M. (2004). Mental health needs and mental health care in jails: The past, the present, and hope for the future. *American Jails, 28,* 9–18.

Sharp, B. (2006). *Changing criminal thinking: A treatment program.* Alexandria, VA: American Correctional Association.

Sharp, P., & Hancock, B. (1995). *Juvenile delinquency: Historical, theoretical, and societal reactions to youth.* Englewood Cliffs, NJ: Prentice–Hall.

Sherman, L. (2005). The use and usefulness of criminology, 1751–2005: Enlightened justice and its failures. *Annals of the American Academy of Political and Social Science, 600,* 115–135.

Shilton, M. (2003). *Increasing public safety through halfway houses.* Washington, DC: Center for Community Corrections. Retrieved July 28, 2008, from http://www.communitycorrectionsworks.org

Sickmund, M., Sladky, T. J., & Kang, W. (2005). *Census of juveniles in residential placement databook.* Retrieved February 28, 2008, from http://ncjrs.org

Sickmund, M., & Wan, Y. (2003). *Census of juveniles in residential placement: 2003 databook.* Washington, DC: Office of Juvenile Justice and Delinquency Prevention.

Sipes, L., & Young, B. (2006, Fall/Winter). The core mission: Partnerships for public safety. *In-Sites Magazine* [Online]. Washington, DC: Community Capacity Development Office, Office of Justice Programs, U.S. Department of Justice. Retrieved March 29, 2007, from http://www.ncjrs.gov/ccdo/in-sites/winter2006/pfv.html

Smith, P., & Smith, W. A. (2005). Experiencing community through the eyes of young female offenders. *Journal of contemporary criminal justice. 21,* 364–385.

Smykla, J., & Williams, J. (1996). Co-corrections in the United States of America, 1970–1990: Two decades of disadvantages for women prisoners. *Women & Criminal Justice, 8,* 61–76.

Snell, T. (2006, December). *Capital punishment, 2005* (NCJ 215083). Washington, DC: Bureau of Justice Statistics.

Snyder, H. (2007). *Juvenile arrests: 2005.* Washington, DC: Department of Justice, Office of Juvenile Justice and Delinquency Prevention.

Snyder, H., Espiritu, R., Huizinga, D., Loeber, R., & Petechuck, D. (2003). *Prevalence and development of child delinquency* (Child Delinquency Bulletin Series). Washington, DC: U.S. Department of Justice, Office of Juvenile Justice and Delinquency Prevention.

Solomon, A., Dedel Johnson, K., Travis, J., & McBride E. (2004). *From prison to work: The employment dimensions of prisoner reentry.* Washington, DC: Urban Institute, Justice Policy Center.

Spelman, W. (2000). The limited importance of prison expansion. In A. Blumstein & J. Wallman (Eds.), *The crime drop in America* (pp. 97–129). Cambridge: Cambridge University Press.

Spohn, C. (1990). Decision making in sexual assault cases: Do black and female judges make a difference? *Women & Criminal Justice, 2,* 83–101.

Springer, C. (1987). *Justice for juveniles.* Washington, DC: Office of Juvenile Justice and Delinquency Prevention.

Stannard, D. E. (1992). *American holocaust: Columbus and the conquest of the new world.* New York: Oxford University Press.

Steiner, B., & Hemmens, C. (2003). Juvenile waiver 2003: Where are we now? *Juvenile and Family Court Journal, 54,* 1–24.

Stinchcomb, J. B. (2006). Visionary leadership: Making proactive, data-driven decisions. *Corrections Today, 68,* 78–80.

Stohr, M. K. (2000). Women and the law. In A. Walsh & C. Hemmens (Eds.), *From law to order: Theory and practice of law and justice* (pp. 269–295). Maryland: American Correctional Association.

Stohr, M. K. (2006). Yes, I've paid the price, but look how much I gained! In C. M. Renzetti, L. Goodstein, & S. L. Miller (Eds.), *Rethinking gender, crime, and justice* (pp. 262–277). Los Angeles, CA: Roxbury Publishing Company.

Stohr, M. K. (2008). Women and the law. In A. Walsh & C. Hemmens (Eds.), *Law, justice, and society: A sociological introduction* (pp. 269–294). New York: Oxford University Press.

Stohr, M. K., & Collins, P. A. (2008-Forthcoming). *Criminal justice management: Theory and practice in justice centered organizations.* New York: Oxford University Press.

Stohr, M. K., & Cooper, J. (2007, May/June). Historic Irish gaols: Cork City and Kilmainham (Dublin). *American Jails.*

Stohr, M. K., Lovrich, N. P., & Mays, G. L. (1997). Service v. security focus in training assessments: Testing gender differences among women's jail correctional officers. *Women & Criminal Justice, 9*, 65–85.

Stohr, M. K., Lovrich, N. P., Menke, B. A., & Zupan, L. L. (1994). Staff management in correctional institutions: Comparing DiIulio's "control model" and "employee investment model" outcomes in five jail settings. *Justice Quarterly, 11*, 471–497.

Stohr, M. K., Lovrich, N. P., & Wilson, G. L. (1994). Staff stress in contemporary jails: Assessing problem severity and the payoff of progressive personnel practices. *Journal of Criminal Justice, 22*, 313–327.

Stohr, M.K., Lovrich, N. P., & Wood, M. (1996). Service v. security concerns in contemporary jails: Testing behavior differences in training topic assessments. *Journal of Criminal Justice, 24*, 437–448.

Stohr, M. K., & Mays, L. G. (1993). *Women's jails: An investigation of offenders, staff, administration and programming* (Final report for grant number 92J04GHP5). Washington, DC: National Institute of Corrections, Jails Division. Retrieved July 29, 2008, from http://www.nicic.org/Library/008747

Streib, V. (2003). *The juvenile death penalty today: Death sentences and executions for juvenile crimes, January 1, 1973—June 30, 2003.* Retrieved July 29, 2008, from Ohio Northern University Web site: http://www.law.onu.edu/faculty_staff/faculty_prof iles/coursematerials/streib/juvdeath.pdf

Sturgess, A., & Macher, A. (2005). Issues in correctional HIV care: Progressive disseminated histoplasmosis. *American Jails, 19*, 54–56.

Sullivan, L. E. (1990). *The prison reform movement: Forlorn hope.* Boston: Twayne Publishers.

Sykes, G. M. (1958). *The society of captives: A study of a maximum security prison.* Princeton, New Jersey: Princeton University Press.

Talbot, T., Gilligan, L., Carter, M., & Matson, S. (2002). *An overview of sex offender management.* Washington, DC: Center for Sex Offender Management.

Tartaro, C. (2002). Examining implementation issues with new generation jails. *Criminal Justice Policy Review, 13*, 219–237.

Tartaro, C. (2006). Watered down: Partial implementation of the new generation jail philosophy. *The Prison Journal, 86*, 284–300.

Tartaro, C., & Ruddell, R. (2006). Trouble in Mayberry: A national analysis of suicides and attempts in small jails. *American Journal of Criminal Justice, 31*, 81–101.

Taylor, R., Fritsch, E., & Caeti, T. (2007). *Juvenile justice: Policies, programs, and practices.* New York: McGraw-Hill.

Thompson, E. P. (1975). *Whigs and hunters: The origin of the black act.* New York: Pantheon Books.

Thompson, J. A., & Mays, G. L. (1991). Paying the piper but changing the tune: Policy changes and initiatives for the American jail. J. A. Thompson & G. L. Mays (Eds.), *American jails: Public policy issues* (pp. 240–246). Chicago: Nelson-Hall Publishers.

Tocqueville, A. de. (1956). *Democracy in America* (H. Hefner, Ed.). New York: Norton Books. (Original work published 1838)

Travis, J. (2000). *But they all come back: Rethinking prisoner reentry* (Sentencing & Corrections: Issues for the 21st Century No. 7). Washington, DC: U.S. Department of Justice, National Institute of Justice.

Travis, J., & Lawrence, S. (2002). *Beyond the prison gates: The state of parole in America.* Washington, DC: The Urban Institute Justice Policy Center.

Tutu, D. (1999). *No future without forgiveness.* New York: Doubleday.

Umbreit, M. (1994). *Victim meets offender: The impact of restorative justice and mediation.* Monsey, NY: Criminal Justice Press.

United State Census Bureau. (2004). *Statistical abstracts of the United States.* Washington, DC: Author. Retrieved July 26, 2008, http//:www.census.gov/prod/2004pubs/03statab/vitstat.pdf.

Van Vooris, P., Braswell, M., & Lester, D. (2000). *Correctional counseling and rehabilitation.* Cincinnati, OH: Anderson.

Vanstone, M. (2000). Cognitive-behavioural work with offenders in the U.K: A history of influential endeavour. *The Howard Journal, 39*, 171–183.

Vaughn, M. S., & Carroll, L. (1998). Separate and unequal: Prison versus free-world medical care. *Justice Quarterly, 15*, 3–10.

Vaughn, M. S., & Smith, L. G. (1999). Practicing penal harm medicine in the United States: Prisoners' voices from jail. *Justice Quarterly, 16*, 175–231.

Vito, G., Allen, H., & Farmer, G. (1981). Shock probation in Ohio: A comparison of outcomes. *International Journal of Offender Therapy and Comparative Criminology, 25*, 70–76.

Vogel, R. (2004). Silencing the cells: Mass incarceration and legal repression in U.S. prisons. *Monthly Review, 56*, 1–8.

Walker, E. (2002). Adolescent neurodevelopment and psychopathology. *Current Directions in Psychological Science, 11*, 24–28.

Walsh, A. (2000). Evolutionary psychology and the origins of justice. *Justice Quarterly, 17,* 841–864.

Walsh, A. (2006). *Correctional assessment, casework, and counseling* (4th ed.). Alexandria, VA: American Correctional Association.

Walsh, A., & Ellis, L. (2007). *Criminology: An interdisciplinary approach.* Thousand Oaks, CA: Sage.

Walsh, A., & Hemmens, C. (2000). *From law to order: The theory and practice of law and justice.* Lanham, MD: American Correctional Association.

Welch, M. (2004). *Corrections: A critical approach* (2nd ed.). Boston: McGraw-Hill

Welch, M. (2005). *Ironies of imprisonment.* Thousand Oaks, CA: Sage Publications.

Welsh, W. (1995). *Counties in court: Jail overcrowding and court-ordered reform.* Philadelphia: Temple University Press.

Wener, R. (2005). The invention of direct supervision. *Corrections Compendium, 30,* 4–7, 32–34.

Wener, R. (2006). Effectiveness of the direct supervision system of correctional design and management: A review of the literature. *Criminal Justice and Behavior, 33,* 392–410.

Western, B. (2003). *Incarceration, employment, and public policy.* New Jersey: New Jersey Institute for Social Justice. Retrieved July 31, 2008, from http://www.njisj.org/reports/western_report.html

Weyr, T. (1988). *Hispanic U.S.A.* New York: Harper & Row.

White, A. (2004). *Substance use and the adolescent brain: An overview with the focus on alcohol.* Durham, NC: Duke University Medical Center.

Whitehead, J., & Lab, S. (1996). *Juvenile justice: An introduction.* Cincinnati, OH: Anderson Publishing.

Whitman, J. Q. (2003). *Harsh justice: Criminal punishment and the widening divide between America and Europe.* New York: Oxford University Press.

Whittick, A. (1979). *Woman into citizen.* Santa Barbara, CA: ABC-Clio.

Williams, N. H. (2007). Prison health and the health of the public: Ties that bind. *Journal of Correctional Health Care, 13,* 80–92.

Wilson, J. Q. (1975). *Thinking about crime.* New York: Basic Books.

Winfree, L. T., & Wooldredge, J. D. (1991). Exploring suicides and deaths by natural causes in America's large jails: A panel study of institutional change, 1978 and 1983. In J. Thompson & G. L. Mays (Eds.), *American jails: Public policy issues* (pp. 63–78). Chicago: Nelson-Hall Publishers.

Winter, M. M. (2003). County jail suicides in a Midwestern state: Moving beyond the use of profiles. *The Prison Journal, 83,* 130–148.

Wright, K. (1989). Race and economic marginality in explaining prison adjustment. *Journal of Research in Crime and Delinquency, 26,* 67–89.

Wright, R. (1999). The evidence in favor of prisons. In F. Scarpitti & A. Nielson (Eds.), *Crime and criminals: Contemporary and classic readings in criminology* (pp. 483–493). Los Angeles: Roxbury.

Wright, R., & Decker, S. (1994). *Burglars on the job: Streetlife and residential break-ins.* Boston: Northeastern University Press.

Yates, H. M. (2002). Margaret Moore: African American feminist leader in corrections. *Women & Criminal Justice, 13,* 9–26.

Yochelson, S., & Samenow, S. (1976). *The criminal personality: A profile for change.* Livingston, NJ: Jason Aronson.

Young, V. D. (1994). Race and gender in the establishment of juvenile institutions: The case of the South. *The Prison Journal, 74,* 244–265.

Young, V. D. (2004). All the women in the Maryland state penitentiary: 1812–1869. In M. K. Stohr & C. Hemmens (Eds.), *The inmate prison experience* (pp. 253–268). Upper Saddle River, NJ: Prentice Hall.

Young, V. D., & Adams-Fuller, T. (2006). In C. M. Renzetti, L. Goodstein, & S. L. Miller (Eds.), *Rethinking gender, crime, and justice* (pp.185–199). Los Angeles: Roxbury Publishing Company.

Zhang, Z. (2004). *Drug and alcohol use and related matters among arrestees, 2003.* Washington, DC: National Institute of Justice. Retrieved October 7, 2008, from http://www.ncjrs.gov/nij/adam/ADAM 2003.pdf

Zimmer, L. (1986). *Women guarding men.* Chicago: University of Chicago Press.

Zimmer, L. (1989). Solving women's employment problems in corrections: Shifting the burden to administrators. *Women & Criminal Justice, 1,* 55–79.

Zimring, F. E., & Hawkins, G. (1991). *The scale of imprisonment.* Chicago: University of Chicago Press.

Zimring, F. E., & Hawkins, G. (1995). *Incapacitation: Penal confinement and the restraint of crime.* New York: Oxford University Press.

Zimring, F. E., Hawkins, G., & Kamin, S. (2001). *Punishment and democracy: Three strikes and you're out in California.* Oxford: Oxford University Press.

Zupan, L. L. (1991). *Jails: Reform and the new generation philosophy.* Cincinnati, OH: Anderson Publishing Company.

Zupan, L. L. (1992). The progress of women correctional officers in all-male prisons. In I. L. Moyer (Ed.), *The changing roles of women in the criminal justice system: Offenders, victims, and professionals* (2nd ed., pp. 323–343). Prospect Heights, IL: Waveland Press.

Zupan, L. L., & Menke, B. (1988). Implementing organizational change: From traditional to new generation jail operations. *Policy Studies Review, 7,* 615–625.

Cases

Atkins v. Virginia, 536 U.S. 304 (2002)

Baze v. Rees, 533 U.S.— (2008)

Breed v. Jones, 421 U.S. 517 (1975)

Coker v. Georgia, 433 U.S. 584 (1977)

Cooper v. Pate, 378 U.S. 546 (1964)

Dred Scott v. Sanford, 60 U.S. 393 (1856)

Eddings v. Oklahoma, 445 U.S. 104 (1982)

Estelle v. Gamble, 429 U.S. 97 (1976)

Ex parte Crouse, 4 Wharton (Pa.) 9 (1838)

Ex parte Hull, 312 U.S. 546 (1941)

Ex parte United States, 242 U.S. 27 (1916)

Furman v. Georgia, 408 U.S. 238 (1972)

Gagnon v. Scarpelli, 411 U.S. 778 (1973)

Gregg v. Georgia, 428 U.S. 153 (1976)

Haley v. Ohio, 332 U.S. 596 (1948)

Holt v. Sarver, 309 F. Supp. 362 (1969)

Hudson v. McMillian, 503 U.S. 1 (1992)

Hudson v. Palmer, 468 U.S. 517 (1984)

In re Gault, 387 U.S. 1 (1967)

In re Winship, 397 U.S. 358 (1970)

Jones v. Cunningham, 371 U.S. 236 (1963)

Jordan v. Gardner, 986 F.2d 1521 (9th Cir. 1993)

Kansas v. Hendricks, 521 U.S. 346 (1997)

Kent v. United States, 383 U.S. 541 (1966)

Madyun v. Franzen, 704 F.2d 954 (7th Cir. 1983)

McClesky v. Kemp, 481 U.S. 279 (1987)

McKeiver v. Pennsylvania, 402 U.S. 528 (1971)

Mempa v. Rhay, 389 U.S. 128 (1967)

Morrissey v. Brewer, 408 U.S. 471 (1972)

Penry v. Lynaugh, 492 U.S. 302 (1989)

Pervear v. Massachusetts, 72 U.S. 475 (1866)

Pulido v. State of California, Marin County Superior Court, California (1994)

Roper v. Simmons, 112 S.W. 3rd 397 (2005)

Ruffin v. Commonwealth, 62 Va. 790, 796 (1871)

Salazar v. City of Espanola, New Mexico District Court (2004)

Sandin v. Conner, 515 U.S. 472 (1995)

Schall v. Martin, 104 U.S. 2403 (1984)

Somerset v. Stewart, 1 Lofft 1, 98 ER 499 (KB 1772)

Stanford v. Kentucky, 492 U.S. 361 (1989)

Thompson v. Oklahoma, 487 U.S. (1988)

Turner v. Safley, 482 US 78 (1987)

United States v. Knights, 534 U.S. 112 (2001)

Wilson v. Seiter, 501 U.S 294 (1991)

Wolff v. McDonnell, 418 U.S. 539 (1974)

Woodson v. North Carolina, 428 U.S. 280 (1976)

Index

About the Authors

Mary K. Stohr is Professor in the Department of Criminal Justice at Boise State University and a former chair of her department. She received her PhD from Washington State University in 1990. She has practical experience in the field as a correctional counselor and officer. She has taught management, corrections, and gender and environmental crime. She is a co-founder of the Corrections Section of the Academy of Criminal Justice Sciences (ACJS). She has published two books and numerous articles dealing with corrections and management issues in criminal justice.

Anthony Walsh is Professor in the Department of Criminal Justice at Boise State University. He received his PhD from Bowling Green State University. A former police officer and probation officer, his special interests include the biological bases of human behavior, and his teaching interests are criminology, statistics, legal philosophy, and criminal justice assessment and counseling. He is the author of 20 books, including *Biosocial Criminology* (Anderson, 2002) and *Criminology: A Global Perspective* (Allyn & Bacon, 2000). He has also authored over 100 articles, presented papers at international criminology meetings, and is a consulting editor for both the *Journal of Genetic Psychology* and the *Quarterly Journal of Ideology*.

Craig Hemmens is Director of the Honors College and Professor in the Department of Criminal Justice at Boise State University. He holds a JD from North Carolina Central University School of Law and a PhD in Criminal Justice from Sam Houston State University. He has previously served as Academic Director of the Paralegal Studies Program and Chair of the Department of Criminal Justice. Professor Hemmens has published 15 books and more than 100 articles on a variety of criminal justice–related topics. He has served as the editor of the *Journal of Criminal Justice Education*. His publications have appeared in *Justice Quarterly*, the *Journal of Criminal Justice, Crime and Delinquency*, the *Criminal Law Bulletin*, and *The Prison Journal*.